THE PICTURE OF THE TAOIST GENII PRINTED ON THE COVER OF this book is part of a painted temple scroll, recent but traditional, given to Mr Brian Harland in Ssu-Chhuan province (1946). Concerning these four divinities, of respectable rank in the Taoist bureaucracy, the following particulars have been handed down. The title of the first of the four signifies 'Heavenly Prince', that of the other three 'Mysterious Commander'.

At the top, on the left, is Liu *Thien Chün*, Comptroller-General of Crops and Weather. Before his deification (so it was said) he was a rain-making magician and weather forecaster named Liu Chün, born in the Chin dynasty about +340. Among his attributes may be seen the sun and moon, and a measuring-rod or carpenter's square. The two great luminaries imply the making of the calendar, so important for a primarily agricultural society, the efforts, ever renewed, to reconcile celestial periodicities. The carpenter's square is no ordinary tool, but the gnomon for measuring the lengths of the sun's solstitial shadows. The Comptroller-General also carries a bell because in ancient and mediaeval times there was thought to be a close connection between calendrical calculations and the arithmetical acoustics of bells and pitch-pipes.

At the top, on the right, is Wên *Yüan Shuai*, Intendant of the Spiritual Officials of the Sacred Mountain, Thai Shan. He was taken to be an incarnation of one of the Hour-Presidents (*Chia Shên*), i.e. tutelary deities of the twelve cyclical characters (see Vol. 4, pt 2, p. 440). During his earthly pilgrimage his name was Huan Tzu-Yü and he was a scholar and astronomer in the Later Han (b. +142). He is seen holding an armillary ring.

Below, on the left, is Kou *Yüan Shuai*, Assistant Secretary of State in the Ministry of Thunder. He is therefore a later emanation of a very ancient god, Lei Kung. Before he became deified he was Hsin Hsing, a poor woodcutter, but no doubt an incarnation of the spirit of the constellation Kou-Chhên (the Angular Arranger), part of the group of stars which we know as Ursa Minor. He is equipped with hammer and chisel.

Below, on the right, is Pi *Yüan Shuai*, Commander of the Lightning, with his flashing sword, a deity with distinct alchemical and cosmological interests. According to tradition, in his earthly life he was a countryman whose name was Thien Hua. Together with the colleague on his right, he controlled the Spirits of the Five Directions.

Such is the legendary folklore of common men canonised by popular acclamation. An interesting scroll, of no great artistic merit, destined to decorate a temple wall, to be looked upon by humble people, it symbolises something which this book has to say. Chinese art and literature have been so profuse, Chinese mythological imagery so fertile, that the West has often missed other aspects, perhaps more important, of Chinese civilisation. Here the graduated scale of Liu Chün, at first sight unexpected in this setting, reminds us of the ever-present theme of quantitative measurement in Chinese culture; there were rain-gauges already in the Sung (+12 century) and sliding calipers in the Han (+1st). The armillary ring of Huan Tzu-Yü bears witness that Naburiannu and Hipparchus, al-Naqqās and Tycho, had worthy counterparts in China. The tools of Hsin Hsing symbolise that great empirical tradition which informed the work of Chinese artisans and technicians all through the ages.

SCIENCE AND CIVILISATION IN CHINA

Joseph Needham
(1900–1995)

'Certain it is that no people or group of peoples has had a monopoly in contributing to the development of Science. Their achievements should be mutually recognised and freely celebrated with the joined hands of universal brotherhood.'

Science and Civilisation in China VOLUME I, PREFACE

*

Joseph Needham directly supervised the publication of seventeen books in the *Science and Civilisation in China* series, from the first volume, which appeared in 1954, through to Volume VI.3, which was in press at the time of his death in March 1995.

The planning and preparation of further volumes will continue. Responsibility for the commissioning and approval of work for publication in the series is now taken by the Publications Board of the Needham Research Institute in Cambridge, under the chairmanship of Dr Christopher Cullen, who acts as general editor of the series.

SCIENCE AND CIVILISATION IN CHINA

In the most ancient times . . . The people lived on fruits, berries, mussels and clams – things rank and evil-smelling that hurt their bellies, so that many of them fell ill. Then a sage appeared who drilled with sticks and produced fire with which to transform the rank and putrid foods. The people were delighted and made him ruler of the world, calling him the Sui Jen (Drill Man). Han Fei Tzu, −3rd century, 'The Five Vermin', tr. Watson (1963), p. 96.

For this reason the ancients seem to me to have sought for nourishment that harmonised with their constitution, and to have discovered that which we use now. So from wheat, by winnowing, grinding, sifting, steeping, kneading, and baking it, they produced bread, and from barley they produced cake. Experimenting with food they boiled or baked, they mixed and mingled, putting strong pure foods with weaker, until they adapted them to the power and constitution of man. For they thought that from foods which are too strong for the human consitution to assimilate will come pain, disease, and death, while from such as can be assimilated will come nourishment, growth and health. Hippocratic Corpus, c. −4th century, 'On Ancient Medicine', cited in Farrington (1955), p. 68.

Just as the universe has the five elemental phases, man has the five viscera, and food has the five flavours. The liver imitates wood, the heart fire, the spleen earth, the lung metal, and the kidney water. Sourness goes to the liver, bitterness to the heart, sweetness to the spleen, pungency to the lungs and saltiness to the kidney. Wood produces fire, fire produces earth, earth produces metal, metal produces water and water produces wood. Wood controls earth, earth controls water, water controls fire, fire controls metal and metal controls wood. So during the four seasons do not eat too much of that which is abundant at the time, nor of that which is then in control. If you do, it will harm the corresponding organs. You should select the flavours which are mutually helpful to make the *chhi* abundant. If the five viscera are not harmed, the five *chhi* are increased, the food and drink are consumed in moderation, and the cold and hot seasons respected, then all diseases will be kept away and life will naturally be prolonged. Pao Sêng Yao Lu *by Phu Chhien-Kuan (Sung Dynasty)*, Shuo Fu. ch. 84, p. 2b, tr. Lu Gwei-Djen, mod. auct.

中國科學技術史

李約瑟著

萬朝鼎

JOSEPH NEEDHAM

SCIENCE AND CIVILISATION IN CHINA

VOLUME 6

BIOLOGY AND
BIOLOGICAL TECHNOLOGY

PART V: FERMENTATIONS AND FOOD SCIENCE

BY

H. T. HUANG, D.PHIL

(HUANG HSING-TSUNG)

FORMERLY DEPUTY DIRECTOR, THE NEEDHAM RESEARCH INSTITUTE, CAMBRIDGE

PUBLISHED BY THE PRESS SYNDICATE OF THE UNIVERSITY OF CAMBRIDGE
The Pitt Building, Trumpington Street, Cambridge, United Kingdom

CAMBRIDGE UNIVERSITY PRESS
The Edinburgh Building, Cambridge CB2 2RU, UK   http://www.cup.cam.ac.uk
40 West 20th Street, New York NY 1011-4211, USA   http://www.cup.org
10 Stamford Road, Oakleigh, Melbourne 3166, Australia

© Cambridge University Press 2000

This book is in copyright. Subject to statutory exception
and to the provisions of relevant collective licensing agreements,
no reproduction of any part may take place without
the written permission of Cambridge University Press.

First published 2000

Printed in the United Kingdom at the University Press, Cambridge

Typeset in Baskerville 11.25/13pt, in QuarkXpress™   [GR]

A catalogue record for this book is available from the British Library

ISBN 521 65270 7 hardback

This book is dedicated to the memory of

SHIH SHÊNG-HAN

Master Expositor of the *Chhi Min Yao Shu*

SHINODA OSAMU

Tireless Chronicler of the Food Canons of China

and

LU GWEI-DJEN

Patient Explorer of the History of Food and Nutrition in China

Without whose pioneering works this contribution would not have been possible.

CONTENTS

List of Illustrations . . . . *page* xiii

List of Tables xxi

List of Abbreviations xxiii

List of Acknowledgements xxv

Series Editor's Preface xxvii

Author's Note 1

40 FERMENTATIONS AND FOOD SCIENCE . . 14

  (*a*) Introduction, *p.* 14
    (1) Food resources in ancient China, *p.* 17
      i Grains, *p.* 17
      ii Oilseeds, *p.* 28
      iii Vegetables, *p.* 32
      iv Fruits, *p.* 43
      v Land animals, *p.* 55
      vi Aquatic animals, *p.* 61
    (2) Ancient Chinese culinary system, *p.* 66
      i Food preparation and cooking: methods and utensils, *p.* 67
      ii Seasonings and spices: what they are and how they are used, *p.* 91
      iii Eating and drinking: dining vessels, implements and furniture, *p.* 96

  (*b*) Literature and sources, *p.* 116
    (1) Ancient classical texts, *p.* 116
    (2) Food canons and recipes, *p.* 121
      i The *Chhi Min Yao Shu* or Essential Arts for the People's Welfare, *p.* 123
      ii Early mediaeval food canons, *p.* 124
      iii Late mediaeval food canons, *p.* 126
      iv Premodern food canons, *p.* 129
      v Works on wine technology, *p.* 132
    (3) Classics of *Materia dietetica*, *p.* 134
    (4) Supplementary sources, *p.* 140
    (5) Secondary sources, *p.* 146

  (*c*) Fermentation and evolution of alcoholic drinks, *p.* 149
    (1) Origin of wine fermentations in China, *p.* 150
    (2) Preparation of *ferments* and wines in early mediaeval China, *p.* 168
    (3) Wine fermentations in late mediaeval and premodern China, *p.* 181

(4) Development of red *ferment* and red wine, p. 192
(5) Evolution of distilled wines in China, p. 203
 i Summary of earlier evidence, p. 204
 ii Recent archaeological discoveries, p. 208
 iii Distilled spirit in the Thang and Sung, p. 221
 iv Development of distilled wine in China, p. 226
(6) Medicated wines in China, p. 232
(7) Wines from fruits, honey and milk, p. 239
 i Wine from grape and other fruits, p. 240
 ii Wine from honey, p. 246
 iii Koumiss and other fermented milk products, p. 248
(8) Alcoholic fermentations, East and West, p. 257
 i Origin of *li* and *chiu* in China, p. 259
 ii Origin of beer in the West, p. 263
 iii Grain fermentations, East and West, p. 272
 iv Special features of the Chinese system, p. 278
(9) Production of vinegar, p. 283

(d) Soybean processing and fermentation, p. 292
 (1) Soybean sprouts, p. 295
 (2) Soybean curd and related products, p. 299
  i The origin of bean curd, p. 302
  ii Transmission of *tou fu* to Japan, p. 317
  iii Products associated with *tou fu*, p. 319
  iv Making of fermented *tou fu*, p. 326
  v Comparison of *tou fu* and cheese, p. 328
  vi Addendum, p. 331
 (3) Fermented soybeans, soy paste and soy sauce, p. 333
  i *Ferments* for food processing, p. 335
  ii Fermented soybeans, *shih*, p. 336
  iii Fermented soy paste, *chiang*, p. 346
  iv Fermented soy sauce, *chiang yu*, p. 358
  v Soy fermentations, China and Japan, p. 374

(e) Food processing and preservation, p. 379
 (1) Fermented condiments, pickles and preserves, p. 379
  i Fermented meat and fish products, p. 380
  ii Fermented fish products of East Asia, p. 392
  iii Fermented fish sauce in ancient Greece and Rome, p. 398
  iv Pickled vegetables and other victuals, p. 402
 (2) Chemical and physical methods of food preservation, p. 415
  i Salting and pickling of vegetables, p. 416

LIST OF CONTENTS

    ii *Fu* and *hsi*, dried meat and fish, *p.* 419
    iii Preservation of fruits, *p.* 424
    iv Cold storage, *p.* 429
  (3) Production of oils, malt sugar and starch, *p.* 436
    i Oil from oilseeds, *p.* 436
    ii Malt sugar from cereals, *p.* 457
    iii Preparation of starch, *p.* 461
  (4) Processing of wheat flour, *p.* 462
    i Pasta and filamentous noodles, *p.* 466
    ii Origin of *man-thou* (steamed bun), *p.* 475
    iii Origin of *hun-thun* (wonton), *p.* 478
    iv Origin and development of noodles, *p.* 480
    v Dissemination of noodles in Asia, *p.* 491
    vi Noodles and Marco Polo, *p.* 493
    vii Production and usage of gluten, *p.* 497

(*f*) Tea processing and utilisation, *p.* 503
  (1) Etymology and literature of tea, *p.* 507
  (2) Processing of tea, *p.* 519
    i Tea processing during the Thang, *p.* 519
    ii Processing of Sung cake teas, *p.* 523
    iii Processing of loose tea during the Yuan and Ming, *p.* 528
    iv Origin and processing of oolong tea, *p.* 535
    v Red tea (*hung chha*) and the black tea of maritime trade, *p.* 541
    vi White, yellow, dark, compressed and scented teas, *p.* 550
  (3) Tea drinking and health, *p.* 554
    i Making and drinking tea, *p.* 555
    ii Effects of tea on health, *p.* 562

(*g*) Food and nutritional deficiency diseases, *p.* 571
  (1) Goitre (*Ying* 癭), *p.* 573
  (2) Beriberi (*Chiao-chhi* 腳氣), *p.* 578
  (3) Night blindness (*Chhüeh mu* 雀目), *p.* 586
  (4) Rickets (*Kou lou* 佝僂), *p.* 588

(*h*) Reflections and epilogue, *p.* 592
  (1) The wonderful world of the grain moulds, *p.* 592
  (2) The uneven flow of food processing innovations, *p.* 595
  (3) The evolution of food technology in China, *p.* 597
  (4) Nature, technology and human intervention, *p.* 601
  (5) Effects of processed foods on health and nutrition, *p.* 603
 Epilogue, *p.* 605

BIBLIOGRAPHIES 609
    Abbreviations, *p.* 610
    A. Chinese and Japanese books before 1800, *p.* 611
    B. Chinese and Japanese books and journal articles since 1800, *p.* 629
    C. Books and journal articles in Western languages, *p.* 657

GENERAL INDEX 675
Table of Chinese Dynasties 734
Romanisation Conversion Table 735

ILLUSTRATIONS

1 Three Legendary Emperors, Fu Hsi, Shên Nung and Huang Ti, who taught the people Fishing, Agriculture and the Art of Healing. From a Japanese scroll by Seibe Wake, +1798, Veith (1972), cover page. . *page* 15

2 The staple grains of ancient China: Panicum millet, Setaria millet, Rice, Barley, Wheat, Soybean and Hemp, from *Chin Shih Khun Chhung Tshao Mu Chuang* (*CSKCTMC*) 金石昆蟲草木狀: (a) *Shu*, Panicum millet; (b) *Chi*, Setaria millet; (c) *Tao*, Rice; (d) *Hsiao Mai*, Wheat; (e) *Shu*, Soybean; (f) *Ma*, Hemp. 20

3 The Nine Provinces of Ancient China as outlined in the *Chou Li* (Lin Yin ed. pp. 344–50). . 22

4 Food remains discovered in Han Tomb No. 1 at Ma-wang-tui. *Hunan sheng po-wu-kuan* (1973) II, Pl. 11, p. 9, lower part only. . . 25

5 Examples of food items recorded on bamboo slips found in Han Tomb No. 1 at Ma-wang-tui. *Hunan shêng po-wu-kuan* (1973) II, p. 243, Pl. 290. . 26

6 *Chi Ma* or *hu ma*, sesame; *CSKCTMC*. . 30

7 *Kua*, melon; *CSKCTMC*. . 34

8 *Hu*, gourd; *CSKCTMC*. . 34

9 *Fêng* or *Wu ching*, Chinese turnip; *CSKCTMC*. . 35

10 *Fei* or *Lai fu*, Chinese radish; *CSKCTMC*. . 35

11 *Khuei*, the mallow; *Mōshi Himbutsu Zukō* 毛詩品物圖考, 1/23. . 37

12 Five Liliaceae; from *CSKCTMC*: (a) *Chiu*, leek; (b) *Tshung*, scallion; (c) *Hsieh*, shallot; (d) *Suan* or *hsiao suan*, rocambole; (e) *Hu* or *ta suan*, garlic. . 41

13 *Yü*, taro. Painting by Shih Thao, c. +1697; from Fu & Fong (1973), *The Wilderness colors of Tao Chi*. . 42

14 *Thao*, peach flowers; *Mōshi Himbutsu Zukō* 毛詩品物圖攷, 3/1. . 45

15 *Li*, plum blossoms; *Mōshi Himbutsu Zukō*, 3/4. . 46

16 *Mei*, Chinese apricot; *Mōshi Himbutsu Zukō*, 3/2. . 47

17 *Li*, Chinese chestnut; *Mōshi Himbutsu Zukō*, 3/6. . 50

18 *Chên*, Chinese hazelnut, *Mōshi Himbutsu Zukō*, 3/5. . 51

19 *Phi-pha*, loquat. *CSKCTMC*. . 53

20 Carp, *Mōshi Himbutsu Zukō*, 7/3a. . 64

21 Turtle, *Mōshi Himbutsu Zukō*, 7/5b. . 65

22 Shang Dynasty pottery *yen* or *hsien* 甗 which consists of a *tsêng* 甑 (steamer) surmounted on a *li*, 鬲 (boiler). From Anderssen (1947), Pl. 90. (1a) *Tsêng* Steamer (1b) Perforated bottom of Steamer; (2) *Li* Boiler with three hollowed legs; (3) *Hsien* assembled from *tseng* and *li*. 77

23 Steamer from Pan-pho, Shensi: (a) Crude steamer with boiler and cover: *Sian Pan-pho po-wu-kuan* (1972); (b) Display of steamer assembly at the Pan-pho Museum, c. −4000; photograph H. T. Huang, 1979. 78

24 Steamer from Ho-mu-tu, Chekiang: (a) Assembly of steamer, boiler and stove, Ho-mu-tu, c. −5000; *CKPJ*, **1989** (9), cover; (b) View of steamer and boiler, *Che-chiang shêng wen-kuan-hui*, (*1978*), p. 76, Pl. 3, no. 3. . 79

25 Pottery tripod cauldron, *ting*, 鼎, from Pan-pho. San Francisco Asian Art Museum (1975), Fig. 57, height 15 cm. 79

26 Traditional Chinese stoves, *tsao*: (a) Clay model from Han tomb. *Yunmeng hsien wên-wu*, (*1981*), Pl. 10, Fig. 6; (b) Traditional *tsao* seen in Taipei, Taiwan, photograph by H. T. Huang. 81

27 Three-pronged skewers for *chih* roasting from Chia yu kuan. Chih Tzu (*1987*), p. 7. 86

28 Eastern Han Kitchen scene from Chuchhêng, Shantung. Jên Jih-Hsin (*1981*). 87

29 Chinese bamboo steamers: (a) Mural with bamboo steamers in a Southern Sung tomb in Kansu. Chhên Hsien-Ju (*1955*); (b) Stove with bamboo steamers in Urumqi, Sinkiang, photograph by Kenneth Hui, c. 1985. 90

30 Bronze vessel types for serving food and drink. After Buchanan, Fitzgerald and Ronan (1981), pp. 158–9. 99

31 Lacquer tray containing five *phan* plates, two *chih* mugs, one winged-cup and one pair of chopsticks from Han Tomb No. 1 at Ma-wang-tui. *Hunan shêng po-wu-kuan* (1973), II, p. 151, item 160. . 101

32 Enlarged section from a mural depicting the Banquet at Hung-Men in a Western Han tomb at Loyang. To the left is a guest holding a horn drinking cup. To the right is a chef roasting a fowl on a large fork over a charcoal fire. Fontein & Wu eds. (1976), Tomb 61, p. 23. . . 103

33 Gold bowl and ladle from the tomb of Marquis of I. Qian, Chen & Ru (1981), p. 45, Fig. 58. 106

34 Changing styles of spoons from Shang to Yuan, after Chih Tzu (*1986b*), p. 22: (a) Shang; (b) Western Chou; (c) Spring and Autumn; (d) Warring States; (e) Han; (f) North & South Dynasties; (g) Sui–Thang; (h) Sung–Yuan. 107

LIST OF ILLUSTRATIONS   XV

35 Styles of forks, prehistoric to Yuan, after Chih Tzu (*1986a*), p. 17.
    For the origin of the forks, cf. Table 10.   108
36 Forks in Eastern Han tombs in Sui-tê, Northern Shensi: (1) Mural,
    Shensi shêng po-wu-kuan (*1958*), Fig. 66; (2) Stove model, Tai & Li
    (*1983*), Fig. 4–1.   110
37 Eastern Han brick painting of a dining or drinking scene from
    Chengtu, Szechuan. After Wilma Fairbank (1972).   113
38 Dining scenes from Eastern Han tombs at Liao-yang, Liaoning:
    (a) From Tomb No. 1, Li Wên-Hsin (*1955*), Fig. 18; (b) From Tomb
    No. 2, Wang Tsêng-Hsin (*1960*), Fig. 3–6.   114
39 First two pages of the Introduction to *CSKCTMC*, Calligraphy by
    Chao Chün, from National Central Library, Taipei.   142
40 Vessels excavated from site 14 at Thai-hsi, Kaochhêng, Hopei.
    Hopei sheng wen-wu yen-chiu so (*1985*), p. 31, Fig. 20.   151
41 Wine fermentation vessels from Thai-hsi, Kaochhêng, Hopei:
    (a) Large *wêng* jar containing 8.5 kg residual mash. Pl. 31 (14:58);
    (b) Vessel resembling 'General's helmet' for heating or fermenting
    mash. Pl. 30, no. 3; (14:42); (c) Funnel for transferring or filtering
    fermentation mash. Pl. 33, no. 2 (14:50).   152
42 Flow diagram for the preparation of superior *ferment*, after Miao
    Chhi-Yü ed. (*1982*), p. 371.   172
43 Preparation of red *ferment*: washing of rice in a flowing stream.
    TKKW p. 288.   198
44 Preparation of red *ferment*: Inoculated cooked rice dispersed on
    mats and trays *TKKW* p. 289.   199
45 Preparation of red *ferment*: Incubation of rice on trays to form red
    ferment *TKKW* p. 290.   200
46 Grains of red *ferment* from Foochow, photograph Chen Jia-Hua.   201
47 Chhêng-teh still, from *Chhing-lung hsien*, **1976** (9), Figs. 1 & 2. (a) Chin
    still assembled. Height 41.5 cm, diameter at widest point, 36 cm.
    (b) Bottom: vessel serving as boiler and steamer. Top: condenser
    with convex bottom.   209
48 Cross-section of Chhêng-teh still, Lin Yung-Kuei, **1980** (1), Fig. 5.   210
49 Shanghai still, parts (a), (b), (c) photograph Ma Chengyuan:
    (a) Complete still; (b) Boiler; (c) Steamer, note the performated bottom.   211
50 Shanghai still, cross-section. Ma Cheng-yuan (*1992*), 174–83, Fig. 1.
    First presented at 6ICHSC, Cambridge, 1990.   212
51 Chhu-chou still, cross-section, Li Chih-Chhao & Kuan Tsêng-Chien
    (*1982*), Fig. 1.   213

xvi   LIST OF ILLUSTRATIONS

52 Diagram of Hellenistic and Chinese stills; adapted from *SCC* Vol. v,
    Pt 4, p. 81, Fig. 1454, (d), (e) and (c^{lll}), (d^{l}).   215
53 Mongolian and Chinese stills (after Hommel) reproduced from
    SCC Vol. v, Pt 4, pp. 62–3: (upper) Mongolian still; (lower) Chinese still.   216
54 Early 20th-century 'Hellenistic' or vase type still, Fang Hsin-Fang
    (*1987*), Fig. 1.   217
55 Early 20th-century 'Chinese' or pot type still, Fang Hsin-Fang (*1987*),
    Fig. 3.   218
56 Brick mural from Szechuan, possibly depicting the distillation of wine,
    photograph from Rawson ed. (1996), p. 199.   219
57 Traditional Szechuan still in the 1940s, Wang Yu-Phêng (*1987*), p. 22.   220
58 Diagram of still described in the *CCPY*, reproduced from *SCC* Vol. v,
    Pt 4, p. 113.   228
59 Arabic Retort, photo from the Science Museum, London, +10th
    century to +12th century.   228
60 Definition of *li* from *Ho Han San Tshai Thu-Huei* (+1711).   264
61 Wooden models of Egyptian bread and beer making, British Museum:
    (a) #45196, model showing female servant stirring mash in the
    preparation of beer. Jars holding the strained fermenting liquid stand
    in the foreground. From Asuyt, Middle Kingdom, c. −1900. (b) #40915
    model of a large workshop for the preparation of bread and beer.
    Female servants shown are grinding barley in the background. Others
    knead dough and some shape dough into loaves. Lightly baked loaves
    are broken up, soaked with water and fermented in beer making. From
    Deir el-Bahri. XIth dynasty, c. −2050.   268
62 Origin of wine and beer in the West.   273
63 Origin of *Chhü, Nieh and Chiu* (*Ferment*, sprouted grain and wine).   273
64 Sumerian and Chinese depiction of fermentation vats: (a) Pictograph
    of a Sumerian fermenting vat, after Forbes (1954), p. 280; (b) Oracle
    inscription for *yu* 酉 (fermenting); after Hsü Tsung-Shu (*1981*), p. 563.   276
65 An amphora of the Yangshao culture from Pan-pho, Sian.
    San Francisco Asian Art Museum (1975), Fig. 28, height 43 cm.   276
66 Brick *ferment* from the Chiangnan Brewery, Shanghai, photograph
    HTH.   281
67 Traditional process for making *tou fu*, from Hung Kuang-Chu (*1984a*),
    pp. 58–60: (a) Soaking soybeans, Fig. 1; (b) Grinding soybeans in a
    rotary mill, Fig. 2; (c) Filtering soybean milk, Fig. 3; (d) Cooking
    soybean milk, Fig. 4; (e) Coagulating soybean milk, Fig. 5;
    (f) Pressing of bean curd and processed products, Fig. 6.   304

LIST OF ILLUSTRATIONS

68 Photograph of engraved mural depicting the making of *tou fu* in Han tomb in Ta-hu-thing, Mi-hsien, Honan, photograph H. T. Huang. 306

69 Drawing of engraved scene on a stone slab from Tomb No. 1, Ta-hu-thing, Mi-hsien, Honan, Chhên Wên-Hua, *NYKK (1991)*, p. 248. 307

70 Photograph of a *tou fu* pressing box seen in the countryside near Mi-hsien, Honan in the 1980s, photograph Chhên Wên-hua. . . 309

71 Experimental coagulation of soy milk; left: with boiling, right: without heating, at the Needham Research Institute, 1991, photograph HTH. 311

72 Cake from precipitate from coagulation of unheated soy milk, photograph HTH. 312

73 Flow diagram of soybean curd process, adapted from Ichino & Takei (*1975*), p. 135. 321

74 Drawing of a hawker selling *tou chiang* by Yao Chih-Han (Chhing), from the National Palace Museum, Taipei, Taiwan. . . 323

75 Rubbing of the complete mural referred to on pp. 302–10. From *Honan wên-wu yen-chiu-so* (*1993*), Pl. 34. 331

76 Revised sketch of segment 3 of the lower panel. Chhên Wên-Hua (*1998*), p. 288. 333

77 Preparation of *tempeh* in the countryside in Bali, photograph HTH: (a) Boiling of soybeans in oildrums; (b) Drained, dehulled soybeans awaiting inoculation with a commercial spore powder; (c) Inoculated soybeans dispensed in flat plastic bags; (d) Bags incubated on trays on a large rack. 343–4

78 Flow diagram for process of making *chiang* (fermented soy paste) in the *Chhi Min Yao Shu*. 350

79 Flow diagram of process for making *chiang* (fermented soy paste) in the *Nung San I Shih Tso Yao*. 353

80 Flow diagram for the process of making *chiang yu* (soy sauce) in the *Hsing Yuan Lu*. 364

81 A traditional soy sauce process – 2nd fermentation showing urns and bamboo covers. Drawing by Hung Kuang-Chu (*1984a*), p. 108. . 365

82 Soy sauce incubating urns and covers – Foochow, 1987, photo HTH. 366

83 Traditional soy sauce process – collecting soy sauce using a cylindrical bamboo filter: (a) Drawing by Hung Kuang-Chu (*1984a*), p. 108; (b) Urn showing cylindrical bamboo filter, Foochow, 1996, photograph HTH. 367

xviii   LIST OF ILLUSTRATIONS

84 Traditional soy sauce process in early 20th century – boiler to cook soybeans. Groff (1919), Pl. 1. 368
85 Traditional soy sauce process in early 20th century – incubation trays for culture of fungi. Groff (1919), Pl. 2, Fig. 2. 369
86 Traditional soy sauce process in early 20th century – incubating jars with bamboo cover. Groff (1919), Pl. 3, Fig. 2. 370
87 Traditional soy sauce process in early 20th century – first drawing of product by siphoning. Groff (1919), Pl. 3, Fig. 1. . . . 370
88 Distribution of fermented condiments in East Asia. The northern area is based on soybean products and the southern area on fish products. After Ishige (1993), Fig. 6. p. 22. 393
89 Distribution of fish sauce in East Asia. After Ishige (1993), Fig. 4, p. 20. 394
90 Bronze cooler from the tomb of Marquis of I (c. −400), Hupei Provincial Museum Qian, Chen & Ru (1981), p. 58. . . . 431
91 Design of an ice house in Fêng-hsiang, Shensi, Spring and Autumn period (c. −500). After Tan Hsien-Chin (*1989*), p. 297: (a) Diagram of foundation; (b) Reconstruction. 432
92 Cross-section of an ice storage well in Chi-nan, Capital of the Chhu Kingdom, c. −4th Century. *Hupei shêng po-wu-kuan* (*1980*), p. 43. Scale: 30 cm. Identity and relative positions of the objects found in the well: 1, 2, wooden logs; 3, wooden base for other objects; 4–6, 8, 9 a pottery *tou* (plates with a pedestal); 7, a broken *tou*; 10, bowl; 11, broken plough; 12, pottery jar. 433
93 Sketch of a 19th-century Chinese ice house in Ningpo. Fortune (1853) 1, p. 82. 436
94 Beam press: decoration on Greek Vase −6th century, after Forbes (1956), p. 113. 438
95 Catonian beam press described in Pliny, after Forbes (1956), p. 114. . 438
96 Oil press from the *Wang Chên Nung Shu*, ch. 16, p. 126. . . . 442
97 Oil press from the *Nung Chêng Chhüan Shu*, ch. 23, p. 576. . . 443
98 Steamer and frying pan used in the processing of oilseeds, *TKKW*, tr. Li Chiao-Phing *et al.* (1980), Fig. 12–4, p. 320. . . . 444
99 Trip hammer used to pound shells of nuts to obtain kernels, *TKKW*, tr. Li Chiao-Phing *et al.* (1980), Fig. 12–3, p. 319. . . . 445
100 Grinding of nuts to break shells and obtain kernels, *TKKW*, tr. Li Chiao-Phing *et al.* (1980), Fig. 12–2, p. 318. . . . 446
101 Southern style oil press, *TKKW*, tr. Li Chiao-Phing *et al.* (1980), Fig. 12–1, p. 17. 447
102 Roller-mill powered by two donkeys for pulverising oilseeds. *TKKW*, p. 97. 448

LIST OF ILLUSTRATIONS

103	Rustic oil press carved out of a large tree trunk, photo Hommel (1937), p. 91.	449
104	Modern oil press based on traditional design. Ministry of Grains (1956), p. 12. (1) Wedge press; (2) Overhead beam connecting the wooden structure; (3) Rolling bar on which the mallet is hung; (4) Lever for turning the rolling bar; (5) Flying mallet; (6) Rod holding the mallet; (7) Additional support for mallet on bar; (8) Movable plank to pull the rope holding the lever; (9) Rope to pull the rolling bar; (10) Lever to lower or raise the movable plank; (11) Wooden support for the operator; (12) Jar for collecting oil; (13) Lower wedge; (14) Upper wedge; (15) Oil cakes to be pressed.	451
105	European wedge press for making rapeseed oil in the 19th century, after Albrecht (1825), pp. 60–2.	453
106	Japanese oil press from Okura Nagatsune (1836), tr. and ed. Carter Litchfield (1974), Fig. 2.	453
107	Ghani – the traditional oilmill of India; from Achaya (1993), cover & p. 42: (a) photograph of a Ghani; (b) cross-section of pit and pestle.	454
108	Neolithic saddle quern in China, from Sha-wo-li, Hsin-chêng, Honan. *Chung-kuo shê-huei kho-hsueh yuan* (1983), Pl. 2, Fig. 5.	463
109	Rotary stone mill, Warring States: *Shensi shêng wen-wu kuan-li hui* (1966), Pl. 8.	464
110	Rotary quern, Western Han *Chung-kuo shê-huei kho-hsueh yuan* etc. (1980), Mancheng Report, Vol. II, Pl. 106, nos. 2 & 3 (cf. description in Vol. I, p. 143).	464
111	Rotary quern, Eastern Han, Rawson ed. (1996), Fig. 84. Model of a man operating a rotary quern.	465
112	Dehydrated *chiao-tzu* (a) and *hun-thun* (b) found in a Thang tomb near Turfan in 1959. Adapted from Than Chhi-Kuang (1987), *CKPJ* (11), p. 12.	479
113	Sliced noodle, *chhieh mien*, photograph S. T. Liu.	486
114	Hung noodle, *kua mien*, photograph S. T. Liu.	486
115	Pull noodle, *la mien*, photograph Wang Yusheng.	487
116	Rope noodle, *so mien*, photograph Wang Yusheng.	488
117	Press noodle, *ya mien*, photograph Wang Yusheng.	489
118	Family of Chinese Pasta Foods (after Ishige).	490
119	Marco Polo tastes spaghetti at the court of Kublai Khan. From Julia della Croce, *Pasta Classica*, Chronicle Books (1987), p. 11.	494
120	A will from the Archives in Genoa, dated February 4, 1279, by soldier Ponvio Bastone in which he bequeathed a basket of 'macheroni', to a relative. Courtesy of the Spaghetti Museum, Pontedassio, Imperia, Italy.	495

121	*Camellia sinensis (L) O. Kuntze*, from Ukers (1935) I, front page.	504
122	First two pages of Lu Yü's *Chha Ching* (Classic of Tea) from a Ming edition?	514
123	Design of seals on Sung tribute teas, Chhên & Chu (*1981*), pp. 70–1. 1a, 1b.	526
124	Picture of a Northern Sung tea house, from *Chhing-ming shang ho thu* 清明上河圖 (A City of Cathay), Pl., Scroll Section v, National Palace Museum, Taipei (1980).	528
125	Rolling tea leaves with feet in China, from Ukers (1935) I, p. 468.	535
126	Map of Southeast China showing the *black tea* country of Wu-I in Fukien (A) and the green tea country of Sing-lo in Anhui (B). In the 18th and early 19th century, tea produced in these regions was transported southwards through Kiangsi all the way to Canton for shipment in the overseas trade.	545
127	Chinese style implements for processing tea in India, from Ukers (1935) I, p. 464.	550
128	The genealogy of the major teas of China.	551
129	Tea utensils from *Chha Chü Thu Tsan* 茶具圖贊, pp. 5–27; a–l, +1269. Altogether twelve pictures, from Chhên & Chu (*1987*), pp. 96–119; Cf. also Ukers (1935) I, pp. 16–18. (1) Bamboo basket for drying tea; (2) Wood anvil and iron mallet to mould the tea into cakes; (3) Iron grinding boat; (4) Stone grinding mill; (5) Gourd as ladle for measuring water; (6) Sieve to separate coarse from fine tea; (7) Brush for removal of dust; (8) Lacquer cup with holder; (9) Porcelain tea bowl; (10) Porcelain tea pot; (11) Bamboo brush for washing pots; (12) Towel for cleaning cups.	558
130	Preparation of *kung-fu* tea, Taipei, photograph HTH.	562
131	Chemical formulae for tea catechins and oxidised catechins.	567
132	Trip hammers activated by human foot for polishing rice. *TKKW*, p. 91.	583
133	A bank of trip hammers activated by water power. *TKKW*, p. 92.	584
134	The Wonderful World of the Grain Moulds. Processed foods related to the *Mould Ferment – Chhü* 麴.	594
135	Evolutionary history of the rotary quern and its applications.	598
136	Evolutionary history of the *Mould ferment* and its applications.	600

TABLES

1	The staple grains and major livestock of ancient China	page 21
2	Vegetables in ancient China cited in the *Shih Ching* and *Li Chi* and found among archaeological remains in Han tombs	33
3	Fruits in ancient China cited in the *Shih Ching* and *Li Chi* and found among archaeological remains in Han tombs	44
4	Support staff for keepers of various livestock as described in the *Chou Li*	56
5	Species of fish cited in the *Shih Ching*	62
6	Methods of cooking revealed in Chou and Han classics	70
7	Comparison of ancient and modern Chinese methods of cooking	88
8	Seasonings used in ancient and modern Chinese cookery	95
9	Serving vessels for food and drink in ancient China	100
10	Archaeological finds of forks in ancient China	108
11	Ancient classics as sources of culinary information	117
12	Contents of the *Chhi Min Yao Shu*	123
13	Food canons and recipe books in late mediaeval China	127
14	Food canons and recipe books in premodern China	129
15	Wine classics of the Sung and Ming dynasties	133
16	Mediaeval and premodern works on *Materia Dietetica*	135
17	Fermentation of a Chou wine and modern Shao Hsing wine	164
18	Names of wines produced by the Chou Chinese	165
19	Types of *ferments* made in Han China	167
20	Methods for preparing *ferments* in the *Chhi Min Yao Shu*	170
21	Yield of alcohol in wine fermentations in the *Chhi Min Yao Shu*	180
22	*Ferments* described in the *Pei Shan Chiu Ching*	185
23	Fermentation of wine in the *Pei Shan Chiu Ching*	186
24	Dimensions of the Chin bronze steamer-still	210
25	Common medicated wines noted in the Food and Drug literature	236
26	References to dairy products in the Chinese food literature	256
27	Comparison of ancient fermentation processes for making alcoholic drinks	275
28	Types of traditional *Chhü ferments* made in China today	280
29	Family of products related to soybean curd	322

30	Preparation of *tou fu* and cheese	329
31	Nutrient value of *tou fu* versus cheese	330
32	Processes for making various types of *chiang*	356
33	Usage of soy condiments in food recipes from the Han to the Chhing dynasties	372
34	Basics of Chinese and Japanese soyfood fermentations	378
35	*Chiang* from animal products in the *Chhi Min Yao Shu*	381
36	Making of fish and meat preserves, *Cha*, in the *Chhi Min Yao Shu*	385
37	Products from fermentation of salted fish in East Asia	395
38	Making of pickled vegetables, *Tsu*, in the *Chhi Min Yao Shu*	406
39	Methods of food preservation involving microbial action	409–10
40	Salting of vegetables from the Sung to the Chhing dynasties	418
41	References to vegetable oils in the *Chhi Min Yao Shu*	440
42	Comparison of European, Chinese and Japanese processes for the pressing of vegetable oils	455
43	Recipes for making *ping* (pasta) in the *Chhi Min Yao Shu*	470
44	Recipes for cooking gluten from the Yuan to the Chhing dynasties	502
45	World production of tea in 1950 and 1988	505
46	World consumption of tea *per capita*, 1984–1987	505
47	The tea books (*Chha Shu*) of China	516
48	Implements used in the processing of tea listed in the *Chha Ching*	521
49	Processing of Sung tribute tea at Pei Yuan	524
50	References to the processing of tea in Ming tea books	530
51	Ming system for processing tea	533
52	Comparing the traditional processing of red, oolong and green teas	537
53	Origin of the major red teas in the China trade in the late 19th century	544
54	Implements used in making tea in Lu Yü's tea classic	556
55	Tea utensils used by Emperor Hui Tsung	559
56	Traditional Chinese views on the effects of tea	565
57	Garway's First Tea Broadside (+1660)	566
58	Comparison of flavanol compounds in the three major teas	568
59	Symbolic correlations of *Wu-Hsing* categories in heaven, earth and man	572
60	The development interval of selected processed foods initiated during the Han	595
61	Summary of processed foods derived from grains	604

ABBREVIATIONS

The following abbreviations are used in the footnotes. For abbreviations used in the bibliographies see, p. 610.

C&C	Chhên Chu-Kuei & Chu Tzu-Chên ed. (1981) *Chung-kuo chha-yeh li-shih tzu-liao hdusn-chi (Selected Historiographic Materials on the History of Tea in China).*
CCIF	*Chhien Chin I Fang*
CCYF	*Chhien Chin Yao Fang*
CCPY	*Chü Chia Pi Yung Shih Lei Chhüan Chi* (Food and Drink section only).
CHPCF	*Chou Hou Pei Chi Fang*
CKCC	Chhên Tsung-Mou ed. (1992). *Chung-kuo Chha Ching* (The Chinese Classic of Tea)
CKYL	*Chin Kuei Yao Lueh*
CKPJPKCS	Chung-kuo Phêng-jên Pai-kho Chhüan-shu pien-wei huei (*1992*), *Chung-kuo Phêng-jên Pai-kho Chhüan-shu* (The Encyclopedia of Chinese Cuisine).
CLPT	*Chêng Lei Pên Tshao*
CMYS	*Chhi Min Yao Shu*
CPYHL	*Chu Ping Yuan Hou Lun*
CSHK	Yen Kho-Chün ed. (*1836*) *Chhüan Shang-ku San-tai Chhin-Han San-kuo Liu-chao wen* (Complete Collection of prose literature from remote antiquity through the Chhin and Han dynasties, the Three Kingdoms, and the Six Dynasties).
FSCS	*Fang Shêng Chih Shu*
HHPT	*Hsin Hsiu Pên Tshao*
HTNCSW	*Huang Ti Nei Ching, Su Wên*
HWTS	*Han Wei Tshung Shu*
HYL	*Hsing Yuan Lu*
IYII	*I Ya I I*
KHTS	*Kuei Hsin Tsa Shih*
LYCLC	*Loyang Chieh-lan Chi*
MCPT	*Mêng Chhi Pi than*
MIPL	*Ming I Pieh Lu*
MLL	*Mêng Liang Lu*
NSISTY	*Nung Sang I Shih Tsho Yao*
PMTT	*Pien Min Thu Tsuan*
PTKM	*Pên Tshao Kang Mu*
PTPHCY	*Pên Tshao Phin Hui Ching Yao*
PTSI	*Pên Tshao Shih I*
PSCC	*Pei Shan Chiu Ching*
SCCK	*Shan Chia Chhing Kung*

SHHM	Shih Hsien Hung Mi
SKCS	Ssu Khu Chhüan Shu
SLKC	Shih Lin Kuang Chi
SLPT	Shih Liao Pên Tshao
SMYL	Ssy Min Yüeh Ling
SNPTC	Shên Nung Pên Tshao Ching, also abbreviated to *Pên Ching* in the text.
SPTK	Ssu Pu Tshung Khan
SSTY	Ssu Shih Tsuan Yao
SWCT	Shuo Wên Chieh Tzu, also abbreviated as *Shuo Wên* in the text
SYST	Sui Yuan Shih Tan
TCMHL	Tung Ching Mêng Hua Lu
TKKW	Thien Kung Khai Wu
TNPS	To Nêng Pi Shih
TPYL	Thai Phing Yü Lan
TTC	Thiao Ting Chi
WSCKL	Wu Shih Chung Khuei Lu
YCFSC	Yin Chuan Fu Shih Chien
WLCS	Wu Lin Chiu Shih
YCFSC	Yin Chuan Fu Shih Chien
YHL	Yang Hsiao Lu
YLT	Yün Lin Thang Yin-shih Chih-tu Chi
YSCY	Yin Shan Chêng Yao
YSHC	Yin Shih Hsü Chih

ACKNOWLEDGEMENTS

We thank the following institutions for permission to reproduce the figures as indicated below.

British Museum   61
Buck's County Historical Museum   103
China Cultural Relics Promotion Center, Beijing   56, 111
Commerce Publishers, Beijing   67 (a–f), 81, 83a
Cultural Relics Publishing House, Beijing   4, 5, 31, 40, 41, 47, 48, 75
National Central Library, Taipei   2 (a–f); 7, 8, 9, 10, 12 (a–e), 19.
Östasiatisca Museet, Stockholm   22
Science Museum, London   59
University of California Press   1

SERIES EDITOR'S PREFACE

This volume sees the ultimate fruition of a long relationship of friendship and support for one of the twentieth century's more extraordinary and often demanding personalities, and for one of that century's most significant intellectual initiatives.

When Joseph Needham went to China to found the Sino-British Science Cooperation Office in wartime Chongqing, he had already made the most meaningful Chinese acquaintance of his life in Lu Gwei-djen, whose presence in Cambridge was one of the principal 'evoking factors' that ultimately called forth *Science and Civilisation in China*. He had set for himself the demanding task of maintaining contact between China's beleaguered scientists in flight from the Japanese invasion, and their sources of information and materials in the wider scientific community beyond the Himalayas. But once in China Needham realised that like so many westerners before him his effectiveness would depend on the quality of Chinese assistance he could obtain. His first task was therefore to find a Chinese secretary with appropriate talents – including excellent English, a scientific training and the sheer intrepid perseverance needed to survive many a rough mile of travel along dangerous roads through war-torn countryside.

On 15 May 1943 an important meeting took place, as we may read in one of Needham's letters to his wife Dophi (Dorothy Moyle Needham, FRS) back in England:

'Before I was up, about half a dozen people were waiting in the parlor to see me . . . Among them was a very nice young man Huang Hsing-Tsung, a science graduate from Hong Kong whom I'm taking on as my secretary.'

Clearly this was the ideal person for the job, as a letter of 20 May confirms:

'This morning I worked the first time with my secretary, Huang Hsing-Tsung, a most charming and very wide-awake young man. He has a good science degree from Hong Kong, and did 6 months' research on phthalocyanines before the Japs came in. He's bright.

And those who knew Needham's working habits in his later years will recognise his relief in finding someone congenial who could work in the informal style he preferred:

'Huang Hsing-Tsung, my secretary, is a charming companion as well as awfully efficient – I never dictate letters, we discuss what we want to say, and he writes it straight down on the typewriter and I sign it. He should learn a lot of general science in this job, is already a good organic chemist.'

Like Needham, the new secretary was a man of literary as well as scientific tastes and education, as an entry of 14 April 1943 confirms:

'Lunch in compartment very nice . . . Read Borrow, HT read Tang poems.'

HT (as he was usually called by Needham, as well as by others) was, however, soon to find that there was more to being Needham's secretary than typing letters and sharing poetry. A diary entry of Thursday 1 June 1944 records one of many similar incidents in Needham's interminable lorry journeys in West China:

'Departed Nanfeng. A most unpleasant time. No repair shop nearer than Ningtu. Gasket of pump proved apparently not the trouble, so the spare diaphragm was put on. But that didn't do the trick either, stopping and starting, feebly going, a most nerve-wracking morning. At last, around noon I (I think it was I) had a brainwave to short-circuit the faulty feed system by conveying alcohol from our drums direct to the engine through our long pipe. It worked, KW drove, and we got to Kuangchang at 1:30.'

As Needham's appended sketch records, HT played an essential role in holding up the alcohol pipe throughout the rest of the journey. All this, and much more by way of experiences both hazardous and uncomfortable, HT survived, his personal respect and affection for Needham growing as time went by. In September 1944 HT went to Oxford to take a D.Phil in chemistry, and subsequently he worked in the food processing and pharmaceutical industries in the United States. But his contacts with Needham and with his work never lapsed. In 1980, when he was a Program Director with the National Science Foundation in Washington, he contributed a section to volume VI part I of *Science and Civilisation in China* on the use of natural plant products and biological agents for the control of insect pests. A few years later, at Needham's request, he began work on the present volume, at an age when most people would have been contemplating retirement. Not only that, but he spent a considerable part of the time until Needham's death in 1995 in Cambridge, serving as a Deputy Director of the Needham Research Institute, and for a period taking day-to-day charge of the *Science and Civilisation in China* project itself.

After HT's return to the USA – formally to retirement at last, but in reality to intensive research and writing – it fell to me as general editor of the *SCC* series to see his manuscript through the process of revision and finally through the Press. This could easily have been a difficult relationship: who was I to say to someone in HT's position how a volume in Needham's own series should be shaped, given that he had known *SCC* since it first began to sprout as a concept for a work 'in no more than one volume' as Cambridge University Press had specified at the outset? But it is a tribute to the character that attracted Needham to his young secretary so many years ago that the relationship was one of the easiest and most gracious I have had with an *SCC* author so far. I am delighted to welcome to the series this fascinating book on a topic whose centrality in Chinese culture and to human welfare in general is ample justification of the breadth and depth of the discussion that its author has set out for us.

AUTHOR'S NOTE

This book may be said to have had its genesis in two memorable events that I experienced in China more than fifty years ago. The first was an enforced holiday I had for several months in the fall of 1942 in my ancestral village, Hothang 鶴塘, and the second a delightful encounter in the spring of 1943 with the eminent scholar, Shih Shêng-Han 石聲漢. Hothang is a tiny village about seventy kilometres north of Foochow 福州. It seemed a world away from the bustling city of Hong Kong where I was a research student less than a year earlier. My comfortable life there was shattered when the Japanese suddenly attacked on the morning of 8 December 1941. After Hong Kong fell on Christmas day I was cut off from my family in Malaya and left without any means of support. I realised that the best thing for me to do was to leave for China as soon as possible. The opportunity came early in February, 1942, when I made my way, with the help of friends, through the New Territories and crossed the border into Free China. My original plan was to travel to Hothang to visit my paternal grandmother, but the plan encountered a delay along the way. I stopped by at Amoy University in Changting 長汀 to deliver a message to Professor Arthur Lee from his niece in Hong Kong who was a classmate of mine.[1] He kindly persuaded the University to offer me a job as an instructor in the Chemistry Department, which I gratefully accepted.

After the semester was over I continued my journey to Hothang, where I had a joyous reunion with my grandmother and other relatives. While in Changting I had met a couple of field officers of the Chinese Industrial Cooperatives (CIC) who indicated to me that they were in need of technical personnel in Szechuan and Kansu. I expressed my interest to serve in their organisation. In July I went to Foochow to visit the Anglican Bishop of Fukien, C. B. R. Sargent, who was the teacher of a close friend in Hong Kong.[2] When I returned to the village there was a letter from CIC headquarters asking me if I would be willing to accept an appointment as a research technician in Chengtu. I was delighted and replied that I would. I immediately resigned from Amoy University, fully expecting to be on my way to Szechuan within a couple of months. But the wheels of bureaucracy grinded ever so slowly. I waited and waited as months went by.

Hothang was a small village with probably no more than a couple of a hundred inhabitants. It had one narrow street paved with rectangular blocks of stone, flanked by hills on one side and a rivulet on the other. A series of steps down the street from my grandmother's house would lead to the market square in the centre

---

[1] Arthur Lee was a Chinese Australian who came to Amoy to teach English. He married one of his students and stayed on to be Professor of English at the University. I am deeply grateful to him and Mrs Lee for their hospitality during my stay in Changting.

[2] Before he was consecrated Bishop of Fukien, Christopher Sargent was the Headmaster of the Diocesan Boy's School in Hong Kong. Unfortunately, he died from pneumonic plague on 8 August 1943.

of the village. Going up the hill in the other direction lies the Ancestral Temple, a building in traditional style with elegant sloping roofs. The only other public building in the village was the Christian Church on the far side of the market square.

Having virtually nothing to do, I spent a great deal of time observing the numerous food processing and culinary activities involved in the preparation of the meals I ate everyday. The primary staple in the village was, of course, rice, coarse white rice and red rice. For breakfast, rice was boiled gently in plenty of water and allowed to simmer into a congee[3] in one of the two large woks that sat on the stove in the kitchen.[4] For lunch and dinner, rice was boiled and the semi-cooked grains steamed in a bamboo steamer.[5] The wash water was usually fed to the pigs. To go with the congee we had roasted peanuts, pickled vegetables, salted duck eggs, fermented soybeans, fermented beancurd, and as a special treat deep fried crullers (*yu thiao* 油條) when they were available from the market. The steamed rice would be accompanied by beancurd, salted fish of various kinds, salted or pickled vegetables, fresh leafy vegetables or beans, dried seaweeds, and, on rare occasions, bacon, sausages, eggs, pork, chicken or fish. Vegetables and meat were usually stir-fried with lard or peanut oil, flavoured with soy sauce, fish sauce, salt, rice wine, vinegar and sesame oil. We drank tea during the day, and occasionally wine in the evening.

For fresh produce we relied on what we could get daily in the market at the centre of the village, which served as a gathering place for the people of several neighbouring villages as well as Hothang. I remember stalls selling various kinds of vegetables, fruits, bean sprouts, peanuts, chickens and occasionally fish. The latter two would be sold live. Depending on the season, I might also find hawkers peddling soft beancurd custard (*tou fu hua* 豆腐花), malt syrup candy (*mai ya thang* 麥芽糖), deep fried crullers (*yu tsa kui* 油炸鬼), and pastries made of glutinous rice (*no mi kao* 糯米糕). Among the shops that lined the sides of the square, several were trading in food products such as rice, wheat flour, salt, loose brown sugar, oil, wine, vinegar, soy sauce, fish sauce and brown sugar blocks. Three were of particular interest to me: the butcher, the beancurd shop and the noodle maker. The butcher probably slaughtered a pig every day. Sections of the animal would be hung on hooks so that the customer could easily decide which part of the pig and how much of it he wanted. The parts in greatest demand were the tenderloin, spareribs, liver, brain and kidney. Occasionally, goat meat might also be on sale, but no beef was available during my stay.

I must have spent hours watching the processing of soybeans into curds which were then pressed into blocks of *tou fu*. The process was identical to that shown in the Eastern Han mural from Ta-hu-thing, Mi-hsien down to the shape of the rotary quern for grinding the beans and the square wooden press for pressing the curds.[6]

---

[3] The significance of congee in the origin of grain fermentations is discussed on p. 260.
[4] The traditional Chinese stove is described on p. 80, and shown in Figs. 26a,b. Quite often chunks of sweet potato would be cooked together with rice in the congee.
[5] For a description of the steamer cf. pp. 76–82 and Figs. 23, 24, 26, 29.
[6] The Eastern Han mural is described on pp. 86–7.

Most of the beancurds produced were sold as blocks of fresh *tou fu*. Portions were pressed further and salted as *tou-fu-kan* 豆腐干 (dried beancurd). In the summer some of the fresh curds would be mixed with brown sugar water and sold as a delicious soupy custard (*tou-fu-hua*). The butcher and the beancurd shop were the busiest places in the village. They were usually sold out by midday. The noodle maker, however, remained open until early evening. He kneaded his dough with a long rolling pin until it was a thin sheet about two feet wide and several feet long. The sheet would be folded into a block and then sliced with a big cleaver. The long filaments of sliced noodle (*chhieh mien* 切麵) were boiled, strained, cooled on a round mat and sold.[7] The shop also served as a mini-restaurant. I often indulged myself with an afternoon snack of stir-fried noodles, noodle soup and on rare occasions wonton and noodles.

There was another shop making wheat flour foods closer to our house. This one made very fine noodles, called *kua-mien* 掛麵 (hung noodle) that is still a specialty of Fukien. The dough is pulled into banks of fine threads and then hung on wooden racks in the open to dry. Poles with racks were installed on the vacant lot next to the shop. On fine days we would see a forest of hung noodles being dried between the racks.[8] It was quite an amazing sight. The dried noodles were folded into bunches, and sold. This same shop also made a bun called *kuang-ping* 光餅 (bright bun) unique to northern Fukien. It is round and has a hole in the middle. It looks and tastes just like a bagel, except that it is smaller. What seemed to me especially interesting is the oven used to bake it. It is simply a large urn enclosed in a large block of clay. Charcoal is burnt at the bottom and the pieces of dough baked on the side of the urn.[9] Considerable skill is needed to collect the bun as soon as it is baked so that it does not fall into the fire. This shop also made moon cakes for the Autumn Festival.

The two most engrossing food processing operations could be seen taking place right next door. The family ran a small workshop making wine from steamed rice with the aid of the red *ferment* (*hung chhü* 紅麴).[10] The *ferment* was purchased locally, and the rice was the same as that we ate for lunch or dinner. There were urns containing wine fermentations at varying stages of maturity. When ready the mash was placed in a cloth bag and pressed in a square box under a block of stone. The red liquid was allowed to settle and then decanted into little urns and sealed. There were two major uses for the red residual mash. It was a popular flavouring agent for cooking chicken, pork and fish. It imparted a brilliantly red colour and a delicious flavour to the food. It was also extensively used as a preservative and pickling agent for meat, fish and vegetables, such as Chinese cabbage, turnips and ginger.[11] The young ginger root pickled in this way was absolutely delectable.

[7] The making of sliced noodles is discussed on p. 484, and illustrated in Fig. 113.
[8] The drying of *kua mien* is shown in Fig. 112.
[9] This is an example of the tandoor oven, seen all across Central Asia from Iran to the West and Sinkiang to the East.
[10] The making of red *ferment* and red wine is discussed on pp. 192–202.
[11] The use of the red wine residues as a preservative is discussed on pp. 302, 411, 413.

Next to the 'winery' is another workshop which made fermented soybeans (*shih* 豉) and soy sauce (*shih yu* 豉油).[12] To prepare *shih* soybeans were boiled, steamed, cooled, mixed with a small amount of previously moulded beans, spread out in an urn and allowed to become mouldy. The moulded beans were then incubated in a minimal amount of brine until it became dark brown. The *shih* was used mainly as a relish for breakfast. To make soy sauce, the cooked beans were mixed with flour and allowed to become mouldy as before. They were then incubated with a liberal amount of brine, and this time, it was the liquid that was collected. The solids were practically all disintegrated. The residue was fed to the pigs. Both the wine and the soyfoods prepared were sold through local shops.

Up the hill just below the Ancestral Temple was a large building called the Tea Trade Centre (*chha hang* 茶行). In it were a series of stoves with large woks for stir-frying tea leaves. In the late 19th and early 20th century this region used to be a flourishing centre for producing black tea (in Chinese red tea or *hung chha* 紅茶) for export.[13] But the industry declined when India replaced China as the world's major producer of black tea. Now the Trade Centre just processed a small amount of green tea for local consumption. The most impressive piece of equipment there was a giant wedge press made out of a single large tea trunk.[14] It was being used to press oil from tea seeds.

Soon it was October. The rice plants in the fields were ready for harvest. I went out several times at dawn to the fields owned by the family to watch how the plants were cut, threshed, and the grains collected. They were carried back to the village, dried on large mats, decorticated in in a clay quern (*lung* 礱), and winnowed in a machine to separate the kernels from the chaff. The kernels were carried to a mill by a stream nearby and polished in a series of trip hammers powered by a large water wheel just like the one shown in the *Thien Kung Khai Wu* 天工開物.[15]

I was impressed by the ingenuity displayed in the processes I saw. It seemed to me they all had a rational, scientific basis. The cooking of the rice allowed the starch to swell and become easily digestible not only to humans but also to microorganisms. The kneading of the wheat flour dough generated gluten which endowed it with flexibility and plasticity. But I marvelled at the extent to which a piece of dough could be stretched lengthwise until it reached almost silken dimensions. The grinding of soybeans in water to form a milk-like emulsion was presumably a natural consequence of the properties of the proteins and fats in the bean. What intrigued me most were the fermentations of grains into wine and soybeans into soy sauce. They were rather complicated processes that required a high level of understanding and technical skill. What is the scientific basis of these processes? How did they come

---

[12] The preparation of fermented soybeans and soy sauce is treated on pp. 336–74. Soy sauce is called *shih yu* (sauce from fermented soybeans) in Fukien and Kuangtung but *chiang yu* (sauce from fermented soybean paste) in most parts of China.
[13] The origin of black tea, called red tea in Chinese, is discussed on pp. 541–9.
[14] The Chinese wedge press is discussed on pp. 441–51. Cf. also *SCC* Vol. IV, Pt 2, p. 206 and Fig. 463.
[15] *TKKW*, pp. 79–92. For a discussion of the machines cf. *SCC* Vol. IV, Pt 2, pp. 151–5; 176–95.

about? What were their origins? How long ago were they discovered? When I asked the people doing the work, the answer was always that they had been around a long time, or that they were the legacy of Shên Nung, the legendary ruler who discovered Agriculture and Medicine.

As it turned out, the answer to some of the questions came sooner than I would have thought possible. By mid-November I had received all the necessary travel documents and a travel advance from the CIC. My grandmother made me a batch of malt candy from barley (or wheat) malt and steamed rice as a going away present. I left Hothang in early December[16] and travelled through Fukien, Kiangsi, Kuangtung, Kuangsi, Kweichow and Szechuan, eventually reaching Chengtu in early February, 1943. But before I had a chance to settle down in my new position, I received a letter in April from Joseph Needham who had recently arrived from England and was setting up a Sino-British Science Cooperation Office in Chungking, asking me if I would be interested in joining his organisation as his secretary and interpreter. After suitable negotiations with the CIC I was hired as his secretary in May and we started on our first peregrination together. From Chengtu we drove to Loshan 樂山 where our host was Wuhan University. There we met Shih Shêng-Han 石聲漢, Professor of Plant Physiology, who had ingeniously built all kinds of apparatus out of the simplest materials available both for research and teaching purposes.

After a week in Loshan we went on to Wu-tung-chhiao 五通橋, the centre of a chemical industry complex where we visited the Huang Hai 黃海 Research Laboratory, which had a programme on improving the strains of fungi used in the saccharification of grains for conversion to alcohol. It was there that I had my first view, under a microscope, of the myceliae of *Aspergillus* species isolated from the Chinese *ferment* (*chhü* 麴). Our next stop was Lichuang 李莊 which could only be reached by boat. We were to sail first on a salt transport boat down the river to Iping 宜賓, and then by steamer to Lichuang. To ensure that there would be no hitch, Shih Shêng-Han decided to come along as our guide. It turned out that his presence was invaluable, since the salt boat developed unexpected trouble after the very first day. Shih negotiated with the owner of a small boat, and we hired him to take us to Iping. But we missed the steamer there and continued the trip in the small boat all the way Lichuang.[17]

---

[16] I did not see Hotang again until January 1996, fifty-five years later. To my amazement it has become a bustling little town with several paved streets, motor vehicles moving to and fro, multistorey houses built of concrete and electric lights in the buildings. Our ancestral house where I stayed in 1942 still stands. It happens to be located in the small section of the old street that has been marked for preservation. The Ancestral Temple remains in good condition. It now doubles as a nursery school and kindergarten. The old Tea Trade Centre has been torn down. The picturesque mill with its huge water wheel has been dismantled. A new town hall stands on the site of the old Christian Church, which has been rebuilt on a location outside of town. The most striking impression I got during my brief visit is that everyone I met (which means practically everyone younger than I) spoke very good Mandarin (at least compared to me), whereas in 1942 few people spoke any Mandarin at all. Coupled with my experience in Taiwan, it is clear that there the search for a common spoken language for all China has been successfully supported by both the Communists in China and the Nationalists in Taiwan, cf. Ramsey, S. Robert (1987).

[17] For details of our boat trip see H. T. Huang (1982), pp. 44–6.

Shih Shêng-Han had received his doctorate from the Imperial College, London and spoke English fluently. He had a wry sense of humour and soon he and Needham were exchanging jokes in English. Conversation flowed easily amd continuously during the two days that we were together cooped up in a small space. We talked about all sorts of things but the topic that received the most attention was the History of Science and Technology in China. Shih seemed to be a fountain of information on the origin of the traditional agricultural and food processing technologies of China. I quickly seized the opportunity and plied him with questions on the science and the history of the food processes that I had seen and pondered on in Hotang half a year ago. I learned that, indeed, many of them had had a long history. In fact, detailed descriptions about most of them can be found in a +6th century compendium called the *Chhi Min Yao Shu* 齊民要術 (Important Arts for the People's Welfare).

Shih Shêng-Han stayed on with us in Lichuang for two days. He and I shared a bedroom in the guest house of Tungchi 同濟 University. Our conversations on traditional Chinese food processing continued deep into the night. He patiently answered all my questions. By then I had learned that he was not only a competent scientist, a noted scholar of Chinese classics but also a renowned calligrapher.[18] Obligingly he wrote down the two poems he had introduced us to during our memorable boat trip on two small scrolls, which I later mounted and framed. They have adorned my study for many years and remained a constant source of inspiration as I laboured in the myriad tasks involved in the writing of this book.

Now, as I look back across a span of half a century, I realise how fortunate I was to have been a principal participant in these memorable events which have in recent years assumed a renewed importance as I laboured to complete the present volume. For my sojourn in Hothang enabled me to witness the practice of traditional Chinese food processing methods in the context of daily living in a small village before the onslaught of modern technology. And my encounter with Shih Shêng-Han gave me an opportunity to discuss the scientific basis and historical background of this technology with a foremost scholar of the field. It was the memory of these events at the back of my mind[19] that encouraged me to accept, with little hesitation, Joseph Needham's invitation in late 1984 to be the collaborator responsible for the writing of Section 40, Biochemical Technology, of his *Science and Civilisation in China* (*SCC*) series.

As Needham reiterated in his invitation, the focus of this Section should be the scientific basis and historical background of the fermentations and food processing

---

[18] An example of Shih's calligraphy still hangs in the Needham Research Institute, Cambridge.

[19] Actually, the memories of those two events were never too deeply buried at the back of my mind. I spent many years in my professional life as a research scientist and research administrator in the fermentation and food industry in the US. I was familiar with the production and application of fungal enzymes used in food processing in the US including amylases, proteases, pectinases, lipases and microbial rennet. What is not generally known is that most of the organisms involved, such as *Aspergillus*, *Rhizopus* and *Mucor* species, were first isolated from the ancient *ferment* 麴 (Chinese *chhü* or Japanese *koji*), the principal agent used in Chinese and Japanese fermentation processes.

technologies that are the mainstay of the Chinese dietary system. Although much has been written about Chinese cuisine and food culture, very little has been written by European scholars on the technology of Chinese processed foods. As a result, the translation of Chinese food terms into English is often highly misleading.[20] He hoped also, that along the way, we would be able to satisfy his personal curiosity about the origin and development of a number of unusual food products that he had encountered during his travels in China in the early 1940s, such as the red *ferment* that colours chicken and fish a brilliant red in Foochow, the *fu ju* (fermented bean curd) that endows the Buddhist's vegetarian stew with a unique flavour in Kuangtung, the delectable aroma of the distilled wine in Kweichow, the soy milk that we consumed every morning at breakfast in the Northwest, and the 'cream' that made possible the delicious 'creamed cauliflower 奶油菜花' that soothed our spirit as we struggled with endless calamities suffered by our truck along the panhandle in Kansu.

As originally conceived in the master-plan for *SCC* published in 1954,[21] and revised in 1979, Section 40 was to consist of one major subject, Fermentation (i.e. the conversion of grains to alcoholic drinks), and two minor ones, Food Technology (i.e. the production of processed foods from soybeans and grains), and Nutrition (with emphasis on nutritional deficiency diseases). Two revisions of the plan were discussed and adopted in the first few years after I started to work on the project. Firstly, Food Technology was greatly expanded since it became clear, based on the wealth of material that Needham and Lu Gwei-Djen had already collected, the space allotted to this subject in 1979 was woefully inadequate. Secondly, processing and utilisation of tea were transferred from Section 42 into this Section.[22] After all, tea and wine are the two principal beverages of the Chinese dietary system; thus the processing of tea should be discussed in the company of the technology of wine. Additional adjustments and revisions proved to be necessary in response to the recommendations of anonymous readers who evaluated the preliminary draft on behalf of the Publications Board of the Needham Research Institute. As it turns out, the book that finally emerges is an enlargement of the 1954 master-plan, that is to say, it contains a major section on Fermentations and a minor section on Nutrition, except that another major section, that on Food Science is added.

The book covers almost all aspects of the traditional food processing technologies that I had witnessed in 1942. It also includes the processing of foods that I had encountered as part of my diet in Hothang, but that were not produced

---

[20] A sentiment shared by David Knechtges (1986), p. 63, who points out that 'exacting philology and careful science' are needed to render a Chinese food term into English.

[21] See *Science and Civilisation in China* (*SCC* hereafter) Vol. 1 (1954), pp. xxxv–xxxvi. This plan was revised in the report on *Status of the Project*, (1979), Cambridge University Press, p. 32.

[22] Tea was originally a topic in Section 42, Agro-Industries, which was the responsibility of Christian Daniels. Under the reorganisation Tea was split into two parts. Processing and utilisation became a chapter in this work. Horticulture and genetics were assigned to the Section 38 as a continuation of the volume on Botany, which is being prepared by Georges Metailie. I regret the inconvenience this must have caused Professor Daniels, whose excellent contribution to Section 42 was published in 1996.

locally. These were imported from neighbouring counties, for example, fish sauce, pickled fish, salted fish, salted pork and cured meat products. As I began the project I was excited by the prospect of rediscovering or refining answers to the many questions on Chinese processed foods that I had asked myself years ago and which were reiterated by Joseph Needham. How did they come about? What were their origins? What are the scientific bases of the technologies? How do they compare with processed foods developed in the West? But new questions had already entered my mind even before the work began. To what extent were Chinese processed foods transmitted to her neighbours? Did Chinese food technology have any influence on the development of the food systems of the West? What are the nutritional value of the processed foods? What is the nutritional efficacy of a traditional Chinese diet?

To provide an adequate background for the consideration of these questions, the book begins, in the Introduction, with a survey of the food resources in ancient China, and an account of how the food materials were prepared, cooked and presented for consumption. This is followed by a review of the Literature and Sources used in this study. From there we begin our exploration of traditional Chinese food processing technology. Our first topic is Fermentation technology, the production of alcoholic drinks in their various manifestations, wines from grains, red wine, distilled wine, medicated wines, and wines from fruits, honey and milk. Included also is a comparison of the very different technologies for converting grains into alcoholic drinks in East Asia and in the West, and an explanation of the reason for this divergence. The account ends with the production of vinegar from wine. In the next topic, we go on to the processes by which soybeans are converted by biological, physical, chemical or microbial methods into palatable and nutritious food products. Perhaps the most striking impression one gets from these accounts is the remarkable role a culture of common grain moulds, of the families *Aspergillus*, *Rhizopus* and *Mucor*, known as *chhü* 麴 has played in the processing of foods in China, a situation without parallel among the food cultures of the world. The influence of the grain moulds is seen even in the next chapter when we consider various technologies for the Processing and Preservation of a variety of foodstuffs, such as the making of pickled meat and vegetables, fermented fish sauces, salted fish and meat, fruit preserves, vegetable oil, malt sugar, starch, noodles and other pasta foods, gluten and the use of cold storage in food preservation. In some of the examples, the role of the grain moulds is augmented by the activity of lactic acid bacteria, which are, of course, well known in the food technology of the West. The next topic is the processing of tea, which has undergone a series of changes since it was first prepared as a drink before or during the Han Dynasty. What came as a complete surprise is the discovery that the tea most widely consumed today, i.e. fully fermented tea, known as 'black tea' in the West but 'red tea' (*hung chha* 紅茶) in China, did not exist until about 1840. What was called 'black tea' in maritime trade by tea merchants of the East India Company from about +1720 to 1840 was actually a partly fermented or *oolong* tea. The discussion on tea ends with a consideration of the effects of tea drinking on health as understood by the Chinese, some of which have turned out to be quite acceptable on the basis of modern scientific studies. The chapter on Nutrition is focussed on the natural history of nutritional deficiency diseases in China and how they were treated by dietary means. The volume concludes with a number of Reflections on the overall process of discovery, development and utilisation in food technology in China, its failures as well as its triumphs, and how the grain moulds have surreptitiously crept their way into the technology used today in the manufacture of many familiar processed foods seen on modern grocery shelves around the world.

Thus, in a nutshell, this work deals mainly with developments in Fermentations and Food Science with a brief foray into Nutrition during the historical period in China from antiquity to the 19th century. A great deal of information is available in the classical and mediaeval literature, although much of it is widely scattered and not easily accessible. But for two issues which Needham had raised in our discussions, I have had to explore developments far back into the prehistoric past. For these I had no choice but to rely entirely on the archaeological record which has shifted significantly through new finds even during the span of time that this work was undergoing preparation. The first is the fermentation of cereal grains into alcoholic drinks, East and West. Why did the Chinese replace sprouted grains by a culture of grain moulds as the saccharifying agent in the conversion of grains to sugar? Why did the West fail to discover the bountiful saccharifying activity of the grain moulds? It is impossible to discuss this issue without considering how grains were processed in China and the West, respectively, in the early Neolithic age at the dawn of the agricultural revolution. The second is why milk and milk products did not became a staple part of the diet in China even though dairy animals were reared by the Neolithic Chinese? It is impossible to discuss this problem without going back to the early stage of animal domestication and the origin of a nomadic pastoral way of life, West and East. Although I have included an answer for only the first of the two issues in this work, the forays into the distant past have reinforced the notion that many of our cherished dietary practices, habits and attitudes may have had a longer history than is generally recognised.

This volume is dedicated to the memory of three pioneers in the study of Chinese food science, culture and nutrition. My debt to Shih Shêng-Han needs no reiteration. It is my misfortune that I never had a chance to meet Shinoda Osamu 筱田統 while he was alive. The depth and breadth of his scholarship and the freshness of his ideas never failed to impress me as I delved into his wide-ranging contributions to the field. A translation of a collection of his essays, under the title, *Chung-kuo shih-wu shih yen-chiu* 中國食物史研究 (Studies on the History of Chinese Foods), has been a constant companion soon after it was published. I sorely missed the support and counsel of Lu Gwei-Djen that I had enjoyed during the last few years of her life. She took great pains to give me useful answers to the many questions I had put to her, and continued to feed me material that she thought would be of interest for this Section even in the last few months before she passed away.

Many friends and colleagues have contributed to the substance of the present volume. Those who have read parts of it in draft form and provided comments and advice and steered me to publications that I had missed are listed at the end of this note. But I do need to single out for special thanks several colleagues from this list who have put in a large amount of time and effort to help me in this endeavour. The late Wu Te-To 吳德鐸 of Shanghai went out of his way to provide me with a valuable array of pertinent publications from China, including all the newly annotated editions of the classics of food literature published in the last two decades by Commerce Publishers, Peking. From 1984 to his untimely death in 1992 we had exchanged a steady stream of letters on many issues, including the origin of distilled wine, relevant to the content of this volume. Ishige Naomichi 石毛直道 of Osaka generously sent me many of his books and publications on food culture and technology in East Asia as well as reprints of articles from the Shinoda Collection at the National Museum of Ethnology. He was my principal source of information and advice on fermented fish products and filamentous noodles. He graciously answered all my questions, some of which must have required a considerable amount of his time. William Shurtleff of Lafayette, California, was my principal consultant on the processing of soybeans. He has collected probably the world's largest data base on soyfoods, and was ever ready to search through it for information I needed. Many useful publications were the gift of Hung Kuang-Chu 洪光注 of Peking and Françoise Sabban of Paris. Finally, E. N. Anderson of Riverside, California, kindly read through the entire draft with a fine toothcomb, and offered hundreds of ameliorations and corrections.

Many other colleagues who had not read any part of the draft had also provided me with valuable publications and helpful advice. Foremost among them is Hu Daojing 胡道靜 of Shanghai. It was through his good offices that I obtained a copy of Miao Chhi-Yü's admirable edition of the *Chhi Min Yao Shu* from the Agricultural History Centre of Nanking University. Others in this group are Li Jingwei 李經緯, Xi Zezong 席澤宗, Zheng Jinsheng 鄭金生, Zhong Xiangju 鐘香駒 and Zhou Jiahua 周嘉華 of Peking; Cao Tianqin 曹天欽, Xie Xide 謝希德, Ma Chengyuan 馬承源, Ma Boying 馬伯英 and Qian Wen 錢雯 of Shanghai; Zhong Xiangchong 鐘香崇 of Loyang; Chen Wenhua 陳文華 of Nanchang; Chen Jiahua 陳家驊 of Fuzhou; Y. C. Kong 江潤祥 of Hong Kong; Tanaka Tan 田中談 of Osaka; Ho Peng Yoke 何丙郁 and Delwen Samuel of Cambridge, England; Ti Li Loo 陸迪利 and Ann Gunter of Washington, DC; Tai-Loi Ma 馬泰來 of Chicago; Hui-Lin Li, Naomi Miller and Carmen Lee of Philadelphia; Frank Hole and Anne Underhill of New Haven; David Heber of Los Angeles, Joseph Chang of Simi Valley and Ida Yu of San Leandro, George Amelagos of Gainsville; and Cyril Robinson of Carbondale, Illinois. To this list I should add the colleagues who have helped me with translations from the Japanese, namely Rao Pingfan, Chen Jiahua, Ushiyama Terui 牛山代輝, Lowell Skar, Ishige Naomichi and Ueda Seinosuke. Jeon San Woon performed the same service for Korean. To all the the scholars mentioned here I offer my heartfelt thanks.

I take pleasure in acknowledging the debt I owe to the Library of the National Science Foundation and the Burke Branch of the Alexandria Library, which, over the years have enabled me to borrow hundreds of books from libraries all over the United States on inter-library loan. In particular I wish to thank John Moffett, Librarian of the East Asian History of Science Library at the Needham Research Institute, his colleagues Gao Chuan, Tracy Austin and Sally Church, and his predecessors Liang Lien-Chu and Carmen Lee for the time and effort they have expended to locate and copy the materials I needed.

Although I searched for establishments that still practice the art of food processing with traditional equipment during my travels in China in the 1980s and 90s, I was unsuccessful in locating any of them in the cities I visited. The best I could find were factories that follow traditional processing methods but carry them out with modern equipment. These I visited with pleasure and profit. They include:

Shanghai Fermentation Plant No. 6 上海釀造六廠
    Soy sauce and soy paste
Shanghai Chinese Wine Fermentation Factory 上海中國釀酒廠
    Rice wines, fruit wines, distilled wines
Chiang-nan Beer Factory, Shanghai 江南啤酒廠
    Beer, (Brick Ferment), distilled wine
Tou fu Production Workshop, at the Fukien Agricultural University, Foochow
    Tou fu is made fresh everyday
Foochow Winery No. 1 福州第一酒廠
    Yellow wine, using the red *ferment*
Foochow Fermentation Plant 福州釀造廠
    Soy sauce and soy paste, fish sauce
Ku-thien Red *Ferment* Production Plant 古田紅麴廠
    Red *Ferment*, supplied to all manufacturers in Fukien

My thanks go to Wu Te-To for arranging all the visits in Shanghai, to Chen Jiahua for the *tou fu* workshop and Foochow Winery No. 1, and to Chen Jiahua and Rao Pingfan for the Fermentation Plant in Foochow and the Red *ferment* plant in Ku-thien and to the personnel of these facilities who patiently explained to me the details of the processes every step of the way. These visits have enhanced my understanding of how the efficiency of the traditional processes have been increased by the application of modern science and technology. In this connection I wish also to thank Chen Wenhua for organising the memorable trip we took together to Ta-huting, Mi-hsien to see the remarkable Eastern Han wall mural which describes the preparation of *tou fu* in great detail.

When this project began in 1985 I was a Program Director for Biochemistry at the National Science Foundation (NSF), Washington, DC. I spent all the time I could spare on it, in the evenings, on weekends and on holidays, as well as 10 per cent of my official time on weekdays with the blessing of the Foundation. In 1988 I received a six month sabbatical for the project. I wish to take this opportunity to thank the NSF for the support I enjoyed until my retirement in 1990. Progress should have been much faster after my retirement, but a new obligation intervened. From 1990 to 1994 I served as a part-time Deputy Director of the Needham Research Institute in Cambridge. While it allowed me to take advantage of the facilities of the Library there, it also meant that a great deal of my time was diverted away from the writing of this book, and progress remained slow. It was not until the end of 1995 that a preliminary draft of the whole work was completed. Unfortunately, in early 1996 disaster struck. I had to undergo open heart surgery which effectively incapacitated me for more than four months. But fortunately, my recovery was uneventful and soon I was able to continue with the work. It is, therefore, with a profound sense of joy and relief that I realise the project, which has dominated my life for more than ten years, is at long last drawing to a close.

Those who have kindly read sections of this book in draft form and offered many useful suggestions are:

Anderson, E. N.   *Riverside, CA*
Aronson, Sheldon   *Queens, NY*
Bedini, Silvio   *Washington, DC*
Blue, Gregory   *Victoria, BC*
Bray, Francesca   *Santa Barbara, CA*
Chang, K. C.   *Cambridge, MA*
Cullen, Christopher   *Cambridge, UK*
Chou, Marilyn   *Yorktown, NY*
Engelhardt, Ute   *München, Germany*
Golas, Peter   *Denver, CO*
Guo Fu   *Peking, China*
Hong Guangzhu   *Peking, China*
Hsu, Cho-yun   *Pittsburgh, PA and Hong Kong*
Huang Shijian   *Hangzhou, China*
Ishigi, Naomichi   *Osaka, Japan*
Liu Zuwei   *Shanghai, China*
McGovern, Patrick   *Philadelphia, PA*
Métailié, Georges   *Paris, France*
Needham, Joseph[23]   *Cambridge, UK*
Powell, Marvin   *DeKalb, IL*
Rao, Pingfan   *Foochow, China*
Robertson, William van   *Monterey, CA*
Sabban, Françoise   *Paris, France*
Shurtleff, William   *Lafayette, CA*
Simoons, Frederick   *Spokane, WA*
Tsien, Tsuen-Hsuen   *Chicago, IL*

---

[23] Deceased, 24 March 1995.

Ueda, Seinosuke   *Kumamoto, Japan*
Wagner, Donald   *Cambridge, UK*
Wu, Te-To[24]   *Shanghai, China*

While their advice and counsel have been invaluable, I alone am responsible for any deficiencies and errors that remain in the text.[25]

Finally, my deepest gratitude goes to the late Joseph Needham, who gave me the inspiration and encouragement to embark on this adventure, and to my wife Rita, whose loving care and constant support have helped to make it come true.

H. T. Huang
Alexandria, VA

---

[24] Deceased, 29 February 1992.
[25] Unless stated otherwise, all English versions of Chinese passages cited in the text are translated by me; I alone am responsible for any errors in the translations.

40. BIOCHEMICAL TECHNOLOGY
(Fermentations and Food Science)

(a) INTRODUCTION

In other parts of this series we have traced the origin and development of agriculture, forestry and animal husbandry in China.[1] In this Section, we shall extend our exploration of the realm of biological applications into the neighbouring domain of food science and technology.[2] We shall consider how some of the commodities generated by the agricultural system are processed and converted into specialised articles of food and drink which we have come to regard as characteristic components of the Chinese diet. Among such processes, the most interesting biochemically are the fermentations used in the making of rice wine, vinegar, soy paste, soy sauce and related fermented products. They are, indeed, ancient prototypes of modern biotechnology. Others may require simple mechanical, environmental or chemical manipulations such as those employed in the preparation of bean sprouts, bean curd and wheat gluten. An intermediate case is provided by tea,[3] the preparation of which may or may not include a 'fermentation'. These products supply either basic nutrients needed for growth and maintenance of health or adjuncts that inject flavour and spice into a diet that would otherwise be unexciting and bland. We shall delineate the origin and development of these products, analyse the scientific basis of the processes involved in their preparation, and compare, where possible, the Chinese technology with the equivalent system extant in the West and in other cultures.

We shall also touch briefly on how the Chinese have utilised foods as a means to control and cure diseases. As indicated by the following quotations from the ancient classic of internal medicine (Figure 1), *Huang Ti Nei Ching, Su Wên* 黃帝內經素問 (The Yellow Emperor's Manual of Corporeal Medicine, Questions and Answers), c. −2nd century, they have recognised early on that an unbalanced diet leads to disease and poor health, and found that certain types of food could be used to cure specific ailments:[4]

> When the liver suffers from an acute attack one should quickly eat sweet food in order to calm it down.

Unless stated otherwise, all English versions of Chinese passages cited in the text are translated by the author; he alone is responsible for any errors.

[1] Cf. Bray, *SCC* Section 41, Vol. VI, Pt 2 (Agriculture); and Daniels and Menzies, *SCC* Section 42, Vol. VI, Pt 3 (Agricultural Technology and Forestry) and the forthcoming volume on Zoology and Animal Husbandry.

[2] The close relationship between 'food science' and 'nutrition' is formally acknowledged by many modern universities, especially in the USA. Thus, we find the two subjects lumped together in one single academic department, namely 'Food Science and Nutrition', in such well-known institutions as Purdue University, Massachusetts Institute of Technology, the Universities of Arizona, Florida, Minnesota, etc.

[3] For a delightful account of the aesthetic side of the story of Chinese tea, cf. Blofeld (1985).

[4] Veith (1972), pp. 199–200, mod. auct. The quotation is found in *HTNCSW*, ch. 22, p. 191.

Fig 1. Three Legendary Emperors, Fu Hsi, Shên Nung and Huang Ti, who taught the people Fishing, Agriculture and the Art of Healing. From a Japanese scroll by Seibe Wake, +1798, Veith (1972), cover page.

When the heart suffers from tardiness, one should quickly eat sour food which has an astringent effect.

When the spleen suffers from dampness one should quickly eat 'salty [bitter]' food which has a drying effect.[5]

When the lungs suffer from obstruction of the upper respiratory tract one should quickly eat bitter food which will disperse the obstruction and restore the flow.

When the kidneys suffer from dryness one should quickly eat pungent food which will moisturise them.

Thus, many common food materials and products such as wheat, barley, millet, cow's milk, malt sugar, rice wine, bean sprouts, bean curd, ginger, scallion, grapes and a host of fruits and vegetables have found their way into the pharmacopoeias, where they are accorded as much respectability as many well-established drugs whose therapeutic properties have been authenticated and affirmed through hundreds of years of clinical practice.[6] But before we embark on an examination of the written records and begin to deal with the issues outlined earlier, it would be useful for us to consider the milieu in which these particular processing technologies and nutritional ideas have germinated and grown to become major factors in the evolution of the Chinese diet and culinary system, or simply, the Chinese cuisine.

A cuisine is the distinctive way a people select, prepare, cook, serve and eat food. As defined by modern food anthropologists, it is made up of four components.[7] The first component includes the raw materials available from the native agricultural and animal husbandry system. The second involves the methods used to prepare and cook them. The third comprises the type of spices that are added during or after the cooking to flavour the food. The fourth component embraces the rules that govern the event of eating, such as when to eat, where to eat, who to eat with, how to serve the food, etc. The status of each of the four components in ancient China will be briefly surveyed, especially in terms of their effect on the development of the technologies and nutritional ideas which are to be reviewed and discussed in the main body of this Section.

For the purpose of this survey, ancient China will denote the period from the start of the Chou dynasty to the end of the Han dynasty, that is, from about −1000 to +200, although we may occasionally have need to refer to archaeological evidence dating from the Shang dynasty or earlier and to consult material collected after +200 in the *Chhi Min Yao Shu* 齊民要術 (Important Arts for the People's Welfare), +544. An overview of this period should provide us with a useful frame of reference for our exploration, since by the end of the Han dynasty much of the raw materials and methods of cooking and presenting the food, that would later characterise the full grown Chinese cuisine, were already in place. To facilitate discussion of subsequent

---

[5] In the original statement 'bitter' food is recommended for the spleen. The commentary points out that there are two problems with the text. Firstly, 'bitter' is used twice, while 'salty' is not used at all. Secondly, 'bitter' contradicts a later statement (p. 199) in this chapter which says, 'the colour which corresponds to the spleen is yellow; its proper food is *salty*' (Veith (*ibid.*), p. 205). It would seem that in this case 'bitter' is a copying error and should be replaced by 'salty'.

[6] Cf. for example, *PTKM* chs. 22, 23, 24; *CCIF*, chs. 3 and 4.   [7] Farb & Armelagos (1980), p. 185.

developments we shall divide the remaining centuries of Chinese history into two periods: a mediaeval period, which lasts from the end of the Han to the end of the Yuan (c. +200 to +1368), and a premodern period, which starts from the Ming (+1368) and ends at about +1800.[8] To arrive at a quick concordance of Chinese historical periods with the Western calendar, the reader will find it helpful to consult the Table of Chinese Dynasties, reprinted at the end of the book between the Index and the Romanisation Conversion Tables.

(1) Food Resources in Ancient China

Tell me what thou eatest, and I will tell thee what thou art.[9]

The four basic categories of foodstuffs recognised in ancient Chinese literature, i.e. grains, vegetables, fruits and meat (or animal products) are essentially similar to those used by nutritionists to classify foodstuffs in our own modern age. For example, in the *Huang Ti Nei Ching, Su Wên* (Yellow Emperor's Manual of Corporeal Medicine, Questions and Answers), we are advised that a balanced consumption of the four foodstuffs, grains (*ku* 穀), meat (*ju* 肉), fruits (*kuo* 果) and vegetables (*tshai* 菜) is required to provide all the nourishment that our body needs.[10] This rudimentary scheme was, however, clearly inadequate in designing the fare for the table of a king or a feudal lord. Thus, the legendary master chef, I Yin 伊尹 of the Shang dynasty, added fish (*yü* 魚), harmonising agents (*ho* 和 i.e. spices), and water to the above four basic foodstuffs, as separate categories of food materials in his catalogue of gastronomic provisions for the king's kitchen to be procured from the far-flung regions of the empire.[11] For our survey of food resources, we will be mindful of both the nutritional and gastronomic traditions of China. We will begin with foodstuffs derived from plants, i.e. grains, oilseeds, vegetables, and fruits (including legumes and spices), and end with those taken from animals, i.e. mammals and fowls followed by fish and other aquatic creatures.

(i) *Grains*

Abundant is the year, with much millet, much rice;
But we have tall granaries,
To hold myriads, many myriads and millions of grain.[12]

---

[8] Our division of Chinese history into ancient, mediaeval and premodern periods is similar to that proposed by Hsü Cho-yun (*1991*), pp. 2–6, which is, in turn, based on the division used by the early 20th century Chinese savant Liang Chhi-Chhao (*1925*), ch. 34, p. 25. Briefly, they regard ancient China as the period when Chinese civilisation established itself in the heartland of China, mediaeval China when it interacted vigorously with its neighbours in Asia, and modern China when it interacted with the world as a whole.

[9] Brillat-Savarin (1926), p. xxxiii.; cf. the German proverb 'Mann ist was Mann isst.'

[10] *HTNCSW*, ch. 70, p. 199.   [11] *LSCC*, ch. 14, *Pên Wei* 本味; Lin Phin-Shih ed. (*1985*), pp. 370–4.

[12] The Book of Odes, *Shih Ching*; Mao 279, Waley 156, tr. Waley (1), p. 161. For the convenience of readers who may wish to look up the original Chinese text, all the poems quoted are identified by number according to Mao's traditional sequence (M 000) as well as Waley's numbering system used in *The Book of Songs* (W 000).

Grains or cereals were (and still are) the major food resource in the Chinese diet. The Chinese term for grain, *ku* 穀, however, was applied not only to cereals (Gramineae) but also to such field crops as *shu* 菽, soybean (Leguminosae) and *ma* 麻, generally interpreted as hemp (Moraceae), which were also cultivated for their seeds. Archaeological evidence and oracle bone inscriptions and textual records[13] clearly show that the major grains grown during the Shang dynasty and Western Chou period were *chi* 稷, the foxtail millet, *Setaria italica* (L.) Beauv, *shu* 黍, the broomcorn millet, *Panicum miliaceum* (L.) Beauv., *tao* 稻, rice, *Oryza sativa* L., *mai* 麥 (barley or wheat),[14] and hemp, *ma* 麻, *Cannabis sativa* L. Written characters found on bronze vessels further indicate that the soybean, *shu*, *Glycine max* (L.) Merrill, was already under cultivation in the early years of Western Chou.[15] All six crops are mentioned, some more frequently than others, in the *Shih Ching* 詩經 (the Book of Odes), an anthology of folk songs and ceremonial odes which contain a wealth of information on plant and animal life in China from the −11th to −7th centuries.[16] Indeed, the harvesting of all six crops is conveniently celebrated in one poem, the 'Ode to the Seventh Month', *Chhi Yüeh* 七月, though not in the same stanza. In the sixth stanza, we find the lines:[17]

> In the sixth month we eat wild plums and berries,
> In the seventh month we boil mallows and beans.
> In the eighth month we dry the dates
> In the tenth month we take the rice
> To make with it the spring wine,
> So that we may be granted long life.

and in the next stanza,

> In the ninth month we make ready the stackyards,
> In the tenth month we bring in the harvest,
> Millet for wine, millet for cooking, the early and the late,
> Paddy and hemp, beans and wheat.

In both stanzas 'beans' in the original is *shu* 菽, i.e. soybeans. Millet for wine is *shu* 黍, and millet for cooking, *chi* 稷. In the last line, 'paddy' is actually a translation

---

[13] Chang, K. C. (1977), pp. 26–7 and (1980), pp. 146–9; Li Hui-Lin (1983); Hsu & Linduff (1988), pp. 345–51. For an account of the food resources and methods of cookery in China during the Neolithic Age cf. Yang Ya-Chhang (1994).

[14] *Mai* can be interpreted as either wheat or barley. The true identity of *mai* will be discussed further on p. 27.

[15] Hu Tao-Ching (1963). The proposal of Yü Hsing-Wu (1957) that the Shang oracle bone inscription 荳 is an ancient version of *shu*, has not been accepted by most scholars. For the phylogenetic relationship of the soybean with other species of the genus *Glycine* see Kollipara, Singh & Hymowitz (1997).

[16] Based on textual examination, the eminent scholar Liang Chhi-Chhao (1955), pp. 109–17, concluded that the poems in the *Shih Ching* were composed during a 500 year span ranging from the early days of the Chou dynasty down to the period before the birth of Confucius, i.e. from −11th to −6th centuries. From linguistic evidence, Dobson (1964) has independently arrived at roughly the same conclusion, and dated the *Shih Ching* around −11th to −7th centuries. These dates are accepted by most modern scholars; cf. for example, Yü Guanying (1983), p. 97 and Kêng Hsüan (1974), p. 395.

[17] M 154, (W 159); tr. Waley (1937), mod. auct.

of *ho* 禾, a word denoting grains in general. Since both millets, hemp, soybean and wheat are all examples of grain crops, it would hardly make sense to translate *ho* as 'grain' in the current context. 'Paddy' would appear to be a reasonable rendition since rice is the logical *ho* to fit into this group of crops.

We have so far interpreted *mai* as wheat. This is a convenient simplification, since *mai* is, strictly speaking, a collective term for both wheat (*Triticum turgidum*) and barley (*Hordeum vulgare*), wheat being the 'lesser' or *hsiao mai* 小麥 and barley the 'greater' or *ta mai* 大麥. Thus, we are not absolutely sure whether the *mai* of the poem *Chhi Yüeh* is actually wheat or barley.[18] However, in another poem the *Shih Ching* has used the word *lai* 萊 for wheat and *mou* 牟 for barley, indicating that a clear distinction between wheat and barley was already recognised.[19] We may now conclude with some assurance that panicum millet, setaria millet, rice, wheat (and barley), soybean and hemp were the principal grain crops grown in China during the centuries when the poems of the *Shih Ching* were collected (Figure 2).

For the origin and history of these crops, the reader is referred to the detailed account given by Bray in Section 41.[20] But for our current discussion it will be interesting to know whether the list of principal grain crops garnered from the *Shih Ching* still remains valid by the time of the Han dynasty. To answer this question, we have summarised in Table 1 the information recorded in the *Chou Li* 周禮 (The Rites of the Chou), former Han, on the staple grains (and major livestock) raised in the Nine Provinces (Figure 3) of ancient China.[21] In the original text panicum and setaria millets, rice and wheat are directly listed by name while soybean is indicated by inference, according to Chêng Hsüan's (鄭玄 +2nd century) annotation, as constituents of the cluster of 'five grains' said to have been cultivated in Yü-chou and Ping-chou. Actually, a special term, *wu ku* 五穀 (five grains), has been used frequently in the classical literature to denote figuratively the staple grain crops of the land.[22] The earliest textual reference to *wu ku* 五穀 is found in the Analects of Confucius, *Lun Yü* 論語 (−5th century), when Tze Lu 子路, falling behind his main

---

[18] *Mai* is mentioned in five poems and must, therefore, be a crop of some importance in the early Chou period. In their translations of the *Shih Ching* the classical Sinologists, Legge, Karlgren and Waley have all followed the traditional interpretation and rendered *mai* as wheat. Some modern scholars such as Shinoda Osamu (*1951*, *1987a*) Amano Motonosuke (*1979*), Yü Ching-Jang (*1956a,b*) and Kêng Hsüan (*1974*) have, however, taken the view that wheat could not have been introduced into China before the −2nd or −3rd century. We tend to agree with the traditional position while conceding that wheat did not become a major crop until about −300, see Bray, SCC Section 41, Vol. VI, Pt 2, pp. 461–3. This problem will be discussed further on pp. 462–3. The wheat of ancient China is most probably a tetraploid durum wheat (*Triticum turgidum*), which was identified as the species found among the grains at Ma-wang-tui.

[19] Cf. W 153, 154 (M 275, 276).

[20] Cf. Vol. VI, Pt 2, pp. 434ff. and Li Fan (*1984*), pp. 22–65. Cf. also Hsu Cho-yun (*1980*) for the state of agriculture during the Han.

[21] *Chou Li*, ch. 8, *Chi Fan Shih* 職方氏 (Director of Regions), p. 344.

[22] The ancient Chinese seemed to have made a fetish of endowing a group of similar entities, concrete or abstract, with a numerical designation. Thus we find in the *Shih Ching* references to *pai ku* 百穀, hundred grains (M 154, 212, 277; W 159, 155, 162), which presumably is a collective term for all cultivated crops, including grains, fruits and vegetables. The *Chou Li* often mentions *chiu ku* 九穀, nine grains (ch. 1, *Ta Tsai* 大宰; ch. 4, *Lin Jên* 廩人, *Tshang Jên* 倉人), which according to Chêng Hsüan's annotation mean *shu* (panicum millet), *chi* (setaria millet), *liang* (another variety of millet?), *tao* (rice), *ma* (hemp), *ta tou* (soybean), *hsiao tou* (small beans), *hsiao mai*

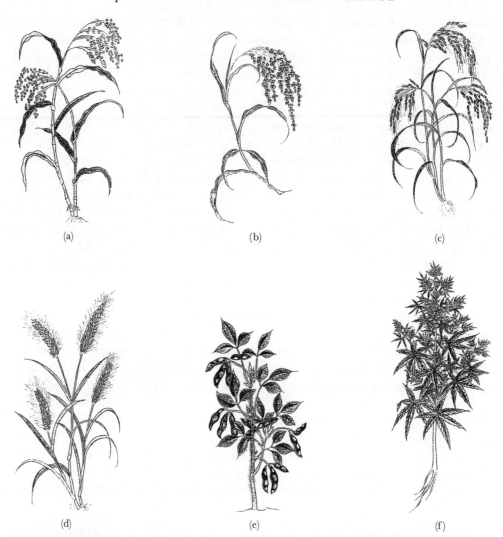

Fig 2. The staple grains of ancient China: Panicum millet, Setaria millet, Rice, Barley, Wheat, Soybean and Hemp, from *Chin Shih Khun Chhung Tshao Mu Chuang (CSKCTMC)* 金石昆蟲草木狀: (a) *Shu*, Panicum millet; (b) *Chi*, Setaria millet; (c) *Tao*, Rice; (d) *Hsiao Mai*, Wheat; (e) *Shu*, Soybean; (f) *Ma*, Hemp.

(wheat), and *khu* (wild rice, grain from *Zizania caduciflora*). In his *Chiu Ku Khao* 九穀考 (Study of the Nine Grains), Chhêng Yao-Thien (c. 1805) has concurred with this view except for *chi* 稷, which he thought was not millet, but rather sorghum, well known in later ages in China as *kao-liang* 高粱. More recently, Hsia Wei-Ying (*1979*), pp. 126–33, re-evaluated the problem and interpreted the ancient nine grains as *shu* millet, *chi* millet, rice, hemp, soybeans, small beans, wheat, barley and wild rice. We also find references to *liu ku*, six grains (*Chou Li*, ch. 1, *Shang Fu* 膳夫), and *pa ku*, eight grains (*Hsing Ching*, 星經, Star Manual; see Vol. 3, p. 703). But by far the most popular term for staple grains is *wu ku*, five grains, presumably because of the special significance of the number 'five' in the Chinese lexicon. Shinoda Osamu (*1987a*), p. 7, has found that the term *wu ku* is used no less than seventy-five times in the pre-Han literature, indicating that it was a common expression in ancient China. But the popularity of 'five' apparently knew no bounds. Besides five grains, in this Section we will come across other familiar clusters of five, such as five elements (or phases), five senses, five flavours, five organs, five sounds, five colours etc.

Table 1. *The staple grains and major livestock of ancient China*[a]

Province	Location	Staple grains	Livestock
Yang-chou 揚州	Lower Yangtze & South	Rice	Bird, beast
Ching-chou 荊州	Middle Yangtze & South	Rice	Bird, beast
Yü-chou 豫州	Honan & Huai valley (?)	Five grains[b]	Six beasts[c]
Chhing-chou 青州	E. Shantung	Rice, wheat	Chicken, dog
Yen-chou 兗州	N. Honan, W. Shantung, S. Hopei	Four grains[d]	Six beasts[c]
Yung-chou 雍州	Shensi, E. Kansu	Panicum & setaria millets	Ox, horse
Yu-chou 幽州	S. Liaoning, N. Shantung, N. Hopei	Three grains[e]	Four beasts[f]
Chi-chou 冀州	S. Shansi	Panicum & setaria millets	Ox, sheep
Ping-chou 并州	N. Shansi, N. Hopei	Five grains[b]	Six beasts[g]

[a] Based on the account given in the *Chou Li* 周禮, Chapter 8, *Hsia Kuan Hsia* 夏官下 (Summer Official, lower section) on the duties of the *Chih Fang Shih* 職方士 (Director of Regions), pp. 344–50. The nine provinces are listed in the order they occur in the text. The key geographical feature is the bend formed by the Huang Ho as it flows south and then turns east towards the China Sea. According to the text, Chi-chou lies within the bend (*ho nei* 河內), Yen-chou, to the east (*ho tung* 河東), and Yü-chou, to the south (*ho nan* 河南) of the river. Yang-chou is south-east (*tung-nan* 東南) of this core region, Ching-chou, directly south (*chêng nan* 正南), Chhing-chou, directly east (*chêng tung* 正東), Ping-chou, directly north (*chêng pei* 正北), and Yu-chou, to the north-east (*tung-pei* 東北). Lastly, Yung-chou is directly west (*chêng hsi* 正西) of the River. Their locations are identified further in Lin Yin's footnotes to the text (pp. 346–50).
  According to the annotations of Chêng Hsuan 鄭玄 (+127–200), the identity of the five grains, six beasts etc. are:
[b] Five grains: panicum and setaria millets, soybean, wheat and rice.
[c] Six beasts: horse, ox, sheep, pig, dog and chicken.
[d] Four grains: panicum and setaria millets, rice and wheat.
[e] Three grains: panicum and setaria millets, and rice.
[f] Four beasts: horse, ox, sheep and pig.
[g] Five beasts: horse, ox, sheep, dog, and pig.
N.B. it should be noted that in this context wheat can mean either wheat or barley.

party, asked an old man whether he had seen his master, Confucius. The old man replied, 'Your four limbs are unaccustomed to toil; you cannot distinguish the five kinds of grain (*wu ku*): who is your master?'[23] In the −3rd century Mencius stated that during the time of the great flood, 'the five grains (*wu ku*) did not ripen' and that 'Hou Chi 后稷 taught the people how to cultivate land and the five kinds of grain (*wu ku*).'[24] Thus, we see that *wu ku* was already a common expression during the times of Confucius and Mencius. Since then it has become synonymous with 'staple grains' in the Chinese language, and has persisted as a popular idiom through the centuries among the common people as well as the literati.[25]

[23] *Lun Yü*, Legge, tr. (*1861*), p. 335. [24] *Meng Tzu*, Legge, tr. (*1895*), pp. 250–1.
[25] Li Chhang-Nien (*1982*) has pointed out that even in China today, after a particularly bountiful harvest, we may still see the local peasantry celebrate the occasion by parading banners with the well-worn couplet, *Wu ku fêng têng, Liu chhu hsing wang* 五穀豐登，六畜興旺, 'the five grains reaching new peaks; the six livestock fat and prosperous'. The couplet may still be displayed regardless of whether three, four or six grains have actually been harvested, or even when one of the crops is not a grain, but a tuber such as the sweet potato.

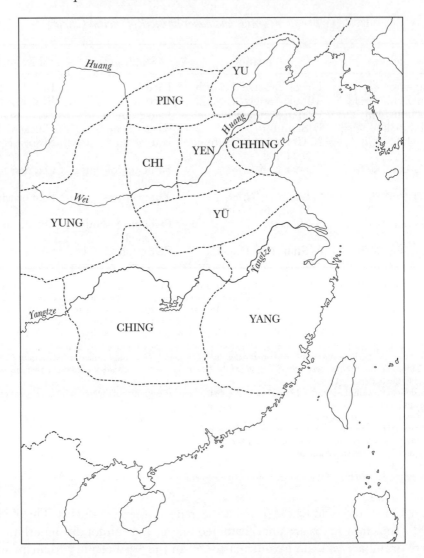

Fig 3. The Nine Provinces of Ancient China as outlined in the *Chou Li* (Lin Yin ed. pp. 344–50).

But what about *ma*, or hemp? Was it also regarded as a member of the group of 'five grains' or 'staple grains'? The evidence suggests that it was. In fact, the earliest statement extant on the identity of *wu ku*, found in the *Fan Tzu Chi Jan* 范子計然 (The Book of Master Chi Ni), −4th century, says:[26] 'The grains that thrive in the East are wheat (or barley) and rice, in the West, *ma* (hemp), in the North, soybean, and in the Centre, *ho* 禾 (presumably, millets).' Even Chêng Hsüan himself, in

---

[26] Cf. Li Chhang-Nien (*1982*); *Chhu Hsüeh Chi* 初學記, ch. 27.

annotating another segment of the *Chou Li*, recognised *wu ku* as hemp, panicum and setaria millets, wheat and beans (*ma, shu, chi, mai, tou*).[27] In this case hemp takes the place of rice, which he had already identified as one of the 'five grains' elsewhere in the same book (see Table 1). While we may be puzzled by Chêng Hsüan's lack of consistency, we must realise that he could have been thinking of different parts of the country or different literary citations when he wrote these particular pieces of commentary.[28] In addition, Wang I 王逸, in his annotation (+2nd century) of the *Chhu Tzhu* 楚詞 (Elegies of Chhu State), c. −300, considered *wu ku* to be rice, setaria millet, wheat, beans and hemp (*tao, chi, mai, tou, ma*).[29] These references leave no doubt that hemp was regarded, at least by some Han scholars, as one of the staple grains of ancient China, provided that hemp is the correct identification for *ma* in the context of *wu ku*.

We cannot end this brief discussion without mentioning yet another ancient definition of *wu ku* found in the classical literature. In the *I Chou Shu* 逸周書 (Lost Records of the Chou Dynasty), c. −245, it is stated that:[30] 'Wheat resides in the East, panicum millet in the South, rice in the Centre, *su* 粟 (millet) in the West and soybean in the North.' This interpretation introduces us to a new word for millet, *su*, which had apparently, by the end of the Han dynasty, acquired a level of importance rivalling that of the ancient *shu* and *chi*.

Actually, *su* as well as another grain, *liang* 梁 are also mentioned in the *Shih Ching*,[31] and occasionally in other pre-Han classics.[32] *Liang*, and, by association, *su*, have been interpreted as the newer, higher quality *Setaria italica* Beauv. var. *maxima*, which had emerged during the Warring States period and surpassed the smaller eared, lower yield *chi*, *Setaria italica* var. *germanica* Trin., in importance by the beginning of the Han dynasty.[33] The status of *liang* remains equivocal; some scholars believe that it may be an ancient name for sorghum, commonly known as *kao liang* 高梁.[34] But there is no question that after the Han dynasty, for whatever reason, *su* gradually evolved into the preferred name for setaria millet in China and has

---

[27] *Chou Li*, ch. 2, *Chi I* 疾醫, p. 47.
[28] This inconsistency is not restricted only to the *Chou Li*. In the *HTNCSW* the five staple grains (*wu ku*) are described in chapter 4 as *mai, shu, chi, tao* and *tou*, i.e. wheat, panicum millet, setaria millet, rice and soybean, but in chapter 5 as *mai, ma, chi, tao* and *tou*, i.e. wheat, hemp, setaria millet, rice and soybean.
[29] *Chhu Tzhu*, ch. 10, *Ta Chao* (The Great Summons). This interpretation of the identity of the five grains is accepted by modern annotators such as Fu Hsi-Jen (*1976*), p. 174 and Tung Chhu-Phing (*1986*), p. 273.
[30] *I Chou Shu*, cited in *Chhu Hsüeh Chi*, ch. 27. See Hu Hsi-Wên (*1981*), p. 25.
[31] For *su* cf. M 187 (W 103); *liang*, M 187, 121, 211 (W 103, 161, 151).
[32] Cf. *Mêng Tzu*, Legge, tr. (*1895*), p. 463; *Kuan Tzu*, ch. 5 (*Chung Ling* 重令), p. 53; *Mo Tzu*, ch. 9 (*Shang Hsien Chung* 尚賢中), p. 46.
[33] Yü Ching-Jang (*1956a,b*), Chao Kang (*1988a,b*), p. 132; cf. also K. C. Chang (1977), p. 26.
[34] Yü Ching-Jang *ibid.*; The fact that carbonised remains of sorghum have been found in a number of Chou and early Han tombs have, however, led some scholars, such as, Li Chhang-Nien (*1982*), Li Fan (*1984*) p. 65, and Li Yü-Fang (*1986*) pp. 267ff. to claim that *liang* is not just a variety of setaria millet, but rather the ancient name for *kao liang*, *Sorghum vulgare* Pers. even though sorghum did not become a popular cereal crop in China until after the Yuan dynasty (see *SCC* Vol. VI, Pt 2, pp. 449–51). This view has been refuted by Yü Ching-Jang (*1956a*), Miao Chhi-Yü (*1984*) and Huang Chhi-Hsü (*1983a*), but the issue remains unclear, see Ho Ping-Ti (1975), pp. 380–4. The case for *kao liang*, however, has been strengthened recently by the discovery by Li Fan *et al.* (*1989*) of fossilised grains of sorghum in a Neolithic site in Kansu, dated at about −3000.

remained so until the present day. To add to the confusion, several other terms, *shu* 秫, *chi* 穄, *chu* 秅, *phei* 秠 etc. were also used to indicate different varieties of millets. The bewildering terminology for Chinese millets has been valiantly tabulated and explained by Bray.[35] Fortunately, for our purpose it will not be necessary for us to steer our way through this maze of millet cultivars. For us, it will suffice to recognise that two major types of millets, *shu* (panicum) and *su* (setaria) were cultivated in ancient China.[36]

The above sketchy account of the cultivated grains of ancient China have recently received vivid corroboration from archaeological finds acquired in a series of Han dynasty tombs excavated in the last two decades. Of these the most spectacular and significant discovery is undoubtedly that which is now known as 'Han Tomb No. 1 at Ma-wang-tui 馬王堆' on the eastern outskirts of Chhangsha in Hunan. The owner of the tomb has been identified as the wife of Li Tshang 利蒼, the first Marquis of Tai 軑, who reigned from −193 to −186. She died a few years after −168. The lady has won considerable posthumous fame as a medical celebrity because her body was found to be remarkably well preserved when her coffin was opened after an interment of twenty-one centuries. But from our vantage point she deserves our special gratitude for arranging to have buried with her the richest storehouse, in terms of both quantity and variety, of ancient foodstuffs that have ever been discovered in China. Moreover, her burial remains include a collection of inscriptions on bamboo slips which list the names of additional foodstuffs not buried with her, and, more importantly, provide valuable information on the methods that were

---

[35] Bray, *SCC* Vol. VI, Pt 2, pp. 437–41.

[36] More than a millennium had elapsed from the early years of the Chou to the end of the Han dynasty. As we have emphasised in a previous section (*SCC* Vol. VI, Pt 1, pp. 463–71), by the +3rd and +4th centuries, many of the names of plants (and animals) mentioned in the *Shih Ching* and other early works had become quite obscure. The situation is compounded by the tendency of ancient Chinese writers to coin new names for different varieties of the same crop. Thus, it has remained ever since a continuing 'task of scholarship to clarify and elucidate the original meanings' of such plants (*ibid.*, p. 463). Surely one of the most vexing and intractable issues in this task is the ancient meaning of *shu* 黍, *chi* 稷, and their relationship to *su* 粟, (cf. *SCC* Vol. VI, Pt 2, pp. 438–41). Archaeological evidence clearly shows that two species of millets were major crops in ancient China, a *Panicum* and a *Setaria*, cf. Huang Chhi-Hsü (*1983a*). Mediaeval and modern scholars have always agreed that *shu* is panicum and *su*, setaria. But no unanimity has existed concerning the identity of *chi*.

What, then, is *chi*? One school holds that *chi* is simply a non-glutinous type of *shu* (panicum). This has been the view of eminent pharmacist–scholars, starting from Thao Hung-Ching (Liang) through the centuries to Su Kung (Thang) and Li Shi-Chên (Ming). The other school regards *chi* as synonymous with *su* (setaria). This is the position found in standard agricultural treatises, starting with *Chhi Min Yao Shu* (N. Chhi), down to *Nung Sang I Shih Tsho Yao* (Yüan) and *Nung Chêng Chhuan Shu* (Ming). As pointed out by Chang Tê-Tzu (1983), modern scholars, both in China and abroad, are still divided on the issue. For example, Chhi Ssu-Ho (*1981*), Yü Ching-Jang (*1956a,b*), Tsou Shu-Wên (*1960*), Hsia Wei-Ying (*1979*), Tsan Wei-Lien (*1982*), Chao Kang (*1988a*) etc. are of the opinion that *chi* is an older name for *su*, while Liu Yü-Chhuan (*1960*), Kêng Hsüan (*1974*), Ping-Ti Ho (*1975*), Chhên Wên-Hua (*1981*), Wang Yü-Hu (*1981a*), Hui-Lin Li (*1983*), etc., accept the idea that *chi* is simply a non-glutinous variety of *shu*. In the hope of breaking the impasse, Yü Hsiu-Ling (*1984*) made a comprehensive and thorough analysis of all the relevant literature on the subject. He has shown, we believe convincingly, that on balance the evidence is in favour of the view that *chi* and *su* both denote the same species of millet, *Setaria italica*, a conclusion with which we can comfortably agree.

In passing, it may be worth mentioning that another view of *chi* is that it is synonymous with *liang* 粱, *liang* being the ancient name for sorghum, *kao liang* 高粱. This notion was first advanced by Wu Jui of the Yüan dynasty (*PTKM*, ch. 23, on *chi*), and later amplified by Chhêng Yao-Thien (*1805*) during the Chhing dynasty. It is still being considered seriously by some scholars today, for example, Yang Ching-Shêng (*1980*).

Fig 4. Food remains discovered in Han Tomb No. 1 at Ma-wang-tui. *Hunan sheng po-wu-kuan* (1973) II, Pl. 11, p. 9, lower part only.

used for preparing and cooking the raw food (Figures 4 and 5). More will be said later about the foodstuffs stored or described on bamboo slips. For now let us take a look at the grains that have been identified in Han Tomb No.1 at Ma-wang-tui:[37]

Shu 黍, broomcorn millet, *Panicum milliaceum*
Chi 稷, foxtail millet, *Setaria italica*
Hsiao mai 小麥, wheat, *Triticum turgidum*
Ta mai 大麥, barley, *Hordeum vulgare*
Tao 稻, rice, *Oryza sativa*
Shu 菽, soybean, *Glycine max*
Ma 麻, hemp, *Cannabis sativa*

This list certainly agrees fully with everything we have said so far about the identity of the staple grains in ancient China.

Although millets were the earliest grains domesticated in Neolithic times in northern China, their importance, relative to wheat, rice and soybeans, had probably declined by the start of the Han dynasty. Nevertheless, their special status in

---

[37] A detailed account of the studies conducted on the food remains found in Han Tomb No. 1 at Ma-wang-tui was published by the Hunan Agricultural Academy, cf. *Hunan nung-hsüeh-yuan (1978)*.

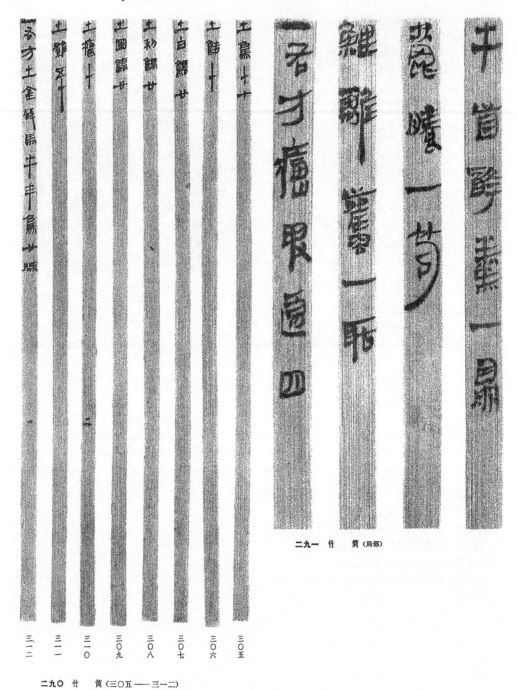

Fig 5. Examples of food items recorded on bamboo slips found in Han Tomb No. 1 at Ma-wang-tui. *Hunan shêng po-wu-kuan* (1973) II, p. 243, Pl. 290.

the history of Chinese agriculture was enshrined by scholars such as Chêng Hsüan, who, in the listing of the five grains often placed *shu* and *chi* ahead of wheat, soybean and rice. After the Han dynasty the setaria millet (*su* 粟 or *chi*) increased its importance at the expense of panicum, and today only setaria millet (popularly called *ku tzu* 谷子) is grown on a significant scale in China.[38] Thus, when we refer to the processing of millet in later chapters, we shall probably be talking of setaria and varieties thereof rather than panicum.

As stated earlier we have interpreted the character *mai* as wheat although it is a term applied to both wheat and barley, wheat being *hsiao* 小 (lesser) *mai* and barley, *ta* 大 (greater) *mai*. Both are autumn-sown and spring-harvested cereals and may have come together to China from the Near East at the end of the Neolithic period.[39] It is curious that in the pre-Han literature we often come across the word *mai*, but seldom *hsiao mai* or *ta mai*. The only pre-Han reference to *ta mai* that has been found is in the *Lü Shih Chhun Chhiu* 呂氏春秋 (Master Lü's Spring and Autumn Annals), −3rd century.[40] The earliest reference to *hsiao mai* is in the *Fan Shêng-Chih Shu* 氾勝之書 (The Book of Fan Shêng-Chih), −1st century.[41] Yet the archaeological evidence from Ma-wang-tui shows conclusively that wheat and barley were both important grains in the early years of the Han dynasty.[42] The role of wheat in food processing will be treated in Chapters (*c*), (*d*) and (*e*), and barley in Chapters (*c*) and (*e*).

Rice, the most important grain crop in China today, was the first cereal domesticated by the Neolithic Chinese.[43] Its special role in the development of the Chinese culinary system will be discussed later in this chapter. Its use in the production of rice wine will be treated in Chapter (*c*).

Millets, wheat (and barley) and rice are all cereals which supply mostly carbohydrates and a lesser amount of protein to the diet. The next grain, soybean, is not a cereal but a legume. It is an excellent source of protein and fat. It was probably domesticated in China at the beginning of the Chou dynasty (about −1000).[44] We shall have a good deal to say about the processing and nutritive value of soybean

---

[38] *SCC* Vol. VI, Pt 2, p. 443. Recent information from China, see *Shansi nung-yeh kho-hsüeh yuan* (*1977*), shows that in 1971 *ku tzu* was grown on about 5% of the total land area devoted to crop production in China. But in North China it could account for as much as 15–20% of the acreage under cultivation.

[39] Bray, *SCC* Vol. VI, Pt 2, pp. 459–64.

[40] *LSCC*, *Jên Ti* 任地 (ch. 26, segment 4), p. 844.; cf. Hsia Wei-Ying (*1956*), p. 47.

[41] *FSCS*, cf. Shih Shêng-Han (*1959*), p. 10.   [42] *Hunan nung-hsüeh-yuan* (*1978*), pp. 4–5.

[43] Results of the recent archaeological excavations at Chia-hu 賈湖 in Honan, Phêng-thou-shan 彭頭山, Pa-shih-tang 八十壋 and Yü-chhan-yen 玉蟾岩 in Hunan and other sites have revolutionised our ideas about the origin of agriculture in China. They show clearly the first cereal cultivated is rice, and that its domestication most likely occurred in the middle reaches of the Yangtze valley at about 11,500 BP, see *Honan shêng wên-wu yen-chiu-so* (*1989*), *Honan shêng wên-wu khao-ku yen-chiu-so* (*1990*, *1996*), Hsieh Chhung-An (*1991*), Hsiang An-Chhiang (*1993*), Liu Chih-I (*1994*, *1996*), Chang Wên-Hsü & Phei An-Phing (*1997*), Yen Wên-Ming (*1997*), Bruce Smith (*1995*), ch. 6 and the news report of Normile (*1997*) for additional details and references. Thus, the older view, see Bray, *SCC* Vol. VI, Pt 2, pp. 481–9, that the cultivation of rice originated in the piedmont zone of Southeast Asia is no longer tenable. In fact, the cultivation of rice along the Yangtze is considerably older than the cultivation of the millets in North China which date back about seven to eight thousand years. It may also have been contemporaneous with the domestication of barley and wheat in the Near East ten thousand years ago.

[44] Bray, *ibid*. pp. 510–14; Hu Tao-Ching (*1963*).

later. Because of its very special role in the Chinese dietary system, soybean processing will command a chapter all by itself (d).

The last of the staple grains, hemp, *ma*, *Cannabis sativa* L, is neither a cereal nor a legume. It has been cultivated since Neolithic times in China, both as a fibre and a food crop. The Chinese recognised early that it is a dioecious plant. The male form, *hsi* 枲, was grown for fibre, while the female form, *chü* 苴, was grown for seed. It is believed that hemp seed was used in China as food until about +600 when it was superseded by other grains.[45] For our purpose, hemp seed is of special interest in that it contains about 30 per cent fats and is useful as a source of vegetable oil. We will, therefore, continue its discussion under the next topic of oilseeds.

(ii) *Oilseeds*

Simmer minced meat well and add it to cooked dryland grown rice, then blend in melted fat; this is called the Rich Fry. Simmer minced meat well and add it to cooked millet, then blend in melted fat; this is called the Similar Fry.[46]

Very little is known about the early history of the extraction of oil from seeds in China.[47] This may appear, at first sight, surprising since edible fat or oil is an indispensable ingredient in the style of Chinese cookery as we know it today, and two of the crops well known in ancient China, hemp and soybean, are excellent sources of oil. Indeed, soybean is one of the principal sources of vegetable oil in the world today, but there is hardly any mention of its being extracted for oil in the ancient Chinese literature.[48] Nor, for that matter, is there any record of the pressing of hemp seeds or any other oilseed for oil in the pre-Han classics.

Yet as we can see from the two recipes from the *Li Chi* 禮記 (Record of Rites), c. +1st century, quoted above, fat must have been used liberally in cooking in ancient China. But it was animal fat and not vegetable oil, and as indicated in the passage about the duties of the *Phao Jên* 庖人 (Keeper of Victuals) in the *Chou Li*, fats from a much wider range of livestock than those we are accustomed to see in modern Western cuisine were deemed suitable. Moreover, for the most important ceremonial or social occasions each type of meat was considered to have been properly prepared only when it was paired with its own particular brand of fat:[49]

The *Phao Jên* (Keeper of Victuals) is in charge of the six livestock, six animals and six fowls for slaughter and processing ... In the spring he presents young lamb and suckling pig, which are cooked in beef tallow; in the summer he presents dried pheasant and fish, which are cooked in canine fat; in the autumn he presents calf and baby venison, which are cooked in lard; and in the winter, he presents fresh fish and goose, which are cooked in lamb tallow.

---

[45] Bray (1984), pp. 532–55; Li Hui-Lin (1974).
[46] *Li Chi*, *Nei Tsê* 內則, p. 467; tr. auct; adjuv. Legge (1985a), p. 468.
[47] In contrast, the invention and development of oil pressing machinery is well documented in the West; cf. *SCC* Vol. IV, Pt 2, p. 206. A discussion of the history of oil pressing will be presented in Chapter (e).
[48] Bray, *SCC* Vol. VI, Pt 2, p. 519.   [49] *Chou Li*, ch. 1, *Phao Jên* 庖人, p. 36.

Thus, all indications are that the practice of extracting oil from seeds was simply unknown before the Han period. Why? We will discuss this issue later in Chapter (e) when we consider the question of the development of oilseed pressing technology. We may point out, at this time, that the earliest record of the recognition that there is fat in a vegetable seed is found in the *Ssu Min Yüeh Ling* 四民月令 (Monthly Ordinances for the Four Peoples), +160, which states that seeds of the *chü* hemp are 'pounded' (*tao chih* 擣治) to make *chu* 燭 (candles).[50] Presumably, after pounding the thick greasy residue would have the consistency of wax. There is no indication of how the 'pounding' was done. And yet by the 6th century, when the *Chhi Min Yao Shu* (Important Arts for the People's Welfare), +544, was completed, oil pressing houses (*ya yu chia* 壓油家) were already in routine operation to receive oilseeds for processing on a commercial scale.[51]

Hemp may still be the earliest seed processed for oil in China, but no direct evidence for it has been found. There is no question that hemp is the *ma* 麻 that was widely cultivated in ancient China as a fibre crop. But there is now less confidence that it is the very same *ma* used as a food crop and recognised by the Han annotators as one of the 'five staple grains' of the realm. The prevailing view is that hemp seed was used as a food grain in ancient China until the Han dynasty, after which the practice gradually declined and eventually died out.[52] However, there is another *ma* (Figure 6), sesame (*Sesamum orientale* L.), originally called *hu ma* 胡麻 (i.e. foreign *ma*) which was also grown in ancient China and whose seed was far superior a food to hemp seed. It was none other than Master Sung himself, the author of the celebrated *Thien Kung Khai Wu* 天工開物 (The Exploitation of the Works of Nature), +1637, who first drew attention to the inherent inconsistency of identifying hemp as an ancient food grain. He says:[53]

There are only two kinds of *ma* that can be used as grain or for oil, *huo ma* 火麻 (hemp) and *hu ma* 胡麻 (sesame). *Hu ma* is also called *chih ma* 芝麻, and was reputed to have been brought to China from the West in the former Han dynasty. In ancient times *ma* was recognised as one of the five staple grains, yet this is certainly inappropriate if hemp itself is meant... After all, the yield of oil from *huo ma* seed is poor, and the cloth woven from its bark is of little worth.

On the other hand, he continues:

Sesame is both delicious and nutritious. It would not be an exaggeration to say that it is the king of all grains. A few handfuls are sufficient to quell one's hunger for a long time. Cakes,

---

[50] *SMYL*, 2nd month, p. 25. The passage reads *Chü ma tzu hei, yu shih erh chung, tao chih tso chu* 苴麻子黑，又實而重，擣治作燭. Although *chu* 燭 nowadays explicitly means candle, it is more appropriately translated in this context as flambeau, since candle, as we know it, was probably unknown at the time when the *SMYL* was written. *Chu* in those days denoted a flammable torch made by binding strips of plant fibre, and soaking the bundle in a slow burning grease, such as the tallow from an animal, or the oily residue obtained by pounding hemp seeds. For a detailed discussion of this viewpoint see Miao Chhi-Yü (*1981b*) ed., pp. 32–3. The *SMYL* further notes in the 7th month, p. 76, that seeds of *hsi erh*, 枲耳 *Xanthium strumarium* (cf. *SCC* Vol. vi, Pt 1, p. 480) and in the 8th month, p. 84, that seeds of the gourd, *Legenaria leucantha*, were also used in the making of flambeaux. Indeed according to the *Fan Shêng Chih Shu* the seeds of the gourd were particularly suitable for incorporation into flambeaux since they burned with a very bright flame, cf. Shih Shêng-Han (1959), p. 24.

[51] *CMYS*, ch. 18, p. 133. [52] Li Hui-Lin (1974). [53] *TKKW*, ch. 1, p. 7; Sun & Sun (1966), p. 24.

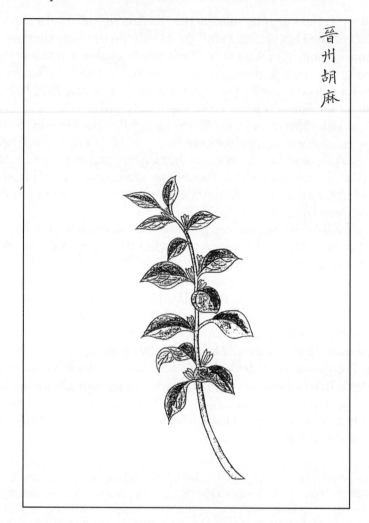

Fig 6. *Chi Ma* or *hu ma*, sesame; *CSKCTMC*.

breads and sweetmeats, when sprinkled with a few sesame seeds, will have their flavours improved and their values increased. When made into oil sesame can enrich the hair, benefit the intestines, make the strong smelling meat savoury and counteract poisonous elements.

This theme has been taken up and expounded in recent years by Li Fan who argues that sesame could have been either indigenous or brought to China during the Neolithic Age.[54] Three types of evidence were cited in support of his thesis. The first is archaeological. Sesame seeds have been recovered from two Neolithic sites in

---

[54] Li Fan (*1984*), pp. 81ff., 240ff.

Chekiang province in the late 1950s.[55] Thus, sesame must have had a much longer history in China than indicated by the legend of its importation during the former Han which has been shown to be highly suspect.[56] The second is botanical. Wild types of sesame are quite common in certain localities in Yünnan.[57] The seeds are gathered by the minority people and pressed into oil or used as food, as the ancient Chinese might have done. The third is textual. *Hu ma* and hemp fruit, (*ma fen* 麻蕡) are described in the *Shên Nung Pên Tshao Ching* 神農本草經 (Pharmacopoeia of the Heavenly Husbandman), +2nd century, and the commentary by Thao Hung-Ching states that of all the eight staple grains, sesame is the most valuable. Moreover, the *Fan Shêng-Chih Shu* (The Book of Fan Shêng-Chih), −1st century, has a brief reference to sesame (*hu ma*) as one of the crops recommended by the legendary I Yin of the Shang dynasty to plant by the '*ou thien* 區田' (shallow pits) system in times of drought.[58] While we find these arguments persuasive we need to take into account the fact that it was hemp seeds and not sesame that were discovered in the Han tombs at Ma-wang-tui. Nevertheless, we cannot dismiss the possibility that sesame was already a significant crop in China by the beginning of the Han dynasty, and that sesame seed could have had just as long a history as hemp as a source of edible oil.

Perhaps of equal antiquity to hemp and sesame as an oilseed crop is *wu ching* 蕪菁 or *man ching* 蔓菁, the Chinese turnip or colza, *Brassica rapa* L., known in the *Shih Ching* (Book of Odes) as *fêng* 葑. It was probably the most important Brassica crop of ancient China, and was grown for its root as well as its leaf and seed. We shall encounter it again in our discussion on vegetables. Its position as a source of oil, however, had started to erode by the time of the Northern and Southern dynasties. Foremost among its competitors were *yun thai* 雲薹, *shu chieh* 蜀芥, *chieh tzu* 芥子, and *sung* 菘 from which various cultivars of *yu tshai* 油菜, oil-bearing vegetables, were later developed. The cultivation of all four crops are described in the *Chhi Min Yao Shu* (Important Arts for the People's Welfare), +544.[59] They were probably the major sources of vegetable oil in mediaeval China. The nomenclature of the Chinese Brassicas is extremely confusing, but according to the recent analysis of Miao Chhi-Yü, we can distinguish between three broad types of Brassicas grown for oilseed in China today.[60] The first is the *chieh tshai* 芥菜 (mustard) type, derived from *shu chieh* 蜀芥, Szechuan mustard, *Brassica juncea* Coss, or *chieh tzu* 芥子, Chinese mustard, *Brassica cernua* Hemsl. The second is the *pai tshai* 白菜, Chinese cabbage, type. It can be related either to *yun thai* 雲薹, *Brassica campestris* L. var. *oleifera*, or to *pai tshai* 白菜, *Brassica chinensis* var. *oleifera*. The third broad type is derived from *kan lan* 甘藍, *Brassica oleracea* L., which was probably imported into China from Europe through Sinkiang during the Yuan dynasty.

---

[55] *Chê-chiang shêng po-wu-kuan* (*1960*) and (*1978*).   [56] Laufer (1919), p. 293; cf. *SCC* Vol. VI, Pt I, pp. 172–3.
[57] Li Fan (*1984*), p. 84.   [58] Shih Shêng-Han (1959), p. 41.
[59] *CMYS*, chs. 18 and 23, pp. 132–6, 146–8.   [60] Miao Chhi-Yü (*1982*), pp. 147–8.

(iii) *Vegetables*

In the seventh month we eat melons,
In the eighth month we cut the gourds.
....
He who plucks turnips or radish,
Must not ignore the lower parts.[61]

The Chinese word for vegetables, *tshai* 菜, is made up of two parts, the radical *tshao* 艸 (grasses) and the word *tshai* 采 (to gather), indicating that vegetables were originally collected from the wild. The word *tshai* 菜, however, has also evolved into a term denoting viands, in opposition to *fan* 飯 or grain food at a meal. For the moment we shall deal with *tshai* only in its role as vegetables. Although in the *Shih Ching* the character *tshai* (to gather) is often found associated with a vegetable, for example, *tshai fêng*, *tshai fei*, *tshai chhi* 采苢, *tshai wei* 采薇 etc.,[62] this does not necessarily mean that the vegetable mentioned had to be exclusively gathered from the wild. For the cultivation of vegetables, as well as fruits, must have started very early in China, presumably during Neolithic times. In the Shang oracle bones were found the pictographs *fu* and *yu*, which certainly suggest a picture of little plants grown within an enclosure.[63]

甫 *fu*　　　　　囿 *yu*　　　　　圃 *phu*

Later the two characters were merged to give us the word *phu*, meaning a vegetable garden or an orchard. Thus, it is likely that by the early Chou period, horticulture had started to develop into a significant sector of the agricultural economy.

The *Shih Ching* mentions no less than forty-six plants which could have been eaten as vegetables.[64] Some were gathered from the wild, others cultivated. Several vegetables were also mentioned in the *Li Chi* which contains a considerable amount

---

[61] The first quotation is from M 154 (W 159), tr. Waley. The second is from M 35 (W 108); tr. auct.; adjuv. Waley, Karlgren (1950), Legge (1871). This couplet provides a sobering example of the pitfalls in interpreting and translating ancient Chinese verse into a foreign language. The original reads:

Tshai fêng tsahi fei　　采葑采菲
Wu i hsia thi　　　　　無以下體

We now know that both *fêng*, turnip and *fei*, radish were important vegetables in ancient China. When one realises that in each case the roots were valued at least as highly as the leafy parts of the plant, the meaning of the couplet then becomes clear, that is, in plucking turnip or radish, one must not forget the edible roots, cf. the *Shih Ching Pai Shu* (+1695) of *Shih Ching Pai Shu* and Chiang Ying-Hsiang, ed. (*1934*) ch. 2, pp. 17–21. The versions by the three eminent Sinologists are as follows:

    Legge   When we gather the mustard plant, and earth melons
                  We do not reject them because of their roots.
    Karlgren One gathers the *fêng* plant, one gathers the *fei* plant
                  Without regard to their lower part.
    Waley   He who plucks greens, pluck cabbage
                  Does not judge by the lower parts.

Clearly, all three translators had missed the key point in the second line.

[62] *Shih Ching*, *tshai chhi* M 178 (W 134); *tshai wei* M 167 (W 131).
[63] Lin Nai-Shên (*1957*), p. 132.　　[64] Lu Wên-Yü (*1957*); cf. Chang, K. C. (1977), p. 28.

Table 2. *Vegetables in ancient China cited in the* Shih Ching *and* Li Chi *and found among archaeological remains in Han tombs*

Chinese name	English name	Latin name	W[a]	L[b]	H[c]
Kua 瓜	Melon	*Cucumis melo*	159	N	M
Hu 瓠	Gourd	*Lagenaria leucantha*	159	—	M
Fêng 葑	Chinese turnip	*Brassica rapa*	50	N	M
Fei 菲	Radish	*Raphanus sativus*	108	—	O
Phu 蒲	Cattail	*Typha latifolia*	144	—	—
Chhin 芹	Oriental celery	*Oenanthe javanica*	250	—	O
Ho 荷	Lotus	*Nelumbo nucifera*	37	—	M
Sun 筍	Bamboo shoot	*Phyllostachys* spp.	144	—	M
Khuei 葵	Mallow	*Malva verticillata*	159	—	M
Liao 蓼	Smartweeds	*Polygonum hydropiper*	230	N	—
Wei 薇	Wild bean	*Vicia augustifolia*	131	—	—
Chhi 苣	Lettuce	*Lactuca denticulata*	134	—	—
Thu 荼	Sowthistle	*Sonchus arvensis*	108	—	—
Chiu 韭	Chinese leek	*Allium ramosum*	159	N	O
Tshung 蔥	Spring onion	*Allium fistulosum*	—	N	O
Hsieh 薤	Chinese shallot	*Allium bakeri*	—	N	—
Chiang 薑	Ginger	*Zingiber officinale*	—	N	M
Hsiao tou 小豆	Lesser bean	*Phaseolus calcaratus*	—	N	M
Ling chiao 菱角	Water caltrop	*Trapa bicornis*	—	N	M
Chih erh 芝耳	Mushrooms	*Basidiomycetes*	—	N	—

[a] *W* Waley's numbering of poems from the *Shih Ching*.
[b] *L* Li Chi; N, *Nei Tse* 內則, ch. 10, *Pattern of the Family*.
[c] *H* Huang Chan-Yüeh (1982), Summary of Food Remnants Found in Han Tombs, pp. 77–9; M, Ma-wang-tui, O, other Han tombs.

of information on food materials and the way they were prepared and served. We have collected from the two classics the names of the vegetables that we believe were cultivated or used extensively in ancient China. They are listed in Table 2, with appropriate references for those whose remains have been found in Ma-wang-tui and other Han tombs.

The two cucurbits shown in Table 2, the melon and the gourd (Figures 7 and 8), are the most important fruit vegetables in ancient China. Their young leaves could also be eaten as vegetables. *Kua* 瓜, melon, *Cucumis melo* L., sometimes known as *thien kua* 甜瓜 (musk melon) is eaten as a fruit. It has gained a certain degree of notoriety as the fruit which contributed to the untimely demise of the lady of Han Tomb No. 1 at Ma-wang-tui. More than 138 melon seeds were recovered from her esophagus, stomach and intestines, indicating that she must have eaten the melon so fast that she swallowed the seeds without realising what she had done.[65] *Hu* 瓠,

[65] *Hunan shêng po-wu-kuan* (1973), I, pp. 35–6; *Hunan nung-hsueh-yuan* (1978), p. 9.

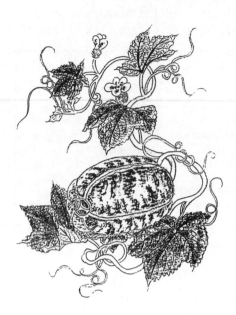

Fig 7. *Kua*, melon; *CSKCTMC*.

Fig 8. *Hu*, gourd; *CSKCTMC*.

the gourd, *Lagenaria leucantha* Standl. was particularly popular since the shells of the fruits made excellent utensils and receptacles.[66] Although not noted as a source of oil, the *Fan Shêng-Chih Shu* (−1st century), states that the seeds produce a bright flame when incorporated in a flambeau.[67] Both crops are described in the *Chhi Min Yao Shu* of the 6th century.[68] They are moderately important crops in China today.

The two cruciferous plants, *fêng* 葑 and *fei* 菲 (Figures 9 and 10) are also described in the *Chhi Min Yao Shu*. We have already encountered *fêng*, the Chinese turnip, as *man ching* or *wu ching*, as well as several other Brassica species, earlier in the segment on oilseeds. *Man ching* was highly prized as a vegetable in ancient China. In the *Lü Shih Chhun Chhiu* (Master Lü's Spring and Autumn Annals), −239, it was named as one of the most delectable vegetables of the land.[69] It remained popular during the mediaeval period.[70] Although it is still being cultivated today, its importance has

---

[66] Li Hui-Lin (1969), p. 256; idem. (1983), p. 48; *Chechiang shêng wên-kuan-huei, po-wu-kuan* (1978).
[67] Shih Shêng-Han (1982), p. 25.   [68] *CMYS*, ch. 14, ch. 15, p. 110 and pp. 118ff.
[69] *LSCC*, ch. 14, *Pên Wei* segment states that a most delectable vegetable is the (*man*) *ching* from Chü Chhü 具區, the area near Lake Thai (Thai Hu 太湖) in present day-Kiangsu province; cf. also Harper (1984).
[70] Legend has it that Chu-Ko Liang 諸葛亮, the great general and strategist of the Three Kingdom (+221−265) period was an energetic booster of *man ching*, so much so that it later became known as the *chu-ko tshai* 諸葛菜 (Chu-Ko vegetable). The famous poet of the Northern Sung, Su Tung-Pho 蘇東坡 was also known for his fondness for *man ching*. Indeed, he devised a recipe for making a man-ching *kêng* (stew). The Southern Sung poet, Chu Pien 朱弁, celebrated *man ching*'s association with these two famous historical figures in a couplet which says:

Shou chê chu-ko tshai   手折諸葛菜
Tzu chu tung-po kêng   自煮東坡羹

To pluck by hand Chu-Ko's vegetable
So as to cook Tung-Po's favourite stew.

Fig 9. *Fêng* or *Wu ching*, Chinese turnip; *CSKCTMC*.

Fig 10. *Fei* or *Lai fu*, Chinese radish; *CSKCTMC*.

declined greatly[71] in the face of competition from newer Brassica cultivars such as *pai tshai* (*B. chinensis* L.), *yu tshai*, (*B. campestris* L.), *chieh lan*, (the Cantonese *kai lan*), *B. alboglabra* etc. Nevertheless, *man ching* deserves recognition as the progenitor of the vibrant group of mustards which constitute the dominant source of leafy vegetables in the Chinese diet today.

The other crucifer, *fei*, the radish, better known as *lo po* 蘿蔔, *lu fu* 蘆菔 or *lai fu* 萊菔 is *Raphanus sativus* L. It is valued for its root. In the *Erh Ya* (The Literary Expositor), −300, it is called *thu* 葖 or *lu pha* 蘆萉. In spite of the profusion of names, it was apparently not considered a major vegetable in ancient China. It was mentioned briefly in the *Chhi Min Yao Shu* practically as an afterthought to the chapter on *man ching*.[72] Ironically, its popularity grew steadily thereafter. Today it is one of the most important vegetables of China.

Next, we have a set of three aquatic plants which were common in the wetlands of the Yangtze basin.[73] The first is the sweet flag (or the cattail), *phu* 蒲, *Acorus calamus* L. (or *Typha latifolia* L.), family Araceae, the tender shoots of which were popular as a vegetable. It is still cultivated in parts of the country today. Then we have the umbelliferous leafy vegetable *chhin* 芹, the oriental celery, *Oenanthe javanica* (Bl.) DC. Under the name *chhi* 芑, the celery from Yün-mêng 雲夢 was judged by the master chef I Yin, in the *Lü Shih Chhun Chhiu* (Master Lü's Spring and Autumn Annals) of the −3rd century, as one of the most delectable vegetables of the empire.[74] It is still cultivated as a leafy vegetable in paddy fields as *sui chhin* 水芹, or wetland celery; but

---

[71] Li Hui-Lin (1969), p. 259. [72] *CMYS*, ch. 18, p. 133. [73] Li Hui-Lin (1983), pp. 43–4.
[74] *LSCC*, ch. 14, *Pên Wei* 本味, p. 371; Yun-mêng is in present-day Hupei province.

it has largely been replaced by the *han chhin* 旱芹 or dryland celery, *Apium graveolens* var. *dulce*, which might have been introduced into China during the Han dynasty. The last entry, *ho* 荷 or *lien* 蓮, the Indian lotus, *Nelumbo nucifera* Gaertn., family Nymphaeaceae, was one of the earliest plants cultivated in China, lotus seeds having been discovered in pottery jars at a Neolithic site near Chêngchou.[75] Both the seed and the tuberous rhizome, known as *lien ou* 蓮藕, were and still are used as food. Lotus seed and root have become well known outside China as food materials that are distinctly Chinese.[76]

Another typically Chinese vegetable is *sun* 筍, the bamboo shoot, probably from a species of the genus *Phyllostachys* or *Bamboosa*. Both the *Erh Ya* (The Literary Expositor), −300, and the *Shuo Wên Chieh Tzu* 說文解字 (Analytical Dictionary of Characters), +121, explain that *sun* is the young shoot of bamboo.[77] In the poem from the *Shih Ching* cited in Table 2, bamboo shoot was served as a course during an important feast:[78]

> Hsien-fu gave the farewell party—
> A hundred cups of clear wine.
> And what were the meats?
> Roast turtle and fresh fish.
> And what were the vegetables?
> Bamboo shoots and reed shoots.

Obviously, bamboo shoot (as well as reed shoot) was so highly regarded in ancient China that it was considered fit to be served at a feast in honour of an important lord. Very little is known about its cultivation during or before the Han dynasty. The earliest account we have on bamboo as a crop is found in the *Chhi Min Yao Shu*.[79]

The next plant, *khuei* 葵 (Figure 11), the mallow, *Malva verticillata* L. of the family Malvaceae, was probably the most popular leafy vegetable in ancient China. Its cultivation is briefly mentioned in the *Ssu Min Yüeh Ling* (+160),[80] and described in considerable detail in the *Chhi Min Yao Shu* (+544).[81] Several varieties were known. But its importance gradually declined after the Thang dynasty.[82] By the Ming

---

[75] *Chêngchou shih po-wu-kuan* (1973) and Li Hui-Lin (1983), p. 43.
[76] See, for example, Gloria Bley Miller (1972), pp. 858–9. In addition, the Chinese 'Indian' lotus is one of the most beloved flowers in the Chinese garden. It has long been a favourite subject of Chinese poets and painters, cf. *SCC* Vol. VI, Pt I, pp. 133–6. It is, moreover, a symbol of purity and holiness and occupies a special place in the mythology of Buddhism, since the Buddha is usually depicted as sitting on a throne formed by the lotus flower.
[77] *Erh Ya*, ch. 13; *SWCT*, ch. 5; cf. also *SCC* Vol. VI, Pt I, pp. 377–94.
[78] *Shih Ching* W 144 (M 261), tr. Waley. [79] *CMYS*, ch. 51, p. 259.
[80] *SMYL*, 1st month, p. 2. Cf. Hsu Cho-yun (1980), p. 217. [81] *CMYS*, ch. 17; pp. 126–31.
[82] That the mallow was still a common vegetable in Thang times is indicated succinctly in the poem, entitled *Phêng Khuei* 烹葵 (Cooking Mallows), by Pai Chü-I, p. 65, who in describing his life style in less fortunate days, lamented:

> Phin chhu ho so yu　　　貧廚何所有
> Chhui tao phêng chhiu khuei　炊稻烹秋葵

which may be translated as follows:

> In my bare kitchen, what do I own?
> But steamed rice and fall mallow stew.

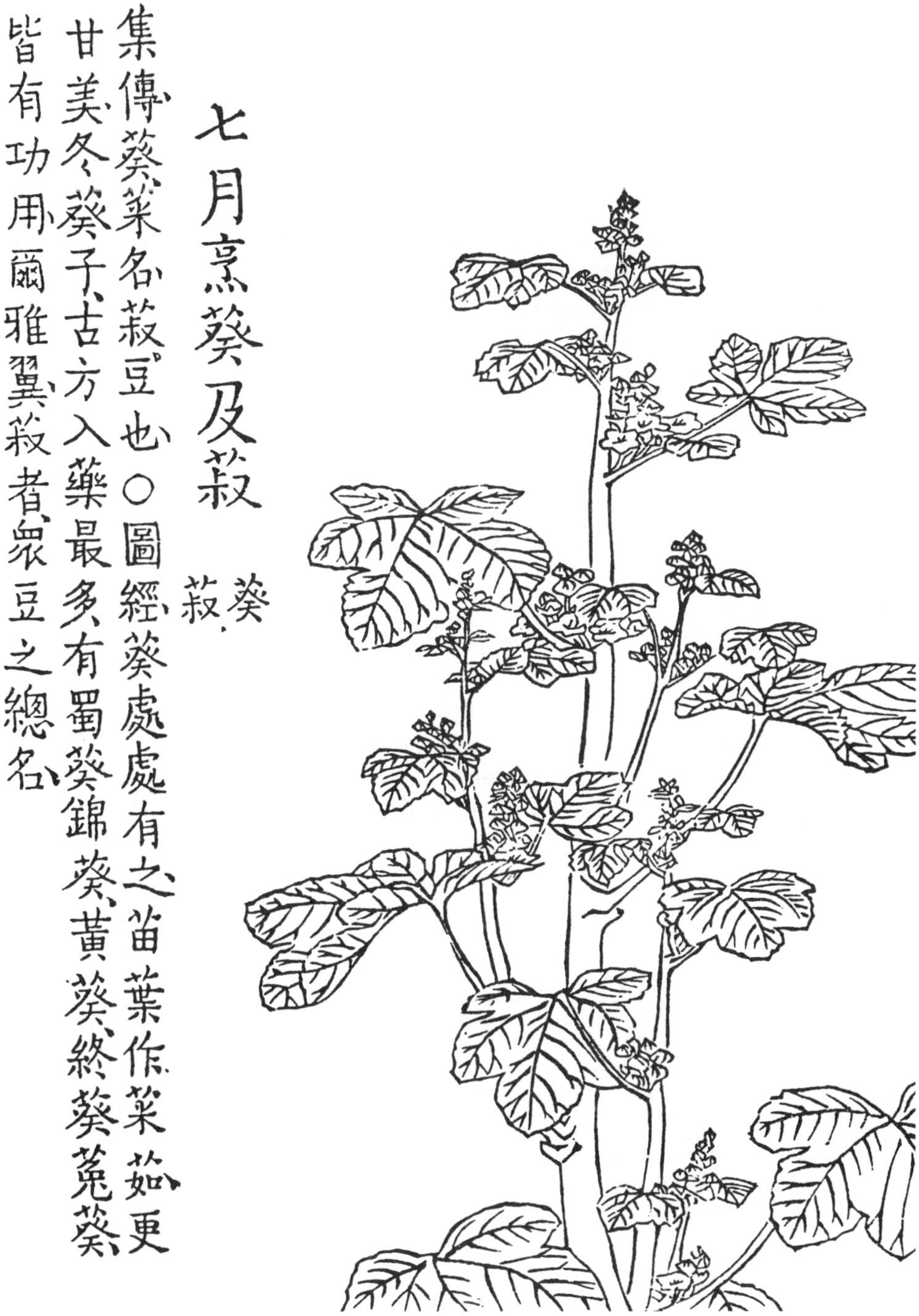

七月烹葵及菽

葵 菽

集傳葵菜名菽豆也〇圖經葵處處有之苗葉作菜如更甘美冬葵子古方入藥最多有蜀葵錦葵黃葵終葵菟葵皆有功用爾雅翼菽者眾豆之總名

Fig 11. *Khuei*, the mallow; *Mōshi Himbutsu Zukō* 毛詩品物圖考, 1/23.

dynasty Li Shih-Chên was quite unfamiliar with it as a vegetable and placed it in the herb class in his great pharmacopoeia, *Pên Tshao Kang Mu* 本草綱目 (+1596).[83] The decline of the mallow is coupled with the rise of the Chinese cabbage, *Brassica chinensis* as the leading leafy vegetable of China. It was most likely caused by a change in food technology and food habits. According to Li Hui-Lin,[84] vegetable oils were not available in ancient China and mucilaginous leaves such as the mallow which helped to blend the flavours and thicken the sauce when cooked with other food ingredients became a highly favoured component of the diet. After the extraction of oil from seeds became practical on a large scale, mucilaginous vegetables were no longer needed for this special function, and the mallow was gradually displaced by other more easily cultivated leafy crops.

The mallow is followed by four other plants which were used as vegetables in ancient China, but are no longer cultivated today. *Liao* 蓼, the smartweed or water pepper, *Polygonium hydropiper* L., family Polygonaceae, was mentioned in the *Shên Nung Pên Tshao Ching* (Pharmacopoeia of the Heavenly Husbandman)[85] and described as a cultivated crop in the *Ssu Min Yüeh Ling* (+160).[86] *Wei* 薇, the wild bean, *Vicia* spp., family Leguminosae, is probably the same as *chhao tshai* 巢菜 which has been occasionally cultivated as a fodder or green manure.[87] *Chhi* 苣 or *Chü* 苣, probably the Chinese lettuce, *Lactuca denticulata* Maxim., family Compositae, is noted in passing as a cultivated crop in the *Chhi Min Yao Shu*.[88] Finally, *thu* 荼 or *khu tshai* 苦菜 has been identified as the sowthistle,[89] *Sonchus arvensis* L., also of the Compositae family. Our interest in *thu* lies mainly in the fact that the word *thu* may possibly be an ancient form of *chha* 茶, tea, a subject which we will discuss in detail in Chapter (*f*).

We now come to three favourite liliaceous plants noted for their pungent odour and piquant flavour. The first is *chiu* 韭, the Chinese leek, *Allium odorum* L. The whole plant including the green shoot and fleshy bulb was consumed as a vegetable. It was highly regarded by the ancient Chinese, who paired it with lamb as offerings in sacrificial ceremonies.[90] The next plant, *tshung* 蔥, scallion or spring onion, *Allium*

Khuei continued to be mentioned occasionally by Sung writers such as Su Tung-Po and Lu Yu, cf. Nie Fêng-Chhiao (*1985*). It was even placed at the head of the list of vegetables given in the celebrated Yuan nutritional treatise *Yin Shan Chêng Yao* 飲膳正要 (p. 160) but its use steadily declined. The mallow is so unfamiliar today, that *khuei* is frequently mistaken for 'sunflower', *Helianthis annus*, an import from the Americas, for which the Chinese translation is *hsiang jih khuei* 向日葵, i.e. sun-facing *khuei*, cf. Wang Yü-Hu (*1982*), p. 44. For example, in a passage from a recent English translation by Irving Y. Lo of the poem 'The Seventh Month' from the *Shih Ching*, which we have quoted on p. 18, *khuei* is rendered as 'sunflower'. It occurs (on p. 10) in the title of the splendid anthology of Chinese poems edited by Liu & Lo (1975), *Sunflower Splendour*. It is an unfortunate title since most of the poems included, were written long before the sunflower made its appearance in the Middle Kingdom. For a similar interpretation see Ho Suan-Chhuan (*1986*) on the identity of *khuei* as described on wooden tablets found in Han tombs at Chü-yen (cf. Tsien Tsuen-Hsuin, *SCC* Vol. v, Pt 1, p. 30).

[83] *PTKM*, ch. 16, p. 1038.
[84] Li Hui-Lin (1969), p. 256. As noted in the *PTKM* (*ibid.*), *khuei* was also known as *hua tshai* 滑菜, the slippery vegetable, because of the mucilaginous character of its leaves.
[85] *SNPTC*, in the first-class herb section. Cf. also *SCC* Vol. VI, Pt 1, p. 486.
[86] *SMYL*, 1st month, p. 2; cf. Cho-yun Hsu (1980), p. 217. [87] Cf. *Tzhu Hai* 辭海 (1979), pp. 613, 1192.
[88] *CMYS*, ch. 28, p. 158; Li Hui-Lin (1969), p. 259. [89] Lu Wên-Yü (*1957*), p. 22.
[90] *Ibid.*, p. 89. The reference pairing the leek with the lamb in sacrificial offerings occurs in the poem, 'The Seventh Month' in the Books of Odes (w 159, M 154), from which we have already cited a passage earlier (p. 18).

fistulosum L. is probably the most widely used seasoning vegetable in Chinese cookery. Two varieties were known, *hsiao tshung* 小蔥, little *tshung*, which is harvested in the summer, and *ta tshung* 大蔥, large *tshung*, which is harvested in the winter.[91] The third and less familiar plant is *hsieh* 薤, the Chinese shallot, *Allium bakeri* Regal. Not mentioned in the *Shih Ching* or *Li Chi* are two other plants of the lily family which were also important in ancient China. One is *hsiao suan* 小蒜, or simply *suan*, *Allium scorodoprasum*, rocambole, believed to be a native Chinese plant.[92] But probably far more important is *ta suan*, 大蒜 garlic, *Allium sativum* L., also known as *hu* 葫 which is assumed to have come from Central Asia. It is surely one of the most extensively

---

In the eighth stanza, we see the phrase *hsien kao chi chiu* 獻羔祭韭, that is, to offer lamb and leek as sacrificial gifts to the ancestors. Waley has translated *chiu* as garlic, Karlgren (1950) as onion and Legge (1871) as scallion. At least all three are in the right genus.

[91] *SMYL*, 1st month, p. 2; cf. Hsu Cho-yun (1980), p. 217.

[92] The identity of *ta suan* and *hsiao suan* has proved to be another vexing problem in our task in 'the elucidation of the ancient' botanical lexicon. We first encountered the problem when we reviewed the pesticidal plants described in the Chinese pharmacopoeia for *SCC* Vol. VI, Pt 1 (pp. 478–508), and decided tentatively that *ta suan* is *Allium scorodoprasum* L. and *hsiao suan*, *Allium sativum* L. We now believe that these assignments are incorrect. Perhaps the oldest use of the word *suan* is found in the *Erh Ya*, ch. 13, where it states that *li* 蒚 is *shan suan*, mountain *suan*. Thus, it is possible that a kind of *suan* was already known in China by the late Chou dynasty. The earliest record of *ta suan* and/or *hsiao suan* occurs in the *Ssu Min Yüeh Ling* (+160). *Hsiao suan* is mentioned in the 6th, 7th and 8th month segments, while *ta suan* is mentioned only in the 8th month. These indications suggest that *hsiao suan* is the older, indigenous plant and *ta suan* the more recent import from abroad. Indeed, Li Shi-Chên stated (*PTKM*, ch. 26) that *suan* or *hsiao suan* is a native Chinese plant while *ta suan* or *hu* was brought back to China, allegedly by Chang Chhien, from the western regions. Shih Shêng-Han (*1963*), pp. 20–1, has examined the early literature and is in basic agreement with this position.

What then are *hsiao suan* and *ta suan*? Bretschneider was probably the first western botanist to assign modern binomial names to the two plants. He concluded in the late 19th century that *hsiao suan* or *suan* is *Allium sativum* L. and *ta suan* or *hu* is *Allium scorodoprasum* L. (B, III, pp. 392–3). Similar conclusions were arrived at by Porter Smith & Stuart (pp. 27–8) at about 1910 and by Read (pp. 218–19) in 1937. Just a few years later, in 1936, Chia & Chia (*1955*), p. 1051, indicated that *hu* or *ta suan* is *Allium scorodoprasum*. More recently, in Vol. 5 of the *Iconographia Cormophytorum Sinicorum* (Institute of Botany), published in 1976, *suan*, which by ancient convention is synonymous with *hsiao suan*, was assigned the botanical name, *Allium sativum* L. Thus, based on these authorities one would naturally be led to conclude that *ta suan* is *A. scorodoprasm*, rocambole, and *hsiao suan* is *A. sativum*, garlic.

But in spite of these apparently authoritative references this conclusion is now suspect. Today, *ta suan* is widely cultivated and extensively used in cookery in China, while *hsiao suan*, usually collected from the wild, is rarely seen in a kitchen. *Ta suan*, as a common agricultural commodity, is readily available for scientific examination. Recent taxonomic studies of authentic specimens, such as those carried out in support of the compilation of the *Chung-kuo thu-nung yao-chih* 中國土農藥誌 (Repertorium of Plants used in Chinese Agricultural Chemistry), 1959, and the *Chung I ta tzhu tien* 中醫大辭典 (Cyclopedia of Chinese Traditional Drugs), 1985, have demonstrated unequivocally that *ta suan*, as we now know it, is, without a doubt, *Allium sativum* L., the venerable garlic known to the ancient Egyptians and Israelites, as described by Moldenke & Moldenke (1952), p. 32, and Michael Zohary (1982), p. 80. It is identified in the *Chung-kuo thu-nung yao-chih*, p. 176, as *Allium sativum* L. var. pekinese (Prokh) Maekawa, and simply as *Allium sativum* L. in the *Chung I ta tzhu tien*. *Hsiao suan*, as indicated in the latter, is identified as rocambole or *Allium scorodoprasm*.

But how did this discrepancy arise? We suspect it is a case of a mistaken identity. We know that today, *suan*, also popularly known as *suan tou*, usually refers to *ta suan*, cf. *Chung-kuo thu-nung yao-chih*, p. 176. But in the ancient texts, starting with *Min I Pie Lu, Hsin Hsiu Pên Tshao* down to *Pên Tshao Kang Mu*, *suan* invariably refers to *hsiao suan*, and *hu* to *ta suan*. It is possible that originally *hsiao suan* did denote garlic, which, according to Rombauer & Becker (1981), p. 584, is also known as oriental garlic or ail. Its Asian origin is indisputable. It could have been indigenous to China or brought there very early, before or during Shang or Chou times. On the other hand, rocambole, also called giant or topping garlic, is a European plant and could have been the *hu* supposedly introduced into China by Chang Chhien. Thus, it is possible that Bretschneider and Porter Smith and Stuart correctly identified *hsiao suan* as garlic, *Allium sativum*, and *ta suan* as rocambole, *Allium scorodoprasm*. However, during the 20th century, as garlic gained in importance and popularity, it became known as *suan* or *ta suan*, while rocambole, no longer under cultivation, was relegated to the status of *hsiao suan*.

used vegetable condiments produced in the world today. Planting times for all five species of *Allium*, Chinese leek, spring onion (both small and large varieties), rocambole, shallot and garlic were listed in the *Ssu Min Yüeh Ling*, indicating that all five vegetables (Figure 12) were already grown widely during the Han dynasty.[93] From the detailed instructions given in the *Chhi Min Yao Shu* for the cultivation of these crops,[94] we may presume that there was no visible abatement of their popularity between the +2nd and +6th centuries.

The next plant listed in Table 2 is *chiang* 薑, ginger, *Zingiber officinale* Rosc., family Zingiberaceae, certainly one of the best known condiments used in Chinese cooking. It is grown mostly in south China, since in the north the roots cannot attain an acceptable size. Ginger was one of the highly prized harmonising agents (*ho* 和) for the king's kitchen named by the legendary I Yin in the *Lü Shih Chhun Chhiu*, −239.[95] The *Ssu Min Yüeh Ling*, +160, recommends that ginger be planted in the fourth month after sprouts begin to appear on the rhizome.[96] It also mentions the cultivation of ginger's close relative *jang ho* 蘘荷, *Zingiber mioga* Rosc.[97] Although apparently a fairly important vegetable in ancient China, we had first encountered it as *chia tshao* 嘉草, a plant with insecticidal activity, in Part 1 of Volume VI of this series. Ginger is today as important as ever, but *jang ho* is no longer grown as a cultivated crop.

After soybean, *hsiao tou* 小豆, or *chih hsiao tou* 赤小豆, the rice bean or red lesser bean, is probably the oldest cultivated legume in China. It is *Vigna (formerly Phaseolus) calcaratus*.[98] In addition to the *Li Chi* (Table 2), it is mentioned in the *Fan Shêng-Chih Shu* and *Ssu Min Yüeh Ling*.[99] The *Ssu Min Yüeh Ling* notes briefly the planting of *wan tou* 豌豆, *Pisum sativum* L. (peas) which was also called *hu tou* 胡豆 since it was reputed to have been an introduction from central Asia.[100]

Table 2 ends with two familiar products which may be considered as characteristically Chinese. The first is *ling-chiao* 菱角, water caltrop, the fruit of *Trapa bicornis* L. or *Trapa natans* L., family Onagraceae, which is still extensively cultivated today in the lower Yangtze basin.[101] It has also been called 'water chestnut'.[102] However, outside China 'water chestnut' is invariably the name applied to *wu-yü* 烏芋 (dark yam) or *ti-li* 地栗 (ground chestnuts), the tuber of *Eleocharis tuberosa* Roxb. The last item, *chih-erh* 芝栭, is a class name for mushrooms. Mushrooms growing on soft ground are *chih* 芝, while those growing on wood are *erh* 栭. We have no idea what specific mushrooms are meant in the citation from *Li Chi*, but six types of *chih* and

---

[93] *SMYL*, 1st month, 6th month, 7th month and 8th month, pp. 2, 76, and 84.
[94] *CMYS*, chs. 19, 20, 21 and 22; pp. 137, 141, 143 and 144.
[95] *LSCC*, ch. 14, *Pên Wei*; the best ginger is said to be from Yang-phu, Szechuan.
[96] *SMYL*, 4th month, p. 47.   [97] *SMYL*, 9th month; cf. *SCC* Vol. VI, Pt 1, pp. 473–4.
[98] Li Hui-Lin (1983), p. 47. Formerly identified as a *Phaseolus*, which is now considered a strictly New World genus. Species misclassified under this name have recently been shown to be all *Vigna*. This is not to be confused with the *chhih tou* 赤豆, the adzuki bean, a cultigen, and thus *Vigna angularis*.
[99] *FSCS*, cf. Shih Shêng-Han (1959), p. 21; *SMYL*, 2nd month, 4th month, pp. 26, 47.
[100] *SMYL*, 1st & 3rd month; pp. 2, 37; Tai Fan-Chin (*1985*), pp. 1–10.
[101] Li Hui-Lin (1983), p. 43.   [102] Smith & Stuart (1973), p. 440.

Fig 12. Five Liliaceae; from *CSKCTMC*: (a) *Chiu*, leek; (b) *Tshung*, scallion; (c) *Hsieh*, shallot; (d) *Suan* or *hsiao suan*, rocambole; (e) *Hu* or *ta suan*, garlic.

Fig 13. *Yü*, taro. Painting by Shih Thao, c. +1697; from Fu & Fong (1973), *The Wilderness colors of Tao Chi*.

the well-known *mu erh* 木耳 (*Auricularia auriculae* (L. ex Hook) Underw.) are listed in the *Shên Nung Pên Tshao Ching*.[103]

We cannot leave this survey of vegetables in ancient China without a few words on *yü* 芋, the taro (Figure 13), *Colocasia esculentum* Schott. of the Araceae which though not mentioned in the *Shih Ching* or *Li Chi*, is named in the *Kuan Tzu* 管子 (The Book of Master Kuan), late −4th century, as an important crop.[104] According

---

[103] *SNPTC*, upper-class and middle-class drugs, cf. Tshao Yuan-Yu (*1987*), pp. 61–2; 246–7.
[104] *Kuan Tzu*, ch. 23, p. 222.

to the *Fan Shêng-Chih Shu*, *yü* was one of the crops recommended by the legendary I Yin of the Shang dynasty for planting in shallow pits in times of drought.[105] Its cultivation is reported in the *Ssu Min Yüeh Ling* and described in detail in the *Chhi Min Yao Shu*, which also mentions *shu* 藷, the Chinese yam, *Discorea opposita* as an exotic tuber of south China.[106] Both tuber crops were useful adjuncts to grain in times of famine and they were grown extensively in mediaeval China. Today, the taro is still being grown as a vegetable, but the dominant tuber in the Chinese diet is now the sweet potato, *Ipomeia batatas* (L.) Lam, introduced to China from America in the 16th century.

From the foregoing account it is clear that a rich variety of vegetable resources were available to the ancient Chinese. As indicated in Table 2 archaeological remains or records of nine of the twenty vegetables listed have been identified in the Han tombs at Ma-wang-tui. They are melon, gourd, turnip, lotus, bamboo shoot, mallow, red bean, ginger and water caltrop. In addition, records of two plants not listed, wild ginger (*jang ho* 蘘荷), *Zingiber mioga* L., and taro (*yü*), *Colocasia esculentum* S. were also found. Remains of four additional vegetables, Chinese leek, radish, spring onion and possibly water celery were discovered in other Han tombs.[107] These corroborations certainly serve to enhance the credibility of Table 2 as a guide to the vegetables grown and consumed in ancient China.

(iv) *Fruits*

> Fairest of trees under heaven and on earth,
> The orange came to be acclimated here,
> Thus destined, it will not live elsewhere,
> But grows in our southern land.[108]

As we have suggested earlier, fruit and vegetable gardens were already an important sector of the economy by the beginning of the Chou dynasty.[109] Indeed, from the occurrence of the ancient form of the character *kuo*, meaning fruit,

Ancient 果   Modern 果

in oracle bones, we can presume that fruits were recognised as a separate category of food commodity as early as the Shang dynasty. At least fifteen fruit trees are mentioned in the *Shih Ching* and several additional ones cited in the *Li Chi*. The more important trees from the two works are listed in Table 3, together with appropriate references to archaeological evidence collected from recent excavations of Han tombs.

---

[105] Shih Shêng-Han (1959), p. 41.
[106] *SMYL*, 1st month, p. 2. *CMYS*, ch. 16, cf. also Bk 10, No. 27; pp. 121, 592.
[107] Huang Chan-Yüeh (1982), pp. 78–9.
[108] *Chhu Tzhu*, ch. 4, *Chü Sung* 橘頌 (Ode to the Orange), tr. Wu-Chi Liu, in Liu & Lo eds. (1975), p. 15; also cf. David Hawkes (1985), p. 178.
[109] Cf. p. 32; Lin Nai-Shên (1957), p. 132.

Table 3. *Fruits in ancient China cited in the* Shih Ching *and* Li Chi *and found among archaeological remains in Han tombs*

Chinese name	English name	Latin name	W[a]	L[b]	H[c]
Thao 桃	Peach	*Prunus persica*	84	N	—
Li 李	Chinese plum	*Prunus salicina*	84	N	O
Mei 梅	Chinese apricot	*Prunus mume*	17	N	M
Hsing 杏	Apricot	*Prunus armeniaca*	—	N	M
Tsao 棗	Chinese date	*Zizyphus jujuba*	159	N	M
Chên 榛	Chinese hazelnut	*Corylus heterophylla*	165	N	O
Li 栗	Chinese chestnut	*Castenea mollissima*	125	N	M
Mu kua 木瓜	Chinese quince	*Chaenomeles sinensis*	18	—	O
Chih chü 枳椇	Raisin tree	*Hovenia dulcis*	170	N	—
Li 梨	Pear	*Pyrus bretschneideri*	—	N	M
Cha 楂	Chinese hawthorn	*Crataegus pinnatifida*	—	N	—
Shih 柿	Persimmon	*Diospyros kaki*	—	N	M
Ying-thao 櫻桃	Chinese cherry	*Prunus pseudocerasus*	—	Y	—
Hua-chiao 花椒	Fagara	*Xanthoxylum simulans*	8	—	M
Kuei 桂	Cassia	*Cinnamomum cassia*	—	N	M

[a] *W* Waley's Numbering of the Poems of the *Shih Ching*.
[b] *L Li Chih*; N, *Nei Tse* 內則, ch. 10, The Pattern of the Family; Y, *Yueh Ling* 月令, ch. 8, The Monthly Ordinances.
[c] *H* Huang Chan-Yueh (1982), summary of food remains found in Han tombs, pp. 77–9; M, Ma-wang-tui; O, other Han tombs.

The peach, *thao* 桃 (Figure 14), *Prunus persica* Batsch. of the Rosaceae, is a fruit of great antiquity in China. Peach stones have been identified at Ho-mu-tu (−5000) and other Neolithic sites in Chekiang.[110] As indicated by its frequent mention in the *Shih Ching* and other classics it was undoubtedly a major fruit crop in ancient China.[111] In popular mythology it is said to have been the food of the immortals (*hsien thao* 仙桃) and it has had a long tradition in Chinese culture as a symbol of longevity.[112] De Candolle in 1884 placed China as the original home of the peach.[113] According to Laufer it was introduced in the −1st or −2nd century from China to Persia (Iran), from where it made its way to Europe and later to the New World.[114] It is cultivated widely in China and throughout the temperate regions of the world today.

[110] Li Fan (1985), p. 165; *Chê-chiang shêng po wu kuan* (1960), *Chê-chiang shêng wên kuan huei* (1960), *Chê-chiang shêng wên kuan huei, po wu kuan* (1978).
[111] Cited in Hsin Shu-Chih (1983), pp. 53–4.
[112] See for example, the story of the monkey king stealing peaches of longevity from the garden of the Celestial Queen Mother as described in the famous novel *Hsi Yu Chi* (*Journey to the West*), ch. 5.
[113] Cf. de Candolle (1884).   [114] Cf. Laufer (1919).

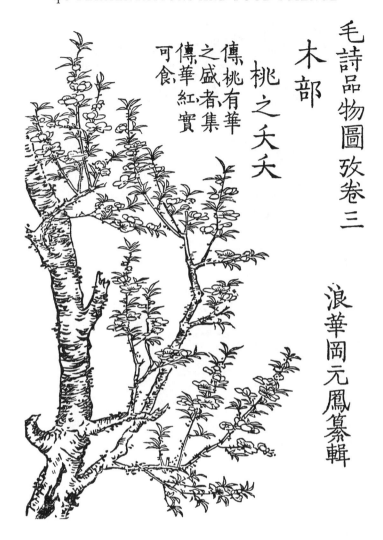

Fig 14. *Thao*, peach flowers; *Mōshi Himbutsu Zukō* 毛詩品物圖攷, 3/1.

Often cited in literary works together with *thao* is *li* 李, the Chinese (Japanese) plum, *Prunus salicina* Lindl (Figure 15). Both trees were cultivated for their ornamental flowers as well as their edible fruits, as, for example, reflected in the following lines from the *Shih Ching*:[115]

> Gorgeous in her beauty,
> As flower of peach or plum.

---

[115] Book of Odes, w 84 (M 24), tr. Waley.

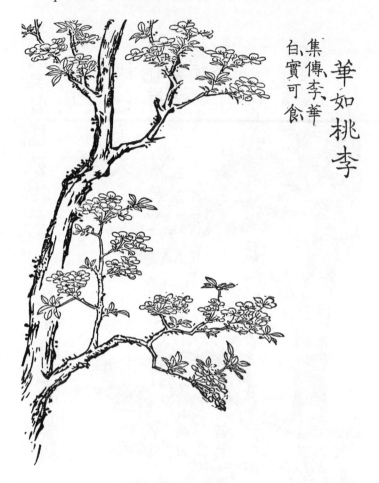

Fig 15. *Li*, plum blossoms; *Mōshi Himbutsu Zukō*, 3/4.

Archaeological remains of *li*, the Chinese plum, have been discovered in several Han tombs in Kiangsu, Hupei, Kuangtung and Kuangsi.[116] They suggest that *li* was an important fruit crop in ancient China. Wild varieties of *li* have been found in Szechuan and eastern Tibet, indicating that these could have been the progenitors of the modern *li* cultigen. Today it is grown very widely in China, from Heilungkiang in the north, all the way to Kuangtung and Kuangsi in the south, and Tibet in the west.

Even more highly prized for the beauty of its flowers than *thao* and *li* is *mei* 梅, the Chinese apricot (or dark plum), *Prunus mume* Sieb. and Zucc. (Figure 16). We have already noted earlier[117] its popularity among poets and painters as a symbol of strength, steadfastness and purity, since its blooms open in early spring to grace a

---

[116] Huang Chan-Yüeh (*1982*), pp. 78–9; Li Fan (*1985*), p. 169.
[117] Cf. Needham *et al.*, *SCC* Vol. VI, Pt I, p. 420.

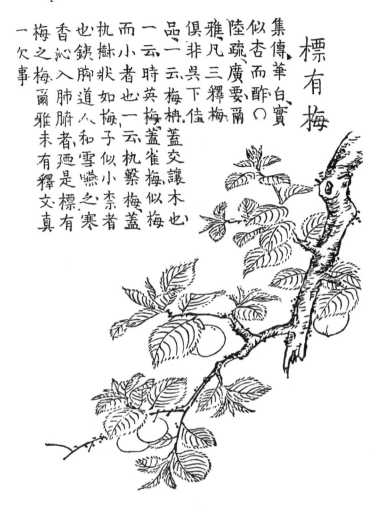

Fig 16. *Mei*, Chinese apricot; *Mōshi Himbutsu Zukō*, 3/2.

still drab and frosty landscape.[118] *Mei* stones have been identified in a tomb in Chiang-ling, Hupeh, of the Warring States period (−480 to −221) and at Ma-wang-tui and other Han tombs.[119] It was presumably a major fruit crop in ancient North and Central China. According to the *Shu Ching* (Book of Documents), −5th century, 'to make a well harmonised soup, one needs salt and *mei* fruit',[120] which indicates that *mei* was used primarily as an acidulant to impart a tart flavour to food. Today, *mei* is grown largely south of the Yangtze River. Hardly any is found in the lower Yellow River valley, the home of the peoples who composed the poems collected in

---

[118] Cf. Bray, *SCC* Vol. VI, Pt 2, p. 550.   [119] Li Fan (*1985*), p. 175.
[120] *Shu Ching* (Book of Documents), *Shuo Ming Hsia* 說命下: *Jo Tso Ho Kêng, Erh Wei Yen Mei* 若作和羹，爾惟鹽梅, tr. auct.; adjuv. Legge (1879), p. 260.

the *Shih Ching*. There are two possible reasons for this discrepancy. First, the climate in North China was considerably warmer during the Shang–Chou period than it is today.[121] By Thang and Sung times the climate had already become too cold for growing *mei*, and its cultivation had to be restricted to regions south of the Yangtze River. Second, *mei* was successfully adapted to grow in the north, but in poor yield, when it was widely used as a food acidulant.[122] Later, as vinegar became readily available, there was no longer a compelling reason to force the cultivation of *mei* in North China and the practice gradually died out.

The next Prunus in Table 3 is *hsing* 杏, the apricot, *Prunus armeniaca* L. Wild varieties of the apricot are found throughout North, Northwest and Northeast China indicating that North China is its original habitat as suggested by de Candolle.[123] Yet there is no reference to it in the *Shih Ching* (Book of Odes). It is curious that while *mei*, a southern plant, is noted five times in the *Shih Ching*,[124] *hsing*, a genuine northern plant of economic importance, is not mentioned at all. It must have been domesticated early since it was named in the ancient text of *Hsia Hsiao Chêng* 夏小正 (Lesser Annuary of the Hsia Dynasty), a work believed to have been compiled between −7th and −4th century, which states that 'in the fourth month, *hsing* "flowers" are seen in the garden'.[125] Hsin Shu-Chi has postulated that the fruits of the early varieties of *hsing* had a rather indifferent taste, and, therefore, it was not deemed worthy of mention in the *Shih Ching*.[126] Later, as improved cultivars became available, *hsing* was accepted as a fruit crop. Certainly by the beginning of the Han dynasty the apricot was already a well-known cultivated fruit. Remains of apricot stones have been found in several Han tombs, including that at Ma-wang-tui.[127]

The Chinese jujube (or Chinese date), *tsao* 棗, *Zizyphus jujuba* Mill., family Rhamnaceae, is another fruit of great antiquity, jujube stones having been identified at a Neolithic site (−5000) at Phei-li-kang in Honan.[128] It is mentioned once (as date) in the *Shih Ching* in the poem 'Ode to the Seventh Month' quoted earlier.[129] But a related wild spiny shrub, *chi* 棘, *Z. spinosa* Hu, which produces a smaller and sour fruit, is mentioned no less than eight times (in six poems).[130] It is likely that *chi* is the wild ancestral form of *tsao* and in some of the poems the two terms were considered synonymous. The close relation between the two is reflected in the fact the both words are made up of the same two characters, *tzhu* 朿. In *tsao*, the jujube, they are placed one on top of the other and in *chi*, the wild jujube, they are placed side by side. Remains of the jujube have been found at Ma-wang-tui and several other Han

---

[121] Chang Kwang-Chih (1980a), p. 144; Chu Kho-Chên (1973).
[122] Hsin Shu-Chih (1983), pp. 11–12. [123] Cf. de Candolle (1884).
[124] *Shih Ching*, M 20, 130, 141, 152 and 204 (W 17, 181, 69, 231 and 165). According to Khung Ying-Ta's commentary (*Mao Shih Chêng I*, +642) on the *Shih Ching*, the *mei* in M 130 and 204 may actually refer to the *nan* 柟 tree, *Phoebe nanmu* Gamble, of the Laurel family rather than to the apricot tree. For a discussion on this issue see Hsin Shu-Chih (1983), pp. 9–12, and Lu Wên-Yü (1957), pp. 73 and 79. However, even accepting Khung's interpretation this still leaves three valid references to *mei* in the *Shih Ching*.
[125] Hsin Shu-Chi (1983), p. 14. [126] Ibid. [127] Huang Chan-Yüeh (1982), pp. 78–9.
[128] Li Fan (1985), pp. 196–7. [129] *Shih Ching*, M 154 (W 159).
[130] *Shih Ching*, M 32, 109, 131, 141, 152 and 219 (W 78, 275, 278, 69, 165 and 287).

tombs.[131] It is still extensively and intensely cultivated throughout China today, although the best varieties appear to come from North China. It is rich in vitamin C and its composition is similar to that of the fig in terms of edible matter, sugar and acid.[132] When dried or preserved in honey it is said to keep for a very long time.

Next we have two nut trees, *chên* 榛, the hazelnut, *Corylus heterophylla* Fisch., family Corylaceae and *li* 栗, the chestnut, *Castanea mollissima* Bl., family Fagaceae. Archaeological remains of the chestnut have been identified at Neolithic sites at Phei-li-kang, Honan and Ho-mu-tu, Chekiang.[133] Carbonised hazelnut found at Pan-pho near Sian has been carbon dated at about −4300.[134] Both are indigenous Chinese plants and were important food nuts in ancient China (Figures 17 and 18). According to the *Chou Li*, the chestnut, hazelnut and jujube were used in sacrificial offerings together with dried peach and *mei* fruit.[135] Furthermore, in the *Tso Chuan* (Master Tso's Commentaries on the Spring and Autumn Annals), the jujube, chestnut and hazelnut were recommended as suitable gifts for presentation by a prospective bride to a prospective groom.[136] Both nuts are cultivated in China today, but the chestnut much more widely than the hazelnut.

Mu kua 木瓜, the Chinese quince, *Chaenomeles sinensis* Koehne of the Rose family, is the fruit of the *chu* 杼 tree, which is found mostly North of the Yangtze river.[137] It was presumably of some importance in ancient China since its cultivation merited a separate chapter in the *Chhi Min Yao Shu*.[138] In South China today, however, the term *mu kua* is more often reserved for the papaya, *Carica papaya*. Related to the Chinese quince is the Chinese hawthorn, *cha* 樝 or *shan cha* 山樝, *Crataegus pinnatifida* Bge., another native plant of China. In the *Erh Ya* (−300) it is called *chhiu* 棣. The dark red berry has a pleasantly sour flavour similar to that of the native American cranberry, *Vaccinium macrocarpus*, and makes a delicious jam.[139] *Chü* 枳 or *chih chü* 枳椇, the raisin tree, *Hovena dulcis* Thunb., family Rhamnaceae, is an unusual fruit tree endemic to North and Central China. The edible part is the swollen peduncle which is filled with a pleasant, yellowish, pear-like pulp, rather than the small fruit itself. It is mentioned in both the *Shih Ching* and the *Li Chi* (Table 3).

Although the Chinese word for pear, *li* 梨, is not cited, several ancient names for wild species of pear are mentioned in the *Shih Ching*. They include *tu* 杜 (*Pyrus phaeocarpa*),[140] *kan tang* 甘棠 (*P. betulaefolia*)[141] and *suei* 檖 (*Pyrus* sp.),[142] which are believed to be the progenitors of the familiar Chinese sand pear, *sha li* 沙梨, *Pyrus pyrifolia* Nakai

---

[131] Cf. notes 127 and 128 above.
[132] Church (1924), cited in Li Hui-Lin (1983). [133] Li Fan (1985), p. 199.
[134] Li Hui-Lin (1983), p. 35; *Chung-kuo kho-hsüeh yuan khao-ku yen-chiu so shih yen shih* (1972).
[135] *Chou Li*, ch. 2, *Pien Jên* 籩人, p. 54.
[136] *Tso Chuan*, ch. 3, *Chuang Kung*, 莊公, 24th year, p. 70; cf. Legge (1872), p. 108.
[137] Cf. Smith & Stuart (1973), pp. 362–3, where it is listed as *Pyrus cathayensis* or *Cydonia sinensis*, and also *SCC* Vol. VI, Pt I, p. 423. For a detailed discussion on the identity of *mou* as *mu kua* see Hsin Shu-Chih (1983), pp. 20–6.
[138] *CMTS*, ch. 42, p. 223.
[139] Under the name *Shan-cha-kao* 山樝糕; cf. Porter-Smith & Stuart (1973) repr, p. 130. This jam or jelly is similar in colour, texture and flavour to the cranberry jelly which normally accompanies the ever-popular roast turkey served at Thanksgiving in the United States and Christmas dinners in the United States and Britain.
[140] *Shih Ching* M 169 (W 145). [141] *Shih Ching* M 16 (W 138). [142] *Shih Ching* M 119, 132 (W 277, 80).

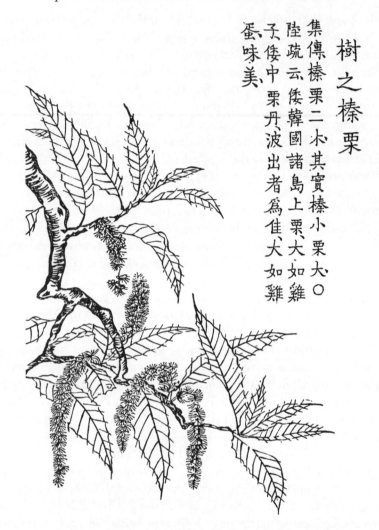

樹之榛栗

集傳榛栗二木其實榛小栗大〇
陸疏云倭韓國諸島上栗大如雞
子倭中栗丹波出者為佳大如雞
蚕味美

Fig 17. *Li*, Chinese chestnut; *Mōshi Himbutsu Zukō*, 3/6.

and white pear, *pai li* 白梨, *Pyrus bretschneideri* Rehd.[143] Indeed, remains of the *sha li* (*P. pyrifolia*) were found in Han Tomb No. 1 at Ma-wang-tui (Table 3). Thus, we may infer that the domesticated varieties of the Chinese pear were probably developed from wild species between the early years of the Chou dynasty and the period of the Warring States. They are Eastern representatives of the enormous pear clan and are different from the Western pears which are derived from *Pyrus communis* L.

Perhaps better known in the West is *shih* 柿, the persimmon, *Diospyros kaki* L., family Ebenaceae, which is widely grown in North and Central China today. It is an

---

[143] Li Fan (*1985*), pp. 183–5.

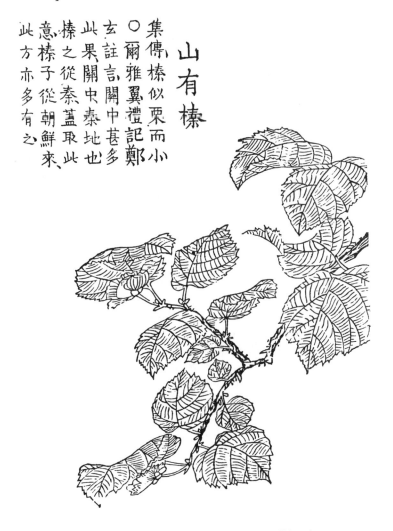

Fig 18. *Chên*, Chinese hazelnut, *Mōshi Himbutsu Zukō*, 3/5.

indigenous Chinese plant. The wild prototype *D. kaki* var. *sylvestris* and a related species *D. lotus* have both contributed to the domestication of this fruit,[144] which probably occurred during the Warring States period. By the beginning of the Han dynasty it had become sufficiently established as to be included for burial in one of the Han tombs at Ma-wang-tui.[145]

The last fruit on this list is *ying thao* 櫻桃, the Chinese cherry, *Prunus pseudocerasus* Lindl., family Rosaceae. It blooms early in spring and already bears fruits before the arrival of summer. In the *Erh Ya* it is known as *hsieh* 楔 or *ching thao* 荊桃. In the

---

[144] Li Hui-Lin (1983), pp. 34–5; Li Fan (1985), pp. 202–4. [145] Huang Chan-Yüeh (1982).

Yüeh Ling 月令 chapter of the *Li Chi* it is called *han thao* 含桃, and is recommended as an appropriate ceremonial offering to one's ancestors. Its remains have been found in the Neolithic site at Phei-li-kan, and at a Warring States tomb in Chiang-ling, Hupeh.[146]

Table 3 ends with two entries which are not fruits but spices. They are included here for convenience since the products are also harvested from a shrub or tree. The first is *hua chiao* 花椒, fagara, the fruit of *Zanthoxylum* species (of the family Rutaceae), of which two kinds were known, the *shu chiao* 蜀椒 from Szechuan (*Z. piperitum*) and the *chhin chiao* 秦椒 from Shensi (*Z. bungei*).[147] It was the primary 'hot and pungent' spice of China until the coming of *hu chiao* 胡椒, black pepper, *Piper nigrum* from India during the Han dynasty.[148] The second is *kuei* 桂, cassia (or Chinese cinnamon), made from the bark of the cassia tree, *Cinnamomum cassia* Blume, of the Lauraceae family, which was an important spice in ancient China.[149] The cassia from *Chao-yao* 招搖 was considered by the legendary I Yin of the Shang dynasty to be a delectable spice fit for the king's kitchen.[150] Remains of both fagara and cassia have been identified in the Han tombs at Ma-wang-tui (Table 3).

There is one other ancient Chinese fruit which deserves to be noted here although it is not included in Table 3. We mean the Chinese gooseberry, *Actinidia chinensis*, Planch, family Dillenaceae, which is now well known in the West as the *kiwi* fruit.[151] It is mentioned in the *Shih Ching* as *chhang chhu* 萇楚 and is referred to variously as *yao i* 銚弋 and *yang thao* 羊桃 in the *Erh Ya*, *Pên Ching* and *Shuo Wên* and as *mi hou thao* 彌猴桃 in the pharmacopoeias since the *Shih Liao Pên Tshao*, +670 and the *Khai Pao Pên Tshao* of +973.[152] Even though it has been a familiar drug since antiquity, it never found favour as an article of food in China. The development of the obscure Chinese gooseberry into the *kiwi* fruit of international commerce is a remarkable story of the vicissitudes of the transmission of an economic plant from one cultural sphere to another.[153]

So far all the fruits we have discussed are natives of North China, but many important fruits have also originated in Central and South China. Several of these

---

[146] Li Fan (*1985*), p. 176.

[147] Porter-Smith & Stuart (1973), pp. 462–3. Mention should also be made of a related spice, *chu yü* 茱萸, *Zanthoxylum ailanthoides*, Sieb. et Zucc., the fruits of which had properties similar to those of fagara. Remains of *chu yü* were found in the Han tombs at Ma-wang-tui. Its cultivation is described in the *CMYS*, ch. 44, p. 226.

[148] The earliest written record of *hu chiao* occurs in the account of the Western regions (*Hsi Yü Chuan* 西域傳) in the *Hou Han Shu* (History of the Later Han Dynasty, +25–220). According to Li Hui-Lin (1979), pp. 46–53, the *chü chiang* 蒟醬 described in the *Nan Fang Tshao Mu Chuang* 南方草木狀 is actually *Piper nigrum*, although later scholars tended to confuse it with betle pepper, *Piper betle*.

[149] Different types of *kuei* are described in the *Nan Fang Tshao Mu Chuang*; cf. Li Hui-Lin (1979), pp. 83–4.

[150] *LSCC*, ch. 14, *Pên Wei*, p. 371.

[151] We thank Hsü Cho-yun for drawing our attention to the connection between *chhang chhu* of the *Shih Ching* and the modern *kiwi* fruit of New Zealand. For a discussion of the identity of *chhang chhu* see Lu Wên-Yü (*1957*), pp. 80–1. Porter Smith & Stuart (1973), pp. 14–15, reminds us that the term 'Chinese gooseberry' has also been applied by Europeans to another fruit in China, that of *Averrhoa carambola*. *Chhang chhu* is translated by Legge, (1871) Bk XIII, Ode 3, p. 217, as the 'carambola' and by Waley (1937), p. 21, as the goat's peach.

[152] *Shih Ching*, (W 2 M 148); *Erh Ya*, ch. 13, p. 28a; *SWCT*, p. 17 (ch. 1, II, 5); *SLPT*, p. 102; *Khai Pao Pên Tshao*, cited in *PTKM*, pp. 1887–8.

[153] Schroeder & Fletcher (1967).

Fig 19. *Phi-pha*, loquat. *CSKCTMC*.

were presumably well known by the Warring States period and certainly by the early Han dynasty.[154] The first is *phi pha* 枇杷, the loquat (Figure 19), *Eriobotrya japonica* (Thumb.) Lindl., a member of the family Rosaceae, which has contributed so many well-known fruits in the North. The *phi pha* was mentioned in the *Shang Lin Fu* 上林賦 (Ode to the Imperial Forest), −2nd century, and Chêng Hsüan (+2nd century) identified loquat and grape as the rare delicacies denoted in the *Chhang Jên*

---

[154] Certainly based on evidence uncovered in excavations of Han tombs; cf. Huang Chan-Yüeh (*1982*).

場人 (Horticultural Officer) passage of the *Chou Li*.[155] The second is *yang mei* 楊梅, sometimes called the Chinese strawberry, *Myrica rubra* Sieb. et. Zucc., of the family Myricaceae, which was also mentioned in the *Shang Lin Fu*. It was cited in *Nan-Yüeh Hsing Chi* 南越行記 (Travels in Southern Yüeh), c. −175 and *Lin-I Chi* 林邑記 (Records of Champa), −2nd century, two works which are no longer extant. Both fruits have been recognised among the archaeological remains found in the Han tombs at Ma-wang-tui.[156]

Undoubtedly much better known to the West is *chü* 橘, the sweet orange or tangerine, *Citrus reticulata* Blanco (or *Citrus nobilis* Lour.), and its congeners, *kan* 柑, *chhêng* 橙, and particularly *yu* 柚, the pomelo, *Citrus grandis* Osbeck. They are probably the oldest domesticated fruits in South China. We have already recounted in some detail the origin and history of oranges and other citrus fruits in China in Section 38.[157] The reader's attention is directed particularly to the delightful story from the *Yen Tzu Chhun Chhiu* 晏子春秋 (Master Yen's Spring and Autumn Annals) −4th century, about the transformation of *chü* 橘 in the South to *chih* in the North.[158] For the present we wish merely to point out the antiquity of oranges and pomelos as fruits, and the high regard with which they were held in ancient China. As noted in the *Yü Kung* (Tribute of Yu) chapter of the *Shu Ching* 書經 (The Book of Documents), dated to −5th century or earlier, the Doomsday provinces of Yang-chou and Ching-chou (Figure 3) furnished oranges and pomelos as part of their tribute presented to the Chou court. In the *Lü Shih Chhun Chhiu*, −3rd century, I Yin recommended the *chü* (orange) of Chiang-phu 江浦 and the *yu* 柚 (pomelo) of Yün-mêng 雲夢 as the most valued of fruits for the kings's table.[159] Finally, records of oranges and pomelos have been identified in the Han tombs at Ma-wang-tui.[160] These accounts leave no

---

[155] *Chou Li*, ch. 4; p. 177. The name *phi-pha* is said to be derived from the shape of its leaves which resemble those of the Chinese musical instrument, *phi-pha* 琵琶. Porter Smith & Stuart (1973), p. 164, point out that 'the term loquat is a transliteration of the Cantonese sound of 櫨橘 (*lu-chü*), which is another name for cumquat (*chin-chü* 金橘) or golden orange. Just how this name came to be applied to the fruit of the *Eriobotrya* is uncertain, as the Chinese books do not indicate such use. However, it seems that this term has gained currency in California, where this fruit is now extensively grown.' Actually, Chinese books are, in fact, the very source of this confusion. Both *lu-chü* and *phi-pha* are mentioned in the *Shang Lin Fu*, and annotators of the *Pên Wei* segment in the *Lü Shih Chhun Chhiu* have suggested that *lu-chü* and *phi-pha* are alternate names for *kan-cha* 甘櫨, cf. Chhiu Phang-Thung, Wang Li-Chhi & Wang Chêng Min (1983), pp. 63–5, one of delicacies mentioned by I Yin. Since then there have been two schools of thought about the identity of *lu chü* in China. One school, as advocated by Li Shih-Chên (*PTKM*, ch. 30, p. 1796), holds that *lu chü* is another name for *chin chü* (pronounced *loquat* and *kumquat* respectively in Cantonese), a small citrus fruit of the genus *Fortunella* (*F. margarita*, cf. *SCC* Vol. VI, Pt 1, p. 374). The other school, of which a leading proponent is the famous Sung poet Su Tung-Pho, assumes that *lu-chü* is identical to *phi-pha*. The literary references in support of each point of view have been summarised recently by Kho Chi-Chhêng (1985). Although historically Li Shih-Chên is probably correct, Su Tung-Pho's view, nevertheless, has continued to receive wide acceptance until the present day. In fact, the identity of *lu-chü* and *phi-pha* was recently reaffirmed in the botanical literature, cf. Yu Tê-Tsun (1979), p. 310. Thus, it is likely that *lu-chü* was, as it is today, a common name for *phi-pha* in Kuangtung, and this name for the fruit was naturally used by the Cantonese immigrants who introduced it into California in the 19th century, cf. Sucheng Chan (1986). Porter-Smith & Stuart are, however, mistaken in mixing it up with the golden orange, *chin-chü*, which was also introduced by Chinese into California, where it is very well known as kumquat (or cumquat).

[156] *Hunan nung-hsüeh-yuan* (1978), p. 12.

[157] Needham *et al.*, *SCC* Vol. VI, Pt 1, pp. 363ff. [158] *Ibid*. pp. 106ff.

[159] *LSCC*, *Pên Wei*; Chhiu Phang-Thung, Wang Li-Chhi & Wang Chêng Min (1983), p. 11.

[160] Huang Chan-Yüeh (1982), p. 79.

doubt that oranges and pomelos were certainly well known if not readily available in the northern heartland of Chinese civilisation.

(v) *Land animals*

> Who says you have no sheep?
> Three hundred is the flock.
> Who says you have no cattle?
> Ninety are the black-lips.[161]

We have seen in Table 1 (p. 21) that six domestic animals were reared as livestock in the nine doomsday provinces of ancient China: horse (*ma* 馬), ox or cattle (*niu* 牛), sheep (*yang* 羊), pig (*shih* 豕), dog (*chhüan* 犬 or *kou* 狗), and chicken (*chi* 雞). The importance of these animals in the economy of the realm is signified by their collective designation as the *liu chhu* 六畜, six livestock, in tandem with *wu ku* 五穀, five grains, in the classical literature. There is now overwhelming archaeological evidence to show that the six livestock listed were already firmly domesticated by the Shang era. Major remains of pigs, dogs, horses, cattle, sheep and chickens were identified in the burial plots excavated at Anyang.[162] Of these, all except horses were also found in garbage dumps indicating that they were consumed as food. The horse, presumably, was used almost exclusively for pulling chariots.

Although listed as the first of the *liu chhu*, there was probably hardly ever a time when the horse (*ma*), *Equus caballus* (family Equidae) was reared as a food animal. It was esteemed for a quite different reason. From the dawn of history until the late Middle Ages the horse was valued throughout the civilisations of Asia and Europe as an instrument of mobility and engine of war, and only incidentally as a source of meat. It was probably the nomadic peoples living on the margins of the Asian steppes who first domesticated the horse and brought it into China in the late Neolithic Age.[163] However, legend credits Hsiang Thu 相土, one of the predynastic lords of the Shang dynasty, with the introduction and utilisation of horses in China.[164] Apparently, they soon acquired a status far above all other forms of livestock because of their military value. Horses are mentioned in no less than forty-two poems in the *Shih Ching* (The Book of Odes), but always in reference to their use in pulling chariots for transportation, hunting and warfare, and none to their consumption as meat.[165] Although the *Chou Li* (Rites of the Chou) describes no less than seven officials involved in various aspects of the rearing, care and management of

---

[161] *Shih Ching* w 160 (M 190); tr. Waley.
[162] Chang Kwang-Chih (1980), p. 143; Chhên Chih-Ta (*1985*), Hsieh Chhung-An (*1985*). For an overview of the domestic animals of China today, see Epstein (1969).
[163] Harris, Marvin (1985), p. 90; Diamond (1993), pp. 266–70 and Mallory, J. P. (1989).
[164] Chang, Kwang-Chih (1980), p. 9; *Shih Pên Pa Chung* 世本八種, p. 358.
[165] For example, M 53, 79, 223 (W 179, 118, 268).

Table 4. *Support staff for keepers of various livestock as described in the* Chou Li

Keeper of Title	Cattle Niu Jên 牛人	Chicken Chi Jên 雞人	Sheep[a] Yang Jên 羊人	Dog Chhuan Jên 犬人
Officers:				
Second class	2	0	0 (0)	0
Third class	4	1	2 (2)	2
Secretaries	2	0	0 (0)	1
Clerks	4	1	1 (1)	2
Assistants	28	0	2 (2)	4
Labourers	200	4	8 (8)	16
Total[b]	240	6	13 (11)	25

[a] In addition to the Sheep Keeper, *Yang Jên*, there is a parallel official, called *Hsiao Tzu* 小子 (Lesser Master), who is responsible for cutting up and cooking the sheep after it is slaughtered. The numbers in parentheses indicate the staff of the *Hsiao Tzu*.
[b] The primacy of the Cattle Keeper is indicated by the size of his staff which is larger than the staff of all the other offices combined.

horses, none had the specific duty of preparing them for sacrificial slaughter.[166] In contrast, there was only one keeper assigned to each of four other livestock, the ox, sheep, dog and chicken, and his duty was, in every case, to collect, maintain and, if needed, prepare the animal for ritual slaughter and processing as food.[167] The relative importance of these offices may be gauged by the size of the staff assigned to them, as shown in Table 4.

Nevertheless, there was no official taboo against eating horseflesh, and in a society where meat was never plentiful, it would be difficult to imagine that horseflesh was not eaten when available. In fact, it was deemed quite acceptable for presentation even at the royal table. A passage in the *Chou Li* on the duties of the *Nei Yung* 內饔 (the Internal Culinary Supervisor), even admonishes that 'when a horse is dark along its spine and its fore-legs covered with piebald marks, its meat is likely to be tinged with the unpleasant odour of insects' and thus unfit for consumption.[168] A similar passage also occurs in the *Li Chi*.[169] But these are rare and isolated references. They do not affect our overall impression that the horse was not regarded as a food animal in ancient China.

---

[166] *Chou Li*, chs. 7 and 8. The seven officials are: *Ma Chih* 馬質 *Hsiao Jên* 校人, *Chhih Ma* 趣馬, *Wu Ma* 巫馬, *Yü Jen* 庾人, *Yü Shih* 圉師, and *Yü Jên* 圉人. Their duties are described on pp. 306, 338–43, and further explained by Hsia Wei-Ying (*1979*), pp. 63–72. Altogether they command a staff of 198 persons.

[167] Keepers of cattle, chicken, sheep and dogs are described on the following pages of Lin Yin's edition of the *Chou Li*: *Niu Jên*, 128; *Chih Jên*, 211; *Yang Jên*, 309; and *Chhüan Jên*, 384. It is curious that there is no keeper of pigs, *Shih Jên* 豕人. In fact, the *Chou Li* only mentions the pig once, in connection with the duties of the *Fêng Jên* 封人 (Superintendent of Enclosures) who has the incidental job of preparing roast pork when needed.

[168] *Chou Li*, ch. 1, *Nei Yung*, p. 38.

[169] *Li Chi*, *Nei Tsê*, Legge tr. (1885), p. 463. This is simply a repetition of the *Chou Li* passage, *ibid*.

On the other hand, the ox (*niu*), *Bos taurus domesticus* Gmelin, family Bovidae, was well known both as a draught animal and a food animal. Another predynastic ancestor of the Shang, Wang Hai 王亥, is given the credit of being the originator of cattle breeding in China.[170] Cattle were apparently reared in large herds during the Shang dynasty. According to oracle bone records, astonishingly large numbers of cattle were used in Shang rituals, for example, 1,000 once, 500 once, 400 once, 300 thrice, and 100 nine times.[171] They continued to be a major livestock in the Chou era. The ox is mentioned in thirteen poems in the *Shih Ching*. In eight poems it is mentioned in association with sheep,[172] mostly in connection with the consumption of the meat of both animals in rituals and banquets. In five poems it is cited by itself, in one as a draught animal to pull carts and in the others as a highly regarded source of meat.[173] There are many references to cattle in the *Chou Li*, usually in reference to the use of beef, veal or beef tallow in sacrificial offerings and for the dining pleasure of the king and his nobles.[174] Occasionally, cattle are mentioned as draught animals for carrying provisions and pulling carts in military operations.[175] Beef is also mentioned frequently as a meat staple in the *Li Chi*, and is listed as an ingredient in the preparation of several of the celebrated *Pa Chên* 八珍 or 'eight delicacies'.[176]

The next animal on the list of six livestock (*liu chhu*) is *yang* (sheep), *Ovis aries* Linne, also family Bovidae. As noted earlier, sheep are mentioned in association with cattle in eight poems in the *Shih Ching*, indicating that their close relationship, both being ruminants, was recognised by the early Chinese. The sheep is mentioned by itself in seven additional poems, of which three refer to the use of its meat in rituals, and the other four deal with the use of lamb's skin for the fur coats of the well-to-do.[177] In five of the poems we are introduced to the word *kao* 羔, signifying a young lamb, which was presumably prized for its skin as well as for its meat, as it is today. *Yang* and *kao* are mentioned in several segments of the *Chou Li*, usually in reference to their use in rituals and the feeding of the high-born.[178] Similarly, sheep are cited in the *Yüeh Ling* and *Nei Tsê* chapters of the *Li Chi*, and mutton is indicated as an ingredient in four of the eight delicacies of the Chou cuisine.[179]

---

[170] Chang Kwang-Chih (1980), p. 9; cf. also pp. 32, 139 and 143 for references to the importance of the water buffalo in Shang society.
[171] Hu Hou-Hsüan (*1944*), pp. 5–6. These numbers are indeed impressive. They are much larger than those reportedly used on ritual occasions in Homer's Iliad and Odyssey, cf. Homer (1985) (1 & 2). However, they pale into insignificance when compared with those employed in a major religious ceremony by another people at about the time of the founding of the Chou dynasty. In Kings I 8: 63, we read that Solomon, and all Israel with him, offered to the Lord twenty-two thousand oxen and a hundred and twenty thousand sheep. The world has probably not seen a ceremonial bloodbath of animals on this scale ever since.
[172] *Shih Ching* M 66, 165, 190, 209, 245, 246, 272, 292 (W 100, 195, 160, 199, 238, 197, 220, 233).
[173] The ox-cart is mentioned in M 227 (W 135). The other poems are M 291, 212, 210, 300 (W 158, 162, 200, 251).
[174] For example, ch. 1, *Phao Jên* (Keeper of Vituals), *Nei Yung* (Internal Culinary Supervisor) and *Shih I* (Nutritionist–Physician); cf. Lin Yin (*1985*), pp. 36, 38, 45.
[175] *Chou Li*, ch. 3, *Niu Jên* (Keeper of Cattle) and ch. 10, *Chou Jên* 輈人 (Yoke-harness Superintendent); Lin Yin ibid., pp. 128, 435.
[176] *Li Chi*, pp. 467–70.
[177] As food, W 159, 161, 195 (M 154, 211, 165); as fur, W 6, 38, 119, 264 (M 18, 146, 80, 120).
[178] *Chou Li*, pp. 36, 38, 46, 308, 309.   [179] Legge (1885) tr., pp. 459, 461, 463, 468–9.

To those of our readers brought up in Europe or America where our exposure to Chinese food is weighted heavily in favour of the Southern, particularly Cantonese, style of cooking, it may come as a surprise that lamb or mutton was one of the principal meats of ancient China. Today, lamb is eaten mostly in North China, near the grasslands where sheep are reared in large numbers, although it has occasionally found its way to the provincial cuisines of Hunan and Szechuan. Both ox and lamb were the preferred animals used in sacrificial rituals and in social entertainment, but the ox was rated higher than the lamb. This attitude is neatly illustrated by the story about King Hsuan of Chhi in the *Mëng Tzu* (Book of Mencius), c. −290:[180]

The King was sitting in the upper part of the hall and someone led an ox through the lower part. The King noticed this and said 'Where is the ox going?' 'The blood of the ox is to be used for consecrating a new bell.' 'Spare it, I cannot bear to see it shrieking with fear, like an innocent man going to the place of execution.' 'In that case, should the ceremony be abandoned?' 'That is out of the question, use a lamb instead.'

The people were displeased.[181] They thought the King miserly in substituting a lamb for an ox. Mencius pointed out that if he was pained by seeing an innocent animal being led to execution, what was there to choose between an ox and a lamb?

There is little doubt that the pig, *Sus scrofa domestica* Brisson of the boar family, and the dog, *Canis familiaris* L. of the wolf family, were the earliest animals domesticated in China. Remains of both animals were found in the oldest Neolithic settlements of the Yang-shao (−4000) and Ho-mu-tu (−5000) cultures.[182] Thus, they already had a long history of domestication behind them when the Shang arrived on the scene. Yet in the *Shih Ching* (The Book of Odes), −11th to −7th centuries, there are only two references to the pig and two to the dog (as hunting hound).[183] Our modern dietary prejudices may incline us to explain this strange omission by supposing that during the Chou era pigs and dogs had become so common and familiar that they were considered somewhat uncouth and thus unworthy of participation in the ritual and festive events celebrated in the poems of the *Shih Ching*. But this view does not hold water. Dogs and pigs were mentioned as perfectly respectable meat animals in *Mo Tzu* (The Book of Master Mo), c. −400, *Mêng Tzu* (the Book of Mencius), c. −290, and *Hsün Tzu* (the Book of Master Hsün), c. −240.[184] In fact, the *Li Chi* states that the Son of Heaven himself was to be served wheat and lamb in the Spring, soybean and chicken in the Summer, hemp seed and dogflesh in the Autumn and millet and pork in the Winter.[185] Surely, dogflesh and pork would not have been served to the Emperor unless pigs and dogs were esteemed at least as highly as sheep and chicken as sources of meat in ancient China.

---

[180] Book of Mencius, Book 1, Pt A; tr. D. C. Lau (1970), pp. 54–5.
[181] And well they might be. After the ceremony the carcass was presumably cut up and the meat distributed to the people. There was, of course, much more meat to be had in an ox than in a lamb.
[182] Chang, Kwang-Chih (1977), pp. 95, 513.
[183] For pigs, M 250 (W 239) and for dogs, M 103, 127 (W 258, 260).
[184] *Mo Tzu*, ch. 49, *Lu Wên* 魯問, p. 373. Book of Mencius, Book 1, Pt A, Lau (1970), tr. p. 51. *Hsün Tzu*, ch. 4, *Yung Ju Pien* 榮辱篇; p. 61.
[185] *Li Chi, Yueh Ling*, pp. 257, 272, 285, 296.

We have quoted earlier (p. 28) a passage from the *Chou Li* on the duties of the *Phao Jên* (Keeper of Victuals) who presents to the royal household dried pheasant and fish cooked in canine suet in the summer, and veal and baby venison cooked in lard in the autumn.[186] An almost identical passage occurs in the *Li Chi*.[187] Evidently canine suet and lard were used routinely as shortening for cooking in the ancient kitchen. The *Chou Li* and *Li Chi* also recommend that 'pork be served with *chi* millet and dogflesh with *liang* millet'[188] and warn that the 'flesh of a dog which was uneasy and with the inside of its thighs red, would be coarse, and the pork from a pig which looked upwards and closed its eyes, would be measly'.[189] These references leave no doubt that pigs and dogs were indeed major sources of meat in the late Chou and early Han dynasties.

The popularity of pig and dog as meat animals apparently reached a high point during the Han dynasty. The popularity of the pig has been maintained through the centuries until the present day. The Chinese have been and are a pork-eating people. Today pork accounts for 94 per cent of the red (mammalian) meat consumed in China each year. It is estimated that in a given year there are altogether 800 million pigs on earth, of which 300 million or 38 per cent are in China.[190]

The story of the dog is quite different. We suspect the dog was originally domesticated in China as a meat animal and only later as a house guard and hunting hound. Of the seven poems classified as dealing with the subject of hunting by Arthur Waley in the *Book of Songs*, the hound was mentioned in just two poems.[191] As we have seen, most of the literature references to the dog were concerned with its use in ceremonial rituals and as food. Indeed, the slaughtering of dogs had acquired so much prestige that it was regarded as a separate profession. Dog butchers, *kou thu* 狗屠, were common during the late Chou and early Han period. Several famous personages of that time, for example, Nieh Chêng 聶政, Kao Chan-Li 高漸離 and Fan Khuai 樊噲 had begun life as dog butchers.[192] Nevertheless, it was during the Han dynasty that the non-food virtues of the dog, as hunting companion, reliable guard and faithful pet began to outweigh its value as food. After the Han, references to the eating of dogflesh all but disappeared from the literature.[193] By the Thang dynasty the very idea of eating dogflesh had become so repugnant that in his annotation of the *Chhien Han Shu* 前漢書 (History of the Former Han Dynasty), Yen Shih-Ku 顏師古 had to explain that 'at that time dogflesh, just like mutton and

---

[186] *Chou Li, Phao Jên*; p. 36.   [187] *Li Chi, Nei Tsê*; Legge (1885) p. 461.
[188] *Chou Li, Shih I* (Nutritionist–Physician), p. 46. *Li Chi, Nei Tsê*, Legge (1885), p. 461.
[189] *Chou Li, Nei Yung*, p. 38; *Li Chi, Nei Tsê*, Legge (1885), p. 462.
[190] Wittwer *et al.* (1987), pp. 309–11.   [191] *Shih Ching* M 103, 127 (W 258, 260).
[192] Nieh Chêng and Kao Chan-Li: Shih Chi, ch. 86; Yang & Yang (1979), pp. 389–402. Fang Khuai: *Han Shu*, ch. 41, p. 1.
[193] Except in popular novels such as the *Shui Hu Chuan* 水滸傳, (late Yüan to early Ming), in which some of the romantic heroes, not bound by the rules of polite society, readily indulge in the eating of dogflesh (for example, ch. 4). Actually, Marco Polo has noted that the people of Kinsai 'eat all sorts of flesh, including that of dogs and other brute beasts', cf. Latham (1958), p. 220. Today, dogflesh is served openly as a delicacy in restaurants in Beijing and Canton. According to K. C. Chang (private communication) one such Beijing restaurant is located next to the Institute of Archaeology. See also Marvin Harris (1985), p. 180.

pork, was eaten by people, so there were dog butchers who slaughtered dogs and sold their meat'.[194]

The last of the six livestock is the chicken (*chi*), *Gallus gallus domesticus* Brisson, of the pheasant family, the first fowl to be domesticated in China.[195] It was valued both as a source of meat and as an alarm to wake people up early in the morning, a function of no mean significance in a rural agricultural economy. Of the four poems in the *Shih Ching* in which chicken is cited, three are in reference to the crowing of the cock, but none to its use as food.[196] However, the *Chou Li* describes a Chicken Keeper, *Chi Jên* 雞人 who is responsible for providing chickens for ritual offerings and there are several references in the *Li Chi* to the use of chicken as food.[197] Chicken and dog are probably the most common animals of ancient China, found even in the humblest of surroundings. They are the only two domestic animals mentioned in the *Tao Tê Ching* (Canon of the Virtue of the Tao). In a crisp commentary on the idyllic life of the people in a simple village, Lao Tzu says:

> They are happy in their ways.
> Though they live within sight of their neighbours,
> And crowing cocks and barking dogs are heard across the way,
> Yet they leave each other in peace while they grow old and die.[198]

In addition to the *liu chhu*, six livestock, the *Chou Li* also uses the term *liu shêng* 六牲, six food animals for slaughter, which have traditionally been taken to mean cattle, sheep, swine, dog, goose (*yan* 雁) and fish (*yü* 魚).[199] In view of what we have said earlier (p. 55), the omission of the horse in the *liu shêng* is not surprising, but it is curious that chicken should have been replaced by goose. Perhaps it is because the goose was rated more highly as meat than chicken. The inclusion of the goose and fish in the *liu shêng* suggests that hunting and fishing were important activities in the economy of early historic China. There is no evidence that the goose, *Anser* sp. (family Anatidae), had been domesticated during the Shang and Chou dynasties. It is mentioned in four poems in the *Shih Ching*, always in a context which makes it clear that the bird was obtained by hunting rather than husbandry.[200] Other game fowls and mammals also mentioned in the *Shih Ching* include the quail (once), pheasant (once), wild duck (twice), rabbit (four times) and deer (six times).[201] Except for the duck all the animals are cited in the *Chou Li* and or *Li Chi* indicating that hunting was still an important means for the acquisition of meat (i.e. animal flesh, *ju* 肉) during Chou and Han times.[202]

---

[194] Annotation to the *Han Shu* (漢書注), ch. 41, p. 1.
[195] Chang Chung-Ko (*1986*); Hsieh Chhung-An (*1985*). [196] *Shih Ching* M 82, 96, 90, 66 (W 25, 26, 91, 100).
[197] *Chou Li*, ch. 5, p. 211. *Li Chi*, *Yüeh Ling* and *Nei Tsê*; pp. 271, 303, 457, 460.
[198] Tr. Fêng & English (1972), *Tao Tê-Ching*, ch. 80.
[199] *Chou Li*, ch. 1, *Shan Fu* 膳夫 (Grand Chef), and ch. 2, *Shih I* 食醫 (Dietician–Physician), pp. 34, 45.
[200] *Shih Ching* M 82, 78, 159, 181 (W 25, 31, 29, 126).
[201] *Shih Ching*: Quail, W 259 (M 112); pheasant W 274 (70); wild deer W 274 (70); rabbit W 117, 189, 274, 297 (M 7, 231, 70, 197) and deer M 125, 183, 244, 262, 297, 302 (M 156, 161, 242, 180, 197, 257).
[202] *Li Chi*, *Wang Chih* 王制 states that the Son of Heaven conducted three hunts a year. The first was to provide appropriate dried meat to fill sacrificial vessels. The second was to furnish delicacies for entertaining guests. The third was to replenish the supply of meat for the daily use of the royal family. Thus, hunting was a serious way to increase the variety and quantity of the meat supply. Cf., Wang Mêng-Ou ed. (*1984*), p. 220.

Han Tomb No. 1 at Ma-wang-tui near Changsha has also yielded invaluable information on the food animals slaughtered at about the year −170. From the food containers, the remains of animals identified include those of:[203] Mammals: cattle, sheep, swine, dog, deer and rabbit, Birds: chicken, duck, goose, pheasant, sparrow, crane.

The results confirm what we have said all along, that the dog was an important food animal but the horse was not. The presence of deer, goose, pheasant, and other exotic birds may simply be indicative of the owner's high station and does not imply that these were important food animals. The remains of several fish were also found, but this is a topic that properly belongs to the next portion of our survey.

(vi) *Aquatic animals*

> Oh, the Chhi and the Chu
> In their warrens have many fish
> Sturgeons and snout-fish,
> Long fish, yellow-jaws, mud-fish and carp,
> For us to offer, to present
> And gain great blessings.[204]

Fish has always been a highly regarded component of the diet in China. Abundant fishing implements (spears, harpoons and hooks), fish bones and stylised fish designs on pottery were recovered from the early Neolithic communities in the lower Huang Ho valley.[205] Fishing remained an important sector of the economy during the Shang and Chou dynasties. Six species of fish, representing varieties that are still commonly eaten in northern Honan today, were identified among the bones discovered in the Shang ruins of Anyang.[206] References to fish or fishing are found in at least eighteen poems in the *Shih Ching* (Book of Odes). No less than thirteen individual fishes are mentioned. Their original names, English equivalents according to Waley's translation, and scientific designations, as determined recently by Chhêng Chhing-Thai, are presented in Table 5.

The large number of species recognised by the authors of the poems in the *Shih Ching* is indicative of the wealth of fish resources available and the importance of fishing as a means of acquiring food in the lower Huang Ho valley by the Shang and Chou communities. Six families are represented by the species listed in Table 5. Eight species, carp (*li* 鯉), lucky fish (*chia yü* 嘉魚), roach (*kuan* 鰥), rudd (*tsun* 鱒), bream (*fang* 魴), mud fish (*yen* 鰋), long fish (*thiao* 鰷) and tench (*hsü* 鱮) are from one family alone, the Cyprinidae, a clear evidence of the special significance of the carp

---

[203] *Hunan nung-hsüeh-yuan* (*1978*), pp. 47–74.
[204] *Shih Ching*: M 282 (W 223) tr. Waley. The *Chhi* 漆 and the *Chu* 沮 are rivers in the ancestral homeland of the Chou tribe in Shensi. Their location is no longer known.
[205] Chang, Kwang-Chih (1977), p. 97, Chhiu Fêng (*1982*) and Wu Shih-Chhi (*1987*).
[206] Wu Hsien-Wên (*1949*). The actual species identified are: *Pelteobargrus fulvidraco* (Richardson), *Cyprinus carpio* L., *Mylopharyngodon aethiops* (Basil), *Ctenopharyngodon idellus* C. & V., *Squaliobarbus curriculus* Richard and a *Mugil* sp. All are well-known fresh water fish, except the *Mugil*, which is found normally along the southeastern coast of China.

Table 5. *Species of fish cited in the* Shih Ching

Chinese name		English name[a]	Latin name[b]	Family	Waley
Chan	鱣	Sturgeon	*Acipenser sinensis*	Acipenseridae	86
Wei	鮪	Snout fish	*Psephurus gladius*	Psephuridae	86
Li	鯉	Carp	*Cyprinus carpio*	Cyprinidae	168
Chia yu 嘉魚		Lucky fish	*Varicorhinus simus*	Cyprinidae	169
Kuan	鰥	Roach	*Elopichthys bambusa*	Cyprinidae	85
Tsun	鱒	Rudd	*Squaliobarbus curriculus*	Cyprinidae	29
Fang	魴	Bream	*Megalobrama terminalis*	Cyprinidae	29
Yen	鰋	Mud fish	*Culter erythroopterus*	Cyprinidae	168
Thiao	鰷	Long fish	*Hemiculter leucisculus*	Cyprinidae	223
Hsu	鱮	Tench	*Hypophthamichthys molitrix*	Cyprinidae	59
Chhang	鱨	Yellow jaw	*Pseudobagrus fulvidraco*	Leiocassisidae	168
Sha	鯊	Eel	*Rhinogobius giurnius*	Gobiidae	168
Li	鱧	Tench?	*Ophiocephalus argus*	Ophiocephalidae	168

[a] After Waley's translation.
[b] According to Chhêng Chhing-Thai (*1981*).

family in Chinese life. The remaining five species are from five families, the Acipenseridae (sturgeon), Psephuridae (snout fish or swordbill fish), Leiocassisidae (yellow jaw), Gobiidae (eel?) and Ophiocephalidae (tench?). The sturgeon, *chan* 鱣, and its relative the swordbill sturgeon, *wei* 鮪, are now commonly known as *hsin* 鱘 and *pai hsin* 白鱘, and remain among the most highly prized fish of China.

In addition to names of fish, the *Shih Ching* also describes various methods used to catch them, for example:

Fishing pole: 'How it tapered, the bamboo rod,
With which you fished in the Chhi!'[207]
Fish nets: 'Fish nets we spread;
A wild goose got tangled in them.'[208]
Fish traps: 'Do not break my dam,
Do not break my fish traps.'[209]

The existence of dams or dikes along which bamboo traps are laid across appropriate openings for catching fish is suggestive of the beginning of aquaculture in China. Surely it required no great leap of imagination to go from building a dike along a river or lake to enclose a body of water to the excavation of an artificial pond. Indeed, the *Shih Ching* itself may have provided the earliest example of aquaculture on record. In the poem *Ling Thai* 靈臺 (Magic Tower) it is said that when the King visited the Magic Park, the 'fish sprang so lithe' in the Magic Pond by the

---

[207] *Shih Ching* w 43 (M 59).   [208] *Shih Ching* w 77 (M 43).   [209] *Shih Ching* w 108 (M 35).

Magic Tower.²¹⁰ The magic pond may be construed as a prototype of a fish rearing pond.

The importance of fishing and a primitive form of aquaculture in the life of the Chou community is further illustrated by the description of the duties of the *Yü Jên* 漁人 (Fishery Superintendent) in the *Chou Li*. The *Yü Jên* directs a staff of 14 assistants, 30 clerks and 300 labourers 'to maintain the basket traps along dikes so as to supply fish according to season. In the Spring he presents the *wei* sturgeon and other kinds of fish, either fresh or in a dried form, for the King's table. He provides fresh or dried fish as needed for ceremonial offerings, for entertaining guests and for funerary sacrifices. He administers official ordinances regulating fishing and collects taxes from fisherman for the King's treasury.'²¹¹

As we have noted earlier, fish was esteemed as a separate category of food (p. 17) by the legendary chef I Yin in the *Lü Shih Chhun Chhiu*, which proclaimed *fu* 鮬 from the Tungting Lake and *er* 鯉 from the Eastern Sea as the most delectable fish of the realm. The exact meaning of *fu* is uncertain. It is also written as *fu* 鮒, and is probably the aquatic mammal, the river dolphin (or river pig), *chiang thun* 江豚, *Neomeris phocaenoides*.²¹² *Er* may originally have been the name of a fish, but the *Shuo Wên Chieh Tzu* (+121) says it means fish roe, which would be an admirable interpretation if it applies to the roe of the large sturgeons of the lower Yangtze.

Six species of fish have been identified from the array of fish bones found among the food remains discovered in Han Tomb No. 1 at Ma-wang-tui.²¹³ They are *li* 鯉 (Figure 20), *Cyprinus carpio* Linne, *chi* 鯽, *Carassius auratus* (Linne), *chhi pien* 刺鯿, *Acanthobromo simoni* Bleeker, *yin ku* 銀鮕, *Xenocypris argentea* Gunther, *kan* 鱤, *Elopichthys bambusa* (Richardson) and *kuei* 鱖, *Siniperca* sp. The first five are from the carp family, and the sixth is a member of the Serranidae. We have come across *li* and *kan* (a more modern name for *kuan* 鰥) earlier as fish cited in the *Shih Ching* (Table 5). The presence of these fish in the tomb serves to underscore the high status that fish enjoyed as a gastronomic resource in ancient China. The prestige of fish continued to grow in the succeeding centuries. Indeed, during the mediaeval period, as the economic centre of gravity moved southward, the most prosperous part of the country such as the lower Yangtze valley, became the proverbial 'land teeming with fish and rice' (*yü mi chih hsiang* 魚米之鄉), an expression equivalent to the biblical 'land flowing with milk and honey' so familiar in the West.²¹⁴

Fish, however, are not the only aquatic animals used as food. A special favourite in ancient China is the turtle. The *Yü Kung* 禹貢 chapter of the *Shu Ching*, Book of

---

²¹⁰ *Shi Ching* w 244 (M 242). Chou Su-Phing (*1985*), *NYKK* 1985/2, 64–70, argues that the magic pond itself is already an example of aquaculture in practice. The earliest work on aquaculture in China is the *Yang Yü Ching* 養魚經 (Manual of Fish Culture) attributed to Thao Chu-Kung of the Chou Dynasty. Actually, its author is unknown. It is incorporated in ch. 6 of the *Chhi Min Yao Shu* (+544).
²¹¹ *Chou Li*, ch. 1, p. 42.
²¹² Chhiu Phang-Thung, Wang Li-Chhi & Wang Chêng Min (*1983*) p. 51. *Chiang thun* is commonly found at the mouth of the Yangtze River, but is known to swim upstream to as far as Ichang and often sojourns at Tungting Lake.
²¹³ *Hunan nung-hsüeh-yuan* (*1978*), pp. 74–82. ²¹⁴ Exodus 3: 17; 13: 5.

必河之鯉

Fig 20. Carp, *Mōshi Himbutsu Zukō*, 7/3a.

Documents (c. −600), records the presentation of giant turtles (*ta kuei* 大龜) as tribute from the doomsday province of Ching-Chou to the Emperor's court. The *Shih Ching* mentions roast turtle (*pieh* 鱉), (Figure 21) as a delicacy served together with fish at a feast in honour of a visiting Lord, referred to earlier on p. 36, and again at a drinking party to celebrate a victorious campaign.[215] This is the soft-shelled fresh water turtle, *Trionyx sinensis* (*Amygda sinensis*), family Trionychidae.[216] Stewed *pieh* turtle is one the delectables presented in the *Chhu Tzhu* (Elegies of the South) to entice a lost soul to return to its old home.[217] The *Chhu Tzhu* also mentions *hsi* 蠵, *Caretta*

---

[215] *Shih Ching* w 144, 133 (M 261, 177). [216] Schafer, Edward H. (1962), also (1967), pp. 214–16.
[217] *Chhu Tzhu, Chao Hun*; Hawkes (1985), p. 228.

南有嘉魚

箋南方水中有善魚
傳嘉魚鯉質鱒鰓
出於丙穴
嚴緝下文樛木非
則嘉魚亦非
名木
魚

○肌
集
名
苞鼈龜鱠鯉

Fig 21. Turtle, *Mōshi Himbutsu Zukō*, 7/5b.

caretta olivacea, family Cheloniidae, a giant sea turtle of South China.[218] Another prized turtle, *yüan* 黿, the large soft-shell *Pelochelys bibroni*, also family Trionychidae, was cited in the *Tso Chuan* (Master Tso's Enlargement of the Spring and Autumn Annals), c. −605.[219] All these turtles are collectively and popularly known as *chia yü* 甲魚 (shelled or armoured fish). They are esteemed highly in China today. The importance of turtle in ancient Chinese cuisine is also reflected in the *Chou Li*, which

---

[218] *Chhu Tzhu, Ta Chao*; Hawkes (1985), p. 234, translates it as turtle.
[219] *Tso Chuan*, Hsuan Kung 宣公, 4th year, p. 184.

lists a *Pieh Jên* 鱉人 (Shellfish Keeper) who was responsible for providing the King's kitchen with tortoise (*kuei* 龜), and turtle (*pieh* 鱉), as well as clam (*chhên* 蜃), oyster (*pang* 蚌), snail (*luo* 蠃) and ant eggs (*chhi* 蚳).[220] It is curious that two well-known shellfish, shrimp and crab, are not included, although both are mentioned in the *Shên Nung Pên Tshao Ching* (Pharmacopoeia of the Heavenly Husbandman).[221] The Shellfish Keeper, however, clearly occupied a less prestigious position than the Fishery Superintendant, since he was supported by only eight assistants and sixteen labourers.[222]

(2) Ancient Chinese Culinary System

Having completed our survey of the food resources in ancient China we are now ready to consider the remaining components of the culinary system, that is to say, how the raw food was prepared and cooked, what flavouring agents were used, and how the finished products were served and eaten. But before we move on to these matters, let us briefly recapitulate what we have found to be the principal food resources in China from the Chou to the Han dynasties:

Grains: the principal grains were millets (*setaria* and *panicum*) in the North and rice in the South, with wheat, soybean and hemp seed as minor grains. By the end of the Han, millets had lost ground to wheat in the North while soybean continued to grow in importance.

Fats and oil: before the Han most of the fat used in cooking was animal fat (lard, beef, sheep or dog tallow). Large-scale processing of vegetable oil from the Chinese turnip, sesame and hemp seeds probably started during the Han.

Vegetables: major vegetables were melon, gourd, Chinese turnip, radish, oriental celery, lotus root, bamboo shoot, mallow, smartweed, lettuce, leek, spring onion, shallot, rocambole, garlic, ginger, peas, red bean, taro, water caltrop, and mushrooms.

Fruits: major fruits were peach, apricot, *mei* apricot, Chinese cherry, pear, persimmon, loquat, orange, pomelo, jujubes, hazelnut, chestnut, quince, fagara and cassia.

Animals: land animals reared as food were cattle, sheep, pigs, dogs and hens. Wild animals hunted as food included deer, rabbit, goose, duck, pheasant and quail. Fish and turtles were highly prized; clams, oyster, snails and other shellfish were also eaten.

Overall, the list is an impressive one, indicative of the wealth of food resources, particularly those of plant origin, that were domesticated, cultivated or harvested from the wild by the ancient Chinese. Only a few items were imports from the West, for example, wheat, sesame (possibly), and garlic (or rocambole), but the majority, an astonishing number, are indigenous to China. Indeed, as noted by Valvilov 'In wealth of its endemic species and in the extent of its genus and species potential of

---

[220] *Pieh Jên*, ch. 1, pp. 2, 43. The *Chou Li* also lists a *Kuei Jên* 龜人, Tortoise Keeper, who supervised the collection and care of tortoises whose shells were to be used for divination. The *Kuei Jên* had a staff of six assistants, eight clerks and forty labourers; cf. pp. 185, 254.
[221] *SNPTC*, crab in Middle Class and shrimp in Lower Class of drugs. [222] *Chou Li*, p. 2.

its cultivated plants, China is conspicuous among other centres of origin of plant forms. Moreover, the species are usually represented by enormous numbers of botanical varieties and hereditary forms.'[1] Most of the products listed can still be seen today in the food markets of China and in many other parts of the world. A number have gone out of fashion, or been replaced by new varieties bred in China or introduced from abroad, such as the mallow, smartweed or oriental celery. One would be hard pressed to have a meal in China or anywhere else today with panicum millet as the cereal, mallow as a vegetable, and dogflesh as part of a main course. Many new crops have entered China during or since the Han dynasty and have established themselves firmly on the Chinese food scene;[2] their effect on the Chinese diet has been enormous.[3] We may have to touch upon specific instances of such introductions in later parts of this study, but for the present, we need to proceed to our next task and consider how the ancient Chinese turned the resources available to them into actual articles of food and drink to be partaken at a meal, in other words, how they prepared and cooked the raw food.

(i) *Food preparation and cooking: methods and utensils*

> We pound the grain, we bale it out,
> We sift, we tread,
> We wash it, we soak it,
> We steam it through and through.[4]

The *Shih Ching* 詩經 (Book of Odes), −1100 to −600, has given us the earliest literary record of methods used for preparing and cooking food in ancient China. In addition to the case of 'steaming' (*chêng* 蒸) mentioned in the verse from the poem *Shêng Min* 生民 (Origin of the People) quoted above, other methods of cookery such as boiling, baking, roasting and pickling are also mentioned. We can find references to the same methods or variations thereof in later classics, such the *Chhu Tzhu* 楚詞 (Elegies of Chhu), c. −300, *Chou Li* 周禮 (Rites of the Chou), *Li Chi* 禮記 (Record of Rites) and in two documents discovered in Han tombs at Ma-wang-tui 馬王堆 near Changsha, firstly, a set of bamboo slips inventorying the foods that were stored (*MWT Inventory*) in Tomb No. 1 and secondly, a manuscript written on silk, now known as the *Wu Shih Erh Ping Fang* 五十二病方 (Prescriptions for Fifty-two Ailments or *52 Ping-fang*) from Tomb No. 3. Although the extant text of the *Chou Li* and *Li Chi* were reconstituted in the Western and Eastern Han respectively, much of the

---

[1] Vavilov (1949/50), p. 26. He goes on to say in a footnote that 'If we take into account the enormous number of wild plants, besides the cultivated ones used for food in China, we may better understand how hundreds of millions of people managed to exist on its soil.'
[2] For example, grape, alfalfa, black pepper, chili pepper, sweet potatoes and corn.
[3] Anderson (1988), pp. 112–48; Simoons (1991). For discussions on Chinese cookery in antiquity see Hu Chih-Hsiang (*1994*) and Liu Chün-Shê (*1994a,b* and *1995*).
[4] w, 238 (M 245), mod. auct. Waley renders the last line as, 'We boil it all steamy', while the original text leaves no doubt that the grains were 'steamed', presumably in a steamer, cf. Karlgren (1950), p. 201. Steaming of grain is also implied, though not explicitly stated, in M 251 (W 173).

material they contain must have originally belonged to the late Chou. In addition, they include valuable commentaries provided by scholars in the Eastern Han period. From the date of Han Tomb No. 1 at Ma-wang-tui (−168) we may assume that the information recorded in the *MWT Inventory* and the *52 Ping-fang* would be valid for the year of about −200. Thus, the five documents together should provide us with a reasonable résumé of the state of cookery in China from early Chou to late Han. What we propose to do now is firstly, to list and review the methods of cooking depicted in the five documents, and secondly, relate them to the utensils available for cooking during this period as indicated by the archaeological record, and thirdly, to identify the type of foods prepared by the methods described in terms of the practices of modern Chinese cookery.

But first we need to have a look at how the raw foods were prepared before they were cooked. We have seen, in the quotation from the poem *Shêng Min*, how the grain, presumably millet or rice, was washed and soaked before it was cooked by steaming to become *fan* 飯 (cooked grain). It is generally conceded that wheat is too hard to be treated this way.[5] It has to be milled into *mien* 麵 (flour) before it can be processed into edible products. The processing of wheat grain into flour, and the flour into a whole range of breads and pastas is a subject unto itself, and will be treated separately in Chapter (*e*). This technology began to take off in earnest in the Han dynasty and eventually led to the division of China into two dietary zones: the South (with pockets in the North) where grain is eaten as cooked granules, *li shih* 粒食; and the North which thrives largely on products from wheat flour, *mien shih* 麵食.[6] But for most of the period we are concerned with, all China may be considered as a *li shih*, granule eating, zone.

Fruits were usually eaten raw. Vegetables were presumably cut into pieces and cooked. It is in regard to meat, especially that from large animals, that preparation before cooking took on a new dimension in the culinary process. After the animal was killed and flayed,[7] it had to be cut into sections small enough to be further processed in the kitchen. This required the service of a skilled professional such as the butcher of King Hui of Liang. We read in the *Chuang Tzu* 莊子 (The Book of Master Chuang), c. −290:[8]

Ting, the butcher of King Hui, was cutting up a bullock. Every blow of his hand, every heave of his shoulder, every tread of his foot, every thrust of his knee, every sound of the rending flesh, and every note of the movement of the chopper, were in perfect harmony-rhythmical like the *Mulberry Grove* dance, harmonious like the chords of the *Ching Shou* music.

---

[5] Cf. Bray, *SCC* Vol. VI, Pt 2, p. 461; Shinoda Osamu (*1987a*), p. 20.

[6] Shinoda Osamu (*1987a*), p. 51. For a discussion of the development of *mien shih* during the Han, cf. Hsu Cho-Yun (*1993*). The concept of *li shih* was noted in the late Chou by *Mo Tzu*, p. 190, and in the *Li Chi*, p. 230. Indications are that *mien shih* was already a part of the dietary experience at the time of the Warring States.

[7] The flaying of an ox or sheep is mentioned in the *Shih Ching*, M 209, line 15, which reads, *Huo po huo phêng* 或剝或烹. It is translated literally by Karlgren as, 'Some flay, some boil' but the line is bypassed by Waley (w 199) who renders it as 'Now baking, now boiling.'

[8] *Chuang Tzu*, ch. 3, tr. Legge (1891), p. 198 cited in Needham, *SCC* Vol. II, p. 45, cf. also Lin Yutang (1948), p. 216, and Watson (1968), p. 50.

In the kitchen, the sections were then cut into large chunks, *tzu* 胾, thin fine slices or strips, *khuai* 膾, large slices, *hsüan* 軒, or chopped into minced meat suitable for pickling into *hai* 醢. Fish in thin slices were also *khuai*; in large pieces they were called *hu* 膴.[9] The cutting process itself was considered an integral part of cookery. Indeed, in the *Chou Li*[10] and in the *Mêng Tzu* (Book of Mencius), c. −290,[11] culinary art is referred to as *ko phêng* 割烹, that is, cutting and cooking, an expression that was later transported to Japan where it is still in use today. In the preparation of *tzu* 漬, one of the Eight Delicacies, the *Li Chi* admonishes that the meat be cut across the grain, which would make the slices more tender and more easily digestible.[12] This is an important point gastronomically as well as nutritionally since meat was sometimes eaten raw, especially in the *khuai* 膾 form.[13] In a familiar passage in the *Lun Yü* (The Analects), c. −450, about his alleged fussiness on food, it is said of Confucius that 'He did not dislike to have his rice finely cleaned, nor to have his *minced meat* cut quite small.'[14] and that 'He did not eat meat which was *not cut properly*, nor what was served without its proper sauce.'[15] We submit that the translation does not do justice to the Master's original intent. In these quotations the expression used for *minced meat* was *khuai* 膾 and for *not cut properly*, *ko pu chêng* 割不正, which literally means 'cut not straight'. If *khuai* had meant minced meat it would be difficult to conceive how it

---

[9] These interpretations are based on *Li Chi*, *Nei Tsê*, pp. 456, 459, 463, and *Chou Li*, *Pien Jên*, p. 54.
[10] *Chou Li*, *Nei Yün*, p. 38.   [11] *Mêng Tzu*, Book v, Pt 1, ch. 7, verse 1; Legge (1895), p. 361.
[12] The relevant passage occurs in *Nei Tsê*, p. 469. It says, 'For the *steeped delicacy* take beef from a freshly killed animal and cut it into thin slices. It is imperative that the *meat be cut across the grain*. Steep the slices in good wine for a couple of days and season them with meat sauce, plum juice or vinegar before eating.' For us, the key point here is that the meat must be cut across the grain. The original phrase is: *pi chuêh chhi li* 必絕其理, which has been translated literally by Legge (1885), p. 469, as 'to obliterate all the lines in it'. The culinary reason for this admonition is explained by Howard Hillman (1981), p. 52. In advising how one should cut flank steak for London broil, he says, '*To cut across the grain* means to slice the connective tissue at right angles. Obviously, the thinner the slices, the shorter the resulting connective tissue segments. The shorter the segments, the less chewy the meat, and therefore the more tender it will be to the eater.' One might add, that since the connective tissue is more resistant to enzymic degradation than muscle, the meat would be digested more readily than if it had been cut with the grain.
[13] The above passage also shows that the meat was eaten virtually raw. Yet to many a competent scholar today it seems inconceivable that meat was sometimes eaten raw in China at the time of Confucius. Indeed, in his edition of the *Li Chi*, Wang Mêng-Ou (p. 469) translated into modern Chinese the relevant passage in *Nei Tsê* on the preparation of *tzu* 漬, the *steeped delicacy*, as follows: 'Cut the meat across the length of the muscle fibres into thin slices. *Cook thoroughly* and steep the slices in good wine.' The step, *cook thoroughly*, is found nowhere in the text. Evidently, the notion that during the Chou period meat could have been steeped and eaten without first cooking it is so preposterous to our learned annotator that he regarded the omission of the cooking step as simply an oversight on the part of the original author, and felt compelled to correct the error in his translation. See also Sun Chung-En (*1985*) for a discussion of this passage.
  Although the practice of eating meat prepared without cooking seems to have declined during the Han, it did not entirely disappear even after several centuries. The *Chhi Min Yao Shu*, p. 422, describes the preparation of *Shêng Shan* 生脠, as follows: 'Take one catty of mutton, four ounces of pork and marinate in clear soy sauce. Cut the meat into fine threads. Dress it with ginger, chicken egg and, in Spring and Autumn, also with perilla and smartweed.' The dish is certainly reminiscent of 'steak tartare,' still a favourite among the gastronomic delights of the modern West.
[14] *Lun Yü*, *Hsiang Thang*, Bk x, ch. 8, verse 1, Legge (1861), p. 232. The significance of this statement in relation to the food habits of Confucius has been extensively discussed by Chao Yung-Kuang (*1990*), p. 20 and Ma Chien-Ying (*1991*), p. 19.
[15] *Ibid.* verse 3, Legge (1861), p. 232. Although the original text merely says, *ko pu chêng pu shih*, 'what is not cut straight is not eaten,' we agree with Legge that meat must be one of the foods to be considered in this context.

Table 6. *Methods of cooking revealed in Chou and Han classics*

Document	Shih Ching Mao	Chou Li Lin	Chhu Tzhu	Li Chi Wang	MWT Inventory	52 Ping-fang
Chêng 蒸	245	PJ54	TC	NT460	125	+
Phêng 烹	149	NY38	IIWJ	LY367	—	+
Chu 煮	—	PHJ40	—	—	—	++
Chien 煎	—	NY38	TC	NT467	126	+
Ao 熬	—	SJ178	—	NT469	+	+
Fan 燔	231	—	—	LY366	—	++
Chih 炙	231	—	TC	LY456	+	+
Phao 炮	231	FJ124	CH	NT468	?	+
Chuo 濯	—	—	TC	—	+	—
Tsu 漬	—	(+)	—	NT468	—	+
Khuai 膾	177	(+)	TC	NT456	+	—

Shih Ching Book of Odes: the numbering of poems is based on Mao's sequence.
Chou Li, Rites of the Chou: PJ, *Pien Jên* 籩人; NY, *Nei Yung* 內饔; PHJ, *Phêng Jên*, 烹人; SJ, *Shê Jên*, 舍人; FJ, *Fêng Jên* 封人. The numbers are pages in the Lin Yin (*1984*) edition of Chou Li.
Chhu Tzhu, Elegies of Chhu: TC, *Ta Chao* 大招; CH, *Chao Hung* 招魂; HWJ, *Hsi Wang Jih* 惜往日.
Li Chi, Record of Rites: NT, *Nei Tse*, 內則; LY, *Li Yün*, 禮運; the numbers are pages in the Wang Mêng-Ou (*1970*) edition of *Li Chi*.
MWT Inventory is based on the bamboo slips found in Han Tomb No. 1 at Ma-wang-tui. The numbering follows that listed in the MWT Report (*1973*). A '+' means that more than one slip was found for the type of food.
52 Ping-fang refers to the Prescriptions for 52 Ailments found in Han Tomb No. 3 at Ma-wang-tui, which applies many methods of cookery to the preparation of ointments and decoctions. A '+' means that the method occurs more than once, and a '++' means that it occurs more than twenty times.

could have been 'not cut properly'. However, if *khuai* had meant small thin slices or threads of meat, the whole passage becomes clear. Meat that was not sliced properly would be meat cut in the wrong direction, that is, with the grain rather than across the grain. The slices or threads when eaten raw would be difficult to chew and digest. The proper sauce was, of course, essential in masking the raw taste of meat, which might otherwise be considered as less than attractive to a refined palate.[16] Thus the passage is an indication of the Master's knowledge of the culinary arts and good nutritional practice, rather than an unreasonable fussiness about food. As far as we know, the practice of cutting meat or fish into *khuai* pieces is unique to China among ancient cultures. It may be an important factor in the evolution of the Chinese cuisine.

Table 6 lists the methods of cooking found in the Chou and Han classics that we have referred to earlier. The first is *chêng* 蒸, steaming, which is mentioned in the

---

[16] *Li Chi, Nei Tsê*, gives several examples of the proper kind of sauce to complement a particular *khuai*, for example, meat sauce with beef *khuai*, mustard sauce with fish *khuai* and meat sauce with fresh venison, cf. Wang Mêng-Ou ed. (*1984*), pp. 456, 457.

Shih Ching for cooking grain to *fan* 飯 (see p. 67). *Fan* may well have been the most important food prepared by steaming, but a whole range of products were also cooked in the same way: cakes or pastry (*er* 餌 in the *Chou Li*), duck (*Chhu Tzhu*), fish (*Li Chi*), young chicken (*MWT Inventory*) and herbs (*52 Ping-fang*).[17] In addition, according to the Book of Mencius a steamed suckling pig was once presented to Confucius when he was away from home.[18]

The next method is *phêng* 烹, boiling (or heating in water), certainly a most basic and versatile technique of cooking. The word *phêng*, of which the archaic form is *phêng* 亨, is also used to indicate cooking in general. Indeed, of the Chou references on *phêng* listed, only the *Shih Ching* cites its use on a specific food, fish or mallow; all the others, from the *Chou Li*, *Chhu Tzhu*, and *Li Chi* relate to *phêng* as a method of cooking in general. Although *phêng* is not mentioned in the *MWT Inventory*, it is cited in the *52 Ping-fang* for the boiling of several decoctions.[19] We also find *phêng* used in the *I Ching* 易經 (The Book of Changes), in relation to the function of the *ting* 鼎, the cauldron,[20] and in the *Tao Tê Ching* (the Taoist Canon), where it says that 'ruling the country is like *cooking* (*phêng*) a small fish'.[21] It is the word used to describe the process of *cooking* a fish soup (*kêng* 羹) in the *Tso Chuan Chhun Chhiu* 左傳春秋 (Master Tso's *Spring and Autumn Annals*).[22]

To modern readers a more familiar word that denotes cooking with water is *chu* 煮. *Chu* is said to be derived from the ancient form *chu* 鬻 which is listed in the *Shuo Wên Chieh Tzu* (Analytical Dictionary of Characters), +121.[23] It is not found in the *Shih Ching*, *Chhu Tzhu* or *Li Chih*, but it is mentioned in the *Chou Li* in connection with the duties of the *Phêng Jen* 烹人, the Grand Chef.[24] It is not mentioned in the *MWT Inventory*, but it is the most frequently cited word seen in the *52 Pingfang* to indicate the process for making decoctions and soups.[25] Presumably, *chu* had acquired common currency by the beginning of the Han dynasty, and it has remained the accepted term to describe cooking in the presence of water until today. Indeed, the best-known poem in the Chinese literature on cooking starts with the word, *chu*. In the late Three-Kingdoms period when the Dynasty of Wei was first founded, the emperor Tshao Phi 曹丕 was jealous of his younger brother Tshao Chih 曹植, who was much admired and beloved as a poet. One day the emperor commanded Tshao Chih, at the risk of his life, to compose, within the time he took to pace seven steps, a poem on the subject 'cooking beans'. To the pleasant surprise of the nervous

---

[17] *Chou Li*, p. 54; *Chhu Tzhu*, p. 172, *Li Chi*, p. 460; *MWT Inventory*, no. 125 and *52 Pingfang*, no. 24, 110, 128, 184, 266, 269. For an account of the methods of cookery available during the Neolithic Period see Yang Ya-Chhang (*1994*).
[18] *Book of Mencius*, Bk III, Part 2, Ch. 7, verse 3; Legge tr. (1895), p. 277; D. C. Lau (1970), p. 112.
[19] *Shih Ching*, *pheng* used on fish, M 149 (W 149); on mallow, M 154 (W 159). *52 Pingfang*, nos. 57, 97, 98, 120.
[20] *I Ching*, Hexagram 50, line 2, tr. Legge (1973) edn, Causeway Books, p. 170.
[21] *Tao Tê Ching*, ch. 60, line 1.
[22] *CCTC*, Chao Kung, 20th year, p. 403. In Legge (1872), p. 684, *phêng* is translated as 'cook'.
[23] *SWCT*, p. 63 (chs. 3b, 6).   [24] *Chou Li*, *Phêng Jên*, p. 40.
[25] *52 Pingfang* shows the word *chu* used in twenty-four recipes, e.g. 19, 31, 35, 38, 40 etc.

onlookers, Tshao Chih did indeed compose a poem, which may be translated as follows:[26]

> Outside, the bean bush burns roaringly hot,
> While beans weep bitterly inside the pot.
> Sir, we're all born of the very same seed,
> Why then press us with such unseemly speed?

The poem made obvious and pointed reference to the fact that both Tshao Phi and Tshao Chih had the same mother, and implied that Tshao Phi's treatment of his brother was unduly vindictive in terms of the standards of Confucian ethics. The emperor felt ashamed and let his brother go free.

In the last line of the poem we are introduced to the word *chien* 煎 which is translated here as 'press'. *Chien* does mean to 'press' or 'harass' but in culinary terms it means to cook in the presence of a little oil, that is to shallow fry or grill, and it is the next term listed in Table 6. The *Fan Yen*, Dictionary of Local Expressions, explains that *chien* means to cook until the food is dry.[27] *Chien* is not found in the *Shih Ching*, but it occurs in the *Chou Li* (as a duty of the Internal Culinary Officer), *Chhu Tzhu* (fried bream and fried flesh of the giant crane), *Li Chi*, (fried minced meat for the Rich Fry), *MWT Inventory* (fried chick) and in several prescriptions of the *52 Pingfang*.[28] Since the Han dynasty *chien* has remained as the generally accepted word for shallow frying in the Chinese culinary vocabulary.

The next word is *ao* 熬 (ancient *ao* 敖), which according to the *Shuo Wên* means to fry without adding fat (*kan chien* 干煎).[29] The *Fan Yen* says *ao* is to dry by heat (*huo kan* 火干) as grains were often dried.[30] It is not found in the *Shih Ching* or used in a culinary sense in the *Chhu Tzhu*.[31] Its meaning in the *Li Chi* is somewhat ambiguous. In the recipe for *ao* (the Grill), one of the Eight Delicacies (*pa chen* 八珍), the meat was simply air dried. On the other hand, in the recipe for *chhun ao* 淳熬 (the Rich Fry), quoted earlier on p. 28, frying with heat was evidently involved.[32] Thus we cannot be absolutely clear as to what is meant when the food was cooked by *ao* except that no water and probably no fat was involved. In another passage the *Li Chi* states

---

[26] The original reads:

Chu tou jang tou chi	煮豆燃豆箕
Tou tsai fu chung chhi	豆在釜中泣
Pên shih tung kên shêng	本是同根生
Hsiang chien ho thai chi	相煎何太急

This is an abridged version of a poem recorded in the *Shih Shuo Hsin Yü*, p. 60. The authentic version has two more lines, but the meaning remains the same. Cf. Gary Lee (1974), p. 122.

[27] *Fan Yen*, ch. 7, p. 7b.

[28] *Chou Li, Nei Yung*, p. 38; *Chhu Tzhu, Ta Chao*, p. 172; *Li Chi, Nei Tsê*, p. 467; *MWT Inventory*, item 126; and *52 Pingfang*, nos. 11, 12, 20, 23, 25, 176 etc.

[29] *SWCT*, p. 208 (chs. 10a, 20).   [30] *Fan Yen*, ch. 7, pp. 7a, 7b.

[31] *Chien-Ao* occur together in a non-culinary context in ch. 17, *Chiu Ssu* (Nine Longings), *Yuan Shang* (Resentment against the Ruler), p. 254; see David Hawkes (1985), p. 309 (line 15), 'My heart within is scorched and burning.'

[32] *Li Chi, Nei Tsê*, 467, 469; cf. Legge (1885), pp. 468–9, who translates *ao* as 'grill'.

that baskets of *ao* processed grains are placed next to the casket at the funerals of important personages.[33] Chêng Hsuan's commentary says that *ao* means *kan chien* 干煎. *Ao* is referred to in the *52 Pingfang*,[34] where its meaning is also equivocal. But in spite of these ambiguities, ten bamboo cases of different animal *ao* meats ranging from rabbit, goose, chicken, crane to sparrow were listed by the *MWT Inventory*,[35] indicating that *ao* was a method of cooking of some importance in early Han.

So far, in the methods described, the food being cooked is shielded from the fire by some type of vessel. We now come to three ancient techniques in which no cooking vessels are used. The first is *fan* 燔, roasting, where a large chunk of meat or a whole small animal is hung directly over the fire, in the manner of the roasting of Peking duck. *Fan* is mentioned several times in the *Shih Ching*[36] and *Li Chi*,[37] and frequently in the *52 Pingfang*. The second is *chih* 炙, which is to place small pieces of meat on a skewer and roast or broil over a charcoal fire, similar to the shish kebab or shashlik of West Asia. It is often mentioned together with *fan* in the *Shih Ching*.[38] We encounter such *chih* meats as roast crane in the *Chhu Tzhu*, roast beef, lamb and pork in the *Li Chi*, and eight kinds of roast meats (three beef parts, two dog parts, pig, deer and a chicken) in the *MWT Inventory*.[39] The *52 Pingfang* also cite *chih* in a number of prescriptions.[40] The last technique is *phao* 炮, which means to wrap the meat up in vegetable leaves, paste clay all around it and bake it directly in the fire. When it is cooked, the clay crust is removed and the meat is cut up and served. *Phao* is cited in the *Shih Ching* and *Chhu Tzhu* (roast kid).[41] It is the method used to prepare one of the Eight Delicacies listed in the *Li Chi* (roast suckling pig).[42] It is not mentioned in the *MWT Inventory*, and it is used in the *52 Pingfang* only once. However, this prescription, recommended for treating haemorrhoids, is of such unusual interest that it is worth quoting in full,[43] 'Choke a yellow hen with soy paste and let it die. Wrap it in miscanthus leaves. Daub wet clay over it and bake it in the fire. When the crust is dry (and well done), eat the chicken. Use the feathers to fumigate the perineum.' We do not know how efficacious the prescription is, but the reader may recognise it as the world's oldest recipe for clay-wrapped chicken, better known today as 'beggar's chicken', which one can still find on the menu of some gourmet Chinese restauraunts in East Asia. In fact, *phao*, clay-wrapped baking, has always remained a rather exotic method of cookery, presumably because of the inordinate amount of work that would have to be done in order to execute it in an ordinary kitchen.[44]

---

[33] *Ibid. Sang ta chi* 喪大記, p. 732. According to Chêng Hsüan the fragrant aroma of the *ao*-cooked grains attract ants away from the corpse.
[34] *52 Pingfang*, e.g. nos. 30, 132, 164, 178, 185, 203.   [35] *Hunan shêng po wu kuan* (1973), I, pp. 115–16.
[36] *Shih Ching*, M 231, 238, 245, 246, 248 (W 189, 249, 238, 197, 203).
[37] *Li Chi, Li Yung*, pp. 366, 367, 369.   [38] *Shih Ching*, M 231, 246, 248 (W 189, 197, 203).
[39] *Chhu Tzhu, Ta Chao*, p. 172; *Li Chi, Nei Tsê*, p. 456; *MWT Inventory, Hunan shêng po wu kuan* (1973), I, p. 134.
[40] *52 Pingfang*, nos. 36, 103, 122, 138, 201, 203 etc.   [41] *Shih Ching* M 231 (W 189); *Chhu Tzhu, Chao Hun*, p. 269.
[42] *Li Chi, Nei Tsê*, p. 468.   [43] *52 Pingfang*, no. 149 (line 258), tr. Harper (1982) mod. auct.
[44] Harper (1984), pp. 46–7. There are few examples of *phao* cookery in the Chinese food literature. One interesting variation is recorded in the *CMYS* (ch. 77, pp. 479–80) in which finely sliced threads of lamb dressed with appropriate seasonings encased in sheep's intestine is directly roasted in hot ashes. The product is said to be exceptionally delicious.

The next method listed in Table 6 is *chuo* 濯 or *chhien* 燖, poaching in water or broth. *Chuo* 濯 is derived from *yao* 鬻, which the *Shuo Wên* says 'is to gently cook in hot meat or vegetable soup'.⁴⁵ It is not mentioned in the four Chou classics or the *MWT Prescriptions*, but four containers of *chuo* meats are listed in the *MWT Inventory* (two beef organs, pig and chicken).⁴⁶ Furthermore, the word *chhien* 煔, a variant form of *chhien* 燖 that is applied in the *Chhu Tzhu* to a quail,⁴⁷ is best interpreted as having been cooked by poaching. Thus poaching was probably a significant method of cooking in ancient China.⁴⁸

Even more gentle than poaching is steeping, *tzu* 漬, which we have cited earlier from the *Li Chi* in regard to the preparation of the 'Steeped Delicacy', one the Eight Delicacies of Ancient China (p. 4, n. 12). It is thus indirectly referred to in the *Chou Li*.⁴⁹ The most important form of steeping is one done in an acid environment, in which case the process is known as *tsu* 菹, pickling. Pickled foods were well known in ancient China. The *Shih Ching* refers to a *tsu* made from a melon, and the *Li Chi* notes three types of *tsu*, prepared from deer, elaphure and fish respectively.⁵⁰ In the *Chou Li* seven pickled (*tsu*) vegetables and seven pickled (*hai*) meat viands are listed as standard fare for the royal family.⁵¹ Three jars of *tsu*, from wild ginger, bamboo shoot and melon, are included in the *MWT Inventory*. *Tzu* 漬 but not *tsu* 菹 is mentioned as a method of making decoctions in the *52 Pingfang*.⁵² We will have more to say on the technology of *tsu* in Chapter (*e*).

Table 6 ends with *khuai* 膾, which we have seen earlier (p. 69) could be interpreted as thin slices or silky strips of meat or simply minced meat. The word is used both as a noun and a verb. The *Shuo Wên* defines *khuai* as 'to finely slice meat'.⁵³ The *Li Chi* states that *khuai* is 'finely sliced fresh meat'.⁵⁴ In translations of the Chinese classics sinologists have consistently rendered *khuai* as mince or minced meat.⁵⁵ While we recognise that 'minced meat' is a form of finely cut meat, it is by no means the most satisfactory interpretation of *khuai*. The only description we have on how *khuai* meat was prepared in ancient China occurs in the *Li Chi, Shao I*, which says: 'Take fresh

---

⁴⁵ *SWCT*, p. 63a (ch. 3b, p. 6).   ⁴⁶ *MWT Inventory*, nos. 51–4, p. 135.
⁴⁷ *Chhu Tzhu, Ta chao*, p. 172. Hawkes (1985) interprets *chhien* as boil.
⁴⁸ It is possible that poached meats were often used in sacrificial ceremonies. Shên Kua in *MCPT*, ch. 3, p. 41 says, 'There are three types of sacrificial offerings (from animals): *hsing* 腥 (raw), *chhien* 燖 (lightly cooked) and *shu* 熟 (fully cooked).'
⁴⁹ *Chou Li*, pp. 34, 45 refer to the Eight Delicacies.   ⁵⁰ *Shih Ching*, M 210; *Li Chi*, p. 463.
⁵¹ *Chou Li*, pp. 55, 57. In *tsu* the vegetables are cut in lengths of about four inches before pickling. Listed also are five additional types of pickled vegetables known as *chhi* 齏, in which the vegetables are processed in smaller pieces.
⁵² *MWT Inventory*, Hunan shêng po wu kuan (1973), I, p. 142; *52 Pingfang*, nos. 3, 20, 73, 99, 120, 138 etc.
⁵³ *SWCT*, p. 90a (ch. 4b, p. 14b), *Khuai, shih chhieh ju yeh* 膾，細切肉也.
⁵⁴ *Li Chi, Nei Tsê*, p. 463, *Jou hsing shih chhieh wei khuai* 肉腥細切為膾.
⁵⁵ For example, on Confucius eating *khuai* meat, in *Lun Yü*, Book x, ch. 8; see Legge (1861a), p. 232 and Lau (1979), p. 103. On comparing *khuai* with jujubes, in *Mêng Tzu*, Bk vii, Pt 2, ch. 36; cf. Legge (1861b), p. 497 and Lau (1970), p. 202. For references to *khuai* in *Li Chi, Nei Tsê* cited earlier, cf. Legge (1885a), pp. 459–60, 462–3, 463. *Khuai* is used as a verb in the *Shih Ching*, M 177 (W 133), cf. sixth verse, *Phao pieh khuai li* 炮鱉膾鯉 which in both Waley and Karlgren is rendered as 'roast turtle and minced carp'. See also, *Chhu Tzhu, Ta Chao*, verse 8, *khuai chü pho chih* 膾苴蓴只 which is translated by Hawkes (1985), p. 234, as 'ginger flavoured mince'.

beef, lamb or fish. Slice it in (large) thin sections and cut it further to give *khuai*.'[56] The statement affirms that *khuai* was prepared from raw meat, but, unfortunately, it does not tell us how the large thin sections were cut further. It is impossible to say whether the further cutting leads to small thin slices, thin threads or mince. On the whole we think it would make more culinary sense to consider *khuai* as thin slices or thin threads rather than mince. The traditional Chinese way to make minced meat is by 'chopping first in one direction, then another, crosshatching continually until the ingredients look almost machine-ground'.[57] This is far easier and less laborious than the method described in the *Li Chi, Shao Yi* passage. Why go through the trouble of cutting thin sections if in the end the meat will all be slashed down to a hash? Indeed, modern Chinese scholars such as Wang Mêng-Ou, Fan Shu-Yün, Fu Hsi-Jên, and Shih Shêng-Han all tend to support this point of view.[58] The *khuai* meats of antiquity, then, are counterparts of today's *jou phien* 肉片 and *jou ssu* 肉絲 that we see as common ingredients in many stir-fried dishes. Unfortunately, the eating of raw meat in the form of *khuai* began to lose favour towards the end of the Chou. No additional information on how it was prepared can be found in the Han or the post-Han literature.[59] However, *khuai* from fish has remained popular through the ages until the present day.[60] In fact, the word *khuai* itself is now taken to mean small thin slices or threads of fish to be eaten raw.[61]

---

[56] *Li Chi*, p. 588.
[57] Miller, G. B. (1972), p. 42. She further points out that minced meat is often made today using a mechanical grinder. 'This tends to squeeze the connective tissue fibres, press out some of the juices, and toughen the food. This is why hand-minced porkballs are invariably lighter.' It is curious that in the pre-Han classics, the word *chhieh* (slicing) is always used for the cutting of meat or fish. There is no specific word for 'chopping' or 'mincing' in the culinary vocabulary. But by the +6th century 'slicing' and 'mincing' had become clearly distinguishable. In the *Chhi Min Yao Shu*, *chhieh* is still used to indicate 'slicing', but a new word *tsho* 剉 has appeared to denote 'mincing'. Cf. Miao Chhi-Yü ed. (*1982*), pp. 420–1, 494; Shih Shêng-Han (*1984*), p. 76.
[58] Wang Mêng-Ou ed. (*1984*), *Li Chi*, p. 463, explains that *khuai* is meat cut into strips until they are as fine as silk. Fan Shu-Yün ed. (*1986*), *Shih Ching*, M 177, p. 271, note 37, on *khuai* of carp, says '*khuai* is thinly sliced fish, and especially, thin slices of raw fish for eating'. Fu Hsi-Jên ed. (*1976*), *Chhu Tzhu*, p. 172, interpretes *khuai tzu pho chih* as 'thin sliced wild ginger', in contrast to 'ginger flavoured mince' of Hawkes, as cited in note 46 above. Shih Shêng-Han (*1984*) in the abbreviated edition of the *CMYS*, p. 77, defines *khuai* as very thin slices or very fine threads of meat.

Based on this interpretation, a revised translation of the passage from the Book of Mencius cited in note 55 would seem to be in order. Mencius was asked which he would prefer, *khuai chih yü yang tsao* 膾炙與羊棗. Legge renders this as 'minced meat and broiled meat or sheep dates', and D. C. Lau, as 'mince and roast or jujubes'. This is an odd comparison. Suppose Mencius had loved minced meat, disliked jujubes but detested roast, how would he have answered the question? It would make more sense to use *khuai* as a verb in this context. Mencius, then, would have to choose between 'thinly sliced roast' and 'jujubes'. His answer would certainly have been thinly sliced roast. Indeed, thinly sliced roast would have been considered delicious by almost everybody. This would illuminate the derivation of the celebrated idiom, *khuai chih jên kou* 膾炙人口, which means something that is in everybody's mouth, hence, widely quoted or highly popular.

[59] Meat *khuai* was eaten raw until as late as the 6th century. The *CMYS*, p. 422, describes the preparation of a dish consisting of finely sliced raw lamb and pork; and comments, p. 450, on the relish that accompanies a meat *khuai*. In later references *khuai* would invariably indicate sliced fresh meat for cooking, or thinly sliced cooked meat, cf. Wang Jên-Hsing ed. (*1987*), pp. 310 (beef, Ming: *Sung Shih Chun Shêng*); 348 (pig skin aspic, Yuan: *Chü Chia Pi Yung*); 275 (sheep head, Yuan: *Yin Shan Chêng Yao*). For a brief account of the evolution of *khuai* in the Chinese diet see I & Chün (*1984*).
[60] For example, *CMYS*, pp. 421, 450; Wang Jên-Hsing (*1987*), pp. 156 (Thang: *Ta yeh shih I chi* 大業拾遺記, from *Thai Phing Kuang Chi* 太平廣記, ch. 234), 366 (Yuan: *YSCY* 89) and 381 (Ming: *Tuo Neng Phi Shih* 多能鄙事).
[61] *Tzhu Hai*, *1979*, miniaturised version, p. 1,510.

During the late Chou and early Han raw *khuai* was probably eaten either directly or after steeping in wine, seasoned by the appropriate sauces. The *MWT Inventory* lists four containers of *khuai* meats, beef, lamb, venison, and fish.[62] It does not say whether the *khuai* had or had not been treated in any way. In any case, the evidence is strong that *khuai* meats were an important component of the foods served at ritual offerings and festive occasions presided over by the king or the nobility.

Other methods of cookery that were also important in ancient China include the making of *fu* 脯 and *hsi* 腊, dry meat, *hai* 醢, pickled minced meat and *chhi* 齏, pickled chopped vegetables. These, together with *tsu* 菹, pickled vegetables, will be considered in detail in chapter (*e*) as aspects of food processing and preservation.

Although, to the untrained eye, the shape of pottery wares made by early cultures all over the world, ranging from East Asia, the Near East, North Africa and the Americas, may look remarkably similar, there is one form of ancient pottery that can immediately be recognised as distinctively Chinese, and that is the *yên* or *hsien* 甗 steamer, which consists of a *li* 鬲, a cooking pot with three bulging legs, surmounted by another pot called *tsêng* 甑 that has a perforated bottom, like a colander (Figure 22).[63] Steam from boiling water in the lower pot rises through the holes and cooks the food in the upper pot. Pottery steamers were already familiar vessels in the Neolithic communities of Pan-pho (Figure 23) and Ho-mu-tu (Figure 24) six to seven thousand years ago, indicating that steaming is one of the oldest methods of cookery in China. As we shall see later the invention and continued use of the steamer has had a major influence on the course of development of Chinese cookery and food technology.

When a *li* is made with solid rather than hollow legs, we have the familiar *ting* 鼎, cauldron (Figure 25), and a *ting* without legs is *huo* 鑊 or *fu* 釜 cooking pot. Beautiful versions of these vessels and variations thereof made in bronze during the Shang and Chou periods now adorn the great museums of the world.[64] But the pottery versions were, presumably, the utensils that were actually used for cooking food in everyday life.[65] Unfortunately, references to cooking vessels in the Chou classics are few and far between,[66] but from the wealth of archaeological evidence now available, it may be inferred that *li*, *yen*, *tsêng*, *ting*, and *huo* or *fu* were the principal cooking vessels during the Shang and Chou periods.[67] One drawback of the three

---

[62] *MWT Inventory*, Hunan shêng po wu kuan (1972), I, p. 135.

[63] The *yên* (or *hsien*) and its constituent parts, the *li* and the *tsêng* are described in Vol. 5, Pt 4, pp. 26–32. The uniquely Chinese character of the *li* is commented on by Creel (1937), p. 44. A similar type of pot with three tapered hollow legs was discovered in the Neolithic culture in Ban Kao, Thailand, cf. Flon (1985), p. 252. As far as we know, no other culture has made a *hsien*-type steamer, which shows that the Chinese were the only people who used steaming as a means of cooking food on a large scale.

[64] For an introduction to the grandeur and beauty of ancient Chinese bronzes, cf. Fong, Wen ed. (1980), and for a detailed study of the Chinese bronzes cf. Ma Cheng-Yuan (1986) and *idem*, ed. (1988).

[65] The bronzes we see in the museums were ritual vessels and symbols of power and prestige. Indeed, as pointed out by K. C. Chang (1980b), p. 45, 'at the highest levels certain bronzes were made to symbolize the dynastic rule of a state, beginning with the Xia, and when dynasties changed hands, so did the vessels.' They were clearly out of the reach of the common people.

[66] *Shih Ching*: *ting* M 292 (W 233); *fu*, M 15, M 149 (W 76, W 149); *Chou Li*: *ting* p. 34, p. 39; *Chhu Tzhu*: *ting* in *Ta Chao*; *Li Chi*, *huo* p. 468.

[67] Cf. K. C. Chang (1973); see also Max Loehr (1968) and Ma Chhêng-Yuan ed. (1988).

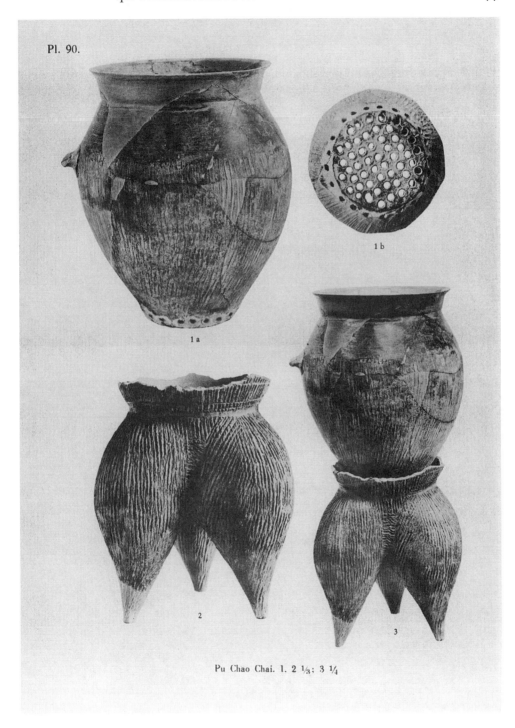

Fig 22. Shang Dynasty pottery *yen* or *hsien* 甗 which consists of a *tsêng* 甑 (steamer) surmounted on a *li*, 鬲 (boiler). From Anderssen (1947), Pl. 90. (1a) *Tsêng* Steamer (1b) Perforated bottom of Steamer; (2) *Li* Boiler with three hollowed legs; (3) *Hsien* assembled from *tseng* and *li*.

(a)

(b)

Fig 23. Steamer from Pan-pho, Shensi: (a) Crude steamer with boiler and cover: *Sian Pan-pho po-wu-kuan* (1972); (b) Display of steamer assembly at the Pan-pho Museum, c. −4000; photograph H. T. Huang, 1979.

(a)   (b)

Fig 24. Steamer from Ho-mu-tu, Chekiang: (a) Assembly of steamer, boiler and stove, Ho-mu-tu, c. −5000; *CKPJ*, **1989** (9), cover; (b) View of steamer and boiler, *Che-chiang shêng wen-kuan-hui*, (*1978*), p. 76, Pl. 3, no. 3.

Fig 25. Pottery tripod cauldron, *ting*, 鼎, from Pan-pho. San Francisco Asian Art Museum (1975), Fig. 57, height 15 cm.

legged vessels *li*, *yen* and *ting* is that much of the heat given by the burning fuel under the pot is dissipated. To contain the heat, insulation in the form of brick or adobe, was built around the vessel and refinement of this process eventually led to the construction of a stove, *tsao* 灶 (or cooking range). The prototype stove was called *tshuan* 爨 in the *Shih Ching*.[68] We are not sure what it looked like, but the stove had evolved into its traditional form by the early part of the Han dynasty. Pottery models of *tsao* abound in burial objects found in Han tombs. A typical example is shown in Figure 26, where it is compared with traditional stoves still in use in China today. It is clear that there has been little change in the basic design in two thousand years. The *tsao* is a rectangular platform about two and a half to three feet high, two and a half feet deep and four to five feet wide. The hollow interior serves as a fuel burning chamber. There is a stoke hole in front and a chimney at the back. On top are one or two large circular openings on which cooking pots such as *huo* and *fu* can be tightly fitted. In most Han models we find one of the pots surmounted by a steamer, *tsêng*, indicative of the importance of steaming as a cooking process in the ancient kitchen. There is evidence that *huo* 鑊 or *fu* 釜 pots made of iron were already fairly common by the Han dynasty.[69] They are probably similar in shape to the *ting* shown in Figure 25, but with the legs removed. Being more durable than pottery and less expensive than bronze, the iron pots eventually became the preferred utensils to be placed on a stove. The *tsao* 灶 is a wonderful invention. It is so very well insulated that little heat is lost. Furthermore, high or low heat could be generated at will by adjusting the amount of fuel in the burning chamber. Indeed, the control of burning in the *tsao* had become such an integral part of the art of cookery that the word *chhui* 炊 meaning literally to blow fire, was often used to denote the entire process of preparing a meal, for example, as in *chhui fan* 炊飯, to cook rice (or another cooked grain).

As we have noted earlier, steaming was used to cook *fan* 飯 (rice or millet grain). The *I Chou Shu* (Lost Records of the Chou) credits the legendary Yellow Emperor, Huang Ti, as the originator of steaming as a way to cook grain.[70] *Fan* was also known as *shih* 食. Thus when Confucius said that 'with coarse *shih* to eat, and water to drink, and my bended arm for a pillow – I have still joy in the midst of these things',[71] he was referring to steamed rice or millet. Since *fan* was the major component of a Chinese meal,[72] the preparation of *fan* may be assumed to be a preeminent function of the steamer. One of the characteristics of steamed rice (or millet) is that it is grainy and loose, so that it could be conveniently carried in a bamboo container, *tan* 箪, as was the custom at the time of Confucius and Mencius.[73] Steaming was also

---

[68] *Shih Ching* M 209.
[69] *Chiang-hsi shêng po-wu-kuan* (1978), p. 158; *Nanching po-wu-kuan* (1979), p. 424; *Li-chou i-chih lien-ho-khao-ku fa-chüeh tui* (1980), p. 1. *Yang-chou shih po-wu-kuan* (1980), p. 5.
[70] *I Chou Shu*, cited in *TPYL yin-shi pu*, p. 328.
[71] *Lun Yü*, Bk VII, ch. 20, Verse 14; tr. Legge (2), p. 200. For other passages in which grain food occurs as *shih* see Chang, K. C. ed. (1977a), p. 40.
[72] The other being viands, *shan* 膳 or dishes, i.e. *tshai* 菜 in modern language.
[73] *Lun Yü*, Legge tr. (1861), p. 188. *Mêng Tzu*, Legge tr. (1895), p. 170.

Fig 26. Traditional Chinese stoves, *tsao*: (a) Clay model from Han tomb. *Yunmeng hsien wên-wu*, (*1981*), Pl. 10, Fig. 6; (b) Traditional *tsao* seen in Taipei, Taiwan, photograph by H. T. Huang.

used for cooking meat and fish dishes. Its importance further increased when the establishment of the milling of wheat to flour stimulated the development of a great variety of bread and pasta products during and after the Han dynasty.[74] Today, steaming remains a distinctive feature of Chinese cookery.

Li 鬲, *ting* 鼎, and *huo* 鑊 (or *fu* 釜) were the vessels used for cooking food by boiling or variations thereof, that is, poaching, simmering–stewing, and pot-roasting. *Li* and *ting* were important in the early part of Chou but *huo* and *fu* increased in usage as the stove, *tsao*, became established.[75] Today the large pot that sits on top of the stove, *tsao*, is known as *kuo* 鍋 in most parts of China. But some remnants of ancient forms still persist in certain dialects. For example, in Cantonese the large pot, *huo* 鑊, is pronounced *wok*, by which name it has become familiar in many kitchens of Europe and America. In Fukien (both Foochow and Amoy) dialects, it is still known as *ting* 鼎 but is pronounced *tiang*.

The legendary Huang Ti was also credited with the preparation of *chu* 粥 gruel or congee, that is grain boiled in plenty of water until soft.[76] This could have been an earlier method of cooking grain than steaming. During the Neolithic Age, the techniques for hulling grain were rather crude. The grains obtained might not give a smooth product when steamed. With prolonged boiling, however, even the hardest grain could be softened to give an acceptable gruel. The ancient form of *chu* 粥 is *chu* 鬻 indicating that it was probably first prepared in a *li* vessel. It was also known as *mi* 糜 which suggests that the grains were so well softened that they started to fall apart. When the *chu* was thick, it was called *chan* 饘, when thin, simply *chu* 粥. The *Li Chi* has several references to eating *chu* (congee), for example, as suitable nourishment for the elderly and the first food permitted after a fast mourning the passing of a parent.[77] A *chu* prepared from a *liang* 粱 millet is mentioned as a remedy for a snake bite in the *52 Pingfang*.[78] Thus, the boiling of grains in water to make *chu* or congee must have been a major use of cooking pots in the Chou–Han period.

Cooking pots were undoubtedly used in the poaching (*chuo* 濯) method for cooking meats discussed earlier (Table 6). But the best-known ancient dishes prepared by cooking in water in such pots were *kêng* 羹, a word which today is usually taken to mean 'soup'. Indeed, it is so construed in the translations of the three poems in which *kêng* is mentioned in the *Shih Ching*.[79] Tu Yu's commentary in the *Tso Chuan* 左傳 (Master Tso's Enlargement of the Spring and Autumn Annals) on the passage which tells us that the 'grand *kêng* is not flavoured with condiments (*ta kêng pu chih* 大羹不致)', explains that *kêng* is the broth from meat, that is, soup.[80] On the other

---

[74] This topic will be discussed in detail later in Chapter (*e*). We may note in passing that steaming became such an integral part of the culinary culture of the Chinese, that it gave rise to the common expression *chêng chêng ji shang* 蒸蒸日上 (literally, rising like steam day by day') to denote a condition of rapid progress or increase in prosperity.

[75] Large free-standing *ting*'s were still in use for cooking in Eastern Han and as late as Northern Wei. Cf. Tanaka Tan (*1985*), pp. 258 and 273.

[76] *I Chou Shu*, *TPYL yin-shih pu*, p. 450.

[77] *Li Chi*, pp. 88, 120, 289, 718, 918, 1015.   [78] *52 Pingfang*, line 92 no. 56.

[79] *Shih Ching*, M 255, M 300, M 302, cf. translations by Waley, Karlgren and Legge.

[80] *Tso Chuan*, Duke Huan 桓公, 2nd year. Tu Yu's 杜預 commentary says, '大羹' 肉汁也.

hand, the *Erh Ya* (The Literary Expositer), c. −300, declares simply that 'meat is *kêng* (*rou wei chih kêng* 肉謂之羹)',[81] and the *I Li* (Personal Conduct Ritual), observes that 'juice from the grand *kêng* is on the stove (*ta kêng chhi tsai tshuan* 大羹湆在爨)'.[82] Thus *kêng* in ancient China can mean anything ranging from a clear broth, thin soup, thin stew, thick stew to plain meat. It might originally have been a general term to denote meat cooked (*phêng* 烹 or *chu* 煮), with or without vegetables, in varying amounts of water, that is viands as opposed to grain food. In a quotation from the *Tso Chuan* about Yellow Springs (*huang chhuan* 黃泉) as a concept of the netherworld, we have previously noted the story of the estrangement between Prince Chuang of the State of Chêng and his mother in the year −721:[83]

> Chuang then shut up Wu Chiang (his mother) in the city of Ying, and swore that he would never look upon her again until they both come to the Yellow Springs. Afterwards he repented of this oath ... One Khao Shu who was Warden of the Marches in the Valley of Ying heard of this and went to court to offer some presents to the prince. The prince made him stay to dinner, and noticed that he had laid aside some of the meat (ju 肉) so he asked him why. 'Your servant', answered Khao Shu, 'still has a mother, and she likes a taste of the best things that I am given to eat. She has never tried this princely dish (*kêng* 羹); pray take no offence if I keep a little for her.'

We know the rest of the story. At Khao Shu's suggestion, the prince arranged to have a tunnel dug in the earth going down to a spring, and had a happy reunion with his mother there. Thus, he was able to meet her without violating his oath. For our purpose, the important point about this quotation is that Khao Shu had put aside meat from the 'princely *kêng*'. One is left with no doubt that there was meat, and perhaps other ingredients, in the *kêng*, which in this case can best be construed as a stew. But it could also have been a general term for 'dishes'.

The *Li Chi* says that '*kêng* (soup or stew) and *shih* 食 (cooked grain) were used by all, from the princes down to the common people, without distinction of degree'.[84] Thus *kêng* must have been a familiar and routine food eaten as an accompaniment to grain. A great variety of meats have been prepared in the form of *kêng*. The *Li Chi* lists pheasant, dried meat, chicken, dogflesh, and rabbits.[85] The juice was often thickened with rice (grain) crumbs, but without the addition of the smartweed, *liao* 蓼. The *Chhu Tzhu* mentions orioles, pigeons, geese, and jackal's meat.[86] *Kêng* was also made from vegetables only. Presumably, this was the kind of *kêng* that the common people ate.[87] The *Chou Li* says that two kinds of *kêng* were used in ritual sacrifices and entertainment of officials: the *thai kêng* 太羹 (grand soup) and *hsing kêng* 鉶羹 (mixed soup).[88] According to Chêng Hsuan, after the meat or fish is well

---

[81] *Erh Ya Chu Shu*, ch. 6, p. 77.
[82] *I Li Chu Shu*, ch. 4, p. 42. See also ch. 25, p. 302, which states that 'no condiments are added to the juice of the grand *kêng*'.
[83] Cf. *SCC* Vol. v, Pt 2, p. 84. Originally from *Tso Chuan*, Duke Yin, 1st year, tr. Needham.
[84] *Li Chi*, p. 463, Legge tr. (1885), p. 464.   [85] *Li Chi*, p. 457.   [86] *Chhu Tzhu, Ta Chao*, verse 7.
[87] *I Li Chu Shu*, ch. 42, p. 500, ch. 46, p. 547. Cf. Steele (1917) II, pp. 118, 154, for reference to a vegetable soup made with sow thistle, vetch, mallows and celery.
[88] *Chou Li*, p. 40.

cooked in a pot, the broth (*chhi* 湇) is removed and placed in a serving vessel and this is called the ancient grand soup (*thai ku chih kêng* 太古之羹).[89] No condiment or vegetables are used in order to preserve the essential purity of the flavour of the meat. When condiments and vegetables are added to the broth the product is called *hsing kêng* 鉶羹 (mixed soup).[90] That *kêng* was a leading viand in ancient China is supported by the fact that *kêng*'s were at the head of the list of foods recorded in the *MWT Inventory*. There were twenty-four cauldrons (*ting* 鼎) of *kêng*, under three separate categories:

(1) Nine Grand Soups, *Yü Kêng* 酐羹: oxhead, mutton, dogflesh, suckling pig, pig, pheasant, duck, chicken and venison with fish.[91] (See Figure 5)
(2) Seven White Soups, *Pai Kêng* 白羹: these are called *pai kêng*, because they were thickened with rice (grain) groats or crumbs, which were sometimes called *pai* 白. They should have the colour and consistency of congee. The materials used were beef, venison with salted fish and bamboo shoots, venison with taro, venison with small bean, chicken and gourd, fresh sturgeon, and crucian carp with lotus root and salted fish.
(3) Eight Vegetable-Meat Soups: These presumably were more like stews. There were three using celery (with dogflesh, goose and carp respectively); three using turnips (with beef, beef and pork respectively) and two using sonchus leaves (with dogflesh and beef respectively).[92]

Another dish that was prepared by using pots was *erh* 胹 (also 濡) or *huo* 臛, braising or pot-roasting, in which meat in large chunks was simmered in its own juice. The *Li Chi* lists the following braised meats: suckling pig, chicken, fish and turtle. According to commentaries by Chêng Hsuan and Khung Ying-Ta, the suckling pig was stuffed with smart-weed and wrapped in sonchus leaves, and then cooked to give the potted roast.[93] The whole animal with the juice was presented at the table. The *Chhu Tzhu* mentions *erh* and *huo* in the preparation of a beef rib, turtle, tortoise, and magpie, and a *chuan* 臇 duck, which was cooked in very little juice and could be construed as a potted duck.[94]

We may assume that pots were also used for shallow-frying, *chien* 煎 and for dry-frying or parching, *ao* 熬. The *Chou Li* refers to the parching (*ao*) of cooked grain to

[89] *I Li Chu Shu*, ch. 25, p. 302, see Chêng Hsuan's comment on juice from *kêng*.

[90] For a discussion of the significance of *kêng* in Chinese history, see Sung Yü-Kho (*1984*) and Wang Hsüeh-Thai (*1985*).

[91] The character 酹 describing the first group of *kêng* on the original bamboo strips is clearly written in Fig. 5, slip 1. The official report of the MWT Excavation, *Hunan sêng po-wu-kuan* (*1973*), p. 130, interprets this character as *yü* 酐. On the other hand, Thang Lan (*1980*), p. 9, thinks it should be *kan* 䵚. But both sides agree that this series of *kêng* were probably 'grand soups', that is, clear broth from meat as indicated in the *Li Chi*, *Chou Li* and *I Li*. There is, however, some doubt that this interpretation may be oversimplified, as shown by Chu & Chhiu (*1980*), p. 61.

[92] The *I Li* recommends a specific vegetable with a particular meat in the making of *kêng*: young bean leaves with beef, sow-thistle leaves with mutton and vetch vines with pork. Cf. *I Li Chu Shu*, ch. 26, p. 314 and Steele (1917) 1, p. 258.

[93] *Li Chi*, pp. 457–8; Legge (1885), p. 460.

[94] *Chhu Tzhu*, Fu Hsi-Jên (ed.), pp. 161, 172; David Hawkes (1985), 'casseroled duck' 鳧 pp. 227–8, 234.

make *chhiu* 糗.⁹⁵ Since the ancient Chinese did not have ovens, one way to parch the grain would be to stir it in a heated pot. The stirring was probably done with a spoon or ladle called *pi* 匕. The earliest reference to this instrument is found in the *Shih Ching*, which states, 'Messy is the stew in the pot; Bent is the thornwood spoon.'⁹⁶ *Pi*, which may be written as 匕, 枇 or 畢, came in various sizes and shapes and were used as both cooking and serving utensils. The *Li Chi* records a *pi* 枇 used in rituals, and presumably also in entertaining, that was three to five feet long.⁹⁷ It also mentions a different, but equally long, *pi* 畢 which is a two-pronged fork. It was so named because its shape is similar to that of the constellation *pi* 畢, now identified as *Epsilon Tauri*. We will have more to say about both the spoon and fork later.

The *pi* fork is of some interest in connection with *chih* 炙 roasting, one of the three processes in which the food was cooked by direct exposure to the fire. We know very little how *fan* 燔 roasting was carried out, and there is not much one can say about *phao* 炮. But *chih* is another matter. The earliest record we have on how *chih* roasting was done is found in a story in the *Han Fei Tzu* about Duke Wên of Chin, *chih* roast and his chef.⁹⁸

During his reign (−636 to −628), Duke Wên was served a dish of *chih* roast that had some hair around it. The Duke summoned the chef, and rebuked him, 'Do you wish to choke me to death? Why is there hair around the roast meat?' The chef knelt and kowtowed before the Duke, saying, 'My Lord! Evidently, I have committed three heinous crimes, each punishable by death. My knife was sharpened on the grindstone until it was as sharp as the keenest sword. It cut the meat with ease, yet it did not cut the hair. This is my first crime. When I carefully pierced the meat piece by piece on the wood skewer, I found no trace of hair anywhere. This is my second crime. When I placed the skewers on the stove, the charcoal was red hot. The meat quickly cooked through, but the hair was unsinged. This is my third crime. Is it not possible that there is someone at the court who is trying to falsely incriminate me?' 'Quite so!' The Duke said. He called the attendants and cross-examined them. He found the culprit and had him killed.

Thus, we are told that in *chih* roasting a wooden skewer was used to pierce a series of cubes of meat, which were then broiled on burning charcoal. This is exactly how shish kebab or shashlik, an immensely popular dish in Eastern Mediterranean countries, is prepared today. Tomb paintings from Chia-yü-kuan in the Wei-Chin period show unmistakably that a three-pronged fork was also used to make a *chih* roast. One example is given in Figure 27. However, the traditional single piece skewer appears to be the most popular way to carry out *chih* roasting in Han times,

---

⁹⁵ Cf. Sun I-Jang's commentary, *Chou Li*, (ed.) p. 53. *Chhiu* is parched, cooked grain, which when mixed with water can be eaten directly. It may be compared to a modern dry cereal.

⁹⁶ *Shih Ching*, w 203 (M 203), 1st verse. Waley translates *pi* as spoon. But in this context it may simply be a stirring stick. Later *pi* apparently evolved into a fork.

⁹⁷ *Li Chi*, p. 667.

⁹⁸ *Han Fei Tzu*, ch. 10, *Nei Chhu Shuo, Hsia*, pp. 86–7, tr. auct.; cf. also Liu Chhang-Jun (*1986*), pp. 169–70. Actually, the text gives two versions of the same story. We have followed the ending in the first version. This story is also discussed in Li Ho (*1986*).

Fig 27. Three-pronged skewers for *chih* roasting from Chia yu kuan. Chih Tzu (*1987*), p. 7.

according to a marvellous kitchen scene incised on a stone slab in an Eastern Han tomb in Chu-Chhêng county, Shantung Province (Figure 28).[99]

On the top of the picture is a row of large hooks on which are hung carcasses and parts of food animals ready for further processing. We can recognise a turtle, a deer, two hangers of fish, a boar's head and shoulders or flanks of ox, sheep or pig. In the next row we see a rack of large trays on the left side and attendants moving towards them carrying large plates. To the right is a chef cutting up fish and another cutting a chunk of meat from a large shoulder hung from above. On the third row near the centre are three chefs busily slicing meat, presumably for *chih* roasting. To their right are two attendants turning skewers of meat resting on the sides of a rectangular hibachi type grill. One of them is simultaneously waving a fan over the burning charcoal. Below them is an attendant piercing cubed meat into a skewer, while another watched for a chance to place a new skewer on the grill. Going down on the right side, we meet the five domestic food animals being led to slaughter; chickens (kept in a cage), a goat (in lieu of sheep), an ox, a pig and a dog. One person is hitting the ox with a heavy (metal) ball. Going down on the left side, the picture shows a servant drawing water from a well, a wood burning stove (*tsao*) on which sits a huge steamer, a helper inserting firewood into the stoke hole, and attendants filtering fermented mash to clear wine using a pottery filter or cloth bag. Down the centre we see a chicken coop, and attendants washing a bird, mixing food in a large basin, and cutting firewood with an axe. Further down is a foreman waving a large spoon at a maid who was sitting down with her cap and gown in disarray and obviously not doing her share of the work. Behind him is another person who appears to be ready to hit her with a stick. At the bottom are a set of fermentation jars and four large storage urns for containing water or wine. At the far left are two small urns of the type that might have been used for steeping meat and vegetables, a topic that will be treated in detail in Chapter (*e*).

[99] Pictures with the three-pronged forks are reproduced in Tanaka Tan (*1985*), figs. 32 and 33. See. Jên Jih-Hsin (*1981*) for a description of the layout of the tomb in Chu-chhêng, Shantung. Thirteen stone slabs with drawings on them were discovered. They offer a rich tapestry of daily life in the Eastern Han. The scene in question was incised on a stone slab measuring 152 cm wide, 76 cm high and 23 cm thick. It has been analysed in detail by Pirazzoli-t'Serstevens (1985) and reproduced in Tanaka Tan (*1985*), fig. 31, p. 269 and in Flon (1985), p. 271.

Fig 28. Eastern Han Kitchen scene from Chuchhêng, Shantung. Jên Jih-Hsin (*1981*).

Table 7. *Comparison of ancient and modern Chinese methods of cooking*

Heat Transfer Agent	Ancient Method	Ancient Products	Modern Method	Modern Products
Steam				
Steaming	Chêng 蒸	Rice, viands, bread, pasta, pastry	Chêng 蒸	Rice, viands, bread, pasta, pastry
Water				
Boiling	Phêng 烹	Plain congee, savoury congee,	Chu 煮	Plain congee, savoury congee,
Simmering	Er 胹	Soups, stews	Tun 炖	Soups, stews
Braising	Huo 臛	Pot roast,	Lu 鹵	Pot roast
Poaching	Chuo 濯	Poached viands	Tshuan 汆	Poached viands
Fat or oil				
Shallow frying	Chien 煎	Viands	Chien 煎	Viands
Deep frying	—	—	Cha 炸	Viands, pastry
Stir-frying	—	—	Chhao 炒	Viands
Direct heat				
Dry parching	Ao 熬	Grain, meat	Po 爆	Grain
Roasting or broiling	Fan 燔	Meats, fowl	Khao 烤	Meats, fowl
Skewer roasting	Chih 炙	Meats	?	Meats
Baking in clay	Phao 炮	Meats, fowl	?	Fowl
No heat				
Steeping	Tzu 渍	Meats, Vegetables	Phao 泡	Vegetables

How do the ancient methods compare with modern practices in Chinese cookery? We have tabulated in Table 7 both the ancient and modern methods of cooking side by side so that the similarities and differences between them can be readily discerned. For the modern methods we have generally followed the nomenclature given in Buwei Yang Chao's celebrated cookbook, which identifies twenty different methods of cooking Chinese food.[100] In principle, cooking involves the transfer of external heat to the raw food. There are only four ways by which this can be done. These have been used as a basis to classify the different methods of cooking shown in Table 7. In the first case, heat is applied to boil water, and the steam generated rises to cook the food. Steaming was a very important method of cookery in ancient China. It still is today. The durability of steaming is in no small measure due to the invention of first the wooden steamer and then the bamboo steaming basket, *chêng lung* 蒸籠, some time before the Sung dynasty (+960 to +1279).[101] Because of its light

---

[100] Chao, Buwei Yang (1963), pp. 39–47.
[101] The use of wooden steamers is mentioned in Lu Yü's famous *Chha Ching* (The Classic of Tea), +770 for steaming tea leaves and in the *I Chien Chi*, +1170, for steaming wine, cf. p. 230). Tanaka Tan (1985), p. 257, figs. 12, 13a, 13b, has reproduced pictures from brick murals showing stacks of bamboo steamers on a stove in Southern Sung tombs, cf. Chhên Hsien-Ju (1955) and *Ssu-chhuan shêng po-wu-kuan* (1982).

weight and ease of handling, the bamboo steamer eventually displaced the heavy pottery and wooden steamers, and greatly expanded the versatility and convenience of steaming cookery. It remains today as a distinguishing feature of the Chinese cuisine (Figure 29).

In the second case, heat is applied to water and the hot water cooks the food *in situ*. By controlling the amount of heat supplied to the water, high (fast boiling), medium (slow boiling or simmering) and low (no boiling), and varying the cooking time, a wide range of results can be achieved. The basic methods in this class described earlier are still being used today, but usually under different names. Thus, instead of varieties of *kêng* 羹 we have *thang* 湯 (hot broth) for soup, *hung shao* 紅燒 (red cook) for stew, *lu* 鹵 for pot roast and *tshuan* 汆 for poaching.[102] It is likely that most modern dishes listed under this category, or reasonable facsimiles thereof, could have been produced in ancient China with the materials and equipment then available.

This, however, is certainly not the case with the third category of methods, that is, frying, which utilises heated fat or oil to transfer heat to the food. Today, three methods of frying are used: *chien* 煎, shallow frying, *cha* 炸, deep frying and *chhao* 炒, stir-frying. Of the three only shallow frying was known to the ancients, who would not have been able to produce any of the deep-fried or stir-fried dishes that are so familiar to us today. Stir-frying may be regarded as a combination of shallow-frying and dry-parching, a rather ill-defined process which probably involved rapid stirring of food ingredients in a heated pot.

Dry-parching, *ao* 熬 is the first of the methods listed under the fourth category of methods in which heat is transferred directly to the food without the mediation of steam, water or fat. The modern version is *po* (or *pao*) 爆. It is used in the preparation of roast chestnuts and presumably for the making of puff grains such as the rice used in puff rice cakes. Roasting *fan* 燔 is now known as *khao* 烤, a good example of which is the famous Peking duck (*pei ching khao ya*) 北京烤鴨. *Phao* 炮, baking in clay, in the guise of 'beggar's chicken' is a rare dish that may only be seen in an exotic restauraunt setting. *Chih* 炙, skewer roasting, which was apparently highly popular in the Han dynasty, has almost all disappeared. One may still encounter it in far Northwest China in the cuisine of the Uighurs or other national minorities, but most Chinese in America or Europe today would be surprised to learn that it was a native Chinese dish.

The last method, steeping, is considered as a method of cooking even though no heat is applied. It is not as important in modern cuisine as it was in ancient China. Steeping was practised for both meat and vegetables. Today, only vegetables are steeped.

In summary, we may say that in the use of steaming and boiling techniques ancient and modern Chinese cooking are basically the same. But in the practice of frying and roasting enormous changes have taken place. Skewer roasting was an important cooking technique in ancient China. It no longer is. Deep-frying and

---

[102] *Tshuan* is not listed in Mrs Chao's cookbook, but she calls the procedure 'plunging' and 'rinsing'.

(a)

(b)

Fig 29. Chinese bamboo steamers: (a) Mural with bamboo steamers in a Southern Sung tomb in Kansu. Chhên Hsien-Ju (1955); (b) Stove with bamboo steamers in Urumqi, Sinkiang, photograph by Kenneth Hui, c. 1985.

stir-frying were unknown in Chou and Han cookery, yet they are among the most widely used methods today. Indeed, stir-frying has been called 'the most characteristic method of cooking in Chinese',[103] and many Western observers will readily agree. Because of the overwhelming importance of stir-frying in Chinese cookery one may even question whether ancient Chinese cooking is distinguishably Chinese. To answer this question we will need to consider the next component in the culinary system, that is, the kind of seasonings available and how they were used to flavour the food.

(ii) *Seasonings and spices: what they are and how they are used*

O that we had meat to eat! We remember the fish we ate in Egypt for nothing, the cucumbers, the melons, the leeks, the onions and the garlic.[104]

We can all sympathise with this lament of the ancient Israelites. We too, remember well the leeks, the onions and the garlic in our diet! While the way our food smells and tastes depends a good deal on the type of raw materials we have and the manner we cook them, it is the seasoning and spice we use that give it its characteristic flavour. They help to endow the food with a harmonious blend of the flavours derived from individual ingredients and were, therefore, called *ho* 和, harmonizing agents. In the *Lü Shih Chhun Chhiu* 呂氏春秋, the legendary chef I Yin 伊尹 tells us that the most delectable harmonising agents were ginger (*chiang* 姜) from Yang Phu (in Szechuan), cassia (*kuei* 桂) from Chao Yao (in Hunan), mushroom (*chün* 菌) from Yueh Lo (probably in South China), fish paste made from sturgeon, and salt from Ta Hsia (in Northwest China).[105] It is pertinent to note that while four of the harmonising agents listed are natural products, one, fish paste, is itself made by the blending of natural ingredients. Seasonings made from fish, meats, soybean, and vegetables were, indeed, extremely important in Chinese cuisine, and the processes for making them will be major topics for discussion in Chapters (*d*) and (*e*).

Harmonising agents are carriers of the five basic flavours that we encounter in the food and drink we consume. In the Chinese lexicon the five flavours are defined as bitter (*khu* 苦), sour (*suan* 酸), sweet (*kan* 甘), pungent (*hsin* 辛) and salty (*hsien* 鹹).[106] The kind of flavour that would contribute to good health was supposed to vary according to the season of the year. Both the *Chou Li* and *Li Chi* advise that in aiming at a balance of flavours in cookery, one should favour 'sour' in the spring, 'bitter' in the summer, 'pungent' in the autumn and 'salty' in the winter.[107] In spite of this injunction the ancient texts do not tell us much about specific seasonings that are bitter. Apart from wine, which is regarded as a bitter commodity in later pharmacopoeias, the only products we can identify are two plants in the *Shih Ching*, *thu* 荼,

---

[103] Chao, Buwei Yang (1963), p. 43. [104] The Bible, Numbers 11:4.
[105] *LSCC*, ch. 14, *Hsiao Hsing Lan* 孝行覽, *Pen Wei* 本味, pp. 370–1. Cf. commentary by Chhiu Phang-Thung, Wang Li-Chhi & Wang Chêng Min (*1983*), pp. 58–9.
[106] *HTNCSW*, ch. 4, pp. 86–7; Veith (1949), pp. 112–13. [107] *Chou Li*, p. 46; *Li Chi*, p. 458.

the bitter herb, and *liao* 蓼, the smartweed (Table 2), which is both bitter and pungent. *Liao* was apparently the favourite vegetable of the compilers of the *Li Chi*, who mentioned *liao* more frequently than any other vegetable.[108]

We have stated that the *mei* 梅 (pp. 46–7) fruit was used to impart a tart flavour to food, but its role as a acidulant was soon displaced by *hsi* 醯, a form of vinegar produced by microbial oxidation of alcohol in the wines that the ancient Chinese apparently delighted in producing in great quantities and numerous varieties. The *Chou Li* lists a *Hsi Jên* (Keeper of Vinaigrette Viands), whose duty it was to see that the king was well supplied with *hsi* viands for ritual and entertainment purposes.[109] He had a staff of two secretaries, twenty female assistants and forty labourers. The *Li Chi* describes the use of *hsi* for the steeping of scallions and shallots and as a sauce to accompany two of the Eight Delicacies, the Bake (suckling pig baked in clay) and the Steeped Delicacy (steeped thin slices of beef).[110]

Two types of sweeteners were known in ancient China, malt sugar, *i* 飴, and honey, *mi* 蜜. The *Shih Ching* refers to malt sugar as well as to sweetness as a flavour.[111] Sweet foods were recommended by the *Chou Li* to promote the healing of muscle in external ailments.[112] The *Li Chi* recommends malt sugar and honey as confection for presentation to the elderly.[113] The *MWT Inventory* includes both malt sugar and honey among the burial remains of Han Tomb No. 1.[114] We may infer from these references that sweeteners were available and used as condiments in cookery. The *MWT Inventory* also lists salt, *yen* 鹽, which must, of course, have been an indispensable ingredient of the condiments in the deceased lady's kitchen.[115] The *Chou Li* describes a *Yen Jen* 鹽人, the Salt Administrator, who also had a staff of two secretaries, twenty lady assistants and forty labourers.[116] The *Li Chi* mentions spreading salt over a chunk of beef in the recipe for the Grill, another one of the official Eight Delicacies.[117]

Unlike bitter, sour, sweet and salty the last of the five flavours we will consider here, *hsin* 辛, pungent, is not a well-defined single flavour but rather a group of associated flavours, piquant, pungent, acrid, peppery, spicy, hot etc. They all elicit unpleasant sensations in our bodies, burning throat, watery eyes and a runny nose. Yet in the right amount they impart a zest and excitement to food that can make them almost irresistable. The spices in this group that we have already mentioned earlier include the vegetables (Table 2), Chinese leek, scallion, Chinese shallot, garlic, rocambole, ginger and mushroom, and the fruits (Table 3) fagara and Chinese cinnamon or cassia. The recipe for the Grill in the *Li Chi* also describes the sprinkling of shredded cassia and ginger on the beef before the application of salt. It further advises that 'To go with finely sliced meat, the accompanying relish should be scallion in the spring, but a mustard in the autumn. To go with suckling pig, it should be Chinese leek in the spring but smartweed in the autumn. With lard,

---

[108] *Li Chi*, pp. 457, 460, 470.   [109] *Chou Li*, p. 57.   [110] *Li Chi*, pp. 463, 469.
[111] *Shih Ching* w 240, 108, 169 (M 237, 35, 171).   [112] *Chou Li*, p. 47.   [113] *Li Chi*, p. 444.
[114] *MWT Inventory*, malt sugar, p. 138, item 97; honey, p. 140, item 114.
[115] *MWT Inventory*, p. 139, item 104.   [116] *Chou Li*, p. 58.   [117] *Li Chi*, p. 469.

scallion should be used, with beef tallow, shallot. For the three slaughtered animals, fagara should be the spice.'[118] Remains of ginger, cassia, and two types of fagara were identified in Tomb No. 1 at Ma-wang-tui and records of Chinese leek and scallion were found in a Han tomb at Yun-mêng.[119] Thus, we may infer that the *hsin* group of condiments were widely used and constituted a characteristic feature of ancient Chinese cookery.

In addition, it was recognised that each type of animal fat (and later, vegetable oil) had its own unique aroma and flavour. Accordingly, beef tallow was said to be *hsiang* 鄉, dog fat *sao* 臊, lard *hsing* 腥 and lamb tallow *shan* 羶. These seemingly unflattering qualities, however, could be used to advantage in cooking: young lamb and suckling pig were fried in beef tallow, dried pheasant and fish in dog fat, calf and baby venison in lard, and fresh fish and goose in lamb tallow.[120] It appears that during the cooking process, the blending of the natural taste and aroma of the fats with ingredients in the meat (and possibly with the right amount and combination of spices) created new flavour qualities that rendered the finished dish inviting to the palate and pleasing to the senses. Thus, in terms of mixing of flavours the ancient Chinese evidently understood the dictum: the whole is more than the sum of its parts.

But there is another aspect of harmonisation that we have not yet touched on, and that is the texture of the food. This was particularly true in the case of *kêng* 羹 soups. The broth from meat tends to be thin. To give it more body and thus a richer feel in the mouth, the *Li Chi* advises the addition of rice (grain) crumbs, *san* 糁, to thicken the soup. The process is called *ho san* 和糁 'harmonise with rice crumbs'.[121] In fact, Chêng Hsuan's commentary states that 'to achieve the perfect soup we need to harmonise it with five flavours and thicken it with rice crumbs'. Another way to thicken the soup but at the same time make it more slippery, *hua* 滑, to the tongue is to add a mucilagenous vegetable, such as the mallow, *khuei* 葵, or smartweed, *liao* 蓼. The *I Li* says that *hsing kêng* 鉶羹 requires the addition of '*hua*', which improves the texture of a well-tempered soup.[122]

Thus, it may be said that the whole objective of the cooking process was to blend the characteristic flavours and texture of the individual ingredients into a harmonious whole, an ideal of the culinary arts that was often articulated in the Chou classics. For example, the *kêng* soup was evoked as a metaphor for the concept of harmony by the Chou philosopher Yen Tzu 晏子:[123]

Harmony may be illustrated by soup. You have the water and fire, vinegar, pickle, salt and plums, with which to cook fish and meat. It is made to boil by the firewood, and then the cook mixes the ingredients, harmoniously equalizing the several flavours, so as to supply what is deficient and carry off whatever is in excess.

---

[118] *Li Chi*, p. 460.
[119] Huang Chan-Yueh (1982), pp. 78–9. For Ma-wang-tui cf. *Hunan nung-hsüeh-yuan* (1978), pp. 28–33. For Yun-mêng, cf. *Hupei shêng po-wu-kuan* (1981), p. 15.
[120] *Chou Li*, p. 36; *Li Chi*, p. 459. cf. Legge (1985), p. 462.
[121] *Li Chi*, p. 457; cf. *Li Chi Chu Shu* for Cheng Hsuan's commentary.   [122] *I Li Chu Shu*, ch. 26a, p. 314.
[123] *Tso Chuan, Chao Kung* 昭公 20th year, Hung Yeh *et al.*, (1983), p. 403.

The same idea was even more eloquently expressed by the words of the legendary chef, I Yin:[124]

In the business of harmonious blending, one must make use of the sweet, sour, bitter, pungent and salty. Whether things are to be added earlier or later and in what amounts – their balancing is very subtle and each thing has its own characteristic. The transformation which occurs in the cauldron is quintessential and wondrous, subtle and delicate. The mouth cannot express it in words; the mind cannot fix upon an analogy. It is like the subtlety of archery and horsemanship, the transformation of Yin and Yang, or the revolution of the four seasons. Thus it is long-lasting yet does not spoil; thoroughly cooked yet not mushy; sweet yet not cloying; sour yet not corrosive; salty yet not deadening; pungent yet not acrid; mild yet not insipid; oily-smooth yet not greasy.

This statement is certainly in consonance with the view of a modern commentator on Chinese cuisine, Lin Yutang, who said that 'the whole culinary art of China depends on the art of mixture'.[125] In this sense, modern Chinese cookery has kept and enlarged on the tradition established by the desciples of I Yin twenty-five centuries ago.

How successful were the ancient Chinese in bringing I Yin's ideal into practical fruition? The best evidence for their culinary achievements is found in two poems in the *Chhu Tzhu* (Songs of the South), *Chao Hun* 招魂 (Summons of the Soul) and *Ta Chao* 大招 (The Great Summons), designed to summon a deceased soul to return to life by offering it the greatest pleasures that this world can provide. They contain the most vivid and inviting descriptions of foods and viands that can be found in any ancient literature. In *Chao Hun* the soul is tempted with:[126]

> Rice, broom-corn, early wheat, mixed all with yellow millet;
> Bitter, salt, sour, hot and sweet – there are dishes of all flavours,
> Ribs of the fatted ox cooked tender and succulent;
> Sour and bitter blended in the soup of Wu;
> Stewed turtle and roast kid, served up with yam sauce;
> Geese cooked in sour sauce, casseroled duck, fried flesh of the great crane;
> Braised chicken, seethed tortoise, high-seasoned but not to spoil the taste;
> Fried honey-cakes of rice flour and malt sugar sweetmeats;
> Jadelike wine, honey-flavoured, fills the winged cups;
> Ice-cooled liquor, strained of impurities, clear wine, cool and refreshing;
> Here are laid out the patterned ladles, and here is sparkling wine.

In the *Ta Chao* the delicacies were described thus:[127]

> The five grains are heaped up six ells high, and the corn of zizana.
> Cauldrons seethe to their brims, wafting a fragrance of well blended flavours
> Plump orioles, pigeons, and geese flavoured with broth of jackal's meat;
> O soul, come back! Indulge your appetite!
> Fresh turtle, succulent chicken, dressed with the sauce of Chhu;

---

[124] *LSCC*, ch. 14, *Pen Wei*, p. 370; tr. Harper (1984), p. 41; cf. also the translation by Knechtges (1986), p. 53.
[125] Lin Yutang (1939), p. 340.   [126] David Hawkes (1985), pp. 227–8.   [127] *Ibid.*, pp. 234–5.

Table 8. *Seasonings used in ancient and modern Chinese cookery*

Flavour	Ancient	Modern
BITTER *Khu* 苦	Sonchus weed, smartweed, wine	Bitter melon,[a] wine
SOUR *Suan* 酸	*Mei* plum, vinegar	Vinegar, lemon
SWEET *Kan* 甘	Malt sugar, honey	Cane sugar
PUNGENT *Hsin* 辛	Alliums, ginger, cassia, fagara	Alliums, ginger, sesame, fagara,[b] cassia, anise, fennel, clove, black pepper,[c] chilli pepper
SALTY *Hsien* 鹹	Salt	Salt
SAVOURY *Chiang* 醬	Meat, fish, soy pastes & pickles	Soy sauce and paste, fish, shrimp, oyster sauces, MSG, pickles.

[a] The balsam pear, *khu kua* 苦瓜, *Momordica charantia*, contains quinine, but is mint-like and refreshing. It is used as a vegetable.
[b] It is known as Szechuan peppercorn, and is used mostly in Szechuan cooking.
[c] A mixture of cassia, anise, fennel, clove and black pepper is also available as *wu hsiang* 五香 (five spice).

> Pickled pork, dog cooked in bitter herbs, and zingiber-flavoured mince,
> And sour Wu salad of artemisia, not too wet or tasteless.
> O soul, come back! Indulge in your own choice!
> Roast crane is next served, steamed duck and boiled quails,
> Fried bream, stewed magpies, and green goose, broiled.
> O soul, come back! Choice things are spread before you.
> The four kinds of wine have been subtly blended, not rasping to the throat:
> Clear, fragrant, ice-cooled liquor, not for base men to drink;
> And white yeast is mixed with must of Wu to make the clear Chhu wine.
> O soul, come back and do not be afraid.

Surely, this long list of delectables is indicative of the high level of sophistication in the culinary arts that the Chinese had achieved in the Warring States period (c. −300). From the references to steaming-cookery and the stress on blending of flavours, as exemplified by the use of bitter herbs, sour sauces, ginger and pickles etc. seen in the *Chhu Tzhu, Li Chi* and the *MWT Inventory*, it would appear that the character of the cookery in the Chou–Han period is already distinguishably Chinese.

How does the use of harmonising agents in ancient China compare with modern practice in Chinese cookery? For thickening soups and sauces, the answer is simple. Purified starch has now completely replaced rice crumbs and mucilaginous vegetables. For the flavouring agents, significant changes are noticeable as indicated by the data tabulated in Table 8. Today, bitter spices or vegetables are rarely seen while wine remains an important condiment. Vinegar is still the mainstay for the sour or tart taste. To impart sweetness cane sugar has overtaken malt sugar or honey. For saltiness salt remains king. But salt was also added indirectly in the form of a great variety of savoury sauces and pastes, *chiang* 醬 and *hai* 醢 etc., described in ancient China. In contrast, modern cookery depends on only a few savoury condiments, soy, plum, fish, shrimp, and oyster sauces, and MSG (monosodium glutamate), which are produced under much better quality control than their ancient counterparts. Soy sauce, by far the most important of this group of condiments, was probably unknown during the Han. The history of its development will be a major topic in Chapter (*d*). The other savoury sauces will be considered in Chapter (*e*). It is in the *hsin* 辛, pungent, category that the greatest change is visible. The onion group, cassia and ginger are still popular, fagara (Szechuan peppercorn) less so. New spices such as sesame, anise, fennel, clove, and black pepper have entered and stayed on the scene. But it is the introduction of the chilli or *Capsicum* pepper from the New World that has had the greatest impact on Chinese cuisine. It is probably eaten in larger amounts and by more people than any other spice in China, or for that matter, all over the world. The popular Szechuan and Hunan cuisines that we know of today would be mere provincial curiosities without the intervention of chilli pepper. But we have said enough about spices, seasoning and harmonising agents. It is time for us to move on and ascertain how the harmonised products were presented, served and consumed.

(iii) *Eating and drinking: dining vessels, implements and furniture*

> Spread out the mats for them,
> Offer them stools.
> Spread the mats and over-mats,
> Offer the stools with shuffling steps.[128]

We have now reviewed the principal methods and equipment used to prepare grain, meat, and vegetable foods eaten at a meal in ancient China. But we have not yet said anything about the drinks that must have been consumed in company with the foods. Even the simplest meal conceivable, that is described in the Analects by Confucius, 'With coarse cooked grain to eat, with water to drink, and my bended arm for a pillow – I have still joy in the midst of these things'[129], there is a drink to go with the food. The best-known drink is, of course, *chiu* 酒, 'wine', not the grape wine of the West, but a wine made from grains by a unique, ancient process. Since the subject of 'wine' will be almost a complete chapter unto itself (*c*), we shall not say anything more about it now, except in connection with its use as part of a meal.

The *Chou Li* tells us that six types of drinks were prepared for the royal family:[130]

---

[128] *Shih Ching* w 197 (M 246), tr. Waley, mod. auct.   [129] *Lun Yü*, Book VII, Legge (1861), p. 200.
[130] *Chou Li*, pp. 34, 52, and *Chou Li Chu Shu*, p. 79. Knechtges (1986), p. 50, note 11, has interpreted *chiang* as vinegar. We prefer the traditional interpretation of *chiang* as the water drained from the grain after it is boiled and strained for steaming. Upon standing it often becomes sour from the action of lactic acid bacteria. It is then called *suan chiang* 酸漿, cf. *CMYS*, p. 509. It may also become slightly fermented. *Chiang* is commonly used as part of the incubation medium in wine fermentations during the Sung, cf. *PSCC*, ch. 3.

(1) *Shui* 水, plain water.
(2) *Chiang* 漿, water drained after boiling grain to ready it for steaming.
(3) *Li* 醴, sweet weak wine after a one day fermentation.
(4) *Liang* 涼, a tea made from parched grain.
(5) *I* 醫, a drink made from plum juice.
(6) *I* 酏, a very thin congee.

The *Li Chi* mentions the same six beverages, but also includes fully fermented wine, *chiu* 酒, filtered and unfiltered.[131] Water (1), the first item on the list is, of course, the oldest of all drinks. For this reason, it was sometimes referred to honorifically as *hsüan chiu* 玄酒, the primordial wine.[132] The fact that the early Chinese drank water is to be expected. What is of greater interest to us is the question of when they started to drink boiled water, a practice of great importance in the maintenance of public health. Its origin is obscure. There is no information in the pre-Han, or even pre-Thang literature on the drinking of boiled water. In fact, the earliest reference we have to it is in the Southern Sung. In the *Chi Lê Pien* 雞肋編 (Miscellaneous Random Notes), +1133, there is a passage which says, 'Even when the common people are travelling they take care only to drink boiled water.'[133] This does not mean, of course, that the practice of drinking boiled water started during the Sung. We suspect the practice had existed long before the Sung. It may have started in early Han as a byproduct of the process of boiling water to make tea. It may also have been simply a logical extension of the ancient concept that cooked food is healthier than raw food.[134] *Chiang* (2), *Liang* (4) and *I* (6) are all variants of drinks derived from grain. *I* (5) is rarely seen in the Han literature. *Li* (3) was apparently much favoured during the Shang dynasty.[135] But *chiu* (wine) was the preferred drink of the Chou people. The *Shih Ching* is replete with references to wine in connection with feasting and entertainment, indicating that it was an important element in the culture and cuisine of Chou society.[136] Both literary and archaeological evidence indicate that its prestige and popularity remained unabated during the Han.

---

[131] *Li Chi*, p. 457. In another context, p. 484, it refers to 'five drinks 五飲', by leaving out item 5, I 醫.
[132] *Li Chi*, p. 531, which states that the 'The Hsia favoured clear water (*shui*), the Shang sweet wine (*li*), and the Chou clear wine (*chiu*).' Cf. pp. 368–9 and 973 for references to the use of water as wine.
[133] Tr. by Needham *et al.* (1970), p. 362 from a passage in Fan Hsing-Chun (*1954*), although they were unable to locate its original source. We have now traced it to the *Chi Lê Pien*, +1130, by Chuang Chi-Yü 莊季裕, Pt 1, p. 8a. We suspect the reason why the practice of drinking boiled water is not mentioned in the post-Han and pre-Sung literature is because the practice was so familiar to everyone that no one deemed it of interest to record it in writing. The same reason may be cited to explain why there is no recipe for cooking rice in all the works of Chinese culinary literature.
[134] The notion that cooking with fire cleanses food is of great antiquity. The *Li Wei Han Wen Chia* 禮緯含文嘉 (Apocryphal Treatise on Rites; Excellences of Cherished Literature), an early Han work based on ancient material, says: 'It was Sui Jen who first drilled wood to obtain fire and taught the people to cook food from raw materials in order that they might suffer no diseases of the stomach, and raise them above the level of the beasts', tr. Needham (1970), p. 364, who notes that, 'It was from these roots that there comes forth the proverbial saying *pai fei wu tu* 百沸無毒 (anything that is boiled cannot be poisonous).
[135] *Li Chi*, p. 531, cf. note 132 above.
[136] *Shih Ching*, see for example, w 30, 75, 95, 144, 156, 157, 159, 169, 193, 194, 195, 196, 197, 199, 200, 203, 204, 212, 233, 242, 250, 262.

Thus, a meal by definition consists of two parts: food, *shih* 食 and drink, *yin* 飲. Both *shih* and *yin* are words that can be used as a noun or as a verb, so that *shih* can also mean to eat and *yin* to drink. In the *Li Chi* food is further divided into *fan* 飯, grain food, *shan* 膳, viands and *hsiu* 羞, confections or supplemental delicacies.[137] We shall have more to say about these categories later but for now the stage is set to allow us to consider the utensils and accessories, that is, all the physical equipment needed to turn the partaking of food and drink into a cooperative, social event. These are:

(1) Vessels for serving food and drink.
(2) Serving implements to bring food from the vessel to the diner.
(3) Furniture to place the vessels within reach of the diner.

Vessels for serving food and drink

Of all the arts of the ancient Chinese none has evoked more admiration than the Shang and Chou bronzes. These strikingly beautiful objects have been studied extensively not only for their archaeological importance and artistic merit but also for their supposed use as utensils for cooking and serving food and drink. The classification of bronze vessels by archaeologists in terms of their culinary uses has been reviewed and evaluated by K. C. Chang. The results have given us a very good idea of the type and shape of vessels involved in the Chou culinary system.[138] From the numerous types of Shang and Chou bronze vessels summarised by him, we have already picked out those that were used in cooking the raw food, *li* 鬲, *ting* 鼎, *yen* 甗, *tseng* 甑, *huo* 鑊 and *fu* 釜 (Figures 22, 23, 24). We are now concerned with those which appear to have been used as vehicles for serving food and drink. The most important of these vessel types are shown in Figure 30. In many cases, versions of the same designs have also been fabricated from pottery, wood and occasionally bamboo. A résumé of the different versions of vessel types known and their presumed end uses is presented in Table 9.

The first item listed is the best known of all Chinese bronzes, *ting* 鼎, the tripod cauldron, which we have already encountered as a cooking vessel. Big and little cauldrons are mentioned in the *Shih Ching*, and the *Chou Li* states that the king was served twelve cauldrons of viands.[139] The *MWT Inventory* describes twenty-four cauldrons of *kêng* 羹 soups and stews, but only eleven cauldrons (seven lacquer and four pottery) were recovered in the tomb.[140] These references provide ample evidence that *ting* was an important implement both for cooking and for serving food in ancient China. The next two vessels, *kuei* 簋 and *yü* 盂, are similar in shape, and apparently also similar in function. *Kuei* is cited in the *Shih Ching*, *Chou Li* and *Li Chi*,

---

[137] *Li Chi*, pp. 456–7.
[138] Chang, K. C. (1973), pp. 503–4; (1977a) pp. 34–5; (1977b) pp. 366–70; (1980) pp. 23–7. For further information on Chinese bronzes see Mizuno (1959), Fong (1980), Ma Cheng-Yuan (1986) and Ma Chêng-Yuan ed. (1988).
[139] *Shih Ching* w 233 (M 292); *Chou Li*, p. 34.
[140] Hunan shêng po-wu-kuan, *MWT Inventory*, (1973), I, pp. 95 and 127.

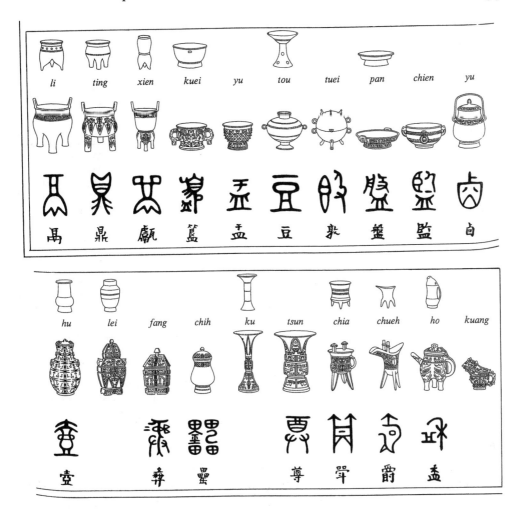

Fig 30. Bronze vessel types for serving food and drink. After Buchanan, Fitzgerald and Ronan (1981), pp. 158–9.

and is said to be a container for grain food.[141] Since the word *kuei* is topped by a bamboo radical, it might originally have been made of bamboo basketry and may be considered as related to *tan* 箪, the bamboo basket for carrying cooked grain mentioned by Confucius and Mencius.[142] The *Shuo Wên* says that *yü* is a container for *fan* (cooked rice or millet). It also says that *wan* 盌, which is a variant form of *wan* 碗, is a small *yü*. This means that *yü* is the venerable progenitor of the modern Chinese bowl, *wan*, of which the best known example is the rice bowl, *fan wan* 飯碗, probably the most important, at least metaphorically, of all food utensils in China today.

[141] *Shih Ching*, w 279, 195 (M 135, 165); *Chou Li*, pp. 54, 178; *Li Chi*, p. 369.
[142] *Lun Yu* ch. 6, v. 9, Legge (1861), p. 188; *Mêng Tzu* Bk 1 (2), ch. 11, v. 3, Legge (1895) p. 171.

Table 9. *Serving vessels for food and drink in ancient China*

Type		Pottery	Bronze	Wood/Lacquer	End use
Ting	鼎	S,C,H	S,C	H	Soup, stew
Kuei	簋	S,C,H	S,C	—	Rice,[a] congee
Yü	盂	S,C,H	S,C	H	Rice, congee
Tou[b]	豆	S,C,H	C	S,C,H	Viands
Phan	盤	S,C,H	S,C	H	Water, viands
I	匜	H	S,C	H	Water
Yu	卣	S,C	S,C	—	Wine
Hu[c]	壺	S,C,H	S,C,H	H	Wine, drinks
Tsun	尊	S,C,H	S,C	H	Wine, drinks
Ku	觚	S,C	S,C	—	Wine
Chia	斝	S,C	S,C	—	Wine
Chüeh	爵	S,C	S,C	—	Wine
Ho	盉	S,C	S,C	—	Wine
Kung[d]	觥	S,C	S,C	—	Wine
Chih[e]	卮	H?	C	H	Wine, drinks
Erh Pei	耳杯	H?	—	H	Wine, soup?

s = Shang, c = Chou and h = Han
[a] Rice in this case means cooked grain in granules. It was usually served in bamboo baskets such as *tan* 簞 or *chuan* 簹.
[b] *Tou* was usually made of wood, which was probably treated in some way to make it impervious to water. A very popular form of *tou* made of bamboo basketry was called *pien* 籩. A pottery *tou* was called *têng* 登.
[c] A square form of *hu* was called *fang* 鈁, and one with a large top and small bottom, *lei* 罍.
[d] Also pronounced *kuang* or *huang*. Originally this was a rhinoceros (or buffalo) horn; simulated versions in bronze were made later.
[e] The earliest drinking vessel is probably the *phiao* 瓢 which is simply a gourd split in half. It was still in use during the Chou.

Almost as familiar as the *ting* cauldron is *tou* 豆, which may be described as a shallow bowl or a deep plate sitting on a pedestal. It was usually made of wood. Versions of *tou* made of bamboo basketry were called *pien* 籩; those made of pottery, *têng* 登. *Tou* and *têng* were the preferred vessels for conveying wet dishes of meat or vegetables, while *pien* was used for dry food. *Tou* and *pien* are cited, often together, in the *Shih Ching*, *Chou Li* and especially *Li Chi*.[143] Mencius had mentioned *tou* as a container for a bean *kêng* stew.[144] The popularity of *tou* waned after the Han dynasty, although it never actually died out. In fact, we can still occasionally see specimens of *tou* vessels, made of porcelain or pewter, being used in Cantonese restauraunts in England or America today.

Perhaps the decline of *tou* was due, at least in part, to the rise of popularity of *phan* 盤, originally designed as a receptacle for the waste water which had flowed from an *i* 匜 pitcher to rinse the diner's fingers. The *phan* is, in effect, a *tou* without the pedestal. We begin to see, by the Warring States period, lacquer *phan*'s that were too shallow to serve as a water receptacle but eminently suitable for serving meat and

[143] For example, *Shih Ching*, w 72, 238 (M 158, 245); *Chou Li*, pp. 54, 55; *Li Chi*, pp. 369, 394, 409, 414, 428, 482.
[144] *Meng Tzu*, Bk 7 (I), ch. 35, Legge (1895) p. 469.

Fig 31. Lacquer tray containing five *phan* plates, two *chih* mugs, one winged-cup and one pair of chopsticks from Han Tomb No. 1 at Ma-wang-tui. *Hunan shêng po-wu-kuan* (1973), II, p. 151, item 160.

vegetable viands. Many well-preserved lacquer *phan* with beautiful designs (Figure 31) were discovered in Han Tomb No. 1 at Ma-wang-tui. Some of these actually had food remains on them (Figure 4), indicating, without a doubt, that *phan* had become, by early Han, the equivalence of modern dining plates for serving viands at a meal.

Thus, the major vessels used for serving food in ancient China are: *ting* (three-legged bowl or tureen) for soup and stews, *kuei* (or *tan*) and *yü* (rice bowl) for cooked grain, and *tou* or *phan* (plates) for viands. The remaining entries deal with the serving of drinks. The first of these, the *i* 匜 water pitcher, however, is not a drinking vessel. It is, nevertheless, an integral component of the dining scene, since it is used in conjunction with the *phan* for the ablution of the fingers of the distinguished guests before they can touch their food. The importance of this ritual will become self-evident later when we consider the way food is actually eaten. Beautiful specimens of lacquer *i* pitchers were found in early Han tombs at Ma-wang-tui and Chiang-ling, indicating that this practice was very much in vogue at least during the former Han.[145]

---

[145] *Hunan shêng po-wu-kuan* (1973) I, p. 88, Fig. 82, II, p. 159, Pl. 167; *Chiangling Fêng-huang-shan 167 etc.* (1976); and *Chi-nan chhêng Fêng-huang-shan 168 etc.* (1975).

The next three entries, *yu* 卣, *hu* 壺 and *tsun* 尊, are vessels used for the storage of drinks, particularly wine. *Yu* and *hu* are pots shaped like a vase. *Yu* is a *hu* with a handle, while *tsun* is a vase shaped like a spitoon (Figure 30). The ancient word *hu* has retained its original meaning to this day, as, for example, in the familiar *chha hu* 茶壺 (teapot) of which we shall hear more of in Chapter (*f*). The *Shih Ching* refers to a feast in which a hundred *hu* pots of wine were served,[146] and in the *Book of Mencius hu* is described as a container for *chiang* or rice water.[147] A slightly different model of *hu* is *lei* 罍, which has the shape of a Grecian urn.[148] *Hu* itself was usually round, but a square form, appropriately called *fang* 鈁 was also known. Specimens of *hu* and *fang* in lacquerware and bronze were discovered in Han tombs at Ma-wang-tui, Chiang-ling and Yun-meng.[149] Both *hu* and *tsun* are cited in the *Li Chi* as containers for wine (and other drinks).[150] Although *tsun* is also cited prominently in the *Chou Li* as a wine vessel, it is no longer used in this sense in the Chinese language today.[151]

To serve the ample store of wine on hand, the ancient Shang and Chou people designed a variety of wine decanters or drinking cups. The most popular ones, listed next in Table 9, are *ku* 觚, *chia* 斝, *chüeh* 爵, *ho* 盉 and *kung* 觥 (Figure 30). The first four of these elaborately crafted objects may also have been used for warming wine. Based on their substantial size and weight, the beautiful bronze specimens of these vessels that we see in the museums today were probably more of ceremonial than culinary significance. Pottery versions of some of these were presumably the utensils seen in everyday life. There is, however, a fair degree of uncertainty in regard to the specific function of these vessels. The *Shih Ching* cites *chia* and *chüeh* as important wine vessels in a context which suggests that they were used directly as drinking vessels.[152] The one reference in the *Tso Chuan* to *chüeh* also implies that it was used for drinking.[153] The *Li Chi* mentions *chia* and *chüeh* in several passages, but the function of these vessels is not clearly defined.[154] All three could function as decanting, warming as well as drinking vessels. *Ho* has not appeared in the Chou classics that were consulted but *kung* is the bronze version of *szu kung* 兕觥, the buffalo or rhinoceros horn, which is featured in the *Shih Ching* as a goblet or cup for drinking wine.[155]

The horn cup apparently remained a popular goblet for drinking wine during the Han period. In a mural depicting the famous 'Banquet at Hung-Men', discovered in a Western Han tomb near Loyang in 1957, one of the participants is shown clearly

---

[146] *Shih Ching* w 144 (M 261). Waley, however, translates *hu* as cup.
[147] *Meng Tzu* Book I (2), ch. 10, 11; Legge tr. (1895), pp. 170, 171.
[148] *Shih Ching* w 40 (M 3), for an illustration of its shape see Chang, K. C. (1977a), p. 174.
[149] *Hunan shêng po-wu-kuan* (1973) I, p. 95, II, Pl. 158; *Chi-nan Fêng-huang-shan 168 etc.* (1975); *Yun-meng Suei-hu-ti Chhin-mu* (1981), pp. 57 and 64. Cf. also *KK* (*1981*) *I* pp. 38, 42, 43.
[150] *Li Chi, hu* p. 395; *tsun* pp. 395, 406, 495. [151] *Chou Li*, pp. 49, 212. [152] *Shih Ching* w 267, 197 (M 220, 246).
[153] *Tso Chuan*, Chao Kung, 5th year p. 358; Chuang Kung 21st year p. 66; Ai Kung, 15th year, p. 492; and Suan Kung, 2nd year, p. 181.
[154] *Li Chi*, pp. 395, 439, 494, 585. On p. 395 three other drinking vessels are mentioned *san* 散, *chih* 觶, and *chiao* 角. Chêng Hsüan's commentary says, 'The container for one *shêng* is *chüeh*, for two, *ku*, for three, *chih*, for four *chiao*, and for five, *tsun*.'
[155] *Shih Ching* w 40, 159 (M 3, 154).

Fig 32. Enlarged section from a mural depicting the Banquet at Hung-Men in a Western Han tomb at Loyang. To the left is a guest holding a horn drinking cup. To the right is a chef roasting a fowl on a large fork over a charcoal fire. Fontein & Wu eds. (1976), Tomb 61, p. 23.

to be holding a horn-shaped wine goblet in his right hand (Figure 32).[156] Perhaps the most elegant of all the ancient Chinese bronzes listed in Table 9 is *ku* 觚. There is no question that this is a drinking vessel, and must have featured prominently in the scenes of sacrificial rituals, feasting and entertainment that adorned the life of the Chou nobility. However, for the common people, the most important drinking vessel during the Chou dynasty was probably *phiao* 瓢, the receptacle that results when a gourd is cut in half. *Phiao* is made famous by Confucius in his statement praising the virtue of Huei: 'With a single bamboo dish of grain and a gourd dish (*phiao*) of drink ... he did not allow his joy to be affected by it.'[157]

Towards the late Chou period two new types of drinking vessels emerged with designs quite different from the Shang and early Chou bronze prototypes. These are shown in Figure 31. The first is the *chih* 卮 stoup, a cylindrical vessel reminiscent of a beer tankard or a coffee mug. It sometimes comes with a cover. In the account of the 'Banquet at Hung-Men' in the *Shih Chi*, Historical Record (c. −90), a *chih* 卮 stoup of wine was given to the bodyguard of the principal guest.[158] Well-preserved lacquered *chih* stoups have been found in Han tombs at Ma-wang-tui, Chiang-ling and Yun-meng.[159] The second has been given the name of *erh pei* 耳杯, cup-with-ears or 'winged cup'. It is a shallow, oval-shaped bowl with two narrow, flat handles at the mouth (Figure 31). It could easily have been used for soup as well as wine. It is surprising that no special name had been coined for this rather uniquely shaped vessel. From the large number of specimens of lacquer *erh pei* that have been discovered in early Han tombs, e.g. eighty at Tomb No. 1 at Ma-wang-tui, and eighty-four

---

[156] Fontein, J. and Wu Tung ed. (1976), Tomb 61, p. 23. [157] *Lun Yu*, Bk 7, ch. 9; Legge tr. (1861), p. 188.
[158] *Shih Chi*, ch. 7, pp. 312–13; Yang & Yang (1979), p. 219.
[159] For Ma-wang-tui cf. *Hunan shêng po-wu-kuan* (1973) 1, p. 95; Chiang-ling, cf. *Chhang-chiang liu yü etc.* (1974), p. 47 and *Chi-nan chhêng Fêng-huan shan etc.* (1975), p. 5; and Yun-mêng, cf. *Yün-mêng hsien wên-wu kung-tso tui* (1981), p. 44.

in Tomb M8 in Chiang-ling, it is obvious that this was the major vessel used for drinks in the late Chou, Chhin and Han era.[160]

Implements for serving food

We have seen that for the drink, *yin* 飲, portion of a meal, we have *hu* 壺 and *fang* 鈁 pots to hold the liquid, and mugs (*chih* 卮), winged cups (*erh pei* 耳杯) or one of the fancy bronze vessels from which to drink it. But for the food, *shih* 食, portion we still need a utensil to bring the food into the mouth. From textual and archaeological information, three types of serving and eating utensils are known to have been in use in ancient China: chopsticks, spoons and ladles, and forks. Chopsticks are, of course, the best-known and most characteristically Chinese eating implement. According to the *Shih Chih* 史記 (Records of the Historian), −90, by Szuma Chhien, King Chou 紂, the last ruler of the Shang dynasty, was the first person to have made a pair of chopsticks out of ivory, which implies that chopsticks must have been known before King Chou's time, that is, before −1100.[161] This statement is often quoted as one example of King Chou's profligate ways which lost him his empire. There is, however, no textual or archaeological evidence in support of Szuma Chhien's account. The earliest archaeological record of chopsticks was found in a tomb at Ta-pho-na, southeast of Tali, Yunnan. Among the bronze artifacts discovered there were two pairs of chopsticks and several spoons. The tomb was dated at sometime in the middle Spring and Autumn period (c. −600).[162]

Chopsticks, formally called *chu* 箸 or *chia* 挾, however, must have become well known by the time of the Warring States (−500 to −300). The *Li Chi* has two references in chapter 1 in regard to how chopsticks should or should not be used. The first occurs in a section dealing with good table manners:[163]

When eating viands with others from the same dishes, do not eat (hastily) to satiety. When eating rice (i.e. cooked grain) with others from the same bowl be sure that your hands are clean.

'Do not roll the rice (or cooked grain) into a ball (with your fingers); do not put rice (you have already picked up) back into the bowl; do not gulp down your drink; do not make a noise in eating . . .

. . . Do not fan your (bowl of) rice (to cool); do not use chopsticks (*chia* 挾) to pick up rice . . .

The key to understanding these admonitions is that the proper way to pick up rice, which would have been steamed and served as loose granules, was to use one's fingers.[164] Chopsticks were meant only for viands. It was, therefore, imperative that when eating rice with others from the same bowl one should have clean hands. It is

---

[160] *Hu-nan shêng po-wu-kuan* (*1973*) I, p. 95; *Chhang-chiang liu yü etc.* (*1974*), p. 52.
[161] *Shih Chi*, ch. 38. The earliest textual reference to chopsticks is probably in the *Han Fei Tzu* (early −3rd century), ch. 7, *Yü Lao* 喻老, p. 57, and ch. 8, *Shuo Lin Shang* 說林上, p. 62. These references were presumably used later by Ssuma Chhien.
[162] K. C. Chang (*1977b*), p. 456; cf., *Yunnan shêng wên wu kung-tso tui* (*1964*).
[163] *Li Chi*, ch. 1, *Chhü Li* 曲禮, pp. 29, 30.
[164] Wang Jên-Hsiang (*1985*) and Shên Thao (*1987*). For further details of the history of chopsticks and their role in Chinese food culture cf. Chao Yung-Kuang (*1997*) and Hsing Hsiang-Chhên (*1997*).

also understandable that one should not put rice that had been picked up back into the common bowl. This explains why in polite society, it was necessary to have *i* 匜 and *phan* 盤 vessels ready for the ablution of the fingers of the diners. The significance of cleaning one's hands is mentioned in another part of *Li Chi*. In chapter 22, in connection with the diet allowed during a period of morning, we read that after the fast is over, one may first eat congee, and then advance to coarse rice and water. Then follows the interesting admonition,[165] 'To eat congee from a bowl (*sheng* 盛) it is not necessary to wash one's hands. To eat rice (grain) from a basket (*chuan* 簹) it is necessary to wash one's hands.' Obviously, clean hands are not essential if one eats congee with a spoon, but they are if one eats rice with one's fingers. It is interesting to note that even the *Li Chi*, a guide to proper behaviour for the aristocracy and well born, mentions a vessel made of bamboo basketry for serving rice at a meal. Basketry could have provided the most widely used vessels for serving cooked grains in ancient China, and we shall have more to say about it later.

The second passage in the *Li Chi* occurs later in chapter 1 but is also concerned with table manners,[166] 'Use chopsticks (*chia* 梜) when there is vegetable in the soup (*kêng* 羹); do not use chopsticks [but use a spoon] when there is no vegetable in the soup.' The implication again is that chopsticks were used only to pick up the solid food in viands. These references suggest that chopsticks were already well known towards the end of the Warring States period.

Chopsticks were common among the remains recovered in early Han tombs at Ma-wang-tui and Chiang-ling. Most significantly a pair of bamboo chopsticks was found together with four plates, a ear-cup and two wine stoups on a serving tray in Tomb No. 1 at Ma-wang-tui (Figure 31), and a bamboo case containing twenty-one bamboo chopsticks was recovered from Tomb No. 167 in Chiang-ling.[167] The chopsticks were from twenty to twenty-five centimetres (six to ten inches) long, well within the range of modern samples. The usage of chopsticks expanded greatly after the Han. As rice bowls became more common, the practice of shovelling rice from a bowl into the mouth began to gain favour and the habit of picking up rice with fingers soon died out. However, the current, familiar expression for chopsticks, *khuai tzu* 筷子, probably did not come into use until later. In any case, it did not find its way into the written literature until early Ming (fifteenth century).[168]

We have already referred to the long spoon *pi* 匕 as an instrument to stir food while it was being cooked in a pot (p. 85). A small *pi* would be the equivalent of the modern spoon, which is used universally for conveying soup from a container (cauldron or bowl) into one's mouth. A more specialized type of spoon is the ladle, *shuo* 勺, which is designed to distribute liquid, such as soup or wine, from a large container to several smaller ones. A wide range of ancient spoons and ladles have been discovered in China in recent years. The most celebrated archaeological find

---

[165] *Li Chi*, ch. 22, *Sang Ta Chi* 喪大記, p. 718.   [166] *Li Chi*, ch. 1, p. 33.
[167] For Ma-wang-tui, *Hu-nan shêng po-wu-kuan* (1973) II, Fig. 160, p. 151; for Chiang-ling, *Fêng-huang Shan 167 etc.* (1976), p. 37.
[168] *Shu Yuan Tsa Chi* 菽園雜記.

Fig 33. Gold bowl and ladle from the tomb of Marquis of I. Qian, Chen & Ru (1981), p. 45, Fig. 58.

of this genre is undoubtedly the exquisitely crafted gold ladle recovered from the inside of a gold bowl (Figure 33) in the tomb of Marquis I of the State of Tsêng in Sui-Hsien, Hupei, which was excavated in 1968.[169] It is presumably a miniature model, being only five inches long. The working size lacquered ladles found at Tomb. No. 1 at Ma-wang-tui are from sixteen to twenty-four inches long.

The *pi* (later called *shih* 匙) spoon has had a long history in China, having been known since the Shang dynasty. A summary of the changing styles of spoons in different periods from Shang to Han is shown in Figure 34.[170] The oldest specimens were made of bone. The early bronze types have a sharp point which suggest that they might have also been used to cut meat. They started to take a rounder form in the Spring and Autumn period, and by the time of the Warring States, lacquered wooden spoons began to appear on the scene. During the Han, lacquer spoons reached the height of popularity and much care was taken to decorate them with beautiful designs. The early spoons (Shang and Western Chou) tend to be quite long, twenty-five to thirty centimetres (ten to twelve inches), while the later ones were shorter, ten to twenty centimetres (four to ten inches), just the length convenient for a dining utensil. The formal Chinese word for a spoon is *chhang shih* 餐匙 but because it is used to mix a soup while flavouring agents are added, it is almost always known today as *thiao kêng* 調羹, that is, the article used to fine tune the soup.

One does not normally associate the fork with utensils used in a Chinese meal, but recent archaeological finds tend to suggest that it may have had a culinary role in ancient China. We have earlier (p. 85) referred to the long, two pronged fork,

[169] Qian Hao, Chen Heyi and Ru Suichu (1981), p. 45.   [170] Chih Tzu (1986b).

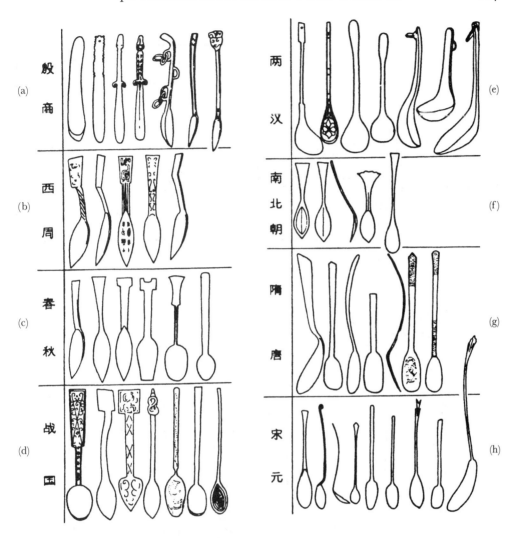

Fig 34. Changing styles of spoons from Shang to Yuan, after Chih Tzu (1986b), p. 22: (a) Shang; (b) Western Chou; (c) Spring and Autumn; (d) Warring States; (e) Han; (f) North & South Dynasties; (g) Sui–Thang; (h) Sung–Yuan.

pi 畢, mentioned in the *Li Chi*, and the three pronged forks used as skewers for broiling *chih* 炙 meat (Figure 27). But now we are talking about smaller versions that are similar to those set on a Western dining table. The archaeological evidence for the fork in China is summarised in Table 10, and the shape of the specimens are shown in Figure 35.[171] Thus, there is no question that forks did exist in ancient China, but were they used as food serving implements? Documentation on the *pi* fork is scarce. There is a reference in *I Li* to the effect that the '*Tsung Jen* 宗人 official

[171] Chih Tzu (1986a).

Table 10. *Archaeological finds of forks in ancient China*

Period	Location	Number	Material	Length cm[a] total	Length cm[a] tooth	Number in fig. 35	References
Chhi-chia[b] culture	Wuwei, Kansu	1	Bone	?	?	8	1
Shang[c]	Chêngchou, Honan	1	Bone	8.7	2.5	7	2
Warring States	Houma, Shansi	2	Bone	?	?	6	3
	Houma, Shansi	1	Bone	?	?	5	4
Warring States	Loyang, Honan	1	Bone	18.2	4	4	5
	Loyang, Honan	51	Bone	12.1	4	1&3	6
Eastern Han	Chiuchhuan, Kansu	2	Bronze	26.3	7.5	9	7
Eastern Chin	Shihhsing, Kuangtung	4	Iron	15	4	not shown	8

[a] All are between 12 and 20 cm in length.
[b] Three-pronged, all others are two-pronged.
[c] Damaged.

References: 1 *Kansu shêng po-wu-kuan* (*1960*), Pl. 4, no. 6.
2 *Honan Shêng wên-hua-chü* (*1956*), Fig. 18.
3 *Shansi shêng wên-kuan hui* (*1960*), Pl. on p. 10, no. 4.
4 *Shansi shêng wên-kuan hui* (*1959*), Fig. 4, no. 7.
5 *Loyang po-wu-kuan* (*1980*), Pl. 5, no. 15.
6 *Chung-kuo kho-hsüeh yuan khao-ku yen-chiu so* (*1959*), Fig. 98, Pl. 72.
7 *Kansu shêng wên-wu kuan-li wei-yuan-hui* (*1959*), Fig. 17.
8 *Kuangtung shêng po-wu-kuan* (*1982*), Fig. 13 (7).

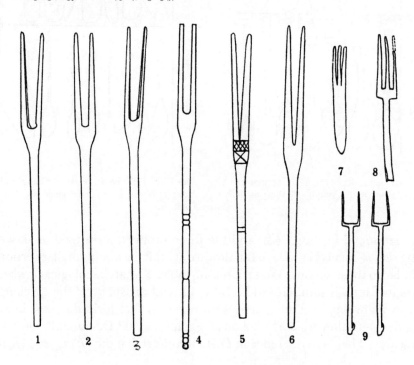

Fig 35. Styles of forks, prehistoric to Yuan, after Chih Tzu (*1986a*), p. 17. For the origin of the forks, cf. Table 10.

enters first with the *pi* fork',¹⁷² but we have no idea of how large it was or what it was for. However, in an Eastern Han tomb in Ta-kua-liang, Sui-Tê county, northern Shensi, there is a stone carving of a dining scene in which three forks are shown hanging near the head of the diner (Figure 36(1)). Such forks also appear in another tomb in a scene of the top of a stove together with a hook, a ladle, a gourd and a whisk (Figure 36(2)).¹⁷³ These pictures are intriguing, and at least suggestive that at one time forks might have been used to serve food in ancient China.

Dining furniture and accessories

There is very little information in the classical literature on the furniture used at a meal in Chou China. One thing we can be sure of is that there were no tables or chairs. People sat on mats on the floor. The earliest record of the physical setup of a meal is given in the *Shih Ching*. In a poem on a clan feast, 'The Weedy Path', we read:¹⁷⁴

> Spread out the *mats* (of bamboo) for them,
> Offer them *stools*.
> Spread the mats and *over-mats* (of reed),
> Offer stools with shuffling step.

The bamboo mats were called *yen* 筵, the reed over-mats, *shih* 席, and stools, *chi* 几. The finer reed mats were placed over the coarse bamboo mats. Presumably, the guests sat on the heels of their feet in a kneeling position. There is no mention of any furniture on which the food vessels were placed. The *chi* is usually taken to mean a low stool for one to lean on, and not for placing dishes of food. In a passage in the *Tso Chuan* on proper conduct between states the King of Chhu was asked, 'why setup a *chi* stool if you do not lean on it, why fill a *chüeh* 爵 vessel if you do not drink from it?'.¹⁷⁵ There is an account in the Book of Mencius, in which the sage is said to have fallen asleep leaning against a low table (*chi*).¹⁷⁶ The *chi* was apparently a regular piece of furniture in a household. In the *Nei Tsê* chapter of the *Li Chi*, *chi* is mentioned twice as part of the equipment and accessories that had to be moved when young members of the family attended to the comforts of the elderly.¹⁷⁷ The *Chou Li* cites a special official, *Szu Chi Yen* 司几筵 (Superintendent of Stools and Mats), whose duty it was to provide the stools and mats appropriate to each occasion as needed by the royal household.¹⁷⁸ In another poem in the *Shih Ching* about a sacrifice offered by the stalwart Duke Liu, we are told:¹⁷⁹

> Walking deftly and in due order
> The people spread out (*yen*) mats and (*chi*) stools,
> He stepped up on the mat and leaned upon a stool.

---

¹⁷² *I Li Chu Shu*, ch. 45, p. 528.
¹⁷³ *Shensi sheng po-wu-kuan (1958)*, Fig. 66, and Tai Ying-Hsin & Li Chung-Hsüan *(1983)*, p. 236.
¹⁷⁴ *Shih Ching*, w 197 (M 246).   ¹⁷⁵ *Tso Chuan*, Chao Kung, 5th year, p. 358.
¹⁷⁶ *Mêng Tsu*, Bk II, Pt I, Legge tr. (1895), p. 228; Lau tr. (1970), p. 93.
¹⁷⁷ *Li Chi, Nei Tsê*, pp. 446–7; Legge tr. (1885), pp. 452–3.
¹⁷⁸ *Chou Li*, p. 214.   ¹⁷⁹ *Shih Ching*, w 239 (M 250).

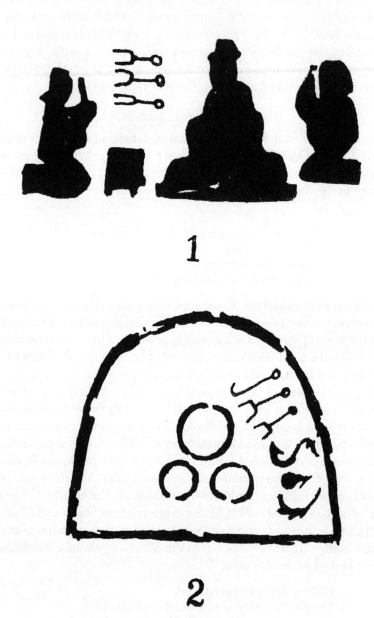

Fig 36. Forks in Eastern Han tombs in Sui-tê, Northern Shensi: (1) Mural, *Shensi shêng po-wu-kuan (1958)*, Fig. 66; (2) Stove model, Tai & Li (1983), Fig. 4–1.

Thus, all indications are that *chi* is a stool for one to lean on. It may be especially needed in helping an elderly person to get up from a kneeling position. In yet another poem, the *Shih Ching* says:[180]

> The guests are taking their seats (*yen* mats);
> To left, to right, they arrange themselves.
> The food baskets and dishes are in their rows,
> With dainties and kernels displayed.

There is no explanation of where exactly the food baskets and dishes were placed 'in their rows' during the feast. Presumably, they were placed on the mats or on the floor, in front or on the side of the guest. Since the vessels had to be on the floor, it would make sense to have them raised as high as possible so that the food is within easier reach of the diner. That is precisely the case. We have *tou* for viands and *ting* for soup. Both are raised above ground. *Tou* stands on a pedestal and *ting* stands on three legs.

Yet, one cannot help wondering why the food was not placed on the *chi* stools, since they were obviously available. It would seem a perfectly natural and convenient thing to do unless the early stools were too narrow for anything to be placed on them. But this point was beginning to be appreciated by the ancient Chinese. There is one passage, in the *Chhü Li* chapter of the *Li Chi*, which suggests that changes had already started to take place in the style of dining by the time of the Warring States,[181] 'When the host leads on the guests to present an offering (to the father of cookery) they will begin with the dishes which were first brought in. Going on from the meat cooked on the bones, they will offer all (the other dishes).'

According to the early Thang commentator Khung Ying-Ta, to perform the offering ceremony, the ancients placed a small sample of the food on the 'table' (*chuo* 桌) between the dishes. Tables were already in common use by Kung's time, but it is not at all clear what kind of 'table' he thought the ancients had. However, he clearly implied that the dishes were placed on a sort of 'table' and not on the mat or on the floor.

In any case, elaborate instructions are given in the *Li Chi* regarding the physical setup of a typical Chou meal.[182] Basically, a gentleman diner kneels on an individual mat. Food and drink vessels are placed on the mat near him in a prescribed order. The other diners are seated in positions relative to the host according to their respective rank and station. For details of the specific arrangements the reader is referred to the excellent summary provided by K. C. Chang.[183] Unfortunately, we do not have any archaeological data to corroborate the account in the *Li Chi*. But, fortunately, we do have a considerable amount of information on the physical setup of a Han meal derived from tomb murals and archaeological finds.

---

[180] *Ibid.* w 267 (M 220).   [181] *Li Chi*, p. 29.
[182] *Li Chi*, cf. *Chhü Li* (pp. 28–9), *Nei Tsê* (pp. 463–5), and *Shao I* (pp. 884–8).
[183] Chang, K. C. (1977), pp. 37–9.

Sometime in the Warring States period or perhaps earlier, trays with low support, *an* 案, rather like a modern cocktail table with very short legs, began to appear on the scene.[184] The earliest type is no more than a tray with a raised foundation, such as the lacquer *an* from Han Tomb No. 1 at Ma-wang-tui shown in Figure 31. The *chi* 几 stool was also made, but it tends to be narrower though higher than the *an* as indicated by the specimens discovered in the Western Han tomb at Chiang-ling, Hupei.[185] A variety of sizes and shapes of *an* articles, made of wood, lacquer, bronze, stone and pottery have been discovered in Han tombs in recent years. These have been reviewed and discussed by Chhên Tsêng-Pi.[186] The most interesting find from our point of view was a group of three bronze *an* specimens recovered from an Eastern Han tomb near Canton. One was rectangular and the other two round. When the tomb was unearthed, two pairs of chopsticks, chicken and pork bones, and rust marks marking the spots where winged cups had been placed were found on top of one of the round trays indicating without a doubt that these *an* trays were used for the serving of food and drink.[187]

Many scenes of food preparation, cooking, feasting and dining have been discovered in Han tombs. We have already referred to the large kitchen scene copied from a mural in a tomb in Chu-chhêng, Shantung (Figure 28), where a stack of seven rectangular *an* trays is shown near the top left-hand corner. The trays appear to be larger than the articles we find in the tombs, but they are similar to those seen in murals from several other Han tombs, notably Liao-yang, Liaoning; Chi-nan, Shantung; Golingal, Inner Mongolia and Phêng-hsien, Szechuan.[188] That round trays were also popular is indicated in the picture from Liao-yang which shows a stack of five round trays.

During the Han dynasty the *an* 案 gradually became established as the serving tray in the food system. In the *Shih Chi* (Historical Record) it is recorded that when the Emperor Kao Tsu visited the principality of Tsao, King Chang Ao himself carried the *an* dinner tray to the Emperor as a gesture of obeisance.[189] In the meantime the *chi* 几 stool also grew in size and importance and evolved into a low long table well suited for a person sitting on a mat on the floor to use for reading and writing. It became known as a *shu chi* 書几 (stool for books) or a *shu an* 書案 (tray for books) and pictures of it, with its characteristically curved legs, are seen often in murals from Han tombs.[190] But its literary function apparently did not detract from its culinary usefulness, for smaller food trays could still be placed on the larger, long low tables. Thus, in a Han meal we are liable to encounter a *chi* low table, a large *an* tray-table or a small *an* tray.

There is, however, another *an* which also evolved into a part of the dining environment. The *Chou Li* lists a *Chiang Tzhu* 掌次 (Housing Superintendent) whose duty

---

[184] See for example, Qian Hao *et al.* (1981), p. 48.   [185] *Chhang-chiang liu-yu etc.* (1974), pp. 41–61.
[186] Chhên Tsêng-Pi (1982), concludes that tables were unknown during the Han.
[187] *Kuangchou shi wen-kuan-hui* (1961).   [188] Tanaka Tan (1985), cf. Figs. 7, 16, 17 and 39.
[189] *Shih Chi*, ch. 104, Biography of Thien-Shu 田叔列傳. The story has given rise to the idiom *Chü an chhi mei* 舉案齊眉 (presenting the tray at one's eye level) as a mark of respect and harmony.
[190] Chhên Tsêng-Pi (1982), pp. 92–3, and Fig. 4.

40 FERMENTATIONS AND FOOD SCIENCE

Fig 37. Eastern Han brick painting of a dining or drinking scene from Chengtu, Szechuan. After Wilma Fairbank (1972).

it was to provide *an* 案 and *di* 邸 for the king and his retinue while they were travelling away from the royal palace.[191] In this context, *an* is a low, narrow platform bed and *di* a screen placed along the sides of the bed. This is presumably the type of bed, called *chhuang* 床, that is referred to in the *Shih Ching* and *Li Chi*.[192] It was also used to sit on, and is presumably the forerunner of the low couch-bed, *tha* 榻, defined in the *Shih Ming*.

What then was the furniture setup in a Han meal? All the information we have is derived from pictures of dining and feasting scenes in murals from Eastern Han tombs. One example, that has been reproduced often, is a dining scene from a brick mural uncovered near Chengtu, Szechuan.[193] It is reproduced in Figure 37. It depicts seven diners sitting in groups on three mats. In front of them are two *an* trays, one rectangular and one square. Standing on the floor, at the bottom of the picture, is a large bowl (*yü* 盂) of soup or drink with a ladle in it, and an ear-cup nearby. On the upper mat one diner is holding a plate (*phan* 盤), and the other two a *chih* 卮 mug each. The diner on the upper part of the lower left mat appears to be

[191] *Chou Li*, ch. 2, p. 61.   [192] *Li Chi, Nei Tsê*, p. 446; *Shih Ching*, W 159, 257 (M 154, 189).
[193] *Chung-chhing shih po-wu-kuan* (1957), p. 28; reproduced in Fairbank (1972), p. 179, K. C. Chang, (1977), Fig. 23, and Tanaka Tan (*1985*), Fig. 53, p. 285.

Fig 38. Dining scenes from Eastern Han tombs at Liao-yang, Liaoning: (a) From Tomb No. 1, Li Wên-Hsin (*1955*), Fig. 18; (b) From Tomb No. 2, Wang Tsêng-Hsin (*1960*), Fig. 3–6.

holding a winged-cup. It is difficult to see to what extent the elaborate rules governing the seating of the guests and placing of the victuals decreed in the *Li Chi* could have been followed in this scene. The seating arrangement seems much too informal. Perhaps it was a drinking rather than a dining party. A somewhat more formal representation is shown in a mural from Liao-yang, Liaoning, which portrays a wealthy couple having a meal (Figure 38a).[194] The man sits on a raised platform (or low bed), *tha* 榻, which is flanked by a standing screen. In front of him is a low narrow *chi* table of the same length as the *tha*. To the far right of the table is an upright

[194] Tanaka Tan (*1985*), Fig. 60, copied from Li Wên-Hsin (*1955*), p. 30, Fig. 18.

instrument which looks like a writing brush. The woman facing him also sits on a low *tha* flanked by a standing screen. Her bed and screen are, however, decidedly lower than his bed and screen. She has a round tray of cups in front of her, but it is placed directly on the floor. In another picture from Liao-yang (Figure 38b) we have a scene of two men dining together each sitting on a *tha* bed.[195] They share a long low table between them. The table and the beds are of the same length. One round tray of food is placed on the table. In both scenes from Liao-yang there is a contingent of attendants bringing food and drink, removing empty utensils or fanning the diners to ensure that they stay comfortably cool.

It is clear from the foregoing discussion that in terms of the physical setup a Chou or Han meal is a very different affair from a modern Chinese dinner. In fact, the seating and furniture arrangement may strike an observer today as being more Japanese than Chinese. This is not surprising since many aspects of ancient Chinese culture transported to Japan have undergone less change there through the centuries than in their original homeland. But now we need to move on and wind up this phase of our survey by addressing the remaining two questions regarding the environment of 'eating and drinking' in ancient China. Firstly, who did the people eat with? The answer is simple, they ate with their families and friends. This is amply demonstrated by the instructions in the *Li Chi* on how family and guests should comport themselves during a meal. Secondly, when did they eat? The answer is also simple, they had three meals a day. In a commentary on the *Shan Fu* 膳夫, the Chief Steward, in the *Chou Li*, Chêng Hsuan says that the king had three meals a day, a morning meal, a midday meal and an evening meal.[196] In the *Nei Tsê* chapter of the *Li Chi* 'gruel, thick or thin, spirits or must, and soup with vegetables' were listed as breakfast fare, and 'beans, wheat, spinach, rice, millet and glutinous millet' as food for lunch and dinner.[197]

In summary, we can say that in terms of the food resources available, techniques of food preparation, methods of cookery, types of spices, use of chopsticks and spoons, frequency of meals, and who one partakes a meal with, the ancient culinary system is quite recognisably Chinese. However, in terms of the types of vessels used for serving food and drink and the physical setup of a meal the ancient system differs considerably from modern Chinese practice. But taken as a whole, it was definitely a Chinese cuisine *in statu nascendi*, and we hope some of the factors that help it to evolve into the mature Chinese cuisine will reveal themselves as we delve into the main body of this work.

[195] *Ibid.*, Fig. 62, taken from Wang Tsêng-Hsin (*1960*), Figs. 3–6.   [196] *Chou Li*, p. 34.
[197] *Li Chi*, p. 444; cf. Legge (1885), p. 451. Although Hu Hsin-Shêng (*1991*), agrees with our position, he is of the opinion that for most of the people the custom was to have two meals a day. On the other hand, Huang Chin-Kuei (*1993a,b*) is of the opinion that during the pre-Chhin period most people only had two meals a day. The midday meal is simply a snack.

(b) LITERATURE AND SOURCES

The patient reader who has come this far with us on our exploration will have already become familiar with the names of several ancient Chinese texts that were referred to repeatedly in the preceding chapter. We will undoubtedly revisit these works from time to time as we find ourselves immersed in a host of specialised treatises that deal with fermentations, food processing, and ideas on health and nutrition. These may exist as independent publications but are often found only as parts of well-known *Tshung Shu* 叢書 collections. Thus, before we go on any further, it will be worth our while to pause for a moment and take a bird's eye view of the scope and character of the principal sources that have been or will be consulted for the tasks ahead.[1] For this purpose it is convenient to divide the relevant literature into four categories (1) ancient classical texts that touch on aspects of the dietary system (2) books dealing with the processing and preparation of food and drink, the so called 'food canons' (*Shih Ching* 食經) (3) books which stress the relationship between nutrition and health, such as the pandects of 'diet therapy' (*Shih Liao Pên Tshao* 食療本草) and (4) supplementary documentation from philosophical texts, dynastic histories, encyclopedias, literary works, special monographs on plants and animals, etc. We shall briefly describe each category in turn and conclude with a listing of the secondary sources that have served us well as guides to the primary literature.

(1) Ancient Classical Texts

The most important ancient Chinese sources that provide useful information related to food and drink are shown in Table 11. Heading the list is the *Shih Ching* 詩經 (Book of Odes), a collection of poems, folk songs and ceremonial odes composed during a period spanning 500 years from the founding of the Chou dynasty to the time before the birth of Confucius, that is from about −1100 to −600. The 305 poems touch on almost every aspect of life and daily living, for example, courtship, marriage, warriors and battles, agriculture, feasting, sacrificial rituals, music and dancing, dynastic legends, hunting, friendship, moral admonitions, etc. In many of the events described, food and drink are of necessity a prominent part of the action. Although quoted often by Confucius and Mencius, the text of the *Shih Ching* that has survived to this day as one of the five Confucian classics is the version attributed to Mao Hêng recovered in the early Han. As we have seen from the previous chapter, it is a veritable fountainhead of information on the food resources and methods of cookery in early Chou China.

---

[1] For previous guides to the literature in this series relevant to the present study, cf. *SCC* Vol. I, pp. 47–51, Vol. VI, Pt 1, pp. 182–321, Vol. VI, Pt 2, pp. 47–93 and Vol. VI, Pt 3, pp. 45–51 and 544–47.

Table 11. *Ancient classics as sources of culinary information*

Chinese title	English title	Date approx.	Culinary content
Shih Ching 詩經	Book of Odes	−1100 to −600	Food resources, methods of cookery
Chhu Tzhu 楚詞	Elegies of Chhu	−300	List of viands
Chou Li 周禮	Rites of the Chou	−300	Food resources, methods of cookery, utensils
Li Chi 禮記	Record of Rites	−450 to +100	Food resources, methods of cookery, utensils
Lü Shih Chhun Chhiu 呂氏春秋	Master Lu's Spring & Autumn Annals	−240	Theory of cookery, list of delectables
Fang Shêng-Chi Shu 汜勝之書	Book of Fan Shêng-Chih	−10	Food resources
Ssu Min Yüeh Ling 四民月令	Monthly Ordinances for the 4 Peoples	+160	Food resources & products
Huang Ti Nei Ching Su Wên 黃帝內經素問	The Yellow Emperor's Manual of Corporeal Medicine	−2nd century	Diet & health
Mawangtui Chu Chien 馬王堆竹簡	MWT Inventory	−200	Food resources, cookery, viands, utensils
Wu Shih Erh Ping-fang 五十二病方	52 Pingfang	−200	Food resources, methods of cookery, decoctions

The *Shih Ching* has a well-deserved reputation of being a trustworthy observer of the events it records. A particularly telling illustration of its reliability as a historical document is provided by the opening lines of the poem *Shih Yüeh Chih Chiao* 十月之交 (Conjunction in the Tenth Month):[2]

> At the conjunction (of sun and moon) in the tenth month,
> On the first day of the moon, the day *hsin mao* 辛卯,
> The sun was eclipsed,
> This was an ugly omen.

It is now known that the day *hsin mao* coincident with the first day of the tenth month occurred in the sixth year of the reign of King *Yu* 幽 of the Chou dynasty. The date can be shown to correspond specifically to August 29, in the year −776 in the

---

[2] M 193, tr. Karlgren. Although listed as W 293 in the index, it is not included in the 1960 edition of Waley's 'The Book of Songs'.

Western calendar. Chinese and European astronomers have found that an eclipse had indeed occurred on that very day.[3]

There are two excellent translations of the *Shih Ching* in English. The first is Arthur Waley's 'The Book of Songs'. It is a remarkable synthesis of poetry and scholarship. Waley has somehow managed to transfer much of the poetic charm in the original into a different language. The second is Bernard Karlgren's literal, word for word rendition. Both versions have been indispensable in our study of the *Shih Ching* as a source of illumination on the life and times of early Chou China. The two translations by James Legge published in the last century are still useful for consultation when a need arises for a third opinion. Of the numerous elucidations and commentaries on the *Shih Ching* in Chinese,[4] four were found to be particularly useful in this work. The first, *Shih Ching Tzhu Tien* 詩經詞典 (A Cyclopedia of Words in the *Shih Ching*), compiled by Hsiang Hsi (1986), is an invaluable dictionary of the archaic words found in the *Shih Ching*. It has the additional virtue of being the first Chinese publication we have come across that numbers the poems in numerical order as they occur in Mao Hêng's traditional arrangement. The second, Lu Wên-Yu's *Shih Tshao Mu Ching Shih* 詩草木今釋 (A Modern Elucidation of the Flora mentioned in *Shih Ching*), 1957, is a convenient guide to the botanical names of the plants mentioned in the poems. The third, *Shih Ching Chhüan I Chu* 詩經全譯注 (The *Shih Ching*: A Complete Translation [into Modern Chinese] with Annotations), by Fan Shu-Yün, 1984, gives a polished rendition of the poems in modern Chinese verse, with learned annotations that draw heavily on all the traditional commentaries. Finally, the well-known *Shih Ching I Chu* 詩經譯注 (The *Shih Ching*: Translation [into Modern Chinese] and Annotations), by Chiang Ying-Hsiang (1934), is a useful companion to the more recent translation of Fan Shu-Yün.

The *Chhu Tzhu* 楚詞 (Elegies of the Chhu State), or 'Songs of the South', is the other ancient book of poetry that has given us valuable information. It is a collection of sixty-seven poems that were composed by several authors towards the last phase of the Warring States period, that is at about −300. In addition to being a later work than the *Shih Ching*, it also differs from it in being strictly a southern Chinese creation. The style is more free flowing and expansive than that which characterises the northern *Shih Ching*. The principal author, Chhü Yuan 屈原 (c. −340 to −278), is the most famous poet of ancient China. He was a patriot of the State of Chhu, a talented poet, and an outspoken official whose advice was disregarded by the king. When Chhu finally fell to Chhin he was so grief stricken that on the fifth day of the fifth month he threw himself into the river and drowned. Since then, his death has

---

[3] Cf. *SCC* Vol. III, pp. 409–10. Actually, there are two possibilities for the day *hsin mao* to fall on the first day in the 10th month during the period when the *Shih Ching* was compiled. The year −776 is favoured by Chu Wên-Hsin (*1934*), Fotheringham (1921), Hartner (1935) and Waley (1936). On the other hand, Hirayama & Ogura (1915) point out that in North China, the eclipse of −776 would only cover one degree of the sun, and would not be visible to the naked eye. They prefer the year −734. Xi Zezong (private communication, 1989), however, using the traditional method of observing the image of the sun in a basin of water during the eclipse of January, 1980 in Yunnan, has found that even such a small eclipse can be readily detected. He concurs with the year −776.

[4] There are 197 such works dated before 1900, and 41 after 1900 listed in Hsiang Hsi (*1986*), pp. 923–34.

been commemorated every year by the Dragon Boat (*Tuan Wu* 端午) Festival. Legend has it that the Chhu people used to throw bamboo tubes containing rice into the river in order to entice the dragons away from Chhü Yuan's body. This is said to be the origin of the special rice dumpling wrapped in Zinzania leaves called *tsung tzu* 粽子, that is now as much a part of the festival as the Dragon Boat race itself.

The *Chhu Tzhu* has been rendered into English by David Hawkes. As the reader can see, the segments of the two poems, the Great Summons and Summons of the Soul, that we quoted earlier on pp. 94–5 are beautifully translated. All the wonder and excitement that characterise the original descriptions of the delectable dishes seem to have been successfully conveyed in the English version. For the Chinese text and annotations we have relied heavily on the recent *Chhu Tzhu I Chu* (Translation and Annotation of *Chhu Tzhu*) by Tung Chhu-Ping, 1986 and *Chhu Tzhu Tu Pên* by Fu Hsi-Jên, 1976.

Next are two books that deal with rites and proper behaviour, the *Chou Li* 周禮 (Rites of the Chou) and the *Li Chi* 禮記 (Record of Rites). There is also a third *Li* classic, the *I Li* 儀禮 (Ritual of Personal Conduct), which, however, does not give as much information of the type we need as the other two *Li*'s. The three books are known collectively as the *Three Li*'s. The *Chou Li* (Rites of the Chou), also known as *Chou Kuan* 周官 (Officials of the Chou), is one of the classics of the Confucian school and has been considered traditionally as the work of Chou Kung 周公, the Duke of Chou, regent of the third Emperor of the Chou dynasty, whom Confucius never tired of citing as the ideal official who serves the land without regard to personal gain. Recent research indicates that the book was probably compiled during the Warring States period under the influence of Tsou Yen 騶衍, the founder of the Yin–Yang school of philosophy.[5] This would place the date of its compilation at about −300. It describes the positions of the principal officials in an idealised Chou state and may well contain material applicable to the situation that existed earlier during the Western Chou. Many of the duties described pertain to activities of interest to us in this study, such as agriculture, animal husbandry, food processing, fermentation of wine and sauces, cookery, food therapy and the presentation of food for rituals and entertainment.[6] There is as yet no English translation of the Chou Li, although Biot's French translation of 1851 is available for consultation. For this work we have been helped greatly by the modern edition of the *Chou Li* of 1985, with a translation (into modern Chinese) and annotations by Lin Yin.

The other book is the *Li Chi* (Record of Rites), compiled by Tai Shêng during the late Western Han from material on public and private rites and personal behaviour current in the Confucian school during the Warring States period. Thus one may assume that the customs and practices described in it apply to the social conditions that existed during the late Eastern Chou. The author is known as Tai the Younger since his father also made a similar compilation which is now known as *Ta Tai Li Chi*,

---

[5] Hsia Wei-Ying (*1979*), pp. 3–12.
[6] These activities are summarised and analysed in Shinoda Osamu (*1987a*), pp. 47–56.

(Record of Rites of Tai the Elder). The *Li Chi* of Tai the Younger or *Hsiao Tai Li Chi* is the one that concerns us. It contains forty-seven chapters of which several, such as *Chhü Li* 曲禮 (Rules of Ceremony), *Wang Chih* 王制 (Royal Regulations), *Yüeh Ling* 月令 (Monthly Ordinances), *Nei Tsê* 內則 (Pattern of the Family), *Shao I* 少儀 (Lesser Rules of Conduct) and *Sang Ta Chi* 桑大紀 (Record of Mourning Rites), have passages that relate to some aspect of food and drink. The *Nei Tsê*, in particular, is a rich source of information on the food and drink that were prepared and served in a gentlemanly household. It describes the recipes for the Eight Delicacies, the only recipes found in an ancient Chinese text. There is an English translation by Legge (1885) and a useful Yin-Tê index. The *Li Chi* version used routinely is the 1970 edition with annotations and translation (into modern Chinese) by Wang Mêng Ou.

The next book is the *Lü Shih Chhun Chhiu* 呂氏春秋 (Master Liu's Spring and Autumn Annals), a compendium of topics on natural philosophy written by a group of scholars gathered by Lu Pu-Wei, the first prime minister to Chhin Shih Huang 秦始皇, the first Emperor of a unified China. It was probably completed in about −240, and contains essays reflecting the different schools of thought in the pre-Chhin period. There are twenty-six chapters. The part that concerns us most is the *Pên Wei* 本味 (Root of Taste) segment in chapter 14 which contains a brilliant discourse on the theory of cookery. We have also made use of information in the four segments dealing with agriculture in chapter 26. There is a 1985 edition with translation and annotations by Ling Phing-Shih, but for our purpose the most useful compilation has been the *Lu Shih Chhun Chhiu Pên Wei Phien* 呂氏春秋本味篇, that is the 'Root of Taste' chapter itself, published in 1983 as a separate volume, with copious annotations and commentaries by Chhiu Phang-Thung *et al*.

Next on the list are two works on agriculture, *Fan Shêng-Chih Shu* 氾勝之書 (The Book of Fan Shêng-Chih), c. −10, and *Ssu Min Yüeh Ling* 四民月令 (Monthy Ordinances for the Four Peoples), +160. The English translation of the extant remnants of the *Fan Shêng-Chih Shu* by Shih Shêng-Han has been an invaluable guide. For the *Ssu Min Yüeh Ling* our major references are the English translation by Hsü Cho-Yun and the annotated edition by Miao Chhi-Yü of 1981.

We have cited a passage from the *Huang Ti Nei Ching, Su Wên* 黃帝內經素問 (The Yellow Emperor's Manual of Corporeal Medicine, Questions and Answers), on the very first page of this work, and will have many occasions to refer to it in the future. The original text was supposed to have been composed by the legendary Yellow Emperor himself. Modern scholarship, however, tends to place its earliest compilation at about the time of the Warring States.[7] It is generally regarded as the oldest of the Chinese medical classics. The version that has come down to us is the one systematised by Wang Pin in +762. It consists of eighty-one chapters arranged in twenty-four volumes. There are several sections that touch on the influence of diet and the environment on health and disease. The work may thus be regarded as the progenitor of the *Shih Liao* 食療 or 'Diet Therapy' tradition, and the corpus of food

---

[7] Li Ching-Wei & Chhêng Chih-Fan, eds. (*1987*), p. 165. See also Veith (1949), pp. 4–8.

therapeutics literature that it has inspired. There is a partial English translation by Ilza Veith and a complete French translation by A. Chamfrault and Ung Kang-Sam. For our study our principal reference has been *Huang Ti Nei Ching Su Wên, I Shih* 黃帝內經素問譯釋 (Translation and Annotation of the Yellow Emperor's Manual of Corporeal Medicine, Questions and Answers), published by the Nanking College of Chinese Medicine, 1981.

Lastly, there are two valuable works found among the burial remains discovered in Western Han tombs at Ma-wang-tui. We have already met in the preceding chapter, the *MWT Inventory* (from Tomb No. 1),[8] and the *52 Pingfang* (from Tomb No. 3).[9] The former consists of 312 bamboo slips of which 164 dealt with food products, 59 with utensils and furniture and the remainder with musical instruments and household effects. Some of the food products have been discussed earlier and we will refer to them again in Chapter (*e*) when we deal with the topic of food processing. The latter, a manuscript written on silk now known as the *Wu Shih Erh Ping Fang* 五十二病方 (Prescriptions for Fifty-two Ailments), which for convenience is referred to as the *52 Pingfang*, is the oldest book of medical prescriptions extant in Chinese. The most interesting feature of the *52 Pingfang* is that many of the concoctions used to cure ailments can easily pass off as dishes eaten at a routine meal. In fact, about a quarter of the 'medicaments' mentioned are food materials or food products. Thus the book serves as a good example of the concept of *Shih Liao*, or diet therapy in action. An English translation of the *52 Pingfang* with a Prolegomena by Donald Harper was published in 1982.

(2) Food Canons and Recipes

We have intimated that there are two types of classical Chinese books on food; one which deals with the technology and mechanics of food preparation in the *Shih Ching* 食經 (Food Canons) tradition[10] and the other which deals with the nutritive and therapeutic properties of food such as the *Shih Liao* (Diet Therapy) component of the *Pên Tshao* 本草 (Pandects of Natural History) compilations. Books of both genres have been considered by some scholars as parts of the overall *pên tshao* corpus.[11] Thus, the excellent *Pên Tshao* catalogue of Lung Po-Chien lists many of the extant books dealing with food and nutrition under the general heading of *Shih Wu Pên Tshao* 食物本草 (Natural History of Foods), which is then subdivided

---

[8] *MWT Inventory*, see *Hu-nan shêng po-wu-kuan* (*1973*), Vol. 1, ch. 10, pp. 130–55. For another discussion and interpretation of the bamboo slips, cf. Thang Lan (*1980*).
[9] This is found in *Ma-wang-tui Han-mu po-shu*, Vol. 4, pp. 1–82.
[10] Starting with the legendary *Shên Nung Shih Ching* 神農食經 (Food Canon of the Heavenly Husbandman), which is cited in Lu Yü's *Classic of Tea*, p. 42 (cf. Chapter (*f*)), books on culinary matters have been called *Shih Ching* 食經 in the Chinese literature, perhaps as an indication of the high regard with which food and drink are held. Shinoda has expanded this tradition further, in Shinoda & Seiichi (*1973*), to include more general works in which food and drink form though a small but yet notable part of the whole, such as the *Chhi Min Yao Shu, Nan Fang Tshao Mu Chuang, Yu-Yang Tsa Tsu* etc. We have followed this tradition in our enumeration of the Food Canons of China.
[11] Cf. *SCC* Vol. VI, Pt 1, pp. 220ff.

into two categories: one on *Phêng Chih Fang Fa* 烹製方法 (Methods of Cookery) which includes the books in the Food Canon tradition, and the other on *Shih Wu* 食物 (Food Resources) which includes the books on diet therapy.[12] The Methods of Cookery category lists thirteen titles, and the Food Resources, thirty-two. However, books dealing with drinks, that is, wine and tea are not included, and Lung's division is by no means accepted by all scholars. For example, Shinoda and Tanaka, in their *Chung Kuo Shih Ching Tshung Shu* 中國食經叢書 (Collection of Chinese Food Canons), list thirty-six pre-1800 titles, of which, however, at least three are in the diet therapy tradition. In his now classic studies of the Shih Ching 食經 (Food Canons) of mediaeval and premodern China, Shinoda discusses more than a hundred texts, without even touching on any of the specialised monographs on wine or tea.[13] If we leave out the books whose major emphasis is on agriculture, geography or literature, we are left with about seventy titles that conform to the food canon and diet therapy traditions. Another useful collection, the special issues on Food and Drink in the *Min Shu Tshung Shu* 民俗叢書 (Series on Folk Literature), edited by Lou Tzu-Khuang 婁子匡 published in Taipei, simply classifies the books according to type of food resource, such as tea, wine, vegetables, fruits etc. It contains seventy-four titles of which twenty-three are concerned with tea, the literature of which we will treat separately in Chapter (*f*). We are fortunate that since about 1984, Commerce Publishers (*Shang Yeh Chhu Pan Shê* 商業出版社), Peking, has been issuing new editions of selected works listed in the above bibliographies under the general heading of *Chung Kuo Phêng Jên Ku Chih Chhung Khan* 中國烹飪古籍叢刊 (Chinese Culinary Classics Series) with updated annotations and often translations into modern Chinese. More than twenty titles have been published so far, including a most useful selection of passages pertinent to food and drink in the pre-Chhin classics edited by Thao Wên-Thai *et al*. For our study their appearance could not have been more timely.[14]

Thus, after putting aside the specialised monographs on tea and wine, we are left with a literature totalling about seventy titles. Many are, however, quite slim, numbering no more than a few pages. There is also a good deal of repetition among them, so that it will be exceedingly tedious to review them all. We will only cover the most important works from the point of view of their significance in terms of food processing and nutrition. With these preliminaries out of the way, we are now ready to consider the category of 'Food Canons' and the recipe books (*Shih Phu* 食譜) that they have helped to generate.

[12] Lung Po-Chien (*1957*), pp. 100–16.   [13] Shinoda Osamu (*1987a*), pp. 99–230.
[14] We were, however, unable to get a complete set of the whole series published by Commerce Publishers. Among those that we missed are two bibliographic surveys of Chinese culinary literature, namely, *Chung-kuo phêng-jen wen-hsien thi-yao* 中國烹飪文獻提要 by Thao Chen-Kang & Chang Lien-Ming (*1986*) and *Chung-kuo phêng-jen wen-hsien kai shu* 中國烹飪古籍概述 by Chhiu Phang-Thung (*1989c*). Apparently, they were in such high demand that they were sold out promptly after each printing was released. Unfortunately, they were not available either in the Cambridge University Library or the Library of Congress in Washington, DC. It was only after this draft was completed that we had a chance to see the table of contents of these works. Although we have missed a few of the lesser known compilations we do not think that any substantive changes in this chapter are called for.

Table 12. *Contents of the* Chhi Min Yao Shu

Book	Chapter	Topic
I	1	Reclamation of land
	2	Collection of seed grain
	3	Cultivation of setaria millet
II	4–13	Cultivation of cereals, beans, hemp & sesame
	14–16	Cultivation of melons, gourds & taro
III	17–29	Cultivation of vegetables
	30	Monthly calendar
IV	31	Planting hedges
	32	Transplanting trees
	33–44	Fruit trees & Szechuan pepper
V	45	Mulberry trees & sericulture
	46–51	Shade & ornamental trees, bamboos
	52–4	Dye plants
	55	Tree felling
VI	56–61	Animal husbandry (including poultry & fish)
VII	62	Commerce and trade
	63	Curing of earthenware
	64–7	Ferment cultures and fermentation of wine
VIII	68–79	Soy and meat sauces, vinegars, & meat preserves
IX	80–9	Roasts, noodles & dumplings, pickles, candies etc.
	90–1	Glue, ink, brushes
X	92	Plants not indigenous to (North) China

(i) *The* Chhi Min Yao Shu *or Essential Arts for the People's Welfare*

Although the *Chhi Min Yao Shu* (c. +540), as shown in *SCC* Vol. VI, Pt 2, is best known as a treatise in agriculture, it is, in fact, a comprehensive work on all aspects of daily labour on a large 5th to 6th century estate in North China.[15] Indeed, in the Preface to his book, the author, Chia Ssu-Hsieh says:[16]

I have collected material from classical and contemporary books, proverbs and folk-songs; I have enquired for information from experts, and learned myself from practical experience. From ploughing and cultivation, down to the making of fermented sauces and vinegar, there is no useful art in supporting daily life that I have not described in detail. Thus, I call my book 'Essential Arts for the People's Welfare'. In all, it comprises 92 chapters divided into ten books.

The author's claim that there is no useful art in supporting daily life that he has not described in detail is generally substantiated by the contents of the book as presented in Table 12. The topics covered include agriculture (chs. 1–16), horticulture

---

[15] The content of the agricultural sections is discussed in Bray, *SCC* Vol. II, Pt 2, pp. 55–9, and the food and drink sections in Shih Shêng-Han (1958), Thao Wên-Thai (*1986b*), Chhiu Phang-Thung (*1986a*) and Sabban (1988b and 1990b).
[16] *CMYS*, Miao Chhi-Yü (*1982*), p. 5, tr. Bray, *SCC* Vol. VI, Pt 1, p. 56.

(chs. 17–34), sericulture (ch. 45), animal husbandry and aquaculture (chs. 56–61), timber and bamboo (chs. 46–51), dye plants (chs. 52–4) and fermentations, food processing and cookery (chs. 64–89). Although the references we cited in Chapter (*a*) are those dealing with horticulture (vegetables and fruits), the chapters of greatest importance to us in this study are 64–89 (fermentations, food processing and cookery as shown in Table 12), which we will make extensive use of in Chapters (*c*), (*d*) and (*e*). They contain some of the most comprehensive and detailed descriptions of culinary processes in the Chinese literature, and represent a valuable record of the state of the art in these areas against which all subsequent achievements have to be measured. Thus, from the perspective of the history of fermentations and food processing, the importance of the *Chhi Min Yao Shu* can hardly be overstated.

Shih Shêng-Han has provided an admirable survey in English of the contents of the *Chhi Min Yao Shu* and an analysis of the source books from which materials for it were drawn. For the original Chinese text we have relied mainly on two recent editions. The first is Shih Shêng-Han's *Chhi Min Yao Shu Ching Shih* 齊民要術今釋 (Annotation and Translation [into modern Chinese] of the *Chhi Min Yao Shu*) of 1957. The other is the *Chhi Min Yao Shu Chiao Shih* 齊民要術校釋 (Annotated Edition of the *Chhi Min Yao Shu*) of Miao Chhi-Yü published in 1982.[17]

Chapters 64–89, with Shih Shêng-Han's modern translation and annotations, have been issued as a separate volume by the Commerce Publishers, Peking, in 1984 as part of the *Chung Kuo Phêng Jên Ku Chih Chhung Khan* (Chinese Culinary Classics Series). We have found it a most convenient vehicle for routine use, but it does not include the descriptions for making dairy products from cow or ewe milk found in chapter 57 (on sheep husbandry). Thus, by depending only on this publication one may be led to perpetuate the erroneous notion that dairy products were never a significant feature of the diet of the Chinese people.[18]

(ii) *Early mediaeval food canons*

For our purpose the mediaeval period is defined as the era from the start of the Three Kingdoms until the end of the Yuan dynasty, that is from about +200 to +1350. It may be subdivided into an early mediaeval period, i.e. from about +200 to +900, and a late mediaeval period, i.e. from about +900 to +1350. Although many books under the heading of Food Canons were written in the early mediaeval period, almost all of them are now lost. The *Han Shu I Wên Chi* 漢書藝文集 (Bibliography to the History of the former Han) lists a *Shêng Nung Shih Ching* 神農食經 (Food Canon of the Heavenly Husbandman) and the *Sui Shu Ching Chi Chih* 隋書經籍志 (Bibliography to the History of the Sui) lists nine books in the food canon genre. One of

---

[17] In the preparation of his edition, Miao Chhi-Yü (*1982*) has meticulously examined all the extant versions of the *Chhi Min Yao Shu* from the Sung until the present day. He has corrected textual errors and elucidated the meaning of words which have long remained obscure. His critiques of the previous editions are presented in two appendices, pp. 733–858.

[18] The history of the role of dairy products in the diet of the Chinese people has been reviewed by Sabban (1986a). The technology of dairy foods in China is treated on pp. 248–56.

these is the *Tshui Shih Shih Ching* 崔氏食經 (Madam Tshui's Food Canon)[19] which is reputed to have been written by the mother of Tshui Hao 崔浩, a famous scholar and a prime minister in the early part of the Northern Wei (Thopa) Dynasty. However, as Shinoda has pointed out, there is also a food canon (*shih ching*) attributed to Tsui Hao himself, in nine volumes, listed in the *Wei Shu* 魏書 (History of the Northern Wei). All that is left of it is Tsui Hao's preface in which he says:[20]

When I was growing up, I loved to watch my mother and aunts go about their household duties. They were expert in the culinary arts. They faithfully served their elders and provided for the ancestral offerings each season. Although they had plenty of domestic help they would often prefer to do some of the cookery themselves. But then for more than ten years, the country suffered turmoil and famine. We barely had enough to eat to stay alive, and my mother had no chance to prepare a proper meal. Noting how ignorant the younger generation had become in such matters, she decided to record in writing her considerable experience and skill in the culinary arts for their benefit. The result is nine chapters of elegant and systematic descriptions that form this book.

From this account it is difficult to determine whether Tshui Hao himself or his mother actually wrote the book. In fact, it may have been the very same book that is listed in the Bibliography to the History of the Sui. But it is clear that the contents did not come from bookish information; it was based on the personal experience of a skilled practitioner. In any case, the book itself, whether by Tshui Hao or his mother, is of great interest to us since it is most likely the *Shih Ching* (Food Canon) that features so prominently as the source of quotations cited in the fermentation and food processing chapters of the *Chhi Min Yao Shu*. Shih Shêng-Han counted thirty-three, but Shinoda Osamu, thirty-seven such quotations. Their subject matter falls into four areas (1) food preservation (2) food processing, including wines and pickles (3) cookery and (4) horticulture, which deals with cuttings for the propagation of trees bearing delicious fruits.

Unfortunately, Tshui Hao had a falling-out with the Wei Emperor in his later years, and he and his whole family were put to death in +450 for high treason. He was thus regarded as a traitor in the remaining hundred years of the Northern Wei. Shinoda has suggested that this is the probable reason why Tshui Hao's name is never mentioned and his *Shih Ching* not specifically identified in the *Chhi Min Yao Shu*, since the author Chia Ssu-Hsieh was himself an official under the Northern Chhi.[21]

At any rate, there were apparently a large number of food canons in circulation during the Thang dynasty. The *I Wên Lo* 藝文略 (Bibliography) of the *Thung Chih* 通志 (Historical Collections), published in the Sung dynasty (c. +1150), lists forty-one books (comprising 366 chapters) in the food canon category. Of these, five were included in the Sui bibliography, and only a few we can identify as Sung works. Thus, the majority are presumed to be Thang contributions. Unfortunately, most of the books

---

[19] Shih Shêng-Han (1958), p. 28.
[20] Quoted by Shinoda Osamu (*1987a*) tr. Kao *et al.*, p. 101, from the *Wei Shu* (History of the Wei Dynasty).
[21] *Ibid.*, p. 104.

are now lost. Only remnants of three early mediaeval texts in the food canon category have survived to this day. The oldest is the Sui *Shih Ching* 食經 (Food Canon) of Hsieh Fêng 謝楓 (c. +600), followed by the *Shih Phu* 食譜 (List of Victuals) of Wei Chü-Yuan 韋巨源 (c. +710) and the *Shan Fu Ching Shou Lu* 膳夫經手錄 (The Chef's Handbook) of Yang Yeh 楊曄 (c. +857). All three remnants are short documents, no more than two to eight pages each. The first two are lists of exotic dishes and food products, and the last records interesting facts about a number of food resources.[22] There is virtually no information on cookery or food technology in them. Nevertheless, they are useful signposts in the development of the Chinese cuisine in that they are the only source of information we have on the kind of dishes and processed foods, such as *thien hua pi lo* 天花饆饠 (a mushroom pilaf), *chien fêng hsiao* 見風消 (fritters of glutinous rice), *po lo mêng chhing kao mien* 婆羅門輕高麵 (steamed Brahmin light wheat cake), that were part of the culinary experience of the upper classes during the Thang dynasty.[23]

(iii) *Late mediaeval food canons*

Fortunately, the survival rate of books on food and drink during the Sung and Yuan is a considerable improvement over the early mediaeval period. The books extant that are consulted frequently in this study are listed in Table 13. The first entry, *Chhing I Lu* 清異錄 (Anecdotes, Simple and Exotic) is a compilation of anecdotes of little known facts and unusual events associated with a wide range of topics current in the Sui, Thang and Five Dynasties era. The entire book consists of 648 anecdotes, of which 238 deal with matters relating to food and drink. The latter selections have been published, under the same title, as a separate volume in the Chinese Culinary Classics Series, arranged under eight headings: delicacies, vegetables, fish, fowl, mammals, wine, tea and fruits. Although it has little to say about food processing, some of the observations provide fascinating clues to the evolution of Chinese cuisine. One useful feature is that it cites the complete extant text of the *Shih Ching* of Hsieh Fêng and the *Shih Phu* of Wei Chü-Yuan.[24] The second entry is not usually considered as a food classic. It is a compilation of the food and drink sections of the *Thai Phing Yü Lan*, the Imperial Speculum of the Thai-Phing reign period, which were published as a separate volume in the Chinese Culinary Classics Series in 1993 under the title of *Thai Phing Yu Lan: Yin-shih Pu* 太平御覽飲食部 (The Food and Drink Sections of the *Thai Phing Yü Lan*). It contains a wealth of information on matters relating to food and drink from the beginning of recorded history until the early Sung. For our purpose, it may be regarded as a Food Canon. Many of the works cited are no longer extant. Of the next four entries, *Shih Chêng Lu* 食珍錄, *Shan Fu Lu* 膳夫錄 and *Yü Shih Phi* 玉食批 are short collections of menus,

---

[22] See comments by Chhiu Phang-Thung (*1986b*).
[23] Some of the lost Food Canons are cited in Japan in the 醫心方 of 丹波康賴, +984, cf. Shinoda Osamu (*1987a*) tr., pp. 109–12.
[24] *Chhing I Lu* (*CKPJKCTK*), pp. 5–16.

Table 13. *Food canons and recipe books in late mediaeval China*

Title of book	Pages	Author	Date	Subject matter
Chhing I Lu 清異錄 Simple and Exotic Anecdotes	147	Thao Ku 陶穀	+960	Anecdotes on food
Thai Phing Yü Lan 太平御覽 Emperor's Daily Readings	761	Li Fang 李昉	+983	Food & Drink sections of Imperial Encyclopedia
Shih Chêng Lu 食珍錄 Menu of Delectables	3	Yu Chhung 虞悰	Northern Sung	List of dishes
Shan Fu Lu 膳夫錄 Chef's Manual	7	Chêng Wan 鄭望	Southern Sung	Anecdotes
Yü Shih Phi 玉食批 Imperial Food List	5	Anon.	Southern Sung	List of dishes
Pên Hsin Chai Shu Shih Phu 本心齋疏食譜 Vegetarian Recipes from the Pure Heart Studio	13	Chhên Ta-Sou 陳達叟	Sung	Anecdotes
Nêng Kai Chai Man Lu 能改齋漫錄 Random Records from the Corrigible Studio	148	Wu Cheng 吳曾	Southern Sung	Anecdotes
Shan Chia Chhing Kung 山家清供 Basic Needs for Rustic Living	112	Lin Hung 林洪	Southern Sung	Anecdotes, recipes
Wu Shih Chung Khuei Lu 吳氏中饋錄 Madam Wu's Recipe Book	28	Madam Wu 吳氏	Sung	Recipes
Chü Chia Pi Yung 居家必用 Compendium of Essential Arts for Family Living	139	Anon.	Yuan	Food processing, recipes
Shih Lin Kuang Chi 事林廣記 Record of Miscellanies	30	Chhên Yuan-Ching 陳元靚	+1280	Food processing, recipes
Yun Lin Thang Yin Shih Chih Tu Chi 雲林堂飲食制度集 Dietary System of the Cloud Forest Studio	45	Ni Tsan 倪瓚	+1360	Food processing, recipes
I Ya I I 易牙遺意 Remnant Notions from *I Ya*	26	Han I 韓奕	Yuan	Food processing, recipes

recipes and anecdotes of exotic and unusual dishes, while the fourth one, Madam Wu's recipe book (*Wu Shih Chung Khuei Lu* 吳氏中饋錄), is noteworthy as the first Chinese cookbook in which exact measurements are given.[25] They show that culinary art had reached a high level of sophistication by the time of the Sung.

[25] Tradition has it that this is the work of a Madame Wu of Kiangsu who lived during the Sung dynasty. It first appeared in *Shuo Fu*, the collection published by Thao Tsung-I of the Yuan. The use of exact measurements was first noted by Lin Yutang in his Preface to Lin & Lin (1969). See also the Preface to *CKPJKCTK* edition of *Chhing I Lu*.

Indeed, the search for gastronomic excellence and rare delectables appeared to know no bounds. The results sometimes border on the grotesque as was noted by the anonymous author of the *Yü Shi Phi* (Imperial Food List):[26]

> Alas! Those who reap the bounties of the earth should first seek to assuage the sorrows of the needy. Otherwise, we are unworthy to sample the simplest fares, far less the excessive refinements of the wealthy. Take for example, the practice of using only the cheeks of the lamb, the jowls of the fish, the legs of the crab, and for wonton or whole melon soup, only the meat of its claws. The rest is discarded with the comment that it is not fit for the nobleman's table. If someone picks up the food, he is called a dog.

It is not surprising that such excesses should evoke a reaction. Thus we see the next entry, the *Pên Hsin Tsai Shu Shih Phu* 本心齋疏食譜 (Vegetarian Recipes from the Pure Heart Studio), seek to reverse this trend by promoting the simplicity and elegance of vegetarian fare. This move against the exotic and bizarre received strong support from the *Shan Chia Chhing Kung* 山家清供 (Basic Needs for Rustic Living) which presents ingenious recipes based on a wide range of vegetarian materials, with interesting historical notes about their origin. In between the two works we have the *Nêng Kai Chai Man Lu* 能改齋漫錄 (Random Notes from the Corrigible Studio), another collection of anecdotes of unusual events associated with various aspects of living spanning the Wei–Chin (+220 to +420) to the Thang–Sung (+618 to +1279) period. Those dealing with food and drink have been collated and published as a separate volume in the Chinese Culinary Classics Series. The remaining four books, *Chü Chia Pi Yung* 居家必用 (Essential Arts for Family Living), *Shih Lin Kuang Chi* 事林廣記 (Record of Miscellanies), *Yün Lin Thang Yin Shih Chih Tu Chi* 雲林堂飲食制度集 (Dietary System of the Cloud Forest Studio),[27] and *I Ya I I* 易牙遺壹 (Remnant Notions from *I Ya*)[28] represent a maturing of the tradition of the food canon genre and the format they established was continued into the Ming and Chhing periods. All contain recipes that deal with fermentations and food processing. The full title of *Chü Chia Pi Yung* is *Chü Chia Pi Yung Shih Lei Chhuan Chi* 居家必用事類全集 (Complete Compendium of Essential Arts for Family Living). What we list in Table 13 is just the Food and Drink section of it, which has been published as a separate volume in the Chinese Culinary Classics Series. The *Yün Lin Thang Yin Shih Chih Tu Chi* deserves a mention as the work of the famous Yuan landscape painter Ni Tsan 倪瓚. For an analysis and critique of the recipes and food products described in these treatises the reader is referred to the comprehensive articles on mediaeval and premodern Chinese food canons by Shinoda Osamu, as well as the overviews of the Chinese culinary literature by Thao Chên-Kang and Chang Lien-Ming and Chhiu Phang-Thung.[29]

---

[26] *Yü Shih Phi* in *Wu Shih Chung Khuei Lu* (*CKPJKCTK*) p. 76, tr. Lin & Lin (1969), p. 37, mod. auct.
[27] For further information on this work see Chhiu Phang-Thung (*1986c*). [28] Cf. *idem.* (*1986d*).
[29] Shinoda, Osamu (*1987a*), pp. 121–52, Thao & Chang (*1986*) and Chhiu Phang-Thung (*1989*). Unfortunately, we did not manage to see a copy of Thao & Chang's work until our manuscript was virtually completed, and our efforts to obtain a copy of Chhiu's work are still unsuccessful. The latter was evidently so popular that it went out of print shortly after it was published. Cf. also Thao Wên-Thai (*1986a*), Huang Chu-Liang (*1986*) and Tai Yün (*1994*).

Table 14. *Food canons and recipe books in premodern China*

Title of book	Pages	Author	Date approx.	Subject matter
To Nêng Phi Shih 多能鄙事 Routine Chores Made Easy	159	Liu Chi 劉基	+1370	Fermentations, food processing, recipes
Chhü Hsien Shên Yin Shu 臞仙神隱書 Book of the Slender Hermit	200	Chu Chhüan 朱權	+1440	Food processing
Pien Min Thu Tsuan 便民圖纂 Everyman's Handy Illustrated Compendium		Kuang Fan 鄺璠	+1502	Food processing
Sung Shih Tsun Shêng 宋氏尊生 Sung Family's Guide to Living	132	Sung Hsü 宋詡	+1504	Fermentations, food processing, recipes
Yin Chuan Fu Shih Chien 飲饌服食牋 Compendium of Food and Drink	192	Kao Lien 高濂	+1591	Tea, fermentations, recipes
Yeh Tshai Po Lu 野菜博錄 Monograph on Edible, Wild Vegetables	149	Pao Shan 鮑山	+1597	Uncultivated edible plants
Hsien Chhin Ngo Chi 閒情偶寄 Random Notes from a Leisurely Life	60	Li Yü 李漁	+1670	Properties of food, gastronomy
Shih Hsien Hung Mi 食憲鴻秘 Guide to the Mysteries of Cookery	162	Chu I-Tsun 朱彝尊	+1680	Fermentations, food processing, recipes
Yang Hsiao Lu 養小錄 Guide to Nurturing Life	98	Ku Chung 顧仲	+1698	Fermentations, food processing, recipes
Hsing Yuan Lu 醒園錄 Memoir from the Garden of Awareness	62	Li Hua-Nan 李化楠	+1750	Food processing
Sui Yuan Shih Tan 隨園食單 Recipes from the Sui Garden	150	Yuan Mei 袁枚	+1790	Cookery, nutrition, wine, tea, recipes
Thiao Ting Chi 調鼎集 The Harmonious Cauldron	871	Anon.	+1760 to +1860	Fermentations, wine, tea, recipes
Chung Khuei Lu 中饋錄 Book of Viands	18	Cheng I 曾懿	+1870	Food processing

(iv) *Premodern food canons*

By premodern we mean the period starting from the Ming through to the last part of the Chhing, that is, from +1350 to about 1850. Major works in the food canon tradition published during this period are listed in Table 14. The first is *To Nêng Phi Shih* 多能鄙事 (Routine Chores Made Easy) of Liu Chi 劉基, who was a distinguished official of the early Ming. The second, *Chhü Hsien Shên Yin Shu* 臞仙神隱書 (Notes of the Slender Hermit) was compiled by Chu Chhüan 朱權, a son of the Prince of Chou, Chu Hsiao 朱橚, whom we have come across in an earlier part of this series.[30] The next is the *Sung Shih Tsun Shêng* 宋氏尊生 (Sung Family Guide to Cultivation of Life), which is the section dealing with food and drink from a larger work known as *Chu Yü Shan Fang Tsa Pu* 竹嶼山房雜部 (Miscellanies from the Bamboo Islet Studio) written by members of the Sung family.[31] This is followed by *Pien Min Thu Tsuan* 便民圖纂 (Everyman's Handy Illustrated Compendium), which, although usually classified as an agricultural treatise, contains a considerable amount of material on food processing. All four works appeared to have appropriated many passages, usually without attribution, from earlier compendia such as the *Chü Chia Pi Yung* (Essential Arts for Family Living) and *Shih Lin Kuang Chih* (Record of Miscellaneous Matters) listed in Table 13. The next book, *Yin Chuan Fu Shih Chien* 飲饌服食牋 (Memoirs on Food and Drink) consists of the culinary section of Kao Lien's *Tsun Shêng Pa Chien* 尊生八牋 (Eight Memoirs on Cultivation of Life). It, too, borrows heavily from previous food canons such as *Wu Shih Chung Khuei Lu* (Madam Wu's Recipe Book), *I Ya I I*, (Remnant Notes from I Ya), and *To Nêng Phi Shih* (Routine Chores Made Easy). But it also contains significant amounts of new material, and is probably the most important publication in this genre in the Ming dynasty. It had itself become the sauce material for later works such as *Shih Hsien Hung Mi* 食憲鴻秘 and *Yang Hsiao Lu* 養小錄. Kao Lien stresses the nutritive value of foods and avoids the excesses of those who search only for gastronomic thrills. In his introduction he says:[32]

In the food we eat in order to sustain life, we should promote the simple and the wholesome. We should not allow what we eat to injure our body, nor let the five flavours do battle with our five organs. This is the basic rule for the cultivation of life. In this volume, I first write about tea and water, then congee and vegetables, followed by brief descriptions of preserved food, wines, noodles, cakes and fruits. I take what is practical, and eschew the exotic and bizzare.

It is quite possible that these sections were at one time circulated separately as independent pamphlets. Wylie, in his well-known bibliography of Chinese books, credits Kao Lien with a series of 'minor treatises' on food, dealing with such topics as soups, congee, noodles and pasta, preserved meats etc., which look remarkably similar in content to the sections in the *Yin Chuan Fu Shih Chien*, indicating that Kao Lien's works were highly regarded by the culinary community.[33]

The *Yeh Tshai Po Lu* 野菜博錄 (Comprehensive Account of Edible Wild Plants) lists 435 uncultivated edible plants and is the largest compendium of its kind. It is included

---

[30] Needham *et al.*, *SCC* Vol. VI, Pt 1, pp. 332–3.
[31] There is another section of this work, the *Yang-sheng pu* 養生部 (Section on Nourishing Life) which also contains useful information for our study.
[32] *YCFSC*, p. 1. [33] Wylie (1867), pp. 153–4.

as a prime example of the literature inspired by the esculentist movement started in early Ming by Prince Chu Hsiao and described in detail elsewhere in this series.[34]

The next work is the celebrated *Hsien Chhin Ngo Chi* 閒情偶寄 (Random Notes from a Leisurely Life) by Li Yü 李漁. It consists of six chapters, each dealing with one aspect of civilised living. The one on food and drink has been issued as a separate booklet in the Chinese Culinary Classics Series. It is a scholarly discourse on the culinary virtues of three major categories of food, in the order of importance as perceived by the author, that is, vegetables first, grains second and meat and fish last.[35] A book with a more practical bent is the *Shih Hsien Hung Mi* 食憲鴻秘 (Guide to the Mysteries of Cookery) by Chu I-Tsun 朱彝尊, which discusses diet and health and describes 450 recipes, including a significant number for food processing and preservation, for example, pickling of vegetables, preservation of fruits, making of fermented sauces, vinegar etc. Some recipes deal in exotic materials, such as bear's paw and tendons of deer; others are unusual in another way such as the vegetarian or mock turtle and vegetarian or mock meat balls. The next two books, *Yang Hsiao Lu* 養小錄 (Nurturing Life Booklet)[36] and *Hsing Yuan Lu* 醒園錄 (Memoir from the Garden of Awareness) both deal with food processing and cookery. Although the *Yang Hsiao Lu* is the longer work, the directions in the *Hsing Yuan Lu* are in greater detail and easier to follow. This book was edited by the well-known scholar and theatre critic Li Thiao-Yuan 李調元 based on the manuscript prepared by his father, Li Hua-Nan 李化楠. In the introduction, Li Thiao-Yuan says:[37]

When my late father was alive, he was a firm believer in doing all things in moderation. He ate mostly simple vegetarian dishes. But when he entertained his parents, he made every effort to provide them with the the best and most delicious viands. On his many tours of duty as a government official, usually in the Kiangsu-Chekiang area, he often had the opportunity to enjoy an exceptionally well-prepared and tasty dish. He would immediately visit the chef, write down the recipe and and try it out himself in the kitchen. In this way, after many years, he had accumulated quite a collection of excellent recipes.

Thus, we see that the recipes had been validated by the author himself who was evidently an able writer and a skilful chef. This is no doubt the reason for the clarity and thoroughness that characterise the recipes in this book. The next entry in Table 14 is the *Sui Yuan Shih Tan* 隨園食單 (Recipes from the Sui Garden) of Yuan Mei 袁枚, probably the most famous of all classical Chinese texts on the culinary arts. It starts off with a discussion on what one should or should not do in cookery, and then launches into a narrative describing 327 recipes in twelve categories of dishes. His recipes have been tested by Lin Ju-Lin and Lin Tsuifeng in a modern kitchen and form the basis of the material in their excellent cookbook.[38] There is, however, little of importance in terms of food processing in the book and for us it is less useful than some of the other books listed in Table 14, such as the *Hsing Yuan Lu* (Memoir

---

[34] Cf. n. 30 above, pp. 331–48. [35] Cf. Liu Sung & Yeh Ting-Kuo (*1986*).
[36] Chhiu Phang-Thung (*1986e*). [37] *Hsing Yuan Lu*, pp. 1–4.
[38] Lin & Lin (1969), cf. also Chou San-Chin (*1986*).

from the Garden of Awareness). The most comprehensive work in this genre in the Chhing dynasty is, however, the *Thiao Ting Chi* 調鼎集 (The Harmonious Cauldron), a massive tome covering 871 pages and containing 2,700 recipes, including about 250 entries on fermentations and food processing. Unfortunately, the author is unknown and there is a good deal of uncertainty as to when the book was completed.[39] Parts of it were said to be in circulation before +1765, but other parts perhaps no earlier than 1860. There is considerable interest in the origin of this book among food historians in China, as many of the recipes in it are identical to those presented in the *Sui Yuan Shih Tan*. The usual questions arise. Which book is earlier? Who is copying whom? But we must leave behind this controversy and point out only that for our study it is assuredly not a book that we can ignore. We end this Table with another slim recipe book, *Chung Khuei Lu* 中饋錄, this time by a Madam Chêng (Chêng I 曾懿).[40] It deals exclusively with food processing and provides a useful summary of the state of the art towards the end of the Chhing dynasty.

(v) *Works on wine technology*

Although many of the food canons that we have mentioned include sections, sometimes quite substantial ones, on the production of 'wines' by fermentation,[41] there has also arisen a distinct and separate literary genre under the specialised term of *Chiu Ching* 酒經 (Wine Canons or Wine Classics). The Food and Drink volumes of the *Min Su Tshung Shu* 民俗叢書 (Series on Folk Literature) include sixteen titles in this category but half of these deal with historical and literary aspects of drinking as an epicurean pastime. The other eight works are listed in Table 15. The first on the list is the *Chiu Ching* 酒經 (Wine Canon) of Su Tung-Pho, the foremost literary figure of the Sung dynasty. It is a succinct account of the process of wine making, indicating that the author probably had had considerable personal experience with what he was writing about. The *Chiu Phu* 酒譜 (Wine Menu), another Sung work is a general, historical and literary treatise on wine with a minimum of technological content. By far the most important entry on this list is the *Pei Shan Chiu Ching* 北山酒經 (Wine Canon of North Hill) by Chu Kung 朱肱 (c. +1117), which contains the most complete description of wine making processes since the appearance of the *Chhi Min Yao Shu* in +540. The next work, *Hsin Fêng Chiu Fa* 新豐酒法 (*Hsin Fêng* Wine Process) is the last chapter of *Shan Chia Chhing Kung* (Basic Needs for Rustic Living) by Lin Hung, Southern Sung (Table 13), and presents the recipe for the making of a famous wine. The *Kuang Chi Chu* 觥記注 (Notes on Wine Vessels) by Chêng Hsiai 鄭獬 is a list of different types of wine vessels from antiquity to the Sung. Next follows the *Chhü Pên Tshao* 麴本草 (Pandects of Ferment Cultures), which, in spite of its title, has little to do with ferments, but provides an invaluable inventory of wines extant in

---

[39] The authorship of the *TTC* is discussed in Thao Wên-Thai (*1986c*) and Chhiu Phang-Thung (*1986f*).
[40] The author is described in Wang Chu-Lou (*1986*).
[41] For example, *CCPY*, pp. 36–47; *YSFSC*, pp. 123–33.

40 FERMENTATIONS AND FOOD SCIENCE

Table 15. *Wine classics of the Sung and Ming dynasties*

Title	Pages	Author	Date approx.	Contents
Chiu Ching 酒經 Wine Canon	4	Su Shih 蘇軾	+1090	Fermentation process
Chiu Phu 酒譜 Wine Menu	38	Tou Phing 竇平	Sung	History, anecdotes
Pei Shan Chiu Ching 北山酒經 Wine Canon of North Hill	26	Chu Kung 朱肱	+1117	Fermentation process
Hsin Fêng Chiu Fa 新豐酒法 *Hsin Fêng* Wine Process	2	Lin Hung 林洪	Sung	Fermentation process
Kuang Chi Chu 觥記注 Notes on Wine Vessels	12	Cheng Hsiai 鄭獬	Sung	Wine vessels
Chhü Pên Tshao 麴本草 Pandects of Ferment Cultures	6	Thien Hsi 田錫	Sung	Starter cultures
Chiu Hsiao Shih 酒小史 Mini-History of Wine	8	Sung Po-Jen 宋伯仁	Sung	List of wines
Chiu Shih 酒史 History of Wine	56	Fêng Shih-Hua 馮時化	Ming	Historical and literary notes

the early Sung.[42] The *Chiu Hsiao Shih* 酒小史 (Mini-History of Wine) is simply a list of names of wines presumably current during the Sung dynasty. Among the names is a grape wine from the Western region. The *Chiu Shih* 酒史 (History of Wine) is a collection of biographies, anecdotes, quotations from the classics and poems about wine. It has little to tell us about the technology of wine making, but serves as a useful guide to literary allusions to wine.

The last work we need to mention in this segment is not usually considered as part of the Food Canon genre, but it really needs to be included in this group in view of its importance to fermentations and food processing. It is the celebrated *Thien Kung Khai Wu* 天工開物 (Exploitation of the Works of Nature) of Sung Yin-Hsin 孫應星, +1637. It contains several chapters of interest to us (1) Cultivation of Grains (4) Processing of Grains (6) Manufacture of Sweeteners (12) Oilseed Processing and (17) Ferments for Wines. We have already quoted a passage from it on the subject of oilseed processing on pp. 29–30. There are two translations of *Thien Kung Khai Wu* in English, one by E. T. Z. Sun and S. C. Sun, and the other by Li Chiao-Phing *et al.* It will be a major reference work for Chapters (*c*) and (*e*). Two other works in this category are the *Wu Li Hsiao Shih* 物理小識 (Mini-Encyclopedia of the Principle

---

[42] Traditionally thought to have been written c. +985. Recent work by Liu Kuang-Ting (personal communication) suggests that it was written later.

of Things), +1664 by Fang I-Chih, and *Kuangtung Hsin Yü* 廣東新語 (New Talks about Kuangtung), +1690 by Chhü Ta-Chü.

(3) CLASSICS OF *MATERIA DIETETICA*

Legend has it that the emperor Shên Nung was the inventor of agriculture as well as herbal medicine in ancient China. According to the *Huai Nan Tzu* 淮南子 (The Book of the Prince of Huai-Nan), c. −120, it came about thus:[43]

> Anciently, the people lived on plants and drank water, collecting the wild fruits from the trees and eating the flesh of grubs and mussels. They often got ill and were hurt by poisonous things. So Shên Nung, the emperor, began to teach them how to sow (and reap) the five staple grains ... He tested the properties of the hundred plants, and the quality of the water, whether sweet or bitter; and thus he caused the people to know what to avoid and what to accept. At that time in a single day they met with seventy plants with active principles.

Presumably he discovered which plants were suitable as food, and which as medicine. Hence, in their medical writings the ancient Chinese were inclined to follow the dictum, *I Shih Thung Yuan* 醫食同源, that is, food and medicine have a common origin. Thus, it was believed that some foods could also serve as medicine (and vice versa), and, as mentioned earlier (p. 3), many food materials and products have found their way into the pharmacopoeias, or pandects of natural history. For example, about a quarter of the medicaments listed in the *52 Pingfang* (Table 11), recently discovered in Han Tomb. No. 3 at Ma-wang-tui, are food materials or products. This tradition is visible in the *Shên Nung Pên Tshao Ching* 神農本草經 (Classical Pharmacopoeia of the Heavenly Husbandman) +2nd century. It is continued and enlarged through a whole series of *pên tshao* compilations down to the great pharmacopoeia, *Pên Tshao Kang Mu* of +1596. The botanical significance of these works has been dealt with at length in Volume VI, Part 1 of this series. They will be discussed further from a medical and pharmaceutical perspective in later parts of Vol. VI. A number of them, however, have a strong emphasis on diet therapy, and may be classified separately as components of a *Materia Dietetica*. They are listed in Table 16.

The first book is the *Chou Hou Fang* 肘後方 by Ko Hung 葛洪, the famous alchemist. The full title is *Chou Hou Pei Chi Fang* 肘後備急方 (Handbook of Medicines for Emergencies). The version we now have has been amended and enlarged by Thao Hung-Ching (c. +500) and later by Yang Yung-Tao (c. +1000, Chin dynasty). It consists of 73 chapters in eight volumes. Although famous as a treatise of medical remedies it also deals with such topics as nutritional deficiency diseases and food incompatibilities. The *Hsin Hsiu Pên Tshao* 新修本草 (New Improved Pharmacopoeia), edited by Su Ching 蘇敬 was the first official pharmacopoeia compiled by imperial decree and the first to have been richly illustrated. The chequered story of this great

---

[43] *Huai Nan Tzu*, ch. 19, p. 1a. tr. Needham, *SCC* Vol. VI, Pt 1, p. 237.

Table 16. *Mediaeval and premodern works on* Materia Dietetica

Name	Author	Date approx.	Content
Chou Hou Fang 肘後方 Prescriptions for Emergencies	Ko Hung 葛洪	+340	Diet & disease, food hygiene
Hsin Hsiu Pên Tshao 新修本草 New Pharmacopoeia	Su Ching 蘇敬	+659	Medicinal properties of foods
Chhien Chin Shih Chih 千金食治 Golden Dietary Remedies	Sun Ssu-Mao 孫思邈	+655	Medicinal properties of foods
Shih Liao Pên Tshao 食療本草 Pandects of Diet Therapy	Mêng Shên 孟詵	+670	Medicinal properties of foods
Wai Thai Mi Yao 外台秘要 Important Prescriptions from a Distant Post	Wang Thao 王燾	+752	Food incompatibilities
Shih I Hsin Chien 食醫心鑑 Candid Views of a Nutritionist–Physician	Tsan Yin 咎殷	Late Thang	Food prescriptions
Shih Shih Wu Kuan 食物五觀 Five Aspects of Nutrition	Huang Thing-Chien 黃庭堅	+1090	Advice on food & diet
Shou Chhing Yang Lao Hsin Shu 壽親養老新書 New Handbook on Care of the Elderly	Chhen Chih Tsou Hsuan 陳直，鄒鉉	+1080 +1307	Diet therapy for the aged
Chêng Lei Pên Tshao 証類本草 Reorganised Pharmacopoeia	Thang Shen-Wei 唐慎微	+1082	Medicinal properties of foods
Yin Shan Chêng Yao 飲膳正要 Principles of Correct Diet	Hu Ssu-Hui 忽思慧	+1330	Food prescriptions
Yin Shih Hsü Chih 飲食須知 Essentials of Food and Drink	Chia Ming 賈銘	+1350	Food antagonisms and incompatibilities
Shih Wu Pên Tshao 食物本草 Natural History of Foods	Li Kao, Li Shih-Chên 李杲，李時珍	+1000 +1641?	Medicinal properties of foods
Shih Wu Pên Tshao Hui Tsuan 食物本草會纂 New Compilation of the Natural History of Foods	Shên Li-Lung 沈李龍	+1691	Medicinal properties of foods
Lao Lao Hêng Yen: Chu Phu 老老恒言：粥譜 Remarks of an Elder: Congee Menu	Tshao Thing-Tung 曹庭棟	+1750	Health-giving congee

work has been told in *SCC* Volume VI, Part 1.⁴⁴ The pictures in the original editions are all lost. Fortunately, ten of the twenty chapters have survived in Japan, of which two versions in manuscript form are known. The one recovered by Fu Yun-Lung 傅雲龍 and printed by him in fascimile in 1889, was reprinted in 1935. The other, in the collection of Mori Tateyaki and brought back to China by Lo Chên-Yü 羅板玉 in 1891, was finally reproduced by the Classics Press in Shanghai in 1981.⁴⁵ But what is even more encouraging is that not only the table of contents, but a substantial part of the text of the *Hsin Hsiu Pên Tshao* have, in fact, been preserved in chapters 2 to 4 of the *Chhien Ching I Fang* 千金翼方, the well-known supplement to the *Chhien Ching Yao Fang* 千金要方 (Thousand Golden Remedies) of Sun Ssu-Mao 孫思邈, the famous Thang physician. Thus, on the basis of these remnants, the Tun-Huang manuscripts, and the quotations trickled through numerous other *Pên Tshao* treatises, it was possible for Shang Chih-Chun 尚志鈞 to reconstruct what would appear to be the complete ancient text of the *Hsin Hsiu Pen Tshao* and publish it in 1981.⁴⁶ In contrast to the fate of *Hsin Hsiu Pên Tshao*, both *Chhien Ching Yao Fang* (thirty chapters) and its supplement, *Chhien Ching I Fang* (thirty chapters) have survived in China in relatively good condition. The two together form a veritable encyclopedia of the medical and pharmaceutical arts of Thang China. The *Chhien Chin Yao Fang* itself is also of particular interest to us in that there is a whole chapter in it (chapter 26) devoted to the topic of diet therapy (*Shih Chih* 食治).⁴⁷ This chapter has been reprinted as a separate volume under the title *Chhien Chin Shih Chih* 千金食治 (Table 16) as part of the Chinese Culinary Classics Series.

The next entry, *Shih Liao Pên Tshao* 食療本草 (Pandects of Diet Therapy), was the first *pên tshao* compilation devoted to diet therapy. According to the Japanese *I Hsin Fang* (Ishinhō) 醫心方 (Collection of Essential Prescriptions) of Tamba no Yasuyori 丹波康賴, +984, the book was written by Mêng Shên 孟詵, a student of Sun Ssu-Mao, and elaborated by Chang Ting 張鼎 as the *Pu Yang Fang* 補養方.⁴⁸ It was said to contain descriptions of 227 medical food products, divided into three volumes. Although considered lost for centuries, quotations from it have survived in the *Chêng Lei Pên Tshao* 證類本草 and fragments of it were found among the documents recovered in Tunhuang in the early twentieth century. The current version available is a reconsitution based on remnants in the *Chêng Lei Pên Tshao* and the Tunhuang manuscript.⁴⁹ Fate has been kinder to the *Wai Thai Mi Yao* 外台秘要 (Important Prescriptions from a Distant Post) by Wang Thao 王燾, which has survived without mishap to this day. It is an invaluable compilation since it is the repository of much information recorded in many pre-Thang medical books, which were lost soon

---

⁴⁴ Needham *et al.*, *SCC* Vol. VI, Pt 1, pp. 264–71. Su Ching later changed his name to Su Kung.
⁴⁵ The facsimile was reproduced in traditional folios in 1981. It was reprinted in book form with an introduction by Wu Tê-To in 1985.
⁴⁶ By Anhui Science & Technology Press, 安徽科學技術出版社 (*1981*).
⁴⁷ For a discussion of the dietary segments of the *CCYF* and *CCIF* cf. Ho I & Ting Wang-Phing (*1986*).
⁴⁸ For a discussion of this issue, cf. Shinoda Osamu (*1987a*), p. 110.
⁴⁹ Edited by Fan Feng-Yuan 范鳳源, based on a Japanese version annotated by Nakao Manzo 中尾萬三, and reprinted in Taipei in 1976.

after the advent of the Sung dynasty. It contains segments dealing with food incompatibilities and diet therapy.

The *Shih I Hsin Chien* 食醫心鑑 (Candid Views of a Nutritionist–Physician) by Tsan Yin 咎殷 was probably compiled in the middle of the ninth century. *Shih I* 食醫 is the official title of the Nutritionist–Physician in attendance at the royal household described in the *Chou Li* (Rites of the Chou).[50] The book stresses the use of foods and food products to cure various diseases. It was still current in the Sung but became lost shortly thereafter. The present version was reconstructed from material in the Japanese edition of *I Fang Lei Chi* 醫方類集, an old Korean medical encyclopedia (+1443), which contains passages from a number of ancient Chinese medical texts no longer extant in China.[51]

The first Sung dynasty work listed is the *Shih Shih Wu Kuan* 食時五觀 (Five Aspects of Nutrition) by the famous poet and calligrapher Huang Thing-Chien 黃庭堅. It is an essay on the proper attitude one should take towards food, and looks at food from the nutritional rather than the gastronomic point of view. It has been published in a collection of short food classics in the Chinese Culinary Classics Series.[52] A more substantial book is the *Shou Chhing Yang Lao Hsin Shu* 授親養老新書 (New Handbook on Care of the Elderly), which comprises four parts. Part one was compiled by Chhên Chih 陳直 during the Sung dynasty and was known independently under the title *Yang Lao Fêng Chhing Shu* 養老奉親書 (New Handbook on Care of the Elderly). Parts 2 to 4 were later added by Tsou Hsüan 鄒鉉 of the Yuan, based on material extant in older texts. It is a major contribution to the *Yang Shêng* 養生 (Cultivation of Life) literature. It contains advice on daily living, medical prescriptions, food recipes and anecdotes all pertaining to the feeding and care of the elderly. This work is now available as a publication under the *Chung I Chi Chhu Chhung Shu* 中醫基礎叢書 (Basic Chinese Medical Classics) Series.

We now come to the celebrated *Chêng Lei Pên Tshao* 證類本草 (Reorganised Pharmacopoeia) for which the full title is the *Chhung-Hsiu Chêng-Ho Ching-Shih Chêng Lei Pei-Chi Pên Tshao* 重修政和經史政類備急本草 (Classified and Consolidated Emergency Pharmacopoeia), +1204, the major work in the *pên tshao* genre of the Sung dynasty. Usually referred to as the *Chêng Lei Pên Tshao*, it is noteworthy from the perspective of diet therapy in that in it are preserved many passages from such lost works as *Shih Liao Pên Tshao* 食療本草, *Shih Hsing Pên Tshao* 食性本草 and *Shih I Hsin Chien* 食醫心鑑. The historical background of the *Chêng Lei Pên Tshao* and the various modifications that it engendered during the Sung have been treated in *SCC* Vol. VI, Part 1.[53]

The next two entries are probably the best known and most important works on diet therapy during the late mediaeval period, the *Yin Shan Chêng Yao* 飲膳正要 (Principles of Correct Diet) of the Yuan by Hu Ssu-Hui 忽思慧 (Hoshoi), and the

---

[50] *Chou Li*, p. 45.
[51] Based on the Japanese edition of 1861. The current Chinese version was reconstructed by the Tung Fang Hsueh Huei 東方學會 in 1924.
[52] Together with *WSCKL (1987)* in *CKPJKCTK* series.   [53] Needham *et al.*, *SCC* Vol. VI, Pt I, pp. 281–94.

Yin Shih Hsü Chi 飲食須知 (Essentials of Food and Drink) of late Yuan to early Ming by Chia Ming 賈銘. Hu Ssu-Hui was the Grand Dietician (*Yin Shan Thai I* 飲膳太醫) of the Mongol court. The *Yin Shan Chêng Yao* was thus the official nutritional guide for the royal household. It comprises three chapters; one deals with delicacies and specialities, two, common viands and drinks and prescriptions for various ailments, and three, nutritional and medicinal properties of foods. Its significance in the history of diet therapy in China has been discussed for Western readers by Lu and Needham.[54] Fortunately for us, the reproduction of a Ming edition preserved in Japan and published in Shanghai by Chang Yuan-Chi 張元濟 in 1920 is now available to us under a new impression in the Chinese Medical Classics Series. The pictures in this edition have served well as illustrations for articles on aspects of cookery and diet therapy in China.[55] The sixty-one food prescriptions for various ailments in chapter 2, with copious annotations, have been published separately under the title *Shih Liao Fang* 食療方 (Diet Therapy Prescriptions) in the Chinese Culinary Classics Series.[56]

In contrast to the *Yin Shan Chêng Yao* which tends to stress the curative functions of food, the *Yin Shih Hsü Chi* 飲食須知 (Essentials of Food and Drink) is concerned mainly with the preventive aspects of therapy. It has also been introduced to Western readers in an article on Food and Drink in the Yuan and Ming dynasties by Mote.[57] The author, Chia Ming, had reached the super-venerable age of a hundred years in +1368 when the Ming dynasty was founded. He died six years later. Thus he was a living testimony to the effectiveness of the injunctions he preached. In his preface he says:

> Drink and food are relied upon to nourish life. Yet if one does not know which substances, by their respective natures, are antagonistic to and incompatible with each other, and consumes them indiscriminately, at the least the five viscera will be thrown out of harmony, and at the most, disastrous consequences will immediately ensue. Thus it is that nourishing foods may at times actually be injurious to life. I have examined the texts, annotations and commentaries of the major pharmacopoeias, and found that each of the substances listed may give rise to harmful as well as beneficial effects, leaving the reader at a loss as to what to do. In compiling this work, I have, therefore, made a point of selecting those facts pertaining to antagonism and incompatibility among foods, so that those who wish to attain good health will be able to apply this information in their daily consumption of drink and food.

The book is divided into eight chapters with contents as follows (1) water and fire (2) grains (3) vegetables (4) fruits (5) flavourings (6) fish and shellfish (7) fowl and (8) mammals. The quality is uneven. It is a repository of much useful information and common sense, but also a good deal of superstitious misconceptions, that have been uncritically copied from older texts. With its emphasis on what is poisonous and to be avoided and what are incompatible with each other, the *Yin Shih Hsü Chi* tends to

---

[54] Lu & Needham (1951). For a discussion of the culinary aspects of this work see Sabban (1986b), Anderson (1994) and Buell (1986 & 1990).
[55] For example, Lu & Needham (1951), Harper (1984).
[56] In the same book as *Chhien Chin Shih Chih*. A new edition, with copious annotations and a translation into modern Chinese by Li Chhun-Fang 李春芳 was also published in 1988.
[57] Mote (1977), pp. 227–33; preface tr. Mote, mod. auct., pp. 228–9.

be somewhat alarmist in tone, and could give the impression that eating was a highly hazardous occupation. It is now available as a separate publication with annotations in the Chinese Culinary Classics Series.

There are several works that are known to have existed under the title *Shih Wu Pên Tshao* 食物本草 (Natural History of Foods). The earliest one is attributed to Li Kao 李杲 (Li Tung-Hêng 李東恒) who lived under the Jurchen/Chin, but the version now extant, allegedly edited by Li Shih-Chên and printed by Yao Ko-Chêng 姚可成, dates from +1638. The same title has also appeared with Lu Ho 盧和 as the author in c. +1570, and with Wang Ying 汪穎 in +1620. The bibliography of this work in its several manifestations and under various authorships is extremely complex, and quite beyond the scope of our study to untangle at this time.[58] For our purpose, the version that we have consulted is the one attributed to Li Kao/Li Shih-Chên which includes 58 chapters and 2,000 entries. It contains the most comprehensive survey of the quality of spring water ever recorded in China.

The format of the next entry, the *Shih Wu Pên Tshao Hui Tsuan* 食物本草會纂 (New Compilation of the Natural History of Foods), by Shên Li-Lung 沈李龍 in early Chhing, is reminiscent of the *Yin Shih Hsü Chi*, with chapters on water, fire, grains, vegetables, fruits, fish, shellfish, fowl and mammals. But there are also two additional chapters, one on everyday household arts, such as emergency foods, dietary incompatibilities and poisonous foods and antidotes, and the other on pulse reading. It represents a summation of the major observations and discoveries in diet therapy from the Thang, Sung, Yuan and Ming texts that we have so far discussed and is an indicator of the state of the art in China before the arrival of new nutritional ideas from the West in the 19th century.

We close Table 16 with a specialised book devoted entirely to congee, the *Chu Phu* 粥譜 (Congee Menu) compiled by Tshao Thing-Tung 曹庭棟 in +1750. It is actually the fifth volume of a five-volume work known as *Lao Lao Hêng Yên* 老老恒言 (Remarks of an Elder), written by the author at age seventy-five as a guide to all aspects of daily living for the elderly. The fifth volume has been issued as one of two parts in a separate publication, under the title *Chu Phu*, in the Chinese Culinary Classics Series. Throughout Chinese history, congee has been regarded as a particularly desirable fare for the elderly, and the author, who died in his nineties, has collected in one volume what he regarded as the most beneficial types of congee in the Food Canon (*Shih Ching*) and Diet Therapy (*Shih Liao*) literature. After a brief discourse on the principles for making good congee, we are presented with one hundred descriptions of the therapeutic and nutritive properties of one hundred types of congee, divided into upper, middle and lower classes. It is not a recipe book since hardly any directions for making a specific congee are given. The slant is strongly vegetarian. Of the one hundred entries, more than eighty contain only ingredients of plant origin. Of the thirty-six entries in the upper class, only two, deer tail and bird's nest congee, include the use of animal products.

[58] Needham *et al.*, *SCC* Vol. VI, Pt 1, p. 585; cf. Lung Po-Chien (*1957*), pp. 104–6.

(4) Supplementary Sources

It has been said, 'All animals feed, but humans alone eat (*and drink*).'[59] Eating and drinking are, of course, essential ingredients of civilised living. The way a people eat and drink becomes one of the distinguishing features of their culture and is ensconced in their literature and art. Thus it is not surprising that the ancient classical texts of China have provided valuable information on the culinary practices of their day. The most important and useful of these have been listed in Table 11 and discussed in Section (*a*)(1) of this chapter. But these are by no means the only texts that are relevant to our study. Indeed, we have already had the occasion to refer to passages from the most sacred books of ancient China, such as the *I Ching* 易經 (Book of Changes), *Shu Ching* 書經 (the Book of Documents), *Tao Tê Ching* 道德經 (Canon of the Tao and its Virtue), *Chuang Tzu* 莊子 (the Book of Master Chuang), *Lun Yu* 論語 (the Analects of Confucius) and *Mêng Tzu* 孟子 (the Book of Mencius) which are generally thought of as discourses in philosophy rather than guides to daily living. Additional ancient classics that have been or will be consulted include the *I Li* 儀禮 (Ritual for Personal Conduct), *Tso Chuan* 左傳 (Master Tso's Enlargement of the Spring and Autumn Annals), *Kuan Tzu* 管子 (The Book of Master Kuan), *Mo Tzu* 墨子 (The Book of Master Mo), *Han Fei Tzu* 韓非子 (the Book of Han Fei), and *Huai Nan Tzu* 淮南子 (the Book of the Prince of Huai Nan). The wide range of the philosophical schools represented in these works is indicative of the pervasive role of the culinary system in the life of the ancient community. In many cases they provide invaluable corroborative evidence to support the views formulated from the information found in the principal sources listed in Table 11.

One of the recurring problems we encounter in this study, as we have in other parts of this series, is the the meaning of ancient words.[60] We have dwelled on the case of the millets *chi* 稷 and *su* 粟 on page 24, and touched on the case of the stools *chi* 几 and *an* 案 on pages 109 and 112 in the preceding chapter.[61] In such situations consultation with lexicographic and encyclopedic works is *de rigueur*. This class of literature, including the *Erh Ya* 爾雅 (Literary Expositor), *Shuo Wên Chien Tzu* 説文解字 (Analytical Dictionary of Characters), *Fang Yen* 方言 (Dictionary of Local Expressions), *Shih Ming* 釋名 (Explanation of Names) down to the *Khang Hsi Tzu Tien* 康熙字典 have been discussed briefly in *SCC* Volume I and in detail in Volume VI, Part 1.[62] Since the 1970s several excellent dictionaries have appeared on the scene, such as the *Chung Wên Ta Tzhu Tien* 中文大字典 (The Encyclopedic Dictionary of the Chinese Language) from Taipei, *Han Yü Ta Tzu Tien* 漢語大字典 (The Great Chinese Dictionary) from Szechuan and *Han Yü Ta Tzhu Tien* 漢語大詞典 (The Great Idiomatic Chinese Dictionary) from Shanghai.[63]

---

[59] Farb & Armelagos (1980), p. 3. [60] Needham *et al.*, *SCC* Vol. VI, Pt 1, pp. 463–71.
[61] Cf. pp. 109–12 above. [62] Needham *et al.*, *SCC* Vol. I, pp. 47–51; Vol. VI, Pt 1, pp. 182–6.
[63] These have superseded older works such as the *Tzhu Yuan* and *Tzhu Hai* mentioned earlier, cf. Needham Vol. I, p. 49.

The dynastic histories are, of course, always a valuable source of information on aspects of daily life, sometimes including food and drink, at the specific period under consideration. The forerunner and type species of this genre is the *Shi Chi* 史記 (Historical Records) of Ssuma Chhien 司馬遷, who recorded the events that marked the development of Chinese Society as a national and cultural unit from legendary times to the founding of the Han dynasty. The format provided by the *Shi Chi* is followed by the histories of the Former Han (*Chhien Han Shu* 前漢書) and the Later Han (*Hou Han Shu* 後漢書), and a continuing series of histories for each of the succeeding dynasties. The Economic Affairs (*Shih Huo Chih* 食貨紀) sections, and sometimes the main narratives, often offer nuggets of useful information on the contemporary culinary scene. Perhaps of greater interest to us, however, are unofficial notes and memoirs which describe the conditions of life and culture at a particular place and time, such as the *Hsi Ching Tsa Chi* 西京雜記 (Miscellaneous Records of the Western Capital), mid +6th century, for the city of Sian during the Former Han. The best known works of this type are the *Tung Ching Mêng Hua Lu* 東京夢華錄 (Dreams of the Glories of the Eastern Capital) by Mêng Yuan-Lao 孟元老 (+1148), the *Mêng Liang Lu* 夢粱錄 (Dreaming of the City of Abundance), by Wu Tzu-Mu 吳自牧 (+1334) and the *Wu Lin Chiu Shih* 武林舊事 (Institutions and Customs of the Old Capital) by Chou Mi 周密 (+1270). From the former we can learn a good deal about the culinary environment in the Northern Sung Capital of Khaifeng before it fell to the Chin army, and from the latter two, that in the Southern Sung capital, Hangchou, in a time of maximum peace and prosperity. Additional impressions of life in the Southern Sung capital are found in the *Tu Chhêng Chi Shêng* 都城紀勝 (Wonders of the Capital [Hangchou]), +1235 and *Hsi Hu Lao Jên Fan Shêng Lu* 西湖老人繁勝錄 (Memoir of the Old Man of West Lake), Southern Sung. Another work that is worth noting is the *Loyang Chieh Lan Chi* 洛陽伽籃記 by Yuang Hsuan-Chih 楊衒之 (+547), which describes the buddhist temples and monasteries in Loyang (and what were grown in the gardens) during the Northern Wei.

Although we have paid special attention to the *Shih Liao* 食療 (Diet Therapy) component of the *Pen Tshao* 本草 corpus, it does not mean that the traditional works on *Materia Medica* are not of value in our study. We still need to regularly consult such standard pharmacopoeias as the *Shên Nung Pên Tshao Ching* 神農本草經 (Classical Pharmacopoeia of the Heavenly Husbandman) of the −1st century, the *Pên Tshao Kang Mu* 本草綱目 (The Great Pharmacopoeia) of the +16th century, and the modern *Chung Yao Ta Tzhu Tien* 中葯大詞典 (Cyclopedia of Chinese Traditional Drugs) published in 1978. The evolution of the *pên tshao* literature has been treated at length in Volume 6, Part 1, in which the fascinating story of another great, but lesser known, pharmacopoeia of the +16th century was also reviewed.[64] This is the *Pen Tshao Phin Hui Ching Yao* 本草品彙精要 (Essentials of the Pharmacopoeia Ranked according to Nature and Efficacy), which was commissioned by the Ming emperor

---

[64] Needham *et al.*, *SCC* Vol. VI, Pt 1, pp. 302–8.

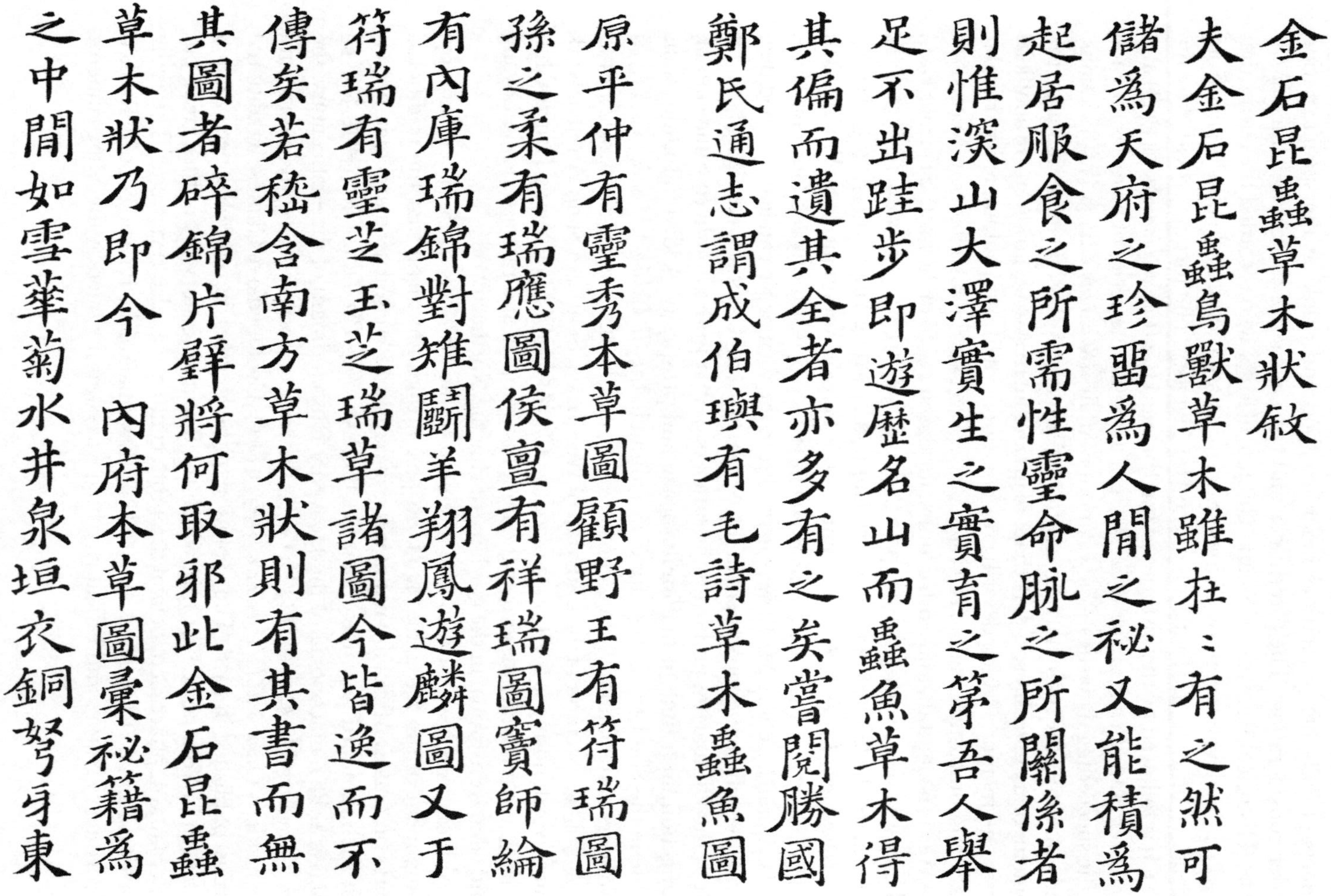
金石昆蟲草木狀敘
夫金石昆蟲鳥獸草木雖往往有之然可
儲爲天府之珍畱爲人閒之祕又能積爲
起居服食之所需性靈命脈之所關係者
則惟溪山大澤實生之實育之第吾人舉
足不出跬步即遊歷名山而蟲魚草木得
其偏而遺其全者亦多有之矣嘗閱勝國
鄭氏通志謂成伯璵有毛詩草木蟲魚圖
原平仲有靈秀本草圖顧野王有符瑞圖
孫之柔有瑞應圖侯亶有祥瑞圖竇師綸
有內庫瑞錦對雉鬬羊翔鳳遊麟圖又于
符瑞有靈芝玉芝瑞草諸圖今皆逸而不
傳矣若嵇含南方草木狀則有其書而無
其圖者碎錦片璧將何取邪此金石昆蟲
草木狀乃即今 內府本草圖彙祕籍爲
之中閒如雪華菊水井泉垣衣銅弩牙東

Fig 39. First two pages of the Introduction to *CSKCTMC*, Calligraphy by Chao Chün, from National Central Library, Taipei.

Hsiao Tsung and completed in +1505. It was kept in the Imperial Academy of Medicine in manuscript form and never published until our own time. The remarkable feature about this MS is that the text is built around the illustrations which were beautifully executed in colour. Several Ming and Chhing manuscripts have survived, but none of the illustrations have been reproduced.[65] Although we have no idea of how it came about, the illustrations have also been copied by the famous Ming album painter Wên Shu 文俶, from +1617 to +1620, and collected as a bound folio under the title *Chin Shih Khun Chhung Tshao Mu Chuan* 金石昆蟲草木狀 (Illustrations of Minerals, Insects, Herbs and Trees), with an introduction written in exceptionally fine calligraphy by her husband, the well-known poet and bibliophile, Chao Chun 趙均. There is no text but the content and order of the paintings leave no doubt that they were intended as illustrations to the *Pên Tshao Phin Hui Ching Yao*. This folio is now preserved in the National Central Library, Taipei, Taiwan.[66] Several of Wên Shu's illustrations are reproduced in Figure 2 in Chapter (*a*), but alas not in colour, and we now present the first two pages of the introduction in Figure 39.

Similarly, we need to acknowledge our debt to the standard agricultural treatises and almanacs other than the *Fang Shêng-Chi Shu*, *Ssu Min Yüeh Ling* and *Chhi Min Yao Shu*, which also contain valuable information pertaining to the preparation of food and drink. These agricultural works have been described in detail in Volume 6, Part 2.[67] Among the ones we have found most useful are the late Thang *Ssu Shih Tsuan Yao* 四時纂要 (Important Rules for the Four Seasons) of late Thang and the *Pien Min Thu Tsuan* 便民圖纂 (Everyman's Handy Illustrated Compendium) of the Ming, which we have mentioned earlier (Table 14). Often consulted also are the three well-known texts of the Yuan dynasty: the *Nung Sang Chi Yao* 農桑輯要 (Fundamental of Agriculture and Sericulture) by the Imperial Board of Agriculture (+1273), *Nung Sang I Shih Tsho Yao* 農桑衣食撮要 (Selected Essentials of Agriculture, Sericulture, Clothing and Food) by Lu Ming-Shan 魯明善 (+1314) and the celebrated *Nung Shu* 農書 (Agricultural Treatise) of Wang Chên 王楨 (+1313). Lu Ming-Shan is an unusual figure in Chinese history, being one of very few Uighurs who made a name for himself as a high-ranking official in central China and as a successful author in the Chinese language. Today he is considered a folk hero among the Uighurs in Sinkiang. Although he and Wang Chên were contemporaries with similar interests, there is no evidence that the two ever met. The major work on agriculture in the Ming dynasty is the *Nung Chêng Chhüan Shu* 農政全書 (Complete Treatise on Agricultural Administration) by Hsü Kuang-Chhi 徐光啟 (+1639). For

---

[65] The original illustrated manuscript for the *Pen Tshao Phin Hui Ching Yao* is now believed to be at the Tanabe Pharmaceutical Co. in Osaka. Complete copies are known to be at the National Library, Peking, Biblioteca Nazionale Centrale Vittorio Emanuele II, Rome and the Staats Preussicher Kulturbesitz, Berlin, cf. Needham *et al. ibid.* p. 302 and Bertuccioli (1956).

[66] For a comparison of this folio with the known illustrated manuscripts of the *Pen Tshao Phin Hui Ching Yao*, cf. Li Chhing-Chi (*1979*).

[67] Bray, *SCC* Vol. vi, Pt 1, pp. 55–72. For a listing of all the agricultural treatises of China see Wang Yü-Hu (1979), originally published in 1964.

our purpose, it may be particularly memorable for the esculentist message it conveys by the inclusion of the entire text of the *Chiu Huang Pên Tshao* 救荒本草 (Treatise on Wild Food Plant for use in Emergencies) within its covers. It also contains a good deal of material, largely as quotations from previous works, pertinent to the processing of vegetables and fruits. The last work we need to mention is the *Shou Shih Thung Khao* 授時通考 (Compendium of Work Days) compiled by imperial order under the direction of Ou Erh-Thai 鄂爾泰 (+1742). It is of value chiefly as a repository of quotations from older texts that would otherwise be no longer available.

We now come to the group of specialised monographs or tractates on specific plants or animals that are such a prominent feature of the literature of natural history in China. We have introduced this group in *SCC* Volume VI, Part 1, and discussed at some length the monographs on a number of plants, mostly ornamentals.[68] Two of these represent plants of interest as food resources, the *Chü Lu* 橘錄 (Monograph on Citrus) of Han Ye-Chih 韓彥直 (+1178), and the *Sun Phu* 筍譜 (Treatise on Bamboo Shoots) by the monk Tsan Ning 贊寧 (+970). Both works contain materials related to food processing and preservation that are pertinent for Chapter (*e*).

The favourite topic for writers of specialised monographs is undoubtedly *Li Chi* 荔枝 or litchi, *Litchi sinensis*, certainly one of the most delicious fruits of China. We have already met with two works on this plant previously.[69] They are the *Li-Chih Phu* 荔枝譜 (Monograph on Litchi) written by Tshai Hsiang 蔡襄 in about +1059 when he was governor of Fukien and the *Tsêng Chhêng Li Chi Phu* 增城荔枝譜 (Treatise on the Litchis of Tsêng-Chhêng [in Kuangtung]) by Chang Tsung-Min 張宗閔 in +1076 which is no longer extant.[70] Another work about litchis in Kuangtung is the more recent *Lingnan Li Chih Phu* 嶺南荔枝譜 (Litchis South of the Range) written by Wu Ying-Ta 吳應達 of the Chhing Dynasty. Altogether no less than fifteen monographs on litchi are known to have existed, and six of them, including Tshai Hsiang's work, share the title *Li-Chih Phu*.[71] Two of these also contain significant information on processing and preservation, one by Hsü Po 徐燉 and the other by Sung Chüeh 宋珏, both of the Ming dynasty. Detailed discussion of the litchi story must, however, await the remainder of Section 38 (Botany).

Not nearly as popular is the case of mushrooms, an important component in the Chinese cuisine. Three monographs are known. The oldest is the *Chün Phu* 菌譜 (Treatise on Fungi) of Chhên Jên-Yü 陳仁玉 (+1245), which describes eleven types of edible mushrooms, including *sung chün* 松菌 (pine mushroom) which has been identified as *Armillaria matsutake* Ito et Imai. The second is the *Kuang Chün Phu* 廣菌譜 (Extensive Treatise on Fungi) by Phan Chih-Heng 潘之恒 (c. +1550) which describes twenty fungi, including several that would be familiar to a modern Chinese kitchen. For example, we see seven variants of *mu erh* 木耳 (wood-ear or

---

[68] Needham *et al.*, *SCC* Vol. VI, Pt 1, pp. 355ff.
[69] *Ibid.*, pp. 359 and 361.   [70] Wang Yü-Hu (*1979*), pp. 70 and 73.
[71] Thirteen are listed in *ibid.*, pp. 313–14; two others are included in the Chinese Folk Literature Series (*Min chu tshung shu* 民俗叢書), section on Fruits and Vegetables (*Kuo Su* 果蔬). Besides Tshai Hsiang, Hsü Po and Sung Chüeh, the other authors of a *Li Chih Phu* are Tshao Fan 曹蕃, Chhên Ting 陳鼎 and Têng Tao-Hsieh 鄧道協.

cloud-ear), *Auricularia auricula* (L. ex Hook), the *hsiang chün* 香菌, the modern favourite shitake, *Lentinus edodes* (Berk) Sing., and the common mushroom of the Western cuisine, *mo ku chün* 蘑菇菌, *Agaricus campestris* L. ex. Fr. The first two are distinguishing ingredients of Chinese cookery and can be found, in dried forms, in Chinese groceries everywhere. The last work is the *Wu Chün Phu* 吳菌譜 (Mushrooms in the Wu Region) by Wu Lin 吳林 (+1683) which records the types of mushrooms available in the vicinity of Soochou during the Khang Hsi reign. Although we can identify only a few of the fungi listed in terms of modern mycological classification, the range of organisms described in detail is certainly impressive and indicative of the sophistication the Chinese had achieved in recognising and utilising wild edible fungi.

Several other monographs on food plants are worth mentioning. The *Ta Tsao Phu* 打棗譜 (Monograph on Jujubes) by Liu Kuan 柳貫 (Yuan dynasty) provides useful information on an important fruit, with a few notes on matters related to processing. The *Yu Ching* 芋經 (Book of Taro) by Huang Shêng-Tsheng 黃省曾 (Ming dynasty) describes the planting and usage of this indigenous tuber crop. The nutritional and culinary value of melons and a few vegetables are discussed in the *Kua Shu Shu* 瓜蔬書 (Book of Melons and Vegetables), another Ming work by Wang Shih-Mou 王世懋, who also treated fruits in a similar tractate called *Kuo Shu* 果書 (Discourse on Fruits). But perhaps the ultimate monograph in this subgroup is a compendium of monographs, such as the *Chhün Fang Phu* 群芳譜 (Monographs on Cultivated Plants), first written by Wang Hsiang-Chin 王象晉 (+1630), and later enlarged to the *Kuang Chhün Fang Phu* 廣群芳譜 (Monographs on Cultivated Plants, Enlarged) by Wang Hao (+1708) 王灝. The latter work, which includes the text of the former, consists of eleven *phu*'s (monographs) on such topics as vegetables (five chapters), tea (four chapters), fruits (fourteen chapters), grains, flowers etc. with descriptions on the processing, preservation and utilisation of many of the products.

In contrast to plants, there are few monographs on animal products except for seafood. The oldest is probably the *Hsiai Phu* 蟹譜 (Monograph on Crabs) by Fu Kung 傅肱 (+1060), which was followed by a similar work entitled *Hsiai Lo* 蟹略 (Discourse on Crabs) by Kao Shih-Sun 高似孫 at about +1180. Both contain useful cooking and processing recipes and testify to the high regard this crustacean enjoys in the Chinese cuisine. Less interesting is a Chhing dynasty *Hsü Hsiai Phu* 續蟹譜 (A Continuation Monograph on Crabs) by Chhu Jen-Huo 褚人獲 which contains mainly anecdotes about eating crabs.

The other seafood monographs deal largely with fish. The *Yang Yü Ching* 養魚經 (Monograph on Fish) is a Ming work by Huang Shêng-Tshêng 黃省曾. It describes eighteen kinds of fish ranging from the highly valued sturgeon to the delectable yet hazardous *ho thun* 河豚 (puffer fish), *Fuga vermicularis*, with instructions of how the latter should be prepared so that it could be eaten safely. The contemporary *Yü Phin* 魚品 (Types of Edible Fish) by *Tun Yuan Chü Shih* 豚園居士 (Resident of the Hidden Garden) discusses twenty kinds of fish in the Yangtze delta. A more extensive work is the *Min Chung Hai Tsho Shu* 閩中海錯疏 (Seafoods of the Fukien Region) of +1593, compiled by Thu Pên-Tsun 屠本畯, and enlarged by Hsü Po 徐𤐌. It lists 246 seafood products of Fukien province, and mentions, for the first time, the bird's nest, that has become a prestigious and inevitable feature of the cuisine at important Chinese banquets. Yet the author, who was well versed in aquatic natural history of southeast China, had only a hazy idea about the origin of this product. He relates the story that the nest was carried in the beak of the swallow as it flew across the South China Sea. When it became too tired it would drop its load on to the surface of the sea and rest on it until it regained enough strength to resume the flight.[72] Thu Pên-Tsun also wrote a *Hai Wei So Yin* 海外索隱 (Guide to the Delicacies of the Sea) which is a collection of anecdotes about different types of seafoods, and mentions the edible jellyfish *cha* 鮓, which is today often served as part of the appetiser dish at a Chinese meal.[73]

Finally, it should be noted that a most valuable source of information lies in the wealth of archaeological finds uncovered from burial sites in China in the last thirty-five years. These include the remains of food materials and products, cooking and serving implements, containers for food and drink, dining furniture, manuscripts on bamboo slips and silk, and murals of scenes of cooking and feasting. Reports of these discoveries are available in Chinese archaeological journals such as *Khao Ku* 考古 (Archaeology), *Khao Ku Hsüeh Pao* 考古學報 (Archaeological Reports), *Wên Wu* 文物 (Reference Materials for History and Archaeology) etc. Articles on the significance of these finds in terms of food resources, technology and culture have appeared in journals such as *Nung Yeh Khao Ku* 農業考古 (Agricultural Archaeology). Special monographs have also been published to commemorate the discoveries from especially noteworthy excavations, such as those from the Han tombs at Ma-wang-tui, Yun-mêng and Man-chhêng.[74] These have provided us with valuable insights into the culinary system in situations where no written records exist or where the written records are sketchy and incomplete, as we shall see in the discussions on several key issues in food processing in Chapters (*d*) and (*e*).

(5) SECONDARY SOURCES

Tracing the paths to the myriad pieces of pertinent information in the primary sources would have been immensely onerous had it not been for the steady hand of a number of excellent secondary sources to guide the way. In addition to the numerous scholarly articles cited in the text, there are several sources of special interest for our exploration that deserve to be mentioned before we move on to the next chapter. First, there are the pioneering studies of scholars outside China on the history of Chinese food culture and the culinary arts. Foremost among these are the works of Shinoda Osamu. A selection of his numerous papers on this subject were collected

---

[72] *Min Chung Hai Tsho Shu*, in *SKCS*, **590**, p. 524. For an account of the history and usage of bird's nest see Chiang & Kuan (*1991*), pp. 98–113. Cf. also Needham *et al.*, *SCC* Vol. IV, Pt 3, p. 537.
[73] This dish is particularly popular in Fukien although it is becoming quite common in other parts of China as well as overseas. It is called *tha* in Foochow dialect.
[74] *Hu-nan shêng po-wu-kuan* (*1973*), *Yun-mêng Shui-hu-ti* (*1981*) and *Chung-kuo shê-huei kho-hsueh-yuan* (*1980*).

in a book entitled *Chugoku Shokumotsu Shi* 中國食物史 (A History of Food in China) published in 1974. Eight of the eleven papers there have been translated into Chinese by Kao Kuei-Lin, Hsieh Lai-Yung and Sun Yin and published under the title *Chung Kuo Shih Wu Shih Yen Chiu* 中國食物史研究 (Studies on the History of Food in China). The papers on the *Shih Ching* 食經 (Food Canons or Classics) of mediaeval and premodern China have been particularly valuable in guiding our coverage of the literature.[75] Shinoda Osamu and Tanaka Seiichi have also compiled a *Chung Kuo Shih Ching Tshung Shu* 中國食經叢書 (A Collection of Chinese Food Classics) in which are reproduced forty of the most important treatises among about 150 titles in this genre that they have identified and located in Japanese collections. Shinoda's research has been continued and expanded by Ishige Naomichi and his collaborators to include comparative studies of the food cultures of Japan, China, Korea, Vietnam and the other countries of Southeast Asia. We are pleased to mention the work of Françoise Sabban of France, probably the only Western scholar who specialises in the history of food culture in China, whose publications have furnished not only many useful references to the primary literature but also new perspectives on several of the problems we have wrestled with in the course of our investigation.

Next are a group of publications from China that deal with historical aspects of Chinese food culture and technology and provide a valuable overview of the field. They include 'A Short History on Food, Culture and the Development of Food Processing Industries in China', *Chung Kuo Yin Shih Wên Hua ho Shih Phin Kung Yeh Fa Chan Chien Shih* 中國飲食文化和食品工業發展簡史, by Yang Wên-Chhi (*1983*); 'A Brief History of Chinese Cookery', *Chung Kuo Pheng Jen Shih Lo* 中國烹飪史略, by Thao Wên-Thai (*1983*), 'Draft of a History of Food Science and Technology in China, Part I', *Chung Kuo Shih Phin Kho Chih Shih Kao, Shang* 中國食品科技史稿，上 by Hung Kuang-Chu (*1984*), 'A Brief History of the Development of Food Processing Industries in China', *Chung Kuo Shih Phin Kung Yeh Fa Chan Chien Shih*. 中國食品工業發展簡史, by Wang Shan-Tien (*1987*), 'Investigation of Food and Drink during the Han-Wei Era', *Han Wei Yin-Shih Khao* 漢魏飲食考, by Chang Mêng-Lun, and 'History of Food and Drink in China' Vol. 1, *Chung Kuo Yin Chuan Shih ti-i-Chüan*. 中國飲饌史第一卷 by Tsêng Chhung-Yeh (*1988*). The 'History of the Fermentation Industry in China', *Chung Kuo Wei Sheng Wu Kung Yeh Fa Chan Shih* 中國微生物工業發展史, by Chhên Thao-Shêng (*1978*) contains excellent treatments of the history of traditional Chinese fermentation processes. There are also collection of essays on food and drink that include discussions on the historical aspects of the subject. One is *Tsin Tsin Yu Wei Than* 津津有味譚 (Talks about Tasty Viands) by Chhên Tshun-Jên 陳存仁 (*1978*) and the other *Chung Kuo Yin Shih Than Ku* 中國飲食談古 (Talks on Food and Drink in Ancient China) by Wang Jên-Hsing 王仁興 (*1985*). The essays are highly anecdotal in content and obviously intended for the general

---

[75] For other useful bibliographies and commentaries cf. Thao & Chang (*1986*), Chhiu Phang-Thung (*1989c*), *CKPJPKCS*, pp. 423–6, 540–3 and Tai Yün (*1994*).

reader rather than the specialist, so that when literature references are mentioned, if at all, exact citations are usually lacking. Much more academic and useful to us are many of the wide-ranging short articles on historical aspects of the culinary arts by Chinese food historians such as Lin Nai-Shen, Chhiu Phang-Thung, Thao Wên-Thai, Chih Tzu, Kuo Po-Nan, Wang Jen-Hsiang etc. that are a standard feature of the popular magazine *Chung Kuo Phêng Jên* 中國烹飪 (Chinese Cuisine). The articles that appeared before 1985 are conveniently reissued in a collection entitled *Phêng Jên Shih Hua* 烹飪史話 (Talks on the History of Cuisine). These have provided us with many invaluable leads that would otherwise have been extremely difficult to come by.

Finally, we come to books in English on Chinese cuisine and culture which we have read with pleasure and profit in this study. The first is Buwei Yang Chao's delightful cookbook *How to Cook and Eat in Chinese* (1945, revised 1963). Her introductory chapters on meal systems, eating and cooking materials, methods of readying and cooking are masterpieces of 'analysis and synthesis'[76] and have served as a friendly guide to our reconstruction of the cuisine of ancient China. An unusual cookbook with a scholarly bent is *Chinese Gastronomy* by Hsiang-Ju Lin and Tsuifeng Lin (1969) which relates modern dishes to ancient and classical cookery, and contains discerning discussions on theoretical aspects of the character of Chinese cuisine such as the criteria of excellence, blending of flavours, variation of texture etc. A valuable source of scholarly information and references to primary works is *Food in Chinese Culture* edited by K. C. Chang (1978), containing essays on food and culture, with anthropological and historical perspectives, from antiquity to the modern era. It includes separate essays by different authors on Ancient China, Han, Thang, Sung, Yuan and Ming, Chhing, Modern China: North, and Modern China: South. The last works to be mentioned are *The Food of China* by E. N. Anderson (1988), and *Food in China: A Cultural and Historical Inquiry* by Frederick Simoons (1991). Anderson describes the development of the Chinese dietary system from antiquity to the present day. Simoons discusses the history, science and culture of different classes of foods. Both works contain excellent bibliographies of materials related to Chinese foods in Western languages.

---

[76] Hu Shih, Forward to Mrs Chao's cookbook (1963), p. xvi.

(c) FERMENTATION AND EVOLUTION OF ALCOHOLIC DRINKS

Since prehistoric times the Chinese have prepared an alcoholic beverage from cooked grains called *chiu* 酒, which has held a dominant position in the drink part of the *yin shih* 飲食 (drink and food) combination through the centuries until the present day. There is no exact English equivalent to the word *chiu*. It has traditionally been translated as 'wine'.[1] As early as +1664 Robert Boyle, the sceptical chymist himself, recognised that the Chinese produced a 'wine'. He said:[2]

In that vast Region of *China*, which is enriched with so fertil a Soil, and comprizeth such variety of Geographical Parallels, they make not (as *Samedo* informs us) their Wine of Grapes, but of Barley; and, in the Northern parts, of Rice, where they make it also of Apples: but in the Southern parts, of Rice onely; yet not of ordinary Rice, but of a certain kind peculiar to them, which serves onely to make this Liquor, being used in diverse manners.

By 'wine', Boyle naturally meant *chiu*. But strictly speaking, wine is the product of the fermentation of grape juice. Thus, to call *chiu* 'wine' is linguistically less than satisfactory. Since it is made from grains, *chiu* can be said to be more akin to ale or beer than wine. Indeed, we ourselves as well as others have suggested that it be called beer or ale.[3] Yet the processes for making the two products differ so significantly that *chiu* can hardly ever be mistaken for beer or ale. In fact, *chiu* actually resembles wine more than beer in terms of its alcohol content (greater than 10 per cent) and its overall organoleptic character.[4] And so, we are faced with a dilemma. How shall we translate the word *chiu*? Wine or beer? Simply to say 'alcoholic drink' would be too cumbersome. After a good deal of deliberation on the pros and cons, we finally decided to stick to tradition and call it 'wine'. Our reasons are threefold.

The first is gastronomic. *Chiu* is used widely in Chinese cooking and dining in a manner analogous to wine in European cuisines. While beer or ale may also be served at meals, it is rarely seen at formal dinners and banquets. It is seldom used in cookery. Indeed, gastronomically speaking, *chiu* in East Asia is the nearest equivalence of wine in Europe and the Neo-Europes.[5]

[1] In English renditions of the Chinese classics, Sinologists have always translated *chiu* as wine. See, for example, translations of the *Shih Ching* (The Book of Odes) by Legge (1871), Waley (1937) and Karlgren (1950).
[2] Boyle (1664), Pt 2, Sec. 1, p. 88. Italics are in the original.
[3] Temple (1986), pp. 77–8; Anderson (1988), p. 21; Simoons (1991), p. 448.
[4] Godley (1986), pp. 385–6, relates that European travellers to China in the 19th century have often remarked on the similarity between *chiu* and grape wine. Davis (1836), pp. 307–8, 330, declared that the local wine 'which is fermented from rice ... nevertheless resembles some of our weaker white wines in both colour and flavour'. The resemblance is, in fact, sometimes so great that even an expert cannot tell them apart, as shown by the trick that Huc (1855) II, p. 322, played in the 1830s on an unsuspecting Englishman in Macao. He said, 'we took it into our heads to fill some bottles with it [Chinese wine], and having first sealed them with great care, we offered them to an English connoisseur in wine. He tasted and not only found it excellent but discovered that it was the product of some celebrated vintage in Spain.'
[5] Crosby (1986), pp. 2–3, defines Neo-Europes as areas in the world which Europeans have successfully colonised, i.e. North and South America, South Africa, Australia and New Zealand.

The second is religious and ceremonial. *Chiu* 酒 was the drink presented to the gods and ancestors at ritual offerings that we read about so often in the *Shih Ching* (Book of Odes), the *Chou Li* (Rites of the Chou) and the *Li Chi* (Record of Rites). Wine played a similar role in ancient Greece and Rome. It was, and still is, the drink consecrated in the sacred Christian ritual of the Holy Eucharist. And for toasts on formal occasions *chiu* is the preferred drink in China, just as wine is in the West.

The last is aesthetic and sensual. *Chiu* or rather the drinking of it had became so embedded into the aesthetic and sensual experiences of the Chinese that it was often noted in their arts and literature, particularly in their poetry. References to the pleasures (and evils) of the cup can be found in the works of successive generations of poets, from the folksongs in the *Shih Ching*, through the poetry of the great Thang and Sung periods to the present day.[6] In the West, the poetic equivalence of *chiu* is wine from the grape, not beer or ale. We see wine mentioned in the masterpieces of ancient Greece (Homer) and Rome (Virgil), and of England from Shakespeare to Edward Fitzgerald. Take the popular lines from Ben Johnson:[7]

> Drink to me only with thine eyes,
> And I will pledge with mine.
> Or leave a kiss but in the cup
> And I'll not look for wine.

Can one imagine replacing beer (or ale) for wine? Certainly not. So, in terms of its total cultural connotation, we believe *chiu* is best rendered in English as 'wine' and it will be so designated in the remainder of this work. When a distinction between *chiu* and European wine (i.e. grape wine) is necessary we may have to specify it as Chinese wine or revert to the word *chiu*. Now that the issue is settled, we can put it aside and move on to consider how the process for making wine originated in China.

(1) Origin of Wine Fermentations in China

He built a pond filled with wine, and a forest of racks hung with meat, so that they can revel and cavort naked through the night.[8]

This quotation from the *Shih Chi* 史記 (Historical Records), c. −90, describes yet another one of King Chou's (紂) decadent ways that lost him his kingdom to the Chou (周) people in about −1050. There may be a slight exaggeration in the manner in which the wine was allegedly stored and served, but the passage certainly leaves

---

[6] Many poems in the Book of Odes deal with drinking and feasting. Wine drinking is a favourite topic of the great Thang and Sung poets such as Li Pai, Tu Fu and Su Shih.
[7] From *To Celia*, in *The Oxford Book of English Verse, 1250–1918*, edited by Sir Arthur Quiller-Couch, p. 221.
[8] *Shih Chi* (Historical Records), ch. 3, History of Yin.

Fig 40. Vessels excavated from site 14 at Thai-hsi, Kaochhêng, Hopei. *Hopei sheng wen-wu yen-chiu so (1985)*, p. 31, Fig. 20. Scale 25cm.

one with the impression that the preparation of wine was already practised on a large scale by the time of the Shang. The importance of wine in the social and ceremonial life of the Shang has long been represented by the existence of the rich array of bronze drinking and storage vessels that we have discussed earlier on pages 98–100 and Figure 30. More direct evidence for the production of wine was obtained in 1974 when the remains of a winery of the mid-Shang period was excavated at Thai-hsi 台西, near Kao-chhêng 蒿城 in Honan.[9] Among the pottery vessels (Figures 40 and 41) found is a large jar with a wide mouth (*wêng* 瓮), Figure 41a, containing 8.5 kg of a greyish white residue which was analysed by Fang Hsin-Fang and shown to contain cell walls of dead yeast.[10] Another interesting type of vessel is the so-called general's helmet (*chiang chün khuei* 將軍盔), Figure 41b, which is believed to be used as a fermentation vessel.[11] A third type of notable vessel is the pottery

---

[9] Hsing Jung-Chhuan (*1982*), Xing & Tang (1984), *Hopei shêng khao-ku yen-chiu so* (1985), Yün Fêng (*1989*) and Chang Tê-Shui (*1994*). Forty-six pottery vessels were recovered from the site including *wêng* 瓮, *kuan* 罐, *lei* 罍, *tsun* 尊 and *hu* 壺. The main features of the winery are described by Underhill (forthcoming).

[10] *Hopei shêng khao-ku yen-chiu so* (*1985*), p. 204. Fang Hsin-Fang has not published a detailed report of the analysis. According to Yün Fêng (*1989*), p. 261, the material probably represents the residues left from a typical fermentation such as that obtained from the making of yellow wine (*huang chiu* 黃酒). The source of this information is not given.

[11] There is soot on the outside of the 'general's helmet'. It may also have been used as a pot or boiler to cook or steam the grain substrate, cf. Xing & Tang and Yün Fêng above. A similar type of vessel, i.e. a *kang* 缸 with a tapered pointed bottom, was also seen at a site near Chi-nan in Shantung, cf. *Shantung ta-hsüeh li-shih hsi.* (*1995*), pp. 19, 21.

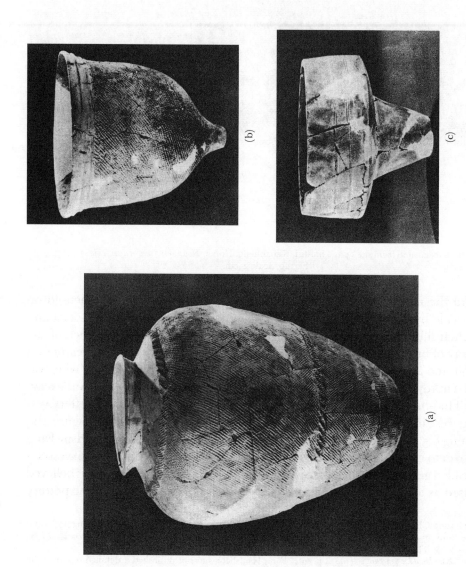

Fig 41. Wine fermentation vessels from Thai-hsi, Kaochhêng, Hopei: (a) Large *wêng* jar containing 8.5 kg residual mash. Pl. 31 (14:58); (b) Vessel resembling 'General's helmet' for heating or fermenting mash. Pl. 30, no. 3; (14:42); (c) Funnel for transferring or filtering fermentation mash. Pl. 33, no. 2 (14:50).

funnel (*lou tou* 漏斗), Figure 41c, which was presumably used for the transfer or filtration of the fermented mash. Pits of peaches, plums and jujubes as well as seeds of hemp and jasmine were found in several urns (*kuan* 罐) suggesting that some kind of fruit wine was being made and that some of the wines were flavoured with herbs. But the principal substrate used in the fermentation appears to be setaria millet, since a large amount of carbonised grains of it was discovered in a storage pit next to the winery during excavations in 1985.[12] Furthermore, in 1980, at a late Shang site at Thien-hu-tshun 天湖村, Lo-shan 羅山 County, Honan, two bronze *yu* 卣 containers with tightly sealed lids were discovered that still had liquid in them.[13] The liquid was shown to contain alcohol.[14]

What were the processes used and when and how were they actually discovered? This is a mystery in the history of food science and technology in China because the traditional process for making Chinese wine (*chiu* 酒) is quite different from that developed for making wine or beer in the ancient Mediterranean world.

The making of wine from grape (and from other fruit juices) needs no inventiveness. It is, in fact, practically unavoidable.[15] Fruit juices contain glucose and fructose which are fermented to alcohol easily by the action of yeasts that are always found, lying in wait as it were, on the skins of the fruits.[16] When the juice is expressed from the fruit, the yeasts would immediately go to work and convert the sugars to alcohol and carbon dioxide. Thus, a wine would be produced spontaneously unless something is done, such as heating the juice, to kill the action of the yeast. There is no mystery as to how the process for making grape wine originated. It happened the first time grapes were pressed and the juice allowed to stand for some length of time. This probably took place in Mesopotamia or Iran sometime before −5000.[17]

The production of an alcoholic drink from a cereal is a much more sophisticated affair. It does not happen naturally in a way that would catch the attention of an observer. The process requires the linking of two separate biochemical steps

---

[12] Yün Fêng (*1989*), p. 262. [13] *Ho-nan shêng Hsin-yang ti-chhü wên kuan hui* (*1986*), p. 164.
[14] Ou Than-Sheng (*1987*) and Li Yang-Sung (*1993*), p. 541. Details of the analyses done at Peking University are not available. This sample is about eight hundred years earlier than the celebrated wine remnants discovered in the Chung Shan tomb of the Warring States period described in Qian, Chen & Ru (1981), p. 43 (Figs. 55 and 56).
[15] This point is made by Tenny L. Davis (1945), p. 25. It is illustrated by the story given below (p. 245) on the ease with which the monkeys in the Huang Shan mountains collect fruits and flowers in a pool and allow them to ferment into wine. But the juice should not be overly exposed to air so that the wine would be oxidised to vinegar (cf. pp. 283–91). A general discussion of the history and scientific basis of the fermentative production of alcoholic beverages is given in Rose (1977*a*), in Rose, ed. (1977), Vol. 1, pp. 1–41. For detailed accounts of the grape wine fermentation see Kunkee & Goswell (1977) and Benda (1982); and on the brewing of beer see MacLeod (1977) and Hilbert (1982).
[16] The basic reaction in a fermentation process is the conversion of a fermentable sugar, normally glucose or fructose, to alcohol and carbon dioxide:

$$C_6H_{12}O_6 \rightarrow 2C_2H_5OH + 2CO_2$$

The reaction is catalysed by a series of enzymes in yeast, usually a strain of *Saccharomyces cerevisiae* which is a common airborne organism and abundant on the skin of fruits.
[17] McGovern *et al.* (1996); cf. also Badler (1995), McGovern & Michel (1995).

(1) saccharification: hydrolysis of the starch in the cereal to fermentable sugars and (2) fermentation: conversion of the sugars by yeasts to ethyl alcohol and carbon dioxide. Since yeast is ubiquitous in the atmosphere, the key step is the hydrolysis of the starch to sugar.[18] One way of doing this is to utilise the amylases in sprouted grains. This is the well-known approach used in the brewing of beer.

Another way of doing this is by utilising the amylases in human saliva. Anyone who has chewed starchy food, such as bread, rice or potato for a short time would notice that it gradually becomes sweet. This fact was evidently known to and used by the primitive peoples of East Asia to prepare alcoholic drinks. Tribal women would chew starchy food until it is thoroughly mixed with saliva and then expel it into a vessel where it would continue to hydrolyse the starch into sugars which would in time be fermented into alcohol by the wild yeasts from the atmosphere. It is said that the Jyomon (繩文) people in Japan in −5th century learned how to ferment wine in this way from the 'non-Chinese people who crossed the East China Sea to Japan from the southern part of China'.[19] Although this method of making wine was also used by the Koreans and Pacific Islanders, there is no record of its having been employed by the Chinese people.[20]

The traditional Chinese process for making wine from grains since the Han Dynasty (−221 to +207) is, however, based on an entirely different principle. The cooked grain is incubated in the presence of water with a microbial culture preparation called *chhü* 麴, which contains a mixture of fungal enzymes, spores and myceliae and yeast cells. There is also no exact English equivalent for *chhü*. It has been translated variously as barm, leaven, yeast, and starter.[21] None is entirely satisfactory but *ferment* is probably is the best we can do since *chhü* contains both enzymes and live organisms. During incubation the enzymes hydrolyse the starch from the grain, the spores germinate, and the myceliae proliferate to produce more amylases. The yeasts grow in number and ferment the sugars generated *in situ* to

---

[18] Starch is a complex polymer of glucose. In the making of beer or Chinese wine from starch the key issue is its hydrolysis to glucose.

$$(C_6H_{10}O_5)_n + nH_2O \rightarrow nC_6H_{12}O_6$$

The glucose liberated is then fermented by yeast to alcohol.

[19] Ueda Seinosuke (*1992*); p. 133; cf. also Shinoda Osamu (*1967*).

[20] *Wei Shu*, ch. 100, Biography of the *Wu Chi* 勿吉, p. 2220; *Sui Shu*, ch. 81, Biography of the *Mo Ho* 靺鞨, p. 1821, Lee & Hyo-gee (1996). Ling Shun-Sheng (1957), describes the use of chewing as a means of brewing alcoholic drinks among the tribal people of Okinawa, Taiwan and other Pacific Islands. McGee (1984), pp. 428–9 relates the account given by Girolamo Benzoni in the 16th century of how Peruvian women in South America chewed corn to make an alcoholic drink.

[21] *Chhü* is translated as leaven in Legge tr. (1885), p. 303 and Li Chiao-Phing *et al.* (1980), p. 427; as yeast in Sun & Sun (1966), p. 289, as *ferment* in Huang & Chao (1945), pp. 24ff., and as starter in Shih Sheng-Han (1958), p. 79. Shih mentions that it has also been translated as barm. In this work we are following Huang & Chao to render *chhü* as 'ferment', but to show that the word has a specific meaning it is printed in italics as *ferment*. A *ferment* may be identified further by various attributes, such as, brick *ferment*, loose *ferment*, wheat *ferment*, rice *ferment*, superior *ferment*, common *ferment*, wine *ferment*, etc. We may also use the term mould *ferment* when it is helpful to emphasise the nature of the organisms that are responsible for the activity of the *ferment*. In other words, *ferment* may be used in the same way as *koji* which is the Japanese pronunciation for 麴. Thus, we say rice *koji*, barley *koji*, soybean *koji* etc.

alcohol. The procedure has been called an Amylomyces or simply Amylo process.[22] How it was discovered is a fascinating problem on which we hope to shed some light later in this chapter.

But before we consider this problem, we need to get an idea about the antiquity of the Chinese amylo process as indicated in the classical record. According to the *Shih Pên* 世本 (Book of Origins), 'I Ti 儀狄 first prepared alcoholic drinks (*chiu lao* 酒醪) at the command of the daughter of the emperor; he achieved drinks with five different flavours. Shao Khang 少康 later prepared wine from millet.'[23] The emperor in question was Ta Yü 大禹, the founder of the legendary Hsia dynasty (c. −2000 to −1520). Shao Khang or Tu Khang 杜康 was the sixth occupant of the Hsia throne. The notion that I Ti and Tu Khang were involved in the invention of wine has been circulated in the pre-Han and Han classics such as the *Chan Kuo Tshê* 戰國策 (Records of the Warring States), Chhin or earlier, and the *Shuo Wên Chieh Tzu* 説文解字 (Analytical Dictionary of Characters), +121.[24] What it implies is that by the time of the founding of the Hsia dynasty (c. −2000) a process for making wine was already known. Thus, the process itself must have been discovered some time before −2000; but how much earlier is difficult to say. The problem has generated extensive speculation and discussion in China since the 1950–60s.[25] The occurrence of pottery prototypes of well-known Shang bronze drinking vessels such as *tsun* 尊, *chia* 斝 and *ho* 盉, and large jars for fermenting and filtering wine, among the artifacts excavated from burial sites of the Lungshan cultural period (c. −2600 to −1900), has led Chang Tzu-Kao and Fang Yang to postulate that the process for making wine from grains was a Lungshanoid discovery.[26] On the other hand Yuan Han-Chhing and Li Yang-Sung held the view that the invention of the fermentation process itself must have taken place long before the appearance of vessels dedicated to the drinking of wine, and surmised that it had occurred during the earlier Yangshao period (c. −5000 to −3000).[27] The results of archaeological excavations in the 1970–80s in the Ta-wên-khou (−4300 to −2600) cultural sphere have tended to support the latter view. Several vessels that can be interpreted as utensils for fermenting and drinking wine were unearthed at burial sites in Shantung.[28] Furthermore, vessels suitable for fermenting and filtering wine were identified among the Yangshao

---

[22] Cf. Lafar (1903), pp. 94–5.   [23] *Shih Pên*, Book of Origins, ch. 1.
[24] *Chan Kuo Tshê*, ch. on *Wei Tshê* (Record of the Wei State); *SWCT*, p. 311 (ch. 14b, p. 15b).
[25] Yüan Han-Chhing (1956), p. 99; Chang Tzu-Kao (1960); Li Yang-Sung (1962); Fang Yang (1964); Lo Chi-Thêng (1978); Tshao Yuan-Yu (1979), p. 233; Chhên Thao-Shêng (1979), p. 29; Tsêng Tsung-Yeh (1986a,b,c); Wang Shu-Ming (1987c); Tsêng Tsung-Yeh (1988), pp. 70–3.
[26] Chang Tzu-Kao (1960) and Fang Yang (1964).
[27] Yüan Han-Chhing (1956) and Li Yang-Sung (1962). As pointed out by Fang Hsin-Fang (1980), p. 141, what we are concerned with here is the origin of the fermentation and not its extensive utilisation. The interval between these two events could have been measured in centuries or even millennia.
[28] Li Chien-Min (1984), Yün Fêng (1987), Wang Shu-Ming (1987a,b,c), and Li Yang-Sung (1993). Results of more recent excavations are reported in Su *et al.* (1989) and *Shantung shêng wen-wu khao-ku yen-chiu so* (1991 & 1995). For additional reports on Neolithic pottery vessels suitable for the fermenting and drinking of wine, and their role in mortuary rituals cf. Anne Underhill (forthcoming).

pottery artifacts recovered in Pan-pho in Shensi.[29] Thus, it would seem likely that the invention of the process for making an alcoholic drink from grain in China is almost as old as the parallel inventions of wine (from grape) and beer in the Near East.

What kind of wine did the Neolithic Chinese actually make during the prehistoric period? Perhaps it would be more accurate to say 'wines', since the earliest records indicate that the Shang and Hsia Chinese, and presumably their ancestors also, prepared at least four types of wines (*chiu* 酒), i.e. *li* 醴, *lo* 酪, *lao* 醪 and *chhang* 鬯.[30] We have already met *li* on p. 97, as one of six drinks mentioned in the *Chou Li* 周禮 (Rites of the Chou). The *Li Chi* 禮記 (Record of Rites) states that it was the preferred drink of the Yin (Shang) people for their sacrificial ceremonies.[31] It is cited often in the Shang oracle bones.[32] We find it noted also in the *Shih Ching* 詩經 (Book of Odes), −1100 to −600, *Chhu Tzhü* 楚詞 (Songs of the South), c. −300, and *Lü Shih Chhun Chhiu* 呂氏春秋 (Master Lu's Spring and Autumn Annals), c. −240.[33] It is usually interpreted as a sweet wine with a low alcohol content. We suspect it was originally made by using a sprouted grain as the saccharifying agent, but by the late Han it was more often prepared by fermenting the must with a regular microbial *ferment* for a short one night incubation.[34] The identity of *lo* is less certain. It is coupled with *li* in two passages in the *Li Chi* and is presumed to be a fermented drink of great antiquity.[35] It may denote a wine made from fruits, grain or milk. The *Ku Shih Khao* 古史考 (Investigation of Ancient History), +250, states, 'In ancient times, there were *li* and *lo*; it was not until after the advent of Yü 禹 that *chiu* 酒 was invented.'[36] This quotation suggests that *li* and *lo* are considerably more ancient than *chiu* itself. We have just encountered *lao* 醪 in the quotation from the *Shih Pên* 世本 (Book of Origins) given in the preceding paragraph. It is simply wine mixed with its lees. A thin alcoholic porridge called *lao* is still being made by the aborigines in Taiwan today. The way it is done may serve as a model to help us understand how the amylo fermentation might have been discovered. *Chhang* 鬯 is a ritual wine made by using a *ferment* prepared with the aid of a plant material. It is mentioned often in the Shang oracle bones and once in the *I Ching* 易經 (Book of Changes) of

---

[29] Wang Chih-Tsun (*1994*). Based on the character of the pottery uncovered, Yün Feng (1987) has pushed the argument further to suggest that wine was already being made during the earlier Phei-li-kang (−6500 to −5500) period.

[30] The identity of *chiu* 酒, *li* 醴, *lo* 酪, *lao* 醪 and *chhang* 鬯 is discussed in detail in Ling Shun-Shêng (*1958*); and briefly by Fang Hsin-Fang (*1980 & 1989*) and Chang Tê-Shui (*1994*). *Chiu* 酒, *li* 醴 and *chhang* 鬯 are seen often in oracle inscriptions, but *chiu* is sometimes written with the water radical on the right of the *yu* 酉 radical, e.g. 酒 as well as on the left.

[31] *Li Chi, Ming Thang Wei* 明堂位 (Hall of Distinction), ch. 14, p. 531; Legge (1885), ch. 12.

[32] Ling Shun-Shêng (*1958*), p. 890, lists 4 occurrences of *li* in oracle bones where it is written as 豊, i.e. without the *yu* 酉 radical. This suggests that *li* was widely used in sacrificial offerings.

[33] *Shih Ching*: M 180, 279 and 290 [W 262, 156 and 157]; Waley translates *li* as 'heavy wine' in W 262, but as 'sweet liquor' in W 156 and 157. *Chhu Tzhu: Ta Chao*; Hawkes (1985), p. 235, translates *Wu li* 吳醴 as the 'must of Wu'. *LSCC*: ch. 1, pt 3, *Chung I* 重己, Lin Phin-Shih ed. (*1985*), p. 23.

[34] *SWCT*, on *li*, p. 312 (ch. 14B, p. 16); *Shih Ming*, ch. 13, p. 8b.

[35] *Li Chi, Li Yun*, ch. 9, 367; *Chi I*, ch. 24, p. 759. The identity of *lo* will be discussed on pp. 250–4.

[36] *Ku Shih Khao*, cited in Ling Shun-Shêng (*1958*), p. 885. The original reads, 古有醴酪，禹時儀狄作酒.

the Chou.³⁷ A *chhang* made from a black panicum millet known as *chü chhang* 秬鬯 is seen in the *Shih Ching*, *Shu Ching* 書經 (Book of Documents), *Chou Li* and *Li Chi*.³⁸ It must, therefore, have been a drink of some importance. The ancient art of making *chhang* is also still alive among aborigines in Taiwan today.³⁹

What kind of wines then were first made by the Neolithic Chinese? Based on the sketchy information shown above we suspect the earliest 'wines' made were *li* and *lo*, the *li* being derived from cooked rice or millet with sprouted grain as the saccharifying agent, and the *lo* from ground fresh fruits mixed with water. In each case wild yeasts in the air would be attracted to and multiply in the sweet solution to convert the sugar to alcohol. But what about the amylo process used to make the standard wine, *chiu* 酒, mentioned earlier? It probably came upon the scene some time later. This new technology was so powerful that it eventually overtook the sprouted grain process to became, by the −2nd century, the dominant feature of the Chinese wine fermentation. The agent of the amylo process is, as we have stated, the *ferment chhü* 麴, Therefore, the key issue we face is this: how was *chhü* discovered? The earliest reference to *chhü* occurs in the *Shu Ching* (Book of Documents), dated before −500, which says:⁴⁰

To make wine or sweet liquor, you need to have *chhü* 麴 *nieh* 櫱
Jo tso chiu li, erh wei chhü nieh. 若作酒醴，爾惟麴櫱。

In this quotation wine is *chiu* 酒 and sweet liquor is *li* 醴. We have purposely left *chhü nieh* untranslated because its meaning has been a matter of some controversy. In fact, there is no agreement as to whether it was a single entity called *chhü nieh*, or two separate entities, *chhü* and *nieh*. As recently reviewed by Fang Hsin-Fang three ways of interpreting the nature of *chhü nieh* are known in the Chinese literature.⁴¹ They are:

1. *Chhü* 麴 and *nieh* 櫱 are two separate entities: *chhü* is a preparation of *ferments* and *nieh* a sprouted cereal grain. This is the traditional view supported by the Chou classics and recently reiterated by Yüan Han-Chhing.⁴² *Chhü* is mentioned in the *Tso Chuan* (Master Tso's Spring and Autumn Annals), and *nieh* is mentioned in the *Chhu Tzhu* (Songs of the South) and the *Li Chi* (Record of Rites).⁴³ The *Shuo Wên* (Analytical Dictionary of Characters), states that '*nieh* is sprouted grain', and the

---

³⁷ Shang oracle bones: cf. Hu Hou-Hsüan (*1944*) and Lin Shun-Shêng (*1958*), pp. 896–7. *I Ching*: cf. Hexagram 51, *Chên Kua* 震卦; Legge (1899), Causeway edn (1973), p. 173.
³⁸ *Shih Ching*: M 262 (W 136, where it is translated as black mead). *Shu Ching*; *Lo Kao* 洛誥, cf. Chhü Wan-Li ed. (*1969*), p. 128. *Chou Li*: cf. *Yü Jên* 鬱人 and *Chhang Jên* 鬯人, pp. 209, 210. *Li Chi*: ch. 10, *Li Chhi* 禮器 (Utensils of Rites); p. 394.
³⁹ Ling Shun-Shêng (*1958*), pp. 903–4.
⁴⁰ *Shu Ching*: *Shuo Ming Hsia* 說命下. In the three English translations of the *Shu Ching*, *chhü nieh* is rendered as '*yeast* and *malt*' by Legge (1879), p. 260; as '*fermenting* and *saccharine* ingredients' by Medhurst (1846), p. 173; and as 'ferment of sugar' by Old (1904), p. 128.
⁴¹ Fang Hsin-Fang (*1980*), p. 141, who also cites the proposal of Yamasaki Hiyachi (*1945*), p. 20, that *chhü* is cake *ferment* and *nieh* is loose ferment in granules.
⁴² Yüan Han-Chhing (*1956*), p. 81; cf. also Tshao Yuan-Yu (*1979*), p. 231.
⁴³ *Tso Chuan*, *Hsüan Kung* 宣公 12th year. *Chhu Tzhu*, *Ta Chao* 大招, translated as yeast by Hawkes (1985), p. 235. *Li Chi*, *Li Yun* 禮運 (Ceremonial Usages), p. 383.

Shih Ming (Exposition of Names), +2nd century, explains that '*nieh* is obtained by soaking *mai* 麥 (barley or wheat) seeds in water until they sprout'.[44] Presumably, since a sprouted grain by itself can only saccharify the grain and cannot ferment the sugar, it has to be used in conjunction with *chhü*.

2. *Chhü* and *nieh* are two entities: *chhü* is the agent for making *chiu* 酒, and *nieh* that for making *li* 醴. This is the view held by Sung Ying-Hsing. In his *Thien Kung Khai Wu* (The Exploitation of the Works of Nature) of +1637, he said, 'In antiquity, *chhü* was used for making wine (*chiu*), and *nieh* for sweet liquor (*li*). In later times, the manufacture of sweet liquor was discontinued because its flavour was thought to be too weak, and "the art of making *nieh* (*nieh fa* 糵法)" was consequently lost.'[45] In both the English versions of *Thien Kung Khai Wu* now available, *nieh* is translated as malt and *nieh fa* as the art of making *nieh* in accordance with the traditional definitions. But there is a serious problem with this interpretation. Sung is alleged to have said that the art of making *nieh* had been lost (*nieh fa i wang* 糵法亦亡), whereas there is no evidence whatsoever that the Chinese ever ceased to use malted grain to produce malt sugar from antiquity down to the modern age. Malt sugar, *i* 飴 or *thang* 餳, is mentioned in the *Shih Ching* (Book of Odes) and the *Chhu Tzhü* (Songs of the South) indicating that the Chou people knew how to sprout grains and use them to produce malt sugar (maltose).[46] The preparation of *nieh* (sprouted wheat) and *i* (malt sugar) are described in detail in the *Chhi Min Yao Shu* (Important Arts for the People's Welfare), +544, and both products are standard items in the Thang and Sung pharmacopoeias and in the great *Pên Tshao Kang Mu* of Sung's own time.[47] Thus, Sung Ying-Hsing's statement, as traditionally interpreted, simply does not make sense.

Fang Hsin-Fang has suggested that the *nieh* that Sung had in mind was not a sprouted cereal, but rather a weaker type of *ferment*, that is a weak *chhü*, that was developed specifically by the late Neolithic or Shang people for making sweet liquor.[48] When sweet liquor fell from favour, there was no longer any reason to continue making this *nieh*. We feel this explanation is much too contrived. There is, in fact, a more direct and simpler explanation. Sung Ying-Hsing had said '*nieh fa i wang*

---

[44] *SWCT*, p. 147 (ch. 7a, p. 21); *Shih Ming*, ch. 13, p. 8b. Although the *Shih Ming* specifically states that *nieh* is sprouted wheat or barley, *nieh* can mean any kind of sprouted grain, such as sprouted millet, rice, wheat or barley. It is unfortunate that *nieh* has been usually translated as 'malt', since in the brewing literature malt is almost synonymous with sprouted barley.

[45] *TKKW*, ch. 17, p. 285; tr. adapted auct. from Sun & Sun (1966), p. 289 and Li *et al*. (1980), pp. 427–8. In each case *nieh* is translated as malt. While in English the word 'malt' means a sprouted grain in general, it is often taken to denote barley malt in particular, especially when one talks about the brewing of beer. Because of this ambiguity, in the present discussion we have refrained from using the word 'malt' as far as possible. When it is used in connection with beer brewing, it would mean a sprouted grain of wheat or barley.

[46] For *i*, see *Shih Ching*, M 237, line 13 (translated by Waley in 240 as rice-cake). For *thang* see *Chhu Tzhü*, *Chao Hung*, translated by Hawkes (1985), p. 228, correctly as malt sugar.

[47] The *PTKM*, ch. 25, p. 1548, lists sprouted panicum millet, setaria millet, rice, wheat (or barley) and beans. Cf. Smith & Stuart (1973), p. 256. 'The grain is moistened and left to sprout and when the process has gone on for a sufficient length of time, it is dried in the sun. The sprouts and husk are rubbed off and the grain is ground into flour for making into cakes or bread.'

[48] Fang Hsin-Fang (*1980*), p. 140.

蘖法亦亡', which is taken to mean 'the art of *making* sprouted grain was also lost'. As we have shown, this statement is quite incorrect. However, *nieh fa i wang* can also be interpreted as 'the art of *using* sprouted grain was lost'.[49] Sprouted grain, *nieh*, was presumably discovered when cereal seeds were accidentally stored under damp, wet conditions. It was soon found to have the ability of liquifying cooked grain and producing a sweet tasting liquor. The early sprouted grain was, however, most likely contaminated with yeast (and other microorganisms). The yeast would, if the right conditions prevailed, convert the sweet material into an alcoholic liquid. So the prehistoric sprouted grain could be used for making *li* 醴 all by itself. In the *Lü Shih Chhun Chhiu* (Master Lu's Spring and Autumn Annals), −239, there is a reference to a clear wine, *i* 酏 and sweet liquor, *li* 醴. Kao Hsiu's commentary tells us that *li* was made from *nieh* and *shu* 黍, that is sprouted grain and [steamed] panicum millet, without the help of *chhü*.[50] Since *li* was supposed to result from just a short fermentation, much of the sugar would remain intact. Thus, *li* would be sweet. But as the process for making sprouted grain improved with the availability of cleaner grains and more sanitary conditions, the level of fermentative organisms in the product declined sharply, until eventually the sprouted grain that was made would have had hardly any fermentative activity. The *li* that could be made from it would, of course, be so weak as to be no longer acceptable to the drinking public. With a short incubation there would not have been time for wild yeasts to contaminate the solution and start a meaningful fermentation. Thus, this problem was thought to exist only because later scholars misread Sung's statement. There is really no discrepancy between the traditional view and Sung Ying-Hsing's view.

3. *Chhü nieh* was a single entity in late prehistoric China, that is before the advent of the Hsia dynasty. It was the term used for a *chhü* prepared from sprouted grain; that is to say, it was sprouted grain that had turned mouldy as well as yeasty.[51] The *chhü* in *chhü nieh* is an adjective and the two characters together would mean a 'fermenting' type of sprouted grain. It was, in fact, simply the yeasty *nieh* postulated in (2) further contaminated with moulds. But by the start of the Chou (c. −1100) *chhü nieh* had developed into two entities, *chhü*, the mould *ferment*, and *nieh*, the sprouted grain. The term *chhü nieh* was still used in the sense of a combined amylolytic and fermentative agent in the *Chhi Min Yao Shu* (Important Arts for the People's Welfare) of +544. In describing the process for making wine (chapter 64) from *Ho Tung chhü* 河東麴, it states that 'cooked grain was added at intervals in amounts commensurate with the *chhü nieh* 麴蘖 (amylo-fermentative activity) in the fermentor'.[52]

This hypothesis, while interesting, is far from convincing. There is no evidence for the existence for the entity *chhü nieh* in the Shang oracle bones. All the available

---

[49] This is also the view of Chhên Thao-Shêng (*1979*), p. 38.
[50] *LSCC*, ch. 1, p. 23; cf. 1989 edn Shanghai Classics Press, p. 14 for Kao's commentary.
[51] Fang Hsin-Fang (*1980*), p. 140.
[52] *CMYS*, ch. 64, p. 365. *Ho Tung* is the name of a prefecture in southern Shansi. Chapter 129, Economic Affairs, p. 3274, of the *Shih Chi* (Historical Record), has a reference to *nieh chhü* 蘖麴 which Shih Shêng-Han (1957) has interpreted as sprouted grain and *ferment*, two entities, but Chhên Thao-Shêng (*1979*), p. 35, has regarded it simply as another expression for amylo-fermentative activity.

Chou documentation indicates that *chhü* and *nieh* were two separate entities. Indeed, Fang himself concedes that *chhü nieh* had already split into *chhü* and *nieh* at the start of the Chou. The two soon developed different functions; *chhü*, ferment, was used for making wine, and *nieh*, sprouted grain, for making malt sugar, *i* 飴, the major sweetener of ancient China.⁵³ The existence of *nieh* (sprouted grain) and *i* (malt sugar) is well documented in the Chou classics. There is evidence that during the Chou *nieh* was also used for making wine without the aid of *chhü*. The *Li Chi* states that 'propriety (*li* 禮), is to people, what sprouted grain (*nieh* 櫱), is to wine (*chiu* 酒)'⁵⁴ and the *Kuan Tzu* (c. −4th century) says that '*nieh* 櫱 may be used to make wine'.⁵⁵ Both quotations imply that sprouted grain by itself could be used to ferment a wine.⁵⁶ There is no need to invoke a new entity *chhü nieh* to explain the statement about wine making quoted earlier from the Book of Documents, that is:

Jo cho chiu li, erh wei chhü nieh 若作酒醴，爾惟麴櫱.

If *chhü* and *nieh* are two different entities as we are inclined to believe, the quotation can be interpreted in two ways:

(1) 'To make wine (*chiu*) we need a *ferment* (*chhü*),
   To make sweet liquor (*li*) we need a sprouted grain (*nieh*),'
(2) 'To make wine or sweet liquor (*chiu* or *li*), we need *ferment* and sprouted grain (*chhü* and *nieh*).'

Indeed, as we shall soon see, both *chhü* and *nieh* were at times used together to make wine during the late Chou. The main difference between *chiu* (wine) and *li* (sweet wine) was in the length of fermentation. Yet the concept of *chhü* and *nieh* acting in concert had become so embedded in the consciousness of the writers that *chhü nieh*, as a figurative term denoting an amylo-fermentative agent or activity, persisted in the literature long after the acceptance of *chhü* and *nieh* as two different entities.⁵⁷ In fact, as late as the close of the Ming dynasty, Sung Ying-Hsing entitled chapter 17 of the *Thien Kung Khai Wu*, which deals with the subject of *ferments*, as *Chhü Nieh* (麴櫱) even though a sprouted grain is not mentioned anywhere in the text.⁵⁸

If the prehistoric or Neolithic sprouted grain (*nieh*) had usually been contaminated with yeasts, then the earliest alcoholic drink (*li* 醴) made by the Neolithic

---

⁵³ The making of malt sugar is described in Chapter (e), pp. 457–60 below.
⁵⁴ *Li Chi, Li Yün* 禮運 (Conveyance of Rites), ch. 10, p. 383. The original states, 故禮之於人也 • 猶酒之有櫱也.
⁵⁵ *Kuan Tzu*, ch. 17, p. 165, 以魚為牲，以櫱為酒. Literally, this means 'to use fish as animal sacrifice, and sprouted grain as wine.' The term *nieh chiu* 櫱酒 (wine from sprouted grain) is also seen in the *Chhien Han Shu* (History of the Former Han), c. +100, cf. note 77 below.
⁵⁶ For reasons to be explained later on pp. 278–9, the 'wine' prepared by using sprouted grain as the saccharifying agent would have to be a *li* 醴, i.e. a wine with an alcoholic content no higher than that of beer (i.e. less than 3%).
⁵⁷ The expression *chhü nieh* is so used in the *Nan Fang Tshao Mu Chuang* (+304), *Thou Huang Tsa Lu* (+835), and *Thai-Phing Huan Yu Chi* (c. +980) as mentioned on p. 183; in poems by Pai Chü-I (Thang) and Su Tung-Pho (Northern Sung) as noted by Shinoda (1973), p. 328 and p. 298; and also in the *Pei Shan Chiu Ching* (+1117), p. 1.
⁵⁸ *TKKW*, ch. 17, p. 285. In this context *chhü nieh* is translated as 'yeast' by Sun & Sun (1966), p. 289 and as 'Leaven and Malt' by Li *et al.* (1980), p. 427.

Chinese would have been a type of ale or beer. Was *chhü* then derived from yeasty *nieh*? Maybe. But the traditional view holds that *chhü* had originated in another way. We have mentioned earlier (p. 156) that one of the ancient alcoholic drinks of China is called *lao* 醪. In fact, the *Shih Pên* (Book of Origins) mentions *chiu* and *lao* as the alcoholic drinks prepared by the legendary inventors I Ti and Tu Khang.[59] *Lao* and *li* were also the two alcoholic drinks associated with soups and decoctions recorded in the *Huang Ti Nei Ching, Su Wên* (The Yellow Emperor's Book of Corporeal Medicine, Questions and Answers).[60] *Lao* is usually interpreted as wine mixed with its lees. How then was wine (*chiu* or *lao*) first prepared? Chiang Thung 江統 of the Tsin dynasty (c. +300) in his *Chiu Kao* 酒誥 (Wine Edict) had this to say:[61]

> Wine (*chiu* 酒) was developed at the command of ancient kings. Legend says it was through the effort of princess I Ti or prince Tu Khang. But most likely, left-over cooked grain (*fan* 飯) was left in the open (in Khung-Sang 空桑). Soon it was covered with a verdant (*microbial*) growth, and upon further storage a fragrant liquor ensued. This is how wine originated. There is no mystery to it.

This view was reiterated later by Chu Kung 朱肱, author of the *Pei Shan Chiu Ching* (Wine Canon of North Hill), +1117, who said:[62]

> According to the ancients, when contaminated cooked grain (*fan*) was fermented with [*cooked*] millet or wheat, a rich liquor (*lao* 醪) would result. This is the beginning of wine. The *Shuo Wên* says, 'muddy wine (*chiu pai* 酒白) is called *sou* 醙'. *Sou* means bad (i.e. contaminated with moulds) *fan* 飯; it also means old. When *fan* is old it becomes bad. If the *fan* is not bad, the wine would not be sweet.

The next step would be for the ancient Chinese to find out that the bad (i.e. mouldy) *fan* could be dried and stored without losing its amylo-ative activity. Thus, *chhü*, the mould *ferment*, was born. As we shall soon see, this was basically how *chhü* was made as described in the earliest fermentation recipes in the *Chhi Min Yao Shu*, +544. In fact, the amylo-process of making wine starting from scratch, as it were, is still practised among the aboriginal tribes in Taiwan. As reported by Ling Shun-Shêng, the process begins with the making of a *ferment* inoculum.[63] Left-over cooked millet or rice is spread at the bottom of a basket, covered with another basket, hung on a stand and kept warm. After three to four days the cooked grains are

---

[59] *Shih Pên*, ch. 1. [60] *HTNCSW*, ch. 14; Veith (1972), p. 151.
[61] *Chiu Kao*, which reads as follows: 酒之所興，肇自上皇。或云儀狄，一曰杜康。有飯之盡，委之空桑。郁積成味，久蓄氣芳。本出于此，不由奇方。
[62] *PSCC*, pt 1, p. 4. The original reads: 古語有之，空桑穢飯，醞以稷麥，以成醇醪，酒之始也。說文酒白謂之醙，醙者壞飯也。醙者老也。飯老即壞，飯不壞則酒不甜。
[63] Ling Shun-Shêng (*1958*), p. 895. It is interesting to compare the making of wine from cooked rice *de novo* by the aborigines in Taiwan with the making of *bouza* beer from bread by the peasants in Egypt and the Sudan. Both are prehistoric procedures that have apparently survived the passage of time and are still in use today. In fact, according to a Japanese local history cited by Kondo (1984), p. 13, this was how Japanese rice *koji* (*ferment*) was discovered. '... once a cask of steamed rice was left uncovered. The owner found to his horror that the rice had moulded. Negligent man that he had already proven himself to be, he failed to dispose of it. Several days later he discovered that the cask of spoiled rice had been transformed into a cask of delicious saké.' This story shows how easily steamed rice can be moulded and fermented if left alone under the proper environmental conditions.

covered with a film of fungal myceliae, ranging from one to three centimetres in length. The best results are obtained when the myceliae are about one centimetre long. If the inoculum is allowed to incubate further it would turn partly black, yellow and red. The wine made from it would be bitter. The next step is to steam a batch of millet or rice, cool the soft granules and mix in the starter culture inoculum. The mixture is placed in a rattan basket lined with banana leaves, and covered with more banana leaves. After three to four days the liquid mash (with the consistency of yoghurt), called *lao* in the native language would be ready for eating as is. But more often than not it is mixed with water in a pottery jar, with banana leaves tightly tied over its opening. After four days the must is filtered through a rattan trough to give a wine, called *sarano-lao*, and dregs called *lakare-no-lao*. The filtered wine is stored and served to guests when the occasion arises. The dregs are mixed with hot water and used as a drink, but according to Ling the flavour is weak and the taste, slightly sour.

In a variation of this procedure, the aborigines also made a *tshao chhü* 草麴, herbal *ferment*, by mixing the leftover cooked grain with extracts of plants, especially leaves from such families as the Rutaceae, Leguminosae, Compositae and Chenopdiaceae. The wine made with this *ferment* would be analogous to the ritual *chhang* 鬯 of ancient China. We will have more to say about herbal ferments later (pp. 183, 185–6).

The ease with which the modern aborigines in Taiwan could prepare an inoculum and make wine from cooked millet or rice suggests strongly that this is probably the way the ancient Chinese discovered the mould *ferment chhü* 麴 and the wine *chiu* 酒 from millet and rice, the principal constituent of their diet. We may say that there were two routes leading to the origin of alcoholic drinks in China: first, through contaminated sprouted grain to make *li* 醴 (sweet liquor), and second, through contaminated *fan* (steamed grain) to make *chiu* 酒 (wine). But the ancient Chinese never learned how to produce a sprouted grain with just the right amount of yeast contamination, and the process eventually died out. On the other hand, they steadily increased their skill in controlling the microbial activity of the *chhü* from cooked grain, and thus, the manufacture of wines of good quality, which had already become a necessary accoutrement of daily living and social intercourse by the time of the early Chou.

The importance of wine in the life of the Chou is shown by the fact that it is mentioned in no less than thirty poems in the *Shih Ching* (Book of Odes). It was already an article of commerce as indicated by the lines:[64]

> When we have got wine we strain it, we!
> When we have none, we buy it, we!
> And taking this opportunity
> Of drinking clear wine.

But Confucius was distrustful of the quality of the wine (and dried meat) that could be bought from the market, indicating that quality control of the wine sold was

---

[64] *Shih Ching*, W 195 (M 165).

poor.[65] Nevertheless, the wine that was produced could have been quite potent, as drunkenness had already become a problem. In another poem in the *Shih Ching*, one proper lady groaned:[66]

> When guests are drunk
> They howl and bawl,
> Upset my baskets and dishes,
> Cut capers, lilt and lurch.
> For when people are drunk
> They do not know what blunders they commit.

Unfortunately, there is little information in the Chou classics about how the wine was actually made. The one and only description we have is from the *Yüeh Ling* 月令 (Monthly Ordinances) chapter of the *Li Chi* (Record of Rites). It states:[67] 'In the second month of Winter orders are given to the Superintendent of Wine to see that the millet and rice are complete, the *ferment* and sprouted grain in season, the soaking and heating cleanly conducted, the water fragrant, the pottery vessels in good condition; and the fire properly regulated.' These are the six requisites for making good wine. Although the statements do give us an idea of the principles underlying the process, they tell us next to nothing about how it was actually carried out, for example, how much materials were used, how they were prepared, how long the fermentation lasted, what temperature the vessels were held at, etc. To get a better understanding of the ancient procedure Wang Chin has compared it with the one used in the early part of the 20th century for the making of the famous Shao Hsing 紹興 wine of Chekiang.[68] The results are tabulated in Table 17. In a superficial way one can say that the two processes are fairly similar, except that in addition to the *ferment chhü* the Chou process also used a sprouted grain (*nieh* 糵) and the Shao Hsing process, a medicated *ferment* called *chiu yao* 酒藥 (wine medicament). We do not know the identity of the sprouted grain employed in the Chou process.[69] The cooked grains, *ferment*, sprouted grain and water were presumably all incubated together. The sprouted grain could give a head start to the liberation of sugar so that the yeast present could begin proliferating and fermenting before the arrival of fresh fungal enzymes in quantity. In the Shao Hsing process, the *chiu yao* (wine medicament) is preincubated with substrates in a separate step to give a rich yeast inoculum, which is then mixed with cooked grain and *chhü* in the main fermentor. This preincubation step is not shown in Table 17.

---

[65] *Lun Yü*, tr. Legge (1893), pp. 222–3. The original reads, 沽酒市脯不食.   [66] *Shih Ching*. w 267 (M 220).

[67] *Li Chi*, p. 300, tr. auct. adjv. Legge (1885), p. 303. In this passage, *chhü nieh* is translated by Legge as leaven cakes, and interpreted by Wang Meng-Ou (*1984*) as yeast.

[68] Wang Chin (*1921*), p. 274.

[69] Nowhere in the *Li Chi* or in any relevant work in the pre-Han literature is there any indication of how *chhü* (*ferment*) or *nieh* (sprouted grain) were prepared. In the case of *nieh* i.e. sprouted grain, we would have to assume that in ancient China it could have been obtained by sprouting any of the cereal grains cultivated, namely, the millets, wheat or barley, or rice. We do not know what kinds of wine were prepared from sprouted millets or rice; but it is interesting to note that sprouted rice was used to ferment wine in early Japan and that efforts have been made recently to reproduce this ancient drink in Kumamoto Prefecture, cf. Ueda (*1995*) and Teramoto *et al.* (*1995*). We thank Sheldon Aronson for drawing our attention to the recent papers by Ueda and his coworkers.

Table 17. *Fermentation of a Chou wine and modern Shao Hsing wine*

Component	Modern Shao Hsing process	Ancient Chou process
Timing	8th moon, start *ferment*	9th moon, start *ferment*
Raw Materials	Glutinous rice, *ferment*, wine medicament, clear spring water	Millet or rice, *ferment*, sprouted grain, fragrant water
Vessel	Fermentor vats	Pottery jars
Ingredients in fermentor	Steamed rice, *fan* 飯, *ferment* (*chhü* 麴), mother of wine, *chiu mu*[a] 酒母, water	Steamed grain, *fan* 飯, *ferment* (*chhü* 麴), sprouted grain, *nieh* 糱, water
Temperature control	Mixing rakes	Regulate fire
Contamination Prevention	Pasteurisation	Clean environment

[a] Prepared by incubating *ferment*, wine medicament and substrate until a culture rich in yeasts is obtained.

We can learn a little more about the Chou process by examining the names of the wines that were produced. The *Chou Li* (Rites of the Chou), lists a *Chiu Chêng* 酒正 (Wine Superintendent) who controls the supply of wines and a *Chiu Jên* 酒人 (Chief Wine Master) who is responsible for the production of wines. From the duties of these officials we can identify five *chhi* 齊 wines and three *chiu* 酒 wines, that were required by the royal household for sacrificial offerings, domestic consumption and entertainment. Actually, the *chhi* wines were not finished products but rather intermediates in the fermentation process. But they were apparently the appropriate and necessary drinks for certain specific ceremonial occasions. Wang Chin has compared the characters of the five *chhi* wines and three *chiu* wines, as interpreted by the Han commentators, with those of their modern counterparts in the Shao Hsing process, and his tabulation is presented in Table 18. It suggests that the Chou vintners were quite sophisticated in recognising the different stages the fermentation had to undergo before a finished wine was obtained. *Li* 醴 was drawn from the fermentation as 'sweet liquor' for ritual purposes and 'clear wine' was obtained presumably by decantation or filtration.

The *Yüeh Ling* (Monthly Ordinances) chapter of the *Li Chi* also tells us that in the first month of summer 'the son of Heaven entertains (his ministers and princes) with strong drink (*chou* 酎)'. According to Han and Thang commentators *chou* is wine obtained by fermenting three successive batches of cooked grain in the same medium, and should have a high content of alcohol.[70] The statement suggests that the technique of adding fresh cooked grain to the fermentation when the original substrate had become exhausted was already an established part of the Chou wine

---

[70] *Li Chi*, p. 274; Legge tr. (1885), p. 271. The interpretation was given by Chêng Hsüan (+127–200) and Yen Shih-Ku (+581–645). The successive addition of substrate permits the build-up of alcohol to levels higher than those achieved in the fermentation of grape juice. This aspect of the technology is discussed further on p. 279.

Table 18. *Names of wines produced by the Chou Chinese*

Name	Usage	Han commentary	Modern counterpart	Step in process
Yüan chiu 元酒	Grand Ritual	Plain water	Water	Raw material
Five chhi 五齊				
1. *Fa chhi* 乏齊	Ritual	Gas and solids rise to surface	Saccharification and fermentation begin	Step 1
2. *Li chhi* 醴齊	Ritual	Liquor thickens and seethes	Saccharification and fermentation continue	Step 2
3. *Ang chhi* 盎齊	Ritual	Liquor seethes, flavour intensifies	Fermentation continues	Step 3
4. *Thi chhi* 緹齊	Ritual	Liquor seethes, turns red	Fermentation continues	Step 4
5. *Chhên chhi* 沉齊	Ritual	Lees settle	Fermentation ends, dregs settle	Last step
Three chiu 三酒				
1. *Shih chiu* 事酒	Drink	Decant as soon as dregs well settled	New wine	New wine
2. *Hsi chiu* 昔酒	Drink	Stored wine	Old wine	Old wine
3. *Chhing chiu* 清酒	Drink	Clear wine	Old, stored	Old wine

technology. It was also during the Chou reign that the art of making the wine *ferment*, *chhü* 麴, was stabilised. The *Yüeh Ling* says that in the last month of spring, 'the son of Heaven presents robes yellow as the leaves of the mulberry tree to the divine ruler'.[71] The actual words used here to denote the yellow robe is *chü i* 鞠衣. The word *chü* used here is also written as *chü* 蘜 in the *Erh Ya* (Literary Expositor), c. −300; and *chhü* 麴 is written as 𮕵 (possibly a variant of 蘜) in the *Shuo Wên*. (Analytical Dictionary of Characters), +121.[72] These relationships suggest that the colour referred to is derived from the colour of the dust (spores) from the culture(s) in the *ferment*, probably those of *Aspergillus oryzae*. The predominance of *A. oryzae* would suggest that the *ferment* was in the form of loose granules.[73]

Records of both *ferments* and wines are found among the burial remains unearthed in Han Tomb No. 1 at Ma-wang-tui. The *MWT Inventory* lists two sacks of *ferment* (*chhü* 麴) and eight jars (*tzu* 資) of wine, of which two are *pai chiu* 白酒 (mature, muddy

---

[71] *Li Chi*, p. 267; Legge tr. (1885), p. 263. The *chhü i* is also mentioned in the *Chou Li*, *Nei Shih Fu* 內司服 (Internal Wardrobe Mistress), Lin Yin (1985), p. 81.
[72] *Erh Ya*, ch. 13, *Shih Tshao* 釋草, p. 26b; *SWCT*, pp. 20 (ch. 1b, p. 12b) and 147 (ch. 7a, p. 21b).
[73] Fang Hsin-Fang (*1980*), p. 143.

wine), two, *yün chiu* 醞酒 (wine fermented with multiple batches of substrate), two, *lei chiu* 肋酒 (filtered wine) and two, *mi chiu* 米酒 (perhaps *li* 醴, light sweet wine).[74] The list also includes eight lacquer vases said to contain *pai chiu*, *yün chiu* and *mi chiu*.[75] Unfortunately, most of the sacks found in the tomb had disintegrated during the long years of interment and remains of the *ferments* could not be identified, but from what is left in some of the pottery jars and lacquer vessels one can readily infer that a liquid such as wine must have originally been stored in these containers.[76] In addition, seven *chi* 卮 goblets and forty wine cups (*erh pei* 耳杯) were found among the lacquer wares buried in the tomb. These artifacts leave little doubt that wine played an important role in the life of the nobility during the early years of the Han dynasty.

Sprouted grain was still used to make *li* (sweet liquor), i.e. *nieh chiu* 糵酒, during the Western Han. The *Chhien Han Shu* (History of the Former Han Dynasty, −206 to +24) notes that sprouted grain kernels (*mi nieh* 米糵) and silk were sent as gifts to the northern pastoral Hsiung Nu 匈奴 tribes who were in a continuous state of contention with the Han Empire.[77] On one occasion, in a proposal to marry a Han court lady, the Hsiung Nu chieftain demanded, among other presents, an annual dowry of 'ten thousand piculs of sprouted grain wine (*nieh chiu* 糵酒)'.[78] It was, in fact, during the late Han that the process was transmitted, probably through Korea, to Japan, where it soon displaced the chewing method of wine making that had been in operation since the −5th century.[79] From the +2nd to +4th centuries a Japanese sweet liquor (醴 *lai* or *rai*) was prepared by mixing sprouted rice with steamed rice. Apparently the sprouted rice (糵 *getsu*) in use was also contaminated with yeasts just as the early *nieh* was in prehistoric China. Recently, Ueda and his collaborators have duplicated the *lai* fermentation using sprouted rice as the saccharifying agent.[80] He is of the opinion that the Japanese *ferment* (麴 *koji*) was derived from the accidental addition of moulded steamed rice to the sprouted rice (*getsu*). This could have happened when *getsu* was used to treat steamed rice that had been inadvertently put aside and started to become mouldy.[81] The result was so outstanding that the brewers realised that there was something special about

---

[74] *Hu-nan shêng po-wu-kuan (1973)* 1, p. 139; bamboo slips 108–11. The report considers *mi chiu* possibly as *li*, but the evidence cited is by no means convincing.

[75] *Ibid.*, p. 143, bamboo slips 168–77.

[76] *Ibid.*, lacquer, p. 95, pottery, p. 127. See also Thang Lan *(1980)*, pp. 22, 29–30.

[77] *Chhien Han Shu*, *Hsiung Nu Chuan* (Biography of the Hsiung Nu), cf. p. 3760 on *mi nieh* 米糵 (sprouted grain kernels) and p. 3763 on *shu nieh* 秫糵 (sprouted millet).

[78] *Ibid.*, p. 3780. In addition the letter demands an annual gift of five thousand *hu* (each *hu* is about 35 litres) of setaria millet kernels and ten thousand bolts of silk. The original reads, 取漢女為妻，歲給遺我糵酒萬石，稷米五千斛，雜繒萬匹.

[79] Ueda Seinosuke *(1992)*, pp. 134–5.

[80] Teramoto *et al.* (1993). They had to add modern yeast in order to complete the fermentation.

[81] Shinoda Osamu *(1967)*, presumably influenced by the theory that rice cultivation originated in the piedmont Southeast Asia, is of the opinion that the use of moulded rice originated in Indo-China and was transmitted via the Yangtze Valley and East China to Japan. In the light of archaeological discoveries in the 1980s and 90s, which show that rice was domesticated in the Yangtze valley as early as 11,500 BP, such a view is no longer tenable. But a South China contribution to the development of the mould *ferment* as we know it today is highly likely.

Table 19. *Types of* ferments *made in Han China*[a]

Name	Made from	Shape and form	Surface feature
Mou 麰	Barley		
Chia 䴷	Wheat		
Kuo 麨	Wheat		
Phi 麱	Wheat?	Cake	
Hua 䴭	Wheat?	Cake	
Chhai 䴳	Wheat?	Cake	
Khu 麯	Wheat?	Cake	
Chhü 麴		Granular	
Mêng 䴷			With mycelial coat
Wu I Chhu 無衣麴			Without mycelial coat

[a] Based on Fang Hsin-Fang (*1980*), *Fang Yen* and *Kuang Yün*.

the steamed rice. Continuing experimentation resulted in the emergence of rice *koji* as the primary agent used in the brewing of Japanese wine (*saké* 酒) even after the advent of modern microbiology in the nineteenth century.[82]

Further progress in the making of mould *ferment* preparations during the Han dynasty can be gauged from the names for different kinds of *chhü* 麴 listed in dictionaries such as the *Shuo Wên*, *Shih Ming*, and *Fang Yen* (Dictionary of Local Expressions), c. −15. Fang Hsin-Fang has classified them according to three criteria (a) raw materials used for their preparation (b) physical state of the products and (c) amount of mycelial growth on the surface.[83] The results are shown in Table 19.

It is interesting to note that practically all the names contain a *mai* 麥 radical, indicating the importance of wheat or barley as the major raw material for making *ferments* since the Han. Wheat is the chief ingredient in most of the *ferments* used in making Chinese wines today. A second point of interest is the emergence of *pin chhü* 餅麴 (cake or brick) which has, since the Han, become the preferred form of *ferments* in the Chinese wine process. Conditions in the interior of the cake tend to favour the growth of *Rhizopus* species, while those on the surface, of *Aspergillus* species. This development is a milestone in the evolution of the Chinese wine fermentation. It explains why in most Chinese processes *Rhizopus* species have remained the predominant fungi responsible for the amylolytic activity of the culture and have provided the wines with their inimitable, characteristic flavours.

Unfortunately, there is no information available on how the different types of *ferments* were prepared. Nor is there any information on the details of the wine process as practised during the Han. But some scattered semi-quantitative data on the process are available. The *Chhien Han Shu* (History of the Former Han), +100, tells us

---

[82] A technical description of the modern Japanese *saké* brewing process is given in Kodama & Yoshizawa (1977). For a popular account of *saké* brewing and culture, see the highly readable and beautifully illustrated book by Hiroshi Kondo (1984).
[83] Fang Hsin-Fang (*1980*), p. 144.

that, 'from 2 *hu* 斛 (20 *tou* 斗) of crude grain, 1 *hu* (10 *tou*) of *ferment*, 6 *hu* and 6 *tou* (i.e. 66 *tou*) of finished wine are obtained'.[84] In other words, 30 volumes of grain solids (crude grain plus) yielded 66 volumes of wine. The large amount of *ferments* added served as a source of enzymes as well as an inoculum of fungi and yeast. In the *Hou Han Shu* (History of the Later Han Dynasty), +450, a commentary states that 'one *tou* 斗 of rice grain gives one *tou* of wine of superior quality; one *tou* of setaria millet gives one *tou* of wine of medium quality; one *tou* of panicum millet gives one *tou* of wine of ordinary quality'.[85] Towards the end of the Later Han, Tshao Tshao, founder of the Wei Kingdom, described a Nine Stage Fermentation Process for wine. 'Use 30 katties (*chin* 斤) of *ferments* and 5 *shih* 石 (or *hu* 斛) of flowing water. Soak the *ferments* early in the twelfth month, and begin the fermentation with the first *hu* of rice (*tao mi* 稻米) in the first month. Add a second *hu* of (cooked) rice after three days, and continue the procedure every three days until nine *hu*'s of rice are used up ... If the product tastes bitter, a tenth batch of rice [may be added]. The wine will then be sweet and more pleasant to drink.'[86] The last statement indicates that after the ninth stage of the fermentation, the alcohol content must have reached a level inhibitory to the action of the yeast, although not to the action of the fungal amylases. The tenth addition of grain generated more sugars, but the yeast was no longer able to convert them to alcohol. In this case, the amount of *ferment*, 30 catties, is less than 5 per cent of the total grain added. It must have functioned mainly as an inoculum of microorganisms. From these reports we may surmise that at least some of the wines made during the Han period involved the fermentation of successive batches of substrate in the broth and that their alcoholic content was comparable to that of the contemporary grape wines of the West.[87]

(2) Preparation of *Ferments* and Wines in Early Mediaeval China

Wine is the gift of the gods. The Ruler uses it to nourish all under Heaven, offer sacrifice, pray for prosperity, support the weak, and heal the sick. For the hundred blessings to occur, wine is indispensable.[1]

This remarkably romantic and incredibly naive view of the benefits of wine is testimony to the extraordinary regard with which it was held by scholars of the Han dynasty. Contrast this view with that of Pliny the Elder who, half-way across the world, bemoaned the fact that 'so much toil and labour and outlay is paid to the price of a thing that perverts men's minds and produces madness'.[2] In fact, in ancient Rome, women were not allowed to drink wine. Infringement of this rule

---

[84] *Chhien Han Shu, Shih Huo Chi* 食貨志 (Economic Affairs Section), ch. 24, p. 1182. Cf. Also Swann (1950), pp. 345–6.
[85] *Hou Han Shu*, ch. 22, *Liu Lung Chuan* 劉隆傳, p. 781.   [86] Quoted in *CMYS*, ch. 66, p. 393.
[87] For a discussion of wine yields during the Han, cf. Yü & Chang (*1980*).
[1] *Han Shu*, ch. 24, *Economic Affairs*, p. 1182; tr. auct. adjv. Swann (1950), p. 344.
[2] Pliny, *Natural History*, tr. Rackham, Book XIV, p. 277.

was punishable by death.³ Furthermore, Pliny tells us that according to Cato, 'the reason why women are kissed by their male relatives is to know whether they smell of "tipple" – that was then the word denoting wine, and also the word "tipsy" comes from it.'⁴ Who would have thought that such a chivalrous gesture could have had so demeaning an origin?

Other quotations from the Han classics reinforce the notion that both the *chhü* 麴 *ferment* and the *chiu* 酒 wine were important commodities in the economy of the realm.⁵ Yet we know very little about how either was prepared other than the few fragments of information presented in the previous subchapter. The popularity of wine continued to grow from the end of the Han (+220), through the turmoil of the Three Kingdoms, the Chin, and through the Northern and Southern dynasties, but little of interest that might have been added to the literature of wine technology in the intervening three hundred years has survived to our day. Fortunately, at the end of this period, in the year +544, a most significant event in the history of food science and technology in China occurred in the appearance of the *Chhi Min Yao Shu* 齊民要術 (Essential Arts for the People's Welfare), the first of an estimable series of comprehensive treatises on agriculture that would be compiled and updated for a number of the dynasties that follow (see Chapter (*b*)).

Four chapters (64–7) of the *Chhi Min Yao Shu* deal with the preparation of *ferments* and wines. The author, Chia Ssu-Hsieh, describes in detail the methods for making nine varieties of *ferments*, and records the brewing of thirty-seven kinds of wines.⁶ The nine *ferment* varieties are listed in Table 20. Actually there are only four major types, the *shên chhü* 神麴, superior *ferment*, the *pai lao chhü* 白醪麴, moderate *ferment*, the *pên chhü* 笨麴, common *ferment* which includes the *i chhü* 頤麴, common *ferment* for autumn wine, and the *pai tuo chhü* 白墮麴, a special cake *ferment*. Some differences between them are immediately apparent from Table 20. Wheat is used for all the *ferments* except the last one, No. 9, which is made from millet.⁷ But even for the wheat-based *ferments*, the way the raw material is treated varies considerably with the type of *ferment* to be prepared. For the superior and moderate *ferments*, the wheat is divided into three parts; one part is roasted (actually stir-fried), one steamed and one kept raw. For the common *ferment*, all the wheat is roasted. They also differ in the shape in which they are made. The superior and moderate *ferments* are round cakes while the common *ferments*, square cakes or bricks. To illustrate the process for making the *ferments*, we will now quote in full the text from the *Chhi Min Yao Shu* for the preparation of two representative types of *ferments*. The first example is the first

---

³ *Ibid.*, p. 247, line 89, tells us that 'the wife of Maetennus was clubbed to death by her husband for drinking wine from the vat, and that Romulus acquitted him on the charge of murder.'
⁴ *Ibid.*, p. 247, line 90.
⁵ Chang Mêng-Lun (*1988*), ch. 8. The evils associated with overindulgence in wine are described in ch. 9. See also Chhên Thao-Shêng (*1979*), p. 32.
⁶ Plus two medicated wines prepared by soaking herbs in finished wine; cf. Miao Chhi-Yü (*1982*), p. 393, with *Acanthopanax spinosa* Miq., and p. 395 with the *shen* 檣 plant, the identity of which is uncertain.
⁷ Huang Tzu-Chhing & Chao Yün-Tshung (*1945*), p. 36, in Ferment Method 4, render *hsiao mai* 小麦 (wheat) as rye. No reason is given for this curious translation.

Table 20. *Methods for preparing* ferments *in the* Chhi Min Yao Shu

No. *ferment*	Substrate steam:roast:raw	Medicament	Dimensions cm (diameter/height)
1. Superior	wheat 1:1:1	none	5.5/2.0
2. Superior	wheat 1:1:1	none	11.0/3.3
3. Superior	wheat 1:1:1	none	6.6/4.4
4. Superior	wheat 1:1:1	yes[d]	unknown
5. *Ho-Tung*[a] superior	wheat 6:3:1	yes[e]	unknown
6. *Pai Lao*[b] medium	wheat 1:1:1	yes[d]	11.0/2.2
7. Common, spring	wheat 0:1:0	yes[f]	(22×22×4.4)[g]
8. Common, autumn	wheat 0:1:0	yes[f]	unknown
9. Common, *Pai Tuo*[c]	setaria millet 2:0:1	yes[e]	square cakes

[a] 河東, is in southwestern Shansi.
[b] 白醪, the preferred *ferment* for making wine from glutinous rice.
[c] 白墮, named after Liu Pai-Tuo 劉白墮, a famous wine maker of the Northern Wei.
[d] Cocklebur leaves?
[e] Decoction of mulberry, cocklebur, artemisia and pepper (probably *Zanthoxylum ailanthoides*) leaves in a ratio of 5:1:1:1.
[f] Cakes are laid on leaves of artemisia for incubation.
[g] Based on 1ft 尺 equals 22 cm and 1 in 寸 equals 2.2 cm.

of five methods described for preparing a *shên chhü* 神麴, superior *ferment*. It is No. 1 in Table 20.

Preparation of superior ferment *from three hu* 斛 *of wheat*[8]

One *hu* each of steamed (*chêng* 蒸), stir-roasted (*chhao* 炒) and raw wheat (untreated) are prepared.[9] The roasting of the wheat should be stopped at the point when the wheat turns yellow, and care should be taken not to char it. The raw wheat should be selected from the finest kinds. Each of the three preparations of wheat is ground separately to a fine state. After the grinding they are mixed together.

Before sunrise on the first day of the seventh moon (usually August), a boy in dark clothing is sent to draw twenty *hu* of water while facing west.[10] No one else is allowed to touch this water. If there is too much water, the unused portion may be discarded, but it must not be used by anyone. The ground grain and water are mixed to a firm consistency[11] by workers who face west. For the caking of the *ferment*, only young boys are employed and they too must work facing west. No dirty person is to be employed, and the work is done away from the living quarters.

[8] *CMYS*, ch. 64, tr. ibid., mod. auct. For the original text see Miao Chhi-Yü ed. (*1982*), pp. 358–9.
[9] According to Wu Chhêng-Lo (*1957*), p. 58, at the time of the *CMYS* one *hu* i.e. 10 *tou* 斗 should equal 39.6 litres. The meaning of *chêng* is clear, but *chhao* at this time signifies the technique of pan-roasting i.e. to stir in a hot pan or pot until the kernels are roasted, cf. the description in the next example on pp. 171–2. Nowadays *chhao* is used for the well-known technique of stir-frying. When *chhao* 炒 became synonymous with stir-frying will be discussed in a paper under preparation.
[10] The original term for 'West' is *sha ti* 殺地, the place of death. In classical Chinese tradition 'East' is the place of birth, and 'West' the place of death. The same notion was held by the ancient Egyptians. In ancient Luxor the living city was on the East bank and the tombs of the dead Pharoahs were located on the West bank.
[11] According to modern practice the raw *ferment* may contain anywhere from 18% to 24% moisture, cf. Miao Chhi-Yü ed. (*1982*), p. 372.

40 FERMENTATIONS AND FOOD SCIENCE

The raw *ferment* cakes are fabricated in a hut roofed with thatch (straw), not with tiles.[12] The earth must be clean (firmly packed) and free from loose dirt. There must be no damp spots. The floor is divided into squares with footpaths so that four alleys are formed. Figures are sculptured from the raw *ferment* mix. Some of them are setup as '*ferment* kings' and there are five kings. The *ferment* cakes are then placed on the floor in rows side by side along the footpaths. Then a person from the owner's family is chosen as the master of ceremonies. Neither a servant nor a guest may serve as the master. Offerings of wine and dried meat are made to the '*ferment* kings' in this way: The king's hands are moistened and used as bowls to receive gifts of wine, dried meat and boiled pastry.

The master of ceremonies reads the sacrificial incantation three times.[13] Everyone kneels twice to the deities. The wooden door to the hut is then closed and sealed with mud to prevent the ingress of wind. After seven days the door is opened, the *ferment* cakes turned over, and the door again closed and sealed with mud. After another seven days, the cakes are piled up on top of each other and the door sealed to prevent wind from coming in. After a third seven days, the cakes are put in an earthen jar and the mouth covered and sealed with mud. After a fourth seven days, a hole is made in each of the cakes and they are strung together. They are hung in sunlight until dry, and then taken in and stored. Each cake is about two and a half *tshun* 寸 (5.5 cm) in diameter and a nineteenth of a *tshun* (2.0 cm) in thickness.

This example shows that, as might be expected, a considerable amount of mystery and superstition is associated with the procedure. But the author, Chia Ssu-Hsieh, was evidently a true experimentalist and kept an open mind on the efficacy of the traditional religious trappings. In methods 2 and 4 he reports that variations in the procedure just quoted in which some of the superstitious elements such as the fabrication of cakes by male children, the making of footpaths and the offering of sacrifices to the *ferment* kings, are eliminated, did not have any adverse effect on the potency of the products. In fact, even the thatch roofed hut with the door opening to the East is not essential. He says, 'It has been found that there is not much difference whether the sacrifice with dried meat and wine is offered or not, therefore, the practice is diminishing.'[14] To facilitate further discussion, a flow diagram of the process is shown in Figure 42. But for now we need to move on to the second example, a process for making a type of *pên chhü* 笨麴, common *ferment*, intended for the brewing of spring wine. It is No. 7 in Table 20.

Preparation of Chhin-chou 秦州 *common* ferment *from wheat*[15]

'The process should start in the seventh moon. If the solar term is early, it should start early in the month.[16] If the solar term comes late, it should start later in the month. Take good, clean wheat free from insects and stir-roast (*chhao* 炒) it in a large *huo* 鑊 pot. The method of stir-frying is as follows: A beam should be installed above the stove, and a large spoon with a

---

[12] A straw (thatched) roof would retain heat better than a tiled one.
[13] Prayers and incantations are said to the kings of the east, west, north, south and centre to ward off vermin and insects. These are not translated; cf. Miao Chhi-Yü ed. (*1982*), pp. 359–60. It is possible that such rituals have always been a part of the fermentation process since antiquity.
[14] Tr. Huang Tzu-Chhing & Chao Yün-Tshung, p. 37; *CMTS*, ch. 64, p. 362.   [15] *CMTS*, ch. 66, p. 387.
[16] In the Chinese calendar, the year is divided into twenty-four solar terms or subseasons called *chieh chhi* 節氣. Cf. Shih-Shêng-Han (1959), p. 68.

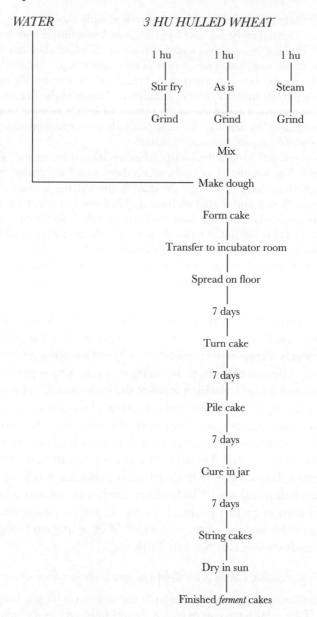

Fig 42. Flow diagram for the preparation of superior *ferment*, after Miao Chhi-Yü ed. (*1982*), p. 371.

long handle is hung on to it with a rope. The fire should be slow as the wheat is stirred rapidly with a rocking motion. The stirring should not be stopped even for a moment. Otherwise the grains will not be cooked evenly. The wheat is roasted until it turns yellow and has a fragrant smell; it should not be singed. When done, it is winnowed, and any extraneous matter removed. The wheat is then ground but not too fine or too coarse. If too fine it will be difficult to filter the wine. If too coarse it will be difficult to press the must.

A few days before the start of the process, *Artemisia* plants are collected and cleaned. The weed free artemisia is dried in the sun until the water content is low. The raw *ferment* is prepared by mixing the ground wheat evenly with water. It should be firm, dry and not sticky to the touch. When done, it is stored overnight and then kneaded further on the next morning until the right consistency is achieved. A wooden mould is used to press the cakes. Each cake is one foot (*chhih* 尺) square and 2 inches (*tshun* 寸) thick.[17] Strong young men are employed to press the cakes firmly. After pressing a hole is pierced through the centre of each cake.

Shelves of bamboo slats (similar to those used for rearing silkworms) are setup on a wooden frame. Dried artemisia leaves are laid on top of the shelves. The raw *ferment* cakes are laid on top of the artemisia, and then covered with a layer of artemisia. The layer of artemisia on the bottom should be thicker than that on top of the *ferment* cakes. The door and windows (of the hut) are then closed and sealed. The incubation is carried out as before. After the third seven-day period (total of twenty-one days) the *ferment* cakes should be fully cured. The door is opened and the *ferment* cakes examined. If there are five coloured myceliae (*i* 衣) on the surface, the cakes are taken out and dried in the sun. If no myceliae are seen, the door is resealed and the cakes allowed to incubate for another three to five days. They are then dried in the sun. During drying the cakes are turned over several times. When fully dried, they are stacked on racks and stored until use. One *tou* 斗 of this type of *ferment* will digest seven *tou* of grain (after it is cooked).

By placing the raw *ferment* cakes on shelves, instead of on the floor, many more cakes can be incubated in the same hut, so that the process can be carried out on a manufacturing scale. The two methods differ only in the way the wheat is treated and in the size of *ferment* cakes. Otherwise, they are basically the same. In a nutshell, partially cooked wheat is ground, mixed with water and pressed into cakes. The cakes are incubated in a closed chamber to serve as substrates for the growth of fungi and yeast that are naturally present. When optimal growth is obtained the cakes are harvested, dried and stored until needed.

Chia Ssu-Hsieh was, of course, unaware of what was really going on during the incubation, except that when it was finished the *ferment* cakes acquired the magical property of being able to digest cooked grain and convert it to wine. He did, however, note the presence of multicoloured coating (*i* 衣) on the surface of the cake, and indicated that this was a sign of the maturity of the *ferment*. In the light of what we know today we can see that his directions are, on the whole, scientifically sound. Mixing of the flour from steamed, roasted and raw wheat with water gave raw cakes that had enough moisture to allow *Aspergillus* spp., *Rhizopus* spp. and yeasts to proliferate, but not enough to turn the cake into a syrup. Thus, the final product remained firm and dry. Although the last step of drying in the sun may kill some of the surface myceliae, it does little harm to the organisms within. In this dry state no further microbial growth can occur and the *ferment* cake can be kept for as long as three years and still remain viable. By sealing the crevices on the walls with mud and providing a thatched roof, the hut becomes a well-insulated airtight chamber in which incubation can take place at a temperature and humidity conducive to the

---

[17] According to Wu Chhêng-Lo (1957), at the time of the *CMTS*, one foot (*chhih*) equals about 24 cm and one inch (*tshun*), about 2.4 cm.

growth of the microorganisms. By choosing the right season, autumn, the hut was ensured not to become too hot or too cold. And by using the same hut over and over again, the same contingent of spores of fungi and yeast cells that had become established in it would infect every fresh batch of raw *ferment* cakes and thus endow them at the end of the incubation with the same level and type of microbial and enzymatic activity as the preceding batches.

The *Chhi Min Yao Shu* tells us in quantitative terms the activity of each type of *ferment*. The measurement used is the number of *tou* 斗 of grain that can be digested by one *tou* of the *ferment*. The exact word used is *sha* 殺 (kill) for digest. The grain, after being cooked into *fan* 飯, would liquify in an aqueous suspension of the *ferment* as it is digested (killed) by the amylases originally present and the additional amylases elaborated by the growing myceliae. The number of *tou* of grain (as *fan*) digested by one *tou* of the major four types of *ferments* are:[18]

Superior, *shên chhü* 神麴	20–30	
Moderate, *pai lao chhü* 白醪麴	10	
Common, *pen chhü* 笨麴	7	
Common, *pai tuo chhü* 白墮麴	15	

The overall digestive and fermentative power during the brewing process is termed *chhü shih* 麴勢 (*ferment* power), which is manifested by the ability of the broth to dissolve the added cooked grain and keep itself bubbling with gas.

With active *ferments* on hand the actual brewing of the wine is a relatively easy matter. Thirty-eight wines are recorded in the *Chhi Min Yao Shu*, based on rice, glutinous rice, panicum and setaria millets as the substrates. We will cite four examples of wine fermentation in full from the text. The first involves the use of a *shên chhü* 神麴, superior *ferment* No. 5 listed in Table 20.

Preparation of wine with a superior ferment[19] *(Example 1)*

Panicum millet is the substrate. One *tou* 斗 of *ferment* is used to digest one *tan* 石 (i.e. ten *tou*) of grain.[20] Glutinous millet will only give a weak wine and is not often used. In preparing the *ferment* cakes, all the surfaces including the sides and the inside of the holes must be scraped clean. Then they are crushed into pieces the size of a date or chestnut, and dried thoroughly in the sun. Each *tou* of *ferment* is soaked in one *tou* and five *shêng* 升 (i.e. 1.5 *tou*) of water.

The best time to collect water for the process is in the tenth month (November) when frost first occurs and the mulberry leaves fall. To make good spring wine, water is collected on the last day of the first month (usually February–March). South of the Yellow River where it is warm, fermentations are started in the second month (March). North of the river, where it is cold, they are started in the third month (April). In general, any time near the Chhing-Ming 清明 Festival[21] would be suitable. The time of the first freezing is close to the end of the year

---

[18] *CMYS*, chs. 64, 65, 66 and 67; pp. 361, 383, 387 and 401.
[19] Tr. Huang Tzu-Chhing & Chao-Yün-Tshung, pp. 41–4, mod. auct.; *CMYS*, ch. 64, pp. 365–6.
[20] In this example one *tou* of *ferment* digests only one *shih* (10 *tou*) of grain. Other examples show the *ferment* digesting 3 to 4 *shih* of grain.
[21] One of the solar terms. It usually occurs around 5 April, near Easter.

when the streams begin to solidify. This water may be collected for spring wine. Any water collected during the other months must be boiled five times. After cooling, it is used to soak the *ferment*. Any deviation from this procedure will disturb the wine. The weather at the first freeze in the tenth month (November) is still mild, and there is no need to cover the earthern jars (in which the water is stored). In the eleventh and twelfth months, the earthen jar must be covered with millet straw.

The *ferment* is soaked for ten days in the winter or seven days in the spring. As soon as bubbles start to form and fragrant foams appear, the *ferment* is ready for brewing the grain. In winter when the cold is intense, the water in the *ferment* broth may freeze, even though the jar has been covered with straw every day. When that happens, the ice crystals are strained out, placed in a pot and warmed until the ice melts. The water should not be heated; once the ice is in liquid form all is returned to the jar. Cooked millet is then added to the broth. If this procedure is not followed, the charge will be injured by the cold.

If the earthen jar has a capacity for five *tan* 石 of grain, the first charge will consist of only one *tan* of the grain. The grain must be washed thoroughly until the wash water is clear. It is then cooked into a *fen* 饙 (partially cooked grain), which is in turn transferred into an empty jar and flooded with boiling water. The flooding will stop when there is a layer of water about one *tshun* over the surface of the *fen*. The mouth of the jar is covered with a basin and allowed to stand. When the water is all absorbed, the *fen* is soft and ready for use. The *fen* granules are spread out on a mat to cool. Any juice is kept in the basin. All of the lumps of millet are broken up by hand. After cooling the grain is transferred into the jar [containing the ferment suspension], and the mixture stirred with a wine paddle. The same procedure is repeated for subsequent charges.

However, in the eleventh and twelfth months (December and January), when the weather is cold and water freezes, the cooked millet is added when it is as warm as the human body. For the mulberry-fall wine or spring wine, all the charges should be added cold. If the starting charge is added cold, subsequent charges should also be cold. If the starting charge is warm, subsequent charges will also be warm. Warm and cold charges should not be intermixed in the same fermentation.

Usually the second charge will consist of eight *tou*, and the third charge, seven *tou* of grain [before cooking], the actual amount will depend on the *ferment* power (*chhü shih* 麴勢) in the must. There is no absolute quantity. In general the grain is divided into two equal parts. The first half is used to make the hot water-flooded *fen* 饙, while the other half is resteamed to make *fan* 飯. If all the grain is added in the form of *fen*, the wine will be dull. If, however, resteamed grain is incorporated, the wine will be light and fragrant. Therefore, it is advisable to divide the grain.

Winter fermentations will be finished after six to seven charges, while eight to nine charges may be appropriate in the spring. In the winter one may need to warm the fermenting broth, in the spring, to cool it. If too much grain is added at one time, too much heat will be generated and the wine will be injured; it will not keep well. In the spring the earthern jar is covered with a single layer of cloth, and in the winter with a straw mat. In early winter when the first charge is added, burning charcoal may be plunged into the jar and a sword laid at its mouth. These will be removed after the wine is ready.

Wine that is prepared in winter will be ready after fifteen days of fermentation. Those fermented in the spring will take ten days. In the middle of the fifth month (June), a bowlful of wine is removed from the jar and exposed to sunlight. If the wine is of good quality, it will remain unchanged. It is inferior its colour will change. Those which change colour should be drunk early while the good ones may be kept through the summer. But after a while, the finished mix will settle and the wine is decanted. It should last till the season of mulberry-fall. If the wine is stored in a cellar, it will pick up an earthy flavour. It is best kept in a thatched hut with several layers of straw roof. A house with tiled roof will be too hot. In the making and soaking of the *ferment*, in the steaming of the grain, and in the fermentation itself, river water should be employed throughout. Families without enough helping hands, may use sweet well-water instead.

Two points are noteworthy. Firstly, the soaking of the *ferment* really constitutes an inoculum stage for the process, during which the enzymes present can immediately start to hydrolyse the protein and starch in the grain. The nutrients liberated support the growth of yeasts and promote the proliferation of fungi. After several days there will be an abundant level of enzymes and yeasts in the medium. The conditions will be ripe to receive a fresh charge of cooked grain and begin the first stage of the fermentation process. The second point is that the author was fully aware of the importance of temperature for the success of the process. A considerable amount of care was taken to ensure that the fermentor jar was not too warm and not too cold. For this reason, it was only advisable to prepare the *ferment* and make fine wines in the autumn or spring. Some superior wine could be made in winter, but it was a struggle north of the Yellow River. We now move on to the next example of the making of wine in which a *pên chhü* 笨麴, common *ferment* (No. 7, Table 20) is used.

Preparation of spring wine with a common ferment[22] *(Example 2)*

The *ferment* cakes are cleaned, broken into small pieces and exposed to the sun until dry. Water is collected on the last day of the first month (February–March). River water is preferred. Well water is too salty; it should not be used for washing or steaming the grain. Generally one *tou* of *ferment* will digest about seven *tou* of grain when suspended in four *tou* of water. Use this ratio to estimate the amount of river water needed for the operation. A seventeen *shih* jar will be suitable for the fermentation of only ten *tan* of grain. If too much grain is added, the jar will not be able to contain the foam. Based on the size of the jar, the right amount of grain to be used can be calculated.

When the *ferment* is soaked in water for seven to eight days it will start to ferment vigorously. The process can then begin. Based on a total addition of 10 *tan* of panicum millet, the first charge will consist of two *tan* of grain, which is steamed twice and spread on a clean mat to cool. The large lumps are broken up before they are added to the fermentation. They should be allowed to float on the surface undisturbed until the next day when they will easily disperse when stirred with a wine paddle. If they are stirred too early the wine will be muddy. After the charge is added, the jar would be covered with a mat.

Afterwards, a new charge of millet grain is added to the fermentor every other day. Each batch is prepared in the same way as the first charge. The second charge consists of one *tan* and seven *tou* of grain, the third charge, one *tan* and four *tou*, the fourth charge, one *tan* and one *tou*, the fifth charge one *tan*, the sixth and seventh charges nine *tou* each. After nine *tan* of grain has been added, the wine should be examined and tasted. If the flavour is robust, no more additions are made. If not, yet another charge of three or four *tou* is made. The wine is tasted again a few days later. If it is still inadequate, another charge of two to three *tou* is added. If the *ferment* power remains strong, the wine still bitter, the usage of grain may

---

[22] *CMYS*, ch. 66, p. 387.

exceed ten *tan*. The important point is to achieve a good flavour, not to aim at reaching a goal of ten *tan*. But one should also take care to avoid too much grain being added; otherwise the wine will be too sweet. Before the seventh charge, each time a charge is made the fermenting broth should be examined with care. If the broth bubbles vigorously, the ferment power is still strong and more grain than the recommended amount should be added, but each charge should never exceed in quantity the preceding charge. If the ferment power appears weak, the charge can be reduced by three *tou*. If the charge is not increased when the *ferment* power is strong, the fermentation will not develop maximum flavour. If the charge is not decreased when the *ferment* power is weak, there will be undigested grain remaining in the broth.

If five or more jars are operated at one time, it is important to have enough grain cooked and cooled so that additions can be made to each fermentor. If only one jar can be charged in time, the other fermentors will miss the proper timing for a successful fermentation. Ideally, the second charge should be made before the day of Han Shih 寒食 (day before Chhing Ming). After that date the process will be a little late. If due to unexpected circumstances the process could not be started in time, river water in spring, though carrying undesirable odour, will have to be used. The grain for digestion must be rinsed until thoroughly clean. The workers' hands must be washed frequently and the nails kept scrupulously clean. There should not be a trace of saltiness on the hands; otherwise, the wine will spoil and will be unable to last through the summer.

From the two examples given we may summarise the wine making process as follows. *Chhü ferment* is soaked in clean water[23] for several (7–10) days until an active fermentation ensues. A charge of cooked grain (panicum millet or rice) is added to the fermentor and allowed to react until all the grain is liquified (1–2 days). A second charge is then added, and the cycle repeated until the *ferment* power is exhausted (7–10 charges). This point is reached when the grain is hydrolysed to sugar but not fermented to alcohol. Thus, the wines would taste a little sweet. By adjusting the amount of charge towards the end of the fermentation, a skilled wine maker can ensure that the finished wine contains just a little bit of sugar and does not taste too sweet. Since amylase enzymes are more tolerant of alcohol than yeasts, this means that a successful fermentation will end naturally when the alcohol content reaches a level that completely inhibits the fermentative action of yeasts (approx. 15 per cent alcohol). The broth is allowed to settle and the wine is decanted as needed, or it is pressed and the wine clarified. Or the broth is consumed as is, i.e. *ho phei* 合醅 (without separation of the residues by filtration).

Care is taken to ensure that during the process the fermentor is not too cold nor too warm. This is aided by carrying out the fermentation in the cool seasons. The means available to control temperature are, however, rather limited. To keep the fermentor warm the approach is to retain the heat evolved during the fermentation as much as possible. They can be covered with mats, raised on bricks, and wrapped with straw. The charges can be added warm and as a desperate measure, burning

---

[23] In the first example, cf. pp. 174–5, it is pointed out that water collected just before freezing may be used for making wine. Water collected at other times has to be boiled five times. The author of the *CMTS* was evidently well aware that clean water free from contaminants is needed to make good wine.

charcoal plunged into the broth. To cool them, covers can be removed and the fermentation slowed down by adding the charge in loose lumps instead of fine granules.

The common *ferment* may be used in a quick fermentation to make a summer wine, as shown in example 3.

Preparation of the 'cock crowing' summer wine[24] (Example 3)

Take two *tou* 斗 of setaria millet and steam it thoroughly.[25] Stir into it two *chin* 斤 of finely ground common ferment,[26] and mix thoroughly with five *tou* of water. The mouth of the jar is covered. When it is prepared one day, the wine will be ready by the time the cock crows on the next morning.

In this case the wine is made in one batch without the addition of subsequent charges of grain. It would be a very weak wine for immediate consumption only. The last example is the preparation of a *fa chiu* 法酒 (regulation wine), in which each charge consists of twice the amount of grain used in the preceding charge.

Preparation of a regulation wine with rice[27] (Example 4)

Glutinous rice would be particularly desirable. In the early morning of the third day of the third month, draw 3.3 *tou* of well water. Grind 3.3 *tou* of common *ferment* and pass it through a fine mesh. Measure out 3.3 *tou* of paddy rice – if that is not available, dryland rice may be used. The rice is steamed twice and spread out to cool. Add the *ferment* to the water followed by the cooked rice. After seven days, add a second charge of cooked rice, prepared from 6.6 *tou* of grain. After another seven days add a third charge made from 13.2 *tou* of grain. After a third seven days add a fourth charge with 26.4 *tou* of grain. The fermentation is then allowed to proceed to completion. The wine should achieve full strength. The wine may be consumed *ho phei* 合醅 (without filtration) and the jar is not sealed. If a clear wine is desired, the jar is sealed with mud. After standing for seven days the wine is clear. Decant the clear supernatant and press the remainder.

There are several points of interest in this procedure. A preliminary incubation (soaking) of the *ferment* in water is again dispensed with. The ratio of the total charge of grain to amount of the *ferment* (49.5:3.3) is the highest of all four examples. So is the ratio of grain to the initial amount of water (49.5:3.3). The mash must have been so thick that it should qualify as a semi-solid fermentation.

Although a wealth of detail on the procedure of wine making is given in these and other examples in the *Chhi Min Yao Shu*, the author has neglected to tell us a most important quantitative aspect of the process, and that is, the yield of finished wine from a typical operation. Perhaps it is because he was interested in wine making mainly as a domestic activity to support the needs of the household or estate and not as an enterprise to manufacture a product for sale. But he has given us enough

---

[24] *CMYS*, ch. 66, p. 395. The meaning of 'cock crowing' is self-evident.
[25] The text says that the grain is steamed into *mi* 糜 which normally means a congee. But in winery parlance, especially in Shantung, it denotes a *chiu fan* 酒飯, i.e. a cooked grain designed for brewing. The rice is continually stirred during steaming so that it becomes softer than ordinary *fan*.
[26] A *chin* 斤 is one-tenth of a *tou* 斗.   [27] *CMYS*, ch. 67, p. 408, method for *kêng* rice.

information for us to make an educated estimate of the final weight (and volume) of the fermented broth and its theoretical alcohol content for the four examples quoted based on the following assumptions:

(1) Each litre of grain or *ferment* weighs 0.8 kg.[28]
(2) Each *kg* of raw grain gives 2.5 kg of cooked grain with a 70% water content.[29]
(3) The *ferment*, as is, contains 60% starch.[30]
(4) The substrate (millet or rice), as is, contains 70% starch.[31]
(5) 80% of the starch is hydrolysed and converted to ethanol and carbon dioxide.[32] The carbon dioxide is lost.

The results are summarised in Table 21. It should be emphasised that the numbers are estimates only and are subject to serious uncertainties. The major ones are the amount of water absorbed when the grain is steamed, the weight lost when starch is converted to carbon dioxide and the efficiency of the process. But the figures arrived at do look reasonable, and are probably within the correct range. All the wine fermentations using superior *ferments* in chapter 64 have been checked in the same way and the final alcohol concentration in the broth ranged from 10 to 13 per cent. Another serious uncertainty is that there is no information at all on the quantity of wine residues or dregs (*tsao* 糟) that were actually obtained after decantation or pressing. The one example with relevant data, in chapter 66, indicates that each *shih* 石 of substrate gives rise to one *tou* 斗 of *tsao*, which based on modern experience, is definitely too low.[33] In any case, we may assume that the alcohol content of the finished wine is probably similar to that of the final broth, i.e. 6 to 11 per cent. Another way of gauging the percentage of alcohol is by checking the stability of the wine. In chapter 67 of the *Chhi Min Yao Shu*, the last wine mentioned is the

---

[28] Based on measurements done on commercial samples of polished rice.
[29] We can only guess at the water content of the cooked rice used as substrate by the early mediaeval vintners of China, since it can vary considerably depending on the conditions of the cooking process. For the purpose of this exercise we have adopted the value published in the *Chung-yang Wei-sheng Yen-chiu Yuan* (*1952*), p. 10 which states that ordinary steamed rice (served at a meal) contains about 70% water. But this assumption also implies that about 25% of the original solid in the grain was lost during the washing and steaming, as is the case found in modern Chinese winery practice recorded in modern publications on wine fermentations, such as that on wine technology by the *Wu-hsi chhing-kung-yeh hsüeh-yuan* (*1964*), pp. 78–9, 176, 190; on the famous Shao-Hsing wine by the *Che-chiang shêng kung-yeh chhang* (*1958*), p. 40, and on red wine by Hsü Hung-Shun (*1961*), pp. 22–3. In modern practice, the rinse water and steam condensate are collected and returned to the fermentation, so that much of the starch washed away is recovered and used as substrate. This accounts, in part, for the excellent yields of alcohol that are achieved in the processes described in the references given above. However, this was not the case for the four examples from the *CMYS* cited in the text, where any material washed away would simply be lost.
[30] Cf. *Wu-hsi chhing kung-yeh hsüeh-yuan* (*1964*), p. 57; Wang Chin (*1921*), p. 273; Chou Hsin-Chhun (*1977*), p. 32.
[31] Cf. *Wu-hsi chhing kung-yeh hsüeh-yuan* (*1964*) p. 66; Wang Chin (*1921*), p. 273; *Che-chiang shêng kung-yeh chhang* (*1958*), p. 8.
[32] This is pure conjecture. There are no data available on which one can make an educated guess. Modern fermentations described in *Wu-hsi chhing kung-yeh hsüeh-yuan* (*1964*), pp. 176, 190, and in Hsü Hung-Shun (*1961*), p. 23, report yields of >95% of alcohol. On the other hand, according to Chhên Thao-Shêng (*1979*), pp. 65, 69, from studies conducted in the 1930s at Nankai University, and by Sun Ying-Chhüan & Fang Hsin-Fang (*1934*), only 50% of the starch is utilised in a fermentation of sorghum. Starch in sorghum, with its content of tannin, may be harder to digest than rice, so the assumption of 80% yield may not be too much out of line.
[33] *CMYS*, ch. 66. According to Miao Chhi-Yü (*1982*), p. 403, fn. 39, the amount of residue in modern fermentations varies from 20% to 40%. *Wu-hsi chhing kung-yeh hsüeh-yuan* (*1964*) gives 35% to 40% residue for Shao-Hsing wine, p. 103, and 17% to 23% for an Amoy wine, p. 190.

Table 21. *Yield of alcohol in wine fermentations in the* Chhi Min Yao Shu

Process	Solids used Ferment grain (kg (litres))		Water added (kg)	Water in grain (kg)	Ethanol produced (kg)	Final weight (kg)	Alcohol content (per cent)
Superior Ferment No. 1	4 (5)	40 (50)	7.5	60	10.6	101.3	10.5
Common Ferment No. 2	11.4 (14.3)	80 (100)	57.2	120	22.2	247.4	9.0
Common Ferment No. 3	0.2	1.6 (2)	5	2.4	0.44	8.8	5.0
Common Ferment No. 4	2.6 (3.3)	39.6 (49.5)	3.3	59.4	10.2	95.2	10.7

Example of calculation: No. 1.
Total weight: 4 (*ferment*), 7.5 (water), 100 (cooked grain) = 111.5 kg.
Weight of starch: $0.6 \times 4 + 100 \times 0.3 \times 0.7 = 23.4$ kg.
Ethanol produced: 80% of $23.4 \times 0.568 = 10.6$ kg.
CO_2 produced: 80% of $23.4 \times 0.543 = 10.2$ kg.
Final weight: $111.5 - 10.2 = 101.3$ kg.
Therefore, percentage of alcohol is $(10.6/101.3) \times 100 = 10.5$.

mulberry-fall wine made using a *pai tuo* 白墮 *ferment*.[34] The fame of the producer of this *ferment*, Liu Pai-Tuo 劉白墮 is recorded in the *Shui Ching Chu* 水經注 (Commentary on the Waterways Classic), a work of the late +5th century.[35] The *Lo Yang Chieh Lan Chi* 洛陽伽籃記 (Buddhist Temples in Lo-yang), +547, also states that Liu Pai-Tuo was a skilful wine maker.[36] The wines he made were extremely stable and could travel more than a thousand *li* 里 without spoiling. This suggests that the alcoholic content of the wines he made, as well as the fine wines described in the *Chhi Min Yao Shu* must have been greater than 11 per cent.

But we can only speculate on these matters since there are no hard facts about how much wine was actually obtained, from the fermented mash or broth either by decantation or pressing. In spite of the numerous references to wine in poems written by such eminent poets as Li Pai, Tu Fu and Pai Chü-I, there is hardly any useful information on the technology of wine that has come down to us from the Thang dynasty.[37] Wang Chi 王績 in early Thang (c. 7th century) is said to have written a *Chiu Ching* 酒經, Wine Canon and a *Chiu Phu* 酒譜, Wine List. The *Chhing I Lu* 清異錄 (Anecdotes, Simple and Exotic), +965, Table 13, states that Wang Chin 王雍 (Thang?) wrote a *Kan Lu Chiu Ching* 甘露酒經 (Sweet Dew Wine Canon) and that his methods were much admired and followed in his day.[38] But all these books are

---

[34] The mulberry-fall wine is one of the well-known wines of early mediaeval China. It was often quoted in poetry. It is named after the 'Mulberry Fall' River, the exact location of which is no longer known, cf. *Nêng Kai Chai Man Lu* (Table 13), p. 38.
[35] *Shui Ching Chu* Chi Shi Ting Ngo, ch. 4, p. 18b (*SKCS* **574**, p. 70).
[36] *Lo Yang Chia Lan Chi, Fa Yün Shi* 法雲寺, p. 15b.
[37] Shinoda Osamu (*1972a*), 'Wines in Mediaeval China', in Shinoda & Seiichi (*1972*), pp. 541–90.
[38] *Chhing I Lu*, p. 104.

long lost. And so for the next relevant work on wine fermentations we will have to wait for the literature that emerged more than five hundred years later in the Sung dynasty.

(3) WINE FERMENTATIONS IN LATE MEDIAEVAL AND PREMODERN CHINA

> Flower paths have not been swept for any guest;
> My thatched gate for the first time opens to you.
> For food – the market's far – no wealth of flavours;
> For wine – my house is poor – only old muddy brew.
> If you don't mind drinking with the old man next door,
> I'll call across the hedge and we can finish off what's left.[1]

This excerpt from a poem by the famous Thang poet Tu Fu indicates that brewing of wine was a familiar domestic chore during his time. The word for 'muddy brew' in the original is *phei* 醅, which we have encountered earlier in *ho phei* 合醅 (p. 177), that is, unfiltered fermented mash. Presumably, every household had a few jars of brew in progress at varying stages of maturity, and a drink could be drawn off at any stage as the occasion required.[2] There is an anecdote in the *Chhing I Lu* (Anecdotes, Simple and Exotic), +960, which illustrates this point well.[3]

'I have always heard that the poet Li Thai-Pai (i.e. Li Pai) was partial to the drink *yü fou liang* 玉浮梁 (pearly floating logs), but never found out what it really is. Recently, I hired a servant from the Wu region and had her ferment some wine. After a while I asked her how the process was coming along. She replied, "It's not ready yet; it's only in the *fou liang* 浮梁 (floating log) stage." She showed me a cup of the brew. It was covered with foam like floating ants. Then it suddenly dawned on me that what Li Pai drank was simply the broth taken while it was still in the floating log stage.'

The *fou liang* (floating log) stage was also known as *fou i* 浮蟻 (floating ant) stage. It is probably similar to the *fa chhi* 乏齊 or *li chhi* 醴齊 steps given in the Chou Li (Table 18). The floating ant as a symbol of wine in progress is a popular metaphor in the poetry of the post-Han and Thang periods. We quote one example from a poem by another well-known Thang poet Pai Chu-I:[4]

---

[1] Tr. Watson (1984), *The Columbia Book of Chinese Poetry*, p. 231 'A Guest Arrives'; cf. *Thang Shih San Pai Shou, Kho chih* 客至, p. 201. The original reads:

花徑不曾緣客掃，蓬門今始為君開。
盤飧市遠無兼味，樽酒家貧只舊醅。

[2] This is in spite of the strict government control on wine making in effect during the Thang, cf. Wang Chin (1921), p. 271.

[3] *Chhing I Lu*, p. 105. Wu is present-day Kiangsu.

[4] *Wên Liu Shih Chiu* 問劉十九 (A Question for Liu, No. 19); cf. *Thang Shih San Pai Shou*, p. 251. 'Green ants' is the name of a wine. The original reads: 綠蟻新醅酒，紅泥小火爐。晚上天欲雪，能飲一杯無？ For other examples of 'floating ants' as wine see Shinoda (*1973a*), p. 326, and *Nêng Kai Tsai Man Lu*, p. 54.

> Green ants – a fresh unfiltered wine,
> Red clay – a burning charcoal stove.
> It looks that it will snow tonight,
> Why not join me and drink a cup?

The presence of dregs in the drink clearly did not deter the Thang poets from indulging in their delight of wine. The process of wine making was evidently practised on a wide scale throughout the realm during the Thang era. Refinements and variations on the basic process as described in the *Chhi Min Yao Shu* must have been made, but as we have said before (pp. 180–1), no records about them have survived to our day. Perhaps some of these are incorporated in the *Chiu Ching* 酒經 (Wine Canon) written by the eminent Sung poet Su Tung-Pho at about +1090:[5]

In the South, rice (glutinous or non-glutinous) is mixed with herbs to form a cake, *ping* 餅 [which is cured to become the wine medicament]. The best types are fragrant in smell and pungent in flavour, firm yet light. The wheat flour is 'fertilised'[6] and mixed with an extract of ginger [to form ferment blocks]. The blocks are incubated [in the *ferment* hut] until streaks are seen on the surface. They are strung on a string and dried in air. The longer it is dried the more potent it becomes.

When making wine, measure out five *tou* 斗 of rice and divide it into five parts. The first part consists of three *tou*, and the rest half a *tou* each. The three *tou* of rice is the initial substrate, and the half *tou* batches, charges for subsequent additions. Three additions are made, and a half *tou* of rice is left over. To start, take four ounces of medicated *ferment*, three ounces of regular *ferment* and disperse them in a small volume of water. Then mix in the cooked rice, transfer the mix to a jar and submerge the contents in a suitable amount of water. After three days, fermentation begins and frothing is visible. The early brew is rather flat and tastes slightly bitter. Three charges of substrate must be added before a good wine is obtained. The medicated ferment is vigorous while the regular ferment gentle in action. It is advisable to taste the broth frequently after a charge is made, and allow the flavour of the broth to determine whether the amount of the next charge should be adjusted up or down. Normally, the first addition is made after three days, and all three additions completed after nine days. After fifteen days the fermentation is ended, and one *tou* of water is added. This water must have been boiled and then cooled. Because the weather is warm in the South, every addition of (boiled?) water or cooked rice must be cooled first. The broth is allowed to stand for five days and filtered. Three and a half *tou* of wine is obtained. This is the main batch. The filtration takes half a day. The remaining unused half *tou* of rice is cooked with three parts of water, and then mixed with four ounces each of wine medicament and ferment, and the residual mash from the filtration. The mash is allowed to ferment further for five days. One and a half *tou* wine is obtained. This is the secondary batch. Altogether the yield of wine is five *tou*. The combined wine is held for five days. It is robust and rich and flavour. The dregs or residual mash from the filtration should be used without delay for fermentation with the last charge of cooked grain. Otherwise it will become dry and parched and the wine made from it, sick. If the fermentation is carried out leisurely, the wine will be rich and the yield high. The results will be the opposite if the fermentation is rushed. So we usually allow thirty days for the whole process.

---

[5] The original is quoted in many works, for example, in the *Chiu Shih* of the Wine Section of the *Min Shu Tshung Shu*, and Chhên Thao-Shêng (*1979*), p. 49, who has translated the account into modern Chinese.

[6] The original is *chhi fei chih* 起肥之, literally, to make fertile, that is to inoculate with a previously made preparation of *ferment*.

At last we are given a definite yield, five *tou* of wine from five *tou* of rice. But any sense of elation soon gives way to exasperation, as we find that Su Tung-Po has not bothered to say how much water is used in the process. Nor does he tell us how much residual mash or *tsao* 糟 is left behind after the filtration. Thus, there is no way one can estimate the final volume or weight of the mash with any degree of confidence. The most interesting part of this account, however, is that a medicated *ping* 餅 as well as a regular *ferment, chhü* 麴, are used in the brew. The medicated *ping* is, in effect, the *chiu yao* 酒藥, wine medicament, that is routinely used today in the brewing of regular, undistilled wine, or *huang chiu* 黃酒, such as the famous Shao Hsing wine. It uses rice instead of wheat as the substrate and is incubated differently from those *chhü* which contain herbal extracts as described in the *Chhi Min Yao Shu*. Actually, the *chiu yao*, wine medicament, is simply a variant form of *ferment* and has had a long history in China. It was presumably used in the making of the fragrant wine *chhang* 鬯 during Shang and Chou times. The earliest description of the preparation of a herbal *ferment* with rice occurs in the *Nan Fang Tshao Mu Chuang* 南方草木狀 (Herbs and Trees of the South), +304. Item number 14 says:[7]

In Nan-Hai herbal *ferments* (*tshao chhü* 草麴) are common. Wines are not made with regular *ferments* (*chhü nieh* 麴糵), but with a special product made by pounding rice flour mixed with many kinds of herb leaves and soaked in the juice of *yeh-ko* 冶葛 (probably *Gelsemium elegans* Benth.). The dough, as big as an egg, is left in dense bushes under the shade. After a month it is done, and is mixed with glutinous rice to ferment wine.

A similar account is given in other Thang works such as the *Thou Huang Tsa Lu* 投荒雜錄 (Miscellaneous Jottings Far from Home), by Fang Chhien-Li, c. +835,[8] and the *Ling Piao Lu I* 嶺表錄異 (Strange Southern Ways of Men and Things), by Liu Hsun, c. +895.[9] It is also mentioned in the early Sung work *Thai Phing Huan Yü Chi* 太平寰宇記 (Geographical Record of the Thai-Phing reign-period) by Yueh Shi, c. +980.[10] It is interesting to note that except for the *Ling Piao Lu I*, the references cited all use the term *chhü nieh* 麴糵 to denote a *ferment*, suggesting either that *chhü nieh* had become an archaic, literary expression for a fermenting agent, or that sprouted grain, at least in the minds of these writers, was still used in north China, in association with wheat *ferment*, for the making of wine. But in the south where rice was the main substrate it is possible that ripened herbal cakes of rice had always been the microbial agent of choice used to ferment wine. This view is corroborated by the prominent role assumed by herbs as constitutents of the *chhü* described in the most important work on the technology of wines in the Sung dynasty, the *Pei Shan Chiu Ching* 北山酒經 (Wine Canon of North Hill) by Chu Kung 朱肱, +1117.

The author, also known as Chu I-Chung 朱翼中, had operated a winery in Hangchou and was experienced in the art of fermentation, presumably leaning towards the southern style. The book is divided into three parts. Part 1 discusses the

---

[7] Tr. Li Hui-Lin (1979), p. 59, mod. auct.   [8] *Thou Huang Tsa Lu*.   [9] *Ling Piao Lu I*, p. 5.
[10] *TPHYC*, ch. 169, items on *Tan-Chou* 儋州 and *Chhiung-Chou* 瓊州. All three references are cited in Ling Shun-Shêng (*1958*), p. 904.

historical background of wine making and covers some of the same material that we have cited from the classics in the preceding subchapters. The most interesting segment in this part is where the author introduces us to the use of a yeast (*chiao* 酵) preparation in the making of wine.[11] He says:[12]

Northerners do not use *chiao* 酵, they only use wash water and call it *hsin shui* 信水 (trusty water?). But *hsin shui* is not *chiao*. Winemakers use *chiao* to liven the brew when the weather is cold. A fermentation is slow to start when no *chiao* is present. If it starts fermenting too late, the first stage (of the process) will not be right. The best way (to get *chiao*) is to take a sample of an actively fermenting culture. If this is not possible, skim the surface of a broth when it is fermenting vigorously, dry the solids with a towel, disperse the dregs in some *ferment* powder and air the mixture on a hanging straw basket. The product is called *kan chiao* 乾酵 (dry yeast), and may be kept and used as needed. The amount of it to use varies with the season. When it is cold, use more. When it is warm use less. As a rule, in the winter the wine maker uses more *chiao* and less *chhü*, in the summer less *chiao* and more *chhü*.

Chu's comment suggests that *chiao* was widely used in wine fermentations in the south, perhaps because the *chhü ferment* made in the north was a richer source of yeast than the types made in the south. In part 2 of his wine canon he describes the preparation of thirteen medicated *chhü ferments*, divided into three categories. These are listed in Table 22. The general procedure is as follows:[13]

The best time to make regulation *chhü* 麴 ferment is in the sixth month when the weather is hot. First prepare the steep liquor. Add a hundred catties of cocklebur, twenty catties each of hops and smartweed, finely chopped, into three *tan* 石 and five *tou* 斗 of water in an earthern jar. Steep for five to seven days, if the weather is cloudy, for ten days. Cover the jar with a basin. Each day stir the contents with a paddle and filter enough liquor to mix with wheat flour to make the *ferment*. This is the recommended procedure for a large-scale operation. For smaller-scale operations, for example, with a batch of three to five hundred catties of flour, take the [right amount of] three herbs, grind them in fresh well water, and use the aqueous extract (to mix with the flour). The herbs will endow the wine with a tart pungent flavour ... Each catty of flour will become 1.4 catties of raw cake. When they are done and dried, the weight will fall back to one catty.

The cakes should be packed firm. If there are hollow pockets within the *ferment* will be poorly balanced. If there is too much water the centre may become sweet and soft. If the water is unevenly distributed the interior will turn black and green. If injured by heat, the centre will be red, if by cold, the growth will be poor and the cake heavy. Only those which are light, white or yellow inside, or with flowery streaks [of myceliae] visible on the surface, are *ferment* cakes of good quality. From the fabrication of raw cakes to the end of incubation, it takes about a month. The cakes are stacked in an airy place and held for more than ten days. They are examined, and if no wet spots are seen inside, dried in the sun. When cooled they are stored in a dry place high above the floor, free from insects and rats. After forty-nine days the cake is ready for use.

---

[11] *Chiao* is first mentioned in the *CMYS*, ch. 82, p. 509, as a leaven to make a raised bread, but this is the earliest reference we have of its use in wine making.
[12] *PSCC*, Pt 1; Hsin Hsing (Taipei, 1975) ed. pp. 1228–30; *SKCS* **844**, pp. 816–17.
[13] *PSCC*, Pt 2; Hsin Hsing, p. 1231; *SKCS* **844**, pp. 817–18.

Table 22. Ferments *described in the* Pei Shan Chiu Ching

No.	Type	Substrate	Number of herbs[a]	Incubation
1.	Yên Chhü 罨麴	Wheat	7	In closed dark room
2.	—	Wheat	4	—
3.	—	Wheat	6	—
4.	—	Wheat	apricot seed	—
5.	Fêng chhü 風麴	Wheat:rice (G) 6:4	15	In airy area, wrapped in leaves in paper bag
6.	—	Wheat:rice (G) 1:1	8	—
7.	—	Wheat:rice (G) 1:1	8	—
8.	—	Wheat:red bean 7:1	4[b]	—
9.	Pao chhü 醪麴	Rice (G)	6	In airy area
10.	—	Rice (G):rice (K) 1:3	6[c]	—
11.	—	Rice (G)	9	—
12.	—	Wheat	ginger	—
13.	—	Rice (G):wheat 2:3	ginger	—

[a] Includes spices such as clove, nutmeg, pepper and cassia.
[b] Includes malt.
[c] Includes a *ferment* inoculum.
G, glutinous rice (*no mi* 糯米); K, kêng rice (*kêng mi* 粳米).

Cocklebur and smartweed are by now familiar to us, but this is the first time we have encountered hops, *shê ma* 蛇麻, *Humulus lupulus*, as an ingredient in a Chinese wine. As shown in Table 22 the *ferments* are divided into three groups according to the way the cakes are incubated. The four products in the first group, *yen chhü* 罨麴 (covered *ferment*) are all made from wheat flour. They are incubated in a dark room presumably in the same way as those described in the *Chhi Min Yao Shu* (pp. 170–1). The second group, the *fêng chhü* 風麴 (aired *ferment*), also consists of four products. Three are made with wheat and glutinous rice (Nos. 5, 6 & 7), and one with wheat and the small red bean (No. 8). For incubation, the cakes are wrapped in leaves (e.g. mulberry) and hung in paper bags in an airy place. The last group, the *pao chhü* 醪麴 (exposed *ferment*), includes five products. Three are made of rice only (Nos. 9, 10, 11), one of rice and wheat (No. 13) and one of wheat only (No. 12). All are incubated on straw or leaves in an airy environment. In fact, the way the *fêng chhü* and *pao chhü* are incubated is similar to the manner the *tshao chhü* 草麴 (herbal *ferment*) is prepared in the *Nan Fang Tshao Mu Chuang* described earlier (p. 183). It is also reminiscent of the method used until recently by the aborigines in Taiwan to prepare *lao* 醪 and the *chhang* 鬯 drinks today.[14]

Herbs are used in all the preparations; some have only one (Nos. 4, 12 and 13), but others can have as many as fifteen (No. 5). They include well-known medicinals

[14] Ling Shun-Shêng (*1958*), pp. 895, 903–5.

Table 23. *Fermentation of wine in the* Pei Shan Chiu Ching

Number	Task
1.	*Wo chiang* 臥漿. Prepare steep water from wheat; allow it to turn sour.
2.	*Thao mi* 淘糜. Wash glutinous rice with running water till clean.
3.	*Chien chiang* 煎漿. Boil steep water until the right acidity and consistency is obtained.
4.	*Thang mi* 湯米. Soak cleaned rice in hot steep water for four to five days; drain water.
5.	*Chêng chhu mi* 蒸醋糜. Steam sour rice, spread on table to cool.
6.	*Yung chhü* 用麴. Prepare inoculum with *ferment* and steamed rice.
7.	*Ho chiao* 合酵. Take leaven (yeast) from an active fermentation or use predried material.
8.	*Thu mi* 酘米. Prepare fermenting batch: mix sour steamed rice with malt, yeast and *ferment* in jar and layer in some rice steep water.
9.	*Chêng thien mi* 蒸甜糜. Steam clean, prewashed sweet rice, and cool on table.
10.	*Thou ju* 投醹. Add additional charges of steamed rice as needed depending on season.
11.	*Chiu chhi* 酒器. Prepare pottery jars by washing and curing with oil.
12.	*Shang chhao* 上槽. Transfer fermented mash on trough for pressing.
13.	*Shou chiu* 收酒. Drain, press and collect wine in jars.
14.	*Chu chiu* 煮酒. Steam jars to stabilise the wine, cool slowly (and seal).
15.	*Huo pho chiu* 火迫酒. Fire-pressure wine: wine in sealed jar is incubated in heated chamber.

such as *jen shên* 人參 (ginseng, *Panax ginseng*), *chhuan hsiung* 川芎 (hemlock parsley, *Ligusticum wallichii*), *pai fu tzu* 白附子 (sweet cassava, *Jatropha janipha*), *fang-fêng* 防風 (*Ledebouriella seseloides*) etc. as well as the spices, betel nut, cassia, clove, ginger, pepper and nutmeg. In No. 8, *mai nieh* 麥蘖 (malt, from either wheat or barley) is listed as one of the herbs. No. 10 is noteworthy in that some powdered *ferment* prepared previously is included as a constituent. In the light of what we know today, it is surprising that the idea of using an inoculum was not practised more widely.[15] It is not clear what the functions of the numerous herbs are.[16] Some may help to cut down the growth of undesirable organisms; others may endow the *ferment* with a distinctive flavour that is passed on to the wine.

Part 3 of *Pei Shan Chiu Ching* deals with the fermentation process itself. Directions are given for each of the fourteen tasks involved from start to finish. These are listed in Table 23. A few comments are necessary to highlight the differences between this procedure and those recorded in the *Chhi Min Yao Shu*.

---

[15] The use of an inoculum is implied in the Wine Canon of Su Tung-Pho.
[16] Chhü Chia-Wei & Fang Hsin-Fang (*1935*) have studied the effect of eleven herbs commonly used in the making of *ferments* on the wine process. They found that (1) some herbs tend to retard the growth of microorganisms (2) none exert an effect on the rate of starch digestion and (3) cinnamon bark and tangerine peel enhance the rate of fermentation at low concentration but become inhibitory at high concentrations. Cited in Chhen Thao-Sheng (*1979*), p. 54.

In the first place the preparation of the initial substrate charge (tasks 1 to 5) to start off the fermentation is now considerably more complex. A thin decoction made by steeping wheat grains in hot water called *chiang* 漿 is allowed to turn sour naturally and then concentrated (tasks 1 and 3). The concentrate is diluted with hot water and used to soak the first batch of rice grain (tasks 2 and 4). Much of the acid is absorbed by the rice. The rice is then drained and then steamed until soft. Oil is added to the boiling water to prevent it from frothing over. This rice is called *tshu mi* 醋糜 (sour steamed rice).[17] It is cooled on a table. The *chiang* from the soaking and steaming is collected for further use. The use of *chiang* in a wine fermentation is recorded in only one example in the *Chhi Min Yao Shu*. In the other examples, the washings from the grain are not used.

In the *Chhi Min Yao Shu* the *ferment* is first soaked in water until a vigorous fermentation ensues before the cooked grain is added. Now the sour steamed rice is mixed directly with malt, the *ferment* (*chhü* 麴), and some yeast (*chiao* 酵), and layered with *chiang* (steep water) in the fermentor jar to start the process (tasks 6, 7 and 8). The inclusion of malt is no surprise, but the use of yeast represents a real innovation and advance in technology. The subsequent charges (tasks 9 and 10) are not first soaked in *chiang* before steaming and they are called *thien mi* 甜糜 (sweet steamed rice). Charges may be added every three to four days up to a total of six to seven charges, depending on the weather.[18]

When the fermentation is completed (from 7–8 to 24–5 days) the wine is pressed (task 12), collected in jars (task 13) and heated (task 14). In task 12 the fermented mash is transferred to a trough (*tshao* 槽), spread on a bamboo mat or screen and covered with a board on top. A heavy object is placed on the board to force the liquid to flow out of the mash. This description implies that the fermented mash is in a semi-solid state. It is probably also in a thixotropic condition. In other words the broth does not separate easily from it unless it is heavily pressed. Unfortunately, there is no illustration or description of the shape and dimensions of the trough available. The juice is allowed to stand for three to five days when a clear wine can be decanted into a collection of storage jars (task 13). From a technological viewpoint, the most interesting task is, however, that of heating the wine (task 14). Two pieces of beeswax, five slices of bamboo leaves, and half a pill of serrated arum are added to a jar of newly pressed wine,[19] which is covered with mulberry leaves. It is placed on a steamer and heated until the wine inside begins to boil. The wax acts as a defoamer and prevents the wine from frothing over. When the fire dies down the jar is removed, placed in in a heap of lime, and allowed to cool very slowly. The wine must be clear before it can be heated, and the mulberry leaf cover should not be disturbed while the jar is being steamed. This operation is probably the world's first pasteurisation process in action in the wine and spirit industry. Another way of

---

[17] *Mi* 糜 or 穈 is also called *chiu fan* 酒飯 (wine rice), cf. Miao Chhi-Yü ed. (*1982*), p. 406, n. 67.
[18] This step, called *thou ju* 投醹, is not listed separately in the *SKCS* edition, presumably due to a copying omission.
[19] Serrated arum is *Arisaema consanguineum*.

heating the wine (task 15) is to let sealed jars of clear wine stand in an insulated room heated by several burning charcoal braziers. When it is hot enough the room is sealed, and allowed to cool slowly. Seven days later the room is opened and the wine removed.[20]

The tasks are followed by examples of the fermentation of ten specific types of wines and the preparation of two *chhü* 麴 *ferments*. It is curious that *ferments* are included here since they clearly belong to part 2. The directions for the tasks are presented in a manner suitable for unit processes. There is no attempt, and no obvious intention, of describing a complete operation from beginning to end. It is disappointing that no exact quantities, especially in terms of amounts of water used, are given either in the tasks or in the specific examples. Thus, in spite of the wealth of details provided it is impossible to estimate the yield of wine in a typical operation as depicted in the *Pei Shan Chiu Ching*. Nevertheless, even with this blemish it still deserves to be regarded as a most significant document in the history of wine making in China. It details the method of using soured grain steep water in the initial fermentation, a procedure that is an integral part of the process for making Shao Hsing wine today. The acidity in the mash favours the growth of yeasts even though the action of the amylases may be slightly retarded. This avoids the need for pre-incubating the *ferment* before the initial batch of substrate is added, as practised in the examples in the *Chhi Min Yao Shu*. It describes the use of a *chiao* 酵, yeast, in wine making for the first time in the Chinese literature. *Chiao* has, of course, been known since the late Han period as a leaven for making cakes and steamed buns.[21] This indicates that the winemakers in Sung times were aware of the difference between the saccharifying and fermenting activities in the ferment and in the brew. Furthermore, they knew that the fermenting activity could be skimmed off the top of the brewing medium and made into a dry preparation for later use. And lastly, it introduces the concept of heating the wine to stabilise it for long-term storage. This is a true invention and its discovery is particularly intriguing, since it is by no means obvious and could not have easily happened by chance. Perhaps, it is related to the custom of drinking the wine warm. We have referred to the wine serving and warming vessels among the the bronzes of the Shang and Chou eras.[22] Since antiquity the Chinese have always warmed a serving pot in a hot water bath before pouring and drinking the wine. Perhaps, one day the water in the bath was overly heated and became hot enough to pasteurise the contents of the pot, and for some reason the pot was put aside, covered and forgotten. Days, perhaps weeks later someone found and tasted the forgotten wine and realised that it was still good. After a few trials, the concept of heating the wine to stabilise it was reduced to practice and became an integral part of the wine making process.

The use of *chiao* 酵 (yeast) in the wine fermentation is reiterated in the *Hsin Fêng Chiu Fa* 新豐酒法 (*Hsin Fêng* Wine Method) by Lin Hung 林洪 of the Southern Sung

---

[20] Cf. *SCC* Vol. v, Pt 4, p. 146. [21] *CMYS*, ch. 82, p. 509.
[22] Cf. pp. 102–3. The practice may have also been influenced by the observation that cooked food keeps longer.

(Table B5). This is a small segment of a larger compilation entitled *Shan Chia Chhing Kung* 山家清供 (Basic Needs for Rustic Living), which is listed in Table 13. It says:[23]

First take one *tou* 斗 of wheat flour, three *shêng* 升 of rice vinegar and two *tan* 桓 of water. Heat to boiling and add one ?? sesame oil, Szechuan pepper and scallion. Use the decoction (*chiang* 漿) to soak one *shih* 石 of rice. After three days drain the rice and steam it to make *fan* 飯 (cooked rice). Heat the original decoction until it boils, skim off the froth, add more oil and pepper, and pour it into a jar (to cool). When the decoction is ready stir in about one *tou* of cooked rice (*fan*), ten catties of *ferment*[24] powder and half a catty of yeast (*chiao mu* 酵母). The next morning distribute the remainder of the *fan* in several jars and add a suitable quantity of the fermenting mash into each jar. Distribute and add two *tan* of water and two catties of flour to the jars. Mix the contents and cover the jars. After a day stir the mash in the jar. The brew will be ready in three, four to five days. . . . Fresh yeast (*chiao mu*) can be skimmed from the fermenting mash for use in the next cycle.

It is interesting to see, for the first time, the use of the term *chiao mu* 酵母, mother of *ferment*, which is now the scientific as well as the common name for yeast in China. In spite of this apparent elevation of the status of *chiao* 酵, it was not adopted universally as an ingredient in all subsequent wine making operations. It is strange that in part 3 of the *Pei Shan Chiu Ching*, where its virtue is extolled, *chiao* is not used in any of the ten examples of wine making described. Perhaps the examples were added to part 3 as an afterthought, or they were interpolations by a later writer. In any case, one would be curious to know how extensive *chiao* or *chiao mu* were used in wine fermentations after their appearance in the *Pei Shan Chiu Ching*. To find the answer we need to look at the relevant literature, which, in this case, comprises the food canons of late mediaeval China listed in Table 13 and those of premodern China listed in 14. The late mediaeval works we have to concern ourselves with are the last four entries from the Yuan period in Table 13. The most informative of the four is the *Chü Chia Pi Yung* 居家必用 (Essential Arts for Family Living), late Yuan, the author of which is unknown. Of the eight methods given for the wine fermentation, only one, for *chi ming chiu* 雞鳴酒 (cock crow wine) employs a yeast preparation. This procedure also calls for the use of malt in addition to *chhü ferment*.[25] The extra boost given by the yeast and malt no doubt enables the fermentation to be completed quickly, two days in the summer, three in spring and five in winter. The *Shih Lin Kuang Chi* 事林廣記 (Record of Miscellanies), +1283, describes eight wine fermentations from grain; only one employs a yeast preparation.[26] None is used in the single example of a fermentation in the *Yün Lin Thang Yin Shih Chih Tu Chi* 雲林堂飲食制度集 (Dietary System in the Cloud-Forest Studio), +1360.[27] Of the seven fermentations in the *I Ya I I* 易牙遺意 (Remnant Notions from *I Ya*), late Yuan, yeast (*chiao*)

---

[23] *SCCK*, pp. 111–12.
[24] The original text reads *mien mo* 麵末 (wheat grits), which is probably a copying error for *chhü mo* 麴末 (ground *ferment*).
[25] *CCPY*, p. 44. The same procedure is also given in two Ming works, *Tuo Nêng Phi Shih* and *Sung Shih Tsun Shêng*, cf. Shinoda & Tanaka (*1973*), pp. 369 and 461.
[26] *Shih Lin Kuang Chi*, ch. 2, cf. Shinoda & Tanaka (*1973*), pp. 62–3.
[27] *Yün Lin Thang Yin Shih Chih Tu Chi*, pp. 20–1.

is used in only two.²⁸ A review of the fermentations described in the food canons of premodern China, listed in Table 14, gave similar results. Thus, it would appear that even though the value of using a yeast preparation in a fermentation is recognised, the advantage is not great enough to make much of a difference to the final result. There is usually sufficient yeast present in a well-made mould *ferment* to enable it to attain a useful level to carry out the fermentation to completion. Added yeast is only needed in special situations such as the brewing of cock crow wine (*chi ming chiu* 雞鳴酒).

Standard *ferments* for wine continued to be manufactured in the traditional way through Ming and Chhing China. The status of the technology during this period can be gauged from this excerpt from the *Thien Kung Khai Wu* 天工開物 (The Exploitation of the Works of Nature) of +1637.²⁹

Wine *ferments* can be made from wheat, rice or [wheat] flour, depending on local conditions and differences between North and South. The principles are, however, the same. For making 'ferment' either wheat or barley may be used. The wheat grains in the hulls are washed with well water and dried in the summer sun. They are ground into a meal, mixed with spent wash water and shaped into cakes. The cakes are wrapped and bound in paper-mulberry (*Broussonetia papyrifera*) and hung in the air. Or they may be covered with rice stalks to encourage it to turn yellow (i.e. encourage the yellow mould to grow). After curing for forty-nine days the *ferment* is ready for use.

To make 'flour *ferment*' take five catties of white [wheat] flour and boil five pints of soybean in the juice of smartweed (*liao* 蓼) until the beans are soft.³⁰ Then they are mixed together with five ounces of powdered smartweed and ten ounces of almond paste and pressed into cakes. As before, the cakes are then either wrapped in mulberry leaves and hung or covered with rice stalks to promote the growth of the yellow mould. Another type [of *ferment*] is made culturing the moulds in cakes of glutinous rice flour mixed with the juice of *polygonum*; neither the method nor length of incubation differ from those described before.

Ferments can be made from a large number of primary and supplementary raw materials, flavoured with herbs that range from a few to as many as a hundred. The ingredients vary according to region and are too numerous to describe in detail. In recent years, in Peking, the seeds of Job's tears, *i* 薏 (*Coix lachryma*), have been used as the main substrate to make the *ferment* for the Job's tear wine. In Ningpo and Shaohsing, Chekiang province, mungbeans are the chief substrate in the *ferment* to make the bean wine. These are the two leading wines of the country (as described in the *wine canons* or *classics*).³¹

If at the wine workshop, the culturing of moulds is inadequate and the process poorly tended, or if the implements are not properly cleaned, then a few pellets of defective *ferment* can ruin bushels of [the winemaker's] grain. For this reason one should always buy *ferments* from reputable manufacturers who have established a good record for products of high quality. The *ferment* for making the yellow wine (*huang chiu* 黃酒) in Hopei and Shantung

---

²⁸ *I Ya II*, pp. 1–7.   ²⁹ *TKKW*, ch. 17, tr. Sun & Sun, and Li Chhiao-Phing *et al.*, mod. auct.

³⁰ Both Sun & Sun (1966) and Li *et al.* (1980) have translated this sentence so as to imply that the flour and beans are boiled together in the smartweed liquor. This can hardly be technically feasible since the flour would gelatinise in the water to a paste and the beans would not get cooked.

³¹ It is not clear which specific wines Sung Ying-Hsing has in mind, but we do find a Yenching (i.e. Peking) regulation wine and a Chin-hua (a city in Chekiang near Shao-hsing) wine among the wines listed in the *Chiu Hsiao Shih* of +1235.

comes from Anhui and is shipped to the North by boat or cart. The 'red wine' (*hung chiu* 紅酒) produced in the South is made with the same kind of *ferment* as that from Anhui.[32] It is called 'fire ferment' (*huo chhü* 火麴) in general, except that the kind sold in Anhui is shaped in square blocks, while that in the South is made into round cakes or balls.

Smartweed is the spirit of *ferment*; and grains are the body. [To make a *ferment*], some old residual mash (*tsao* 糟) must be added so as to bring the spirit and body together. It is not known when the first wine mash originated – but the process is similar to the use of old vitriol in the calcination of the alum.[33]

This procedure is similar to those described in the *Pei Shan Chiu Ching* (Wine Canon of North Hill) of +1117. The high regard accorded to smartweed may be misplaced since perfectly good *ferments* can be made with wheat flour without the use of smartweed. The most interesting part of this excerpt is the statement that 'some old residual mash must be added to bring the body and spirit together'. This is a clear indication that the use of an inoculum to make *chhü ferment* was probably quite common in the Ming dynasty. We have noted that the use of an inoculum is mentioned in one example of the making of a *ferment* in the *Pei Shan Chiu Ching*. Yet one would be hard pressed to find it mentioned in the recipes for making *ferments* given in the food canons of the late mediaeval and premodern China listed in Tables 13 and 14. Perhaps, the practice was already more widespread than intimated by Chu Kung, since the advantage of using an inoculum would be obvious to any practitioner after he has tested the concept. It just did not get into the literature because the authors of the food canons tended to copy copiously from previous works and were not particularly interested in the technological aspects of the wine making process.

Although the standard, traditional fermentation as delineated in the *Chhi Min Yao Shu* and modified in the *Pei Shan Chiu Ching* remained more or less unchanged throughout the Ming and Chhing dynasties, other developments had taken place in the Sung period that greatly expanded the scope of wine technology in China. A good way of discerning the changes is to look at the status of wine making as described in the *Chü Chia Pi Yung* (Yuan, c. +1300), listed in Table 13, and compare it with that in the *Pei Shan Chiu Ching* (+1117), listed in Table 15. We are struck immediately by two major innovations that had become well established by the Yüan dynasty. The first is the discovery of the red *ferment*, *hung chhü* 紅麴, caused by the predominance of red fungi of the genus *Monascus* in the cake. This led to the fermentation of a new type of wine, the so-called red wine or *hung chiu* 紅酒. The second is the commercialisation of distilled wine, commonly called *shao chiu* 燒酒 (burnt wine) or *pai chiu* 白酒 (white wine). Thus, during the late Sung the universe of Chinese wine divided itself into two major constellations, the fermented wine, commonly known as *huang chiu* 黃酒, yellow wine (of which the *hung chiu*, red wine is a variant) and the distilled wine, *shao chiu* or *pai chiu*. We can also discern an

---

[32] This is a curious statement since the author knows full well that the 'red wine' is made using a 'red *ferment*' as described in a later paragraph in the same chapter of *TKKW* (ch. 17).
[33] Which is described in *TKKW*, ch. 11.

evolutionary change in the growing importance of medicated wines. For example we find a wine named after the herb, *wu chia phi* 五加皮, bark of *Acanthopanax spinosum*, in the *Chiu Hsiao Shih* (Mini-History of Wine), +1235, and a recipe for incorporating it in a medicated wine in the food canon *Chü Chia Pi Yung*.[34] These topics will be covered in the next three subchapters. We shall discuss first the discovery and development of the red *ferment*, next, the evolution of distilled wine, and finally, the medicated wines of China.

(4) Development of Red *Ferment* and Red Wine

During my travels in the Western region, I often needed a place to spend the night and a simple dinner. Once I was served a dish cooked with a red ferment from Kua-chou. It was soft, smooth, rich and succulent, and seemed to dissolve in the mouth.[1]

The above quotation from a poem by Wang Tshan 王粲 who lived in the late Han dynasty (+177–217) is reputed to contain the earliest reference to the red *ferment*, *hung chhü* 紅麴, in the Chinese literature. But it is a claim that is difficult to substantiate. In the first place, *hung chhü* (red *ferment*) itself is not mentioned again until seven hundred years later in the Five Dynasties period although there is a reference to a red coloured wine in the *Lo Yang Chia Lan Chi* 洛陽伽藍記 (Description of Buddhist Temples in Loyang) in +547,[2] and another in the *Yu Yang Tsa Tsu* 酉陽雜俎 (Miscellany of the Yu-yang Mountain) of +860.[3] But from the respective contexts it is doubtful that in either case the red wine in question could have been prepared *via* the use of a red *ferment*. Secondly, the red *ferment* has always been made with rice as the substrate, and the Kua-chou area in the far Northwest would seem to be an unlikely place where its discovery would have been first noted.[4] There may, however, be an authentic reference to a red wine from a poem by the Thang poet Li Ho 李賀 (+791 to +817).[5]

> In the crystal cup, the (sparkling) amber mash is thick,
> And the wine drops from the little trough are pearly red.

The word for trough here is *tshao* 槽 which, as we have seen earlier in the *Pei Shan Chiu Ching* (Wine Canon of North Hill), is the equipment used to separate wine from

---

[34] *CCPY*, p. 43.

[1] *Chhu Hsüeh Chi* 初學記, ch. 26, paragraph 12, p. 24. Kua-Chou 瓜州 is usually taken to mean present-day Tun-Huang 敦煌.

[2] *Lo Yang Chia Lan Chi*, ch. 3, p. 11b; cited by Hung Kuang-Chu (*1984a*), p. 80, and Shinoda Osamu (*1973a*), p. 331.

[3] *Yu Yang Tsa Tsu*, ch. 7, mentions the preparation of a red wine by servants of a Chia family during the Wei (+220 to +264) period using water collected midstream in the Yellow River, cf. Shinoda, *ibid*.

[4] There is, however, another Kua-chou 瓜州 in Kiangsu, located at the confluence of the Grand Canal and the Yangtze River. It would not be unreasonable to find the red *ferment* in this southern Kua-chou. However, Wang Tshan was quite explicit that the Kua-chou in his poem was located in the Western regions.

[5] From *Chhang Ku Chi* 昌谷集, cited in *Chiu Shih*, p. 35; tr. auct.; adjuv. Needham, *SCC* Vol. v, Pt 4, p. 144. We now prefer the translation of *tshao* as 'trough' (of a wine press) rather than 'channel' (side-arm of a distillation apparatus), as formerly interpreted.

the residual, fermented mash (p. 187). If this quotation does indeed signify a true 'red wine', then a 'red *ferment*' must have already been known before Li Ho's time. But the next reference we have on the red *ferment* did not occur until sometime later, in the *Chhing I Lu* (Anecdotes, Simple and Exotic) of +960 in an entry entitled, *Fermented Mash: Bone of Wine*:[6]

> The Director of Victuals for the *Mêng Shu* 孟蜀 kingdom has a collection of a hundred books on cookery.[7] Among the recipes there is one for a red pot-roast lamb. The method is to simmer the meat with red *ferment*, and make sure that it is thoroughly soaked in the bone of the wine. When ready, cut in thin slices and serve. Note: bone of wine is the fermented mash after the wine is pressed, i.e. *tsao* 糟.

Presumably, when the dry *ferment* is simmered, it would absorb water and assume the consistency of fermented mash (*tsao*). The passage shows not only that the red *ferment* is an agent for the making of wine, but also that the red residual mash is used in cooking, an application that we will discuss further later on. When we encounter the red *ferment*, *hung chhü* 紅麴 again, it is in a poem by the great poet and gastronome of the Sung dynasty, Su Tung-Pho (+1036–1101). Writing when he was in South China, he says:[8]

> In search of local products for an old friend,
> I sent him post haste 'cured mash' and 'red ferment'.

In another poem, written while viewing the moon and thinking of a friend, he soliloquises:[9]

> Last year we shared a plate of tender alfalfa,
> Tonight I pour a *Min* 閩 wine, red as cinnabar.

Min is a synonym for Fukien province. Thus, these verses indicate that 'red *ferment*' and 'red wine' were very much specialities of the South, which they still are today. They may also explain why in the *Pei Shan Chiu Ching* (+1117) there is no mention of either red *ferment* or red wine. The author, Chu Kung, presumably gained his experience of the art in the North and his treatise is based largely on the traditional style of wine making. He did make the observation however, in the general directions for making *ferments* in part 2, that when the *ferment* cake is injured by heat the centre will be red. This suggests that the cake is contaminated with a red organism, and provides a clue as to how the red *ferment* might originally have been discovered. Someone, probably sometime before or during the Thang period, noticed the red growth in the interior of a *ferment* cake, decided that it was something that had emerged during the incubation and wondered whether it could grow further if it

---

[6] *Chhing I Lu*, pp. 31–2.
[7] *Shu* is one of the ten kingdoms of the Five Dynasties period founded by the *Mêng* family.
[8] Poem No. 2 dedicated to Chiang Ying-Shu 蔣穎叔; cf. Shinoda Osamu (*1973a*), p. 300. The original reads: 剩興故人尋土物，臘糟紅麴寄跎蹄.
[9] 'Looking at the moon on the tenth day of the eighth month', in *Collected Poems of Su Shih*, p. 375. The original reads: 去年舉君苜蓿盆，夜傾閩酒赤如丹. Young alfalfa was used as a vegetable; cf. *CMYS*, ch. 29, p. 162.

was mixed with fresh *ferment* substrate and incubated under similar conditions. He soon found that the red stuff would indeed propagate, and the new growth could be used to inoculate yet another fresh batch of substrate, and the process continued on and on. Furthermore, the red product so obtained was able to ferment cooked grain to wine just like a traditional *ferment*. So a new kind of *ferment* was born. This was a true invention. It was triggered by the observation of an unusual phenomenon, and nurtured to fruition by a curiosity about what it meant, and a compulsion to test the consequences with appropriate experiments. This is probably the earliest example of the selection of a desired microbial culture from a mixed population of organisms, a technique that has been applied on a massive scale in modern times in the fermentation industry.

The key to the success of this selection is the ability of the red mould, *Monascus* sp., to withstand a higher temperature (and also a lower pH) than the other grain moulds, *Aspergillus* sp., *Rhizopus* sp., and yeast etc. that comprise the microflora of a typical *ferment*. Under normal conditions, these more abundant and faster growing species would quickly take over all available space in the *ferment* cake and completely crowd out any chance the red mould might have had to occupy a significant part of the territory. But when the temperature is higher than usual, growth of the normal microflora would be inhibited and the red mould would be able to overtake them to become the dominant species in the *ferment*. To prepare a product with consistent properties it would make sense to inoculate each new batch of substrate with an old culture of the red mould, so that it could have a headstart over the other fungi that are naturally present in the environment.

A large number of wines in circulation commercially during the Sung period are named in two of the books presented in Table 15, the *Chhü Pên Tshao* 麴本草 (Natural History of Ferments) which lists fourteen wines and the *Chiu Hsiao Shih* 酒小史 (Mini-History of Wine), which lists one hundred and two. In addition, the *Chhü Wei Chiu Wên* 曲洧舊聞 (Old News from the Winding Wei River), +1127, lists more than two hundred types of wines made by prominant families, localities and catering establishments.[10] But no red wine or *ferment* is mentioned in any of the three works. By the Yuan dynasty, however, red *ferment* and red wine had become such familiar commodities that they had found their way into the food literature. Red *ferment* is cited as an ingredient in the preparation of rice vinegar in the *Nung Sang I Shih Tsho Yao* 農桑衣食撮要 (Essentials of Agriculture, Sericulture, Clothing and Food), +1314.[11] Both red *ferment* and red wine are described in the *Chü Chia Pi Yung* 居家必用 (Essential Arts for Family Living) and the *I Ya I I* 易牙遺意 (Remnant Notions from *I Ya*), two of the Yüan food canons listed in Table 13.

The *Chü Chia Pi Yung* contains probably the earliest account of the process for making a red *ferment* and a red wine. But before a *ferment* can be made, one needs first to prepare an inoculum called a *chhü mu* 麴母, mother of *ferment*. The procedures for making the mother of *ferment* and the red *ferment* are as follows:[12]

---

[10] *Chhü Wei Chiu Wên*, ch. 7, SKCS **861**, pp. 327–9. [11] *Nung Sang I Shih Tsho Yao*, p. 92. [12] *CCPY*, pp. 38–9.

Preparation of mother of ferment

To make *chhü mu* (mother of *ferment*) wash well one *tou* 斗 of glutinous rice and steam it to make *fan* 飯 (cooked rice). Add enough water as if it was a wine fermentation. Mix in two catties of superior red *ferment* and transfer the mash into a jar. The fermentation will be complete in seven days in the winter, three days in the summer, and five days in the spring or fall. Place the content in a large bowl and stir the mash to a thick paste. To make red *ferment*, use two *shêng* 升 of paste for evey *tou* of *kêng* 粳 rice. The whole batch will make fifteen *tou* of superior red *ferment*.

Preparation of red ferment

To make *hung chhü* 紅麴 (red *ferment*), take fifteen *tou* of regular *kêng* rice. Wash thoroughly and soak the rice in water overnight. On the next day steam the rice until it is about 80 per cent cooked through and divide it into fifteen separate portions. Mix by hand two *shêng* of the above mother ferment with each portion of cooked rice until the ferment is evenly distributed. Place all the portions in one big pile and cover it with a piece of cloth. The top should be secured with a mat and the floor covered with straw. It is important to see that the pile is not too cold and does not become too hot. If too hot, the batch will be ruined. When the pile feels hot to the touch, the cover should be taken off and the pile separated into smaller piles. If the surface is barely warm, the smaller piles are recombined and covered as in the original pile. If the temperature is about right do not disturb the pile. Someone has to stay awake all through the night to monitor the pile. On the next day, at midday the material is separated into three piles, an hour later into five piles, in another hour or two, recombined into one pile, and in another hour or two, separated into fifteen piles. In other words, the material is separated and recombined every one to two hours as needed, depending on the level of heat in the material. The procedure is repeated again and again.

On the third day, the *ferments in nascendi* are distributed into five to six bamboo baskets, and rinsed with freshly drawn well water. It is then combined into one pile, and separated and recombined every one to three hours as needed, depending on the level of heat. On the fourth day, the *ferment* grains are again placed in bamboo baskets and rinsed with freshly drawn well water. If all the grains do not float on the water, they should be piled together again and treated in the same way as before for another day. On the fifth day, the grains are rinsed once more. They all float on the water. They are then dried in the sun and ready for use in a wine fermentation.

This is a much more tedious and elaborate procedure than the ones employed in the making of regular *ferment* as described in the *Chhi Min Yao Shu* and the *Pei Shan Chiu Ching*. The substrate is not pressed into cakes. It remains as loose granules. The process calls for a tremendous amount of patience. The continuing cycle of making large piles, separating into little piles, followed again by large piles, little piles etc. can easily drive a workman up the wall. The key to the success of the operation is, of course, to keep the material on the hot (*je* 熱) side. Since the microbial growth generates heat, the whole point of the exercise of piling and separating is to dissipate just enough heat to keep the culture at the desired temperature. It required considerable skill, since the Chinese did not have an objective way of measuring the temperature, and the workers had to gauge what the correct 'hotness' was through experience. With the red *ferment* on hand, the *Chü Chia Pi Yung* then proceeds to describe the fermentation of a red wine.[13]

Preparation of wine with the red ferment

The raw materials are two pints (*shêng*) of red *ferment* and one *tou* of glutinous rice. One and a half to two ounces (*liang* 兩) of regular *ferment* may also be included. Wash the rice thoroughly. Boil five *shêng* of water with a handful of rice. Cool and soak the rice in this water, for two nights in cold weather and for one night in warm weather. Strain the rice and steam it until it is cooked through. Wash the red *ferment* and grind it to a meal. Soak it in warm water until it starts to ferment, then use it in the fermenting jar. The regular *ferment* needs no pre-incubation. It is finely ground and mixed thoroughly with the rice substrate, which is then transferred to the jar. The mixture is suspended in spent wash water and the lumps are broken up. If lumps remain the wine may become too sour. If desired some water may also be added. After two days the mash is stirred. After three days the wine may be pressed. Or wait for another one to two days to see if the flavour improves. After the wine is pressed, the residual mash may be returned to the jar and mixed with a gruel made by cooking one *shêng* of glutinous rice in three *shêng* of water. The fermentation is allowed to proceed for another two days, before the mash is pressed again. The wine so obtained may be blended with the main batch for drinking. But if the intent is to store the wine until next year, then it is best not to blend them. If necessary, the residual mash from the second pressing may be used to ferment a third batch of wine.

It is noted that some regular *ferment*, in addition to the red *ferment*, is employed in the process. A regular *ferment*, as well as yeast, *chiao* 酵, is also used in a recipe for making the red wine of Chien-chhang, *Chien-chhang hung chiu* 建昌紅酒, as recorded in the other Yuan food canon, *I Ya I I*.[14] The same recipe is given in the *Yin Chuan Fu Shih Chien* 飲饌服食牋 (Compendium of Food and Drink) by Kao Lien late in the Ming dynasty (Table 14).[15] Unfortunately, none of the above examples provide enough information to enable us to estimate the yield of wine obtained. But modern experience suggests that the red *ferments* of the Sung and Yuan periods were probably just as effective in converting starch from grains to alcohol as the traditional cake *ferments* of the same era.

The red *ferment* is mentioned in the celebrated Yuan dietary treatise *Yin Shan Chêng Yao* 飲膳正要 and several Ming works. Among these are *Pien Min Thu Tsuan* 便民圖纂 (The Farmer's Manual), +1502; *Sung Shih Tsun Shêng* 宋氏尊生 (Sung Family's Guide to Living), +1504; *Mo Ngo Hsiao Lu* 墨鵝小錄 (A Secretary's Commonplace book), +1571; *Pên Tshao Kang Mu* (Pandects of Natural History), +1596,[16] and *Thien Kung Khai Wu* 天工開物 (Exploitation of the Works of Nature), +1637. The procedures for making the red *ferment* described in the *Mo Ngo Hsiao Lu* and the *Pên Tshao Kang Mu* are virtually the same as that quoted earlier from the *Chu Chia Pi Yung*, that is to say, they require piling and separating the *ferment* granules at regular intervals

---

[13] *CCPY*, p. 44. The same recipe is given in *Sung Shih Tsun Shêng* ch. 3; cf. Shinoda & Tanaka (*1972*), p. 461.
[14] *I Ya I I*, p. 4. Chien-Chhang is located in southwestern Szechuan.
[15] *YCFSC*, pp. 125–6. Neither *I Ya I I* nor *YCFSC*, however, gives a recipe for making the red *ferment*.
[16] *YSCY*, ch. 3, p. 176. *Pien Min Thu Tsuan*, ch. 14, p. 8b. *Sung Shih Tsun Shêng*, ch. 3 in Shinoda & Tanaka (*1973*), p. 461. *Mo Ngo Hsiao Lu*, quoted in Hung Kuang-Chu (*1984*), p. 82. *PTKM*, ch. 25, p. 1547.

during the incubation period. The *Mo Ngo Hsiao Lu*, however, does include two new recommendations. The first is the addition of vinegar (and three herbs) to the cooked grain. This should lower the pH of the grain and further discourage the proliferation of undesirable organisms. The second is to advise the operator to keep the incubating cooked grains at a level of heat equal to his own body, which, as it turns out is close to the optimum growth temperature of about 38°c for *Monascus* sp.

The procedure for making red *ferment*, called *tan chhü* 丹麴 (cinnabar *ferment*), by Sung Ying-Hsing in the *Thien Kung Khai Wu* represents a significant departure from the older method, and deserves quoting in full.[17]

Preparation of red ferment in the Thien Kung Khai Wu

The method for making red *ferment* (*tan chhü*) is an innovation of more recent times. Its principle is to extract wondrous powers out of the rank and putrid. Its practice is to transform and distill the essences [of grains]. When thinly spread on fish or meat – the things in the world most easily spoilt – the red *ferment* will enable them to retain their fresh qualities even in the height of summer. For as long as ten days no fly will come near them and their colour and flavour will remain as fresh as before. It is indeed a marvellous medicament.

Grains of the common (non-glutinous) *hsien* 秈 rice, either early or late, are pounded and hulled to the most excellent whiteness and then soaked in water for seven days.[18] When the stench has become unbearable, the grains are taken to a river and rinsed clean with free flowing water. Only the water from mountain streams should be used. The water from large rivers will not do. [Figure 43.]

After washing, some odour may still remain, but when the material is steamed in a steamer, it will be transformed into a fragrant aroma. When the rice is about half-way cooked through, the steaming is stopped. It should not be fully cooked. The half-cooked rice is immediately immersed in cold water. When the rice has cooled off, it is steamed again. This time it is cooked through.

The cooked rice is heaped together, several *tan* 石 to a pile, for the addition of mother of *ferment* or *hsin* 信. The mother of ferment must be made from the residual mash of the best red wine at a proportion of one *tou* 斗 of mash to three *shêng* 升 of the natural juice of smartweed mixed in alum water. Two catties of the mother of *ferment* are added to every *tan* of steamed rice while it is still hot, then mixed by several pairs of hands until the rice is cooled. The pile is kept under observation in order to look for signs of fermenting activity. When fermentation starts, the pile will become slightly warmer. The mixture of rice and *ferment* is placed in several large baskets and rinsed once with alum water. It is then distributed among separate woven bamboo trays, and placed on shelves in an airy room in order to catch the breeze. From now on air would be the determining factor [in the process]; fire and water would have little effect.

Each bamboo tray contains about five *shêng* of steamed rice. The room [in which the trays are shelved] should be large and have a high ceiling so that the heat from the tile roof would not become oppressive. The room should face south, never west. The material is turned over three times every hour. For seven days the workers tending the *ferments* will sleep near the trays. They should never allow themselves to sleep soundly and should be prepared to get up several times during the night.

[17] *TKKW*, ch. 17; tr. Sun & Sun (1966), pp. 292–3, mod. auct., adjuv. Li *et al.* (1980).
[18] *Chü Chia Pi Yung* uses *kêng* rice. The non-glutinous *hsien* rice probably makes it easier to obtain the final product in discrete grains.

Fig 43. Preparation of red *ferment*: washing of rice in a flowing stream. *TKKW* p. 288.

At first the rice is snowy white, but after one or two days the colour turns pitch black. From black it turns to brown, from brown to rust, from rust to red and at its deepest red the colour changes to a light yellow. With the help of air currents, we can see these changes occurring in front of our very eyes. The process is called 'cultivation of yellow *ferment*' (*shêng huang chhü* 生黃麴). The *ferment* produced through this process is twice as potent and priced twice as high as ordinary *ferments*. The grains are washed once with water between the black and brown stages, and again between the brown and red stages. After it has turned red it is not washed again. In making this *ferment*, it is essential that the operators' hands, and the mats and trays used, are scrupulously clean. The slightest amount of dirt will bring the entire operation to ruin.

Fig 44. Preparation of red *ferment*: Inoculated cooked rice dispersed on mats and trays *TKKW* p. 289.

The main innovation is to spread the loose granules on bamboo trays and stack them on shelves [as shown in Figures 44, 45]. By adjusting the amount of material on each tray and the extent of airiness in the room, one can achieve, by experience, a balance of heat generation and dissipation so that the overall temperature remains at the optimum level. This procedure is much more convenient than the original method of piling and separating the grains at stated intervals day and night. It should be considerably easier on the workers even though they still have to look after the trays during the night.

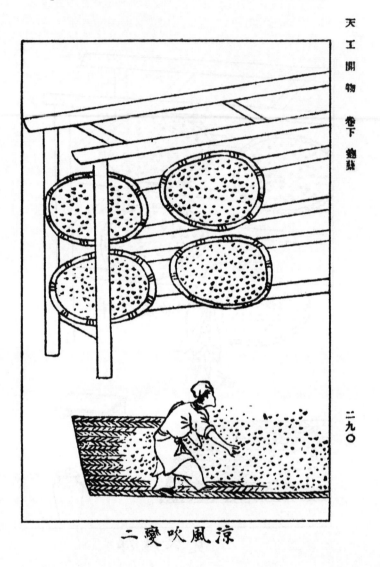

Fig 45. Preparation of red *ferment*: Incubation of rice on trays to form red *ferment TKKW* p. 290.

From the four descriptions of the process of making red *ferments* available, that is, those from *Chü Chia Pi Yung, Mo Ngo Hsiao Lu, Pên Tshao Kang Mu,* and *Thien Kung Khai Wu* we can discern the factors that were regarded as critical by the practitioners in order to achieve a successful operation.

(1) The use of an inoculum, a mother of *ferment, chhü mu* 麴母 or *hsin* 信, prepared with material from a previous successful fermentation.
(2) All utensils, materials and workers' hands that come into contact with them should be scrupulously clean.

(a)

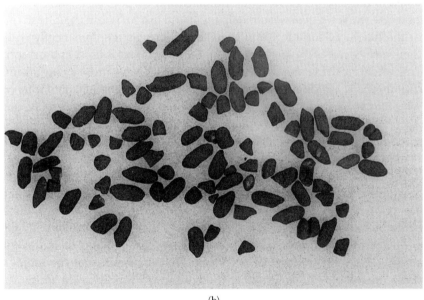

(b)

Fig 46. Grains of red *ferment* from Foochow, photograph Chen Jia-Hua.

(3) The *ferment* grains should be kept near body temperature during incubation.
(4) They should be kept slightly acid, using vinegar or alum.
(5) They should not be allowed to dry out; hence the washing of the grains during the incubation.

All these directions constitute good manufacturing practice in the light of our current understanding of the nature of the process. They provide the desired inoculum and optimal growth conditions in terms of temperature, acidity and humidity that favour the propagation of *Monascus* sp. and yeast and discourage that of other fungi. Even so, the procedure is still significantly more demanding than that needed for the making of regular *ferments*, which, according to the food canons and pharmacopoeias can be readily made without the use of preformed microbial inoculum. Thus, while the regular *ferment* has been made (and is being made) all over China, the red *ferment* has remained (and still remains) largely a speciality of South China, particularly the province of Fukien, the Min region that gave its name to Su Tung-Pho's red wine. According to the *Tan Chhi Pu I* 丹溪補遺, the best red *ferments* have come from northern Fukien, from Sung-Chhi 松溪, Chêng-Ho 政和, Chien-Ou 建甌 and Ku-Thien 古田 counties. Today, Ku-Thien remains the home for red *ferment* of the highest quality, but it is now made in other parts of China, including Taiwan, as well as overseas where there are heavy populations of Chinese from Fukien such as Malaysia, Singapore and the Philippines.

Almost as important economically as the red *ferment* itself is the red fermented mash, *hung tsao* 紅糟, that is left after the wine is pressed. This is the 'bone of wine' mentioned in the *Chhing I Lu* that we quoted earlier on p. 193. As Sung Ying-Hsing had observed, the red *ferment* would help meat and fish to 'retain their fresh qualities even at the height of summer' and that 'for as long as ten days no fly will come near them'. *Hung tsao* is the preferred way the red *ferment* is used as a preservative, colouring and flavouring agent for foods, and we will have more to say about these applications when we discuss food processing later in Chapter (*e*). However, the term *hung chiu* 紅酒 (red wine) gradually went out of fashion, and today, the wine made using the red *ferment* is also called *huang chiu* 黃酒, yellow wine.

The red *ferment* (*hung chhü* 紅麴) or red yeast was first listed as a therapeutic agent in the *Pên Tshao Kang Mu* of +1596.[19] Since then it has remained as one of the more obscure drugs in the Chinese pharmacopoeia. This situation, however, may be ready for a dramatic change. Recent studies in Japan and China show that the red yeast contains a class of compounds called monacolins which have the ability to lower blood cholesterol in animal models of hypercholesterolemia.[20] Clinical trials in China and the United States now indicate that the red yeast is also effective in lowering the serum cholesterol in humans.[21] It turns out that the major active ingredient, monacolin K, is identical to lovastatin,[22] the first of a series of statin

---

[19] *PTKM*, ch. 25, pp. 1547–8. It is said to improve digestion and enhance circulation.
[20] Endo (1979); Zhu, Li & Wang (1995, 1998).
[21] Cf. Heber, *et al*. (1999) for the results of the US trial and a summary of the clinical trials in China.
[22] Lovastatin was originally isolated from a strain of *Aspergillus terreus*. Its chemical name is [1S-[1α(R*),3α,7β, 8β(2S*,4S*),8aβ]]-1,2,3,7,8,8q-hexahydro-3,7-dimethyl-8-[2-(tetrahyrdo-4-hydroxy-6-oxo-2H-pyran-2-yl)ethyl]-1-naphthalenyl 2-methyl-butanoate. After ingestion this lactone is hydrolysed to the corresponding β-hydroxy acid, a potent inhibitor of 3-hydroxy-3-methyl-glutaryl coenzyme A (HMG-CoA) reductase, the enzyme which catalyses the rate limiting step in the synthesis of cholesterol. All the statin drugs act in a similar manner.

cholesterol-lowering drugs introduced in the recent decade. These exciting discoveries suggest that within a few years the red yeast may become a highly popular dietary supplement used to reduce the incidence of coronary and other circulatory diseases in the general population. At the same time, this development may also generate a new level of interest in the fermentative and culinary applications of the red *ferment* that would propel the wines fermented with it and the foods prepared with the wine residues (*hung tsao* 紅糟), from regional specialities into products consumed and enjoyed throughout China.

(5) Evolution of Distilled Wines in China

The litchis are newly ripe, the colour of a cock's crown,
One catches the first whiff of amber from the burnt-wine.
How one would like to pluck a branch, and to drink a cup!
But there's no other guest here with whom to share this treat.[1]

Distilled wine or spirit is known in China today as *shao-chiu* 燒酒 (burnt-wine or wine that burns) in the South and *pai chiu* 白酒 (white wine) in the North. The poem by Pai Chü-I (+772 to +846) quoted above is one of several references in which 'burnt-wine' is mentioned in the Thang literature. Is this 'burnt-wine' distilled wine? Or is it merely wine that has been warmed or mulled, which is the way the Chinese have traditionally drunk wine since antiquity. The answer has an important bearing on the invention of the distillation of alcohol, a problem of major interest in the history of chemistry and chemical technology. Did distilled spirit first appear in China as *shao chiu* (+7th to +9th century), or in Europe as *aqua ardens* or *aqua vitae* (mid +12th century)? This question has been examined at considerable length in Volume 5, Part 4 of this series more than a decade ago,[2] but in order to bring the discussion up-to-date and to provide an adequate background for assessing the newer developments, it will be necessary to recapitulate some of the salient points presented in the previous disquisition. The subject had elicited much debate among historians of science in China, such as Yüan Han-Chhing, Tshao Yüan-Yü and Wu Te-To, in the 1950s and 60s, and the debate has been revived with increasing intensity in recent years.[3] Indeed, it is probably the most challenging unsolved problem in the history of chemistry and food science in China.

---

[1] From *Li-chi Lou Tui Chiu* 荔枝樓對酒 (Drinking at the Litchi Pavilion), tr. Needham, *SCC* Vol. v, Pt 4, p. 143, mod.; cf. *Pai Chü-I Chi Chuan*, Collected Works of Pai Chü-I, p. 393. The original reads:

荔枝新熟雞冠色，燒酒初開琥泊香。
欲摘一枝傾一盞，西樓無客共誰嘗。

[2] *SCC* Vol. v, Pt 4, pp. 121–58.
[3] Cf. Yüan Han-Chhing (*1956*), Tshao Yüan-Yü (*1963*, *1979*), Wu Tê-To (*1966*, *1982*, *1988a,b*), Hsing Jung-Chhuan (*1982*), Hung Kuang-Chu (*1984a*), pp. 148–51, Mêng Nai-Chhang (*1985*), Fang Hsin-Fang (*1987*), Liu Kuang-Ting (*1987* and an undated MS), Wang Yu-Phêng (*1987*), Chang Hou-Yung (*1987*), Huang Shih-Chien (*1988* and *1996*), Li Pin (*1992*) and Li Hua-Jui (*1990*, *1995b*).

(i) *Summary of earlier evidence*

The controversy about the origin of distilled wines in China centres on two seemingly contradictory accounts given by the preeminent Ming naturalist Li Shih-Chên in the *Pên Tshao Kang Mu* (The Great Pharmacopoeia) of +1596. The first deals with the origin of *shao chiu*, also called *huo chiu* 火酒 (fire wine) or *a-la-chi chiu* 阿剌吉酒 (Mongol *araki*). He says:[4]

The making of burnt-wine was not an ancient art. The technique (*fa* 法) was first developed in Yüan times (+1280 to +1367). Strong wine (*nung chiu* 濃酒) is mixed with the fermentation residues (*tsao* 糟) and put into a steamer (*tsêng* 甑). On steaming (*tsêng* 蒸) the vapour is made to rise and a vessel (*chhi* 器) is used to collect the condensing drops (*ti lu* 滴露). All sorts of wine that have turned sour can be used for distilling (*chêng shao* 蒸燒). Nowadays, in general, glutinous rice (*no-mi* 糯米), or ordinary rice (*kêng mi* 粳米) or glutinous millet (*shu* 黍) or the other variety of glutinous millet (*shu* 秫) or barley (*ta mai* 大麥) are first cooked by steaming (*chêng shu* 蒸熟), then mixed with *ferment* (*chhü* 麴) and allowed to brew (*niang* 釀) in vats (*yung* 甕) for seven days before being distilled (*i tsêng chêng chhü* 以甑蒸取). [The product] is as clear as water and its taste is extremely strong. This is distilled spirits (*chiu lu* 酒露 or dew of wine).

Many Chinese and Western scholars were only too happy to accept the words of so eminent an authority;[5] but overlooked is an equally important passage in the same chapter of the *Pên Tshao Kang Mu*, which describes the nature of grape wine in these words:[6]

There are, in fact, two sorts of grape wine, that obtained by fermentation which has an elegant taste, and that made like *shao chiu* (*ju shao chiu fa* 如燒酒法), which has a powerful action (*ta tu* 大毒). The makers mix the juice with *ferment* (*chhü* 麴), just as in the ordinary way for the fermentation of glutinous rice. Dried raisins ground up can also be used in place of the juice. This is what the emperor Wên of the Wei Dynasty meant when he said that the wine made from grapes was better than that from *chhü* and rice, because the intoxicating effect of the former fades away more quickly. In the distillation method many dozens of catties of grape are first treated with the 'great *ferment*' *ta chhü* 大麴, such as used to make vinegar, and then put into a steamer and steamed (*ju tsêng chêng chih* 入甑蒸之). A receiver (*chhi* 器) is used to collect the distillate (*ti lu* 滴露). This is a beautiful pink colour. Anciently, such a brandy was made in the Western region (*hsi yü* 西域). It was only when Kao-chhang 高昌 (i.e. Turfan in modern Sinkiang) was captured during the Thang period that this technique was obtained.

Kao-chhang was captured by the Thang forces in +640.[7] Thus, according to this passage the art of steaming fermented mash to obtain distilled spirit was already known to the Chinese by the 7th century. Whether the technique came eastwards from Turfan, or that it just developed in China as a result of the stimulus from

---

[4] *PTKM*, ch. 25, p. 1567; tr. Needham *et al.*, *ibid.*, p. 135, mod. auct.
[5] For example, in *San Tshai Thu Hui*, +1609, *Wu Li Hsiao Shih*, +1664, Laufer (1919), p. 238, and Shinoda Osamu (*1957a*).
[6] *PTKM*, ch. 25, p. 1568; tr. Needham *et al.*, *SCC* Vol. v, Pt 4, p. 136.
[7] Actually, Kao-chhang has been under the control of the Chinese before the Thang conquest at various times since the Han Dynasty, cf. Huang Shih-Chien (*1994*), p. 90.

Turfan is immaterial at this point. What is important for us is that this passage cannot be reconciled with Li Shih-Chên's earlier statement that the 'The technique [of making burnt-wine] (*chhi fa* 其法) was first developed in Yüan times.' We shall attempt to explain the nature of this discrepancy after we have reviewed the current status of the problem.

Unfortunately, no records about the distillation of either grape wine or grain wine in the Thang dynasty have survived to our day.[8] There are, however, a number of references to 'burnt-wine (*shao chiu* 燒酒)', or similar expressions, in the Thang literature which have been collected and discussed earlier and are now listed below:[9]

Author	Date	Topic
Mêng Shên 孟詵	+670	Steam fermented mash many times.[10]
Li Chao 李肇	+810	Burnt spring wine, *shao chhung chiu* 燒春酒.
Pai Chu-I 白居易	+820	Burnt wine, *shao chiu* 燒酒.
Fang Chhien-Li 房千里	+830	After-burnt wine, *chi shao chiu* 既燒酒.
Yung Thao 雍陶	+840	Burnt wine, *shao chiu* 燒酒.
Liu Hsün 劉恂	+880	To burn wine; wine not burnt is *chhing chiu* 清酒, clear wine.[11]

In addition, several other references of interest belonging to the Sung period were also noted. The first is a reference dated at the beginning of the Sung dynasty that appears to directly denote the existence of a distilled spirit. In the *Chhü Pên Tshao* 麴本草, Pandects of Ferment Cultures, +990 (listed in Table 15), there is a description of a twice 'burnt' (*shao* 燒) Siamese toddy, which was so potent that it would overcome even the most capacious drinker.[12] The second is a passage in the *Meng Liang Lu*, Southern Sung, which says that at the dawn market one could buy 'crystal burnt-wine, red and white (*shui ching hung pai shao chiu* 水晶紅白燒酒) which had a gentle flavour and evaporated as soon as it entered the mouth'.[13] Neither citation, however, clearly indicates a relationship between 'burnt-wine' and a distillation process. The last reference occurs in a rhapsodic ode about a wine called

---

[8] Tshao Yuan-Yu (*1979*), p. 551 has made a thorough search of the Thang histories and has not come up with any material to corroborate Li Shih-Chên's remark on distilled grape wine. Wu Tê-To (*1988*), p. 85, found a similar reference to the making of grape wine during the Thang in the *Thai Phing Yü Lan*, ch. 844 and in *Nan Pu Hsin Shu* 南部新書, ch. 22, by Chhien I 錢易 of the Sung. Unfortunately, neither passage describes the method that was obtained from Kao-Chhang. Li Shih-Chên's original source for this passage is presumably lost.

[9] *SCC* Vol. v, Pt 4, pp. 141–4. The quotation from Li Ho's poem is not included since it may refer to red wine, rather than distilled wine.

[10] *SLPT*, item 178, pp. 111–12. The term *chêng* 蒸 is actually used, but it is not sure that it means steaming in this context, cf. Liu Kuang-Ting (unpublished MS). The meaning of *chêng* will be discussed later.

[11] *SCC* Vol. v, Pt 4, p. 144. This statement is from the *Ling Piao Lu I* (Strange Things Noted in the South), +890, which states that 'some wines that are not burnt, are used for making "clear" wine'. It is not seen in the extant editions of the *Ling Piao Lu I*. It is found in the passage quoted in the *Thai Phing Yu Lan*, ch. 845. The original reads, *I yu bu shao che, wei chhing chiu ye* 亦有不燒者，為清酒也。

[12] *SCC* cf. Vol. v, Pt 4, pp. 144–5. There is, however, some question as to whether Thien Hsi (+940–1003) is actually the author of the *Chhü Pên Tshao* cf. Liu Kuang-Ting (undated MS). In the *Min Su Chhung Shu* 民俗叢書 (Compendium of Folk Literature), Section II, published in Taipei, it is listed as a work of the Ming dynasty.

[13] *SCC* ibid., p. 147, n. f. The original occurs in *MLL*, ch. 13, p. 108.

Tung-Thing Chhun Sê 洞庭春色 (Spring Colours by the Tung-thing Lake), a product made by the fermentation of orange juice, in which the Sung poet–scholar Su Tung-Pho (+1036 to +1101) said:[14]

> Suddenly the cloud vapour condenses like melting ice,
> Whereupon tears come forth dripping down like liquid pearls.

In spite of the poetic imagery, one is tempted to conclude that the process described in this couplet cannot be anything but a distillation. If so, distilled wines were certainly not unknown in Su Tung-Pho's time. It is curious that there is no mention of a burnt-wine in the *Pei Shan Chiu Ching* (+1117). As for the significance of the passage on 'fire-pressured wine', *huo pho chiu* 火迫酒, we are now of the opinion that it refers to the pasteurisation of wine and has nothing to do with the process of distillation.[15]

Furthermore, it was noted that the use of steam distillation for the preparation of perfumed water, i.e. that extraction of essential oils, was well known to the Sung iatro-chemists.[16] Three references are particularly relevant. The first is the *Thieh Wei Shan Tshung Than* 鐵圍山叢談 (Collected Conversations at Iron-Fence Mountain), +1115 by Tshai Thao 蔡絛 on rose-water. The second is the *Yu Huan Chi Wên* 游宦紀文 (Things Seen and Heard on Official Travel), +1233, of Chang Shih-Nan 張世南, on oil of citrus. The third is the *Chu Fan Chih* 諸番記 (Records of Foreign Peoples and their Trade), +1225, by Chao Ju-Kua 趙汝适, also on rose water. The distillation vessels, i.e. steamers, are said to be made of white metal or tin. Since the Sung technologists knew how to prepare perfumed water by steam distillation, there is no reason why they would have refrained from using the same equipment and methodology to prepare distilled spirits from fermented mash.

These then are the major pieces of evidence examined earlier on the existence of distilled spirit and the art of distillation in the Thang and early Sung. It has been said earlier that none 'is quite decisive; some are relatively convincing, others less so. But there are times when probabilities accumulate to such an extent as to change quantity into quality and justify a circumstantial conclusion.'[17] Before we attempt to re-evaluate this opinion in the light of recent debates on these references we need to mention two new discoveries that have been made since the above words were written. The first is another quotation attributed again to Su Tung-Pho in the *Wu Lei Hsiang Kan Chi* 物類相感集 (On the Mutual Response of Things according to their Categories), c. +980:[18] 'When the wine catches fire, smother it with a piece of blue cloth.' There is no way that a wine that burns can be prepared except by distillation. Second, there is a story in the *I Chien Chih* 夷堅志 (+1185) which relates the tale of a winery attendant who died in a fatal accident while steaming wine (*chêng*

---

[14] Tr. Needham, adjv. Hagerty, *SCC* Vol. v, Pt 4, p. 145; cf. *TSCC*, *Tshao Mu Tien*, ch. 226 and *Chiu Shih*, ch. 1, p. 8. The original reads: 忽雲蒸而冰解，旋珠零而涕潛.

[15] The possibility that *huo pho chiu* denotes distillation was first suggested by Wu Tê-To (*1966*), p. 54; the idea was discussed in *SCC* Vol. v, Pt 4, p. 146.

[16] *SCC* Vol. v, Pt 4, pp. 158–62.   [17] *Ibid.*, p. 146.

[18] Tr. auct.; cf. *Wu Lei Hsiang Kan Chi*, p. 6. The original reads, 酒中火焰，遺青布拂之自滅. According to *Sung Shih* (History of the Sung), it was actually written by the monk Tsan Ning, a contemporary of Su Tung-Pho. This quotation was first cited by Tshao Yüan-Yü (*1979*) (repr. in Chao Khuang-Hua (*1985*), pp. 550–6).

chiu 蒸酒).¹⁹ The term *chêng chiu* may also mean to 'distil wine'. Wu Têo-To and Li Hua-Jui have both interpreted the anecdote as evidence that the preparation of distilled wine was already practised on a significant scale during the Southern Sung.²⁰

But before we proceed further we ought to take note of the types of distillation equipment designed and fabricated by the mediaeval Chinese alchemists that are described in the *Tao Tsang* (Taoist Patrology) and reviewed previously in *SCC* Volume v, Part 4.²¹ The only distillation apparatus as setup that is ready for use is the mercury still from the *Tan Fang Hsü Chih* 丹房須知 (Indispensable Knowledge of the Chymical Elaboratory) of +1163.²² Although the text was first published in +1163, the material in it can be traced as far back as the +3rd century to the time of Ko Hsuan, the great-uncle of Ko Hung.²³ There is no mention in the alchemical literature about this apparatus being employed for the distillation of wine, but it is clear that it could have easily been used for that purpose. In other words, based on the Taoist literature the capability for carrying out a distillation has probably existed in China in the alchemical community since the +3rd century. With their penchant for secrecy, it is possible they had actually prepared a distilled wine but did not spread the word except to those within their own fraternity. Unfortunately, at the time when *SCC* Volume v, Part 4 was completed no working example or model of such a still has been found among the remains that have been uncovered in the archaeological sites of mediaeval China.

Documentary evidence from the alchemical literature has, so far, been equivocal. The simplest type of still for converting cinnabar to mercury is the *ambix* or 'pomegranate' flask (*shih liu kuan* 石榴鑵) as shown in *Tao Tsang* 907, *Chin Hua Chhung*

---

¹⁹ There are two passages on this incident. The first is in *I Chien Ting Chih* 夷堅丁志, ch. 4, p. 569, which simply says, *I chiu chiang yin chêng chiu hui huo chung* 一酒匠因蒸酒墮火中. The second is in *I Chien San Chih* 夷堅三志, 壬卷 ch. 9, pp. 1526–7, which tells the story of one Yang Ssu 楊四 who loved to eat chickens but also delighted in killing them in a most sadistic way. He used to place a couple of chickens in a cage and scald them with boiling water. The chickens would run amok until all the feathers fell off. He would then kill and eat them at leisure. In thirty years he must have consumed well over ten thousand chickens. One day while he steamed a batch of wine (*chêng chiu* 蒸酒) he fell asleep by the side of the stove. The steamer he used consisted of a wooden barrel large enough to accommodate over ten jars of wine. The stove was heated by burning rice straw. Suddenly the steamer barrel collapsed splashing boiling water, steam and hot wine from the broken jars all over Yang's body. He jumped about just like his scalded victims. In two days he died.

Although we cannot rule out the possibility that Yang Ssu was actually distilling a wine, the second anecdote also suggests that the *chêng chiu* 蒸酒 (steaming of wine) may in this case be the same as the *chu chiu* 煮酒 (cooking the wine) step in task 14, used to stabilise the wine.

²⁰ Wu Tê-To (*1988b*), p. 91, and Li Hua-Jui (*1995b*), p. 51. Two other Sung references on burnt wine have been cited in recent reports by other authors from China. The first is from *Thai Phing Hui Ming Ho Chi Chü Fang* (Standard Formularies of the People's Pharmacy in the *Thai-Phing* reign-period), +1178, p. 5, which has been cited by Chu Chhêng (*1987*), to indicate that 'burnt wine' was used to break up particles to make *hsiang mo* 香墨 (fragrant ink). However, Wu Tê-To (private communication) has found that the original text actually says *shao chhu* 燒醋 (burnt vinegar) and not *shao chiu*. Presumably, the text that Chu Chhêng used had a copying error.

The second reference, first cited by Tshao Yüan-Yü (*1985,1979*), occurs in a treatise on forensic medicine, the *Hsi Yüan Lu* (The Washing Away of Wrongs), +1247, in which a mouthful of *shao chiu* is used to extract the toxin from the wound inflicted by a snake bite. The passage, however, is not in the original text of *Hsi Yüan Lu*. It is found only in an amended work, *Pu Chu Hsi Yüan Lu Chi Chêng* (the *Hsi Yüan Lu* with Amendments and Annotations) +1796, ch. 4, p. 12, and is, therefore, of no value in evaluating the date of discovery of wine distillation in China.

²¹ *SCC* Vol. v, Pt 4, pp. 68–80 on 'The stills of the Chinese alchemists', and Fig. 1452, p. 77.
²² Ibid., p. 77, Fig. 1452.   ²³ The author of *Pao Phu Tzu*.

Pi Tan Ching Pi Chih 金華沖碧丹經秘旨, +1225. The flask is placed over a crucible and heat is applied from above for *destillatio per descensum*.[24] In 1972 several round flasks were seen among the medicinal artifacts discovered at a Thang tomb at Ho Chia Chhun 何家村, near Sian.[25] These were identified by Wu Te-To as 'pomegranate' flasks described in *Tao Tsang* 907. We do not share Wu Te-To's thesis that this type of still could have been used for the preparation of distilled wine, but the discovery certainly reinforced the notion that distillation as a process was well known to the alchemists in the Thang.

Nevertheless, other archaeological studies in recent decades have proved to be more rewarding. As indicated earlier, Needham *et al.* have noted a wine-distilling scene from the frescoes at the cave-temples of Wang-fu-hsia 萬佛峽, Yu Lin 榆林 Cave near Ansi, Kansu, dating from the Hsi-Hsia period (+1032 to +1227).[26] Since we do not have a drawing or description of the internal structure of the still we cannot tell whether it has a side tube or not. If it does, it will be of Chinese design, otherwise it will be Mongolian. But the significance of the Yu Lin frescoes is now dwarfed by several recent exciting archaeological discoveries.

(ii) *Recent archaeological discoveries*

The first is the discovery, in 1975, of a bronze-steamer still in a Chin 金 (+1115 to +1234) tomb near Chhêng-teh 承德, Hopei province.[27] It is really a typical steamer (*tseng* 甑) surmounted by an open pot with a convex bottom which serves as a cooling device. Thus, it consists of three parts, the boiler (*fu* 釜 or *kuo* 鍋), the container or steamer (*tsêng* 甑) and a condenser (Figure 47). A disposable grating, probably made of bamboo, separates the boiler from the container. A cross-section of the fully assembled vessel is shown in Figure 48, and its dimensions are listed in Table 24.[28] From Figure 48 we can see that the bottom of the condenser now forms a dome over the steamer, and fits snugly over the annular gutter with a side tube at the mouth of the steamer. To work it, water is placed in the boiler at the bottom, fermented mash on the grating in the middle, and cold running water on top in the condenser. As the steam from the boiler rises, it heats the mash and carries alcohol vapour with it to the cooled, domed ceiling. The steam and alcohol vapour condense, the liquid runs down the sides of the dome to the annular gutter where it is carried by the side tube to a collecting bottle.

The apparatus has been tested using a batch of fermented mash and shown to work well.[29] That it was actually used for distilling spirit from fermented mash is

---

[24] *SCC* Vol. v, Pt 4, p. 58.   [25] Wu Tê-To (*1982*).

[26] Cf. *SCC* Vol. v, Pt 4, pp. 64–7, particularly Fig. 1443, p. 66. The paintings referred to are copies made by Tuan Wên-Chieh (*1957*) at Yü Lin Khu published in *The Frescoes of Yü Lin Khu*, Tunhuang Research Institute. Liu Kuang-Ting and others (undated MS) have suggested that the particular fresco was painted during the Yuan period.

[27] In Hopei province; cf. *Chhing-lung hsien ching chan tzu ta-tui ko-wei hui* (*1976*) and Lin Yung-Kuei (*1980*).

[28] Lin Yung-Kuei 林榮貴 (*1980*), pp. 66–71.

[29] *Chhing-lung hsien ching chan tzu ta-tui ko-wei hui* (*1976*). Two experiments were carried out. Volume of mash: volume of water = 1:1. Yield of distillate was 0.9 catty from 8 catties of mash, and 0.56 catty from 6 catties of mash. The distillate contained approx. 9% alcohol. Since the content of alcohol in the mash is not reported, it is difficult to say how successful the distillation was.

(a)   (b)

Fig 47. Chhêng-teh still, from *Chhing-lung hsien*, **1976** (9), Figs. 1 & 2. (a) Chin still assembled. Height 41.5 cm, diameter at widest point, 36 cm. (b) Bottom: vessel serving as boiler and steamer. Top: condenser with convex bottom.

strongly indicated by the water marks on the walls of the boiler. From the bottom to a height of 6 cm, the wall is very dark. This is the zone of contact with the boiling solution as it continues to accept more and more dissolved matter from the mash as the steaming progresses. Next there is a grey zone of about 10 cm which covers most of the space that contains the fermented mash. Finally, there is a zone of another 10 cm which shows considerable signs of corrosion presumably because this is where steam and air together made contact with the metal. The marks would be quite different if the vessel had been used for the preparation of mercury or extraction of essential oils from plants.

Even more exciting is the discovery of a very similar type of bronze vessel in the collection of the Shanghai Museum,[30] and another one at the Archaeological Bureau of Chhu-chou, Anhui, near Nanking.[31] Both vessels have been dated at about the Eastern Han period, more than 800 years before the Northern Sung. A picture of the Shanghai specimen fully assembled is shown in Figure 49, and a cross-section of it in Figure 50. The vessel consists of two parts. The top part is a traditional steamer (*tsêng* 甑) with a built-in grating as its bottom. The bottom part is the boiler (*fu* 釜). The cover or condenser to be placed on top of the steamer is lost.

---

[30] First reported by Wu Tê-To in 1986 at the 4th International Conference on the History of Science in China, May 1986, Sydney, Australia. A detailed report was presented by Ma Chhêng-Yüan at the 6th International Conference on the History of Science in China, Cambridge, *1990*, and published in Shanghai (*1992*).

[31] Li Chih-Chhao & Kuang Tsêng-Chien (*1986*).

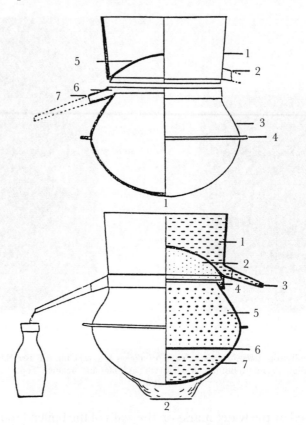

(1) 1. Condenser pot, 2. Drain pipe, 3. Steamer, 4. External flat ring 5. Convex bottom, 6. Circular gutter, 7. Drain tube.

(2) 1. Cold water, 2. Wine vapour, 3. Stopper. 4. Condensed wine, 5. Fermented mash, 6. Perforated grid, 7. Boiling water.

Fig 48. Cross-section of Chhêng-teh still, Lin Yung-Kuei, **1980** (1), Fig. 5.

Table 24. *Dimensions of the Chin bronze steamer-still: total height 41.5 (all dimensions are given in centimetres)*

Boiler and steamer	Condenser
Total height 26.0	Total height 16
Height of neck 2.6	Height of dome 7.0
Diameter at mouth 28.0	Diameter at mouth 31.0
Maximum diameter of boiler 36.0	Diameter at bottom 26.0
Annular gutter, width 1.2, depth 1.0	Drainage spout about 2.0 (damaged)
External rim, width 2.0	
Drainage spout, length 20	

Fig 49. Shanghai still, parts (a), (b), (c) photograph Ma Chengyuan: (a) Complete still; (b) Boiler; (c) Steamer, note the performated bottom.

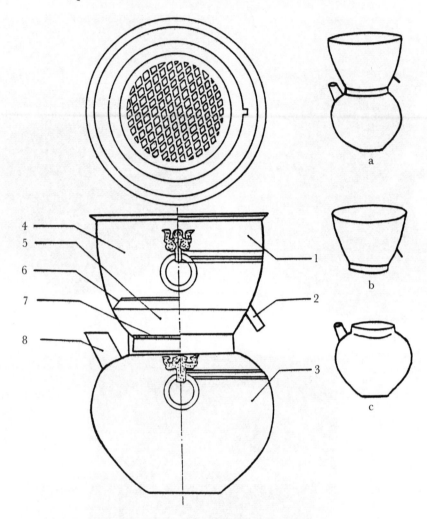

a. Still, assembled   b. Steamer   c. Boiler

1. Steamer, 2. Drain tube, 3. Boiler, 4. Condensing chamber, 5. Fermented mash, 6. Circular gutter, 7. Grid, 8. Return vent.

Fig 50. Shanghai still, cross-section. Ma Cheng-yuan (*1992*), 174–83, Fig. 1. Total height 45.5 cm. First presented at 6ICHSC, Cambridge, 1990.

The key to its function is an annular gutter placed above the grating around the inner wall of the steamer (Figure 49c). Connected to the gutter is a spout to allow any liquid collected in the gutter to flow out into a collecting jar (Figure 49a). One curious feature is a vent tube in the upper part of the boiler (Figure 49b) whose function has elicited a considerable amount of speculation. To prepare distilled wine, the fermented mash is placed on top of the grating and heated by the steam rising

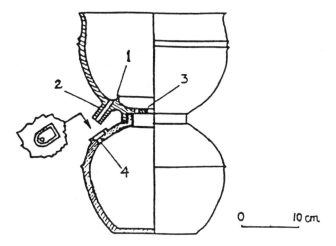

1. Circular gutter   2. Collecting tube
3. Perforated grid   4. Return vent

Fig 51. Chhu-chou still, cross-section, Li Chih-Chhao & Kuan Tsêng-Chien (*1982*), Fig. 1.

from the boiler. Presumably, when in use, the boiler vent is plugged and the steamer is surmounted by a cover or some type of condenser (similar to the one found in the Chhing-lung still). The condensate flows down the sides of the vessel into the annular gutter and is led away by the spout into a collecting vessel.

As shown by its cross-section (Figure 51), the construction and dimensions of the Chhu-Chou 除州 specimen is similar to that of the Shanghai model. The cover is also lost. Instead of a vent on the boiler we now have an opening which presumably could serve the same function as the vent on the Shanghai boiler (Figure 49b). The geometry of the Shanghai still suggests that if the spout on the steamer and the vent on the boiler are placed on the same plane adjacent to each other they can be easily connected so that the condensate from the annular gutter would flow back into the boiler. In effect the still would function as a rudimentary Soxhlet extraction apparatus for continuous steam distillation of the substrate placed on the grid of the steamer. A similar connection can be arranged for the Chhu-Chou still.

Ma Chhêng-Yüan and his colleagues at the Shanghai Museum have tested the vessel in several ways.[32] Four series of experiments were conducted. Except for the first series all the experiments were carried out with a bronze replica of the original parts.

---

[32] Ma Chhêng-Yüan (*1992*), p. 181. Many specimens of covers of ancient Chinese bronze steamers are now known. The cover that was fabricated for the steamer was based on the designs of known specimens. In the experiments with the bronze replica a wet cheesecloth was placed over the cover to lower the temperature of the condenser.

1. Steam distillation of fermented mash. The original steamer was used, together with an aluminium pot as the boiler and a specially fabricated cover as the condenser. The substrate was a fermented mash prepared from glutinous rice used by the Shanghai Brewery for production of *chhi pao ta chhü* 七寶大麴 brand of distilled spirit. The alcohol content of the distillate was 20–27%.

2. Direct distillation of wine. The apparatus was used as a simple still. In one test the alcohol content was raised from 51.1% to 79.4% and in another test from 15.5% to 42.5%.

3. Role of packing materials in the steamer. It was thought that the packing materials could act to fractionate the ascending liquid and increase the alcohol content of the condensate. Fermented wine was the substrate. Use of fennel, cotton gauze and gourd sponge in the steamer all increased the alcohol content of the distillate.

4. Steam distillation of scented ingredients. The spout from the steamer and the vent on the boiler are connected. Cassia or fennel was loaded on the grid. After continuous steam distillation a cassia oil (b.p. 240–60°C) or a fennel oil (b.p. 160–220°C) was obtained.

The results of these experiments demonstrate that the technology for making distilled spirits from either clear wine or fermented mash has existed in China since the Eastern Han. Furthermore, this equipment can be adapted for the continuous steam distillation of essences of perfumes and spices. But from the viewpoint of the history of distillation the most interesting aspect of these discoveries is that both the Han and Chin stills utilise an annular gutter to collect the condensate and thus conform to the 'Hellenistic' rather than the 'Chinese' model of a still as delineated in *SCC* Volume V, Part 4.[33] It would appear that a reassessment of the origin of the stills used in traditional Chinese distilleries is in order. To facilitate discussion, we have reproduced the diagrams for the two types of stills in Figure 52. Unfortunately, there is no drawing or description available from Chinese sources of a commercial still in use during the Yuan, Ming and Chhing dynasties. Foreign observers have filled only part of the gap. The Jesuit Cibot did see a still in action in about +1780, but he found the Chinese alembics so rustic that he 'would not dare to give a description of them.'[34] There is, however, a description by Guppy of the equipment used for the distillation of *samshu* (i.e. *san shao* 三燒) in the 1880s which can be interpreted to denote a central catch bowl and side tube of the Chinese type.[35] On the other hand, the Japanese *rangaku* (a pharmaceutical extractor-still) of the +18th

---

[33] *SCC* Vol. V, Pt 4, pp. 80–121, cf. p. 81, Fig. 1454. The still-head with an annular gutter is believed to be derived from the Mesopotamian rim-pot stills of the −3rd and −4th millennia. It was in use by the Greek alchemists of the +1st to +4th centuries as shown by Taylor (1945). It was, therefore, designated as 'Hellenistic'.

[34] *SCC* Vol. V, Pt 4, p. 149.

[35] *Ibid.*, p. 114; Guppy, H. B. (1884) says, 'When the fermented millet is taken out of the jars, it is placed in a large wooden vat or tub, the bottom of which is made of kind of grating; and beneath this vat is placed a large boiler of water which is heated by an adjacent furnace. The steam ascending through the grating and passing through the fermented millet finally comes into contact with a cylinder of cold water; it is there condensed, and trickling off in a little gutter finds its way out through a long spout in a clear stream of veritable *samshu*.'

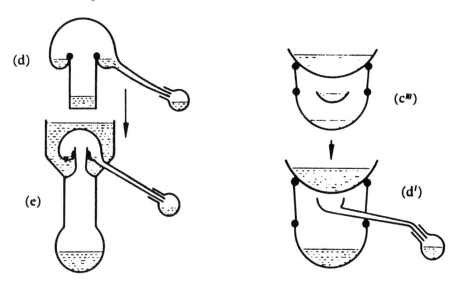

Fig 52. Diagram of Hellenistic and Chinese stills; adapted from *SCC* Vol. v, Pt 4, p. 81, Fig. 1454, (d), (e) and (c^{III}), (d^I).

century, which has previously been thought of as an apparatus produced under Western influence, is now seen to be identical in design to the Chin bronze still shown in Figures 47 and 48.[36] It may be presumed to be of Chinese origin. Two pictures of wine stills in Chhing China drawn by Europeans have been reproduced in *SCC* Volume v, Part 4, one from Mason (1800) and the other, Gray (1880), but without a diagram of the cross-section of each still, it is difficult to tell whether they belong to the 'Hellenistic' or 'Chinese model'.[37] These may be compared to the drawings and photos of the stills in early 20th century China from Hommel (1937) which were also presented.[38] Hommel's diagrams of the Mongolian and Chinese stills are reproduced here in Figure 53. Recently, Fang Hsin-Fang published diagrams of the 'Hellenistic' and 'Chinese' styles of industrial equipment he had found in use in the early 1930s for the distillation of wine in different parts of China.[39] Since he saw them well before any plans to modernise the plants got underway, they may be regarded as representative of the traditional stills in operation during the Chhing and Ming dynasties (at least in overall design, if not in detailed construction). Both types are derived from the ancient Chinese steamer, *tseng* 甑. To avoid confusion we shall adopt Fang's nomenclature and call the Hellenistic the *vase* (*hu* 壺) type and the Chinese the *pot* (*kuo* 鍋) type. An example of the vase (*hu* 壺) type

---

[36] *SCC* Vol. v, Pt 4, p. 114, Fig. 1488a, 1488b. [37] *Ibid.*, p. 67, Fig. 1444, and p. 68, Fig. 1445.
[38] *Ibid.*, pp. 62–5, Figs. 1436–40; cf. Hommel (1937), p. 143, notes that the Mongols employed a still with a central collecting bowl to distil a liquor called *arrihae* from *airak* which is soured mare's milk (similar to koumiss). Although Hommel's book, *China at Work*, was published in 1937, the photographs were taken during 1921–6 and 1928–30.
[39] Fang Hsin-Fang (*1987*). Cf. also Li Chhiao-Phing (*1955*), p. 209.

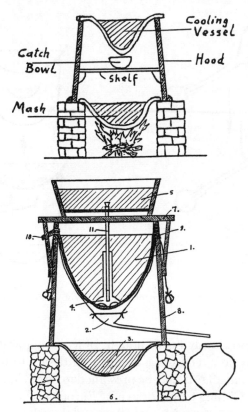

(Upper) Mongolian Still. (Lower) Chinese Still. (1) Condensing or cooling vessel. (2) Pewter catch basin with conveyer pipe. (3) Cast iron bowl with mash. (4) Pewter funnel. (5) Shallow wooden tub. (6) Fire box. (7) Wooden frame supporting shallow tub. (8) Barrel-shaped hood. (9) Gasket of sewn cloth, filled with sand. (10) Overflow pipe. (11) Wooden pipe with wooden stopper for letting cold water run down into the condensing vessel.

Fig 53. Mongolian and Chinese stills (after Hommel) reproduced from *SCC* Vol. v, Pt 4, pp. 62–3: (upper) Mongolian still; (lower) Chinese still.

still, found in a distillery in Tangshan 唐山, Hopei, is shown in Figure 54. It is, in fact, a direct descendant of the Han or Chin bronze still, except that the annular gutter is now part of the condenser rather than the steamer. We have no idea when this critical transformation took place. It is critical since the arrangement allows the steamer itself to be made larger and lighter by using wood for its fabrication, an essential step in the scaling-up of the process. An example of the pot (*kuo* 鍋) type, used in Fenghsien, near Thaiyuan in Shansi, is shown in Figure 55. The main difference between the two types is that in the 'vase' type the cooling surface is concave, and in the 'pot' type, convex. In the latter case the annular gutter is replaced by a central bowl (with a side-tube) below the condenser. But at what date this simple and elegant invention was made still remains at present beyond

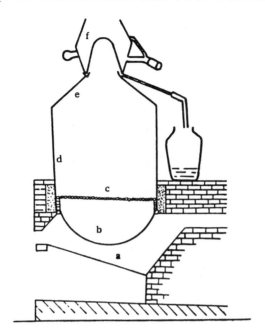

(a) Stove, (b) Boiler, (c) Perforated grid, on which fermented mash is placed,

(d) Steamer, (e) Cover, containing circular gutter with drain tube,

(f) Condenser, with convex cooling surface.

Fig 54. Early 20th century 'Hellenistic' or vase type still, Fang Hsin-Fang (*1987*), Fig. 1.

conjecture.[40] The 'vase' type occurred mainly in Hopei and the Northeast, and the 'pot' type, in other parts of China in the famous distilleries of Shansi, Szechuan, Kweichow and Sinkiang. To what extent the shape of the vase type condenser (Figure 54) which is made of tin, has been influenced by the Moor's head of the West, is a topic of considerable interest that deserves further elaboration.[41]

But we must now return to our main question, and that is, when was the technology, as exemplified by the specimens of late Han stills, first used for the preparation of distilled wine? This brings us to the third archaeological discovery in the form of two Han brick murals from Szechuan, one found in Phêng-hsien 彭縣 in 1955, and

---

[40] Cf. *SCC* Vol. v, Pt 4, p. 97.
[41] Hung Kuang-Chu (*1984a*), pp. 144 and 150, based on the dimensions of the Chin still discovered at Chhing-lung (Figure 47), thinks it cannot be very efficient when used with solid fermentation mash. He suggests that it was probably employed mainly for distilling koumiss rather than making Chinese distilled wine. The significance of the Shanghai, Chhu-chhêng and Chhêng-teh stills has recently been discussed by Huang Shih-Chien (*1996*).

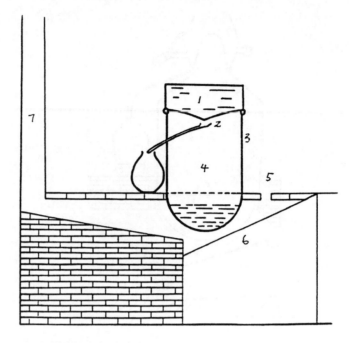

1. Cold water,   2. Collecting bowl and tube,   3. Steamer,
4. Fermented mash,   5. Stove opening,   6. Fuel,   7. Chimney.

Fig 55. Early 20th-century 'Chinese' or pot type still, Fang Hsin-Fang (*1987*), Fig. 3.

the other in Hsin-tu 新都 in 1979.[42] The two murals are almost identical. A photograph of the mural from Hsin-tu is shown in Figure 56.[43] For almost three decades it was thought that the murals depicted the process of the fermentation of wine. Indeed, this was the interpretation adopted in the commentaries that accompanied the photographs of the murals that were published. For example, the 1983 commentary says:[44]

---

[42] Liu Chih-Yuan *et al.* ed. (*1958, 1983*).

[43] Courtesy of the British Museum. The brick mural from Hsin-tu is 50 cm wide and 28 cm high. Rubbings and photographs of the murals have been published in China, namely Liu Chih-Yuan *et al.* ed. (*1958*), Pl. 9 and Liu Chih-Yuan *et al.* ed. (*1983*), Fig. 49, Wang Yu-Phêng (*1987*), Pl. 1 and in Japan, Tanaka Tan (*1985*), p. 283, Fig. 51. The quality of the reproductions is, however, less than satisfactory. We would probably have to be content with one of these reproductions for Figure 56 had fate not intervened in our favour. It was by chance that I was in London in mid-September 1996. I took the opportunity to visit the British Museum to see an exhibit of recent archaeological finds from China entitled, 'Mysteries of Ancient China'. Imagine my delight and surprise when among the wonderful artifacts I browsed through, I suddenly came upon the very object of this discussion, the Han brick mural itself in glorious relief and beauty. The lighting was just perfect to examine the scene in detail. It is listed as item no. 103, p. 199, in the catalogue edited by Jessica Rawson (1996), and was on loan from the Szechuan Provincial Museum, Chengtu.

[44] Liu Chih-Yuan *et al.* ed. (*1958*), pp. 4–5 and Liu Chih-Yuan *et al.* eds. (*1983*), pp. 50–1. The passage is a combination of the 1958 and 1983 commentaries.

Fig 56. Brick mural from Szechuan, possibly depicting the distillation of wine, photograph from Rawson ed. (1996), p. 199.

The mural depicts the fermentation of wine. In the right [centre] is a large mixing pan [on top of a stove]. A woman holds on to the rim with her left hand, and is stirring the content with her right hand, presumably mixing the ferment with the substrate. To her left is a man who appears to be helping her. In front of the pan is a processing counter which shelters three fermentation jars. The jars are connected by a spiral tube to an opening at the top of the counter, through which additional substrates may be added to the jar as needed.

The main problem with this interpretation is that the 'fermentation jars' under the counter are simply too small. In fact, the scene includes a labourer carrying two such jars using a regular carrying pole. There are many Han murals depicting fermentation processes, such as the famous kitchen scene from Chuchhêng, Shantung, shown in Figure 28. The fermenting jars shown are much larger. It would be impossible for one person to carry two such vessels. The discovery of the Shanghai Han still has stimulated a re-evaluation of the scene and led Wang Yu-Phêng to postulate that these Szechuan brick murals actually represent the preparation of distilled spirit from fermented mash.[45] The large pan near the centre of the scene is actually the open top of the condenser which sits on top of the steamer and the boiler assembly below. The steamer and boiler are hidden from view. The woman by the pan is actually stirring the cooling water which has to be changed from time to time. The condensate that forms under the pan is collected (either by a central bowl or an

---

[45] Wang Yu-Phêng (*1987/1989*), pp. 277–82.

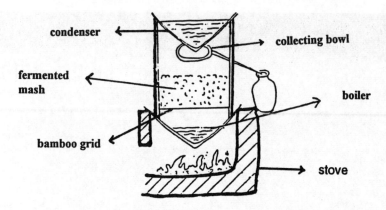

Fig 57. Traditional Szechuan still in the 1940s, Wang Yu-Phêng (*1987*), p. 22.

annular gutter) and allowed to flow through a tube via the openings on the counter into the small jars below. Several jars are needed to collect the fractions of the condensate that run off from the still. The earlier fraction would have a higher alcohol content than a later fraction. Wang Yu-Phêng found that the setup shown in the Hsin-tu mural is quite similar to that seen in the countryside in Szechuan in the 1940s for the making of distilled spirit or *shao chiu*. Such a small distillery is called a *shao fan* 燒坊. A cross-section of a typical still is depicted in Figure 57.[46]

This provocative interpretation, however, has been refuted by Yü Ming-Hsien who considers the scene to represent the pressing of the fermented mash, that is task 13 in Table 23, as described in the wine making process in the *Pei Shan Chiu Ching*.[47] Instead of stirring the cooling water of the condenser, the woman is using a heavy implement to press on a round board sitting on top of the mash in a collecting trough below. The construction of the trough itself is hidden from view. If so, the fermented mash must have been in a solid or semi-solid condition. If it had been in a liquid condition, the mash would have been simply filtered as indicated in the mural from Chuchhêng, Shangtung shown in Figure 28. It is curious that the board for pressing the mash should have been round. Presses usually are square or rectangular. Furthermore, the implement she holds in her right hand does not look heavy enough to act as a weight to press on the mash.

Unfortunately, the mural does not show enough detail to allow us decide between the two interpretations. Futher information is needed before we can decipher the true meaning of these murals. For the present, we may just have to say that the mural probably depicts a type of fermentation process, in which a wine of high alcoholic content (perhaps, 15–20%) was produced. This could explain the small

---

[46] This sketch was published by in *ibid.*, p. 22. It is reproduced in Rawson ed. (1996), p. 199. The reader will recognise it as a rustic form of the 'Chinese' or 'pot' still described earlier, pp. 214–17.

[47] Yü Ming-Hsien (*1993*).

jars used in storing the finished product. But we cannot completely dismiss the possibility of a distilled wine during the Han.[48] There are, two anecdotes in the Han literature that suggest the existence of a distilled spirit. The first occurs in the *Hou Han Shu* (History of the Later Han),[49] which tells us that one Chao Ping 趙炳 'climbed up to the thached roof of his host's house, and proceeded to light a fire to cook a meal. His host was greatly concerned. But Chou Ping smiled and asked him not to worry. He finished his cookery with no injury to the house.' The fuel he used is presumably distilled spirit with a high content of alcohol. The second anecdote is found in the *Shên Hsien Chuan* (Lives of Holy Immortals), +4th century, of Ko Hung, which relates that during the reign of the Emperor Hêng-Ti 恆帝 (+147 to +167), a Taoist holy man, Wang Yuan 王遠 told his followers,[50] 'I am giving you a very special wine. It is made in the celestial kitchen. Its flavour is rich and robust. Ordinary mortals should not drink it directly. It will cause the bowels to rot. Please dilute it with water before you drink it.' Evidently, this Taoist adept had the means to prepare distilled spirits.

Based on the evidence presented above it is clear that the technology for distilling wine already existed during the Han and the possibility of the technology being used to prepare distilled spirit certainly cannot be ignored. The same type of equipment shown in the Han still of the Shanghai Museum was still being used during the Chin 金 (+1115 to +1234) period.

(iii) *Distilled spirit in the Thang and Sung*

We have summarised on pp. 204–8 the scattered references to *shao chiu* 燒酒 or burnt-wine in the Thang literature. The heart of the controversy is the issue of whether the term *shao chiu* as seen in the Thang texts meant distilled wine or simply wine that has been warmed or mulled. This issue has been the cause of heated debate in recent years.[51] Unfortunately, there is no direct documentation on the actual making of a *shao chiu* or distilled wine in the Thang literature. Usually, wine is warmed in a hot water bath, so to apply the term *shao* 燒 which means to burn, to the process would seem to be rather inappropriate. However, the recent discovery of a Sung wine-warmer that uses charcoal to heat the wine directly has been cited by Li Pin as proof that the Thang *shao chiu* should be interpreted as 'mulled wine'.[52] But there is a problem. In China all wines are drunk warm, including distilled wine.

---

[48] Even though we do not agree with the notion among some historians that distillation of wine was practised widely during the Han Dynasty, e.g. Huang, Ray (1988), p. 45 and Rawson ed. (1996), pp. 199–200.
[49] *Hou Han Shu, Fang Shu Lieh Chuan* 方術列傳, Chung-Hua ed. (1965), p. 2742.
[50] *Ku Chin Thu Shu Chi Chhêng*, Chengtu (1985) edn, section 4, *Po-wu hui-pien* 博物彙編, Sub-section 2, *Shen I tien* 神異典, ch. 232 (or ch. 9 of *Shên hsien lieh chuan* 神仙列傳), Biography of Wang Yuan, pp. 62, 139–40.
[51] In particular by Wu Tê-To (1988), Mêng Nai-Chhang (1985), Liu Kuang-Ting (1987) and an undated MS, Chang Hou-Yung (1987), Huang Shih-Chien (1988 & 1996), Li Hua-Jui (1990) and Li Pin (1992).
[52] Li Pin (1992), p. 79. For the original excavation report see *Ssue-chhuan shêng wên-wu khao-ku etc.* (1990), pp. 123–30. Reference to the charcoal wine warmer is on p. 128.

If warming the wine before drinking is the normal practice, why was it necessary to give the warmed wine a special name?[53]

Furthermore, there are two perfectly good words used to denote warming or mulling a wine in the Thang literature.[54] One is *wên* 溫 as seen in a poem by Yuan Chieh 元結 in the line,[55] 'Burning (*shao* 燒) wood to warm (*wên* 溫) the wine.' Another is *nuan* 暖 which is seen in three poems, two by Pai Chü-I in,[56] 'To warm (*nuan* 暖) wine in the forest, burn (*shao* 燒) red leaves,' and 'Gently warm (*nuan* 暖) the shallot spiced wine,' and the other by Li Hao in, 'If it is not warmed (*pu nuan* 不暖) the colour of the wine will be slow to appear.' In these quotations, the meaning of either *wên* or *nuan* is crystal clear. The word *shao* also appears in two of the quotations, and in each case there is no doubt that it means 'to burn'. These references, which have so far been neglected by those involved in the controversy, tend to reduce the possibility that the term *shao chiu* 燒酒 could have been used to denote warmed or mulled wine.

Another discrepancy is found in the reference to Liu Hsün (on p. 205) in the *Ling Piao Lu I*, which says,[57] 'When the wine is ripe, it is stored in earthen jars and heated (*shao* 燒) with charcoal fire. When the wine is not heated, it is used as clear wine (*chhing chiu* 清酒).' In this case, it would be difficult to interpret the word *shao* as merely to 'warm' since the 'clear wine' is itself also 'warmed' before it is drunk. However, Li Pin has ingeniously proposed that *shao* in this context means to heat the wine to a pasteurising temperature.[58] The statement can then be translated as 'when the wine is not pasteurised, it is used as clear wine'. He goes on to suggest that the term *chi shao* as in *chi shao chiu* 既燒酒 (p. 205) maybe taken to mean merely wine that has been pasteurised. In other words, *shao* and *chi shao* are synonymous with the term *huo pho* 火迫 (fire-pressured) as in the *huo pho chiu* that we encountered in the *Pei Shan Chiu Ching* earlier on p. 187. One might ask: if *shao* were an accepted term for heating the wine for pasteurisation, why then was it necessary to coin the term *huo pho* for the same operation? It is pertinent to note that in the example given (Table 23, task 15) it is 'clear wine' that was subjected to the heat treatment. This would suggest that 'clear wine or *chhing chiu*' was also pasteurised. The citation in the *Ling Piao Lu I* then implies that 'when the wine is not heated,

---

[53] Indeed, *shao chiu* (distilled wine) itself was also drunk warm. Hommel (1937), p. 146, describes a warmer for distilled spirit. Interestingly, the fuel used is the spirit itself. As far as we know, no one has called mulled distilled wine, *shao shao chiu*. In the West, grape wine is usually drunk chilled. Calling mulled Chinese wine *shao chiu* would be akin to calling chilled grape wine 'chilled wine' or 'frosty wine', and listing it as a separate category of wine. There is simply no need to do this.

[54] We are indebted to Shinoda Osamu (*1972a*) 'Wine in Medieval China', p. 337, for three of the four pertinent references on *wên* and *nuan*.

[55] The original reads, 燒柴為溫酒. It occurs in a poem entitled, 'Remembering Mêng Wu-Chhang in the snow' 雪中懷孟武昌.

[56] For the first two references see note 54 above. In Pai Chü-I's first poem the original line reads, 林間暖酒燒紅葉. In Li Hao's poem, the line is 不暖酒色上來遲. The second Pai Chü-I's poem is the *Chhun Han* 春寒 (*Chhüan Thang Shi*, p. 5124), which contains the line 酥暖薤白酒.

[57] *LPLI*, from *TPYL*, *yin shi pu*, p. 120. The original reads, *Chi shu, chu I wa yung, yung fên sua huo, shao chih*; *I yu pu shao chê, wei chhing chiu yê* 既熟，貯以瓦甕，用糞掃火，燒之（亦有不燒者，為清酒也）. The statement in parenthesis occurs in a footnote to the text.

[58] Li Pin (*1992*), pp. 80–1.

it is used as clear wine [which is also heated]'. The statement is self-contradictory. Unfortunately, there is no definite documentation in the Thang literature that would relate *shao chiu* directly to distilled wine, pasteurised wine or simply mulled wine.

But fortunately for us, there are a number of references in the Thang literature which may throw additional light on the problem. These indicate that there were two kinds of wine in circulation, one with regular strength and one that is unusually potent. The first indication of the existence of a high-potency wine is the account of the 'frozen-out wine' as reviewed earlier in *SCC* Volume v, Part 4.[59] Distilling is not the only way to obtain a liquor with a high alcoholic content from fermented wine. Freezing can achieve the same purpose. Evidence is unequivocal that frozen-out wine had been sent from the Western region to the Central kingdom during the Thang, so that those in positions of authority, including writers and poets, had a chance to taste the richness of its flavour. This potent wine was apparently an extremely rare commodity.

The second reference concerns the wine cups used during the Thang dynasty. Chang Hou-Yung has studied two wine cups discovered in a Thang tomb in Han Sen Tsai 韓森寨 near Sian in 1958.[60] They are made of porcelain and their dimensions are: diameter of mouth, 3.5, 3.4 cm; diameter of round foot, 2.0, 1.5 cm; and height, 2.7, 2.3 cm. These diminutive cups are similar in size and shape to those used to serve burnt-wine during the Ming–Chhing periods as well as in China today. They are obviously too small for serving fermented wine. The existence of these small wine cups implied something potent was to be served in them. What could it be, if not a distilled spirit?

Thirdly, as pointed out by Mêng Nai-Chhang, apparently two types of drinking vessels were used for wines during the Thang, a small one and a large one.[61] In a poem in the 'Drinking to the Eight Immortals' series, Tu Fu said of Li Pai, 'Give him one *tou* 斗 of wine, and he will spout forth a hundred poems.'[62] And yet in another poem, on how he missed his friend, he declares that from Li Pai, 'Quickly a thousand poems, float from one cup of wine.'[63] No matter how one looks at it a *tou* has to be many, many times larger than a cup. How then is it possible that in one case one *tou* can elicit a hundred poems, and in another a mere cup can inspire a thousand? Tu Fu gives us further examples in 'Drinking to the Eight Immortals' of how the effect of a cup of wine in one case can be far more pronounced than a *tou* or two of another wine. But we should let Li Pai speak for himself. In the second of the two poems on 'Drinking Alone Beneath the Moon' he tells us, 'Three cups penetrate the Great Truth, one *tou* accords with Nature's laws.'[64] If we look at the

---

[59] *SCC* Vol. v, Pt 4., pp. 151–4. [60] Chang Hou-Yung (*1987*). [61] Mêng Nai-Chhang (*1985*).

[62] *Li Pai i tou shih pai phien* 李白一斗詩百篇。

[63] *Min chieh shih chhien shou* 敏捷詩千首
 Phiao ling chiu i pei 瓢零酒一杯

[64] *San pei thung ta tao* 三杯同大道
 I tou ho tzu jan 一斗和自然

original, we see that the 'Great Truth', *ta tao* 大道, is more or less equivalent to 'Nature's Laws', *tzu jang* 自然, so that taken literally the effect of three cups is equal to that of one *tou*. This makes no sense unless the three cups contain a potent spirit (i.e. distilled or frozen-out wine) and the one *tou*, regular, fermented wine.

Finally, there is the discrepancy between the prices for wines recorded in the Thang texts. For example, Li Pai says in one poem,[65] 'A golden goblet for the beauteous wine, ten thousand cash a *tou*.' Yet Tu Fu, a contemporary, remarked in another poem,[66] 'Come quickly and join me to partake of a *tou*, I just happen to have three hundred copper coins.' Li Pai's wine appeared to cost more than thirty times per *tou* than Tu Fu's. How can it be? Unless Li Pai was talking about distilled wine, and Tu Fu, regular fermented wine.

From these considerations we may infer that in addition to regular, fermented wine there was in existence a high-potency wine which was probably prepared by distillation. This could be the *shao chiu* 燒酒, burnt-wine, mentioned by the Thang poets. It was potent, very expensive and available only in limited quantities. The only evidence that such a wine could have been prepared in a still is Li Shih-Chên's account of the steaming of fermented grape mash cited earlier. We lack direct documentation of such a process in the Thang literature. The situation presumably did not change significantly during the Sung. The relevant Sung references which support the presence of a distilled wine cited earlier (p. 205) are also a matter of controversy. We may summarise the arguments as follows:

1. The *Chhü Pên Tshao* 麴本草, +990, cites a twice 'burnt' (i.e. distilled) wine from Siam. This is now considered an early Ming work.[67] It is thus no longer relevant to our discussion.

2. We find in the *Meng Liang Lu*, Southern Sung, the statement that one could buy[68] 'crystal burnt wine, red and white'. The interpretation of this *shao chiu* as distilled wine has been questioned on two counts. First, the burnt-wine is said to be either red or white, whereas distilled wine is usually regarded as white. This is not a serious objection since it is quite normal for some of the colour in the original wine to be splashed over during the distillation. Second, the text goes on to say that the wine[69] 'had a gentle and fragrant taste, and evaporated as soon as it entered the mouth'. Since distilled spirit is said to have a fiery mouth feel, Huang Shih-Chien has argued that anything that can be described as 'gentle and fragrant' can hardly

---

[65] *Chin chun chhing chiu tou shih chhien*　　金樽清酒斗十千

[66] *Su lai hsiang chiu yin i tou*　　速來相就飲一斗
Chhia yu chhing thung san pai chhien　　恰有青銅三百錢

[67] As pointed out by Liu Kuang-Ting (undated MS), pp. 1–2 and Huang Shih-Chien (*1988*), p. 162 there are two problems with this reference. Firstly, there is no record of a *Chhü Pên Tshao* by Tien Hsi in the Sung Bibliographies. Secondly, the term used for Siam, Hsien-Lo 暹羅, in this work did not exist until the year +1349. The significance of this passage in terms of the prevalence of distilled spirit is, therefore, suspect.

[68] *MLL*, p. 108, the original text reads, *Shui ching hung pai shao chiu* 水晶紅白燒酒.

[69] *Ibid.*, the original text states, *chhi wei hsiang nuan, ji khou pien hsiao* 其味香軟，入口便消.

be construed as a distilled wine.[70] Actually, to those who are connoisseurs of distilled spirits, this is exactly how a good specimen should be characterised.[71]

3. The rhapsody on the *Tung-Thing Chhun Sê* 洞庭春色 remains intriguing. It is difficult to interpret the poem other than a distillation process.

4. The statement on wine catching fire in the *Wu Lei Hsiang Kan Chi* 物類相感集 would indicate that the wine in question is a distilled wine. Li Pin, however, has suggested that what we are dealing with here is the alcohol vapour at the mouth of the bottle of wine catching fire when it is warmed by a charcoal fire.[72] This is certainly possible, but quite unlikely since the amount of charcoal in a warming pan is hardly adequate to cause the wine to boil.

5. The story from the *I Chien Chih* 夷堅志 (+1185) about the untimely death of the winery attendant who was scalded while *steaming wine* (*chêng chiu* 蒸酒) has been interpreted by Wu Tê-To and Li Hua-Jui as evidence that a distilled wine was being made on a substantial scale.[73] But if we examine carefully the story in the *I Chien Chih* we would notice two interesting facts: first, the steamer consisted of a large wooden barrel, and second, over ten jars of wine were placed in the steamer [presumably on the grating] to be steamed.[74] Thus, while it is possible that the 'steaming' in this context actually denotes a distillation, it could also be interpreted as a process to 'pasteurise' the wine in the manner of the *chu chiu* 煮酒 step as described in the *Pei Shan Chiu Ching* cited earlier.[75]

On the other hand, a large wooden barrel could easily have served as a vehicle to distil wine provided it was fitted with an appropriate distillation head, a topic that we shall discuss later. In fact, Li Hua-Jui has found two poems in the Sung literature in which the term *chêng chiu* 蒸酒 is used in the sense of a product rather than a process, one by Su Shun-Chhing 蘇舜欽 and the other by Chhin Kuan 秦觀.[76] Since

---

[70] Huang Shih-Chien (*1988*), pp. 160–1. Shinoda Osamu (*1972a*), pp. 360–1, was the first to draw attention to the passage from *MLL* as evidence for the existence of a distilled spirit.

[71] For example, see Wên Ching-Ming & Liu Ching-An (*1989*). In extolling the virtues of the *fen chiu* 汾酒, a famous distilled spirit, the wine is said to be 'soft in the mouth, and gives a sweet taste when swallowed' (入口綿，落口甜). Similar accolades have been accorded to other famous distilled spirits of China, cf. Li Hua-Jui (*1995b*), p. 56.

[72] Li Pin (*1994*), p. 80. It should be pointed out that the Arab chemist Jabir (12th to 13th century) had noticed that fires could burn at the mouth of bottles when the wine was salted and boiled, cf. al-Hassan & Hill (1986), p. 141.

[73] Cf. note 20 above.

[74] *I Chien San Chih, Jên Chuan*, p. 1527; the original text reads, *yung ta thung tsuo chêng, kho yung chiu than shih yü* 用大桶作甑，可容酒壇十餘. Cf. n. 19 above.

[75] Cf. pp. 187–8, Table 23, task 14, *PSCC, SKCS* **844**, p. 830, which says that in *chu chiu* 煮酒 (boiling wine), the jar of wine is placed on the grating of a steamer and steamed.

[76] Li Hua-Jui (*1995b*), p. 51. For Su Shun-Chhing 蘇舜欽 cf. *Su Hsüeh Shih Chi*, ch. 6, a farewell poem to Chhên Chin-Shi 陳進士：

時有飄梅應得句，苦無蒸酒可沾巾 。

For Chhin Kuan cf. *Huai Hai Chi* 淮海集, ch. 2. In a poem remembering Liu Chhüan-Mei 劉全美 he says:

素冠長跪蒸酒歠，云是劉郎字全美。

In these passages *chêng chiu* 蒸酒 is, in our opinion, preferably interpreted as a distilled wine, a product in its own right, rather than wine that has been steamed.

pasteurisation or stabilisation by 'steaming' or direct heat was an integral component of the wine-making process, it would seem to us to make more sense to denote *chêng chiu* as 'distilled wine' rather than 'steamed wine'.

From these observations we would regard (1) as invalid (2) and (4) as supportive (3) and (5) as mildly supportive. But we still do not have documentary evidence that relates a distillation process to a high potency wine.

(iv) *Development of distilled wine in China*

Taking into account all of the evidence so far presented, what can we say about the origin of distilled wine in China? Was it invented or prepared for the first time during the Han, the Thang or the Sung? Or was it developed in China or introduced from abroad during the Yuan? Before we attempt to reach a conclusion, let us briefly summarise what we would consider as the key features of the evidence.

Firstly, two bronze stills dated to the Eastern Han (+25 to +220) and one to the Chin (+1115 to +1234) have been discovered. Their existence indicates that the technology of making distilled wine has been known in China since the Eastern Han. The construction of these stills is uniquely Chinese.[77] In fact, they are directly derived from the Chinese bronze steamer (*tsêng* 甑) by the insertion of a circular gutter at the mouth of the steamer. By having a vent in the boiler that connects to the outlet tube of the steamer, the still can also be used for continuous steam distillation in the manner reminiscent of the Soxhlet extractor.

The size of the Han and Chin stills indicate they are models for experimentation rather than full-scale equipment for production. There is no convincing evidence that distilled wines were produced for consumption during the Han, although it is entirely possible that small amounts were prepared and tasted by the alchemists.

Secondly, there is considerable indirect evidence to indicate the existence of two types of wines during the Thang, a fermented wine of regular strength and a highly potent wine. The latter is probably the burnt-wine (*shao chiu* 燒酒) cited by the Thang poets. Furthermore, distilled water (*chêng chhi shui* 甑氣水) was listed as a medicament in the *Pên Tshao Shih I* (Supplement to the Natural Histories), +725.[78] The use of steam-distillation to extract ingredients from drugs is cited in the *Wai Thai Mi Yao* (Secret Formulary from the Outer Terrace), +752.[79] Both operations are easily carried out in the type of Han still that has been discussed at some length above. These references show clearly that during the Thang, the Chinese had the technology to prepare distilled wine either from fermented mash or from finished wine.

---

[77] Although in the terminology proposed in *SCC* Vol. v, Pt 4, p. 81 they would be classified as Hellenistic stills, there is no evidence that their development was in any way due to influence from the West.

[78] *PTSI*, cited in *PTKM*, ch. 5, p. 409. The text says, *I chhi chhêng chhü* 以器乘取, that is, to collect [the water] with a receiver. This was first noted in *SCC* Vol. v, Pt 4, p. 62.

[79] *WTMY*, ch. 29, on 集驗去黑子及贅方, *SKCS* **737**, p. 229. The term used for steam distillation is *chêng ling chhi liu* 蒸令氣溜, which means literally, 'steam and allow the vapour to condense'. This reference was first cited by Liu Kuang-Ting (undated).

Thirdly, although we do no have a description of a wine distillation process in the Sung literature, the evidence suggests that burnt-wine, i.e. *shao chiu*, was already sold in the market of the capital and that it would catch fire when ignited. Furthermore, the steam distillation of flower essences was also an art well known to the Sung iatro-chemists.[80]

So much then for a review of the state of the art from the Han to the Sung. What about the Yuan, the dynasty during which, according to Li Shih-Chên, the art of making distilled wine was developed (see p. 204)? There are two references in the Yuan literature that are pertinent to our discussion. The first is a passage from the *Yin Shan Chêng Yao* (Principles of Correct Diet), +1330, of Hu Ssu-Hui, the Dietician at the Yuan Court:[81] '*A la chi* 阿剌吉 wine: Sweet and pungent in flavour, strongly heating and toxic. Relieves cold obstructions and counteracts chilling vapours. Heat superior wine to boiling point and collect the dew. This is *a la chi* wine.' *A la chi* (araki in Mongolian) wine is, thus, clearly prepared by direct distillation of finished, fermented wine. How was this distillation carried out? This brings us to the second reference, a recipe entitled *Nan fan shao chiu fa* 南番燒酒法 (The burnt-wine method of the Southern Tribal folk) from the compendium *Chü chia pi yung shih lei chhuan chi* 居家必用事類全集 (Essential Arts and Assorted Techniques for Family Living), c. +1300.[82] A note to the title says that the foreign name of the product is *a-li chhi chiu* 阿里乞酒 i.e. *a-ra-qi* wine. Although it has been quoted earlier, it is so central to our discussion that we need to quote it again.

Take any kind of substandard wine, be it sweet, sour, insipid or light. Pour it into a pot until it is about 80 per cent full. Place another pot on top of the first one, with the two mouths facing each other at a slight angle. Bore a hole on the side of the empty bottle and attach to it a hollow bamboo tube to act as a drain pipe. The other end of the tube is placed in the mouth of a second empty pot which acts as a receiver. Fill the space around the mouths of the two pots with pieces of porcelain or pottery and then seal with a lute made of a mixture of paper fibre and lime [so that the still will be airtight]. The lower bottle is firmly set on a large urn filled with the ashes of burnt-paper. Place two to three catties of hot, burning charcoal in the ashes around the bottle. The wine in the bottle soon begins to boil. The vapour rises into the empty bottle and condenses along its sides. The condensate flows into the bamboo tube and is collected in the receiver. The product is colourless just like pure water. The distillate from sour wine is tangy and sweet, while that from weak wine tends to be sweet. About one third of the original volume can be recovered as good [distilled] wine.

A graphic representation of the apparatus is shown in Figure 58. Even a cursory inspection of this setup will tell us that this still is a rudimentary version of the Arabic retort, a specimen of which is shown in Figure 59. Retorts are designed for simple distillation. Unlike the Han and Chin bronze stills that we discussed earlier, they cannot be used for the steaming of fermented mash and continuous steam

---

[80] Cf. *SCC* Vol. v, Pt 4, pp. 158–62.   [81] *YSCY*, ch. 3, p. 122.
[82] *CCPY*, p. 47. tr. auct. adjuv. Needham *et al.*, cf. previous quotation in *SCC* Vol. v, Pt 4, pp. 112–13. This is an example of a 'pot and gun barrel' still. Actually, it may also be regarded as a rudimentary retort. Since it came to China through the southern tribes, it was probably transmitted from the Indian culture area.

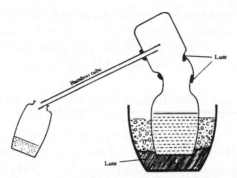

Reconstruction of the 'pot and gun-barrel' still described in the *Chü Chia Pi Yung Shih Lei Chhüan Chi* of +1301. An Indian or Western design with no cooling for either the helm or the side-tube.

Fig 58. Diagram of still described in the *CCPY*, reproduced from *SCC* Vol. v, Pt 4, p. 113.

Fig 59. Arabic Retort, photo from the Science Museum, London, +10th century to +12th century.

extraction of essential oils. Since the distilled spirits of China are normally made by steaming fermented mash, and seldom by the distillation of finished wine, this primitive and inelegant apparatus for distilling *a-ra-qi* wine could hardly have been of much value to the Chinese distillers.

There is actually one other reference in the Yuan literature on the making of distilled wine. It is the 'Ode to Araki Wine' (*Ya-la-chi chiu fu* 軋賴機酒賦) written by Chu Têh-Jun 朱德潤 in +1344 after he had received a present of this wine from a high official.[83] But it is loaded with such poetic imagery that we are unable to decipher how the still described in the poem is actually constructed. It is possible, as interpreted by Huang Shih-Chien, that the distillate is collected in a circular gutter.[84] If so, the still is a descendant of the Han and Chin bronze stills described earlier, and can hardly be considered as a Yuan invention. In this case the araki spirit was distilled from a finished wine and not from fermented mash.

We can only surmise that during the Yuan, Ming and Chhing dynasties the stills that were used to steam-distil fermented mash were either of the 'vase' or *hu* 壺 type, or of the 'pot' or *kuo* 鍋 type, in other words stills derived from the traditional Chinese steamer *tsêng* 甑, and not even remotely related to retorts used by the Arabs. Based on the information recorded by foreign observers in the 18th and 19th centuries and Fang Hsin-Fang's observation in the early 20th century, it would seem that the stills in use throughout China since the Yuan were of the 'pot' type rather than the 'vase' type. But there is no question that distilled spirits gained popularity during and after the Yuan.

After examining all the available evidence we are inclined to propose the following scenario for the development of distilled wines in China. The steaming of fermented mash and collecting the condensate was probably first accomplished during the Eastern Han in the type of bronze still found in the Shanghai Museum. Experimental amounts of spirits were prepared, but the art remained a secret held tightly by Taoist alchemists. During the Thang, limited amounts of distilled spirit were prepared and distributed. It was rare and extremely expensive. It was known to connoisseurs as burnt-wine (*shao chiu*), but the nomenclature of the product remained in a state of flux. Improvements in the process were made during the Thang and Sung. By late Sung or early Yuan a major breakthrough took place in the technology of wine distillation which turned distilled wine into an affordable article of commerce. It was then produced and consumed on a large scale. In other words, distilled spirit was invented during the Han, developed through the Thang and Sung and reached fruition as a commercial success in the Yuan. This may seem to the reader an unduly long process for the development of a food product. But in fact, as we shall show in the succeeding chapters, this is by no means an unusual situation when compared with the history of the development of other processed foods

---

[83] The poem is cited and discussed by Lo Chih-Thêng (*1985*), Chu Chhêng (Shêng)(*1987*), Huang Shih-Chien (*1988*) and Li Hua-Jui (*1995b*), p. 62.

[84] Huang Shih-Chien (*1988*). There is no mention of a grating in the poem. The araki wine was distilled from finished wine and not from fermented mash.

in China. We shall have more to say on this subject as part of our Reflections in Chapter (h).

How does this scenario measure up against the two apparently conflicting statements made by Li Shih-Chên in the *Pên Tshao Kang Mu*? Firstly, it takes into account Li's assertion that a high potency wine was prepared by steaming (*chêng* 蒸) fermented grape mash in a steamer (*tsêng* 甑) during the Thang dynasty. Secondly, it provides a rationale for Li's claim that the art of making *araqi* style distilled wine was developed during the Yuan. To explain what we mean we need to digress for a moment and briefly review the evolution of the still in China.

From the Han to the Sung the only still we know of is of the *vase* or Hellenistic type as exemplified by the Han and Chin bronze stills. These stills are characterised by an annular gutter built into the inner rim of the steamer (e.g. Figure 50). The only specimens we have of such stills are small models for experimental use. Larger versions for preparative purposes were presumably fabricated using bronze or pottery. The former would be costly to make and the latter cumbersome to use. And so distilled spirits remained a rare and expensive commodity during the Thang and early Sung. But something highly significant took place during the Sung which changed the state of the art of distillation by the beginning of the Yuan. A new type of still came into being; it was widely adopted and used for the production of distilled wine on a commercial scale. How did this change come about?

We have noted earlier that the Han and Chin bronze stills are descendants of the steamers for culinary use which are such a common feature of Chinese kitchens. As shown earlier, large pottery steamers sitting on pottery boilers are a common sight on murals and models of Han stoves.[85] Major changes, however, took place in the design of kitchen equipment between the Han and the Sung. Large iron pots in the shape of *wok*s replaced pottery pots as the boiler that sat on the stove. Pottery steamers for cooking were replaced by shallow wooden or bamboo steamers that were light and capable of being stacked on top of each other. A mural of a Sung stove, with a boiler and a stack of short bamboo steamers is shown in Figure 29a. Such steamers were a common sight in Chinese kitchens until recent times but they were ill-suited for conversion into stills.

Another approach was to replace the large pottery steamer with an open-ended wooden barrel. One opening is placed on a large *wok* or boiler. A grating made of wood and bamboo could be installed near the lower opening to house the materials to be steamed. The *I Chien Chih* 夷堅志 tells us that this type of steamer was used to steam jars of wine during the Sung (p. 225). According to the famous *Classic of Tea* (*Chha Ching* 茶經) of Lu Yü wooden steamers were used during the Thang to steam fresh leaves in the processing of tea.[86] To convert such a steamer into a still the challenge was to provide a condenser with a collecting device on top of the

---

[85] Cf. pp. 81 and 87 (Figures 26a and 28).
[86] Cf. note 75 above. The use of wooden steamers for steaming fresh tea leaves is mentioned in the *Chha Ching* (Classic of Tea), ch. 2, p. 7. Cf. also p. 521.

barrel. We suspect that this was the major problem that taxed the ingenuity and inventiveness of the Chinese wine technologists during the Thang and the Sung. One solution was to incorporate the annular gutter into a concave condenser. The result would be a device similar to the Moor's head of the West that we see in the *vase* type still described by Fang Hsin-Fang in the early 20th century (Figure 54). This was eventually accomplished, but we have no idea when. In the meantime a simpler solution was developed, and that was to place a central collecting bowl under a convex condenser. This arrangement is known today as the Mongolian still, which according to Dr MacGowan was used to prepare mutton wine and according to Hommel (Figure 53a) for the distillation of koumiss.[87] The main drawback of this still was that each time the central bowl was full, the condenser on top had to be removed before the distillate could be collected. This deficiency was rectified when a drain tube was attached to the central bowl to allow the condensate to flow into a collecting vessel outside the steamer. We have then the so-called 'Chinese' still which was noted by foreign observers in the 18th and 19th centuries (Figure 53b).[88] Such a still was easy and inexpensive to fabricate. We suspect the first 'Mongolian' style wooden still was built in late Thang or early Sung, and the first 'Chinese' style wooden still, in late Sung. It was this 'Chinese' still that made possible the commercial production of distilled spirit during the Yuan. Thus, when Li Shih-Chên said that 'the technique (for making burnt-wine) was first developed in Yuan times',[89] he was, we presume, referring not to the general method of distillation but rather specifically to the Chinese still, which was widely used during the Yuan and which, of course, was unknown to the ancients.

This Chinese style still, which Fang Hsin-Fang described as the pot or *kuo* type, is an indigenous (or Sino-mongolian) invention. It could not have been transmitted from the West, although it could easily have been used for distilling the *a-la-chi* wine mentioned in the *Yin Shan Chêng Yao*. Thus, we may say that Li Shih-Chên was correct in both his statements as quoted on p. 204. Firstly, distilled wine was known during the Thang, even though it was not necessarily prepared only from grape wine. Secondly, a new and convenient technique for wine distillation, as exemplified by the Chinese still, was introduced commercially during the Yuan. This 'art' (i.e. method) was not known to the ancients. It remained the method of choice for the commercial production of distilled wine through the Yuan, Ming and Chhing dynasties.

---

[87] Hommel (1937), p. 143. For the account by Dr MacGowan of the construction of a Mongolian still in the 19th century see the preparation of mutton wine on pp. 237–8. In spite of its name, the Mongolian still was not necessarily invented by the Mongols. It was used widely by the Mongols for the distillation of *koumiss* or *airak* (*soured mare's milk*). The distilled wine is called *arihae*. For a discussion of the geographical distribution of still types, cf. *SCC* Vol. v, Pt 4, pp. 103–21.
[88] See discussion and note, pp. 214–18.
[89] *PTKM*, p. 1567. The original words are, '*chi Yuan shih shih chhuang chhi fa* 自元時始創其法' i.e. the technique was developed during the Yuan, presumably by Chinese technicians. It does not say that the method came from abroad.

(6) MEDICATED WINES IN CHINA

Shên Shu-Chan 申叔展 asked Huan Wu-Shê 還無射, 'Do you have any wheat *ferment* (*mai chhü* 麥麴)?' He answered, 'No!' 'Do you have any hemlock parsley (*shan chü chhiung* 山鞠窮)?' 'No!' 'What then, do you do when you are sick in the stomach?'[1]

This quotation from the *Tso Chuan* or *Chhun Chhiu Tso Chuan* 春秋左傳 (Master Tso's Enlargement of the Spring and Autumn Annals), contains the earliest reference to the use of a *ferment* as a medicament. One may infer that wine too must have also been used very early in the treatment of illnesses. Indeed, the *Chou Li* (Rites of the Chou) indicates that the very word *i* 醫, for the medical arts, is itself the name of a wine made by the fermentation of a thin congee.[2] But the tradition of adding herbs or other drugs to wines probably started with the ritual wine *chhang* 鬯, which as we have already seen is one of the earliest wines made in ancient China (p. 156). *Chhang* is seen frequently on oracle bones of the Shang as well as bronze inscriptions of the Chou. On the bronzes it is often associated with the word *yü* 鬱, later identified as the herb, *yü chin tshao* 鬱金草, *Curcuma aromatica* Salib. of the ginger family.[3] In the *Shu Ching*, *Shih Ching* and the *Tso Chuan* we also see the term *chü chhang* 秬鬯, indicating that the *chhang* was made from the black panicum millet *chü* 秬.[4] The *Chou Li* lists a *Chhang Jen* 鬯人, Superintendent of Ritual Wines, whose duty it was to provide the wines and the accessories used on all ritual and ceremonial occasions, as well as a *Yü Jên* 鬱人, Superintendent of Ritual Herbs, who had the responsibility for blending the *chhang* wine with *yü chin tshao* herb and ensuring that the proper vessels were available to contain the resultant *yü chhang* 鬱鬯 wine.[5] The *Li Chi* states that the Son of Heaven alone had the prestige to present *chhang* wines as gifts, and mentions *yü chhang* 鬱鬯 several times as the fragrant wine for use in both rituals and entertainment.[6]

The textual evidence available suggests that *chhang* was the earliest fragrant wine, next came *chü chhang* and shortly thereafter *yü chhang*. The Chou literature also mentions a *chhang tshao* 鬯草, fragrant wine herb,[7] that was presumably used in the making of a herbal *ferment* (*tshao chhü* 草麴) in the manner later described in the *Nan Fang Tshao Mu Chuang* (A Flora of the South), +304, *Lin Piao Lu I* (Strange Occurrences South of the Range), c. +900 and *Thai Phing Huan Yü Chi* (The State of the World during the Thai-Phing reign period), c. +980. The nature of this *chhang tshao* is not known but it could have meant any kind of plant material incorporated in a *ferment*. At least some of *chhang tshao* wine herbs had come from South China or some

---

[1] *Tso Chuan, Hsüan Kung* 宣公 12th year (−597), p. 200. The *Tso Chuan* deals with events taking place between −722 & −453.
[2] *Chou Li, Chiu Chêng* 酒正 (Wine Administrator), p. 49 and *Chiang Jên* 漿人 (Superintendant of Beverages), p. 52.
[3] Ling Shun-Shêng (*1958*), pp. 896–7; cf. *Chiang-su hsin i-hsüeh yuan* (*1986*), *The Cyclopedia of Chinese Drugs*, pp. 1316 and 1735.
[4] *Shu Ching, Lo Kao* 洛誥; cf. Chhü Wan-Li ed. (*1969*), p. 187. *Shih Ching*, w 136 (m 262) translates it as black mead. *Tso Chuan, Hsi Kung* 僖公 28th year, p. 133 and *Chao Kung* 昭公 15th year, p. 389.
[5] *Chou Li*, pp. 209–10.   [6] *Li Chi*, pp. 76, 219, 394, 435, 758.   [7] Ling Shun-Shêng (*1958*), p. 900.

other far-off, exotic place.⁸ According to Ling Shun-Shêng many types of plant material were, until recently, used by the aborigines in Taiwan to make *ferments* and *chhang* wine directly from cooked grain without the aid of any pre-formed source of cultures. As we have noted earlier (pp. 161–2) the aborigines were also able to make *ferments* and a wine called *lao* from cooked grain even without the addition of plant material. The ease with which *chhang* and *lao* could be made by a primitive people with the simplest type of equipment is strongly indicative of the probability that one of the first *ferments*, either herbal or non-herbal, had originated in South China.

The therapeutic value of wine itself was recognised early in China. In fact the word *i*, for medicine, is in part derived from the radical *yu*, for wine.

i 醫   *yu* 酉

According to the *Huang Ti Nei Ching, Su Wên* (The Yellow Emperor's Manual of Corporeal Medicine, Questions and Answers):⁹

The ancient sages prepared decoctions and wines just to have them ready in case they were needed. They were there but hardly ever used. But in mediaeval times, people strayed away from the correct mode of living, and ill winds soon began to come among them. By imbibing the wines and decoctions, they were able to stay healthy.

Wine generates heat when taken internally. It stimulates the bodily functions and dispels lethargy and coldness. The *Po Wu Chi* 博物志 (Records of the Investigation of Things), +290, relates the story of three men taking a long walk on a cold misty morning.¹⁰ One had neither food nor drink, one had a light meal and one had drunk wine. The weather proved to be more treacherous than they had expected. The one on an empty stomach died. The one who had a light meal became ill. The one who had drunk wine survived unscathed. Pan Ku 班固, author of the History of the Former Han (*Chhien Han Shu* 前漢書), was so impressed as to say that wine is the first of a hundred medicaments (*pai yao chih chang* 百葯之長).¹¹ Besides being a drug in its own right, a wine may also serve as a useful adjunct to help the absorption of other drugs. The *Shên Nung Pen Tshao Ching* (Classical Pharmacopoeia of the Heavenly Husbandman) says that,¹² 'depending on its nature, a drug may be taken as a pill, as loose powder, as a decoction after boiling in water or soaking in wine, or as a fried paste'. This means that wines are a good vehicle to help carry other drugs into the vital system. What then could be more natural than to incorporate various herbs in wine and thus produce a medicated wine?

Apart from the fragrant ritual wine *yü chhang* the oldest reference to a herbal wine is the cinnamon or cassia wine from the *Chhu Tzhu* (Songs of the South), c. −300.¹³ But the earliest prescriptions we have for the making of medicated wines occur

---

⁸ *SWCT*, p. 106.   ⁹ *HTNCSW*, ch. 14, p. 115; cf. Veith (1972), p. 152.
¹⁰ *Po Wu Chi*, ch. 10 (*HWTS*), p. 12.   ¹¹ *Chhien Han Shu*, p. 1183; cf. Swann (1950), p. 348.
¹² *SNPTC*, Tshao Yuan-Yu ed. (*1987*), p. 11.
¹³ *Chhu Tzhu*, Tung Chhu-Phing ed. (*1986*), p. 44; Hawkes, tr. (1985), p. 102.

in two silk manuscripts discovered in Han Tomb No. 3 at Ma-wang-tui, the *Yang-Shêng-Fang* 養生方 (Manual for Nurturing Vitality) and the *Tsa-Liao-Fang* 雜療方 (Miscellaneous Prescriptions) which may be dated at about −200. The former lists six prescriptions in which herbs are fermented with panicum or seteria millet or rice to produced medicated wines. The latter lists one prescription. The herbs used include such familiar items as *niu hsi* 牛膝 (*Achyranthes bidentata*) and *kao pên* 蒿本 (*Lingusticum sinense*), but other ancient names mentioned still remain to be deciphered. The manuscripts are so badly damaged that many critical parts of the description of the processes are lost. One prescription is, however, in good enough condition to allow us to reconstruct the major steps involved. Since it represents the earliest record of a specific fermentation process in China (c. −200), it is worthwhile to quote it in full:[14]

To make *lao* 醪 (a muddy wine): take one *tou* 斗 each of *tsê chhi* 澤漆 and *ti tsieh* 地節, cut them up fine and soak them in five *tou* of water. Filter and discard the residue. Boil *tzu wei* 紫葳 in the juice, filter again to get the broth. Disperse one *tou* each of (x) *chhü* 麴 (*ferment*) and *mai chhü* 麥麴 (*ferment* from wheat or barley) in the water; allow the suspension to sit overnight and filter. Take one *tou* each of (x) millet and (x) rice and cook separately to give cooked grain. Combine the grains and mix them with the fermenting liquor, like adding soup to cooked rice, and allow the fermentation to proceed. Grind three pieces of *wu thou* 烏頭, five pieces of ginger and (x) pieces of *chiao mu* 焦牡 and mix them together. Place the mixture at the bottom of the jar and then add the fermenting mash. Pour in the filtered herbal broth and mix the contents well. Finally, pour in ten *tou* of finished fine wine, and do this three times. [Allow the fermentation to proceed until completion.] Drink one cup [of the medicated wine] each afternoon between 3 and 5 o'clock. If one itches, a massage will be helpful. After continuing the treatment for a hundred days, the [patient's] eyesight will be clearer, the hearing sharper and the four limbs stronger and more agile. In addition, (x) ailment will be cured.

Even though the specific ailment that is cured is missing, the overall benefits as stated are pretty much what are touted for the imbibition of medicated wines today. A number of examples of the use of wines to treat specific illnesses are, however, recorded in the literature. One story concerns the famous physician Shun-Yü I 淳于意 of the Western Han.[15] A concubine of the King of Tzu-Chhuan 菑川 was in labour but unable to deliver the baby. Shun-Yu I examined her and prescribed a dose of *lang tang* 莨菪 (probably henbane, *Hyoscymus niger* L.) taken with a good shot of wine. Soon afterwards she gave birth to a healthy baby. Another story concerns the Thang physician Sun Ssu-Mo who was asked to treat a monk suffering from a nervous disorder.[16] He first had the patient eat some very salty food before going to

---

[14] Text reproduced in Ma Chi-Hsing (*1985*), p. 571. The spaces occupied by (x) indicate missing text in the original manuscript. Tentatively, *tsê chhi* is identified as *Euphorbia helioscopia* L., *ti tsieh*, *Polygonatum odoratum* (Mill) Druce, and *tzu wei*, *Campsis grandiflora* (Thunb.) Loisel.

[15] *Shih Chi*, ch. 43, Biography of the physician *Pien Chhüeh* 扁鵲 who lived during the Spring and Autumn period, but most of the chapter actually deals with the exploits of the Western Han physician Shun-Yü I.

[16] The story is quoted in Sun Wên-Chhi 孫文奇 and Chu Chun-Po 朱君波 (*1985*), p. 4. According to Porter-Smith and Stuart (1973), p. 71, and *Chiang-su hsin I-hsueh yuan* (*1986*), p. 1379, *ju hsiang* is probably *Boswellia carterii* Birdw.

bed. In the middle of the night the latter became exceedingly thirsty, and woke up Sun to ask for help. Sun prescribed a pint of wine taken with a dose of cinnabar, seeds of Chinese jujubes, and *ju hsiang* 乳香 (Bombay mastic). Later he prescribed another half a pint of wine with a smaller dose of the drugs. Thereupon, the patient fell into a deep sleep. Two days later he awoke and was completely cured.

But tradition has it that wines not only cure ailments, they also ward off evil spirits that give rise to epidemics. The best known and possibly the oldest example of such a wine is the *thu su* 屠蘇 wine, said to have been invented by the famous physician Hua Tho 華跎 of the late Eastern Han and Three Kingdoms period (+208 to ?). It is made by soaking a number of herbs in wine. The exact formula might have changed from time to time, but Li Shih-Chên's consists of the following: *ta huang* 大黃, *kuei chih* 桂枝, *chieh keng* 桔梗, *fang fêng* 防風, *shu chiao* 蜀椒, *pa chhi* 菝契, *wu thou* 烏頭 and small red beans.[17] It is drunk by the whole family on New Year's day so that everyone can stay healthy for the rest of the year. Other protective wines are the realgar wine drunk on the 5th day of the 5th month, the *tuan wu* (dragon boat race) festival and the *chu yu* 茱萸 wine for the *chhung yang* 重陽 day (9th day of 9th month).

One of the best known herbal wines is the chrysanthemum wine, made with parts of the plant, *Chrysanthemum morifolium* Ramat. (formerly *C. sinensis*), which is first mentioned in the *Hsi Ching Tsa Chi* 西京雜記 (Miscellaneous Records of the Western Capital), c. + mid 6th century. The relevant passage reads:[18]

When the chrysanthemums are in full bloom, collect the flowers, stems and leaves and mix them with cooked millet to carry out a fermentation. The mash will be ripe on the 9th day of the 9th month of the following year. The wine is then ready for drinking. So, it is called chrysanthemum wine.

But as in the case of the ancient *yü chhang* 鬱鬯 (p. 156) chrysanthemum wine may also be made by steeping the plant parts in pre-formed wine as shown in the *Chhien Ching I Fang* (Supplement to the Thousand Golden Remedies), +660. Indeed, this would appear to be a far more convenient way of preparing a herbal wine than a *de novo* fermentation. The soaking of herbs in a finished wine is indicated for the preparation of all three medicated wines described in the *Chhi Min Yao Shu*, +540. In one the herb is the well-known *wu chia phi* 五加皮, *Acanthopanax spinosum*, probably the earliest reference to its use as a wine ingredient in the Chinese literature. In another, the herbs are ginger, black pepper, and the juice of pomegranate, and in the third, ginger, black pepper, long pepper and clove.[19] Of the twenty prescriptions for medicated wines given in the *Chhien Ching I Fang*, only two are prepared by a *de novo* fermentation process. All the others involve steeping or soaking a collection of herbs in wine, the largest number of ingredients being forty-five for the *tan shên* 丹參

---

[17] *PTKM*, ch. 25, p. 1561. The identity of the herbs are: *ta huang*, *Rheum officinale*; *kuei chih*, *Cinnamomium cassia*; *chieh kêng*, *Platycodon grandiflorum*; *fang fêng*, *Siler divaricatum*; *shu chiao*, *Zanthoxylum* sp.; *pa chhi*, *Smilex china*; and *wu thou*, *Aconitum fischeri*.

[18] *Hsi Ching Tsa Chi* (*HWTS*), ch. 3, p. x.

[19] *CMYS*, ch. 66, p. 393, wine soaked in herb (浸藥酒法); p. 394, pepper wine (胡椒酒法); and p. 395, mixed herbal wine (和酒法).

Table 25. *Common medicated wines noted in the Food and Drug literature*

Source	CMYS	CCIF	SLPT	CPT	SCYL	PSCC	CHS	YSCY	CCPY	PTKM
Date +	540	655	670	985	1080	1117	1235	1330	Yuan	1596
Chü hua[1] 菊花	−	+	−	+	+	+	−	−	+	+
Niu hsi[2] 牛膝	−	+	+	−	+	+	−	+	+	+
Ti huang[3] 地黃	−	−	+	+	−	−	−	−	−	+
Chiang[4] 薑	+	−	+	+	−	−	−	−	−	+
Hu ku[5] 虎骨	−	−	+	−	+	−	−	+	−	+
Kou tzu[6] 枸子	−	+	+	+	+	−	−	+	−	+
Kuei[7] 桂	−	−	−	−	−	−	+	−	−	−
Pai yang[8] 白羊	−	−	−	+	−	−	+	+	+	+
Thien men tung[9] 天門冬	−	−	−	−	−	+	+	+	+	−
Wu chia phi[10] 五加皮	+	−	−	−	−	−	−	−	+	−

[1] *Chrysanthemum morifolium*. [2] *Achyranthes bidentata* (ox knee). [3] *Rehmannia glutinosa*. [4] *Zingiber officinale* (ginger). [5] *Panthera tigris* (tiger bone).
[6] *Lycium chinense* (matrimony vine). [7] *Cinnamomum cassia* (cassia). [8] *Ovis aries* (lamb). [9] *Asparagus lucidus*. [10] *Acanthopanax spinosum*.

CMYS   *Chhi Min Yao Shu*.   PSCC   *Pei Shan Chiu Ching*.
CCIF   *Chhien Ching I Fang*.   CHS   *Chiu Hsiao Shih*.
SLPT   *Shih Liao Pen Tshao*.   YSCY   *Yin Shan Cheng Yao*.
CPT   *Chhü Pen Tshao*.   CCPY   *Chü Chia Pi Yung*.
SCYL   *Shou Chhing Yang Lao*.   PTKM   *Pên Tshao Kang Mu*.

A '+' sign means that the medication is mentioned in the work listed in the heading.

wine.[20] The *Shih Liao Pên Tshao* (Pandects of Diet Therapy), +670, lists thirteen medicated wines, each one named after its principal herbal component, but does not describe how the wines are made, whether by steeping or fermenting.[21] Medicated products are among the many wines mentioned in both food canons and diet therapy classics of the Sung, Yuan and Ming, as for example, in the *Chhü Pên Tshao*, +985, *Shou Chhing Yang Lao Hsin Shu*, +1080, *Chêng Lei Pên Tshao*, +1082, *Pei Shan Chiu Ching*, +1117, *Chiu Hsiao Shih*, +1235, *Yin Shan Chêng Yao*, +1330, *Chü Chia Pi Yung*, late Yuan, and *Pên Tshao Kang Mu*, +1596.

Many of the wines listed contain more than one herbal component. The record number of ingredients (forty-five) in the *tan shên* wine of the Thang has probably never been equalled. It does appear to carry a good thing to excess. The same familiar herbs and spices incorporated in the *ferments* described in the *Pei Shan Chiu Ching* (Wine Canon of North Hill), for example, ginseng, hemlock parsley, sweet cassava, cassia, ginger and nutmeg etc. are all used in the making of medicated wines (pp. 185–6). But most of the known wines are named after their principal ingredient and usually contain only one or two (sometimes three) herbs. Ten such wines, which have remained popular through many centuries are given in Table 25. Each wine has been mentioned in at least three of the classical works listed. The only exception is the wine spiced with cinnamon (or cassia), which was first noted in the *Chhu Tzhu* (Songs of the South) and later celebrated in poems by Su Tung-Pho.[22] It is an ingredient in numerous wines, including the ancient *thu su* 屠蘇 wine mentioned above. Most of the entries are based on familiar drugs, such as *niu hsi* 牛膝, *ti huang* 地黃, and *kou tzu* 枸杞 and the ever popular *wu chia phi* 五加皮. Two wines with animal components are included, the venerable tiger-bone wine and the estimable mutton wine. The tiger-bone is still going strong, but the lamb appears to have lost its appeal.[23]

Although the Chinese prepared mutton wine by including mutton as an ingredient to the fermentation, the Mongols prepared it in a process that includes a distillation. An account of the procedure was published in 1873 by Dr MacGowan. We are quoting it in full because it contains the best description we have of how a Mongolian still was constructed in the 19th century.[24]

The following were the ingredients: 1 sheep, 40 catties of cow's milk whisky, 1 pint of skimmed milk, soured and curdled, 8 ounces of brown sugar, 4 ounces of honey, 4 ounces of fruit of dimocarpus, 1 catty of raisins, and a half a dozen drugs weighing in all about one catty. The sheep must be two years old, neither more nor less, a male, castrated.

Plant necessary for distillation – 1 large pot (cast iron), 1 wooden half-barrel opened at the bottom,[25] 1 smaller pot (cast iron), 1 earthenware jar, felt belts, cow dung, fire.

---

[20] *CCIF*, ch. 16, pp. 181–4.   [21] *SLPT*, pp. 108–9.   [22] Chou Chia-Hua (*1988*), pp. 86–7.
[23] Sun Wên-Chhi & Chu Chun-Po (*1985*) list one lamb wine (p. 14), but eight tiger-bone wines (pp. 34ff).
[24] Dr MacGowan (1871–2), p. 239.
[25] *Ibid.* footnote on p. 239, 'This is about 2 feet high (English): tapers. At the bottom it is large enough to sit on the rim of the big pot; at the top it is small enough to let the small pot sit in it without falling through. It is called a *Boorher*.' The Mongolian still as described is basically the same as that shown by Hommel (Figure 53a).

Process – Set the *boorher* on the large pot, calk the joining first with paper, then daub the outside with cow dung and ashes. Make the *boorher* airtight by plastering it all over outside with cow dung.

Pour in the wine, add half the raisins (i.e. 8 oz) cut or crushed, half the black sugar, the pint of airik, and the bones of the sheep's legs from the knee downwards after breaking them open.

From the other bones strip all the fat and most of the flesh, leaving them fleshy. Hang them head and all inside the *boorher* high enough to be beyond the reach of the whisky, and low enough to be out of the reach of the pot above. Break up the medicines into small pieces (do not pound them) and put them into the earthenware pot. Into that pot put also the honey, white sugar, dragon's eye and the remaining half of the black sugar and raisins. Suspend the earthenware pot in the centre of the *boorher*, put on the pot above, make the joining air-tight by paper, cloth and felt bands. Apply fire to the great pot. When the upper pot feels warm to the touch, fill it with cold water and stir. When the water becomes too hot to touch, ladle it out and fill it up with cold water. When this second potful of water becomes too hot for the hand, slacken the fire, take off the upper pot, and the earthenware pot will be seen full of a dirty brown liquor boiling furiously. Take out the earthenware, pour off the liquid, replace the earthenware pot, replace the upper pot, fill with cold water. When this potful of water becomes hot, the whole thing is over. The earthenware pot is again about half-filled, pour it off and let it cool. When reasonably cold put it up in jars and close them with the membrane of ox or sheep bladders.[26]

The liquor thus prepared is said to have a 'very strong odour of mutton; it is sweetish and unctuous'. The alcohol content is about 9 per cent.

It should be noted that the entries in Table 25 merely represent some of the most prominent members of the medicated wine fraternity. Its membership has grown steadily from antiquity to the modern era. The Ma-wang-tui silk manuscripts, c. −200, list seven medicated wines, the *Chhien Ching I Fang*, +655, twenty, the *Pên Tshao Kang Mu*, +1596, sixty-seven, and the modern *Yao Chiu Yên Fang Hsüan*, 1985, by Sun & Chu, three hundred and sixty one. Yet, the role of herbs and spices in the making of Chinese wines may actually be even greater than indicated by these numbers. We have noted earlier that herbs and spices are used extensively, if not invariably, in the preparation of the *ferments* that provide the essential amylofermentative activity in a fermentation. This is true of the cases described in the *Chhi Min Yao Shu* and *Pei Shan Chiu Ching*, as it is in commercial practice in China today.[27] Thus, it is inevitable that these accessory ingredients in the ferment will be carried into the finished products and contribute significantly to the enormous range of subtle flavours available to the vintner in the creation of many distinctive varieties of Chinese wine.

The use of spices and herbs in alcoholic beverages is however, by no means a unique Chinese phenomenon. The ancient Greeks added cassia leaves and sesame

---

[26] The description continues: '*Remarks* – The great bulk of the flesh of the sheep is not used, nor any of the fat. All the marrow bones are broken open. The skull is not broken open nor the tongue extracted from the head. At the end of the process the mutton on the bones is cooked, but tastes badly. The *hoieu nood* (dragon's eye) which was put in black came out white. The quantity of cow's milk wine in the [lower] pot is not much diminished, but the strength is gone and what remains is good for throwing away only.'

[27] *Wu-hsi chhing-kung-yeh hsüeh yuan* (*1964*), pp. 42–52.

oil to date wines, and Egyptian physicians created special brands by the addition of extracts of lupin, skirret, rue, mandrake etc. to beer.[28] Pliny the Elder described wines flavoured with a host of herbs and spices, including myrrh, asparagus, parsley-seed, southernwood, cardamom, flowers of cinnamon, saffron, sweet rush, etc.[29] Of greatest interest is his report that hops were an ingredient of the Iberian beer of his day, but in the classical world this plant was used mainly in medicine and the young shoots were eaten as a vegetable.[30] According to Forbes this most important herb incorporated into an alcoholic drink in the West did not come into general use until late in the Middle Ages, probably in the +13th to +14th centuries.[31] Hops were introduced as a preservative and a flavouring agent. Today they are best known as the ingredient that imparts the characteristic bitter flavour to beer. It would be unthinkable to have a beer without hops. The word 'hoppe' did not appear in the English language until about +1400. The timing of its arrival prompted the ditty:

> The Bible and Puritans, Hops and Beer,
> Came into England, All in one year.

Presumably, the disapproval of the puritans did not impede the popularisation of beer (and ale) in England in the ensuing centuries.

So much then for 'medicated wines,' but the story does not actually end here. In the Chinese lexicon any wine that is not made from grains is considered a 'medicated' wine. Indeed, wines made from fruits (such as grapes), honey and milk are usually listed among medicated wines in the Chinese literature. These products are of considerable historical and technological interest and warrant a separate discussion, which we shall now provide on pages 239–57.

(7) WINES FROM FRUITS, HONEY AND MILK

> Beautiful grape wine glittering in a white jade goblet;
> I was about to drink it, when the lute hastened me to mount my horse.
> Should I lie drunk on the battlefield, please do not laugh at me;
> For how often have men returned safe from war since the ancient days?[1]

The above poem, entitled 'Song of Liang-chou' (*Liang-chou tzhu* 涼州詞) by the Thang poet Wang Han (c. +713) is probably the best known drinking song in all Chinese literature. Yet curiously it deals with grape wine, and not the more common wine from grains, indicating that grape wine was already an established gastronomic delicacy during the Thang dynasty. In fact, by then grape wine had been known in China for eight hundred years.[2] But grape was not the only substrate

---

[28] Forbes (1954), pp. 277, 281. [29] Pliny the Elder (1938), Vol. IV, Bk 14, pp. 249, 257.
[30] Forbes (1956), p. 141; cf. also Pliny *ibid*. Vol. VI, Bk 21, p. 222. [31] *Ibid*.
[1] Tr. Hsiung Deh-Ta (1978), p. 172, mod. auct.; cf. *Thang Shih San Pai Shou* (Three Hundred Thang Poems), p. 263. The original, titled *Liang Chou Tzhu* 涼州詞, reads as follows: 葡萄美酒夜光杯，欲飲枇杷馬上催，醉臥沙場君莫笑，古來征戰幾人回。
[2] Cf. *SCC* Vol. v, Pt 4, pp. 136–41, 151–2.

other than grains used to make alcoholic drinks. Wines were also made from honey and milk. We shall now consider the history of the fermentation of wines derived from these non-conventional substrates, first from grape, to be followed by honey and milk.

(i) *Wine from grape and other fruits*

There is passage in the *Shih Chi* (Historical Records), c. −90, which notes that in Ta-Yüan 大宛 (Ferghana) and vicinity 'wine is made from grapes, the wealthier inhabitants keeping as many as ten thousand or more *tan* 石 stored away. It can be kept for as long as twenty or thirty years without spoiling. The people love their wine just as their horses love their alfalfa.'[3] Seeds of the wine grape, *phu thao* 葡萄 (*Vitis vinifera* L.), was brought back to China from Central Asia at about −126 by the envoy Chang Chhien, and grown near the capital as a fruit and a medicament.[4] Apparently it was most successfully cultivated in parts of Kansu, namely Lung-hsi 隴西, Wu-yüan 五原 and Tun-huang 敦煌.[5] The *Shên Nung Pên Tshao Ching* (Pharmacopoeia of the Heavenly Husbandman), a Han compilation, notes that wine can be made from it,[6] and during the Three Kingdom's period, c. +220, the Emperor of Wei, Tshao Phei 曹丕, wrote:[7] 'Grapes can be fermented (*niang* 釀) to make wine. It is sweeter than the wine made (from cereals) using *ferments* and sprouted grain (*chhü nieh* 麴蘖). One recovers from it more easily when one has taken too much.'

This terse statement does not tell us how the grape wine was made. For reasons to be discussed later, we may presume that it was brewed with the aid of the same kind of *ferment* as that used in the making of wines from grain.[8] The *Lo-Yang Chieh Lan Chi* (Buddhist Temples of Loyang), +547, describes grapes as large as jujubes grown in a temple garden,[9] and the History of the Northern Chhi Dynasty (*Pei Chhi Shu*), +550−577, records the presentation, at about +550, of a plate of grape wine by Li Yüan-Tsung 李元忠 to the emperor.[10] But during the Han and until the start of the Thang (c. −100 to +600) the grape remained an unfamiliar delectable and its wine a

---

[3] Tr. Needham, *SCC* Vol. v, Pt 4, p. 152; cf. *Shih Chi*, ch. 123, p. 15a. A similar statement is found in the *Po Wu Chi*, p. 7b. Chêng Hsuan's commentary on the *Chhang Jên* 場人 (Superintendent of the Imperial Gardens) in the *Chou Li*, p. 177), lists the grape as one of the delicacies grown for the royal household. But there is also an indigenous wild grape in China, called *ying yü* 蘡薁, *Vitis thunbergii*, which we have translated as 'berry' in the poem Seventh Month from the *Shih Ching* (w 159, M 154) quoted on p. 18. According to the *Pên Tshao Phin Hui Ching Yao*, +1505 (ch. 32, p. 772), *ying yü* can also be used to make wine.
[4] Cf. *SCC* Vol. I, pp. 174ff. The characters for grape, *phu thao*, have gone through a series of changes. They are written in the *SNPTC* as 蒲桃, *Hsin Hsiu Pên Tshao* as 蒲陶, *I Wên Lei Chi* as 蒲萄, *Chhien Chin Yao Fang* as 蒲桃 and *Chhien Chin I Fang* as 葡萄.
[5] *MIPL*, Shang Chih-Chün (1981), p. 87; it is cited in the *HHPT*, ch. 17, p. 225 and *PTKM*, ch. 33, p. 1885.
[6] *SNPTC*, Tshao Yuan-Yu (1987), p. 317.
[7] *CSHK* (San Kuo Section), ch. 6, p. 4a; tr. Needham mod. auct; cf. *SCC* Vol. v, Pt 4, p. 138, n. c for other references.
[8] Cf. Yüan Han-Chhing (1956), p. 99.   [9] *LYCLC*, *HWTS* edn ch. 4, p. 14b; cf. Jenner, tr. (1981), p. 232.
[10] *Pei Chhi Shu*, book 22, ch. 14, p. 3. It is curious that the text should say a plate of grape wine (*phu thao chiu i phan* 蒲桃酒一盆) unless what is meant is a tray of cups of grape wine. However, the word *chiu* (wine) is left out in a similar passage in the *TPYL*, ch. 72, p. 3a, and the *Yüan Chien Lei Han*, ch. 403, 5c.

rare, exotic drink in China. It is noteworthy that the grape is mentioned only as an addendum to the chapter on the peach in the *Chhi Min Yao Shu* of +540, even though a whole chapter in it is devoted to the cultivaton of alfalfa, a crop that was reputedly brought to China at the same time as the grape from the Western regions by Chang Chhien.[11]

Grapes and grape wine received a new boost in popularity during the Thang with the conquest of Kao-chhang (Turfan) in +640, as has been alluded to by Li Shih-Chên in the passage quoted earlier.[12] There is another account of this event recorded in the *Thai Phing Yü Lan*, +983, which reads as follows:[13]

Grape wine was always a great thing in the Western countries. Formerly they sometimes presented it (as tribute) but it was not until the capture of Kao-chhang that the seeds of the 'mare teat' grape (*ma ju phu thao* 馬乳蒲桃) were obtained, and planted in the imperial gardens. The method of wine making (*chiu fa* 酒法) was also obtained, and the emperor himself took a hand in preparing it. When finished it came in eight colours, with strong perfumes like those of springtime itself; some types tasted like a kind of whey (*thi ang* 醍盎). (Bottles of it) were given as presents to many officials, so people at the capital got to appreciate the taste of it.

It is difficult to say whether this account corroborates Li Shih-Chen's claim that a distilled wine was also made from grape at this time, since we cannot be sure of what is meant by the 'method of wine making (*chiu fa* 酒法)', stated in the text. One possibility is that it refers to the procedure of carrying out the fermentation with the yeasts that occur naturally on the skin of the grapes. In any case, the passage tells us that grape wine was becoming well known during the Thang. The famous 'mare teat' grape is elongated in shape in contrast to a spherical type called 'grass dragon pearl (*tshao lung chu* 草龍珠)'. In time vineyards growing these varieties became established outside the imperial gardens, and we find the poet Han Yü, +768 to +824, lamenting the fate of a neglected vineyard in these words:[14]

> The new twigs aren't yet everywhere – half are still withered;
> The tall trellis is dismembered – here overturned, there uplifted.
>
> If you want a full dish, heaped with 'mare teats',
> Don't decline to add some bamboos, and insert some 'dragon pearls'.

Vines continued to flourish in parts of arid Kansu which gave rise to the grape wine of the Western Region (Liangchou 涼州) celebrated in the poem by Wang Han. Vineyards of the mare teat apparently did well in the region around Thai-yuan in Shansi which became a major centre for the manufacture of grape wine in Thang China.[15] The 'mare teat' wines of Thai-yüan was acclaimed in a poem by

---

[11] *CMYS*, ch. 34, pp. 191–2 on grape, and ch. 29, pp. 161–3 on alfalfa.
[12] Cf. p. 204. [13] Tr. Needham *et al.*, *SCC* Vol. v, Pt 4, p. 139; *TPYL*, ch. 844, p. 8a.
[14] Tr. Schafer (1963), p. 143; for the original cf. *Han Chhang-Li Chi*, 韓昌黎集, ch. 9, p. 29.
[15] Thaiyüan is today the capital of Shansi province. Schafer (1963), p. 144, notes the submission of grape wine as tribute from Thaiyüan to the Thang court.

Liu Yü-Hsi 劉禹錫 (+772 to +842), which contains a description of how the vine is grown in a typical vineyard:[16]

> And now the vine is planted out,
> It climbs the wooden frame about,
> The lattice shades with tender green,
> And forms a pleasant terrace screen.

And it ends with a paean in praise of the 'mare teat' wine:

> We men of Tsin, such grapes so fair,
> Do cultivate as gems most rare;
> Of these delicious wine we make,
> For which men ne'er their thirst can shake.
> Take but a measure of this wine,
> And Liang-chou's rule is surely thine.

From these scattered accounts we may conclude that *vinifera* grape and grape wine had been known in China since the Western Han. For centuries grape wine was brought to China from Central Asia as tribute or as an article of commerce. Grape was cultivated as a crop in limited areas in Kansu, and some wine was made from it by the use of traditional Chinese *ferment*. But the Chinese-made grape wines apparently could not compare in quality with those from Central Asia, and could not compete with the indigenous wines made from grain.[17] Grape wine remained a rare, exotic drink. The wine industry received a boost in early Thang due to the introduction of the 'mare teat' grape from Kaochhang and a 'new fermentation process' which utilised the wild yeasts that occur naturally on the fruit. This variety apparently did quite well in the Thaiyüan region and the wine from it gained a significant degree of popularity. It was during the Thang that we first see the lure of grape wine celebrated in poetry, and the recognition that it could be made without the use of the traditional Chinese *ferment*. The *Hsin Hsiu Pên Tshao*, the New Pharmacopoeia of +650, states that 'unlike wines from grain, grape wine and mead do not require *chhü* 麴', and 'to make grape wine, just allow the juice to stand and it will naturally ferment into wine'.[18] Nevertheless, this boost in popularity proved to be short-lived and grape wine remained an exotic commodity and a medicament in China in the ensuing centuries until modern times. Why?

We can see two possible reasons. Firstly, the *vinifera* grape from Central Asia probably never attained its full potential of quality and yield under cultivation in the heartland of China. For a *vinifera* plant to flourish and the fruit to develop its full flavour, it needs a mild winter and a long, hot dry summer such as found in

---

[16] Tr. Sampson, Theos. quoted in Schafer (1963), pp. 144–5. The poem is titled *Phu thao ko* 葡萄 歌. For the original text cf. *Liu Mêng Tê Wên Chi*, ch. 9, pp. 5a–5b. *Tsin* 晉 is equivalent to today's Shansi. The *CMYS*, ch. 31, p. 192, notes that the grape is a vine and needs to be grown on a wooden support. The 'pleasant terrace screen' could easily have been extended to form an arbour, as practised by the ancient Egyptians (Forbes (1954), p. 290, fig. 285) and by the Uighers in Turfan, Sinkiang today.

[17] *Hou Han Shu*, ch. 118 notes that the grape wine from *Li-I*, near Samarkand, is particularly good.

[18] *HHPT*, ch. 19, p. 287 on wine; ch. 17, p. 225 on grape.

Mediterranean West Asia and Europe. In North China, however, the winter is harsh and the rains come in the summer. Thus, while we do read of occasional successes in the cultivation of grape, it did not become a crop of major economic importance in ancient and mediaeval China.

Secondly, the art of fermenting grape juice to wine was poorly developed in China. We have said that the making of wine from grapes is practically inevitable. That may be so. But to make a really 'good' wine is altogether a different matter. It involves attention to a thousand details that can only be learned by example or experience. Initially, the Chinese made wine from grape juice through the use of *chhü* (*ferment*), that is, following the traditional way of making wine from grains. It was a case of the right agent but the wrong substrate. The mouldy flavours introduced by the fungi in the *ferment* could have done considerable violence to the taste of the final product. Moreover, growth of the fungi would have consumed some of the sugars which would otherwise have been available for conversion to alcohol, thus causing a reduction in yield. After the conquest of Kao-chhang in +640 it is said that a new 'method of making (grape) wine was also obtained'. What this new method is or how it was obtained are not explained.[19] But by +650 the *Hsin Hsiu Pên Tshao* was telling us that wine can be made by just allowing the grape juice to ferment spontaneously. Presumably, this was the new method mentioned in the quotation from *Thai Phing Yü Lan* shown above. It no doubt works, but without further refinements it could not have given rise to a wine of good quality with any degree of consistency.[20]

Although there are other references to grape wine in the poetry of the Thang, there is virtually no information on the details of how the wines were made.[21] The situation is no better in the Sung and Yuan. The famous poet Su Tung-Pho[22] drank a grape wine, and there is a description of *phu thao chiu* 葡萄酒 in the *Pei Shan Chiu Ching* (Wine Canon of North Hill) of +1117.[23] Only it turned out that this *phu thao chiu* is not really grape wine at all, but regular rice wine fermented with the addition of some grape juice to the must. This so called 'grape wine' was placed by the author in the category of medicated wines, among which it is often listed even today.[24]

According to Marco Polo, the Thaiyüan region remained a major centre of viticulture and the manufacture of grape wine during the Yuan dynasty.[25] There are several references to grape wine in the official histories of the Yuan dynasty. The *Yüan Shih* (History of the Yuan) states that a decree of +1251 discontinued the tribute of grape wine from An-i 安邑, and another in +1296 did the same for Thaiyüan 太原

---

[19] Cf. *SCC* Vol. v, Pt 4, pp. 136–7.
[20] Spontaneous fermentation, depending entirely on the wild yeasts present on the fruit at harvest, was, of course, the way that grape wine had originally been made in Mesopotamia and Egypt, cf. Forbes (1954), p. 282. But as we shall see in the next subchapter the use of a yeast starter to start a fermentation had probably become the norm long before the grape and grape wine came to Greece and established themselves there.
[21] Cf. Shinoda (*1972a*), pp. 333–4 for references to poems by Li Pai, Wang Chi and Pai Chü-I.
[22] Shinoda Osamu (*1972b*), p. 301.   [23] *PSCC*, ch. 3.   [24] Sun & Chu (*1986*), p. 28.
[25] Marco Polo, *The Travels*, Latham, R. tr., (1956), p. 165. Regarding Thaiyüan, Polo says, 'It has many fine vineyards, producing wine in great abundance. Wine is produced nowhere else in the province, except in this district, and from here it is exported throughout the whole province.'

and Phingyang 平陽.²⁶ It records the sending of grape wine in +1331 as tribute to the Yuan court by kingdoms in the West, and that one convoy of grape wine to the capital was pillaged by barbarian bandits.²⁷ According to the Economic Affairs Section of the *Hsin Yüan Shih* (New History of the Yuan), the government Wine and Vinegar Bureau explains in +1231 that since grape wine is fermented without the use of grain *chhü* 麴, it should not be taxed at as high a rate as regular wine.²⁸

Some of the grape wine was apparently distilled to give brandy. According to the *Yin Shan Chêng Yao* (Principles of Correct Diet), +1330, 'there are many different grades of grape wine. The strongest comes from *ha-la-huo*, the next from the Western tribes, and the next from Phingyuan and Thaiyuan.'²⁹ It is pertinent to note that the Mongols called distilled liquor *arrihae*. *Ha-la-huo* could be a corruption of *ar-ri-hae*. We suspect distilled grape wine was a popular drink at the Mongol court. But we still have no description of the distillation of grape wine in the Thang, Sung or Yuan literature.

In fact, so far we have searched in vain for any description of how the grape juice was fermented with or without the use of *chhü*. The only recipe we have been able to find for the making of grape wine is in the *Yin Chuan Fu Shih Chien* (Compendium of Food and Drink) of +1591, and it says:³⁰ 'Take one *tou* of pressed juice from grapes, and blend in four ounces of *chhu* 麴 *ferment* in a jar. Cover the mouth and [the juice] will turn into wine as indicated by the presence of an exotic fragrance.'

What is most curious is the fact that even by this late date there is no indication that the Chinese had used a yeast preparation, *chiao* 酵, to make grape wine. Yet *chiao* is utilised in at least three recipes for rice wines in the very same work, the *Yin Chuan Fu Shih Chien*. Why did the Chinese fail to make the simple connection between the use of *chiao* (yeast) and the conversion of fermentable sugar into alcohol, a connection that had been well known in the West since the ancient civilisations of Egypt and Mesopotamia? We will discuss this anomaly in the concluding chapter.

Interest in grape wine apparently declined during the Ming and Chhing.³¹ But grape as a fruit had grown in popularity. Foreign observers in the 18th and 19th centuries commented favourably on the quality of the grapes found in the Peking–Tientsin region.³² Perhaps, through hundreds of years of selection, the varieties of

---

²⁶ *Yüan Shih*, ch. 4, p. 14b; ch. 19, p. 3a. Phing-yang is a prefecture southwest of Thai-yüan, and An-i is a town in Phing-yang.
²⁷ *Yüan Shih*, ch. 35, 19b, ch. 41, 12a.   ²⁸ *Hsin Yüan Shih*, ch. 72, p. 1.
²⁹ This version is quoted in *PTKM*, ch. 25, p. 1568; cf. *SCC* Vol. v, Pt 4, p. 137, tr. Needham, mod. auct. The version in the *YSCY*, ch. 3, p. 122 is slightly different in that it simply says that the flavour of the other wines is inferior to *ha-la-huo*.
³⁰ *YSFSC*, p. 127; cf. also pp. 125, 126 and 127 for recipes in which yeast (*chiao*) is used in the fermentation.
³¹ Huc (1855), I, p. 322, a Frenchman who travelled in China from 1838 to 1852, observed that 'grape wine was in use under every dynasty, and every reign to the 15th century ... The Chinese of our day, however, do not cultivate the vine on a large scale, and do not make wine of grapes; the fruit is only gathered for eating either fresh or dried.'
³² John Bell, a Scotsman who in +1719 to 1722 travelled from St Petersburg to Peking, where he found the grapes to be 'fine, and of agreeable taste' (p. 142). Charles Gutzlaff, while sailing along the China coast in the 1830s, was impressed by the quality of the grapes he found at the tip of the Shantung peninsula and in the region of Tientsin. He was surprised that 'no wine is extracted from the excellent grapes which grow abundantly on the banks of the Pei-ho and constitute the choicest fruit of the country' (p. 102). We are indebted to Michael Godley (1986), pp. 383ff. for these interesting references.

vinifera cultivated there had become well adapted to the soil and climate of North China. But there is no indication that these acclimatised varieties had played any role in the rise of a modern grape wine industry in the Shantung peninsula in the twentieth century.[33]

Wines were also made from other fruits. One is the 'orange' wine made famous by the Sung poet-scholar Su Tung-Pho in the poem *Tung-thing Chhun Sê* 洞亭春色.[34] Its excellent flavour is said to rival that of grape wine. Nothing, however, is said about the way the orange wine was prepared. Another is a wine described in the *Kuei Hsin Tsa Shih* 癸辛雜識 (Miscellanies from the Kuei-Hsin Street), c. +1300, when the author, Chou Mi stored a hundred pears in a sealed jar and found, six months later, that it had turned into an alcoholic drink.[35] But these are really mere curiosities. Their status in China never reached that achieved by cider or perry in Europe. Wines from the nectar of flowers were also known. The *Chiu Thang Shu* (Old History of the Thang Dynasty), relates the fermenting of the juice of coconut flowers into wine.[36] More astounding is the story, first told in the *Phêng Lung Yeh Hua* 蓬攏夜話 of the Ming, of how the monkeys of the famous Huang Shan 黃山 brewed a fragrant drink:[37]

There are many monkeys in Huang Shan. In the spring and summer they collect miscellaneous flowers and fruits and store them in a rocky crevice. In time they would ferment into a wine (*chiu* 酒). The fragrant aroma would be detectable hundreds of steps away. A woodcutter venturing deep into the woods may come upon it. But he should not drink too much of the wine lest the monkeys discover the reduction in the amount of fluid left. If so, they would lie in wait for the thief and playfully torture him to death.

According to the experience of recent travellers, the descendants of the Ming monkeys of Huang Shan have apparently not forgotten this remarkable accomplishment of their ancestors.[38] The phenomenon suggests that the fermentation of the sugars in flowers and fruits to alcohol must have been known to man very early in the pre-historic era. It is interesting to note that the oldest of all such sources of sugar is probably honey, a concentrate of nectar collected by bees. When honey is

---

[33] For the early history of the modern grape wine industry in China, cf. Godley (1986).
[34] We have already referred to an ode to this wine on p. 206. If our interpretation is correct, the product could have been an orange liqueur similar to cointreau. Su Tung-Pho was a great connoisseur of wines. For a discussion on all the wines described and mentioned in his works, cf. Chou Chia-Hua (*1988*), reprinted in Li Yin (*1989*), pp. 175–90.
[35] *KHTS*, First Addendum, p. 14, which also says that grape is fermented by 'Moslems' without the use of *chhü ferment*. Fruit wines extant during the Ming mentioned in Hsieh Kuo-Chên (*1980*), Pt 1, pp. 186–8, include pear, jujubes, coconut milk, litchi, water melon and persimmon. Today, wine is also fermented from pineapples and papaya, cf. Nakayama (1983).
[36] *Chiu Thang Shu*, ch. 197. E. N. Anderson (1997), private communication, points out that 'the nectar of coconut flowers is not abundant and is not worth the trouble. What is obviously intended here is toddy, which is made from the sap of the stalk of coconut flowers. You cut the flower stalk and let the sap run into a jar.' This is how Ceylon toddy is obtained which is then distilled into arrack, cf. Nathanael (1954).
[37] *Phêng Lung Yeh Hua*, ch. 148, p. 4b.
[38] Tai Yung-Hsia (*1985*). More recently Lin Yen-Ching (*1987*), records that he and a companion had actually discovered such a crevice among the rocks while exploring the remote forests of Huang Shan. They could smell the alcohol several feet away. They drank some of the brew and found it delicious. The phenomenon is apparently well known to the local residents, who dubbed the drink as 'monkey wine'.

diluted with water it is easily fermented by yeasts, which are ubiquitous in the environment, into alcohol; giving rise to the ancient drink we know today as mead.

(ii) *Wine from honey*

Honey, *mi* 蜜, may well be the only fermentation substrate that goes back beyond Neolithic times.[39] But its history in China is a fairly recent one. It is mentioned as a sweetening agent in the *Chhu Tzhü* (The Songs of the South), c. −300,[40] and as a medicament in the *Prescriptions for 52 Ailments*, −200.[41] The *Shên Nung Pên Tshao Ching* (Western Han) lists it as *shih mi* 石蜜 (stone or wild honey).[42] The earliest reference to mead, *mi chiu* 蜜酒, however, did not occur until the Thang dynasty. The *Shih Liao Pên Tshao*, +670, states that *mi chiu* is good for exanthematous illnesses[43] and the *Hsin Hsiu Pên Tshao*, +659, tells us that honey can be fermented into wine without the use of a *chhü ferment*.[44] Unfortunately, the latter does not say what was used, if not *chhü*, to promote the fermentation.

For the best known reference in Chinese literature on mead, we are again indebted to the Sung poet–gastronome Su Tung-Pho, who in the year +1080 learned how to make it from honey from a friend and wrote a poem in its honour. In the introduction to the poem, he says: 'The Taoist priest, Yang Shih-Chhang 楊世昌, from western Szechuan, is an expert in making mead of great excellence. I have obtained the recipe from him and this poem is in commemoration of the occasion.' But the poem actually had little to say about the detailed method of fermentation, which, fortunately, is preserved in a compilation by Chang Pang-Chhi 張邦基 called *Mo Chuang Man Lu* 墨莊漫錄 (Random Notes from Scholar's Cottage), +1131, and reads as follows:[45]

Su Tung-Pho often made wine from honey in Huangchou 黃州. He composed a poem on mead, but his technique was known only to a few people. [His method is as follows.] Heat four catties of honey and blend in hot water to give one *tou* 斗 of solution. Then grind two ounces of good wheat *ferment* together with one and a half ounces of southern rice *ferment*, place the resulting powder in a thin silk bag and suspend the bag in the honey solution in a sealed jar. The vessel is cooled when the weather is very hot, warmed when it is cool, and

---

[39] Forbes (1954), p. 275.
[40] Cf. Hawkes (1959/1985), p. 228, 'Fried honey cakes of rice flour' and 'jade-like wine, honey flavoured'.
[41] Chung I-Yen & Ling Hsiang (*1975*), p. 60.
[42] *SNPTC*, Upper Class Drugs, cf. Tshao Yüan-Yü (*1987*), p. 275. Thao Hung-Ching's commentary says it is found among rocky cliffs, hence the name stone honey. The hint of a rudimentary apiculture in China is probably noted in the *Po Wu Chi* (c. +190), ch. 10, p. 12a, considerably later than in Europe where apiculture was already a well-established industry by the time Virgil finished his epic account of the rearing of bees in Pt 4 of the Georgics, c. −28.
[43] *SLPT*, p. 108.
[44] *HHPT*, p. 287. Incidentally, readers of Waley's *The Book of Songs* will find 'mead' mentioned in w 136 (M 262), and perhaps come to the erroneous conclusion that mead was known in China during the time of the Chou. Actually, in this particular case, mead is an incorrect translation of *chü chhang* which may be rendered as 'black millet wine'.
[45] Tr. Lu Gwei-Djen, mod. auct., *Mo Chuang Man Lu*, ch. 5, p. 15a, is quoted and discussed in full in Chou Chia-Hua (*1988*), pp. 82–3.

heated when it is very cold. After a day or two, the fermentation begins and continues for several days until the bubbling stops. The wine will be clear and be ready to drink. At first it will retain some of the taste of honey but after half a month the flavour will be rich and strong. When it first begins to bubble, the addition of another half a catty of cold honey, as is, will give superior results. I have tried it myself and found the wine sweet and intoxicating, but the heavy drinker may think it is not potent enough.

Assuming that about 80% of the sugars present are converted to alcohol it is possible to estimate the alcoholic content of the finished mead, using the same type of calculation employed earlier for the wines from grain listed in Table 21.[46] The figure we arrive at is about 13.1%, which may represent a maximum obtainable yield. The recipe looks simple enough. Yet it has been reported that at one time Su Tung-Pho abandoned the procedure after several unsuccessful trials.[47] Based on our current knowledge of fermentation technology, we can see two problems. One is to maintain the process at near or below 30°C, since the honey water would spoil readily if the temperature should rise much above this optimum value.[48] The other is caused, as in the case of grape juice, by the use of *chhü* 麴, which, even when it is of the highest quality, may introduce undesirable mouldy flavours into the fermented product.

Nevertheless, the interest of Su Tung-Pho undoubtedly made mead a better known alcoholic drink in China. Mead is mentioned in two Southern Sung works, *Hsü Pei Shan Chiu Ching* 續北山酒經 (Continuation to the Wine Canon of North Hill), and the *Chiu Hsiao Shih* 酒小史 (Mini-History of Wine).[49] Recipes for a making mead found their way into the Ming food canon and pharmaceutical literature. One is given in the *Yin Chuan Fu Shih Chieh*, +1591, which says:[50]

Heat three catties of honey with one *tou* of water and transfer the solution to a bottle to cool. Mix in two ounces of ground white *ferment* (*chhü* 麴) and two ounces of yeast (*chiao* 酵) and seal the bottle with wet paper. Allow the bottle to stand in a clean place, for five days in the spring and autumn, three days in the summer and seven days in the winter. Wine of good quality would be formed.

In this case yeast (*chiao*) is actually used but together with *chhü ferment*. However, we find *chhü ferment* used again by itself in another recipe, given in the *Pên Tshao Kang Mu*, +1596, based on a method attributed to the eminent Thang pharmacist Sun Szu-Mao. It states, 'Mix one catty of sandy honey, one *shêng* 升 of steamed glutinous rice, five ounces of wheat *ferment* with five *sheng* of hot water. Incubate in a sealed jar for seven days.'[51]

---

[46] According to Wu Chhêng-Lo (*1937*), revised by Chhêng Li-Chün (*1957*), pp. 58–60: one Sung *tou* = 10 *shêng* = 6.64 litres; 1 Sung catty = 596 g. Total volume is 6.64 litres and total weight of honey added = 4.5 catties or 2.68 kg. At 80% sugar, total substrate is 2.14 kg which at 80% conversion should give 1.09 × 0.8 = 0.87 kg of alcohol. The product, therefore, contains about (0.87/6.64) × 100 = 13.1% alcohol.
[47] *Pi Shu Lu Hua* 避暑錄話, quoted in Chou Chia-Hua (*1988*), p. 83.
[48] For a brief discussion on the mead fermentation, cf. Chou Chia-Hua (*1988*), p. 83. A more extensive account is given by Huang Wên-Chhêng (*1985*).
[49] *Hsü Pei Shan Chiu Ching* by Li Pao 李保; *Chiu Hsiao Shih* by Sung Po-Jen 宋伯仁, +1235, p. 4.
[50] *YCFSC*, p. 127 under grape wine. [51] *PTKM*, ch. 25, p. 1564.

The incorporation of glutinous rice would, of course, bring the taste of the mead closer to that of regular rice wine. In this case, honey is considered as an additive, and the product, naturally, another medicated wine.[52] Mead in China, however, never attained the degree of popularity that it enjoyed in mediaeval and modern Europe.

(iii) *Koumiss and other fermented milk products*

A more complex and interesting process technologically is the preparation of koumiss or kumiss (transliterated as *hu mi ssu* 忽迷思), also called *ma thung* 馬湩, made by the fermentation of mare's milk. The importance of mare's milk in the life of the pastoral peoples of Central Asia can hardly be exaggerated.[53] Apparently, the making of koumiss from milk is an ancient art. Li Shih-Chên states that during the Han a wine (*chiu* 酒) was made from mare's milk (*ma ju* 馬乳), but he says nothing about the way the fermentation was carried out.[54] The *Chhien Han Shu* (History of the Former Han Dynasty) lists a *Tung Ma* 挏馬 (Prefect of the Mare Milkers), who, according to Ying Shao's annotation, was responsible for collecting mare's milk and processing it into mare's milk-wine (*thung ma chiu* 挏馬酒) and other products for the Imperial Household.[55] But there is no clear description of how koumiss was prepared. Except for one reference to a *ju chiu* 乳酒 (milk-wine) in a poem by the famous Thang poet Tu Fu,[56] koumiss is hardly mentioned in the Chinese literature between the end of the Han and the beginning of the Sung even though the *Chhi Min Yao Shu*, +544, and the Thang Diet Therapy classics describe several other products derived from milk such as *lo* 酪, *su* 酥 and *thi hu* 醍醐,[57] the making of which will be dealt with later. Koumiss did not become well known in China until the coming of the Mongols and it relapsed into obscurity when they left. But it was the favourite alcoholic drink of the Mongols and over-indulgence in it is said to have been one of the reasons that hastened the demise of the Yuan dynasty.[58]

The making of an alcoholic drink from milk is a more complex process than the fermentation of fruit juice or honey. Mare's milk contains less fats, less proteins but more lactose than cow's milk. It is low in casein and does not curdle in acid like cow's milk. During the process some of the fats are separated as cream, the proteins

---

[52] Sun & Chu (*1986*), p. 136.

[53] Tannahill (1988), pp. 131–2. Mare's milk is a particularly good source of vitamin C, which would otherwise be deficient in the diet of a pastoral people.

[54] *PTKM*, ch. 50, p. 2769.

[55] *Chhien Han Shu*, ch. 79, p. 7a; cf. Bielenstein (1980), p. 34, on the Han official *Thung ma ling* (Prefect of the Mare Milkers) among whose duties was the fermentation of mare's milk to kumiss. Yin Shao 應劭 is the author of *Fêng Su Thung I* 風俗通義. Li Chhi's (李奇) annotation states that mare's wine is made by hitting or punching the milk (撞挏乃成也), cf. also Hsü Chu-Wên (*1990*).

[56] The original line is: 山瓶乳酒下清雲, written in commemoration of the gift of a bottle of milk-wine by a Taoist priest, quoted in Shinoda Osamu (*1973a*), p. 334.

[57] Schafer (1977), p. 106, has interpreted *lo* as kumiss, *su* as clotted cream and *thi-hu* as clarified butter. On the other hand Pulleyblank (1963), pp. 248–56, considers *lo* as 'boiled or soured milk', *su* as 'butter' and *thi-hu* as 'clarified butter'. As the reader will soon see, our point of view on these products is closer to that of Pulleyblank than to Schafer's.

[58] Yüan Kuo-Fan (*1967*).

remain unchanged and the lactose is either converted to lactic acid or fermented to alcohol. Thus, we have two fermentative processes in operation simultaneously. The most vivid description we have on the making of koumiss is provided by a Western traveller to China, William of Rubruck, c. +1253, who reports:[59]

Kumiss is the customary drink of Mongols and other pastoral peoples of Asia. The method of making it is as follows. Take a large horse skin bag and a long bat that is hollow inside. Wash the bag clean and fill it with mare's milk. Add a little sour milk [as inoculum]. As soon as it begins to froth beat it with the bat and continue doing so until the fermentation stops. Every visitor to the tent is required to hit the bag several times when he enters the tent. The kumiss will be ready to drink in about three to four days.

Presumably the inoculum, after generations of selection, contains yeast as well as lactic acid bacteria. By filling the bag near to the top the milk is exposed only to a small amount of air. The lactobacilli hydrolyse the lactose and convert some of the glucose released into lactic acid, which lowers the pH, but the proteins from mare's milk do not curdle and remain in solution. The milk is left in a relatively anaerobic condition. The yeast is encouraged to proliferate, hydrolyse and ferment the lactose to alcohol. By hitting the bag the milk is churned and a soft mass of cream floats to the top. When done the cream is skimmed off leaving behind a slightly milky liquid – koumiss.[60] It is said to have a 'tangy taste similar to freshly fermented grape wine. After a man has taken a draught it leaves a taste behind like that of almond milk, and it makes the inner man most joyful, intoxicating weak heads and provoking the flow of urine.'[61]

It is curious that koumiss should have had the reputation of being an intoxicating drink. The sugar (lactose) content of modern mare's milk, though higher than cow's milk, is only 6.2%. If it is totally hydrolysed and fermented to alcohol, we would have a drink with an alcohol content of only about 3.3%, that is about the strength of beer. This would represent the maximum potency attainable. Most specimens would undoubtedly contain less alcohol since some of the lactose is converted to lactic acid. Koumiss is tame stuff compared with *chiu*, grape wine or mead. But perhaps the Mongols did drink copious amounts of koumiss, so much so that they managed to get drunk on it.[62]

[59] Tr. auct. from Yüan Kuo-Fan (*1967*), p. 14, which is taken from the 'The Journey of William of Rubruck to the Eastern Parts of the World, 1253–55,' Rockhill (1900), p. 66 and J. Pinkerton, 'The Remarkable Travels of William de Rubriquis . . . into Tartary and China, 1253' in *A General Collection of Voyages and Travels* (1808–14 Vol. VII, p. 49 (quoted in Tannahill (1988), p. 123). A similiar procedure is described in *Hei Tha Shih Lu* 黑鞯事略 (Brief Notes on the Black Tartars), c. +1237; cf. *Hei Ta Shin Lüeh Chien Cheng*, p. 93.

[60] Koumiss is still made on a large scale today in the Soviet Union. The method employed is similar to that used by the mediaeval Mongols. According to Steinkraus *et al.* (1983), p. 276, the milk is placed in a wooden vessel, mixed with a small portion of old koumiss, and held for fifteen to twenty-four hours. 'Additional heat and agitation is applied if necessary to stimulate the fermentation. The fermentation is complete when the milk is thoroughly sour and sends a thick mass to its surface. It is then beaten and stirred until the "curd" is thoroughly broken and forms a thick liquid. It is again covered and fermented for an additional twenty-four hours or longer, and blended until perfectly smooth.' For further information on how koumiss is made from cow's milk see Kosikowski (1977), pp. 42–6, Prescott & Dunn, 4th edn (1982), pp. 158–9, and Campbell-Platt (1987), pp. 108–9.

[61] Rockhill (1900), p. 67.

[62] According to de Rubruck, the Mongols had four types of alcoholic drink: grape wine, wine from grains, koumiss and mead, cited in Yüan Kuo-Fan (*1967*), p. 143.

At any rate koumiss, or more accurately milk wine, did not completely disappear from the Chinese scene. In the nineteenth century when the Europeans began to inquire into the Chinese pharmacopoeias they found koumiss and other wines from milk among the ingredients for some of the medicinal wines. These were loosely translated as 'cow's milk whisky'.[63] But today it is not even listed in works on medicinal wines, although similar products are assuredly being made and consumed by Mongolian, Tibetan and other minorities of northwestern China.

There is, however, another fermented drink from milk that is mentioned far more often than milk wine (*ju chiu* 乳酒) in the Chinese literature, namely *lo* 酪.[64] The word *lo* has had a chequered history in China. It probably started out as a term for a sort of alcoholic drink prepared from fruits such as the *lo* in *li lo* 醴酪 seen in the *Li Chi* (Record of Rites),[65] but it soon began to acquire two other meanings, one, a drink derived from grains, and the other, a drink from milk.[66] Thus, since antiquity there have been three types of *lo* in China, a *kuo lo* 果酪 (*lo* from fruits), *ju lo* 乳酪 (*lo* from milk) and a *mi lo* 米酪 (*lo* from grains). But by the time of the Thang *lo* had become a word that was applied almost exclusively to a fermented product obtained from milk.

The word *lo* is mentioned in the *Chhu Tzhü* (Elegies of the South) in the verse, 'Fresh turtle, succulent chicken, dressed with the *sauce* (*lo*) of Chhu'.[67] The commentary of Chu Hsi says that '*lo* is the juice of milk (*lo, ju chiang ye* 酪，乳漿也)'.[68] *Lo* was already a well-known product during the Western Han. The *Shih Chi* (Historical Records), −90, relates that Chung Hang admonished the Hsiungnu thus:[69] 'All the foods of China were not as convenient and satisfactory as milk and milk products (*thung lo* 湩酪).' The *Hou Han Shu* (History of the Later Han Dynasty), +450, says of the Wu Hêng (烏恒) tribe, that they 'eat meat and drink *lo* (*shih jou yin lo* 食肉飲酪).'[70] The *Shuo Wên* defines *lo* as 'juice from milk (*ju chiang yê* 乳漿也).' The *Shih Ming* says '*lo* is fattening. It is made from fluid milk. It makes one fat and adds a glow [to one's complexion].'[71] From these remarks we may infer that *lo* was a drink derived from milk, but, as yet, we have no idea of what it tasted like or how it was made.

---

[63] MacGowan (1871–2), p. 238.

[64] In modern dictionaries the character 酪 is pronounced as *lao*. In the *Shuo Wên*, and in the great *Khang Hsi Tzu Tien*, it is pronounced as *lo*. This suggests that throughout most of Chinese history *lo* is the preferred pronunciation. We have adopted *lo* in this account. Other Western historians of Chinese food culture, such as Sabban (1986a), in her excellent treatment on the history of dairy products in China, pp. 31–65, have preferred to romanise 酪 as *lao*. The reader is urged to consult her article for further information on the cultural and culinary aspects of dairy foods in China.

[65] *Li Chi*, ch. 9, *Li Yung*, p. 367, *Chi I* 祭義, ch. 24, p. 759.

[66] For a discussion on this issue cf. Ling Shun-Shêng (*1958*), pp. 888–9.

[67] Tr. Hawkes (1959/1985), p. 234. The original, *Ta Chao*, Tung Chhu-Phing (*1986*), p. 274, reads, *hsien hsi kan chi, ho chhu lo chih* 鮮蠵甘雞，和楚酪只. *Lo* is translated here as 'sauce'. It is uncertain whether this *lo* is a milk product.

[68] Cited in *Han Yu Ta Tzhu Tien*, p. 1403; cf. also Fu Hsi-Jên (*1976*), p. 172, note 28.

[69] Tr. Lattimore (1940/1988), p. 488; cf. *Shih Chi*, ch. 110, History of the Hsiungnu, pp. 2898–9. Chung Hang 中行 was a eunuch of high position at the Han court. He was forced to escort the princess who was given in propitiatory marriage to the Hsiungnu chief. Because he had been sent against his will, he declared that 'he would yet become a calamity to Han'. The passage shows that milk and *lo* were the primary foods of the Hsiungnu people.

[70] *Hou Han Shu*, ch. 90. Biography of Wu Hêng 烏恒傳, p. 1979.

[71] *Shih Ming*, ch. 13, p. 8b, *lo, chê yê, ju chih so tso shih jên fei chê yê* 酪澤也，乳汁所作使人肥澤也.

By the time of the Northern and Southern Dynasties (+420–580) *lo* had become a major beverage in the Northern part of China under the reign of dynasties established by invading nomadic tribes.[72] Presumably for this reason, the method for making *lo* from the milk of cows or ewes was well known to Chia Ssu-Hsieh who recorded it in the *Chhi Min Yao Shu*, +544. The description starts with a number of general admonitions on the management of the herd and when and how the milking should be conducted.[73] It then goes on to say:[74]

The milk collected is placed in a pot and heated with a gentle fire. If the fire is too strong, some of the milk solidifies at the bottom of the pot will char. It is advisable to collect in the 1st and 2nd month an ample supply of cow or sheep dung and dry it. The dung makes the best fuel to heat milk. A grass fire will cause dust to fall into the milk, and a wood fire will easily cause the milk to char. Dry dung burns gently, and is free from these defects. Keep stirring the milk with a [wooden] ladle. Do not allow it to froth over. After a while, stir the milk by moving the ladle up and down against the side of the pot. Do not move the ladle in a circular motion. It will inhibit the curdling. Do not blow on the milk. This will cause the milk to separate.[75] After boiling four to five times, stop the heating and pour the milk into a shallow vessel. Do not disturb the milk. When it has cooled somewhat, peel off the skin that forms on the surface and place it in another container. It will be used to make *su* 酥.

Bend a tree stem into a circle and use it to hold open a round silk bag. Filter the warm milk through the bag and collect it in an earthenware bottle for incubation. A new bottle can be used without prior heat treatment. If it is an old bottle, it should first be heated in hot ashes so that all the water embedded in the wall is expelled. Be sure to turn the bottle in the ashes so that the walls are heated evenly. If any moisture is left it will cause the *lo* to break (*tuan* 斷)[76] and to be unable to form. During incubation, the temperature must be controlled. Ideally, it should be warm, slightly warmer than human body temperature. If it is too hot, the *lo* will turn sour; if too cold, no *lo* will form. [Before incubation] add to the filtered hot milk pre-formed sweet *lo* as the starter (*chiao* 酵). For every pint of milk add half a spoon of starter. The starter is placed in the large ladle and stirred with a small spoon to an even consistency. It is then added to the filtered milk and mixed thoroughly by using the ladle. The bottle is then wrapped with a woollen or silk blanket to keep it warm. After a while, the blanket is replaced by a piece of cloth. By the following morning, the *lo* will be ready.

---

[72] See the comparison between *lo chiang* 酪漿 (juice of *lo* translated as 'yoghurt drink') and tea in Chapter (*h*), p. 511 below.

[73] *CMYS*, ch. 57, on Rearing Sheep, p. 315, gives the following admonitions: (1) When milking it is important that about a third of the mother's milk should be reserved for the calves or kids. (2) Collection of milk for making *lo* should start towards the end of the 3rd month or early in the 4th month, when there is rich pasture to feed the milking animals. The process should be slowed down by the end of the 8th month and eventually stopped altogether. (3) The mothers and their offspring are kept apart at night. They are sent out early in the morning to forage on the grass when they are laden with fresh dew. The animals are taken home, and milked. After milking, the animals are kept together and sent back to feed again in the pasture. This way, recovery of milk will be maximal and both mothers and their young are well fed and healthy.

[74] *Ibid.* pp. 315–18.

[75] The original text says, 吹則解. The meaning of this sentence is obscure. It has also baffled Shih Shêng-Han (*1958*), p. 407.

[76] The meaning of *tuan* in this context is also obscure. The text continues, 'if after repeated heat treatment of the bottle, the *lo* still breaks, it is likely that the workshop is infested with snakes or frogs. In that case, burn some human hair or horns of cattle or sheep in the premises. The odour will drive the pests away.' Since we are dealing with a lactic acid fermentation, *tuan* could mean a lack of production of acid and thus a failure to form a curd. Lactic acid bateria are subject to infection by bacteriophage, which would lyse the bacteria present. Cf. Barnett (*1953*), cf. pp. 77–8.

If one is too far from the city and no preformed *lo* is available, soured rice water[77] may be used as the starter. Use one spoon of sour water for each *tou* 斗 of milk, and mix the starter evenly in the milk. When the starter is sour, the product is also sour. When the starter is sweet, using too much of it will also make the product sour.

The text continues with several interesting recipes.

Preparation of dry *lo* (*kan lo* 乾酪): Do this in the 7th and 8th month. Expose the *lo* to the sun. When a skin of fat forms on top, remove it. Repeat the process until all the fat is removed and no more skin forms. When about a *tou* of defatted *lo* is collected, heat the material in a frying pan and pour it on a shallow dish. Allow the sun to dry it further, until it reaches a consistency suitable for rolling into balls about the size of pears. Dry the balls in the sun until dry. It can be kept for years and is an ideal food to take on long journeys.

Preparation of strained *lo* (*lu lo* 漉酪): Place a batch of rich *lo* of good quality in a cloth bag and suspend it. Allow the water to drip out of the bag. When all the whey has dripped out, heat the residue gently on a pan and place it on a plate. Dry it in the sun. When it no longer feels wet, roll it into balls the size of pears. It will also keep for years. Shavings from it may be added to congee or soup. The flavour is superior than *kan lo*. Although heating the product is slightly detrimental to its flavour, which is not as good as that of raw *lo*, it will not keep as well unless it is heated. Without heating the *lo* will be infested with insects and will not survive the summer.

Preparation of a starter for making *lo* from mare's milk (*ma lo chiao* 馬酪酵): Mix 2–3 pints of donkey's milk with mare's milk (amount can be varied). Incubate the milk to form *lo*. Take the sediment from the *lo*, roll it into balls and dry them. These can be used as a starter for making *lo* in the following year.

In these passages we have intentionally refrained from translating the word *lo* 酪 since we are not at all sure if there exists an equivalent product among the dairy foods of the West. *Lo* has been translated variously as cheese, soured milk, koumiss, yoghurt or simply fermented milk.[78] Based on the directions on how it is made in the *Chhi Min Yao Shu*, it is clear that regular *lo* is formed by a lactic acid fermentation. Except for *lo* from mare's milk, i.e. *ma lo* 馬酪, there is no indication that it was an intoxicating drink. The milk is first heated and cooled. During cooling a good deal of the fat separates and is removed. As more acid is produced some of the protein, particularly casein, is coagulated. It is a short overnight fermentation. There is no description of what *lo* looks like. It has to be a fluid since it is consumed as a drink. Thus, it probably resembles buttermilk or soured milk more than yogurt. It contains suspended solids which can be separated from the whey by filtration in a silk bag to give a *lu lo* 漉酪 (drained *lo*). It can also be dried by evaporating the water in the sun to give a *kan lo* 乾酪 (dried *lo*). These products are, in fact, 'lactic cheeses'.[79] The

[77] The original is *chhu sun* 醋飧. *Sun* is cooked grain soaked in water. The water often turns sour due to the action of lactic acid bacteria. The *CMYS* devotes a whole chapter to this topic under the title *sun fan* 飧飯, cf. ch. 86, p. 524.

[78] *Cheese*: Shih Shêng-Han (1958), pp. 88–9; *Sour milk*: Pulleyblank (1963), pp. 248–56; *Koumiss*: Schafer (1977), p. 106; *Yoghurt*: Anderson (1988), pp. 50, 55, although he did not directly refer to *lo* as yoghurt and Sabban (1986a), pp. 36–7.

[79] Cf. n. 76 above, Barnett (1953), p. 61.

flavour and consistency of *lo* is largely determined by the type of organisms present, presumably *Lactobacillus* and *Streptococcus* spp. There was probably very little yeast in the *lo* starter, but some of the lactose present could have been hydrolysed in the process. Indications are that there was a sweet *lo* (*thien lo* 甜酪) and a sour *lo* (*suan lo* 酸酪).

From the above discussion it appears likely that *lo* is a sort of defatted soured milk, liquid yogurt or buttermilk in which the protein is only slightly coagulated.[80] We suspect that most of the *lo* mentioned in the Thang and Sung literature, unless it is prepared from mare's milk, probably contained very little alcohol, if any. In addition to *lo* there are three other milk products that are frequently mentioned in the Chinese literature, that is, *su* 酥, *thi hu* 醍醐 and *ju fu* 乳腐. *Lo*, *su* and *thi hu* are often mentioned together as a metaphorical hierarchy representing the stages in the development of the soul.[81] *Su* is listed in the *Ming I Pieh Lu* (Informal Records of Famous Physicians) +510, together with the milk from cows, ewes and mares. The *Hsin Hsiu Pên Tshao* (New Improved Pharmacopoeia), +659, quoting Buddhist scripture, says that milk gives rise to *lo*, *lo* to *su*, and *su* to *thi hu*.[82] Milk and its three derivatives are listed in the *Shih Liao Pên Tshao* (Handbook of Diet Therapy),[83] +670, and in all the major pharmacopoeias of the ensuing dynasties. The *Chhi Min Yao Shu* has advised us that when cooling the heated milk to make *lo*, the skin that forms on the surface is removed and used to make *su*. But that is not the only source of *su*. The *Chhi Min Yao Su* also gives us a recipe for preparing *su* 酥 from *lo* 酪:[84]

Method for Extracting *Su* from *Lo* (*phêng su fa* 抨酥法). First make a churning stick from elm wood. Take an elm wooden bowl and cut off the top half. Bore four holes about an inch in diameter on the sides of the lower plate. Attach a long stick to the centre of the bowl – to form a sort of rake similar to that used in mixing wine mash. Both sweet and sour types of *lo* may serve as the substrate. Even batches that are highly soured may be used. Place the yoghurt drink in a jar of suitable size and expose it to the sun. Start early in the morning and wait until the sun is at the southwestern corner (i.e. later afternoon). Then agitate the liquid with the churning stick, making sure that the plate often reaches the bottom of the jar. After a 'meal time' heat up some water until it is just hot enough to touch with one's hands. Pour the hot water into the jar. The volume should be about half that of the substrate. Agitate further until the oily *su* begins to separate. Add the same volume of cold water and agitate some more. At this time it is not necessary to have the plate reach the bottom of the jar, since the *su* floats to the top. Add another batch of cold water. Stop the churning when the oily *su* starts to solidify.

Place a bowl of cold water next to the jar. Move the congealed *su* by hand from the jar to the bowl. Eventually all of the *su* will be transferred. The residual defatted, diluted soured milk is used for mixing with cold cooked grains or congee. When the water in the bowl is cold, all the floating pieces of *su* will be solidified. They are dried (dripping off the water), rolled by hand into balls and kept in a bronze or a water-proofed earthenware pot. When enough material is collected in about ten days, the balls are heated by a slow-burning cow or sheep dung fire. The procedure is similar to the method for extraction of fragrance from

---

[80] This translation follows Pulleyblank (1963). Another translation that we shall use following Jenner (1981), p. 215, is 'yoghurt drink'. The original text in the *Loyang Chieh Lan Chi* uses the term *lo chiang* 酪漿 which could be construed as a liquid form of *lo*, which is, in fact, regular *lo*.
[81] Schafer (1977), p. 106. [82] *HHPT*, ch. 15, p. 171. [83] *SLPT*, pp. 29–30. [84] *CMYS*, ch. 57, pp. 317–18.

flowers and spices shown elsewhere [in the *Chhi Min Yao Shu*].[85] As the oil heats up the residual moisture in it will evaporate explosively and make a noise like heavy raindrops falling on water. When all is quiet the *su* is ready.

The text continues to say that the creamy skin collected when the heated milk is cooled during the *lo* process is also heated in the same way to give *su*. In reading the above passage the astute reader may have noticed that there is an ambiguity associated with usage of the word *su*. *Su* is applied to the creamy material that separates from the soured milk after churning. It is also used for the final product obtained after the crude *su* is heated in a pot. From these accounts it is likely that *su* is simply butter. What then is *thi-hu*? Surprisingly, there is very little information about its preparation. All we have is a statement from The *Chêng Lei Pên Tshao* (Reorganised Pharmacopoeia), +1108, which quotes Su Kung (c. +670) in saying that '*thi hu* is the esssence of *su*. Each *tan* 石 of good quality *su* contains 3–4 pints of *thi-hu*. Heat *su*, churn it and refine it (*je, phêng, lien* 熱，抨，煉). Allow it to sit in a jar until it solidifies. Poke the soft solid and *thi hu* would separate and it can be poured out.'

This statement indicates that *thi hu* is simply clarified butter or butter oil. It is the familiar *samna* of Middle-Eastern cookery and the *ghee* of Indian cookery. In our view *lo*, *su*, and *thi-hu* are soured milk, butter and clarified butter oil respectively. It follows the interpretation adopted by Pulleybank and Sabban[86] and differs from that given by Schafer who considers *lo* as koumiss, *su* as clotted cream and *thi-hu* as butter.[87] Based on the method described in the *Chhi Min Yao Shu* we would have to conclude that the conditions of the process favour a lactic acid fermentation rather than a combination of lactic acid and alcohol production.[88] In general it would appear that the *lo* of ancient and mediaeval China was closer to soured milk than to koumiss.

Abbreviated versions of the recipes for making *lo* and *su* in the *Chhi Min Yao Shu* are presented in the *Chhü Hsien Shên Yin Shu* (Notes of the Slender Hermit), +1440 and repeated in the *Pên Tshao Kang Mu* of +1596.[89] A more extensive version of the recipe for *lo* is found in *Chü Chia Pi Yung* of the Yuan, but in the latter work the method for *su* now includes sheep tallow or lard as ingredients.[90] This suggests that after the Yuan, the word *su* may no longer indicate pure butter, but simply a shortening based on animal fat.

What about *ju fu* 乳腐? It emerged as a dairy food during the Thang dynasty. It is listed in the *Shih Ching* 食經 (Food Canon), c. +600, of Hsieh Fêng and in the

---

[85] *Ibid.*, ch. 52, pp. 263–4. [86] Sabban (1986a), pp. 36–40; Pulleyblank (1963), cf. n. 78 above.
[87] Schafer (1977). Schafer's interpretation of *lo* as koumiss is followed by Freeman (1977), p. 156.
[88] According to current knowledge, to produce koumiss from either mare's or cow's milk requires the use of large inoculum (10–39%) containing actively growing lactic acid bacteria as well as *Torula* or *Sachcharomyces* yeast at an incubation temperature of 28°C. For a lactic acid fermentation to make yoghurt or soured milk a 5% inoculum of culture is used and the incubating temperature is 38°C or higher. See, for example, Kosikowski (1977), pp. 45–9. In the *CMYS*, the temperature recommended is slightly warmer than body temperature, which is close to 38°C and would favour the lactic acid process.
[89] *PTKM*, ch. 50, pp. 2788–90.
[90] *CCPY*, method for *lo*, p. 136; method for *su*, p. 135. This warns us that in recipes containing *su* in recipes in subsequent works, one cannot assume the *su* used is automatically butter.

Shih Liao Pên Tshao (Compendium of Diet Therapy), +670.⁹¹ In the biography of Mu Ning 穆寧 in the Thang histories⁹² his four highly successful and virtuous sons are each compared to one of the four dairy products, the oldest to *lo*, the next to *su*, the third to *thi hu* and the youngest to *ju fu*. Although *ju fu* also known as *ju ping* 乳餅 is mentioned often in the subsequent food literature, the earliest record of its preparation is not seen until the *Chü Chia Pi Yung* (Essential Arts for Family Living) of the Yuan (+1271 to +1368).⁹³

To make *ju fu*: Filter one *tou* 斗 of cow's milk through a silk bag into a pot. Boil it three to five times. Dilute with a little water. Coagulate the milk by adding vinegar, just like the making of *tou fu* 豆腐 (bean curd). When the curds are fully formed, drain it in a silk bag, and press it with a piece of stone. Add salt and store the curds at the bottom of a jar.

This passage is adjacent to another recipe which describes the making of milk curds *ju thuan* 乳團 from soured milk (*lo* 酪) by a similar procedure.

From this account it would seem that there are two types of Chinese milk curd, one *ju fu* 乳腐 or *ju ping* 乳餅, is curdled with vinegar, and the other *ju thuan* 乳團 is curdled by naturally formed lactic acid. They are versions of the Western cottage cheese, and represent the dairy equivalent of *tou fu* (pressed bean curd) which will be discussed in Chapter (d). Since the *Chhi Min Yao Shu* (+544) processed milk products are often mentioned in agricultural treatises⁹⁴ as well as food canons and recipes. They were apparently familiar food products available in the markets of the capitals of the Sung.⁹⁵ To give us an idea of their scope of usage, the major references to dairy products in the Chinese food literature are summarised in Table 26. We have included several works of *Materia Dietetica* in the Table to serve as a baseline.⁹⁶ In these cases milk and its products are regarded as medication as well as food. In the other works, *ju* (milk) is only entered if it is used as an ingredient in cookery. The table shows that dairy foods were a significant, if minor, factor in the diet of the Chinese during the Thang, Sung and Yuan periods. Butter (*su* 酥) was particularly popular; it was the shortening of choice for many rolls and cakes. But the role of butter began to erode by the Yuan dynasty as shown by the adulteration of *su* with tallow and lard in the *Chü Chia Pi Yung*, noted earlier. By the beginning of the Chhing even *lo* (soured milk) had shifted its meaning. As shown in the *Shih Hsien Hung Mi*,

---

⁹¹ *Shih Ching* in *Chhing I Lu*, p. 15; *SLPT*, enlarged by Chang Ting 張鼎, Renmin Weisheng ed. (1984) p. 62.

⁹² *Chiu Thang Shu*, ch. 150, pp. 4116–17; *Hsin Thang Shu*, ch. 88, p. 5016. Mu Ning lived at the time of the An Lu-Shan Rebellion in the 8th century.

⁹³ *CCPY*, p. 137. The same recipe is reproduced in the *Chhü Hsien Shên Yin Shu* and cited in the *PTKM*, ch. 50, p. 2792.

⁹⁴ *SSTY*, 7th month, p. 177; *NSISTY*, 4th month, p. 79, 5th month pp. 85–6.

⁹⁵ *TCMHL*, pp. 17, 43, 61. A Chang 張 family in the southern capital of Hangchou was noted as specialists in the making of cheese from milk (*ju lo* 乳酪); *MLL*, pp. 135, 137. The statement by Freeman (1977), p. 156, that 'kumiss, fermented mare's milk' is among the beverages commonly consumed during the Sung is most likely mistaken. He has apparently followed Schafer's interpretation and regards *lo* as synonymous with koumiss. The Northern Sung encyclopedia he cites, the *Shih Wu Chi Yüan* 事物紀原 (The Origin of Things), c. +1080, ch. 9, pp. 7a, 7b, about the origin of koumiss is actually about the origin of *li lo* 醴酪. Thus, what he says about the popularity of koumiss is correct if we replace koumiss with 'processed milk' products.

⁹⁶ *SLPT*, *YSCY* amd *YSHC*. Table 26 is adapted from a more extensive tabulation given by Sabban (1986), p. 65.

Table 26. *References to dairy products in the Chinese food literature*[a]

Reference	Ju 乳	Lo 酪	Su 酥	Thi-hu 醍醐	Ju fu[b] 乳腐
Chhi Min Yao Shu +544	+	+	+	—	—
Shih Ching +600	—	—	+	—	+
Shih Liao Pên Tshao +660	+	+	+	+	+
Shih Phu +700	+	—	+	—	—
Tung Ching Mêng Hua Lu +1148	+	+	+	—	+
Mêng Liang Lu +1334	+	+	+	+	+
Yin Shan Chêng Yao +1330	+	+	+	+	+
Yün Lin Thang Yin Shih Chih Tu Chi +1360	—	—	—	—	+
Chü Chia Pi Yung +1360	—	+	+	—	+
Yin Shih Hsü Chih +1368	—	+	+	—	+
Chhü Hsien Shên Yin Shu c. +1440	+	+	+	—	—
Yin Chuan Fu Shih Chien +1591	+	+	+	—	+
Shih Hsien Hung Mi +1680	+	+?	+	—	—
Yang Hsiao Lu +1698	+	+?	—	—	—
Hsing Yuan Lu +1750	+	—	+	—	+

[a] Based on Sabban (1986), p. 65.
[b] Includes *ju ping* 乳餅.

+1680 and *Yang Hsiao Lu*, +1698, *ju lo* 乳酪 had been turned from a soured milk into a type of milk pudding.[97] And, as if to add insult to injury, even the name *ju fu* 乳腐, traditionally reserved for dairy milk curds, is absconded to denote fermented *tou fu* in the *Shih Hsien Hung Mi* and in the *Sui Yuan Shih Tan* of +1790.[98] One would have to admit that by the time of the Chhing, milk products were no longer a significant factor in the diet of the people of China.[99]

After reviewing all the information presented above we are left with two surprising observations. Firstly, although milk (from the mare, cow or the ewe) has been a component of the *Materia Medica* and *Materia Dietetica* since the 6th century, it is the fermented product (as koumiss or soured milk) that was consumed in mediaeval and premodern China as a food or beverage rather than milk itself. In fact, we do not find a recommendation for drinking milk in the food literature until the *Shou Chhin Yang Lao Hsin Shu* 授親養老新書 (Nurturing of the Elderly), +1080 to +1307 (Sung to Yuan), which says that the drinking of cow's milk is particularly beneficial to the

---

[97] *SHHM*, p. 13 and *YHL*, p. 7 give the identical recipe for making *ju lo*: 'Add half a cup of water to a bowl of cow's (or ewe's) milk. Blend in three spoons of wheat flour and filter. Heat in a pot until the liquid boils. Add powdered sugar and beat vigorously with a wooden spoon as the milk boils. Filter into a bowl and serve.' This product can hardly qualify as a true *lo*.

[98] *SHHM*, p. 163; *SYST*, p. 119.

[99] Milk products, however, did not completely disappear, as shown by the recipes for making clotted cream in sheets (*niu phi* 牛皮) and milk curds (*ju ping* 乳餅) which are included in the *Hsing Yuan Lu* (p. 39), and reproduced in the *Thiao Thing Chi* of late Chhing, pp. 237–8, 262–3. In the latter case, it would seem that the author was simply copying the recipes from the *Hsing Yuan Lu*.

elderly.¹⁰⁰ The recommendation is reiterated in the *Yin Chuan Fu Shih Chien*, +1591, of the Ming Dynasty.¹⁰¹ This discrepancy has long been regarded as a peculiar quirk of Chinese food culture. We now suspect the real reason is that the Chinese and their nomadic neighbours to the North were, and still are, lactose malabsorbers. They cannot digest the lactose in fresh milk, but they can tolerate fermented milk in which the lactose has been converted into lactic acid (or alcohol).

Secondly, it is remarkable that the Chinese and particularly their nomadic neighbours of the Asian steppes, for whom milk was a major food resource, never learned to use rennet to curdle milk, and were thus excluded from experiencing the wonders of true *cheese*, i.e. rennet cheese, an indispensable component of the great cuisines of Europe. Considering the substantial trade that moved along the silk road between China and Europe from the Han to the Ming it is remarkable that, somehow, this secret of the West never found its way to Central and East Asia.¹⁰²

(8) Alcoholic Fermentations, East and West

And Noah began to be an husbandman, and he planted a vineyard. And he drank of the wine and was drunken, and he was uncovered within his tent.¹

According to the above passage from Genesis, Judaic tradition tells us that wine was first made shortly after the great floods in West Asia. In China 'wine' is said to have been invented at the direction of the daughter of Emperor Ta Yü, who tamed the great floods and founded the Hsia dynasty.² These legendary accounts, if not taken too literally, are consistent with modern archaeological evidence which indicates that alcoholic drinks were already known at the dawn of history in both West and East Asia, having presumably been discovered independently in each region during the Neolithic Age. But in spite of the similarity in the basic chemistries involved, the details of the fermentation technology in the two cultures have developed along quite different lines. What are the reasons for this discrepancy? We believe the

---

¹⁰⁰ *Shou Chhing Yang Lao Hsin Shu*, ch. 1, p. 30a. Although the work was compiled in +1080, it was significantly enlarged in +1307. The exact date of the recommendation is thus sometime between the two dates. However, milk was used as a decoction mixed with other diet therapy ingredients during the Thang. Schafer (1977), p. 105, cites a poem by Pai Chü-I, +772 to +846, *Chhun Han* 春寒 (Spring Chill), *Chhüan Thang Shih*, ch. 30, p. 4b, stating that 'nothing could please him more on a cool morning than rehmannia (*ti huang* 地黃) taken with milk'. The *Shou Chhing Yang Lao Shu*, p. 29b, also recommends a drink made with milk, black pepper (*pi ba* 蓽茇) and water.

¹⁰¹ *YCFSC*, p. 171.

¹⁰² Two hypotheses for the discovery of rennet by the people of the prehistoric Near East are discussed by Mair-Waldberg (1974). One is that it happened when milk was inadvertently stored in a pouch made from a calf's stomach. Another, is that it was the result of the slaughter of a young animal soon after it had nursed its mother. In any case it was a rare accident unlikely to be repeated elsewhere. It was not known to the pastoral nomads of East and Central Asian steppes. Its use was already common knowledge by the early years of the Roman Empire, cf. Loeb edition of Columella (1968), p. 285.

¹ Genesis, 10. 20. For a discussion of the culture of grape wine in the Bible cf. Sasson (1994).

² Cf. p. 161, the *Chiu Kao* 酒誥 (Wine Edict), text quoted in Hu Shan-Yüan (*1939*), p. 266, Yang Wên-Chhi (*1983*), p. 7, and Chang Yuan-Fang (*1991*), p. 360. Judaic chronology indicates that Noah lived at about −2000. Chinese records hold that the Hsia dynasty was founded in about −2000. Alcoholic drinks were probably discovered very much earlier, cf. pp. 155–6. Our present knowledge indicates that alcoholic drinks were known long before −2000.

answer lies in the differences between the specific natural and man-made environments that existed in the respective societies when the process for converting fermentable sugars to alcohol was first discovered. Among these, the most important are, firstly, the kind of raw materials available as substrates for the fermentation, and secondly, the culinary techniques for handling such materials that were in place when the invention occurred.

We have reviewed and analysed the historical evidence on the development of alcoholic drinks from grains in China (pp. 154–62). In the West two types of beverages were produced in the early societies of Mesopotamia and Egypt, wine from grape, and ale or beer from grain.[3] The grape we are concerned with here is the wine grape, *Vitis vinifera*, believed to have been domesticated in the region south of Caspian Sea.[4] The grains involved are barley and wheat, which were first domesticated somewhere in Southwestern Asia.[5] The conversion of grape juice to wine requires little or no inventive intervention by man. Indeed, the process is virtually inevitable.[6] The oldest written records in Mesopotamia, dated near −2000, take note of the existence of beer as well as grape wine.[7] But all indications are that both types of fermented beverages must have been developed earlier in the Neolithic Age, and that wine appeared before beer.[8] The fermentation of grape juice to wine requires no heating and could have been carried out in in the simplest type of container such as that made of animal skin, wood or stone. A grape fermentation could have been observed even before the domestication of *Vitis vinifera* which probably began at about −8000.[9]

---

[3] Partington (1935), p. 303.

[4] Bianchini & Corbetta (1976), p. 160. For more recent discussions of the issue cf. Zohary (1995), Olmo (1995), Badler (1995) and McGovern & Michel (1995). For further information see other articles in McGovern, Fleming & Katz eds. (1995).

[5] Harlan (1981) and Feldman, Lupton & Miller (1995) on wheat, and Harlan (1995) on barley. For the domestication of plants in the Near East, see Roaf (1990); Naomi Miller (1991), Bar-Josef (1992); Naomi Miller (1992); Zohary & Hopf (1993); Valla (1995); Grigson (1995) and Hopkins (1997). The origin of agriculture is discussed in Smith, Bruce (1995), Miller (1991), pp. 135–7 and Hawkes (1983), pp. 27–46.

[6] The wine grape is truly the fermentation substrate *par excellence*. It has a firm but thin skin that can be crushed easily to give an abundance of juice. The juice contains 18%–22% fermentable sugar, glucose and fructose, much higher than any other fruit known. The result is a wine with an alcohol content of about 10%–11%, sufficient to give it prolonged stability when properly stored. The mild acidity of the juice (pH 3.1–3.7) is optimal for the conversion of sugar to alcohol by the fermenting yeast and still somewhat inhibitory to the growth of contaminating bacteria. In addition, wild yeasts occur in profusion on the skin of the grape so that they are already in place and ready to act as soon as the grape is crushed, cf. Benda (1982), p. 311.

For a comparison of the sugar content of various fruits see Fowles (1989), 'The Compleat Home Wine-maker', *New Scientist*, 2 September 1989, p. 39, Table 1. Most fruits other than grapes contain 5 to 10 per cent sugar, the only exception being ripe bananas, whose sugar content may reach 18 per cent. Indeed, banana wine is made locally in many parts of the tropics (E. N. Anderson, private communication).

[7] Partington (1935), cites several references to beer in Egyptian papyri, pp. 195–7, and from Sumerian sources, 303–4. One can argue about whether the ancient fermented drink made from grains should be called 'beer', since beer today means a beverage prepared from sprouted barley flavoured with hops. Moreover, ancient beer was also prepared from wheat. But for convenience we shall stick with 'beer' since it is the term that is commonly used by archaeologists and food historians.

[8] More recently, Michel, McGovern & Badler (1992) have found chemical evidence of the brewing of barley beer in Godin Tepe as early as the −3500 to −3000. Furthermore, McGovern *et al.* (1996) have shown that grape wine was already being made on a substantial scale in the Northern Zagros mountains by −5400 to −5000.

[9] Olmo (1995), p. 487 and *ibid.* (1996), p. 36. On the other hand Zohary (1995) maintains that definitive evidence of cultivation is not available until the Chalcolithic period (c. −3700 to −3200).

On the other hand the preparation of fermented drinks, i.e. ale or beer, from wheat or barley is a matter of considerable complexity. The problem is analogous to the preparation of *chiu* 酒 (wine) and *li* 醴 (sweet wine) from millets and rice in China. Yet the process that was eventually adopted for making Chinese wine turned out to be very different from that established in the West for the making of beer. Why? For an answer we need to go beyond the historical record and try to ascertain how the fermentation of grains to alcohol could have originated in the prehistoric past in China and in the West.

(i) *Origin of* li *and* chiu *in China*

When we reflect upon the information that we have gathered and examined on pages 154–62 we are led to the conclusion that the development of the 'wine' fermentation in China is greatly influenced, if not determined, by the nature of the grains available and the way they were processed and cooked for human consumption. The native starch in grain kernels occurs in granules that are relatively resistant to hydrolysis by digestive enzymes. When heated in water the granules break down, the starch is gelatinised and rendered susceptible to enzymatic hydrolysis. How the Chinese of the early Neolithic period made their grains digestible is suggested by a familiar passage in the *Li Chi* (The Record of Rites), which says in part:[10]

At the first use of ceremonies, they began with food and drink. They roasted millet and meat; they excavated the ground in the shape of a jar, and scooped water from it with their two hands; they fashioned a drumstick of reeds and used the earth as a drum . . . Formerly the ancient kings had no houses. In winter they lived in caves which they had excavated, and in summer in nests which they had framed. They knew not the transforming power of fire, but ate the fruits of plants and trees, and the flesh of birds and beasts, drinking their blood and swallowing their hair and feathers . . . The later sages then arose, and men learned to take advantage of the benefits of fire. They moulded metals [into tools] and fashioned clay [into pottery] so as to rear towers with structures on them, and houses with windows and doors. They baked (*phao* 炮), broiled (*fan* 燔), boiled (*phêng* 烹) and roasted (*chih* 炙). They produced sweet liquor or wine (*li* 醴) and fruit wine (*lo* 酪).

We agree with Yuan Han-Chhing's suggestion that a rudimentary fruit wine (*lo*) was probably the earliest fermented beverage prepared in Neolithic China.[11] But, unlike the *vinifera* grape, the fruits native to China were not well suited for wine making, and in time, interest in *lo* waned as *li* 醴 and later *chiu* 酒 arose to become the favourite drinks of the Shang and Chou people. How did this happen? Legend has it that the Yellow Emperor (Huang Ti) 'invented earthenware pots and steamers', and 'steamed *ku* 穀 (hulled grains) to make *fan* 飯 (steamed granules), and

---

[10] *Li Chi*, *Li Yung*, p. 366, tr. Legge (1885), pp. 368–9, adjuv. auct. The full passage is cited in K. C. Chang (1977), pp. 44–5 and in Anderson (1988), p. 34.
[11] Yuan Han-Chhing (*1956*), p. 79. Indeed, as noted above (p. 153) the presence of pits of peach, plum and jujubes in pottery jars in the ruins of an ancient 'winery' at Thai-hsi 台西 near Chengchou 鄭州, suggest that fruit wine was still being made during the mid-Shang, cf. Xing & Tang (1984) and Underhill (forthcoming).

boiled *ku* to make *chu* 粥 (congee)'.[12] Extensive archaeological excavations in China have shown that tripod cooking pots are a common sight among the pottery remains of the Phei-li-kang and related cultures of the −6th to −7th millennia, while pottery steamers are well known from sites of the Yangshao and Ho-mu-tu cultures (−4th to −6th millennia).[13] Since the archaeological evidence shows that boiling pots arrived at least a millennium before steamers, we would have to presume that *chu* was made long before *fan*.

It is fortunate for us that both rice and millet have soft kernels and husks that are easily removed.[14] This means that both grains could be gently boiled to give congee or steamed to give cooked granules so that the prehistory of the culinary arts in North and South China developed along similar lines.[15] In ancient China sprouted grain (*nieh* 糱) was used to make *li* (sweet liquor) and microbial *ferment* (*chhü* 麴) to make *chiu* (wine). Tradition holds that the *ferment chhü* was derived from contaminated *fan*. But nothing is known about the origin of *nieh*. How was the saccharifying activity of *nieh* discovered? We suspect it might have something to do with the making of congee. Consider, for example, the following scenario. Some of the grains were stored in a damp and wet place, and sprouted before they were dehusked.[16] If when their kernels were heated in water to make congee, the fire accidentally died out and the suspension was inadvertently left in a semi-heated condition for some length of time, part of the starch would be gelatinised and hydrolysed by the amylases present into malt sugar.[17] The result would have been a broth with a deliciously sweet taste. Upon further standing the dispersion would have developed a fragrant alcoholic flavour which the Chinese has characterised as *chhun* 醇.[18] This may be how *nieh* 糱, *nieh mi* 糱米 (sprouted grain kernel), *i* 飴 (malt sugar) and *li* 醴

---

[12] *Ku Shih Khao*, cited in Thao Wên-Thai (*1983*), p. 10, says that 'Huang Ti made boilers and steamers (黃帝作釜甑)', and, 'Huang Ti was the first person to steam grains to make *fan*, and boil grains to make *chu*' (黃帝始蒸穀為飯，烹穀為粥).

[13] Cf. Chang, K. C. (1986), Fig. 54, p. 99, for archaeological finds of cooking pots associated with the Phei-li-kang culture and our discussion above (p. 76) on Pan-pho and Ho-mu-tu pottery steamers. Creel (1937), p. 44, pointed out that the *li* 鬲 tripod and the *hsien* 甗 steamer are among the most distinctive types of pottery found in Shang China. These pottery vessels are unique to Chinese culture. For a discussion of the diet of the Chinese in the Paleolithic and Neolithic periods see Tsêng Tshung-Yeh (*1988*), ch. 1, pp. 1–92.

[14] Rice husks are found in great abundance at the prehistoric site at Ho-mu-tu dated at about −5000, cf. *Che-chiang shêng po-wu-kuan* (*1978*), p. 107, Yu Hsiu-Ling (*1976*), and Chang, K. C. (1986), p. 210. This certainly suggests that in contrast to the tightly glumed wheat and barley grains of the early Neolithic West, the rice cultivars of China, by −5000, had husks that were easily removed.

[15] Today, rice is still cooked throughout China in the same two ways; gently boiled to give *chu* 粥 (congee), or steamed to give *fan* 飯 (cooked granules). In North China, *chu* is often called *hsi fan* 稀飯 (thin or diluted *fan*). In Chinese restaurants in Europe or America *fan* is usually called 'steamed rice', even though a steamer is not necessarily used for the steaming.

[16] This is a situation that would be all too common in the warm humid spring and summers of central and south China.

[17] The condition would be similar to that used in the mashing of malt (i.e. sprouted barley or wheat) in the making of beer. Mashing is often carried out at 70–80°C. The cereal amylases are among the most heat resistant enzymes known and would remain active at this temperature In fact, in the early years of the 20th century they were used for the desizing of textiles at >80°C.

[18] Thus, the original partially liquified, fermented congee may be regarded as the progenitor of the ancient *lao* 醪 congee and its modern counterparts such as the *lao* of Taiwan, *lao chao* of China, *doburoku* of Japan, the *tapeh* of Indonesia and the *gasi* of the Philippines, cf. Frake (1980), p. 167. As an example of how such products are prepared today see Wang & Hesseltine (1970), pp. 574–5.

(sweet liquor) were discovered. Further testing would reveal that *li* could be more conveniently made by incubating *fan* (steamed granules) with *nieh*.

That *nieh* could have been used to make *li* from *fan* indicates that it was contaminated with yeasts. The prehistoric Chinese were, of course, unaware of this possibility and probably thought that the ability to convert the malt sugar solution to a liquid with an alcoholic (*chhun* 醇) aroma was an inherent property of the *nieh*. Since the level of yeast was uncertain, the extent of fermentation achieved was also uncertain. Nor did they discover that they could bypass the problem by adding a sample of must from a previous fermentation to the incubation mixture. However, by the time they realised that their *nieh* gave inconsistent results when used to make *li*, the problem was already made moot by the appearance of the *ferment chhü* 麴 to the culinary scene. *Chhü* was a source of yeasts as well as moulds. This could be the main reason *nieh* and *chhü* were often used together in the Chou processes for making wine.

As noted earlier (pages 161–2) the *ferment chhü* arose when someone inadvertently left a basket of steamed rice in the open for several days. Drying the mouldy mass probably provided the earliest prototype of a *ferment* preparation. Eventually, it would be found that flour from grains could also support mould growth if it was mixed with water to form a cake.[19] This type of rice *ferment* eventually became known in North China as *chhü* and in South China as *chiu yao* 酒藥 (wine medicament).[20] How then did the original rice (or millet) *ferment* of antiquity acquire the character *chhü* 麴, written with a wheat (*mai* 麥) radical, during the Han, and why in the later ages was it usually made with wheat as the raw material rather than rice?

Actually, the character for the *ferment chhü*, according to the *Shuo Wên* (Analytical Dictionary of Characters), +121, was originally written as 鞠, which is constructed by placing a bamboo radical on top of the word denoting a chrysanthesum.[21] In fact, it is written simply as 籟 in the Ma-wang-tui Inventory suggesting that this was the form of the character at the beginning of the Han.[22] The etymology of *chhü* is thus consistent with the notion that the product was first formed when steamed rice

---

[19] The moulds evidently grow very slowly on the kernels of uncooked grains, even when they are presprouted. A recipe for a *ferment* made from steamed rice, *nü chhü* 女麴 (glutinous rice *ferment*), is given in the *Chhi Min Yao Shu*, +544, p. 534 cf. p. 336. In this case the rice is steamed, pressed into little cakes, and then incubated in between layers of artemisia leaves. This may be regarded as an alternative method for preparing the wine medicament of antiquity.

[20] The earliest description of how a *ferment* was made from rice flour is in the *Nan Fang Tshao Mu Chuang* (Flora of the South), +304, cf. Li Hui-Lin (1979), p. 59, on 'Herb leaven'. In this recipe the dough is made by mixing rice flour with the juice of various herbs and left in dense bushes in the shade. After a month it is done (presumably covered with myceliae). The procedure suggests that it took a long time for the rice flour (uncooked) cake to become infected by fungi. It was probably not easily reproducible in areas outside South China. A raw rice flour cake, known as *shitogi*, has, in fact, been traditionally used as an offering to various shrines in Japan. Ueda Seinosuke (*1996*) recently examined *shitogi* samples left in the Ono shrine in Shiga-cho for several days but failed to find any relevant fungal growth. To hasten the process, an inoculum is used in the making of the wine medicament for fermenting the Shao Hsing wine. Rice flour is mixed with smartweed powder and preformed wine medicament before the cake is incubated, cf. *Che-chiang shêng kung-yeh chhang* (1958), pp. 10–11.

[21] *SWCT*, p. 147 (ch. 7A, 21), see also p. 61 (ch. 3B, 2) which says that 鞠 is equivalent to 籟.

[22] MWT Inventory, item 146, *Hu-nan shêng po-wu-kuan* (1973) I, p. 142, items 146 and 147. The editors compare 鞠 with inscriptions in a Western Han tomb in Loyang and conclude that it is equivalent to 鞠 which is itself a variant of 麴 the character finally adopted for the mould *ferment*.

granules were exposed to air in a bamboo basket and that at some time it would acquire the colour of the yellow chrysanthemum. Although the character 麴 is found in the *Fan Yen* (Dictionary of Local Expressions), c. −15, it is not listed as a separate entry in the *Shuo Wên*.[23] This suggests that 麴 did not displace the older form of the character until after the Han.

Thus, it is quite possible that in the −2nd millennium the *ferment chhü* 麯 used in North China was prepared from steamed or powdered rice (or millet) in the same way as the wine medicament of the South. The situation began to change with the introduction of wheat (*mai* 麥) and barley (*mou* 麰) as cultivated crops at about the middle of the −2nd millennium.[24] The grains of both crops have kernels that are too hard to be conveniently cooked into congee or steamed granules. For this reason they were not fully integrated into the Chinese dietary system until the advent, at about −300, of the rotary quern which facilitated the milling of grains into flour, a topic that we shall consider in detail in a later chapter (pages 466–70). It is likely that some time during the Chou, sprouted wheat and barley were prepared and tested against steamed rice or millet for the preparation of malt syrup (*i* 飴) and sweet wine (*li* 醴). The preparation of malt syrup was a resounding success, but the making of *li* was probably less successful.[25] Soon barley or wheat malt became the preferred agent used for the preparation of malt syrup.

We suspect the success achieved with malted wheat and barley in making malt sugar may have suggested to the Chou chefs that there was something special in these grains that would facilitate the making of wine. They might have first tried to make a *ferment* from steamed granules of wheat or barley with discouraging results. They then decided to grind the kernels into flour and to mix the four lightly with water to form little cakes. These might also have been steamed before they were exposed to air. After many days the cakes would be covered with tiny threads and a layer of yellow powder. This was the origin of the wheat or barley *ferment* (*mai chhü* 麥麴) described in the *Chhi Min Yao Shu*, of +544. It was found that when the *ferment* prepared in this way was dried, it could maintain its activity for several years. In time wheat became the preferred substrate for making the *ferment* in North China. As indicated by the presence of the *mai* 麥 radical in the names given to them, all the different Han *ferments* listed in Table 19 (p. 167), except one, may be presumed to have been prepared from wheat. The ascendancy of wheat in the making of the mould *ferment* is probably also a reflection of the growing importance of wheat as a crop, and the primacy of North China as the political and cultural centre of Chinese civilisation.

[23] *Fang Yen*, ch. 13, p. 12.
[24] How wheat and barley arrived to China is still a matter of controversy. For a discussion of when the cultivation of wheat and barley began in China see Ho Ping-Ti (1975), pp. 73–6; and Bray (1984), Vol. 5, Pt 2, pp. 459–65. The possibility remains open that barley was domesticated independently in China, cf. T.-T. Chang (1983), p. 78, whereas wheat was most likely transmitted from West Asia.
[25] Teramoto *et al.* (1993), show that the amylase activity of barley malt is greater than that of sprouted rice. However, the flavour of the rice wine made from barley malt is inferior to that from sprouted rice. We suspect the amylase activity of wheat malt is also greater than that of sprouted rice.

We may summarise our speculative excursion by saying that the invention of alcoholic drinks in prehistoric China is guided by two factors: firstly, the nature of the founder crops, rice and millet, which yield grains with kernels that are relatively soft, and secondly, the development of pottery cooking pots, followed by pottery steamers. Gentle heating of grains that had been inadvertently sprouted, i.e. *nieh* 糵, in water led to the discovery first of malt sugar (*i* 飴), and later of sweet wine (*li* 醴). Accidental exposure of steamed grains in air led to the discovery of the mould *ferment chhü* (麯 later 麴) which was used to prepare the wine *chiu* 酒. In time, the mould *ferment* completely displaced the use of sprouted grain in the making of wines. The sweet wine *li* itself gradually lost favour and disappeared from the culinary scene in China after the demise of the Han Dynasty (+220).

But *li* did not die out. It was during the Han or a little later that the technology of making *li* from *nieh* (as sprouted rice) migrated via Korea to Japan, where *nieh* was known as *getsu* 糵 and *li* as *rai* 醴.[26] *Getsu* was also displaced in Japan in the mediaeval period when the Japanese learned how to prepare rice *koji* 麴 and to utilise it for the making of *saké* 酒 wine. Nevertheless, *rai* continued to be produced as a speciality beverage by employing *koji* in a short, one-night incubation. It was a ritual drink presented to the gods in sacrificial ceremonies (Figure 60). One can still enjoy it today in tea houses in Japan under the name *amazaké* 甘酒 (sweet wine), although the modern product contains little or no alcohol.[27]

This, however, is not the end of the story. *Amazakè* was transmitted to the United States as part of the complement of health foods introduced by the macrobiotic movement in the latter part of the 20th century. For many years it was a dietary speciality of interest only to members of the macrobiotic community. Since 1980 it has gained widening circles of acceptance among devotees of natural and organic foods as a non-dairy drink derived from rice. It is now commercially produced and can be readily seen in health food stores and supermarkets that supply organically grown fruits and vegetables.[28] One cannot help but marvel at the twists and turns of cultural transmission that brought a reincarnation of an obscure and forgotten beverage of ancient China to the grocery shelves of a modern technological society so far removed in time and space from its ancestral home.

(ii) *Origin of beer in the West*

The brewing of beer also depends on the enzymatic activity of sprouted grains to hydrolyse the starch in grains into fermentable sugars. Thus, it is analogous to the

---

[26] Since the Japanese language does not have an L sound, *lai* is pronounced *rai*.   [27] Ueda Seinosuke (*1992*).
[28] The story of *amazakè* in America is told in Shurtleff & Ayoyagi (1988b). A typical Amazake product such as 'Rice Dream' is a milky liquid consisting of dextrins, maltose and other solubilised constituents of rice. It is also made into an ice-cream-like dessert. The latter version of *amazaké* is promoted in Germany, cf. personal communication from Ute Engelhardt. In addition to *amazaké*, the ancient Chinese malt sugar *i* 飴 apparently accompanied sprouted rice when it came to Japan. Malt sugar, known as *amé* and written as 飴, in the form of a rice candy or sugar syrup, is still being made by allowing barley malt to act on steamed rice. The syrupy form of *amé*, known as *midzu-amé*, is a popular confection parallel to the *mai ya thang* 麥芽糖 of China, cf. p. 457 below, as well as Oshima, Kintaro (1905), pp. 22–3.

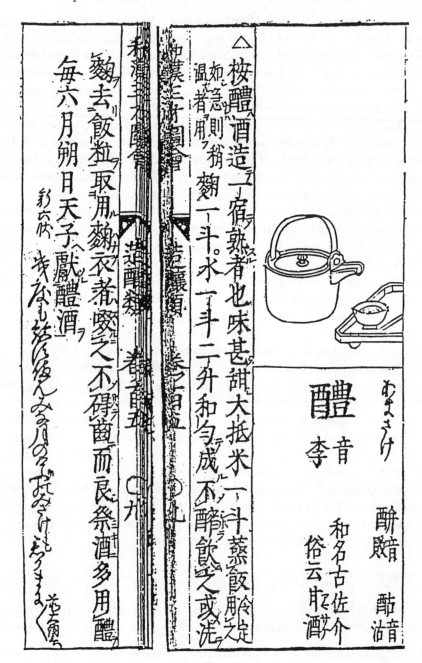

Fig 60. Definition of *li* from *Ho Han San Tshai Thu-Huei* (+1711).

making of *li* in prehistoric and ancient China. Because of the hardness of their kernel, barley and wheat were seldom cooked in the form of whole grains. The grains have to be ground into meal or flour[29] before they can be processed into digestible foods. Furthermore, the earliest forms of wheat and barley that were domesticated at about −8000 were covered with tight glumes or husks that were troublesome to separate from the edible kernels.[30] The ears were usually toasted or parched to loosen the chaff before it could be separated from the kernels by rubbing between two stones and the kernels ground to a meal or flour.[31] The heating would denature the starch granules in the grain and make it more digestible. But it would also have inactivated the enzymes generated by any presprouted grains that might have been present, and denatured the gluten precursors in the grains thus rendering them less suitable for making bread.

When mixed with water the coarse meal would form a paste or porridge, which would be the forerunner of the later Greek *maza* and the Roman *puls* as well as the still later Tibetan *tsamba*. When mixed with water the flour would form a dough which could be baked on a hearthstone to give a flat bread, the forerunner of the flat breads that are still widely made in the Near East and Central Asia. It is unlikely that under these conditions a fermented beverage could have evolved from either porridge or bread. Yet beer is so seductive a commodity in the daily life of modern man in the West that some scholars have been tempted to attribute the interest in the domestication of grains during the early Neolithic Age to their importance as a source of beer rather than food.[32] This appealing notion implies that beer existed before agriculture. It does not take into account the tricky problem of how beer could have been brewed without heat-resistant water-proof containers which did not come into being until after the advent of pottery.

---

[29] According to Moritz (1958), p. xxv, 'The milling of grain is probably the oldest of all human arts: it almost certainly goes back to a time when grain was still being gathered from plants growing wild, and for long ages thereafter the "daily grind", in a very literal sense, was prominent among the less pleasant household tasks.' In fact, mortar and pestle and grinding mills are among the earliest stone implements recovered from Paleolithic and early Neolithic sites, cf. Cohen (1977), p. 134, Howe (1983), p. 62, Hole (1989), p. 102, and Valla (1995), p. 173.

[30] For an account of the origin and evolution of cultivated species of wheat and barley in the Neolithic Age in the Near East cf. Zohary & Hopf (1993), pp. 18–64.

[31] A popular account of how primitive wheat and barley gathered from the wild were 'parched' to liberate the kernels from the glumes is given in Tannahill (1988), pp. 22–6. For a more scholarly discourse on the subject see Hillman, G. (1984 & 1985) who studied the processing of archaic wheat as practised recently in Turkey that may duplicate the operations featured in Sumerian texts. Such a technique was still used by minority tribes in Yunnan in the 1960s – they collected wild wheat, roasted and pounded the grains in a mortar to remove the glume from the kernels, which were then processed further, cf. Chin Shan-Pao, cited in Li Fan (*1984*), p. 42. The roasting of new wheat by a similar procedure was also noted by Chao Yung-Kuang (*1995*), p. 255, in the Chinese countryside during the Cultural Revolution. In fact, in antiquity rice was also roasted or burnt to aid removal of the husk as indicated in the definition of the word *pien* 煸 in the *Kuang Yün*, +1011, as 'to burn husked rice to obtain the kernels (*shao tao tso mi* 燒稻作米).'

[32] Braidwood *et al.* (1953) and Katz (1987). It is interesting to note that a similar idea was put forth by the eminent Chinese historian Wu Chhi-Chhang in 1935, cited by Yuan Han-Chhing (*1956*), p. 82. Wu said 'When our ancestors domesticated millets and rice, their principal objective was to ferment wine not make *fan* (steamed granules) . . . the custom of eating steamed rice arose from the drinking of wine.' The problem with all these imaginative flights of fancy is that it is immensely laborious to make beer without a waterproof container in which the sprouted grain can be heated. Thus, the making of beer does not appear feasible until the arrival of heat resistant pottery cooking vessels, i.e. about −6000 in the Near East and probably earlier in China.

But by about −6000 these obstacles had for all practical purposes disappeared. New varieties and species of wheat and barley were developed in which the husks could be removed from the kernels without preroasting.[33] The invention of pottery gave the Neolithic chefs a new dimension in which to exercise their skill and imagination. They now had pots to boil and simmer in and urns that could be used as ovens.[34] These technological advances made it possible to prepare meals and flour, and thus porridge and bread, from grains that had not been pre-roasted. The stage was then set for the evolution of processes that would convert grains into fermented beverages.

We have no direct knowledge of how these processes evolved. It has been suggested that it probably happened through an accident in the normal procedure for making either a porridge or a bread. Let us first consider the case of porridge.[35] One way to make a porridge was to heat raw barley or wheat meal slowly in water. If the meal used had been prepared from grains that had sprouted,[36] we would have a situation parallel to that of the proposed discovery of *i* (malt sugar) and *li* (sweet wine) in prehistoric China.[37] Another way of making porridge was to mix barley meal, prepared from pre-roasted grain, with water or wine.[38] If some meal derived from raw, pre-sprouted grain, were accidentally mixed with roasted meal and hot water, and left to stand for days, the result would also be a fermented broth similar to the *lao* and *li* of prehistoric China.

It is not possible to decide which scenario is the more probable one. But once it was realised that the fermented broth exhibited an aroma and flavour reminiscent of those of grape wine we can easily visualise what further developments might follow. For example, the fermentation would start more quickly and proceed more

---

[33] Eventually the hulled types would be largely replaced by free-threshing varieties. For a brief discussion of the evolution of new species and varieties of domesticated wheat and barley, cf. Hopkins (1997) and for further details, cf. Harlan (1995), Feldman, Lupton & Miller (1995), Zohary & Hopf (1993), pp. 18–82, Miller (1992) and Harlan (1981).

[34] We suspect one of the early forms of the oven is the *tandoor* which is simply a large earthen urn partially submerged in the ground or encased in a large block of mud. Charcoal is allowed to burn at the bottom of the urn; the dough is plastered on the walls and collected as soon as it is baked. This is how the flat breads in the Near East and throughout Central Asia are still being baked. Such an oven is depicted in an Egyptian tomb mural of Dynasty XVIII, cf. Forbes (1954), p. 278, Fig. 180, c. −1500, and other murals and wooden models of the same period.

[35] The *porridge* origin of beer is mentioned in McGee (1984), p. 420 and James and Thorpe (1994), p. 332. Brothwell & Brothwell (1969), p. 166, appear to suggest that beer could have originated from either bread or gruel.

[36] As pointed out by Forbes (1954), p. 278, 'Malting was not invented for brewing, and is older than the baking of loaves of bread: it was intended to make cereal and other seeds or fruits more palatable. Such foodstuffs can be made pleasanter and more digestible by the germination induced on prolonged soaking in water.' The same view is emphasised by Corran (1975), p. 16, who points out that, 'The biochemical explosion that accompanies germination gives rise to the production of vitamins and other nutritional desirable substances, and also makes the grain, apart from the husk, more assimilable by the human body.'

[37] There is no direct evidence that a porridge was ever made in this way in the prehistoric Near East (Marvin Powell, private communication). This possibility, however, cannot be ruled out.

[38] This was apparently the way barley porridge was prepared in classical times in the Mediterranean world. For example, in the Iliad the lady Hecamede made a pottage with barley meal and wine, flavoured with honey, onion and cheese, cf. Iliad, tr. E. V. Rieu, p. 214. It is interesting to note that Tibetan *tsampa* is prepared in a similar manner, cf. Tannahill (1988), p. 25.

vigorously if an active grape wine must were added to the incubating mixture. Eventually, the chef would discover that sprouted barley or wheat (i.e. malt) itself, husk and all, when heated in water could produce a fermentable broth.[39] This would be the precursor of the modern beer process in which malt alone is used as both substrate and catalyst to generate the sugar that is later fermented by yeast to alcohol.

Although the scenario sounds quite plausible, we do see a serious difficulty with it. If this was the way beer or ale originated, then malt sugar would have been a key intermediate in its evolution. Yet, unlike what happened in China, there is no mention of a sweet product being obtained by treating grains with malt in the early Sumerian or Egyptian documents or even in the classical Greek or Roman literature. One possible explanation may be that the prehistoric Near East was so well endowed with other sweeteners such as honey, dates and grapes, that the existence of another sweetener was not a matter that attracted much attention.

What about an origin of beer or ale based on bread? Bread is often mentioned as the precursor of beer as indicated in ancient Egyptian and Sumerian documents summarised by Partington (1935) and Forbes (1954).[40] It was presumed that beer was probably first prepared from bread, just as *bouza* is still being made in the countryside in Egypt and the Sudan today. The process is identical to that depicted in numerous murals and wooden models found in ancient Egyptian tombs,[41] an example of which is shown in Figure 61. It is described by Partington as follows:[42]

The grain was moistened and allowed to sprout, then ground in a mortar, moistened with water, formed into lumps with leaven and baked superficially into a sort of bread, the inside being raw. This bread was then broken up and allowed to ferment in a tub (of water) for about a day, and the liquor pressed through a strainer.

Was this the way beer originated? If so, we would be talking mainly about beer in Egypt, long regarded as the original home of raised bread.[43] The situation might have been slightly different in Sumeria where the bread was prepared largely from barley flour.[44] Since barley is a poor source of gluten, barley flour would be a poor substrate for raised bread. It is not likely that a leaven would have been added intentionally to a barley dough. But as long as the flour was prepared from sprouted

---

[39] In the modern beer process, the malt is prepared from husked barley and is used without removing the husk. We suspect that the barley malt of antiquity was also husked. It is likely that free threshing barley was preferred for making barley meal or flour when it became available. In the early Neolithic period emmer was probably the most common wheat of the New East. Sprouted emmer would need to be dehusked before the kernels could be processed into meal or flour.

[40] Partington (1935), pp. 195–7, 303–5; and Forbes (1954), pp. 277–81. Cf. Corran (1975), pp. 16–17; Darby, Ghalioungi & Grivetti (1977) II, pp. 529–50; MacCleod (1977), p. 45 and Tannahill (1988), pp. 48–9.

[41] Tenny Davies (1945), pp. 26–7, cites a mural in the sacrificial chamber of Achethetep-her of the Old Kingdom (−2686−2181); for other examples see Corran (1975), p. 33, Darby *et al.* (1977), p. 536 and Wilson (1988), pp. 10 and 18.

[42] Partington (1935), p. 196.

[43] See for example, Jacob (1944), pp. 26–34 and 37, cf. Exodus 12:34, which says. 'And the people took their dough before it was leavened, their kneading troughs being bound up in their clothes upon their shoulders.' This suggests that the Israelites learned how to make leavened bread during their sojourn in Egypt.

[44] Early literature in Mesopotamia suggests that barley was a more important crop than wheat. As noted by Harlan (1995), p. 143, 'The Sumerians had a god for barley, but not one for wheat.'

Fig 61. Wooden models of Egyptian bread and beer making, British Museum: (a) #45196, model showing female servant stirring mash in the preparation of beer. Jars holding the strained fermenting liquid stand in the foreground. From Asuyt, Middle Kingdom, c. −1900. (b) #40915 model of a large workshop for the preparation of bread and beer. Female servants shown are grinding barley in the background. Others knead dough and some shape dough into loaves. Lightly baked loaves are broken up, soaked with water and fermented in beer making. From Deir el-Bahri. XIth dynasty, c. −2050.

grain, contamination of the dough with yeast could occur easily, especially if the milling and baking operations were carried out in close proximity to vats where grape juice was being fermented to wine. When the dough was baked lightly, the bread broken up and then soaked in water, the starch would be hydrolysed to sugar and the sugar fermented to alcohol. In time the Sumerians would have found that the addition of malt and of fermenting wine to the incubation would enhance the conversion of bread to beer.

This scenario explains why *bappir* (usually interpreted as beer bread) and malt were used together in early Sumerian recipes for brewing beer.[45] For example, one recipe from c. −2600 calls for the use of grain, *bappir* and *buglu* (presumably malt). For 100 sila (40 litres) of beer, 72 sila of emmer wheat without husk, 12 sila of *bappir*, and 96 sila of *boglu* were needed.[46] They would also explain why in a later recipe, encoded in a Hymn to the goddess Ninkasi (c. −1800), malt and *bappir* were mashed together and the 'wort' separated and fermented with wine must as a source of yeast.[47]

In ancient Egypt husked emmer was the principal source of wheat flour.[48] When the dough prepared from yeasty, malted emmer flour was allowed to stand it would rise or expand as the carbon dioxide liberated by the fermentation of sugar was trapped in the gluten matrix. The attending Neolithic cook would certainly have noticed this amazing phenomenon. She would be delighted to find that the 'raised bread' baked from it was not only delicious but also smoother and softer in texture than 'ordinary' bread.[49] When a lightly baked raised bread of this type was suspended in water, it would ferment gently to give a 'beer'.

---

[45] The interpretation of *bappir* as 'beer bread' has recently been criticised by Stol (1994) and M. A. Powell (1994). However, they do agree that *bappir* is a product of malt, i.e. sprouted grain, and that it is probably cooked in an oven. Thus, for our purpose we may consider *bappir* as a putative 'beer bread' and it would perform the same function as that of a true 'beer bread' as traditionally interpreted in the Sumerian recipes for making beer.

[46] Cited in Partington (1935), p. 304. For further information on the culture and economy of beer in ancient Sumeria cf. papers by P. Michalowski, M. A. Powell, M. Stol, H. Neumann and C. Zaccagnini in a Symposium volume edited by Milano (1944).

[47] A translation of the hymn by Civil (1964) is reproduced in Katz & Maytag (1991) who prepared a batch of Ninkasi beer in a modern brewery (Anchor Brewing Co.) in San Francisco in 1989. The *bappir* used was a barley bread baked twice. This would have killed any residual yeast that might have been present. Thus in contrast to the earlier Sumerian recipe an external source of yeast had to be added to the fermentation. Briefly, the process they used was as follows:

(1) Preparation of *bappir* (beer bread), which was probably made from barley and the preparation of malt (sprouted barley)
(2) Mashing of *bappir* and malt in water
(3) Separation of 'wort' from spent grains and chaff
(4) Fermentation of wort by the addition of yeast

If their interpretation is correct, the Sumerians in −1800 already knew that two stages were involved: firstly, saccharification of the gelatinised starch in the bread and secondly, fermentation of the sweet solution to an alcoholic drink. The procedure is also in agreement with the glossary of Mesopotamian brewing terms given by Hartman & Oppenheim (1950), such as *dark beer, turbid beer, beer with a head, beer bread, clarified mash* and *green malt*.

[48] Harlan (1981), p. 7. According to Feldman *et al.* (1995), p. 189, 'emmer was the most prominent cereal in the early farming villages of the Near East'.

[49] This scenario is supported by the recent work of Samuel (1996b). She examined bread samples from Egyptian tombs dated at between −1200 to −2000 by scanning electron microscopy and concluded that 'deliberately germinated' emmer wheat flour was used to bake some types of leavened bread. It is possible that by about −2000 this type of bread already had had a long history and that it could have served as an early substrate used in the making of beer in prehistoric times. For further details on how wheat flour and bread were prepared in ancient Egypt cf. Samuel (1989).

However, as in the case of the *li* fermentation in China the level of yeast contamination in the malted flour was variable and uncertain. As the technology of processing malted emmer wheat improved, yeast contamination in the flour would decline to a negligible level. By this time the chef would have learned that a piece of dough from a previous fermenting batch could be used as a leaven to initiate the rising of a new batch of dough. Soon she would find out how easily a piece of active dough could be kept alive for use in the baking of the next round of bread. The cool, moist environment used for storage would promote the growth of lactobacilli as well as yeast. This was probably the way the famous sourdough was discovered.[50] If by chance the activity of the sourdough died out, a new dough could always be restarted by using a fermenting wine must as the leaven.[51]

The making of beer by soaking raised bread in water is a simple and convenient procedure, especially when the baking of bread was already a daily household chore. The simplicity of the operation may explain why this prehistoric procedure has persisted in the Egyptian countryside for the making of *booza* until recent times.[52] For a large-scale operation, however, it was a cumbersome and laborious process, since the kernels of the sprouted grain had to be separated from the chaff and the dough allowed to rise and then moulded and baked individually. It did not take long for the ancient brewer to realise that what was important was not the bread itself but rather the ingredients behind it, that is to say, heated or cooked grain (as meal or flour), malt (sprouted grain) and yeast (the froth from a wine or beer fermentation). It was also recognised that there were two stages in the process: one, saccharification (solubilisation of grain solids and generation of a sweet solution), and two, fermentation (bubbling of the solution to give a drink with a wine-like flavour). Furthermore, the rate of saccharification could be increased by the addition of fresh malt to the mashing mixture, and the rate of fermentation by the addition of fresh yeast. These observations would lead to the development of a two-stage process. A clue to how this development might have taken place is provided by the pioneering work of Delwen Samuel, who studied ancient Egyptian bread and beer residues by scanning electron microscopy.[53]

---

[50] Bread in the Old World was raised by sourdough from antiquity until the modern age when fresh yeast became available commercially, cf. Wood, ed. (1996a), p. 5. Wood also (1996b) showed how easy it was to prepare a sourdough starter by exposing a bowl of flour in water to air.

[51] Pappas (1975), p. 3, suggests that the bread of the Egyptians was 'leavened with yeast-rich foam scooped from the top of fermenting wines'. Tannahill (1988), p. 52, proposed that raised bread arose when ale instead of water was used to mix the dough. The proposal, however, contradicts her earlier assertion (p. 48) that ale arose from bread-making.

An alternative theory on the origin of bread is given by McGee (1984), p. 273: 'Before bread, grain was either simply parched on hot stones or boiled into a paste or gruel. Two discoveries ushered in a new age, first, that pastes too could be parched – hence flat bread – and second, that paste set aside for a few days would ferment and become aerated – hence, raised bread.' Cf. Panati (1987), p. 400. The main problem with this hypothesis is that when such a paste is exposed to air for days, it is likely to be covered with fungal myceliae rather than yeasts.

[52] Bread beer is also brewed in Russia (as a type of *kvas*) and in Finland (as *kalja*). We thank E. N. Anderson for this information. According to Smith & Christian (1984), p. 74, 'Home-made *kvas* is nowadays fermented from malted rye, barley or wheat and rye, wheat or buckwheat flour or from pastry, bread or rusks; sugar honey and various fruits, berries and herbs are used to flavour it.' The similarity to the Egyptian *booza* process is unmistakable.

[53] Samuel (1989, 1993, 1995, 1996a & 1996b).

Based on the results of her examination of beer residues from Deir el-Medina (−1550 to −1307) and Amarna (c. −1307) Samuel deduced that in mid −2nd millennium beer was made in Egypt by a two-step process as follows: 'First cereal grains were malted and heated to provide sugar and flavour, and these grains were then mixed with sprouted, unheated grains in water. The resulting water and starch solution was then decanted and fermented [with yeast] to make the beer.'[54] A team at the Scottish and Newcastle brewery in Newcastle tested Samuel's recipe and came up with a surprisingly attractive product.

At first sight it would seem that Samuel's reconstructed brewing recipe had nothing to do with a bread-derived origin of beer. Yet if we examine her procedure and reflect on how it could have come about, we can see that it is actually a logical extension of the original raised bread approach. When we compare the three beer processes we have discussed (the *booza* process, the *Ninkasi* process, and the *Samuel* process), it is clear that each process can be divided into three distinct reactions, *viz.*

(1) Denaturation or gelatinisation of native starch by heat.
(2) Saccharification of starch by malt enzymes to dextrins and maltose.
(3) Fermentation of sugars by yeast to alcohol and carbon dioxide.[55]

In the *booza* process, starch is denatured during the baking of beer bread, and reactions (2) and (3) are carried out simultaneously in one operation. The process is probably rather inefficient since some of the malt enzymes and yeasts in the dough must have been inactivated by the heat in reaction (1). In the *Ninkasi* process, starch is also denatured during the baking of *bappir*, but reactions (2) and (3) are carried out separately, with (2) catalysed by malt enzymes and (3) catalysed by yeast. Thus, the *Ninkasi* may be regarded as an extension of the *booza* process. In the *Samuel* process starch is denatured by heating sprouted grain, but reactions (2) and (3) are identical to those in the *Ninkasi* process. It is, in fact, the *Ninkasi* process with the *bappir* replaced by heated, sprouted wheat meal.[56]

How did the Egyptian brewers at about −1500 heat this sprouted wheat meal for beer making? We suspect it could have been kneaded with a little water, placed in a

---

[54] News report in *Science*, (1996), **273**, p. 432. This conclusion is based on the observation that most of the beer residues Samuel, (1996b), pp. 489–90, examined contained no trace of yeast. She found that 'Large pieces of chaff and coarse fragments of grain predominate. The microstructure is composed of morphologically unaltered starch granules, pitted and channeled but undistorted starch granules, and fused starch. This would be characteristic of a mixture of coarsely ground cooked and uncooked malt. The high proportion of chaff and the lack of yeast suggest that these contents are spent grain, that is the residues that are left after rinsing, sugars, dextrins and starch from processed malt.' 'These findings suggest that fermentation was initiated in the rinsed sugar- and starch-rich liquid obtained after straining out the bulk of cereal husk.'

[55] Yeasts were well known in Egypt by the −2nd millennium. The Ebers Papyrus (XII to XIII dynasties, −1991 to −1633) mentions wine yeast, beer yeast, mesta-yeast, growing yeast, bottom yeast, yeast juice and yeast water, cf. Partington (1935), p. 197.

[56] The sprouted wheat (i.e. wheat malt) was presumably ground without separating the kernel from the chaff. Samuel (1996a) has made a thorough study of the processing of sprouted and unsprouted emmer wheat and found that the processing of wheat or barley malt into a refined chaff-free flour was a laborious and time consuming task.

clay mould, and baked in hot ashes.[57] The 'loaves' that result could be mashed with fresh malt as required in reaction (2). This interpretation may explain why loaves of putative 'bread' (actually bread moulds) continue to feature prominently in murals and wooden models that appear to depict the process of brewing beer in Egypt in the −2nd millennium.[58]

From these accounts we may say that beer originated in the West from the accidental manipulation of either porridge or bread. Our preference is for bread, for three reasons. Firstly, it explains why malt sugar solution was not recognised as an intermediate in the fermentation and concentrated for use as a sweetener. Secondly, it explains how the two known recipes for making beer in the −2nd millennium, the *Ninkasi* process as interpreted by Katz and Maytag, and the Egyptian process as reconstructed by Samuel, could have evolved. Finally, it provides an answer to a puzzling question. Since malt was already an ingredient in the mashing step in both the *Ninkasi* and *Samuel* recipes of the −2nd millennium, why did the ancient brewers fail to realise that a wort could be obtained simply by mashing malt itself in hot water? The answer is, we submit, that the mashing of *bappir* (beer bread) in water at ambient temperature had been practised for so long that it had become ingrained in the mind-set of the brewers of the −2nd millennium. When malt was introduced as an ingredient in the saccharification step, the brewers simply followed tradition and added malt to the 'bread' in the mashing operation. It took perhaps another millennium before anyone realised that mashing malt all by itself could provide a satisfactory wort.

(iii) *Grain fermentations, East and West*

Based on the above discussion we may represent the origin of wine and beer in the West by the scheme shown in Figure 62. The one ingredient that connects wine to beer is yeast. Since grape wine preceded beer, perhaps by as much as one to two millennia, we suspect the yeast used in the first beer fermentation probably descended from wine. Wine was already produced on a large scale by −5000 in the Zagros mountains.[59] By the time beer arose, yeast spores from fermenting wine vats would have become plentiful in the air over the culinary establishments of the Near East. As suggested in Figure 63, beer was discovered when a piece of bread made from sprouted grains and leavened with yeast was soaked in water. In

---

[57] Wood (1996), cites the −3rd millennium Egyptian kitchen discovered by Mark Lehrner in which breads were baked in clay moulds placed in hot ashes. The process he describes is very similar to that presented by Corran (1975), p. 16, who says, 'Ground malt can be mixed into a dough, provided too much water is not used. Cakes of this dough, if baked at a temperature of 50°C, slowly dry off; a brown outer skin forms and the inside part of the cake becomes dry enough to allow the cake to be transported in comfort.' This cake (or beer bread) can be crumbled and mixed with fresh malt and warm water to make a mash.

[58] Darby, Ghalioungui & Grivetti (1977), **2**, pp. 537, 542, Forbes (1954), pp. 270–1, 278–9. The close connection between baking, brewing and fermenting of the grape-must is discussed by Weinhold (1988).

[59] McGovern *et al.* (1996). The antiquity of the wine fermentation easily explains why the Egyptians already had relatively pure preparations of yeasts by −1500, cf. Partington (1935), p. 197.

40 FERMENTATIONS AND FOOD SCIENCE

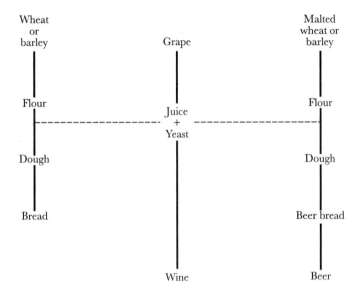

Fig 62. Origin of wine and beer in the West.

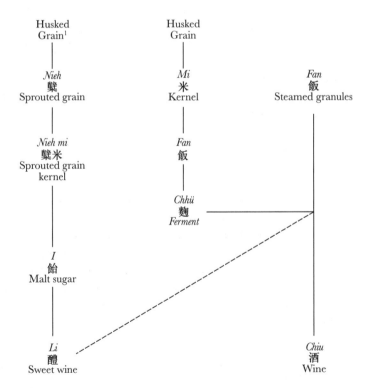

[1] Rice or millet with husk

Fig 63. Origin of *Chhü, Nieh* and *Chiu* (*Ferment*, sprouted grain and wine).

due course of time the process was greatly simplified. The intermediate steps of sprouting, milling, dough making, leavening and baking were in varying degrees cut out. Finally malt was simply mashed in hot water and yeast added to the filtered wort.

In the case of China, two types of fermented drinks, *li* (sweet liquor) and *chiu* (wine) were also made in antiquity. But they were both prepared from grains, i.e. millets and rice. Their origin is represented in the scheme shown in Figure 63. Initially, *li* was prepared by manipulating the kernels (*mi* 米) of sprouted grain *nieh* 櫱 that had been contaminated with yeast. The intermediate malt sugar *i* 飴 was isolated and used as a sweetener. This process was discontinued during the Han and *li* was made by using *chhü* in a one-night fermentation. Eventually *li* itself ceased to be made, and *chiu* (wine) became the principal alcoholic drink of China until the present day.

The key element in Figure 63 is the mould *ferment chhü* 麴. At first *chhü* was used together with *nieh* but during the late Chou (−3rd to −6th century) it was found that *chhü* alone was able to saccharify the starch and ferment the sugar into wine. We have shown earlier (pages 161–2) how this unique invention could have originated from accidental exposure of steamed grains (*fan* 飯) to the air.[60] After several days, the granules began to emit a fragrance that attracted the attention of the attending cook. The realisation that the mouldy myceliae produced a pleasant effect probably prevented the Neolithic observer from throwing the contaminated granules away. The ease with which airborne fungi can grow on steamed millet and rice paved the way for the discovery of *chhü*.

We can see the same factors at work in the origin and development of alcoholic fermentations in the West. The raw materials involved are wine grapes on the one hand and wheat and barley on the other. The inevitability of the production of wine from grapes needs no further comment. More importantly, the grape wine fermentation provided a readily available source of yeast that was fairly free from contaminating microorganisms. Again, the course of events was directed by the nature of the starting raw material, in this case wheat and barley. It is the hard kernel of wheat and barley that prompted the pre-sprouting of the grains before they were dehusked and ground into flour or meal. When yeast was added to the flour, it was the gluten content in the wheat that enabled the dough to rise by trapping the carbon dioxide evolved as the residual sugars present were fermented. Adoption of the baking of leavened bread as a routine culinary practice thus paved the way for the making of 'beer bread' which, when soaked in water would in time produce an alcoholic drink.

---

[60] Since the Neolithic Age both millet and rice have been steamed to give tender, fluffy granules that are highly palatable and easily digestible. But steamed granules are also an excellent substrate for the growth of airborne fungi. Firstly, they provide a large surface area for spores to alight on and germinate. Secondly, the starch within the grain is already gelatinised and rendered highly susceptible to hydrolysis by the enzymes elaborated by the growing fungi. The sugars liberated induced the growth of yeasts which in turn fermented some of the sugars to alcohol.

Table 27. *Comparison of ancient fermentation processes for making alcoholic drinks*

Product	Western		Eastern	
	Wine (grape)	Beer	Li (sweet liquor)	Chiu (Chinese wine)
Raw material	wine grape	wheat, barley	sprouted grain	rice, millet
Culinary background	—	bread (porridge)	i 飴	fan 飯
Saccharifying agent	—	malt amylase	grain amylase	fungal amylase
Fermenting agent	yeast	yeast	yeast	yeast
Microbial culture	single	single	mixed	mixed
Process	submerged	submerged	submerged	semi-solid

The characteristics of each type of ancient fermentation, East and West, are compared in Table 27. Grape wine is a special case that has no parallel in China. The preparation of *li*, sweet liquor, from malt was discontinued during the Han. Thus, for this comparison the products of interest are beer in the West and *chiu* in China. We may say that beer probably arose as an extension of the art of bread making, and Chinese wine arose as a result of the steaming of dehusked grain. In China saccharification of starch to sugar and fermentation of sugar to alcohol take place in tandem within one vessel. While the same situation might have held in the Neolithic West, by the 2nd millennium in both Sumeria and Egypt, saccharification and fermentation were carried out sequentially as two separate operations. In the West yeast was the only microorganism involved in the making of wine and beer; in China a mixed culture of fungi and yeast in the form of *chhü* was the essential microbial ingredient for the fermentation of Chinese wine.

While all indications are that the fermentation of grains into alcoholic drinks arose independently in China and in the West, there is one curious coincidence that suggests the possibility of a transmission of ideas between the two cultures. We refer to the use of the amphora, a conical pottery jar with a pointed bottom, in the making and storage of wines and beer. It is well known that amphoras were the containers that carried wine, olive oil and garum in the maritime trade in the Mediterranean in classical times. They were used for the brewing and storage of beer in ancient Mesopotamia and Egypt. Amphora-shaped vessels are often seen in Egyptian murals and models of brewing activities.[61] In fact, the Sumerian pictograph (Figure 64a) representing the brewer's vat is in the shape of an amphora. What is not so well known is that the pottery amphora (Figure 65) was also fabricated

---

[61] Forbes (1954), pp. 283, 290; Forbes (1956), p. 132; Corran (1975), pp. 23, 33; Darby *et al.* (1977), pp. 541, 558; Wilson (1988), pp. 10, 57.

Fig 64. Sumerian and Chinese depiction of fermentation vats: (a) Pictograph of a Sumerian fermenting vat, after Forbes (1954), p. 280; (b) Oracle inscription for *yu* 酉 (fermenting); after Hsü Tsung-Shu (*1981*), p. 563.

Fig 65. An amphora of the Yangshao culture from Pan-pho, Sian. San Francisco Asian Art Museum (1975), Fig. 28, height 43 cm.

and used in the Yangshao (−5000 to −3000) culture of Neolithic China.[62] The function of this uniquely shaped vessel has, however, received little attention. Most Chinese archaeologists would probably agree with Tsêng Tsung-Yeh that it was used for collecting water from a pond or the river.[63] But amphoras were apparently involved in the brewing of wines from grains. The radical for fermentation, *yu* 酉, was first written in the shape of an amphora (Figure 64b) as seen in oracle-bone inscriptions. What is particularly interesting to us is that the amphora apparently died out in China during the late Ta-wên-khou and Lungshan period (c. −3500 to −2000), and was replaced by conical vessels with a wide mouth.[64] The oracle-bone inscription for *yu* 酉, which would be dated at about −1500, was based on the shape of an amphora or vessels derived from it.[65] This suggests that the origin of fermented drinks from grain must have extended well beyond the Lungshan culture into the Yangshao era. Did the amphora travel from East to West or from West to East? Or is it just a coincidence that both China and the Near East both developed it at about the same time? The answer will have to remain a mystery until further information on the origin of the amphora in the East and West is unearthed.

We have shown that while the mould *ferment* originated from the accidental handling of steamed rice or millet, it has been made primarily with wheat since the Han. In fact, according to the *Chhi Min Yao Shu*, to make a *ferment* wheat was roasted or steamed, ground into a meal and moulded into cakes, which were then exposed to the air and allowed to become infected by fungi and yeasts (pages 170–4). In other words, wheat was manipulated in a way not unlike the methods used by the ancient Mesopotamians and Egyptians in the making of porridge or bread. Why then was there no parallel invention of the mould *ferment* in the West? We think the answer may rest on two factors: the first is nature and the second technology.

By nature we mean the character of the grains cultivated and the environment in which they were utilised. In China the grains, millet and rice, contain soft kernels that can be directly boiled or steamed. The climate is hot and humid in summer. Plant residues of both grains are hospitable to the growth of fungi of the genus *Aspergillus*, *Rhizopus* or *Mucor*. As a result, the air over the Neolithic communities in China was probably loaded with spores of these fungi. The conditions were thus favourable for spores to find a suitable spot on the steamed granules to alight on,

---

[62] Chang, K. C. (1977), p. 123; *ibid.* (1986), pp. 126, 129–31 and Ho Ping-ti (1975), p. 130. Neither author has commented on the function of the amphoras.

[63] Tsêng Tsung-Yeh (*1988*), p. 64.

[64] These may be regarded as amphoras with the upper part cut off. Examples of such vessels are seen in K. C. Chang (1986), p. 171, Su *et al.* (*1989*), p. 108, Fig. 11, nos. 4 and 11, and illustrated by Underhill (forthcoming), Fig. 3. The 'general's helmet' noted on p. 151, Figure 41(b) may be regarded as an adaptation of the ampohora design. It should be noted that a vessel with a pointed bottom is basically unstable, and needs a support for it to stand upright. It must have served a highly desirable and unique function for it to have been made at all.

[65] Hsü Tsung-Shu (*1980*), p. 563; and Fang Hsin-Fang (*1980*), p. 142. The amphora is ideal for separating clear wine from the fermented mash, since the residues could settle and be packed tightly at the conical bottom with a minimum of surface exposure. This is presumably the reason why, in spite of the extra labour required to keep it upright, it was used widely for the storage of wine in the Classical West. On the other hand, after the Shang, when good methods of filtration were developed, the need for vessels with pointed bottoms ceased to be compelling and they disappeared from the culinary scene.

germinate and grow. On the other hand in ancient Sumeria and Egypt the grains were wheat and barley. Their hard kernels had to be ground into meal or flour before they could be cooked. A smooth paste prepared from flour would have a limited surface to attract airborne fungal spores. However, a wet meal could have provided a suitable substrate for the growth of fungi. In fact, the type of spent grain residues from the processing of beer that Samuel examined would have supported fungal growth quite nicely, yet she found no significant trace of fungal contamination.[66] We suspect that in the dry climate of the Near East the air over the Neolithic communities was probably deficient in the desired type of fungi.[67] But there could be another factor, and that is technology.

Even if luxuriant fungal growth had been found in a batch of wet meal or spent beer grain in the ancient West, an observer would probably be repelled by the contamination and throw it away. He would not have seen any value in the moulded grain. It is fortuitous that in Neolithic China steaming produced kernels of rice or millet that were not only soft and fluffy but also swollen with an enormous amount of water. Because of their high water content, once the granules were over-run with fungal myceliae they would begin to liquify and attract the growth of yeasts. An observer could see not just the mouldy growth and the liquification of the grain, but more importantly, smell the fragrant aroma emitted by the liquid. Thus the original discovery depends not only on nature, i.e. the softness of the grain and the presence of the right fungi in the environment, but also on technology, i.e. the prior art of fabricating a pottery steamer and using it to prepare granules heavily laden with water.

(iv) *Special features of the Chinese system*

One interesting effect of the use of the *chhü ferment* allowed the Chinese vintners to exploit fully an advantage inherent in the use of fungal amylases over the cereal enzymes. The reason is as follows. Malt contains an alpha and a beta amylase. Both enzymes are involved in the hydrolysis of starch. The α-enzyme cuts the linear polymer into smaller oligomers, thus exposing a host of fresh non-reducing ends.[68] The β-enzyme splits off the second glucose residue from the non-reducing end and liberates a molecule of the disaccharide, maltose. The hydrolysis continues until the linear polymer is completely degraded. The maltase enzyme in yeast hydrolyses maltose to two glucose units which are then fermented to alcohol. Maltose, however, is an inhibitor of β-amylase, so that once its concentration reaches 7 per cent,

---

[66] Samuel (1996a), p. 7.

[67] The author has tried to duplicate the procedure that the aborigines in Taiwan routinely use to make wine *de novo* from rice. Steamed brown rice was left in the open on a dish (a bamboo basket not being available) in the balcony of his apartment in Alexandria, Virginia during the hot humid summer. While in Taiwan mycelial growth was evident after three to four days (cf. pp. 161–2), he had yet to see any significant growth even after a week. What usually happens is that the rice granules eventually become so dry that no self-respecting fungi care to grow on them. The local air is apparently too free from fungal spores and too dry to duplicate what happens in Taiwan.

[68] For an account of the chemistry of the hydrolysis of starch by amylases see Kulp (1975), pp. 62–81.

the rate of maltose production is drastically reduced. Thus under normal conditions, the concentration of alcohol in beer stays at about 3 to 4 per cent. The fungal population in *chhü* produces an α-amylase and an amyloglucosidase. The latter attacks the glucose residue at the non-reducing end and liberates a glucose molecule which is directly fermented by yeast. Glucose does not inhibit the action of amyloglucosidase so that high concentrations of it can be accumulated in the fermenting medium. Amyloglucosidase is also less sensitive to alcohol than is cereal β-amylase. Thus, by repeatedly recharging the medium with fresh substrate i.e. cooked rice or millet granules as shown in the examples on pages 174–6, it was possible to build up the concentration of glucose and hence alcohol in the must to rather high levels. The only limiting factor then is the sensitivity of the yeast to alcohol. In general, the fermentative activity of yeast stops when the alcohol in the medium reaches 11 to 12 per cent, the normal alcohol content of grape wine. As we have shown in Table 21, p. 180, several of the wines described in the *Chhi Min Yao Shu* could have rivalled the alcohol content of contemporary grape wines of Central Asia in terms of alcohol content.

By repeatedly challenging the yeasts in the *ferment* through the centuries with increasing amounts of substrate, the vintners had unwittingly selected strains of yeast that would remain active even after the alcohol content reaches 15 to 16 per cent. This is the average alcohol content of the celebrated Shao Hsing wine[69] and the Fukien Yellow wine[70] (as well as for Japanese saké[71]). The selection must have occurred through yeasts carried in *ferments* used as seed to inoculate a new batch of raw *chhü* before it is incubated, even though some Chinese vintners have continued to prepare *chhü* from the raw ingredients *de novo* without the benefit of an inoculum.[72] In general, the processes for making *chhü ferment* as described in the *Chhi Min Yao Shu* have undergone only minor changes to evolve into the four major types of *ferments* made in the present day (Table 28). Of these the one truly significant innovation is the invention of the red *chhü ferment* before the advent of the Sung.[73]

The ancient and mediaeval Chinese, of course, had no idea of what was in the *chhü* that gave it its *chhü shih* 麴勢 (strength of *ferment*) or saccharifying-cum-fermenting power.[74] But we now know that these activities are due to the presence of fungi and

---

[69] *Che-chiang sheng kung-yeh chhang* (*1958*), pp. 47–9, reports 16.7%, 17.4%, and 19.4% alcohol for three grades of Shao Hsing wine.

[70] According to Hsü Hung-Shun (*1961*), p. 23, Fukien yellow wine contains from 14.5% to 16.1% alcohol.

[71] Kodama & Yoshizawa (1977), p. 460, state that the average alcohol content of saké is 15.0%, v/v. But during the fermentation, the alcohol content of the mash could reach 20%.

[72] According to Chou Hsin-Chhun (*1977*), pp. 1–5, the traditional method for making brick *chhü ferment*, similar to that described in the *CMYS*, was used in the early 1950s to start a modern factory for making distilled wine from sorghum on Quemoy Island. Wheat was the principal ingredient and no inoculum was added before incubation. In time, the microorganisms in the *ferment* were separated and identified. The most desirable fungi and yeasts were selected for use to inoculate future batches of *ferment*, although the specific organisms were not revealed. No preformed inoculum is used in the making of *chhü* for the manufacture of Shao Hsing wine as described in *Che-chiang shêng kung-yeh chhang* (*1958*), pp. 14–17, and *Wu-hsi chhing-kung-yeh* (*1964*), pp. 54–6. The author has seen a similar procedure followed in making *chhü* bricks without a preselected inoculum at the Chiangnan Beer Factory in Shanghai in Sept. 1987. This factory makes traditional Chinese wines as well as beer.

[73] Cf. pp. 192–202.   [74] *CMYS*, ch. 66, p. 388.

Table 28. *Types of traditional* Chhü ferments *made in China today*[a]

Type	Raw material	Form	Major organisms	Uses
Great *Ferment* 大麴	Wheat, barley, peas	Brick	*Rhizopus* spp., *Aspergillus* spp., yeasts	Distilled wine
Little *Ferment*[b] 小麴	Rice, rice bran, herbs	Granule	*Rhizopus* spp., *Mucor* spp., yeasts	Table wine
Wheat *Ferment* 麥麴	Wheat, barley	Cake	*Aspergillus oryzae*, *Rhizopus* spp., yeasts	Table wine
Red *Ferment* 紅麴	Rice	Granule	*Monascus purpureus*, yeasts	Table wine

[a] Adapted from Chou Hêng-Kang (*1964*), pp. 75–88, Hung Kuang-Chu (*1984*), p. 89, and Chin Chiu-Kung (*1981*), pp. 65–95.
[b] *Hsiao chhü* is also called *chiu yao* 酒藥, wine medicament.

yeasts in the product. Through microbiological studies carried out in the 1930s and 1950s, the organisms present in Chinese *ferments* have been identified as follows:[75]

Moulds: *Aspergillus oryzae, A. niger, A. flavus,* etc.
Rhizopus oryzae, R. delemar, R. japonicus etc.
Mucor mucedo, M. rouxii, M. javanicus etc.
Monascus purpureus, Penicillium, Absidia spp.
Yeasts: *Saccharomyces cerevisiae, S. mandshuricus*
Pichia, Willia and *Monilia* spp.
Bacteria: *Acetobacter, Lactobacillus* and *Clostridium* spp.

The principal traditional types of *chhü ferments* produced in China today are listed in Table 28. We have already seen a picture of granules of red *ferments* in Figure 46. A picture of brick *ferments* is now shown in Figure 66. The population in each case is highly complex. Moulds are always the most numerous; next come yeasts and lastly bacteria. The moulds elaborate amylases that hydrolyse the starch, the yeasts ferment the sugars to alcohol, and the bacteria produce trace compounds that affect the aroma and taste of the wine. The distribution of organisms within each type may vary considerably according to the special conditions that prevail in the location where the *chhü* is made. These variations no doubt contribute to the unique character of each variety of the celebrated table wines and distilled wines that are currently being manufactured and consumed in China.

Although pure cultures of the moulds and yeasts isolated from the *ferments* have been tested and used in making *ferments* and wine, there is every indication that most

[75] Chhen Thao-Shêng (*1979*), pp. 51–3; Chou Hsin-Chhun (*1977*), pp. 9–24; Chou Hêng-Kang (*1964*), pp. 76–88; *Wu-hsi chhing-kung-yeh* (*1964*) pp. 49, 57–9.

Fig 66. Brick *ferment* from the Chiangnan Brewery, Shanghai, photograph HTH.

vintners in China still tend to favour the old fashioned mixed cultured products.[76] But it would be incorrect to say that the ancient procedures described in the *Chhi Min Yao Shu* (pages 169–78) and *Pei Shan Chiu Ching* (pages 183–8) have remained virtually unchanged. In the making of *ferments* it is now common practice to inoculate the cooked substrates with *ferment* powder before incubation. Another notable innovation is the use of a pre-incubated culture rich in yeasts in the main fermentation.[77] For example, as indicated earlier on Table 17, p. 164, in the Shao Hsing wine process, three components are mixed in an aqueous medium in the fermenting vat: first, the substrate, i.e. steamed glutinous rice, second, *chhü* 麴 ferment, and third, a yeast culture known as *chiu mu* 酒母 or *chiu niang* 酒娘 (mother of wine),[78] which is itself prepared by incubating a batch of substrate with *chhü* (ferment) and *chiu yao* 酒藥 (wine medicament) and allowing the saccharification and fermentation to

---

[76] *Wu-hsi chhing-kung-yeh* (*1964*), p. 57; Hsü Hung-Shun (*1961*), pp. 8. 11–12; Hung Kuang Chu (*1984a*), p. 89. and *Che-chiang sheng kung-yeh chhang* (*1958*), pp. 14 and 16.

[77] It is not clear when this innovation was adopted in the making of table wines, such as the Shao Hsing wine. It may have been a Chhing dynasty development. But we cannot rule out the possibility that it may have come to China by transmission from Japan towards the end of the 19th, or beginning of the 20th century. Modern microbiological studies on the making of wine from grains had started much earlier in Japan (late 19th century) than in China (1930s). The incorporation of Western beer-brewing technology was attempted in the Meiji era, leading no doubt, to an appreciation of the primary role of yeasts in the fermentation of sugars to alcohol, cf. Kodama & Yoshizawa (1977), pp. 45–9.

[78] *Chiu niang* as an ingredient in the wine fermentation, is listed in the Chhing compilation *Thiao Ting Chih* (The Harmonious Cauldron), ch. 8, p. 656. There is, however, no mention of how the *chiu niang* was prepared.

proceed until a high population of yeasts is achieved. The addition of yeasts greatly reduces the lag time incurred in the production of alcohol in the main fermentor. This basic procedure is also utilised in Japan for the making of *saké* or Japanese *chiu*. The same three components are employed in the main fermentor: first, steamed rice, second, *koji* 麴 i.e. *chhü ferment* and third, a yeast culture called *moto* 酛 or *chiu yüan* 酒元. But while China has continued to depend on mixed cultures to produce *chhü ferment* and *chiu mu* (yeast culture), Japan has fully adopted the use of pure cultures to make *koji* (from *Aspergillus oryzae*) and *moto* (from *Saccharomyces cerevisiae*).[79]

Thus, we see that in China (and Japan) the ancient practice of carrying out both the saccharification of starch and the fermentation of sugars simultaneously in one vessel has remained unchanged. But in the West, perhaps as early as the latter part of the Old Kingdom in Egypt, it was realised that the process actually consisted of two distinct phases, saccharification and fermentation, and the two steps were performed separately. Later, it was found that malt could, by itself, be mashed to give a substrate suitable for use in the second, fermentation step. This is probably how beer has been made since the early Middle Ages.[80]

In the East, however, the distinction between saccharification and fermentation in the making of wine from grains was never clearly understood until the arrival of modern science from the West. This is curious since the making of malt sugar (*i* 飴) from sprouted grain has been practised at least since the Chou, and the use of yeast (*chiao* 酵) as a leaven for making steamed bread has been known since the Han.[81] It was noted that *chiao* could be skimmed off the foam from an actively fermenting mash. Indeed, the addition of *chiao* to hasten a fermentation is recommended in the *Pei Shan Chiu Ching* (+1117)[82] and alluded to in several wine recipes recorded in the subsequent literature (pages 188–90). Moreover, it was well known that diluted honey and grape juice (pages 243, 246) could be fermented by itself without the intervention of *chhü ferment*. Yet no one ever thought of adding *chiao* to one of these fermentable substrates to see if it would convert them to wine. This is one of the missed opportunities that we shall consider in the concluding chapter.[83]

---

[79] For details of the process, cf. Kodama & Yoshizawa (1977).

[80] For details of the beer process, cf. MacLeod (1977). The main problem with the one step process is that the optimum temperatures for saccharification and fermentation are not the same. Mashing is best performed at is 60° to 70°C, while the optimum temperature for the fermentation is about 20°C for top yeasts and 10°C for bottom yeasts. In the one step process, a compromise has to be struck. Both saccharification and fermentation have to be carried out at a temperature that is optimal for neither reaction.

[81] It is interesting to note that in China the very first raised bread, though steamed and not baked, which is now called *man thou*, was made precisely in the same way by adding a fermenting must to the wheat flour dough, cf. p. 469 and n. 26.

[82] *PSCC*, p. 8, recommends mixing the foam from an actively bubbling mash with ferment particles and drying the mixture to give a dry yeast powder. This product was preincubated with *ferment* and substrate to give a rich yeast culture for use in the main fermentation. Presumably, this is the earliest record of a 'mother of wine'.

Actually, even in a *ferment* preparation in which yeasts are relatively few in number it would have been extremely difficult to separate the saccharifying activity from the fermenting activity, since most of the fungi present can by themselves ferment sugars to alcohol, albeit at a low rate.

[83] As a parting note the reader may be interested to see an example of what the Chinese thought of the nature of the active ingredient in wine in the following quotation from the *Li Hai Chi* 蠡海集, c. +1400, pp. 731–2.

(9) PRODUCTION OF VINEGAR

The *Hsi Jên* 醯人, Supervisor of Vinaigrette Viands, is in charge of the seven steeped vegetables, five pickled vegetables and other vinegary dishes that are needed in ritual offerings and entertainment of guests.[1]

When wine is exposed to air, it is susceptible to oxidation to vinegar by acetic acid bacteria, usually of the *Acetobacter* species, which are ubiquitous in the environment. The overall chemical reaction is:[2]

$$C_2H_5OH + O_2 \rightarrow CH_3COOH + H_2O$$

Thus, once wine is known, the discovery of vinegar is only a matter of time. The first vinegar was probably simply wine that had been left around and turned sour. Indeed, the English word vinegar is derived from the French words *vin* (wine) and *aigre* (sour). So one would have thought that vinegar, *tshu* 醋, should be almost as old as wine itself in China and in the West. This may well have been the case, but surprisingly there is hardly any mention of the word *tshu* in the pre-Han literature,[3] even though *suan* 酸, meaning sour, was well known as one of the five flavours, *wu wei* 五味, of ancient Chinese cuisine.[4] The earliest condiment to impart sourness is the

---

Someone said, 'Wine is brewed following the same processes as are used for preparing poisonous drugs such as wolf's bane (*wu thou, Aconitum fishcheri*); therefore it makes people drunk.' Another guest refuted him, saying, 'Not at all; how can *wu-thou* be said to make people drunk? It is because wine is made from rice and microbial ferment (usually derived from wheat) which are contradictory to each other. The rice plant flowers bloom in the daytime while the wheat plant flowers bloom at night. Thus they are as opposite as midnight and noon. This why wine can make people drunk.' But I was not satisfied and made difficulties, saying, 'In the south, people almost always make vinegar from rice and wheat; why then does vinegar not make people drunk? Moreover, in the north there are grape wine, perry, jujube wine, kumiss, while in the south there is mead, wines made from the sap of trees and from coconuts – all these can make people drunk. How then can these facts be explained from the opposition of rice and wheat?' So both of the two speakers look rather discouraged.

Then I said, 'Wine tastes both pungent and sweet; it comes from the respective essences of wheat and rice. It is because it has such pure and condensed Yang that it can go straight through the pulses (vessels) and enter the organs. When the wine enters the mouth it does not stop in the stomach, but immediately circulates all over the hundred pulses, which is why when one is drunk the respiration is faster and deeper, and the complexion becomes red. Those who are experienced can drink buckets of it and still have room for more. If all of it had remained in the stomach, the stomach would hardly contain it. Vinegar cannot make people drunk, because with its sour taste it belongs to the category of Yin things and has a stopping and closing nature. Not only do people not get drunk on it, but they do not want to drink too much of it. As for all the other wines, although they are not made from rice and wheat, they all belong to the category of pure Yang, and all have a pungent and sweet taste; hence they make the people drunk. Among the various sources from which wines can be made, none can themselves make people drunk except the coconut water and the tree sap, which already have something of that quality naturally in them.'

[1] *Chou Li*, ch. 2, p. 57.   [2] For a discussion of the technology of vinegar production, see Nickel (1979).
[3] *Tshu* is mentioned once in the commentary to the *Shu Ching* (Book of Documents), 孔氏傳, Legge (1879), p. 128, but it seems to have been used as an adjective for sour and not as a noun, cf. Hung Kuang-Chu (*1984a*), p. 114. Moreover, it occurs in a passage that is regarded as a forgery, so that its authenticity is in doubt. It is odd that vinegar is also hardly mentioned in the ancient Egyptian and Sumerian records. Darby, Ghalioungui & Grivetti (1977) II, p. 617, lament that 'It is a frustrating fact that although the Egyptians could not have ignored vinegar, they did not leave a single mention of this relish in any text we know.' See also Partington (1935), p. 197.
[4] The earliest mention of five flavours is probably in the *Tao Tê Ching*, c. -450, ch. 12. The *Chhu Tzhu* (Songs of the South), c. -300, tr. Hawkes (1985), notes a sour sauce (p. 228) and a sour Wu salad (p. 234); *Han Fei Tzu* (c. -3rd century), ch. 18, p. 149, mentions sour, sweet, salty and bland flavours; *HTNCSW*, tr. Veith, often lists *suan* as one of the five flavours, e.g. ch. 5, p. 118; ch. 9, p. 139; ch. 22, p. 206.

mei 梅 apricot, which we have alluded to earlier on pages 000 and 000, but its use seemed to have waned by the Spring and Autumn period. What took its place? Most scholars believe that it was supplanted by *hsi* 醯, apparently the chief acidulant used during the Warring States and Western Han period. In fact, *hsi* is considered as the archaic form of *tshu* 醋, the word for vinegar in currency since +6th century. The earliest mention of *hsi* is in the Analects of Confucius,[5] and it occurs in the *Chou Li*, as quoted above, and in the *Li Chi*.[6] More significantly, it is cited as an ingredient in eleven recipes in the silk manuscript, *Wu Shih Erh Ping Fang* (Prescriptions for Fifty-two Ailments) discovered in Han Tomb No. 3 at Ma-wang-tui.[7] This shows that by early Han *hsi* was already a well-defined product that was used as a condiment as well as a carrier for drugs. But none of these references can tell us how *hsi* was made.

The *Shuo Wên* (+121) says that '*hsi* tastes sour and is made from wine (*chiu* 酒) prepared from congee (*chu* 粥)'.[8] This seems clear enough. But there are other references in the literature which suggest that *hsi* 醯 was a general term for meat sauces that had turned sour.[9] To becloud the issue further two additional words for sour condiments, *tso* 酢 and *tai* 酨, are also included in the *Shuo Wên*.[10] *Tso* is mentioned in the *Chi Chiu Phien* (Handy Primer for Urgent Use), c. −40, and the *Ssu Min Yüeh Ling*, (Monthly Ordinances for the Four Peoples), +160.[11] *Tai* is used in four recipes in the *Prescriptions for Fifty-two Ailments*, c. −200.[12] The latter also prescribes a 'bitter wine', *khu chiu* 苦酒, in one recipe.[13] Moreover, the *Shuo Wên* lists the word *tshu* 醋 but explains it as a form of toast that a guest drinks to honour the host.[14] Thus, during the Han *tshu* did not mean vinegar, and vinegar was represented by at least four different expressions, *hsi*, *tso*, *tai* and *khu chiu*. One may wonder why there should have been so many names coined for the same condiment, but it is possible that each expression did at one time actually denote a specific product prepared in a particular way. Unfortunately, any knowledge of the fine distinctions between them has long since perished.

In any case, indications are clear that an artificial flavouring agent to impart a sour taste to food has been known in China since the time of Confucius (c. −6th century). Although called by several names during the late Chou and Han periods, it was primarily a type of vinegar derived from the further processing of wine

---

[5] *Lun Yü*, ch. 5, v. 23; Legge tr. (1861a), p. 182, translates *hsi* as vinegar.
[6] *Li Chi*, ch. 12 *Nei Tsê*, pp. 460, 463, 469; Legge tr. (1885a), interprets *hsi* as pickle on p. 462, brine on p. 463 and vinegar on p. 469.
[7] Harper (1982), recipes 29, 30, 94, 112, 121, 132, 140, 145, 158, 185, 260. For the reproduction of manuscript and original commentaries see *Mawangtui Han-mu po-shu* (*1985*).
[8] *SWCT*, p. 104 (ch. 5A, p. 20).
[9] For example, the *Shih Ming* says a pickled mince meat with much liquid is called *hsi*. For a discussion of this issue see Hung Kuang-Chu (*1984a*), pp. 112–17.
[10] *SWCT*, p. 313 (ch. 14B, p. 18) for both words.
[11] *Chi Chiu Phien*, p. 31a; *SMYL*, 4th and 5th months (*1981*) edn, pp. 47, 53.
[12] Harper (1982), recipes 8, 208, 210 and 253.
[13] *Ibid.*, recipe 197. Thao Hung-Ching, c. +510, notes that because vinegar has a bitter taste, it is also known as bitter wine, cf. *PTKM*, ch. 25, p. 1554.
[14] *SWCT*, p. 312 (ch. 14B, p. 17a).

fermented from grain. The ancient literature had little to say about the details of this process. In fact, the earliest record of the making of vinegar did not appear until the 6th century with the publication of that estimable compendium, the *Chhi Min Yao Shu*, (Important Arts of the People's Welfare), +544, by which time vinegar had probably been known in China for at least a thousand years.

The *Chhi Min Yao Shu* describes twenty-four recipes for the preparation of *tso* 酢, which the author noted was commonly known in his time as *tshu* 醋.[15] All the recipes, except one, are from chapter 71.[16] In eight of the recipes the term *khu chiu* 苦酒 (bitter wine) is used in the title, and in six recipes the end products are called *tshu* 醋. The starting materials include steamed cereal grain (*fan* 飯) made from millets, rice, wheat and barley, cooked bran, baked bread, spoiled wine, regular wine, residue from wine fermentations, honey and over-ripe peaches. Chapter 71 is a remarkable document, packed with informative details and discerning observations on the science and art of vinegar making in the 6th century. To facilitate discussion of its contents, we have numbered the recipes in the order they occur in the text. They can be divided into four categories depending on the nature of the raw materials used (1) those that start with cooked grain (2) those that start with finished or unfiltered wine (3) those that are the same as (2) but with a herb or adjuvant added, and (4) those which start with directly fermentable substrates, such as honey. Selected recipes representing each of the four categories and containing information of unusual interest are translated and reproduced in full below:

1. Preparation of vinegar from barley[17] *(Recipe 7)*

Begin the process on the seventh day of the seventh month. If this is not possible, collect and store all the raw materials, including the water drawn on the seventh day, so that everything will be ready to go on the fifteenth day of the month. Except for these two days, all other days are inauspicious for this process.

Vinegar urns should be placed inside the house near the main door. As an example, for each *shih* 石 of *hung* 䴷 (a granular wheat *chhü* 麴 *ferment*), measure out three *shih* of water, and one *shih* of coarse grain barley (called *tsao* 造).[18] Since the grains are not intended for making *fan* (cooked grain for eating), it is preferable to use them partially polished. They are first winnowed to remove the chaff, washed well and steamed thoroughly. Spread the hot granules on a suitable surface to cool. When they reach body temperature add them to the fermenting urn and mix them well with the other ingredients using a paddle. Cover the mouth of the urn with a piece of gauze.

---

[15] *CMYS*, ch. 71, Shih Shêng-Han (*1984*), p. 86; this statement is left out in Miao Chhi-Yü ed. (*1982*), p. 429, who suspects that it is a Southern Sung interpolation.

[16] The lone exception, from ch. 34, describes the conversion of over-riped peaches to vinegar; cf. Miao Chhi-Yü ed. (*1982*), p. 191.

[17] *CMYS*, ch. 71, p. 431, recipe 7.

[18] *Hung* is a granular *ferment* rich in *Aspergillus oryzae* and *A. flavus*. For the preparation of *hung* see the recipe for yellow coat *ferment* on p. 335 (*CMYS*, ch. 70, p. 414). One *shih* equals 10 *tou* or 100 *shêng*; each *shêng* = 396.3 ml during the Northern Wei, cf. Wu Chhêng-Lo (*1957*), p. 58. *Tsao* is partially polished grain, cf. Miao Chhi-Yü ed. (*1982*), p. 439, note 11. Since barley is not easily polished, a considerable amount of labour is saved by using partially polished *tsao* rather than fully polished *mi* 米.

After three days the contents begin to seethe and froth. As soon as this happens, stir the mash from time to time. If the mash is not stirred white pellicles will form on the surface of the liquid and the product will be poor.[19] Stir the mash thoroughly with the stem of a wild jujube tree. If human hair should fall into the urn, it will also ruin the vinegar. The mash will recover when the hair is removed.

On the sixth or seventh day, take five *shêng* 升 of setaria millet – the grains should not be too highly polished – wash them well and steam them thoroughly. Spread the cooked grains out to cool. When they reach human body temperature, add them to the fermenting urn, and cover the mouth with a piece of gauze.

Inspect the urns after three to four days. If the grains are all digested, stir the mash and taste a sample. If the taste is sweet, then leave the urn alone. If the sample tastes bitter, steam another two to three *shêng* of setaria millet and add the batch to the fermentation. Use your judgement to see if more grain should be added. After two weeks, the vinegar may be edible but only after three weeks will it attain its full body and flavour. Mix one cup of vinegar with one bowl of water to make the product drinkable. In the middle of the eighth month decant the clear liquid and transfer it to a clean urn. Cover the mouth and seal it with clay. The vinegar will keep for several years.

During the fermentation spray cold water on the outside of the urn every two to three days as needed to dissipate the heat, but do not allow raw water to leak into the urn. One may also use panicum or glutinous grains as well as white or yellow varieties of setaria as the substrate.

2a. Preparation of Vinegar from Unfiltered Wine[20] (Recipe 9)

All sorts of wine of poor quality such as those that are too tart or mash that started well but turned sour before filtration, may be used as substrate for making vinegar. For example, mix five *shih* of unfiltered must, one *tou* of powdery *chhü ferment*, one *tou* of wheat *hung ferment*, and one *shih* of clear well water in a fermentor urn. Feed in two *shih* of steamed millet after the grains have been cooled to body tempaerature and stir well. Cover the mouth with gauze. Stir again twice each day. The mash will mature in about seven days in the spring and summer, but it will take longer in the fall and winter. The product is fragrant and tasty, and settles well. After a month the clear liquid is separated and stored in another urn.

2b. Preparation of vinegar from clear wine[21] (Recipe 10)

Filtered spring wine that is too acid to drink can be used to make vinegar. For example, mix one *tou* of wine with three *tou* of water in an urn. Allow the urn to sit in the sun. When it rains cover the mouth with a plate so that no rain water can get in. When it is sunny remove the cover.

After seven days the solution may develop a foul odour and a coat or film (*i* 衣), may form on the surface. Do not worry. Above all, do not move the urn or stir its contents. After several tens of days, the conversion to vinegar will be complete. The film will have settled, and the product will be fragrant and delicious.

3. Preparation of vinegar flavoured with lesser beans[22] (Recipe 17)

Thoroughly soak five *tou* 斗 of lesser beans, i.e. *hsiao tou* 小豆 or adzuki beans, and place them in the fermenting urn. Add some steamed panicum millet grains to cover the beans

---

[19] Cf. Shih Shêng-Han (1958) 2nd edn, p. 82. [20] *CMYS*, ch. 71, recipe 9; p. 432.
[21] *Ibid.* ch. 71, recipe 10; p. 432. [22] *Ibid.* ch. 71, recipe 17; p. 434.

and top with three *tan* 石 of wine. Cover the mouth with gauze. After twenty days vinegar will be formed.

4. *Preparation of vinegar from honey*[23] *(Recipe 22)*

Dissolve one *tou* of honey in one *tan* of water. Cover the mouth of the urn with a piece of gauze and allow it to stand in the sun. In twenty days vinegar will be formed.

From these examples we can see that the processes actually fall into three basic types depending on the nature of the substrate used. In the first type (example 1, recipe 7) the substrate is grain and the process involves three steps (a) saccharification of starch to sugars (b) fermentation of sugars to alcohol and (c) oxidation of alcohol to acetic acid, *viz*.

$$\text{Starch} \underset{(a)}{\rightarrow} \text{Sugars} \underset{(b)}{\rightarrow} \text{Alcohol} \underset{(c)}{\rightarrow} \text{Acetic acid}$$

In the second type (example 4, recipe 22) the substrate is sugar and only steps (b) and (c) are needed. It is remarkable that neither *ferment* (*chhü* 麴) nor starter (*chiao* 酵) is added to the substrate. Presumably, when the honey is diluted the yeasts resident in it would be able to proliferate immediately and ferment the sugar to alcohol, which would, in turn be oxidised by *Acetobacter* into acetic acid. If a starter had been added the process would have been much faster. In the third type (examples 2a and 2b; recipes 10 and 17), the substrates include finished or unfinished wine and the process is simply step (c). In this case, herbs or spices may be added with the wine and the products represent the first 'medicated vinegars' made in China.

The recipes in chapter 71 indicate that the early medieval Chinese vintners had a good appreciation of the technological factors that influence the conversion of wine to vinegar. For example, they were aware of the importance of temperature, substrate concentration and aeration in achieving a successful operation. We know today that the optimum temperature for the action of *Acetobacter* strains in traditional Chinese vinegar processes is about 25–35°C, but in some traditional operations the temperature may reach as high as 45°C.[24] The *Chhi Min Yao Shu* recommends two ways to control the temperature. The first is to start the process in the spring or Autumn, and allow the air in the room to cool the fermentor.[25] The second is to spray the urn with cold water if it gets too hot, and wrap the urn with felt or spray it with hot water if it gets too cold.[26]

---

[23] *Ibid*.
[24] The optimum temperature may vary widely depending on the nature of the process. Hung Kuang-Chu (*1984*), p. 113, says that in a trickling filter-type process, the optimum temperature is 25–35°C, although Chhü Chih-Hsin (*1959*), p. 27, tells us that in a traditional vinegar process with grains as the starting substrate, the temperature in the oxidation step may be allowed to go as high as 44–5°C. On the other hand, Nickel (1979), p. 165, states that in the modern submerged process 'the temperature of operation is generally about about 30°C . . . but most strains (of *Acetobacter*) will die above 38°C'. These observations suggest that the modern Chinese *Acetobacter* strains may be more heat resistant than those currently in use in the West.
[25] The seventh day of the seventh moon (approx. first to third week of August) or the fifth day of the fifth moon (approx. first to third week of June) appear to be the favourite dates.
[26] *CMYS*, ch. 71. recipe 14 concludes with the statement: 'To make vinegar in the summer, it is necessary to spray [the urn] with cold water. In spring and fall, it is necessary to keep (the urn) warm. Wrap it in cloth or spray with hot water, as needed.'

Another factor that affects the process is the alcohol concentration in the medium. Alcohol is somewhat inhibitory to *Acetobacter*, and it is best to keep its concentration to under 7% if the oxidation is to proceed at a reasonable rate.[27] Although Chia Ssu-Hsieh, the author of *Chhi Min Yao Shu*, could not have known what alcohol is or how it could be measured, he was apparently aware of the fact that if the wine was too strong the conversion to vinegar would be slow. And so in example 2b (recipe 10), he recommends that one part of wine be mixed first with three parts of water before the medium is set out in the open air to become exposed to the action of airborne *Acetobacter* organisms. Furthermore, he gives a higher water:solid ratio to the recipes for making vinegar from grain than those for the fermentation of wine as described in chapters 64–7.[28] This would, of course, lower the alcohol concentration of the wines fermented *in situ* and thus facilitate their oxidation to vinegar.

A third critical factor is aeration. Oxygen from the air is needed to support the growth of *Acetobacter* and to react with alcohol to give acetic acid. At least three of the practices recommended in the recipes are conducive to retention of oxygen in the system. The first is simply to stir the medium. This will allow gaseous oxygen to diffuse into and dissolve in the solution. But in processes which begin with grains as the substrate, saccharification and fermentation have to take place first before any alcohol becomes available for oxidation. Neither step requires oxygen. Thus, there is no good reason to stir the medium until after the fermentation (as shown by gas evolution) has started to calm down. Surprisingly perhaps, this is precisely the procedure adopted in several of the vinegar recipes in the *Chhi Min Yao Shu*. After charging the substrates the urns are allowed to stand undisturbed for several days before their content is stirred on a regular schedule.[29]

A second way to increase the availability of oxygen to the medium is not to cover the fermentor-urns tightly. The recipes very sensibly recommend that the mouth of the fermentor-urn be covered with a piece of gauze.[30] This will prevent dust from falling into the urn but gas exchange between the air and the medium would be unimpaired. A third practice that favours oxygen retention is to add bran to the mash. This is particularly effective when thick fermented dregs are included in the medium.[31] The tiny air bubbles trapped in the porous bran particles will increase the availability of oxygen in the medium. This practice is still being used in some wine processes in China today.[32]

From examples 2b and 3 (recipes 10 and 17) we may infer that the early mediaeval Chinese were aware of the fact that wine could be converted to vinegar without the assistance of *chhü ferments*. Though they did not speculate on the nature of the agent

---

[27] Hung Kuang-Chu (*1984a*), p. 113.   [28] Mêng Ta-Phêng (*1976*), p. 100.

[29] *CMYS*, ch. 71. In recipe 2 the mash is stirred after standing for seven days, and in recipe 7, three days. But the attitude towards stirring is not consistent. In recipe 6 the mash is stirred right from the first day, and in recipe 10, where the substrate is simply diluted wine, no stirring is recommended. One suspects that in each case the author was merely copying down the instructions given to him by an experienced vintner, and took little notice of the conflicting advice offered in these four recipes.

[30] *Ibid.* recipes 1, 2, 3, 7, 9 etc.   [31] *Ibid.* recipe 13.   [32] Chhü Chih-Hsin (*1959*), pp. 37–9.

that causes this conversion, they did realise that it was something pervasive in the environment. It was not an entity that had to be transmitted from one operation to another. Thus, in none of the recipes in the *Chhi Min Yao Shu* is there an appreciation that a previous batch of vinegar can be used to hasten the conversion of the next batch of substrate. But the vintners were aware that something extraordinary was happening while the conversion took place.

Example 2b (recipe 10) describes concurrent appearance of foul odour and a film or coat, *i* 衣, on the surface of the diluted wine after it had been exposed to air for seven days.[33] But this should cause no alarm. Eventually, the coat will settle and good fragrant vinegar will result. What is this *i*? We agree with Mêng Ta-Phêng that in this case it probably respresents a film of *Acetobacter* bacteria.[34] Since the medium is stationary, it is only on the surface that the bacteria can have ready access to the oxygen in the air to grow and to oxidise the substrate. It would also explain why this process is so slow. The text says that it takes several tens of days. If the recipe had recommended stirring of the medium, the process would probably have finished much sooner.[35]

Recipes 7 (example 1) and 12 describe the formation of white pellicles, *pai pu* 白醭 on the surface of the grain mash after it has been incubated for 2 to 3 days. The presence of pellicles indicates that the vinegar would be spoilt. The remedy is to remove them (recipe 12) or stir the medium. Obviously, *pai pu* must look quite different from the *i* film discussed in the preceding paragraph. What then is *pai pu*? Shih Shêng-Han suggests that it is a film of *Acetobacter*.[36] But we have just said that *i* is *Acetobacter*. Mêng Ta-Phêng has proposed that *pai pu* is a film of yeast.[37] This is not satisfactory either. In the first three days of the process the primary reactions are saccharification and fermentation. There should be plenty of sugars in the medium for the yeast to ferment to alcohol. Yeast cells would be carried in the froth and foam of the fermentation rather than be constrained to form a pellicle on the surface of the liquid. Another possibility is that *pai pu* is formed by a special type of *Acetobacter*, such as *A. xylinum*, a well-known slime former which often coats the packings of trickling vinegar generators and renders them inoperative.[38] Thus, the presence of *pai pu* would greatly inhibit the action of the good strains of *Acetobacter*. But the description provided is too sketchy to permit a definitive identification. The only way to resolve the issue is to duplicate the conditions of the processes shown in examples 1 and 10

---

[33] The word *i* 衣 is usually used to denote the coating of myceliae and spores that form on the surface of *ferments*. For example, the loose granules of *ferment* employed in the making of soy sauce is known as *huang i* 黃衣 yellow coat; Cf. p. 335.
[34] Mêng Ta-Phêng (*1979*), p. 99. The recipe does not tell us how extensive this film is, but it could have covered the entire surface of the medium. Under the film, the environment would be somewhat short of oxygen and thus conducive to the production of foul-smelling metabolites.
[35] According to Nickol (1979), p. 165, in a modern submerged process with efficient stirring, the cycle time for a batch of 12% vinegar is only thirty-five hours.
[36] Shih Shêng-Han (1958), pp. 82–3. [37] Mêng Ta-Phêng (*1976*), p. 99.
[38] Nickel (1979), states on p. 158 that 'Over 100 species, subspecies and varieties of *Acetobacter* have been classified. Many of these species probably would be acceptable in a vinegar generator with the exception of *A. xylinum*, a heavy slime former.'; see also p. 163.

(recipes 7 and 10), allow *pai pu* and *i* to form and then examine them with the methods of modern microbiology.

Continuing the tradition seen in the *Prescriptions for Fifty-two Ailments*, vinegar has been mentioned regularly as a medicament in standard pharmacopoeias such as the *Hsin Hsiu Pên Tshao* (+659), *Shih Liao Pên Tshao* (+670), *Chêng Lei Pên Tshao* (+1082) and *Pên Tshao Kang Mu* (+1596). Although vinegar was known to have been made from a variety of substrates such as grains (millet, wheat barley and rice), fermented wine residues, wine, honey, peaches, grapes, jujubes, cherries etc., the preferred product for medicinal use is that derived from rice.[39] But except for a few brief notes in the *Pêng Tshao Kang Mu* the pharmacopoeias have had little to say on the methods of vinegar preparation. The recipes described in the *Chhi Min Yao Shu* were evidently so well accepted that they remained as standard practices of vinegar making throughout the Sui, Thang and Sung dynasties. It was not until the Yuan that recipes for making vinegar appeared again in the literature. The topic is treated in the *Chü Chia Pi Yung*, (ten recipes), *Shih Lin Kuang Chi*, (three recipes) and the *I Ya I I* (three recipes) as well as in the agricultural work *Nung Sang I Shih Tsho Yao* (three recipes).[40]

The recipes are, however, largely a rehash of those given in the *Chhi Min Yao Shu*. There are two, however, that deserve more than a passing notice. The first is recipe 7 from the *Chü Chia Pi Yung*.[41] It describes the preparation of vinegar from malt sugar. A small amount of *pai chhü* 白麴 (white *ferment*) is mixed with the aqueous malt sugar solution in an urn, covered tightly with paper and the must allowed to stand in the sun. After twenty days white pellicles, *pai pu* 白醭, may be seen on the surface. The recipe warns that the pellicles should not be disturbed. When they eventually sink into the bottom of the urn, the vinegar is done. The second is recipe 1 from the *Nung Sang I Shih Tsho Yao*.[42] It describes the making of aged vinegar from rice. After mixing the steamed rice with loose *ferment* and a red *ferment* in an urn, the mash is allowed to stand without stirring. After twenty days a white coat, *pai i* 白衣, may be seen on the surface. It may sink by itself. If it does not, it could be removed when the aqueous phase is strongly sour. It would seem that both the *pai pu* and the *pai i* behave in the manner reserved for the *i* of recipe 10 (example 2b) of the *Chhi Min Yao Shu* and unlike the *pai pu* of recipe 7 (example 1).[43]

Of the Ming and Chhing food canons listed in Table 14 (p. 129) vinegar recipes are found only in the *Tuo Nêng Phi Shih*, +1370 (fourteen recipes), *Sung Shih Tsun Shêng*, +1504 (twenty-one recipes), *Shih Hsien Hung Mi*, +1680 (five recipes), *Yang Hsiao Lu*,

---

[39] Cf. *HHPT*, p. 291; *SLPT*, p. 119; *CLPT*, p. xx; and *PTKM*, pp. 1554–5 which states a preference for rice vinegar that has been aged for two to three years.

[40] *Shih Lin Kuang Chi* is dated at about +1280, and *Nung Sang I Shih Tsho Yao*, +1314. The other two are usually regarded as late Yuan works. Their exact dates of publication are uncertain, cf. their respective prefaces to editions published in the *CKPJKCTK* series.

[41] *CCPY*, p. 51. [42] *NSISTY*, p. 92.

[43] *CMYS*, ch. 71, pp. 431–2 (recipes 7 and 10). The confusion is not surprising, since the terms were never clearly defined. What is one man's *i* may be another man's *pu*. Again we note that the situation calls for experimental study.

+1698 (three recipes), *Hsing Yuan Lu*, +1750 (four recipes, and the *Thio Ting Chih*, c. +1760 to 1860 (eighteen recipes). There is nothing new or of unusual interest in these recipes which are largely copied from previous compilations, which are themselves based on the *Chhi Min Yao Shu*. None of the authors or compilers had tried to improve the processes or understand the phenomena underlying the conversions.[44]

Thus, we find that the traditional methods recorded in the *Chhi Min Yao Shu* in +560 for making vinegar remained virtually unchanged until the coming of modern microbiology to China in the 19th century. In fact, the basic procedures are still in use today to produce the famous vinegars of China, for example, the *lao chhên tshu* 老陳醋 (well-tempered vinegar) of Shansi, the *hsiang tshu* 香醋 (fragrant vinegar) of Chêngchiang, Kiangsu and the *Paoning tshu* 保寧醋 (Pao-ning vinegar) of Szechuan.[45] In each case the raw material, sorghum, glutinous rice or wheat have to undergo saccharification (by fungal enzymes), fermentation (by yeasts) and oxidation (by bacteria). The fungi and yeasts needed are supplied by the specially made *ferment* and the bacteria originated from the airborne population.[46] The steps may proceed simultaneously or follow each other consecutively. It is a long and laborious process. Its chief virtue is that the trace metabolites generated in the each step remain in the mash and are carried over into the finished product. It is the blend of these metabolites with the acetic acid that gives the vinegar the characteristic flavour and fragrance that have been so highly prized in the gourmet kitchens and dining-rooms of China for hundreds of years.

---

[44] The best example of this copying is the case of the so called *wu mei tshu* 烏梅醋, dark *mei* vinegar, an anhydrous sour condiment. It is made by soaking the flesh of *mei* apricot (cf. p. 48) in strong vinegar, drying it and grinding it into grits, which can be rehydrated with water to give a 'vinegar.' The procedure is first described in the *CMYS* (ch. 71, recipe 21), and repeated, sometimes verbatim sometimes with embellishments, in the *Chü Chia Pi Yung*, p. 51, *Shih Lin Kuang Chi*, Shinoda & Tanaka (*1973*) I, p. 267, *Tuo Nêng Pi Shi* (*idem*, p. 372), *Sung Shih Tsun Shêng* (*idem*, p. 471) and *Thiao Ting Chi*, p. 30.

[45] For an account of the manufacture of these vinegars, see Hung Kuang-Chu (*1984a*), pp. 120–8.

[46] Although most *ferments* contain some *Acetobacter* strains when fresh, cf. Chou Hêng-Kang (*1964*), pp. 159, 173, they tend to die off during storage. Thus, the bacteria that convert alcohol to acetic acid are presumably of airborne origin.

(d) SOYBEAN PROCESSING AND FERMENTATION

Of the five staple grains of ancient China, soybean is the one that has had the shortest history of cultivation in the Western world.[1] While millets, wheat, barley and rice were in varying degrees part of the established food system in the ancient and mediaeval West,[2] soybean did not become a cultivated crop outside East Asia until the present century. It was first introduced into Europe in the early 19th century, where it remained as a horticultural curiosity in the Jardin des Plantes in Paris and the Kew Gardens in London. It came to the New World in 1765 to the estate of Samuel Bowen, near Savannah, Georgia,[3] and was promoted as a crop for forage and green manure, but for more than a century it attracted little interest. It was not until World War I when the shortage of edible oil drew attention to soybean as a potential source of oil that it was taken seriously as a cultivated crop. Since the 1930s it has steadily grown in importance and is now, after corn and wheat, the third largest crop in the USA.[4] Upon harvest the beans are processed to extract the oil and the defatted meal is used mainly as a protein supplement in animal feeds.

The remarkable aspect of the meteoric rise of the soybean as an economic crop in the 20th century is that it has very little to do with the way soybean has been traditionally utilised for thousands of years in China and other parts of East Asia. In the modern West soybean serves primarily as a source of edible oil and a protein-rich animal feed. In the East it has always been, and still is, regarded as the source for a number of processed soyfoods that are a regular and important component of the human diet. Attempts to introduce the traditional East Asian soy foods to the West have met with only indifferent success and the experience provides an interesting case study of why the food products and habits of one culture cannot be simply transplanted carte blanche into another culinary environment.

For the traditional Chinese, traditional soybean foods were evolved in response to a need to turn soybean into palatable products that fitted into the Chinese culinary system. Although it has been praised in such glowing terms as 'that miracle, that noblest of crops, that wondrous plant, the soybean', by enthusiastic boosters in

---

[1] Cf. pp. 27–8. For an account of the history of soybean in China cf. Bray, *SCC* Vol. VI, Pt 2, pp. 510–14 and Hymowitz (1970 & 1976).
[2] Wheat, barley, millet and rice were known in the ancient Mediterranean world, cf. Darby, Ghalioungui & Grivetti (1977), II, pp. 457–99.
[3] Hymowitz & Harlan (1983). Cf. also Wittwer, *et al.* (1987), p. 184.
[4] Cater, Cravens, Horan *et al.* (1978), pp. 278–9. Since World War II China has lost its eminence as the principal producer of soybeans. Of the world's total production in the early 1980s, estimated at about 80 million metric tons annually, 50 million are contributed by the US, 15 million by Brazil, 10 million by China, and smaller amounts are contributed by Mexico, Indonesia and Argentina.

the West,[5] to the ancient Chinese 'the chief virtue of the soybean was that it produced good crops even on poor land, that it did not deplete the soil, and that it guaranteed good yields even in poor years, so that it made a useful famine crop'.[6] For in spite of its admirable nutrient content,[7] the soybean is far from being an ideal food. Apart from the immature beans which can be eaten directly as a vegetable, the raw mature beans as harvested and stored suffer from several serious defects when used as food. Firstly, the soy proteins are difficult to digest. The beans contain proteins, such as trypsin inhibitor, which suppress the action of proteases in the human digestive system. Unless they are thoroughly cooked to inactivate the inhibitors, the proteins would be poorly digested resulting in growth inhibition, pancreatic hypertrophy and hyperplasia.[8] Secondly, the carbohydrate component contains raffinose and stachyose, two α-galactosides which are not degraded by human digestive enzymes. They pass into the colon where they are metabolised by anaerobic bacteria, leading to the generation of gas and flatulence.[9] To be sure, both defects are shared by other legumes, but the undesirable effects may be more pronounced in soybeans – in ancient China these were eaten as grains in much larger amounts than other legumes, which were consumed as vegetables. Finally, soybean contains a so called unpleasant 'beany' flavour which was one of the major obstacles that had to be overcome before soybean oil could be commercialised on a large scale. It still remains as an impediment to the integration of soy proteins in Western style food products.[10]

---

[5] Anderson & Anderson (1977), p. 346. See particularly Platt (1956) who quotes the following encomium on the soybean:

>   Little Soybean who are you   From far off China where you grew?
>   I am wheels to steer your car,   I make cups that hold cigars.
>   I make doggies nice and fat   And glue feathers in your hat.
>   I am very good to eat,   I am cheese and milk and meat.
>   I am soap to wash your dishes,   I am oil to fry your fishes,
>   I am paint to trim your houses,   I am buttons on your blouses.
>   You can eat me from the pod,   I put pep back in your sod.
>   If by chance you've diabetic   The things I do are just prophetic.
>   I'm most everything you've seen   And still I'm just a little bean.

[6] Bray, *SCC* Vol. VI, Pt 2, p. 514, and Yü Ying-Shih (1977), p. 73. The *Fan Shêng-Chih Shu*, –1st century, states that 'From soya beans a good crop can be easily secured even in adverse years, therefore it is quite natural for the ancient people to grow soya as a provision against famine. Calculate the acreage to be covered by soya beans for members of the whole family according to the rate of 5 *mou* per capita. This should be looked as 'the basic' for farming.' (tr. Shih Shêng-Han (1959), pp. 18–21.)

[7] On a dry weight basis the soybean contains approximately 40% protein, 20% oil, 20–30% carbohydrate, 3–6% fibre and 3–6% minerals. It has twice as much protein as meat as well as most other beans and nuts, and is a good source of thiamine, riboflavin, vitamin A, iron and calcium. Cf. Witwer *et al.* (1987), p. 188, Anderson & Anderson (1977), pp. 347–8, and Platt (1956).

[8] For a review of legume toxins in relation to protein digestibility in the rat, cf. Rackis, Gumbmann & Liener (1985). According to Liener (1976), however, not all the adverse effects can be attributed to the trypsin inhibitor. The undenatured storage protein is somehow inherently refractory to the action of trypsin.

[9] Raffinose is α-D-galactopyranosyl-(1→6)-α-D-glucopyronosyl-β-D-fructofuranoside, and stachyose is α-D-galactopyranosyl-α-D-raffinose. The relation of these oligosaccharides to flatus production has been shown by Calloway *et al.* (1971), and Rackis (1975, 1981). For an analysis of the content of these saccharides in various legumes cf. Sosulski *et al.* (1982).

[10] Cater *et al.* (1978), p. 287. Cf. also Platt (1956), p. 835. The undesirable flavour and odour are produced when the polyunsaturated oils in the beans are oxidised through the action of lipoxidase. In the intact bean the enzyme is separated from its substrate, but they can easily come into contact with each other when the structure of the bean is damaged.

Through experience the ancient Chinese found that soybeans could be made more palatable and better tolerated when they were cooked in water for a long time, leading to a *tou chu* 豆粥, bean congee or gruel.[11] This was the way soybean was often prepared and served. It is interesting to note that this treatment does indeed help to inactivate the protein toxins, limit the generation of the beany flavour and reduce the level of flatus causing saccharides, thus turning the raw bean into a wholesome and digestible food.[12] The soybean gruel must have been extremely smooth and thin, since in the pre-Han references the ingestion of soybean is usually depicted by the expression *chho shu* 啜菽, which means literally to 'sip or suck soybean'. For example, the *Li Chi* 禮記 (Record of Rites) talks about *chho shu yin sui* 啜菽飲水 (sipping soybean and drinking water), while in the *Lieh Tzu* 列子 (the Book of Master Lieh) we find the phrase *chho shu ju huo* 啜菽茹藿 (sipping soybean and chewing the leaves).[13] *Chho* 啜 is an unusual word in the food lexicon. It is a manner of ingestion that is half-way between eat, *shih* 食 and drink, *yin* 飲. Soybean appears to have been the only grain that was prepared in such a manner that it could be sipped or sucked. *Chho* is not used even for a thin rice or millet congee the ingestion of which is described by the word *shih* (to eat) in the *Li Chi*.[14]

But soybeans were also steamed to make *tou fan* 豆飯 (cooked bean granules) in the manner applied to millet and rice. For example, the *Chan Kuo Tshê* 戰國策 (Records of the Warring States), Chhin, and the *Shi Chi* 史記 (Historical Records), −90, describe *tou fan* 豆飯 and *huo kêng* 藿羹 (soybean leaf soup) as food for the common people.[15] The results, however, were less than entirely satisfactory, and cooked soybean was for a long time regarded as a coarse and inferior food.[16] In the famous 'Contract between a Servant and His Master', c. −59, one of the hardships the employee was supposed to accept was 'to eat only cooked beans (*fan tou* 飯豆) and drink only water'. Upon hearing this and other indignities that were to be imposed on him under the contract, the poor man broke down and wept bitterly.[17] Moreover, it was recognised that eating too much soybean could be hazardous to

---

[11] The Biography of Fêng I 馮異 in the *Tung Kuan Han Chi*, and in *Lieh Chuan* 7 of the *Hou Han Shu*, relates how he prepared bean congee for the Emperor Kuang Wu, founder of the Eastern Han dynasty, when the imperial troops were cold and hungry during a strenuous campaign. The making of bean congee is mentioned in the Biography of Shih Chhung 石崇, *Lieh Chuan* 3, p. 12b, *Chin Shu*, who found it to be a time-consuming job. The *Ching Chhu Sui Shih Chih*, p. 2a, records that bean congee was prepared as part of the New Year ritual offerings made on the fifteenth of the first month. The above references are all cited in Li Chhang-Nien (*1958*), pp. 58–9, 64, 69.

[12] Liener (1976) for inactivation of protein toxins and Pires-Bianchi et al. (1983) for reduction of flatulent saccharides.

[13] *Li Chi*, ch. 2, *Than Kung Hsia*, p. 179; and *Lieh Tzu*, ch. 7, *Yang Chu*, in *Lieh Tzu I Chu*, p. 187. The term '*chho shu*' is also found in the Biography of Ming Kung 閔貢, *Tung Kuang Han Chi*, and in Chang Han's *Tou Kêng Fu* 豆羹賦 (Ode to the Soybean Stew), c. +3rd century, and *Erh Ya I* 爾雅翼, +1174, chuan on *shu* 菽 (soybean). Cf. citations in Li Chhang-Nien ed. (*1958*), pp. 58, 63, 81.

[14] *Li Chi*, for example, ch. 4, *Than Kung Hsia*, p. 166, and ch. 47, *Sang Fu Szu Chih*, p. 1015.

[15] *Chan Kuo Tshê*, ch. 8, *Han Kuo Tshê* 韓國策 (Record of the Han State), p. 3; *Shih Chi*, ch. 70, *Chang I Lieh Chuan* 張儀; cited in Li Chhang-Nien ed. (*1958*), pp. 38, 48.

[16] *Lun Hêng* 論衡, ch. 8, *I Chêng* 藝增, p. 86, considers both wheat and soybean (when cooked as granules) as coarse and unpleasant. The original reads, 食豆麥者，皆謂糲而不甘. Cf. Li Chhang-Nien ed. (*1958*), p. 55.

[17] *Thung Yüeh* 僮約, tr. Hsü Cho-Yün (1980), p. 233 in which eat *fan tou* is translated as 'eat beans'. The relevant passage is cited in Li Chhang-Nien *ibid.*, p. 49.

one's health. The *Po Wu Chih* 博物志 (Notes on the Investigation of Things), c. +190, warns that after eating soybeans for three years, the abdomen would feel heavy (bloated?) and one's movement would be impaired.[18] The sentiment is echoed in the *Yang Shêng Lun* 養生論 (Discourse on Nurturing Life), c. +260.[19] In addition, the *Tsin Shu* (History of the Tsin Dynasty, +265 to +419), says that soybeans are notoriously difficult to cook, *tou chih nan chu* 豆至難煮,[20] presumably meaning that it takes a long time to cook them to a condition of acceptable palatability.

To circumvent these problems the Chinese developed ways of processing soybeans to produce foods that are wholesome, attractive and nutritious. In the centuries following the Han one sees a gradual decline in importance in soybean cooked as either steamed granules, *tou fan* 豆飯 or thin gruel, *tou chu* 豆粥. Soybeans were increasingly processed into (1) soybean sprouts (2) soybean milk, soybean curd and related products and (3) fermented whole beans, soybean paste and soy sauce. The origin and development of these soy foods will be the topics for discussion in this chapter.

(1) Soybean Sprouts

Yellow Curls from the Soybean. The flavour is sweet and mild. It is used to treat numbness in the joints, muscle and knee.[21]

The above quotation from the *Shên Nung Pên Tshao Ching* is generally regarded as the earliest reference to soybean sprout in the ancient literature. The text calls it *ta tou huang chüan* 大豆黃卷 which, translated literally, means yellow curls from the great bean. Since sprouted grain, *nieh* 櫱, such as that from rice, millets or barley had been known and used since early historic times for making malt sugar, it is not surprising that soybean should have been treated in the same way.[22] But it is difficult to say when soybean sprouts were first prepared and used as medicine or food. Although the *Pên Ching* is supposed to be based on Chou and Chhin material, its compilation was not completed until the end of the Han Dynasty, c. +200.[23] There is no way of knowing which items in it were in currency before the Han and which were introduced during the Han. On the other hand we can be sure that all the products cited in the *Prescriptions for 52 Ailments*, a document redacted in the early −2nd century, were available at about −200. It mentions soybean, *shu* 菽, in four recipes but soybean sprouts, *ta tou huang chüan*, in none.[24] Yet it mentions *nieh mi* 櫱米, sprouted millet as well as *shu mi* 秫米, dehusked millet grain.[25] Taken together, these two documents tend to suggest to us that soybean sprouts probably first came into medicinal use during the Han dynasty.

---

[18] *Po Wu Chih* (*HWTS*), ch. 4 on food intolerance, p. 6b. The idea is repeated in *Ko Wu Chhu Than* 格物麤談 (Simple Discourses on the Investigation of Things), +980, Pt 2, p. 22.
[19] *Yang Shêng Lun*, by Chi (or Hsi) Khang 稽康, of the Three Kingdoms period; cited in Li Chhang-Nien ed. (*1958*), p. 61.
[20] *Chin Shu, Shih Chhung Chuan, Lieh Chuang* 3, p. 12b; cited in Li Chhang-Nien *ibid.*, p. 64.
[21] *SNPTC*, middle-class drug, p. 337. [22] See the discussion on *nieh* and *chhü*, pp. 157–61.
[23] Following Chinese tradition *Shên Nung Pên Tshao Ching* is abbreviated in the text as *Pên Ching*.
[24] *52 Pingfang*, recipes 39, 211, 279, 281. [25] Chung I-Yen and Ling Hsiang, (*1979*), p. 58.

Since then, *ta tou huang chüan* has been cited regularly as a medicament in the pharmaceutical literature. The *Wu Shih Pên Tshao* 吳氏本草 (Wu's Pharmaceutical Natural History), c. +235, explains that *ta tou huang chüan* is simply the young sprout obtained when soybeans are allowed to germinate.[26] Thao Hung-Ching, c. +510, further states that *huang chüan* 黃卷 is made by drying the sprout of the black soybean when it reaches a length of half an inch.[27] Since then *Ta tou huang chüan* is described in all the standard pharmacopoeias, starting with the *Ming I Pieh Lu*, c. +510, through the *Hsin Hsiu Pên Tshao*, +659, *Chhien Chin Yao Fang*, +660, *Shih Liao Pên Tshao*, +670, *Chêng Lei Pên Tshao*, +1082, *Pên Tshao Phing Hui Ching Yao*, +1505, *Pên Tshao Kang Mu*, +1596, to the present day *Chung Yao Ta Tshu Tien* (Cyclopedia of Chinese Traditional Drugs). Its effect is said to be laxative, resolvent and beneficial to the hair and skin.[28]

The earliest documentation we have on the preparation and use of soybean sprouts as food is in the *Shan Chia Chhing Kung* (Basic Needs for Rustic Living), of the Southern Sung, where they are listed under the title, *Ngo Huang Tou Shêng* 鵝黃豆生 (Swan-like Yellow Bean Sprout).

A few days before the 15th day of the 7th moon, the people soak black soybeans in water to allow them to sprout. They spread grain chaff on a tray, top the layer with sand, plant the beans in the medium and press the surface with a wooden board. When the beans are sprouted [remove the board and] cover the tray with an [inverted] pail. Each morning remove the pail and allow the sprouts to be exposed briefly to the sun. This will enable the sprouts to grow evenly and be protected from the sun and wind. On the 15th day display the tray in front of the ancestral tablet. After three days, remove the tray and wash the sprouts clean. Grill the sprouts with oil, salt, vinegar and spices to give a savoury dish. It is especially good when rolled in a sesame pancake. Because the sprouts are lightly yellow, and are shaped like the neck of a swan, they are called *Ngo Huang Tou Shêng* (Swan-like Yellow Bean Sprouts).[29]

Bean sprouts were apparently a common article of food during the Sung dynasty. In the *Tung Ching Mêng Hua Lu* 東京夢華錄 (Memories of the Eastern Capital), +1149, Mêng Yüan-Lao recalls bean sprout, *ya tou* 牙豆 as one of the edible staples found in the common market in the Northern Sung capital, Khaifêng.[30] But by this time the sprouts of other beans had also entered the culinary scene. In fact, Mêng Yüan-Lao himself mentions the sprouting of mung beans and the lesser beans for sale by the street hawker. The *Pen Tshao Thu Ching* 本草圖經 (Illustrated Pharmacopoeia), +1061, states even that the 'white sprouts from the mung bean,

---

[26] *Wu Shih Pên Tshao* (also known as *Wu Phu Pên Tshao*), section on grain, p. 83.
[27] Cited by Li Shih-Chên in *PTKM*, ch. 24, p. 1507.
[28] *SNPTC* says that it is used to alleviate suffering in cases of oedema with loss of sensation, cramps and pains at kneecaps, and the *MIPL* recommends it for curing digestive disorders, restoration of skin and hair blemishes and strengthening of the vital energy, *chhi*. Cf. Porter-Smith & Stewart, (1973), p. 190.
[29] *Shan Chia Chhing Kung*, pp. 83–4. The fifteenth day of the seventh month is the Summer Festival *Chung yuan* 中原 when offerings are made to the ancestral spirits. A long soybean sprout with a yellow tip may resemble the neck of a swan with a yellow beak, hence the name Swan-like bean sprout. Except for the step of exposing the germinating beans to sunlight, this recipe is basically similar to those given in later works, in which the beans are always kept in the dark to prevent the generation of chlorophyll and the emergence of a bitter taste.
[30] *TCMHL*, ch. 3, p. 25 and ch. 8, p. 54.

lu tou 綠豆, are considered a delicacy among vegetables',[31] suggesting that they were superior to the sprouts derived from the soybean. Indeed, in the *Chü Chia Pi Yung* 居家必用 (Essential Arts for Family Living), a Yüan dynasty work, when we first encounter the term *tou ya* 豆牙, the modern expression for bean sprouts, it is the mung bean that is chosen as the starting material for making bean sprouts.[32]

Clean mung beans and soak them in water for two nights. When the beans start to swell, rinse them with fresh water and dry them thoroughly. Choose a clean surface, wet it with water and spread several sheets of paper over it. Sprinkle the beans over the paper and cover them with a plate. Rinse the tray with fresh water twice a day. When the sprout is about an inch long, discard the husk. Blanch the sprouts in boiling water, and season with ginger, vinegar, oil and salt to give a delicious dish.

Virtually the same recipe is given in the *I Ya II* 易牙遺意 (Remnant Notions from *I Ya*) another Yüan work, except that it also recommends frying the sprouts with diced meat.[33] The procedure is repeated in the *Yin Chuan Fu Shih Chien* 飲饌服食牋 (Compendium of Food and Drink), +1597, in which sprouts from the broad bean, *han tou* 寒豆 and soybean, *ta huang tou* 大黃豆, are also mentioned.[34] But bean sprouts were considered generally as a poor man's food. They were neglected by most of the Ming and Chhing food canons and recipe books listed in Table 14. They are not even mentioned in the massive Chhing recipe book, *Thiao Ting Chi*, the Harmonius Cauldron, +1760–1860. They are, however, held in high esteem in the *Hsien Chhin Ngo Chi* 閒情偶記 (Random Notes from a Leisurely Life), +1670, which states that 'among vegetables, the purest are bamboo shoots, mushrooms and bean sprouts (*tou ya* 豆牙)'.[35] The famous gastronome of the Chhing dynasty, Yüan Mei, in the *Sui Yüan Shih Tan* 隨園食單 (Recipes from the *Sui* Garden), +1790, had this to say about bean sprout, *tou ya*, 'Though common, yet it's value is inestimable. Those who disdain it do not realise that often it is the humble hermit who is most worthy to keep company with emperor and king.'[36]

From this sketchy account we may surmise that the pharmaceutical soybean sprout, *ta tou huang chüan* 大豆黃卷, probably came into general use in medicine no later than the Han dynasty. Yet that there is no evidence of the use of bean sprouts in food until the Sung. We believe the reason is that *ta tou huang chüan* was always considered as a medicine and hardly, if ever, as food. The drug was, and still is, prepared by allowing the bean to germinate until the sprout is about half an inch long.[37] This means that the bean is still largely intact and firmly bound to the

---

[31] *Pên Tshao Thu Ching*, cited in Li Chhang-Nien (*1958*), p. 272. The same sentiment is echoed by Li Shih-Chen, in *PTKM*, ch. 24, p. 1516.
[32] *CCPY*, p. 73.   [33] *I Ya II*, Pt 1, p. 29.
[34] *YCFSC*, pp. 89, 122. The identity of *han tou* is uncertain; it could be either *chhan tou* 蠶豆 (broad bean, *Vigna faba*) or *wan tou* 豌豆 (peas, *Pisum sativum*).
[35] *Hsien Chhin Ngo Chi*, p. 10.
[36] *SYST*, p. 104. The last sentence is a paraphrase of the original, which says literally that 'it is *Chhao* 巢 and *Yu* 由 who are worthy of the company of *Yao* 堯 and *Shun* 舜'. *Yao* and *Shun* are legendary emperors of China who ruled before *Yü* 禹, the founder of the Hsia dynasty. *Yao* offered the throne to the hermit *Chhao* who declined it, and *Shun* offered his throne to *Yu* with the same result.
[37] Cf. Thao Hung-Ching, cited in *PTKM*, ch. 24, p. 1507; cf. *SLPT*, p. 106.

husk.³⁸ In other words, the pharmaceutical soybean sprout of Chinese medicine is really soybean that has just begun to sprout. It is mostly bean and very little sprout (especially after it is dried). Gastronomically, it probably still retains the disadvantages of raw soybean without much of the advantages of the sprout. It should, therefore, more accurately be called a sprouted bean. Indeed, in appearance it is quite unlike the sprouts, *tou ya* 豆牙, that are a common vegetable in Chinese cookery today. The latter are largely sprouts and only a little bit of bean.³⁹

The transition from the sprouted soybean, *ta tou huang chüan* 大豆黃卷, to soybean sprouts, *huang tou ya* 黃豆牙, may have occurred some time during the Thang or early Sung, in response to the discovery that the mung bean, *lü tou* 綠豆, gave a sprout that is both delicious and easily digestible. The mung bean, *Vignna mungo* var. *radiata*, a cultigen, may have come to China from India or Southeast Asia during the late Han. It is mentioned in the *Chhi Min Yao Shu* 齊民要術 (Important Arts for the People's Welfare), +544 and in the *Shih Liao Pên Tshao* 食療本草 (Manual of Diet Therapy), +670.⁴⁰ As we have noted earlier (pp. 296–7) mung bean sprouts were praised highly as a vegetable in the *Thu Ching Pêng Tshao* of +1061, indicating that they were already well known as a culinary staple in the early Sung. It stands to reason that the procedure for making mung bean sprouts should have also been applied to other beans, including the soybean. By the end of the Sung, all these dietary bean sprouts were called *tou ya* 豆牙, to distinguish them from the pharmaceutical sprouted bean, which continued to be known as *ta tou huang chüan* until the present day.⁴¹

Soybean and mung bean are still the most popular starting materials for making bean sprouts today, the soybean more prevalent in the North and the mung bean in the South. The general name for bean sprouts is *tou ya* 豆牙, but mung bean sprouts are often called *ya tshai* 牙菜 (*nga chhoi* in Cantonese), sprouted vegetable, in the South, especially in Kuangtung.⁴² The making of sprouts represents a simple and convenient way of turning the recalcitrant soybean into a a crunchy, nutritious, and delicious vegetable. The sprouting process destroys toxic proteins and flatus inducing oligosaccharides, and at the same time increases the content of ascorbic acid, riboflavin and nicotinic acid.⁴³ The sprouts can be made at all times of the year, and are thus particularly valuable in the winter when fresh vegetables are in short supply. They can be eaten in cooked dishes or raw as in a salad. Indeed, it is as an ingredient for salads that bean sprouts are beginning to gain acceptance in the cuisines of North America and Europe.

---

[38] For an illustration of *ta tou huang chüan*, the herbal drug, see *Chiang-su hsin i-hsüeh yuan* (*1985*), p. 147.

[39] The soybean sprouts we see in Oriental food markets today can have up to three-inch sprouts. For mung bean sprouts the length is about one inch. For an excellent description of the growing of bean sprouts see Miller, G. B. (1966), pp. 822–3.

[40] *CMYS*, ch. 3; p. 43. *SLPT*, p. 118. Tai Fan-Chin (*1985*) has suggested that it came to China in the early Thang dynasty.

[41] Chinese food historians, e.g. Chhêng Chien-Hua (*1984*), p. 376, have often confused *ta tou huang chüan* with *tou ya*, regarding the two as synonymous. As we have shown, this is not necessarily the case.

[42] Anderson & Anderson (1977), p. 327.

[43] Platt (1956). For a discussion of germination on the changes in nutrient value of cereals and legumes cf., Finney (1983), Khader (1983), and Suparmo & Markakis (1987).

(2) SOYBEAN CURD AND RELATED PRODUCTS

> Planting soybeans, the seedlings are sparse,
> My strength is spent, my heart is weary.
> If only I had known of Huai-Nan's art,
> I could just sit idly and reap my gains.[1]

The second approach to enhancing the nutritional value of soybeans is through mechanical and chemical treatment. The beans are ground in the presence of water and the resultant soybean milk (*tou chiang* 豆漿) is coagulated to give soybean curd, i.e. *tou fu* 豆腐, from which a host of products are obtained by further processing, such as pressing, drying, smoking, frying and fermenting.

The centre piece and the best known member of this array of soy products is undoubtedly *tou fu* or pressed bean curd. To a modern consumer who has become aware of the healthful benefits of high quality vegetable proteins, *tou fu* must surely rank as the most significant gift to the world from the Chinese food system. Yet we are not at all sure as to when and how soybean curd was prepared for the first time. Indeed, the origin of *tou fu* is today regarded as a major unsolved problem in the history of food science and nutrition in China. But this was not always the case. As shown in the above quotation from the Sung philosopher Chu Hsi, tradition had long held that *tou fu* was invented by Liu An 劉安, the Prince of Huai-Nan 淮南, who lived in the Western Han from −179 to −122. This view was echoed in a number of Ming works, for example, in the *Wu Yüan* 物原 (The Origin of Things), +15th century, of Lo Chhi and the *Tshao Mu Tzu* 草木子 (Herbs and Plants), +1378, of Yeh Tzu-Chhi.[2] And Li Shih-Chên says categorically in the *Pên Tshao Kang Mu*, that 'the art of making *tou fu* originated with Liu An, the Prince of Huai-Nan'.[3] Thus, throughout the Sung, Ming, Chhing and the early years of the Republic, it was generally accepted that *tou fu* was invented during the Western Han.[4] This view has been reinforced by the fact that the famous '*Tou fu* of the Mount of Eight Venerables (*pa kung shan tou fu* 八公山豆腐)' has been and still is produced, allegedly according to Liu An's original procedure, in the Huai-Nan region in Anhui Province.[5] *Pa kung shan*, Mount of the Eight Venerables, is said to be the site

---

[1] Cf. *Wu Li Hsiao Shih* 物理小識 (Mini-encyclopedia of the Principle of Things), +1643, ch. 6, p. 13a, section on food and drink, quoted from one of the poems extolling the virtues of a vegetarian diet by Chu Hsi (+1130 to +1200). The original reads:

Chung tou tou miao hsi,	種豆豆苗稀
Li chieh hsin i fu.	力竭心已腐
Chao chih Huai Nan shu,	早知淮南術
An tsuo huo chhuan pu.	安坐獲泉布

In his commentary Chu Hsi said, 'according to tradition, *tou fu* was invented by the Prince of Huai-Nan'. Chu Hsi may be somewhat naive in implying that in making *tou fu* one could simply sit back and relax, but it is probably true that even in Sung times that food processing was more lucrative than farming as it certainly is today.

[2] *Wu Yüan*, p. 26. *Tshao Mu Tzu*, ch. 3b, p. 81. [3] *PTKM*, ch. 25, p. 1532.
[4] Li Chhiao-Phing (*1955*), p. 154. [5] Kuo Po-Nan (*1987a*), 373–7.

where Liu An and his fellow practitioners of the art of alchemy together achieved immortality.[6]

Indeed, it was not until the early 1950s that doubts were first raised about the validity of this traditional view. Yüan Han-Chhing thoroughly combed the *Huai Nan Tzu* (Book of the Prince of Huai-Nan), c. −2nd century, and could find no mention of the word *tou fu* 豆腐, or its alternative names *li chhi* 黎祁 or *lai chhi* 來其 in the text.[7] Furthermore, he scrutinised a host of pre-Sung publications and could find no evidence at all of the existence of *tou fu* in any Thang or pre-Thang work. It is not mentioned in standard texts such as the *Shuo Wên Chieh Tzu*, *Shih Ming*, *Fang Yên*, *Chhi Min Yao Shu*, *Thai Ping Yü Lan*, etc. The only relevant reference to *tou fu* he came across was in the *Pên Tshao Yen I* 本草衍義 (Dilations of the Pharmacopoeia) of Khou Tsung-Hsi of the late eleventh century, which contains the statement, 'grow soybean; mill it to make a curd. It can be eaten'.[8] Yüan Han-Chhing thus concluded that *tou fu* was already in common use during the Sung, and that it was probably invented in the early years of the eleventh century but definitely not before the demise of the Thang.

In 1968, in an article on the origin of *tou fu*,[9] Shinoda Osamu pointed out that in the *Chhing I Lu* 清異錄 (Anecdotes, Simple and Exotic), a compilation made by Thao Ku during the Five Dynasties (+907 to +960), there is a passage which says:[10] 'When Shih Chi 時戢 was the magistrate of Chhing Yang 青陽, he stressed the virtue of frugality among the people, and discouraged the consumption of meat. Instead, he promoted the sale of *tou fu*, which gained the sobriquet, little lamb chops.'

Consequently, Shinoda inferred that *tou fu* was already produced and marketed commercially by the latter part of the Thang dynasty. This piece of evidence pushed back the date for the invention of *tou fu* by at least another hundred years. There the matter seems to have been laid to rest, and for the next two decades, few scholars took seriously the notion that the Prince of Huai-Nan had anything to do with the invention of *tou fu*.[11] But by the 1980s some food historians began to have second thoughts on the matter and felt that perhaps the traditional view had been too hastily dismissed. The issue was reopened by Hung Kuang-Chu who made an exhaustive search of a wide range of pre-Thang literature, including dictionaries,

---

[6] Lu Shu (*1987*) cites another legend for the origin of the name *Pa Kung Shan*. It is said that after a long and fruitless day of alchemical experimentation, Liu An felt tense and weary. He took a walk up a hill, and met eight elderly men coming down the slope. By their vigorous stride and energetic demeanour, he knew they were clearly no ordinary mortals. He took the opportunity to ask them what he should do in order to achieve immortality. The elderly men immediately taught him how to grind soybeans to make soy milk, and how to curdle the milk to make bean curd. And so the place was called Pa Kung Shan in honour of the eight celestial visitors. This legend suggests that Taoist alchemists could have played a key role in the original invention, subsequent development and eventual adoption of *tou fu* as a processed food.

[7] Yüan Han-Chhing (*1954*) wrote in response to a 1953 article by Li Thao 李濤 and Liu Szu-Chih (*1953*), who reiterated the view that *tou fu* was invented by the Prince of Huai-Nan. See also Yüan Han-Chhing (*1981*).

[8] *Pên Tshao Yen I*, ch. 20, p. 3.

[9] Shinoda Osamu (*1963*), tr. Chinese, Yu Ching-Jan (*1971*); cf. Shinoda (*1968*).

[10] *Chhing I Lu*, *Shuo Fu*, vol. 61, p. 5. Chhing-yang is in southern Anhui, not far from present-day Huai-Nan. It was established as a county in +742. We have not be able to determine the year when Shi Chi served as a magistrate there.

[11] Cf. Tshao Yüan-Yü (*1985b*).

agricultural compendia, pharmacopoeias, food canons, histories, miscellaneous compositions, and the literary output of such eminent poets and writers as Li Pai, Tu Fu, Pai Chü-I, Han Yü, Liu Tsung-Yüan etc., for signs indicative of the existence of *tou fu* during and before the Thang.[12]

He too failed to find a reference older than the one in the *Chhing I Lu* 清異錄. But he noted that during the Sung there was a dramatic increase in the number of times that the term *tou fu* had occurred in the literature. In addition to the poem by Chu Hsi and the statement from the *Pên Tshao Yen I* quoted earlier, there are numerous references which signify that *tou fu* was indeed a familiar and important article of food for the Sung people. For example, the *Wu Lei Hsiang Kan Chi* 物類相感集 (On the Mutual Response of Things according to their Categories), c. +980, says, 'frying *tou fu* in soybean oil produces a flavourful dish'.[13] The poet Lu Yu of the Southern Sung, in the *Lao Hsüeh An Pi Chi* 老學庵筆記 (Notes from the Learned Old Age Studio), +1190, states that 'Tung-Pho was fond of eating honey with *tou fu*, wheat gluten and cow's milk,' and that a man from Chia-Hsing 嘉興 'had opened a *tou fu* soup shop next to his bookstore'.[14] The *Mêng Liang Lu* 夢梁錄 (Dreams of the Southern Capital), +1334, of Wu Tzu-Mu, notes that in Lin An 臨安 (present-day Hangchou), the capital, *tou fu* soup and grilled *tou fu* were sold in a wine shop, and a food stall specialising in vegetarian stews also sold grilled *tou fu*.[15] Lin Hung's *Shan Chia Chhing Kung* (Basic Needs for Rustic Living), Southern Sung, describes two dishes containing *tou fu*: one called Hsüeh Hsia Kêng 雪霞羹 (Snow and Red Cloud Soup), prepared by cooking *tou fu* with hibiscus flowers, and the other, *Tung-Pho Tou Fu* 東坡豆腐 (*tou fu* à la Tung-Pho), made from yew nuts, scallion, oil and soy sauce.[16] Chhên Sou-Ta's *Pên Hsin Tsai Su Shih Phu* 本心齋素食譜 (Vegetarian Recipes from the Pure Heart Studio), a Southern Sung work, notes that *tou fu* may be cut in strips and flavoured with condiments.[17] But what impressed Hung Kuang-Chu most is a playful essay in the *Chhêng Tsai Chi* 誠齋集 (The Devout Vegetarian) of Yang Wan-Li 楊萬里 (+1127 to +1206), entitled, *Tou Lu Tzu Jou Chuan* 豆盧子柔傳 (Biography of *Jou* the son of *Tou Lu*), and subtitled '*Tou Fu*'. It should be noted that *Tou Lu* 豆盧 was a well-regarded surname during the Thang and Five Dynasties period, and *jou* 柔 is simply a metaphor for *fu* 鮒 or *fu* 腐. Thus, '*Tou-Lu Tzu Jou*' is just another name for *tou fu*. The piece is full of puns that are difficult to translate, but in a nutshell we can say it purports to describe the life story of a certain Tou-Lu Fu, the son of Tou-Lu.[18] His ancestral home is in Wai-Huang 外黃 county, in northeastern Honan. His name is synonymous with finely ground (or cooked) soybean. His colour is pure white and he exudes a flavour reminiscent of the ancient 'grand soup (*ta kêng* 大羹)' or 'primordial wine (*hsüan chiu* 玄酒)'. His texture resembles that of the dairy products *thi hu, shu* and *lo* 醍醐，酥，酪 i.e. butter oil, butter and soured milk, derived from milk. He first came upon the scene towards the end of the Han,

---

[12] Hung Kuang-Chu (*1984a*), p. 48.   [13] *Wu Lei Hsiang Kan Chi*, p. 12.
[14] *Lao Hsüeh An Pi Chi*, ch. 7, p. 3b; ch. 1, p. 8.   [15] *MLL*, ch. 16, Wine Shop, p. 131, and Noodle Shop, p. 136.
[16] *SCCK*, pp. 83, 94.   [17] *Pên Hsin Tsai Su Shih Phu*, p. 36.
[18] *Chhêng Tsai Chi*, ch. 117, pp. 3–5; *SKCS*, **1161**, 487–8; quoted in Hung Kuang-Chu (*1984a*), pp. 49–50.

then retired as a hermit for many years and reappeared in public at the late Wei period (c. +530 to 550).

In the story, Tou-Lu Fu the man and *tou fu* the product are inextricably woven. To Hung Kuang-Chu, this account not only reaffirms the notion that *tou fu* was first developed during the Han, but also indicates that the invention was not disseminated to a significant extent until after the late Wei.[19] He believes that the author, Yang Wan-Li did not invent the story out of thin air, but rather based it on some evidence available to him at the time. The lack of documentary support for this view is troubling but not necessarily fatal. In the first place, throughout Chinese history thousands upon thousands of books were written, copied, circulated for a time, and lost. The texts now extant and accessible to us are a mere fraction of what must have been in existence at an earlier age. There may have been many references to bean curd between Han and Thang in the literature but we would never know it. Secondly, according to the standard dictionaries, from the *Shuo Wên*, *Shih Ming* to the *Khang Hsi Tzu Tien* the word *fu* 腐 normally means 'decayed' or 'rotten', hardly an appellation fit for a food product. In fact, the term *tou fu* did not come into common usage until after the tenth century. If bean curd had been invented during the Han dynasty, it probably would not have been called a *fu* 腐. If it had been known by a completely different name bean curd could easily have eluded the scrutiny of the surviving literature conducted by modern scholars.

In a search for alternative names for *tou fu* in the Thang and pre-Thang literature, Hung Kuang-Chu was equally unsuccessful. He did, however, find, in the biography of Mu Ning 穆寧 in the *Thang Shu* 唐書 (History of the Thang) a reference to *ju fu* 乳腐 (literally decayed milk), a type of acid coagulated milk curd or cheese, that had been mentioned earlier in the Food Canon of Hsieh Fêng in the Sui dynasty (+589 to 618) and in the *Shih Liao Pên Tshao* in early Thang (+670).[20] These citations indicate that by the Thang era it had become respectable to use the term *fu* in the designation of a food product. It was only then that *tou fu* became an acceptable name for bean curd. Before this date it is likely that bean curd would have been called by some name other than *tou fu*. Thus the traditional view that a bean curd had originated in the Western Han may well be correct but we have not yet been able to find out what it was called in the pre-Thang literature.

(i) *The origin of bean curd*

It turned out that just as Hung Kuang-Chu embarked on his investigation, a new and decisive piece of evidence was being uncovered from archaeological excavations carried out in China in the late 1960s. Before we discuss the significance of this

---

[19] Hung Kuang-Chu (*1984a*), p. 50.
[20] Cf. The Food Canon of Hsieh Fêng as reproduced in the *Chhing I Lu* (Food Section) in the *CKPJKCCK* series, p. 15, and the *Shih Liao Pên Tshao*, p. 62. In *Thang Shu*, folio 28, ch. 105, p. 3, the four sons of Mu Ning are compared to the following products: *lo*, soured milk; *su*, butter; *ti hu*, clarified butter; and *ju fu*, milk curds or cheese. *Ju fu* is to dairy milk what *tou fu* is to soybean milk. For further information on these products, cf. p. 254.

finding it will be necessary to consider first what the literature tells us about the traditional process used in the making of *tou fu*.

The earliest intimation of the process probably occurs in a poem by Lu Yu (+1125–1210), which contains the verse: 'Test the quern to turn it well. Wash the pot and cook the curd.'[21] The term used for bean curd here is *li chhi* 黎祁 which Lu Yu notes is what the people in Szechuan call *tou fu*. This brief sketch is later augmented by another verse from a poem by the Yüan poetess, Chêng Hsiung-Tuan 鄭允端: 'Turn the quern to let the jade liquid flow. Cook the milk and let the clear fluid glow.'[22] While these verses hardly constitute a systematic description of the process, they do tell us that in order to make *tou fu* we need to (1) grind the beans in water to give a jade liquid (milk) and (2) cook and coagulate the milk and then allow the clear fluid to separate from the curd. But it was not until the late Ming that we are given a clear recipe of the process by Li Shih-Chên, who says:[23]

The method for making *tou fu* originates with the Prince of Huai-Nan, Liu An. Black bean, soy bean, white bean, mud bean, peas, mung bean etc. can all be used. The steps are:

1. Soak the beans [in water].
2. Grind the beans [to give a bean milk].
3. Filter the milk [to remove coarse residues].
4. Cook the milk [to homogeneity].
5. Add *yen lu* 鹽鹵 (bittern), leaf of *shan fan* 山礬 (mountain alum), or vinegar to curdle the milk.
6. Collect the curd.

Another way to form the curd is to mix the hot milk in a jar with gypsum powder. Various salty, bitter, sour or pungent materials may also be used to coagulate the milk. If a film should form on the surface of the milk, it should be collected and dried to give a *tou fu phi* 豆腐皮 (bean curd skin), which is itself a delicious food ingredient.

The steps outlined above are basically the unit operations used in the making of *tou fu* today.[24] Although no details are given in the *Pên Tshao Kang Mu*, we can surmise that the procedure used in Li Shih-Chên's time is very similar to a traditional *tou fu* process as practised today in the Chinese countryside. A detailed description of such a process, illustrated with drawings, has been published by Hung Kuang-Chu.[25] The drawings are reproduced in Figures 67a–f. The same curdling agents or coagulants

---

[21] This verse has been quoted often by writers about the origin of *tou fu*, for example in Hung Kuang-Chu (*1987*), p. 13. The original reads:

Shih phêng tui nien chan 　　試盤碓碾轉
Hsi fu chu li chhi. 　　　　　　洗釜煮黎祁

[22] Cf. Hung Kuang-Chu (*1984a*), p. 54 and Ang Tian-Se (*1983*), p. 32. The original reads:

Mo lung liu yü ju, 　　　　　　磨礱流玉乳
Chien chu chieh chhing chhüan. 　煎煮結清泉

[23] *PTKM*, ch. 25, p. 1532.
[24] Hung Kuang-Chu (*1984a*), pp. 57–61 and (*1987*), pp. 1–7; Shurtleff & Aoyagi (1975), pp. 76–112; and Ichino Naoko and Takei Emiko (*1985*).
[25] Hung Kuang-Chu (*1984a*), pp. 58–60. Figure 67f however, is not necessarily a picture of a typical *tou fu* press. The ones I saw in China in 1942–4 and in 1987 and 1981 all use a square or rectangular wooden box to contain the fresh curd as it was pressed. According to Shurtleff & Aoyagi (1983), pp. 100, 304, wooden boxes are also used as *tou fu* presses in the countryside in Japan.

Fig 67. Traditional process for making *tou fu*, from Hung Kuang-Chu (*1984*a), pp. 58–60: (a) Soaking soybeans, Fig. 1; (b) Grinding soybeans in a rotary mill, Fig. 2; (c) Filtering soybean milk, Fig. 3; (d) Cooking soybean milk, Fig. 4; (e) Coagulating soybean milk, Fig. 5; (f) Pressing of bean curd and processed products, Fig. 6.

listed by Li Shih-Chên, i.e., bittern, mountain alum, vinegar, and gypsum are still in use today.[26] *Yen-lu*, bittern or bitterns, is a common byproduct of the salt industry, being the mother liquor that remains behind when sodium chloride is crystallised from brine. Its major ingredients are usually magnesium chloride, magnesium sulphate and sodium chloride.[27] *Shan fan*, mountain alum or China box, is *Murraya exotica* L. which we have already come across previously as an insectidal plant.[28]

With this background we are now ready to evaluate the significance of the recent evidence on the origin of *tou fu* that has come to light in China. During 1959–60 a team of archaeologists from Honan Province discovered two Eastern Han tombs at Ta-hu-thing 打虎亭 village, Mi-hsien 蜜縣 county.[29] Both tombs had been cleaned out by grave robbers. Few burial objects of interest remain except for the pictures engraved on stone slabs in tomb No. 1 and murals painted on the walls of tomb No. 2. A report of the excavation was published in 1972.[30] It is richly illustrated with reproductions of geometric decorations and engravings of people engaged in various household, kitchen and dining activities. But none of the illustrations in the report has anything to do with *tou fu*. In 1981, however, in a volume commemorating thirty years of archaeological discoveries in Honan, it was disclosed that one of the unpublished scenes carved on the stone slabs appears to be a depiction of the making of *tou fu*.[31] This revelation soon generated a great deal of interest among food historians of China since it constituted definitive evidence that *tou fu* was already being made during the Later Han. The claim has been repeated by Chhêng Pu-Chi and Huang Chan-Yüeh, queried by Kuo Po-Nan and included in a table of major events in the history of Chinese food culture compiled in Japan.[32] But since no photographs, sketches or detailed descriptions of the original scene had been published, it was impossible for any interested party to examine the evidence and assess the merit of this claim. This situation was finally rectified in 1990 when Chhên Wên-Hua published a line drawing as well as photographs of the scene and discussed in detail the significance of each of the individual segments in the picture.[33]

---

[26] *Wu Li Hsiao Shih*, ch. 6, p. 12 recommends the use of gypsum to curdle the soy milk; cf. also *Hu Ya* 湖雅 by Wang Jih-Chên (1813–81). Kuo Po-Nan (*1987a*), p. 337, mentions the recent introduction of glucono-δ-lactone by the Japanese as a precipitating agent which, according to Lu Shu (*1987*), p. 380, has been adopted in the making of the finest *pa kung shan tou fu* in the region corresponding to the ancient principality of Huai-nan.

[27] *Yen lu* is known as *nigari* in Japanese. In ancient China, *yen lu* may also mean *shan yen* 山鹽, i.e. mountain salt (or rock salt). The Economics Affairs chapter of the *Shih Chi*, states, 'East of the mountains, the people consume sea salt (*hai yen* 海鹽); west of the mountain, they consume mountain salt (*shan yen*).'

[28] *SCC* Vol. VI, Pt 1, p. 500. Hung Kuang-Chu (*1984a*), p. 55, has interpreted leaf of *shan fan* as alum such as $Al_2(SO_4)_3 K_2SO_4 \cdot 24H_2O$.

[29] Mi-hsien lies about forty kilometres southwest of Chêngchou, capital of Honan Province. Ta-hu-thing is six kilometres west of Mi-hsien.

[30] The report of the excavation by An Chin-Huai & Wang Yü-Kang (*1972*) includes the kitchen scene (fig. 11) and dining scene (fig. 12) discussed and reproduced in Yü Ying-Shih (1977).

[31] Suggested by Chia O. in *Honan shêng po wu kuan* (*1981*), p. 284.

[32] Huang Chan-Yüeh (*1982*), p. 72; Chhêng Pu-Khuei (*1983*); Kuo Po-Nan (*1987a*), pp. 375–6; Nakayama Tokiko ed. (*1988*), p. 560.

[33] Chhên Wên-Hua (*1990*), paper presented at the 6th International History of Science in China, Aug. 2–7, Cambridge, England, and published in *NYKK 1991* (1), pp. 245–8. I thank Prof. Chhên for accompanying me on a visit to Han Tomb No. 1 at Ta-hu-thing, Mi-hsien in September, 1991. Based on my own examination of the wall pictures, I am satisfied that Figure 69 is a faithful representation of the scene depicted on the mural.

Fig 68. Photograph of engraved mural depicting the making of *tou fu* in Han tomb in Ta-hu-thing, Mi-hsien, Honan, photograph H. T. Huang.

Tomb No. 1 is the larger of the two tombs at Ta-hu-thing, Mi-hsien. It is 26.46 metres long and 20.68 metres at its widest point. The maximum height of the ceiling is 6.32 metres. The interior consists of an entry hall, a corridor, a fore chamber, a middle or main chamber, a rear chamber, a south side chamber, an east side chamber and a north side chamber.[34] Our main concern is with the east side chamber. On its ceiling are carvings of domestic animals and fowls and a variety of geometric designs. On the walls are engraved drawings depicting various activities involved in the preparation of food, for example, the slaughter of chickens and steer, the hanging of sides of meat, cooking operations, fermentation of wine, the making of *tou fu* etc. In addition, the pictures also show all types of kitchen implements and utensils, such as stoves, stools, trays, low tables, cabinets, bowls, plates, urns, jars, pots, boxes etc.

On the south wall of the east side chamber two murals are engraved on stone slabs. The mural on the east depicts a variety of kitchen activities. The mural on the west may be divided into two panels. The upper broad panel deals with the fermentation of wine. The lower narrow panel contains the scene that appears to represent the making of *tou fu*. It is 130 centimetres long and 40 centimetres high. As shown in the photograph (Figure 68) and the line drawing (Figure 69) the scene is divided

---

[34] For details of the construction of the tomb, cf. An Chin-Huai & Wang Yü-Kang (*1972*), pp. 49–50. Its considerable size and extensive decorations suggest that the owner was someone with very substantial means.

河南省密县打虎亭漢墓東耳室画像石製豆腐图

Fig 69. Drawing of engraved scene on a stone slab from Tomb No. 1, Ta-hu-thing, Mi-hsien, Honan, Chhên Wên-Hua, *NYKK* (*1991*), p. 248.

into five segments which, in order to facilitate discussion, are numbered (1) to (5) going from left to right. Briefly, the segments show:

(1) Two persons standing behind a large basin.
(2) A person stretching his right hand holding a ladle towards a stone rotary quern.
(3) Two persons holding a net or cloth over a very large basin as if filtering coarse material from a liquid. A large ladle floats near the top of the basin. A third person stands on the left side as if he is directing the operation.
(4) A person stirring the contents of a small basin with a rod. To his right is a jar presumably containing an ingredient needed for the operation.
(5) A large box standing on legs with a platform on top. The platform is pressed down by a pole anchored to the box at one end and a heavy weight attached to the other. On the left side of the box is an outlet tube at the bottom to carry the liquid into a collecting jar.

Since no titles are carved on the slab, and the picture in question occurs on the same slab as another picture depicting a wine fermentation, it is necessary, first of all, to ascertain whether the segments shown are actually related to the fermentation of wine. But the traditional Chinese wine process does not require the use of a mill or a cloth filter in the preparation of the mash for fermentation. So the scene cannot possibly be a portrayal of the wine making process. Nor does it bear a similarity to the process for producing vinegar or soy sauce. On the other hand, it does fit in extraordinarily well with the operations involved in the making of *tou fu* as described by Li Shih-Chên and as practised in the countryside in China today (Figures 67a–f). The steps shown in the tomb scene are analysed by Chhên Wên-Hua and compared with the traditional *tou fu* process as follows:[35]

---

[35] Chhên Wên-Hua (*1991*).

1. *Soaking, Chin tou* 浸豆 (Figure 67a): Soybeans (5 kg) are soaked in water (15 kg) for 5–7 hours (5 hours at 25°C, 7 hours at 15°C). No condiments are added. Soak until the concave edge of the bean is flat. The purpose of the soaking is to soften the protein matrix in the seed to facilitate further processing. Segment (1) of the tomb scene fits this operation very well.

2. *Milling, Mo tou* 磨豆 (Figure 67b): The soaked beans are milled to give soy milk, *tou chiang*. To each 5 kg of beans another 30 kg of water is added. The water must be mixed in evenly. The milling breaks up the structure of the seed and releases the protein into the solution. Segment (2) is consistent with this description. The stone mill shown is presumably turned by hand. In the tomb drawing the right hand of the attendant points towards segment (1) as if he is ready to scoop up more soaked beans for milling. The shape of the mill is similar to that seen in the countryside near Ta-hu-thing today.

3. *Filtering, Kuo lü* 過濾 (Figure 67c): The ground bean puree from the milling is strained through a piece of cloth to remove insoluble particles. Segment (3) depicts this operation perfectly. Two persons are holding the cloth containing the coarse residues and are about ready to remove them from the large basin. The floating ladle near the top suggests that the basin is nearly full. The third person on the left of the basin appears to be supervising the operation. Next to him by the side of the basin stands a container on a tall pedestal. It could be a lamp. Or it could contain a material needed in this operation. We know that in the Honan countryside it is customary to add some edible oil to the milled bean suspension in order to cut down the amount of foam. It is possible that the material in the container is the edible oil to be used as a defoamer if needed.

4. *Cooking, Chu chiang* 煮漿 (Figure 67d): The filtered milk is heated in a pot to boiling to homogenise the milk and to remove the beany taste associated with raw soy bean. This operation is not shown in the picture from Tomb No. 1 at Ta-hu-thing. The significance of this omission will be discussed later.

5. *Coagulating, Tien chiang* 點漿 (Figure 67e): After boiling for 2–3 minutes the fire is extinguished or the milk is transferred to a jar. The emulsion is allowed to cool slowly. When the temperature reaches 75–80°C, a coagulating agent is added and the milk stirred gently in a prescribed manner. If particles are seen to form the stirring is stopped. The jar is covered to slow the rate of cooling. The particles will settle upon standing. Segment (4) shows an attendant stirring the curd. A jar to his left may be the container for the coagulant.

6. *Pressing, Chên ya* 鎮壓 (Figure 67f): The soft curd is collected in a piece of cloth, placed in a wooden box and pressed to remove the whey. The pressed curd is then cut into smaller pieces and sold as pressed bean curd or *tou fu*. Segment (5) in the tomb picture shows clearly a box of curd being pressed and whey being collected. It is interesting to note that the complement of box, rod and hanging counter-weight is almost identical to the equipment in use for pressing *tou fu* in the countryside in the vicinity of Ta-hu-thing in the 1980s,[36] as can be seen from the photograph shown in Figure 70.

Fig 70. Photograph of a *tou fu* pressing box seen in the countryside near Mi-hsien, Honan in the 1980s, photograph Chhên Wên-Hua.

From this account, one has to admit that on the whole, the tomb drawing agrees remarkably well with the operations for making *tou fu* as described by Li Shih-Chên and as currently practised in the countryside in China. Segments (1), (2) and (3) correspond to steps 1, 2, and 3, and segments (4) and (5), steps 5 and 6. The only jarring note is that there is no step 4, cooking the milk (*chu chiang* 煮漿). Why? One possible explanation is that in Han times no special vessel was reserved for cooking the milk. In the background of this scene and in the adjacent picture, kitchen implements and activities are already displayed in such profusion that it was deemed superfluous to add yet another common culinary operation to this scene.[36a] Everyone knows that soy milk has to be cooked before it is coagulated. But this is not entirely satisfactory. After all, the artist was presumably commissioned to depict the whole *tou fu* making process on the slab. If cooking the milk had been recognised as an integral part of the procedure, it would be highly irresponsible of him to leave it out and thus render the scene incomplete.

Several reasons for excluding the segment on cooking the soy milk from the scene have been considered. The first is that there is not enough space on the slab to include the entire process. Something had to be left out, and the cooking operation became the victim. But this is clearly not the case. There is, in fact, so much room on the slab that a lot of extraneous articles had to be included to fill in the

---

[36, 36a] Chhên Wên-Hua, private communication (1990).

background. A second is that the proprietor wanted to keep the process a secret and intentionally left a key step out of the mural. A third is that during the Han, when the process was first developed, step 4, cooking soy milk (*chu chiang*), was not yet a part of the procedure. In other words, the earliest *tou fu* was made without heating the milk. This hypothesis is, of course, based on the tacit assumption that it is possible to make *tou fu* without first cooking the milk. It is, however, impossible to tell from available documentary evidence whether this assumption is valid. All known accounts of ancient or modern processes for the making of *tou fu* include cooking the milk as an integral step in the procedure. There is no record of anyone ever having tried to produce a curd from soy milk that has not been cooked.[37] Prevailing scientific opinion suggests that heating is essential to put the soy milk in the proper physical state ready for coagulation.[38] But experts with extensive experience in the art of making *tou fu* have given us conflicting opinions. Some thought uncooked milk can be used to make *tou fu* while others held the opposite point of view.[39] To resolve this uncertainty in our own mind, we decided to conduct a series of experiments ourselves in 1991 to ascertain whether uncooked soy milk could be coagulated to make *tou fu*. The detailed procedure and the results are summarised in the footnote.[40]

---

[37] In the *Shih Hsien Hung Mi* (Guide to the Mysteries of Cookery), c. +1680, p. 60, cooking the milk is left out in the description given for the making of *tou fu*. But the passage also admonishes that the skin (which presumably forms on the surface of the milk when it is heated) should not be removed, indicating that in this case the heating step was involved though not explicitly stated.

[38] The reason for heating the soy milk in terms of our current understanding of the chemistry of soybean milk has been discussed by Yang Shu-Ai *et al.* (*1989*), pp. 70–1 and Chiang Ju-Tao (*1991*), pp. 21–2. They suggest that the soy proteins are held in solution by interaction between the ionic groups and by extensive hydrogen bonding between their hydrophilic groups and water. Heating to near boiling breaks down some of the hydrogen bonds, exposes the free sulphhydryl groups and allows the protein molecules to interact with each other. The milk is thus placed in a latent-coagulating condition. The coagulants expose the hydrophobic groups, reduce the solubility of the protein matrix and force them to curdle out of the aqueous phase in the form of a gel. Without the heat treatment the proteins would simply precipitate out of the solution as a flocculant.

[39] Chhên Wên-Hua consulted *tou fu* makers in China while the author consulted experts outside China. Particularly helpful were William Shurtleff and Akiko Aoyagi, soyfood populariners extraordinary and authors of the celebrated work, *The Book of Tofu* (1975), Kao Chhuan 高川, a colleague at the Needham Research Institute, who had made *tou fu* routinely in her kitchen for several years in Cambridge, and Tzeng Mao-Chi 曾茂吉, the proprietor of *Mrs Cheng's Soybean Products*, Honolulu, HI.

[40] The first series of experiments were carried out in the spring of 1991 at 2B Sylvester Rd, Cambridge, England. The help of Kao Chhuan in the design and execution of the experiments, using the implements and appliances normally available in an English kitchen, is gratefully acknowledged. The general procedure is as follows:

(1) *Soaking*. Soybeans 125 g are soaked in 600–800 ml of water for 14–16 hrs at room temperature, i.e. 17–19°C.
(2) *Milling*. The soaked beans are rinsed and divided into two lots. Each lot is mixed with 400 ml water and blended in a Waring blender (1 quart size) for 1 min. and 15 secs. to a finely dispersed bean puree.
(3) *Filtering*. The two batches of bean paste are combined and squeezed through a piece of kitchen towel. About 400 ml of soy milk is collected.
(4) *Cooking*. The milk is divided into two parts, each about 200 ml.
 (a) Part A is heated in a pot with constant stirring until the milk froths and begins to boil (temperature of milk about 94°C). It is then poured into a glass container, and allowed to cool to about 75°C.
 (b) Part B is poured into a glass container and kept at room temperature.
(5) *Coagulating*. About 0.5 g gypsum is blended with 1 ml water; half the gypsum mixture is added to Part A and half to Part B. In each case the milk is stirred with a teaspoon for about a minute. The glass container is covered with transparent plastic wrap and left to stand. In about half an hour the milk in Part A begins to jellify, and a fluffy precipitate starts to form in Part B. After 4 hrs a photograph (Figure 71) of the two glasses is taken; by now in Part A the gel completely fills the glass, and in B the fluffy precipitate settles to occupy half the volume of the glass.

40 FERMENTATIONS AND FOOD SCIENCE 311

Fig 71. Experimental coagulation of soy milk; left: with boiling, right: without heating, at the Needham Research Institute, 1991, photograph HTH.

Basically, we found that it was quite easy to make *tou fu* from soy beans by the classical Chinese procedure using the implements available in a modern kitchen. But the soy milk must first be cooked to boiling point. Without cooking, or heating the milk without boiling it, the addition of a coagulant yields a fluffy precipitate rather than a firm, resilient gel. This is illustrated pictorally in Figure 71. Although the fluffy precipitate can be pressed to give a bean cake, the product does not have the elasticity or firmness that is typical of a piece of *tou fu* (see Figure 72). This proto *tou fu* retains a raw, beany taste and is in every way a less palatable product than normal *tou fu*. The experiment was repeated in a commercial soyfood workshop and the results were the same.[41] Evidently, cooking the milk had a profound impact on the

(6) *Pressing*. The gel from part A is firm and resilient and has the texture of standard soft *tou fu*. After washing with water, it can be cut into slices and pressed further, or it can be consumed directly, usually by grilling on a frying pan and spiced with a bit of hot sesame oil. The golden whey in Part B is decanted and the precipitate is collected in a piece of cloth and pressed to give a wet soy 'cake'. As shown in the photograph (Figure 72), the cake looks like a block of *tou fu*, but it does not have the texture or the elasticity of *tou fu*. It crumbles when pressed between the fingers. It can be grilled but it does give the smooth feel typical of *tou fu* and leaves a strong beany taste in the mouth. In every way it is not as attractive a product as normal *tou fu*.

[41] A second series of experiments was carried out at the workshop of *Mrs Cheng's Soy Products* in Honolulu, HI in the winter of 1991. The company made soy milk and *tou fu* from organically grown soybeans. After the milling operation a sample of the puree was put aside and used for our experiments. The results were the same as those seen in the first series of experiments. We thank Mr Tzeng Mao-Chi, the proprietor for his assistance and collaboration.

Fig 72. Cake from precipitate from coagulation of unheated soy milk, photograph HTH.

properties of micelles that enable the insoluble polymeric proteins and fat to stay in the water as an emulsion. In the uncooked state the micelles have little interaction with each other. In the presence of the coagulating agents they are destabilised and the insoluble proteins simply flocculate out of the aqueous phase. After heating, the micelles greatly increase their interaction with each other. When the micelles are destabilised the proteins retain their tertiary links to each other and separate *en masse* as a matrix out of the aqueous phase resulting in a firm gel. Furthermore, the boiling destroys the lipoxidase activity in the raw milk and hence much of the beany flavour that would otherwise develop in the soy milk itself or the product that was made from it later.[42]

The results of these experiments may be interpreted as in partial support of the notion that at the time when Tomb No. 1 at Ta-hu-thing was built, cooking the soy milk to boiling point was not yet an integral part of the process depicted in the picture shown in Figures 68 and 69. The prototypic *tou fu* that was made without boiling the milk could not help but be inferior to the product that we now know as *tou fu* in terms of both its physical and organoleptic properties. Even if the milk had been heated but not boiled, the results would have been variable and inconsistent.[43]

---

[42] For a discussion of the chemical and physical changes which occur during the heating of soy milk, see Yang Shu-Ai *et al.* (*1989*), pp. 70–1 and Chiang Ju-Tao (*1991*), pp. 21–2.

[43] According to Yang Shu-Ai *et al.*, *ibid.*, pp. 72–2, the results of coagulating the soy milk after heating it to various temperatures are as follows: to 70°C, no curd, to 80°C a very soft curd, to 90°C for 20 mins., a curd with a beany odour, and to 100°C a typical *tou fu* curd.

As a result, it never gained acceptance as a standard article of food. Indeed, this proto *tou fu* lingered on as an exotic rarity outside the mainstream of the culinary system for several hundred years after its invention during the Han, until it was discovered that a consistently delicious, superior product could be made from soy milk as long as it was first cooked to its boiling point. Today, it is easy for us to think that this is but a minor amelioration of the overall process. But in practice it could represent a considerable innovation. Perhaps the staff had not yet learned how to heat the milk to boiling without causing it to froth over. While cooking milk to boiling point may sound straightforward in theory, it is not at all that easy in practice. As every cook who has boiled cow's milk knows, the greatest care has to be exercised to ensure that the milk does not froth over. So it is with soy milk. To heat a large pot of soy milk can be a very tricky operation indeed, unless a defoamer is on hand to cut down the froth. At present we have no idea as to whether the Han Chinese knew that certain substances have the ability to break up foams and froths. We have earlier noted the presence of a pedestal to the left of the large basin in segment 3 of the picture (Figure 69), and wondered whether it was a lamp or a container for an antifoam such as vegetable oil. If it were a lamp, and not a container, then it would suggest that the Chinese had not yet learned about the existence of antifoams and how they could be used to contain froth. In any case, whatever the reason, 'cooking the milk to boiling point' was not then a part of the procedure, and thus excluded from Figure 68. Unfortunately, without the boiling step, the process would not yield a product that is attractive in appearance and pleasant to the taste. It was not until the late Thang that the technique for cooking the milk was mastered and incorporated into the process.[44] After that *tou fu* quickly took off as a widely available, common article of food accessible to both the rich and the poor.

If this version of the history of *tou fu* is correct then it becomes relatively easy to explain two rather inconvenient facts not readily reconcilable with the notion that *tou fu* was invented during the Han. The first is that out of more than three dozen murals and stone carvings of culinary scenes discovered in Later Han, Wei and Chin (+20 to +400) tombs in the last three decades,[45] the picture under discussion is the only one that deals with the making of *tou fu*. It was obviously an art known to and practised in only a few select culinary establishments. The second fact is the lack of any mention of the making of bean milk or a bean curd in that inestimable classic, the *Chhi Min Yao Shu* (Important Arts for the People's Welfare) of +544, for which the author had claimed that 'From ploughing and cultivation, down to the

---

[44] Hung Kuang-Chu (*1987*), p. 3, warns that it is important to prevent deposition of a precipitate on the wall of the pot and to control the froth while the milk is being heated. It is the practice of the *tou fu* makers in the countryside to clean the pot thoroughly and to line the walls with a film of vegetable oil before it is used to cook soy milk. This procedure is not at all obvious and could have taken centuries of experimentation for it to be worked out. It is likely that Taoist adepts and Buddhist devotees who had a special interest in vegetarian substitutes of meat and fish helped to bring the development to a successful conclusion.
[45] Tanaka Tan (*1985*) compiled and reproduced forty pictures from nineteen tombs of this period located in nine provinces ranging from Ninghsia and Kansu in the west to Shantung in the east. None includes a scene indicative of the making of *tou fu*.

making of fermented sauces and vinegar, there is no useful art in supporting daily life that I have not described in detail.'[46] Surely this indicates that *tou fu* making had not yet gained the stature of one of the 'useful arts' for the people's welfare in the +6th century.

As we have seen earlier, in ancient China the soybean was usually consumed as cooked grains, *tou fan* 豆飯, or bean congee, *tou chu* 豆粥. Neither product is gastronomically appealing. The former is not easily digestible and it takes an inordinately long time to cook the latter. Thus, the ancient chefs were well aware of the need for new methods to turn this valuable resource into a palatable and attractive food for the table. It was against this background that soybean milk and soybean curd were invented. That the invention should have occurred during the Han is due, we believe, to the convergence of two factors, one technological and one cultural. Firstly, it was the time when the technology for grinding grains such as wheat was becoming widely available. There is now ample archaeological evidence to show that stone rotary mills were already in general use by the beginning of the Western Han (−206).[47] This development was no doubt instrumental in stimulating the rise of wheat as the major grain in north China. As noted earlier (pages 68, 265) wheat grain is hard to cook; it has to be milled into flour, *mien* 麵, before it can be made into palatable products. Once the grinding of wheat in rotary mills had become commonplace, it was only a matter of time before someone applied the same technology to soybean, another grain that is hard to cook. Unfortunately in this case the result proved to be rather disappointing in that the milled product turned out to be flakes rather than flour.[48] But the matter did not stop there. It stands to reason that someone would have tried to cook the flakes in water in the same way that whole soybeans were boiled in order to make *tou chu*, bean congee. If he had been a keen observer he would have noticed that when the milled flakes were mixed with water, part of the bean solids would merge into the liquid to give a milky emulsion. This would have prompted him to try grinding the beans directly in water in order to get most of the bean solids into the aqueous phase. Thus, soybean milk, *tou chiang* 豆漿, was born.

In our view, the preparation of soy milk by grinding soybeans in water is the key invention in the whole process. While it is but one application of a well-established technology, the result is unique and totally unexpected. No other grain cultivated in

---

[46] *CMYS*, Introduction; Miao Chhi-Yü ed. (*1982*), p. 5.

[47] A host of stone and clay models of rotary mills have been found in Western and Eastern Han tombs in China in the last three decades, cf. Li Fa-Lin (*1986*), pp. 146–67, Lu Chao-Yin & Chang Hsiao-Kuang (*1982*), pp. 90–6, and Chhên Wên-Hua (*1987*), pp. 112–13. The evidence leaves no doubt that these implements were used on a wide scale by the early Han. This topic was discussed in *SCC* Vol. IV, Pt 2, pp. 185–92. The use of rotary mills to grind wheat flour will be treated on pages 463–5.

[48] When raw soybean is ground, the cell structure is disrupted, but the matrix of protein, fat and carbohydrate, because of its high fat content, can still stick together to give a viscous mass. The result is a spongy flake. In the modern process for soybean meal the bean is first roasted at a high temperature before it is ground to a soybean meal. The heat denatures the matrix of constituents in the seed so that they can be broken up into fine particles by the grinding action of the mill. In the presence of water, however, the constituents released by the damaged cell structure will dissolve or become emulsified in the liquid to give a soy milk.

Han China, wheat, barley, rice, and millets when treated in the same way would have yielded a milky emulsion. Indeed, the product has the same pearly appearance and smooth, silky texture that are the hallmark of human and animal milk. Poets have waxed lyrical about its loveliness. Yet because of the unpleasant beany taste generated when the raw bean was ground, soy milk, *tou chiang*, did not become an immediate hit as a drink among the well-to-do in Eastern Han society. Furthermore, just like regular milk, *tou chiang* could not stay fresh for long. If left to its own devices it would spoil and undergo changes that render it unfit for human consumption. What then could be done to make soy milk more palatable or process it into a product with a longer shelf life?

This is where the second factor came into play. During the Han there was a conspicuous increase in contact between the people of the Middle Kingdom and the pastoral nomads of the northern steppes. The Chinese thus became familiar with milk from domestic animals as a significant article of food, and with the products derived therefrom, such as soured milk, cream, butter and koumiss. The Biography of the Hsiung Nu in the *Shih Chi* states that they 'look with disdain at Han foods and regard them as inferior to milk and yogurt drink (*thung* 潼 *lo* 酪)'.[49] Yen Shih-Ku's annotation says that *thung* is milk (*ju chih* 乳汁).[50] While the Han Chinese never adopted such dairy foods whole-heartedly into their own dietary system, they could easily have learned how some of these products were prepared. Thus, once it was realised that a 'milk' could be obtained by grinding soybeans in water, it would be natural for a resourceful chef to try to process it into products similar to those derived from livestock milk. Perhaps the first soybean curd was obtained accidentally when he or she added various condiments, such as vinegar or crude salt to improve the flavour of the milk. As pointed out by Li Shih-Chên[51] several types of salty, bitter, sour and pungent materials were effective as curdling agents. In fact, the first curd might even have happened spontaneously.[52] Just as the first yoghurt probably arose from the action of indigenous lactic acid bacteria in dairy milk, so the first bean curd could have resulted from the action of adventitious organisms associated with soy milk. But these are all speculations. The important point is that the Han Chinese learned that soybean milk can sometimes form a curd similar if not identical in appearance to those obtained from livestock milk. This discovery

---

[49] *Shih Chi*, *Hsiung Nu Lieh Chuan*, ch. 110, p. 2899. Perhaps, an earlier reference to *thung* as milk is found in the *Lieh Tzu* (Book of Master Lieh), *Li Ming* 力命, p. 158. In the *Han Shu*, Biography of Yang Hsiung (p. 3561), there is a reference to *mi li* 爓蠡 which according to Chang Yen 張晏, is *kan lo* 乾酪 (dried yoghurt or cheese). The meaning of *lo* and other products derived from milk was discussed on pages 248–57.

[50] The earliest occurrence of the term *ju chih* 乳汁 is probably in the *Wu Shih Erh Ping Fang* (Prescriptions for 52 Ailments), recipe 180, an early Han manuscript discovered in Tomb No. 1 at Ma-wang-tui; cf. Harper (1988), p. 479. It is next mentioned in the *Min I Pieh Lu*, (Informal Records of Famous Physicians), c. +510, as recorded in *CLPT*, ch. 15, 3b.

[51] *PTKM*, ch. 25, p. 1532.

[52] Hung Kuang-Chu (*1984b*), *Tou fu shên shih khao* 豆腐身世考 has pointed out that after cooking, the soy milk can coagulate spontaneously, but the process is very slow. Even without cooking, through the oxidation of the sugars to acids by indigenous bacteria, one may surmise that the milk when kept at room temperature will eventually coagulate, but the process will be even slower and more difficult to control.

provided them with an incentive to improve the reliability of the process and the palatability of the product. But interest in this product, just as interest in dairy products, was at first only peripheral, and progress was slow. It was not obvious that heating the milk would meet both objectives, since in the parallel processes used by the pastoral peoples for the making of dairy products such as kumiss, cream, butter, and possibly yoghurt drink, heating of the milk was not necessarily involved. It may be pertinent that by the 6th century boiling had become a part of the procedure for making *lo* 酪 from milk.[53] From then on it would be natural for someone to try boiling the soy milk first before curdling it with a coagulant. At any rate, several hundreds of years would have to elapse before a reliable procedure was achieved. Unfortunately, we have no knowledge at all of the detailed evolutionary refinements that had to take place in order to turn the original invention into a viable, economic enterprise.

We do know, however, that from the end of the Han dynasty to the end of the Thang, the place of soybean in the Chinese food system had undergone a major change. In Han China soybean was still considered a grain, that is a primary food, *chu shih* 主食. It was cooked to give granules, *tou fan* 豆飯, or congee, *tou chu* 豆粥, which formed the principal component of a meal. But by the time of the Thang it had ceased to be a primary food, and was well on the way to becoming a supplemental food, *fu shih* 副食, albeit one of singular and extraordinary importance in the Chinese food system. *Tou fan* and *tou chu* practically disappeared from the literature after the 6th century, as did the expression *chho shu* 啜菽, sipping soybeans, a manner of ingestion unique for *tou chu*.[54] These changes suggest that during the turbulent centuries between the end of the Han and the rise of the Thang, while soybeans were gradually being displaced by wheat as a primary food, they steadily gained ground as a valuable raw material to be processed into supplemental foods with exceptional nutritional and gastronomic qualities, such as soybean sprouts, soybean curds, soybean paste and soybean sauce. Of particular interest to us is that once soybean curd was firmly established, it would itself become the starting point for the creation of a series of processed foods that have added high nutritional value, excitement and variety to the Chinese diet. At the same time the technology of bean curd was transported eastwards across the China sea to Japan, where it not only became a part and parcel of the native food system but also generated new varieties that rival in delicacy and gastronomic appeal the best products of its homeland. In the next three sections we shall first address the question of how *tou fu* came to be transmitted from China to Japan, then describe the evolution of various soy foods derived from *tou fu*, including the development of fermented *tou fu*, and finally we shall compare the role of soybean curd products in the dietary system of China with that of cheese in the West.

---

[53] Cf. p. 251; *CMTS*, ch. 57, p. 316.

[54] Li Chhang-Nien ed. (*1958*), p. 69. The latest reference to *tou fan* and *tou chu* are in the *Ching Chhu Sui Shih Chi* (New Year Customs in the Chin-Chhu Region), mid 6th century, and to *chho shu* in the *Liang Shu* (History of the Liang), +635. Their occurrence in later documents are usually citations of earlier works.

(ii) *Transmission of* tou fu *to Japan*

It is now well known in the West that *tou fu* is no less important in the dietary system in Japan as it is in China. There is a general consensus that it came originally from China, but considerable uncertainty as to when and how. The subject has elicited a good deal of interest among food historians in China in recent years.[55] The reason for this is in part prompted by the hope that a solution to this problem may help to illuminate the question of the origin of *tou fu* itself in China. Two theories have been advanced. The first follows popular tradition in Japan which holds that *tou fu* was brought from China by the delegation of Buddhist monks headed by the master Kanshin 鑑真 (Chien Chên) in +754.[56] The second subscribes to the view that *tou fu* came to Japan with the Zen master, Ingen 隱元 (Yin Yuan), in +1654.[57] It is likely that both theories are unfounded. There is no question that *tou fu* was already a common article of food when Ingen arrived in Japan. In fact, he was surprised to find types of *tou fu* that were quite different from those he was accustomed to seeing in China. After he founded the celebrated Mampuku-ji temple south of Kyoto, he taught the monks and local *tou fu* masters how to make the Chinese-style pressed *tou fu*. According to Shurtleff and Aoyagi, 'although this very firm variety became quite popular in Japan in the next century, it is now made by only one *tou fu* shop in Japan and is featured in Zen Temple Cuisine only in restauraunts near Mampuku-ji'.[58] It is possible that because of Ingen's involvement in the promotion of Chinese style *tou fu*, later Chinese scholars misunderstood its meaning and thought he was the transmitter of *tou fu* as a generic product.[59]

While the traditional view that *tou fu* was brought to Japan by Kanshin is certainly plausible, there is as yet no documentary evidence to support this claim. Shinoda Osamu has made a careful study of the Japanese literature and he was unable to find any reference to bean curd in works contemporaneous with the Thang period.[60] The earliest reference he saw is in a diary of a court official, Nakatsomi Sukeshige 神主中臣祐重, who in the year +1183 noted that the imperial menu for the day included *tang fu* 唐腐 (Chinese *fu*, i.e. *tou fu*). About fifty years later, in +1239, a letter of thanks from the famous Buddhist priest Nichiren Shonin 日蓮上人 uses the word *suridofu*, which was presumably a kind of *tou fu*, although its exact nature is unknown. No additional references were found in the 13th century, but there was a notable increase in the number of times bean curd was cited in the 14th century. Shinoda tabulated twelve such references dated from +1338 to +1534. Bean curd

---

[55] Kêng Chien-Thing & Kêng Liu-Thung (*1980*), Hung Po-Jên (*1983*), Hu Chia-Phêng (*1986*), and Kuo Po-Nan (*1987b*).

[56] Kêng & Kêng, *ibid.*, point out that in 1963 in a ceremony at Nara commemorating the 1200th anniversary of Kanshin's death, the largest group in attendance were processors of soybean, particularly makers of *tou fu*, indicating the respect he is held as the transmitter of the art to Japan. See also Hu *ibid.*, and Kuo, *ibid.* For a brief layman's account, in English, of the history of *tou fu* in Japan cf. Shurtleff & Aoyagi (1975), rev. edn (1979) pp. 114–18.

[57] Cf. Hung Pu-Jên, *ibid.* [58] Shurtleff & Aoyagi (1975), p. 115. [59] Hung Pu-Jên (*1983*).

[60] Shinoda Osamu (*1968*), pp. 35–6. This section of Shinoda's paper has been translated into Chinese and published in Hung Kuang-Chu (*1984a*), pp. 65–8. The discussion which follows is largely based on Shinoda's work.

was known by several names in this period, such as *thang fu* 唐腐, *thang pu* 唐布 (both pronounced as *tofu*), *mao li* 毛立 (*mota*) and *pai pi* 白壁 (*haku*). The term *tou fu* 豆腐 was not used until +1489. After the 16th century bean curd is mentioned so often that it is obvious it had become a common, easily accessible article of food in Japan.

Shinoda made two interesting observations regarding the making of *tou fu* in mediaeval Japan. He noted that in the 15th century *tou fu* was most often prepared in winter. For example, in the *Oyudononoue no nikki* 御湯殿上日記 (Daily Log of the Yü Thang Palace), the making of *tou fu* was recorded forty-four times [*sic*] during the decade from +1477 to +1488. Of these, ten batches were prepared in the 10th month, four in the 12th month and thirty-two in the 11th month. No bean curd was apparently made in the summer. The same preference for making *tou fu* in the winter is also revealed in other works. Thus, we may surmise that the stability of the product that was produced probably left much to be desired. To ensure that it would stay fresh for a reasonable length of time, the best thing to do was to make it only in winter.[61] The second observation is that Buddist monasteries and temples were the most active establishments involved in the making of *tou fu* in mediaeval Japan. This is not surprising if, as is generally believed, *tou fu* was originally transmitted from China and disseminated in Japan by Buddhist monks. It is to be expected that monasteries should have become centres of *tou fu* technology. With their commitment to abstinence from eating the meat of animals, the monks were particularly in need of a good vegetarian source of protein in their diet.[62] *Tou fu* fits this bill admirably. Thus it is only natural that they should have played a leading role in promoting this nutritious, high protein food of plant origin as a substitute for meat among all strata of society in Japan.

From this account one would have to conclude that *tou fu* probably came to Japan sometime in late Thang or early Sung, allowing at least a hundred years between the firm establishment of the technology in China and its successful transfer to a new geographical and cultural environment. It was probably brought over by Buddhist monks as part of the religious and cultural intercourse between the two countries.[63]

---

[61] Shinoda Osamu, *ibid*. p. 36, has shown that the same conclusion is also implied in a painting from the middle years of the Muromachi period (+1392 to +1568) depicting a group of women in Kyoto selling *tou fu*. The written description says that the *tou fu* had came from Nara, thirty kilometres away. This suggests that the event must have taken place in winter, since *tou fu* could not possibly have stayed fresh throughout the journey if the weather had been hot.

[62] The special place of *tou fu* in the life of the Buddhist monks is illustrated by a proverb composed by Ingen in its honour, cf. Abe & Tsuji (*1974*), p. 4. It reads:

Mame de   まめで
Shikaku de   四角で
Yawaraka de   軟で

Each line has a double meaning, and the poem has been translated in two ways by Shurtleff & Aoyagi (1975), p. 94 as follows:

Made of soybeans,   or   Practising diligence,
Square, cleanly cut,   Being proper and honest,
And soft.   And having a kind heart.

[63] It was probably at about the same time that *tou fu* was transmitted to Korea where it also became an important part of the Korean dietary system.

The date of Kanshin's arrival, +754, may be a bit too early for the process, then still in a developing state, to have been selected as part of the culinary accoutrement of the Buddhist delegation. But we simply do not know. Without solid evidence we can neither prove or disprove whether Kanshin had anything to do with this transmission.

What we do know is that in the 14th and 15th centuries bean curd evolved into types of *tou fu* that acquired a distinctly Japanese character. They were softer, whiter and more delicate in flavour than those normally produced in *tou fu*'s original homeland. But the story of *tou fu* did not remain static in China, where, as we shall see, much ingenuity went into the development of a series of new foods based on bean curd technology that have further extended the importance of *tou fu* in the dietary system of the Chinese people. Indeed, the variety of products derived from *tou fu* was already intimated in +1665 by Friar Domingo Fernandez-Navarrete, probably the first European to encounter the soybean products of China. In his journal he noted:[64]

Before I proceed to the next chapter . . . I will here briefly mention the most usual, common and cheap sort of food all China abounds in, and which all men in the empire eat, from the emperor to the meanest Chinese, the emperor and great men as a dainty, the common sort as necessary sustenance. It is called *t'eu fu*, that is paste of kidney beans . . . They draw the milk out of the kidney-beans, and turning it, make great cakes of it like cheeses, as big as a large sieve, and five to six fingers thick. All the mass is as white as the very snow, to look to nothing can be finer. It is eaten raw, but generally boil'd and dressed with herbs, fish and other things. Alone it is insipid, but very good so dressed and excellent fry'd in butter. They have it also dry'd and smok'd, and mixed with caraway seeds, which is best of all. It is incredible what vast quantities of it are consum'd in China, and very hard to conceive there should be such abundance of kidney beans.

Except for mistaking kidney-beans for soybeans, the observation is reasonably reliable.

(iii) *Products associated with* tou fu

By the 19th century a host of products derived from soybean milk have been developed. The most complete enumeration of these is that given in the *Hu Ya* (Lakeside Elegance), c. 1850, of Wang Jih-Chên. His notes on the principal products are listed as follows:[65]

Tou fu 豆腐 is prepared by grinding soybeans finely [in water], cooking the milk in a pot, and then coagulating it with gypsum or bittern.
  The milk before coagulation is called *tou fu chiang* 豆腐漿.
  The curds are wrapped in a piece of cloth and placed in a wooden box to drain off excess water.

---

[64] F. Dominick Fernandez Navarette (*1665*), pp. 251–2.
[65] *Hu Ya*, in Shinoda and Tanaka (*1973*), II, pp. 497–515. The passage is translated and reproduced in Hung Kuang-Chu (*1984a*), pp. 23–4. A similar enumeration is also given in Wang Shih-Hsiung (*1861*), p. 63.

The soft product is called watery *tou fu*.

The softest product, which cannot be recovered as blocks is called *tou fu hua* 豆腐花 or *tou fu nao* 豆腐腦.

The curds when placed in layers between sheets of cloth and then pressed are known as *chhien chang* 千張 or *pai yeh* 百頁.

When a film forms on the surface of the heated milk, it can be peeled off to give *tou fu i* 豆腐衣, or *tou fu phi* 豆腐皮 as noted in the *Pên Tshao Kang Mu*.

When small blocks of *tou fu* are deep fried, they form an outer coat with an empty interior, called *yu tou fu* 油豆腐.

When pressed and dried *tou fu* is cut in small blocks and then simmered in soy sauce, the product is known as *tou fu kan* 豆腐干. When cooked with 5 spices it is called *wu hsiang tou fu kan* 五香豆腐干.

The plain dried *tou fu* is known as *pai tou fu kan* 白豆腐干.

When dried *tou fu* is smoked with smoke from wood shavings it becomes *hsün tou fu* 薰豆腐.

When dried *tou fu* is soaked in brine and fermented, the product is called *chhou tou fu kan* 臭豆腐干.

These products are still being made in China today under more or less the same names. To facilitate discussion, a flow diagram for the making of *tou fu* and products associated with it is presented in Figure 73.[66] Incorporated in the diagram is a variation from the traditional procedure that has gained acceptance in China in recent years. It is the so-called 'hot extraction' method in which the freshly ground puree (crude milk including residual material), sometimes called *hsi chiang* 細漿, is immediately heated before it is filtered. In contrast, in the traditional 'cold extraction' process, the crude milk is filtered first before it is cooked.[67] Both types of treatment are used in China today.[68] The flow diagram delineates the derivation of soy milk (*tou chiang*), bean curd skin (*tou fu phi*), bean curd flower (*tou fu hua*) or bean curd brain (*tou fu nao*), pressed bean curd, i.e. *tou fu* itself, if the pressed curd is in the form of blocks, or *chhien chang* (thousand sheets) if it is in the form of sheets. This may be as good a place as any to point out that although *tou fu* is usually translated as bean curd, strictly speaking, it should be called 'pressed bean curd', or still more accurately, 'pressed bean curd in blocks'. In fact, there is no simple or elegant way to translate the term *tou fu* into English, but the word has now come into such widespread use in the English language that soon, we hope, no translation will be needed.[69]

After pressing, the large blocks of *tou fu* are usually cut into smaller blocks which are kept in cold water before they are sold. As shown in Figure 73, the smaller pieces

---

[66] After Ichino Naoko and Takei Emiko (*1985*), p. 135.

[67] For a discussion on the extent of use of both types of extraction in China, Japan and Korea see Ichino & Takei, *ibid.* p. 141.

[68] Hung Kuang-Chu (*1987*), p. 46, notes that in 'hot extraction' the deleterious enzymes in the bean solids are inactivated more quickly than in the 'cold extraction' process. The resultant milk will have a less beany flavour and become more pleasant to the human palate. Moreover, the heated puree from the 'hot extraction' process will be easier to filter than the cold puree. On the other hand, more energy will be consumed to heat the crude puree than the filtered milk.

[69] Shurtleff & Aoyagi (1979), p. 112 point out that the rendering of *tou fu* as soybean cheese is not satisfactory either. True, dairy cheese is made from dairy milk curds just as *tou fu* is from soybean curds, but it is salted, fermented and ripened, whereas *tou fu* is simply the pressed bean curd without further treatment.

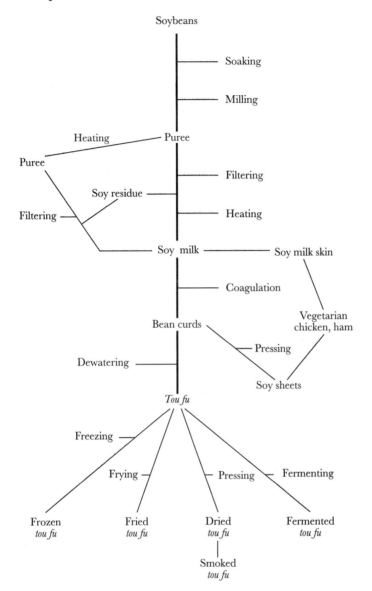

Fig 73. Flow diagram of soybean curd process, adapted from Ichino & Takei (*1975*), p. 135.

may be processed to give *tou fu kan* 豆腐干 by drying, *hsün tou fu* 薰豆腐 by smoking, *tung tou fu* 凍豆腐 by freezing, *tou fu phao* 豆腐泡 by deep frying and *fu ju* 腐乳 by fermenting. The whole gamut of bean curd associated foods is listed in Table 29. The first item is soy milk, *tou chiang* 豆漿 or *tou fu chiang* 豆腐漿 which must be regarded as the senior member of this collection, since it had to have been available before any *tou fu* could be made. Yet except for its role as an intermediate for *tou fu* there was for a long time apparently very little interest in *tou chiang* itself as a food or drink.

Table 29. *Family of products related to soybean curd*

Product	Description	Method of production
Tou chiang 豆漿	Soybean milk	Soybean ground in water, filtered, cooked
Tou fu phi 豆腐皮	Soy milk skin	Film skimmed off heated soy milk
Tou hua 豆花	Bean curd	Soymilk coagulated, washed?
Tou fu 豆腐	Pressed bean curd in blocks	Bean curd wrapped in cloth, pressed
Chhien chang 千張	Pressed bean curd in sheets	Bean curd placed in layers of cloth sheets, pressed
Tung tou fu 凍豆腐	Frozen *tou fu*	*Tou fu* frozen, then thawed
Tou fu phao 豆腐泡	Deep fried *tou fu*	Small blocks of *tou fu* deep fried
Tou fu kan 豆腐干	Dried *tou fu*	Blocks of *tou fu* dried in air (sun)
Hsün tou fu 薰豆腐	Smoked *tou fu*	Dried *tou fu*, smoked
Fu ju 豆腐乳	Fermented *tou fu*	*Tou fu*, fermented

In fact, the first reference to the word *tou chiang* did not appear in the literature until the late Yüan, in the *I Ya I I* (Remnent Notions from *I Ya*), where it is mentioned in passing in the preparation of a thick puree (*fu* 腐) of poppy seed and mung bean powder.[70] It is said that the poppy seed extract is to be cooked in a way similar to *tou fu chiang*. The statement implies, of course, that *tou fu chiang* or *tou chiang* must have been around for quite some time. It further implies that although *tou chiang* had been readily available, it was not itself of much interest as a food product. It probably did not become popular until the Chhing dynasty, when the sale of *tou chiang* by hawkers became a sight common enough as to induce an artist to record the scene in a drawing (Figure 74) for posterity.

To those who have enjoyed the ubiquitous *tou chiang* 豆漿 and *yu thiao* 油條 (fried crullers) at breakfast in Chinese communities in East Asia, it may come as a surprise that *tou chiang* was not immediately in demand as a food after it was discovered. But there may have been very good reasons for its lack of popularity. Firstly, there is the unpleasant beany taste caused by the oxidation of indigenous fats by the enzyme lipoxidase released when the bean is crushed. Secondly, there is the presence of protease inhibitors and recalcitrant oligosaccharides which render the milk less than easily digestible. Cooking the milk cures a large part of the problem by inactivating the lipoxidase as well as the protease inhibitors. But the indigestible oligosaccharides still remain.[71] The situation is reminiscent of that of dairy milk which as itself never became a recognised part of the Chinese food system because of its lactose content.

---

[70] *I Ya I I*, p. 52. The next reference in the literature is in the *Pên Tshao Kang Mu Shih I*, (Supplemental Amplification of the *Pen Tshao Kang Mu*), +1765, p. 365.

[71] After prolonged heating, up to ninety minutes, substantial amounts of raffinose and stachyose, are removed, cf. Pires-Bianchi *et al.* (1983). It was, perhaps, during the Chhing that the Chinese found out that prolonged heating of the soy milk would render it more easily digestible. This paved the way for *tou chiang* to be accepted as an article of food.

Fig 74. Drawing of a hawker selling *tou chiang* by Yao Chih-Han (Chhing), from the National Palace Museum, Taipei, Taiwan.

The advantage of making *tou fu* from the cooked milk is that the bulk of these oligosaccharides are left behind when the bean curd is pressed, drained and washed.

The next item in the Table is *tou fu phi* 豆腐皮, bean curd skin, which today is commonly known as *fu chu* 腐竹 (bean curd bamboo) presumably because in shape it resembles a strip of bamboo. The finished product comes in long wrinkled strips with a light yellow colour. As mentioned earlier (page 303), the *Pên Tshao Kang Mu*, +1596, says that as a skin is formed on the surface of the milk when it is heated, it can be peeled off and dried to give *tou fu phi*. One may assume that the product had been made for at least a couple of hundred years before Li Shih-Chên's time. It is mentioned in the *Pên Tshao Kang Mu Shih I* (Supplemental Amplification of the *Pên Tshao Kang Mu*), c. +1765, and in several literary works of the Chhing dynasty.[72] As can be seen from Table 29, *fu chu* is the product in this series that contains the least amount of water. It can be kept for a long time without refrigeration, and has been a popular ingredient for many Buddhist-style vegetarian dishes.[73]

Tou hua 豆花 or *tou fu hua* 豆腐花 (bean flower) is simply the freshly coagulated curd without additional treatment. It is also called *tou fu nao* 豆腐腦 (bean curd brain). The earliest reference we have to bean flower is in the *Sui Yüan Shih Tan* 隨園食單

---

[72] *Pen Tshao Kang Mu Shih I*, p. 365. One of the poems on *tou fu* by Li Thiao-Yüan (c. +1778) describes the froth on the surface of cooked soy milk folding like a cloth. *Tou fu phi* is directly mentioned as a dish in the popular Chhing novels *Yü Lin Wai Shih* by Wu Ching-Tzu (+1701 to 1754), ch. 22 and *Ching Hua Yüan* by Li Ju-Chên (+1763–1830), ch. 23, and as a skin for dumplings in the *Hung Lou Mêng* by Tshao Hsüeh-Chhin, ch. 8. All four references are cited in Hung Kuang-Chu (*1987*), pp. 30–2.

[73] *Fu chu* is also translated as dried bean curd sticks, as it is so called in the basic Buddhist vegetarian recipe given in Miller, G. B. (1966), p. 624.

(Recipes from the *Sui* Garden) written by the famous Chhing gastronome Yüan Mei at about +1790. A recipe for *fu yung tou fu* 芙蓉豆腐 (hibiscus tou fu) says,[74] 'Place *fu nao* 腐腦 in well water and heat to boiling point three times to remove the beany flavour. Suspend the curd in chicken soup and heat to boiling point. Before serving, garnish with laver and pieces of shrimp.' It is noteworthy that the *fu nao* is first boiled to remove the beany flavour as well as to dilute the indigestible oligosaccharides in the whey, indicating that the raw *tou hua* needed treatment to make it wholly attractive. Today *tou fu hua* is usually drained as much as possible to remove the whey, and is frequently suspended in a sugar syrup and served as a refreshing cold desert.

Of *tou fu* itself little need to be said here, except that after the Sung it became a regular feature in many of the standard food canons of the Yüan, Ming and Chhing dynasties such as *Yin Shih Hsü Chi* (+1350), *I Ya I I* (Yüan), *Yin Chuan Fu Shih Chien* (+1591), *Shih Hsien Hung Mi* (+1680), *Yang Hsiao Lu* (+1698), *Hsing Yuan Lu* (+1750) and *Sui Yüan Shih Tan* (1790).[75] Little is known about the history of pressed bean curd sheets called *chhien chang* 千張 or *pai yeh* 百頁. They were first mentioned in the 19th century in the *Hu Ya* (Lakeside Elegance) of Wang Jih-Chên and the *Sui Hsi Chü Yin Shih Phu* (Food and Drink Menu of the Random Resthouse) of Wang Shih-Hsiung. They are often cut into thin strips or used as wrappers, in the same manner as wheat dough skin, for making pastries.

The next item *tung tou fu* 凍豆腐 is simply frozen *tou fu*. It was first mentioned in the *Shih Hsien Hung Mi* (Guide to the Mysteries of Cookery), +1680, which states that:[76]

In the depth of winter expose a cake of *tou fu* in a basin of water outdoors overnight. The water will freeze even though the *tou fu* itself may not. But the beany taste will be washed off and the flavour of the *tou fu* much improved. Another way is to freeze the *tou fu* itself without the water. When thawed it will look like a little beehive. Wash it well; heat it in a soup base or fry it in oil. Regardless of how it is cooked it will be an unusual dish.

The same recipe is repeated in the *Yang Hsiao Lu*, +1698. The *Hsing Yüan Lu*, +1750, suggests allowing a whole batch of frozen *tou fu* squares to thaw out slowly and then storing them in a cool place so that they can be used in the summer. The recipe in the *Sui Yüan Shih Tan*, +1790, calls for boiling the thawed frozen *tou fu* in water to remove the remaining beany flavour before it is simmered in a soup base.[77] This suggests that in Yüan Mei's time even the best made *tou fu* might still retain a perceptible beany taste. In North China freezing the *tou fu* in winter provides an easy way to make a novel type of *tou fu* with an interesting, sponge-like texture that could be used with profit by the sophisticated vegetarian chef.[78]

---

[74] *SYST*, p. 100.
[75] *YSHC*, p. 51; *IYII*, p. 52; *YCFSC*, p. 104; *SHHM*, pp. 60–4; *YHL*, pp. 20–1; *HYL*, pp. 13–16; *SYST*, pp. 86, 99–101. *Tou fu* is mentioned in popular novels such as the *Shui Hu Chuan* (Yüan-Ming), ch. 13; *Hsi Yu Chi* (Ming), ch. 55, ch. 61; *Jü Lin Wai Shih* (Chhing), chs. 2, 16, 19, 20, 21 etc. and *Hung Lou Mêng* (Chhing), chs. 8, 41, 61.
[76] *SHHM*, p. 63.   [77] *YHL*, p. 20; *HYL*, p. 16 and *SYST*, p. 101.
[78] According to Hung Kuang-Chu (*1987*), p. 48, and Ichino and Takei (1985), p. 136, frozen *tou fu* is also made in China and Japan today using an industrial-scale deep freezer.

Although *tou fu phao* 豆腐泡 (*tou fu* puff) or *yu tou fu* 油豆腐 (fried *tou fu*), the next item, must have been known for a long time, it was not mentioned in the literature until the arrival of the *Hu Ya* of the 19th century. As noted earlier, *tou fu phao* is made by deep frying small blocks of *tou fu*.[79] While the art of shallow frying, *chien* 煎, has been known since antiquity, deep frying, *cha* 炸, did not become popular until a later age, perhaps as late as the Sung. Deep fried foods are mentioned often in the *Tung Ching Mêng Hua Lu* 東京夢華錄 (Dreams of the Glories of the Eastern Capital), +1148, and other works of the same genre.[80] We have not encountered deep fried *tou fu* in the standard Ming and Chhing food canons, but the deep frying of a *tou fu* product is, as we shall soon see, clearly indicated in the *Shih Hsien Hung Mi*, +1680, in the making of deep fried 'stinky' dewatered *tou fu*.

Much more is known about the history of the next product, *tou fu kan* 豆腐干, which translated literally means dried *tou fu*. Actually it may still contain as much as 70% water (Table 29) even though it is made by pressing the curd as hard as possible and then air-drying the extra firm product. Thus, it is more accurately dewatered or semi-dried *tou fu*. It is first mentioned indirectly in the *Shih Hsien Hung Mi*, in connection with the making of *hsün tou fu* 薰豆腐 (smoked *tou fu*), which states:[81] 'Press soybean curd until it is as dry as possible. Soak it in brine, wash well and dry it in the sun [to give *tou fu kan*]. Spread sesame oil over the surface and then smoke it. Another method: Soak *tou fu* in brine, wash well and dry it in the sun. Boil it in soup base, then smoke it.'

Both *tou fu kan* 豆腐干 and *hsün tou fu* 薰豆腐 are cited in the *Yang Hsiao Lu*, +1698, and *tou fu kan* again in the *Sui Yüan Shih Tan*, +1790.[82] Another type of *tou fu kan* is the *wu hsiang tou fu* 五香豆腐 (five spiced *tou fu*) made by simmering *tou fu kan* in a sauce seasoned with five spice. The advantage of *tou fu phao*, *tou fu kan* and varieties thereof is that they have a longer shelf life than *tou fu* itself. All are used as starting materials in recipes of many vegetarian dishes.

(iv) *Making of fermented* tou fu

The last item in Table 29 is, *fu ju* 腐乳, or fermented *tou fu*, long renowned as one of those special gastronomic delectables that is often irresistible to the aficionado but obnoxious to the uninitiated. Technologically it is the most interesting of all the products derived from *tou fu*, and thus deserves a more extensive treatment than the others. The earliest reference to a fermented *tou fu* is found in the *Phêng Lun Yeh Hua* 蓬櫳夜話 (Night Discourses by the Pêng Lun Mountain) of Li Jih-Hua (+1565 to +1635) which contains the following passage:[83]

---

[79] See quotation on p. 320. Hung Kuang-Chu (*1987*), p. 51, points out that today, to make *tou fu* puffs, some soda (sodium carbonate) is usually added to the milk before it is curdled. This helps the *tou fu* cubes to puff up more quickly when they are deep fried.
[80] *TCMHL*, p. 17; *Tu Chhêng Chi Shêng* (Wonders of the Capital), +1235, p. 7; *Hsi Hu Lao Jên Shêng Lu* (Memoir of the Old Man of the West Lake), Southern Sung, pp. 7, 17; *MLL*, pp. 133–7.
[81] *SHHM*, p. 62.   [82] *YHL*, pp. 20–1; *SYST*, p. 118.
[83] *Pêng Lun Yeh Hua*, p. 5, quoted by Hung Kuang-Chu (*1987*), pp. 84–5.

The people from the *i* 黟 District love to ferment (*hai* 醢) *tou fu* in the fall.[84] They wait until it changes colour and is covered with a hairy coat. The hair is carefully wiped off and the cake gently dried. It is then deep fried in hot oil, just like the making of *san* 饊 pastry. The oil is drained off and the cake cooked together with other food materials. It is said that the product has the flavour of *yu* 鮋 fish.

A more detailed account of the making of *tou fu ju* is found in the *Shih Hsien Hung Mi*, +1680, which says:[85]

To make Fukien style *tou fu ju*, press *tou fu* until it is as dry as possible or wrap it in fine cotton paper and dessicate it in fresh ashes. Cut the block into thick square pieces and place them in rows on the floor of a bamboo steamer pad. After all the tiers are filled cover the steamer. [The optimum time for making it] is in the second or third month in spring or ninth or tenth month in autumn. Put the steamer in an airy place. After five to six days the surfaces [of the *tou fu*] will be covered with a hairy growth, which may gradually turn black or a greenish red. Take the squares and wipe off the hair with a piece of paper. Save it and make sure not to damage the skin. For [the *tou fu* from] each *tou* of beans, prepare three catties of soy sauce and one catty of fried salt. (If soy sauce is not available, use five catties of salt.) Grind eight ounces of fresh red ferment spiced with clean peppercorn, fennel and liquorice, and mix the powder with salt and wine. Place the *tou fu* in a jar and add the wine sauce mixture. Then seal the mouth of the urn with clay. Allow the urn to stand for six months and and an excellent flavour will ensue.

Two other passages in the *Shih Hsien Hung Mi* are also of interest in this connection. One describes a method for making *tsao ju fu* 糟乳腐, i.e. *fu ju* 腐乳 aged with a fermented mash.[86] 'Transfer ripened *ju fu* or samples that are too salty in layers to a new vessel. Line fermented mash or residue between the layers and allow the material to age. A product with a unique flavour is obtained.' The other deals with the making of *tou fu fu* 豆腐脯 a deep fried 'stinky' dewatered *tou fu*.[87] 'Take good quality [pieces of] *tou fu* and grill in oil. Then cover it with a cloth screen to keep out flies and other insects. When a "stinky" odour is developed fry the pieces again in hot "boiling" oil. The flavour [of the product] is excellent.'

Two points of interest emerge from these passages. The first is that by the time that *Shih Hsien Hung Mi* (+1680) was published, *fu ju* and *ju fu* had apparently become synonyms for fermented *tou fu*. The word *fu* now could mean a gel or custard made

---

[84] The actual term used here is *hai fu* 醢腐, where *hai* is the ancient word for making a sauce by auto-digestion. The hairy growth obviously refers to fungal myceliae. The i district lies in southern Anhui.

[85] *SHHM*, pp. 61–2. The original passage is interspersed with notes which describe a variation of the process as it was practised in Chekiang. These notes make the text a bit confusing to read. They were left out in the translation. However, from these notes we may reconstruct the Chekiang process follows: 'After the steamer is filled with *tou fu* squares it is steamed [to cook the *tou fu*]. While still hot the steamer is placed on a bed of rice straw and covered completely with rice husk. This should be done in a spot with little movement of air. After five to six days the *tou fu* squares are taken out. The hairy growth is pressed down and flattened. [This will help to keep the product fresh.] The squares are then layered in a jar. Sprinkle a pinch of salt on each piece of *tou fu* until all the surfaces are evenly salted. For each layer of *tou fu* there is a layer of salt. When the salt is dissolved, each piece is heated in the sun by day and marinated in the sauce mixture [as indicated in the Fukien process]. Continue sunning and marinating until all the sauce is exhausted. Soak the layered *tou fu* in the jar with wine. Then seal the mouth of the urn with clay. Allow the urn to stand for six months and an excellent flavour will ensue.'

[86] *SHHM*, p. 63.   [87] *Ibid.*, p. 64.

from any edible suspension or emulsion of food material, and *ju* any type of dairy or soy milk derived product. The second is that although the word *cha* 炸 was not used, there is no question that the frying in 'boiling' oil shown in the second passage indicates that deep frying was a common method of cookery during the Ming.

By the middle of the Chhing dynasty local varieties of *fu ju* had begun to win national fame, such as the *fu ju* of Suchow and the white *fu ju* of Kuangsi. The *Sui Yüan Shih Tan* (+1790) says:[88]

Ju fu: The ones from the [shops] near the front of the Temple of General Wên in Suchow are particularly good. The colour is black, and the flavour is clean. There are two types, a wet and a dry. The product with some shrimp paste in it is also attractive, but may have a slight fishy taste. The white *fu ju* from Kuangsi is also outstanding, especially that made by the family of the offical Wang Khu 王庫.

Technically, the most interesting accounts of the making of *fu ju* are found in the *Hsing Yüan Lu* (+1698). Five recipes are presented, representing two types of methodology. One uses ground wheat *ferment* as shown in the translation given below:[89]

First prepare yellow wheat *ferment* as previously described and comminute it to a fine powder. Take ten catties of fresh *tou fu* and two catties of salt. Cut the *tou fu* into thin rectangular pieces. Sprinkle a layer of salt over a layer of *tou fu*. Allow the *tou fu* to soak in the brine [that is generated]. After five or six days remove the *tou fu* but keep the juice for later use. Arrange the *tou fu* pieces neatly in a steamer and steam until they are well cooked. Hang the steamer with its contents in an empty room for half a month when the *tou fu* becomes covered with luxuriant fungal growth. After scraping off the hairy surface the pieces are air dried. Now treat the *tou fu* with dry yellow *ferment* as follows. Decant the salty juice from the soaking step and mix in dried *ferment* to form a paste. Spread a layer of *tou fu* over a layer of *ferment* paste and cover with a layer of fragrant (i.e. sesame) oil. Add a few whole pieces of fagara. Place the stacks in a jar and seal the mouth securely with mud. Warm the jar in the sun during the day. After a month the product will be ready for the table.

The other methodology uses the mash left from the fermentation of wine from grains.[90] In general, we may say two types of processes were developed to make *fu ju*. In the first type, a two-step procedure is followed (1) the *tou fu*, cut into small blocks, is steamed, cooled and exposed to the action of air-borne fungi (2) they are then aged using wine ferments. The product from step (1) is quite bland; most of the flavour is developed in step (2), the aging step. In the second case, the *tou fu* blocks were fermented and aged concurrently using red *ferment* (and some white *ferment*). Variations of these methodologies can give rise to a variety of flavours. The organisms responsible for converting *tou fu* to *fu ju* in various Chinese processes have been isolated and studied. The most important ones appear to be *Mucor sufu*, *Mucor rouxanus*, *Mucor wutuongkiao*, *Mucor racemosus*,[91] *Mucor sinensis* (later identified as *Actinomucor elegans*) and *Mucor dispersus*.[92] The products are usually packaged in squares,

---

[88] *SYST*, p.119. [89] *HYL*, p. 14. [90] *Ibid.* p. 15. [91] Hung Kuang-Chu (*1987*), p. 78.
[92] Wai, N. (1929), *ibid.* (1964); Hesseltine, (*1965*), *ibid.* (*1983*); Wang & Hesseltine (1970).

about three inches wide and three-quarters of an inch thick. It has the consistency of a soft cheese such as Brie. Because of its high salt content *fu ju* is used either as a relish to accompany bland foods, such as rice congee, or as a condiment to flavour other cooked dishes. Today, *fu ju* is still made in China using natural airborne organisms, but many factories have switched to the use of pure cultures as the inoculum.[93] The Japanese make a similar product called *tofu no misozukè*, tofu pickled in miso. Limited amounts of *fu ju* are also made in the Philippines, Vietnam, Indonesia and the United States.[94]

(v) *Comparison of* tou fu *and cheese*

When we talk of 'cheese' today we usually mean rennet cheese, which is made by curdling the milk with the enzyme rennet. In this sense, there is no evidence that cheese was ever made in China, even though dairy products were a significant component of the Chinese diet during the North and South, Thang and Sung Dynasties.[95] When cheese was introduced to China in the 19th century it was a new and exotic food product that was alien to the Chinese culinary system. It still has not yet made its way into the Chinese kitchen.[96] Similarly, as Friar Navarrete had observed in +1665, the Europeans in Manila never tasted *tou fu*, even though it was readily available in the Chinese community there.[97] There was no easy way to incorporate *tou fu* into the well-established cuisines of the West. It was only in the latter part of the present century, when the virtues of a vegetarian source of protein received serious attention that *tou fu* began to gain acceptance as a palatable and nutritious food in Europe and America.[98]

Tou fu is often translated as 'soybean curd' or 'soybean cheese'. Neither term is correct. True, there is a similarity between *tou fu* and dairy cheese.[99] *Tou fu* is made

---

[93] Hung Kuang-Chu (*1987*), pp. 79–84. Five basic types of *fu ju* are listed on p. 105.

[94] Shurtleff & Ayoyagi (1975), pp. 358ff. In the Philippines it is known as *tahuri*, in Vietnam as *chao* and in Indonesia *taokoan* or *takoa*. All five basic types of Chinese *fu ju* are made in the US.

[95] Cf. pages 248–56 for an account of Chinese dairy products. Tannahill (1988), p. 126 mentions 'curd' cheese in the Thang, but what she means by 'curd' cheese is not defined.

[96] Although dairy products such as milk, cream and butter are no longer strangers to the Chinese diet, cheese still awaits acceptance. Even today Anderson (1988), p. 146 heard it described as 'the mucous discharge of some old cow's guts, allowed to putrefy'.

[97] F. Dominick Fernandez Navarette (1665).

[98] Cf. Shurtleff & Aoyagi (1979) 'The Book of Tofu,' which has been influential in making *tou fu* better known to the peoples of America and Europe. In the 1950s one would have to seek out an oriental food market to buy *tou fu* in London or New York. Today, *tou fu* is sold in many food supermarkets in the US. In 1991 the author was very pleasantly surprised to find *tou fu* cubes cooked with mushroom and zucchini on the menu in the students' cafeteria at Robinson College, Cambridge, England and to learn that many of the students were vegetarians.

[99] Indeed, *tou fu* was often interpreted as 'cheese' by European observers. For example, Captain John Saris, in +1613, was reported to have said, 'Of *cheese* they have plentie. Butter they make none, neither will they eat any milke, because they hold it to bee as bloud.' For other references to *tou fu* as *cheese*, cf. Franklin (1770), pp. 245–6, *The Indian Agriculturist (Calcutta)*, **1882**, Dec. 1, pp. 454–5 and Church, Arthur H. (1886), pp. 140–4; as *fromage*, cf. Baron de Montgaudry (1855), pp. 20–2, Duméril (1859), p. 106 and Champion, Paul (1866); or as *kaese*, cf. Stoeckhardt & Senff (1872) and Dr Ritter (1874). We thank Bill Shurtleff for extracting these interesting references from his soybean database.

Table 30. *Preparation of* tou fu *and cheese*

Process	*Tou fu*	Cheese
Raw material	Soy milk made from soybean and water	Milk from dairy animals
Coagulant	Chemical, c.g. $CaSO_4$, $MgSO_4$ etc	Rennet enzyme
Pressing of curds	Heavy pressure	Light pressure
Further processing	No, except for derived products cf. Table 29	Salting and curing with microorganisms 4–6 months or longer
Shelf life	Poor	Good

from curds obtained by the coagulation of soybean milk, just as cheese is made from curds obtained by coagulation of dairy milk. But there are important differences. Soybean curds are formed when a chemical (calcium sulphate, bittern, acids etc.) destabilises the micelles in the milk and allows the proteins to bond with each other to form a gel. On the other hand, dairy milk curds are formed when an enzyme (rennet, extracted from the abomasum of a suckling calf) hydrolyses the casein into a less soluble protein, para-casein. *Tou fu* is simply pressed soybean curd. It is bland and quite perishable. On the other hand, cheese is pressed dairy curd that has been salted and ripened with microorganisms. It has a salty yet distinctive flavour, and a long shelf life. The preparation of the two products are compared in Table 30.

There are cheeses that are simply curds obtained by treating milk with acid, such as cottage cheese or ricotta. They are the dairy equivalents of fresh *tou fu*. There are also *tou fu* products that are salted and cured with microorganisms, that is, types of fermented *tou fu* or *fu ju*. These are the true soybean equivalents of cheese. But in terms of their importance as processed foods, we can say that what *tou fu* is to the Chinese diet is comparable to what cheeses are to Western diets.

In Chinese cuisine *tou fu* is used in the same way as meat or fish. It can be eaten by itself, when properly seasoned. But usually it is cooked with other food ingredients using all the standard methods of Chinese cookery. Although it is an inexpensive substitute for meat, it is highly prized by emperors and peasants alike.[100] Cheese may be eaten by itself as an appetiser, dessert or filling for sandwiches. It is also used extensively in cooking, not only as a food ingredient but also as a versatile flavouring agent, e.g. in the Italian lasagna, Greek moussaka or French quiche. In Italy it is

---

[100] Cf. p. 300 for reference to *tou fu* as a substitute for lamb, *Chhing I Lu*, ch. 6, p. 5a in *Shuo Fu*, ch. 61. Lai Hsin-Hsia (*1991*) relates stories of how the founding emperor of the Ming dynasty, and two Chhing emperors, Khang Hsi and Chhien Lung, had the highest regard for *tou fu* as food. *Tou fu* was among the gifts a regional commander presented to Chhoe Pu, a Korean official who was stranded in China in +1488, cf. 'Chhoe Pu's Diary' tr. Meskill (1965), p. 73.

Table 31. *Nutrient value of* tou fu *versus cheese*

Product per 100 g	*Tou fu* regular	Fermented *tou fu*	Cottage cheese	Cheddar cheese
Proximate g				
Water	84.5	70	80	37
Energy kcal	76.0	116	85	403
Protein	8.1	8.2	17.3	25
Carbohydrate	1.9	5.1	1.85	1.28
Lipid	4.8	8.0	0.4	33
Crude fibre	0.1	0.31	0.0	0.0
Ash	0.7	8.7	0.7	3.9
Minerals mg				
Calcium	105	46	32	721
Iron	5	2.0	0.2	0.7
Magnesium	103	52	4.0	28
Phosphorus	97	73	104	512
Potassium	121	75	32	98
Sodium	7	2,873	13	620
Lipid g				
Saturated	0.69	1.16	0.27	21.1
Monounsaturated	1.06	1.77	0.11	9.4
Polyunsaturated	2.7	4.52	0.02	0.94
Cholesterol mg	0	0	7	105

Source: USDA Handbook: Composition of Foods. No. 816, Dec. 1986, pp. 148, 152. No. 8–1, Nov. 1976, 01–009, 010014.

treated as a table condiment such as salt or pepper. One can hardly have a meal in an Italian restauraunt without a heavy sprinkling of grated Romano or Provolone on the pasta.

The nutrient composition of *tou fu* and cheese (cheddar) are compared in Table 31. We see immediately that the ratio of lipid to protein is much lower in *tou fu* (4.8/8.1 = 0.57) than it is in cheese (32/25 = 1.28). Furthermore, 60% of the lipid in *tou fu* is polyunsaturated; only 15% is saturated. There is no polyunsaturated lipid in cheese and 73% of the fat is saturated. In addition, there is no cholesterol in *tou fu*, while there is 105 mg of cholesterol per 100 g of cheese. Finally, the sodium content is low in *tou fu* but high in cheese. When all these factors are taken into consideration, for those who need to control their blood lipid and cholesterol levels, there is no question that *tou fu* is a more healthy source of protein than cheese. It will surely grow in importance in Europe and America as methods for incorporating *tou fu* into the culinary systems of the West are developed and popularised.[101]

[101] Shurtleff & Aoyagi (1979), pp. 143–98, have developed many recipes for incorporating *tou fu* into Western style dishes, e.g. dressings, spreads, dips, salads, soups, souffles, etc.

Fig 75. Rubbing of the complete mural referred to on pp. 306–9. From *Honan wên-wu yen-chin-so* (1993), Pl. 34.

(vi) *Addendum*

The question of whether the Eastern Han stone mural described above (pp. 306–9) is, indeed, a depiction of the making of *tou fu* has been hotly debated recently in China. The controversy was prompted by the publication of the official report on the Han tombs at Ta-hu-thing in which a rubbing, presented here as Figure 75, and a line tracing of the whole panel were reproduced.[102] Although they show that a

---

[102] *Honan wên-wu yen-chiu-so* (*1993*), pp. 133, 134 and Pl. 34. The picture of the rubbing on Pl. 34 is particularly valuable. For the first time one can see detailed incised lines on the mural showing the outlines of the rotary mill, the wooden grater and decorations on the basins. This is impossible in my photograph (Figure 68). Even when I was at the tomb in person and could examine the panel itself at close range, the lighting in the chamber was so poor that I could not make out any of the details so clearly delineated on Pl. 34. In their comments on this mural the editors simply say, pp. 128 and 132, that it has been interpreted as operations associated with a wine fermentation by some people or as a depiction of the making of *tou fu* by others.

few details in Chhên Wên-Hua's sketch (Figure 69) of the lower panel were in error, these were not considered serious and attracted little attention. However, in 1996 Sun Chi[103] took strong exception to Chhên's interpretation and insisted that the scene actually represents stages in a wine fermentation. He raised four major objections. Firstly, in the second segment the object regarded as a mill is really a large bowl. Secondly, in the third segment what is seen across the top of the basin is a wooden plank and not a cloth filter. Thirdly, the press shown in the last segment cannot be the type used for pressing bean curd since it does not conform to style of the *tou fu* press drawn by Hung Kuang-Chu (Figure 67e). Finally, the critical act of heating the soy milk is nowhere to be seen. According to him, these segments actually depict stages in a wine fermentation as described in the *Chhi Min Yao Shu*. Segment 1 simply shows the act of adding cooked grain to the fermentor; segment 2, a bowl containing ground *ferment*; segment 3, the breaking up of lumps of cooked grain added to continue the fermentation; segment 4, the stirring of the fermentor after the addition of a feed; and segment 5, the pressing of the must to give a clear wine. This view was given wide circulation in a National Newspaper by two reporters in 1997, who chastised Chhên for careless scholarship and disseminating misleading information.[104] These assaults elicited spirited responses from Chhên Wên-Hua and Chia O,[105] which were published, together with reprints of the original critiques, in late 1998.[106]

After examining the illustrations in the report and re-examining the mural itself, both Chhên and Chia[107] remain convinced that segment 2 does, indeed, depict a rotary mill and not a large bowl. Chhên admits that his original sketch of segment 3 is in error. His corrected version (Figure 76)[108] shows one person holding a long rectangular grater, and the other holding a cloth bag over the basin. The scene still suggests that some sort of filtration is going on. They both present photographs to show that the traditional *tou fu* presses seen in small workshops in China conform to the Ta-hu-thing model rather than to Hung Kuang-Chu's drawing. Above all, they point out that the *Chhi Min Yao Shu* clearly specifies the vessels used for wine fermentations as narrow mouthed jars called *wêng* 甕[109] whereas all the vessels shown in the

---

[103] Sun Chi (*1996*). [104] Tung & Wên (*1997*).
[105] Chia O was the first person to suggest that the lower panel of this mural might be a scene depicting the making of *tou fu*, cf. *Honan shêng po-wu kuan* (*1981*), p. 284.
[106] Chia O (1998), Chhên Wên-Hua (1998), Sun Chi (*1998*) and Tung & Wên (*1998*).
[107] Chhên Wên-Hua made another visit to Ta-hu-thing and was able to examine the mural under good lighting, which had greatly improved since our visit there in 1991. Chia O is a research fellow of the Honan Provincial Institute of Archaeology, and had easy access to the tomb.
[108] Chhên Wên-Hua (*1998*), p. 288.
[109] *CMYS*, e.g. pp. 363, 364 & 365, recommends that the fermentor be insulated in cold weather with straw or husks and be cooled by dousing cold water on the vessel in hot weather. This can be done easily with a *wêng* in which a must will have a small, exposed surface area relative to its volume. Evaporation of the alcohol produced would be minimal. On the other hand, a *kang* would be difficult to insulate. In it a liquid would have a large surface area relative to its volume. The alcohol formed could easily evaporate. In the famous kitchen scene from Chu-chhêng (Figure 28), the fermentors are *wêng* vessels, while a *kang* is used to collect the wine filtered from the must through a cloth filter. As we shall see in the next subchapter, *kang* type vessels are used in soy paste and soy sauce fermentations where the substrate is semi-solid and evaporation of water is not a problem, or even encouraged.

Fig 76. Revised sketch of segment 3 of the lower panel. Chhên Wên-Hua (*1998*), p. 288.

lower panel are wide mouthed basins called *kang* 缸, which are quite unsuitable for use as wine fermentors. Thus, Sun's claim that the vessels in the lower panel are fermentors is fatally flawed. We agree with Chhên and Chia that, all things considered, the lower panel is best interpreted as a representation of a process for making *tou fu*. The lack of a segment dealing with the boiling of soy milk is troubling, but it may indicate that what we are seeing is not a fully mature process but rather one that was still undergoing development.

(3) Fermented Soybeans, Soy Paste and Soy Sauce

> Sauces and pickles are brought
> For the roast meat, for the broiled,
> And blessed viands, tripe and cheek;
> There is singing and beating of drums.[1]

The above quatrain is taken from a poem celebrating a clan feast in the *Shih Ching* (Book of Odes). The characters for 'sauce' and 'pickle' in the Chinese original are *than* 醓 and *hai* 醢 respectively. These are words no longer in use. What then were *than* and *hai*? According to Chêng Hsüan's commentary, '*hai* is *chiang* 醬 made from meat. When the *chiang* is thin, it is called *hai than* 醓醢, and *than* is the juice from *hai*.'[2]

---

[1] *Shih Ching*, w 197 (M 246), fourth stanza. The original reads:

Than hai i chien	醓醢以薦
huo fan huo chih	或燔或炙
chia yao phi chüeh	嘉殽脾臄
huo ko huo o.	或歌或咢

[2] *Mao Shih Chêng I* 毛詩正義 (Elucidation of the Book of Odes), c. +640, M 246. The commentary says, '*than hai* 醓醢, juice of meat'. See also *Chou Li*, ch. 2, *Hai Jên* 醢人, Lin Yin ed. (*1985*), pp. 55–6.

The *Shih Ming* (Expositor of Names), +2nd century, says, 'when *hai* is soupy, it becomes *than*'.[3] Thus, *hai* can be construed as a thick meat sauce (or a paste), and *than* a thin one. But what about *chiang*? The *Shuo Wên* tells us, '*chiang* is *hai*; it is made by blending meat with wine'.[4] From such a spotty description we can only have a vague idea of what *hai* actually was or how it was prepared. Fortunately, we do have other more specific information on how *hai* or meat *chiang* was made in antiquity. In his commentary on the *Hai Jên* 醢人, Superintendent of Meat Sauces, in the *Chou Li*, Chêng Hsüan says,[5] 'To make *hai*, cut meat in thin slices, dry them in the sun, slice them in thin strips, mix them with salt and mould *ferment* (*chhü* 麴) in a jar and soak the mixture in fine wine. After a hundred days *hai* is formed.'

It is unlikely that the fungi and yeast contained in the *ferment* could multiply to any extent in the salty, alcoholic, sugarless medium, under the somewhat anaerobic condition of the incubation. It is, however, likely that the protease and amylase enzymes resident in the *ferment*, as well as any residual enzymes left in the meat itself, would remain active and partially hydrolyse the meat proteins to soluble peptides or amino acids, the elements responsible for the flavour of the sauce. The same passage in the *Chou Li* mentions *hai* prepared from venison, snail, oyster, ant's eggs, fish, rabbit, and goose, as well as a *hai than*. The *Chou Li* further states that the *Shan Fu* 膳夫 (Chief Steward) stores 120 jars of *chiang* 醬 in the royal household and that the *Shih I* 食醫 (Grand Dietician) is responsible for upholding the quality of 100 types of *chiang* needed by the King.[6]

Hai 醢 made from frog, rabbit and fish are mentioned in a passage in the *Nei Tsê* chapter of the *Li Chi*. Also mentioned are *chiang* from fish roe and *chiang* from mustard, as well as a *hai chiang* which presumably simply means a *hai* type of *chiang*.[7] In other passages the *Li Chi* tells us a *hsi chiang* 醯醬 should be eaten with vegetable dishes, although it should be avoided when one is in mourning.[8] Since, as we have seen in Chapter (*c*), pages 283–5, *hsi* 醯 is the prototype for vinegar, we can surmise that *hsi chiang* probably indicates an acid type of *chiang* possibly made by blending *chiang* with vinegar.[9] These records indicate that *chiang* is a general term for fermented or pickled condiments prepared from finely divided food materials; *hai* is *chiang* made from animal products, and *than* is juice drained from mature *hai*.

When Confucius said that he would not eat a food without its proper *chiang* 醬 (sauce),[10] we may assume that what he was talking about was a sort of liquid or colloidal condiment containing bits of finely digested meat or other edible products.

---

[3] *Shih Ming*, ch. 13, *Shih Yin Shih* (On Food and Drink).   [4] *SWCT*, p. 313 (ch. 14B, p. 19).

[5] *Chou Li*, ch. 2, *Hai Jên*, pp. 55–6. The commentary further notes that *than hai* is juice from meat. Hung Kuang-Chu (*1984a*), pp. 91–2, has tested Chêng Hsüan's procedure in three separate experiments using finely cut lean pork and confirmed that a meat sauce with good flavour and aroma is produced.

[6] *Chou Li*, ch. 1; *Shan Fu*, p. 34; *Shih I*, p. 45.

[7] *Li Chi*, *Nei Tsê*, p. 457. In the case of the mustard *chiang*, as in today's mustard sauce, it could also mean a finely ground material dispersed in a watery or oily base.

[8] *Li Chi*, ibid.; *Sang Ta Chi*, p. 718; and *Hsien Chuan*, p. 918.

[9] *I Li Chu Shu*, p. 42. Chêng Hsüan's commentary says that *hsi chiang* is a mixture of *chiang* and vinegar; cf. Hayashi Minao (*1975*), pp. 58–9.

[10] *Lun Yü*, Book 10, ch. 8, verse 3; Legge (1861a), Dover ed. (1971), p. 232.

These types of *chiang* sauces were the principal savoury condiments of pre-Han China. Their production was continued during the Han and probably for centuries afterwards. We shall have more to say about the technology of their production in Chapter (*e*), but their culinary importance was soon overtaken by savoury fermented condiments processed from the soybean, namely, fermented beans (*shih* 豉), soy paste (*chiang* 醬), and soy sauce (*shih yu* 豉油 or *chiang yu* 醬油).

The processing of soybean into these fermented condiments is the third method the Chinese developed to turn the soybean into palatable and attractive food products. The concept of making condiments out of soybeans undoubtedly had its inspiration in the meat sauces (*chiang*) of antiquity. Its realisation, however, would not have been possible without the invention of the mould *ferment* – *chhü* 麴, which, by the end of the Chou dynasty, had become a stable and readily available ingredient in the food system. It is *chhü* technology that provides the conceptual framework for the three major soyfood fermentations. Fermented soybean, *shih*, was probably produced first, followed quickly by soy paste, *chiang*, and much more leisurely, by soy sauce, *shih yu* or *chiang yu*. But the application of *chhü* technology to non-alcoholic foods also stimulated the development of variants of mould *ferments* that were useful agents in facilitating the conversion of soybeans into fermented soyfoods. And so before we delve into the history of these three soyfoods, we need first to acquaint ourselves with those specialised *ferments* which we shall encounter not only in the fermentative conversion of soybeans but also in many other operations in food processing and preservation that are to be treated in Chapter (*e*).

(i) *Ferments for food processing*

The *Chhi Min Yao Shu*, +540, lists three mould *ferments* of this type, known as *huang i* 黃衣 (yellow coat *ferment*) or *mai hun* 麥䴷 (moulded wheat), *huang chêng* 黃蒸 (yellow mould *ferment*), and *nü chhü* 女麴 (glutinous rice *ferment*). Methods for making them are described as follows:

1. *Huang i* 黃衣 (yellow coat *ferment*):[11] In the sixth month take a batch of wheat kernels and wash them thoroughly. Soak them in water in an urn, until the water develops a sour taste. Drain off the water, and steam the wheat kernels until they are well cooked. Spread the wheat granules on a mat placed on a wooden frame to a thickness of about two inches. Cover the layer thinly with young reeds of *Miscanthus sacchariflorus* collected the day before. If miscanthus reeds are not available use cocklebur (*Xanthum strumarium*), but be sure all weeds are cleaned out and no dew is left on the leaves. Cover the granules when they are cooled.

After seven days, when a luxuriant yellow coat forms on the granules, the product is ready. Dry the granules in the sun. Be sure to discard only the cocklebur. Do not winnow the yellow granules. The people of Chhi often winnow the granules in the wind to get rid of the yellow coat. This is a big mistake. When the wheat *ferment* (also called *mai hun* 麥䴷) granules are used in a fermentation, *it is the yellow coating that provides the bulk of their fermentative activity*. If the yellow coat is winnowed away, the activity will be severely impaired.

---

[11] *CMYS*, ch. 68, p. 414.

2. *Huang chêng* 黃蒸 (yellow mould *ferment*):[12] In the sixth or seventh month take raw wheat grain and grind it into flour. Mix with water and steam until the flour is cooked. Spread on a mat to cool. Cover with miscanthus leaves and proceed as for *mai hun* (i.e. *huang i*). When it is ready, do not winnow and damage its fermentative activity.

3. *Nü chhü* 女麴 (glutinous rice *ferment*):[13] Clean thoroughly three *tou* of glutinous rice, and steam them until the grains are soft. Allow the rice to cool completely. Press the soft grains in a *chhü* 麴 mould to form cakes. Place the cakes over a bed and under a layer of artemisia leaves. Incubate the trays in the same way as in the production of wheat *ferment*. After three times seven (21) days, open the covering and see if a yellow coat has formed around the cakes. Then the process is finished. If after three times seven days no coat is visible, continue the incubation until a rich coat is achieved. Remove the cakes and dry them in the sun. They are then ready for use.

From these recipes we can see that *huang i* and *nü chhü* consist of granules or cakes covered by spores of the fungi, whereas *huang chêng* is more like a powder. It is interesting to note that the author, Chia Ssu-Hsieh, was well aware that the fermentative activity resides in the yellow dust that covers the *ferment* particles. And yet he never went a step further to separate the dust out and study its nature and properties. If he had, he would have found that the dust was an ideal inoculum to seed the medium in the preparation of a new batch of ferment. We shall have more to say about this observation later, but it is now time for us to proceed to the history of *shih* (fermented soybean), *chiang* (soy paste), and *chiang yu* (soy sauce) in detail.

(ii) *Fermented soybeans,* Shih

Shih 豉 (also read as *chhih*) is among the foods stored in pottery jars and listed in the bamboo slips discovered in Han Tomb No. 1 at Ma-wang-tui.[14] On this slip *shih* is written as 枝, which the *Shuo Wên* says is prepared by incubating soybeans with salt.[15] There are no references to *shih* in the pre-Chhin (−221 to −209) literature. But there is no doubt that by late Western Han, *shih* had become a major commodity in the economy of the realm. The *Huo Chih Lieh Chuan* 貨殖列傳 (Economic Affairs) chapter of the *Shih Chi* 史記 (Historical Records), c. −90, refers to *nieh-chhü yen-shih chhien ho* 糵麴鹽豉千合, 'thousands of jars of saccharifying mould *ferment* and salty fermented soybeans', as articles of commerce.[16] The *Chi Chiu Phien* 急就篇 (Primer

---

[12] *Ibid.* The terms *huang i* and *huang chêng* are difficult to translate. To avoid confusion, we have rendered them as 'yellow coat *ferment*' and 'yellow mould *ferment*'.

[13] *CMYS*, ch. 88, p. 534. Literally, *nü chhü* means 'female *ferment*'. The reason for this name is obscure. To indicate its origin, we have simply translated it as 'glutinous rice *ferment*'.

[14] *Shih* was identified as the material stored in jars 126 and 301, cf. *Hunan shêng po-wu-kuan* (1973) I, p. 127. The bamboo slip in question is listed as no. 101 by the Hunan Provincial Museum Ma-wang-tui research team, *ibid.* p. 138. In Thang Lan (1980), p. 25, the same slip is listed as no. 139. The MWT research team interprets the character written on the slip as *shih*. This view is shared by Hayashi Minao (1974), p. 58. But Thang Lan thinks it means a *chiang* made from a kind of fish. We favour the interpretation of the MWT team.

[15] *SWCT*, p. 149 (ch. 7B, p. 1b).

[16] *Shih Chi*, ch. 129, *Huo Chih Lieh Chuan*, p. 3274. The statement can be interpreted in two ways. Firstly, as four products, that is: *nieh* (sprouted grain), *chhü* (*ferment*), *yen* (salt), and *shih* (fermented bean), or secondly as two products: *nieh chhü* (saccharifying ferment) and *yen shih* (salty fermented soybeans). We prefer the second interpretation.

for Urgent Use), c. −40, mentions *wu i yen shih hsi chho chiang* 蕪荑鹽豉醯酢醬, 'fetid elm seed, salty fermented soybeans, pickles and vinegar, and *chiang*'.[17] The *Chhien Han Shu* 前漢書 (History of the Former Han Dynasty) relates that two of the seven wealthiest merchants of the realm had accumulated their fortunes by trading in *shih*.[18]

The *Shih Ming* (Expositor of Names), early +2nd century, defines *shih* 豉 as *shih* 嗜, that is, something delectable and highly desirable.[19] It was a popular food product. In fact, the extensive trade in *shih* during the Han–Wei period leaves no doubt that it was considered a culinary necessity of daily living. When the Prince of Huainan (the legendary inventor of *tou fu*) was exiled for fomenting rebellion (in −173) against his brother the Han emperor Wên-ti, he and his retinue were, nevertheless, provided with such necessities of life as 'firewood, rice, salt, *shih* and cooking utensils'.[20] *Shih* was also used as a drug. It is listed as a medication in the *Wu Phu Pên Tshao* 吳普本草 (Wu's Pharmaceutical Natural History), c. +210 to +240, and as a middle-class drug in the *Min I Pieh Lu* 名醫別錄 (Informal Records of Famous Physicians), c. +510, which says 'it is bitter in taste, cold and non-toxic'.[21] None of these references, however, tells us how *shih* was actually made. For the earliest description of a process, we have to wait until the appearance of the *Chhi Min Yao Shu*, +544, in which a whole chapter (72) is devoted to the preparation of *shih*.

Although chapter 72 describes four methods for making *shih*, one of them actually deals with the making of a *shih* from wheat, a product that is of minor interest in the food system when compared to the *shih* obtained from soybeans. Of the remaining three methods, two (i.e. methods (1) and (3)) deal with the making of *tan shih* 淡豉, a bland product used in medicine as well as in food, and one (method (2)) with *yen shih* 鹽豉, a salty product used to flavour food. Methods (1) and (2) will be described in detail. But before we do that, we need to take note of a number of conditions which the author thinks should be followed if a successful product is to be achieved.[22]

It is imperative that the incubation be carried out in a shady, warm hut with a thoroughly clean floor. For method (1) a pit about two to three feet deep should be dug on one side of the room. The hut should have a thatched roof. A tiled roof would not be acceptable. All windows and the door should be tightly sealed with

---

[17] *Chi Chiu Phien*, p. 31. The commentary says that *shih* is made by incubating soybeans. *Wu i* is listed in the *Pên Ching* as a middle-class drug. It is the seed of the fetid elm, *Ulmus macrocarpa* Hance, cf. Bernard E. Read (1936), p. 195, item 607, Porter-Smith & Stewart (1973), p. 448 and Tshao Yüan-Yü (*1987*), p. 245.
[18] *Chhien Han Shu*, ch. 61, *Huo Chih Chuan* (Record of Economic Affairs), p. 3694, reports that two of the seven wealthiest merchants of the realm traded in *shih*.
[19] *Shih Ming*, ch. 13.
[20] *Shih Chi*, c. −90, *Huai Nan Hêng Shan Lieh Chuan* 淮南衡山列傳 (Biography of Prince Huai Nan), ch. 118, p. 3079. Other accounts have been cited by Chang Mêng-Lun (*1988*), pp. 108–9, e.g. *San Kuo Chi*, c. +290, *Tshao Chen Chuan* 曹真傳 (Biography of Tshao Chen), *San Fu Chüeh Lu* 三輔決錄, c. +153, *Shih Shuo Hsin Yü* 世說新語, c. +440, p. 21, and *Hou Han Shu*, c. +450, *Tung Chuo Chuan* (Biography of Tung Chuo).
[21] *Wu Phu Pên Tshao*, p. 84, says it benefits *chhi* 氣 (vital energy) and the *MIPL*, p. 204, recommends it for 'colds, headaches, chills and fever, malaria, noxious effluvia, irritability, melancholy, decline, difficult breathing, painful and cold feet, and for the destruction of poisons in pregnant domestic animals'. tr. Porter-Smith & Stuart (1973), p. 193.
[22] *CMYS*, p. 441, edited and condensed by auct.

mud so that no draft, mice or insects can get in. A small opening is cut (through the door). It should be just big enough for one person to pass through. It should be blocked by a thick straw screen when it is in use. Ideally the incubation room should be as warm as a human arm pit. The best time for the process is late spring (4th moon) or early summer (5th moon). Next best is early autumn (7th to 8th moon). It can be carried out at other times, but it should be remembered that when the weather is too hot or too cold it will be difficult to keep the incubating temperature at the desired level.

(1) *Method for making* Tan *(bland)* Shih[23]

Boil 10 *tan* 石 (c. 400 litres) of winnowed, clean soybeans in a large pot. Stop boiling when the bean feels soft between the fingers. Strain and cool.

When the beans are at the right temperature, i.e. warm in winter and cold in summer (when touched by hand), spread them on the floor of the incubation hut in conical piles. Inspect the piles twice a day. Place a hand inside a pile. When the inside feels as warm as one's armpit, the pile should be turned. This means the beans on the outside of the pile should be moved inside the pile and vice versa. After four to five turns, the pile is warm both inside and outside and a white coat (myceliae) begins to appear. The tip of the pile is then pressed down each time it is turned. After four more turns the height of the pile is reduced to six inches. A yellow coat (spores) starts to appear. The beans are then spread to a height of about three inches. They are allowed to incubate undisturbed for three days. A rich yellow coat now covers the beans. The moulded beans are taken outside the hut, winnowed to remove the loose coat of dust, soaked in water, placed in a basket, rinsed thoroughly with water and air dried on a mat.

The pit is lined with straw mats and filled loosely with millet husk to a height of three feet. The beans are place in the pit. A worker will stamp on them until they are as compact as possible. The beans are then covered by a mat on top of which are placed more husks. These are also trampled down.

The *shih* (fermented beans) will be ready in about ten days in summer, twelve days in autumn and fifteen days in winter. If the incubation is too short the colour will be light, if too long, a bitter flavour will develop . . . When dried in the sun the fermented soybeans may keep for up to a year without spoiling.

(2) *Method of making (salty)* Shih[24]

Take 1 *tan* (c. 40 litres) of soybeans, scour them well and soak in water overnight. Steam the beans on the next day. Rub a grain between the fingers. If the skin slips out, the batch is ready.

Spread the beans on the ground to a thickness of two inches – or on a mat if the ground is not suitable. Allow the beans to cool. Cover the layer with rushes also to a thickness of two

---

[23] *Ibid.* pp. 442–3. At the end of this passage, the text continues to say,

It is difficult to carry out this task well and easy to ruin the whole batch. It is important to have the process under control. If the incubation is not properly controlled and becomes too hot, the soybeans will be like rotting mud and stink to high heaven. Even pigs and dogs will not want to touch them. If the incubation becomes too cold, subsequent warming will still result in a *shih* with poor flavour. Therefore, it is essential to keep a close watch over the temperature to ensure that it is within a favourable range. This will not be easy. In fact, it is harder to control than a wine fermentation.

[24] *Ibid.* pp. 443–4, tr. Shih Sheng-Han (1962) 2nd edn, p. 87, mod. & ed. auct. This recipe is taken from the *Shih Ching* 食經. According to Wu Chhên-Lo (1957), p. 58, during the Northern Chhi, l *sheng* = 0.4 litre.

inches. After three days inspect the beans to see if they have turned yellow [i.e. are covered with yellow myceliae]. If they have, remove the rushes and spread the beans further to form a thinner layer. Make grooves along the layer with the finger to divide the beans into 'plots'. Repeat the operation, i.e. reform the thin layer, and divide it into plots, three times each day. Stop further manipulation after three days.

Meanwhile, cook another batch of soybeans to get a thick decoction. Take 5 *shêng* 升 (one *shêng* = 0.4 litre) of a glutinous rice *ferment* (*nü chhü* 女麴) and 5 *shêng* of good table salt. Mix them with the yellow moulded beans and sprinkle them with the bean decoction. Knead them well until some free juice is left between the fingers. Then transfer the mixture into a pottery jar. If the jar is not full, fill the empty space with wild mulberry leaves. Tightly seal the mouth with mud. Place the jar in the middle of the courtyard for twenty-seven days. Then spread the contents out and let them dry in the sun. Steam the fermented beans, and sprinkle on them a decoction of mulberry leaves. Steam them as long as if they were raw beans. When done, spread and sun again. After steaming and sunning three times, the finished product is obtained.

From these two examples we can see that the process for making *shih* involves two main stages (1) surface culture of wild moulds on cooked soybeans, and (2) enzymatic hydrolysis of bean constituents in a tightly enclosed space. In the first stage, the beans are soaked, washed, cooked, cooled and spread out to allow airborne fungal spores to alight and grow on them. This stage is similar to the process described in Chapter (*c*), pages 170–3 for making the *mould ferments* used in the brewing of *chiu* or Chinese wine. In fact, the product may be regarded as a type of soy *ferment* (soy *chhü*). In the second stage, the moulded soybeans are cleaned to remove loose spores, washed and incubated with or without additives (*ferments* and/or salt) under more or less anaerobic conditions. In effect, the soy *ferment* is itself the substrate for the fungal enzymes they contain *in situ*. When the hydrolysis is complete the fermented beans are steamed and dried in the sun so they can be stored for use as needed. During the drying, part of the aromatic amino acids released (or peptides thereof) are oxidised to melanin which endow the *shih* with a dark colour.

In the first stage, as the spores germinate and myceliae proliferate, the fungi produce enzymes which digest the proteins, carbohydrates and fats in the beans. Although such action should render the beans less toxic and more digestible to humans, some of the peptides from the proteins may give the beans a bitter taste. Indeed, the appellation *ta khu* 大苦 (great bitter) had been applied by some Han commentators to *shih* itself.[25] Presumably, it is for this reason that the moulded soybeans are washed thoroughly with water before they are used in the next stage. Nevertheless, the washed beans must still contain significant amounts of residual fungal enzymes and, perhaps, even myceliae. During the second, near-anaerobic incubation, the conditions are not conducive to further fungal growth. The residual enzymes, however, can continue to hydrolyse the protein and carbohydrates in the soybeans, but the extent of hydrolysis, being limited by the amount of water available, is insufficient to damage the shape and internal structure of the beans.

---

[25] Wang I's commentary on the *Chao Hung, Chhu Tshu*, interprets *ta khu* 大苦 (Great Bitterness) as *shih*. Cf. Fu Hsi-Jên ed. (*1976*), p. 161.

Any sugar released would support the growth of lactic acid bacteria, which, in turn, would arrest the emergence of undesirable bacteria. If salt and other condiments are present, they would also have a chance to penetrate and flavour the beans during the incubation. It is curious that for making *tan shih* (bland fermented beans) the second stage is carried out in a pit. To be sure, the pit is exceptionally well insulated to keep the soy ferments at an even temperature. But the shovelling of beans in and out of the pit must have been extremely awkward and hard on the backs of the workers. It did not take long for this method of incubation to disappear from the literature.

Indeed, the use of a jar instead of a pit for the second incubation is already indicated in the very next account we have on the making of bland *shih*, that found in the late Thang *Ssu Shih Tsuan Yao* 四時纂要 (Important Rules for the Four Seasons) of Han O.[26] Other than this change, the methods given in this work for bland and salty *shih* may be regarded as simplified and abbreviated versions of the procedures described in the *Chhi Min Yao Shu*. *Shih* production is mentioned periodically in the Sung, Yuan and Ming literature. Two recipes for making *shih* are given in the *Wu Shih Chung Khuei Lu* 吳氏中饋錄 (Madam Wu's Recipe Book), a Southern Sung work of uncertain date.[27] They describe only the second stage of the process. The key starting material is *huang tzu* 黃子, presumably a preparation of yellow moulded beans, but no information is given on how they were prepared. Apparently, by Sung times *huang tzu* was already a readily available food processing material. Various other ingredients, such as wine, *tsao* 糟 (wine fermentation residues), vegetables and spices are added to the incubation which otherwise follows the traditional procedure closely. Brief recipes for making *shih* are found in *Chü Chia Pi Yung Shih Lei Chhüan Chi* (Complete Compendium of Essential Arts for Family Living), late Yuan, *To Nêng Phi Shih* (Routine Chores made Easy), +1370, and *Yin Chuan Fu Shih Chien* (Compendium on Food and Drink), +1591.[28] But for the next detailed description of the *shih* process we have to await the advent of the Great Pharmacopoeia, *Pên Tshao Kang Mu*, by Li Shih-Chên, +1596. He says,[29]

Shih 豉 can be made from various kinds of soybeans. That from black soybeans is good for medicine. There are two types, a bland type, *tan shih* 淡豉, and a salty type, *yen shih* 鹽豉. For therapy use the decoction from the bland *shih* or the salty type (of *shih*). To make:

Tan shih: Take 2–3 *tou* (20–30 litres) of black soybeans in the sixth month. Winnow them clean and soak them overnight in water. Steam the beans through and spread them on a mat. When almost cool cover them with artemisia leaves. Examine them every three days. When they are covered, but not luxuriantly, with a yellow coat, they are dried in the sun and winnowed clean. Add enough water to wet the beans, so that when scooped by hand, there

---

[26] *SSTY*, 6th month, items 40, 41, 42; Miao Chhi-Yü ed. (*1981*), pp. 161–2, gives three methods for making *shih*: bland *tou shih* 豆豉 (fermented soybeans), *yen shih* 鹽豉 (salty fermented soybeans) and *fu shih* 麩豉 (fermented wheat bran).

[27] *WSCKL*, p. 26. Both incubations are carried out in jars. Salt is included in the incubation indicating that the products are of the salty type.

[28] *CCPY*, pp. 58–9; *TNPS*, p. 372; *YCFSC*, pp. 96, 101.

[29] *PTKM*, ch. 25, p. 1527. At this point Li Shih-Chên explains that by the term *shih hsin* 豉心 (heart of *shih*, cited in some of the prescriptions) he means the product taken from the centre of a batch; he does not mean the naked bean obtained by removing the coat.

will be wetness between the fingers. Place them in a jar and pack them down tightly. Cover them with a three-inch layer of mulberry leaves. Seal the jar's mouth with mud, and let it stay in the sun for seven days. Take the beans out and dry them in the sun for an hour. Then wet them and put them back in the jar. Repeat this treatment seven times. Steam the beans, cool, dry and store them in a jar.

Yen shih: Take 1 *tou* of soybeans, soak them in water for three days, steam them through and spread them on a mat (as before). When the beans are covered with a yellow coat, they are winnowed, soaked in water, strained from the water and dried in the sun. For every four catties of beans mix in one catty of salt, half a catty of fine strips of ginger and season with pepper, orange peel, perilla, fennel and almond in a jar. Add water until it is one inch above the surface of the mixture. Top the contents with leaves and seal the vessel. Sun the jar for a month and the *shih* will be ready.

In principle, Li Shih-Chên's procedures are identical to those described in the *Chhi Min Yao Shu* and reiterated in the *Ssu Shih Tsuan Yao*. Similar procedures are repeated in four recipes given by Li Hua-Nan in the *Hsing Yüan Lu* (Memoir from the Garden of Awareness), +1750.[30] Although the first stage, the production of moulded soybeans (or soy *ferment*), remains unchanged, a variety of additional materials such as fagara, sugar, wine, melon juice, melon meat, melon seeds, liquorice, mint, magnolia bark, fritillary corm etc. have been included in one or more recipes for the second stage incubation. These spices or herbs no doubt gave each of the products its own unique fragrance and flavour.

We have noted earlier (pp. 336–7) that *shih* was an important processed food during the Han–Wei period. Several examples of the use of *shih* as a condiment are found in recipes in the *Chhi Min Yao Shu*.[31] The *Hsin Hsiu Pên Tshao* (The New, Improved Pharmacopoeia), +659, notes that,[32] '*shih* is used extensively as food. During spring and summer when the weather is unsettled, *shih*, either steamed or pan-fried and soaked in wine, is a particularly good relish.' A similar recipe is given in the *Shih Liao Pên Tshao* (Compendium of Diet Therapy), +670.[33] *Shih* is mentioned in the *Chhien Ching Yao Fang* (A Thousand Golden Remedies), +655, *Nêng Kai Tsai Man Lu* (Random Records from the Corrigible Studio), Southern Sung, *Tung Ching Mêng Hua Lu* (Dreams of the Eastern Capital), +1187, *Mêng Liang Lu* (Dreams of the Former Capital), +1334, and *Yin Shan Chêng Yao* (Principles of Correct Diet), +1330, and a *shih hsin chu* 豉心粥 (heart of fermented bean congee) is described in the *Shou Chhin Yang Lao Hsin Shu* (Nutritious Recipes for the Aged), +1080–1307.[34] An interesting recipe in which *shih* is stewed with *tou fu kan* 豆腐干 (dried *tou fu* blocks) and bamboo shoots is given in the *Shih Hsien Hung Mi* (Guide to the Mysteries of Cookery), +1680.[35]

[30] *HYL*, pp. 11–13.
[31] *CMYS*, ch. 76, rabbit stew; ch. 77, steamed chicken, braised lamb, steamed bear meat, steamed fish; ch. 80, roast duck.
[32] *HHPT*, ch. 18, p. 280. In later pharmacopoeias, for example, *Pên Tshao Phing Hui Ching Yao*, ch. 36, p. 833 and *PTKM*, ch. 25, p. 1528, this quotation is attributed to Thao Hung-Ching. It is, however, not found in the newly reconstituted edition of *Min I Pieh Lu* edited by Shan Chih-Chun (*1986*).
[33] *SLPT*, item 187, pp. 117–18.
[34] *CCYF*, ch. 26, in *Chhien Chin Shih Chi*, p. 63. See also *Nêng Kai Tsai Man Lu*, p. 19. *TCMHL*, pp. 22, 41. *MLL*, ch. 16, p. 1389. *YSCY*, ch. 3, p. 240. *Shou Chhin Yang Lao Hsin Shu*, ch. 2, p. 47b.
[35] *SHHM*, p. 162.

Although in the first stage of the *shih* process, cooked soybeans are usually incubated in air in the absence of additives, in some cases they are mixed with wheat flour before the incubation. This may hasten the growth of the airborne natural fungi. The earliest description of this practice is found in the *Nung Sang I Shih Tsho Yao* (Selected Essentials of Agriculture, Sericulture, Clothing and Food), +1314, which says,[36] 'Black soybeans are cleaned and cooked through. They are strained, mixed evenly with some flour and spread on a mat. When cool, they are covered with paper mulberry leaves to make *huang tzu* 黄子 (yellow moulded seeds). When the beans are completely covered with a yellow coat, they are dried in the sun.'

In the second stage of the process sliced melon and eggplant are incubated together with the mould-coated beans. When finished, pickled melon and eggplant are collected as well as the fermented beans. The use of flour is also indicated in a *shih* recipe from the *Yang Hsiao Lu* (Guide to Nurturing Life), +1698 and in one of the four recipes in the *Hsing Yüan Lu*.[37] When the amount of flour is large, the procedure becomes very similar to that developed for the making of soy *chiang* 醬, which is the topic of our next discussion. But before we deal with that, we would like to introduce the reader to another important processed soy food, called *tempeh*, which is prepared by a procedure somewhat similar to that described for the first stage in the making of *shih*. Although *tempeh* is scarcely known in China and Japan it is widely produced and consumed throughout Indonesia. It must be considered as an important soyfood product.

Tempeh is of unusual interest in that it is the only major processed soyfood that did not originate in China or Japan. Its derivation is obscure. It has probably been made in Java for at least several hundred years but there is no information as to when or where it was first made. Its preparation is very simple. Soybeans are soaked in water, dehulled, boiled, cooled, dried and then inoculated with a previous batch of *tempeh*. The soft beans are wrapped in banana or other large leaves and placed in a warm place for twenty-four to forty-eight hours. (Figure 77a–d) When a luxuriant growth of white myceliae binds the beans into a firm cake, the *tempeh* is ready for cooking and consumption.[38] The procedure is somewhat reminiscent of that used in making the *ferment* of both bland and salty *shih*, and *tou huang* 豆黄 (yellow coated bean) as described in the *Pen Tshao Kang Mu*.[39]

---

[36] *NSISTY*, 6th month, p. 93. To carry out the second stage, the text continues, 'Take two catties of sliced melon and eggplant. For each catty (of vegetable), add 1 oz of clean salt, and suitable amounts of finely cut ginger, orange skin, perilla, cumin, pepper, and liquorice. Mix them together and let them stand overnight. On the next day winnow the moulded beans to remove free myceliae. Mix the beans and vegetables in an urn and blend in the free vegetable juice. Top with a layer of bamboo leaves. Press with brick or stone. Seal the mouth with paper and mud. After staying in the sun for a month, the beans, melon and eggplant are removed and exposed to the sun until dry. Collect and store.'

[37] *YHL*, p. 19; *HYL*, p. 11.

[38] For traditional methods of making *tempeh* in the Indonesian countryside see Burkhill (1935), pp. 1080–6 and Stahel (1946). For an updated modern version of the process see Hesseltine (1965), pp. 154–63, Wang and Hesseltine (1979), pp. 115–19, Djien (1986) and Winarno (1986). For information on how *tempeh* is cooked, cf. Soewitao (1986).

[39] *PTKM*, ch. 25, pp. 1527–31.

(a)

(b)

Fig 77. Preparation of *tempeh* in the countryside in Bali, photograph HTH: (a) Boiling of soybeans in oildrums; (b) Drained, dehulled soybeans awaiting inoculation with a commercial spore powder.

(c)

(d)

Fig 77. Preparation of *tempeh* in the countryside in Bali, photograph HTH: (c) Inoculated soybeans dispensed in flat plastic bags; (d) Bags incubated on trays on a large rack.

For us it is intriguing that the organism occurring in most *tempeh* starters is a *Rhizopus oligosporus*, a mould in the same genus as the *Rhizopus oryzae* and variants thereof which are the ubiquitous components of the Chinese wine-making mould *ferment*. This coincidence raises two questions. Firstly, is there a connection between *tempeh* and China? Yoshida has suggested that the *tou huang* (and hence *shih*), is the progenitor of *tempeh* since both products were made with the aid of plant leaves.[40] Ishige has pointed out that, etymologically, the word *tempeh* could have been derived from the Chinese *tou ping* 豆餅 (bean cake).[41] But these suggestions hardly constitute persuasive evidence for a Chinese connection with *tempeh*. The earliest reference to *tempeh* in Javanese literature is in the *Serat Centini*, dated at about 1815, which mentions 'onions and uncooked témpé'.[42] The earliest reference to soybeans in Indonesia is the account by the Dutch botanist Rumphius (+1747), who noted that soybeans were used for food and as green manure.[43] Thus, soybeans had arrived in Indonesia by +1747, but we have no idea as to how the beans were processed for food.

The second question is this: Since the procedure for making *tempeh* is so close to the first stage of the *shih* process, why did the Chinese miss making a *tempeh* type product? The reason, we think, is largely technological. The objective of the Chinese was to make a stable, fermented soybean product that could be used as a drug, relish or condiment. Thus, the *shih* process was designed in such a way as to restrain both the level of fungal growth and enzymatic hydrolysis so that the structural integrity of the beans is preserved. The product could then be steamed, dried and stored for later use. As it happens, the conditions needed for making good *shih* are exactly those which are detrimental to making a *tempeh*. If we compare the *shih* and *tempeh* processes, we will see two significant differences. Firstly, the beans are treated differently after they are soaked. In making *shih*, whole beans, including the seedcoats, are directly steamed and used. But in making *tempeh*, the beans are soaked in water, dehulled, boiled, and the cooking water drained away. Removal of the hull renders the beans more accessible to mycelial growth, and discarding the cooking water removes substances that inhibit the formation of good *tempeh*.[44] Secondly, the conditions of incubation for the growth of fungi are significantly different. In the *shih* process, the beans (with the hull) are piled loosely, so that the environment is strongly aerobic and conducive to sporulation. In the *tempeh* process the cooked naked beans are pressed together tightly, and the cake wrapped in banana leaves. Such a situation discourages sporulation, encourages mycelial growth and allows the growing mycelia to penetrate and bind the entire mass of beans into

---

[40] Yoshida, Shuji (*1985*), pp. 168–9.   [41] Ishige Naomichi, personal communication, 1993.
[42] Shurtleff & Aoyagi (1985), p. 10. The *Serat Centini*, written in verse, tells of the adventures of 'students' wandering in the countryside in search of truth. In the context of a reception, various foods were listed, including 'onions and uncooked *tempe*'.
[43] Rumphius, G. E. (1747), p. 388.
[44] Wang & Hesseltine (1965) found that during boiling, a heat-stable inhibitor of protease synthesis in *R. oligosporus* is extracted by the water. Thus, cooking the beans in excess water, and later discarding the water are essential steps in a successful *tempeh* operation.

tempeh.⁴⁵ But once the growth reaches an optimum level, the product has to be used almost immediately because of its highly perishable nature. Thus, it is highly unlikely for anyone making *shih* to so deviate from the standard procedure as to end up accidentally with a product similar to *tempeh*.⁴⁶ *Tempeh* was invented by a people who had no preconceived notions about how soybeans should be processed, and were free to apply fermentative technology to soybeans in a novel and unconventional way quite unrelated to the Chinese experience.

(iii) *Fermented soy paste,* chiang

Chiang 醬 is one of the food products contained in pottery jars and listed in the bamboo slips found in Han Tomb No. 1 at Ma-wang-tui.⁴⁷ The word also occurs in the silk manuscript *Wu Shih Er Ping Fang* (Prescriptions for Fifty-two Ailments) discovered in Han Tomb No. 3 at Ma-wang-tui.⁴⁸ It appears that by the time of the Western Han, the word *chiang* had already undergone a subtle change from its ancient meaning. It was increasingly used specifically to denote the fermented paste obtained from soybeans. Since then, it is understood that the character *chiang*, unless otherwise indicated, usually means *tou chiang* 豆醬, i.e. *chiang* derived from soybean. Similarly, the character *tou*, unless otherwise specified, usually means the soybean. When the word *chiang* is applied to a specific type of fermented sauce or paste, a prefix is used to indicate the kind of raw material from which the *chiang* was made. Thus, in addition to *chiang* itself, the Ma-wang-tui bamboo slips also list a *ju chiang* 肉醬 (meat paste), a *chhüeh chiang* 雀醬 (sparrow paste), and a *ma chiang* 鯥醬 (a kind of fish paste) as well as a *than* 醓 and a fish *than*.⁴⁹ The types of *chiang* made from animal parts will be addressed separately in Chapter (*e*), but those of plant origin, such as *mai* 麥 (or *mien* 麵) *chiang* 醬 (wheat *chiang*), *yü jên chiang* 榆仁醬 (elm nut *chiang*), and *ta mai chiang* 大麥醬 (barley *chiang*) will be briefly considered later in this account.

The earliest literature reference to *chiang* as a fermented soy paste is found in the *Chi Chiu Phien* (The Handy Primer), c. –40. The commentary says that soybean and wheat flour are mixed to produce *chiang*.⁵⁰ Fermented soy paste, as *shu chiang* 菽醬 is mentioned in the *Wu Shih Erh Ping Fang*,⁵¹ and as *tou chiang* 豆醬, in the *Lun Hêng*

---

⁴⁵ Wang & Hesseltine (1979), p. 118. They found that 'too much aeration will cause spore formation and may also dry up the beans, resulting in poor growth'. Presumably, this means poor mycelial growth.
⁴⁶ Perhaps the incentive for developing a *tempeh* type product never existed in China in the Middle Ages, since a highly palatable, protein rich food had already been made from soybean. This product is *tou fu*. Having eaten both *tempeh* and *tou fu*, I would suspect that gastronomically most East Asians would prefer *tou fu* to *tempeh*.
⁴⁷ *Hunan sheng po-wu-kuan* (1973) I, p. 139, jar 106. In Tang Lan (1980), p. 27, it is jar no. 150.
⁴⁸ *Wu Shih Erh Ping Fang*, recipes 143, 149, Harper (1982), pp. 412, 428.
⁴⁹ Jar nos. 90, 91, 93, 94. No. 98, *ma chiang*, is interpreted by the Ma-wang-tui research team as a horse meat *chiang*, but Tang Lan (1980), p. 25 (list no. 140), shows that the word *ma* should have been written with a fish radical on the left to indicate the *ma* as a fish. Since there was a taboo against eating horse meat, Tang Lan's suggestion would make sense.
⁵⁰ *Chi Chiu Phien*, p. 31; note by Yen Shih-Ku (7th century).
⁵¹ *Wu Shih Erh Ping Fang*, recipe 143 in which the dregs from a *shu chiang* are used in a salve to treat haemorrhoids.

(Discourse in the Balance) of Wang Chhung, +82.[52] The *Ssu Min Yüeh Ling* 四民月令 (Ordinances for the Four Peoples), c. +160, advises the making of fermented paste (or sauce) from soybean or soybean groats in the first month.[53] None of these early references tell us exactly how *chiang* was prepared. For that information we have, once again, to consult the *Chhi Min Yao Shu*, +544.

We have noted earlier (p. 9) that in some recipes for making *shih*, limited amounts of wheat flour are added to hasten microbial growth. To make *chiang*, larger amounts of wheat flour are mixed with the soybean and microbial growth may be hastened even further by inoculating the mash with preformed mould *ferments* such as *huang i*, *huang chêng* and *nü chhü*. This is done in the first recipe in chapter 70 as quoted below.

Method for making Chiang (fermented soy paste)[54]

The best time to make *chiang* is the twelfth or first month, followed by the second month, but no later than the third month. Use jars that are waterproof. If water seeps out from the jar, the *chiang* will spoil. Avoid jars that have previously been used for making vinegar or pickled vegetables. Place the jars in the sun; if necessary raise them on a block of stone. During the summer rains, be sure not to allow the bottom of the jars to stand in water . . .[55]

Use spring planted black soybeans. The spring planted beans are small but even in size. Those planted later are larger but uneven. The dry beans are steamed in a steamer. Allow the steaming to go on for half a day. Unload the beans, and repack them in the steamer. Those that were on top are placed in the bottom, and vice versa. Unless this is done, some beans will be cooked, others will be raw; they will not be steamed evenly. [Continue steaming.] Be sure all the beans are well steamed . . .[56]

Bite a bean and see if it is dark and cooked through. If it is, collect and dry the beans in the sun. At night pile them together and cover them. Do not let them get wet. When it is time to remove the hulls place the beans in a steamer. Steam them vigorously. Take the beans out and dry them in the sun for a day. On the next morning, clean the beans by winnowing. Select the good beans to fill a mortar [and pound them]. Pounding the beans will not crack them. If they had not been treated with a second steaming, the pounding will most likely crack them, and they will be difficult to clean.

---

[52] *Lun Hêng*, ch. 23, *Ssu Wei* (Four Taboos), p. 226. Classics Press, 1990. Wang Chhung tells us that there is a taboo against making soybean paste when thunder is heard, that is, after Spring rains have arrived. The preferred time to make *chiang* is the 1st month, to give the farmers useful employment when it is too cold to work in the field.

[53] *SMYL*, 1st month, 'One can make various sauces (or pastes). In the first ten days of the month, the soybean is roasted; in the second ten days the soybean is boiled. Use the groats to make *mo-tu* 末都 (a type of *chiang*), which, by the conjunction of the sixth and seventh month, is to be used for preserving melons, as well as for making fish sauce, meat sauce and soy sauce.' tr. Hsü Tso-yün (1980), mod. auct. The actual term used here for soy sauce is *chhing chiang* 清醬 (clear paste). We will have more to say about it later (p. 364). *Mo-tu* has also been interpreted as fermented soy paste by Miao Chhi-Yü ed. (*1981*), p. 23 and Hung Kuang-Chu (*1984a*), p. 93.

[54] *CMYS*, ch. 70, pp. 418–20.

[55] At this point the text goes on to say, 'Take a rusted iron nail. With one's back towards the *sui sha* . . . direction, place the nail under the stone supporting the jar. This way, even if the *chiang* is eaten by a pregnant woman, it would not spoil.'

[56] The text continues, 'Cover the fire with ashes and let it smoulder all night without extinguishing. The best fuel is dried cow dung. Pile the dung and leave the center vacant. It burns as well as charcoal. If possible collect an ample supply and use it for cooking. There is no dust and the fire will not be too strong. They are much better than dried straw.'

The beans are winnowed to remove broken pieces and soaked in a large basin of hot water. After a while they are washed to remove the black seed coats ... Strain the beans and steam them.[57] Steam the beans for the same length of time as for cooking rice to make *fan*. Remove the beans [from the steamer], and spread them out on a scrupulously clean mat. Allow them to cool.

In the meantime, take white table salt, *huang chêng* 黃蒸 (yellow mould *ferment*), *chü* 蓾 leaves, and wheat *ferment* (*pên chhü* 笨麴) and dry them thoroughly in the sun.[58] If the colour of the salt is yellow, the *chiang* product will be slightly bitter. If the salt is not dry the *chiang* will easily spoil.[59] Use of 'yellow mould *ferment*' will add a red tinge to the product and improve its flavour. The 'yellow mould *ferment*' and common *ferment* should be ground to a powder and sieved through a horse tail hair screen.

The approximate proportion of the ingredients are: three *tou* of dehulled, cooked beans, one *tou* of ground common *ferment*, one *tou* of ground 'yellow mould *ferment*', five *shêng* of white salt, and as much *chü* leaves as can be picked up with three fingers. If less salt is used, the *chiang* will turn sour; adding the salt later will not help. If superior *ferment* (*shên chhü* 神麴) is used, one *shêng* of it can replace four *shêng* of common *ferment*, because of its greater digestive power. When measuring the beans, heap them liberally above the vessel; do not level the top. On the other hand, pack the salt and *ferments* loosely in the measuring vessel and level the top flat. The measured ingredients are mixed strenuously by hand in a large basin, until everything, including the moisture, is evenly distributed. This operation is done facing the planet Jupiter, to insure that no grubs will form. Place the mixture in a jar and pack the contents tight. The jar should be full; if it is only half full the product will not ripen properly. Cover the jar and seal the mouth with mud, so that the jar is airtight.

When the fermentation is done (after about thirty-five days in the twelfth month, twenty-eight days in the first and second months and twenty-one days in the third month), open the cover. The fermented mash will show large cracks, and shrinkage away from the wall of the jar. A yellow coat (of spores and myceliae) will be visible everywhere. Remove all the chunks and break them into pieces. Divide the *chiang* substrate from two jars into three parts (to fill three jars for the next stage).

Before sunrise, collect water from a well. Dissolve three *tou* of salt in one *tan* (ten *tou*) of water. Stir well; after settling, decant off the clear solution for use. Soak some yellow mould *ferment* (*huang chêng*) in the salt solution in a bowl. Rub the particles until a thick yellow suspension [of spores and myceliae] is obtained. Discard the residue, add more salt solution and pour the suspension into a jar. For example, for one hundred *tou* of moulded bean substrate, provide three *tou* of *huang chêng* and varying amounts of salt solution. The important thing is to achieve a mixture with the consistency of thin congee, since the fermented beans in the mixture can absorb a considerable amount of water.

Allow the urn to stand in the sun with the mouth open. [As the proverb says, 'Drooping mallow and sun dried *chiang*', both are excellent products.] For the first ten days stir the contents from top to bottom several times a day. Then stir once a day until the thirtieth day. Be

---

[57] The text admonishes, 'Water may be added as needed. Do not change the wash water. If the water is replaced, some of the bean flavour will be lost and the *chiang* will be of poor quality ... Use the wash water to boil the broken pieces to dryness, and process the residue to make a side batch of *chiang* for immediate consumption. The wash water is not used for the main batch.

[58] According to Miao Chhi-Yü ed. (*1982*), p. 426, note 6, *chü* is a plant product, but its identity is not known. Note 7, the text says *mai chhü*, wheat *ferment*; but it is interpreted here as *pên chhü* 笨麴, common *ferment*, in opposition to *shêng chhü* 神麴, superior *ferment*, stated in the next paragraph.

[59] It is curious to see salt used in the first fermentation. This may be another reason why the process was so slow. In any case, this practice was not seen again in the subsequent literature.

sure to cover the mouth when it rains. Do not let water get into the jar. Extraneous water will promote the emergence of grubs. Stir thoroughly after each rain. The *chiang* is ready for use in twenty days, but it will take one hundred days for full flavour to be developed.[60]

From this description we see that the process was divided into three stages (a) Preparation of soybean substrate (b) First Fermentation and (c) Second Fermentation. A flow diagram of the three stages is presented in Figure 78. The first stage is devoted to making the soybeans receptive to fungal growth. Unlike the *shih* fermentations discussed earlier, in this process the beans are dehulled as well as steamed. In fact, they have to be steamed three times before the seedcoats are sufficiently loosened for removal by pounding. The procedure is neither efficient nor effective. Hung Kuang-Chu comments that it would only work when the soybeans are freshly harvested. In his experience it would not work well with older (stored) beans.[61]

The first fermentation (stage b) is aimed at producing a suitable 'moulded bean' for use as the substrate in the second fermentation. This moulded bean was later given the name of *chiang huang* 醬黃 (yellow *chiang ferment*).[62] It is analogous to the *huang tzu* 黃子 (yellow seed *ferment*) produced in the first stage of the process for making *shih*. Inoculation of the incubation mixture with two kinds of *ferments* (*huang chêng* and common wine *ferment*) was a useful procedure. It should greatly hasten the rate of microbial growth. Similarly, the wheat flour present in the two *chhü* inoculants for the incubation mixture would enhance the amounts of nutrients available to support microbial growth. In the second fermentation (stage c) the high salt concentration probably inhibits fungal growth, but it is not expected to affect seriously the ability of the fungal enzymes in the *chiang*-substrate to hydrolyse the proteins and carbohydrates present *in situ* to give a savoury, pasty product.

When this process is compared with that described for the making of *shih* (see pages 338–41) we can see several differences. Firstly, the soybeans are dehulled as well as cooked by steaming whereas in the *shih* process they are only steamed. This should make them more receptive to invasion by growing fungi. But the method of dehulling is cumbersome and time consuming. There are surely more efficient ways of achieving the same end.[63] Secondly, the first fermentation is carried out under near-anaerobic conditions, whereas in the *shih* process the conditions are aerobic. By fermenting the beans in a sealed jar filled to the top, oxygen would soon be depleted and growth would be slow. It is no wonder that, even with a heavy dose of inoculum and the availability of wheat flour as additional nutrient, the fermentation took as long as twenty to thirty days. The choice of an anaerobic condition for

---

[60] The author comments: 'If a pregnant woman who may spoil the *chiang* is present, place leaves of jujubes, *Ziziphus spinosus*, in the jar and the *chiang* will recover. Some people use "filial stick" to stir the material, or roast the jar. The *chiang* is saved, but the foetus may be lost.'
[61] Hung Kuang-Chu (*1984a*), p. 96.
[62] For references to *chiang huang* see *TKKW*, ch. 17, p. 286, and *HYL*, p. 7. Consult also Miao Chhi-Yü ed. (*1982*), p. 427, note 11.
[63] The steaming, sunning and dehulling takes four days. Compare, for example, with the way soybeans are dehulled in the traditional *tempeh* process, which would take less than a day.

1. Preparation of substrate

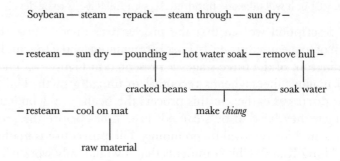

2. First stage fermentation

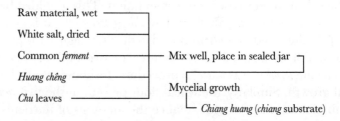

3. Second stage fermentation

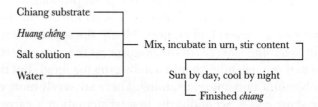

Fig 78. Flow diagram for process of making *chiang* (fermented soy paste) in the *Chhi Min Yao Shu*.

the first fermentation is curious since all the other moulded inoculants described in the *Chhi Min Yao Shu*, such as the *ferments* used in the making of wines (*chhü*) and fermented foods (*huang i, huang chêng* and *nü chhü*),[64] as well as the yellow seed *ferment* (*huang tzu*) are incubated as a surface culture exposed to air. Thirdly, in the second fermentation the incubating mixture is submerged in water and stirred from time to time whereas in the *shih* process little additional water, other than that resident

---

[64] The use of *huang i* and *huang chêng* for making various meat and fish *chiang* is described in *CMYS*, ch. 70; pp. 420–1. These processes will be discussed in Chapter (*e*).

within the beans, is provided and the mash is incubated without stirring. Thus, in the *chiang* process there is plenty of water and mechanical agitation to digest and break the beans down to a paste, while in the *shih* process there is only enough water to cause a limited amount of hydrolysis and no forced mixing to injure the structure of beans. Lastly, it is difficult to see what useful purpose is served by adding a substantial amount of *ferment* to the second fermentation in both the *chiang* process (as *huang chêng*, yellow mould *ferment*, p. 348) and the *yen shih* process (as *nü chhü*, glutinous rice *ferment*, p. 339). There is already an ample supply of enzymes, spores and myceliae in the substrate itself, i.e. *chiang huang* or *huang tzu*. In any case, with the high salt concentration in the aqueous medium, it is doubtful that the spores could germinate and the myceliae proliferate to a significant extent. Hydrolysis of the proteins and carbohydrates in each case would have to rely largely on the action of the enzymes already resident in the substrate.

It is natural that much of the later work would be aimed at correcting the major deficiencies noted above. We find that significant changes were already in place in the recipe for making *chiang* recorded in the *Ssu Shih Tsuan Yao* (Important Rules for the Four Seasons) of Han O, a late Thang work. A method for making *chiang* in ten days is described as follows:[65]

Chiang Substrate. Take one *tou* of yellow soybeans, clean three times (to remove extraneous matter), soak in water, drain off water, steam thoroughly until the beans are tender (*lan* 爛) and collect on a flat surface. Mix the beans with two *tou* and five *shêng* of wheat flour. Be sure all the beans are individually coated with the flour. Steam again until the flour is cooked. Spread and allow to cool to human body temperature. Cover the ground evenly with cereal leaves. Spread the beans on top, and cover them with another layer of leaves. Incubate for three to four days. The beans are wrapped around with a dense yellow coat. Dry them in the sun and store the finished *chiang* substrate.

Chiang Incubation. When it is time to make *chiang*, get ready one *tou* of water for every *tou* of the *chiang* substrate used. Dissolve five *shêng* of table salt in water at body temperature. Mix the filtered salt solution with the *chiang* ferment in a jar. Seal tightly. After seven days stir the contents. Place three ounces of Han pepper in a cloth bag, and hang the bag in the jar. Add one catty of cold, cooked [edible] oil and ten catties of wine. After ten days the *chiang* will be ready.

Although the description is rather sketchy, we can see that the process has been simplified and streamlined. It now follows the same general scheme that had been developed for the brewing of *chiu* (wine) and the making of *shih* (fermented soybeans). As is the case with the *chiu* and *shih* processes, the first stage involves an aerobic surface culture, and the second, a mildly anaerobic submerged incubation. These changes should make the process more efficient. Even without the use of preformed inoculum, the first stage now takes only four days, instead of thirty days. The second fermentation takes only ten days whereas the old method requires more than twenty days. Although there is no indication that the beans are dehulled

---

[65] *Ssu Shih Tsuan Yao*, Miao Chhi-Yü ed. (*1981*), 7th month, item 52, p. 185.

before use, the text says that they are steamed until *lan*, i.e. soft, ripe and tender. This means the internal structure of the beans is already damaged; they are thus rendered easily susceptible to invasion by proliferating fungal myceliae.

The next pertinent description of *chiang* production appeared about four hundred years later in the *Nung Sang I Shih Tsho Yao* (+1314). We shall quote the recipe for making *chiang* in full:[66]

Method for fermenting Chiang

Take one *tan* (100 catties) of soybeans. Stir-fry (*chhao* 炒) until cooked, mill to remove the hulls, boil until soft, and strain off the water. Mix the beans evenly, while still hot, with sixty catties of white wheat flour. Line a low table fully with bamboo leaves. Spread the bean-flour mixture on the leaves to a height equivalent to the thickness of two fingers. When cool, cover with mulberry leaves or cocklebur. Wait until a yellow coat forms around the particles. Remove the leaves and cool for a day. On the next day sun the material [*chiang* substrate] until dry. Break up any chunks and winnow until clean.

Blend (the *chiang* substrate) with about forty catties of salt and two *tan* of rain water. If the mash is too thin, stir-fry some white wheat flour, cool, and add an appropriate amount (to get the desired consistency). If it is too thick, heat licorice in a salt solution. When it is cool, mix the solution with the mash as needed (to achieve the right consistency). On the night of a 'fire' day, load the ingredients (in an urn) for incubation with the aid of light from a lamp. This will discourage the development of grubs. Add cumin, anise, licorice, scallion and pepper. The product will be fragrant and rich in flavour.

We may note a couple of innovations introduced in this process. Firstly, the beans are stir-fried and milled to remove the hull. This should facilitate microbial growth during the first fermentation. Secondly, spices are added to the second fermentation to enhance the flavour of the finished *chiang*. But the author gives no detailed instructions on how the second fermentation should be carried out. We can only surmise that the conditions are similar to those described in the *Chhi Min Yao Shu* and in the *Ssu Shih Tsuan Yao* cited earlier. Based on these assumptions, we can construct a flow diagram of the process as shown in Figure 79.

Variations of this process are shown in many recipes for making *chiang* given in the Yuan, Ming and Chhing literature. The anonymous late Yuan work *Chü Chia Pi Yung* (Essential Arts for Family Living) presents two methods for making *chiang* from soybeans, one for *shu huang chiang* 熟黃醬 (ripe yellow soy paste) and the other for *shêng huang chiang* 生黃醬 (raw yellow soy paste).[67] The names are actually misleading, since the 'ripe' and the 'raw' refer to the manner in which the beans are treated before mixing with wheat flour for the first fermentation and do not refer to the quality of the final product. In the *shu* or 'ripe' case, the beans are roasted and ground into flour; in the *shêng* or 'raw' case the beans are soaked overnight and boiled until soft and tender. The product from the first fermentation is called *huang tzu* 黃子 as well as *chiang huang* 醬黃. Similar methods with minor variations can be

---

[66] *NSISTY*, 6th month, p. 91. A 'fire' is indicated as a 'hot' day on the calendar.   [67] *CCPY*, pp. 54, 56.

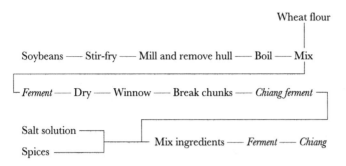

Fig 79. Flow diagram of process for making *chiang* (fermented soy paste) in the *Nung San I Shih Tso Yao*.

found in the *Shih Lin Kuang Chi* 事林廣記 (Record of Miscellanies), +1289;[68] *I Ya I I* (Remnant Notions from I Ya), late Yuan;[69] *To Nêng Pi Shih* 多能鄙事 (Routine Chores Made Easy), +1370;[70] *Chhü Hsien Shên Yin Shu* 臞仙神隱書 (Book of the Slender Adept), c. +1400;[71] *Sung Shih Tsun Shêng* 宋氏尊生 (Sung Family's Guide to Living), +1504;[72] and *Pên Tshao Kang Mu* (The Great Pharmacopeia), +1596.[73] The beans are prepared for fermentation in several ways, roasted and dehulled, dried and dehulled or simply cooked by steaming or boiling until soft and tender. The important thing is that they are more thoroughly cooked, and hence more susceptible to invasion by fungi, than those used in the making of *shih*. It is curious that in all the above cases, the first fermentation is always inoculated by exposing the solid medium to wild airborne organisms when preformed inoculum is readily available. In spite of the fact that the earliest description of a *chiang* process, that recorded in the *Chhi Min Yao Shu*, employs a preformed inoculum in making *chiang huang* (yellow *chiang* substrate), Chinese technologists have, in general, shied away from using a preformed inoculum in making their *ferments* (or starter cultures). They seem to feel that there is a mystique about wild organisms which is missing in preformed cultures. It is only in the twentieth century that manufacturers of fermented soyfoods in China have begun to realise that inoculation with preformed cultures can greatly increase the efficiency of their processes and the quality of their products.[74]

The *Huo Chih Chuan* (Economic Affairs) chapter of the *Han Shu* (History of the former Han Dynasty) states that thousands of jars of *hsi chiang* 醯醬 were traded in the market, and that a merchant had become wealthy by selling *chiang*. Thus *chiang* was undoubtedly an important commodity in Han economy.[75] It was also used as

---

[68] *SLKC*, ch. 4, Shinoda & Seiichi ed. (*1973*), p. 267. [69] *I Ya I I*, p. 9.
[70] *TNPS*, Shinoda & Seiichi (1973), p. 372. [71] *Chhü Hsien Shên Yin Shu*, Shinoda & Seiichi, *ibid.* pp. 428–9.
[72] *Sung Shih Tsun Shêng*, Shinoda & Seiichi *ibid.*, p. 467.
[73] *PTKM*, ch. 25, p. 1552. [74] Chhên Thao-Shêng (*1979*), pp. 77–81.
[75] *Han Shu*, ch. 1, *Huo Chi Chuan*, p. 3687, states that among the commodities traded were thousand jars of *hsi chiang*, which can mean *hsi* and *chiang* (vinegar and sauce) as two products or *hsi chiang* (a tart sauce) as a single product. The latter view is favoured by Hayashi Minao (*1975*), pp. 58–60. On the other hand, according to the *Huo Chi Chuan*, p. 3694, a Mr Chang had become wealthy by trading in *chiang*, indicating that *chiang* was an important economic commodity in its own right.

medicine. It is listed as a lower-class drug in the *Min I Pieh Lu*, c. +510, which says '*Chiang* is salty, sour, and cold. It dispels heat, restrains anxiety, and counteracts toxic side effects of drugs, heated soups and fire.'[76] The *Hsin Hsiu Pên Tshao* (+659) states, '*Chiang* is usually made from soybeans. Lesser amounts are made from wheat. Some are also made from meat and fish; these are called *hai* and are not used in medicine. Others are made from the seeds of the Chinese elm and the fetid elm.'[77] But *chiang* was valued mainly as an agent which enhances and harmonises the flavour of foods. Yen Shih-Ku (7th century) has stated that, '*chiang* is to food, what a general is to an army'.[78] Thao Ku in the *Chhing I Lu*, +960, says, 'Soy paste, *chiang* 醬, is the host of the eight delicacies (*pa chên* 八珍), while vinegar, *chhu* 醋, is their manager,'[79] thus placing the value of *chiang* as a condiment higher than that of vinegar. Nevertheless, we have little specific information on *chiang*'s role in cookery through the Wei, Chin, and Thang periods. The *Chhi Min Yao Shu* mentions a fragrant *chiang* as an ingredient in cooking a melon.[80] A *chiang* is mentioned in Hsieh Fêng's *Shih Ching* 食經 (Food Canon), c. +610.[81] At least nine recipes in which *chiang* is used, including the celebrated *Tung Pho Tou Fu* (*tou fu* a la Su Tung Pho) are presented in the *Shan Chia Chhing Kung* 山家清供 (Basic Needs for Rustic Living), Southern Sung.[82] The *Mêng Liang Lu* (Dreams of the Former Capital), +1334, lists a five flavour crab in wine and *chiang*.[83] Although the references to *chiang* that have survived are by no means plentiful it is clear that *chiang* was a ubiquitous ingredient in the Chinese culinary system from the Han to the Sung. Indeed, in another passage, the *Mêng Liang Lu* states, 'the things that the people cannot do without every day are firewood, rice, oil, salt, *chiang*, vinegar and tea'.[84] Thus, *chiang* was regarded as one of the proverbial 'seven necessities of life' in the welfare of the Sung people, and its importance would remain undiminished for several centuries to come.

Actually, we find that since the declining days of the Southern Sung *chiang* has been mentioned with increasing frequency in the culinary literature. The *Wu Shih Chung Khuei Lu* (Madam Wu's Recipe Book), Southern Sung, gives five recipes in

---

[76] *MIPL*, p. 314.
[77] *HHPT*, p. 292. According to Tshao Yüan-Yü (*1987*), p. 226 and 245 the Chinese elm is *Ulmus campestris* Sm. var.laevis. planch, and the fetid elm *Ulmus macrocarpa* Hance, cf. also Porter-Smith & Stewart pp. 448–9.
[78] Commentary on *chiang* in *Chi Chiu Pien*, p. 31.
[79] *Chhing I Lu*, p. 28. The eight delicacies are described in the *Li Chi, Nei Tsê*; cf. Legge (1885a), pp. 468–70. The term is now used to denote the most highly regarded delicacies of the realm.
[80] *CMYS*, ch. 87, p. 529.
[81] *Shih Ching* (Food Canon) by Hsieh Fêng, quoted in *Chhing I Lu*, p. 15. Cf. also Shinoda & Seiichi (*1973*), p. 116.
[82] *SCCK*, pp. 38, 48, 55, 73, 77, 81, 94, 97, 104 in recipes for rabbit, fish, fowl, beans and vegetables. The exact date of the author, Lin Hung, is unknown, except that he lived during the Southern Sung.
[83] *MLL*, p. 133.
[84] *Ibid.*, p. 139, tr. Freeman (1977), p. 151, mod. auct. Freeman translates *chiang* as soybean sauce, which is probably incorrect, and for our purpose misleading. As we shall see later, *chiang*, in its own right, was more important in Sung cookery than soy sauce. We have preferred to keep it untranslated. According to the *Thung Su Pien* 通俗編 (Origin of Common Expressions), +1751, p. 598, the original text in the *Mêng Liang Lu* read, 'firewood, rice, oil, salt, wine (*chiu*), *chiang*, vinegar and tea'. When the list was popularised in Yuan musical plays, which allowed only seven syllables in each line of the lyric, wine was discarded leaving us with the now proverbial seven necessities.

which *chiang* is used as a condiment,[85] but in a slightly later compilation, the *Chü Chia Pi Yung* (Compendium of Essential Arts for Family Living) of the Yuan period,[86] the number of such recipes has grown to no less than twenty. *Chiang* is mentioned often in such standard works on the culinary arts as the *I Ya II* (Remnant Notions of I Ya), Yuan/Ming,[87] the *Yin Chuan Fu Shih Chien* (Compendium of Food and Drink), +1591,[88] the *Shih Hsien Hung Mi* (Guide to Mysteries of Cookery), c. +1680,[89] the *Hsing Yüan Lu* (Memoir from the Garden of Awareness), +1750,[90] and the *Sui Yüan Shih Tan* (Recipes from the *Sui* Garden), +1790,[91] which probably represents the height of Chinese gastronomy in the premodern age.

Although soy *chiang* has been the preeminent *chiang* condiment in Chinese cookery, *chiang* made from other plant materials have also achieved a degree of recognition. As we have noted earlier, the *Hsin Hsiu Pên Tshao* tells us that *chiang* was also made from wheat and elm nuts. These, too, have found their way into standard cookery recipes. The earliest record on the making of wheat (*mai*) *chiang* and elm nut (*yü jên* 榆仁) *chiang* are found in the *Chhi Min Yao Shu* (+540). We quote:[92]

Wheat Chiang: Soak one *tan* of wheat grain overnight. Cook through and incubate [as described earlier] to convert the wheat into *huang i* (yellow coat *ferment*). Heat to dissolve three *shêng* of table salt in one *tan* and six *tou* (i.e. sixteen *tou*) of water. Transfer eight *tou* of clear salt solution into a jar. Mix the yellow coat *ferment* evenly with the water and incubate in the sun. After ten days the *chiang* will be ready.

Elm Nut Chiang: Take one *shêng* of elm nuts and grind and sift the flour. Mix it evenly with one *shêng* of clear wine and five *shêng* of *chiang*. Incubate for a month and the *chiang* will be ready.

In the latter case elm-nut flour is the substrate, *chiang* (presumably soy *chiang*) is the *ferment* and wine takes the place of water. Since the incubation contains much more *chiang* than substrate, what we end up with, in effect, is a *chiang* with an elm-nut flavour. Unfortunately, there is a gap of seven hundred years before we see the next pieces of relevant information. The preparation of *mien chiang* (wheat flour *chiang*) is described in the *Shih Lin Kuang Chi* of +1280,[93] and recipes for making *chiang* from wheat flour (*mien* 麵), barley, lesser bean (*hsiao tou* 小豆), broad bean (*tao tou* 刀豆) and elm nuts (*yü jên* 榆仁) are found in the *Chü Chia Pi Yung*, c +1350.[94] The *Yin Shan*

---

[85] WSCKL, p. 7 (fish), 11 (meat), 15 (melon), 16 (garlic sauce) and 21 (vegetable).
[86] CCPY, pp. 71, 77, 83 (mien), 84, 88, 88, 89, 96 (yü jên), 99, 100, 101, 102, 105, 106, 101, 107, 119, 121, 122, 132, 132, 133.
[87] I Ya II, pp. 20, 23, 30, 32, 41, 49, 50, 53.
[88] YCFSC, pp. 74, 77, 77, 81, 84, 85, 97, 98, 103, 114, 115, 119, 120, 149, 150.
[89] SHHM, pp. 50, 70, 93, 102, 126, 145, 157; also for sweet *chiang*, pp. 72, 107, 120, 131, 131, 163.
[90] HYL, pp. 21, 28, 31, 37, also sweet *chiang*, pp. 21, 26.
[91] SYST, pp. 53, 58, 62, 76, 77, 86, 87, 91, 92, 94, 104, 106, 108, 120. See also *thien chiang* (sweet *chiang*), pp. 47, 52, 107, 107, 115 and *mien chiang* (wheat *chiang*), pp. 55, 59. For an appreciation of Yüan Mei's contribution to Chinese cuisine, see Lin & Lin (1969).
[92] CMYS, p. 421.
[93] SLKC, ch. 4, Shinoda & Tanaka (1972), p. 267. The recipe given is actually similar to that for making *tou chiang*. It is possible that the author did not have a clear idea of the difference between *mien chiang* and *tou chiang*.
[94] CCPY, pp. 55–6.

Table 32. *Processes for making various types of chiang*

Type of chiang	Soybean *tou* 豆	Wheat *mien* 麵	Sweet flour *thien mien* 甜麵	Lesser bean *hsiao tou* 小豆	Broadbean *tao tou* 刀豆	Barley *ta mai* 大麥	Elm Nut *yü jen* 榆仁
Date Reference	13th century *CCPY*	13th century *CCPY*	16th century *PTKM*	13th century *CCPY*	13th century *CCPY*	13th century *CCPY*	13th century *CCPY*
First fermentation	Growth of airborne fungi						
Primary substrate	Bean, whole or ground	Grain, or flour	Flour	Bean, ground	Bean, whole	Grain, whole	Dehulled nuts and wheat *ferment*
Secondary substrate	Flour	—	—	—	Flour	Flour	
Product	Bean *ferment*	Wheat *ferment*	Wheat *ferment*	Wheat *ferment*	Bean *ferment*	Barley *ferment*	
Second fermentation	Digestion of proteins and carbohydrates						
Substrate	Bean *ferment*	Wheat *ferment*	Wheat *ferment*	Lesser bean *ferment*	Bean *ferment*	Barley *ferment*	
Additive	Table salt Water	Table salt Water	Table salt Water	Table salt Water	Table salt Water	Table salt Water	Table salt Water
Ratio salt: substrate	1:4	1:4	3:10	1:3.2	1:4	1:5	1:5

CCPY   *Chü Chia Pi Yung*.
PTKM   *Pêng T'shao Kang Mu*.

Chêng Yao, +1330, says that in countering the toxicity of foods soy *chiang* is superior to *mien chiang*,[95] indicating that the name *mien chiang* 麵醬 had supplanted the *mai chiang* 麥醬 of the *Chhi Min Yao Shu*.

Similar recipes are reproduced, often verbatim, in the *To Nêng Pi Shih*, +1370,[96] *Chhü Hsien Shên Yin Shu*, +1400,[97] and *Sung Shih Tsun Shêng*, +1504.[98] They are incorporated in the *Pên Tshao Kang Mu*, +1596,[99] which, for the first time introduces us to the *thien mien chiang* 甜麵醬 (sweet wheat flour *chiang*). These processes are summarised in Table 32. In general, the raw materials are either cooked thoroughly or ground into flour, made into dough, steamed and cut into slices. Sometimes they are mixed with wheat flour before further processing. In the first fermentation the substrate is exposed to airborne fungi until good mycelial growth is obtained. In the second fermentation, the partly fermented substrate is allowed to undergo enzymatic hydrolysis in a strong aqueous salt solution under the same conditions described in the *Chhi Min Yao Shu* for making *chiang*. The product usually has the consistency of a thick paste.

Table 32 shows that *chiang* can be made from a variety of legumes and cereals. It is curious that while *mien chiang* and *thien mien chiang* are listed as separate products, they are made practically by the same procedure.[100] Sometimes even the demarcation between *tou chiang* and *mien chiang* is blurred.[101] One gets the impression that several of the authors of the standard culinary works had no practical experience in what they were writing about. Often they were simply copying blindly from a previous work. In any case, both *mien chiang* (fermented wheat paste) and *thien mien chiang* (fermented sweet flour paste, sometimes called *thien chiang* 甜醬) are listed as condiments in the cooking recipes of the Yuan, Ming and Chhing periods.[102] There are few references to the use of *chiang* from the lesser bean, broad beans, barley or elm nut in the food literature and we would have to assume that their role was minor. But while soy *chiang* was undoubtedly the principal savoury condiment in Chinese cookery from the Han–Wei period until the early Chhing, it has ceased to be so since the middle of the 18th century. This honour now belongs to soy sauce, *chiang yu* 醬油, which is the last member of the triad of Chinese soy condiments and the topic of our next discussion.

---

[95] *YSCY*, p. 121.   [96] *TNPS*, Shinoda & Tanaka (*1972*), p. 372.
[97] *Chhü Hsien Shên Yin Shu*, Shinoda & Tanaka, *ibid*. p. 429.
[98] *Sung Shih Tsun Shêng*, Shinoda and Tanaka, *ibid*. p. 469.   [99] *PTKM*, ch. 25, p. 1552.
[100] Indeed, the *PTKM* (*ibid*.) describes the preparation of both *thien mien chiang* (sweet flour *chiang*) and *mien chiang* (wheat flour *chiang*) on the same page. The two procedures are virtually the same except in the way the flour is treated before it is exposed to airborne fungal spores. For *thien mien chiang* the flour is made into dough, cut into slices and steamed; for wheat *mien chiang* the flour is mixed with water and shaped into cakes. But the reader would be misled to think that the former is always made with cooked flour, and the latter with raw flour. In the recipe for making wheat *mien chiang* from the *Chü Chia Pi Yung*, the flour is made into dough, cut into slices and steamed for the first fermentation. It would seem that the name of the product depends on the whim of the reporter.
[101] In the *SLKC*, p. 267, the method described for making *mien chiang* is practically the same as that recorded in the *NSISTY* for making *chiang* (i.e. soy *chiang*), quoted on p. 352. We suspect that *mien chiang* of the premodern era was often made with soybean as a secondary substrate, cf. *SHHM*, p. 50, recipe for making sweet *chiang*.
[102] For recipes using *mien chiang* see: *CCPY*, pp. 83, 96, 101, 123; *TNPS*, p. 385; *SHHM*, pp. 162, 163; *HYL*, pp. 21, 24; and *SYST*, pp. 55, 59. For *thien chiang* or *thien mien chiang* see: *YCFSC*, p. 81; *SHHM*, pp. 72, 107, 120, 131, 131, 163; *YHL*, pp. 83, 83; *HYL*, pp. 21, 26; and *SYST*, pp. 47, 52, 107, 107, 115.

(iv) *Fermented soy sauce*, chiang yu

Soy sauce, *chiang yu* 醬油 (or *shih yu* 豉油), has been called 'the most important flavourer of Chinese food'.[103] It is also the mainstay of the cuisines of Japan, Korea and Vietnam and is, without doubt, the best known and most widely used processed Chinese soyfood in the West today. At first glance it would seem obvious that soy sauce must have evolved from *chiang* or *shih*, and yet, in spite of all that we know about the origin of *chiang* and *shih*, we do not have a firm idea of when or how soy sauce was first prepared and utilised.

The earliest references to the term *chiang yu* occur in two late Sung works. Lin Hung's *Shan Chia Chhing Kung* (Basic Provisions for Rustic Living) gives four recipes in which *chiang yu* is used to flavour various vegetables and seafood.[104] Madam Wu's Recipe Book (*Wu Shih Chung Khuei Lu*) describes the use of *chiang yu* in cooking meat, crab and vegetables.[105] Neither reference tells us how *chiang yu* was prepared. This information does not necessarily mean that soy sauce first came into existence during the Sung. It means rather that by the time of the Sung, *chiang yu* had become the accepted name for the liquid condiment derived from *shih* or *chiang*. Most scholars would agree that soy sauce as an entity is almost as old as soy *chiang* itself but opinions differ as to what it was called before it became known as *chiang yu* during the Sung.[106]

In fact, it is surprising that the sauce made from *chiang* should have been called *chiang yu* 醬油 at all. Throughout the history of the Chinese language *yu* 油 has always meant the oily (or greasy), water insoluble substance derived from animal, vegetable or mineral sources, for example, *chu yu* 豬油 (lard), *niu yu* 牛油 (butter), *chhai yu* 菜油 (rape seed oil), *hsiang yu* 香油 (sesame oil), *thung yu* 桐油 (tung oil), *shih yu* 石油 (petroleum) etc. *Chiang yu* (or *shih yu* 豉油) is the one, notable exception. It is an aqueous solution (or suspension) of a variety of substances, and by no stretch of any imagination can it be construed as a *yu* (oil).[107] Nevertheless, the term *chiang yu* gained acceptance in the Sung and is now firmly ensconced in the language, perhaps an indication of its very special role in the daily life of the Chinese. Thus,

---

[103] Chao, Buwei Yang (1963), p. 27. She goes on to say, 'With soy sauce you can cook an untiring series of Chinese dishes with nothing but those foods that you can get at any American chain market.'

[104] *SCCK*; p. 34, soy sauce, ginger threads and a little vinegar used as a dressing for lightly cooked young chives; p. 47, stir fry of spring bamboo shoots, bracken leaves and fish or shrimp seasoned with soy sauce, sesame oil, salt and pepper; p. 66, soy sauce used with sesame oil, pepper and salt to flavour young bamboo shoots, mushrooms and Chinese wolfberry seeds; p. 98, soy sauce and vinegar used to season blanched shoots of the daylily.

[105] *WSCKL*; p. 8, soy sauce used to marinade sliced raw meat before stir-frying; p. 10, crab cooked in wine, soy sauce, vinegar, wine fermentation lees and sesame oil; p. 21, soy sauce, sesame oil, vinegar and pepper mixed as a dressing for various vegetables.

[106] See for example, Hung Kuang-Chu (*1984a*), p. 94; Chang Lien-Ming (*1987a,b*); Kuo Po-Nan (*1989*).

[107] The *Yung-Lo Ta Tien, I Hsüeh Chi* 永樂大典醫學集, +1408, *1986*, pp. 618–19 lists twenty-six different kinds of oils *yu*. *Chiang yu* 醬油 is not mentioned. However, the example of the term *chiang yu* as soy sauce has encouraged the use of the word *yu* to denote other culinary sauces, e.g. *mai yu* 麥油, a water-based sauce made by fermenting wheat, *HYL*, +1750, pp. 8–9, *sun yu* 筍油, bamboo shoot sauce, and *hsia yu* 蝦油, shrimp sauce, *SYST*, +1790, p. 101. A more recent and relevant example is oyster sauce, *hao yu* 蠔油 (*hor yao* in Cantonese), which is a popular condiment in Cantonese cookery.

we have a surprising situation in which an English translation, 'soy sauce' (sauce prepared from soybean), is a more rational name for a product than its Chinese original, *chiang yu* (oil derived from *chiang*).

There are several products associated with *chiang* and *shih* mentioned in the literature that could be construed as ancient precursors of soy sauce, that is, the *chiang yu* popularised during the Sung. The earliest product in this collection is the *chhing chiang* 清醬 listed in the *Ssu Min Yüeh Ling* (Monthly Ordinances for the Four Peoples), +40, that we have already touched on earlier (page 347, note 53). Several authors are of the opinion that *chhing chiang* is an ancient name for soy sauce.[108] Taken literally *chhing chiang* means 'clarified *chiang*', which is a fitting description of soy sauce. On the other hand, in the passage cited *chhing chiang* is preceded by *yü chiang* 魚醬 (fish paste) and *ju chiang* 肉醬 (meat paste); thus, one can hardly be certain that *chhing chiang* actually refers to clarified soy paste, and not simply another type of savoury paste.[109]

An unexpected piece of evidence favouring the former view has come from the *Wu Shih Erh Ping Fang* (before −168), in which there is a reference to the use of a *shu chiang chih tzu* 菽醬之滓 (dregs from soy *chiang*) in a salve for treating haemorrhoids.[110] If there had been a fraction of dregs, it would surely have to be balanced against a fraction of clarified liquid, from which the dregs were separated. In other words, the very existence of dregs implies the existence of a clarified sauce, which would appropriately be called *chhing chiang*.

The *Chhi Min Yao Shu*, +540, lists in its food processing chapters three seasoning agents that have been considered as potential precursors of soy sauce, *chiang chhing* 醬清, *shih chih* 豉汁, and *shih chhing* 豉清. *Chiang chhing* is *chhing chiang* written backwards and presumably means the same thing. It is used as a condiment in five recipes.[111] Very much more popular is *shih chih* (aqueous extract of fermented soybeans); it occurs in at least twenty-six recipes.[112] Finally, there is *shih chhing* 豉清 (clarified fermented soybeans) which is used in three recipes.[113] Unfortunately, the *Chhi Min Yao Shu* gives no indication on how *chiang chhing* and *shih chhing* were prepared, although in the recipe for a fish and vegetable stew there is, luckily for us, a brief statement on how *shih chih* was obtained by boiling *shih* in water.[114]

---

[108] Hsü Cho-Yün (1980), p. 217; Hung Kuang-Chu (1984a), p. 94; Wang Shan-Tien (1987), p. 468 have interpreted *chhing chiang* as soy sauce.

[109] Miao Chhi-Yü ed. (1981), p. 24. Kuo Po-Nan (1989), considers *chhing chiang* to be a thin soy sauce, without the flavour of true soy sauce.

[110] *Wu Shih Erh Ping Fang*, Prescription 143, cf. Harper (1982), p. 412. There is no indication of how this soy paste, *shu chiang* 菽醬, was prepared or how the dregs were separated.

[111] *CMYS*; *chiang chhing*: ch. 70, p. 422, 422; ch. 76, p. 464; ch. 77, p. 479; ch. 87, p. 529.

[112] *CMYS*; *shih chih*: ch. 76, pp. 463, 463, 463, 464, 464, 464, 465, 466, 466, 467; ch. 77, pp. 478, 478, 478, 479, 479, 480, 480; ch. 80, pp. 494, 494, 494, 496, 496, 496; ch. 82, p. 509; ch. 87, p. 528; ch. 88, p. 532.

[113] *CMYS*; *shih chhing*: ch. 76, pp. 466, 467; ch. 77, p. 478.

[114] *CMYS*; ch. 76, p. 465. The text states, '*Shih chih* 豉汁: Cook [*shih*] in a separate pot of water. Allow it to boil once. Strain off the *shih* beans, and after the soup settles decant the clear solution. Do not stir the water during cooking. Stirring will muddy the decoction. It will not clarify after straining. When cooking *shih chih* stop the process as soon as the water reaches the light brown colour of amber. Do not let it become too dark. Then the juice will be bitter.'

Shih chih was apparently a well-known flavouring agent during the Han. It is referred to, at least 350 years before the *Chhi Min Yao Shu*, in the *Shih Ming* (Exposition of Names), +2nd century, which states:[115] 'Roast-dry meat: marinade meat in sugar, honey and *shih chih* (fermented soybean juice) and roast it until it has the consistency of dried meat.'

Moreover, according to Hung Kuang-Chu and Kuo Po-Nan the preparation of *shih chih* by percolating soup over *shih* is intimated in the famous poem by Tshao Chih (c. +220) on cooking beans which we have alluded to earlier (A-2 p. xx).[116] The first two lines of the original version may be paraphrased thus:[117]

> The soup obtained by cooking beans in water
> When strained over *shih* (fermented soybeans) gives *chih* (juice).

We may call this the percolation method for making *shih chih*. It is somewhat more sophisticated than the direct boiling method described in the *Chhi Min Yao Shu*. In any case, we suspect the juice made by either method would probably have the consistency of a soup rather than a sauce. It could not have had a long shelf life. Thus, in the recipes from the *Chhi Min Yao Shu*, it is probable that *shih chih* had to be made fresh each time it was used. And yet, the percolation method was still in use more than a thousand years after Tshao Chih's poem. The very same method is recounted in the *Chü Chia Pi Yung* (Yuan) and in the *Pên Tshao Kang Mu* (Ming).[118] Taken altogether the evidence tends to support the view that *shih chih* was an aqueous decoction or percolation of *shih*.[119] It could not have been the precursor or prototype of *chiang yu* or soy sauce that we are looking for.

But this is not to be the end of the story. *Shih chih* is mentioned in the Thang work *Shih Liao Pên Tshao* (Nutritional Therapy), +670, which comments on the excellence of the product from the *Shan* prefecture and describes briefly how it was made:[120]

The fermented soybean juice (*shih chih* 豉汁) from *Shan* 陝 prefecture is superior to ordinary *shih*. To make it, ferment soybeans to yellow mould *ferment* (*huang chêng* 黃蒸) stage. For each *tou* [of ferment] add four pints of salt, four ounces of pepper [and incubate in water]. It will

---

[115] *Shih Ming*, ch. 13. [116] Hung Kuang-Chu (*1982*), and in Kuo Po-Nan, (*1989*).
[117] *Shih Shuo Hsin Yü*, p. 60. tr. auct. The original reads:

Chu tou chhih tso kêng   煮豆持作羹
Lu shih i wei chih.   漉豉以為汁

[118] *CCPY*, p. 60. The text reads, 'Chengtu Method for making *shih chih*: it should be carried out after the 9th month but before the 2nd month. Take 3 pecks (*tou*) of good quality soy nuggets. Heat a batch of clear sesame oil until it is well done and smoke free. Disperse a pint of cooked oil in the *shih*, steam in a steamer, spread the nuggets out to cool and let dry in the sun. This steaming and drying is repeated three times. Mix the nuggets evenly with a peck of table salt, and break up any large lumps [that may form]. Percolate through three to four pecks of soup. The filtrate is placed in a clean pot, seasoned with ground Szechuan pepper, pepper, ginger, strips of orange peel (one ounce of each) and scallion, and boiled until one third of the volume is lost. Store in a waterproof porcelain jar. Be sure to use a clear sesame oil. The product has excellent fragrance and flavour.' This recipe is copied virtually verbatim in the *PTKM*, p. 1527.
[119] Hung Kuang-Chu (1982), p. 99 points out that in North China soy sauce is sometimes called 'ling yu', a juice obtained by straining [water over *shih*].
[120] *SLPT*, p. 118. The words in brackets, though not in the original, are added to preserve the sense of the text in translation.

be half done in three days in spring, two days in summer and five days in winter. Add five ounces of raw ginger and it will have a clean and delicate flavour.

This sketchy statement is the only description we have on how *shih chih* was prepared during the Thang. There are two stages. In the first stage the soybeans are said to be fermented into *huang chêng* 黃蒸 (yellow mould *ferment*), which normally denotes, as we have seen on p. 336, the loose *ferment* prepared from wheat. In this case, *huang chêng* could well be a copying error for *huang tzu* 黃子, the substrate used in the making of *shih*. The second stage fits that of a process for making *shih*. But the description is woefully inadequate. If enough water were present, one could visualise, at the end of the incubation the separation of a liquid condiment which could have been called *shih chih*. Such a liquid would be the direct result of a fermentation. It would be quite different from the product obtained by extracting preformed *shih* with water or soup. Furthermore, the *Pên Tshao Shih I* 本草拾遺 (A Supplement for the Pharmacopoeia), +739, notes that the *shih chih* from Shanchou can be kept for years without deterioration.[121] These references suggest that the *shih chih* of Shanchou, and perhaps *shih chih* during the Thang in general, was prepared by fermentation and that it had keeping qualities comparable to those of modern soy sauce. This fermentation-derived *shih chih* would be a fitting precursor to the *chiang yu* of late Sung.

The third product the *Chhi Min Yao Shu* mentions is *shih chhing* 豉清 (clarified fermented soybeans). Nothing is known about the way it was prepared; presumably it would bear the same relationship to *shih* as *chiang chhing* (or *chhing chiang*) is to *chiang*. In going over the recipes in the *Chhi Min Yao Shu* it is easy to get the impression that the three products are used interchangeably. Is it possible that they are merely different names for the same product? The answer is No. There are two cases in which both *chiang chhing* and *shih chih* are used in the same recipe, and another case in which both *shih chih* and *shih chhing* are used to cook the same dish.[122] These examples leave no doubt that these names do represent three separate entities.

The next piece of evidence we need to consider is found in a recipe for making *shih* in the *Ssu Shih Tsuan Yao*, late Thang, which says:[123]

To make salty *shih*: Clean one peck of black soybeans. Discard those with defects. Steam until soft and ripe. Ferment them into yellow coated beans in the usual way. Winnow off the yellow coats, wash with boiled water, and drain until dry. For each peck of beans, prepare five pints of salt, half a catty of ginger cut into thin strips, and one pint of clean pepper. Dissolve the salt in water; be sure the brine is at human body temperature. Place the beans in a jar, add a layer of pepper, and then a layer of ginger. Add the brine to a depth of five to seven inches above the beans. Cover with pepper leaves and seal the mouth of the jar with mud. After twenty-seven days, collect the beans and dry them in the sun. *The residual juice is boiled and stored separately*. It is particularly good for vegetarian fare.

---

[121] Quoted in *CLPT*, ch. 25, p. 17a; *PTPHCY*, +1505, p. 833, and *PTKM*, p. 1527.
[122] *CMYS*, for *chiang chhing* and *shih chih*, pp. 464, 479; for *shih chhing and shih chih*, p. 478.
[123] *Ssu Shih Tsuan Yao*, Miao Chhi-Yü ed. (1981), 6th month, item 41, p. 161.

This *residual juice* should have the properties we now attribute to soy sauce, *chiang yu*. Thus, this passage is the earliest description we have of how a prototype of soy sauce might have been prepared in the pre-Sung literature. Unfortunately, the *Ssu Shih Tsuan Yao* does not tell us what this residual juice was called. Therefore, we cannot relate it specifically to either *shih chih* or *shih chhing*. In a similar manner, a juice could also have separated from the mash in the *chiang* fermentation, which was probably how the *chhing chiang* or *chiang chhing* of antiquity was discovered.

From these accounts we may infer that prototypes of soy sauce were made from both *chiang* and *shih* and called by various names in early mediaeval China. They were liquids separated from *chiang* and *shih* when excess water was used in the respective second stage fermentations. Interest in these byproducts grew slowly but steadily.[124] By the time of the Sung, they had become important culinary ingredients in their own right, and were standardised under the collective name of *chiang yu* (soy sauce), since the product was usually made with *chiang* as the intermediate. We do not know why the term *yu* 油 was introduced to denote the product, in preference to the older names of *chih* 汁 and *chhing* 清.[125] Although the name *chiang yu* soon gained wide acceptance, the fact remains that soy sauce continued to be made for centuries as an extension of both the *chiang* or the *shih* processes. In most of China soy sauce is derived from *chiang*, but in the South, especially in Kwangtung and Fukien, it has always been and still is known as *shih yu* 豉油 in the local dialects, and some of the best soy sauces are produced commercially today as an extension of the process for making *shih*.[126]

The role of *shih* as a precursor to soy sauce is further emphasised in the very first specific reference we have to the making of *chiang yu*. It is found in the *Yün Lin Thang Yin Shih Chih Tu Chi* (Dietary System of the Cloud Forest Studio), +1360 by the famous Yuan painter Ni Tsan; it says:[127] 'For every official peck of yellow bean *ferment* (*huang tzu* 黃子), have ready ten catties of salt and twenty catties of water. On a *fu* 伏 day mix them [in a jar] and incubate.'

The description is too short on details to be of value as a guide for making soy sauce. The point of interest is that the starting material used is *huang tzu*, which usually denotes the substrate for making *shih*, and not *chiang huang*, the substrate for making *chiang*. Thus, we can say that even in Southern Kiangsu, during the Yuan,

---

[124] *Thang Shu, Pai Kuan Chi* 百官志 (List of Officials), says 'The Bureau of Fermentations employs 23 technicians for *chiang*, 12 for vinegar and 12 for *shih*.' This statement shows that at the end of the Thang, *chiang yu*, soy sauce, had not attained the status of an independent condiment which commands a separate crew for its production.

[125] Wang Shan-Tien (*1987*), p. 468, notes that during the Chhing Dynasty, *chiang yu* is often referred to as *chhing chiang*, e.g. *Shun Thien Fu Chi* 順天府志 (Gazetteer of the Shun-Thien Prefecture), Chhing, ch. 50, *Shih Huo Chi*, No. 2, item 36, says '*chhing chiang* is *chiang yu*'. In North China soy sauce is often called '*chhing chiang*' or simply '*chiang*'. In fact, in the recent past in Peking, many restaurants provided two small bottles of condiments on every table, one for vinegar and one for soy sauce. The soy sauce bottle would be marked *chhing chiang* (or *chiang chhing*).

[126] This agrees with the theory of Sakaguchi Kinichiro (*1976*), p. 224, that Japanese *miso* traces its ancestry back to Chinese *chiang*, but Japanese *shoyu* to Chinese *shih* (1979). This issue is discussed in Shurtleff & Aoyagi (1976), p. 224.

[127] *YLT*, p. 1. *Huang tzu* is the moulded soybean used as a substrate for making *shih*. A *fu* day is one of the three *kêng* days during the summer, based on the calendrical system according to 'celestial stems' and 'earthly branches'. What it amounts to is that the operation should be carried out when the weather is warm.

soy sauce was also prepared as a derivative of *shih*. This view is reinforced in another brief recipe for making *chiang yu* recorded in the Yuan work *I Ya I I*:[128]

Clean yellow soybean substrate (*huang tou* 黃豆) free from its coat (spores). Mix one *tou* bean substrate with six catties of salt and more than the usual amount of water (for making *chiang*), and incubate. When the product is ripe, the residual beans will remain at the bottom and the sauce (*yu*) will stay on top.

Again, the substrate used is moulded bean, but flour could have been included in its preparation. In this case undigested beans, i.e. *shih*, settle at the bottom, and the sauce can easily be decanted. A more detailed account for making soy sauce is found in the *Pên Tshao Kang Mu*, +1695, but it is presented as a method for *tou yu* 豆油 (soybean oil), which in this case is obviously a synonym of *chiang yu*:[129] '*Tou yu*. Boil three *tou* of soybeans in water until soft. Blend in twenty-four catties of wheat flour, and incubate until a yellow *chiang* substrate is formed. Mix ten catties of substrate with eight catties of salt and forty catties of well water. Ferment in a jar in the sun until the product is ready.'

Although the passage does not tell us how the soy sauce is separated from the fermenting mash, it is absolutely clear that here we have a process based on *chiang* and the product can rightfully be called *chiang yu*. The next description of a *chiang yu* process occurs about a hundred years later in the *Yang Hsiao Lu*, +1698. One of the three sketchy recipes given states:[130]

Cook yellow (or black) soybeans until soft. Mix the beans with white flour and the decoction, and knead until the dough is solid. Form flat or convex cakes. Cover with artemisia leaves until a good growth of yellow mould is obtained. Grind the yellow-coated cakes. Incubate the meal in brine in a jar. Warm in the sun to achieve [a thin] soy paste (*chiang*). Strain through a tight bamboo sieve over a large-mouthed urn. The soy paste remains on top and *chiang yu*, soy sauce, is collected below.

A more detailed description of the process is found in the *Hsing Yuan Lu*, +1750, but it is presented as a method for making *chhing chiang* 清醬.[131]

Wash one *tou* of clean yellow soy beans. Boil until the beans are soft and the colour turns red. Blend the beans and the cooking water evenly with twenty-four catties of white flour. Spread and pack the lumps firmly on bamboo or willow [leaf] trays and cover them with rice straw. Place the trays in a windless room, and incubate for seven days until good growth of myceliae is obtained. Remove the straw. Place the trays in the sun during the day and keep them indoors at night. Repeat the procedure for fourteen days. If it rains during the day, the trays should be placed in the sun for additional days until the goal of fourteen days is reached. The mouldy bean flour mixture should be very dry. This is the way to make yellow *chiang* ferment (*chiang huang* 醬黃).

[128] *IYII*, p. 10.
[129] *PTKM*, p. 1552, as an appendage to the item on *chiang*. Strictly speaking, *tou yu* is the oil expressed from soybeans, but the term is sometimes, as it is in this case, applied to soy sauce.
[130] *YHL*, pp. 11–12.
[131] *HYL*, pp. 6–8. This is the second of three methods for making *chhing chiang* described in this work. It is chosen for translation because it gives the most detailed account of how the final sauce is recovered. It should be pointed out that the amount of flour used in this example seems rather large when compared with a first recipe (p. 6) in which each *tou* of beans was blended with only 3–5 catties of flour.

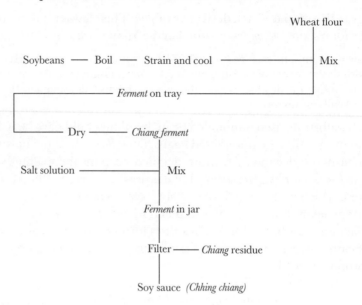

Fig 80. Flow diagram for the process of making *chiang yu* (soy sauce) in the *Hsing Yuan Lu*.

For each *tou* of yellow *chiang ferment*, measure five *tou* of well water in a jar. Place exactly fifteen catties of raw salt in a bamboo basket, and hang the basket in the well water. Allow the salt to dissolve in the water. Discard the residue in the basket. Add the yellow *chiang ferment* to the water and allow the mixture to warm in the sun for three days. On the morning of the fourth day, stir the contents with a wooden paddle (do not do this when the urn is in the sun). After two more days, stir again. Repeat the procedure three to four times. After about twenty days, the *chhing chiang*, i.e. clear *chiang* or soy sauce, should be ready for use.

To separate the soy sauce (*chhing chiang*), we need a finely woven bamboo cylindrical tube that is open at both ends. The southerners call such a device *chiang chhou* 醬篘 (circular sieve for *chiang*). It is readily available in the local markets in the capital (Peking). The same markets will also have for sale covers of varying sizes for the urns. When the sauce is ready, place the circular filter in the centre of the urn. Push it down so that the tube sits securely on the bottom of the urn. Remove the mash inside the tube until the bottom of the urn is clearly visible. Place a brick on top of the tube to prevent it from floating upwards. Clear sauce will flow from the mash into the tube. By the next morning, the tube will be filled with liquid. Transfer the clear sauce slowly with a bowl into a clean jar. Cover the jar with a piece of cloth to prevent flies from falling in. Warm the jar in the sun for half a month. To make more sauce, increase the quantity of raw materials. After the sauce is ready, one can also collect the beans that float to the top of the mash with a sieve. When half-dried, the product makes a delicious *tou shih* 豆豉 (fermented bean relish).

With this passage we are left with no doubt that *chhing chiang* is, indeed, synonymous with *chiang yu*. The process may be represented by the flow diagram shown on Figure 80. According to Hung Kuang-Chu, except for the use of the cooking water to mix the beans and flour, the procedure shown in the diagram is basically the same as that seen by him in the Chinese countryside for the making of soy sauce,

腌酱、晒酱

Fig 81. A traditional soy sauce process – 2nd fermentation showing urns and bamboo covers. Drawing by Hung Kuang-Chu (*1984a*), p. 108.

chiang yu (Figures 81 and 82). Even the rather quaint method of using a cylindrical bamboo tube to recover the liquid sauce is still practised in the production of the famous Kuanthou soy sauce of Fukien (Figure 83a and b).[132] In fact, the *Hsing Yuan Lu* procedure is quite similar to that observed by Groff for the manufacture of soy sauce in Kuangtung in the early years of the 20th century.[133] The steps she observed in the process are:

(1) Boil soybeans until soft, drain off water. Figure 84 (boiler pot).
(2) Mix cooled beans with flour until every bean is covered with flour; the ratio of bean:flour is about 7:6.
(3) Spread bean-flour nuggets on tray; incubate in dark room for one to two weeks until good growth of yellow mould is achieved. Figure 85 (incubation trays).
(4) About seventy catties of mould-coated nuggets are mixed with clarified solution of forty catties of salt in one hundred and fifty catties of water to fill a large wide-mouthed jar. Each jar is provided with a conical bamboo cover. Figure 86 (bamboo cover and jar).

---

[132] Hung Kuang-Chu (*1984a*), pp. 106–9 (cf. Figs. 4 & 5). Kuanthou is a town east of the provincial capital of Foochow. It is the home of a *shih yu* (soy sauce) made by extension of the method for producing *shih*, in which no flour is used in the first fermentation. According to Shurtleff & Aoyagi (1976), p. 222, a similar (though not identical) cylindrical bamboo filter has also been used in small Japanese workshops to collect soy sauce from the incubating mixture. For a description of the traditional *shih yu* process practised in China in recent years, cf. Chhên Thao-Shêng (*1979*), pp. 75–6.

[133] Groff, Elizabeth (1919). The factory was in Hsinan, about fifty miles southwest of Canton.

Fig 82. Soy sauce incubating urns and covers – Foochow, 1987, photo HTH.

(5) The jars are placed in the yard and exposed to the sun for several months.[134] The first drawing of the liquor (by siphoning) is made in three to four months. Figure 87. To the residue is added fresh salt solution and the mixture put in the sun again until it is ready for a second drawing. As many as four drawings may be made.

(6) The drawn sauce is placed in a clean jar and aged in the sun for one to six months.

Because the fermented beans do not disintegrate in the process, a liquid is drawn easily from the mash.[135] For this reason, soy sauce is often known as *chhou yu* 抽油 (drawn sauce) among the Cantonese. The thin sauce from the first drawing is called *shêng chhou* 生抽 (fresh drawn sauce), and the thicker sauce from later drawings is known as *lao chhou* 老抽 (mellow drawn sauce).[136]

[134] Although Groff does not say so, presumably the contents are stirred from time to time during the submerged incubation. Any stirring, however, could not have been very vigorous, since at the first drawing the beans still retain their shape as nuggets, cf. figure 84.

[135] It can be collected by using the cylindrical bamboo filter, by siphoning, or by having a spout on the wall near the bottom of the jar.

[136] *Shêng chhou* and *lao chhou* are terms that the reader is likely to encounter on the labels when purchasing a bottle of soy sauce from China in an oriental food market in America or Europe.

40 FERMENTATIONS AND FOOD SCIENCE

(a)

(b)

Fig 83. Traditional soy sauce process – collecting soy sauce using a cylindrical bamboo filter: (a) Drawing by Hung Kuang-Chu (*1984a*), p. 108; (b) Urn showing cylindrical bamboo filter, Foochow, 1996, photograph HTH.

Fig 84. Traditional soy sauce process in early 20th century – boiler to cook soybeans. Groff (1919), Pl. 1.

Fig 85. Traditional soy sauce process in early 20th century – incubation trays for culture of fungi. Groff (1919), Pl. 2, Fig. 2.

Fig 86. Traditional soy sauce process in early 20th century – incubating jars with bamboo cover. Groff (1919), Pl. 3, Fig. 2.

Fig 87. Traditional soy sauce process in early 20th century – first drawing of product by siphoning. Groff (1919), Pl. 3, Fig. 1.

As we can see from the flow diagram in Figure 80 the traditional Chinese process for making soy sauce consists of two main stages (1) A surface fermentation to produce a mould-coated bean-flour substrate (called *chiang huang* 醬黃 or *huang tzu* 黃子) (2) A submerged incubation in a salt solution to partially digest the ferment substrate from the previous step. The conditions are such that the liquid condiment that forms is easily separated from the residual undissolved nuggets. The main difference between a *chiang* (fermented soy paste) process and a *chiang yu* (fermented soy sauce) process is that in the latter case the second fermentation is carried out in more brine and with less stirring than is optimal for making *chiang*. As a result, many of the residual beans are still intact and can be recovered for use as *shih*.

Although both *shih*, fermented beans, and *chiang*, fermented bean paste, are listed in the pharmacopoeias, soy sauce, *chiang yu*, was never officially listed. There are, however, indications that soy sauce was occasionally used as a therapeutic agent. The *Wai Thai Mi Yao* (Secret Formulary from the Outer Terrace), +752, states, 'For ulcers, apply *chhing chiang* and sulphur powder.'[137] The *Chhien Chin Pao Yao* 千金寶要, +1124, has this to say, 'To treat a person bitten by a dog, daub soy sauce (*tou chiang chhing* 豆醬清) over the afflicted area for three to four days,' and 'For throbbing of fingers and toes, daub with warm soy sauce (*chiang chhing* 醬清) and honey.'[138] The principal use of soy sauce has, of course, always been as a flavouring agent. But when did it become a major condiment in the culinary system, and how important was it relative to *shih* and *chiang* at different stages of Chinese history? To answer these questions, we have tabulated (Table 33) the number of times fermented soybeans (*shih*), soy paste (*chiang*) and soy sauce (under its various names as well as *chiang yu*), occur in cooking recipes in the food literature from the Han to the Chhing. Let us see what the tabulation tells us.

The occurrence of *chhing chiang* in the *Ssu Min Yüeh Ling*, and of *shih chih* in the *Shih Ming*, coupled with the indirect evidence from the *Wu Shih Erh Ping Fang* (page 359), strongly suggest that some kinds of liquid soy condiment had been obtained as a byproduct of *chiang* or *shih* since the Han dynasty. They were known variously as *chhing chiang* 清醬, *chiang chhing* 醬清, *shih chih* 豉汁 and *shih chhing* 豉清. *Shih chih*, however, was a term applied mainly to an aqueous extract of *shih*. It might also have indicated a sauce obtained from an extension of the *shih* fermentation. We have no information on how the other three were made. But for the purpose of Table 33, we shall consider them all as precursors or proto-precursors of soy sauce. Included in the Table are also other synonyms of soy sauce encountered in the Sung and post-Sung literature, that is, *chiang chih* 醬汁 (juice of *chiang*), *chiang sui* 醬水 (water extract of *chiang*), *shih yu* 豉油 (sauce from *shih*), *tou yu* 豆油 (sauce from soybeans), and *chhiu yu* 秋油 (sauce harvested in autumn).

As we have noted earlier (pp. 337, 353) both *shih* and *chiang* were major commodities in the economy of the former Han. They are still important ingredients in Chinese cuisine today, *chiang* more than *shih*. Because much of the relevant literature

---

[137] *Wai Thai Mi Yao* cited in Wang Shang-Tien (1987), p. 468.
[138] *Chhien Chin Pao Yao*, +1134, ch. 3, p. 68 and ch. 5, p. 106.

Table 33. Usage of soy condiments in food recipes from the Han to the Chhing dynasties

Reference	Chiang	Chiang chhing	Chhing chiang	Chiang yu	Chiang chih	Chiang shui	Shih	Shih chih	Shih chhing	Shih yu	Tou yu	Chhiu yu
SMYL +160	—	—	1	—	—	—	—	—	—	—	—	—
CMYS +540	2	5	—	—	—	—	20	26	2	—	—	—
SSCKS/Sung	9	—	—	3	1	—	—	—	—	—	—	—
WSCKLS/Sung	5	—	—	3	—	—	—	—	—	—	—	—
MLL Yuan	2	—	—	—	—	—	1	1	—	—	—	—
SLKC +1280	1	—	—	—	—	—	—	—	—	—	—	—
CCPY Yuan	17	1	—	—	—	—	—	—	—	—	—	—
YLT +1360	3	—	—	1	—	2	—	—	—	—	—	—
IYIY Yuan	10	—	—	1	—	1	—	—	—	—	—	—
TNPS +1370	5	—	—	—	—	—	—	1	—	—	—	—
YCFSC +1591	20	—	—	5	1	1	4	1	—	—	—	—
SHHM +1680	8	—	1	37	2	1	4	2	—	—	—	—
YHL +1698	14	—	—	11	—	—	1	—	—	—	—	—
SYL +1750	4	—	5	1	—	—	—	—	—	—	8	—
SYST +1790	15	—	24	138	1	6	2	—	—	—	—	62
TTC Chhing 3	58	—	4	66	—	2	2	—	—	—	—	—
TTC Chhing 7	33	—	—	—	1	1	1	—	—	—	—	—

SMYL Shih Min Yüeh Ling.
CMYS Chhi Min Yao Shu.
SCCK San Chia Chhing Kung.
WSCKL Wu Shih Chung Khuei Lu.
MLL   Mêng Liang Lu.
SLKC Shih Lin Kuang Chi.
CCPY Chü Chia Pi Yung Shih Lei Chhuan Chi.
YLT   Yün Lin Thang Yin Shih Chih Tu Chi.

IYII   I Ya II.
TNPS   To Nêng Phi Shih.
YCFSC Yin Chuan Fu Shih Chien.
SHHM   Shih Hsien Hung Mi.
YHL   Yang Hsiao Lu.
SYL   Hsing Yuan Lu.
SYST   Sui Yuang Shih Tan.
TTC   Thiao Thing Chi ch. 3 Meats, ch. 7 Vegetables.

was lost, it is difficult to gauge their relative importance during the centuries between Han and Thang. Based on the Table 33, we can say that at the time of the *Chhi Min Yao Shu*, +544, *shih* was used extensively as a relish and a condiment, but its popularity gradually waned. Usage of *shih* has stayed at a low, though still significant, level since the Sung to the preset day. On the other hand, while *chiang* was used in fewer recipes than *shih* in the *Chhi Min Yao Shu*, it soon overtook *shih* to become a most important condiment in Chinese cookery. It is included in recipes in all the cookery books from Sung to Chhing listed in the Table.

During the Sung the various sauces obtained by modification of the *shih* or *chiang* process were unified under the name *chiang yu*, as this type of product began to gain popularity as a savoury condiment in the Chinese diet. Terms such as *shih chih* and *chhing chiang*, however, continued to be seen from time to time in the Ming and Chhing food literature. *Chiang yu* started to rival *chiang* in importance during the Ming, and by the early years of the Chhing had surpassed it in culinary usage. In the last Ming entry, *Yin Chuan Fu Shih Chien*, +1591, *chiang* occurred in twenty recipes while soy sauce in only seven recipes (five for *chiang yu*, one for *chiang chih* and one for *chiang sui*). But in the first Chhing entry, *Shih Hsien Hung Mi*, +1680, the eight citations for *chiang* are overwhelmed by forty-one citations for the soy sauce group.[139] In *Yang Hsiao Lu*, +1698, there were four citations for *chiang* versus eleven citations for *chiang yu*. In *Hsing Yuan Lu*, +1750, we see four citations for *chiang* against fourteen for soy sauce (five *chhing chiang*, one *chiang yu* and eight *tou yu*). By the last entry for the 18th century, the *Sui Yuan Shih Tan*, +1795, soy sauce had reached its present eminence as the principal condiment of Chinese food. The soy sauce group was used ninety-five times, but *chiang* only fifteen times. Apparently Yüan Mei had an aversion to the term *chiang yu*. He preferred to use *chhing chiang* and *chhiu yu* which are deemed more elegant expressions than the common *chiang yu*. According to the *Sui Hsi Chü Yin Shih Phu*, by Wang Shih-Hsiung in mid 19th century, the first harvest in the autumn of soy sauce produced through the summer was called *chhiu yu*,[140] but the term is hardly ever used today. We end the Table with parallel data from two chapters of the massive *Thiao Thing Chi* (The Harmonious Cauldron) which may be regarded as the summation of the state of the culinary arts of China in the 19th century. There are two hundred and twelve entries for soy sauce (*chiang yu*) and associated products *versus* ninety-one for *chiang*. The dominance of soy sauce was reiterated, but *chiang* (fermented soybean paste) remained an important condiment in the food system, as it still is today.[141]

Thus, we see that although some kind of sauce from fermented soybeans was known in the early Han, it took a thousand years of gestation before it gained recognition

---

[139] In fact, in one recipe, pp. 56–7 soy sauce (*chiang yu*) itself was used as an additive in the making of *shih*.
[140] *Sui Hsi Chü Yin Shih Phu* (Dietary System of the Random Rest Studio) by Wang Shih-Hsiung, c. 1850, p. 39.
[141] According to the *CKPJPKCS* (Encyclopedia of Chinese Cuisine), p. 285, the major types of *chiang*, fermented paste, produced in China are *tou chiang* 豆醬, *mien chiang* 麵醬 and *tou pan chiang* 豆辦醬, i.e. fermented soybean paste, wheat flour paste and broad bean paste. Actually, the term *tou pan chiang* is somewhat ambiguous. In a typical oriental food market in the US one will find *tou pan chiang* listed on the label as broad bean paste, but more often it is listed as chunky soybean paste. See also Shurtleff & Aoyagi (1976), p. 245, for types of *chiang* available in such a market.

in the Sung as a culinary ingredient in its own right. Why this development took such a long time will be discussed in Chapter (*h*). During the Sung, methods for its preparation were refined and standardised. Its usage in food recipes grew slowly but steadily in the Yuan and Ming Dynasties. By the early years of the Chhing it had attained the status of the most popular condiment in the Chinese food system. Today, soy sauce is the preeminent condiment not only in China, but also in Japan and other countries in East Asia, which brings us to the next topic of our discussion, that is, the transmission of soybean fermentation technology from China to Japan.

(v) *Soy fermentations, China and Japan*

It is believed that long before active cultural interchange between China and Japan began, the Japanese had independently developed savoury sauces by pickling salted fish, shellfish and meat. These condiments were called *hishio*. When the Japanese adapted the Chinese writing system to their language, *hishio* was written as *chiang* 醬. In a poem by Okimuru Imiki (+686–707) in the *Man'yoshu*, the character for *hishio*, as *chiang*, is mentioned in the following context:[142]

> I want to eat red snapper
> With a dressing of minced garlic and vinegar *hishio* 醬
> So do not offer me a leek soup.

Other kinds of *hishio* mentioned in the *Man'yoshu* are made from crab, wild game and deer. During the Nara Period (+710–794), different types of *hishio* began to appear in the literature, some derived from grains and soybeans for example, *misho* 醬 (*chiang*), *miso* 味醬 (*wei chiang*), rice *hishio* 米醬 (*mi chiang*) and *shih* 豉.[143] What is remarkable is that *chiang* 醬 was apparently obtained in liquid form, suggesting that it was a primitive prototype of soy sauce.[144] These references further suggest that the methods for making *chiang* and *shih*, as recorded in the *Chhi Min Yao Shu*, +544, had been transmitted to Japan and were being applied by the Japanese to prepare similar types of fermented soyfoods from locally available raw materials. How this information was transmitted to Japan is not clear. Some scholars believe that it came directly from China to Japan, while others think that it came to Japan via Korea.[145]

Tradition has it that the earliest *miso*, i.e. a *hishio* made with soybeans was prepared by exposing cooked soybean cakes to wild fungal spores and fermenting the moulded beans in brine, thus tracing its genealogy back to the *shih* (rather than

---

[142] *Man'yoshu*, scroll 16, No. 44, tr. Pierson, Shurtleff & Aoyagi (1976), p. 219. The *Man'yoshu* is a collection of early Japanese songs and poems ranging from +315 to about +760. The complete work, in twenty volumes, has been translated by Pierson (1929–63). According to Ishige, personal communication, *hishio* was also written as 醢 duing the Nara and Heian (+794 to +1185) periods. 醬 was used mainly to denote products made from vegetables. When *hishio* was made from animal materials such as fish and meat, a prefix was added to the word 醬, such as 魚醬 or 鹿醬. The *Engishiki* (+907) says, 肉醬曰醢 ('a meat *chiang* is called *hai*'.) It was understood that 醢 referred only to products made from animal parts.

[143] Kikkoman K. K. (*1968*), pp. 69–70, quoted from *Yoryo Ritsuryo* by Fujiwara Fuhito (+718) and Todaiji Shosoin documents (+730–48).

[144] Sekine, S. (*1969*), p. 193. But there is no indication on how this *hishio* (*chiang*) was prepared.

[145] Tanaka Seiichi, tr. Huo Fêng & I Yung-Wên (1991), pp. 74–5.

the *chiang*) of Han China.¹⁴⁶ Other types of *miso* based on *chiang* technology soon followed. But the Japanese gradually transformed *miso* into versions which were significantly different in texture and flavour from those of their Chinese progenitors. As noted earlier, the Chinese *shih* or *chiang* process consists of two stages. In the first stage, cooked soybeans (with or without flour) are incubated in air to give moulded beans, known as *huang tzu* 黃子 (without flour) or *chiang huang* 醬黃 (with flour). In the second stage, the moulded beans are hydrolysed in brine into the flavouring components of *shih* or *chiang*. The Japanese began to vary the first stage by replacing moulded beans with moulded rice or barley, i.e. rice or barley *chhü* 麴 or *koji* in Japanese. In the second stage, this *koji* was incubated with cooked beans in brine to hydrolyse the proteins and polysaccharides into the flavouring components of *miso*. Now *koji* supplies all the enzymes, but the substrate is provided by both *koji* and cooked beans. The result was a product with a lighter colour and a sweeter taste than the traditional Chinese *chiang*. *Miso* had became distinctively Japanese. Indeed, in the beginning of the Heian period a new way of writing *miso* was introduced, which consisted of two characters, one *wei* 味 (pronounced *mi* in Japanese) meaning flavour, and the other a new word, *so* 噌, which has no Chinese equivalent:¹⁴⁷

MI 味 SO 噌

Thus, *miso* was no longer written as a *chiang*. This development may be looked upon as a declaration of independence of *miso* from its Chinese progenitor. Soon *miso* was widely accepted. In the famous epic of the Heian Period, *The Tale of Genji*, we find both *miso* and *hishio* (meat and fish sauces) featured as condiments in many dishes served at court.¹⁴⁸ *Miso* is listed in the earliest dictionary of the Japanese language, the *Wamyō Ruijūshō* 和名類聚抄 (+903–908), which states that it is produced in the countryside as well as in the capital.¹⁴⁹

During the Kamakura Period (+1185 to +1333) *miso* became a staple in the Japanese diet, favoured by the rich and poor alike. It was in this period that the famous *miso* soup originated.¹⁵⁰ This has become the major way *miso* is used in the Japanese diet, one that is uniquely Japanese and totally unfamiliar to Chinese cuisine even now.¹⁵¹ According to tradition it was in the middle of the Kamakura Period (c. +1255)

---

¹⁴⁶ Shurtleff & Aoyagi (1976), pp. 38b and 218b. Ishige has seen in the Korean countryside blocks of cooked soybeans hung from eaves of houses until they become mouldy. Such a block of moulded beans is called *meju* which could have been transmitted as *miso* into Japanese.

¹⁴⁷ According to Kikkoman K. K. (1968), p. 71, the new written form of *miso* first appeared in the *Fusō Ryakuki* 扶桑略記 of +938. Sekine, S. (1969), p. 202, shows that *miso* is more often written as 末醬 during the Nara Period.

¹⁴⁸ *The Tale of Genji*, cited in Shurtleff & Aoyagi (1976), p. 221.

¹⁴⁹ *Wamyō Ruijūshō* 和名類聚抄, +934.

¹⁵⁰ Kuo Po-Nan (*1989a*), p. 15, states that a red *chiang* (*miso*) soup is mentioned in a poem by the Sung Buddhist monk Huei Hung 惠洪 (+1071 to 1128) in honour of a visiting Japanese monk. This would suggest that *miso* soup was probably invented sometime before the Kamakura Period. The existence of *miso* soup was apparently known to some Buddhist monks in China, but it did not become a part of Chinese cuisine. A similar soup made from salty fermented soybeans, *yen shih thang* 鹽豉湯, is mentioned in the *TCMHL*, +1148, p. 22, and *Tu Chhêng Chi Shên*, +1235, p. 7. This product, too, disappeared from the food scene after the Sung.

¹⁵¹ Shurtleff & Aoyagi (1976), p. 231, estimates that as of 1983 about 80 to 85% of the *miso* consumed in Japan is in the form of *miso* soup.

that the Buddhist monk Kakushin brought back from China a method for making a new type of *miso* which he had learned at the Kinzanji (金山寺 Chin Shan Ssu), Temple of the Gold Mountain, a famous Zen centre in the Sung dynasty. Yuasa, Wakayama prefecture, soon became a centre for making this Kinzanji *miso*, which is nowadays prepared by incubating a barley *ferment* (i.e. barley *koji*) with cooked soybeans flavoured with minced eggplant, ginger root, white *uri* melon, *kombu* (sea vegetable) and burdock root.[152]

Legend has it that the earliest type of Japanese soy sauce, *tamari*, was discovered in Yuasa in the 13th century as a dark fragrant liquid left at the bottom of the jar in a Kinzanji *miso* operation. Initially, the word *tamari* was written with the two Chinese characters, *tou yu* 豆油, meaning oil or sauce from beans. Later *tamari* was given a different character, *liu* 溜, which in Chinese means to flow or drip (but in Japanese means to collect, as water collects in a pond).[153]

tamari 豆油 *tou yu*   *liu* 溜

This story of the origin of *tamari* is reminiscent of the way a liquid soy condiment was left at the bottom of the fermentation jar in the *shih* (soy nugget) process described in the late Thang Dynasty almanac, the *Ssu Shih Tsuan Yao*, which we cited earlier (page 361). Yuasa has remained a centre for making Kinzanji *miso* and *tamari*, but in later centuries it also became a centre for the production of *shoyu* 醬油 (i.e. *chiang yu* in Chinese), the typical soy sauce of Japan.

The earliest reference to *shoyu* in Japan occurs in the *Ekirinhon Setsuyoshu* 易林本節用集 a Japanese dictionary of +1597.[154] The procedure for making *shoyu* may have

---

[152] The Gold Mountain Temple is west of Chênchiang 鎮江 (now the capital of Kiangsu province) by the Yangtzu River. In the *Chü Chia Pi Yung*, p. 58, there is a Chin Shan Ssu (Golden Mountain Temple) recipe for making fermented soybeans, *tou shih*, which says:

Take a convenient amount of yellow soybeans. Soak overnight in water and steam thoroughly. When cool, mix the beans evenly with wheat flour and blend with wheat bran. In a clean room spread the granules on a mat to a thickness of two inches. Cover with a layer of grasses, wheat straw (or artemisia stems) and cockleburr leaves. After five to seven days a yellow coat should develop [around the beans]. Clean and winnow off the wheat bran. Wash the beans with water and dry them in the sun.

For each *tou* of moulded bean substrate add one *tou* of supplemental ingredients, and have a thoroughly cleaned jar ready. The supplemental materials are: fresh melon (cut into two-inch chunks), a fresh eggplant (cut into four pieces), clean orange peel, lotus seeds (softened in water and sliced in halves), fresh ginger (in large slices), Szechuan peppercorn, fennel (slightly roasted), liquorice (chopped), perilla leaves and garlic (in cloves). Mix them evenly. Place one layer of moulded beans at the bottom of the jar. Cover with a layer of supplemental materials, then top with a layer of salt. Repeat the sequence until the jar is full. Press firmly and cover tightly with bamboo leaves. Seal the jar with mud. Heat the jar in the sun. After half a month remove the ingredients, mix them evenly, repack them in the jar and seal with mud. Sun the jar for seven times seven days. Do not add water. Enough water will leach out of the melons and eggplants to form a brine. The amount of salt used should be adjusted [according to taste].

We do not know whether this was the recipe that Kakushin brought back to Japan. According to Ishige (personal communication), although in Japan, Kinzanji is commonly written as 金山寺, this is a mistake. The temple that Kakushin visited was actually 經山寺 in Hangchou, the capital of Chekiang. Be that as it may, Kakushin's recipe would have gone through many changes through the ages.

[153] Shurtleff & Aoyagi (1976), p. 222. The legend is suspect since *tamari* is made with a *koji* prepared from soybeans only, in a manner similar to the Chinese method for making *shih* and *shih yu* (soy sauce from *shih*). *Tamari* today is considered as a soy sauce made without the use of wheat, or other cereal grains. Its manufacture is described in Shurtleff & Aoyagi (1980).

[154] *Ekirinhon Setsuyoshu* 易林本節用集.

been transmitted from China in the preceding century.[155] As a result, the Japanese process for making *shoyu* is practically the same as the Chinese process for making the *chiang* type of *chiang yu*. On the other hand, there is no evidence that the Chinese ever tried to adopt the Japanese method for *miso* to make new styles of *chiang*. Thus, even though Japanese soy sauce, *shoyu*, and Chinese soy sauce, *chiang yu*, are quite similar in organoleptic properties, Japanese *miso* and Chinese *chiang* have continued to remain as distinctively different products.

Through the Muromachi (1336–1568) and Edo (1600–1853) Periods *miso* continued to grow in richness and variety. Basically the principal types of *miso* fall into three categories (1) those made from rice *koji* (2) those made from barley *koji*, and (3) those made from soybean *koji*.[156] At the same time, as its importance grew, soy sauce also diverged into two different types (a) *tamari* type, made largely from soybean with little or no wheat, and (b) *shoyu* type, made from soybean and wheat. We can say that *tamari* traces its lineage through soybean *miso* to *shih*, and *shoyu* through *chiang yu* to *chiang*.[157] The Japanese also made fermented savoury soybeans (i.e. *shih*) under the name *kuki* and *hamanatto*, but the history of this product remains obscure.

Japan is now the world's leader in fermented soyfood production. Because it was the first country in East Asia to become industrialised, the Japanese have been the pioneers in applying modern technology to the traditional processes for making *shoyu* and *miso*.[158] A great deal of research has been carried out on the microbiology and biochemistry of the *shoyu* and *miso* fermentations. Excellent reviews of this work are available in English.[159] Much applied research has also been done on the Chinese *chiang* and *chiang yu* fermentations.[160] From these studies it is clear that the traditional procedures underlying these processes are almost identical. They all involve two consecutive stages:

Stage 1: Cooked whole soybeans (singly or mixed with wheat), rice or barley are used to support a surface culture of *Aspergillus oryzae*.[161]

---

[155] Cf. Sasagawa Rinpu 世川臨風 and Adachi Isamu 足立勇 (*1942*); Hung Kuang-Chu (1982), p. 102, Tanaka Seiichi, tr. Huo Fen & I Yung Wên (*1991*), pp. 74–6, Adachi Iwao 足立嚴 (*1975*) and Tei Daisei 鄭大聲 (*1981*). Tradition also has it that Kakushin brought *shoyu* from China to Japan, cf. Johnstone, (1986) p. 38.

[156] The three types are further classified according to their salt content into sweet, mildly salty and salty groups. Each group is again divided by colour into white, light yellow and red. Shurtleff & Aoyagi (1983), pp. 30–44, quote the production of the three major types as follows: rice *miso*, 81%, barley *miso*, 11% and soybean *miso*, 8%.

[157] See the discussion in Sagakuchi, K. (*1979*), who argues that *miso* is derived from *chiang* and *shoyu* from *shih*. The same view is held by Yokotsuka T. (1986), p. 325. To us, it is clear that *shoyu* is derived from *chiang*, and *tamari* from *shih*, while *miso* owes its origin to both *shih* and *chiang*. Today *shoyu* accounts for more than 97% of the soy sauce production in Japan, and *tamari* less than 3%, cf. Shurtleff & Aoyagi ibid., p. 49.

[158] According to Sumiyoshi, Y. (1987), in the 1980s Japan produced annually 1.2 million tons of *shoyu* and 750,000 tons of *miso*. This means that each person consumed 10.2 litres of *shoyu* and 6.7 kg (c. 15 lbs) of *miso* each year. The amount of *miso* produced in large plants is 580,000 tons, but it is estimated that another 180,000 tons are produced in families and small workshops. Cf. also Wang & Hesseltine (1979) II, pp. 98, 105.

These figures are considerably higher than the Chinese consumption of *chiang yu* and *chiang*. Wang Shan-Tien (*1987*), p. 474, estimates that China produces annually about 2 million tons of soy sauce and 300,000 tons of *chiang*. The annual *per capita* consumption is thus less than 2 litres of soy sauce and 300 g of chiang. These figures may be underestimated, since the production in the countryside in homes and small workshops may not be included in the national statistics. The *per capita* figures are based on a total population of one thousand million.

[159] For example, Wang & Hesseltine (1979); Yong & Wood (1974); and Yokotsuka, T. (1986).

[160] Chiao, J. S. (1981) and Hesseltine & Wang (1986).

[161] As the mould proliferates it produces a host of hydrolytic enzymes. Proteins and carbohydrates in the substrate are hydrolysed to provide nitrogen and energy needed to maintain fungal growth. The procedure is, in principle, the same as that for making the *chhü* or *koji* used in the brewing of *chiu* or *saké*.

Table 34. *Basics of Chinese and Japanese soyfood fermentations*

Soyfood	*Chiang*	*Miso*	*Chiang yu*	*Shoyu*
Stage 1				
Materials	Cooked soybean	Rice or barley	Cooked soybean	Cooked soybean
	Wheat flour	—	Wheat flour	Cracked wheat
Exposure to air	Culture of *A. oryzae*	Culture of *A. oryzae*	Culture of *A. oryzae*	Culture of *A. oryzae*
Product	*Chiang* ferment	Rice or barley *koji*	*Chiang* ferment	Soy wheat *koji*
Stage 2				
Materials	*Chiang* ferment	Rice or barley *koji* cooked soybean	*Chiang* ferment	Soy wheat *koji*
	—		—	—
	Brine	Salt	Brine	Brine
Anaerobic digestion	Mild stirring	Stationary	Mild stirring	Mild stirring

Stage 2: The resultant mould culture, singly or mixed with cooked soybean, is incubated in brine to allow the enzymes to hydrolyse proteins into lower peptides and amino acids, carbohydrates into sugars, and fats into fatty acids and glycerol.[162]

By varying the type and amount of raw materials and the condition of incubation, the same two-stage process has been exploited to give us products with different texture and organoleptic properties, that is, *chiang*, *miso*, *chiang yu* and *shoyu* (Table 34). The production of these delicious and yet inexpensive savoury condiments from soybeans and grains must surely rank as the most imaginative, as well as the most useful application of microbial technology in the premodern age. But the spread of fermented soyfoods did not stop at Japan. We see similar products in Korea (*jang*, *doen jang*), Vietnam, Thailand, Indonesia (*taucho*, *kechap*) and Malaysia (*taucheo*). They all owe their origin to the ancient *ferment*, *chhü* 麴 (or *koji* in Japanese), which was developed initially to support the brewing of alcoholic drinks from grains as shown in Chapter (*c*). The application of fungal technology, however, does not end with soyfoods. *Chhü* also plays a role in the production of *chiang* (paste/sauce) prepared from meat and fish, and in the preservation of vegetable foods. These topics will be considered as a part of food processing technology, and discussed in the next chapter.

---

[162] The sugars are partly converted by lactic acid bacteria (*Pediococcus halophilus*, and *Streptococcus* sp.) to acids, thus lowering the pH from neutrality to 4.5, a level at which alcoholic fermentation by yeasts (*Saccharomyces rouxii* and *Candida* sp.) would be favoured. Some of the fatty acids are esterified into esters of lower alcohols. These may be produced by transesterification between the ethanol fermented by the yeasts and the triglyceride fats. The oil is removed and used in the plastics industry, cf. Wang & Hesseltine, (1979), p. 103 and Shillinglaw (1957). This problem has been eliminated since defatted soybeans have replaced whole soybeans in most Japanese *shoyu* factories. It is interesting that no such accumulation is mentioned in the detailed recipes for making *chiang yu* in the Ming and Chhing literature (cf. pages 363–5). It is possible that because wild cultures were used, the moulded beans were more active in degrading fats than those inoculated with a pure culture of *A. oryzae*.

(e) FOOD PROCESSING AND PRESERVATION

Having completed our survey of the technology of processing of soybeans we may now turn our attention to the processing and preservation of animal products, fruits, vegetables, oilseeds and grains other than soybean. As we have seen earlier, on pages 17–65, the ancient Chinese were blessed with a rich variety of food resources. But materials in their natural state are seldom ideal as food for human consumption. Meat and fish are highly perishable; fruits and vegetables are seasonal products with a limited shelf life; and oilseeds and cereals may require extensive processing before they become desirable articles of food. Thus, the Chinese have expended a great deal of time and effort to devise methods to preserve the raw food materials or to convert them into palatable food products with good keeping qualities. For these purposes, a variety of methodologies were developed; some were microbiological or biochemical, others chemical or physical. The application of these methodologies to the food materials, i.e. animal products (particularly meat and fish), vegetables, fruits, oilseeds and cereals (particularly wheat), is the subject of this chapter. Our discussion will consist of three parts: first, the preparation of fermented meat and fish products and pickled vegetables and fruits, second, the use of physical and chemical methods in food preservation and processing, and third, the conversion of oilseeds and grains into highly desirable and nutritious food products.

(1) FERMENTED CONDIMENTS, PICKLES AND PRESERVES

Fermented or pickled meat, fish, fruits and vegetables were well known in pre-Han China. For example, the passage on the *Hai Jên* 醢人 (Superintendent of Fermented Victuals) in the *Chou Li* (Rites of the Chou) states that 'to take care of the domestic needs of the royal household, the superintendent prepares sixty jars for storing fermented condiments and preserves; he fills them with *wu chhi* 五齊 (five types of finely sliced pickled meat or vegetables), *chhi hai* 七醢 (seven kinds of boneless meat paste), *chhi tsu* 七菹 (seven varieties of coarsely cut pickled vegetables) and *san ni* 三臡 (three kinds of meat paste still mixed with bone).'[1] Chêng Hsüan's commentary, c. +160, explains that the five *chhi* pickles were prepared from the root of the cattail, tripe, clams, pork shoulder and young shoots of the water rush. The seven *hai* pastes were obtained from meat, snail, oyster, frog, fish, rabbit and goose. The seven *tsu* pickles were made from chives, turnip, water mallow, mallow, celery, miniature bamboo shoot, and regular bamboo shoot. And, lastly, the three *ni* pastes with bone were processed from three kinds of venison (from the common deer, the elaphure, and the roebuck). The Ma-wang-tui Inventory lists jars of *than* 醓 (meat

---

[1] *Chou Li, Hai Jên*, pp. 55–7.

and fish sauce or juice), three jars of paste (*ju* 肉, meat; *chhüeh* 雀, sparrow; and *ma* 鮬, probably a kind of fish) and three jars of *tsu* pickles (*jang ho* 蘘荷, wild ginger; *sun* 筍 bamboo shoots and *kua* 瓜, melon). Thus, fermented condiments and pickles made from a wide variety of animal and plant materials were an important feature of the diet of the ancient Chinese.

(i) *Fermented meat and fish products*

The best known and perhaps the most ancient of these products is *hai* 醢 or *chiang* 醬 (paste or thick sauce) from animal parts. As we have noted earlier (page 333), *hai* is mentioned in the *Shih Ching* (Book of Odes) and in the *Li Chi* (Record of Rites).[2] *Hai thun* 醢豚, that is, suckling pig flavoured with *hai*, is one of the delicacies mentioned in the *Chhu Tzhu* (Elegies of the Chhu State), –300,[3] and the *Lü Shih Chhun Chhiu* (Master Lü's Spring and Autumn Annals), –239, lists the *hai* made from the sturgeon and snout fish (*chan* 鱣 and *wei* 鮪) as one of the most delectable seasoning agents of the realm.[4] *Chiang* made from the *tung* 鮦 fish is mentioned in the *Ssu Min Yüeh Ling* (Monthly Ordinances for the Four Peoples), +160.[5] Yet the only information we have on how it was made is the brief description in Chêng Hsüan's commentary on the *Hai Jên* in the *Chou Li* that we have quoted earlier (page 334), which states that to make *hai*, finely cut meat is mixed with salt, *chhü* 麴 (mould *ferment*) and good wine for a hundred days. The most noteworthy feature of the procedure is the addition of the mould *ferment* to the salty incubation medium. This endows the process with a feature that is distinguishably Chinese.

In principle, it is possible to make a salty fermented paste/sauce from meat or fish without the intervention of a preformed mould *ferment* by just allowing the enzymes present *in situ* to act on the proteins in the meat or fish substrate. Such a process, in modern biochemical terms, is known as *autolysis*. Its simplicity makes it likely that the earliest types of *hai* were made in this way. After all, this was how the ancient Greeks and Romans made their favourite condiment, *garum*, and relish, *salsamentum*, from fish, and this was presumably also how the early Japanese made *hishiho* paste/sauce from fish, shellfish and meat before the advent of the mould *ferment* called *chhü* 麴 (*koji* in Japanese). We can see two advantages arising from the addition of an external *ferment* to the incubation. Firstly, it provides powerful fungal enzymes to facilitate the digestion of protein. Secondly, the residual starch in the mould *ferment* preparation would be hydrolysed to glucose, which, in turn, would be converted to lactic acid by lactic acid bacteria. The acid generated would protect the product from contamination by pathogenic and putrefying bacteria. Thus, the use of microbial *ferment* represents a significant advance over the original ancient

---

[2] *Shih Ching*, W 197 (M 246); *Li Chi*, p. 457.
[3] *Chhu Tzhu, Ta Chao* (The Great Summons), Fu Hsi-Jên ed., p. 172; Hawkes (1985), p. 234, line 35 renders *hai thun* as pickled pork.
[4] *LSCC*, ch. 14, *Pen Wei*, p. 371.
[5] *SMYL*, 4th month. The 1st month is noted as a good time for making fish *chiang* and meat *chiang*.

Table 35. Chiang *from animal products in the* Chhi Min Yao Shu

No.	Substrate[a]	*Ferment*[b]	Solvent	Salt	Ratio[c]
1	Finely chopped beef, lamb, venison, rabbit pheasant (10)	Ground *ferment* A (5) B (1)	None	(1)	10:6:0:1
2[d]	Finely chopped beef, lamb, venison, rabbit, fresh fish (10)	Ground *ferment* A (5) B (1)	Wine (10)	(1)	10:6:10:1
3[e]	Finely sliced large fish or whole small fish (10)	*Ferment* B whole (1) ground (2)	None	(2)	10:3:0:2
4[f]	Dry mullet, soak in water, use whole (10)	Ground *ferment* A (4) B (1)	None	(2½)	10:5:0:2.5
5	Finely sliced fish (10)	*Ferment* (5)	Wine (2)	(3)	10:5:2:3
6	Shrimps (10)	None, cooked rice (3)	Water (5)	(2)	10:0:5:2
7	*Chu I* Fish entrails	None	None	+++	
8	Crab soaked in sugar	None	Smartweed decoction	++	

[a] Figures in parentheses indicate the number of *shêng* (pints) of raw material used.
[b] *Ferment* A is wine *ferment*, *ferment* B is yellow mould *ferment*, *huang chêng*. The reason for using two types is not clear.
[c] This is the ratio of number of pints of substrate: *ferment*: solvent: salt.
[d] In this recipe the incubating jar is warmed in ashes, and the product is ready after one day.
[e] This recipe also calls for a pint of dry ginger and an ounce of orange peel.
[f] If yellow mould *ferment* is not available, ground sprouted wheat may be used.
A '++' means that a liberal amount of salt is used and a '+++' means that a large amount of salt is used.

technology. If this view is correct, we are then faced with two questions. One, when did the Chinese first start to use a *ferment* in the process? Two, was the same methodology also applied to the making of the snail, frog, rabbit and fish *hai* mentioned in the *Chou Li* and the *Li Chi*? We can only speculate on the answer to the first question, but we may be able to make an educated guess at the answer to the second.

The relevant information is once again provided by the *Chhi Min Yao Shu* (Important Arts for the People's Welfare), +544. In addition to fermented soy paste (*chiang* 醬), chapter 70 (Methods for *Chiang*) gives eight recipes for the preparation of *chiang* from meat, fish and shellfish. The processes described are summarised in Table 35. As a representative procedure, we shall quote in full recipe 1 for the making of a meat *chiang*.[6]

Procedure for making fermented meat paste (Chiang)

Beef, lamb, roebuck, venison and rabbit can all be used. Take good quality meat from a freshly killed animal, trim off the fat and chop it well. Old meat that has dried out should not be used. If too much fat remains, the meat paste will taste greasy. Dry a batch of microbial *ferment* in the sun, grind it and sift the powder through a fine silk sieve. Mix together approximately one *tou* 斗 (i.e. ten *shêng* 升) of chopped meat, five *shêng* of *ferment* powder, two and

[6] *CMYS*, p. 420.

half *shêng* of white salt, and one *shêng* of yellow mould *ferment* (*huang chêng* 黃蒸). The yellow mould *ferment* has also been sun-dried, ground and sifted through a sieve.

The materials are blended evenly on a tray, then placed in a jar. If there is bone [in the meat] it should be chopped up further before the meat is used to fill the jar. The marrow in the bone tends to be rich in fat. Its presence would make the *chiang* fatty and greasy.[7] [When the jar is filled] the mouth is sealed with mud. It is then exposed to the sun. If the process is carried out in winter, the jar should be buried in a pile of cereal husk. After two times seven days, the jar is opened for inspection. If the odour of the *ferment* has disappeared, the *chiang* should be ready for consumption.

Buy a newly killed pheasant, boil [the dressed bird] thoroughly until all the meat is disintegrated in the soup. Strain off the bone. Allow the soup to cool and then use it to dilute the *chiang* [as obtained above]. Chicken soup may also be used [instead of the pheasant soup]. The important point is not to start the process with old meat, which will give the *chiang* a greasy taste. If no pheasant or chicken soup is available, dilute the *chiang* with a good wine. The diluted *chiang* is further ripened by warming in the sun.

Thus, we can say that this recipe is an enlargement of that described by Chêng Hsüan in the Later Han (page 334). The main difference is that in this case no wine (and no other aqueous liquid) is added to the incubating mixture. The substrate is hydrolysed rather slowly, since the high salt concentration is inhibitory to the action of the proteolytic enzymes and water, a key reactant, is only sparingly available. The product is a thick paste which has to be diluted with chicken soup or wine before it is served. If there is a significant amount of bone in the substrate, the product would be a *ni* 臡 rather than a *hai* 醢. The diluted *chiang* would presumably have the consistency of a thick 'sauce' or puree. The same procedure is applied to fish as the substrate in recipes 2, 3, 4 and 5 with similar results. In recipes 2 and 5 wine is a component of the incubation mixture; the digestion is much faster and the product is said to be ready in a day. These observations suggest that it is certainly possible, or perhaps even likely, that the various types of *chiang* mentioned in the *Chou Li* and *Li Chi* were prepared by the same procedure.

Thus, the hydrolysis of substrate proteins aided by enzymes in the mould *ferment* (*chhü*) is the basic methodology for making fermented paste (*chiang*) from animal parts from the Han to the time of the *Chhi Min Yao Shu*, +544. But that is not the only way that meat and fish *chiang* was made during this period. One variation is shown in recipe 7, used for the preparation of the famous *chu i* 鰒鮧, which literally means 'chasing *i*' *chiang*. Legend has it that,[8]

When the Han emperor Wu-ti (−140 to −88) chased the *I* 荑 barbarians to the sea shore, he smelled a potent, delicious aroma, but could not see where it came from. He sent an

---

[7] There is another recipe for making meat *chiang* in the *CMYS*, but it is presented in ch. 81, p. 505, as a method for making *tzu* 滓. In this case, it is also stated that any bone attached to the meat should be chopped up together with the meat. Thus, the product obtained could be a *ni*, as well as a *hai*. The word *tzu* is found only in this context in the *CMYS*; its origin is obscure.

[8] *CMYS*, p. 422. The *I*'s, in this case, refer to the minority peoples who inhabited the coastal regions of China stretching through the ancient kingdoms of Wu and Yueh up to the Shantung Peninsula before the arrival of the Han people from the central plains. The *I* people of Shantung and the northeast coast were referred to as *Tung I* (Eastern I), and those in the south as *Nan I* (Southern I), cf. Pulleyblank (1983).

emissary to investigate. A fisherman revealed that the source was a ditch in which was piled layer upon layer of fish entrails. The covering of earth could not prevent the aroma from escaping. The emperor tasted a sample of the product and was pleased with the flavour. This *chiang* then became known as *chu i* to commemorate the fact that it was obtained while chasing the I barbarian. It is simply a fermented paste made from fish entrails.

To make *chu i*: take the intestine, stomach and bladder of the yellow fish, shark and mullet, and wash them well. Mix them with a moderate amount of salt and place them in a jar. Seal tightly and incubate in the sun. It will be ready in twenty days in summer, fifty days in spring or fall and a hundred days in winter.

No microbial *ferment* or wine was used. The process depended entirely on autolysis, that is, on the action of the enzymes present in the fish organs on the proteins *in situ*.[9] The presence of high levels of salt would ward off contamination by putrefying bacteria. Recipe 6, however, is by no means an exception to the norm. The use of autolysis is encountered in another recipe in the *Chhi Min Yao Shu*. It is given in chapter 75 (Dried meat and fish) and describes a method for making *i yü* 浥魚, literally a damp salted dried fish.[10]

Method for *i yü*: It can be made in any season. All kinds of live fish can be used, except sheatfish and others without scales. Remove the gills, slit the stomach, slice the fish lengthwise into two parts and clean each part thoroughly. Do not remove the scales. In the summer apply extra salt. In spring, summer and winter, salt according to taste, but the flavour should always be on the salty side. The two parts are reunited to form a whole fish. In the winter the fish may be wrapped in the mat. In the summer the fish is layered in a jar. The mouth is sealed so that no flies can enter it to lay eggs. Holes should be drilled at the bottom and plugged. The fishy juice formed can be drained off as needed.[11] When the flesh turns red, the fish preserve is ready. For eating, the salt is washed off, and the fish boiled, steamed or roasted. Its flavour often surpasses that of fresh fish. The preserved fish may be processed further to make *chiang* (paste) or *cha* 鮓 (pickle); or it may be barbecued or deep fried.

The recipes for *chu i chiang* and *i yü* indicate that autolysis was still practised as an means of processing fish products in the +6th century. In fact, they may represent surviving examples of a technology that was widely applied in early historic China (−1500 to −500). The procedures are based on the same principle as that employed by ancient Greeks and Romans for making *garum* and *salsamentum*,[12] and similar to those current in Southeast Asia for the preparation of a variety of fermented fish products.[13]

Mention of *cha* in the recipe for *i yü* leads us to yet another variation in the technology for meat preservation, one based on the intervention of lactic acid bacteria.

---

[9] The proteases from fish viscera have been found to be particularly effective in hydrolysing fish proteins, cf. Beddows & Ardeshir (1979); Noda *et al.* (1982) and Itoh, Tachi & Kikuchi (1993), p. 179, and Knochel (1993), p. 219.

[10] *CMYS*, p. 460.

[11] This would have provided an excellent device for collecting a fish sauce from the incubation, if the incubation had been allowed to continue.

[12] For an account of the role of fermented fish paste and sauce in the life of the ancient Greeks and Romans, cf. Curtis (1991).

[13] Cf. Lee, Steinkraus & Reilly, eds. (1993).

In Recipe 6, Table 35 instead of a mould *ferment* (*chhü* 麴), cooked rice (*fan* 飯) was added to the incubation. The starch from the rice would be hydrolysed to sugar by the shrimp enzymes and some of the sugar converted to lactic acid, thus lowering the pH of the medium. The acidity would inhibit hydrolysis of the proteins, help to preserve the integrity of the shrimp, and suppress the growth of undesirable bacteria. The same principle also works in recipe (8), which, indeed, bears the title of 'a method for the preservation of crab'. As shown by Shinoda Osamu in his classic studies on the history of *cha*, this was already a major method for the preservation of fish in the 6th century.[14] In fact, the product preserved in this way was important enough to have been given a special name, *cha* 鮓, and a whole chapter, No. 74, is devoted to its preparation in the *Chhi Min Yao Shu*. *Cha* is mentioned in the *Erh Ya* (Literary Expositor), c. −300, as a synonym of *chih* 鮨, a *hai* 醢 made from fish.[15] According to the *Shih Ming* (Expositor of Names), +2nd. century, '*cha* 鮓 is obtained by fermenting [the substrate] with salt and rice (*mi* 米) to give a pickled product suitable for consumption as food'.[16] The character *cha* is formed by combining the radical for fish, *yü* 魚, with the right half of the word *tsho* 乍, one of the ancient progenitors of vinegar suggesting, that *cha* is fish kept in sour medium. The word *cha* is not seen in the *Chou Li* or *Li Chi*. It was probably coined sometime early on in the Han dynasty when the product *cha* became a significant factor in the Chinese diet.

Chapter 74, entitled 'Fish *Cha*', of the *Chhi Min Yao Shu* gives eight recipes for making *cha*. They are summarised in Table 36, which also includes a recipe for making pork *cha* from chapter 80.[17] The most detailed description of the procedure is provided by recipe (1), which we shall quote in full:[18]

Method for fish cha: Take a fresh carp, the larger the better. A lean one would be preferable to a fat one. While a fat fish may have superior flavour, it does not keep as well. If the fish is longer than one and half foot, the skin will be tough and the bones sturdy. Such a fish would not be suitable for slicing into *khuai* 膾, but it can still be used for making *cha* 鮓.[19]

After the scales are removed, the fish is cut into pieces about five inches long, one inch wide and half an inch thick.[20] As a piece is cut, it is placed in a basin of water. Clean off the

---

[14] Shinoda Osamu, papers on 'Investigation of *Cha* (Sushi)' Nos. 1 to 10, 1952–1961. The papers are listed in Ishige ed. (*1989*), 1, pp. 10–11. Of particular interest to us are papers No. 1 (1952), 'Evolution of *Cha* (*sushi*) in China', which was translated into Chinese by Yü Ching-Jang (1957), and No. 9 (1957), 'Historical References to *Sushi* in China.'

[15] *Erh Ya*, ch. 6, Explanation of Implements, states that '*chih* is to fish as *hai* is to meat', indicating that *cha* is a kind of *chiang*.

[16] *Shih Ming*, ch. 13. Shinoda (*1952* tr.*1957*) suggests that the technique of fermenting fish with rice originated in South China, since during the Eastern Han the principal grains in the North were millets, barley and wheat. However, we have to remember that the original word in the text for 'rice', *mi* 米 does not necessarily indicate rice since *mi* can also mean any type of dehusked grain.

[17] Ch. 74 contains seven recipes, six for fish *cha*, and one for pork *cha*. Ch. 80 is on roast meats but it includes one recipe for pork *cha*.

[18] *CMYS*, p. 454.   [19] *Khuai* is thinly sliced fish, cf. pp. 69–74 for a discussion on its identity.

[20] The text continues at this point, 'Each piece should have a bit of skin attached to it. If the pieces are too large, the outside will be overheated, and acid production would be inhibited. Only the central section will be good to eat. The meat too close to the bone tends to be too raw and fishy to be edible. Often only about a third of the whole fish can be consumed with enjoyment. Small pieces which will ripen more evenly are, therefore, preferred. The dimensions of the pieces are to serve as a guide, and should not be followed blindly. The vertebra should be cleaved directly. Where the meat is thick the skin can be narrow. Where the meat is thin the skin should be wider. Every piece should have skin on it. Do not use any piece without skin in the process.'

Table 36. *Making of fish and meat preserves*, Cha, *in the* Chhi Min Yao Shu

Number	Substrate	*Fan* (cooked rice)	Salt	Spices	Incubation[a]
1.	Carp, in large chunks	Yes	Yes	Yes[b]	Layer fish, rice in jar
2.	Chunks of fish	Yes	Yes	Yes/No[c]	Wrap in lotus leaves
3.	Chunks of carp	Yes	Yes	No	Incubate overnight
4.	Carp chunks				
	Stage 1	No	Yes	No	Bury in salt
	Stage 2	Yes	No	No	In jar
5.	Chunks of fish				
	Stage 1	No	Yes	No	Bury in salt
	Stage 2	Yes	No	No	Soak in water
6.	Chunks of fish	Yes	Yes	Yes[d]	Ingredients mixed in jar
7.	Chunks of rehydrated dry fish	Yes	Yes	Yes[e]	Layer ingredients
8.	Chunks of cooked pork	Yes	Yes	Yes	Mix ingredients
9.	Chunks of raw pork	Yes	Yes	—	As in no. 1

[a] Unless other stated, the ingredients are incubated in a sealed jar kept at room temperature or warmed in the sun. All recipes are in ch. 74, except number 9, which is found in chapter 80.
[b] Dogwood seed, orange peel and good wine are mixed with the rice.
[c] If available, dogwood seed and orange peel may be added.
[d] Wine, ginger, orange peel and dogwood seed are used.
[e] Line jar with dogwood leaves, and mix rice with dogwood seed.

blood as the piece is being soaked. When all the pieces are sliced and washed, strain off the water, return the pieces to a basin of fresh water, wash and strain again. The pieces are placed on a plate, mixed evenly with white salt and transferred onto a basket. The basket is placed on a horizontal stone tablet to allow the water in the fish to drain away. We call the salt, 'chasing water salt', since it expels water as it is absorbed by the fish. If the water in the fish is not thoroughly drained out, the shelf life of the *cha* will be poor. It is quite safe to allow the water to drain overnight. When the draining of water is complete, roast a piece to check the saltiness of the fish. If it is not salty enough, put more salt in the *sang* 穄 (i.e. starch substrate in the form of *fan* 飯, cooked rice). If it is too salty, no salt need be added. After the starch substrate is properly applied, place a layer of salt over the incubating medium.

The starch substrate is prepared by cooking non-glutinous rice. The cooled rice should be on the hard side. It should not be too soft. If it is too soft the fish preserve will spoil more easily. Mix into the rice whole dogwood seeds, finely sliced orange peel, and a little good wine. Ideally, the consistency of the rice granules should be such that they would stick to the fish pieces. The function of the spices is to enhance the aroma of the product; thus, only a small amount is needed. If orange peel is not available, one may replace it with *tshao chü tzu* 草橘子 (grass tangerine seeds).[21] Wine is useful to deter contamination. It will improve the quality of the *cha* and hasten its ripening. Usually for every *tou* of fish pieces, use half a *shêng* of wine.

[21] The identity of *tshao chü*, grass orange, is uncertain.

Layer the fish pieces in a jar. One row of fish is placed beside one row of starch substrate, and repeat until the jar is full. Pieces with the most fat should be layered near the top of the jar. They will spoil more easily, and should be consumed first. Extra amounts of starch substrate should be placed on top of the last layer of fish. Criss-cross a row of regular bamboo leaves against a row of broad bamboo leaves to cover the top of the medium. Repeat until eight layers are in position. If no broad bamboo leaves are available, use leaves of wild rice or the water reed. In winter and spring, when there are no leaves, they may be replaced by stems. Finally, cut off strips of bamboo and criss-cross them to form a cover over the mouth of the jar. If bamboo is not available, use strips of bramble. Leave the jar inside the house. Do not expose it to the sun or leave it near the fireplace. Doing so will promote spoilage, and ruin the flavour.[22] In cold weather wrap it liberally with straw. Do not allow it to freeze. When a red liquid oozes over the mat, discard it. When a light liquid comes up and tastes sour, the product is ready. When serving it, tear the pieces by hand. Cutting it with a knife would give it a fishy taste.

Chia Ssu-Hsieh, the author, understands that when fish is salted, water is expelled as salt is absorbed by the tissues. Under these conditions, the extent of autolysis is limited, since the concentration of active water in the fish would be low. The sugar derived from the starch substrate, *sang*, would be converted by lactic acid bacteria to lactic acid, resulting in a lowering of the pH of the medium. In effect, the fish is preserved by pickling in an acid salty medium. It should be noted that the method was not restricted just to fish. Meat could also be treated in the same way. In spite of its title, chapter 74 includes a recipe (No. 8 of Table 36), in which pork is the raw material, although it is first cooked before it is incubated with the starch substrate in the salty medium. But in recipe 9 (from chapter 80), the pork is incubated without cooking. In the later literature the term *cha* was further expanded to include pickles from vegetables as well as fish and meat.

As we have seen, meat and fish pastes (*chiang* 醬) were the principal condiments and relishes in pre-Han China. They continued to be highly regarded in the *Chhi Min Yao Shu* of the 6th century. But their importance steadily declined in the face of rising competition from fermented soy products, which by the time of the Sung had become the dominant flavouring agents in the Chinese culinary system. Yet fermented meat and fish paste still lingered on the scene for several more centuries after the Sung. We find the preparation of meat and fish *chiang* mentioned in the standard food literature from time to time until the late Chhing, for example, in the *Ssu Shih Tsuan Yao* 四時纂要 (Important Rules for the Four Seasons),[23] late Thang, *Wu Shih Chung Khuei Lu* (Madam Wu's Recipe Book),[24] Southern Sung, *Shih Lin Kuang Chi* (Record of Miscellanies),[25] +1283, *Chü Chia Pi Yung* (Compendium of Essential Arts for Family Living),[26] Yuan, *To Nêng Pi Shih* (Routine Chores Made Easy),[27]

---

[22] The author appears ambivalent on this point. In another recipe, Table 36, No. 6, the directions call for placing the incubating jars in the sun.
[23] *SSTY*, 12th month, describes the making of rabbit *chiang* and fish *chiang*.
[24] *WSCKL*, meat *chiang*, p. 11; fish *chiang*, p. 9.
[25] *SLKC*, a fish head, *chiang*, a sliced fish *chiang* and mixed fish and rabbit meat *chiang*, p. 268.
[26] *CCPY*, a meat *chiang*, p. 56, a venison *chiang* (*hai* 醢), p. 57 and a fish *chiang*, p. 81.
[27] *TNPS*, *chiang* made from meat, bird, rabbit, p. 378, venison *hai*, p. 378; fish *chiang*, p. 378.

+1370, *Chhü Hsien Shên Yung Shu* (Notes of the Slender Hermit),[28] +1440, *Yin Chuan Fu Shih Chien* (Compendium of Food and Drink),[29] +1591, and *Shih Hsien Hung Mi* (Guide to the Mysteries of Cookery,[30] +1680. A fish-head paste, *yü thou chiang* 魚頭醬 is mentioned twice in the *Mêng Liang Lu* (Dreams of the Former Capital), +1334.[31] However, none of these works, as well as other food canons from Sung to Chhing, describe their use in cookery. The only examples we have found occur in chapter 80 of the *Chhi Min Yao Shu*, in which fish paste (*yü chiang* 魚醬) is used in one recipe and *yü chiang chih* 魚醬汁 (juice from fermented fish paste) in five recipes.[32] Evidently, fermented meat and fish pastes were seldom used in cooking. They were more often served as a condiment or relish to accompany or complement the eating of main items of food as described in the *Chou Li* and *Li Chi*.[33] After the *Shih Hsien Hung Mi*, the preparation of fermented meat or fish paste is rarely mentioned in the food literature.[34] It would appear that since the early days of the Chhing dynasty, they had ceased to be a significant factor in the Chinese food scene.

Pickled fish (*cha* 鮓) seems to have enjoyed a more enduring popularity than that of meat and fish pastes in the centuries following the *Chhi Min Yao Shu*. There are numerous references to *cha* scattered in the Sui and Thang literature,[35] and at least a dozen types of *cha* are mentioned in the *Mêng Liang Lu* as food articles seen in the markets of the Southern Sung capital.[36] The preparation of *cha* is described in most of the standard food canons of the Sung, Yuan and Ming, such as the *Wu Shih Chung Khuei Lu*,[37] Southern Sung, *Shih Lin Kuang Chi*,[38] +1283, *Chü Chia Pi Yung*,[39] Yuan, *I Ya I I*,[40] Yuan/Ming, *Tuo Nêng Phi Shih*[41] +1370, *Yin Chuan Fu Shih Chien*,[42] +1591, *Shih Hsien Hung Mi*,[43] +1680, and *Yang Hsiao Lu* (Guide to Nurturing Life),[44] +1698. The raw materials used, include pork, lamb, goose, sparrow, various kinds of fish, shrimp, mussel and clams. Several variations on the basic procedure are mentioned. One is

---

[28] *Chhü Hsien Shên Yin Shu*, meat *chiang*, venison *hai*, p. 428.
[29] *YCFSC*, fish *chiang*, p. 79. On p. 84 there is a process for meat *chiang*, in which finely chopped meat is incubated with *chiang*, presumably fresh soy *chiang* which would supply both *ferment* and salt.
[30] *SHHM*, fish *chiang*, p. 101. A fish roe *chiang*, p. 50 and a chicken *hai*, p. 114 are mentioned, but they are not fermented products and thus irrelevant to our discussion.
[31] *MLL*, pp. 134, 139.
[32] *CMYS*, ch. 80, *yü chiang*, p. 496; *yü chiang chih*, three recipes on p. 495, and two on p. 496.
[33] See *Chou Li*, *Hai Jên*, p. 55, and *Li Chi*, *Nei Tsê*, p. 457. For example, the *Li Chi* states that sliced meat strips shoud be eaten with a frog *hai*, a meat soup, a rabbit *hai*, thinly sliced meat, a fish *hai*, finely sliced fish, a mustard *chiang*.
[34] There are no references to meat or fish *chiang* in the three major culinary works of the Chhing dynasty *Yang Hsiao Lu*, +1698, *Hsing Yang Lu*, +1750, and *Sui Yüan Shih Tan*, +1790. However, a fish *chiang* and a shrimp *chiang* are mentioned on pp. 404 and 449 in the *Thiao Thing Chi* (The Harmonious Cauldron).
[35] Shinoda (*1957*), pp. 45–6.
[36] *MLL*, pp. 109, 111, 134, 138–9 for the varieties of *cha* mentioned. Cf. the *Chhing I Lu*, p. 3, for a fish *cha* in the shape of a peony, and Shinoda (*1957*) for additional references.
[37] *WSCKL*, meat, p. 6; yellow sparrow, p. 12.
[38] *SLKC*, p. 265, yellow sparrow, pork, lamb, goose, duck, carp; p. 268, lamb, shrimp.
[39] *CCPY*, pp. 84–6, four kinds of fish, yellow sparrow, clam and goose. [40] *IYII*, p. 15, fish.
[41] *TNPS*, pp. 374–5, five kinds of fish, yellow sparrow, pork, lamb, goose, duck, mutton, shrimp. Many recipes are copied verbatim from *SLKC* or *CCPY*.
[42] *YCFSC*, p. 70, meat, p. 73, fish, pork, lamb, p. 82, musssel, p. 84, yellow sparrow.
[43] *SHHM*, p. 97, fish, p. 102, mussel, p. 113, chicken, pork, lamb.
[44] *Yang Hsiao Lu*, pp. 73–4, two kinds of fish.

to use a mould *ferment* (*chhü* 麴) instead of cooked rice (*fan* 飯) as in the case of the celebrated yellow sparrow *cha* of the *Wu Shih Chung Khuei Lu*.[45] In other recipes such as that for fish *cha* in the *Chü Chia Pi Yung*,[46] both *chhü* and *fan* are added. And still in another case, as in the fish *cha* in the *Yang Hsiao Lu*,[47] neither *chhü* nor *fan* were used. No references to *cha* are seen in the *Hsing Yuan Lu* (Memoir from the Garden of Awareness), +1750, and the *Sui Yuan Shih Tan* (Recipes from the Sui Garden), +1790, although recipes for three types of fish *cha* are given in the *Thiao Thing Chi* (The Harmonious Cauldron), the massive compendium of food recipes collected during the 19th century.[48]

Based on these records, we would infer that both fermented paste (*chiang*) and pickled preserves (*cha*) of animal origin started to recede from the Chinese food scene by the beginning of the 18th century. Indeed, they seem to have almost totally disappeared nowadays. One no longer encounters any of these products in the food markets in China nor reads about their use in contemporary cook books. This conclusion is certainly true with regard to meat *chiang* and *cha*. But what has happened to fermented fish products is not quite so simple. While the terms for fish *chiang* and *cha* are seldom encountered in the modern Chinese culinary vocabulary, fermented fish products, though surviving in China only in limited localities, are alive and flourishing as major components of the diet of millions of people in Southeast Asia and Korea.[49] This situation immediately raises two questions. First, why did fermented meat and fish products lose their appeal? And the second, what is the relationship between the thriving fermented fish products of Southeast Asia and the ancient fish *chiang* and *cha* of China?

The answer to the first question is fairly clear. We have noted that as fermented soy condiments increased in popularity, fermented meat and fish products declined in importance. Eventually the soy condiments drove their meat and fish counterparts off the market. But then another question arises. Based on the amount of space allotted to them in the *Chhi Min Yao Shu*, we may say that fermented soybeans (*shih* 豉) and paste (*chiang* 醬) were already beginning to overtake animal-derived fermented pastes and preserves in the 6th century.[50] Yet it was not until the end of the 17th century, more than 1,200 years later that soy products completely displaced their animal-derived counterparts. Why did it take so long? To find the answer we need to consider the function of the various products being compared. Fermented soybeans and soy paste (*shih* and *chiang*) were used primarily as condiments although the former were also occasionally used as a relish. In the case of meat and fish,

---

[45] *WSCKL*, p. 12. In this case a red *ferment* is also added.
[46] *CCPY*, p. 84. A red *ferment* and cooked rice are used.
[47] *YHL*, p. 73. In effect, this is a procedure for making fermented, salted fish slices, a type of *yên yü* or *shiokara*, cf. pp. 392–3.
[48] *TTC*, pp. 372, 374, 392. Fish *cha* is also mentioned in the *Hu Ya*, Wang Yüeh-Chêng (1872), in Shinoda & Seiichi, p. 510.
[49] Cf. Lee, Steinkraus & Reilly eds. (1993).
[50] Compare the space allotted in the *CMYS* to fermented soy products in chs. 70 and 72 with that allotted to fermented meat and fish products in chs. 70 and 71.

while fermented paste (*chiang*) was used as a condiment, fermented pickle (*cha*) was consumed mainly as a relish. Thus, it would make sense to leave *cha* out of this discussion and compare fermented soybeans and soy paste (*shih* and *chiang*) only with fermented meat and fish pastes (*chiang*).

When we compare soy condiments with fermented meat and fish pastes, one factor immediately stands out, and that is, the difference in cost of the respective raw materials. Soybeans are an inexpensive agricultural commodity, while meat and fish are expensive animal products. Soybeans can be stored indefinitely, while meat and fish are highly perishable and are preferably used when fresh. Thus, it is inevitable that meat and fish paste would be more expensive than soy *chiang*. Moreover, since meat and fish represent a highly heterogenous group of substrates, it would be more difficult to standardise or scale up the process, except in the case of trash fish and shellfish, such as shrimp and oyster, in localities where they occur in great abundance. On the other hand, meat and fish pastes might have possessed unique organoleptic properties which soy paste could not duplicate. It is presumably for this reason that fermented meat and fish pastes continued to be made and used in China during the years from the Sung to the Ming. It achieved a limited but stable share of the market. This situation might have persisted indefinitely had it not been for another factor which came into play. Soy sauce gained recognition as a separate and desirable soy condiment and its popularity rose steadily over the next several centuries. It appeared that there was a swing of culinary preference from semi-solid condiments such as *chiang* (paste) to liquid products, such as *yu* (sauce). By the end of the 17th century, soy sauce (*chiang yu* 醬油) had become the dominant condiment in Chinese cuisine. Soy paste was relegated to a secondary position. The virtues of meat and fish pastes were no longer sufficient to keep them alive. It is no accident that the only product in the animal derived category that survives today is fish sauce (*yü lu* 魚露), which literally means fish dew, a topic to which we shall return shortly.

But before we deal with the history of fish sauce we need first to consider what happened to *cha*. Meat and fish preserves were not competitors of fermented soy paste and soy sauce. Why then did *cha* also lose favour in China at the end of the 18th century? The answer may have nothing to do with the changing preference in fermented soyfoods.[51] *Cha* is preserved meat or fish made by incubating the raw material with salt and cooked rice. But not all preserves were prepared by this procedure. By the beginning of Sung, alternative technologies were already current for making preserves of meat, fish, and vegetables. One is by incubating the substrate

---

[51] Ishige & Ruddle (*1990*), p. 44, have pointed out another factor that has helped to hasten the decline of *cha*, and that is the general tendency of the Chinese through the centuries to move away from raw foods to cooked foods. Thus, the ancient practice of consuming uncooked meat or fish in the form of *jou khuai* 肉膾 and *yu khuai* 魚膾 that we discussed on pp. 69–70, 74–5, have all but disappeared from the Chinese culinary scene. Yü Ching-Jang, in his translation of the Shinoda (*1957*) paper on *cha*, related how he almost vomited when he took his first bite of *narezushi* during an early visit to Japan. Later he received from Shinoda a reprint of his paper on the history of *cha* in China. He was so amazed that the product had a Chinese connection, that he decided to translate it into Chinese for the benefit of his countrymen.

in *tsao* 糟, the dregs or residues from wine fermentations, and another is by incubating the raw material in *chiang*, the finished soy paste.⁵² Apparently, by the end of the 18th century, the use of *tsao* and *chiang* had superseded the methodology of *cha*. The applications of *tsao* and *chiang* for the preservation of foods will be considered in detail later in this chapter. Let us return now to the second question and examine how the technology of fish sauce, paste and preserve (*than* 䤃, *chiang* 醬 and *cha* 鮓) might be related to the modern fermented fish products of Southeast Asia.

Based on the information we have gathered from the Han and pre-Han literature the historical development of fermented fish and shell fish products in China may be depicted as follows:

1. *Autolysis*: Salted fresh fish is incubated alone in an enclosed space to give, depending on the length of treatment, a fermented salted fish, *yên yü* 淹魚, a fish paste, *hai* 醢, and eventually juice or sauce, *than* 䤃. This most ancient technique was probably discovered well before the −10th century. It could have been the way the earliest fermented fish (and meat) foods on record, that is, the *hai* and *than* of the *Shih Ching*, were prepared.⁵³ The methodology is very simple. The only essential ingredient is salt,⁵⁴ which was certainly available when Chinese civilisation began at about −2000. This procedure, however, was largely displaced by the more efficient *chhü fermentation* by the time of the Western Han.

2. *Chhü fermentation*: Chopped fish is incubated with salt and the mould *ferment* (and sometimes with wine) to give a paste (*hai*), and which can be pressed to give a sauce or juice (*than*). This methodology was probably invented some time between −6th and −10th century. To differentiate the fermented paste made by the new technique, from that made by *autolysis*, it was given the name *chiang*, a term quite familiar to the disciples of Confucius in the −6th century. According to Chêng Hsüan, the *chhü* process was used to make the animal-derived *chiang* and *hai* listed in the *Chou Li* and *Li Chi*.⁵⁵

3. *Lactic acid fermentation*: Chunks of fish are incubated with salt and *fan*, cooked rice, to give *cha*, a pickled preserve. We may presume that this was the technology used for making the pickled vegetables, *tsu* 菹, so often mentioned in the *Chou Li* and *Li Chi*.⁵⁶ It was probably applied to meat and fish some time between −1st and

---

⁵² *Chhing I Lu*, pp. 31–2, *tsao* is called the bone of wine; *WSCKL*, pp. 9, 17, 22.
⁵³ The earliest fermented animal-derived paste and sauce were known as *hai* and *than*, as listed in the *Shih Ching*, poem W 197 (M 246), cf. p. 333. This suggests that these products were current between −700 to −1100. If they had been made by autolysis, the process could have been developed well before −1000.
⁵⁴ Salt was an important commodity during the Shang, cf. Chang, Kwang-Chih (1980), p. 258.
⁵⁵ Cf. pp. 333–4. The proposed chronology is based on two assumptions. The first is that the Chinese *ferment* (*chhü*) did not gain recognition as an independent culinary entity until the beginning of the Chou. The justification for this assumption has been discussed on pp. 157–62. Thus, it is unlikely that *chhü* could have been used to make fermented meat paste before the Chou dynasty. The second is that when a new term is introduced, it means that either an entirely new product is discovered, or that a new process is developed to produce an old product. The earliest reference to the word *chiang* is in the Analects of Confucius, who lived in the −6th century. Thus, it is likely that the *chhü* fermentation as applied to animal substrates was developed sometime between the maturing of *chhü* as a food product and the advent of the word *chiang*, i.e. between the −11th and −6th centuries. If so, the *chiang* and *hai* derived from animal substrates listed in the *Li Chi* and *Chou Li*, which are based on material current before −300, were probably made using the *chhü* fermentation process.
⁵⁶ The nature of *tsu* will be discussed later, pp. 402–4.

–3rd century. A new noun *cha* was coined to denote preserved meat or fish prepared in this way.[57]

All three procedures are described in the *Chhi Min Yao Shu*, although the examples of autolysis (method 1) may only be remnants of a methodology that was already on its way out. The manner the substrate is prepared is keyed to the nature of the end-product. If the product is to be a paste, the substrate is finely chopped, as in method 2. If it is to be a preserved fish, the substrate is cut into large chunks as in method 3. However, there was little interest shown in the post-Han food literature on the making of a fermented fish sauce, even though meat or fish sauce, under the name, *than*, is mentioned in the *Chou Li* and listed among the bamboo slips uncovered in Han Tomb No. 1 at Ma-wang-tui.[58] Thus, from the Han to the Sung the situation regarding fermented fish products parallels that which we found in fermented soyfoods.[59] For fish the preference was for fermented preserves (*cha*) and pastes (*chiang*); for soyfoods it was for fermented beans (*shih*) and pastes (*chiang*). We have already delineated (pages 358–62) the emergence of fermented soy sauce as a condiment in its own right during the Sung and its rise during the Chhing as the dominant condiment in Chinese cuisine. Did a similar development take place in regard to fermented fish products? Based on the standard food literature from the Sung to the Chhing 'No' would apparently have to be the answer. Furthermore, there is no mention of a fermented fish sauce (*yü chiang yu* 魚醬油) in the late Chhing culinary works such as the *Sui Hsi Chü Yin Shih Phu* 隨息居飲食譜 (Dietary System of the Random Rest Studio), 1861, by Wang Shih Hsiung, *Chung Khuei Lu* 中饋錄 (Book of Viands) by Chêng I, c. 1870, and the anonymous *Tiao Thing Chi* (19th century).

And yet a fish sauce, *yü lu* 魚露 (literally fish dew), also called *yü chiang yu* or *chhi yu* 鯕油, is today a well-known condiment near certain cities along the Southeastern coast of China. The modern Encyclopedia of Chinese Cuisine (*Chung-kuo Phêng-jên Pai-kho Chhuan Shu* 中國烹飪百科全書), states that *yü lu* is made from trash fish in Chekiang, Fukien, Kwangtung and Kwangsi. A shrimp sauce, *hsia yu* 蝦油, is also well known in the coastal areas of these provinces.[60] They are said to have been in use for hundreds of years. Why then is fish sauce not mentioned at all in the culinary classics of the 18th and 19th centuries? One possibility is that following the example of the emergence of soy sauce from fermented soy nuggets and soy paste, a fish sauce did begin to evolve from fish preserve and fish paste during the Ming and the Chhing. But it was a development necessarily restricted to areas where inexpensive trash fish from the sea was readily available. Fish sauce, therefore, remained as a product of regional interest only, outside the mainstream culinary culture of North and Central China and ignored by the major food writers of the Ming and Chhing. Indeed, Hung Kuang-Chu has expended a considerable amount of effort to trace the lineage of modern fish sauce (*yü lu*) to the fish paste (*yü chiang*) and fish preserves

---

[57] In this case, the new word may signal the application of *tsu* technology to a new category of substrates.
[58] *Chou Li Chu Shu*, *Hai Jên*, ch. 6, p. 88; *MWT Inventory*, item 90, cf. p. 380.
[59] Cf. discussion on pp. 371–3.
[60] *CKPJPKCS* (1992), pp. 707–8. See also Ishige & Ruddle (1985) and (1990).

(*cha*) of ancient and mediaeval China, but was unable to establish any connection between them.[61]

(ii) *Fermented fish products of East Asia*

There is, however, another possible explanation for the origin of modern Chinese fermented fish sauce. The conversion of small fish (as well as shell fish such as shrimps) into fish (and shrimp) sauce may have been transmitted from Southeast Asia into China during the 17th or 18th centuries. Based on his extensive studies of the ethnogeography of fermented condiments in East Asia, Ishige Naomichi has divided East Asia into two culinary regions (Figure 88), a northern region dominated by soy products, and a southern region dominated by fish products.[62] But he also found that, as shown in Figure 89, there are pockets in the soy condiment region, for example coastal parts of Japan, Korea, the Shantung Peninsula and Southeastern China, where fish sauce is still a significant component in the diet of the people.[63] He noted that even within the fish condiment region, fish sauce is found only in restricted localities in Indonesia, Kalimantan and the Philippines. The heartland of fish sauce culture clearly lies in continental Southeast Asia, across Vietnam, Laos, Cambodia, Thailand and Burma. To understand the significance of these observations on the transmission of fermented fish condiments, particularly fish sauce, in East Asia, it would be helpful for us to examine the geographical distribution of these products as summarised in Table 37.[64]

There is a bewildering variety of fermented, salted fish products made and consumed in the countries of East Asia,[65] but they can all be classified into four categories (Table 37): (1) fermented salted fish (2) fish paste (3) fish sauce and (4) fish pickle. In Ishige's papers they are referred to as (1) *shiokara* (*yen hsin* 鹽辛)[66] (2) *shiokara* paste (3) fish sauce and (4) *narezushi* (*cha* 鲊). The methods used for their preparation follow those we have discussed earlier on page 390. Category (1) comprises the products made by limited autolysis, so that the shape of the original fish is retained (see method 1, p. 12). In category (2) the fish is allowed to autolyse more extensively and is then ground to a paste. In category (3), the autolysis is allowed to proceed even further until all the tissues are liquified. In category (4) the fish is incubated

---

[61] Hung Kuang-Chu (*1984a*), pp. 180–7. To us, it is particularly pertinent that fish sauce is commonly known as *yü lu*. If the example of soy paste (*chiang*) to soy sauce (*chiang yu*) is followed, fish sauce should have been called *yü chiang yu*. The very name *yü lu* suggests that the modern Chinese fish sauce is not descended from the traditional fish paste, which was made by incubating chopped fish flesh with salt and wine *ferment*. Our thesis is that it was disseminated from Southeast Asia, most likely Vietnam, in the 19th century.

[62] Ishige (1993), p. 22, fig. 6.   [63] Ishige (1993), p. 20, fig. 4.

[64] Adapted from Ishige & Ruddle (*1985*), pp. 182–3 and updated with information from Lee, S. W. (1993), Mabesa & Babaan (1993), Mohamed Ismail Abdul Karim (1993), Putro (1993), Myo Thant Tyn (1993), Phithakpol (1993), Itoh, Tachi, & Kikuchi (1993), and Cherl Ho Lee (1993). We thank Kim Vogt and Khien Du for a discourse on the various fermented fish products of Vietnam. Most of them are available in Oriental food markets in Alexandria, Virginia.

[65] For example, B. Phithakpol (1993), pp. 157, 161–3, lists twenty-two types of fermented fish and shrimp products in Thailand, and Cherl Ho Lee *ibid.*, p. 190, lists thirty-one types of such products in Korea.

[66] The characters *yen hsin* means salty and pungent.

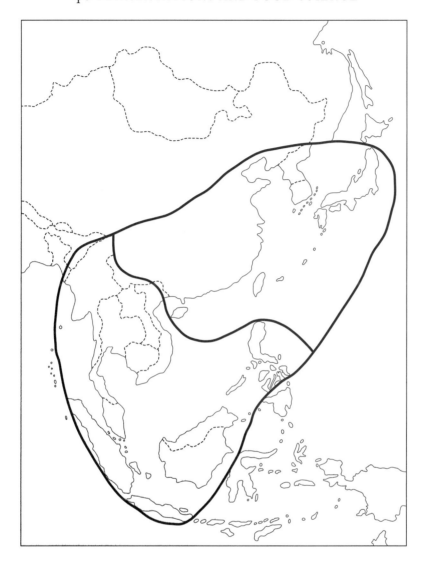

Fig 88. Distribution of fermented condiments in East Asia. The northern area is based on soybean products and the southern area on fish products. After Ishige (1993), Fig. 6. p. 22.

with cooked rice and preserved by the lactic acid produced (method 3, p. 390). It is important to point out that not all the products included in Table 37 are made and consumed today. Some were important in the past but are seldom seen today. For example, fermented salted fish of the *i yü* type (called *yên yü* 腌魚 in China and typified by the Japanese *shiokara*) was last mentioned in China in the *Chhi Min Yao Shu* (+544). It is virtually a forgotten food, except in certain localities where it can be conveniently prepared at home as a byproduct of fish sauce. Fish paste (*chiang*), and

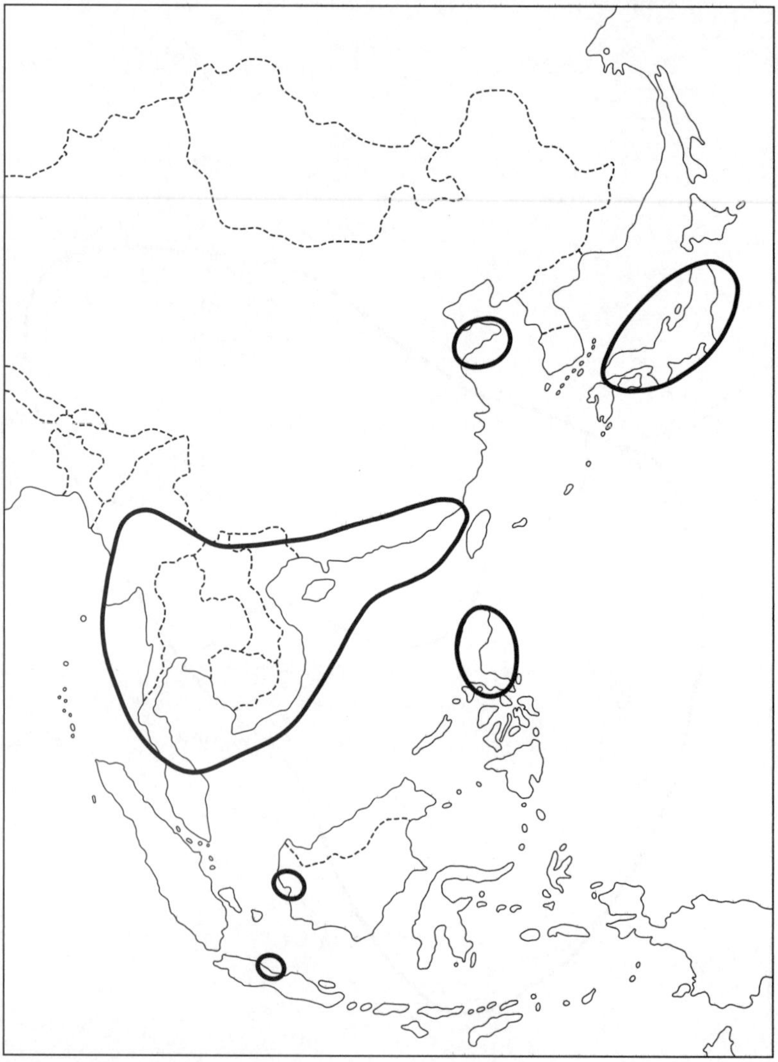

Fig 89. Distribution of fish sauce in East Asia. After Ishige (1993), Fig. 4, p. 20.

pickled fish (*cha*) have virtually vanished from the Chinese culinary scene since the 19th century.[67] *Shiokara* and *narezushi* were formerly important in Japan, but are now recognised only as local specialities with limited appeal.[68] In Korea *jeot-cal* 鯲鮧 and *sik-hae* 食醢 were well known by the 15th century. *Sik-hae* is hardly used today, but

[67] Cf. pp. 10, 11. Although the damp, fermented, salted fish is no longer made as an article of commerce, dried, salted fish has remained enormously popular as a relish or condiment in China today.

[68] Ishige (1993) states that *shiokara* (i.e. *yên yü* in Chinese) is a relic food in China, pp. 17, 29. One of the localities where it is still being made is Foochow and adjacent counties along the coast. According to Chen Jia Hua (private communication), it is made by mixing whole *chhi* fish with 20–30% of its weight of salt, and allowing the mixture to incubate at room temperature for about two months. The meat takes on a pink colour and is easily peeled off from the skin and bone. It is considered a delicacy for those who are brought up on it.

Table 37. *Products from fermentation of salted fish in East Asia*

Country	Products from salted fish			
	Fermented fish	Paste	Sauce	Pickled preserve
China	*i yü*	*yü chiang*	*yü lu chhi yu* *yü-chiang yu*	*cha*
Japan	*shiokara*	*shiokara* paste	*shiojiri ishiru ikanago*	*narezushi*
Korea	*jeot-kal*	*jeot-kal* paste	*jeot-kuk*	*sikhae*
Vietnam	*ca kho mam*	*mam nem*	*nuoc mam*	*mam tom chua*
Cambodia	*prahoc*	*prahoc*	*tuk trey*	*phaak*
Laos	*pa daek*	*pa daek*	*nam pa*	*som pa*
Thailand	*pla deak pla ra*	*kapi*	*nam pla budu*	*pla ra pla som*
Burma	*ngapi kong*	*ngapi*	*ngan-pya-ye*	*nakyang: khying:*
Malaysia	—	—	*budu*	*kasam ikan masin*
Indonesia	*terasi ikan*	—	*kecap ikan*	—
Philippines	*bagoong*	—	*patis*	*burong isda*

jeot-cal has survived largely as an ingredient of the wildly popular *kimchi*.[69] It is pertinent that fish sauce is not mentioned in the older literature in all three countries before 1800. The fish sauce currently made and used in Japan, Korea and China are evidently of recent origin.

In contrast to Northeast Asia, fermented fish products (Table 37) are widely produced and consumed today in the countries of Southeast Asia. Fermented fish (*yen yü* or *shiokara*) is a popular relish in Vietnam, Cambodia, Laos, Thailand, Burma, and Luzon and the Visayas in the Philippines.[70] Pickled fish (*cha* or *narezushi*) is also a major article of food in the same countries.[71] Fish sauce is most popular in continental Southeast Asia with outposts along the Southeastern coast of China, Shantung, Korea and Japan.[72] According to Ishige, microbial ferment (*chhü* or *koji*) is added to the salted fish in the preparation of certain *shiokara* and *narezushi* products, reminiscent of the use of *chhü* in the making of the Chinese fish paste (*chiang*).[73] This shows a definite connection between the fermented fish technology of China and that of Southeast Asia. But it is uncertain as to what this connection actually means in terms of the origin of fermented fish condiments in the countries of East Asia.

Unfortunately, we have very little concrete information about the early history of fermented fish foods in Southeast Asia. It is hoped that food historians in the region will be able to correct this deficiency in the near future. But based on what we do know we may make the following generalisations:

1. The process of fermenting salted fish by autolysis probably arose independently in the countries of East Asia, China, Japan, and Korea in the north and Vietnam, Laos, Cambodia, Thailand and Burma in the south. Fishing is one of the

[69] Lee Sung Woo (1993), pp. 36–7, 38–40. [70] Ishige (1993), p. 17.
[71] Ibid., pp. 15–17. [72] Ibid., p. 19. [73] Ibid., pp. 19, 29.

earliest human activities aimed at gathering food. The technology for salting fish and allowing it to incubate is very simple. Apart from fish, the other key ingredient is salt, which was available from rock deposits and from evaporation of sea water during the Neolithic Age.[74] Salting is a common method for food preservation known to peoples of diverse cultures throughout the ancient world.[75] If salted fish were accidentally allowed to stand around before drying, autolysis would automatically occur. A keen observer would recognise that the salted fish had now acquired a new and appealing flavour. He/she would be tempted to try the method again. In other words, the fermentation of salted fish was an invention that was just waiting to happen. Thus, the fermented fish (腌魚 shiokara) type of product has probably been known in the different ethnic regions of East Asia well before the dawn of recorded history.

2. The technique of making pickled fish (cha or narezushi) with the lactic acid generated by fermenting salted fish with cooked rice was probably developed after the use of autolysis had become well established. Ishige suggests that it arose in the Mekong and associated river basins in Southwest China, and the adjacent highlands of Burma, Thailand and Laos, as a result of the coincident development of irrigated rice cultivation and rice field aquaculture.[76] Attractive though it is, there is no direct corroboration for this hypothesis. At least for the present, we cannot ignore the fact that the earliest mention of pickled fish (cha) is in the Erh Ya, −300,[77] and disregard South-Central China as a possible centre for the origin of this technology.[78] It appears to us unlikely that there had been more than one centre for the origin of this technique. It had to be transmitted from South-Central China to the countries of Southeast Asia or vice versa. In any case, it is probably from China, through commercial and cultural contacts and transmissions of such works as the Chhi Min Yao Shu, that this technique was disseminated to Korea and Japan.

3. The addition of a mould ferment (chhü or koji), in making fermented fish products started in China and spread southwards towards Vietnam, Cambodia and Thailand.[79] This transmission probably took place during the early Han when

---

[74] Ibid., pp. 25–6.

[75] Salting of fish was practised by the ancient Egyptians, cf. Darby, Ghalioungui & Grivetti (1977), p. 369; Tannahill (1973, 1988), pp. 53–4. After salting and drying, the product was stored as salted dry fish.

[76] Ishige (1993), p. 15. The area corresponds to the piedmont zone which is regarded by some scholars as the most likely centre for the domestication of rice, cf. SCC Bray, Vol. VI, Pt 2, p. 486. The idea is reminiscent of that of Shinoda (1970) who had earlier proposed that cha (or narezushi) originated among the Miao tribes of south China and spread to north China during the Han Dynasty. The most recent archaeological information, however, shows that the domestication of rice occurred in about −8500 to −7500 along the middle stretches of the Yangtze River, cf. Bruce Smith (1995), pp. 128–32 and the recent News report by Normile (1997). For further information see the citations on p. 27, n. 43.

[77] Cf. p. 384. Cha is also listed in the Ma-wang-tui inventory of foods that was discovered in Han Tomb No. 1. Furthermore, in China, the use of a lactic acid fermentation for food preservation was first applied to the pickling of vegetables before it was applied to meat and fish. As we have mentioned earlier (p. 379) the pickled vegetables are known as tsu in the Chou Li. Three kinds of tsu are listed in the Ma-wang-tui inventory. But detailed methods for making tsu are not described until much later (in chapter 88 of the CMYS).

[78] Cf. n. 76 above.

[79] In this connection it is interesting to note that angkak, the red ferment (hung chhü), is added to cooked rice in Thailand for making a pickled fish called pla-paen-daeng (daeng meaning red), cf. Phithakpol (1993), p. 160. Angkak is sometimes used in the Philippines for making burong isda (Table 37), cf. Mabesa & Babaan (1993), p. 87. The

Vietnam was part of the Han empire. The same technique was apparently also adopted in Korea for making *sikhae* as shown in the *Eumsikbo* (*yin shih phu* 飲食譜), of +1700s and the *Ojuyeonmunjang jeonsango* (*wu chou hêng, wên chhang kao* 五州衍文長稿), c. 1850.[80] The Koreans, however, have introduced a new twist, the addition of malt, to the process. Today, *koji* is all but forgotten, but malt is still being used to make *sikhae*. It is interesting to note that sprouted wheat was also used in a recipe for making a fish paste in the *Chhi Min Yao Shu* (Table 35, item 4), but this concept was not developed further in China.

4. We have noted earlier the lack of any mention of a fish sauce in China, Japan and Korea before the 19th century. Indeed, according to Ishige 'the deliberate production of fish sauce is a relatively new development in the history of fermented fish products' in East Asia.[81] Just as soy sauce was initially a byproduct of the making of fermented soybeans (*shih*), it is likely that fish sauce was first recovered as a byproduct in the making of fermented, salted fish (*shiokara*). Since the countries of continental Southeast Asia are where the action is regarding fermented fish products, it stands to reason that this would be the region where a dedicated fish sauce process would most likely have originated. In the making of *shiokara*, the fish is gutted, cut up if necessary, salted and incubated. Autolysis is slow and the product usually retains its original shape. The intestines, which contain powerful and salt tolerant proteases, are often salted and incubated separately reminiscent of the *chu i* (fish entrails) paste of the *Chhi Min Yao Shu* (Table 35). But when numerous small fish, such as anchovies, are salted whole and incubated with their visceral organs intact, autolysis is much faster and within a reasonable time all the flesh will be hydrolysed to leave behind a liquid sauce and the bones of the fish. This is how fish sauce is made today in Thailand, Vietnam and pockets along the southeastern coast of China. We do not know when the process was first developed, but from Vietnam it could easily have migrated up the southeastern coast of China to Canton, Hong Kong, Swatow, Amoy and Foochow where fish sauce is known as *yü lu* 魚露 or *chhi yu* 鯖油.[82]

---

latter also reported experiments carried out with with papain, bromelain and enzyme extracts from *Aspergillus oryzae* and *Aspergillus niger* to shorten the length of incubation for *burong isda*. The use of *koji* to speed up the fish sauce fermentation has been successfully demonstrated recently in Japan, cf. Itoh, Tachi & Kikuchi (1993), p. 180.

[80] Lee, Sung Woo (1993), p. 42. Lee further points out that *koji* is used instead of malt for making a sandfish *sikhae* (or *narezushi* in Japanese). *Sikhae*, of course, is made of two ancient Chinese words, *sik* (*shih* 食) for cooked grain, and *hae* (*hai* 醢) for fermented meat paste.

[81] Ishige (1993), p. 30. However, there are historical antecedents to fish sauce seen in the literature, such as the juice from fish paste (*yü chiang chih* 魚醬汁) in the *CMYS* cited on p. 387. Ishige (private communication) has found Japanese references to similar prototypes in the *Engishiki* 正喜式 of the 10th century and the *Honchou shokkan* 本朝食鑑 of +1669.

[82] *Chhi* is the name of a small fish used for making fish sauce in Foochow. The word is not found in all dictionaries. According to Chhên Chia-Hua (private communication) the method of making Foochow style *chhi yu* is as follows. 'Mix cleaned, whole *chhi* fish (butterfish?) with about 25 to 30% of its weight of salt. Layer the salted fish in a jar. Cover it with a bamboo lid. Let it stand in the sun and mix the contents from time to time. All the flesh will be liquified in three to six months. The sauce is drained off and filtered. The residue, consisting almost entirely of bones is collected for use as fertiliser.' The procedure is similar to that cited by Ishige and Ruddle (*1985*), p. 194, on the making of fermented fish sauce in Swatow and Foochow.

The key to success in the fish sauce process apparently lies in the presence of the visceral enzymes during the incubation. Noda *et al.* have shown that 3 alkaline proteases from the pyloric caeca of the sardine still retain 40% of their activity at 20% salt concentration in hydrolysing fish proteins.[83] This is the reason why in the Chinese method for making fish paste, which employs microbial *ferment* to digest the proteins in chopped fish, the extent of hydrolysis was never sufficient to achieve substantial liquification of the substrate. And so we do not see the emergence of a dedicated fish sauce from the Chinese process for making a fermented fish paste. The fish sauce we now see was first made in Southeast Asia and later transmitted to outposts on the Southeastern coast of China, where it is called *yü lu* to show its lack of connection with the traditional fermented fish paste, *yü chiang*, which has itself disappeared from the food system.

(iii) *Fermented fish sauce in ancient Greece and Rome*

The incorporation of visceral enzymes also explains why the ancient Greeks and Romans were successful in making a fermented, salted fish sauce called *garum* or *liquaman* long before such a product became popular in Southeast Asia.[84] The principal commodity produced by the salting of fish was salted fish, called *salsamentum*. Parts of the fish that were considered as trash, such as the innards, blood and gills, as well as whole small fish, were salted and allowed to ferment by autolysis to make a sauce. The Romans produced four separate fermented products, *garum*, *liquamen*, *muria* and *allec*. *Garum* is the choice product, *liquamen* and *muria* represent second and third drawings, while *allec* is the residue that remains after the batch of sauce was collected. Unfortunately, while mentioned frequently in the classical Roman literature, no reliable description of how the fermentation was carried out has survived to this day.[85] The best and most complete description of the process is found in the *Geoponica*, a +10th century Greek work said to have been derived from a +6th century Latin treatise. We quote:[86]

Chhi yu is sometimes referred to as *chhi chih* 鯺汁 (juice of the *chhi* fish), just as *shih yu* 豉油 is sometimes called *shih chih* 豉汁. In Fukien dialects *chhi yu* is pronounced *kieh yu* and *chhi chih* is pronounced *kieh chiap*. The term *kieh chiap* has the distinction of being the precursor of the common English word 'ketchup', which in its modern incarnation is probably the most popular condiment seen in the USA today. According to the *Oxford English Dictionary* ketchup is derived from *kôe-chiap* or *ké-tsiap* (Amoy dialect) which denotes a 'brine of pickled fish or shell-fish'. Apparently, fish sauce under the name *kieh chiap* was a favourite condiment among the Chinese seamen who worked on the foreign ships that sailed along the South China coast in the 18th century. To the ears of the Western sailors on the ship *kieh chiap* sounded very much like *ketchup*. Thus, *ketchup* was born. It has, however, evolved into a completely different product with its own identity.

[83] Beddows & Ardeshir (1979), pp. 603, 613, and Noda *et al.* (1982), pp. 1565–9.

[84] Curtis (1991) points out that the ancient *garum* and *liquamen* of Greece and Rome are counterparts of the fish sauces of Southeast Asia. See pp. 19–26 for a summary of the methods of production of the fish sauces of modern Asia and a discussion of the nutritional value of these products.

[85] *Ibid.*, pp. 9–12, notes that Roman authors Manilius, Pliny the Elder and Columella of the +1st century have provided sketchy descriptions but no precise details of the process for making *garum*.

[86] *Geoponika*, 20, 46, 1–6, tr. Curtis (1991), pp. 12–13. See also the translation from the Greek of Cassianus Bassus by T. Owen, (1806), Vol. II Book 20, ch. 46, pp. 299–300.

The so-called liquamen is made in this manner: the intestines of fish are thrown into a vessel and salted. Small fish, either the best smelt, or small mullet, or sprats, or wolffish, or whatever is deemed to be small, are all salted together, and, shaken frequently, are fermented in the sun.

After it has been reduced in the heat, *garum* is obtained from it in this way. A large strong basket is placed into the vessel of the aforementioned fish, and the garum streams into the basket. In this way, the so called liquamen is strained through the basket when it is taken up. The remaining refuse is alex.

The Bithynians prepare it in this manner. It is best if you take small or large sprats, but if not, wolffish, or horse-mackerel, or mackerel, or even alica, and a mixture of all, and throw these into a baker's kneading trough, in which they are accustomed to knead meal. Tossing into the modius of fish two Italian sextarii of salt, mix up thoroughly in order to strengthen it with salt. After leaving it alone for one night throw it into a vessel and place it without a lid in the sun for one to three months, agitating it with a shaft at intervals. Next take it, cover it and store it away.

Some add to one sextarius of fish, two sextarii of old wine.

Next if you wish to use the garum immediately, that is to say, not ferment it in the sun, but to boil it, you do it this way. When the brine has been tested, so that an egg having been thrown in floats (if it sinks, it is not sufficiently salty), and throwing the fish into the brine in a newly made earthenware pot and adding in some oregano, you place it on a sufficient fire until it is boiled, that is until it begins to reduce a little. Some throw in boiled-down must. Next throwing the cooked liquid into a filter you toss it a second and third time through the filter until it turns out clear. After having covered it store it away.

The best *garum*, the so-called *haimation*, is made in this way: the intestines of tunny along with gills, juice and blood are taken and sufficient salt is sprinkled on. After having left it in the vessel for two months at most, pierce the vessel and the *garum*, called *haimation*, is withdrawn.

There are several points of interest in this description. In the general procedure fish intestines from the larger catch are mixed with the small fish. The enzymes resident in the small fish, the main substrate, are thus fortified by the enzymes from the added intestines. Their combined action is able to completely hydrolyse the proteins in the raw materials and yield a sauce, which is actually a concentrated solution of protein hydrolysate in brine. The valuable role of fish visceral enzymes in the process is again highlighted. In fact, the best *garum*, called *haimation*, was prepared by the use of intestines, gills and blood only, without the flesh which is normally the most desirable part of the fish. In this particular case the product is ready in two months. The process is virtually the same as that described for the *chu i chiang* (paste derived chasing *I* barbarians) discussed on pages 382–3. The ancient Chinese apparently did not incubate the medium long enough to allow the fish entrails to become liquified. Nor were they interested in drawing the liquid that could have separated from the fermented paste, at least at the time of the Western Han. It is quite a coincidence that the product was collected by placing a basket into the liquified medium, similar to the way soy sauce was recovered from the fermented soy paste described in the *Hsing Yüan Lu* (pages 363–4). The ratio of substrate to salt, one modius to two sextarii, is 8:1, somewhat low in salt by East Asian standards.

But since the salted fish is exposed to the sun, it is likely that after evaporation the final concentration of the sauce would be near the level in modern fish sauce, that is, about 20–25%. Thus, we can say that the ancient Roman method for making *garum* and *liquamen* exactly parallels that being used for making fish sauce in East Asia today.

There are three basic requirements for a successful centre for processing salted fish products. Firstly, fish must be plentiful. Secondly, fresh water must be available for cleaning the fish. Thirdly, a ready supply of salt should be nearby. There are many locations along the shores of the Mediterranean where such conditions were met.[87] Being highly perishable, fish had to be processed as soon as possible after it was caught to minimise the proliferation of putrefying bacteria. But even under the best conditions, a certain amount of delay in processing was inevitable with the result that a characteristic odour of decaying fish would permeate the facility and be carried into the product. This characteristic aroma of the sauce is at once its charm and its bane. It has invited the contemptuous witticism of the elite Roman writers of the +1st century. For example, Manilius says, 'There the entire mass of putrid confusion is mixed together' (*illa putris turbae strages confunditur omnis*) when he describes the preparation of salted fish for fermentation.[88] In a similar vein, *garum* is called 'that liquid of putrefying matter' (*illa putrescentium sanies*) by Pliny the Elder and 'the expensive liquid of bad fish' (*pretiosa malorum piscium sanies*) by Seneca.[89]

In spite of these pejorative comments on its characteristic aroma, *garum* (or *liquamen*) was the favourite condiment of the ancient Roman world. In the *De re coquinaria* of Marcus Apicius, the celebrated Roman gastronome of the +1st century, there are listed nearly 350 recipes flavoured with fish sauce but only 31 recipes flavoured with salt.[90] It was a very expensive condiment and was used judiciously and sparingly in cookery. *Garum* was also used as a drug in human and in veterinary medicine.[91] In contrast, no such use has been reported for fish sauce in East Asia.[92]

The Greeks probably made and consumed salted fish and fish sauce as early as the −5th century, but it was between the +1st to +3rd centuries, under the rule of Rome that these products achieved the zenith of their popularity.[93] They were an important commodity in the economy of the Roman Empire, surpassed only by wine, olive oil and grain. But thereafter, their popularity began a slow and steady decline. References to fish sauce were few and far between in the ensuing centuries.

---

[87] Curtis (1991), pp. 46–7. [88] Manilius, *Astronomica*, 5: lines 670–4, cited in Curtis (1991), p. 11.

[89] Pliny, *Historia natura*, 31.93; Seneca, *Epistula*, 95.25, cited in Curtis, *ibid*. Schafer (1977), p. 115, is following a distinguished tradition when he refers to the 'evil smelling fish sauces of modern Indonesia'.

[90] Curtis (1991), p. 29. In his translation of *Apicius de re Coquinaria* (Cookery and Dining in Imperial Rome), Vehling (1977), appears to be uncertain of the nature and identity of *garum* and *liquamen*. They are rendered variously as broth, stock and gravy in the recipes. In his adaption of Apicius for the modern kitchen Edwards (1984), calls the fermented sauces Roman fish-pickle and refers to them as stock or fish stock in his recipes.

[91] Curtis (1991), pp. 27–37; references on this subject are given to the *De materia medica* of Discorides, *De alimentorum facultatibus* of Galen, and the Hippocratic corpus.

[92] Of the three fermented salted fish products mentioned in the food canons of China, only pickled fish (*cha*) is included in the pharmacopoeias, cf. *PTKM*, ch. 44, pp. 2430, 2436, 2442, 2484.

[93] Curtis (1991), pp. 2, 46–8, 64–5, 99, 178.

By the year +1000 it had all but disappeared as a factor in the life of the peoples of the Mediterranean region. The decline and disappearance of *garum* as a condiment after it was held at such high esteem in the +1st and +2nd centuries is a major unsolved problem in the history of food culture.[94]

Curtis suggests that this decline may be traced to the negative attitude of early Christians towards fermented fish sauces as an ingredient in their diet. This is reflected by the proscriptions pronounced by ascetics such as St Pachomius[95] and St Jerome[96] in the +5th century. The Christians were evidently strongly influenced by the Jewish dietary guidelines delineated in the Leviticus. In considering whether a fish sauce such as *garum* may or may not be eaten, they would be reminded of verses such as: 'But anything in the seas and rivers that has not fins and scales . . . is an abomination to you . . . Of their flesh you shall not eat' Leviticus *11*: 10–11. 'You shall not eat the blood of any creature, for the life of every creature is in its blood; whoever eats it shall be cut off' Leviticus *17*: 14.

Since among the raw materials used for making *garum* were fish blood and whole small fish, which were often without scales, it would be easy for a devout Christian to declare this product as 'abominable' and unfit for human consumption. But the attitude of the Christians was sometimes quite ambivalent. Even St Pachomius allowed himself to be served *garum* when he was sick, and other clerics in the +7th century continued to regard *garum* as an acceptable condiment to flavour their food.[97]

Perhaps even more than the rise of Christianity, the fall in the fortune of fermented fish sauces could have been the result of the cataclysmic political and cultural changes that shook the Mediterranean world between the 5th and 10th centuries. Barbarian tribes from the north, Visigoths, Huns, Vandals and Franks, invaded the heartland of the Empire from the 5th to 6th centuries. The ardent followers of a new faith, Islam, from Arabia overran the Middle East, North Africa and Spain in the 7th to 8th centuries. These invasions led to two major changes, one political and the other cultural. With the breakup of the Empire into unstable kingdoms, the network of production and trade in fish sauce that served the needs of the upper-class citizens of Imperial Rome could no longer be sustained. Furthermore, it is hard to imagine that the new masters of the region from the European North and Arabian East could appreciate the flavour of *garum* without being repulsed by its disagreeable odour.[98] Interest in fish sauces thus fell with the disintegration of the

---

[94] The rise of the barrel-pickled herring industry in Northern Europe at the end of the 12th century may be regarded as a reincarnation of the *salsamentum* of the Mediterranean. In the cold North, however, the autolysis of the herring never reached the level needed to produce a paste (*chiang*) or a sauce (*yu*). The product is essentially a *shiokara* (pickled fish). Cf. Knochel (1993), p. 213; and White (1976), p. 19.

[95] Curtis (1991), p. 35 cites St Pachomius, *Regula Sancti Pachomii* 45; but an exception is made for sick monks, as in *Regula Sancti Pachomii* 46.

[96] *Ibid.*, p. 35, cites St Jerome, *Epistula, 108*: 17.   [97] Curtis, *ibid.*, pp. 184–90.

[98] In terms of their effect on the uninitiated, the aroma of fish sauce can be compared to that of well-ripened cheese, irresistible to some but nauseating to others. *De gustibus non est disputandum. Ibid.*, p. 189 cites a zoning regulation in the Byzantine period which prohibits the erection of a fish sauce factory and a cheese factory less than three miles from a town because of their offensive odour.

Roman Empire and it soon became a relic known only to scholars conversant with the classical literature of Greece and Rome.

(iv) *Pickled vegetables and other victuals*

In addition to meat and fish pastes (*hai* 醢 and *ni* 臡), the *Chou Li* lists *tsu* 菹 and *chhi* 齊 as the fermented foods stored in reserve to meet the needs of the royal household.[99] *Tsu* and *chhi* are usually interpreted as pickled foods, especially vegetables.[100] *Tsu* is a word of great antiquity. Its earliest occurrence is in the *Shih Ching*. We quote:[101]

> In the midst of the fields are the huts;
> Along the boundaries and balks are gourds.
> He dries them, *pickles* them,
> And offers them to his great forefathers.

The word for *pickles* in the original text is *tsu* 菹 (also written as *tsu* 俎), which has been traditionally taken to mean *yen chih* 鹽漬 (salt and incubate), i.e. to pickle.[102] The quotation thus implies that the art of pickling was already practised in the early years of the Chou dynasty, i.e. by −1100. But the word *tsu* has meant so many different things that its interpretation as pickle is not so clear cut as to be beyond challenge.[103] We learn from the passage on the duties of the *Hai Jên* 醢人 (Superintendent of Fermented Victuals),[104] that on different ritual occasions, each offering of vegetable *tsu* is paired with its own fermented meat condiment, for example:

---

[99] *Chou Li*, ch. 2, *Hai Jên*, p. 55.
[100] Although all the seven *tsu*'s listed in the *Chou Li*, *Hai Jên*, are made from vegetables, three of the five *chhi*'s are animal products. This suggests that *tsu* may also be applied to animal products.
[101] *Shih Ching*, W 200, M 210.
[102] This view is accepted by standard commentaries such as the *Mao Shih Chêng I* 毛詩正義, by Khung Yin-Ta, Thang, *Shih Chi Chuan* 詩集傳, by Chu Hsi, Sung, and *Shi Chhi* 詩輯 by Yen Tshan, Sung. It is also held by the modern eminent scholar Hsia Wei-Ying (1980). Knechtges (1986), p. 50, concurs, but he, following Biot, renders *chhi* as marinade.
[103] According to the *Han Yü Ta Tzu Tien* (Great Chinese Dictionary) *tsu* has, in various contexts, been used to mean (1) a pickle (2) a meat paste/sauce (3) an ancient form of punishment (4) a dry grass (5) bananas (6) a straw mat and (7) a marshy place.
[104] *Chou Li*, *Hai Jên*, p. 55. The texts reads as follows:

Hai Jên: He is in charge of filling the vessels (*tou* 豆) with the proper victuals offered on each of the four ritual occasions. On the first occasion, *tsu* from chives is paired with meat sauce/paste, *tsu* from cattail root with elaphure meat paste, *tsu* from turnip with venison paste, and *tsu* from water mallow with roebuck meat paste. On the second occasion, *tsu* from mallow is paired with snail paste, *tsu* from beef tripe with oyster paste, *tsu* from frog with ant egg paste, and *tsu* from pork shoulder with fish paste. At the third offering, *tsu* from celery is paired with rabbit meat paste, *tsu* from cattail with meat sauce, *tsu* from spring bamboo shoots with wild goose meat paste and *tsu* from winter bamboo shoots with fish paste. At the last offering, the vessels are filled with light congee or meat congee. These represent the foods presented at regular ritual ceremonies. Similar victuals are offered at funeral rites and at banquets for honoured guests. It is also his duty to see that the king, queen and prince are properly supplied. On major occasions he should provide sixty jars of fermented victuals containing the five *chhi*, seven *hai*, seven *tsu* and three *ni*. For entertainment of guests, he should provide fifty jars of fermented victuals. The needs of other occasions are also his responsibility.

Tou in this context is a shallow bowl on a raised pedestal, made of wood and often lacquered, cf. p. 100.

tsu	*condiment*
chives	meat sauce/paste
cattail root	elaphure meat paste
Chinese turnip	venison paste
water mallow	roebuck meat paste

But it does not tell us how they were paired. Were they mixed together in one vessel or were they placed in separate dishes but eaten together?

In his commentary to this passage Chêng Hsüan explains how the meat condiments (*hai* and *ni*) were made (cf. page 380), but gives no indication as to how *tsu* was prepared. He does, however, tell us that 'the cattail root is cut into four inch segments to make *tsu*',[105] and explains that 'those that are harmonised (*ho* 和) with *hsi chiang* 醯醬 (soured condiments) are called *chhi* 齊 when finely cut, and *tsu* 菹, when coarsely cut or left uncut'.[106] These statements leave *tsu* open to the interpretation that it is simply fresh vegetables cut into large segments. Each type of *tsu* is then flavoured with its complementary paste/sauce before it is presented at the altar or consumed at the table. Indeed, Hung Kuang-Chu has argued with considerable vigour that the word *tsu* in the pre-Han literature has nothing to do with pickles.[107] It simply means any kind of vegetable or meat cut into pieces that are longer than four inches.

This position is bolstered by a statment in the *Li Chi* that in the cutting of meat, 'Elaphure meat and venison are prepared as *tsu*,'[108] which has traditionally been interpreted as 'Elaphure meat and venison are pickled in slices.'[109] But if *tsu* were to denote both pickled meats and vegetables, why did it become a term used exclusively for vegetable pickles after the Han? In Hung's interpretation this difficulty would be removed. The statement now reads 'elaphure meat, venison and fish are sliced into large pieces'. This rendition had, in fact, been adopted independently earlier by Wang Mêng-Ou[110] in his edition of the *Li Chi*.

On the other hand, the *Hai Jên* passage from the *Chou Li* also tells us that there were seven types of *tsu* (as well as five *chhi*'s, seven *hai*'s and three *ni*'s) stored in jars, so that they would always be ready for use in rituals or entertainment.[111] If *tsu* simply

---

[105] Chêng Hsüan's commentary to the above passage, cf. *Chou Li Chu Shu*, p. 88.
[106] Ibid., p. 89. Chêng Hsüan's commentary on the *Hsi Jên* (Superintendent of Acid Condiments) says, 'the five *chhi* and seven *tsu* require vinegar (acidified condiments) to reach full flavour'.
[107] Hung Kuang-Chu (*1984a*), pp. 158–62.
[108] *Li Chi*; the statement occurs in the *Nei Tsê*, p. 463: and *Shao I*, p. 588. The full text of the passages are as follows:

Nei Tsê: 'Fresh beef, lamb and fish are sliced and then cut to make *khuai* 膾. *Elaphure meat and venison are prepared as tsu*, wild boar is prepared as *hsuan* 軒. They are sliced and not cut. Roebuck and rabbit meat are sliced and cut as *khuai*. Scallion and shallot are cut in sour sauce (*hai* 醯) so that they are softened.'
Shou I: 'When fresh meat is sliced and cut finely, it becomes *khuai* (fine strips); when sliced coarsely, it becomes *hsüan* (chunks). *Elaphure meat, venison and fish are prepared as tsu*. Roebuck and rabbit meat are prepared as *khuai*. Wild pigs are prepared as *hsüan*. Scallion and shallot are cut and soaked in sour sauce (*hsi*) so that they are softened.'

[109] This is the way *tsu* is translated in Legge (1885), I, p. 463 and II, p. 80.
[110] Wang Mêng-Ou ed. (1970), pp. 463 and 588. [111] *Chou Li, Hai Jên*, p. 55; cf. n. 104 above.

means fresh vegetables cut into sections larger than four inches, and *chhi* those cut less than four inches, no useful purpose would be served by storing them in jars. They would quickly wilt and deteriorate. There is, however, an alternative explanation. The key to a solution of the discrepancy may well lie in Chêng Hsüan's commentary, which says that each *chhi* and *tsu* is 'harmonised' with *hsi chiang*', which may be interpreted as '*harmonised with vinegar and its complementary meat paste/sauce*,'[112] for example, chives with vinegar and meat sauce, cattail root with vinegar and elaphure meat paste, etc. To *harmonise* with the condiments can mean to flavour the vegetable before it is consumed, in the manner of the tossing of a salad with a seasoning before it is served. But it can also mean to allow a mixture of the freshly cut raw vegetable, vinegar and its paired condiment to stand for a substantial length of time. In effect, this is tantamount to allowing the raw material to be pickled in the preformed paste or sauce. Such a vegetable/vinegar/meat-paste combination can certainly be stored (i.e. incubated) in a jar, and taken out for use as the need arises. In fact, this aging is very likely to improve the quality of the product. Thus, in the *Hai Jên* passage, the pairing of the raw material with a preformed paste can be interpreted as allowing the vegetable segments to pickle in the paired condiment. Regardless of whether the ingredients are freshly mixed or suitably aged, the product would still be called *tsu*.[113]

There are a few other references of the Han period that need to be mentioned. The *Shuo Wên*, +121, says, '*tsu* is pickled vegetable (*chho tshai* 酢菜)'.[114] The *Shih Ming*, +2nd century, tells us, '*tsu* is *tsu* 阻, i.e. to stop. The raw food is fermented so that it will not rot when kept at room temperature.'[115] The *Ssu Min Yüeh Ling*, c. +160, recommends the making of *tsu* from mallows in the ninth month.[116] They are all consistent with the view that the *tsu* of antiquity is pickled vegetable (or animal parts). There is, furthermore, another statement from the *Shih Ming* which actually gives us a good idea about how one type of *tsu* was made. It reads,[117] '*Cha* 鮓 is obtained by fermenting (*niang* 釀) [the substrate] with salt and cooked grain to give a *tsu* (pickled product) suitable for consumption as food.' In other words, *cha* was a type of *tsu* and it was made by fermenting the material with salt and rice. It clearly implies that *tsu* had been made by a lactic acid fermentation.

From these references we can say that vegetables were pickled, that is made into *tsu*, by incubation in an acid medium or by a lactic acid fermentation. We still do not know how the processes were carried out in antiquity. For that information we have once again to wait until the appearance of the *Chhi Min Yao Shu*, +544, which devotes a whole chapter (ch. 88) to 'Methods for making *tsu* and the preservation of raw vegetables' (*Tso tsu, tshang shêng tshai fa* 作葅，藏生菜法).

---

[112] *Chou Li Chu Shu*, p. 89. The original reads, 醯醬所和, which can be interpreted as 'harmonised with vinegar and *chiang*', and also construed as a 'marinade' as rendered by Knechtges (1986), p. 50.

[113] See *Chhi Min Yao Shu*, cf. Table 38; most of the recipes in method 5 would fit this description quite well. Although *tsu* is featured prominently in the *Chhi Min Yao Shu*, the word practically disappeared from the Chinese culinary lexicon after the 6th century. In contrast, *chhi*, remained in use in recipes until the Sung dynasty.

[114] *SWCT*, p. 24 (ch. 1b, p. 20).   [115] *Shih Ming*, ch. 13, gives definitions of *cha* and *tsu*, p. 8b.

[116] *SMYL*, p. 94.   [117] *Shih Ming*, ch. 13.

Chapter 88 contains a wealth of information on how vegetables were pickled or preserved in the +6th century. But before we examine it, we need to warn the reader that there is another chapter in the *Chhi Min Yao Shu* which has the word *tsu* in its title, namely chapter 79, on *Tsu lu* 菹綠 (Pickles and greens).[118] *Tsu*, in this context, means sliced, cooked meat flavoured with spices and vinegar. As for *lü* (green), it is also a type of cooked meat, and more will be said about this later.

The relevant recipes described in chapter 88 are categorised and listed in Table 38. To facilitate discussion, they are numbered in the order they occur in the text. In cases where a recipe covers more than one method of preparation, the individual procedures are identified by a suffix, for example, 34a and 34b. By our reckoning, the chapter contains altogether thirty-six recipes.[119] But not all of them deal with the pickling of vegetables (or fruit). No. 16 is the important procedure for making rice *ferment* (*nü chhü* 女麴) that we have already described on page 336.[120] No. 5 gives a method for preserving vegetables by burying them in the ground. In No. 9, *tsu hsiao* 菹消, a preformed *tsu* is used as a seasoning for cooking mutton and pork.[121] In Nos. 14 and 30 melon is preserved in an aqueous extract of oak bark and dried plum. No. 33 describes *chüeh* 厥 (lettuce) as a delectable vegetable, sweet and crunchy when eaten raw, but no processing is mentioned.[122] The remaining thirty recipes are categorised under eight Methods in the Table according to the ingredients used in the incubation. They include salt, *ferment*, *tsao* 糟 (wine fermentation residues), wine, starch substrate and vinegar, singly or in combinations. Of these, twenty-six actually have the word *tsu* in their title. The other four are represented as methods of preservation (*tshang* 藏) of vegetables, but the procedures employed are virtually the same as those for making *tsu*. The identification numbers of these four recipes are placed in parentheses, namely, (12), (13), (15), (34).

The recipes in Table 38 belong to two general types (A) those that involve microbial activity, i.e. methods *1, 2, 3, 4* and *7*; and (B) those that involve little or no microbial action, i.e. methods *5, 6* and *8*. As an example of a Type A process, Recipe 1 is quoted in full below:[123]

To make salty pickles of mallow, cabbage, colza, and mustard green: Select leaves of good quality, and tie them in bunches with cattail stem. Prepare a solution of salt in water; make sure it is really salty. Wash the vegetables in the salty solution, and place them in a jar. If they are first washed with fresh water, the pickle will quickly deteriorate. After the salty wash water clarifies, pour enough of it into the jar to cover the vegetables. Do not move the vegetables around. At this point the colour of the vegetables is still green. If they are taken out, washed to remove the salt, and then cooked, they will taste as good as fresh vegetables.

---

[118] *CMYS*, pp. 488–9. The title is rather obscure. The meats mentioned include chicken, duck, goose, pork and cicada. *Tsu* is cooked meat, sliced and flavoured with spices and vinegar. *Lu* appears to mean the same thing.

[119] The number of recipes can vary depending on how one defines a 'recipe'. Wang Shan-Tien (1987), p. 235, counts twenty-nine recipes overall in ch. 88 of which twenty-three deal with the making of salty types of *tsu*.

[120] *CMYS*, p. 534.

[121] Compare with the recipe for *tsu hsiao* described in ch. 79. It is virtually the same as the *tsu hsiao* in ch. 88.

[122] *CMYS*, p. 537, see p. 545 note 47, which identifies the plant as *Pteridium aquilinum* Kuhn. The young leaves, known as *chüeh tshai* 蕨菜, are eaten as a vegetable. The root is a source of starch.

[123] *CMYS*, p. 531.

Table 38. *Making of pickled vegetables,* Tsu, *in the* Chhi Min Yao Shu

Method	Vegetable incubated with	Vegetables used	Recipe No.
1	Salt, *ferment*, starch substrate	Mallow, cabbage turnip leaves, mustard greens	1, 3
2	Salt, *tsao*[a]	Whole melon, sliced melon, ginger, bracken, pears	(13), 17,[b] 18, 29[b] (34b)
3	Salt, *ferment*	Cabbage	7b
4	Salt, starch substrate	Mallow, artemisia, shallot, melon	6, 8, (12a), (12b), (15), (34a)
5	Salt, vinegar	Mallow, cabbage, turnip, mustard green, mustard seed, celery seed, melon, turnip and radish roots, bamboo shoot, seaweed, watercress etc.	2, 19, 20, 21, 22, 23, 24, 25, 26, 27, 28 31,[b] 32,[b] 35
6	Salt (brine)	Cabbage	7a
7	Starch substrate	Mallow, cabbage	11
8	Vinegar	Cattail, water mallow[c]	4, 10, 36

[a] *Tsao* is the fermentation residue that remains after the wine is separated. It has the consistency of a thick paste.
[b] Notes on unusual recipes:
Recipe 17, melon is preserved in a fermenting wine mash.
Recipe 29, ginger is incubated in *tsao* without salt, washed, and then stored in honey.
Recipe 31, pears are incubated in water (up to year), then flavoured with honey.
Recipe 32, wood ears mixed with soy nugget juice (*shih chih*) & clarified soy paste (*chiang chhing*) without salt, spiced with ginger and fagara.
[c] Identified as *Nymphoides peltatum*, Miao Chhi-Yü (1982), p. 538. Waley (87) translates the plant as 'water mallow'.
Notes on recipes not included in Table 38:
Recipe 5, vegetables are preserved by burying in earth.
Recipe 9, in this recipe *tsu* is used as condiment for cooking pork and mutton.
Recipe 14, melon is preserved in oak (*yuan*) and plum (*mei*) juice.
Recipe 16, this recipe deals with the making of rice *chhü ferment*.
Recipe 30, same as recipe 14.
Recipe 33, no processing is described in this recipe; the vegetable cited is probably Chinese lettuce, cf. Miao Chhi-Yü (1982), p. 545.

For turnip and mustard greens: after soaking for three days [in salt water], remove the vegetables [from the jar]. Grind *panicum* millet and use the flour to make a congee. Decant the thin decoction. Take wheat *ferment* (*mai hun* 麥䴷), grind it and sift the powder through a silk screen. Spread a layer of vegetables [in a jar], sprinkle on lightly a coating of *ferment* powder, and then cover it with the hot thin congee. Repeat this process until the jar is full. When forming a layer, the bundles of vegetables should be placed side by side in such a way that the head of one is next to the tail of the other, and so on. The original salt solution is then poured into the jar [before it is sealed]. The pickle will have a yellow colour and a pleasant flavour.[124]

[124] This recipe does not describe how the jar is sealed or how long it is to be incubated, presumably for months. Perhaps, by this time the author has described so many processes of this type, cf. chs. 70, 71, 72, 74 that he assumes the reader would by now be quite familiar with the basic procedure involved.

As we can see, the process consists of two stages. Firstly, the vegetable is soaked in brine. Salt would be absorbed and water expelled from the tissue. Secondly, the salted vegetable is incubated with salt, *ferment* and a starch solution. The *ferment* hydrolyses the starch into sugar which is in turn converted by *lactobacilli* to lactic acid. The process is quite similar to that described earlier for the making of fish *cha* on pages 384–5 and further highlights the close relationship between *cha* and *tsu* as indicated in the *Shih Ming*.[125] Many variations of this methodology are shown in recipes in methods *2, 3, 4* and *7*. In method *2*, we have the earliest recipe on the use of *tsao*, wine fermentation residues, for the preservation of foods. This technique has since developed into a major methodology for the preservation of not only vegetables, but also meat and fish. We shall have more to say about it later.

Recipes of the B type, involving little or no microbial action, are listed in methods *5, 6* and *8*. We shall quote one example, Recipe 2 from method *5*:[126]

Method for making blanched pickle: Chinese cabbage and turnip leaves are suitable for this treatment. Select good-quality vegetables. Blanch them in hot boiling water. If the vegetables have already started to wither, wash them with water, drain and allow them to recover their freshness overnight. Then blanch them.

After blanching, plunge the vegetables briefly in cold water. Then mix them with salt and vinegar and season with cured sesame oil. The product will be fragrant and crisp. If a large batch is prepared, it can be kept until spring with no spoilage.

Method *5* was evidently a popular way of preparing vegetables for eating or preservation. It comprises the largest group of recipes listed in Table 38, fourteen out of the total of thirty. Two aspects of the overall procedure are noteworthy. Firstly, the vegetables are blanched in boiling water and cooled in cold water before they are processed further. The exceptions are melon (no. 19), purple laver (nos. 21, 28) and pear (no. 31). This is the way that vegetables are often treated nowadays before they are prepared as salad, or preserved by canning or freezing.[127] Secondly, many of the *tsu* were apparently ready to be served immediately (e.g. nos. 2, 20, 22, 24, 26, 32) after they were seasoned with salt and vinegar. They are therefore salads rather than preserves. It comes as a surprise to us that salads were a popular dish in +6th century China. But their appeal declined steadily after the *Chhi Min Yao Shu*. Enough salt and vinegar were present so that upon keeping, the vegetables would become pickled.

We should take particular note of method *6*. It has just one recipe (no. 7a), in which the vegetable is simply pickled in brine. But it is the first description of the wet salting of vegetables in the Chinese food literature. This technique has grown in the ensuing centuries into a major method of food preservation. Finally, we have three recipes in Category *8*, in which the vegetable is incubated in vinegar without salt. All three type B methods, i.e. *5, 6* and *8*, of preservation are based on chemical and

---

[125] *Shih Ming*, ch. 13, definition of *cha*.   [126] *CMTS*, p. 532.
[127] Cf. for example, Rombauer and Becker (1975), pp. 154, 803, 825. Blanching is particularly helpful for vegetables with thick stalks such as broccoli or cauliflower when they are used in salads.

physical action. Their significance in the history of vegetable preserves will be discussed further in a later segment of this chapter.

Of the type A methods, *1, 3* and *4* (and *7*) are variations of the same theme. Method *1* is, in fact, a combination of methods *3* and *4*. It may well be the oldest method, since it is implied in the passage on *cha* in the *Shih Ming* cited earlier.[128] Method *2* involves the use of *tsao* 糟 (fermentation residues) which has turned out to be an important innovation in the technology of food preservation. It was probably introduced during the late Han. There is another type A method which has, however, arisen from a recipe in the type B group. This is Recipe 32 (method *5*) on the making of *tsu* from the wood fungus or wood ear, *mu erh* 木耳 (*Auricularia auriculae*).[129] After the ears are blanched, cooled and sliced, they are spiced with coriander and scallion, mixed with vinegar, fermented soybean juice (*shih chih* 豉汁) and clarified soy paste (*chiang chhing* 醬清) until the desired taste is achieved, and flavoured with ginger and pepper. It is served immediately, but no doubt upon keeping, it would give a pickled wood ear. This recipe may be regarded as the forerunner of another popular technique of food preservation in the Chinese food system, one in which the vegetables, as well as meat and fish, are incubated in a fermented soy paste (*chiang*).

Thus, from Table 39 we can discern three overall techniques of food preservation practised in the +6th century that involve microbial activity:

(1) Incubation with salt and a starch substrate (with or without a microbial *ferment*) to induce the production of lactic acid.
(2) Incubation in *tsao*, wine fermentation residues.
(3) Incubation in *chiang*, fermented soy paste.

All three techniques may be applied to both vegetables and animal products (meat and fish). Technique (1) has been discussed in connection with the preparation of *cha*, fish and meat preserves (pages 384–5). Our concern now is with its application to vegetables. As shown in Table 38 the use of *tsao*, technique (2), as a medium of preservation is mentioned in several recipes in the *Chhi Min Yao Shu*, but the technique must have been developed sometime earlier. The word *tsao* 糟 occurs as early as c. −300 in the *Chhu Tzhu*, Elegies of *Chhu* (Songs of the South), in the poem Yü Fu (The Fisherman).[130] There is a reference to the preservation of meat in *tsao* in the *Chin Shu*,[131] suggesting that such an application was probably introduced sometime before the Chin (+265 to +419). The use of fermented soy paste (*chiang*) to preserve foods, technique (3), is intimated in the *Chhi Min Yao Shu*. It was probably a later development, starting perhaps during the Thang.

---

[128] *Shih Ming*, ch. 13, *Shih Yin Shih*, definitions of *tsu* and *cha*.   [129] *CMYS*, pp. 536–7.
[130] *Chhu Tzhu*, ed. Tung Chhu-Phing (1986), p. 215, line 5. Hawkes (1985), p. 206, translates *tsao* as 'dregs'. The word *tsao* also occurs in the *Chou Li*, (*Chiu Chêng*, p. 49) and the *Li Chi* (*Nei Tsê*), but in these texts it means simply unfiltered wine rather than the dregs from the wine.
[131] *Chin Shu*, Biography of Kung Chhün, p. 2061.

Table 39. *Methods of food preservation involving microbial action*

Incubation in	Starch substrate		Tsao – wine residues		Chiang – soy paste	
	Vegetable	Animal	Vegetable	Animal	Vegetable	Animal
CMYS +544	mallow, cabbage, turnip, mustard greens	6 recipes, 1 pork, 5 fish cf. Table 36	melon, ginger, bracken	meat	wood ear	—
CIL +965	—	fish	ginger	crab, lamb	melon, leafy vegetables	—
WSCKL Late Sung	radish, zizania stem, carrot	meat, yellow sparrow, mussel	eggplant, radish, ginger, melon	crab, pig's head and leg, fish	melon, mustard, pear, citron, gluten, seaweed	crab
CCPY Yuan	bamboo shoots, zizania stem, carrot	yellow sparrow, 4 fish, goose, clams, crab	melon, leafy vegetables, eggplant, bamboo shoot, ginger	fish, crab, lamb	melon, chopped melon	crab
YCFSC +1591	zizania stem,	fish, meat, pork, lamb, yellow sparrow, mussel	radish, leafy vegetables, bamboo shoot, eggplant, zizania stems, ginger	crab, pig's head, fish	melon, garlic stem, pear, citron, gluten	crab
SHHM +1680	—	fish, mussel, chicken, pork, lamb	melon, leafy vegetables, bamboo shoot, ginger	fish, crab, goose egg	melon, eggplant, mushroom	crab, pork, pork thigh
YHL +1698	—	2 fish	ginger, eggplant, bamboo shoot etc.	fish, crab, duck egg	melon, eggplant, mushroom	crab,
SYST +1790	—	—	leafy vegetables	pork, chicken, fish	melons, ginger,	pork, chicken
TTC 18th to 19th century	zizania stem, carrot, radish	3 fish	bamboo shoots, radish, leafy vegetables, eggplant,	lamb, pork, pig organs, fowls, crab, fish, snails	melons, eggplant, leafy vegetables, ginger	chicken, goose, crab, egg, fish

Table 39 continued. Methods of food preservation involving microbial action – références

Source	Starch substrate		*Tsao* – wine residues		*Chiang* – soy paste	
	Vegetable	Animal	Vegetable	Animal	Vegetable	Animal
CMYS +544	ch. 88, cf. Table 38	ch. 74 cf. Table 38	ch. 88, cf. Table 38	506	ch. 88	—
CIL +965	pp. —	3	32	16, 31	4, 5	—
WSCKL Late Sung	pp. 19, 23	6, 10, 12	17, 18, 22	5, 6, 9, 13	17, 20	13
CCPY Yuan	pp. 69, 70	84, 85, 86	66, 67, 68, 68	80, 81, 82, 83	71	84
YCFSC +1591	p. 98	70, 73, 82, 84	89, 90, 91, 98	72, 79, 83, 85, 99, 102	88, 89	83
SHHM +1680	pp. —	97, 100, 102, 113	70, 78, 84	101, 108, 110, 119, 154	72, 74, 77, 80, 89	109, 131, 145
YHL +1698	pp. —	73, 74	37, 38, 43	77, 82, 83, 91	35, 40, 41	82
SYST +1790	pp. —	—	118	59, 76, 88	121, 122, 123	59, 71
TTC 18th to 19th century	pp. 546, 603	372, 374, 392	90, 523, 537, 538, 557, 599	149, 150, 179, 192, 199, 204, 207, 331, 341, 350, 357, 358, 359, 437, 448, 314, 280, 506	494, 495, 497, 522, 535, 547, 558, 564, 587, 857, 815, 855	149, 281, 282, 299, 313, 314, 315, 341, 394, 438

CMYS   *Chhi Min Yao Shu.*
CCPY   *Chü Chia Pi Yung.*
YHL   *Yang Hsiao Lu.*
CIL   *Chhing I Lu.*
YCFSC   *Yin Chuan Fu Shih Chien.*
SYST   *Sui Yuan Shih Tan.*
WSCKL   *Wu Shih Chung Khuei Lu.*
SHHM   *Shih Hsien Hung Mi.*
TTC   *Thiao Thing Chi.*

How important were these technologies in the Chinese food system in the ensuing centuries after the *Chhi Min Yao Shu*? To answer this question we have collected representative data on the extent of their usage as reflected in the culinary literature from the +6th century to the 19th century, and we summarise them in Table 39. Although the data on technique (3), incubation with salt and starch, with or without a microbial *ferment*, as applied to fish and meat have already been examined earlier in this subchapter, they are, nevertheless, included so that the pattern of usage of all three techniques can be compared together at a glance.

We have left out references to a number of well-known food canons of the Yuan–Ming period, such as the *Shih Lin Kuang Chi*, +1280, *Yün Lin Thang Yin Shih Chih Tu Chi*, +1360, *I Ya I I*, Yuan, *To Nêng Phi Shih*, +1370 etc., since the information they contain usually repeats what are already included in the references listed in Table 39. Their exclusion would not affect the pattern of usage of the technologies shown in the Table. We have included, however, the *Chhing I Lu* (Anecdotes, Simple and Exotic), +965, which contains the earliest references to the use of *tsao* on crab, and of *chiang* on vegetables.

Let us first of all compare the three techniques used for the preservation of vegetables as shown in Table 39. In spite of the prominence given to it in the *Chhi Min Yao Shu*, incubation with a starch substrate (or *ferment*)[132] received only casual acceptance from the Sung to the Ming. It almost entirely faded out of the picture during the Chhing, a situation reminiscent of what had happened to the fish preserve, *cha*, that was prepared in a similar manner.[133] However, it is still alive and is the method used for making the famous *phao tshai* 泡菜 of Szechuan. On the other hand, incubations with *tsao* have maintained a high profile from the Sung to the Chhing even though incubations with *chiang* went through various changes in procedure during the same period.

In the *tsao* process, the wine residues are fortified with salt and often thinned with wine or boiled water. The vegetables are cleaned, cut into the proper size and then immersed in the wine residues in a jar. Some recipes recommend that the raw vegetables be blanched in boiling water and cooled before it is cut.[134] Others recommend that the vegetables be soaked in a cooled solution of lime and alum in water.[135] The product may be ready in 3–7 days, and can be kept for several months to a year. Among the most popular preserves made this way are ginger, eggplant, radish and leafy vegetables.[136]

The situation with the soy paste (*chiang*) process is more complicated. The earliest references in the *Chhing I Lu* do not state how the products were made. Although all the other preserves listed in Table 39 are said to have been prepared from soy paste, several are actually made by a variety of procedures. Some are, indeed, incubated

---

[132] For an example of a detailed procedure cf. Table 38, Recipe 1, and pp. 407–8. [133] Cf. pp. 384–6.
[134] *CCPY*, p. 66, eggplant. [135] *YCFSC*, p. 102, radish, *zizania* stem, bamboo shoot, melon and eggplant.
[136] According to Hung Kuang-Chu (*1984a*), p. 176, vegetables preserved in *tsao* are rarely seen in North China, but highly popular in the Yangtze basin and the South. The *tsao* eggplant from Nanking and *tsao* melon from Yangchou are renowned nationally. From personal experience, the author can say that young ginger preserved in red wine residues is a particularly delicious hors d'œuvre or relish.

in soy paste,[137] some in a medium undergoing a *chiang* fermentation,[138] some in a sweet wheat *chiang* (*thien mien chiang* 甜麵醬),[139] and others simply in soy sauce.[140] These recipes suggest that the process was still undergoing development from the Sung to the Chhing. The most interesting and bizarre example of this procedure is the use of soy paste for the preservation of fruits, pear and two kinds of citron (as well as seaweed and wheat gluten). The very brief recipe is first given in the *Wu Shih Chung Khuei Lu*, late Sung. It is repeated in the *Yin Chuan Fu Shih Chien*, +1591, and again in the *Thiao Thing Chi*, 19th century.[141] The incubation of a sour or sweet fruit in a salty, savoury medium has given rise to a unique group of preserved fruits with an appealing flavour.[142]

Incubations in *tsao* and *chiang* are still important for making preserved vegetables in China today. *Chiang* appears to have caught up and surpassed *tsao* in usage in the 20th century. Its present eminence is reflected by the fact that preserved and pickled vegetables are now known collectively as *chiang yen tshai* 醬腌菜, which means literally vegetables preserved in fermented paste and salt. Actually, the term embraces all three types of pickled vegetables: firstly, *chiang tshai* 醬菜, those preserved in *chiang* and *tsao*; secondly, *yen tshai* 腌菜, those preserved only with salt; and thirdly, *suan tshai* 酸菜, those pickled in acid, either via a lactic acid fermentation, or the addition of vinegar.[143] All these methods can trace their origin to the recipes in chapter 88 of the *Chhi Min Yao Shu* that are listed in Table 39. All the *yen tshai* (salted preserves) and part of the *suan tshai* (soured preserves) involve little or no biochemical or microbial action, and will be considered separately in the segment on 'Chemical and Physical Methods in Food Processing.' We need to turn our attention now to the use of these techniques for the preservation of animal products.

As we have shown earlier on pages 384–5, the technique of incubating fish (or meat) with a starch substrate, such as cooked rice, to make *cha*, was probably developed during the Han dynasty. It was an important method for the processing and preservation of fish and meat in the 6th century. According to Table 39 its importance remained undiminished during the Sung, Yuan and Ming dynasties. But its influence declined rapidly in the 18th century. The method is not mentioned in the *Hsing Yuan Lu* of +1750 or the *Sui Yuan Shih Tan* of +1790. It is represented by only three recipes in the massive *Thiao Thing Chi* of the 19th century. Today, fish and meat *cha* are almost all forgotten in China, although fish *cha* (or *narezushi*) is healthy and well in Southeast Asia.

---

[137] *CCPY*, p. 71, cut melon; *SYST*, p. 121, ginger. [138] *WSCKL*, p. 20, eggplant; *CCPY*, p. 71, eggplant.
[139] *YHL*, p. 35, melon. [140] *SHHM*, p. 77, eggplant, p. 80, mushroom.
[141] *WSCKL*, p. 17; *YCFSC*, p. 89; *TTC*, pp. 815, 855, 857.
[142] Not all fruits are sweet. It depends on their maturity. Françoise Sabban (private communication) points out that preserving 'raw fruits in brine is a common practice in Taiwan (and also on mainland China)'. In Europe the well-known 'mostardadi Cremona' is a kind of traditional mustard, hot and sweet at the same time, with fruits in it. The salted *mei* fruit, under the name *li hing mui*, has become an integral component of the snacks of Hawaii, cf. Laudan (1996), pp. 80–3.
[143] The term *chiang yen tshai* is not found in the *Thiao Thing Chi* or in any of the food canons before the 19th century. It was presumably coined in the 20th century. It is best translated as 'pickled vegetables'. The classification we use here is modified from the system proposed by Than Yao-Hui (*1982*).

According to Table 39 the earliest recipe we have of the preservation of a meat product in *tsao* 糟 (wine residues) is given in the *Chhi Min Yao Shu*, chapter 81 on, 'Preparation of Meat Products.' The procedure is as follows:[144]

Method for Tsao-Preserved Meat: The process can be carried out in all seasons. Dilute the wine residue with water until it has the consistency of congee. Add salt until it is very salty. Immerse a batch of skewer-roasted meat in the medium [in a jar] and incubate it in a cool place in the house. It can be roasted and eaten to accompany wine or cooked rice. Even in the summer it will not spoil in ten days.

This technique appeared to have increased its popularity during the Sui–Thang period. It has enjoyed a steady if not spectacular level of usage from the Sung to the Chhing. The *Mêng Liang Lu*, +1334, lists *tsao* preserves of sheep leg, crab, goose and pig's head on sale in the markets of the Southern Sung capital of Hangchou.[145] One of the favourite *tsao* preserves is crab, which is mentioned in most of the food canons extant since the *Chhing I Lu*, +965.[146] The product is so popular that the recipe is given in verse in the *Chü Chia Pi Yung*:[147]

> Thirty female crabs, no male among them,   [wash clean, wipe dry]
> Five pounds wine residues, four ounces salt,   [blend together]
> Half pint fine vinegar, and half pint wine,   [dilute with, incubate]
> Ripe in seven days, stay fresh till next year.

This verse is repeated in the *Shih Hsien Hung Mi, Yang Hsiao Lu, Thiao Thing Chi* as well as the Ming agricultural handbook *Pien Min Thu Tsuan*, (Everyman's Handy Illustrated Compendium), +1502.[148] Fish and crab are usually incubated raw, while meats such as pig's head and trotters are cooked first, pressed (to remove bones) and then incubated in the wine residue. Both procedures are in use in China today.[149] Most animals products (chicken, duck, pig trotters and internal organs) are also flavoured during incubation in salt with various spices, such as pepper, anise, cassia, and orange peel. Of the materials that are incubated raw, the most famous product is the *tsao* preserved duck egg of Chekiang. Lightly cracked eggs are layered between salt and wine residues and incubated for five to six months. Another interesting application is the use of *tsao* in the making of fermented *tou fu*, i.e. *tou fu ju* 豆腐乳, which has been discussed earlier on pages 327–8. *Tou fu ju* is now, in its own right, an important relish and condiment in Chinese cookery.[150]

Compared to *tsao*, the use of soy paste (*chiang*) to preserve meat and fish was a rather late development (Table 39). The earliest example of a product made this way did not appear in the literature until the Southern Sung, when *chiang*-preserved

---

[144] *CMYS*, p. 506.
[145] *MLL*, pp. 109, 134, 136, 138 on *tsao* preserved sheep trotters, crab, goose and pig's head.
[146] *Chhing I Lu*, p. 16. When the Sui Emperor Yang-ti (reign +605–616) visited Yangchou (in Kiangsu), he was presented with a gift of *tsao* preserved crabs.
[147] *CCPY*, pp. 83–4.
[148] *SHHM*, p. 110; *YHL*, p. 83; *TTC*, p. 438; *PMTT*, ch. 8, p. 9b. The same recipe is also given, though not in verse, in the *To Nêng Phi Shih*, +1370, p. 378 and *Chhü Hsien Shên Yin Shu*, +1440, p. 438.
[149] The details which follow are described in the *CKPJPKCS*, p. 720.   [150] Cf. pp. 326–8.

crab is mentioned in the *Wu Shih Chung Khuei Lu*. A recipe for the process appeared soon afterwards in the *Chü Chia Pi Yung*, Yuan. It is repeated in the *Yin Chuan Fu Shih Chien*, *Shih Hsien Hung Mi*, *Yang Hsiao Lu* and *Thiao Thing Chi*.[151] But as can be seen from the Table, the use of *chiang* for the preservation of meat and seafood never achieved a level of usage approaching that of *tsao*. Although the 19th-century compilation *Thiao Thing Chi* contains several recipes entitled *chiang*-processed meat and fish, some of them actually utilise soy sauce and others fermented wheat paste (*thien mien chiang*) as the incubating medium.[152] In almost all the examples, the materials are incubated raw. When ready, they are dried in air or cooked before storage.

Some time in the present century a major change took place in the methodology of *chiang*-based meat preservation. It has been expanded to include the *lu* 鹵 method of cookery, in which the meat is simmered for eight to nine hours in a stock consisting of soy sauce, wine and various condiments such as sugar, ginger, anise, cassia, fagara etc. In other words, instead of incubating the raw material in a live environment, in which the enzymes and microorganisms in the soy paste/sauce can function, it is now heated in a lifeless medium. This procedure is often referred to as the *chiang-lu* 醬鹵 method. It involves only physical and chemical, but no biochemical intervention. The *chiang-lu* method has virtually displaced the old traditional incubation, except perhaps in the remote countryside where the making of soy sauce is still practised as a household art.

The word *lu* has been translated as pot-stewing; it is red-cooking (*hung shao* 紅燒) done on a large scale.[153] *Lu* first appears in the culinary literature in the *Mêng Liang Lu*, which mentions a *lu hsia* 鹵蝦 (*lu* shrimp) on sale in a salted fish shop.[154] *Lu*-cooked chicken is listed in the *Shih Hsien Hung Mi*, *Yang Hsiao Lu*, *Sui Yuan Shih Tan*, and *Thiao Thing Chi*.[155] *Lu* may be regarded as a stock or marinade.[156] The origin of its use in the context of pot-roasting is obscure. It may be derived from the *lü ju fa* 綠肉法 recipe, chapter 79 of the *Chhi Min Yao Shu*, which is entitled '*Tsu lü* 菹綠', where *tsu* is the term for pickled vegetables (or meat) discussed earlier, and *lu* is the character for the colour 'green'. The recipe for *lü ju fa* is as follows:[157] '*Lü Ju Fa*: cut pork, chicken or duck in one-inch squares. Cook the meat in soy paste juice and salt. Flavour it with scallion, ginger, orange peel, celery and garlic. Add vinegar. The meat when sliced is called *lü ju*.'

Literally, *lü ju* means *green meat*. Generations of scholars have been puzzled by the use of the word *lü* (green) in this context. Examination of this passage shows that the procedure for making the *lü ju* (green meat) of the *Chhi Min Yao Shu* is similar to that

---

[151] Also in *TNPS*, +1370, p. 378, *Chhü Hsien Shên Yin Shu*, c. +1440, p. 438, and *PMTT*, +1502, ch. 8, p. 10a.
[152] *TTC*, p. 149, pork in soy sauce and wine; p. 314, duck in soy sauce; p. 282, chicken in fermented wheat paste, p. 315, duck in wheat paste; p. 341, chicken in wheat paste; p. 394, fish in wheat paste.
[153] Chao, Bu Wei Yang (1963), pp. 40–2, explains the difference between *red cooking* and *lu*.
[154] *MLL*, p. 139.   [155] *SHHM*, p. 113; *YHL*, p. 86; *SYST*, p. 75, also *lu* duck, p. 80; *TTC*, 287.
[156] For example, cf. *WSCKL*, pp. 20, 22, 25 on the making of salted vegetable preserves. When vegetables are steeped in brine, *lu* is the stock left behind after the product is removed. It can then be used to steep a second batch of raw material, and the process repeated.
[157] *CMYS*, ch. 79, pp. 488–9.

for making the *lu ju* (*lu*-cooked meat) of modern times.[158] The *lü* 錄 of the *lü ju fa* recipe is thus an archaic form of the *lu* 滷 of modern *lu*-cooked meat. The *chiang-lu* method used nowadays allows large chunks of beef, lamb, chicken, duck etc. to simmer in the stock. The cooked meat is then sliced as needed. It makes an excellent appetiser or side dish, similar to the delicatessen meats of the West.[159]

(2) Chemical and Physical Methods of Food Preservation

The spoilage of perishable foods is due largely to two factors. The first is contamination by undesirable microbes, and the second, the continued action of enzymes resident in the raw material itself. The objectives of food preservation are to arrest the onset of microbial contamination, and to inhibit or control the activities of the resident enzymes in such a way that the adverse effects are minimised. So far, the methods we have discussed to achieve these objectives all involve some type of microbial intervention. They are largely based on the use of either a mould *ferment* (*chhü* 麴), or products derived therefrom, that is, *tsao* 糟, a byproduct of wine fermentation from grains and *chiang* 醬, a companion product to the soy sauce fermentation from wheat and soybean. In one variation, the *chhü* is replaced by a starch substrate, such as cooked rice, which encourages the proliferation of lactic acid bacteria and the production of lactic acid. All these processes may be regarded as direct or indirect consequences of the invention of *chhü*. We shall have more to say about the role of *chhü* in the development of the technology of food processing and preservation in China later.

But for now, it is time for us to turn our attention to other traditional methods of food preservation that involve no microbial intervention, that is, those that are essentially chemical or physical in nature. These include:

(1) *Salting*: table salt, i.e. sodium chloride, is probably the oldest chemical preservative for raw meat, fish and vegetables known to man. At high concentrations it slows down the action of enzymes in the food materials, and draws water from microbial cells so that they die of dehydration.
(2) *Incubation in vinegar*: the increase in acidity inhibits enzyme activity and destroys undesirable bacteria.
(3) *Drying*: drying also removes water from microbial cells. It is accomplished by exposing the material to air, sunlight, an artificial source of heat, and smoke.
(4) *Incubation in sugar*: the sugar content in a fruit can be raised to such a level that it will draw water from a microbial cell.
(5) *Controlled atmosphere*: by allowing carbon dioxide to accumulate in the air so that it will slow down the metabolic reactions within fruit that contribute to the disintegration of the tissues.
(6) *Cold storage*: all metabolic activity in the food materials is lowered at a lower temperature.

---

[158] *CKPJPKCS*, p. 350.   [159] *Ibid.*, p. 286.

In the preservation of vegetables, fish and meat a combination of methods are often used, for example, salting and incubation in vinegar for pickled vegetables, salting and drying for salted pork, incubation in sugar and drying for preserved fruits etc. Thus, instead of addressing each method in turn, we shall now consider how these techniques have been applied to the preservation of specific foods: firstly, vegetables, secondly, meat and fish, and thirdly, fruits. Finally, we shall describe how the Chinese have exploited cold storage, a method that can be applied to all types of foods.

(i) *Salting and pickling of vegetables*

In the three microbially oriented methods of food preservation we reviewed earlier, viz. the use of wine residues, soy paste, and starch substrate, salt is an indispensable ingredient in each process. Salting is thus an integral component of these procedures. In the use of wine residues and starch substrate, the concentration of salt can be kept quite low since its function is augmented by the alcohol in the wine residues or the lactic acid produced from the starch substrate. Of course, if the concentration of salt is high enough, it could, by itself, deter microbial growth and inhibit enzyme activity, thus protecting the raw food from spoilage. In fact, we have already cited such procedures earlier, in which salt serves as the sole preservative, as on page 383 for preserving fish, and page 407 for preserving a vegetable.

In the case of vegetables, it is quite possible that the earliest pickle, *tsu* 菹, as listed in the *Shih Ching*, −1100 to −600, was made by salting.[1] In fact, a salted vegetable, the simplest type of *tsu*, is referred to by Chêng Hsüan in his commentary on the passage on the *Phêng Jên* (The Grand Chef) in the *Chou Li* (Rites of the Chou), which states, 'The grand soup is not adorned with the five flavours; the sacrificial soup is seasoned with a *salted vegetable* (*yen tshai* 鹽菜).'[2] As we have indicated earlier, the *tsu* of the *Li Chi* (Record of Rites) and *Chou Li* (Rites of the Chou) was made by incubation with vinegar and a savoury condiment, and that of the *Shih Ming* (Exposition of Names), +2nd century, incubation with a starch substrate.[3] Both types of *tsu* may be regarded as refined variants of the originally simple salted vegetable. Examples of methods for making all three products are found in the *Chhi Min Yao Shu* of the 6th century. We are thus inclined towards the view that salting of vegetables was a practice already well established in Chou China.

Recipe 7a, chapter 88, of the *Chhi Min Yao Shu* (Table 38) describes the making of a *tsu* pickle from cabbage by steeping it in a solution of three *shêng* (approx. 1.2 litres) of salt in four *tou* (16 litres) of water.[4] It does not say how long the vegetable is left in

---

[1] According to Chêng Hsüan the *tsu* of the *Chou Li* and *Li Chi* was prepared by incubating the vegetable with *hsi* (vinegar) *chiang* (paste/sauce). Cf. *Chou Li Chu Shu*, p. 89a. Vinegar was not discovered at the time when the poems in the *Shih Ching* were composed. Thus the *tsu* of the *Shih Ching* had to be prepared by another method.

[2] *Chou Li, Phêng Jên*, p. 40. Chêng Hsüan's commentary in *Chou Li Chu Shu*, p. 62b reads, *ta kêng pu chih wu wei yeh; hsing kêng chia yên tshai i* 大羹不致五味也，鉶羹加鹽菜矣.

[3] Cf. pp. 402–4.

[4] *CMYS*, p. 532. The salinity of the solution would depend on the size of the salt crystals. The density of NaCl crystal is 2.17. If the density of the table salt in bulk is >1.5, the 1.2 litres would weigh >1.8 kg. The salinity would be greater than 1.8/40 or 4.5%.

the salt solution, or how the product is collected. More popular than salting alone is the use of salt plus vinegar. There are several recipes for *tsu* based on this procedure listed in Table 38. Thus, pickling in brine or in brine plus vinegar was an accepted method for preserving vegetables in the +6th century.

It is curious that the word *tsu* is rarely seen in the food literature after the *Chhi Min Yao Shu*. The last reference we have to *tsu* as a pickle may be in the *Shih Liao Pên Tshao*, +670, which states, in regard to the *tshao hao* 草蒿 (prince's feather), that 'it is incubated in salt and vinegar to make *tsu* (*chhu yên wei tsu* 醋淹為菹)'.[5] This phrase not only marks the exit of *tsu*, but also the ascendance of the word *yên* 淹, which means to salt, cover with salt or incubate with salt. *Yên*, also written as 醃, occurs quite often in the *Chhi Min Yao Shu*.[6] In the *Shuo Wên Chieh Tzu*, *yên*, written as 腌, is defined as *chih ju* 漬肉 (steeped meat).[7] Apparently *yên* can be written in three ways, 腌, 淹 or 醃. It was probably during the Thang that *yên* evolved into the preferred term to denote the salting process as well as the salted products, such as *yên yü* 腌魚 (salted fish), *yên ju* 腌肉 (salted meat), and *yên tshai* 腌菜 (salted vegetable). Salting reached a high level of acceptance during the Sung[8] and has remained so until the present day.

To give us an overview of the significance of salting in making vegetable pickles from the Sung to the Chhing, we have listed the relevant recipes from the major food canons of this period in Table 40. The data show that all the common vegetables have been processed in this way. The extent of usage has remained high for almost a thousand years, and compares favourably with that achieved by the *tsu* pickles prepared by the three microbially oriented techniques (Table 39). Unlike the decline of pickles (*cha* 鮓) made by incubation with a starch substrate, there was no diminution in interest in salted vegetables (*yên tshai* 腌菜) in the 18th and 19th centuries. Salting alone is the most common method, although it is occasionally augmented by vinegar or vinegar plus wine and flavoured with a variety of spices. As an example of a typical procedure, a Southern Sung recipe from the *Wu Shih Chung Khuei Lu*, called '*Kan pi wêng tshai* 干閉甕菜' (Drying pickling in a sealed jar), is quoted below:[9]

Kan pi wêng tshai: Load 10 catties of vegetable with 40 ounces of roasted salt in a jar. Each layer of vegetables is topped by a layer of salt [and vice versa]. Incubate (*yên*) the salted vegetables for three days. Remove and put the vegetables on a tray. Rub each stalk gently and place it in another jar. Collect the salted liquor (*lu* 鹵) for later use. After another three days, the vegetables are removed, rubbed and placed in a fresh jar. The liquor is also collected for later use. This process is repeated altogether nine times. After the last treatment the vegetables are loaded into a jar. Each layer of vegetables is covered with a film of fagara and aniseed. The layering is continued until the jar is fully and tightly packed. Three bowls of the previously collected salted liquor is added to each jar. The mouth is sealed with mud. The pickles will be ready after the New Year.

---

[5] *SLPT*, p. 12. The word *yên*, written as 醃, also occurs on p. 70, with regard to the white fish which can be 'salted (*yên*) or stored in wine residue'.
[6] *CMYS*, *yen* as 醃 is seen on p. 455, second and third paragraphs and as 淹 on pp. 494, 505, 529, 533, 537.
[7] *SWCT*, p. 90a (ch. 4B, p. 14).
[8] The earliest mention of *yên* in the Sung food literature is in the *SCCK*, p. 100.   [9] *WSCKL*, p. 20.

Table 40. *Salting of vegetables from the Sung to the Chhing dynasties*

Source	Incubation			
	Salt only	Salt + vinegar	Salt + vinegar + wine	Vinegar only
WSCKL Southern Sung	melon, eggplant garlic, leafy vegetables, chives pp. 16, 18, 20, 22, 24, 24	winter melon and garlic, leafy vegetables 19, 23	garlic and melon 15	—
CCPY Yuan	ginger, melon, 2 chives, leafy vegetables pp. 63, 68, 68, 70	eggplant, 2 melon, winter melon 65, 67, 67, 67–8	—	ginger 67
YCFSC +1591	melon, 2 mustard, 2 leafy vegetables, eggplant, chives pp. 87, 88, 96, 98, 99, 100, 102	—	garlic and melon 88	—
SHHM +1680	2 cabbage, leafy vegetables, mustard green, melon pp. 66, 66, 67, 68, 74	ginger and vegetables 68	chopped cabbage 67	—
YHL +1698	4 cabbage, leafy vegetables, mustard green pp. 30, 30, 31, 32, 33, 34	—	sliced mustard, green, radish 31	ginger 38
HYL +1750	melon, plum,[a] apricot, radish, 3 peanuts, mustard green pp. 51, 52, 53, 53, 53, 56, 57	ginger,[b] garlic,[a] mustard green[c] 50, 52, 56	mustard, green[d] cabbage[d] 538	— 57, 59
TTC 19th century	radish, cabbage, mustard greens, celery, lettuce, ginger pp. 46, 538, 539, 550, 556, 568, 570, 574, 580, 581, 587, 596, 598	colza 550	radish 538	ginger 45

[a] Sugar is added to the final incubation.
[b] Sour plum extract is used instead of vinegar; sugar is also added.
[c] In an alternative procedure, vinegar is replaced by wine.
[d] Vinegar is left out in these two recipes.

This recipe gives us a considerable amount of detail on how salting of vegetables was carried out. It is interesting to note that *yên* is used here to indicate incubation with salt, and *lu* to denote the salted liquor collected. The use and reuse of *lu* liquor in the steeping or incubating medium is a practice also seen in other recipes.[10] We have come across *lu* earlier in connection with the pot-roasting of meat in soy sauce (page 415). Apparently by the 12th century *yên* and *lu* have become established terms in the culinary vocabulary. This recipe is reproduced verbatim in several later food canons.[11] In some recipes, the products are flavoured with sugar or spices.[12] In other recipes they are dried and kept as a dry vegetable pickle.[13] There is a considerable lack of consistency in the way the word *yên* is written in the food literature.[14] We suspect these are copying errors. How *yên* is written, 淹, 醃 or 腌, in the old texts, appears to depend on the whim of the copyist. In any case, it is always written as *yên* 腌 in the modern literature. Thus, one may say that the salting technology for vegetables had reached a high level of maturity by the time of the Sung. With minor variations, the basic technology has remained the same throughout the ensuing centuries, and is clearly very much alive in China today.[15]

(ii) Fu *and* hsi, *dried meat and fish*

The making of dried meat (*fu* 脯 or *hsi* 腊) is a very ancient art in China. The earliest reference to the word *fu* is in the *Shih Ching*. In offering various victuals to the Departed Spirit, the suppliants chanted:[16]

> Your wine is well strained,
> Your food well sliced.

In the original poem, 'Your food well sliced' is *erh yao i fu* 爾殽伊脯 which means literally, 'Your viands, sliced dried meat (*fu* 脯).' The earliest reference to the word *hsi* is in the *I Ching* (The Book of Changes), early Chou.[17] Hexagram 21, *Shih Ho* 噬嗑 (Biting through), states on line 3, 'Bites on dried meat (*hsi* 腊), and strikes on something poisonous.' These quotations suggest that *fu* and *hsi* were already part of the common culinary experience of the early Chou people. But a millennium had to pass by before we have a clear indication what *fu* and *hsi* are. From the *Shuo Wên*, +121, we learn that *fu* is dried meat, and that *hsi*, written as 腊, is also dried meat.

---

[10] *WSCKL*, pp. 22, 25; *YCFSC*, pp. 98, 99.
[11] For example, *Yin Chuan Fu Shih Chien*, +1591, p. 96; *Yang Hsiao Lu*, +1698, p. 31; *Hsing Yuan Lu*, +1750, p. 56; and *Thiao Thing Chi*, 19th century, p. 568.
[12] For example, *YCFSC*, p. 87; *HYL*, pp. 50–1, 52.   [13] *YCFSC*, pp. 87, 99.
[14] In the *To Nêng Pi Shih*, +1370, and the *Chhü Hsien Shên Yin Shu*, early +1400, *yên* is written as 淹. In the *Chü Chia Pi Yung* it is written as 醃 in the Shinoda & Tanaka collection (*1972*), and as 腌 in the *CKPJKCTK* edition of 1986. In the *Hsing Yüan Lu*, +1750, pp. 50–2, it is sometimes written as *yên* 醃 in the titles of recipes but as *yên* 腌 in the narratives. For *TNPS*, pp. 376–7 and *CHSYS*, p. 438 cf. Shinoda & Tanaka (*1972*).
[15] Hung Kuang-Chu (*1984a*), p. 167 lists eighteen varieties of salted vegetables in current production. This does not include those that are dried or partially dried. Cf. the *CKPJPKCS*, Encyclopedia of Chinese Cuisine (*1992*), p. 663, which lists the common vegetables that are pickled by salting today.
[16] Tr. Waley (w 203, M 248). Karlgren (1974) translates the line as 'your viands are sliced'.
[17] Wilhelm, tr. (1968), p. 489; Hsieh Ta-Huang, *I Ching Yü Chieh* (*1959*), p. 184.

These definitions reaffirm what we would expect *fu* and *hsi* to be from the way they are used in the ancient literature.

The next reference we have to *fu* is in the *Lun Yü* (The Analects), −5th century, where it is reported of Confucius that 'He did not consume wine or dried meat (*fu*) bought in the market.'[18] This shows that *fu* was already an article of commerce during the time of Confucius. *Fu* occurs several times in the *Chou Li*, for example, in the passages on *Shan Fu* (The Chief Steward), *Wai Yung* (External Culinary Supervisor), *Pien Jên* (Director of Ritual Vessels), and *Hsi Jên* (Supervisor of Dried Viands).[19] The last passage emphasises the importance of *hsi*, that is dried meat or fish, in the ceremonial and private life of the Chou court. Chêng Hsüan's commentary states, 'when the dried meat is thinly sliced, it is called *fu*, when the whole small animal is dried, it is called *hsi*.[20] *Hsi* from fish is listed among the viands under the supervision of the *Wai Yung*. It is also mentioned in the *Chuang Tzu* (The Book of Master Chuang), c. −290, which records that when Prince Jen had landed his fish, he cut it up and made it into *hsi* (dried fish).'[21]

Both *fu* and *hsi* occur in the *Li Chi*. In the *Chhü Li* (Rules of Ceremony), *fu* is shown as a viand served during a meal or at certain ceremonial occasions.[22] The *Nei Tsê* (Pattern of the Family) mentions several types of *fu* in a long list of victuals. It also tells us that at a normal meal for an official, either sliced fresh meat (*khuai* 膾) or dried meat slices (*fu*) may be served but not both in the same meal. If one is served, then there is no place for the other.[23] In the *Sang Ta Chi* (The Greater Record of Mourning Rites) a *yü hsi* 魚腊 (dried fish) is said to be among the foods placed near the coffin at a funeral.[24]

But none of these references actually reveals how *fu* or *hsi* was made. To be sure, we know that the meat was sliced and dried, but we have no idea of the details of the process. For that information we have, once again, to wait for the appearance of the *Chhi Min Yao Shu*, which devotes all of chapter 75 to the making *fu* and *hsi*. Of the seven recipes presented, No. 7, on *i yü* 浥魚 (wet salted fish), has already been quoted in our discussion of fish preserves (page 383). To give us an appreciation of how the process was carried out, recipe 1 is quoted in full below:[25]

Method for making five-flavoured *fu*: The best time for the process is the first, second, ninth or tenth month. Meat from cattle, sheep, the roebuck, deer, wild boar or domestic pig may all be used. The meat is cut into strips or slices. It should be cut along the grain rather than across the grain.[26] When done, they are separated from each other.

Pound the bone (from the cattle, sheep etc.) into pieces. After prolonged boiling decant off the soup. Skim off the residues floating on top and allow the soup to settle and clarify.

---

[18] *The Analects*, tr. D. C. Lau, p. 103.
[19] *Chou Li, Shan Fu*, p. 34; *Wai Yung*, p. 39; *Pien Jên*, p. 54; and *Hsi Jên*, p. 43.   [20] *Chou Li Chu Shu*, p. 65b.
[21] Kuo Chhing-Fan ed. (*1961*), *Chuang Tzu Chi Shih*, ch. 26, *Wai Wu* (External Things), p. 925, and Burton Watson tr. (1968), p. 296, mod. auct.
[22] *Li Chi*, pp. 28, 72, 76.   [23] *Li Chi*, pp. 459, 460.   [24] *Li Chi*, p. 732.   [25] *CMYS*, p. 459.
[26] This advice is the opposite to that given in in the *Li Chi* as discussed on pp. 60–70 for slicing meat into *khuai*. In this case, the meat is cut along the grain so that the muscle fibres remain largely intact. The meat strips or pieces will be more able to survive the extra treatment. In consistency a beef *fu* would be very much like a modern western beef jerky.

Take fermented soybeans (*shih* 豉) of superior quality and wash off any extraneous matter with cold water. Boil the fermented beans in the bone soup. When the soup has acquired the right colour and flavour, it is strained to remove the beans. When cool, add salt to taste. Do not make it too salty. Finely slice and chop white scallion stem. Add the hash, together with ground fagara, ginger and orange peel to the soup. Adjust the amount added according to one's preference. Immerse the meat pieces in the soup. Rub them with the hands until the spices are thoroughly absorbed.

Collect the *fu* slices after incubating them for three nights. As for the *fu* strips, taste the meat from time to time. When the flavour is right remove them. Pierce a string through the meat pieces and hang the line under the eaves on the north side of the building and allow the meat to air dry. Squeeze the strips by hand when they are about half dry. When the *fu* pieces are ready, they are kept in an empty and quiet storage room. If there is smoke around, the *fu* may taste bitter. The pieces should be hung in a paper bag. If they are placed in a jar, they will spoil more easily. It they are not covered with paper, the flies will get at them. The *fu* strips made in the twelfth month are called *chu fu* 瘃脯 (chilled *fu*). They should last until the summer. When taking the dried meat for use, take the fatty ones first. The fat can turn rancid, and thus spoil more easily.

From this description we can see that the process consists of three steps. First, the meat is cut along the grain in slices or strips. Then they are steeped in an aqueous decoction of animal bones and fermented soybeans flavoured with salt and spices. Finally, the pieces are hung on a string to dry in air. The other five recipes may be regarded as variations of this basic theme.

> Recipe No. 2, plain *fu*: Plain salted water is used for steeping, and the pieces when half dry are gently pounded to remove residual water.[27]
>
> Recipe No. 3, sweet *fu*: The sliced meat is directly dried in air; the steeping is omitted. No salt is used.
>
> Recipe No. 4, fish *fu*: Brine is forced into the visceral cavity of the whole fish, which is then dried in air. When the fish is ready it is gutted and soaked in vinegar.
>
> Recipe No. 5, five flavoured *hsi*: Goose, duck, chicken, rabbit, fish etc. are cleaned and processed whole as in Recipe 1.
>
> Recipe No. 6, plain *hsi*: The cleaned animals are cooked in boiling water and then dried in air.

It is recommended that these processes be carried out in the twelfth month, which is commonly known as *la yüeh* 臘月. This is probably the reason why in later centuries cured meat became known popularly as *la jou* 臘肉. This term was first used by Chhên Yüan-Ching of the Southern Sung in his *Sui Shih Kuang Chi* 歲時廣記 (Expanded Records of the New Year Season).[28] Eventually, *la* would replace *hsi* in the culinary literature. Both *hsi* and *fu* are mentioned in the *Ssu Shih Tsuan Yao*, late Thang, the *Wu Lin Chiu Shih* (Institutions and Customs of the Old Capital), +1280,

---

[27] Since fermented soybeans (*shih*) had not been invented in early Chou, it is likely that the earliest *fu* was made by this method, i.e. by using a salt liquor as the incubating medium.

[28] *CKPJPKCS* (Encyclopedia of Chinese Cuisine), p. 316.

and *Mêng Liang Lu* (Dreams of the Abundant Capital), Southern Sung.²⁹ *Fu* is seen in the *Tung Ching Mêng Hua Lu* (Dreams of the Glories of the Eastern Capital), +1147.³⁰ *Hsi* is rarely seen after the Sung, although *fu* is occasionally mentioned until the end of the 18th century.

The *Wu Shih Chung Khuei Lu*, in a section entitled *Fu* and *Cha* (Dried and Preserved Products), gives one recipe on making dried salted meat (*yên jou fa* 鹽肉法) and one air dried salted fish (*fêng yü fa* 風魚法).³¹ The number of relevant recipes increased to about ten in the *Chü Chia Pi Yung*, Yuan, *Yin Chuan Fu Shih Chien*, +1591, *Shih Hsien Hung Mi*, +1680 and *Hsing Yüan Lu*, +1750.³² The processes are identified or characterised by such words as *yên* (incubate in salt), *fu* (dried) or *la* (cured by drying). The basic procedure is to rub salt on the material, incubate it until it is well salted, daub it with spices, wine, wine residues or vinegar, and dry it in air, in the sun, over a fire or in a stream of smoke. Often the salted liquor from a previous batch is added to the incubation. If the material is hard, pressure is applied to exude water from the tissues. The materials processed in this way include, pork, lamb, beef, venison, roebuck flesh, pork tongue, pork thigh, beef tongue, duck, chicken, goose, and various kinds of fish.

One of the most famous products in this category is the ham from Chin-hua (*Chin-hua huo thui* 金華火腿). The technique for making this ham is based on a recipe for cured pork which first appears in the *Chü Chia Pi Yung*, under the title 'The Wu-chou Method for Curing Pork (*Wu-chou la chu fa* 婺州臘豬法),' which we now quote:³³

Three catties of pork would be a suitable batch. Rub one ounce of salt on each catty of meat. Incubate the meat in a jar for several days. Turn them over two to three times each day. Place them in some wine and vinegar and incubate them again for three to five days. Turn them over three to five times each day. Remove the chunks and let them air dry. Prepare a pot of vigorously boiling water and a jug of sesame oil. Separate the chunks of pork. Blanch a chunk in the boiling water and remove it quickly. Brush sesame oil over the pork, and smoke it over a stove outlet. After a day daub incubating liquor and vinegar over the pork and allow it incubate for ten days. Collect the pork chunk and smoke it further over a stove outlet. If normal stove usage does not generate enough smoke, allow rice hulls to smoulder in the stove and generate smoke for ten days. The smoke should be continuous, day and night. The same procedure can be used to cure lamb.

Variations of this recipe dedicated to making ham (*huo thui* 火腿), including the ham of Chin-hua, are described in the *Yin Chuan Fu Shih Chien*, *Shih Hsien Hung Mi*, *Hsing Yüan Lu* and *Thiao Thing Chi*.³⁴ One innovation is that the pork thigh is pressed by a heavy stone over a bamboo mat while it is incubated, no doubt to help the meat to expel water and absorb the salt. Curing ham is nowadays largely a factory operation. There are at least fifteen localities which produce hams of top quality, but the ham of Chin-hua is still rated as the best.³⁵

---

²⁹ *SSTY*, 12th month; *WLCS*, ch. 6, p. 124, ch. 9, p. 172; *MLL*, ch. 13, pp. 109, 111 and ch. 16, p. 134.
³⁰ *TCMHL*, ch. 8, p. 53. ³¹ *WSCKL*, p. 8.
³² *CCPY*, pp. 74, 74, 75, 75, 76, 76, 77, 77, 77, 82; *YCFSC*, pp. 70, 71, 71, 71, 76, 78, 78, 78, 82; *SHHM*, pp. 98, 103, 107, 107, 114, 122, 123, 123, 124, 124, 125; *HYL*, pp. 18, 18, 19, 19, 20, 22, 23, 24, 34.
³³ *CCPY*, pp. 74–5. ³⁴ *YCFSC*, p. 71; *SHHM*, pp. 122–3; *HYL*, pp. 18–9; *TTC*, pp. 216–17.
³⁵ *CKPJPKCS*, p. 265. The ham of Hsüan-wei comes in a close second.

Although we can trace the technology of curing and drying meat from early antiquity to the present day, there are curious lapses which mar the continuity of the record. One such case is in the history of sausage, *hsiang chhang* 香腸, formerly known as *kuan chhang* 灌腸. The making of sausages by filling intestines with minced meat is probably an ancient art,[36] but we have no earlier references to it than the recipe in chapter 80 (Roast Meats) in the *Chhi Min Yao Shu*, +544.[37] Sausages (*kuan chhang*) are listed among the prepared foods available in the markets of the Northern Sung capital of Kaifeng as well as in the Southern Sung capital of Hangchou.[38] Presumably, these were also meat sausages. Yet the only recipe we have on *kuan chhang* during the Sung–Yuan–Ming period, the one in the *Chü Chia Pi Yung*, Yuan, uses sheep's blood to fill the sausage casing. The same recipe is repeated in the *To Neng Pi Shih*, +1370.[39] It was not until four hundred years later that we have another recipe for sausage. It is found in the *Hsing Yuan Lu* of +1750.[40] No blood is mentioned. This time the recipe describes the making of a pork sausage and the drying of it in air. The making of *kuan chhang* from pork is included among the recipes in the *Thiao Thing Chi*, 19th c.[41] The common name for sausages today is *hsiang chhang* (fragrant intestines). They are an important component of the processed meats industry.[42] The cured pork sausage (*la chhang* 臘腸 or *laap cheong* in Cantonese) of Kwangtung is well known all over China as well as in Chinese communities around the world.

Another lapse in the record is the fate of a semi-dried salted fish product known as *hsiang* 鯗. The term is not used in the *Chhi Min Yao Shu* although it could have covered salted fish products such as *i yü* 浥魚 (p. 383). Yet the *Mêng Liang Lu*, +1320, a record of conditions extant in the Southern Sung capital, tells us that there was a special quarter in Hangchou where merchants dealing in dried salted fish (*hsiang*) were wont to congregate.[43] There were between one and two hundred shops supplying *hsiang* in and out of the city. More than sixteen types of *hsiang* as well as all sorts of processed seafoods were on sale. *Hsiang* was obviously a major food product during the Southern Sung. Nevertheless, it is rarely mentioned in the pre-Sung or post-Sung food literature that we have reviewed.

On the other hand, one unique processed food did manage to find its way into the food canons. We refer to a finely shredded meat or fish called *sung* 鬆, a product that has no parallel in Western cuisine. *Sung* literally means loose, lax or relaxed. There is a recipe for chicken *sung* in the *Shih Hsien Hung Mi*, +1680, *Yang Hsiao Lu*, +1698, and *Sui Yuan Shih Tan*, +1792, and recipes for pork *sung* and fish *sung* in the

---

[36] Darby, Ghalioungui & Grivetti (1977), I, p. 161 also lament the dearth of information on the status of sausages in ancient Egypt and the mediaeval Mediterranean world. 'The historical record concerning this food preparation is amazingly blank; no record known to us from the Dynastic Period discusses it. But during the Late Byzantine Period in Egypt, just prior to the Arab conquest in 641, a thriving sausage industry existed in Middle Egypt, perhaps indicative of a longer history than the record immediately shows.'

[37] *CMYS*, p. 494. Chopped and suitably flavoured lamb is used to fill sections of sheep intestine. The raw sausages are then barbecued before they are served or stored away.

[38] *TCMHL*, ch. 3, p. 22; *Wu Lin Chiu Shih*, ch. 6, p. 122.

[39] *CCPY*, p. 97; *TNPS*, in Shinoda & Seiichi, eds. (1973), p. 381.

[40] *HYL*, p. 24, using minced pork, spices and pig intestine and curing the sausage by drying in air.

[41] *TTC*, pp. 198–9. [42] *CKPJPKCS* (1992), p. 634. [43] *MLL*, p. 139.

Hsing Yüan Lu, +1750.⁴⁴ To illustrate a typical procedure, we quote the recipe for pork *sung* from the *Hsing Yüan Lu*:⁴⁵

Take a whole pork loin. Cook it thoroughly [in water] over a hot fire. Cut it diagonally into large pieces. Add fragrant mushrooms and cook the pork further in the original soup until the chunks start to fall apart. Remove the lean meat and shred it into fine fibres by hand. Next add sweet wine, soy sauce, ground aniseed, and a little sugar to the fibres and heat them in the frying pot (*kuo* 鍋) over a slow fire. Keep stirring until all the water is driven off. Remove and keep the product.

The dates of these references indicate that the art of making *sung* was probably perfected in late Ming or early Chhing. The products soon gained wide acceptance, and their popularity has remained high to the present day. They have a long shelf life, and make a particularly welcome relish when served with rice congee at breakfast time.⁴⁶

(iii) *Preservation of fruits*

Three traditional methods have been used in China to preserve or prolong the useful shelf life of fruits: drying, saturating with sugar, and storage in a controlled environment. Drying is the simplest method; it is, no doubt, also the oldest. The *Chou Li* lists dried fruits among the viands presented by the *Pien Jên* (Superintendent of Ritual Viands) at ritual ceremonies at the Chou court.⁴⁷ There are no other references to preserved fruits in the Han or pre-Han literature. However, examples of all three processes can be found in the *Chhi Min Yao Shu*, +544, in chapters (33 to 42) dealing with the cultivation of fruits. Let us take the case of drying first. Chapter 33 provides the following description of a method for drying jujubes (*tsao*):⁴⁸

Sun dried jujubes: First clean the ground. If there are goosefoot the jujubes may acquire a bad odour. Setup a rack to support a bamboo screen. Place the jujubes on the screen. Push them with a toothless rake into a pile, and then spread them out evenly. Repeat this procedure twenty times a day. Leave them spread out during the night. The cool night air would hasten the drying. If it rains cover the fruits with a straw mat. After five to six days, take the red soft fruits [that are almost dry] and put them in the sun on a screen placed on a higher rack. When they are dry they can be piled a foot high without harm. Discard those that show signs of rotting. These will never dry and will only contaminate the good fruits. Those that are not yet dried will be treated in the same way as before.

In a variation of this procedure the jujubes are sliced and then dried to make *tsao fu* (dried jujube slices). A crab apple (*nai* 柰) *fu* is also made in the same way.⁴⁹ Fruits

---

⁴⁴ *SHHM*, p. 114, the recipe is also applicable for making fish, pork and beef *sung*; *YHL*, p. 86; *HYL*, pp. 22, 34; *SYST*, p. 69.
⁴⁵ *HYL*, p. 22.
⁴⁶ This type of shredded meat or fish is often served at breakfast buffets at the better hotels in South China and Taiwan.
⁴⁷ *Chou Li, Pien Jên*, p. 54.
⁴⁸ *CMYS*, p. 183. The procedure is very similar to that practised today in Hopei, China, cf. Miao Chhi-Yü, (*1982*), p. 189, note 22. Dried jujubes, under the name *ta tsao* (great jujube) is listed as an upper-class drug in the *Ming I Pieh Lu*, +510; cf. *PTKM*, ch. 29, p. 1756.
⁴⁹ *CMYS*, p. 215.

can, of course, be dried over a fire. One type of dried persimmon was made by this method.[50] Fruits can also be smoked. A so-called dark dried apricot (*wu mei* 烏梅) was prepared by smoking raw green fruits.[51] It was used largely in medicine. In another variation, the *mei* apricots were soaked in brine overnight and then dried in the sun by day. After repeating the treatment ten times, a dried 'white *mei* fruit' (*pai mei* 白梅) was obtained.[52] A similar method was used to make a dried 'white plum' (*pai li* 白李).[53]

Perhaps because of the simplicity of the technique, drying of fruits is seldom discussed in the standard food canons from the Sung to the Chhing. It is, however, mentioned in the Monograph on Litchi (*Li Chih Phu*), +1059, of Tshai Hsiang, which gives two recipes for making dried litchi,[54] and in the Agricultural Treatise (*Nung Shu*), of Wang Chên, +1313, in which the making of dried persimmon, commonly called persimmon cake (*shih ping* 柿餅) is described.[55]

The second method, saturating fruits with sugar, is analogous to the inundation of vegetables, meat and fish with salt. There are several examples of its application in the *Chhi Min Yao Shu*. For drying grapes we are told to:[56]

Collect fully ripened berries in branches. Cut off the stems with a knife but do not allow any juice to leak out. Mix the grape berries with two parts of honey and one part of *tzu* 脂 (grape juice?) and boil gently five to six times. Strain off the grapes and dry them in air. This treatment not only yields a delicious product, but also one that can be kept over the summer without spoilage.

An interesting variation is to salt the fruits first before soaking them in honey as in the Szechuan method for preserving *mei* apricots:[57] The use of honey as a

---

[50] *CMYS*, p. 218.
[51] *CMYS*, p. 200. An extract of this dark apricot is itself used to preserve melons in Ch. 88, cf. Table 38.
[52] *CMYS*, p. 200.   [53] *Ibid.*, p. 197.
[54] *Li Chih Phu*, p. 3, records two methods:

The first is the Red Salt Method: 'In the countryside people soak red hibiscus flowers in salted plum juice liquor to obtain a red decoction. Litchi fruits are steeped in it. They are then dried in the sun. The fruits acquire a red colour and a sweet–sour flavour. They can be kept three to four years without deterioration, but, they do not taste like natural litchi.'
The second is the White Sun Method: 'Dry the fruits when the sun is strong, until the flesh is hard. Store them in a tightly sealed jar for a hundred days. This is called "sweating it out". After sweating, the litchi will have a long shelf life. Otherwise they will rot within a year.'

A variation of this procedure is given in the *Chhün Fang Phu* (An Assembly of Perfumes), +1630, cited in Wang Shan-Tien (*1987*), p. 204. Li Shih-Chên, *PTKM*, ch. 31, p. 1817, remarked that 'when fresh the flesh of litchi is white; when dry it turns red. It can be dried in the sun or over a fire.'

[55] *Nung Shu*, by Wang Chên, cited in Wang Shan-Tien (*1987*), p. 206, which says:

Method for Dried Persimmon: 'Peel off the skin of fresh persimmon and flatten it. Dry it in the sun. Keep it in a jar. When a frosty powder forms on the surface, it is ready for eating. It is a cooling food.'
According to Wang Shan-Tien (*ibid.*), this is still the way the popular *shih ping* is made in Shensi and Shansi today. The *shih ping*, persimmon cake, is mentioned as a vegetarian delicacy in the popular Ming novel *Hsi Yu Chi* (Journey to the West), cf. ch. 79, p. 908; ch. 100, p. 1126.

[56] *CMYS*, p. 192.
[57] *Ibid.*, p. 200. The text reads:

Method from the Food Canon (*Shih Ching*): Take large *mei* apricots. Remove the skin, [salt] and dry them in air. Do not expose them to the wind. After two nights, wash off the surface salt, and immerse them in honey. Change the honey each month. They will be as good as new in a year.

preservative is extended to litchi in the *Li Chih Phu* (Monograph on Litchi), +1059 and to the orange in the *Chü Lu* (The Orange Record), +1178.⁵⁸ This type of preserved fruits came to be known by the prefix *mi chien* 蜜煎 (cooked in honey)⁵⁹ and later as *mi chien* 蜜餞, which may be translated as 'honeyed sweetmeats'. The term *mi chien* was first used in the History of the Three Kingdoms (*San Kuo Chi*), c. +290, which mentions honeyed lotus seeds, arrowroot slices and winter melon strips.⁶⁰ According to the History of the Liao Dynasty (*Liao Shih*) *mi chien*, i.e. honeyed sweetmeats, were among the presents sent by the Kitan to the Sung emperor.⁶¹ According to the *Tung Ching Mêng Hua Lu*, +1147, honeyed sweetmeats carved in the shape of flowers, *mi chien tiao hua* 蜜煎雕花, were on sale in the market of the Northern Sung capital of Kaifeng.⁶² The *Wu Lin Chiu Shih*, +1280, notes that varieties of *mi chien*, including *mi chien tiao hua*, were served at the imperial court and some were even available for sale in the food markets of Hangchou.⁶³ Taken together these references indicate that honey-preserved fruits were probably known during the Later Han, and were common and popular articles of food during the Sung.

During the Sung we begin to see honey being replaced by cane sugar for the preservation of fruits.⁶⁴ This trend is continued during the Yuan as cane sugar becomes more readily available.⁶⁵ In the *Chü Chia Pi Yung* there are six recipes with *mi chien* (honey cooked) in its title.⁶⁶ One pertains to the making of *mi chien* fruits in general, and the others to the *mi chien* of winter melon, ginger, bamboo shoots, green apricot, and arrowroot respectively. The general recipe recommends the use of crude solve (*phu hsiao* 朴消) to treat the sour fruits before they are boiled in honey.⁶⁷ The 'honeyed' sweetmeats are followed by five recipes for 'sugared' (*thang* 糖) products, namely 'sugared' crispy plum, peppered plum, Chinese strawberry, arrowroot, and Chinese quince. The basic procedure is the same as that used for honeyed sweetmeats, except that the honey is replaced by cane sugar. In the case of the strawberry and of the quince, the sweetmeats are dried in the sun and then stored. Similar recipes for both honeyed and sugared products are found in the food canons of the Ming and Chhing.⁶⁸

---

⁵⁸ *Li Chih Phu*, p. 4; *Chü Lu*, p. 13. *PTKM*, ch. 31, p. 1817 refers to this method as *lu chin mi chien* 鹵浸蜜煎 soak in salt and cook in honey.
⁵⁹ *PMTT*, ch. 14, p. 14a.   ⁶⁰ *San Kuo Chi*, cited in Wang Shan-Tien (*1987*), p. 205.
⁶¹ *Liao Shih*, Section on Geography, cited in Wang Shang-Tien, *ibid*.
⁶² *TCMHL*, ch. 2, p. 15.   ⁶³ *WLCS*, ch. 6, p. 123; ch. 8, p. 163; ch. 9, p. 171.
⁶⁴ Cf. for example, the *hsiang thang kuo tzu* 香糖果子 (fragrant sugared fruits) mentioned in the *TCMHL*, ch. 2, p. 15; *thang tshui mei* 糖脆梅 (sugared crispy apricots) and *wu mei thang* 烏梅糖 (sugared dark apricots) in the *WLCS*, ch. 6, p. 123; and *thang mi kuo shih* 糖蜜果食 (sugar and honey treated fruits) in the *MLL*, ch. 16, p. 137.
⁶⁵ The technology of sugarcane processing is detailed by Christian Daniels in *SCC* Vol. VI, Pt 3, of this series.
⁶⁶ *CCPY*, pp. 30–2.
⁶⁷ *CCPY*, p. 30. Other recipes in this group may recommend that the raw material be blanched in boiling water or pre-treated with solutions of lime, salt or copper carbonate powder. The translation of *phu hsiao* as 'crude solve' follows the nomenclature suggested by Needham *et al.* for saltpetre and related products in *SCC* Vol. V, Pt 7, pp. 100ff.
⁶⁸ For example, *Tuo Nêng Phi Shi*, +1270, ch. 3, Shinoda & Tanaka eds. (*1973*), pp. 388–91; *Chhü Hsien Shên Yin Shu*, +1400, 3rd, 4th, 6th and 7th months, Shinoda and Tanaka, *ibid*. pp. 418, 421, 432; *Pien Min Thu Tsuan*, +1502, ch. 14, p. 14a; *Shih Hsien Hung Mi*, +1680, pp. 91, 92, 95, 96; *Yang Hsiao Lu*, +1698, pp. 64, 65, 66, 69; *Thiao Thing*

The third method of prolonging the shelf life of fruits (and also vegetables) is to store them in a controlled environment. The *Chou Li* tells us that the *Chhang Jên* (Superintendent of Orchards and Gardens) collects valuable fruits according to season and stores them.[69] How they were stored is not specified. We can only surmise that the methods used by the ancients would be similar to those recorded in the *Chhi Min Yao Shu* of +544. This invaluable compendium describes three procedures for the prolonged storage of fruits.

The first is to keep the fruits in a cool enclosed area. It was applied to the storage of grapes (*phu thao* 葡萄):[70]

Storage of grapes: when the grapes are fully ripe, take down the whole vine. Dig a pit under the house. Drill a hole in the wall near the floor. Plant the stem in the hole and pack it tight. Seal the opening to the pit with earth. The grapes should stay fresh throughout the winter.

A similar procedure was used to store pears (*li* 梨):[71]

Storage of Pears: pick the pears immediately after the first frost. If exposed to too much frost, they would not last until the summer. Dig a pit deep under the house. Be sure that the floor is not damp. Place the pears in the pit. Do not cover them. They would keep until summer. When picking pears, handle them carefully. Be sure they are not damaged.

There are two conditions in an underground pit that help to slow down the metabolic activities involved in the decaying of fruits. The first is temperature, which would be near freezing during a typical winter in North China. Lower temperature means lower metabolism. The second is the controlled atmosphere. During storage the fruits would continue to breathe, in other words, to consume oxygen and expel carbon dioxide. But since the fruits are in an enclosed pit, this would lead to a decrease in the oxygen content of the air, and thus a further retardation of metabolic activity. The moisture level would be adequate to prevent the fruits from drying out. Under these conditions the success of the operation would depend a great deal on the integrity of the fruits. If they are bruised they would become susceptible to microbial infection. Thus, the author makes a special point that the pears should not be bruised during handling. In the case of grapes, the individual berries are not touched by hand, since the whole vine is harvested and stored.

The second procedure is to surround the fruit with an inert material as exemplified in the procedure for storing Chinese chestnuts (*li* 栗):[72] 'Storage of fresh

---

Chi, 19th century, pp. 801, 802, 808, 817, 818, 827, 833, 844, 846. Processed fruits are rarely seen in the *Hsing Yüan Lu*, +1750 or the *Sui Yüan Shih Tan*, +1792 which tend to focus on the preparation of savoury dishes. Nevertheless, these products remain a standard feature of Chinese cuisine today, cf. *CKPJPKCS*, p. 376.

[69] *Chou Li, Chhang Jên*, ch. 4, p. 177.
[70] *CMYS*, ch. 34, p. 192. tr. auct. adjuv. Shih Shêng-Han (1958), p. 60. Our interpretation differs somewhat from Shih's in several places. Firstly, in Shih's version the grapes are harvested in separate clusters. We say a whole vine is taken. Secondly, in Shih's version, several holes are made on the side of the pit. We think only one hole is made. Finally, according to Shih, the last step is to, 'fill and cover the pit with earth'. We think it should be to 'fill the opening to the pit with earth'. Cf. comment by Miao Chhi-Yü, (*1982*), p. 195, note 33.
[71] *CMYS*, ch. 37, p. 205 tr. auct. adjuv. Shih Shêng-Han (1958), p. 60.
[72] *CMYS*, ch. 38, p. 210 *ibid*. According to the textual research of Miao Chhi-Yü (*1982*), p. 212, the chestnuts are kept until the second month of the next year, and not the fifth month as shown in Shih's translation.

chestnuts: store in containers with fine sun-dried sand. Cover with a pottery pan. They may all sprout by the second month of the next year, but they will be free from worms.'

Since the chestnuts are inundated with sand their rate of respiration would be very low. But the treatment may fool the seeds into behaving as if they had been planted in the ground, so eventually they would all begin to sprout. In any case, the procedure may be looked upon as an extreme case of atmospheric control.

The third procedure is one that we have already come across in Table 38 (p. 406), as a method for making pear pickle (*li tsu* 梨葅).[73] We are concerned here only with the first part of the recipe, the making of *lan* 㳕, in which pears are steeped in water in a bottle from autumn of one year to spring of the next. Before serving, the pear is peeled, sliced and flavoured with honey.

There are few examples of the application of the first procedure in the subsequent literature. There are, however, emulations of the procedure in which the fruits are stored in sealed urns, rather than in large pits. The *Chü Chia Pi Yung*, Yuan, describes the storage of Chinese olives (*kan lan* 橄欖) in a pewter jar, which after being filled is sealed with paper.[74] The procedure is repeated in the *Pien Min Thu Tsuan*, +1502 and *Shih Hsien Hung Mi*, +1680.[75] In the latter case the mouth of the jar, fabricated in the form of a circular gutter, is covered with a porcelain bowl. The gutter is 80% filled with water, which is replenished continuously as needed. The seal keeps the jar airtight and the humidity generated prevents the fruits from drying out. Another example from the *Chü Chia Pi Yung* is the storage of pomegranates (*shih liu* 石榴) in a pottery jar. The fruit is collected with the stem, placed in the jar and sealed tightly with paper.[76]

The third procedure, steeping in water, is described in the *Chü Chia Pi Yung*. It is particularly effective for storing green fruits.[77] Water collected in the twelfth month (*la shui* 臘水), fortified by a sprinkling of copper carbonate (*thung chhing mo* 銅青末) dust, is recommended. Fruits treated in this way included plums, loquats, apples, jujubes, grapes, water caltrop, sweet melon and olives. This recipe is repeated in the *Pien Min Thu Tsuan* 便民圖纂 (Everyman's Illustrated Compendium).[78] A modified recipe in which the copper carbonate is replaced by menthol and alum is reported in the *Shih Hsien Hung Mi*.[79] The overall procedure is also repeated in the *Thiao Thing Chi*.[80]

It is the second procedure that has received the most attention in the food literature. The *Chü Lu* (Monograph on the Orange), +1178, recommends that after harvest the oranges should be stored in a clean airtight room, surrounded by rice straw.[81] It is important that the room be as far away as possible from any place where wine may be handled. Whenever a bruised fruit is found, it should be removed immediately so that it would not contaminate the remaining healthy

---

[73] *CMYS*, ch. 88, p. 536.   [74] *CCPY*, p. 36.   [75] *PMTT*, ch. 14, p. 12b; *SHHM*, p. 89.
[76] *CCPY*, p. 35; see also *PMTT*, ch. 14, p. 12a.
[77] *CCPY*, p. 35. The copper carbonate presumably acts as a fungicide.   [78] *PMTT*, ch. 14, p. 13b.
[79] *SHHM*, p. 96.   [80] *TTC*, p. 816, reiterates the use of copper carbonate, menthol or alum in the water.
[81] *Chü Lu*, p. 12.

fruits. Three variations of this procedure are described in the *To Nêng Pi Shih*:[82] One is to store the oranges among pine needles. Another is to submerge them in mung beans, and the third is to immerse them in sesame seeds in a pewter vessel. The second and third variants are repeated in the *Pien Min Thu Tsuan*, which adds the admonition that oranges should not be in contact with rice, since rice would hasten the rate of decay.[83] The storage of oranges among pine needles or mung beans is reiterated in the *Yang Hsiao Lu*, *Hsing Yüan Lu* and *Thiao Thing Chi*.[84]

The immersion of chestnuts in sand described in the *Chhi Min Yao Shu* was repeated in the *Chü Chia Pi Yung* and *Pien Min Thu Tsuan* but it was not seen in the later Chhing literature.[85] Two extensions of this method are given in the *Chü Chia Pi Yung*, one the storage of jujubes in millet straw, and the other the storage of pears together with radish roots.[86] Each pear is collected with the stem intact. The stem is then pierced into a radish and wrapped in paper. It is said to be also applicable to oranges.

These references indicate that the aforementioned methods were applied widely from the Sung to the Chhing. But with the introduction of refrigeration technology from the West in the late 19th century, the traditional methods for the storage of fruits have become less and less useful, especially in the cities. In some areas of the countryside, however, such as the loess regions of Shensi, where natural caves abound, the storage of fruits in a controlled environment may still be an economical and convenient way of prolonging the shelf life of fresh fruits.

Although refrigeration is a modern technology, the use of ice to preserve foodstuffs is not. As we shall see, it is actually one of the oldest methods of food preservation known to the Chinese.

(iv) *Cold storage*

The earliest reference we have to ice collection and storage is in the *Shih Ching*. In the well-known poem 'Seventh Month' (*chhi yüeh* 七月), which we have cited earlier (page 18), the first two lines of the last stanza reads as follows:

> In the days of the Second they cut the ice with tingling blows,
> In the days of the Third they bring it into the cold shed.

The words for 'cold shed' in the original text are *ling yin* 凌陰 which according to Chêng Hsüan means 'ice house'.[87] The *Tso Chuan* (Master Tsochhiu's Tradition of

---

[82] *TNPS*, Shinoda & Tanaka (*1973*), p. 395.
[83] *PMTT*, ch. 14, p. 12b. Shih Shêng-Han, cited by Wang Shan-Tien (1987), p. 207, who points out that the post-harvest metabolism of fruits is adversely affected by ethylene and ethanol. The admonition to stay away from wine is thus scientifically sound. The reason hulled rice is deleterious is not due to the rice grain itself but rather to the microbes that reside on the surface of the grain. On the other hand, sesame seeds and mung beans are coated by a film of wax which make them relatively free from microbes growing on the skin.
[84] *YHL*, p. 64; *HYL*, p. 61; *TTC*, p. 843.   [85] *CCPY*, p. 34; *PMTT*, ch. 14, p. 13a.
[86] *CCPY*, p. 35; The same pear recipe is repeated in the *PMTT*, ch. 14, p. 13a.
[87] Waley (w 159, m 154). Day of the Second means the 12th month, and day of the Third, the 1st month. For a discussion of the role of ice storage in ancient China see Tan Hsien-Chin (*1989*). See also Wei Ssu (*1986*), Chhên Hung (*1987*); Wang Sai-Shih (*1988*); Kuo Po-Nan (*1989b*), and Hsing Hsiang-Chhên (*1989*).

the Spring and Autumn Annals), discusses the collection, storage and usage of ice during the reign of Chao Kung, Duke of the State of Lu.[88] That ice was an important commodity in the life of the royal Chou household is indicated by the listing of a special Superintendent of Ice Management in the *Chou Li* (Rites of the Chou) whose duty was as follows:[89]

Ling Jên 凌人: He is in charge of all matters related to ice. Each year during the twelfth month he orders the cutting of ice. Three times of what is estimated to be needed is collected and stored in the ice house. In the spring he prepares the large urns (*chien* 鑑) used for the cold storage of food. All the victuals, wines and other beverages required by the Internal and External Culinary Supervisors are stored in them. The ice cools the foods presented at sacrificial rites, or served to guests at state dinners, or is placed under the caskets during the funeral of the king or queen. In the summer gifts of ice are handed to deserving officials. The *Ling Jên* manages all these activities. In the autumn he cleans the ice house and a new cycle of collection and usage begins.

The *chien* referred to may be construed as an ancient counterpart of the modern ice box. The perishable food in a small covered container is placed in the middle of the *chien* and surrounded with ice. The *chien* is itself stored in the ice chamber. More than twenty bronze *chien* vessels have been excavated from tombs dated to the late Spring and Autumn or Warring States.[90] The most noteworthy is the beautiful *chien* recovered from the tomb of Marquis I (c. −450) of Tsêng which is shown in Figure 90.[91] The collection and storage of ice is also mentioned in the *Li Chi*.[92] To help him carry out all these duties the *Ling Jên* is assisted by two scribes, two supervisors, two assistant supervisors, eight foremen and eighty workers.[93] Ice management at the Chou court was, therefore, a major undertaking that probably no one else beside the king was able to command. What kind of facility was available to store the enormous amount of ice that must have been collected per season? The literature has little to tell us on this subject, but fortunately for us, recent archaeological discoveries in China have been able to throw considerable light on this question.

The remains of an ice house of the Chhin 秦 Kingdom, dated to the Spring and Autumn period (c. −500 to 600) was found in Fêng-hsiang 鳳翔, near Yung-chêng 雍城, Shensi Province.[94] The cellar consists of a large pit in the centre of a platform of packed earth, measuring 16.5 metres East to West, and 17.1 metres North to South. The rectangular pit is 2 metres deep and measures 10 × 11.5 metres at the top and 8.5 × 9 metres at the bottom. The pit is supported by a wall of pressed earth at

---

[88] *Tso Chuan*, Chao Kung, 4th year, 2nd paragraph, shows that in the year −538 ice was collected in the mountains and brought to the capital and stored in ice houses. Ice was used not only for cooling food and drink but also to massage and cool the royal corpses so that they would stay fresh looking for as long as needed to complete the funeral rites.
[89] *Chou Li, Ling Jên*, p. 53.   [90] Chhên Hung (*1987*).
[91] Qian Hao, Chen Heyi and Ru Suichu (1981), p. 58.   [92] *Li Chi, Yüeh Ling*, p. 304.   [93] *Chou Li*, p. 2.
[94] *Shensi shêng Yung-chhêng khao-ku tui (1978)*. See also Ang Chin-Huai (*1992*), pp. 344–5. The construction of this ice house reminds Kuo Po-Nan (*1989b*) of the traditional ice houses that he had seen in Peking. Indeed, it is similar to the ice house in 18th-century America that is still on display today in Thomas Jefferson's estate in Montecello, Virginia.

Fig 90. Bronze cooler from the tomb of Marquis of I (c. −400), Hupei Provincial Museum Qian, Chen & Ru (1981), p. 58.

least 3 metres thick. The floor is lined with stone slabs. A tunnel lined with clay pipes runs from the floor through the west wall to allow melted water to flow out of the cellar into a nearby stream. The cellar is sheltered by a tiled roof. It is estimated that it has a capacity to store 190 cubic metres of ice. Along the walls of the cellar are found large amounts of dried, and decayed plant matter, presumably straws that were used as insulating material. A diagram of the ice house is shown in Figure 91.

The ruins of several ice chambers of a different design have been excavated from sites of Warring States palaces with exciting results. One of the most notable is that at the *Han* 韓 capital at Hsin-chêng 新鄭 in Honan. The chamber is a rectangle room measuring about 28.5 feet north to south, 8.2 feet east to west and about eleven feet in height.[95] Along the eastern wall there are five wells with diameters varying from 30 to 38 inches, and depths from 69 to 97 inches. The wells are formed by placing circular pottery collars on top of each other. The area occupied by the wells is about one-third of the total area of the room. Found in the wells are tiles and bricks and remnants of utensils such as bowls, plates, pots, steamers, pipes and the bones of swine, cattle, sheep, chicken etc. About two-thirds of the artifacts

---

[95] Ma Shih-Chih (*1983*).

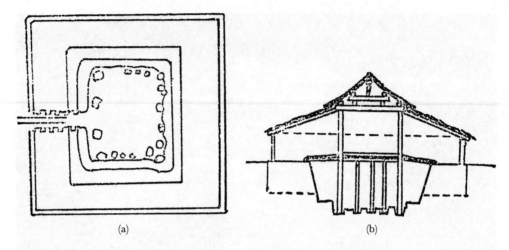

Fig 91. Design of an ice house in Fêng-hsiang, Shensi, Spring and Autumn period (c. −500). After Tan Hsien-Chin (*1989*), p. 297: (a) Diagram of foundation; (b) Reconstruction.

recovered are animal bones. The tiles contain characters such as *kung* 公 and *chu* 朱, which have been interpreted as *kung* 宮 (palace) and *chhu* 廚 (kitchen). Thus *kung chu* 公朱 is an archaic way of writing *kung chhu* 宮廚 (palace chef). These artifacts, therefore, indicate that the wells are not part of the water supply but rather receptacles for storing ice. Seven similar types of ice wells are found in the foundation of Building 1 of the Chhin 秦 palace complex at Hsien-yang 咸陽.[96] Bones of domestic animals also abound in these wells, indicating that they were part of the ice storage system. Three such wells were discovered in the palace in the Yen 燕 capital of Hsia-tu 下都 and eighteen in the Chhu 楚 capital of Chi-nan 紀南.[97] The structure of the wells are all the same, that is, they are formed by fitting a large pottery collar one on top of the other. A cross-section of one of the wells is shown in Figure 92. In one of the wells at Chi-nan there is a large urn at the bottom of the well. When these wells were first discovered, archaeologists were at a loss to explain what they were for. We are now reasonably sure that the function of these wells was to provide a well-insulated cavity for the storage of ice.

Ice chambers (*ping shih* 冰室) or ice wells (*ping ching* 冰井) are mentioned in the *Yeh Chung Chi* 鄴中記 (Record of the *Yeh* region), c. +4th century, the *Shui Ching Chu* 水經注 (Commentary on the Waterways Classic), c. +5th century, the *Loyang Chhieh Lan Chi* (Memoirs of Buddhist Temples of Loyang), +530 and in the *Hsin Thang Shu* (New History of the Thang Dynasty).[98] The *Thai Phing Yü Lan*, +983, refers to 'treasure wells' (*pao ching* 寶井) which were used for the storage of precious goods as well

---

[96] Chang Hou-Yung (*1982*); cf. *Chhin-tu Hsien-yang khao-ku kung-tso chan*, (*1980*).
[97] *Hopei shêng wên-hua chü wên-wu kung-tso tui*, (*1965*), and *Hupei shêng po wu-kuan Chiang-ling Chi-nan chhêng kung-tso chan*, (*1980*).
[98] *Yeh Chung Chi* by Lu Hui 陸翽, p. 1b; *Shui Ching Chu* by Li Tao-Yuan 酈道元, p. xx; *LYCLC*, ch. 1, pp. 2a, 6a; *Hsin Thang Shu*, *Pai Kuan Chi*, p. 1248.

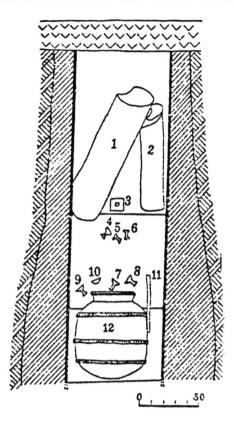

Fig 92. Cross-section of an ice storage well in Chi-nan, Capital of the Chhu Kingdom, c. −4th Century. *Hupei shêng po-wu-kuan* (*1980*), p. 43. Scale: 30 cm. Identity and relative positions of the objects found in the well: 1, 2, wooden logs; 3, wooden base for other objects; 4–6, 8, 9 a pottery *tou* (plates with a pedestal); 7, a broken *tou*; 10, bowl; 11, broken plough; 12, pottery jar.

as for foods.[99] The occurrence of ice wells is seldom mentioned in the literature after the Sung, but the usage of ice continued to grow in the centuries that followed.

Apart from its use in the storage of foodstuffs, ice was probably also directly consumed in iced drinks, which were a luxury much enjoyed by the ruling class. In the *Chhu Tzhu*, −300, the poem *Chao Hun* lists 'Iced cold liquor, strained of impurities, clear wine cool and refreshing,' among the inducements offered to persuade the departed spirit to return to its corporeal body.[100] During the Northern Chhi (+550 to +577), while Prince Kao Jui 高睿 was leading ten thousand men to the Great Wall in the heat of summer, he was presented by the local governor with a cart of ice victuals. Even this princely gift, however, was hardly sufficient to be shared with all his troops. Not willing to enjoy such luxury while his troops suffered, he allowed the ice to melt away without enjoying any of it himself.[101]

---

[99] *TPYL*, *Chü Chhu Pu*, 居處部, ch. 189, no. 17, p. 917.
[100] From *Chhu Tzhu*, *Chao Hun*, tr. Hawkes (1985), p. 228, line 103. For further information on iced drinks, cf. Liu Chên-Ya & Liu Phu-Yü (*1994*).
[101] *Pei Chhi Shu* 北齊書, +636, *Chao Chün Wang* 趙郡王, Chung-Hua ed. (1972), ch. 13, p. 171.

Officials of the Thang considered it a great honour to be awarded a gift of ice by the emperor.[102] In fact, Pai Chü-I, the poet, wrote a poem of thanks after he was so honoured.[103] But ice was no longer the exclusive preserve of the imperial court. A poem by Tu Fu mentions iced water and iced lotus root strips being served during an outing. The situation was similar during the Sung. Ice and snow (*pin shueh* 冰雪) and cold water (*liang shui* 涼水) were sold in the market in the 6th month at the Northern Sung capital at Khaifeng.[104] The *Mêng Liang Lu* records that in the height of summer, officials attending an audience in the Southern Sung capital were also given ice and snow by the court to allow them some relief from the oppressive heat.[105] At the same time, iced *mei* flower wine was served in some of the tea houses of Hangchou.[106] But ice was extremely expensive. In the summer it is said to be worth its weight in gold.[107]

Chilling with ice was apparently applied to dairy foods. An iced butter or cream (*ping su* 冰酥) is mentioned in a poem by Mei Jao-Chhen 梅堯臣 (+1002 to +1060) of the Northern Sung, and an iced yoghurt (*ping lo* 冰酪), in a poem by Yang Wan-Li 楊萬里 (+1127 to +1206) of the Southern Sung.[108] We have no idea of whether *ping su* (iced cream or butter) and *ping lo* (iced yoghurt) are the same or different products, nor how they were prepared. *Ping lo* was a favourite dish at the Yuan court. The Imperial Tutor Chhên Chi 陳基 wrote a poem to celebrate the occasion when he received a dish of *ping lo* as a mark of favour from the Emperor Shun Ti 順帝.[109] The fact that it was popular in exalted circles during the Sung and Yuan may have spawned the legend that the Chinese invented ice-cream and that Marco Polo brought it from China to Italy in the 13th century.[110] After the Yuan iced yoghurt disappeared from the Chinese scene.

The practice of preserving fish in ice is first noted in the *Wu Chün Chi* (Gazetteer of the Wu Region), +1190.[111] This practice was continued during the Ming and ice was used to preserve perishable foodstuffs such as fish when they were shipped along the grand canal from the Yangtze region to the imperial palace in Peking.[112] By the time of the Chhing, ice had become quite a common commodity. Hawkers selling iced drinks and iced concoctions were a common sight in the streets of Peking, and no doubt in other major cities as well.[113] There are numerous references to the existence of ice storage cellars and the storage of fish and other perishables on ice.[114] But the literature has little to say about the mechanics of ice

---

[102] *Thang Yü Lin*, ch. 4. [103] Pai Chü-I, cited in Wang Sai-Shih, (*1988*), p. 17.
[104] *TCMHL*, ch. 8, p. 53. [105] *MLL*, ch. 4, p. 22. [106] *MLL*, ch. 16, p. 130.
[107] *Sui Shih Kuang Chi*, ch. 2, p. 17a, *SKCS* **467**–20. [108] Cited in Kuo Po-Nan (*1989b*).
[109] *Ibid.*, the relevant stanza reads: 色贏金盆分外近，恩兼冰酪賜來初.
[110] Panati, C. (1987), p. 418, cf. also Kuo Po-Nan (*1989b*), Frances Wood (1995), pp. 79–80.
[111] *Wu Chün Chi* 吳郡志, ch. 29, item *thu wu* 土物.
[112] Mote (1977), p. 215, quotes a passage from the *Khe-tso chui-yü* by Ku Chhi-Yuan 顧起元, which describes the procurement system used to regulate the traffic of foodstuffs and other materials along the Grand Canal. It notes that shad (*shih yü* 鰣魚) was shipped under refrigeration from the South to Peking. See also Kuo Po-Nan (*1989b*), p. 11 who quotes a Ming poem celebrating the use of ice for transporting fish.
[113] *Yenching sui shih chi* 燕京歲食記, cited by Wang Sai-Shih (*1988*), p. 18.
[114] Hsing Hsiang-Chhên (*1989*) cites several references from local histories to ice cellars and the use of ice for keeping fish fresh. No information, however, is given on how the ice was made and collected.

collection, storage and distribution. The most detailed information on the subject has, surprisingly, come from a foreign resident. In 1844 the inveterate observer of the Chinese scene, Robert Fortune, wrote to the *Gardeners' Chronicle* from his home in Ningpo as follows:[115]

Ice-Houses in China. A short time before I left England, you published in the *Gardeners' Chronicle* a number of letters and plans for the construction of ice-houses, but, as far as I can remember, nothing at all resembling the Chinese one, which I shall now describe to you. On the left bank of the Ningpo river, proceeding upwards from the town and forts of Chingbai, and in various other parts in the north of China, I have met with these ice-houses. When I inspected them for the first time, last winter (1843), their construction and situation differed so much from what I had been accustomed to consider the essentials of an ice-house at home, that I had great doubts as to their efficiency; but at the present time, which is the end of August, 1844, many of these houses are yet full of ice, and seem to answer the end most admirably. You are probably aware, from my former descriptions of the country, that the town of Ningpo is built in the midst of a level plain, from 20 to 30 miles across. These ice-houses stand on the river sides, in the centre of this plain, completely exposed to the sun – a sun too very different in its effects from what we experience in England – clear, fierce and burning – which would try the efficiency of our best English ice-houses, as well as it does the constitution of an Englishman in China.

The bottom of the ice-house is nearly on a level with the surrounding fields, and is generally 20 yards long by 14 yards broad. The walls, which are built with mud and stone, are very thick, 12 feet in height and are, in fact, a kind of embankment rather than walls, having a door through them on one side and a kind of sloping terrace on the other, by which the ice can be thrown into the house. On the top of the walls or embankment a tall span roof is raised, constructed of bamboos thickly thatched with straw, giving the whole an appearance exactly resembling an English haystack. And this is the simple structure which keeps ice so well during the summer months under the burning sun of China.

The Chinaman, with his characteristic ingenuity, manages to fill his ice-house in a most simple way, and at a very trifling expense. Around the house he has a small flat level field, which he takes care to overflow in winter before the cold weather comes. It then freezes, and furnishes the necessary supply at the door. Again in spring these same fields are ploughed up, and planted with rice; and any water which comes from the bottom of the ice-house is conveyed into them by a drain constructed for the purpose. Of course, here as in England, the ice is carefully covered up with a thick coating of straw when the house is filled. Thus, the Chinaman, with little expense in building his ice-house, and an economical mode of filling it, manages to secure an abundant supply for preserving his fish during the hot summer months. This I believe, is the only, or at least the principal purpose to which it is applied in this country, and never for cooling wine, water, or making ices, as we do in Europe.

We are, indeed, fortunate to have Fortune's perceptive letter. His fascinating account answers, at least in part, our question on the mechanics of ice collection and storage as a commercial activity in the Chhing dynasty. The efficiency of the scheme explains how ice became a readily available and widely used commodity in the cities of China. Accompanying Fortune's letter is a sketch of the ice houses

---

[115] Fortune 1845. The same account is included in Fortune (1853) I, p. 82.

Fig 93. Sketch of a 19th-century Chinese ice house in Ningpo. Fortune (1853) I, p. 82.

which is reproduced here as Figure 93. Fortune's assertion that the ice was never used for cooling wine or making ices is obviously mistaken if it is taken as a general statement, but he could have been correct if he had meant only the situation as it existed in the Ningpo region in his day. In any case it would have been a simple matter to have water for drinking frozen separately and stored in the ice house. We do not know when this ingenious system of ice management was developed, but we do know that within a hundred years of Fortune's letter, it has disappeared from the Chinese food scene, having fallen victim to competition from the mechanical refrigeration technology imported from the West.

(3) Production of Oils, Malt Sugar and Starch

So far, our survey of the technology of Chinese processed foods has covered the fermentation of cereals into wine and vinegar, the conversion of soybeans into versatile soyfoods, and the pickling and preservation of meats, fish, vegetables and fruits. We shall now complete the picture with an examination of the technologies that have been developed to process oilseeds, rice, wheat, and other cereals, to give a variety of nutritious and attractive foods. The topics that remain to be explored are: extraction of oil from oilseeds, preparation of malt sugar and starch from cereals, fabrication of pasta and noodles from wheat, and extraction of gluten from wheat. We shall deal with the first two in this subchapter and the remaining two in the next.

(i) *Oil from oilseeds*

Recalling daily life in the Southern Sung capital of Hangchou, the *Mêng Liang Lu* (Dreams of the Former Capital), +1334, states that, 'the things that people cannot do without every day are firewood, rice, oil, salt, soy paste (*chiang*), vinegar and tea'.[1]

---

[1] *Mêng Liang Lu*, ch. 16, p. 139. See discussion by Freeman (1977), p. 151.

It is clear that by the time of the Sung, edible oil had advanced to the position as one of the proverbial 'seven necessities' of life. And yet, as we have shown earlier (pages 28–9), the pressing of seeds for oil, was unknown in pre-Han China. The fats used in cookery cited in the *Chou Li* and *Li Chi* were animals fats, or in modern parlance, *hun yu* 葷油 (animal oil) as opposed to *su yu* 素油 (vegetable oil). The animal fats were known as *kao* 膏 for the softer type and *chih* 脂 for the harder type.[2] In fact the very word *yu* 油 was not seen in antiquity. The only reference to *yu* in the *Shuo Wên* (Analytical Dictionary of Characters), +121, designates it as the name of a river, *Yu shui* 油水, which originates near Wu-ling.[3] The *Fan Shêng-Chih Shu* (The Book of Fan Shêng-Chi), c. +10, notes that the seeds of the gourd burn brightly when incorporated in a torch.[4] We have already alluded to the pounding of the seeds of the hemp, cocklebur and gourd in the *Ssu Min Yüeh Ling* (Monthly Ordinances for the four Peoples), +160, for use in making flambeaux.[5] The *Shih Ming* (Expositor of Names), +2nd century, also describes the pounding of the seeds of *nai* 柰 (crab-apple) and *hsing* 杏 (almond) and the use of the oil (*yu* 油) to paint on silk.[6] But there is no indication in these references that any fatty liquid was ever separated from the pounded mixtures. The next occurrence of the word *yu* 油, in the sense of an oil, is in the *Po Wu Chi* (Record of the Investigation of Things), c. +290, of the Chin dynasty which mentions the heating of hemp seed oil, *ma yu*, 麻油 to drive off residual water, and the hazard of fire when ten thousand piculs of oil, *yu*, are accumulated in one warehouse.[7]

Thus, the ancient Chinese were well aware of the fact that the seeds of certain plants contained a liquid fat, i.e. an oil. The actual pressing of the seeds to squeeze out the oil, however, was not accomplished until the late Han. This is rather surprising when we consider the fact that the pressing of oil from seeds was already an established fact almost two millennia earlier in the Mediterranean World. Archaeological evidence shows that olive oil was in use in Greece during the Early Minoan and Early Cycladic periods.[8] The earliest remains of an olive oil press (c. −1800 to −1500) were found in Crete. A beam-press for olives (c. −1600 to −1250) was discovered on one of the Cyclades,[9] and a picture of such a press in operation is part of the decoration on a Greek vase of the −6th century (Figure 94).[10] The first written description of a beam press was given by Cato in *De Agri Cultura*, −3rd century (Figure 95).[11] It is one of the presses mentioned by Pliny in his *Natural History*, the others being two types of lever and a screw press. Two lever presses and two screw presses are described by Hero, and there is a painting of a wedge press in the house of Vettii in

---

[2] It should be noted that in the modern vernacular the word *yu* is applied equally to both animal and vegetable fats. For example, we talk of lard as *chu yu* (oil from swine) and butter as *niu yu* (oil from cow), whereas in the classics, lard would be *chu kao* 豬膏 and butter *su* 酥.

[3] *SWCT*, p. 226 (ch. 11A, p. 5). *Wu-ling* prefecture covers the region bordering Hunan, Hupei and Kweichou during the Western Han.

[4] *FSCS*, tr. Shih Shêng-Han (1959), pp. 24–5.   [5] Cf. p. 29, and note 48.   [6] *Shih Ming*, ch. 13, p. 9.

[7] *Po Wu Chi*, ch. 3, p. 6a. A brief account of the pressing of oilseeds has been presented in Daniels (1996), *SCC* Vol. vi, Pt 3, pp. 7–11, of this series as one of the representative agroindustries of China.

[8] Vickery (1936), pp. 51–2.   [9] Forbes (1956), pp. 112–18.   [10] Ibid., p. 113.

[11] Cato, *De Agri Cultura*, x–xiii, (1934), pp. 26ff., 32ff.

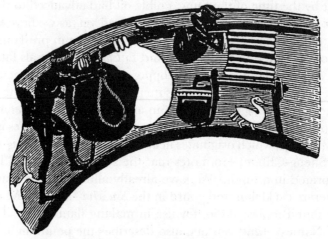

Simple beam-press for extracting oil from olives and juice from grapes. The beam is held in a recess in the wall on the right, not shown. Loaded with weights, human and inanimate, it compresses the rope-bound fruit on the press-bed. The juice is collected in an urn beneath the press-bed. From a Greek vase of the sixth century.

Fig 94. Beam press: decoration on Greek Vase –6th century, after Forbes (1956), p. 113.

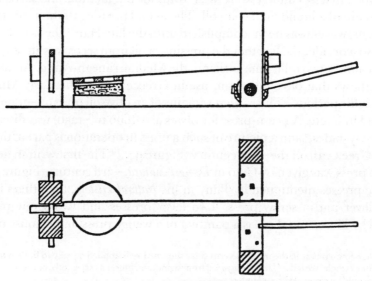

Press described by Cato: side elevation and plan. The left end of the press-beam is caught under a cross-beam between wooden posts. Grapes or crushed olives are packed under this heavy press-beam, and pressure is applied with a rope wound round a drum on the right. The drum is mounted between posts and is rotated by pulling on detachable spokes. Press-beam may be up to 50 ft long.

Fig 95. Catonian beam press described in Pliny, after Forbes (1956), p. 114.

Pompeii.¹² Pliny also describes a roller-mill (*mola olearia*) for crushing the olive fruits.¹³ These accounts show that in terms of mechanical devices for pressing oil from seeds the West was far in advance of the Chinese. Why this should be so is an interesting question, but before we address it we need first to complete our story on the development of oil-pressing technology in China.

At any rate, by the time the *Chhi Min Yao Shu* was completed in +544, the pressing of *man ching* 曼菁 (colza seeds) for oil had already become an established commercial activity, but there is no mention of how the colza oil was used.¹⁴ Fortunately, we are given a little more information on the oil extracted from the perilla seed (*jên* 荏), *Perilla ocymoides* L.:¹⁵

Perilla (*jên*) are harvested and pressed into oil, which is used to pan-fry (*tshao* 炒) cakes. The oil has a lovely green colour and a delightful fragrance, but for frying cakes it is not as good as sesame oil (*hu ma yu* 胡麻油), though superior to hemp seed (*ma tzu* 麻子) fat. The hemp seed fat has a disagreeable odour. Perilla oil should not be used to wet hair; it will cause hair to wilt. When used to flavour soups and stews, perilla (*jên*) seeds are better than hemp seeds (*ma tzu*). They work well when incorporated in flambeaux. . . . The oil is very effective for making oiled silk. In this application it is also superior to hemp oil.¹⁶

This passage provides us with two significant facts: first, that perilla oil was used in cooking, and second, as cooking oil it is inferior to sesame oil but superior to hemp seed oil. The pressing of neither sesame seeds nor hemp seeds is mentioned in the *Chhi Min Yao Shu*, but there are several references to the way the oils derived from them were used. To give us an overview of how extensively the vegetable oils were used, the relevant references are summarised in Table 41. From the table we may infer that vegetable oils were used extensively for cooking, lighting lamps, oiling cloths and salving hair in the +6th century. Sesame oil was rated as the best product for cooking. Next came perilla oil, and last, hemp oil.¹⁷ It is not sure where colza oil, the only one recorded as having been prepared on a large scale, would stand on this scale.

But it would seem that even then, animal grease remained as the preferred fatty ingredient in cookery. For example, in the skewer-roasting of a whole suckling pig, the recipe calls for basting the animal with wine and lard as it is turned over the open fire. It is only when lard is not available, that *ma yu* (hemp seed oil) may be

---

¹² Pliny (1938), vol. v, pp. 387ff. and vol. iv, p. 2901. Hero, *Mechanica*, iii, ii, 13–21. For an interpretation of these sources cf. Drachmann (1932).
¹³ Pliny (1938), vol. ii, p. 640.
¹⁴ *CMYS*, ch. 18, p. 133, which states that 'Two hundred piculs of colza seeds harvested from a hundred *mou* of land are sold to the oil mills for pressing', and comments that this yield is equal in monetary worth to that of 600 piculs of hulled rice or millet.
¹⁵ *CMYS*, ch. 26, p. 154. *Jên* is characterised as *pai su* 白蘇, that is, white perilla. Also mentioned in this chapter are *tzu su* 紫蘇 (purple perilla), *Perilla frutescens* Brit. var. crispa Decne, and *shui su* 水蘇 (water perilla), *Stachys aspera* Michx. var. japonica Maxim. All were used in food.
¹⁶ Although pressing of sesame seeds is not mentioned, ch. 13, p. 108, states that the 'white' type of sesame, *pai hu ma*, gives more oil. For a discussion of the different types of *ma* plants cultivated by the Chinese, such as hemp, sesame, ramie etc. cf. *SCC* Vol. vi, Pt i, p. 171.
¹⁷ Hemp seed oil was such a common article of daily living that its colour was used as a standard to denote the colour of wine, cf. *CMYS*, ch. 65, p. 391.

Table 41. *References to vegetable oils in the* Chhi Min Yao Shu

Oil from seeds of	CMYS Ch.:p.	Pressing mentioned?	Usage
Colza	18:133	yes	—
Perilla	26:155	yes	cooking, lighting, oil cloth, hair salve
	87:529	no	cooking
Almond	36:200	yes	—
Sesame	26:155	no	cooking
	52:264	no	cosmetic
	88:532	no	seasoning on vegetables
Hemp	26:155	no	cooking, oil cloth
	80:494	no	cooking
	87:529	no	cooking
Oil, seed not identified	87:528	no	cooking

used.[18] Again, in the frying of fish and wheat pastries, animal grease, such as *kao yu* 膏油, *kao chih* 膏脂 and the *chih kao* 脂膏 from oxen and sheep (i.e. beef and sheep tallow), rather than vegetable oil is the fat specified in the recipes.[19] Animal grease was also regarded as superior to hemp seed oil for sealing the pores of pottery jars and urns that are a familiar feature of the fermentation processes for making wine and fermented pastes and preserves described earlier in Chapters (c) and (d).[20]

In spite of the references cited in Table 41, we have no indication whatsoever of how the seeds were pressed. Nevertheless, it is clear that vegetable oils were already a significant component of the economy in the +6th century. Their importance probably continued to expand during the Thang. Pressing seeds for oil, *ya yu* 壓油, is listed in the *Ssu Shih Tsuan Yao* (Important Rules for the Four Seasons), of the late Thang, as one of the tasks to be accomplished in the 4th month of the year. By the time of the Sung, vegetable oils had become a major commodity, manufactured and traded throughout the land. This point is illustrated by the following passage from the *Chi Le Pien* (Miscellaneous Random Notes), +1130:[21]

Oil is used everywhere for eating and burning, but the best variety is that made from sesame seeds (*hu ma*), commonly called 'grease hemp' (*chih ma* 脂麻) or 'fragrant hemp' (*chih ma* 芝麻). It is said to have eight perverse characteristics. When the weather is rainy or overcast, the crop is poor, but when it is very dry there is an abundant harvest. Its blossoms droop downwards, while its fruits reach upwards. When the seeds are roasted and pressed, they yield raw oil. When carts are lubricated with it, they run smoothly, but awls and needles (smeared with it) become rough. In Ho-tung, however, people eat hemp oil, *ta ma yu*, which has an offensive smell and like perilla seed oil, is suitable for use in the manufacture of rain-garments. In Shensi, they also consume oil made from almonds, red and blue hibiscus

[18] *CMYS*, ch. 80, p. 494. [19] *CMYS*, ch. 78, p. 485 for fish; ch. 82, pp. 509–10 for various wheat pastries.
[20] The procedure for sealing pottery jars is described in *CMYS*, ch. 83, and in *SSTY*, 4th month.
[21] *Chi Le Pien*, vol. 1, p. 25a; tr. Mark Elvin (1970), in Shiba Yoshinobu, pp. 80–1.

seeds, and rape-turnip seeds, besides using them in their lamps. Tsu Thing 祖珽 [a statesman of the 6th century] lost his sight through the smoke of rape-turnip seed oil in his eyes, but in recent times I have not heard of anyone suffering in this way. In Shantung they also used an oil made from cocklebur seeds, which is useful for curing colds. There is not much sesame oil in Chiang-Hu. Oil from the seeds of the *thung* 桐 tree, *Aleurites fordii*,[22] is much used in lamps, but its thick and filthy smoke is especially to be shunned, if objects such as pictures and statues are nearby. Clothes stained with it can only be cleaned with winter melon. It has a clear colour and a sweet taste, but anyone who eats it by mistake will suffer from vomiting and diarrhoea. Drinking either wine or tea will purge it, for southern wine contains a good quantity of lime. There have been cases of women anointing their hair with it in error. It congealed into a lump, and every method resorted to in the effort to remove it proved unsuccessful, until at length the hair fell out. There is also *phang phi* 旁毗 seed oil, from the plant whose root is the drug *wu yao* 烏藥 (*Lindera strychnifolia*). Villagers use it in torches for lighting but its smoke has a particularly unpleasant smell, and so it is little used in the cities. The oil from [the fruits of] the Chinese tallow tree, *wu chiu* 烏桕 (*Sapium sebiferum*) is hard and can be poured to form candles.[23] It is used everywhere in Kuang-nan... In Ying-chou they also eat fish oil, which has a somewhat rank odour.

The availability of such a wide variety of vegetable oils suggests that the technique of extracting oils from plant seeds had now reached a high level of proficiency. Shên Kua noted that Northerners loved to use hemp oil for frying food.[24] According to the *Mêng Liang Lu*, oil pressing establishments were seen in the southern capital of Hangchou.[25] All indications are that production and commerce in vegetable oils were a significant sector of the Sung economy.[26] And yet we cannot find in the literature of the period any description of how the seeds were processed to yield the oil. For that information, we have to wait for the appearance of the *Nung Shu* (Agricultural Treatise), +1313 by Wang Chên of the Yuan dynasty, which provides the first description of an oil press in the Chinese literature.[27]

Yu Cha 油榨 (Oil Press) – Instrument for producing oil. Take four large sturdy timbers, each measuring more than ten feet long and about five feet in circumference.[28] Stack them together as one large log lying horizontally on the ground. The top part is cut to form a slot and the lower part, supported by a board, is shaped to form a [cylindrical] trench.[29] A circular tunnel is chiselled from the [bottom of the] trench through the wood to connect via a channel to the collecting vessel. To produce oil, stir-fry sesame seeds in a large pan on the stove until they are roasted (Figure 98). Then break up the seeds in a trip hammer (Figure 99) or a grinding mill (Figure 100). After cooking in a steamer, bind the meal with straw to form firm round cakes, which are then placed side by side in the trough [of the press]. Pack the trough tight with blocks of wood. Insert a long wedge between the sections. From a high

---

[22] Cf. *SCC* Vol. VI, Pt 1, pp. 490–2. [23] *Ibid.*, p. 498. [24] *MCPT*, ch. 24, p. 244. [25] *MLL*, ch. 13, p. 105.
[26] For a discussion of the production and commerce of various vegetable oils during the Sung, cf. Shiba Yoshinobu, tr. Mark Elvin (1970). Cf. Chhi Hsieh (*1987*), Sung Economic History, Part 2, pp. 663–5.
[27] *Nung Shu*, ch. 16, Chung Hua (*1956*), ch. 16, p. 296.
[28] This statement is puzzling. With a five foot circumference, each log should have been large enough by itself to scoop out a trough for an oil press (cf. quotation from the *Thien Kung Khai Wu* below).
[29] This sentence is difficult to interpret. The original text reads: *Chhi shang tso tshao, chhi hsia yung hou pang, khan tso ti phan* 其上作槽，其下用厚板. There is no mention of the scooping out of any wood in order to fashion a cylindrical pressing chamber.

Fig 96. Oil press from the *Wang Chên Nung Shu*, ch. 16, p. 126.

40 FERMENTATIONS AND FOOD SCIENCE 443

油榨

Fig 97. Oil press from the *Nung Chêng Chhüan Shu*, ch. 23, p. 576.

position, strike the wedge with a hammer or a mallet into the trough. As the blocks tighten oil begins to flow from the trough. This horizontal press is known as the sleeping trough. When the press is vertical, it is known as a standing trough. In that case, pressure may be applied with either a horizontal wedge or a levered horizontal beam. In that case, recovery of oil is rapid.

It is difficult to tell from this passage how the press is constructed. The drawing that accompanies it, reproduced here as Figure 96, bears no direct relationship to the text. In fact, the very same passage is quoted in the *Nung Cheng Chhüan Shu* (Complete Treatise on Agricultural Administration), by Hsü Kuang-Chhi, +1639, but illustrated by an entirely different drawing (Figure 97).[30] Nevertheless, the passage, augmented by the illustrations, does give us the sense that what the Chinese had been using all along was a wedge press. The principle of the wedge press has been discussed in an earlier volume of this series.[31] The most common Chinese wedge press was fabricated from large tree trunks, slotted and hollowed out, as

---

[30] *NCCS*, ch. 23, p. 576. The passage is copied verbatim from the *Wang Chên Nung Shu* without ascription.
[31] Cf. *SCC* Vol. IV, Pt 2, pp. 206–10. See particularly fig. 463 for a press made from a hollowed-out trunk.

Fig 98. Steamer and frying pan used in the processing of oilseeds, *TKKW*, tr. Li Chiao-Phing *et al.* (1980), Fig. 12-4, p. 320.

shown in the following passage from the *Thien Kung Khai Wu* (Exploitation of the Works of Nature) by Sung Ying-Hsing, +1637.[32]

The large log used for making an oil press must have a circumference equal to the embrace of a man's arms. It is then hollowed in the centre. Camphorwood is best suited for this

---

[32] *TKKW*, ch. 12, p. 210; tr. Sun & Sun, 1966, p. 216, mod. auct.; adjuv. Li Chhiao-Phing (1980), pp. 311–14; cf. also Phan Chi-Hsing (*1993*), pp. 48–9.

Fig 99. Trip hammer used to pound shells of nuts to obtain kernels, *TKKW*, tr. Li Chiao-Phing *et al.* (1980), Fig. 12–3, p. 319.

purpose, followed by sandalwood and alder wood. (A press made of alder wood will rot rapidly if there is moisture on the ground.) The grains of the three woods are circular or spiral, and do not run along the length of the log. The wood, therefore, will not split at the ends even though the central section is subjected to heavy hammering with a pointed wedge. The press should not be made of wood with longitudinal grains. In the Central Plain and North of the [Yangtze] River, where timber of such size is scarce, four pieces of wood can be joined together with iron hoops. The central portion is then hollowed out to serve as the pressing cavity; thus small wood is made to function as large timber.

Fig 100. Grinding of nuts to break shells and obtain kernels, *TKKW*, tr. Li Chiao-Phing *et al.* (1980), Fig. 12–2, p. 318.

The size of the hollowed cavity in the log varies with the dimension of the wood. Its capacity ranges from less than five pecks to over one *tan*. When fashioning the hollow, first cut out a long slot on the surface [of the log], then use a curved chisel to scoop out the wood to form a cylindrical trench. On the bottom of the trench, drill a hole through the wood and connect it to a trough through which the oil expressed can flow into a receptacle. The flat slot [i.e. opening into the cavity] is about three to four feet long and three to four inches wide,

Fig 101. Southern style oil press, *TKKW*, tr. Li Chiao-Phing *et al.* (1980), Fig. 12–1, p. 17.

depending on the general shape of the wood.[33] There are no standard dimensions for it. The pressing wedge and ram are made of sandalwood or oak, for no other wood is suitable. The wedge is cut and shaped with an axe but is not planed. It is desirable that its surface should be rough; a smooth-surfaced wedge may spring back [after the impact of the blow]. The ramming wood and the wedge are both hooped with iron rings to prevent them from breaking.

The press is now setup (Figure 101). Next, materials such as sesame seeds or rape seeds are stir-fried slowly in a pan over moderate heat (tree-grown seeds, such as those of the Chinese

---

[33] Our interpretation of how the press is fabricated differs significantly from those offered by Sun & Sun (1966), p. 218, Li Chhiao-Phing *et al.* (1980), p. 312, and Phan Chi-Hsing (*1993*), pp. 48–9.

Fig 102. Roller-mill powered by two donkeys for pulverising oilseeds. *TKKW*, p. 97.

tallow tree and the *thung* tree, are rolled and steamed without preliminary roasting) until a fragrant scent is given off. Then they are cracked or pulverised [with a trip hammer, Figure 99, a rotary quern, Figure 100, or a roller-mill, Figure 102]³⁴ into fine fragments after which they are steamed. For frying sesame and rape seeds, a flat bottomed pan, no more than six inches deep, should be used. The seeds in the pan must be stirred and turned rapidly. If the pan is too deep or the stirring too slow, the seeds will be parched and the yield

---

³⁴ For a discussion of the history of the trip hammer (*tui* 碓), the rotary quern (*mo* 磨) and the roller-mill (*nien* 碾) cf. *SCC* Vol. IV, Pt 2, pp. 50–2, and 108–206. Cf. also pages 462–6 below.

40 FERMENTATIONS AND FOOD SCIENCE

Fig 103. Rustic oil press carved out of a large tree trunk, photo Hommel (1937), p. 91.

of oil will be reduced. The frying pan is fixed on the stove, but its shape is different from that of the boiling pot for the steamer (Figure 98).

The trough of a [double edged] roller mill is anchored by burying its bottom in the ground. (The surface of a trough made of wood is covered with sheets of iron.) Over the trough is an iron roller suspended from a wooden pole and pushed [along the trough] to and fro by two workers standing facing each other. A man who has large capital can build an ox-pulled roller mill with stones, the work of one ox being equal to that of ten men (Figure 102). Some seeds such as cotton seeds and the like, are ground in a rotary quern (Figure 100) instead of being rolled.

The rolled seeds are screened, the coarser fragments being returned for a second rolling. The fine meal is placed in a steamer and steamed (Figure 98). When it has been steamed sufficiently, it is taken out and wrapped in rice or wheat straw to form thick round cakes which are tied either with iron hoops or split rattan cords. The cakes are of such a dimension that they would fit snuggly into the hollow cavity of the press.

Since the oil is liberated by the action of steam on the meal, inefficient handling of the steamed meal will allow the hot vapour to evaporate, thus diminishing the yield of oil. The skilled worker therefore always pours the meal out quickly, wraps it quickly, and just as quickly ties it up. This is the secret of obtaining a high yield. There are workers who have laboured a lifetime at the job and are still ignorant of this fact. The cakes, when properly packaged, are loaded into the press until it is packed full. The wedge is inserted and struck by the battering ram, forcing out the oil like spring water. The residues left in the press are called pressed cakes (*ku ping* 估餅). The pressed cakes of sesame, colza or rape are all rolled into a fine powder, and then screened to remove the stalks and husks. This cake powder is further treated by steaming, wrapping and pressing [for a second recovery of oil]. This second pressing will yield half as much oil as that obtained from the first pressing. In the case of the seeds from the Chinese tallow and *thung* trees, however, the oil is entirely squeezed out in the first pressing; therefore, the pressed cakes need not be reprocessed.

This description is succinct yet unambiguous. Furthermore, it is illustrated by the drawing reproduced in Figure 101, which shows clearly a southern-style press fabricated from the log of a very large tree. The press is raised above ground by two large wooden frames. We can see the long, narrow opening slot (3 to 4 feet long and 3 to 4 inches wide) in the middle of the log, the oilseed cakes packed next to each other inside the cavity, blocks of wood to fill the space left, two long wedges sticking out above the slot, and a suspended battering ram ready to slam on the top of the wedge and drive it into the press. Finally, we can see oil dripping from a hole at the bottom of the press into a collection vessel. The description also reiterates the pretreatment of the seeds (stir-drying, crushing and steaming etc.), exactly as described in the *Nung Shu* of +1313. Thus, we may say that the method reported by Sung Yin-Hsing in +1637 is practically identical to that described by Wang Chên in +1313, and presumably also the same as that extant in the *Chhi Min Yao Shu* of +544.

A photograph of a rustic oil press made from a log in the early part of the 20th century is shown in Figure 103. It shows the round cakes ringed with bamboo ropes on the right and wooden blocks used to fill the empty space on the left. A wooden wedge is perched on the rim of the press. Such presses could easily be seen in the countryside in China in the 1940s.[35] A more sophisticated version of the same type of a wedge press was in use in Wan-hsien, Szechuan, in the late 1950s (Figure 104).[36]

The *Thien Kung Khai Wu* lists and evaluates the following oils processed in this manner.[37]

For eating, the oils of sesame seeds, turnip seeds, yellow soy beans, and cabbage seeds are the best. Next in quality comes perilla and rape-seed oil; next, tea seed oil; next amaranthus seed oil; the last in quality is hemp seed oil. The best lamp oil is that made from the kernels of the tallow tree seeds, followed by rape-seed oil, linseed oil, cotton-seed oil, and sesame oil (it burns off too quickly in the lamp), while tung oil and mixed tallow-seed oil are the worst. (The poisonous odour of the tung oil is sickening and the mixed tallow-seed oil, prepared from both the tallow hulls and the seeds of the tallow tree, tend to solidify and appear turbid.)

The yield of oil (in catties) per *tan* (picul or 120 catties) of various seeds is summarised below:

---

[35] Cf. *SCC* Vol. IV, Pt 2, p. 206, n. g. The author has seen a similar press in 1942 in a village near Foochow, Fukien used for pressing tea seed oil. With the wave of modernisation in China in recent years, such implements are probably rarely seen today.
[36] *Liang-shi pu yu-chih chü chia kung chhu* 糧食部 etc. (*1958*), vol. I, p. 12.
[37] *TKKW*, pp. 209–10; tr. Sun & Sun (1966), pp. 215–16.

40 FERMENTATIONS AND FOOD SCIENCE

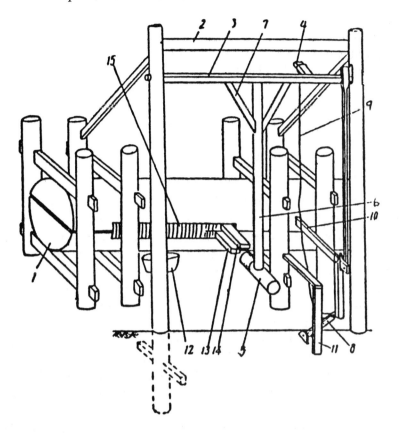

1. 撞榨，2. 木架，3. 滚筒，4. 翘板，5. 飞锤，6. 飞锤杆，
7. 扯牵，8. 脚踏板，9. 绳子，10. 升降杆，11. 保护栏，
12. 油池，13. 下尖，14. 上尖，15. 饼，

Fig 104. Modern oil press based on traditional design. Ministry of Grains (*1956*), p. 12. (1) Wedge press; (2) Overhead beam connecting the wooden structure; (3) Rolling bar on which the mallet is hung; (4) Lever for turning the rolling bar; (5) Flying mallet; (6) Rod holding the mallet; (7) Additional support for mallet on bar; (8) Movable plank to pull the rope holding the lever; (9) Rope to pull the rolling bar; (10) Lever to lower or raise the movable plank; (11) Wooden support for the operator; (12) Jar for collecting oil; (13) Lower wedge; (14) Upper wedge; (15) Oil cakes to be pressed.

Sesame, castor and camphor	40
Turnip	27
Rape	30–40
Tea	15[38]

[38] In the original Ming edition as cited in Phan Chi Hsin (*1992*), pp. 46–7, there is a note on tea which says, 'this tree is over ten feet tall; the seed resembles that of the *chin ying tzu* 金櫻子 (cherokee rose, *Rosa lavigata*), which should be shelled before the oil is released from the kernel'. This note is omitted in the Commercial Press edition, repr. *1987*. In its place is another note which says, 'the flavour of the oil is similar to that of lard . . . It can be used to light a fire or poison fishes'. This suggests that some type of tea oil may be poisonous.

Tung tree	33
Tallow tree (hull)	20
Tallow tree (kernel)	15
Tallow tree (mixed kernel and hulls)[39]	33
Holly	12
Soybean	9
Cabbage	30
Cotton	8
Amaranthus	30
Hemp, linseed	20

The *Thien Kung Khai Wu* also reminds us that,[40] 'Aside from pressing, oils can be extracted from seeds by other processes. Castor oil and perilla oil are prepared by a boiling process.' In another passage, we are told,[41] 'The boiling process requires the use of two pots. The castor or perilla seeds are crushed into meal, put into a pot, and boiled in water. When foamy oil rises to the top, it is skimmed and poured into the other pot, which is dry, and placed over a low fire to evaporate the moisture, thus resulting in a water-free oil. However, the yield of oil from this method is low [as compared to the pressing method].'

The traditional Chinese wedge press described by Sun Yin-Hsing, +1637, and Wang Chên, +1313, remained virtually unchanged until the 20th century.[42] It is quite similar to the wedge press used in Europe for extracting rapeseed oil depicted by Albrecht in the 19th century (Figure 105).[43] There is also a third type of wedge press which is uniquely Japanese in design; it is shown in Figure 106.[44] This press, used for the extraction of rape, sesame and perilla seeds, is described in the *Seiyu Roku* (On Oil Manufacturing) of 1836, which is illustrated with lively woodcut prints. In this case, the material to be pressed is wrapped in a large hemp sack, placed in a metal ring and weighted down with a special block of stone. A special bar sits on top of the stone. It passes through a window in a heavy wooden pillar on each side of the pressing chamber. By ramming a wedge into each window simultaneously, the bar is made to press heavily on the stone and the sack causing the oil to flow from the meal.

We should mention, in passing, yet another type of oil mill which works on quite a different principle from that of the wedge presses of East Asia and Europe, and that is *Ghani – the traditional oilmill of India*.[45] It a basically a glorified mortar and pestle, in which both the crushing and pressing are achieved in one instrument. Its design, construction and operation are illustrated in Figures 107a,b. It is said to have been in use continuously since the −7th century to modern times for extracting oil from sesame, rape-mustard, coconut and castor seeds. It is interesting that this design was never transmitted to India's neighbours.

[39] How the hull tallow, kernel oil and the mixed tallow oil are recovered is described in *TKKW*, pp. 211–12; cf. Sun & Sun (1966), pp. 219–20.
[40] *Ibid.*, p. 210; tr. Sun & Sun (1966), p. 217.   [41] *Ibid.*, p. 211, tr. Sun & Sun, p. 219.
[42] For the introduction of Western oil pressing technology into China in the 19th century see Brown (1979 and 1981).
[43] Albrecht (1825), pp. 60–2.   [44] Okura Nagatsune (1974), fig. 18.   [45] Achaya (1993).

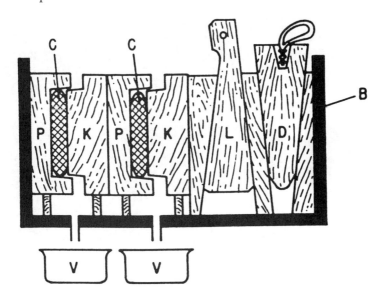

Wedge press of the type described by Albrecht for extracting rapeseed oil. The crushed, roasted seeds were wrapped in horsehair cloths C and then placed between the press plates P (the pan) and K (the lid) in the press box B. Following insertion of the blocks and wedges as shown, the drive-wedge D was driven in with a mallet producing pressure to express the oil from the seeds. After allowing time for the oil to exude and run into the vessels V beneath, the releasing-wedge L was driven in, the press plates were removed, and the oilcake was recovered

Fig 105. European wedge press for making rapeseed oil in the 19th century, after Albrecht (1825), pp. 60–2.

Fig 106. Japanese oil press from Okura Nagatsune (1836), tr. and ed. Carter Litchfield (1974), Fig. 2.

454   40 FERMENTATIONS AND FOOD SCIENCE

(a)

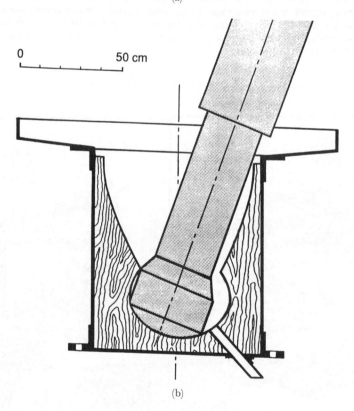

(b)

Fig 107. Ghani – the traditional oilmill of India; from Achaya (1993), cover & p. 42: (a) photograph of a Ghani; (b) cross-section of pit and pestle.

Table 42. *Comparison of European, Chinese and Japanese processes for the pressing of vegetable oils*

Step	European	Chinese	Japanese
Source	H. Albrecht +1825	Sun Yin-Hsing +1637	Okura Nagatsune +1836
Drying[a]	+	(+)	+
Roasting[b]	+	+	+
Crushing[c]	+	+	+
Steaming[d]	?	+	+
Pressing[e]	+	+	+
Refining[f]	+	—	+

[a] Suppresses metabolism and inhibits microbial contamination. In China the seeds are automatically dried before storage although Sun Yin-Hsing did not include it in his account.
[b] Removes more water, coagulates the proteins in the fat-bearing cells, and allows the oil droplets, and allows fine droplets to coalesce into larger drops.
[c] Breaks up cell structure and facilitates the release of oil.
[d] Further coagulates structural proteins.
[e] Squeezes oil from the seed meal.
[f] Addition of ashes and lime neutralises free fatty acids present and converts them into soaps.

The European, Chinese and Japanese processes are compared in Table 42. Except for the difference in the pressing equipment, all three processes are very similar. In fact, the Chinese and Japanese processes are practically identical. Yet one hesitates to conclude that the Japanese process is transmitted from China, since the wedge presses are so different in design. Did the three processes arise independently of each other? We do not have sufficient information to make a judgement. We can only say that in order to release the oil bound in the seeds, certain conditions have to be met. One is that the proteins in the fat-bearing cells should be coagulated, and the other that the internal structure of the seeds should be destroyed. These objectives are achieved by heating (roasting and steaming) the seeds and by pulverising them into a fine meal. The minute droplets of oil released in the cells can then coalesce and separate from the other seed components when the meal is pressed. Since the technologies available to meet these objectives were similar in Europe and in East Asia during the preindustrial age, it is not surprising that the traditional processes developed separately on the two continents should turn out to be so similar. Thus, the design of the process is, in fact, largely determined, if not ordained, by the nature of the raw material. One could ask that in view of the long history of other types of presses used in the West for pressing olive oil,[46] why was it that the wedge press was adopted as the preferred instrument for the extraction of rapeseed oil? The answer is probably technological. It is easier to press oil from crushed olives than it is from crushed rape and other vegetable seeds. The oil in olives (and most fruits) is not as tightly bound to the other cell components as that in vegetable seeds. It takes tremendous pressure to squeeze out the oil in vegetable

[46] Olive oil was known in Egypt at the time of Ramses II, Darby, Ghalioungui & Grivetti (1977), II, p. 784, and in the Aegean in the early Minoan and Cycladic periods, Vickery (1936), p. 51.

seeds, and before the 20th century the wedge press was the most efficient way to achieve that pressure.

We may now return to the question as to why the oil press was so late in its arrival to China. The question could be put in another way. Why were oil presses developed so early in the West? The answer may be both cultural and technological. Olive oil has been called the 'soap, butter and lamp-fuel of antiquity'.[47] It was one of the most important commodities in the economy of early Mediterranean civilisations. But it was more than just a commercial commodity. Winners at the great Olympic games received garlands made of olive leaves.[48] The Kings of Israel were anointed with [olive] oil as a sign of their acceptance by the Lord.[49] Thus, plant oils in general, and olive oil in particular, were symbols of mystical and religious power as well as objects of practical utility. The ancient Greeks and Romans, therefore, had a compelling reason to devise more efficient means of extracting oil from olives without delay. By the time of Christ the simple beam press of the −6th century (pages 437–9) had been replaced by the Catonian press, and several other models, including a screw press and a wedge press, were developed.[50] The situation in ancient China was quite different. Oils served the rather mundane functions of cooking and lighting. Animal fats were apparently plentiful until the Chhin (−206 to −201) to supply the needs of cooking and lighting. In any case, animal fat was able to serve these functions until the Warring States Period (−480 to −220). While it would have been convenient to have been able to extract oil from oilseeds efficiently, it was not considered as a matter of urgency until the Han Dynasty. As a result, by the +1st century, the Mediterranean region was way ahead of China in the technology of oil pressing.

Although the technology of oil pressing remained unchanged from the time of Sung Ying-Hsing to the 20th century, several important changes have taken place in the complement of crops cultivated for the production of vegetable oils in China. The first change is the ascendency of rape as the most important oil crop. The second is the introduction of the peanut from America in the 16th century.[51] The third is the introduction of another American crop via Europe, the sunflower.[52] The American imports have found China a fertile ground on which to prosper. Today, peanut rivals rapeseed as the most important oilseed crop in China, and sunflower has surpassed sesame as a source of oil. In 1985 the production of the four major vegetable oils (in millions of metric tons) were as follows: Rapeseed 5.6; Peanut 5.6, Sunflower 1.9; and Sesame 0.7. After these come flax seed, soybean, hemp seed, perilla seed,cotton seed and castor bean.[53] The domination of peanut and

---

[47] Bowra, C. M., (1965), p. 65.   [48] *Ibid.*, p. 134.
[49] I Samuel *10*: 1; *16*: 13. The practice was followed by the mediaeval Kings of Europe.
[50] For a brief review of the types of presses in use at the time of Pliny cf. Pliny (1938), vol. IV, Pt 2, pp. 206–10.
[51] Ho Ping-ti (1933).
[52] The sunflower originated in North America and was imported into Europe c. +1510. Its earliest record in China is in the *Chhün Fang Phu* of Wang Hsiang-Chin, c. +1621.
[53] Wittwer *et al.* (1987), pp. 241–51. The case of soybean deserves a brief comment. Although it has been a major crop in China for 3,000 years, it has never been regarded as an important source of oil. The reason is that the lipids in soybean are so tightly bound to the cell ingredients that it is very difficult to press the oil out using traditional Chinese oil pressing technology. In fact, the yield of oil from soybean is probably the lowest of all the

sunflower as oilseed crops in China is but one example of the enormous impact the cultivated plants of the New World has had on the welfare of the people in parts of the Old World.

(ii) *Malt sugar from cereals*

The two major sweeteners of ancient China were *i* 飴 (malt sugar or maltose) and *mi* 蜜 (honey). Sugarcane was known, but it was centuries later before the Chinese developed a procedure to extract cane sugar from its juice.[54] The word *i* first occurs in the *Shih Ching* in a poem celebrating the founding of the Chou kingdom,[55]

> The plain of Chou was very fertile,
> Its *chin* and *thu* sweet as malt sugar.

Malt sugar in the original is *i*. It has always been interpreted as the sweet product obtained after saccharifying cooked cereals with sprouted cereal grain (millets, rice, wheat or barley), which is called *nieh* 櫱. In a passage in the *Li Chi* (Record of Rites) sons are admonished, when serving food to their parents, to 'sweeten the dishes with dates, chestnut, honey and malt sugar'.[56] The poem *Chao Hun* 'Summons of the Soul' from the *Chhu Tzhu* (Elegies of the Chhu State), c. −300, mentions honey, malt sugar as well as sugarcane juice (*chê chiang* 蔗漿).[57] But *i* was not the only word used to denote malt sugar. Another ancient word for malt sugar is *thang* 餳 (also pronounced *hsing*). The *Chi Chiu Phien* (Handy Primer), c. −40, says, 'Digest dehulled grain (*mi* 米), separate the juice and cook it. When the product is soft and pliable, it is called *i*, when it is hard and strong, it is called *thang*.'[58] The *Fang Yên* (Dictionary of Local Expressions), c. −15, informs us, '*thang* is known as *chhang huang*, 悵餭' which according to Kuo Phu's commentary is dried *i* (malt sugar). Furthermore, '*I* 飴 is called *thang* 餳', in one region and '*thang* 餳 is called *thang* 糖', in another.[59] Later *thang* written as 糖 became the established word for cane sugar. The preparation of both *thang* and *i* are mentioned in the *Ssu Min Yüeh Ling*, +160.[60] The *Shih Ming*, +2nd

---

seeds listed in the *TKKW* (see p. 452 above). But even Western oil pressing technology in the 19th century was not significantly superior, cf. Brown (1979 and 1981). It is only with the development of the solvent extraction process in the 20th century that it has been possible to extract the oil in soybean efficiently on a commercial scale. Today, soybean is the most important source of vegetable oil in the US. It is still not a major source of oil in China.

[54] For a brief history of sweeteners in China, cf. Sabban (1988).

[55] *Shih Ching*, M 237, W 240, first two lines of the third stanza; tr. auct. adjuv. Waley (1937) and Karlgren (1950). The original reads:

> *Chou yüan wu wu*   周原膴膴
> *Chin thu ju I*   菫荼如飴

Chin is a wild plant, sometimes used as vegetable. Its identity is uncertain. *Thu* has been interpreted variously as the sowthistle, smartweed or the progenitor of tea. Waley renders *i* as 'rice-cakes', and Karlgren as 'honey-cakes'. Neither translation is technically correct.

[56] *Li Chi*, ch. 12, *Nei Tsê*, p. 444, '*Tsao li i mi i kan chih* 棗栗飴蜜以甘之'; cf. Legge (1885), tr. I, p. 451.

[57] *Chhu Tzhu I Chu*, annotated by Tung Chhu-Ping, (*1986*), p. 258. In this verse *mi* is honey, *chhang huang* 悵餭 is malt sugar and *chê chiang* is sugarcane juice, cf. translation by Hawkes (1985), p. 228, who renders *mi* as honey and *chhang huang* as malt sugar, but *chê chiang* is rendered as yam juice.

[58] *Chi Chiu Phien*, p. 32a.   [59] *Fang Yên*, ch. 13.   [60] *SMYL*, 10th month, p. 98.

century, also states that *thang* 餳 is prepared by digesting (i.e. hydrolysing) rice, while *i* is simply a softer form of *thang*. Yet there is another ancient word used to denote malt sugar, *pu* 餔 or 哺, which according to the *Shih Ming* is a cloudy type of *thang*.[61] The *Shuo Wên*, +121, also tells us, '*I* is obtained by incubating hulled grain (*mi* 米) with malt (*nieh* 糵).'[62] From these references we can feel confident that *i*, *thang*, and *phu* are all names used in antiquity to denote malt sugar. We may also infer that by the Eastern Han, malt sugar had been made from grains routinely for more than a thousand years. But we have no idea of the details of the process. For that we have once again to wait for the advent of the *Chhi Min Yao Shu* in +544.

First we learn how sprouted wheat was prepared. The procedure is described in chapter 68. We shall quote it in full:[63]

Method for making sprouted wheat: In the eighth month, soak unhulled wheat grains in a pan, drain off the superfluous water, and expose it to the sun. Flush the grains with water and drain it off at least once a day. When the seeds begin to sprout, spread them on a piece of cloth to a depth of about two inches. Sprinkle with water once a day; stop when the seeds are fully sprouted. Collect the loose sprouts, dry them and do not let them form a cake. Once they cake up, they will lose activity. This type of sprouted wheat is good for making white malt sugar. To make dark malt sugar, wait until the sprouts start to turn green. Form them into a cake, and cut it into pieces to dry. To prepare amber-coloured malt sugar use malted barley instead of wheat.

Having prepared the malt (i.e. sprouted wheat), we are now ready to make the malt sugar. Recipes for this purpose are given in Chapter 89, which is entitled, 'Preparation of *Thang* 餳 and *Pu* 餔'. The first recipe is for the making of 'white' *thang*:[64]

Method for Making white malt sugar: The best result is obtained when white, loose sprouts are used. Those entangled in a cake are unsuitable. Use a well-cured pot; otherwise the malt sugar will be dark. Scrub the pot until it is scrupulously clean, and free from any grease. Place a steamer (*tsêng* 甑) above the pot to contain excessive foaming.

One part of dried malt will digest twenty parts of grain. The grain must be pounded a score of times, throughly cleaned, steamed into *fan* 飯, and spread to cool until lukewarm. Blend the malt with the cooked grain until the components are evenly distributed. Load the mash into an urn with an orifice at the bottom. Do not press the substrate by hand; just allow it to loosely level out. Cover the urn with a quilt to keep it warm. Pack straw around it in winter. After a full day in winter, or half a day in summer, most of the grain should be liquified.

Heat [a large pot of] water until it bubbles. Pour it into the urn until it stands about one inch above the surface of the mash. Then stir to mix the contents of the urn. After about one meal time, loosen the stopper in the orifice, collect the free-flowing syrup and heat it in a pan. When the syrup begins to boil, add two more ladlefuls of the thin syrup. The fire should be low and steady. Too strong a fire chars the syrup and creates a burnt smell. After all the syrup is added to the pot, and when there is no further danger of frothing over, take the

[61] *Shih Ming*, ch. 13, p. 8b. [62] *SWCT*, ch. 5(2), p. 3 (p. 107).
[63] *CMYS*, ch. 68, p. 414; tr. auct., adjuv. Shih Shêng-Han (1958), p. 77.
[64] *Ibid.*, ch. 89, p. 546; tr. auct., adjuv. Shih Sheng-Han (2958), pp. 78–9.

steamer off the boiler. Assign one man to stir the boiling liquid constantly with a ladle. Do not stop even for one moment, for without stirring the syrup [at the bottom of the pot] will char. When it is well done, stop the fire. After it is sufficiently cooled draw out the treacle.[65] If good quality *Seteria* seeds are used, the malt sugar will be crystal clear.[66]

The other recipes included in Chapter 89 are for the making of (1) *hei thang* 黑餳, a dark malt sugar using a malted wheat cake as the saccharifying agent (2) *hu pho thang* 琥珀餳, an amber coloured malt sugar, using malted barley (3) *pu* 餔, a cloudy malt sugar syrup, and (4) *i* 飴, a malt sugar. The raw materials mentioned are *liang mi* 粱米, a large-grained setaria millet, *chi mi* 稷米, a setaria millet, *shu mi* 黍米, a panicum millet and just plain *mi* 米, which can denote rice or any hulled grain.[67] The basic procedure is the same for all the recipes, and consists of three steps:

(a) Incubate steamed grain with malt. The α and β amylases in the malt, hydrolyse the gelatinised starch in the grain into water soluble oligosaccharides and maltose. This will cause the mash to liquify.
(b) Leach out the soluble sugars from the mash by boiling water and collect the extract in a pan.
(c) Heat and stir the extract until a syrup is obtained. Depending on its water content, it can be congealed into soft candy. If air is beaten into the syrup, it will congeal into a hard candy.

The procedure is simple enough.[68] The unique feature is that the malt and the substrate are incubated in a semi-solid environment. No additional water other than that absorbed by the grain is added. The sugar released is thus always held in a concentrated aqueous solution. The high osmotic pressure effectively protects the sugar from being contaminated by stray yeasts from the atomosphere. The sugars are then dissolved in a minimum amount of boiling water, which is then evaporated to give a malt syrup. The process is thus economical in its use of water and fuel, and is well suited for small-scale operation in a home kitchen.

In the *Ssu Shih Tsuan Yao* (Important Tasks for the Four Seasons) of the late Thang, the preparation of *thang* 餳 (malt sugar) is listed as one of the tasks that is to be performed in the third month.[69] However, the importance of malt sugar during the Thang was overshadowed by the advent of the processing of sugarcane into crystalline cane sugar, sucrose.[70] By the Sung dynasty, cane sugar (*thang* 糖) had become

---

[65] At this point the thick syrup should still be warm and pliable. It can be drawn out, sometimes in a spiral, to give candy stick.
[66] *CMYS*, pp. 546–7.
[67] The interpretation of *liang*, *chi* and *shu* follow the Terminology of Chinese millets given by Bray in *SCC* Vol. VI, Pt 2, p. 440.
[68] This step is comparable to the mashing process used in the brewing of modern beer. In this case the substrates and malt are incubated as a suspension in water. The 'wort' or sugar solution obtained this way usually contains about 10% of carbohydrates. It would be expensive to evaporate this relatively dilute solution to a syrup.
[69] *SSTY*, 3rd month, item 40.
[70] The history of the processing of sugarcane to refined cane sugar and its effect on the economy is described in Daniels, C. (1996), *SCC* Vol. VI, Pt 3, Agricultural Technology.

the dominant sweetener in the Chinese diet. Its primacy is affirmed by the publication of Wang Cho's *Thang Suan Phu* (Monograph on Cane Sugar) in +1154. The words *thang* (cane sugar) and *mi* (honey) occur often in the celebrated reminiscences of daily life in the capitals of the Sung Dynasty, namely, the *Tung Ching Mêng Hua Lu* for the Northern Sung, and the *Mêng Liang Lu* and *Wu Ling Chiu Shih* for the Southern Sung.[71] But malt sugar, either as *i* or *thang*, is hardly ever alluded to in such works or in the food canons of the Yuan, Ming and the Chhing dynasties.[72] This does not, however, necessarily mean that malt sugar was completely displaced by cane sugar by the time of the Sung.

In fact, malt sugar, as *i thang* 飴糖, has remained a standard item in the major pharmacopoeias since its first appearance in the *Ming I Pieh Lu* (Informal Records of Famous Physicians), +510.[73] It is included in the *Hsin Hsiu Pên Tshao* (Newly Improved Pharmacopoeia), +659, *Shih Liao Pên Tshao* (Manual of Diet Therapy), +670, *Chêng Lei Pên Tshao* (Reorganised Pharmacopoeia), +1082, *Yin Shih Hsü Chi* (Essentials of Food and Drink), +1350, *Pên Tshao Phin Hui Ching Yao* (Essentials of the Pharmacopoeia ranked according to Nature and Efficacy), +1505, and the *Pên Tshao Kang Mu*, +1596.[74] It is also listed as *i* 飴 in the *Chhien Ching Yao Fang* (A Thousand Golden Remedies), +655, and as *thang* 餳 in the *Yin Shan Chêng Yao* (Principles of Correct Diet), +1330.[75] It is said to be sweet, mildly warming, and non-toxic, and to alleviate general weakness, raise muscle tone, relieve flatulence, and soothes difficulty in swallowing.

That malt sugar continued to be a significant factor in the Chinese diet, is affirmed by its inclusion in the discussion on sweeteners in the *Thien Kung Khai Wu* of +1637. We are told that,[76]

The method of making malt sugar consists of soaking the grains of wheat or rice and the like [in water] and allowing them to sprout. They are then dried in the sun, cooked and treated, and malt sugar is obtained. The best quality is white. A reddish variety, resembling amber, is called jellied malt sugar (*chiao i* 膠飴); it melts in the mouth and is a favourite in the palaces. The confectioners in the south refer to malt sugar as 'little sugar' (*hsiao thang* 小糖) in order to distinguish it from cane sugar.

There are hundreds of ways in which the malt sugar is prepared by sweet makers to please the human palate. They are too many to enumerate here. There is a type called 'nest of silken threads' which is used exclusively by the Imperial household. It is possible that the method of its preparation will be handed down to future generations.

The text makes it clear that cane sugar was the most important sweetener; next came honey; then came malt sugar. The situation has remained unchanged to the present day. Malt sugar, now known as *mai ya thang* 麥芽糖 (sprouted wheat sugar)

---

[71] *TCMHL*, ch. 2, pp. 14, 15, 18, ch. 8, 53; *MLL*, ch. 13, pp. 109, 111; *WLCS*, ch. 6, pp. 122–3; cf. the earlier discussion on the use of sugar to preserve fruits, pp. 425–6.
[72] Ibid.   [73] *MIPL*, p. 98; reconstituted and edited by Shang Chih-Chün (1986).
[74] *SLPT*, No. 171, p. 105; *HHPT*, ch. 17, p. 278; *CLPT*, ch. 24, p. 8b; *YSSC*, ch. 5, p. 51; *PTPHCY*, ch. 35, p. 815; and *PTKM*, ch. 25, p. 1550.
[75] *CCYF* (in *Chhien Chin Shih Chih*) ch. 4, p. 61; and *YSCY*, p. 120.
[76] *TKKW*, ch. 6, p. 130; tr. Sun & Sun (1966), p. 130, mod. auct.

never developed into a commodity traded all over the country. Yet it has survived as a poor man's honey in the countryside, a sweetener that can be prepared in the home kitchen from readily available local materials by a procedure that has proved to be one of the more enduring, yet lesser known, legacies of the *Chhi Min Yao Shu*.[77]

(iii) *Preparation of starch*

We have noted earlier that the *Li Chi*, c. −300, had recommended the use of rice or millet crumbs to thicken soups and stews (page 93). Today, starch would be the favourite ingredient for this purpose. But it is likely to be a starch prepared from potatoes, sweet potatoes or corn, three crops that were unknown to the ancient Chinese. But long before the importation of these New World crops in the 16th century, the Chinese had learned how to prepare a refined starch from cereals. We do not know when the procedure was developed. It is first described in the *Chhi Min Yao Shu*:[78]

Method for making Mi Fên 米粉 (Refined Starch): Setaria is the best, spiked millet, second. (Always use one single kind of cereal, never a mixture.) Pound until it is exceedingly fine. (Discard the broken kernels.) Start out with one type of grain. Mixed grains, glutinous rice, wheat, glutinous and common panicled, millets will not give good results. Mix the powdered grain with water and tramp in a wooden trough for ten turns until the supernatant is clear. Immerse the flour in plenty of cold water in a big jar. There is no need to change the water (for thirty days in spring or autumn, twenty days in summer and sixty days in winter), the more it stinks the better. When the time is up, draw fresh water from the well and pour it into the jar. Stir by hand and discard the sour solution; repeat until all the odour is washed out. Remove the residue and grind it in a coarse earthern mortar. Add water and stir. Transfer the white suspension into a pongee (*chüan* 絹) bag and strain into a receiving jar. The coarse particles in the bag and the mortar are again ground, stirred, mixed with water and strained again as before. When all the particles have been finely ground stir the [whole content of] the receiving jar with a rake vigorously for a long time, and then allow the suspension to settle. Decant off the clear supernatant and draw off the thick residue into a large pan. Stir with a rod for more than 300 rounds, always in the same direction, never the reverse.[79] Allow the pan to stand with a cover to avoid dust. When well settled, lightly ladle out the supernatant water with a large spoon. Apply three layers of fine cloth above the wet residue. Pour millet chaff on the cloth and dry ashes on the chaff. Replace the wet ashes with dry ashes until the newly laid ashes remain dry. [Remove the layers of cloth.] With a knife, peel off the top, bottom and sides of the residual cake. Keep this coarse, non-lustrous material separately for ordinary use.[80] (This material is derived from the exterior layers of

---

[77] In the 1940s malt sugar, *mai ya thang*, was produced locally in many small towns and villages across China. In 1942 the author had the privilege to observe the making of malt sugar from sprouted barley and steamed rice, in the manner described in the *Chhi Min Yao Shu*, in a village not far from Foochow, the capital of Fukien. People still make malt sugar confection even though Fukien is one of major centres of production of sugarcane and cane sugar in the country.

[78] *CMYS*, ch. 52, p. 264; tr. Shih Shêng-Han (1958), pp. 75–6, mod. auct.

[79] This motion would provide a centrifugal force to cause a fractional sedimentation of the solids according to size and weight.

[80] One of the 'ordinary' uses of such starch is probably in the sizing of paper, an application of great importance in the history of Chinese civilisation; cf. Tsien Tsuen-Hsuin, *SCC* Vol. v, Pt 1, p. 73. Cf. Phan Chi-Hsing 9 (*1979*), pp. 61–2.

the original grain and so lacks lustre.) The bowl-shaped central portion, which is smooth, lustrous and as white as the white of an egg, is called *fên-ying* 粉英 (flower of the flour). *Fên-ying* is derived from the core of the original grain; thus, it is lustrous and smooth. On a fine, sunny and calm day, spread the *fên-ying* on a platform; slice it with a knife into pieces in the shape of the teeth of a comb. Dry it in the sun. When fully dry, rub it vigorously. (Vigorous rubbing makes the starch finer. It may still be gritty if it is not rubbed.) Reserve the product for making cakes for honoured guests, and as a base for cosmetics or body powder.

The recipe is reminiscent of the procedure for preparing starch from wheat grains described by the Roman author Cato in *De Agri Cultura*, c. −180, which says:[81]

Recipe for starch: clean hard wheat thoroughly, pour into a trough, and add water twice a day. On the tenth day drain off the water, squeeze thoroughly, mix well in a clean tray until it is of the consistency of wine dregs. Place some of this in a new linen bag, and squeeze out the creamy substance into a new pan or bowl. Treat the whole mass in the same way, and knead again. Place the pan in the sun and let it dry.

The basic principle is the same in both the ancient Roman and early mediaeval Chinese recipes. Bacterial contamination is allowed to break up the cell structure and the starch granules released are collected and dried.[82] The procedure in the *Chhi Min Yao Shu*, however, introduces several refinements. Firstly, the grains are pounded to a powder, which should facilitate the destruction of their internal structure. Secondly, the circular stirring of the suspension enables the particles to be fractionated by centrifugal force. And thirdly, the use of an absorbent to remove water from a highly hydrophilic cake. These innovations probably enabled the Chinese to produce a highly refined starch suitable for use in cosmetics as well as coarser fractions used in paper-making, textiles, adhesive pastes and as a thickener in cookery.[83]

Starch has also been extracted from the water caltrop, the arrowroot, and roots of the water lily and kudzu vine in China.[84] Speciality starches from these crops can still be seen. But practically all the starch traded commercially today is produced from corn, sweet potatoes or potatoes, three American crops introduced into China during the Ming dynasty. In this instance, the American imports have displaced all the native crops to become the dominant source of a vital material in the economy of modern China.

(4) Processing of Wheat Flour

We have noted earlier (pages 18–20, 68) that the primary food of the ancient Chinese was cooked (preferably steamed) grains, that is to say, varieties of millet, rice and

---

[81] Marcus Porcius Cato, reprinted 1979, p. 89.
[82] The same principle is the basis of the giant 'wet milling' industry that produces starch from corn today. The success of the procedure is due to the fact that native starch granules are relatively resistant to enzymatic action, and thus able to survive the microbial contaminants that destroy the internal structure of the grain. It is only when starch is gelatinised (by heat) that it becomes susceptible to the action of animal, cereal or microbial amylases.
[83] For recipes for cosmetic powders, cf. *CMYS*, ch. 52, pp. 264–5.
[84] Wang Shang-Tien, (*1987*), pp. 414–22.

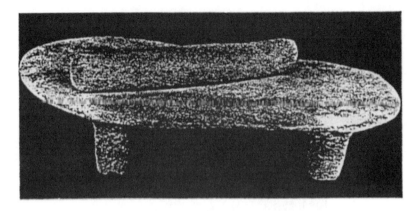

Fig 108. Neolithic saddle quern in China, from Sha-wo-li, Hsin-chêng, Honan. *Chung-kuo shê-huei kho-hsueh yuan (1983)*, Pl. 2, Fig. 5.

perhaps barley. The wheat grain, however, is too hard to be treated in this manner. It has to be milled into flour before it can be processed into a variety of palatable foods, such as pasta, breads, cakes and noodles.[1] Usage of wheat is, therefore, tied to the availability of the rotary stone mills needed to grind the hard kernels efficiently into flour.[2] Although saddle querns were known in China during the Neolithic Age (Figure 108), rotary stone mills did not appear until the Warring States period (Figure 109).[3] They grew in popularity during the Western Han and had become commonplace in the Eastern Han.[4] (Figures 110–111). The emergence and rise of wheat flour (*mien* 麵) as a major component of the diet, parallelled the proliferation of rotary querns during the Han. This development continued through the Three Kingdom's period into the Thang and eventually divided China into two dietary

---

[1] For a discussion on this issue see Bray, Vol. 6, Pt 2, p. 461; Amano Motonosuke (*1962*) and Shinoda Osamu tr. (*1987*), pp. 20–3.

[2] This view is held by Amano and Shinoda, *ibid.* as well as Chinese scholars such as Yü Ying-Shih (*1977*), p. 81, Chhiu Phang-Thung (*1989a*) and (*1989b*), and Chhên Shao-Chün & Wu Chao-Su (*1994*). It should be pointed out, however, that wheat could have been ground to flour before the invention of the rotary stone mill in China. The ancient Egyptians and Mesopotamians made raised bread from flour ground from wheat kernels using a saddle quern, long before the appearance of the rotary quern in the Middle East, which was no earlier than −1000. The low efficiency of the saddle quern was apparently quite adequate to provide the small amounts of millet or rice flour needed to make a variety of pastries consumed as snacks or desserts during the Shang and Chou periods. It was only after wheat became a significant crop that the availability of implements to grind grains into flour became a major factor affecting its utilisation. The invention of the rotary quern (or stone mill) is discussed in detail by Needham *et al.* in *SCC* Vol. iv, Pt 2, pp. 185–92.

[3] At the Chhin capital of Li-yang, cf. *Shensi sheng wen-wu etc* (*1966*), p. 14.

[4] For example, Western Han: Thu Ssu-Hua (*1956*), p. 37, *Loyang khao-ku tui* (*1959*), p. 206, *Loyang fa-chüeh tui* (*1963*), p. 22, *Man-chhêng fa chüeh tui* (*1972*), p. 9, and *Yang-chou po-wu kuan* (*1980*), p. 424; Eastern Han, Huang Chan-Yüeh (*1956*), p. 46, Ko Chia-Chin (*1959*), p. 23 and *Anhui shêng wên-wu etc* (*1959*), p. 46. For additional examples cf. Li Fa-Lin (*1986*) and Chhên Wên-Hua (*1983, 1989* and *1995*). The fact that rotary mills were already known during the Warring States period would refute Shinoda's hypothesis that they were introduced into China as a result of the western expansion of the Han empire, cf. Chhiu Phang-Thung (*1988*) and Tsao Lung-Kung (*1983*).

Fig 109. Rotary stone mill, Warring States: *Shensi shêng wen-wu kuan-li hui* (*1966*), Pl. 8.

Fig 110. Rotary quern, Western Han *Chung-kuo shê-huei kho-hsueh yuan* etc. (*1980*), Mancheng Report, Vol. II, Pl. 106, nos. 2 & 3 (cf. description in Vol. I, p. 143).

Fig 111. Rotary quern, Eastern Han, Rawson ed. (1996), Fig. 84. Model of a man operating a rotary quern.

zones: the south (with pockets in the north), where the primary food is cooked grains, i.e. *li shih* 粒食 (granule food); and the north, where the primary food is processed from wheat flour, i.e. *mien shih* 麵食 (flour food).[5]

Wheat flour contains all the components commonly found in cereal grains, i.e. starch, proteins (including enzymes), lipids and sugars. But its proteins are anything but common. There are four classes of proteins in the flour, albumins and globulins that are water soluble and gliadin and glutenin that are water insoluble. When the flour is mixed with water, the gliadin and glutenin combine to give a remarkable protein polymer called gluten. As the dough is kneaded the gluten forms a matrix which surrounds the starch and water molecules. The dough becomes stiff and acquires two important properties, plasticity and elasticity. When the dough is leavened with yeast, the yeast breaks down the sugars present *in situ* into alcohol and carbon dioxide. It is the delicate balance of plasticity and elasticity that enables the

---

[5] This division is, of course, only partial. It does not mean that there are no pockets of *li shih* in the *mien shih* region or *vice versa*, cf. Shinoda tr. (1987), above. Not all wheat in the world today is, however, consumed as *mien shih* (i.e. as products from flour). Semolina (groats of durum wheat) is steamed to give couscous, a staple food of North Africa and cream of wheat (granules of farina) cooked as a porridge is a common breakfast cereal in North America. Sabban (1990) has called such products 'unformed' wheat foods in contrast to bread and pasta which are 'formed' foods.

dough to trap the carbon dioxide released to give a spongy miracle which we call raised bread or simply bread.[6] One modern Western gastronome has called bread,[7] 'the most fundamentally satisfying of all foods; and good bread with fresh butter, the greatest of all feasts.' Of all the grains, wheat is the only one that produces enough gluten to make this marvellous food possible. Rye proteins also produce a gluten, but it is too weak to be useful. Wheat is therefore the premier grain of Western civilisation.

As is well known, the place of bread in the Chinese dietary system is nowhere as prominent as that in the West. But this does not mean that gluten has had no impact on the development of the Chinese cuisine. In fact, as we shall soon see, it is gluten that endows the dough with such qualities that it can be rolled into sheets, pulled and stretched, and moulded in different ways to give a whole array of Chinese pasta foods. In fact, gluten itself has been isolated from the flour and used as an independent source of protein. Together with *tou fu*, gluten has served as an important source of protein in the traditional Chinese diet. But before we trace the history of gluten, we need to deal with how wheat flour itself is processed into the familiar pasta foods in China.

(i) *Pasta and filamentous noodles*

In terms of the structure of a Chinese meal both *mien shih* (flour food) and *li shih* (granular food) are parts of the *shih* (grain) category as opposed to the *tshai* 菜 (dishes) category.[8] During the Han, *li shih* could be steamed rice or a variety of millet. In either case the cooked product would be called *fan* 飯. Today, *fan* means almost exclusively steamed rice. As a mark of its special importance '*fan*' is also used informally to denote a meal. Thus, *chhih fan* 吃飯, literally to eat steamed rice, is to have a meal.[9] The situation in regard to *mien shih* is more complicated. According to the *Shuo Wên*, +121, *mien* simply means wheat flour.[10] It was mixed with water to form a dough which was cooked in various ways to give products called *ping* 餅. But today, certainly in the southern *li shih* zone, *mien* is simply noodles, or more specifically filamentous noodles.[11] In the northern *mien shih* zone, *mien* could mean filamentous noodles in particular, or foods made from flour in general. As for *ping*, in Han times it was evidently any type of food made from wheat dough, in other

---

[6] For a popular exposition of the scientific basis of gluten formation, dough behaviour and bread making cf. McGee (1984), pp. 273–315. For a similar exposition in Chinese cf. Wu Chih-Sung (*1993*).

[7] Beard (1973), p. xi.

[8] Chang, K. C. (1977), p. 40, and Chao, B. Y. (1972), p. 3, stress the division between *fan* (grain food) and *tshai* (dishes) in a typical Chinese meal. This division was already in evidence in the late Chou in the form of *fan* 飯 and *shan* 膳 shown in the *Li Chi, Nei Tsê*, p. 456, cf. p. 98 above.

[9] Thus one could be invited to a home to *chhih fan* 吃飯 enjoy a delicious meal without ever seeing a grain of *fan* 飯.

[10] *SWCT*, p. 112 (ch. 5B, p. 13).

[11] As pointed out by Hung Kuang-Chi (*1984a*), p. 30, the designation of *mien* as wheat flour today is largely a literary convention. In daily discourse, *mien* almost invariably means *mien thiao*, filamentous noodles. If one were to order a *mien* at a Chinese restaurant (China and elsewhere), one would get a serving of noodles. Similarly, a package of *mien* in an oriental food store would mean a package of noodles.

words a pasta. Today, *ping* means a cake, biscuit or a type of pastry. To avoid confusion, we shall adopt the following translation of *mien* and *ping* in our discussion:

Product	Han and early Medieval	Late Chhing and Modern
Mien 麵	Wheat flour only	Wheat flour or noodle
Ping 餅	Pasta	Cake & pastry

Since the late Chhing *mien* has meant either wheat flour or noodle, or more specifically, filamentous noodles. The word noodle by itself is somewhat ambiguous. The Oxford English Dictionary regards noodle as 'a strip or ball of dough made with wheat flour and egg'. The American dictionaries tend to define 'noodle' as a filamentous (or ribbon-shaped) pasta.[12] For our purpose we shall adopt the Chinese and American usage and consider 'noodle' and 'filamentous noodle' as synonymous. Thus when we talk about the legend of Marco Polo bringing noodles from China to Italy, we mean 'filamentous noodles' and not 'pasta',[13] as Western writers have often assumed.[14] Chinese writers, on the other hand, have always specified that what Marco Polo had brought back to Italy is *mien thiao* 麵條 (filamentous pasta or noodle) and not pasta in general.[15] This is the version of the legend that will be considered later.

Two questions immediately arise. What kinds of pasta or *ping* were made from the wheat flour (*mien*) by the Han Chinese? How and when did the early *mien* (flour) evolve into a *mien* with dual personalities (flour or noodle)? Let us now try to answer the first question.

The earliest occurrence of the word *ping* 餅 is in the Mo Tzu (The Book of Master Mo), −4th century.[16] The context suggests that it is probably some kind of pastry. Cake or pastry made with rice flour were known in the Warring States period. For example, *erh* 餌 and *tzu* 粢 are listed in the *Chou Li* as two of the foods contained in

---

[12] The *Webster's Third New International Dictionary* calls noodle 'a food paste made with egg and shaped typically in ribbon form', and the *Random House Dictionary* defines noodle as a 'narrow strip of unleavened egg dough that has been rolled thin and dried, usually boiled and served in soups, casseroles etc.; a ribbon shaped pasta'. Except for the inclusion of egg, these are acceptable definitions of the Chinese noodle. The Chinese egg noodle will be touched on later.
[13] Having been brought up in the southern Chinese *li shih* 粒食 tradition, it never occurred to me that *mien* could be anything but 'noodles', and 'noodles' anything but 'filamentous noodles'. Thus, when I discussed with Gregory Blue the legend of Marco Polo bringing back noodles from China to Italy, what I had in mind was, of course, filamentous noodles. Gregory became very interested in the subject and spent a considerable amount of time looking up various editions of Marco Polo's book. When he wrote to me about the result of his inquiry, I was surprised to read that he had found no evidence that 'Marco Polo ever brought "pasta" back from China to Italy.' Evidently to him, noodle means pasta. But to me noodle is but one form of pasta. Cf. Gregory Blue (1990 & 1991).
[14] McGee (1984), p. 316, doubts that Marco Polo ever brought back noodles from China, but he seems to treat the noodle as synonymous with pasta. James & Thorpe (1994), p. 311, also tend to use 'pasta' and 'noodle' interchangeably. In Tannahill (1988), p. 234, Root (1971), p. 369, and Anderson & Anderson (1977), p. 338, the question that was considered is whether Marco Polo brought *pasta* back to Italy from China. This is also the way the issue is viewed in a series of papers edited by Sabban *et al.* (1989).
[15] For example, Chhiu Phang-Thung, (1988) p. 16, (1989b), p. 9; Wang Jên-Hsing (1985), p. 66; Yang Wên-Chhi (1983), p. 74 and in our own experience in conversations with numerous Chinese friends. However, Wang Wên-Tsê (1991), p. 8 refers rather loosely to *mien shih* 麵食 (noodle foods) as having been brought back to Italy from China by Marco Polo.
[16] *Mo Tzu*, ch. 46, p. 337.

ritual baskets, *pien* 籩 to be presented at the ancestral altar.[17] According to Chêng Hsüan's commentary they are both prepared from rice or millet flour. When the product is steamed it is called *erh*, when boiled, *tzu*. *Erh* is also mentioned in the *Li Chi* and the *Chhu Tzhu*.[18] Thus, the ancient Chinese already knew how to grind millet and rice grain into flour and process the flour into foods before the Han dynasty.

The word *ping* appears again in the *Chi Chiu Pien* 急就篇 (The Handy Primer), c. −40, in a short list of processed grain foods, that includes also rice cake, steamed wheat/barley grain, and a bean rice porridge.[19] In the *Fang Yen* 方言 (Dictionary of Local Expressions), c. −15, *ping* is called *tho* 飥, and *erh* is called *kao* 糕 or *tzu* 粢.[20] We shall meet with *tho* again later. The *Shuo Wên* (Analytical Dictionary of Characters), +121, says, '*ping* is *tzu* made from wheat flour', '*tzu* is *ping* made from rice flour', and '*kao* is a kind of *erh*'.[21] Somewhat more informative is this passage from the *Shih Ming* (Expositor of Names), +2nd century,[22] '*Ping* is prepared by blending flour with water. *Hu ping* 胡餅 is contoured after the *ta man hu* 大漫沍. Others attribute the name to the *hu ma* 胡麻 (sesame seeds) sprinkled over it.[23] There is *chêng ping* 蒸餅 (steamed pasta), *thang ping* 湯餅 (boiled pasta in soup), *ho ping* 蠍餅 (grub shaped pasta), *sui ping* 髓餅 (bone marrow pasta), *chin ping* 金餅 (gold pasta), and *so ping* 索餅 (rope pasta), often with the name as an indicator of its shape.' If so, *so ping* (rope pasta) would have to be the earliest form of a filamentous noodle known.

During the Han dynasty pasta (*ping*) had become so well accepted that it was sold as a processed food in the streets.[24] But the *Ssu Min Yüeh Ling*, +160, warns,[25] 'By the start of autumn do not eat cooked pasta (*chu ping* 煮餅) or water based pasta (*sou ping* 溲餅).' The author then explains in a note that 'in the summer when one drinks water [with meals], these two pastas stiffen in the water and becomes hard to digest. If one inadvertently eats such a stale pasta, one will get sick. This can be demonstrated experimentally by soaking the two [and other] pastas in water. Only a pasta originally made by mixing flour with wine will disintegrate.' This passage is difficult to interpret, since we have no idea of how the pastas were prepared and stored. Normally, a well cooked pasta would be easily digestible. A problem would only

---

[17] *Chou Li*, ch. 2, p. 54; *Chou Li Chu Shu*, ch. 5, p. 82.
[18] *Li Ch, Nei Tsê*, ch. 12, p. 457; *Chhu Tzhu I Chu, Tsao Hun*, p. 258.
[19] *Chi Chiu Pien*, p. 30b, *Ssu Pu Tshung Khan* edition.
[20] *Fan Yen* (*HWTS*), ch. 13, p. 12.   [21] *SWCT*, pp. 107, 108 (ch. 5B, pp. 3, 6).
[22] *Shih Ming* (*HWTS*), ch. 13, p. 8a. The identity of *chin ping* is unknown.
[23] Chhiu Phang-Tung, (*1992b*), p. 7, points out that the term *hu ping* has been interpreted in three ways. Firstly, it was brought back from the West, i.e. *hu* region; the prefix *hu* being an indication of its origin. Secondly, it was sprinkled with sesame (*hu ma* 胡麻) and so it became *hu ping*. Thirdly, it resembled the leaf of the plant *ta man hu*, which is shaped like a turtle shell. But *ta man hu* can also be interpreted to mean a large flat piece of bread. We suspect *hu ping* is originally a general name for *nang* 饢 or *nan*, a flat bread that is often made in large sheets across Northwest China and Central Asia. *Nan* is usually baked in a tandoor oven. The word *tandoor* or variations thereof is used to denote oven in a host of languages including Hebrew, Farsi, Kurdish, Aramaic, Assyrian, Persian, Turkish and other Central Asian Turkish languages (Kazakh, Uzbek, Kirghiz, Uighur & Turkmen), cf. Alford & Duguid (*1995*), pp. 29–37. It seems likely to us that *hu ping* was transmitted to China during the Western Han dynasty. For further information on *hu ping* cf. Chhên Shao-Chün (*1995*).
[24] *Han Shu*, ch. 99B, p. 4123; *Hou Han Shu*, ch. 63, p. 2085, ch. 64, p. 2122, ch. 82, p. 2737, cf. also Shang Ping-Ho (*1991*), p. 89.
[25] *SMYL*, 5th month; *Miao Chhi-Yü* ed.(*1981*), pp. 54, see notes on pp. 61, 66.

arise if it had been customary to eat these particular pastas cold without further cooking.[26] If they had been allowed to become too dry, they could cause indigestion if they had not been rehydrated by boiling or steaming before they were eaten. The passage is also significant in that it reveals to us that at least one type of pasta was made by mixing wine (and thus yeast) with flour to form the dough,[27] and this could well have been the precursor of the steamed bread and buns that are such a familiar feature of Chinese cuisine today.

The role of pasta in the Chinese diet continued to grow after the Han.[28] Its popularity was so great that the Western Chin (+265 to +317) writer Shu Hsi 束皙 even wrote an 'Ode to Pasta' (*ping fu* 餅賦) in its honour.[29] He recommended that the flour be sifted twice through a screen; it would then be as fine and white as the driven snow. More than a dozen different types of pasta with exotic (and mostly indecipherable) names are mentioned. One recognisable name is *man thou* 曼頭, which is to be enjoyed particularly in the spring, when the weather is neither hot nor cold. Another is *thang ping* 湯餅 (pasta soup). The poem points out that in the harsh winter, when the nose freezes and the mouth is surrounded by frost, there is nothing more welcome than a bowl of hot steaming pasta soup. The popularity of pasta soup is illustrated by the story of Ho Yen, an unusually fair and handsome official at the court of the Wei Emperor Ming-ti, whose reign was +227 to +237.[30] The emperor was suspicious that Ho Yen used powdered cosmetics to whiten his complexion. On a very hot summer day he invited Ho Yen to have a bowl of steaming pasta soup. Poor Ho Yen sweated profusely, and had to wipe his brow and face with his sleeve, but his complexion remained as fair as before.

As yet we still have little or no knowledge of how the different types of pasta (*ping*) were made in early mediaeval China. Once again, we have to wait until the appearance of the *Chhi Min Yao Shu*, +540, which devotes chapter 82 to this subject.[31] To facilitate discussion, the fifteen recipes it contains are summarised in Table 43. We have omitted the last recipe (16) since it does not deal with the making of pasta, but rather the removal of sandy contaminants from finished wheat flour. Of the remaining recipes, neither no. 1 nor no. 2 is directly concerned with the preparation of a specific pasta product. They deal, however, with the preparation of the leaven

---

[26] This is not an unreasonable assumption since *ping* is simply a version of *erh* and *tzu* made from wheat flour. Both *erh* and *tzu* were cooked foods that were eaten without further warming or cooking. Wan Jên-Hsing (*1985*), pp. 62–3, has suggested that in this passage (from the *SMYL*) cooked pasta (*chu ping*) could have been fried in oil, in which case the product would most likely have been eaten without further cooking. Upon standing it could become hardened, rancid and difficult to digest. If so, we can now explain the curious story of the Eastern Han Emperor *Chih ti* who died in +146 after eating a sample of *chu ping*; cf. the passage on *ping*, ch. 860, in the *Thai Ping Yü Lan* and quoted in Miao Chhi-Yü ed. (*1981*), p. 66.

[27] One of the foods mentioned in the *Chou Li, Hai Jên* passage is *i shih* 酏食, which according to Chêng Hsüan's commentary in *Chou Li Chu Shu* (+2nd century), ch. 6, p. 88, is a *ping* made with fermenting wine. This could be the earliest and most rudimentary form of raised pasta (or bread).

[28] Hung Kuang-Chu (*1984a*), pp. 20–46; Wang Jên-Hsing, (*1985*), pp. 52–61; Chhiu Phang-Tung (*1989a*), (*1989b*), (*1992a*), and (*1992b*); Chhên Shao-Chün & Wu Chao-Su (*1994*).

[29] *Ping Fu*, in *Yen Kho-Chü* edn (1836), vol. 2, ch. 87, pp. 1962–3. For a discussion of the different types of *ping* mentioned in this poem, see Knechtges (1986), and Sabban (1990).

[30] *Shih Shuo Hsin Yü*, ch. 5, p. 151; *Ching Chhu Sui Shih Chi*, p. 3b (*HWTS* edn). [31] *CMYS*, pp. 509–11.

Table 43. *Recipes for making* ping *(pasta) in the* Chhi Min Yao Shu

Number	Pasta	Flour	Base	Leaven	Raw mixture	Cooking method
1	*Chiao* 酵 leaven	Wheat	Water	+	Dough	None
2	*Pai mien* 白麵 raised dough	Wheat	Water	+	Dough	Used in other recipes
3	*Shao ping* 燒餅 roast pasta	Wheat	Water	+	Dough + filling	Roast
4	*Sui ping* 髓餅 marrow pasta	Wheat	Water + bone marrow + honey	?	Dough	Bake
5	*Tshan* 粲 loose pasta	Rice	Water + honey	—	Batter	Fry in oil
6	*Kao huan* 膏環 ring pastry	Rice	Water + honey	—	Dough	Fry in oil
7	*Chi tzu ping* 雞子餅 egg pasta	none	none	—	none	Fry in oil
8	*Hsi huan pin* 細環餅 ringlet pasta	Wheat	Water	?	Dough	Fry in fat
9	*Phou thou* 餢鍮 fried pasta	Wheat	Water	+	Dough	Fry in oil
10	*Shui yin* 水引 drawn pasta	Wheat	Water	—	Dough	Boil in water
11	*Po tho* 餺飥 thin drawn pasta	Wheat	Water	—	Dough	Boil in water
12	*Chhi tzu mien* 碁子麵 chessmen pasta	Wheat	Water	—	Dough	Steam
13	*Luo suo* 籪䴹 pasta nuggets	Wheat millet	Water	—	Dough	Steam
14	*Fên ping* 粉餅 rice filaments	Starch	Water	—	Batter	Boil in water
15	*Thun phi ping* 豚皮餅 suckling pig skin	Starch	Water	—	Batter	Spread on hot pan

needed for the making of raised pasta. They are, therefore, of critical interest and deserve to be quoted in full.³²

Recipe 1: To make leaven for *ping* (*ping chiao* 餅酵), take one *tou* 斗 of soured rice water and reduce its volume to seven *shêng* 升 by boiling. Soak one *shêng* of *kêng* 粳 rice in the sour water and then cook it over a slow fire until it has the consistency of congee.³³ For one *tan* 石 of flour, use two *shêng* of this leaven in the summer, and four *shêng* in the winter.

Recipe 2: To make raised dough, *pai ping* 白餅, from one *tan* of flour. Take 7–8 *shêng* of white rice and cook it to give a congee. Add 6–7 *shêng* of white wine as a starter and place the suspension near a warm fire. When large bubbles begin to appear as if a wine is being brewed, decant off the liquid from the congee and mix it with the flour. When the dough rises, it is then ready for making *ping*.

Pai ping is interpreted here as a raised dough since like a blank piece of paper it can be turned into a variety of finished products.³⁴ After putting aside the first two recipes, we are left with recipes for thirteen pastas; in five of them the wheat dough is roasted, baked or fried in oil (nos. 3, 4, 7, 8, 9); in four the wheat dough is boiled or steamed (nos. 10, 11, 12, 13), and in the remaining four, the dough is made of rice flour (nos. 5, 6, 14, 15). Let us consider these groups in turn.

Of the five recipes in the first group, two definitely use raised dough: No. 3 for *shao ping* 燒餅 (meat bun) and no. 9 for *phou thou* 餢飳 (fried pasta cakes). The *shao ping* is made by wrapping raised dough around a filling of cooked [minced] lamb, flavoured with scallion, soy nugget juice and salt, and roasting the raw bun on a pan. The dough from one *tou* of flour will require two catties of lamb meat. The bun sounds rather like a baked version of a modern stuffed *pao-tzu* 包子 and quite different from the *shao ping* (sesame bun) that is so popular in northern Chinese cuisine today.³⁵ The nature of *phou thou* is uncertain; they are probably some kind of fried pastry. It is made by kneading a piece of raised dough and frying it in oil.³⁶ When it is cooked through, the fried pasta would float to the surface. The text says, 'let it float up by itself', which would suggest that plenty of oil was used and thus the product is deep fried. If so, this could be the earliest record of deep frying as a method of cookery in China. How the dough is shaped is not clear, but we could very well have a pasta here that is a precursor of the *yu thiao* 油條, a favourite breakfast food in China today.

Recipe 4 describes a bone marrow pasta (*sui ping* 髓餅). The dough, made by kneading together wheat flour, bone marrow, water and honey, is baked in an oven

---

³² *Ibid.*, p. 509. This recipe is reminiscent of that given by Pliny for the making of leaven in Pliny (1938), Book XVIII, xxvi. 102–xxvii 105 (Rackham translation, p. 255).
³³ Presumably, such a thin congee provides a good medium for wild yeast and perhaps lactobacilli to grow in it, so that after a short time it can serve as a starter for making yeast dough.
³⁴ A blank piece of paper is called *pai tzu* 白紙 (white paper). Some readers may object to calling products from raised dough, pasta, since in the West raised dough is associated with the making of bread. But the distinction is not always clear. Not all bread is leavened (e.g. Norwegian flat bread) and not all pasta is unleavened (e.g. pizza).
³⁵ *Pao-tzu* with pork filling is familiar to many Westerners under its Cantonese name, *char-siu-pao* 叉燒包. *Shao ping* 燒餅 is translated as sesame-sprinkled hot biscuits in Buwei Chao (1972), p. 203.
³⁶ Miao Chhi-Yü ed. (*1982*), p. 514, note 13 identifies *phou thou* as a sort of fried round pastry.

normally used for making *hu ping* 胡餅. Unfortunately, we have no indication of what this oven is like.[37] Recipes 7 and 8 are both incomplete. In each case a vital piece of information is missing. In no. 7 the product is an egg pasta (both chicken and duck). The recipe simply says,[38] 'Break an egg in a bowl; add a little salt and fry in a pan to give a round pasta about 2/10th of an inch in thickness.' This does not make sense, since, if we follow the directions literally, what we get is a fried or scrambled egg that can hardly qualify as a pasta. We suspect that the author, Chia Ssu-Hsieh, simply forgot to mention that the egg is to be mixed with flour or raised dough before it is fried. It is more likely that this step was in the original manuscript, but it was left out inadvertently by the copyist. Recipe 8 is aimed at making of a ringlet pasta (*hsi huan ping* 細環餅), which is also called *han chü* 寒粔 (cold pastry).[39] The text says,[40] 'Mix honey and water with flour to make a dough. If there is no honey, replace it with a decoction of red jujubes. Beef or sheep tallow may also be used. Good results are obtained by mixing cow or goat's milk with flour to make the dough. In this way, the pasta will be tasty and crispy.' It is unfortunate that the recipe does not say how the dough is to be cooked. Presumably, it is one of those cases where the procedure is so well known that the author did not think it necessary to have it set down on paper.[41] Thus, we can only guess that *han chü* was either baked or fried in oil.

In the second group of recipes, two of the products, *shui yin* 水引 and *po tho* 餺飥 (nos. 10 and 11), are boiled in water, and two, *chhi tzu mien* 碁子麵 and *luo suo* 餢䴷 (nos. 12 and 13), are steamed. *Shui yin* is generally regarded as a type of filamentous noodle, while the identity of *po tho* is a matter that will be discussed later. *Chhi tzu mien* (chess pasta) is made by rolling flour dough to the diameter of a chopstick, cutting the rods into small sections, and cooking them in a steamer. Before the pasta is

---

[37] *Hu-ping* is usually interpreted as the ancient form of *shao-ping*, cf. Wang Tung-Fêng ed. (1987), *Chien ming Chung-kuo phêng-jên tzhu tien* (1987), p. 352, and *CKPJPKCS* (1992), p. 96a. Cf. n. 23 above. The oven for making *hu-ping* is, we suspect, simply the tandoor oven seen all across Central Asia from Iran all the way to Sinkiang, cf. Alford & Duguid (1995) and Marks (1996). The tandoor oven could be seen in North China fifty years ago for baking the popular *shao-ping*. Today it is still being used by the Uighurs to make *nang* and *gerdeh* in Sinkiang, and by the Han Chinese to make *kuang-ping* in northern Fukien. The *gerdeh* is a bagel with a hole covered by a thin film of dough, and the *kuang-ping* 光餅 is simply a half-sized bagel. In both cases the dough is steamed before it is baked. They both taste exactly like a bagel. How the bagel reached Sinkiang and Fukien is an intriguing problem. Or was the European bagel transmitted from Central Asia? Having been brought up among Northern Fukien immigrants in Malaya, the *kuang ping* was a familiar part of my culinary experience. I was surprised when I ate a bagel for the first time in America fifty years ago. My immediate reaction was: how did the *kuang-ping* come to America?

[38] *CMYS*, ch. 82, pp. 509–10, tr. auct. In this sentence, Miao's edition says 'use no salt' while Shih Shêng-Han's edition says, 'add a little salt'. On this occasion we have followed Shih's edition.

[39] It is also called *ho tzu* 蝎子, which is the *ho ping* 蝎餅, grub shaped pasta cited in the *Shih Ming*, cf. p. 468. The use of such a name has a parallel in the vermicelli (little worms) of Italy. Similarly, *so ping* 索餅 (string or rope pasta) has its equivalent in the Italian spaghetti (little strings).

[40] *CMYS*, ch. 82, p. 510. This is one of the rare recipes in which milk is an ingredient, and shows that milk was readily available in North China in the 6th century.

[41] *Han chü* 寒粔 has remained an elusive entity in the culinary history of China. The *PTKM*, +1696, ch. 25, p. 1541, says that according to Lin Hung's *Shan Chia Chhing Kung*, *han chü* is prepared by blending wheat and glutinous rice flour, and frying the dough in hemp oil. However, all Lin Hung says is that to make *han chü* it is essential to use honey and oil, cf. *SCCK*, p. 15 in the *CKPJKCTK* series. He does not say how it is cooked. For a further discussion of the identity of *han chü* cf. Knechtges (1986), p. 55.

eaten, it is reboiled and then added to a soup. It is called chess pasta, because the sections are about the size of chess pieces.[42] What is particularly interesting is the name given to the product, *chhi tzu mien*. This is the first time the term *mien* is applied to a specific pasta product. Thus, there was already a margin of uncertainty as to what was *mien* and what was *ping*. The last item in this group is *luo suo* (no. 13). In this recipe, a dough made of wheat flour and cooked millet is pressed *through* a loosely woven bamboo tray, i.e. a woven bamboo sieve. The pieces that emerge are about the size of peas.[43] They are steamed, dried and stored for later use.

The last group of recipes (nos. 5, 6, 14, and 15) all use rice flour or starch as the raw material. In recipe 6,[44] 'a dough is made from glutinous rice flour, honey and water. It should have about the same water content as that of a regular dough for making *thang ping* (pasta soup). Roll a piece of dough to a length of about eight *inches*. Twist it, join the two ends and cook it in a pan of oil.' The product is called *kao huan* 膏環 or *chü nü* 粔籹. The same ingredients are used in recipe 5 to make *tshan* 粲 (loose pastry).[45] But instead of a dough, enough water is added to form a batter, which is then allowed to flow through numerous holes at the bottom of a bamboo cup into a pan of hot oil. Recipe 14 for *fên ping* 粉餅 adopts the same procedure, except that the honey is left out and the batter is made of refined rice starch and water. It is then allowed to flow through holes drilled on a section of an ox horn, into a pan of boiling water.[46] The product goes well with a savoury soup, or with a sweet sesame puree. A thinner version of the same batter is used in the last recipe (15). A copper pan is set to float on a large pot of boiling water. The pan is pushed to rotate quickly as a ladle of batter is dropped on it. The batter spreads by centrifugal force and

---

[42] We are referring here to the ancient Chinese chess game of *wei chhi* 圍碁, which is better known in the West under its Japanese name, *go*. The pieces, either black or white can be quite small.

[43] *CMYS*, p. 511. This is my interpretation of the original text, which actually says, 'Press the dough against a winnowing basket or tray (*po chi* 簸箕) as hard as possible to get little pieces the size of peas.' But there is more to the statement than meets the eye. The key to the process is the nature of *po chi*, which normally means a bamboo winnowing tray or basket, cf. *SCC* Vol. VI, Pt 2, p. 363. The tray may be tightly woven or loosely woven. In the latter case, it becomes a sieve. I tried to understand what the statement means by pressing a dough prepared from cooked millet and wheat flour on a level wooden tray and on a metal colander, a woven bamboo tray or sieve not being available. I find that if the 'tray' is tightly woven, no amount of pressing will yield pasta fragments the size of peas. On the other hand, if the 'tray' is loosely woven the dough could be pressed through the holes to give pasta nuggets with a thickness commensurate with the diameter of the holes. Thus, our interpretation seems to make sense. Both the chess pasta and pasta nuggets are reminiscent of the Italian pasta called orzo or riso, which is about the size of a rice grain and is often used in place of a filamentous pasta at a meal.

As far as we know, this function of the *po chi* (bamboo sieve) is no longer seen in China today in the making of pasta nuggets. But there is more to the life of *luo suo* than I had expected. Recently, I was pleasantly surprised to see in a cooking presentation on TV, a Japanese chef pushing a ball of biscuit dough through a small version (about a foot in diameter) of the winnowing bamboo sieve to give filaments with the thickness of chopsticks and about an inch long. These are fried in oil to give a crunchy, sweet snack. I immediately called our friend Ushiyama Terui in Tokyo and asked her if she knew what the pastry is. She kindly made a special visit to the Ajinomoto Library of Dietary Culture and inquired about this pastry. It turned out to be the popular sweet snack called *karintou* (花林糖), which Ushiyama had consumed all her life but never knew how it was made. Furthermore, the Ajinomoto librarian thinks that it is related to a product described in chapter 82 of the *Chhi Min Yao Shu*. Thus, we have here another example of a mediaeval culinary technique that has died out in China, but survived in Japan.

[44] *CMYS*, p. 509.   [45] *Ibid.*

[46] Recipe 14 is, in principle, similar to the making of spätzle, except that unlike spätzle, there is no egg in the batter. Cf. for example, Rombauer & Becker (1975), p. 204.

eventually covers the whole surface of the pan as a thin pasta film. The pan is removed and the pasta film peeled off. The thin film or pancake has the look of a piece of suckling pig skin.[47]

In spite of its loose organisation and the uneven quality of the recipes, chapter 82 is nevertheless, full of fascinating and revealing details. For example, we learn that both the wheat and rice flour were sifted through a silk screen before use (nos. 5 & 10) and of the existence of an oven used for making *hu ping* (sesame pasta), which is probably the forerunner of the familiar *shao ping*, sesame bun, of today. We learn that receptacles with holes (made with bamboo or ox horn), much like a modern colander, were used to drop batter into hot oil or boiling water to make soft pasta nuggets. In fact, we may have the first indication of deep frying as a method of cookery, even though the word *cha* 炸 is not used. We learn also that flour dough was forced through holes in bamboo sieves to give pasta nuggets.

Indeed, as revealed in this chapter, pasta (*ping*) in 6th-century China covered a wide range of foods made from grains. The dough from wheat flour could be unleavened or raised. It could be fried, baked, steamed or boiled. It could be mixed with marrow, fat, honey and egg to change the texture and flavour of the finished pasta. It could be shaped into various forms from buns to little grubs.[48] It could be used to wrap meat and other fillings. But what is most interesting to us is that in spite of the specific declarations in the *Shuo Wên* and *Shih Ming* that *ping* is made from *mien* (wheat flour), there are four recipes out of the thirteen listed in Table 43, in which rice flour, rather than wheat flour is the raw material. This shows that even as new types of pasta based on wheat flour (*mien*) emerged during the Han, Wei and Chin dynasties, a similar series of foods based on rice flour (*fên*) were also being developed. Thus, while the word *ping* might have been restricted to pasta made from wheat flour in the Han, it had by the 6th century expanded to cover all types of processed foods made from flour obtained from either wheat or rice (or millet).

We have already noted with special interest (page 472) that in recipe 12 the product is called *chhi tzu mien* 棋子麵 (chess pasta). According to the original text another name for it is *chhieh mien chu* 切麵粥 (sliced pasta congee). In this case *mien* actually means *ping*; the raw material has become the product. In fact, the term *chhieh mien* (sliced pasta) later became the name of one of the most popular forms of filamentous noodles in China. This reminds us that *mien* and *ping* were still evolving entities at the time of the *Chhi Min Yao Shu*; and their respective meaning did not become stabilised until several centuries later. What we see in recipe 12 may mark

---

[47] *CMYS*, p. 511. The thin pancake can be cut up and added to a savoury soup or a sweet sesame (or fruit based) soup. Modern versions of this 'suckling pig skin' are still seen in the cuisine of Fukien. In the Foochow area there is a much beloved dish called *ting-pien-hu* 鼎邊糊 (pronounced *diang bian hu* in the local dialect) which is made by spreading a thin layer of rice flour batter along the upper wall of a large cooking *wok*. As the film dries it is scraped and allowed to fall into the hot savoury soup in the centre of the *wok*. In southern Fukien thin round rice pancakes called *po ping* (pronounced *bo biang*) are made on a flat bottomed pan. These are used to wrap chopped meat and vegetables to give spring rolls. When fried they are just like the ubiquitous egg rolls one sees in Chinese restaurants in America except that the skins are thinner than those made with wheat flour.

[48] *Ibid.*, recipe 8, p. 510.

the beginning of a trend which would eventually lead to the designation of *mien* as noodles.

Apart from the anomaly in recipe 7 (on egg pasta) and the lack of a recipe for *hu ping* and *thang ping* which are mentioned in recipes 4 and 6 respectively, there are at least two serious omissions in chapter 82 in the versions of the *Chhi Min Yao Shu* that have survived to our day. The *Pei Hu Lu* 北戶錄 (Records of the Northern Gate), +875, lists several kinds of pasta, including *man-thou ping* 曼頭餅 and *hun-thun ping* 渾吨餅.[49] Other than filamentous noodles, these are two of the most popular and important pasta foods in China today. *Man thou* is the steamed bun that often replaces rice in a meal in the *mien shih* regions of North China, while *hun-thun* is the popular *wonton* that is served in Chinese restaurants everywhere. The commentary to the *Pei Hu Lu* says that both items are described in the *Chhi Min Yao Shu*. This indicates that *man-thou* and *hun-thun* were extant in the Thang editions of the *Chhi Min Yao Shu* but were somehow lost between the Thang and the Sung.[50] This is a highly critical point, since it shows that both *man-thou* and *wonton* were already well known in 6th-century China. We can now answer the first question we posed on page 467 and say that in addition to *man-thou* and *wonton* the kinds of pasta made from the Eastern Han to the Northern Chhi included all the foods mentioned in the recipes listed in Table 43.[51] Because of their unique importance we shall discuss *man-thou* and *wonton* in turn before we take up the story of filamentous noodles.

(ii) *Origin of* man-thou *(steamed bun)*

The earliest form of steamed bun is probably the *chêng ping* (steamed pasta) of the *Shih Ming* (+2nd century). The term *man-thou* 曼頭 first appeared in the *Ping Fu* (Ode to Pasta), c. +300, in which it is recommended that *man-thou* be served on festive occasions in the spring. Legend has it that *man-thou* was invented by the famous general Chuko Liang of the Three Kingdom's period,[52] when he led an army to pacify the barbarians south of the Shu Kingdom (present-day Szechuan). According to local custom it was necessary to offer a human head as sacrifice to the spirits in order to be assured of victory. He declined to do so. Instead, he offered a mock human head made by wrapping raised flour dough around cuts of lamb and pork and painting a human face on the surface of the resultant steam bun. From then on the steamed bun became known popularly as *man-thou* 蠻頭 (barbarian head) which was later changed to the more genteel form of *man thou* 漫頭 (large head).

It is difficult to say how much credence this legend deserves, since the story is not recorded in the official histories of the period. But indications are that the early

---

[49] *Pei Hu Lu*, ch. 2, *Shih Mu* 食目, p. 15b.
[50] The references to the *Chhi Min Yao Shu* are found in the commentary is by Tsui Kuei-Thu 崔龜圖, who also lived during the Thang.
[51] The complete complement of pasta foods, including even cakes and pastries made from rice flour, is nowadays called *mien-tien* 麵點.
[52] Hu Chih-Hsiao (*1984*); Thang Huan (*1988*); Chhên Shao-Chün & Wu Chao-Su (*1994*), p. 221; Chhên Shao-Chün (*1995*).

mediaeval *man thou* was a loaf or a large bun that had a meat filling. Thus, it is really a large version of what we would nowadays call *pao-tzu* 包子 (a little wrap-around). The term *pao-tzu* was first seen in the *Chhing I Lu* (Simple and Exotic Anecdotes), +965.[53] It cannot be construed as just another name for *man-tou* since both *pao-tzu* and *man-tou* are mentioned as distinct entities in the *Tung Ching Mêng Hua Lu*, +1147, the *Wu Lin Chiu Shih*, +1280, and *Mêng Liang Lu*, +1334 of the Sung.[54] It seems likely that *man-thou* represented a large loaf (with or without a filling) and *pao-tzu* a small bun (with a filling). *Pao-tzu* is referred to in the *Chü Chia Pi Yung*, Yuan and the *Yin Shan Chêng Yao*, +1330, but is not mentioned in the food canons of the Ming and Chhing.[55] On the other hand, *man-thou* continued to be described as a product with a filling in the *Yin Shan Chêng Yao*, +1330, *Yun Lin Thang Yin Shih Chih Tu Chi*, +1360, *Chü Chia Pi Yung*, Yuan (repeated in the *Tuo Nêng Phi Shih*, +1370), *I Ya I I*, Yuan, *Shih Hsien Hung Mi*, +1680, and *Sui Yüan Shih Tan*, +1790.[56] Surprisingly, with the possible exception of the *Sui Yüan Shih Tan*, none of these works mentions a *man-thou* without a filling.[57] The first definite description of a *man-thou* without a filling is found in the *Thiao Ting Chi* of the 19th century.[58] Today, *man thou* is unequivocally a steamed bun without a filling. A steamed bun with a filling is always called *pao-tzu*.

Yüan Mei points out in the *Sui Yüan Shih Tan* that the secret to making a good *man-thou* lies in the quality of the leaven (*chiao* 酵).[59] Unfortunately, this is also the ingredient that has received the least amount of attention from the authors of the food canons. For example, a recipe for *man-thou* in the *Chü Chia Pi Yung*, Yuan, says,[60] 'For two and a half catties of flour, use one cup of *chiao* (leaven). Make a hole in the dry flour. Pour into it the liquid *chiao*. Knead until the dough is soft, adding fresh flour [if needed]. Allow the dough to rise in a warm place.'

All this tells us is that the leaven is a liquid or a powder. There is no indication of how it was prepared. When the dough is partly raised it is rolled gently and used to wrap a suitable filling. The raw product is allowed to rise further before it is steamed.

We have noted earlier that two types of leaven for making raised pasta were described in the *Chhi Min Yao Shu* (Table 43, nos. 1 & 2). One is fermenting wine and the other a soured rice decoction. There is no indication as to whether these were the leavens used in making the *man-thou* and *pao-tzu* sold in the capitals of the Northern and Southern Sung.[61] In fact, we do not have another recipe for a leaven until the late Yuan to early Ming. In a recipe for *ta chiao* 大酵 (great leaven) the *I Ya I I* says:[62]

---

[53] *Chhing I Lu*, p. 30.
[54] For *pao-tzu* cf. *TCMHL*, pp. 18 and 28, *MLL*, pp. 134 and 137 and *WLCS*, p. 125. For *man-tou* cf. *TCMHL*, p. 53, *MLL*, pp. 111 and 137, *WLCS*, pp. 121, 125, and 172.
[55] *CCPY*, pp. 120–1; *YSCY*, p. 43.
[56] *YSCY*, ch. 1, p. 42; *YLTYSCTC*, p. 6; *CCPY*, 118; *TNPS*, Shinoda & Seiichi, eds. (1973), p. 384; *IYII*, p. 24; *SHHM*, p. 6; *SYST*, pp. 128, 137, 139.
[57] *SYST*, p. 128, refers to a *chhien tshêng man thou* 千層饅頭 (thousand layer steam bun) which is exactly what we now call a *hua chüan* 花卷 (a flower roll). It has no filling.
[58] *TTC*, pp. 745–7 describes *man thou* with and without filling. [59] *SYST*, p. 139.
[60] *CCPY*, p. 118. [61] Cf. n. 54 above. [62] *IYII*, pp. 33–4.

Have ready 5 *shêng* of white superior glutinous rice, 3 ounces of fine *ferment* (*hsi chhü* 細麴), and 4 ounces of red *ferment* processed into *tsao* (wine residues). Boil the rice in water to give a congee. Break up the *ferment*, dilute the *tsao* in warm water, and mix [both ingredients with the congee] in a porcelain jar. Leave the jar in a warm place or surround it with warm water. After about a week, the mixture is fermenting vigorously. Filter off the residual solids to collect the [liquid] leaven. If the leaven is relatively thick it will be effective. If it is too thin, it should be mixed with the residual solids, warmed and filtered again. In the winter the mixture should be incubated for one and a half weeks.

The procedure suggests that the leaven is available as a liquid. It is actually quite similar to that presented in recipe 2 in the *Chhi Min Yao Shu* (Table 43), except that it is no longer linked to a concurrent wine fermentation. This may well have been the type of leaven used in the making of *man-thou*, *pao-tzu* and other raised pasta under the general title of *ping* mentioned in the Sung literature. A simplified version of this procedure is to mix the wine residues (*tsao* 糟) with flour, dry and then store the dough. A portion of this dried dough may be taken as needed and soaked in water. When a vigorous fermentation ensues, the brew is filtered and the filtrate used as leaven.[63]

The *I Ya I I* also describes the making of a raised dough using *hsiao chiao* 小酵 (lesser leaven).[64] *Hsiao chiao* is simply *chien* 鹼, a crude mixture of alkali and alkaline carbonates, which is often mixed with water and flour when making the dough. As acids are produced in the dough by microorganisms, some of the carbonates are decomposed to give carbon dioxide which would be trapped in the dough matrix. *Chien* is used in a recipe for making *chêng ping* 蒸餅 (steamed pasta) in the *Yin Shan Chêng Yao*, which says:[65] 'Method for *chêng ping*: Take *chiao tzu* 酵子 (little leaven), salt and *chien*, mix with warm water and blend with flour to make a dough. On the second day add more flour and knead to make a larger dough. Each catty of flour makes two large loaves, which are then steamed in a steamer.'

What is this little leaven that is cited here? According to Wang Jên-Hsing, in the countryside in North China, a freshly raised dough is often sliced into little cakes (about an ounce each) and then dried in air.[66] When needed, a cake is broken up in warm water and then used as a leaven to raise a freshly made dough. Such little cakes are called *chiao tzu* 酵子. What is interesting is that, just as in the Yuan dynasty recipe cited above, the *chiao tzu* is always used in conjunction with some salt and alkaline carbonates. The salt helps to strengthen the gluten in the dough and the *chien* carbonates generate carbon dioxide, as acid is formed by microbial activity. Thus, the carbonates in the *chien* augment the action of the yeasts in the leaven.

In summary, we have seen four types of leaven used in the making of raised dough in China. The first, and the earliest, is the must from a fermenting wine medium (Table 43, recipe 2). The second is the sour leaven made from a decoction of rice (Table 43, recipe 1). The third is simply a dried actively rising dough, which is easy

---

[63] *Ibid.*, p. 35.   [64] *Ibid.*, p. 34.
[65] *YSCY*, p. 44. *Chêng ping* is written as 䬓 餅. In the 'Everyman' edition, *chien* is written as in *yen chien* 鹽減.
[66] Wang Jên-Hsing (*1985*), pp. 53–4.

to store and rehydrate before use. Lastly, alkaline salts, rather like baking soda, can serve as a leaven by themselves or in combination with the dried risen dough.

(iii) *Origin of* hun-thun *(wonton)*

It is possible that *hun thun* (or wonton) also has its origin during the Han. The *Fang Yen* says that *ping* may be called *hun* 餛 and the *lao wan* 牢丸 of the *Ping Fu* (Ode to Pasta) may signify a type of wonton.[67] If we accept the Tsui Kuei-Thu's 崔龜圖 commentary on the *Pei Hu Lu* 北戶錄 (+873), then *hun thun ping* 渾吨餅 was already an established variety of pasta at the time of the *Chhi Min Yao Shu*. Tsui further states that it is written in the *Kuang Ya* as *hun-thun* 餛飩, and that according to Yen Chih-Tui 顏之推 *hun-thun* was shaped like a crescent moon and eaten by everybody.[68] Tsui's claim received unexpected corroboration in 1959 from archaeological excavations of a Thang tomb at Asana, near Turfan in Sinkiang. Dehydrated samples of *hun-thun* and *chiao-tzu* 餃子 were discovered, which in spite of their age, have retained their original shape remarkably well (Figure 112).[69] *Hun-thun* was apparently a well-known pasta food during the Thang. The Food Menu (*Shih Phu*) of Wei Chü-Yuan, c. +700, lists a group of twenty-four types of *hun-thun* among the delicacies presented to the emperor, and the *Shih I Hsin Chien* (Candid Views of a Nutritionist–Physician), mid +9th century, uses *hun-thun* in two of its prescriptions.[70]

The origin of the term *hun-thun* 餛飩 is obscure. Some scholars think that it is derived from the *hun-thun* 渾吨 in the *Chuang-Tze*, which is the name of the imaginary emperor of the central region.[71] Literally, it means chaos. When it was utilised as the name of a pasta, the water radical to the left was replaced by a *shih* 食 (food) radical. Presumably, the idea is that it represents 'the nebulous state of confusion when the world began'.[72] The Cantonese name *won-ton* is written as 雲吞 (cloud swallowing), since a bowl of wonton soup is reminiscent of layers of clouds in the sky.[73] A variant of *hun-thun* is *chiao-tzu* 餃子 (or *chiao erh* 角兒). *Hun-thun* is made from a square piece of very thin dough (wonton skin) while *chiao-tzu* is made from a round piece of a thicker dough. In either case, the product can be construed as a derivative of *thang ping* 湯餅. *Hun-thun* is usually boiled in soup; *chiao-tzu* can be steamed, boiled

---

[67] *Fang Yen* (c. −15), p. 12a. We agree with Knechtges (1986), p. 62, who has interpreted *lao wan* 牢丸 of the *Ping Fu* as a kind of dumpling. Thus, a wonton type of pasta was probably already known during the Han. *Lao wan* is mentioned together with *hun thun* 渾吨 in the *Pei Hu Lu*, ch. 2, *Shih Mu* 食目, p. 15b which may be the reason why Than Chhi-Kuang (1987) says that *chiao tzu* was called *lao wan* during the Thang. Than also tells us that *hun thun* was called *pien shih* 扁食 during the Ming. Evidently *hun thun* has assumed a variety of names in Chinese history. In China today it is known as *wonton* 雲吞 in Kuangtung, *chhao shou* 抄手 in Szechuan and *pien shih* 扁食 in Fukien.

[68] *Pei Hu Lu*, ibid. We are unable to locate Tsui's quotation from the edition of *Kuang Ya* in the *Szu Khu Chhuan Shu*.

[69] Than Chhi-Kuang (1987). The *hun thun* is 3 cm long and 1.9 cm wide; the *chiao-tzu* is 5 cm long and 1.5 cm at the broadest point. Presumably both products have shrunk from their original size. It is the *chiao-tzu* that looks like a crescent moon.

[70] *Shih Phu*, cf. *Chhing I Lu*, p. 7; *Shih I Hsin Chien*, pp. 25b, 29a.

[71] *Chuang-tze Chi Shih*, p. 309; cf. Burton Watson tr. (1968), p. 97.

[72] Cf. footnote in Buwei Chao (1970), p. 211; cf. also Phan Chên-Chung (1988) and Ah Ying (1990).

[73] Zee (1990), pp. 70–1, gives an entertaining exposition of the name *wonton*, but he is mistaken to think that *hun-thun* is phonetically derived from *wonton*. It is probably the other way round.

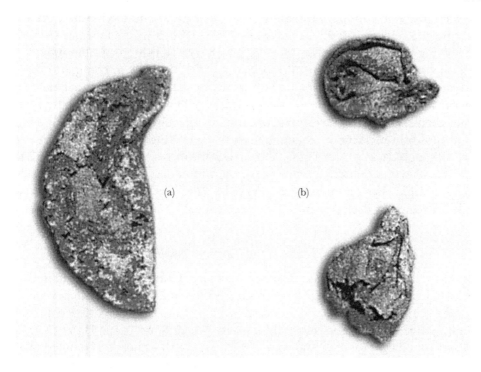

Fig 112. Dehydrated *chiao-tzu* (a) and *hun-thun* (b) found in a Thang tomb near Turfan in 1959. Adapted from Than Chhi-Kuang (*1987*), *CKPJ* (11), p. 12.

or pan-fried, in which case it is called *kuo-thieh* 鍋貼 (pot stickers). The earliest recipe we have for making *hun-thun* is found in the *Wu Shih Chung Khuei Lu*, Southern Sung, which says,[74]

Mix one catty of white flour with 3/10th of an ounce of salt. Knead with water to form a dough. Knead a hundred times and divide the dough into small balls. Flatten the balls with a rolling pin. Use pea flour to prevent sticking. Make sure the edges are thin. Use the skin to wrap the filling. The skin should be firm.

Hun-thun is mentioned in the celebrated reminiscences of life in the Sung capitals of Kaifêng and Hangchow, namely, *Tung Ching Mêng Hua Lu*, *Mêng Liang Lu* and *Wu Lin Chiu Shih* as well as the lesser known *Hsi Hu Lao Jên Fan Shêng Lu*.[75] *Chiao-tzu* as *chiao erh* is seen in the *Wu Lin Chiu Shih*.[76] Recipes for *hun-thun* or *chiao erh* are found in

---

[74] *WSCKL*, p. 31. Compare the procedure with that given in modern cookbooks. 'To make *won-ton*, place the filling in the centre of the square skin (about 4″ x 4″). Fold the skin in half diagonally to form a triangle and seal the edges with beaten egg. Bring the sharp corners together and seal with egg. To make *chiao-tzu* place the filling in the centre of the round dough. Fold to give a half circle and seal the edges.' Cf. Miller, G. B. (1970), pp. 698–701 and Buwei Yang Chao (1972), pp. 211–16. In Western Europe and America *won-ton* skin is readily available in oriental food stores and *chiao-tzu* and *kuo-thieh* are served in most Chinese restaurants.
[75] *TCMHL*, p. 29 (*hun-thun* shops); *MLL*, pp. 108, 136; *WLCS*, p. 51; *HHLJFSL*, p. 6.   [76] *WLCS*, pp. 122, 125.

the standard food canons of the Sung, Yuan, Ming and Chhing.[77] Still it comes as a surprise to find that in the *Sui Yuan Shih Tan*, +1790, a recipe for *chiao tzu* with a meat filling is entitled *tien pu ling* 顛不稜, which is obviously a transliteration of the English word 'dumpling'. Furthermore, Yuan Mei, the author, comments that he had eaten the steamed *tien pu ling* in the port city of Canton and declared it delicious. He observed that the secret lies in the meat filling. This encounter indicates that by +1790 there was already a significant amount of culinary intercourse between China and Britain. Actually dumpling is not a bad translation for *chiao-tzu* or *hun-thun*, although their closest equivalent in European cuisine is the Italian ravioli. Thus, *hun-thun* (i.e. wonton) has been a part of the Chinese food scene for more than 1,500 years and *chiao-tzu* for at least 1,000 years. *Hun-thun* is usually eaten as a snack, while *chiao-tzu* may itself constitute a complete meal.

(iv) *Origin and development of noodles*

Today, one of the most popular dishes one sees in a Chinese restaurant in the West is *hun-thun mien* 餛飩麵, that is wonton soup with noodles. In this case, *mien* means noodles, or specifically filamentous noodles. This brings us back to the second question we posed on page 469. How did the *mien* (wheat flour) of the Han evolve into the *mien* (noodles) of today? But in order to engage this question we need first to deal with the problem of the origin of noodles, which has been a matter of considerable interest among food historians in China and Japan.[78] Most scholars agree that the *so ping* 索餅 of the *Shih Ming*, (+2nd century) is the earliest form of filamentous noodle mentioned in the Chinese literature. Others feel that *thang ping* 湯餅 is actually the precursor of noodles. But the two terms are not mutually exclusive. *So ping* was simply one variety of *thang ping*. Although *thang ping* is usually interpreted as thin slices of dough reminiscent of present-day wonton skin, some of the pieces might have been shaped as long thin filaments and thus qualify as *mien thiao* 麵條 i.e. noodles.[79] The first description we have of the making of a noodle is recipe 10 from the *Chhi Min Yao Shu* (Table 43) for *shui yin* 水引 (wet drawn pasta). It says,[80]

To make *shui yin* (drawn pasta) or *po to* (thin drawn pasta), sift flour through the finest silk screen. Cool the clear decoction from a meat soup and use it to blend with the flour. For *shui yin* rub the dough between the hands until it is no thicker than a chopstick. Cut it to a length of a foot. Fill a pan of water. Place each length of dough in the water and rub it against the side of the pan until it is as thin as a leaf of chive. Then cook it in boiling water.

What we end up with is probably a noodle slightly broader than a fettucini. Presumably, if one retains the long piece of dough with the thickness of chopstick (or rub it to make it thinner) and boil it directly, the result would be a *so ping*. Thus, this

[77] *SCCK*, p. 82; *WSCKL*, p. 31; *CCPY*, pp. 118, 126, 127; *YSCY*, p. 42; *YLTYSCTC*, p. 5; *IYII*, p. 49; *YCFSC*, pp. 146, 148; *SHHM*, p. 36; *YHL*, p. 27 and *SYST*, pp. 127, 137.
[78] Hung Kuang-Chu (*1984*), pp. 30, 32; Wang Jên-Hsing (*1985*), pp. 62–6; Chhiu Phang-Tung (*1988*); Ishige Naomichi *et al.*, *Foodeum*, Summer 1988, no. 1, pp. 4–13.
[79] Wang Jên-Hsing, *ibid.*; Yün Fêng (*1987*); Chhiu Phang-Tung (*1988*), p. 15.   [80] *CMYS*, ch. 82, p. 510.

short recipe, in fact, tells us how to make two types of noodles, a round *so ping* and flat *shui yin*. There is, actually, another method for making *shui yin* in the *Chhi Min Yao Shu* that is implied, though not stated, in recipe 14 for making *fên ping* 粉餅 (rice filaments). A spoon-shaped piece of ox horn is sewn to the bottom of a silk cloth bag. Holes are drilled in the horn. A soft batter of rice flour is poured into the bag and squeezed through the holes into a pot of boiling water. The recipe says that if it is desired to have the product assume the shape of *shui yin*, then slits with the appropriate shape can be cut into the piece of horn. In recipe 13 for making *luo suo* 𪌿𪌎 (pasta nuggets) wheat-millet dough is pushed through a bamboo sieve. Thus, the chefs of the age of the *Chhi Min Yao Shu* already had the knowledge and the means to push a piece of dough through the right type of orifices to make *shui yin* or even *so ping*. The term *shui yin* is seldom seen after the *Chhi Min Yao Shu*, but *so ping* has persisted in the literature. It is mentioned in the *Shan Fu Ching* 膳夫經 (The Chef's Handbook), +857, where it is compared to *pu tho* 不托, a term synonymous with *po tho* and used in this context to denote pasta in general. The text says that '*pu tho* can be thin and broad, or small as a grain of millet. It can be long like a ribbon or square like a leaf; and when it is thick it can be sliced.'[81] Six recipes of cooked *so ping* used in diet therapy prescriptions for various diseases are given in the *Shih I Hsin Chien* 食醫心鑑 (Candid Views of a Nutritionist–Physician), Late Thang,[82] indicating that it was a well-known filamentous pasta in the Thang dynasty. These references suggest that noodles had been known by several names by the time of the Thang, *so ping*, *shui yin*, *thang ping* and *po tho*, and that they were prepared by rubbing a chunk of dough to form ropes or filaments, slicing a large sheet of dough to form long strings, and pushing a soft piece of dough through small orifices.

We have already noted on pages 472–3 that in the *chhi tzu mien* of the *Chhi Min Yao Shu* (Table 43, Recipe no. 12), the meaning of the word *mien* had shifted from 'flour', a raw material to 'pasta', a finished food. Apparently, sometime later the meaning shifted further and *mien* became noodles. When did this shift take place? The *Shih Ching* (Food Canon), of Hsien Fêng, c. +600, lists a dish called *fou phing mien* 浮萍麵 (floating duckweed pasta) and a restaurant menu in the *Chhing I Lu*, +965, lists a *hsüan tshao mien* 宣草麵 (daylily pasta).[83] In both cases, *mien* represents a finished dish; it has to be a pasta. It could have been a noodle, but it could also have included related pastas, such as *chhi tzu mien*. These references, clearly tell us that by the end of the Thang the word *mien* had achieved a dual meaning: wheat flour, or a pasta made from it.

According to the *Tung Ching Mêng Hua Lu* (Dreams of the Glories of the Eastern Capital), +1147, the food shops (*shih tien* 食店) in the Northern Sung capital served

---

[81] *Shan Fu Ching*, p. 5 in Shinoda & Seiichi, eds. (1972), p. 115. The term *pu tho* is derived from the *Fang Yên*, which says that *ping* is called *tho* (*ping wei chih tho* 餅謂之飥). The word *tho* was still used in the sense of a noodle in the *Yin Shan Chêng Yao*, +1330, p. 86.
[82] *Shih I Hsin Chien*, pp. 2a, 10b, 11a, 12b, 12b, 18a, 23a, 25a, 25b. The term *po to* also is used in one prescription, p. 13b.
[83] *Shih Ching* of Hsieh Fêng, in Shinoda & Seiichi, eds. (1972), 116, and *Chhing I Lu*, p. 31.

six kinds of *mien* and one kind of *chhi tzu* (chess pasta).[84] Although the names of the *mien* dishes may sound a bit foreign to us, it is clear that their primary ingredient is a pasta made from flour. This pasta is a separate, distinct entity. It is not *chhi tzu*, chess pasta. It is most likely a noodle. If so, the word *mien*, in the sense of a noodle, would have been used continually in the food canons from the Sung to the present day. The text further says that one of the dishes was prepared by mixing meat and *mien*. 'Formerly, it was eaten with a spoon, but now chopsticks are used.'[85] Of course, it is easier to eat filamentous noodles with chopsticks than with a spoon. That *mien* had become a synonym for noodles is also consistent with the recollections recorded in the *Mêng Liang Lu* (Dreams of the Former Capital), +1334, which states that *mien shih tien* 麵食店 (pasta shops) in the Southern Sung capital served sixteen different types of *mien* dishes and seven types of *chhi tzu* dishes.[86] Among the *mien* dishes are those cooked or topped with thin chicken strips, stir-fried chicken, pork, lamb, fish, eel and bamboo, etc., which could quite credibly pass off as noodle dishes served in Chinese restaurants today. *Chhi tzu* (chess pasta) pieces were also considered as *mien* since they are, in effect, noodles that are extra short. But their popularity waned and they disappeared from the food scene after the Yuan.[87] The word *mien* then became either wheat flour or noodles made from it. The answer to the second question we posed on page 467 is, therefore, that *mien* assumed the mantle of 'noodles' during the Sung.

But *mien* continued to mean wheat flour until today. Even so, the term *ping* as pasta or noodle was still seen occasionally in the late mediaeval period. The *Shan Chia Chhing Kung* (Basic Needs for Rustic Living), Southern Sung, in a recipe entitled *lo fu mien* 蘿菔麵 (radish *mien*), describes mixing the juice from the Chinese radish with flour to form a dough, but calls the product a *ping*. In a recipe entitled *pai ho mien* 百合麵 (lily pasta), the product is still called a *thang ping*.[88] It would appear that *mien* and *ping* were interchangeable. On the other hand, both *so ping* and *thang ping* are products in a recipe entitled *yü yüan so ping* 玉延索餅 (rope pasta from the Chinese yam), indicating that *so ping* and *thang ping* were distinct entities.[89] What is

---

[84] *TCMHL*, ch. 4, p. 29. Among the *mien* foods are *nuan yang mien* 軟羊麵 (soft lamb *mien*) and *chha rou mien* 插肉麵 (blended meat *mien*). For a more detailed discussion of the pasta and noodles of the Sung cf. Sabban (1989a).
[85] *TCMHL*, ch. 4, p. 29. The original reads, *Chiu chih yung shih, chin chieh yung chu* I 舊只用匙，今皆用箸矣.
[86] *MLL*, ch. 16, pp. 135–6.
[87] The identity of *chhi tzu* is now a matter of uncertainty, cf. Hung Kuang-Chu, *CKPJ* (1985) 5, p. 33; Têng Kuang-Ming (1986a, 1986b).
[88] *SCCK*, pp. 73, 23.
[89] *Ibid.*, p. 78. In Thang and Sung literature the term *thang ping* is often used to denote filamentous noodles, which is often served on birthdays as a symbol of longevity. The Thang poet Li Yü-Hsi 劉禹錫 wrote a poem when he was invited to a *thang ping* banquet celebrating the birth of a baby boy of a friend.

憶你懸弧日，余為座上賓。舉箸食湯餅，祝辭添麒麟

At your birth when they first hung out the bow,
I was the most honoured guest at the birthday feast.
Wielding my chopsticks I ate boiled noodles,
And composed a congratulatory poem on a heavenly unicorn. Tr. Knechtges (1986)

Thang ping also occurs in the poems of such famous Sung poets such as Lu Yu and Su Shih; cf. Yün Fêng (1987). The term *thang ping* is still used as a literary expression of noodles served on birthdays.

most interesting in this recipe is that to make *so ping* the flour obtained from the yam [presumably as a thin dough] is poured into a bamboo cylinder and allowed to flow through holes at the bottom into a pan of boiling water. This is the method described in Recipes 5 and 14 (Table 43) in the *Chhi Min Yao Shu* for making loose pastry and rice filaments. It is natural that it would be applied later to the making of noodles.

The earliest recipe we have on the making of a *mien* as noodle is in the *Wu Shih Chung Khuei Lu* (Madam Wu's Recipe Book), Southern Sung. It is entitled *Shui hua mien fa* 水滑麵法 (Method for making wet slippery noodle). We quote:[90]

Take fine white flour. Blend [with water] and knead into dough. One catty [of flour] will give ten lumps of dough. Place them in water and allow the dough to mature (i.e. achieve the correct degree of plasticity and elasticity). Take a lump; pull and stretch it until it is broad and thin. Cook it in boiling water. Flavour [the noodle] with sesame oil, almond oil, salted dry bamboo shoot, (various pickled vegetables)[91] . . . or add cooked meat, which makes a particularly delicious dish.

The words used for pull and stretch are *chhou* 抽 and *chuai* 搜. Presumably, the pieces are shaped by hand and then boiled. They could very well be long strips reminiscent of the *shui yin* of the *Chhi Min Yao Shu*. This recipe is repeated verbatim in the *I Ya I I*, Yuan, and *Yin Chuan Fu Shih Chien*, +1591.[92] A similar, (but not identical) recipe is given in the Yuan compilation *Chü Chia Pi Yung*, which presents seven recipes for the making of *mien*.[93] The first and common step in all of them is to mix wheat flour with water and thoroughly knead the dough. Salt, oil and *chien* 鹼 (alkaline carbonates) are sometimes added to the mixture. The dough is then treated in different ways according to the type of product desired.

1. *Wet slippery noodles* (*shui-hua-mien* 水滑麵): the dough is rolled (with a rolling pin) and kneaded until it is mature. It is manipulated by hand into pieces (or strips) the size of fingers. After soaking in water for two hours, the dough is firm and ready for cooking in boiling water.[94]

2. *Thread noodles* (*so mien* 索麵): lumps of dough are coated with oil and worked by hand to give ropes the thickness of a chopstick. The rope is then placed around two rods and pulled apart as far as possible to give a very long and thin thread.[95] It is then boiled in a pot.

---

[90] WSCKL, p. 31.
[91] These include preserved melon (in fermented soy paste), preserved eggplant (in wine residues), ginger, pickled chives, and silky strips of yellow squash.
[92] IYII, pp. 50–1; YCFSC, p. 146. A thin wet slippery [noodle], *hsi shui hua* 細水滑 is also mentioned in the *Yin Shan Chêng Yao*, p. 32 (no. 41).
[93] CCPY, pp. 113–17. The same recipes are repeated *verbatim* in the *Tuo Nêng Phi Shih* of +1370 (early Ming). There is also a recipe for making *shan yao mien* 山藥麵 (Chinese yam noodle), in which the yam is ground up, made into thin pancakes and sliced into noodles. No wheat flour is used.
[94] This recipe is a bit more informative than that for *shui hua mien* in the *Wu Shih Chung Khuei Lu* quoted above. Even so, the exact manipulations involved and the shape of the product are not absolutely clear.
[95] As pointed out by Hung Kuang-Chu (1984a), p. 43, this is probably a procedure for making *kua mien* 掛麵, because the product is extremely thin. In this process, the dough is rubbed and stretched to form ropes with the diameter of a chopstick. They are joined end-to-end to form a loop and a series of such loops placed around two rods. One rod is kept stationary and the other pulled and slightly twisted until the rods are as far apart as possible. The thin threads are dried in the sun. It is still a common sight today to see such noodles hung on racks to dry in villages near Foochow (Figure 114).

3. *Ribbon noodles* (*chin-tai-mien* 經帶麵): the dough is rolled with a rolling pin and rerolled until it is very thin. The sheet is then sliced [with a knife] to give long ribbons ready for cooking.

4. *Rolled flat noodles* (*tho chiang mien* 托掌麵):[96] the dough is divided into little balls, which are rolled to give flat round sheets as large as the mouth of a shallow cup. They are then boiled.

5. *Silky red noodles* (*hung ssu mien* 紅絲麵): in this case the dough is made with clear shrimp soup rather than water. The dough is rolled thin and then sliced. The noodles will be red in colour.

6. *Silken green noodles* (*tshui lü mien* 翠縷麵): here the juice from tender leaves of the locust tree are used to make the dough. It is rolled and then finely sliced.

7. *Radish noodles* (*lo fu mien* 蘿菔麵):[97] a decoction of radish is used to make the dough, which is rolled to a thin sheet and sliced.

In four out of the seven recipes (3, 5, 6 & 7) the noodles are formed by cutting a thin flat dough with a knife. The general name for such noodles is aptly *chhieh mien* 切麵 (sliced noodles) which were probably the most popular type of noodles made during the Sung.[98] Recipes 1 and 4 do not require cutting, but they could still give filamentous products that resemble *mien*. They may be regarded as more refined extensions of the *so ping* listed in the *Shih Ming* of the +2nd century. The product from recipe 2 is probably the *kua mien* 掛麵 (hung noodle) of today.[99] The term *kua mien* first occurs in the *Yin Shan Chêng Yao*, +1330.[100] Although it is the principal ingredient of a noodle dish, there is no description of how the *kua-mien* is made. There are, however, other interesting bits of information in the *Yin Shan Chêng Yao* that should not be left without notice. For example, we read of the use of white *mien* (flour) to make *mien* (noodles).[101] Such a statement can be confusing if one is not well acquainted with the dual personality of the word *mien*. We also see references to *mien ssu* 麵絲 (silky noodles), and the slicing of white flour [in the form of dough] to give fine noodles.[102] In particular, we see the earliest occurrence of the term *mien thiao* 麵條 (filamentous noodle) in a diet therapy recipe.[103]

One disadvantage in the *kua mien* process is that an inordinate amount of space is required. An interesting variant called *chhên mien* 抻麵 (pull noodle), better known as *la mien*, 拉麵 was later developed. A loop of dough is pulled apart by two hands, coiled back into two loops and then pulled again. The procedure is repeated until

---

[96] The title is difficult to translate. The *tho* 托 here is the phonetic equivalent of *tho* 飥, an archaic word for pasta. As we have noted earlier (note 81) the *Fang Yen*, p. 12, says, *ping wei chih tho* 餅謂只飥 (*ping* is called *tho*).

[97] The original title is *kou mien* 勾麵, which is difficult to interpret. We have followed Hung Kuang-Chu's lead (*1984a*), p. 44, and called it Chinese radish noodle.

[98] There is a recipe for making *chu mien* 煮麵 (cooked noodle) in the *Yün Lin Thang* Dietary System, p. 1. The product is of the sliced noodle type. There are few recipes dealing with the technique of noodle making in the food canons of the Ming and Chhing. We find five noodle recipes in the *Sui Yuan Shih Tan*, pp. 124–5, but they have little to say on how the noodles are made.

[99] See n. 94, above. In Fukien, such thin stretched noodles are also called *hsien mien* 線麵 in the local dialects.

[100] *YSCY*, ch. 1, p. 30, no. 37. For ease of identification we have numbered the ninety-five recipes in ch. 1 in the order that they occur in the text. The culinary aspects of this collection of recipes have been analysed by Sabban (1986b).

[101] *YSCY*, ch. 1, p. 24, no. 18.   [102] *Ibid.*, p. 27, nos. 26, 27; p. 30, nos. 34, 35.

[103] *Ibid.*, ch. 2, p. 87. The term *mien thiao* was not seen again until the *Su Shih Shuo Luo* of the late Chhing.

hundreds of strands of thin noodles are obtained.[104] *Chhên mien* (as *tshe mien* 搓麵) is first mentioned in the *Chu Yü Shan Fang Cha Pu*, c. +1500, by Sung Hsü and his family, which gives a brief description on how it is made.[105]

Based on these accounts we can say that noodles were among the pasta foods developed during the Han. The earliest forms were *so ping* (rope pasta) and *thang ping* (soup pasta), which probably consisted of pasta in small sheets. During the Thang it was found that thin sheets of dough could be sliced with a knife to give long filaments of pasta. This was the beginning of *chhieh mien* (sliced noodles). The discovery that ropes of dough made by hand could be pulled and stretched to give extremely fine threads probably took place in the Sung. This gave rise to *kua mien*, also known as *hsien mien* 線麵 (thread noodle), and a later variant called *chhên mien* (pulled noodle). The method of squeezing a dough through orifices to give filaments is described in the *Chhi Min Yao Shu* (see Table 43, no. 13). The technique was applied to making yam noodle in the Sung, and for making buckwheat as well as wheat flour noodle in the Yuan.

These traditional methods of making noodles are still very much alive in China today. The sliced noodle (*chhieh mien*) remains the most popular product (Figure 113). *Kua mien* is seen mainly in Fukien (Figure 114), and *la mien* in the North (Figure 115). Rope noodles (*so mien*) rubbed by hand are still made in the Northwest (Figure 116). Noodles obtained by allowing the dough or batter to squeeze through orifices, *ho lo mien* 河漏麵 or pressed noodle (*ya mien* 壓麵), are found in the Northeast (Figure 117). The technique was first described in Recipes 5, 13, 14 (Table 43) in the *Chhi Min Yao Shu*.[106] It has been applied mostly in the making of rice and buckwheat noodles.

Many Westerners who live in large cities may have enjoyed the delightful Cantonese tea brunches at which one can savour a whole series of little dumplings, cakes, pastries, and even noodles while sipping interminable cups of tea. It may be regarded as a midday Chinese version of that equally delightful British institution – the afternoon tea. The Cantonese call these brunches *yum cha*, i.e. *yin chha* 飲茶 (drinking tea), and the delectables served *dim-sum*, i.e. *tien-hsin* 點心 (dots of heart's desire).[107] In addition to varieties of *pao-tzu*, *chiao-tzu* and wonton etc. made of wheat flour (*mien-fên* 麵粉), one would also be able to enjoy similar products made from rice flour (*mi-fên* 米粉). It is perhaps not unexpected that just as the *mien* (wheat flour) of antiquity gradually evolved into the modern *mien* (noodles), so the *mi-fên* (rice

---

[104] To see an experienced chef turn a piece of dough into thousands of strands of noodles by hand within a few minutes is akin to witnessing a minor miracle. Having seen the performance several times, I can understand why the art of making of *la mien* has become a popular attraction for tourists in China in recent years.

[105] *Chu Yü Shan Fang Cha Pu* 竹嶼山房雜部, ch. 2, *Yang shêng pu* 養生部, pt 2, cf. *SKCS*, **871**, p. 130. Cf. also Chhiu Phang-Thung (*1988*), p. 16. *Chhên mien* (as *chen thiao mien* 槙條麵) is mentioned in a description of *mien thiao* 麵條 in Hsüeh Pao-Chhên (*1900*), p. 49. But even as the author waxes enthusiastically about the dramatic way it is made, he gives no information on the details of the process. The procedure is difficult to describe, but having seen it once, it is not likely to be forgotten.

[106] In recipe 14 the substrate is *fên ying*, the refined rice starch obtained by the method described earlier, cf. pp. 461–2. It is not mung bean flour as some editions of the *CMYS* seem to indicate, cf. Miao Chhi-Yü ed. (*1982*), p. 513, n. 16.

[107] According to the *Nêng Kai Tsai Man Lu*, +1190, p. 26, the term *tien hsin* was first used in the Thang. It is mentioned often in the *Mêng Liang Lu*, pp. 108, 134, 137. See Mu Kung (*1986*) in *Phêng Jên Shih Hua*, pp. 455–6.

Fig 113. Sliced noodle, *chhieh mien*, photograph S. T. Liu.

Fig 114. Hung noodle, *kua mien*, photograph S. T. Liu.

Fig 115. Pull noodle, *la mien*, photograph Wang Yusheng.

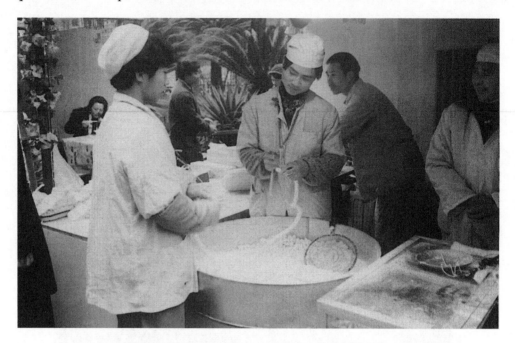

Fig 116. Rope noodle, *so mien*, photograph Wang Yusheng.

flour) of antiquity shifted its meaning to become the *mi-fên* (rice noodles) we know of today. Because of their lack of gluten, the dough made from rice, buckwheat and mung bean lack the plasticity and elasticity exhibited by wheat flour dough. They cannot be rolled into sheets or ropes. There is no way to make the sliced, pulled and hung products characteristic of wheat flour noodles. One remedy is to add gluten to the flour as in the case of the Japanese 'soba' noodles, or by coagulating squeezed filaments of flour dough immediately into boiling water. The sudden heat gelatinises the starch and turns the batter into filaments of a very thick paste.

This type of noodle was called *ho lou mien* 峆咯麵 or *ho lo mien* 河漏麵. It is used to make buckwheat noodles in Shansi and Shensi. The first record of such buckwheat noodles is in Wang Chen's *Nung Shu* (Agricultural Treatise) of +1313.[108] In South China, rice noodles made by this procedure were originally called *mi-lan* 米欖, but they are now known as *mi-hsien* 米線 (rice threads) or *mi-fên* 米粉 (rice noodle).[109] A rice counterpart of sliced noodle (*chhieh mien*) has also been developed. Water-soaked rice grains are ground to give a batter, which is spread thinly on a flat bottom pan and steamed. The thin translucent flat sheet is called *fên phien* 粉片 (rice sheet), which can be cut, shaped and filled to make many kinds of *dim sum* that one sees at a Cantonese tea-brunch. But most interesting for us is that these sheets can be sliced

---

[108] *Nung Shu*, ch. 7, p. 61, on *Chiao Mai* (buckwheat).
[109] Chhiu Phang-Thung (*1987*) and Chu Jui-Hsi (*1994*).

Fig 117. Press noodle, *ya mien*, photograph Wang Yusheng.

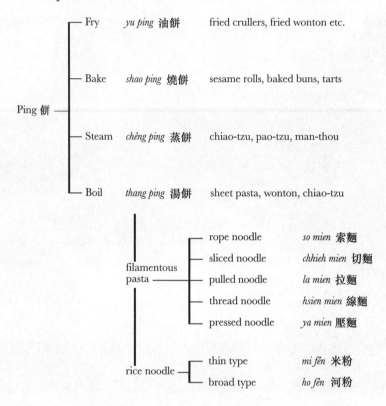

Fig 118. Family of Chinese Pasta Foods (after Ishige).

to give broad filamentous noodles that are called *ho fên* (*hor faan*) 河粉, flat rice noodles in Kuangtung and *kui thiao* 粿條 (rice cake filaments) in Fukien.[110]

The place of noodles in the family of Chinese pasta foods, i.e. the *ping* of antiquity, is represented by the diagram shown in Figure 118. Wheat flour can be processed into four types of pasta: those that are boiled, those that are steamed, those that are baked and those that are fried.[111] This classification follows that proposed by Ishige for his study of the diffusion of noodle culture and technology from China to her neighbouring countries. By Ishige's definition a noodle is a 'liner processed food made from the flour of cereals, beans and potatoes, which must be boiled for serving'. For our purpose, it is a convenient and useful definition.

Pasta sheets and noodles (pasta filaments) can be considered as members of the *thang ping* (boiled pasta) group. As we have seen, the dough can be flattened into small sheets suitable for wrapping fillings to make wonton and *chiao-tzu*. It can be

---

[110] Lin Hao-Jan 林浩然, ed. (1985), p. 72.
[111] The principal 'steamed pasta' is *man-thou* (steamed bun), which we discussed earlier on pp. 475–8. According to the *Tung Ching Mêng Hua Lu*, p. 30, there were shops that specialised in 'baked' pasta (*hu ping*) and 'fried' pasta (*yu ping*) in the Northern sung capital. An example of a modern 'baked' pasta is the sesame bun (*shao ping* 燒餅). The best known example of a modern 'fried' pasta is popular *yu thiao* 油條 (fried twirlers) served often at a Chinese breakfast.

fashioned to give (1) sliced noodles, *chhieh mien* (2) pulled noodles, *la mien* (3) thread noodles, *hsien mien* and (4) pressed noodles, *ya mien* or *ho lo mien*. An egg noodle (*chi tzu mien* 雞子麵) was first mentioned in the *Chu Yü Shan Fang Tsa Pu*, c. +1500. It was made by adding eggs to the dough, which was then processed in the usual way and sliced into filaments.[112] This innovation reduced the blandness of the product and increased the overall popularity of noodles. Finally, methods (4) and (1) have also been applied to make a variety of noodles from rice flour, e.g. *mi fên* and *ho fên*.[113] In view of their versatility and wide appeal, it is not surprising that the technology of noodles should have spread to China's neighbours in East and Central Asia.

(v) *Dissemination of noodles in Asia*

From what we have seen, the origin and development of the various types of noodles in China, except for *ho-fên* (flat rice noodle), are well documented. There is no question that noodles are a Chinese invention. From China they have, at various times, spread into Japan, Korea, Central Asia and Southeast Asia. Unfortunately, other than the case of Japan, very little is known about how and when such disseminations took place.

Chinese pasta and noodles were probably introduced into Japan during the Thang dynasty.[114] *So ping* 索餅 is listed as one of eight *Thang kuo* 唐果 (Chinese snacks) in the *Wamyo-ruizyushu* 和名類聚抄, +930, and as one of the nine Chinese foods in the *Chûjiruki* 廚事類記, c. +1090. Unlike its fate in China the term *so ping* has persisted in Japan until the present day. It was at first a coarse strand with the thickness of a chopstick like the original *so ping* of the *Shih Ming*. It was called *muginawa*. It could be stretched somewhat, but the strand would lose its elasticity as it dries out. The technique of rubbing oil on the dough was introduced from South China in the 14th century. This enabled the Japanese to make *somen* (i.e. *so mien* 索麵) as fine as the contemporary Chinese *hsien mien*.[115]

But *so ping* was not the only precursor of noodles in Japan. Another *Thang* food introduced to Nara was *konton* 餛飩 (i.e. *hun-thun*), which in course of time, was written as 溫飩 and pronounced as *udon*. It was presumably a flat piece of dough. The method of slicing rolled flat dough to make thin noodles was probably introduced from China in the 15th century. The noodles made in this way were then called *udon*. Yet another Thang import to Nara is *po tho* 餺飥 (*hakutaku*), the exact nature of which is not fully known. It is regarded by some scholars as the precursor of *somen* 素麵

---

[112] *Chu yü shan fang tsa pu* 竹嶼山房雜部, ch. 2, *Yang shêng pu* 養生部, pt 2, cf. *SKCS*, **871**, 129. The famous *i fu mien* 伊府麵 (*I* family noodle), prepared from fine egg noodles that were fried in oil after they had been boiled, was probably introduced during the Chhing. Served in soup it is a dish considered distinguished enough to be on the menu of formal banquets, cf. Wang Jên-Hsing (*1985*), p. 163.

[113] Using the starch from mung bean, very fine noodles can be made by pressing the dough through orifices and heating the filaments in boiling water. The product is called *fên ssu* 粉絲 (silky starch threads). Anyone who has not handled it before will be amazed at its remarkable ability to absorb water without falling apart.

[114] Tanaka Seiichi (*1987*) tr. by Huo Fên & I Yung-Wên (*1990*), pp. 59–71 and Ishige *et al.* (*1994*).

[115] As we have noted earlier, these very fine hung noodles are still called *so mien* in the local dialect in Fukien.

(*so mien*) noodles. Noodles made from buckwheat flour fortified with some wheat flour is called '*soba*'. Both *udon* and *soba* have been popular since the 17th century. On the other hand, the steamed bun *man thou* 饅頭 did not arrive in Japan until the Yuan Dynasty. It was brought over in +1349 by Lin Ching-Yin 林淨因 who is still remembered today as the father of *man-thou* in Japan.[116]

According to the *Yin Shih Chih Wei Fang* 飲食知味方, +17th century, pressed noodle (*ho lou mien* 河漏麵), and thread noodle were known in Korea.[117] A buckwheat noodle was prepared by pouring a batter of buckwheat flour and mung bean starch through holes at the bottom of a gourd. The filaments that form are dropped directly into boiling water. Later potato starch was also used in place of mung bean starch. Buckwheat noodles are now called *naeng myon* 冷麵 (cold noodle) in Northern and Eastern Korea. It is not known when thread noodle, *hsien mien*, was introduced to Korea. It is only popular in the southern wheat growing area. Sliced noodle (*chhieh-mien*) is known in Korea as 'kál kuksu'.

La-mien, as well as *chhieh-mien*, is well known in Central Asia, Mongolia and Tibet. These transmissions probably took place during the Chhing Dynasty when these regions enjoyed a good relationship with the Chhing Empire. *La-mien* is considered the progenitor of '*lagman*', a Uyghur word for filamentous noodles that is also used by other peoples of Central Asia. A similar type of *la-mien* is known in Tibet as '*menshhitshi*'. It is curious that the technique for making *la-mien* has diffused only westward. It was not transmitted eastward to Japan or southward to Southeast Asia. In contrast, sliced noodles (*chhieh-mien*) has diffused to the west, east and the south. It is the most widely consumed noodle in Asia.

The emigration of Chinese since the Ming from Fukien and Kuangtung has been largely responsible for the diffusion of noodle technology to Southeast Asia. The immigrants brought with them not only noodles made with rice flour, namely *mi-fên* (rice noodle) and *ho-fên* (flat rice noodle), but also *mien-thiao* (wheat noodles) and *hsien-mien* (thread noodle). They are, of course, part and parcel of the diet of the Chinese population. But noodles have also gained wide acceptance among the local non-Chinese populace.[118] The making of rice noodles (*mi-fên*) has become a part of the local trade as attested by such names as 'Numbanhchock' in Cambodia, 'sen mee' in Thailand and 'bihun' in Malaysia and Indonesia.

The most innovative application of noodle technology in the 20th century, however, did not take place in China, but rather in Japan. We refer to the development of 'instant noodles', in the late 1950s.[119] Today, 'instant noodles' are consumed not only in East Asia but also in Europe, America and other parts of the world.

---

[116] There is a shrine in his honour in Nara, cf. Lin Chêng-Chhiu (*1989*) and Hu Chia-Phêng (*1989*).
[117] This description of the diffusion of noodles from China to other parts of Asia is based largely on Ishige Naomichi (1990) and (1995), ch. 9.
[118] *Ibid.*
[119] Instant noodles are made by freeze-drying noodles and flavouring materials separately and packaging them together. All one needs to do is to add hot, boiling water to the packaged materials, and the noodle will be ready for eating.

(vi) Noodles and Marco Polo

It is a remarkable fact that across the great land mass of Eurasia, noodles are a staple food in China and her neighbours in the Far East, and in one solitary European country, Italy, in the far West, but nowhere in between. Why? What special connection can there possibly be between China and Italy? The one and only obvious answer is Marco Polo, a Venetian merchant who travelled to China in +1271 and returned to Italy in +1292. While in China he served as an advisor to the Yuan Emperor Kublai Khan and visited many parts of the country. It was probably as a prisoner of war in Genoa that he collaborated with Rustichello of Pisa to write an account of his travels under the ambitious title 'Description of the World' (*Divasament dou Monde*) which is best known to English readers as *The Travels of Marco Polo*. Although the book was written in French, it was soon translated into most of the European languages.[120] For almost two centuries it fired the imagination of generations of European explorers. It was one of the books that Christopher Columbus took with him in his epoch-making voyage of 1492 that led to the discovery of the New World.[121]

We do not know when and how the legend of Marco Polo bringing noodles back to Italy from China started. We do know that it has become ingrained in the folklore of both China and the West (Figure 119).[122] It is a very plausible story, but there is no evidence to support it. Leonardo Oelschki has suggested[123] that the legend is based on a passage in the *Il Milione*, the Italian edition of the work by Ramusio, which is quoted below:[124]

With regard to food, there is no deficiency of it, for these people especially the Tartars, Cathaians or inhabitants of the province of *Manji* (i.e. South China), subsist for the most part, upon rice, panicum, and millet; which three grains yield, in their soil, a hundred grains for one. Wheat, indeed, does not yield a similar increase, and bread not being in use with them, it is eaten only in the form of *vermicelli or of pastry*.

In other translations the phrase *in italics* is rendered as:

'vermicelli and pastes of that description'[125]
'macaroni and other viands made of dough'[126]
'noodles and other pasty foods'[127]

These citations do not lend support to Oelschki's hypothesis, but rather indicate that Marco Polo was already familiar with a filamentous pasta before he went to China.

---

[120] For a discussion of the various manuscripts and editions of the book known see Latham (1958), pp. 24–7, and Blue (1990 & 1991). The definitive edition is that of Moule & Pelliot (1938) entitled *The Description of the World*.
[121] Blue (1990), p. 40.
[122] In the West, the legend is fanned by such paintings as the one shown in figure 119. It is today probably of wider currency in China than in the West. It was reiterated in an editorial of the influencial magazine, *Chung Kuo Phêng Jên* (Chinese Cuisine), *1987* (4), p. 3. Gastronomically, the ease with which Chinese noodles blend in with Italian pasta is shown in Teubner, Rizzi & Tan (1996).
[123] Olschki (1960), p. 177 n. 100.   [124] Marsden, William, trs. & eds. (1908), re-ed., Wright, p. 209.
[125] Yule & Cordier, trs. & eds. (1975), p. 438.   [126] Moule & Pelliot, eds. & trs. (1938), p. 244.
[127] Latham, trs. (1958), p. 152.

Fig 119. Marco Polo tastes spaghetti at the court of Kublai Khan. From Julia della Croce, *Pasta Classica*, Chronicle Books (1987), p. 11.

In fact, filamentous pastas were known in Italy long before Marco Polo made his famous journey. In +1154, Idrisi, an Arab traveller, writes that in Trabia, a small village near Palermo, Sicily, 'cibo di farino in forma di fili', i.e. food from flour in the form of strings is made.[128] Thin threads of pasta were also known in the Middle East. They are featured in two Islamic recipes from Bagdad in +1226, one of which is quoted below:[129]

Rishta – Cut fat meat into middling pieces and put into the saucepan, with a covering of water. Add cinnamon-bark, a little salt, a handful of peeled chick-peas, and half a handful of lentils. Boil until cooked; then add more water, and bring thoroughly to the boil. Now add spaghetti (*rishta*, which is made by kneading flour and water well, then rolling over fine and cutting into thin threads four fingers long). Put over the fire and cook until set to a smooth consistency. When it has settled over a gentle fire for an hour, remove.

The method of preparation leaves no doubt that in *rishta* we have a type of filamentous noodle. The other recipe deals with the cooking of *itrīyya*, which is translated as 'macaroni'; but how it is made is not described. Furthermore, *itrīyya* was noted as a food consumed in Bukhara by the great Arab scholar and physician, Abu Ali al-Husein Ibn Sina, better known in the West as Avicenna (+980 to +1037), who said,[130]

---

[128] Idrisi, Abu Abdallah Mohammad ibn Muhammad ibn Idris, *Il diletto di chi appassionato per le perigrinazioni attraverso il mondo*. Cf. also Amédée Jaubert, *Geographie D'édrisi*, Paris (1840), p. 78.

[129] Arberry, A. J. tr. (1939), p. 45. Arberry notes that '*Rishta* is a Persian word meaning "thread", then commonly used for spaghetti. In the original manuscript it is written as *Rāshtā*.'

[130] Cited by Rosenberger, B. (1989), p. 80, which reproduces the translation of L. Leclerc (*Recueil de simples*, vol. I, no. 100) of Ibn Sina's recipe from Ibn al-Baytar's *Gami al-mufridat*, tr. G. Blue, from the French. Rosenberger gives another recipe for *itrīyya*, pp. 81–2 from the anonymous Kitab-al-tabih, which shows that it could be prepared in different ways.

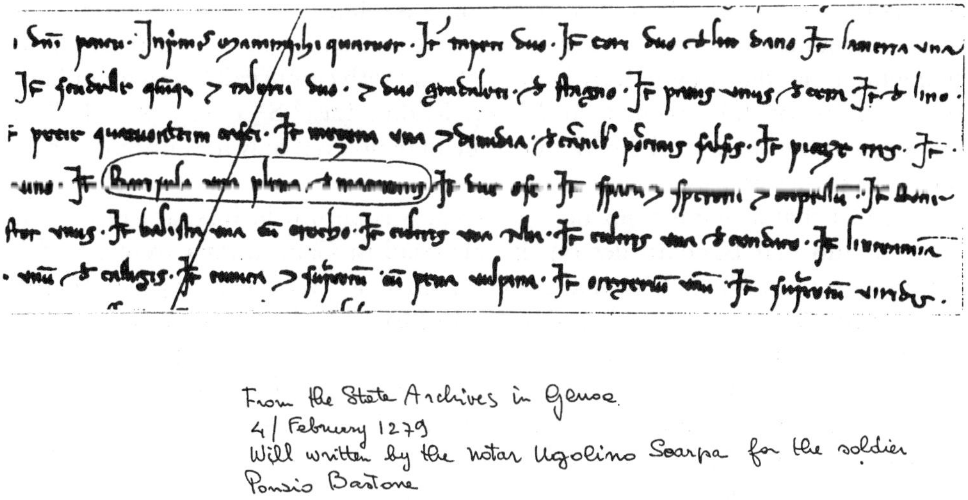

Fig 120. A will from the Archives in Genoa, dated February 4, 1279, by soldier Ponvio Bastone in which he bequeathed a basket of 'macheroni', to a relative. Courtesy of the Spaghetti Museum, Pontedassio, Imperia, Italy.

Itriyya, edible pasta: It is prepared in the form of strips. One makes it with unleavened dough, and cooks it with or without meat. In our country it is called *rishta*. It is warm and very wet. It is slow to digest and heavy on the stomach because it contains no leaven. Cooked without meat it is lighter according to some though they may be mistaken. If one adds pepper and almond oil it is a bit less difficult. When one digests it, it is very nourishing. It frees the lungs of coughing and spitting up of blood, particularly if one cooks it with purselane. It relaxes the chest.

Ibn Sina is known to have been born at Afshana, near Bukhara. His teacher is al-Farabi, the great commentator on Aristotle. They both worked in Bagdad. However, when he says 'in our country' he means, presumably, Iran and the region around Bukhara. *Rishta* is apparently one of many Arabic food names in currency before +1200 that is of Persian origin.[131] Furthermore, there are several documents on file at the Spaghetti Museum, Pontedassio, Imperia, Italy, dated +1244, +1279 and +1284 respectively which show that pasta, maccheroni and vermicelli were known foods before Marco Polo's return to Italy in +1292.[132] One of the documents is reproduced in Figure 120. Thus, the legend that Marco Polo had brought

---

[131] Rodinson (1949), see particularly, pp. 150–1. On p. 138 he also mentions two *rishta* recipes in the *Kitab al-Wusla ila 1-Habib* which is dated at about +1230, well before Marco Polo. In footnote 9, Rodinson states that *rishta* was still being prepared in Lebanon and Syria and often cooked with lentils.

[132] The Spaghetti Museum is mentioned in Waverley Root (1971), pp. 369–70. The documents are (1) a statement of notary Givanuinus de Perdono, dated 1 August 1244, which refers to pasta (2) a will from the National Archives in Genova, dated February 4, 1279, by soldier Ponvio Bastone in which he bequeathes a basket of 'maccheroni' to a relative, and (3) a document dated 13 February 1284 which mentions vermicelli. We thank Dr Patrick Johnson and Mrs Angela Sharpe for a visit to the Spaghetti Museum to obtain copies of these documents, which are now on file at the Needham Research Institute, Cambridge, England. We regret to learn that the Spaghetti Museum is now closed.

filamentous noodles back to Italy from China is totally without foundation.[133] We are left without an explanation for two questions. Firstly, what is the origin of the filamentous noodles seen in Italy in the 12th and 13th centuries? Was it an indigenous invention or did it come from Islamic Middle East? Secondly, why of all the countries in European and the Mediterranean region is it only in Italy that noodles have become a staple food in the diet?

The fact that a filamentous noodle, *rishta*, was known in Bukhara in the 10th century, and that Bukhara lies on the silk road in Central Asia were noted by Rosenberger and Ishigi.[134] We may recall that in Idrisi's report of +1154 cited earlier, the original Arab term for 'cibo di farino in forma di fili (filamentous noodle)' is *itriyya* which, as we have seen, is practically synonymous with the *rishta* in the Bagdad cookbook that was translated as spaghetti. Thus, it is entirely possible that the idea of filamentous noodles may have travelled from China to Europe along the same route as the art of paper-making through Central Asia and the Arab world.[135] Ishige is of the opinion that while paper-making was disseminated by transmission, noodle was disseminated by diffusion. It was presumably the Arabs who brought noodles to Sicily after it fell under their domination. From Sicily it gained entry into the Italian peninsula. This leads us straight to the second question. Why did noodles flourish only in Italy? Apparently the Turks and Arabs never became overly fond of noodles.[136] Nor did the majority of the peoples of the Mediterranean world that they had conquered. Noodles are of only peripheral interest to the countries that lie on the rim of the Mediterranean sea, except for Italy.

A possible answer is suggested by an observation of Gregory Blue, who has drawn attention to a passage in Polo's account of the kingdom of Fansur which he visited on his way back to Italy.[137] Talking about the flour from the sago palm, Polo said, 'And they make of it many cakes of dough which are very good to eat and other eatables of theirs.'[138] Blue found that in the original Latin Z manuscript, the line actually reads:[139] '*& fiunt ex ea lagana & diuerse epule que le pasta fiunt que sunt ualde bone.*' He suggests an alternative translation:[140] 'And they make of it lasagna and other dishes made of pasta which are exceedingly good.

---

[133] Sabban (1989b), pp. 99–100, 'Conclusions'. We agree, of course, with her conclusion that Marco Polo did not bring 'pasta' to Italy from China.

[134] Rosenberger (1989), p. 93, and Ishige Naomichi (1995), pp. 361–72.

[135] Rosenberger, *ibid.* observes that the word *rishta* was borrowed by the Arabs from Sassanid Persia following the Islamic conquest, and that pasta (which would include noodles) could have been transmitted to Persia via the Silk Road through Khurassan. Ishige Naomichi, *ibid.* takes the concept a step further. He stresses the connection between Avicenna and Bhkhara, and the possibility that 'noodles' had travelled the same route as paper making. For the transmission of paper making, cf. Tsien, *SCC* Vol. v, Pt 1, pp. 293ff.

[136] As shown by books on Turkish and Middle Eastern cuisine such as those of Roden (1974), Najor (1981), Arto der Haroutunian (1984) and Algar (1991). Although *rishta* and *itriyya* are mentioned in some of them, they constitute only a minor part of the wealth of recipes presented. Nor are noodles (or pasta) mentioned in Andrew Dalby's history of food and gastronomy in Greece nor in books on Indian cookery such as Chakravarty (1959) and Panjabi (1995). Thus, we are mystified by the assertion of Franke (1970), p. 16, based on a study of the non-Chinese words in the *Yin Shan Chêng Yao* (+1330), that pasta foods such as 'noodles' and 'ravioli' were 'originally non-Chinese, and may have been introduced into China from the Near East in the Mongol period only'. His views were recently reiterated by Wood (1995), p. 79. Such a view contradicts all the facts that we have gathered on the long history of pasta foods in China in this chapter.

[137] Blue (1990), pp. 41–5.   [138] Moule & Pelliot (1938), vol. I, p. 170.
[139] *Ibid.*, p. lxviii.   [140] Blue (1990), p. 44.

Blue goes on to say,

'Let us consider for a moment the history of this *lagana*. It seems to have originated from the ancient Greeks', referring to a pastry made of flour and oil. The dish was apparently introduced in antiquity into Magna Graecia and etymologically it was from this word that the Latin *laganum* and later *lasania*, and the Italian *lasagna* (still *lagana* in Calabrian dialect) derived. Horace told of a dish of leeks, chick peas and *laganum* in one of his satires, and Apicius uses the terms *pasta* and *laganum* several times in his +3rd century cookbook, though as with various ancient culinary terms the semantic content of *laganum* is not really clear. Vehling's gloss in Apicius has it that *laganum* was a small pancake made of flour and oil.[141] According to Buonassisi the ancient Romans served it in strips, the dough having been baked before the other ingredients were added.[142] It was apparently at some unspecified point in mediaeval times that the innovation of boiling the dough rather than baking it was made. The earliest attested use of the term *laganum* in the Italian vernacular seems to have been made by Marco's contemporary, Iacopone da Todi (ca. 1240–1306).

Blue's commentary suggests that there was a separate tradition in Italy of making a type of of baked 'pasta' that had nothing to do with China. This tradition had its genesis in the Imperial days of Rome. Although a known food, 'pasta' was not a major constituent of Roman cuisine. It survived without attracting much attention through the Dark Ages to the Renaissance. When the concept of noodles arrived, it merged easily into the indigenous circle of pasta foods, such as *laganum*. Even so, filamentous noodles did not become a conspicuous part of the food system until the invention of the Tagliatele press in the 15th century. In fact, they did not become popular until the 18th century when spaghetti came into contact with tomato sauce.[143] This was surely one of the most felicitous unions in culinary history. The rich flavour and zest of the new world fruit gave the staid old-world pasta a new lease of life. The popularity of filamentous noodles soared, and spaghetti with tomato sauce has remained the most typical yet celebrated of Italian foods until the present day. We regret to say that Marco Polo had nothing to do with this happy development. It was strictly a marriage between old world staidness and new world charm.

(vii) *Production and usage of gluten*

Westerners visiting China in recent years may have had the experience of enjoying an excellent dinner at a Buddhist Monastery or a gourmet vegetarian restaurant. What is unusual and surprising about such a meal is that all the dishes have names similar to those that one would normally encounter in a regular restaurant, such as *thien suan yü* 甜酸魚 (sweet and sour fish), *mo ku chi phien* 蘑菇雞片 (chicken slices with mushrooms), *chhao hsia jên* 炒蝦仁 (stir fried shrimps) etc. With the eyes closed, one could easily be beguiled into thinking that fish, chicken or shrimp were actually

---

[141] Vehling, ed. tr. (1977), p. 289; cf. also the translation by André (1965), p. 111.
[142] Buonassisi (1977), p. 7. This is also the opinion of Algar (1991), p. 173.
[143] McNeill (1991). The wonderful Italian cuisine that we know today would be impossible without the tomato. McNeill, p. 41, points out that 'It is not accidental that Italy today produces almost half as much tomatoes as all of North America.'

being eaten.[144] And yet no animal products are used in these dishes. Everything served is prepared from materials of vegetable origin. How is this possible? The answer is simple. The secret of these 'mock' fish, chicken or shrimp dishes lies in a skilful blending of *tou fu* or *mien chin* (wheat gluten) with starch and root vegetables such as the Chinese yam, suitably flavoured with soy condiments, mushrooms and spices. Without gluten many of these vegetarian gourmet creations would not have been possible. It is gluten that supplies the texture and chewiness that make the 'mock' pretensions to fish, meat and shrimp believable.

Although gluten is probably well known to all peoples who use flour as a staple food, the Chinese were the only ones who had extracted and used it separately as a substitute for meat and other animal products, and thus raised vegetarian fare to a new height of culinary inventiveness. How did all this happen? What is the origin of gluten? In regard to the latter question, the great naturalist Li Shih-Chên in about +1596, had this to say:[145]

Gluten is sweet, cooling and non-toxic . . . It is made by kneading and washing wheat flour or bran in water. The ancients have no knowledge of it, yet today it is a most important component of vegetarian food. It is very good boiled, but nowadays people often fry it in oil, which would turn it into a 'heating' type of food.

He then quotes Khou Tsung-Shih, c. +1116, who states,[146] 'When white flour is chewed, it becomes gluten. Its stickiness can be used to trap insects and birds.' From these quotations, one might be led to think that gluten was discovered sometime during the Sung. Actually, gluten was probably isolated at least 500 years earlier. The *Ko Chih Ching Yuan* 格致鏡原 (Mirror of the Origin of Things), +1717, quotes a passage from the *Shih Wu Kan Chu* 事物紺珠 (Notes on Miscellaneous Affairs) by Huang I-Chêng, a Ming work no longer extant, to the effect that[147] '*mien-chin* (gluten) was not known in antiquity; it was introduced by the emperor Liang Wu-Ti', who reigned from +502 to +549. Although this claim has not been corroborated by other references, we do believe that the attribution to Liang Wu-Ti has a reasonable chance of being correct. As we shall soon show, the making of gluten is described in the *Chhi Min Yao Shu*, +544, which indicates that it must have been discovered at least fifty to a hundred years earlier.

---

[144] A rather extreme example of the extent to which the vegetarian chefs went to make their creations simulate animal-derived dishes is given in the celebrated Ming novel *Ching Phing Mei* (The Golden Lotus). In ch. 39, p. 294, there is a scene describing several nuns and devout ladies enjoying a vegetarian repast. The dishes resemble their meat counterparts so well that one of the old ladies with failing eyesight had to be convinced that the food was not forbidden.

The Moon Lady put the cakes on small gilded plates and offered them first to Aunt Yang and then to the two nuns. 'Take some more, old lady' she said to Aunt Yang, but the old lady protested that she had had enough. 'There are some bones on the plate', Aunt Yang said. 'Please, sister, take them away. I might put them in my mouth by mistake.' This made everyone roar with laughter. 'Old lady', said the Moon Lady, 'This is vegetarian food made to look like meat. It has come from the temple, and there can't possibly be any harm in eating it.' (tr. Egerton (1939), pp. 179–80.)

[145] *PTKM*, ch. 22, p. 1455.   [146] *Ibid.*; cf. also *CLPT*, ch. 25, p. 14b.
[147] *Ko Chih Ching Yuan*, ch. 25, p. 1b; *SKCS*, **1031**, p. 345.

To understand how it came about, we need to revisit Chapter 82 of the *Chhi Min Yao Shu* and look at recipes 10, *shui yin* 水引 (drawn pasta), and 11, *po tho* 餺飥 (thin drawn pasta) listed in Table 43 (p. 470). Based on the method of preparation, there is little doubt that *shui yin* is a filamentous pasta drawn by hand. The identity of *po tho*, however, has puzzled scholars for centuries. We have translated it as 'thin drawn pasta' since it is made by further processing of the dough used for making drawn pasta (*shui yin*). Recipe 11 actually says:[148]

Po-tho: Shape the dough [for making *shui yin*] until it is as thick as a thumb. Cut it into two-inch sections. Soak them in water. Rub each piece against the side of the pan until it is as thin as possible. Cook it in vigorously boiling water. It not only looks beautifully white and shiny, but also feels luscious and slippery in the mouth.

Let us examine the procedure carefully. When the dough is immersed in water and rubbed against the side of the pan, it is, in effect, being kneaded under water. In this process the starch will be washed away. Eventually the thin film that is left would be almost pure gluten.[149] No wonder Chia Ssu-Hsieh made a special point to call attention to the whiteness of the product and the luscious feel it imparts in the mouth. In fact, we have here the first description for the extraction of gluten from wheat flour dough. This fact, however, was not recognised by most contemporary scholars who continued to regard *po tho* simply as another kind of *ping* or pasta.[150] Its identity is still a matter of controversy today.[151] And yet if someone in the +6th century had continued to make *po tho* in the manner described in the *Chhi Min Yao Shu*, sooner or later he would have realised that the product he got had properties quite different from a soup pasta (*thang ping*) or a rope pasta (*so ping*). The product was probably well known to chefs and those had worked directly in the kitchen. Eventually, it was given the name *mien chin*, i.e. the sinew of flour, since it represents the quintessence of the flour that was left after the weaker parts (the starch) were washed away.

Nevertheless, pasta products construed to be *po tho* continued to be mentioned in the food literature. A *pu tho* made from the drug *ti huang* 地黃 and flour listed in the *Shan Chia Chhing Kung* is said to be a remedy to dispel intestinal worms, but there is no description of how it was prepared.[152] Nor is there any detail on the way a red silken *pu tho* is made in a recipe in the *Shih Lin Kuang Chih*, +1283, except that the flour

---

[148] *CMYS*, ch. 82, p. 510.
[149] This is first pointed out by Khuei Ming (*1988*). I have myself made *po to* in our home kitchen according to directions of Recipe 11 of the *CMYS* (Table 43). There is no question that the product is gluten in the form of thin ribbons.
[150] According to the Sung writer Chhêng Ta-Chhang 程大昌, at one time *po tho* was construed as a thin flat dough shaped by pressing between two palms. Hence, it was also called *chang tho* 掌托 (i.e. held by hand). Later when the rolling-pin and knife became available, there was no need to manipulate the dough by hand, the product was then called *pu tho* 不托 (no holding by hand), cf. *Yen Fan Lu* 演繁露, cited in Miao chhi-Yü ed. (*1982*), p. 515, note 18.
[151] Li Pin (*1991*), suggests that *po tho* was one of the foods transmitted to China from Central Asia. This view is refuted by Wang Lung-Hsüeh (*1991*).
[152] *SCCK*, p. 19. *Ti-huang* is the root of *Rehmannia glutinosa*, a common plant in North China; cf. *PTKM*, ch. 16, p. 1019.

was mixed with a shrimp juice to form the dough.[153] The term *po tho* is still seen in two recipes in the *Chü Chia Pi Yung*, Yuan, one for a yam flour noodle and the other one a flat pasta punctuated with holes.[154] But these are apparently remnants of an obsolete terminology which would soon disappear from the food vocabulary.

The word *mien chin* was already a well-established term in the Northern Sung. We first encounter it in a passage about iron and steel from Shên Kua's *Mêng Chhi Pi Than* (Dream Pool Essays), +1086, which says,[155] 'Steel is to iron as *mien chin* (gluten) is to *mien* (flour). It is only after thoroughly washing the dough that gluten is revealed.' *Mien chin* is seen again in the *Lao Hsüeh An Pi Chi*, c. +1190:[156] 'Bean curd (*tou fu* 豆腐), gluten (*mien chin* 麵筋) and cow's milk (*niu-ju* 牛乳) were soaked in honey. Most of the guests could not appreciate them, but Su Tung-Pho had a sweet tooth, and ate them to his heart's content.'

At the same time, gluten was also being represented by another word, *fu* 麩, which originally meant bran. It is not known when or how *fu* became an alternative word for *mien chin*. The *Thai Phing Yü Lan*, +983, cites *Chhang Chi Chieh Ku* which says, 'Finely refined (washed) wheat flour is called *fu*.'[157] The *I Chien Chih*, +1185, relates that in the northern section of Phing Chiang city, the Chou family sold *fu* and *mien* (pasta) for their livelihood.[158] *Fu* as another name for gluten was apparently widely accepted during the Southern Sung. The *Mêng Liang Lu* lists a casserole of *fu* and lamb, a dragon (shrimp) *fu*, parched *fu* with five flavours and a grilled *fu* among the dishes served in a pasta food (*mien shih*) restaurant.[159] The *Wu Lin Chiu Shih* mentions stalls that sold *fu* and *mien* in the market in the capital, Hangchou.[160]

The first detailed description for preparing gluten is given in the *Shih Lin Kuang Chi*, +1325, under the title *Hsi mien chin* 洗麵筋 (Washing gluten):[161]

Take one *phing* 枰 of white flour and four ounces of salt and mix them with enough warm water until a dough is formed. Allow it to sit for a while. Then place the dough in a basin of cold water and squeeze it (as if washing it) until gluten is obtained. Repeat the process in a fresh basin of cold water until all the starch is washed off. Divide the gluten and shape the pieces into small balls. Then steam them.

During the Sung, gluten had so established itself in the food system that its virtues were celebrated in verse. Wang Yên wrote a poem in praise of *fu chin* 麩筋 with the following lines:[162]

---

[153] *Shih Lin Kuang Chi, Pieh Chi*, ch. 4, by Chhen Yuan-Ching 陳元靚, +1325. cf. Shinoda & Seiichi (*1973*), p. 270.
[154] *CCPY*, pp. 116–17. In the second recipe the flat pasta pieces are impregnated with sheep tallow. When the pasta is boiled, the fat melts and the pieces are left with holes in them. The same recipes are repeated in the early Ming *Tuo Nêng Phi Shih*.
[155] *MCPT*, ch. 3, item 56, p. 42.
[156] *Lao Hsüeh An Pi Chi*, ch. 7, p. 4a. This is one of the few references to cow's milk consumed as food during the Sung.
[157] *TPYL*, Wang Jen-Hsiang ed. (*1993*), p. 397, no. 589.   [158] *I Chien Chih*, cited in Khuei Ming (*1988*).
[159] *MLL*, ch. 16, p. 136.   [160] *WLCS*, ch. 6, p. 136.
[161] *Shih Lin Kuang Chi, Pieh Chi*, ch. 4, p. 24a. The *phing* 枰 is probably a flat *tou* which in Sung times is equal to about 6.6 litres.
[162] Wang Yen 王炎 (+1138–1218), *Shuang Chhi Chi* 雙溪集 (Double Rivulets Collection), cited by Liu Tsün-Ling (*1985*). The original reads: 色澤似乳酪，味勝雞豚佳。

It has the colour of fermented milk,
And a flavour superior to chicken or pork.

And the Taoist monk, Ko Chhang-Kêng said of *fu chin*:[163]

The silken *tou fu* (bean curd) is said to be delectable,
But it is *fu chin* (gluten) that has the cleanest taste.

A considerable amount of poetic license is no doubt invoked in these statements, but they are, nevertheless, evidence of the high regard with which gluten was held among the literati of the Sung. Very little additional information is available on how gluten was prepared after the *Shih Lin Kuang Chih*. The *Yin Shih Hsü Chih*, +1350, says tersely, '*Mien chin* is obtained by washing *fu* 麩 (wheat bran). It is sweet and cooling, but if it is fried in oil, its character changes to heating. It is not easy to digest, and should be avoided by children and those who are ill.'[164] Presumably, this was one of Li Shih-Chên's sources. The next statement on the isolation of gluten after the *Pên Tshao Kang Mu* is in the late Chhing *Sui Hsi Chü Yin Shih Phu*, 1861, which adds an interesting usage for gluten.[165] 'If one should inadvertently swallow a coin, take gluten, roast it without destroying its shape, ground it and take it with boiled water. If the coin is caught in the throat, it will be coughed out. If it is in the stomach, it will be eliminated with the stool.'

It is possible that Marco Polo had heard of gluten during his travels. He is reported to have said:[166]

There is another kind of religious men according to their usage who are called in their tongue 'sensin', who are men of very great abstinence according to their custom and lead their life so very hard and rough as I shall tell you. For you may know quite truly that in all the time of their life they eat nothing but semola and bran, that is the husks that are left from wheat flour. And they prepare it as we prepare it for swine; *for they take that semola, that is bran, and put it in hot water to make it soft and leave it to stay there some time till the whole heart or grain is removed from the husk*, and then they take it out and eat it washed like this without any substantial taste.

Even though the description is somewhat garbled, it does convey the impression that what the ascetic monks ate was gluten. As indicated by references to it in the popular novels *Hsi Yu Chi* 西游記 (Journey to the West), +1570, of the Ming and *Ju Lin Wai Shi* 儒林外史 (The Unofficial History of the Scholars), +1740, of the Chhing, gluten steadily gained acceptance outside the circle of Buddhist devotees.[167] Recipes for cooking gluten are found in the major food canons of the Yuan to the Chhing. They are summarised in Table 44. We see that gluten can be boiled, fried, pickled or shredded. The cooked gluten can be blended with other

---

[163] Ko Chhang-Kêng 葛長庚, Sung, *Chhiung Kuan Hsien Sêng Chi* 瓊琯先生集 (Collection of the Excellent Stone Master), cited by Liu Tsün-Ling, *ibid*., which says: 嫩腐雖云美，麩筋最清純。
[164] *Yin Shih Hsü Chih*, ch. 2, p. 15 .   [165] *Sui Hsi Chü Yin Shih Phu*, p. 29.
[166] Moule & Pelliot (1938), p. 191. Cf. also Yule & Cordier (1903), p. 303, and Latham (1958), p. 111. 'Sensin' may be a transliteration of *shen hsiang* 神仙. This observation would tend to support the idea that Marco Polo actually did go to China, even though he did not mention the Great Wall or the custom of drinking tea.
[167] *Hsi Yu Chi*, ch. 100, p. 1126, ch. 82, p. 937 and ch. 68, p. 774. *Ju Lin Wai Shi*, ch. 2, p. 30.

Table 44. *Recipes for cooking gluten from the Yuan to the Chhing dynasties*

Literature (source)	Gluten (named as)	Cooking (method)	Reference (page[a])
I Ya II 易牙遺意 Yuan	*Fu* 麩	Pickled, pan-fried	52
Chü Chia Pi Yung 居家必用 Yuan	*Mien Chin* 麵筋	Dried venison, mock fermented soybeans, mock fermented soybeans, barbecued meat jerky	132–3
Yin Chuan Fu Shih Chien, +1591 飲饌服食牋	*Fu* 麩	Pickled pan-fried	148–9
Shih Hsien Hung Mi 食憲鴻秘 +1680	*Fu* 麩	Picked, mock meat, deep-fried, smoked	36–7
Yang Hsiao Lu 養小錄 +1698	*Mien chin* 麵筋	Pan-fried, smoked	21
Sui Yuan Shih Tan 隨園食單 +1790	*Mien Chin* 麵筋	Fried, boiled, shredded	107

[a] Page references are for editions of the *CKPJKCTK* series.

culinary ingredients to give a variety of dishes. Some of these dishes are intimated in a vegetarian menu listed in the 19th century compendium *Thiao Ting Chi*[168] and in recipes for cooking gluten in the 20th-century *Su Shih Shuo Luo*.[169]

Even so, until the 1950s, good-quality vegetarian dishes prepared with gluten have been the preserve of Buddhist monasteries and a few gourmet restaurants. But in the latter part of the 20th century a great deal has been done to make such processed foods available to the general populace. With the successful application of modern technology to the mass production of such foods, gluten-based products have become easily available not only in China but also in overseas Chinese communities in Asia and America. Indeed, *mien chin* has come a long way from the prototypic *po tho* of the *Chhi Min Yao Shu* to the refrigerated or canned versions of gluten-based mock abalone, shrimp, chicken etc. sold in oriental markets in America and Europe.[170]

[168] *Thiao Ting Chi*, pp. 119 and 86. [169] *Su Shih Shuo Luo* by Hsüeh Pao-Chhên (c.1900), p. 43.
[170] We have tried several of these products. The refrigerated mock shrimp and tinned mock abalone are particularly good.

(f) TEA PROCESSING AND UTILISATION

We have now traced the history of the major processed foods and drinks that have helped to shape the character and flavour of the Chinese cuisine. Although most of them are well known in East Asia, only a few can be said to have made a noticeable impression on the peoples of the West. But there is one exception – tea. Tea is prepared from the young leaves of the subtropical plant *Camellia sinensis* (L.) O. Kuntze (Figure 121). It is the only product from China in the 'food and drink' (*yin-shih* 飲食) category that has gained wide acceptance throughout the world. Tea is today a truly international beverage, not just a Chinese or East Asian drink. China has long ceased to be the world's largest producer of tea (Table 45). The honour has belonged to India since the early years of the twentieth century.[1] The most fervid drinkers of tea today are not Chinese, but Irish and English (Table 46). It is said that in the city of London alone, twenty million cups of tea are drunk every day.[2]

There is, however, little doubt that the original home of tea is China.[3] But when a decoction or soup of tea leaves was first used as a medication or a drink is a matter shrouded in mystery. The uncertainty about its origin has spawned two legends. According to the first legend, tea was one of the plants selected by *Shên Nung* 神農 (The Heavenly Husbandman), one of the three 'primordial sovereigns' (*san huang* 三皇) of Chinese prehistory, who has been regarded as the 'father' of agriculture, animal husbandry, pharmacy and medicine. We have encountered the version of *Shên Nung*'s exploits given in the *Huai Nan Tzu* (The Book of the Prince of Huai Nan), c. −120, in *SCC* Volume 6, Part 1.[4] There is another version of this story which says, '*Shên Nung* tasted a hundred plants in one day, of which seventy were toxic. He found

---

[1] Cf. Chhên Tsung-Mao ed. (*1992*), *Chung Kuo Chha Ching* (*CKCC*), pp. 726 and 732. However, in 1987 the area under tea cultivation in China, 15.67 million *mou*, was about two and a half times that in India, 6.21 million *mou*. This would suggest that the Indian tea industry is about three times more efficient than its Chinese counterpart.

[2] Smith, Michael (1986), cf. 'Note' in the front of the book. We suspect this may be a slight under-estimation. Smith does not state how he arrived at his figure. According to the *CKCC*, p. 755, the UK imported 162.7 thousand tons of tea in 1988. With a population of 56 million, the consumption per head per year is, therefore, 2.9 kg. It takes about a 2.2 g. of tea leaves to make a cup of tea. This comes to 3.6 cups of tea per person per day. Assuming the population of London to be about 10 million, it may be closer to the truth to say that almost 35 million cups of tea are drunk in London every day.

[3] The discovery of wild tea trees in Assam in the 1820s, cf. Bruce (1840), had led some Western scholars to claim that the original home of tea was not China but India. As pointed out by Gardella (1994), p. 22, this is strictly a non-issue since there is no evidence that 'tea played any role in Indian agriculture and the Indian diet prior to that time'. For further discussions on this isssue cf. Chhên Chhuan (*1984*), pp. 22–44 and (*1994*), pp. 2–14, *CKCC*, pp. 5–7, and Ho Chhang-Hsiang (*1997*). Cf. also Li Kuang-Thao *et al.* (*1997*) and Huang Kuei-Shu (*1997*) for the current status of wild tea trees in Southwestern Yunnan.

[4] *SCC* Vol. VI, Pt 1, p. 237.

Fig 121. *Camellia sinensis (L) O. Kuntze*, from Ukers (1935) 1, front page.

Table 45. *World production of tea in 1950 and 1988*

Country	1950	1988
China, mainland	62.2	545.4
China, Taiwan	9.7	23.6
India	278.5	701.1
Sri Lanka	139.0	228.2
Bangladesh	23.6	42.6
Indonesia	35.4	135.6
Japan	44.1	89.8
Turkey	0.2	153.2
Kenya	6.8	164.0
Soviet Union	—	120.0
Others[a]	14.1	275.3
World total	613.6	2,478.8

Source: CKCC (1992), adapted from table 2, pp. 728 & 732.
All figures are in thousand tons.
[a] Include Iran, Vietnam, Malawi, Tanzania, Luanda, South Africa, Zimbawe, Argentina and Brazil.

Table 46. *World consumption of tea* per capita[a], *1984–1987*

Country or region	Amount (kg)		Country or region	Amount (kg)	
	1984–6	1985–7		1984–6	1985–7
Qatar	3.74	3.21	Afghanistan	0.63	NA
Ireland	3.03	3.09	South Africa	0.56	0.53
United Kingdom	2.94	2.81	Sudan	0.55	0.56
Iraq	2.72	2.51	India	0.55	0.55
Turkey	2.65	2.72	Denmark	0.46	0.45
Kuwait	2.55	2.23	Sweden	0.36	0.35
Tunisia	1.81	1.82	USA	0.36	0.34
New Zealand	1.77	1.71	China	~ 0.35 (1988)	
Hong Kong	1.69	1.63	Switzerland	0.29	0.27
Saudi Arabia	1.69	1.40	Germany, Federal Republic of	0.26	0.25
Egypt	1.54	1.44			
Bahrain	1.52	1.45	Algeria	0.24	0.22
Sri Lanka	1.43	1.41	Norway	0.21	0.22
Syria	1.43	1.26	Tanzania, United Republic of	0.20	0.21
Australia	1.31	1.22			
Jordan	1.12	1.12	Finland	0.17	0.19
Iran	1.05	NA	France	0.17	0.17
Morocco	0.99	0.97	German Democratic Republic	0.16	0.16
Japan	0.94	0.99			
Chile	0.91	0.93	Austria	0.16	0.15
Pakistan	0.90	0.86	Czechoslovakia	0.14	0.14
USSR	0.85	NA	Belgium/Luxembourg	0.13	0.13
Poland	0.81	0.86	Italy	0.06	0.06
Kenya	0.80	0.76	Portugal	0.02	NA
Canada	0.68	0.62	Spain	0.02	NA
Netherlands	0.65	0.65	Thailand	0.01	0.01

[a] From International Tea Committee (1989).
NA, not available.
From IARC Monographs Vol. 51, p. 216.

tea (*chha* 茶) to be an effective antidote.'⁵ This was how tea came to be employed as a medication. The effect of tea was found to be so pleasant that it was later adopted as a beverage. This legend was cited by Lu Yü, author of the famous Tea Classic (*Chha Ching* 茶經), the first book ever written about tea in China.

The second legend was probably fabricated by the followers of Chhan 禪 (better known in the West under its Japanese name, Zen) Branch of Buddhism. It is said that Bodhidharma, an Indian prince who converted to Buddhism and became a Buddhist saint, came to China in about +526 to spread the doctrine that perfection can be achieved through inward meditation. For years he prayed and meditated without sleep. One day, he caught himself on the verge of falling asleep. He was so annoyed with himself that he cut off his eyelids and threw them to the ground. Lo and behold, they immediately took root and grew into the world's first tea plant, whose leaves enabled meditators to remain alert for long periods without sleep.⁶ This charming story has been widely circulated among believers for generations. It has even inspired a modern Chinese poet to write:⁷

> It was told that
> When the monk awoke,
> Like crisscrossing footprints
> On deserted mountains and snowy hills,
> Traces of his dream emerged before his very eyes.
> Feeling repentent and restless,
> He snapped the thick, sleep-tempting eyelashes
> From his thick bearded face.
> Accordingly, in a single night,
> Shrubs of bitter tea began to grow;
> They can restrain the hot tempers of the secular man
> And love-cravings of the religious monk.

Although we cannot take these legends seriously, they do remind us of the antiquity and the religious connotations of tea drinking in Chinese civilisation. Unfortunately, the legend of Shên Nung has become so deeply ingrained in the culture of tea in China that it is often cited in support of the notion that tea was discovered as a drink at about the year −3000.⁸ Perhaps, the legend does contain a

---

⁵ It is said that this passage occurs in the *Shên Nung Pên Tshao Ching*, but no one has been able to find it in the surviving text of this classic. Another alleged source is the *Shên Nung Shih Ching* 神農食經 (Shên Nung's Food Canon), which is cited by no less an authority than Lu Yü in his *Chha Ching*, p. 42. This book is no longer extant. Unfortunately, the fable has been repeated so often in China in recent years that many modern publications on tea have continued to treat it as if it were a matter of fact rather than fancy. Cf. for example Chhên Chhuan (*1984*), p. 2, and Wang Shan-Tien (*1987*), p. 487. Such statements have been criticised by Chou Shu-Pin (*1991*). It is ironic that in the same issue of the *NYKK*, both the articles before and after Chou's paper cite the story of Shên Nung as evidence that tea drinking started 4,700 years ago in China – cf. Chao Ho-Thao (*1991*) and Ou-Yang Hsiao Thao (*1991*). In fact, the legend is still widely circulated in the countryside as shown in Wu Chia-Khuo (*1993*).

⁶ Blofeld, John (1985), p. 1; Maitland (1982), p. 8.   ⁷ Chang Tsho (1987).

⁸ Ukers (1935) II, p. 501, 'A Chronology of Tea', begins with the date −2737 as the year when the legendary Shên Nung discovered tea. Hardy (1979), p. 120, lists −3254 as the year when tea was first planted in China. Evans (1992), p. 10, suggests that the history of tea began in −1600. Cf. also the chronology given in the *CKCC*, p. 677.

grain of truth. If so, it may reveal itself when we examine the early literature on tea in China. Our purpose in the present chapter is threefold: first, to discuss the etymology and literature of tea in China; second, to trace the development of the technology of tea processing through the ages, and third, to examine the Chinese view of the effects of tea drinking on human health. The botanical aspects of tea, its ecogeography, genetics and horticulture as well as its dissemination to the West is being dealt with in the Section on Horticulture in this series currently under preparation by Dr Georges Metailie.

(1) Etymology and Literature of Tea

The word *chha* 茶, which for many centuries has been the term denoting tea, did not become established in the Chinese vocabulary until the Thang dynasty. Before the Thang, many scholars believe, tea was represented by several characters, of which the oldest and most important is *thu* 荼.[9] The character *chha* 茶 was obtained by simply eliminating one stroke from the character *thu* 荼. We have already encountered the word *thu* in the verse from the *Shih Ching* cited on page 457.[10] *Thu* is also mentioned in two of the poems in the *Chhu Tzhu* (The Songs of the South), c. −300.[11]

Other scholars, however, maintain that *thu* had nothing to do with tea, *chha*.[12] The *thu* referred to in the *Shih Ching* (and in the *Chhu Tzhu*) is, in fact, a bitter vegetable (*khu tshai* 苦菜) such as the sowthistle or smartweed. This is the view expounded by Lu Chi in his annotations to *thu* in the *Mao Shih Tshao Mu Niao Shou Chhung Yü Su* (An Elucidation of the Plants, Trees, Birds, Beasts, Insects and Fishes mentioned in the Book of Odes), c. +245.[13] The word *thu* also occurs in the title of an official in the *Chou Li* (Rites of the Chou), namely, *Chang Thu* 掌荼 (Superintendant of *Thu*), whose duty it was to ensure that the *thu* plant was available for the funeral rites at the royal household.[14] There is no indication that tea, as we understand it,

---

[9] Bretschneider (1892), Vol. II, Pt II, pp. 130–1; Ukers (1935) I, pp. 492–3; Bodde (1942); Chhên Chhuan (*1984*), pp. 2–9; Li Fan (*1984*), p. 158.

[10] In the verse cited, i.e. M 237 (W 240) *thu* is translated by Waley as sowthistle. *Thu* also occurs in M 35, in which it is again translated as sowthistle by Waley (W 108):

    Who says that sowthistle is bitter,
    It is sweeter than shepherd's-purse.

But these are not the only poems in the *Shih Ching* that mention *thu*. We find *thu* interpreted as a weed in M 291 (W 158), as rush wool in M 93 (W 36) and as rush flower in M 155 (W 231). These 'lesser' manifestations of *thu* are of no consequence for our current discussion.

[11] *Chhu Tzhu*, Fu Hsi-Jên (1976) ed.; ch. 4, *Pei Huei Fêng* 北回風, p. 122; ch. 17, *Shang Shih* 傷時, p. 260. Hawkes (1985), p. 180, line 13 and p. 315, line 5 translates *thu* in each case as bitter herb.

[12] Dudgeon, John, (1895), p. 4; Blofeld, J. (1985), p. 2; Shih Yeh Thien I 矢野田一, cited in Chhên Chhuan (*1984*), p. 5.

[13] *Mao Shih Tshao Mu Niao Shou Chhung Yü Su*, Pt I, p. 14, Chung Hua, repr. (*1985*).

[14] *Chou Li*, p. 176; cf. Chêng Hsüan's commentary in *Chou Li Chu Su*, p. 249.

was ever an integral part of a funeral ceremony in ancient China. The *thu* in this case could very well have been the *thu* of the *Shih Ching*. In fact, *khu tshai* (bitter vegetable) is listed as a first-class medication in the *Shên Nung Pên Tshao Ching* (The Pharmacopoeia of the Heavenly Husbandman), which was compiled during the Han but was largely based on late Chou material. The text says that *khu tshai* 苦菜 is also called *thu tshai* 荼菜;[15] in other words *thu* is a bitter vegetable.

Actually, both schools of thought are correct. The problem is that the word *thu* was applied in ancient China to at least two types of plants, one a bitter vegetable (*tshai* 菜) such as sowthistle, smartweed etc., and the other a woody plant (*mu* 木) such as tea. A good illustration of this dichotomy is found in the *Erh Ya* (Literary Expositor). Chapter 13 (Herbaceous Plants) says, '*thu* is a bitter vegetable (*thu khu tshai* 荼苦菜)', but chapter 14 (Woody Plants) tells us, '*chia* is bitter *thu* (*chia khu thu* 檟苦荼)'.[16] What is most interesting, however, is Kuo Pho's commentary (c. +300) on the latter statement. Referring to *chia* he says,[17]

> The plant is a small tree like the *chih tzu* 梔子 (*Gardenia jasminoides* Ellis). The leaves grown in the winter may be boiled to make a soup for drinking. Nowadays, those that are gathered early are called *thu* 荼. Those that are gathered late are called *ming* 茗. Another name for them is *chhuan* 荈. The people of *Shu* 蜀 (present Szechuan) call them *khu thu* 苦荼 (bitter *thu*).[18]

This is the first description we have of the tea plant and how its leaves were used. But the earliest record of the word *thu* used unequivocably in the sense of tea is proabably in the *Thung Yüeh* 僮約 (A mock contract between a servant and his master), dated at −59. Among the duties of the slave were to prepare tea (*phêng thu* 烹荼) when guests arrive, and to buy tea from Wu-yang (*Wu-yang mai thu* 武陽買荼).[19] The tradition of serving tea to guests is a hallmark of Chinese culture. To buy tea from Wu-yang, away from the home of the master,[20] would suggest that tea was already a processed food product with a long shelf-life and could be shipped from place to place without deterioration. In this context, *thu* can hardly be anything else but tea.

The reference to tea in the *Thung Yüeh* is corroborated by several passages in the *Hua Yang Kuo Chi* 華陽國志, +347, a historical geography of Szechuan down to

---

[15] *SNPTC*, Tshao Yuan-Yü ed. (*1987*), p. 325.
[16] *Erh Ya*, ch. 13, p. 24b; ch. 14, p. 40b.   [17] *Ibid*. tr. Derk Bodde (1942), mod. auct.
[18] Karlgren (1923), *Analytic Dictionary*, no. 1322, p. 373 gives two ancient pronunciations for the character *thu*: '*d'uo*, sowthistle, bitter herbs, bitter; *d'a*, leaf from which a bitter drink is made, tea plant, tea'.
[19] Cf. Hsu, Cho-Yun (1980), pp. 231–4 for a complete translation of this contract in English. It is discussed by Goodrich and Wilbur (1942), and we have already referred to it in Chapter (*d*), pp. 294–5. Wu-yang is also written as Wu-tu 武都 in some texts.
[20] The contract was drawn up in Chêngtu, Szechuan. According to Chhên Chhuan (*1984*), p. 6, Wu-yang is a tea producing area near Phêng-hsien, which is about 30 km northwest of Chengtu. Chou Wên-Thang (*1995*) has argued that the *thu* in the contract is not tea, but smartweed, a kind of vegetable. This view has been vigorously refuted by Fang Chien (*1996*). We tend to agree with the latter since vegetables, in general, are usually grown and consumed locally. It would not make sense to have to go 30 km away to buy it.

+138, which notes that since antiquity, tea, *chha*, had been a product sent as tribute to the Emperor's court.[21] Among the well-known tea producing localities are Wu-yang 武陽, which we have just met in the *Thung Yüeh*, and Fu-ling 涪陵 which we shall come across again in reference to the *Chhi Min Yao Shu*, +540.

The next reference to tea drinking is in the Biography of Wei Yao in the *San Kuo Chih* (History of the Three Kingdoms), c. +290. Wei Yao was a historian at the court of Sun Hao, King of Wu, who is remembered chiefly for his drunkenness and debauchery. Wei Yao was invited often by the King to day-long feasts at which each guest was expected to drink at least seven *shêng* (pints) of wine. Because of his poor health Wei Yao was hard pressed to keep up with his master. In order to help him, 'someone secretly gave him some tea to take the place of wine' (*huo mi szu chha i tang chiu* 或秘賜茶以當酒).[22]

Another reference of the same period is in the *Po Wu Chih* (Records of the Investigation of Things), c. +290, which informs us that 'the drinking of true tea will cause people to suffer lack of sleep' (*yin chêng thu ling jên shao mien* 飲真茶令人少眠).[23] It is noteworthy that the term *chêng thu* (true *thu*) is used in the quotation, which would indicate that there was more than one kind of *thu* in circulation. Furthermore, the quotation occurs in a paragraph entitled 'Foods to be avoided' (*shih chi* 食忌), suggesting the ability of tea to prevent one from becoming sleepy was at that time hardly regarded as a blessing. In this case, it is again clear that *thu* has to be 'tea'.

Of the above four references, two are for *thu* and two for *chha*. But we have to be wary of the *chha* in the references from the *Hua Yang Kuo Chi* and the anecdote from the *San Kuo Chih*, since it could very easily have been the result of the omission of one stroke in the copying of *thu*. In fact, one often sees *thu* replaced by *chha* in later editions of *Thung Yüeh* and *Po Wu Chih*. The same caution should be exercised when we see the word *chha* mentioned in accounts of events which took place between the rise of the Three Kingdoms and the grand reunification under the Sui (+3rd to +6th centuries), for example, the *chha* allegedly found in the *Kuang Ya* 廣雅 (Enlargement of the *Erh Ya*),[24] +230, *Kuang Chih* 廣志 (Extensive Records of Remarkable Things),[25]

---

[21] Although dated at about +347, *Hua Yang Kuo Chih* (Records of the country south of Mt Hua) records events in the Szechuan region from Chou times +138. In the *HWTS* edition, *chha* (tea) is mentioned on pp. 2b, 4b, 11b, 12a and 16a. Fu-ling 涪陵 occurs on 4b and Wu-yang 武陽 on p. 12a. Some scholars have cited these references as evidence that tea had already been as tribute to the early Chou court, but we agree with Chu Tzu-Chên 朱自振, *CKCC*, p. 12, that this conclusion is unwarranted, and that the era when tea first became a product of commerce was probably the Warring States.

[22] *San Kuo Chih*, *Wu Chih*, ch. 20, Biography of Wei Yao, p. 1462, tr. Bodde (1942), p. 74, who states that this is the earliest passage on tea cited by Pelliot (1922), *T'oung pao*, 21, 436. Although the character *chha* occurs in this quotation, one cannot be sure if it was indeed there in the original text. A simple omission of one stroke in copying would turn *thu* into *chha*. This story is also mentioned in the *Hsü Po Wu Chih*, ch. 5, p. 951.

[23] *Po Wu Chih*, ch. 4, p. 6b; cf. Bodde (1942), p. 75, note 5.

[24] The passage is no longer found in the surviving text in the *Kuang Ya*. It is, fortunately, preserved in the *Thai Phing Yü Lan*, ch. 867, p. 3843, and is quoted below, p. 510.

[25] The original is lost. This is cited by Chhên and Chu (*1981*) from the *TPYL*, ch. 867, but it is not found in the Chung Hua (1960) edition in the NRI library. It says that the leaves are boiled to give a tea (*ming chha*). *Ming* 茗 is another name for tea.

late +4th century, *Shih Shuo Hsin Yü* 世說新語 (New Talk of the Times),[26] +440 to +510, *Wei Wang Tshao Mu Chih* 魏王花木志 (Flora of the King of Wei),[27] +515, and the *Chin Shu* 晉書 (History of the Chin Dynasty),[28] +646. Of the first four references, *Wei Wang Tshao Mu Chi* is long lost, and the pertinent passages that contain the word *chha* in the *Kuang Ya*, *Kuang Chi*, and *Shih Shuo Hsin Yü* are not found in the surviving texts of these works. Instead they occur in quotations collected in the *Thai Phing Yü Lan* (The Emperor's reading book of the Thai-Phing reign), which was compiled in +983, long after the character *chha* had been adopted as the official term for tea. One cannot, therefore, be sure that the character *chha* cited in the relevant passages was actually *chha* and not a corruption of *thu*.

Book 10 of the *Chhi Min Yao Shu*, +544, which is devoted to 'products not native to North-central China' (*fei Chung-kuo wu chhang chieh* 非中國物產者), contains two entries pertinent to our present discussion: No. 53 on 荼 *thu* (bitter vegetable), and No. 95 on 樣, *thu* or *chha* (tea).[29] The addition of the *mu* 木 (wood) radical to the word *thu* was obviously an attempt to differentiate between the *thu* of the herbaceous bitter vegetable, and the *thu* of the woody tea. The text of No. 95 consists of three quotations. We have already seen two; one is the passage from the *Erh Ya* on *chia* 檟, another is the statement on tea and sleep from the *Po Wu Chih*. The third is from the '*Ching Chou Thu Ti Chi* 荊州土地記' (Record of the Ching Chou Region), which says that the '*thu* from *Fu-ling* is the best,' (*Fu-ling thu tsui hao* 浮陵茶最好), probably +4th century. Although the title of entry No. 95 uses the new *thu* 樣 with the *mu* (wood) radical, it is the old *thu* that is found in all three quotations. It would seem that the nomenclature of tea was still in a state of flux. Tea was a product in search of a name in the 6th century.

Next to *thu* the word that was used for tea most often during this period is *ming* 茗. According to the *Chha Ching* the word *ming* first occurs in the *Yen Tzu Chhun Chhiu* (c. −4th century), but this *ming* may have been a copying error for *thai* 苔.[30] The *Shuo Wên*, +121, says, '*thu, khu thu* 荼苦荼 (*thu* is bitter *thu*)' and '*ming, thu ya* 茗荼芽 (*ming* is *thu* sprouts)'.[31] Hsü Hsüan's commentary of the 10th century notes that '*thu* is what

---

[26] Also quoted in *TPYL*, ch. 867, p. 3844. This is the passage quoted by Bretschneider, *Botanicum Sinicum*, Pt II, (1893), pp. 30–1. 'Wang Mang 王濛 (father-in-law of the Chin Emperor Ai-ti) was fond of drinking *thu*. Everyone within his sight was required to drink it. His subordinates were unhappy about his demand. Whenever one had to go and see him, he says "Alas! Today we have to endure *water penance*."' The passage is also quoted in C&C, p. 206.

[27] The original book is no longer extant. The pertinent passage is preserved in the *TPYL*, ch. 867, p. 3845 and says, '*Chha*: the leaves are like those of gardenia. They can be boiled to make a drink. The old leaves are called *chhuan* 荈, and the young buds, *ming* 茗'. Cf. also C&C (*1981*), p. 207.

[28] *Chin Shu*, ch. 77, Biography of Lu Na 陸納, p. 2027; and ch. 98, Biography of Huan Wên 桓溫, p. 2576.

[29] *CMYS*, Bk 10, item 53, p. 645, and item 95, p. 690. The Fu-ling in this passage is probably the same as the Fu-ling in the *Hua Yang Kuo Chih*, n. 20 above.

[30] *Chha Ching*, ch. 7, p. 43, says, 'When Ying (i.e. Yen Ying or Yen Tzu) served as Minister to the Duke Ching of Chhi, *he frequently took a bowl of hulled rice, five eggs and some tea and vegetables*', tr. Ross Carpenter (1974), p. 123. The part in italics reads in the original, 食脫粟之飯，炙三弋，五卵，茗菜而已. The key words are *ming tshai* 茗菜 which is translated as 'tea and vegetables'. But in the surviving editions of the *Yen Tzu Chhun Chhiu*, *ming* is replaced by *thai* 苔. However, it is also possible that both *ming tshai* 茗菜 and *thai tshai* 苔菜 are expressions for 'tea' since some tea trees are known as *thai chha shu* 苔茶樹.

[31] *SWCT*, *thu* p. 26; *ming* p. 27 (ch. 1B, 24 and 26).

we nowadays call *chha*'. In addition to the quotation from the *Wei Wang Hua Mu Chih* referred to earlier, *ming* is mentioned in the *Kuang-chou Chi* 廣州記 (Record of Kuangtung),[32] +4th century, *Nan-Yüeh Chih* 南越誌 (Record of the Southern Yüeh),[33] c. +460, *Thung Chün Yao Lu* 桐君葯錄 (Master Thung's Herbal),[34] probably Han, and *Ming I Pieh Lu* 名醫別錄 (Informal Records of Famous Physicians),[35] c. +510. Except for the *Kuang-chou Chi*, all the references are cited from chapter 867 of the *Thai Phing Yü Lan* since the original works are no longer extant.

There are several interesting anecdotes on tea under the name *ming* recorded in the *Loyang Chhieh Lan Chi* (Buddhist Temples of Loyang), +530.[36] It is said that when Wang Su 王肅 was forced to flee his native Southern Chhi, he came north to serve at the court of the Northern Wei State. At first he could not bear the taste of yoghurt drink (*lo chiang* 酪漿) which was the common beverage of the realm. He was so fond of drinking *ming* decoction (*ming chih* 茗汁), i.e. tea, that he would drink as much as a bucket (*tou* 斗) at a time. His associates nicknamed him the leaky pot, *lou chih* 漏卮.

But years later after he was thoroughly acclimatised to the ways of the North and could drink milk products with ease, the emperor asked him, 'How would you compare tea (*ming chih*) to yogurt drink (*lo chiang*)?' Wang Su replied jokingly, 'Tea does not quite hit the mark; it is but a slave compared to yoghurt drink (*wei ming pu chung, yü lo tso nu* 惟茗不中與酪作奴).' And so the term 'yoghurt's slave (*lo nu* 酪奴)' became a nickname for tea.

The passage goes on to say that although tea (*ming yin* 茗飲) was served at formal functions at the Northern Wei court, most of the notables tended to stay away from it. It only appealed to those who had recently arrived from the South. Indeed, tea was humourously known as 'water penance (*shui o* 水厄)' by the northerners. From these accounts we may infer that at the beginning of the 6th century, tea was still considered as an alien drink in North China, although it was already well accepted in the South.

The *Ching Tien Shih Wên* 經典釋文 (Textual Criticism of the Classics), c. +600, commenting on chapter 14 (Woody Plants) of the *Erh Ya*, lists *thu* 荼, *ming* 茗 and *chhuan* 荈.[37] It notes that *thu* is also written as 檟, and is used as a drink by the people of Szechuan. Using the *fan-chhieh* system, this 檟 is pronounced as *chih* + ch(ia) = *chi-ia*. We now have a phonetic as well as a pictographic distinction between *thu* as bitter vegetable and *thu* as tea.

---

[32] *Kuang-chou Chi*, extant only in quotations in the *Thang Lei Han* 唐類函 (A Thang Compendium), ch. 178, compiled in +1618, gives the name of a Southern type of *ming*, cf. *TPYL* ch. 867, p. 3845, Chung Hua (1960).

[33] *Nan Yüeh Chih*, c. +460, says *ming* is bitter and pungent, and is known locally as *kuo lo* 過羅, cf. *TPYL*, p. 3845.

[34] *Thung Chün Yao Chi* gives names of several localities where *ming* is harvested. The book is lost, but the relevant passage on *ming* is preserved in *TPYL*, ch. 867, p. 3845.

[35] *MIPL* says *ming* tea will lighten the body and change one's bones. The passage is not found in the surviving text of the *MIPL*. It is taken from the *TPYL*, ch. 867.

[36] *LYCLC* (HWTS), ch. 3, *Chêng Chüeh Shih* 正覺寺, pp. 11b–12a. This anecdote occurs on pp. 214–15 in Jenner's translation of 1981.

[37] *Ching Tien Shih Wên*, on *Erh Ya*, ch. 14 (Woody Plants). For an explanation of the *fan-chhieh* system cf. Vol. I, pp. 33–4.

Ming is listed as a middle-class herb in the Woody Plants (*mu pu* 木部) section of the *Hsin Hsiu Pên Tshao* (The Newly Improved Pharmacopoeia), +569.³⁸ The entry says: '*Ming*, also called bitter *thu* 荼. Its taste is sweet and bitter. It is slightly cold but non-toxic. It alleviates boils and sores; promotes urination; dispels phlegm and relieves thirst. It keeps one awake. It is harvested in the spring . . . To make a drink from it add dogwood, scallion and ginger.' There is little doubt that what is being portrayed is tea. Also listed, but in the Vegetable (*tshai pu* 菜部) section is *khu tshai* 苦菜 or *thu tshao* (bitter herb).³⁹ It is also said to lessen the need for sleep. The compiler, in a comment, suspects that *khu tshai* may be just another name for *ming* 茗. This is hardly possible since *ming* is woody and *khu tshai* herbaceous. One may infer from these remarks that there was still a good deal of confusion regarding the identity of bitter vegetable *versus* tea in the early years of the Thang.

The entry on *ming* in the *Shih Liao Pên Tshao* (Nutritional Pharmacopoeia), c. +670 says,⁴⁰ '*Ming* leaves – promote bowel function, dispel heat, dissolve phlegm. Boil to obtain a decoction, which makes a good medium for cooking congee.' It then goes on,⁴¹ '*Chha* 茶 (tea), reduces flatulence and relieves boils and itches. It can be kept overnight. The best tea is made from fresh leaves. However, they can be steamed and pounded and then kept for later use. When it is too old it may promote flatulence. The people then mix it with freshly sprouted leaves of locust or willow.' It is frustrating that the passage does not tell us how *chha* was related to *ming*, but we may have here the earliest statement of how tea was processed. The leaves were steamed and then pounded. We also learn from the two pharmacopoeias that tea was often spiced with scallion and ginger, and that it was made into a congee.

Tea as a medication continued to be referred to as *ming* in the standard pharmacopoeias, through the *Pên Tshao Shih I* (A Supplement to the Pharmacopoeias), c. +725, *Chêng Lei Pên Tshao*, +1108 to the great *Pên Tshao Kang Mu*, +1596.⁴² But as a drink, it shed the names of *thu* and *ming* and became known almost exclusively as *chha* 茶 after the 8th century. How do we know that this was the period when the change in nomenclature took place? The most reliable evidence has come from the inscriptions found on Thang stone tablets. In the *Thang Yün Chêng* 唐韻正 (Thang Rhyming Sounds), +1667, Ku Yen-Wu wrote,

During my visit to Mount Tai (*Thai-shan* 泰山) I had a chance to examine a number of Thang stone tablets with inscriptions of *thu* and *chha*. Those written in +779 (in connection with tea as medication) and +798 (in connection with the use of tea in banquets) contained the word *thu* 荼 . . . there was no change in the character. But those inscribed in +841 and +855 had the word *chha* 茶, obtained by removing one stroke from *thu*. Thus, the change must have taken place after the middle years of the Thang dynasty.⁴³

---

³⁸ *HHPT*, ch. 13, p. 116. Note that *thu* is written as 梌, with the wood (*mu* 木) radical added, but the grass (*tshao* 艹) radical left out. This character is not seen in later works.
³⁹ Ibid., ch. 18, pp. 250–1. ⁴⁰ *SLPT*, item 44, p. 23.
⁴¹ Ibid. The original says *chêng thao ching su yung* 蒸擣經宿用.
⁴² *PTSI*, preserved in *CLPT*, ch. 13; *PTKM*, ch. 32, p. 1870.
⁴³ *Thang Yün Chêng*, cited in C&C, pp. 345–6, which suggests that the change was probably adopted by the Emperor in about +800 to +820.

From these accounts we are inclined to conclude that the use of tea leaves in medicine or as drinks probably began in Szechuan some time before the Han Dynasty. Since tea is not mentioned in the *Shih Ching*, which lists practically all the economically useful plants found in ancient China, one might infer that tea was 'discovered' after the founding of the Chou, c. −1000. On the other hand, Szechuan is isolated from the heartland of Chinese civilisation in the lower Yellow River valley. It should come as no surprise that the people who composed the poems of the *Shih Ching* were not aware of tea's existence.[44] It is likely that tea was originally processed from leaves harvested from trees in the wild.[45] As it increased in popularity, it became necessary to make the leaves more accessible. This led to the cultivation of the plant followed by the development of dwarf bushes which are easily harvested. At any rate the evidence of tea tribute in the *Hua Yang Kuo Chi* cited earlier (pages 508–9) suggests that by the time of the Warring States (−480 to −221) tea was already harvested and processed in Szechuan; and by the −1st century, as shown by the *Thung Yüeh* (Contract with a Slave) of −59, tea drinking was already an established custom in the daily life of the gentry in Szechuan, and tea, as processed tea leaves, was already an article of commerce that could be bought in one city and brought home to another.

In any case, it was during the Han Dynasty that tea was introduced from Szechuan to the population centres of the North China plain and the lower Yangtze Valley. Chou Shih-Yung has made the exciting suggestion that one of the items in the Ma-wang-tui Inventory is *chia* 櫃, an alternative name for tea.[46] If this interpretation is correct, then we can say that tea had already found its way to Hunan in about −200. As it gained acceptance as a drink, its cultivation began to spread eastwards. The story about Wei Yao cited earlier (page 509) shows that tea drinking was well accepted in the Wu court in Nanking in about +260. Tea plantations were established in suitable sites in Hupei, Hunan, Kiangsi, Anhui, Chekiang and Fukien. For centuries, it was considered chiefly as a Southern beverage. As we have seen from such nicknames as *shui ê* 水厄 (water penance) and *lo nu* 酪奴 (slave of milk), it was accepted with considerable reluctance by the Northerners.[47]

---

[44] Szechuan may yet prove to be an independent centre of cultural development in China as indicated by recent archaeological discoveries in Sanxindui, Kuang-han, Szechaun, cf. Bagley (1988 & 1990).

[45] Evans (1992), p. 10, states that the ancient people of Szechuan felled the wild tea trees in order to harvest the leaves. This is not impossible, but one can easily visualise more economical ways of harvesting the young leaves, for example, by someone climbing up to the branches bearing the sprouting buds.

[46] Cf. Chou Shih-Yung (*1979*). This slip is no. 135 in the *MWT Inventory*, *Hunan shêng po wu kuan* (*1973*), I, p. 141. The character in question is 梠. It was originally interpreted as *yü* 奠 (i.e. *yu li* 郁李, a kind of cherry), by the compilers of the Inventory in *Hunan shêng po wu kuan* (*1973*), I, p. 141, and also by Thang Lan (*1980*), p. 20. Later Chu Teh-Hsi and Chiu Hsi-Kuai (*1980*), p. 67, reinterpreted it as *yu* 柚 pomelo. We think that Chou Shih-Yung has made a persuasive case to indicate that it is *chia* 櫃, one of the names for tea. For further discussion on the issue cf. Chou Shih-Yung (*1991*) and Tshao Ching (*1992*). For an opposing view cf. Fang Chien (*1997*) who argues that since it occurs in a group (items 133–6) that contains jujube, pear and *mei* fruit, it should also be a fruit. On the other hand, in the next group (items 137–9) we see bamboo shoots placed together with two types of *mei* fruits. So mixing tea (a plant product) in with fruits (also plant products) could be quite acceptable to the marquise's kitchen staff who had arranged the foodstuffs to be interred with the lady for her enjoyment in the after life.

[47] See n. 34 above. For a brief account of the spread of tea-drinking in the pre-Thang period cf. Ceresa (*1996*).

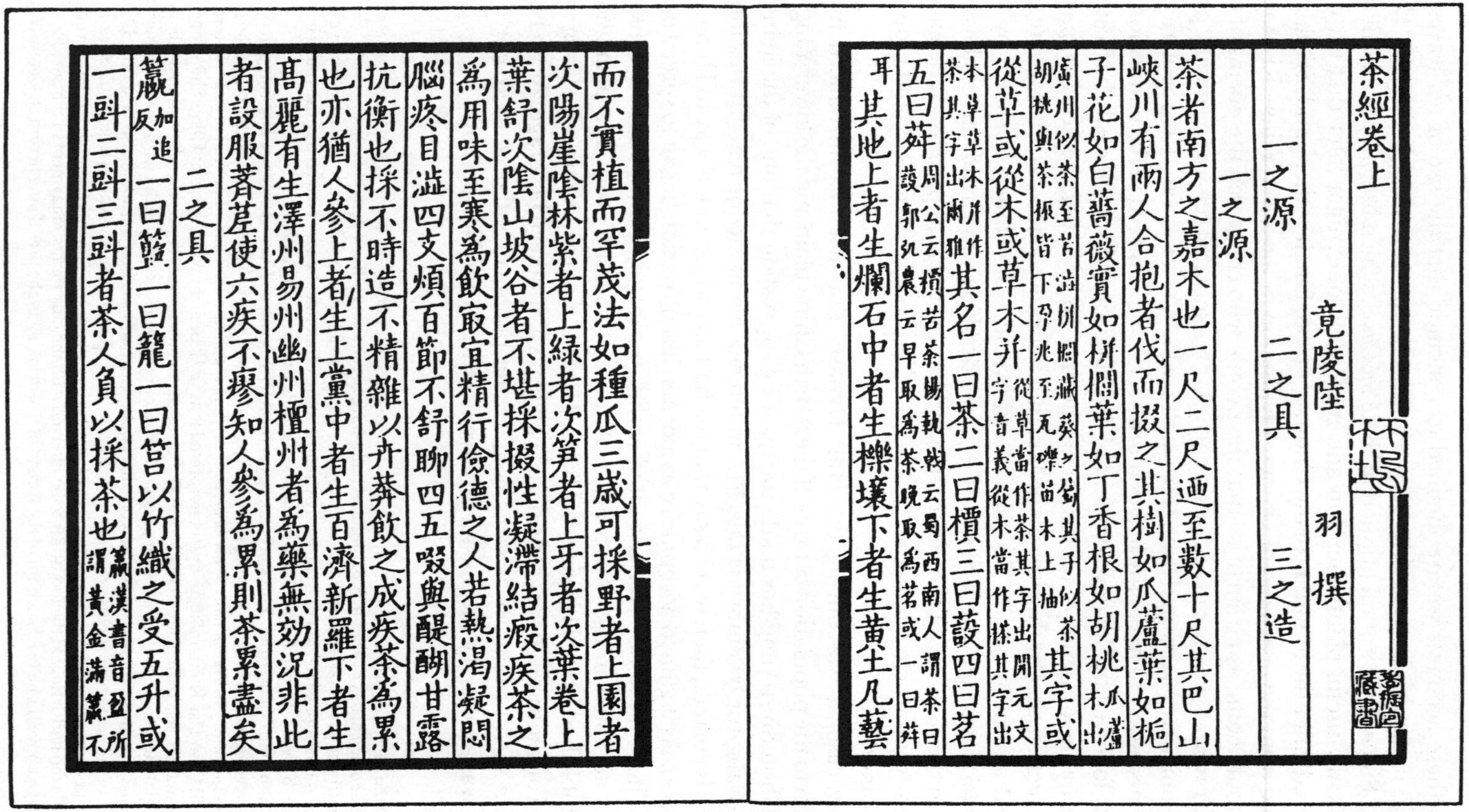

茶經卷上

竟陵陸　羽　撰

一之源　二之具　三之造

一之源

茶者南方之嘉木也一尺二尺迺至數十尺其巴山峽川有兩人合抱者伐而掇之其樹如瓜蘆葉如梔子花如白薔薇實如栟櫚葉如丁香根如胡桃（瓜蘆木出廣州似茶至苦澁栟櫚蒲葵之屬其子似茶胡桃與茶根皆下孕兆至瓦礫苗木上抽）其字或從草或從木或草木并（從草當作茶其字出開元文字音義從木當作搽其字出本草草木并作荼其字出爾雅）其名一曰茶二曰檟三曰蔎四曰茗五曰荈（周公云檟苦荼揚執戟云蜀西南人謂茶曰蔎郭弘農云早取爲茶晚取爲茗或一曰荈耳）其地上者生爛石中者生櫟壤下者生黃土凡藝

而不實植而罕茂法如種瓜三歲可採野者上園者次陽崖陰林紫者上綠者次筍者上牙者次葉卷上葉舒次陰山坡谷者不堪採掇性凝滯結瘕疾茶之爲用味至寒爲飲最宜精行儉德之人若熱渴凝悶腦疼目澁四支煩百節不舒聊四五啜與醍醐甘露抗衡也採不時造不精雜以卉莽飲之成疾茶爲累也亦猶人參上者生上黨中者生百濟新羅下者生高麗有生澤州易州幽州檀州者爲藥無效況非此者設服薺苨使六疾不瘳知人參爲累則茶累盡矣

二之具

籯（加追反）一曰籃一曰籠一曰筥以竹織之受五升或一斗二斗三斗者茶人負以採茶也（籯漢書音盈所謂黃金滿籯不）

Fig 122. First two pages of Lu Yü's *Chha Ching* (Classic of Tea) from a Ming edition?

One factor that encouraged the spread of tea in the North was the increasing popularity of Buddhism, especially the Chhan 禪 (Zen in Japanese) branch that we referred to earlier on page 506. Tea drinking was adopted by the Zen Buddhist monks who found it an invaluable aid to help them stay awake during the long hours that they were required to spend in meditation.[48] Several of the emperors of the Northern Dynasties during the second partition (+479 to +581, see Table of Chinese Dynasties) were devout Buddhists. No doubt their attitude helped to make tea drinking fashionable among members of their courts. As the popularity of tea grew in the North, tea plantations expanded in the South. By the beginning of the Thang, tea was already a significant economic activity of the realm and it was well poised to become a major contributor to the religious, cultural and artistic life of this glorious dynasty.[49]

The coming of age of tea was marked by the publication of the *Chha Ching* 茶經 (The Tea Classic) by Lu Yü in +760.[50] This is the first specialised book about tea ever written (Figure 122). It is also the type specimen of a genre of books in China that have come to be known as *Chha Shu* 茶書 (Tea Books), which have appeared at frequent if irregular intervals from the Thang to the Chhing. They are the counterparts of the *Nung Shu* 農書 (Agricultural Treastises) that have played such an important role in the development of agriculture in China (see Volume 6, Part 2 in this series). It is estimated that about 100 titles of *Chha Shu* were written before 1800, but many of them are no longer extant.[51] They are the principal sources for any study on the historical development of the culture of tea in China, including that of processsing and usage. The ones that have been found most useful in this study are listed in Table 47.[52]

The first item is, of course, Lu Yü's *Chha Ching*.[53] Although a small book with ten brief chapters, it touches on every aspect of the technology and culture of tea known in the 8th century. The chapter headings are:

(1) Origin, characteristics, names and qualities of tea.
(2) Implements for picking and processing leaves.

---

[48] For the influence of Buddhism on tea drinking, cf. *Chha Ching*, pp. 57–8; *Fêng Shih Wên Chien Lu* 封氏聞見記 late +8th century, cited in C&C, p. 211, and Wang Ling (*1991*), who also points out the influence exerted by Taoists during the Tsin and Northern and Southern dynasties (+4th to +6th centuries).

[49] The culture and economics of tea during the Thang is discussed by Ku Fêng (*1991*). The barter of tea for horses with the nomadic tribes of the north during the Thang is first mentioned in the *Fêng Shih Wên Chien Lu*, cited in C&C, p. 212.

[50] For a discussion of the date of publication of *Chha Ching*, cf. Fu & Ouyang (*1983*), p. 91. Other scholars, e.g. *CKCC*, p. 595, and C&C (*1981*), p. 1 prefer +758 as the date of completion.

[51] *CKCC*, p. 595. Fifty-eight titles are listed in C&C.

[52] The full texts or excerpts of the relevant texts of all the books listed are found in C&C. For an account of the major Tea Books of China, cf. *CKCC*, pp. 595–9. Many of the titles in Table 46 are included in the *Chha Shu Chhüan Chi* 茶書全集 (Complete Collection of Tea Books) of Yü Chêng 喻政, +1613.

[53] Lu Yü was held in such high esteem that he was posthumously declared the 'Tea God' (*chha shên* 茶神) or 'Tea Sage' (*chha shêng* 茶聖). For a brief account of his life cf. Blofeld (1985), pp. 2–6; and Fu & Ouyang (*1983*), 75–90, in which are reprinted the full texts of Lu Yü's biography in the *Hsin Thang Shu*, ch. 196 and the *Chhüan Thang Wên*, ch. 433, with annotations. An abbreviated translation of the *Chha Ching* is found in Ukers (1935), I, pp. 13–22. For a complete English translation see Carpenter (1974).

Table 47. *The tea books* (Chha Shu 茶書) *of China*

Title	Author	Date	Content
Chha Ching 茶經 Tea Classic	Lu Yü 陸羽	+760	Horticulture, processing, art of drinking, history and literary references
Chien Chha Shui Chi 煎茶水記 Water for Making Tea	Chang Yu-Hsin 張又新	+825	Sources and quality of water
Shih Liu Thang Phin 十六湯品 Sixteen Tea Decoctions	Su I 蘇廙	+900	Types of boiled water
Chha Phu 茶譜 Tea Compendium	Mao Wên-Hsi 毛文錫	+925	Types of tea produced in different areas
Chhuan Ming Lu 荈茗錄 Tea Record	Thao Ku 陶穀	+960	Anecdotes about tea
Chha Lu 茶錄 Tea Discourse	Tshai Hsiang 蔡襄	+1051	Art of handling, making and drinking tea
Tung Hsi Shih Chha Lu 東溪試茶錄 Tea in Chien-an	Sung Tzu-An 孫子安	+1064	Tea production locations in Chien-an, Fukien
Phin Chha Yao Lu 品茶要錄 Elite Teas Compendium	Huang Ju 黃儒	+1075	Tea processing in Chien-an
Ta Kuan Chha Lun 大觀茶論 Imperial Tea Book	Chao Chi 趙佶	+1107	Overview: horticulture, processing, drinking etc
Hsüan Ho Pei Yuan Kung Chha Lu 宣和北苑貢茶錄 Pei-yuan Tribute Teas	Hsiung Fan 熊蕃 Hsiung Kho 熊克	+1121, +1125	Vintages of tribute tea from Pei-yuan, Chien-an during the Hsüan Ho reign period
Pei Yuan Pieh Lu 北苑別錄 Pei-yuan Tea Record	Chao Ju-Li 趙汝礪	+1186	Tea culture and technology in Pei-yuan, Chien-an
Chha Chü Thu Tsan 茶具圖贊 Tea Utensils with Illustrations	Shên An Lao Jên 審安老人	+1269	Illustrated book of tea utensils
Chha Phu 茶譜 Tea Discourse	Chu Chhüan 朱權	+1440	Art of handling, preparing and drinking tea
Chha Phu 茶譜 Tea Discourse	Chhien Chhun-Nien 錢椿年	+1539	Tea varieties, preparation and drinking
Chu Chhüan Hsiao Phin 煮泉小品 Teas from Various Springs	Thien I-Hêng 田藝蘅	+1554	Tea varieties and drinking
Chha Shuo 茶説 Talks on Tea	Thu Lung 屠隆	+1590	Varieties, processing, storage, drinking
Chha Khao 茶考 Comments on Tea	Chhên Shih 陳師	+1593	General comments, processing, drinking
Chha Lu 茶錄 Tea Records	Chang Yuan 張源	+1595	Processing, storage, and art of drinking
Chha Ching 茶經 Tea Classic	Chang Chhien-Tê 張謙德	+1597	Overview, quality, utensils
Chha Shu 茶疏 Tea Commentary	Hsü Tzhu-Yü 許次紓	+1597	Processing, handling, storage and drinking
Chha Chieh 茶解 Tea Explanations	Lo Lin 羅廩	+1609	Technology, usage and utensils
Chha Chien 茶箋 Notes on Tea	Wên Lung 聞龍	+1630	Processing
Chieh Chha Chien 岕茶牋 Chieh Tea Notes	Fêng Kho-Ping 馮可賓	+1642	Tea from Lo-Chieh, processing, drinking
Hsü Chha Ching 續茶經 Tea Classic Continued	Lu Thing-Tshan 陸延燦	+1734	Encyclopedia of Tea

Compiled with the aid of: Chhên & Chu (*1981*), CKCC (*1992*), *Shuo Fu* 説郛 and *Min Shu Tshung Shu* 民俗叢書.

(3) The picking and processing of tea.
(4) Utensils for boiling and drinking tea.
(5) The boiling of tea for drinking.
(6) The art of drinking tea.
(7) Historical references to tea.
(8) Elite teas of different locations.
(9) Utensils that may be omitted.
(10) How to copy this book on a silk scroll.

Chapter 8 makes it clear that by Lu Yü's time the centre of gravity of tea production had already shifted from Szechuan towards the Eastern coastal provinces. In fact, the finest tea was supposed to have come from *Ku Chu Shan* 顧渚山 near the border between Kiangsu and Chekiang. This was where the first 'tribute tea' was produced for the enjoyment of the Thang emperor and his court.[54] The *Chha Ching* also discusses the places where the purest water for making tea can be found, as well as how the tea should be prepared for drinking. These topics are treated further in the *Chien Chha Shui Chi* 煎茶水記 (Water for Making Tea) and the *Shih Liu Thang Phing* 十六湯品 (Sixteen Tea Decoctions).[55]

Next we come to the *Chha Phu* 茶譜 (Tea Compendium), a description of the types of tea produced in different regions and the *Chhuan Ming Lu* 荈茗錄 (Tea Record), a collection of anecdotes on tea preserved in the *Chhing I Lu*, +960. They are followed by the first Tea Book of the Sung Dynasty, the *Chha Lu* 茶錄 (Tea Discourse) of Tshai Hsiang, who was known as the Tea Commissioner.[56] It tells us how to evaluate, store, handle and make tea for drinking and describes the proper utensils to be employed in these tasks. Tshai Hsiang took particular note of the superior quality of the tea from Chien-an, in Northwestern Fukien, a tea producing area that he thought had been neglected by Lu Yü. Perhaps as a result of his promotional efforts, Chien-an soon became another centre of 'tribute tea' that supplied the needs of the Sung emperor and his court. Tea production in Chien-an is the theme of the next two works, the *Tung Hsi Shih Chha Lu* 東溪試茶錄 (Tea in Chien-an) which describes locations of tea plantations in the region, and the *Phin Chha Yao Lu* 品茶要錄 (Elite Teas Compendium) which deals with the pitfalls in the processing of tea leaves.[57]

The most remarkable book of the series is perhaps the *Ta Kuan Chha Lun* 大觀茶論 (The Imperial Tea Book).[58] It was written by no less a personage than Emperor Hui Tsung, who was the supreme ruler of China from +1101 to +1125. *Ta Kuan* was his

---

[54] The area is west of present-day I-hsing in Kiangsu and Chhang-hsing in Chekiang. It is said that by the end of the 8th century 30,000 labourers and a thousand technicians were involved in the production of 'tribute tea', cf. Chhên Tsung-Mou (*1992*), p. 30 and Blofeld (1985), pp. 6–7.

[55] The original *Shih Liu Thang Phing* is no longer extant. It is preserved as part of *Chhing I Lu*, ch. 4, as is the *Chhuan Ming Lu*, which is also listed in Table 47.

[56] Tshai Hsiang was a famous tea connoisseur. We have met him before, *SCC* Vol. VI, Pt 1, p. 361 as the author of the *Li Chih-Phu* (Monograph on Litchi). For some of his exploits cf. Blofeld (1985), pp. 16–20.

[57] Chien-an is modern Chien-ou. It lies south of the famous Wu-i mountains, which is known in the export tea trade in the 18th and 19th centuries as Bohea, and which Fortune (1848), p. 220, called 'the great black-tea country of Fokien'.

[58] For an English translation of excerpts from the *Ta Kuan Chha Lun* cf. Blofeld (1985), pp. 28–31.

reign title from +1107 to +1111. The book is a brief but masterly exposition of the science and art of tea in the Northern Sung. Even though Hui Tsung is well known as a learned scholar, poet, calligrapher and tea connoisseur, it is extraordinary that a 'Son of Heaven' living in virtual isolation from the populace should have been so knowledgeable about the details of the cultivation, processing and utilisation of tea. We shall return to this work from time to time. The importance of the teas of Chien-an is reiterated in the *Hsüan Ho Pei Yuan Kung Chha Lu* 宣和北苑貢茶錄 (Pei-yuan Tribute Teas), Pei-yüan being one of the best tea plantations in the Chien-an area, and in the *Pei-yüan Pieh Lu* 北苑別錄 (Pei-yüan Tea Record).[59] The vintages of various tribute teas are described in detail and with a level of sophistication and enthusiasm reminiscent of that shown by modern wine connoisseurs in discussing the great vintages of Western grape wines.

The next work, *Chha Chü Thu Tsang* 茶具圖贊 (Tea Utensils with Illustrations) is a description of tea utensils. It is followed by three Ming books dealing with the art of tea preparation and drinking. Two are under the title *Chha Phu* 茶譜 (Tea Discourse), one by Chu Chüan and the other by Chhien Chhun-Nien, and the third is the *Chu Chhuan Hsiao Phin* 煮泉小品 (Teas from Various Springs). More useful to us are the next eight Tea Books, the *Chha Shuo* 茶說, *Chha Khao* 茶考, *Chha Lu* 茶錄, *Chha Ching* 茶經, *Chha Shu* 茶疏, *Chha Chieh* 茶解, and *Chha Chien* 茶箋, and *Chieh Chha Chien* 岕茶箋. They are all comprehensive books about tea in the tradition of Lu Yü's *Chha Ching* and Emperor Hui Tsung's *Chha Lun*, and contain valuable information about the processing of tea leaves during the Ming. The last entry is the *Hsü Chha Ching* 續茶經 (The Tea Classic – Continued), +1734. Although largely a collection of previous works it does contain excerpts from several books that are now lost. The same cannot be said of another Ching collection of the same type, the *Chha Shih* 茶史 (History of Tea), +1669, which we have not included in Table 45. It is curious that no original Tea Books have appeared during the Chhing Dynasty. For any significant new developments during this period other sources will have to be consulted.[60] But before we move on to the next topic we need to acknowledge our thanks to three modern Tea Books that have greatly aided our inquiry. The first is *Chung-kuo chha-yeh li-shih tzu-liao hsuan-chi* 中國茶葉歷史資料選輯 (Selected Historiographic Materials on Tea in China) compiled by Chhên Chu-Kuei and Chu Tzu-Chên (*1981*), an exhaustive compendium of material on the history of tea in the Chinese literature. The second is *Chung-kuo chha ching* 中國茶經 (the Modern Tea Classic) edited by Chhên Tsung-Mou (*1992*). It is an encyclopedia of information on tea in China, including its history, geography, horticulture, processing technology, health effects, and culture. The third is *Chung-kuo ti-fang-chih chha-yeh li-shih tzu-liao hsuan-chi* 中國地方志茶葉歷史資料選輯 (Selected Historiographic Materials on Tea in the local Gazetteers of China), compiled by Wu Chio-Nung (*1990*). It is a valuable companion to the first work, which does not contain material

---

[59] Hsüan Ho is also a reign title of Hui Tsung. It lasts from +1119 to +1125.
[60] For example, local gazetteers.

from the local gazetteers of China. In addition, the volume on Tea of the *Chung-kuo nung-yeh pai-kho chhuan shu* 中國農業百科全書 (The Encyclopedia of Chinese Agriculture) and the special issues on Tea of the journal *Nung Yeh Khao Ku* (Agricultural Archaeology) have also been a useful source of information.

(2) Processing of Tea

When we talk of 'tea', we may have in mind one of two things. The first is 'tea' the product, an article of commerce that we buy at the store and keep at home for use as needed. Today, it is usually available in the West in the form of loose leaves or coarse powder or as tea-bags. We may also mean 'tea' the drink, the 'cup that cheers'. The second is prepared as an infusion of the first. Depending on the type of tea used to make the infusion, we may drink it straight (as in China), mix it with milk and/or sugar (as in Britain) or blend it with lemon and sugar (as in Russia). In this work we are concerned with both the technology of processing the product and the art of making the drink. We shall now trace the technology that has evolved in processing fresh tea leaves harvested from the plant into the 'tea-product' from the Thang Dynasty until the end of the 19th century. We shall touch on the making of the 'tea-drink' later.

(i) *Tea processing during the Thang*

According to the statement in the *Shih Liao Pên Tshao*, +670, cited earlier (page 512), in the very beginning the first 'tea drink' was probably obtained by boiling fresh leaves in water and consumed as a medicinal preparation. In the time-honoured tradition of Chinese herbs, the leaves could have been simply air dried and kept as a herb for later use. But as tea gained favour more as a drink to be enjoyed than as a decoction to be endured, the Chinese people began to look for better ways to process the leaves so that the desirable qualities inherent in the fresh leaves would not only be preserved but perhaps even enhanced. This task has been a continuing goal that has occupied the time and talent of tea technicians from early mediaeval times to the modern age. Indeed, since the Thang, every dynasty has brought significant changes to the technology of processing tea so that each dynasty can almost be defined by the type of tea that is consumed by its populace.

Although a considerable amount of information on the processing of tea must have accumulated through the centuries between the start of the Han (−206) and the founding of the Thang (+618), the only record we have of it is a passage from the *Kuang Ya* (Enlargement of the *Erh Ya*), +230, which says:[1]

---

[1] *Chha Ching*, ch. 7, p. 43, tr. auct, adjv. Carpenter (1974), pp. 122–3, and Smith (1991), p. 60. Although this passage is not found in the surviving text of the *Kuang Ya*, it is preserved in the *TPYL*, ch. 867, which is cited in C&C, p. 203. *Ching* consists of Eastern Szechuan and Western Hupei, while *Pa* is Western Szechuan. The custom of brewing tea with ginger, sesame and other flavouring agents has survived to this day, cf. Blofeld (1985), p. 140, and Tshao Chin (*1991*) who describes the *chiang yen chha* 姜鹽茶 (ginger-salt tea), a brew containing ginger, salt, soybean, sesame and tea, of Hsiang-yin county in Hunan province.

In the States of Ching 荊 and Pa 巴 the people pick the leaves and turn them into cakes. If the leaves are old, the cakes are formed with the aid of a rice paste. Before brewing a drink, they first roast a cake until it assumes a red colour. Then they grind it into a powder, put it in a pottery jar and cover it with boiling water. They mix in scallion, ginger and orange [seeds]. The resulting drink wakens one from the effect of wine and prevents one from falling asleep.

There is no indication how the cakes were prepared from the leaves. Based on the statement in the *Shih Liao Pên Tshao* just cited, they were probably steamed, pounded and then pressed into cakes. In fact, this could have been the way tea leaves were processed during the Western Han and the Three Kingdoms period.[2] Steaming remains an integral step in the manufacture of tea as described in detail in the very first Tea Book of China, Lu Yü's *Chha Ching* (Tea Classic) of +760. At first sight it would appear that even this venerable classic has very little to tell us about the processing of tea in Lu Yü's day. The pertinent paragraph on harvesting and processing in chapter 3 simply says:[3]

Tea leaves should be picked in the second, third or fourth months of the year. In good soil fresh stems should reach a height of four to five inches. They are ready when they assume the appearance of wild bean or bracken. Leaves should be collected with the dew on them. Look for shoots among stems grown closely together. Pick shoots on the healthiest stem in a group of four to five stems. Do not pick on a day when it rains or when it is partly sunny but cloudy. Pick when it is sunny and clear. The leaves are *steamed, pounded, shaped in a mould, dried over low heat, hung in air, and then sealed*.[4] The product is tea.

The short, terse sentence *in italics* is the only statement we have in the *Chha Ching* that actually deals with the processing of fresh tea leaves. But it reveals next to nothing about how each of the steps was carried out. Fortunately, chapter 3 is not the only place where processing is treated in the *Tea Classic*. A considerable amount of useful information is also recorded in chapter 2, which gives the names and explains the functions of the implements used in the harvesting and processing of the leaves. To facilitate discussion, the implements are listed and described briefly in Table 48.[5] Item 1 comprises the bamboo baskets used for carrying the freshly picked leaves from the field to the processing station. They may vary in size from half a *tou* to three *tou*.[6] Items 2 to 15 are the implements used in performing the key processing steps. By examining what the *Chha Ching* says about each item, we can reconstruct the details of each step as follows.[7]

---

[2] As shown in the *Thung Yüeh*, and the Biography of Wei Yao, cf. pp. 508 and 509 respectively.
[3] *Chha Ching*, ch. 3, pp. 12–13. Wild bean and bracken are *wei* 薇 and *chueh* 蕨 in the original text.
[4] The original reads, *chêng chih* 蒸之, *tao chih* 搗之, *pho chih* 拍之, *pei chih* 焙之, *chhuan chih* 穿之, *fêng chih* 封之.
[5] Fu & Ouyang eds. (*1983*), pp. 6–12, cf. also Ukers (1935) I, pp. 16–17 and Carpenter (1974), pp. 62–9.
[6] According to Wu Chhêng-Lo (*1957*), p. 58, a Thang *tou* is 5.94 litres.
[7] A similar reconstruction is given in the *CKCC*, p. 31. Of the implements used, two deserve more than a passing comment, no. 8, the bamboo tray, *phi li* and no. 11, the heating ditch, *pei*. In later ages the tray is often used to dry tea over the heating ditch rather like trays in an oven. This rather primitive system was apparently transmitted to India and used by British planters when they first tried to establish a tea industry in the foothills of the Himalayas. In the early years of this enterprise, tea was processed by the procedure and with the equipment brought over by technicians from China. According to Ukers (1935) I, p. 469, 'Under the then prevalent custom, a

Table 48. *Implements used in the processing of tea listed in the* Chha Ching

1.	*Ying* 籝 or *lan* 籃	Bamboo basket, to collect the leaves as they are picked, size 5 *shêngs* to 3 *tous*
2.	*Tsao* 灶	Stove with a built-in wok (*fu* 釜), preferably one with a brim, for use as a boiler for the steamer
3.	*Chêng* 甑	Steamer made of wood or pottery, using a bamboo basket as the grid (*pi* 箄)
4.	*Chhu chiu* 杵臼	Mortar and pestle
5.	*Kuei* 規 or *chhüan* 棬	Mould, made of iron for pressing the pounded tea into cakes; may be round, square or with designs
6.	*Chhêng* 承 or *chan* 砧	Stone or wood platform anchored to the ground
7.	*Yen* 檐	Oil cloth to cover the platform. The mould is placed over it
8.	*Phi li* 芘莉	Screen or mat formed by weaving bamboo strips across two 3 ft long poles. It is 2.0 ft wide and 2.5 ft long. The unused five inch portions of the poles act as handle to hold the screen
9.	*Chhi* 棨	Awl with a sturdy wooden handle, used to pierce eyes through the tea cakes
10.	*Phu* 扑	Bamboo rod thin enough to string through tea cakes so that they can be moved from place to place
11.	*Pei* 焙	Slow heat furnace for drying tea; made by digging a ditch 2 ft deep, 2.5 ft wide and 10 ft long. It is edged by a 2.0 ft high wall above ground. All surfaces are smoothed out with mud
12.	*Kuan* 貫	Bamboo splints 2.5 ft long; act as skewers for heating wet tea cakes over the furnace
13.	*Phêng* 棚	Rack, made of wood, with two tiers about 1 ft apart; used for drying tea cakes. Skewers of tea cakes are first placed in the lower level, and later to the upper level
14.	*Chhuan* 穿 or 串	Long strips of bamboo used to string tea cakes for final airing
15.	*Yü* 育	Drying chamber framed by wood and covered with bamboo mats. It is divided by a partition. There is a cover for the upper compartment and a floor for the lower. A door opens into the lower compartment in which is placed a very gentle fire

(a) *Steaming*: items 2 and 3 suggest that the stove, boiler and steamer are similar in shape and size to those shown in Han dynasty models and murals that we saw earlier on pages 81 and 87 (Figures 26a and 28). The steamers were made of either pottery or wood. Each steamer could probably hold a large batch (3 *tou*) of leaves. The text does not indicate how long each batch was steamed, but it does say that the leaves should be dispersed (hence cooled) after steaming. Otherwise, much valuable gel (*kao* 膏) could be lost.

single wicker sieve was inserted inside a bamboo frame called "dhool", which was placed over a charcoal fire made in a hole in the ground.' The rolled tea leaves were placed on the sieve to dry. It is strange to see a reincarnation of the *phi li* and the *pei* more than a thousand years after Lu Yü's death in a country so far away from his home.

(b) *Pounding*: item 4 (*chhu chiu* 杵臼) is presumably the mortar and pestle or trip hammer as discussed in *SCC* Volume 4, Part 2, Plate 128, Figures 358, 359 and Table 56. It is used to break up the cell structure of the leaf tissues and render the leaves pliable for pressing.

(c) *Pressing*: the pounded leaves are pressed in an iron mould (item 5, *kuei* 規). There is enough viscous gel from the cells to bind the leaves together and assume the shape dictated by the mould. The text does not say whether the mould has a top or a bottom, except that is is placed on an anvil (item 6) which is covered by a piece of oil cloth (item 7). After pressing, the cakes are spread on a bamboo screen (item 8) to dry.[7] An eye is pierced through each cake with a sharp awl (item 9). They can then be strung on a long bamboo stick (item 10) and moved to the drying furnace.

(d) *Drying*: the furnace (*pei* 焙, item 11) is a rectangular ditch sided by a wall and topped by a wooden rack (*phêng* 棚, item 13). The wet cakes are hung on bamboo sticks (*kuan* 貫, item 12) and placed across the width of the rack to dry. Presumably, a bed of burning or smouldering charcoal at the bottom of the ditch is the source of the heat. The dimensions of the furnace and the rack make it clear that the tea cakes were more than four feet above the burning charcoal. The drying was, therefore, slow and gentle.

(e) *Stringing*: when dry, the tea cakes are hung together in strings (*chhuan* 穿, item 14). In the lower Yangtze basin they are strung with thin strips of bamboo; but in the Szechuan-Hupei region they are strung with ropes made of tree bark. Each string represents a packaged quantity of tea.[8]

(f) *Storage*: the dimensions of the storage chamber (item 15, *yü* 育) are not specified. A charcoal stove is provided during the spring rains to keep the tea dry so that the product does not become mouldy.

The procedure as reconstructed applied only to the processing of cake tea (*ping chha* 餅茶), which was apparently the principal type of tea made during the Thang all across the subtropical provinces of China from Szechuan in the west to Chekiang in the east.[9] But in chapter 6, the *Tea Classic* tells us that tea was also available in the form of 'coarse tea, loose tea, and powdered tea' (*chhu chha, san chha, mo chha*, 粗茶，散茶，末茶).[10] Unfortunately, there is no information on how the Thang loose

---

[8] This step is difficult to interpret. Carpenter (1974), p. 68, renders *chhuan* as 'ties' and Ukers *ibid.*, p. 16, as a package of finished tea. But *chhuan* normally means to thread or string. Thus, in this context it is more likely that it denotes *chhuan* 串, to string together. A *chhuan* may represent different quantities of tea depending on the part of the country. According to the *Chha Ching* in the lower Yangtze a large string weighs one catty, a medium string, half a catty and a small string, 4–5 ounces. But in the Szechuan area, a large string weighs 120 catties, a medium one, 80 catties and a small one, 50 catties. These observations suggest that the cakes of tea were relatively small in the lower Yangtze, and quite large in Szechuan. The word *chhuan* 穿 was formerly written as *chhuan* 釧, meaning a hairpin, cf. Fu & Ouyang (*1983*), p. 11.

[9] Fu & Ouyang, *ibid.*, pp. 68–72; Carpenter (1974), pp. 143–8. By the Thang dynasty, tea was grown all across subtropical China from Szechuan in the west to Chekiang in the east. The famous purple shoot (*tsu shun* 紫筍) was produced in Chekiang in the Ku-chu-shan 顧渚山 area (northwest of present day Chhang-hsing 長興). During the *Chen Yuan* reign period (+785 to +805), it is recorded, cf. C&C, p. 212, that 30,000 labourers were conscripted annually in Ku-chu-shan to process the tea sent as tribute to the imperial court at Changan.

[10] *Chha Ching*, ch. 6, p. 38; Carpenter (1974), p. 116.

tea, coarse tea and powdered tea were produced. The only indication we have is a couplet from a poem by the Thang poet Liu Yü-Hsi 劉禹錫 (772–842), namely,[11]

> I lean on the luscious bushes to pluck the eagle's beak,
> The leaves are *stir fried*, the room is filled with the scent of tea,

which suggests that stir-frying was already a method used to process fresh tea leaves, presumably into loose-leaf tea. The *Chha Phu* of Mao Wên-Hsi, c. +935 (Table 47) mentions several locations in Szechuan where the best loose tea was produced, suggesting that loose tea was already a product of some economic importance during the Thang.[12]

(ii) *Processing of Sung cake teas*

The most elaborate and characteristic implement for the processing of tea described in the *Chha Ching* is the drying furnace (*pei* 焙). In time the word *pei* began to denote not just the drying furnace but the entire work station where all the processing operations were carried out. Some *pei*'s evidently produced better teas than others, and batches of tea were often identified by the *pei* in which they had been processed.[13] The *Tung Hsi Shih Chha Lu*, +1051, counted thirty-two government-owned tea processing stations (*kuan pei*'s 官焙) in the Chien-an 建安 district, and thirty-two such stations in the Chien-hsi 建溪 district in Northwestern Fukien.[14] These were the most prestigious tea stations during the Sung and the home of the renowned tribute teas sent to the Emperor's court.

How were these famous Sung tribute teas produced? Much relevant information on the processing of tea is scattered in the major Tea Books of the Northern Sung (+960 to +1127) listed in Table 47, namely, Tshai Hsiang's *Chha Lu*, the *Tung Hsi Shih Chha Lu*, the *Ta Kuan Chha Lun*, and the *Hsüan Ho Pei Yuan Kung Chha Lu*, but the most complete treatment of this subject is found in the *Pei Yuan Pieh Lu*.[15] The procedure described in this work is summarised in Table 49. It is noted that before processing can begin, the station has to be made ready to for a new season of activity. This operation is called *khai pei* 開焙, which means literally to open up the furnace or work station. Presumably, its purpose is to ensure that all the implements were in working order, the supply of water and firewood (or charcoal) plentiful, and all the needed personnel mobilised. This is usually done three days before the onset of the

---

[11] Cited in C&C, p. 214. The original reads:

| *Tsu phang fang tshung tsê ying chui*, | 自傍芳叢摘鷹觜 |
| *Ssu hsü chhao chhêng man shih hsiang* | 斯須炒成滿室香 |

Eagle beak signifies the fresh buds that unfold to give leaves in the shape of eagle's beaks.
[12] *Chha Phu*, cited in C&C, p. 24.
[13] In the same manner as modern wines are identified by the winery and locality which produced them.
[14] *Tung Hsi Shih Chha Lu*, in C&C, pp. 34–8. We also learn from this work that there were altogether 1,336 tea processing stations in Chien-an of which 32 were government owned. The majority, i.e. 1,302 stations were privately owned. These numbers are indicative of the growing importance of tea in the economy of the Sung empire.
[15] *Pei Yuan Pieh Lu*, in C&C, pp. 84–8.

Table 49. *Processing of Sung tribute tea at Pei Yuan*

1. *Picking* (*Tshai chha* 採茶)
   This should begin before dawn when there is dew on the leaves, and stopped after sunrise. It must be done with the finger-nails, not with the fingers, lest the freshness and purity of the leaf be contaminated with body sweat.

2. *Grading* (*Chien chha* 揀茶)
   The tea leaves are graded as follows: *small bud*, *medium bud* (with a single leaf on the shoot), *purple bud* (with two leaves on the stalk), *a bud with two leaves* and *stem tops*. Only the first two grades are used for making tribute tea.

3. *Steaming* (*Chêng chha* 蒸茶)
   The leaves are washed until scrupulously clean. They are placed in the steamer over vigorously boiling water. The steaming must be just right. If over-steamed, the tea will be yellow and taste flat. If under-steamed, the tea will be green and heavy, and have a grassy taste.

4. *Pressing* (*Cha chha* 榨茶)
   After steaming, the leaves, now called *chha huang* 茶黃 (tea yellow), are cooled by rinsing with water. They are lightly pressed to remove the surface water, and then pressed heavily to express as much juice (*kao* 膏) from the leaf tissue as possible.

5. *Rolling* (*Chiu chha* 研茶)
   The leaves are rolled with a wooden handle on an earthenware plate. They are then dispersed in an appropriate amount of water. This must be done when the leaves are cooked and pressed dry.

6. *Moulding* (*Tsao chha* 造茶)
   The rolled leaves are shaped into cakes in a mould. The insignia of the tea is pressed on the surface by a template. The cakes are placed on a mat to dry.

7. *Roasting* (*Pei chha* 焙茶)
   This step is also known as *kuo huang* 過黃 (firming the tea yellow). The cakes are first toasted over a hot fire and then rinsed with hot water. This is repeated three times. They are then warmed slowly over the fire. The length of time needed is dependent on the thickness of the cake, 6–8 days for thin cakes and 10–15 for thick cakes. They are then cooled by fanning and then stored in an airtight chamber.

excited insects (*ching chih* 驚蟄) period, which falls on about 5 March in the Western calendar, and is the best time for harvesting the types of leaves required for the processing of teas of the highest quality.

Table 49 shows that there are major differences between the Sung tribute tea process and the Lu Yü's Thang process described earlier. First of all, in the Thang *picking* (step 1) was done on a sunny day, but during the Sung, to ensure that the leaf was not in anyway 'dried' by the sun, it had to be completed before sunrise. Quality rather than quantity was emphasised. The *Ta Kuan Chha Lun* admonishes, 'Leaves of whitish colour shaped like a sparrow's tongues or grains of corn are best. One leaf per shoot (*small bud*) is ideal; two leaves per shoot (*medium bud*) is next best; if there are more the product will be inferior. If the leaves have dirt on them, they are immediately dipped into freshly drawn water, which the picker has to carry with him/her.'[16] In fact, picking was a sophisticated team operation that required skilled labour, good organisation and accurate timing.[17]

Step (2), the *grading* of leaves was introduced to ensure that leaves of the same type were used to make the highest quality tribute teas such as Tshai Hsiang's little dragon cake (*hsiao lung thuan* 小龍團). According to Ouyang Hsiu (+1007 to +1072) each catty of this tea is worth two ounces of gold.[18] Step (3), *steaming*, is presumably carried out in the same type of equipment as in the Thang.

Step (4), *pressing*, is also a Sung innovation. The objective is apparently to remove as much juice from the steamed leaves as possible. This stands in sharp contrast to the *pounding* step in the Thang procedure, where the objective is just the opposite, that is, to lose as little juice from the leaf tissues as possible. The difference in treatment may be a reflection of the difference in the nature of the leaves. The Thang process was applied to leaves in areas such as Ku-chu-shan and Lo-chieh-shan (in Kiangsu–Chekiang) which is considerably further north than Fukien. The leaves would not be as succulent and rich in tea components as those grown in the south. The procedure was, therefore, designed to retain as much of leaf gel or juice as possible. The situation is reversed in Chien-an, Fukien during the Sung. The leaves are too rich in tea components. As pointed out in the *Ta Kuan Chha Lun*, 'If [the leaves are] pressed too lightly, the colour will be dark and the tea bitter.'[19] Unfortunately, there is no description of the kind of press that was used for both the light pressing and the heavy pressing.

Step (5) *rolling* (*yen chha* 研茶), is, we believe, quite a different type of treatment from the *pounding* (*tao chha* 擣茶) in the Thang procedure. The motion is probably similar to the kneading and rolling of a dough. The objective appears to be to weaken the internal structure of the leaf without ruining its external shape. Thus, it is possible to disperse the leaves in water to wash away any bitter components that may still remain in the extruded juice on the leaf surface.

Step (6), *moulding* is similar to that practised in the Thang process. The rolled and washed leaves after draining off excess water is pressed in a mould to give a cake of the desired shape. But in the Sung process a template is imprinted on the surface of the cake with an Imperial design which identifies it as a type of tribute tea. The various designs of templates used in the *Pei Yüan* plantation are illustrated in the *Hsüan Ho Pei Yüan Kung Chha Lu*, +1125.[20] Two of the designs are shown in Figure 123. The cakes are then placed on a mat to air dry.

---

[16] *Ta Kuan Chha Lun*, in C&C, p. 44, tr. Blofeld (1985), p. 29, mod. auct. Actually, there is a grade of leaf that is ranked even higher than the *small bud*. According to the *Pei Yuan Pieh Lu*, in some cases when a small bud is steamed and dispersed in water, peeling the outer leaf will yield a fine needle-like heart which is called *shui ya* 水芽 (water sprout). The tea from *shui ya* is probably the finest and the most expensive ever made. The *Chi Le Phien* (Miscellaneous Random Notes), +1139, Pt. 3, p. 9, relates how these special small buds sprouted from tea trees that were more than ten feet in height after a storm.

[17] The *Pei Yuan Pieh Lu*, in C&C, pp. 84–8 notes that 250 pickers were assembled at early dawn in a typical picking operation at the Fêng-huang Shan 鳳皇山 tea plantation. A gong was struck at the fifth hour. The supervising official issues a pass to each picker to enter the plantation. Before sunrise the gong is struck again, and all the pickers have to bring their harvest to the processing station. The pickers are trained to know not only how to pick but also what to pick.

[18] Ouyang Hsiu, *Kui Thien Lu* 歸田錄, c. +1070, cf. C&C, p. 235.

[19] *Ta Kuan Chha Lun*, C&C, p. 44.   [20] *Hsüan Ho Pei Yüan Kung Chha Lu*, C&C, pp. 55–82.

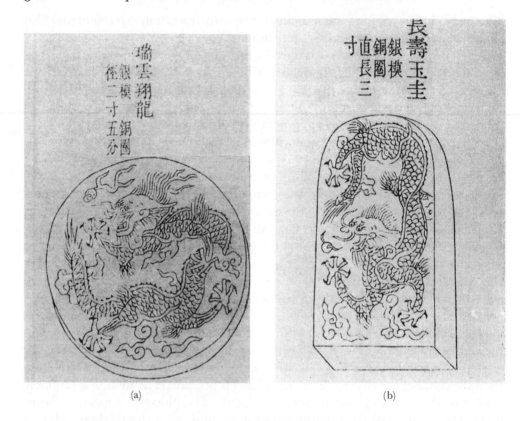

Fig 123. Design of seals on Sung tribute teas, Chhên & Chu (*1981*), pp. 70–1. 1a, 1b.

Step (7), *roasting* (*pei chha* 焙茶), the final step, is the washing and drying (or toasting) of the cakes over a slow fire. But it is done in a manner that would not mar the Imperial insignia on the surface of the cake. Unlike the Thang process, no eyes are pierced through the cakes. They cannot, therefore, be hung through a skewer over the furnace. How was the drying carried out? Unfortunately, the *Pei Yuan Pieh Lu* does not give a description of the furnace, but we know it had to be very different from that given in Lu Yü's *Chha Ching*. The first intimation of a Sung drying furnace occurs in Tshai Hsiang's *Chha Lu* of +1051. In the section on tea implements, we read:[21] '*Chha Pei*: The *chha pei* 茶焙 is [a tray] made of woven bamboo. It is covered with a layer of reeds (*jo* 蒻) to help it to retain heat, and placed at least a foot above the source of the heat. It is kept evenly warm so that the colour and flavour of the tea will be retained.'

Corroboration for this description is found in Emperor Hui Tsung's *Ta Kuan Chha Lun*, +1107. In a passage entitled *Tshang Pei* 藏焙 (Storage of Dried Tea product), we are told:[22]

---

[21] Tshai Hsiang's *Chha Lu*, C&C, p. 31. [22] *Ta Kuan Chha Lun*, C&C, p. 47.

If the tea is over-dried (*shu pei* 數焙), its surface will be wrinkled and the fragrance diminished. If is insufficiently dried (*shih pei* 失焙), the colour will be uneven and the flavour dissipated. Drying of the newly sprouted leaves will drive off undesirable vapours from water, land and dampness. Transfer a burning fire into the furnace (*lu* 爐). Mix seven-tenths of it with light ashes, leaving three parts of the fire exposed. These are also doused with ashes. Place the bamboo drying tray (*pei lou* 焙簍) over the fire to dry off the dampness. Then spread the the tea cakes on it so that every cake will receive the needed amount of heat. If the fire is too strong, cover it with more ashes. The strength of the fire should be adjusted to meet the size of the batch. Place the hand over the heat source. It should feel hot but not scalding.

The two passages indicate that the cakes were dried on bamboo trays placed above the furnace. The heat had to be mild so that the bamboo mat would not itself be burnt. We see that the word *pei* has wandered far from its original meaning as the drying furnace itself. In Sung times it may denote the processing station, the drying process as a whole, or a drying tray. New terms were introduced to indicate more specifically implements used in the *pei* process, such as *lu* for the furnace and *pei lou* for the drying tray. Unfortunately, no dimensions for the *lu* and the *pei lou* are given.

Although the Sung Tea Books have dealt only with the processing of tribute teas, there are strong indications that tea was growing rapidly as a popular drink among the common people. In fact, tea was a commodity that was produced and traded widely during the Northern Sung.[23] Tea cafes (*chha szu* 茶肆) such as the one depicted in the famous Sung scroll painting *Chhing Ming Shang Ho Thu* 清明上河圖 (Figure 124) were a familiar sight in the Northern capital of Kaifeng.[24] According to the *Mêng Liang Lu*, +1334, they were even more numerous in the Southern Sung (+1127 to +1279) capital of Hangchou. By then, tea had already been ranked, together with firewood, rice, oil, salt, fermented soy paste, and vinegar, as one of the seven proverbial necessities of daily living.[25] Two types of commercial teas were available, cake tea (*phien chha* 片茶) and loose tea (*san chha* 散茶). Cake tea includes a regular *phien chha* and a *la chha* 臘茶 (waxed cake tea). According to the Economic Affairs chapter of the *Sung Shih* (History of the Sung Dynasty), +1345, regular *phien chha* was prepared by 'pressing steamed leaves in moulds and stringing the cakes' while *la chha* was prepared in the same way as the tribute cake teas.[26] There is some indication that loose tea might have been prepared by stir-frying (*chhao* 炒) the leaves, a technique that was first mentioned in a poem by Liu Yü-Hsi of the Thang,[27] The details of how the stir-frying was carried out during the Sung are, however, unknown.

---

[23] Chu Chung-Shêng (1985), pp. 131–47, has estimated the annual production of tea during the Southern Sung to be about 37 million catties. According to Wu Chêng-Lo (*1957*), p. 60, a Sung catty is equal to about 0.6 kg. The annual production was thus about 22 million kg. Cf. also Chhên Chhuan (*1984*), pp. 57–66.

[24] *A City of Cathay*, National Palace Museum (1980), Plate 7, Scroll Section v. This is based on a Chhing copy of the Sung original done in +1736.

[25] *Meng Liang Lu*, ch. 16, p. 130 for a description of tea shops, and p. 139 on the seven necessities of daily living.

[26] *Sung Shih*, ch. 183, Economic Affairs, pp. 1795ff. On *phien chha*, the original text reads, *shih chuan mo chung chhuan chih* 實捲模中串之. In the case of *la chha* a coating of waxy oil such as *nao tzu* 腦子 was applied to give the cake a smooth surface so that the impression left by the template would be clearly visible.

[27] Cf. n. 11 above.

Fig 124. Picture of a Northern Sung tea house, from *Chhing-ming shang ho thu* 清明上河圖 (A City of Cathay), Pl., Scroll Section v, National Palace Museum, Taipei (1980).

From the perspective of processing, the principal difference between the Sung and the Thang cake teas is that while in the Sung process the steamed leaves were rolled before they were pressed in a mould into cakes, in the Thang process they were pounded. The leaves in the Sung tea cake still retained their shape, but those in the Thang cake tea were disintegrated beyond recognition. It is the rolling step that paved the way for the processing of loose tea that rose in popularity and displaced the cake tea during the Yuan and Ming.

(iii) *Processing of loose tea during the Yuan and Ming*

Actually, loose tea had begun to overtake cake tea in importance even during the Southern Sung. In fact, by the time of the Yuan Dynasty (+1271 to +1368) loose tea had become the dominant form of tea in the country. The *Nung Shu (Agricultural Treatise)* of Wang Chên, +1313, says that there were three kinds of tea in circulation, a *ming chha* 茗茶, a *mo chha* 末茶 and a *la chha* 臘茶. The first two were probably loose teas, and the last a cake tea. Wang Chên tells us further that of the three, *la chha* was the most expensive and difficult to make. It was mainly reserved for tribute and was seldom seen in the market place.[28] He also gave us the first description of how loose tea was processed.[29]

Tea leaves should be picked early in the morning, preferably before the *chhing ming* 清明 (clear and bright) and *ku yu* 穀雨 (grain rains) periods . . . They are then lightly cooked in a steamer. When done, disperse them on a bamboo tray and roll (*jo* 揉) them while they are still damp. Place the tray in a gently heated chamber (*pei* 焙) to dry the leaves without danger of charring. The tea product is then wrapped in *jo* 箬 (bamboo) leaves so that the heat will dissipate slowly.

The procedure as described is certainly a great deal simpler than those developed for the processing of Thang and Sung cake teas. The production of tribute cake teas was abolished by the first Emperor of the Ming Dynasty (+1368 to +1644) in +1391. Loose tea, called *ya chha* 芽茶 (bud tea) or *yeh chha* 葉茶 (leaf tea), was accepted as tribute.[30] The ostensible purpose of this edict was to lighten the intolerable burden the production of tribute cake teas had placed on the people of the tea-growing regions of the country. The real reason was probably that the quality of loose teas had so improved that the inordinate amount of labour involved in the processing of cake tea could no longer be justified. And yet, the Tea Books of the Ming period have relatively little to say about the processes used to make loose teas. All we have are tidbits of information without any systematic treatment such as that found in the *Chha Ching*, +760 and in the *Pei Yuan Pieh Lu* of +1186.

The first Tea Book of the Ming Dynasty (Table 47), the *Chha Phu* of Chu Chhüan, +1440, does not even mention processing but does have an interesting passage on storage:[31]

Tea should be stored in [containers made of] bamboo leaves. It prefers a warm and dry environment to one that is cold and damp. Warm it an an oven (*pei* 焙). The *pei* is made of wood. The tea is placed on top, and the fire (heat source) below. Cover it with bamboo leaves so that the heat is retained. The warming should be repeated once every three days preferably at human body temperature. This will help to preserve the tea and ward off dampness. The fire should not be too strong lest the tea is burnt. Those that are not regularly warmed in the oven should be stored in a bamboo leaf container placed high above the ground.

---

[28] *WCNS*, ch. 10, p. 113. The passage also mentions in passing how cake teas of the Sung and Thang were prepared. The description is so sketchy that it could not have been considered a topic of great interest by the author.
[29] *Ibid.*, pp. 112–13. The approximate time of arrival of *chhing ming* is 5 April, and of *ku yü* 20 April. A similar procedure is also given in the *Nung Sang I Shih Tsho Yao*, 2nd month, p. 40.
[30] The nomenclature of loose teas in mediaeval China is rather confusing, cf. Chhên Tsung-Mou (*1992*) ed. pp. 26–7. During the Sung, loose tea was known variously as *san chha* 散茶, *tshao chha* 草茶, and *mo chha* 末茶, but during the Ming the preferred names were *ya chha* and *yeh chha*.
[31] *C&C*, p. 122. We have met the author before in *SCC* Vol. VI, Pt I, p. 332 as a younger brother of Chu Hsiao (the Prince of Chou), the author of the celebrated *Chiu Huan Pên Tshao* 救荒本草 (Treatise of Wild Plants for use in Emergencies), +1406.

Table 50. *References to the processing of tea in Ming tea books*

Chhien Chhun-Nien, *Chha Phu* 茶譜, +1539, p. 125
 'Stir-fry and oven-dry to the proper degree' 炒焙適中
Thien I-Hêng, *Chha Phu* 茶譜, +1554, p. 130
 'Sun-dried tea is best; oven-dried is second best' 火作者為次，生晒者為上
Thu Lung, *Chha Shuo* 茶說, +1590, p. 134
 'Sun the tea . . . better than stir-fry over fire' 日晒茶. . . . 勝於火炒
Chang Yüan, *Chha Lu* 茶錄, +1595, p. 140
 'When the *kuo* (wok) is very hot, quickly stir-fry the tea' 候鍋極熱，下茶急炒
Hsü Tzhu-Yü, *Chha Shu* 茶疏, +1597, p. 151
 'As the leaves are picked, they are quickly stir-fried and dried' 旋摘旋炒
 'Tea from Chieh is not stir-fried but steamed' 芥茶不炒，甑中蒸熟
Lo Lin, *Chha Chieh* 茶解, +1609, p. 166
 'To stir-fry, the pan should be hot. To stir-dry, the pan should be warm' 炒茶，鐺宜熱。焙，鐺宜溫
Wên Lung, *Chha Chien* 茶箋, +1630, p. 174
 'While stir-frying, someone should be ready with a fan' 炒時須一人扇之

All page references are to Chhên & Chu (*1981*).

We can understand the need for this 'heat' treatment since the climate in South China is normally cool and extremely damp in the spring. The purpose of this 'heat (*pei*)' treatment is to ward off moisture so that the tea would not become mouldy. Fortunately, many of the remaining Tea Books of the Ming dynasty (Table 47) do have a few pertinent words to say about processing. The references and key phrases from them are presented in Table 50. Let us go over them quickly in turn.

From Chhien Chhun-Nien's *Chha Phu* of +1539, we learn for the first time that stir-frying is now an integral part of the process. The fresh tea leaves are stir-fried (*chhao* 炒) on a hot pan and then dried in an oven or drying chamber (*pei* 焙). The *Chha Phu* by Thien I-Hêng of +1554 informs us that sun-drying is also used; in fact it is superior to stir-frying on a pan. The same sentiment is expressed by Thu Lung in the *Chha Shuo* of +1590. But these entries tell us hardly anything useful about the details of the process. The next entry, Chang Yüan's *Chha Lu*, +1595, is considerably more informative. We quote:[32]

Discard the old leaves and bits of stems from the freshly picked leaves. The cooking pan or wok (*kuo* 鍋) is two feet four inches wide. Stir-fry about half a catty of leaves. The pan should be extremely hot just before the leaves are dropped into it and quickly stirred. The fire should not be diminished. Reduce the fire when the leaves are done. They are then dispersed on a bamboo screen and tossed several times before they are returned into the pan. The fire should then be gradually reduced, and stirring continued until the leaves are thoroughly dry.

---

[32] C&C, pp. 140–1. Before describing the processing, the *Chha Lu* recommends that the leaves be picked five days before the arrival of *grain rains* (about 5 April). Young sprouted purple leaves are the best; those slightly wrinkled are next; followed by those that are somewhat lumpy. Those that are as smooth as bamboo leaves are the least desirable.

In this case, stir-frying has become the backbone of the process. It serves two functions. The first is to wilt the leaves and heat-denature the contents without charring them; and the second is to stir the wilted leaves until they are dry. The pan is presumably a concave-type wok similar in size to that shown Figure oo. The technique of stir-frying is described in greater detail in the *Chha Shu* of Hsü Tzhu-Yü, +1597, who admonishes:[33]

When tea leaves are first picked, their aroma is not fully developed. The power of fire will release their latent fragrance. For stir-frying, the worst utensil is a newly cast iron pan (*tang* 鐺). The raw iron taste will destroy the flavour of the tea. Above all, do not use a pan that is contaminated with any dirt or grease. The effect of grease is even worse than raw iron. Ideally, the pan should be a vessel dedicated to processing only tea. It is not to be used for any other purpose. The pan should be fueled with tree stems, and never with branches or leaves. The latter will burn too quickly and die down just as quickly. The pan should be scrupulously clean. The fresh leaves are stir-fried as soon as they arrive from the field. Stir-fry only four ounces of tea at a time. Start with a low fire to wilt the leaves and then with a high fire to hasten their denaturation. Use a wooden spatula to quickly stir-fry the leaves until they are half-cooked and their fragrance partly released. They are then ready to be transferred to a drying basket (*lung* 籠). The basket is lined with cotton paper before it is placed in the oven. After the tea is dried and cooled they are stored in a bottle. If labour is plentiful several stir-fry pans and several drying baskets may be kept in operation at one time. When labour is short and only one or two pans are in use, four to five drying trays may still be needed, since *stir-frying is fast but oven-drying is slow* (*chhao shu erh pei chhih* 炒速而焙遲).

The last sentence makes it clear that in this case two operations were involved: *chhao* (stir-frying) and *pei* (oven-drying), although it is not so clear how the tea in the baskets (*lung*) was dried. Presumably, it was placed in a *pei* (drying oven or chamber). The *Chha Shu* goes on to say that in the *Chieh* 岕 district[34] the tea is steamed rather than stir-fried before it is dried in an oven, a procedure similar to that described in the *Wang Chên Nung Shu*, +1313 (see p. 529).

We see that in Chang Yüan's procedure both the 'denaturing' and 'drying' of the leaves are carried out on the frying pan, while in Hsü Tzhu-Yü's procedure only the 'denaturing' is done in the frying pan but the 'drying' takes place in an 'oven'. Both types of procedures are described in the *Chha Chieh* of Lo Lin, +1609, which says:[35]

First Recipe:

To stir-fry tea, the pan should be hot. To dry tea, the pan should be warm. The stir-frying should be a quick operation. When the pan is hot enough to burn one's hand, pour the leaves onto it. While the leaves crackle stir them in rapid motion with the hands. Disperse the leaves on a bamboo mat, and cool them immediately by fanning on them. Roll and rub (*jo nuo* 揉挼) the leaves, and repeat the stir-frying. Then transfer them to a pan over a light fire and stir them until they are dry. *The colour of the tea product is a dark green*. If after stir-frying, the leaves are not fanned, the colour of the tea would be different.

---

[33] C&C, p. 151. [34] The *Chieh* (or *Lo Chieh* 羅岕) district is in Chekiang, near the famous *Ku-chu-shan*.
[35] C&C, p. 166.

Second Recipe:

Freshly picked tea leaves are flavourful and succulent. They should be stir-fried over a high heat to release their fragrance. But the fire should not be too strong. The worst thing is to leave the leaves half dry. Do not dry them in the pan. Place them on a basket and dry them over a slow heat. After the leaves are properly cooked [but before they are dried], they should be rolled and rubbed [presumably by hand]. This action helps to mix the gels and liquids in the leaves, so that the flavour will be more easily extracted by hot water.

In both recipes three steps are involved (1) cooking and wilting the leaves by stir-frying (2) rolling and rubbing the wilted leaves and (3) drying the treated leaves. Of particular interest is the fact that step (2), rolling and rubbing, is reinstalled as an integral part of the process. This step was mentioned in Wang Chên's *Nung Shu* (Agricultural Treatise) of +1313, but has been left out by all the previous five Ming Tea Books listed in Table 50. The first recipe also stresses the importance of cooling the stir-fried leaves quickly by fanning before they are rolled and rubbed. The tea will then have the desired dark green colour. The importance of fanning the stir-dried leaves is reiterated by Wên Lung in the *Chha Chien*, +1640, which says:[36]

In stir-frying [the tea leaves] someone should be ready with a fan to quickly dispel the heat. Otherwise, both the colour and fragrance will be harmed. Based on my own experience, the product that has been fanned is a jade green, that has not been fanned is yellow. As soon as the leaves are transferred from the pan to a large porcelain plate, they are immediately fanned. When they are cool to touch they are rolled by hand. Then they are returned to the pan, stir-dried over a light fire, and then further dried in an oven (*ju pei* 入焙). The rolling liberates the tissue essence, so that the fragrance and flavour will be released more easily when an infusion is made. According to Thien I-Hêng, author of the *Chha Phu* of +1554, the best tea is made by exposure to the sun without stir-frying or rolling, but I have not had a chance to try his process.

Wên Lung then goes on to discuss the construction of his *pei* (oven or drying chamber). After repeating the description of the Thang *pei* in the *Chha Ching* he suggests that nowadays it is not necessary to follow Lu Yü's prescription and says:[37]

I have constructed a drying chamber (*pei* 焙). The height does not exceed eight feet; each side of the square is less than ten feet. The walls and ceiling are perpendicular to each other. They are plastered with thick cotton paper, so tightly that no crevices are present. Distribute three to four large fire-urns (*huo kang* 火缸) in the room. Bamboo mats lined with freshly washed new hemp cloth are placed on the fire-urns. Scatter the wilted and rolled leaves on the mat, and seal the chamber to allow the tea leaves to dry. Do not cover the leaves. For the leaves at this stage are still damp, and will turn yellow if they are covered. Allow the drying

---

[36] *Ibid.*, pp. 173–4. The *Chha Chieh* repeats the admonition given in the *Chha Shu* (cf., p. 531) that raw iron utensils should not be used for stir-frying the tea leaves. But it contradicts the *Chha Shu* in recommending the use of pine needles as the fuel to heat the pan for stir-drying. It also advises the use of one's 'hands in stir-frying tea (*tshao chha* 炒茶). The hands will not only distribute the tea leaves evenly but also feel the temperature of the pan.' This requires considerable skill when the aim is to cook or wilt the leaves and the pan is very hot. But it will be easy if the purpose of the stirring is to dry the leaves, and the pan is merely kept warm.

[37] *Ibid.*, p. 174. The drying chamber in Lu Yü's *Chha Ching* is described on p. 522; cf. n. 7 above.

Table 51. *Ming system for processing tea*

Step	Operation		
1. Wilting and denaturing	Stir-frying *chhao* 炒	Steaming *chêng* 蒸	Sunning *shai* 晒
2. Rolling	Rolling *ju nien* 揉撚	—	—
3. Drying	Stir-frying *chhao* 炒	Baking *pei* 焙	Sunning *shai* 晒

to go on for two to three *shih* 時.[38] When they are almost dry, cover them with a bamboo lid [and allow the drying to continue]. When the leaves are completely dry, they are cooled and stored. The same procedure may be used if the tea has to be re-dried.

From these accounts we may summarise the Ming tea processing system in Table 51. Basically, it consists of three steps. The first is wilting-cum-denaturing. The leaves are heated by stir-frying, steaming or exposure to the sun, so that they become pliable and amenable to rolling by hand. The stir-frying and steaming will not only wilt the leaves but also denature and inactivate the enzymes in the leaf tissue. Exposure to the sun will wilt the leaves, but the heat would not be sufficient to inactivate the enzymes. As we shall see, sunning will therefore affect the quality of the tea in a special way. The second step is rolling. The leaves are rolled and rubbed into the desired shape. This step is left out in some of the Tea Books. We suspect the omission is unintentional since it was probably a step so well known to the tea processors that it required no reiteration. The last step is drying, which can be accomplished by stir-drying on a pan over a low heat, baking or sunning. By varying the types of leaves collected and the types the treatment used in steps 1 and 3, a wondrous array of teas can be produced. It is remarkable that basically the same procedure was found to be in operation by Robert Fortune in 1848[39] and by William

---

[38] One Chinese *shih* is equal to two hours. Cf. Vol. III, p. 313. For a discourse on the measurement of time in China cf. Bedini (1994).

[39] Fortune (1852), pp. 276–8, describes the processing of 'green' tea in China in the 1840s. A condensed version of his account by Hardy (1979), p. 60, follows:

The leaves are carried home to the cottage where the operation of drying is performed . . . The drying pans that sit on the stoves are of iron, round and shallow, and, in fact, are the same, or nearly the same as those used by the natives for cooking their rice. A row of these are built into the brick work . . . The pans become hot soon after the warm air has begun to circulate in the flue beneath them. A quantity of leaves from a sieve or basket, are now thrown into the pans, and turned over and shaken up. The leaves are immediately affected by the heat. This part of the process lasts about five minutes, in which time the leaves lose their crispness and become soft and pliable. They are then taken out and thrown upon a table, the upper part of which is made of split bamboo . . . Three or four persons now surround the table and the heap of tea leaves is divided into as many parcels, each person taking as many as he can hold in his hands, and the rolling process commences. I cannot give a better idea of this operation than by comparing it to a baker working and rolling his dough . . . This part of the process also lasts about five minutes, during which time a large portion of green juice is expressed.

When the rolling process is completed the leaves are removed from the table and shaken out for the last time . . . and are exposed to the action of the air. The leaves which are not soft and pliant are again thrown into the drying pans, and the second heating commences. Again one individual takes his post at the furnace, and keeps up a slow and steady fire. Others resume their places at the different drying pans – one at each – and commence stirring and throwing up the leaves, so that they may all have an equal share of the fire, and none get scorched or burned.

Ukers in the early years of the 20th century.[40] Even details such as the crackling sound produced when the fresh leaves touch a hot frying pan, or the fanning of the leaves while they are being stir-dried, were noted in Ukers' account (as well as the rolling of the pan-fried leaves by feet, see Figure 125).

At any rate, the Ming methods were much simpler than those employed in the making of cake teas during the Sung. In fact, some of the Ming authors were quite critical of the techniques favoured by the Sung producers. For example, Hsü Tzhu-Yü, in +1597, complained of the use of only the heart of the sparrow tongue leaf for making the famous No. 1 Pei-yüan tea.[41]

The price of one cake was 400 thousand cash. Yet it would suffice for only a few cups of drink. Why on earth should it be so expensive? The true flavour of the tea would have been lost when the leaf was torn and soaked in water. The external fragrance added would further mask the true flavour of the tea. These doctored products that result cannot possibly compare with those made by our current procedure. *As the leaves are picked they are quickly stir-fried and dried.* Both the colour and fragrance are preserved, and one can taste the true flavour of the tea.

The Ming tea connoisseurs also had a better understanding of how it should be stored. On this subject Wên Lung, +1630, has this to say:[42]

Tea should be stored in a cool and dry environment. If it is damp the tea will lose both flavour and fragrance. If it is hot the tea will turn yellow and become bitter. Tshai Hsiang (author of the Sung *Chha Lu*, cf. Table 47) is mistaken in claiming that tea likes heat. Our current practice is to keep tea in large urns at the upper storey of the building. The mouths should be covered with bamboo leaves. They must not be left open. The cover will protect the leaves from the bad vapours [in the air]. On a sunny and dry day, distribute the tea into smaller containers.

---

[40] The process had changed little in William Ukers' description of the manufacture of green tea in the early part of the 20th century, cf. Ukers (1935), 1, pp. 303–4.

The green tea *kuo* 鍋 is set in the top of a waist-high, brick stove, and about five inches below the level of its surface. The *kuo* itself is about 16 inches in diameter and ten inches deep . . . For steaming the leaf, the *kuo* is made very nearly red hot with a wood fire, and into it are thrown about half a pound of leaves by the firer. These he stirs rapidly about while they produce a crackling sound and a great quantity of steam. The operator frequently raises the leaves a little above the top of the stove and shakes them over the palms of his hands to separate them and allow the steam to penetrate. At length, after two or three brisk throws about the *kuo*, the operator collects the leaves together in a heap, and, with a single deft motion sweeps it into a basket held in readiness by another workman.

From the *kuo* the steamed leaf goes at once to a table covered with matting for rolling . . . After the leaves have been rolled into a ball, they are shaken apart and then twisted between the palms of the hands; the right hand passing over the left with a slight degree of pressure as the hand advances, and relaxing again as it is returned. This twists the leaves regularly and in the same direction. After rolling the leaf is spread out in sieves and allowed to cool for a short time; then it again goes to the *kuo* for the first firing [i.e. stir drying]. This time the fire is considerably diminished, and charcoal, instead of wood, is used for fuel, in order to avoid any smoke; but the pan is still kept so hot that the finger cannot be borne upon it for more than an instant. Close attention is given to regulating the heat; a fireman being constantly employed in this duty, while another fans the leaf throughout the entire operation.

This stir-drying is repeated two more times until it is acceptably dry as indicated by the appearance of a bluish tint. The three stir-dryings take about ten hours. The rolling of leaves by feet in the making of inferior Twankay tea was also cited by Mui & Mui (1984) from Ball (1948), pp. 233ff.

[41] *Chha Shu*, +1597, C&C, p. 150.   [42] *Chha Chieh*, C&C, p. 167.

Fig 125. Rolling tea leaves with feet in China, from Ukers (1935) 1, p. 468.

(iv) *Origin and processing of oolong tea*

The loose tea that we have so far discussed falls under the category of 'green tea', which is still the most popular type of tea consumed in China today.[43] As noted earlier, the process developed during the Ming has remained virtually unchanged from the 16th to early 20th centuries.[44] The continuing dominance of green tea may be the reason why the authors of the standard 'Tea Books' published during the Chhing Dynasty (+1644 to 1911) have focussed on compilations of previous works rather than developments in tea technology of their own times. This is most unfortuante since it was during the Chhing that two new types of tea, oolong or *wu lung*

---

[43] *CKCC* (1992), pp. 132–254, lists 138 varieties of green tea, 10 varieties of red (i.e. black) tea and 13 varieties of Oolong tea being produced in China today. In addition there are 4 varieties of white tea (白茶), 10 varieties of yellow tea (黃茶) and 5 varieties of dark tea (黑茶). Altogether 176 varieties of tea are listed, out of which 138 varieties, i.e. 78% are green teas.

[44] Cf. ns. 39 and 40 above. There is one other Chinese tea that deserves a brief mention. The Phu-erh 普洱 tea from Yunnan consists of a group of unfermented teas prepared from a broad leaf variety of bush, cf. Blofield (1985), pp. 75–7. *CKCC*, pp. 238 and 438, classifies it as a type of non-enzymatically oxidised dark tea (*hei chha* 黑茶) usually in cake or brick form. They are well known for their health-giving properties, and are popular in South China.

烏龍 (dark dragon) tea and *hung chha* 紅茶 (red tea but called 'black tea' in the export trade) joined *lü chha* 綠茶 (green tea) to become the tea triumvirate of maritime commerce. These exciting innovations apparently attracted little attention among the literate tea connoisseurs of the Chhing Dynasty, who had continued the traditional practice of identifying teas according to the localities where they were produced.[45] In fact, the very words *lü chha*, *oolong* and *hung chha* were not seen in the Chinese literature until the latter part of the 19th century, even though the terms 'green tea' and 'black tea' had been familiar expressions in the vocabulary of European tea traders since the late 17th century.

To provide a baseline for further discussion we shall start with an account of the traditional way red tea (*hung chha*) and oolong tea were processed in the early years of the 20th century. For this information we shall again turn to William Ukers whose personal observations are summarised below:

Red tea:[46] After sorting, the leaves are spread out on a large bamboo mat and placed outdoors to wilt in the sun. If the leaves have been gathered in the rain, they may be left on bamboo trays in a drying room fitted with several charcoal fires burning in earthenware pans. While drying, they are stirred occasionally with the hands and tossed into the air to *prevent excessive* fermentation.[47]

After wilting, the leaves are cooled on bamboo trays to check fermentation. When they emit a faint fragrance they are given a light rolling for about ten minutes with the palms of the hands, after which they are again stirred and tossed for thirty minutes. Rolling and stirring are repeated three or four times, until the leaves turn dark in colour and become quite soft.

The procedure then continues in the same way as the process for green tea.

Oolong tea:[48] The leaf is slightly withered before panning [i.e. stir-frying], and during this process, a light fermentation is allowed to develop. For this purpose the leaf is spread three to four inches deep in large bamboo baskets, and placed in the shade, where it is frequently turned over for four to five hours. The temperature of the leaf, during this time, ranges from 83° to 85° F. Finally, it changes in colour and gives off a characteristic apple-like odour, which is the signal for checking any further fermentation by drying.

The process then continues with the same procedure as that for green tea.

During the initial withering or wilting treatment the leaf loses moisture and turgidity. The cell wall becomes more permeable to chemicals. When exposed to oxygen in the air, the catechins in the leaf cells are oxidised enzymatically into theaflavins and thearubigins (Figure 131), which endow the red tea with its characteristic colour. Aroma and flavour components are released. In the early research

---

[45] For example, the *Meng-ting* from Szechuan, *Lung-ching* from Hangchow, *Sung-lo* from Anhui, *Lo-chieh* from Chhang-hsing (Chekiang), Wu-I from Fukien etc., cf. *Pen Tshao Kang Mu Shih I* (+1765), cited in C&C, pp. 374–5.

[46] Ukers (1935) I, pp. 302–3 on the processing of 'black tea'. Cf. also Balfour (1885), vol. III, p. 831, on the manufacture of *congu* and *souchong* teas. The term 'red tea' does not exist in English. *Hung chha* in Chinese is translated as 'black tea' in English.

[47] It seems to us that Ukers is mistaken. The tossing of the leaves would promote rather than impede 'fermentation'.

[48] Ukers (1935) I, pp. 293 and 305. Cf. also pp. 334–7 for the processing of oolong tea in Taiwan (Formosa).

Table 52. *Comparing the traditional processing of red, oolong and green teas*

Tea	Green	Red	Oolong
1. Wilting (and fermenting) *Wei tiao* 萎凋ᵃ	no	yes	partial
2. Rolling *Jo nien* 揉捻ᵇ	no	yes	no
2a. Fermenting *fa hsiao* 發酵	—	yes	—
3. Heat inactivationᶜ *Sha chhing* 殺青	yes	yes	yes
4. Rolling: *Jo nien* 揉捻	yes	yes	yes
5. Drying by heat *Kan tsao* 乾燥	yes	yes	yes

ᵃ Leaves are wilted in the sun and then kept in the shade to allow oxidation of the cell components.
ᵇ If rolling (2) and fermenting (2a) are done thoroughly then step (4) is no longer needed, and steps (3) and (5) may be combined.
ᶜ Usually by stir-frying. In the tea trade this step is known as 'firing'.

on the chemistry of tea it was thought that yeast and other microorganisms were responsible for these reactions.[49] Hence, the process was called 'fermentation'. Although we now know this is a misnomer, the name has been used for so long among tea producers and traders that it has stuck as the accepted term in the tea literature. Rolling ruptures the cells and mixes their contents, thus increasing the rate of fermentation when the leaves are further exposed to air.

The steps involved in the processing of red, oolong and green tea (Table 51) are compared in Table 52. We see that steps (3), (4) and (5) are common to all three products. The differences lie in the way the leaves are treated before step (3) begins. In the case of red tea, the leaves are wilted (step 1), rolled (step 2) and allowed to ferment further until the process is complete (step 2a). Since the leaves are already wilted and rolled, the primary function of step (3), stir-frying or firing, is to denature the enzymes and dry the rolled leaves. Step (4), the second rolling, further exudes the juice left in the leaves and shapes the leaves in their final form. In modern practice, step (4) is often dispensed with and steps (3) and (5) are combined.

In the case of oolong tea, the leaves are wilted (step 1) but not rolled. The fermentation is, therefore, only partial. In the case of green tea, the freshly picked leaves are wilted quickly at high heat. The cell enzymes are immediately denatured and no fermentation is possible. Thus, we can say that green tea is unfermented, red tea is fully fermented and oolong is semi-fermented or partly fermented. The characteristics of oolong are, therefore, in between those of red tea and green tea. By adjusting the processing conditions, oolong teas varying from 15% to 60% fermentation can be made. The main difference between red tea and the other teas is that it is wilted, rolled and allowed to 'ferment' to completion before it is 'fired'.

From these accounts we can see that if during the processing of green tea if the freshly picked leaves were accidentally left to stand in the shade or in the sun for

[49] For a discussion of this phenomenon cf. Yamanishi ed. (1995), pp. 461–7, cf. also Ukers *ibid.*, pp. 526–34 and Chhên Chhuan (*1984*), pp. 266–7.

some length of time, a certain amount of wilting and fermentation would have taken place naturally before the leaves were stir-fried. The resultant tea would have been lightly fermented, and might have acquired sufficient fermented flavour to distinguish it from the green teas processed in the usual way. This is probably how oolong tea originated. The key incident must have occurred somewhere in northwestern Fukien, the original home of partly fermented (oolong) teas in China in the late Sung or early Ming.

Let us recall that the most prestigious Sung tribute cake teas came from Pei-yuan, near Chien-chou (present-day Chien-ou). As loose-leaf tea gained popularity, the centre of production moved north to the Wu-I 武夷 mountains, which became the premier tea producing region of the province. In fact, by +1582 the *Min Ta Chi* 閩大記 (Greater Min Records) was able to declare that the tea from Wu-I was the best in the province and the same sentiment was shared by Hsü Tzhu-Yü in his *Chha Shu* of +1597.[50] By early 17th century Hsü Po had this to say:[51]

In the hills both soil and climate are congenial to tea. Within the nine bends there are several hundred establishments engaged in the manufacture of tea. Each year several tens of thousands of catties of tea were produced. Transported over water and land, the tea was distributed into the four corners of the realm. Within the seas no name was more highly honoured than Wu-I.

Wu-I tea is mentioned and praised in the *Min Hsiao Chi* 閩小記 (Lesser Min Records), +1655, *Kuang Yang Tsa Chi* 廣陽雜記 (Miscellanies about Kuang-Yang), +1695 and the *Tshung Hsün Chhi Yü* 聰訓齊語 (Didactic Notes), +1697.[52] We can infer from these references that by the 17th century Wu-I tea was an important Fukien tea well known throughout China. It was during this century that maritime trade in tea between China and Europe began and Wu-I tea became a major participant in this event. According to Thomas Short (1690–1772), 'the Europeans contracted their first acquaintance with the green tea; then "Bohea" took its place.'[53]

Bo-hea happens to be the way Wu-I 武夷 is pronounced in the Amoy dialect, Amoy being the home of Chinese merchants involved in the export of tea to Europe in the late 17th century. Bohea soon became a household word in the tea trade. Short's observation indicates that the Europeans preferred Bohea (also known as *black tea*) to green tea, and that there was a significant difference between the flavour of green tea and that of Bohea tea. In fact, some Europeans were under the impression that Bohea (i.e. *black*) and green teas were derived from two different species

---

[50] *Min Ta Chi*, cited in C&C, p. 302; and *Chha Shu*, C&C, p. 149. The same sentiment was expressed by the Jesuit scholar du Halde (1736) I, pp. 13–14, cf. Gardella (1994), p. 30.

[51] *Chha Khao*, cited in C&C, p. 317. The nine bends of the river are a distinguishing feature of the landscape in the Wu-I mountains. They were noted by Robert Fortune in his travel to the 'Woo-e-shan' in 1849, cf. Fortune (1852), p. 240. For further information on tea of the Wu-I region, cf. Kung Chih & Yao Yüeh-Ming (*1995*) and Chhên Hsing-I (*1995*).

[52] Cited in C&C: *Min Hsiao Chi*, p. 341; *Kuang Yang Tsa Chi*, p. 354; and *Tshung Hsün Chhi Yü*, p. 355.

[53] Short, Thomas, *Discourses on Tea, Sugar, Milk, Mead, Wines, Spirits, Punch, Tobacco etc. with Plain Rules for Gouty People*, London, 1750; cited in Ukers (1935), I, p. 29.

of plants. Linnaeus was persuaded to assign the name *Thea viridis* to the green tea plant, and *Thea bohea* to the Wu-I tea plant. It is quite possible that the Chinese merchants themselves, who negotiated the transactions with the foreign merchants, had only the vaguest idea of how the two types of tea came about.

Up until the early 18th century there was no record in the Chinese literature of how the processing of Wu-I tea might have differed from that of green tea. The first inkling of the nature of this difference was revealed in the *Sui Chien Lu* 隨見錄 (Record of Second Encounters), +1734 which comments that while the flavour of most teas suffered when exposed to the sun, the Wu-I tea thrived on being in the sun.[54] This statement suggests that at some stage in the process, exposure of the leaves to the sun would be advantageous. Further illumination on this point occurs in the *Wang Tshao Thang Chha Shuo* 王草堂茶說 (Talks on Tea from the 'Wang' Grass Pavilion), +1734. We quote:[55]

Leaves of the Wu-I tea are picked from the 'grain rains' to the 'summer begins' (approx. April 20 to May 20). The tea is picked in intervals of twenty days. At the first picking the leaves are large and robust. At the second and third pickings the leaves are smaller, the flavour thinner and the taste a bit bitter. At the end of summer and beginning of autumn the plants are picked once more. This is called the autumn dew. The fragrance is strong and the taste pleasant, but the yield in the next year will suffer. Therefore, pick with restraint.

After the leaves are picked they are placed on bamboo baskets and exposed to the wind and sun. This step is called *shai chhing* 曬青 (wilting in the sun). When the green colour is a little faded, they are stir-fried and oven dried. Some elite teas such as the Yang-Hsien 陽羨 is steamed (*chêng* 蒸) but not stir-fried (*chhao* 炒); it is oven-dried (*pei* 焙). Others such as Lung-ching 龍井 is stir-fried and stir-dried but not oven-dried. Its colour is thus pure. Only Wu-I is both stir-fried and oven-dried. When done, it is half-green and half-red. Green is the colour of stir-frying, and red, of oven-drying.

Although the author is mistaken in assuming that green is the colour induced by stir-frying and red the colour by oven-drying, the description leaves no doubt that what we have here is an outline of a rudimentary oolong tea process. The *shai chhing* step is what distinguishes Wu-I from the regular green teas of the day. The exposure to sun and wind causes the leaves to wilt and enables the cell components to undergo a 'fermentation' or 'enzymatic oxidation' (i.e. step 1 in Table 52). The change in flavour that ensues was evidently well received by some tea connoisseurs. This is the basis for the remarks of Thien I-Hêng in +1554 and Thu Lung in +1590 on the superiority of sun-processed tea over heat-processed tea that are cited in Table 50. We now know that in any procedure that involves a wilting of the leaves in the sun, a certain amount of fermentation is bound to take place. But we have no idea when exposure to the sun (and to the wind) was first adopted as part of the

---

[54] *Sui Chien Lu*, author unknown, no longer extant, found in Lu Thing-Tshan 陸廷燦 *Hsü Chha Ching* 續茶經 (Continuation to the Tea Classic), +1734, *SKCS* **844**, p. 697, cited in C&C, p. 362.
[55] *Wang Tshao Thang Chha Shuo*, author unknown, no longer extant, found in *Hsü Chha Ching*, +1734, *SKCS* **844**, pp. 695–6, cited in C&C, p. 363.

procedure in Wu-I. It could have occurred sometime towards the end of the Sung, when loose tea began to overtake cake tea in popularity.[56]

That Wu-I tea in the 18th century was significantly different from regular green tea is supported by two pieces of evidence. The *Pên Tshao Kang Mu Shih I* (Supplementary Amplifications to the Great Pharmacopoeia), +1765, says:[57]

Wu-I Tea comes from Chhung-an 崇安 in Fukien. Its colour is black and its taste sour. It helps digestion and lessens wind. It stimulates the spleen and relieves a hangover. According to Tan Tu-Kho all teas are cooling in their effect. Those with weak stomachs have had to stop drinking them. But Wu-I is warming and will not upset the stomach. Those who cannot drink [green] tea may drink it without harm.

Even more striking is this passage from Yuan Mei's *Sui Yuan Shih Tan* (Recipes from the Sui Garden) +1790,[58]

Wu-I Tea: normally I am not fond of this tea. To me it seems heavy and bitter. Drinking it is like drinking medicine. [My mind was changed] in the autumn of 1786, when I toured the Wu-I mountains and visited Man-thing Peak and Thien-yu Temple. The monks there eagerly invited me to taste their teas. The cups were the size of walnuts and the pot no larger than an orange. [They advised me] 'When a sip is taken, do not swallow it immediately. First, smell its fragrance, then test its taste. Mull over its flavour slowly in the mouth, then drink it down. The scent will overwhelm the nose even as the sweetness lingers on the tongue.' After the first cup, I drank one or two more cups. My cares were dispelled and anxieties relieved. A sense of well-being and contentment came over me. It was then that I realised while Lung-ching is noted for its purity and Yang-hsien for its clarity, none can compare in flavour with Wu-I. Wu-I is as different from them as jade is to quartz. Now I know why Wu-I is renowned throughout the land. Its reputation is well deserved. The tea leaves may be infused up to three times, yet their flavour is not depleted.

Readers who are familiar with Chinese tea culture will recognise this account as Yuan Mei's encounter with the art of *kung fu* 功夫 tea, in which tea is brewed as a rich infusion and served in tiny cups.[59] It is a practice that is highly popular among tea votaries in southern Fukien, Taiwan, northeastern Kuangtung and among the Chinese diaspora in Southeast Asia. The infusion, being stronger than the usual

---

[56] In this connection the remarks of Liang Chang-Chü 梁章鉅, *Kui Thien Suo Chi* 歸田瑣記 (*1845*), cited in C&C (*1981*), pp. 401–2, may be pertinent. He reminds us that according to the *Wu I Tsa Chi* 武夷雜記, by Wu Shih 吳拭, a Ming work (cited in C&C, p. 336), Tshai Hsiang, the Tea Commissioner, in about +1050, was already aware of the virtue of Wu-I tea, but his favourable opinion was not shared by later writers. 'This is not because that Wu-I did not produce [good] tea during the Sung, but rather that the technology of processing there was not up to par. Its tea was, therefore, regarded with indifference. It was not until the beginning of the Yuan that two tea officials were assigned to the Wu-I region. There were 202 tea plantations. A new processing station was setup in *Shih Chhü Chhi* 四曲溪 (Four Bend Brooks) . . . In time, Wu-I tea was not only welcome throughout the four corners of the realm, but also shipped via Kuangtung to the countries across the seas.' Liang's remarks would suggest that Wu-I tea started to enjoy a renaissance during the Yuan. The climate would have been favourable for experimentation and innovation. It is quite possible that the adoption of sun-drying as part of the procedure occurred at this time.

[57] *PTKMSI*, p. 252; cf. C&C, p. 374. Compare this statement with that in the *Yin Shang Chêng Yao*, p. 58: 'In general, tea is sweet and bitter, slightly cooling and non-toxic.'

[58] *SYST*, pp. 144–5; cited C&C, p. 383.

[59] For further information on Kung-fu tea cf. Blofeld (1985), pp. 133–40. Kung-fu tea is also described in the *Min Tsa Chi* 閩雜記 (Miscellaneous Notes on Fukien), 1857, by Shih Hung-Pao, cited in C&C, p. 408.

green tea, is savoured like liqueur rather than drunk like wine. Normally, oolong is the tea of choice for preparing this type of infusion. From these references we may therefore infer, that most of the Bohea (i.e. Wu-I) or 'black' teas exported to Europe in the 18th century were actually what we would now call oolong teas. The term oolong (*wu lung* 烏龍), however, was not seen in the written records until 1857, when Shih Hung-Pao 施鴻保 drew attention to an excellent tea called *wu lung* produced in Sha-hsien 沙縣, south of the Wu-I region.[60] *Wu-lung* was mentioned again as the name of a local tea in the *Min Chhan Lu I* 閩產錄異 (Unusual Products of Fukien), 1886.[61]

The use of 'oolong' as a generic term representing the Wu-I (i.e. partly fermented) type of tea, first appeared among tea traders in Formosa (Taiwan) in the 1860s. The Chinese planters who produced it probably came from parts of Fukien where oolong tea was grown and processed. According to Ukers, export of Formosan oolong to the West started in 1869 and it soon became a major factor in the economy of the island.[62] In this connection it is interesting to note that in 1892 a traveller from the mainland had this to say in his diary:[63] 'The hills near Taipei are fully planted with tea. Those of highest quality are known as Wu-lung 烏龍. It is much favoured by westerners.'

This remark suggests that 'oolong' was not yet a term commonly used by the Chinese towards the end of the 19th century. But since the early years of the 20th century oolong has been accepted universally as the generic term for partly fermented teas. The origin of oolong tea is thus quite clear. It was first made when freshly picked leaves were allowed to wilt in the sun before they were subjected to the green tea process. It has been around at least since the early Ming under the name Wu-I. But what about red tea? What is its origin? This has turned out to be the most intriguing question of all.

(v) *Red tea (*hung chha*) and the black tea of maritime trade*

For more than a century China has produced and exported through her maritime trade three types of tea, *lü chha* (green tea), *oolong* (dark dragon tea) and *hung chha* (red tea). Green tea is unfermented, oolong is partly fermented and red tea is fully fermented. But while *lü chha* is translated into English as 'green tea', *hung chha* is rendered into English as *black tea*. In fact, the term *black tea* has been used in the tea trade since the latter part of the 17th century to denote a dark coloured tea from the Wu-I region of northwestern Fukien, which was also known as Bohea tea. Since *black tea* is today the generic name for fermented teas, and *black tea* has been traded since the 17th century, Western[64] as well as Chinese scholars have usually assumed

---

[60] *Min Tsa Chi*, ibid.   [61] *Min Chhan Lu I*, cited in C&C, p. 425.   [62] Ukers (1935) II, p. 230.
[63] Chiang Shih-Chhê 蔣師轍, *Thai Yu Ji Chi* 臺游日記 (Diary of Travels in Taiwan), *1892*, cited in C&C, p. 429.
[64] Gardella (1994), p. 30, states, 'The development of fermented teas in the sixteenth century produced the so-called "black teas" (*hung-chha*) that became Minbei [i.e. Northern Fukien] staple export to the West . . .' Cf. also Daniels, SCC Vol. VI, Pt 3, p. ?? 'Level two technology, which commenced in the Ming, included the preparation of black tea with four more or less distinct unit processes: withering, rolling, oxidation and firing.'

that a fermented tea (i.e. *hung chha*) was already produced in China towards the end of the Ming dynasty. In fact, such eminent Chinese historians of tea as Chhên Chhuan[65] and Wu Chio-Nung[66] have consistently translated the *black tea* of early European documents on the tea trade as *hung chha* 紅茶 in Chinese.

We ourselves also assumed that the *black tea* of the 17th century was the equivalent of 'red tea' or *hung chha* of modern China when we began our examination of the Chinese tea literature. But as we reviewed the literature we were surprised to find that the term *hung chha* is not even seen in print earlier than 1860.[67] When Father Gasper da Cruz reported on his visit to China in +1560, he mentioned that the Chinese served 'a drink called *ch'a*, which is somewhat bitter, red and medicinal'.[68] This report has been cited by Chhên Chhuan as evidence that the Chinese already had 'red tea' in the 16th century.[69] In a similar vein, the *Phien Kho Yü Hsien Chi* 片刻餘閒集 (Notes from Moments of Leisure), c. +1753, states that a tea called Chiang-hsi-wu 江西烏 (Dark Kiangsi) is black in colour but gives a red infusion.[70] We do not think that either of these references is indicative of the existence of a red tea in the 16th and 18th centuries, but rather that some semi-fermented teas, such as Wu-I, which looks black, can give rise to a red infusion, a fact that is thoroughly familiar to all drinkers of Chinese tea.

The first time the term *hung chha* (red tea) occurs in print in Chinese is in the *Chhung Yang Hsien Chi* 崇陽縣志 (Gazetteer of Chhung-yang County), 1866, which states that 'in about 1850, Cantonese merchants came here to buy tea. They require that we pick young buds, wilt them in the sun and then roll them, without resorting to stir-frying [in a pan]. In rainy weather the leaves may be dried over charcoal . . . The tea is exported overseas and is called *hung chha*.'[71] This was soon followed by a report in the *Pa Ling Hsien Chi* 巴陵縣志 (Gazetteer of the Pa-Ling County), 1872, which says that 'In 1843, after the opening of the treaty ports to foreign traders, the Cantonese brought funds to promote the making of *hung chha*. The tea turns red

---

[65] In his history of China's tea exports, Chhên Chhuan (*1993*) consistently uses the term *hung chha* for exports of 'black tea' for the years between +1790 and 1840, cf. pp. 294, 297, 303 and 304.

[66] Wu Chio-Nung (*1987*), p. 91, says that in his translation of Ukers' *All About Tea*, 'black tea' and Bohea tea were each rendered as *hung chha* 紅茶 (i.e. red tea). When both terms occur simultaneously 'black tea' remains as *hung chha*, but 'Bohea tea' becomes *Wu-I chha* 武夷茶.

[67] C&C, p. 23, states that *hung chha* (red tea) is mentioned in two recipes of the early Ming food canon *To Nêng Phi Shih*. They are cited on p. 290. We found that both recipes are reproduced from an earlier Yuan work, *Chü Chia Pi Yung* which by now must have become thoroughly familiar to our readers. We have checked the text of these recipes in the edition of *TNPS* and *CCPY* reproduced in Shinoda & Seiichi (*1972*), pp. 397 and 335, and do not find the term *hung chha* used in either recipe. C&C further claim that a method for processing *hung chha* is found in the *Wu I Shan Chih* 武夷山志 (Gazetteer of the Wu-I Hills) of 1846. We are unable to verify this claim in the segment on tea reproduced in Wu Chio-Nung (*1990*), pp. 324–6.

[68] Cited in Ukers (1935), I, p. 25. Father Gaspar da Cruz was a Portuguese Jesuit missionary who worked in China from +1556 to 1560.

[69] Chhên Chhuan (1984), p. 194.

[70] *Phien Kho Yü Hsien Chi* by Liu Ching 劉靖, c. +1753, cited in C&C, p. 367. The passage is also cited in *CKCC*, p. 114.

[71] *Chhung Yang Hsien Chi*, 1866, cited in Wu Chio-Nung (*1990*), p. 401. Chhung-Yang is in southern Hupei near the border to Kiangsi and Hunan. This passage is also cited by Kung Chih (*1996*) to support his claim that red tea (*hung chha*) was already being made in +1753.

during its exposure to the sun. Hence it is called *hung chha*.'[72] *Hung-chha* is also mentioned in two other gazetteers, the *An Hua Hsien Chi*, 1872,[73] and the *I Ning Chou Chi*, 1873.[74] It is curious that although Fukien is generally acknowledged as the original home of red tea, the term *hung chha* is not seen in the local gazetteers of Fukien of the same period.[75]

Both the name *hung chha* and a method for processing it are found in the *Chung Chha Shuo Shih Thiao* 種茶説十條 (Ten Talks on Tea), c. 1874. We quote:[76]

To process red tea: pick the leaves before the rains begin. Spread the leaves on a mat and sun them. When they are wilted, gather them in a pile. Roll and crush them with the feet to get rid of the bitter juice. After rolling they are sunned again. When they no longer feel sticky when rubbed between the fingers, they are stored tightly in a bag. In about three hours the leaves will feel warm and their colour will turn. This is called *shang han* 上汗 (inducing sweat). After sweating, the leaves are again exposed to the sun until they are completely dry.

The practice of rolling the wilted leaves by feet was still seen by Ukers in the early years of the 20th century (Figure 125).[77] From the above quotations we can see that red tea was already produced on a large scale by about 1870, but we still have no idea as to when and how the practice began. In describing the history of tea in the Thung 峒 region, the *Chhun Phu Sui Pi* 純浦隨筆 (Casual Notes on Local Products), of 1888, had this to say:[78]

Hung-chha processing started in the later years of the Tao-Kuang (1821–51) reign period. Tea brokers from Kiangsi were collecting tea in the I-Ning Prefecture, and so they visited the Thung region. They taught the people how to make red tea. Only one type of tea plant was grown. Leaves picked before the rains were the first lot. They were processed into an experimental oolong. Unpicked plants were reserved for seed. Those picked in the summer were called 'lotus flower', and the last picked of the season were called 'autumn dew'. The leaves were either sunned or steamed. To make red tea, the leaves were exposed to the sun, and while still hot, covered with a piece of cloth. After the leaves turned red, they were dried again in the sun. No fire was used.

---

[72] *Pa Ling Hsien Chi*, 1872, cited in Wu Chio-Nung (*1990*), p. 435. Pa Ling is present-day Yüeh-Yang in northern Hunan.

[73] *An Hua Hsien Chi* 安化縣志 (Gazetteer of the An-hua County), 1872, cited in Wu Chio-Nung (*1990*), pp. 488 and 490, is also in Hunan.

[74] *I Ning Chou Chi* 義寧州志 (Gazetteer of the I-Ning Prefecture), 1873, cited in *CKCC*, p. 215, and in Wu Chio-Nung (*1990*), p. 248. The text reads, 'During the Tao-Kuang (1821–1851) reign period tea prospered in the Ning-chou area. Tea plantings are seen in every village. The products include green tea (*chhing chha* 青茶), red tea (*hung chha*), oolong, white down, flower fragrance and brick tea.' I-ning is present-day Hsiu-shui 修水 in northwestern Kiangsi.

[75] Wu Chio-Nung (*1987*), p. 90.

[76] Tsung Ching-Fan 宗景藩, *Chung Chha Shuo Shih Thiao*, cited in C&C, p. 417. 'Before the rains' probably means before the 'grain rains' period (usually 20 April). *Shang han* is a term the Chinese used in the 19th century to denote 'fermentation'. It is sometimes called *fa han* 發汗. The method as described may not be the most efficient way to promote fermentation (which is actually an oxidation and requires air as a reactant) but it was one that could be easily carried out with simple equipment showing that the writer had little appreciation of the nature of the process.

[77] Ukers (1935) I, p. 468.

[78] *Yeh Jui-Yen* 葉瑞延 (*1888*), *Chhun Phu Sui Pi*, cited in C&C, p. 428. The Thung is presumably the name of the area near present-day Thung-ku. No rolling of the leaves is mentioned in the red tea process. We suspect this was an oversight on the author's part.

Table 53. *Origin of the major red teas in the China trade in the late 19th century*

Type[a]	Trade Name	Locality[b]	Earliest Reference[c]
North China Congous	Keemuns	*Chhi-men* 祁門	1875, technique transferred from Fukien
	Ningchows	*Ning-chou* 寧州	c. 1845 50 technique from Kiangsi traders
	Ichangs	*I-chhang* 宜昌	c. 1840, prompted by traders from Kuangtung
South China Congous	Paklins	*Pai-lin* 白琳	c. 1850s prompted by Kuangtung traders
	Panyongs	*Than-yang* 坦洋	1851 started by local entrepreneurs
	Chingwos	*Chêng-ho* 政和	mid 19th century

[a] The trade names follow Ukers' nomenclature, (1935) I, pp. 228–9. They are a conglomerate of transliterations based on Cantonese, Amoy dialect, and Mandarin, and were used extensively by tea merchants in the China trade. Congu is an anglicised form of *Kung-fu* 工夫.
[b] *Chhi-men* is in south Anhwei, and I-chang is west of Wuhan on the Yangtze River in Hupei. Ning-chou is in northwestern Kiangsi. All three locations which produce North China Congus are, in fact, in Central China. The three South China Congus are all produced in Fukien. Pai-lin is a village south of Fu-ting 福鼎, Than-yang is in Fu-an 福安 county, and Chêngho is in north Fukien.
[c] Based on the information given in Chhên Tsung-Mou, *Chung Kuo Chha Ching*, (1992), pp. 213, 215–19.

It is clear from above passages that *hung chha* (red tea) was processed in a manner consistent with that reported by Ukers cited earlier (pp. 536–7) and that described in Table 52. It is also evident that tea brokers were the driving force behind its development and commercialisation. When and where did this innovation originate? To try to answer this question we have listed in Table 53 names of the most popular red teas in the China trade described by Ukers,[79] and the dates of their earliest appearance as shown in Chinese sources. Several inferences may be drawn from the table. Firstly, red tea took off as a major product sometime between 1850 and 1860, almost simultaneously in several provinces. This means that experimentation of red tea processing must have started earlier, perhaps as early as 1830. Secondly, tea brokers from Kuangtung were instrumental in persuading the producers to shift their processing methods towards the production of red teas. Their action was, no doubt, a response to the demands of the foreign tea buyers, which, in turn, represented the preference of the tea-consuming public in England and other European countries. Red tea was, therefore, a product developed specially for export. Finally, it is interesting to note that the most successful producers of red tea in Fukien were in localities, such as Pai-lin, Than-yang and Chêng-ho, in the northeastern part of Fukien, which were not noted previously as centres for the production of quality green or oolong teas. Thus, the rise of red tea did not occur at the expense of green and oolong teas, and was a boon to the overall economy of Fukien province.

[79] Ukers (1935) I, pp. 228–33 describes the different types of black teas (i.e. red teas or *hung chha*) produced in China during the heyday of the tea trade in the 19th century.

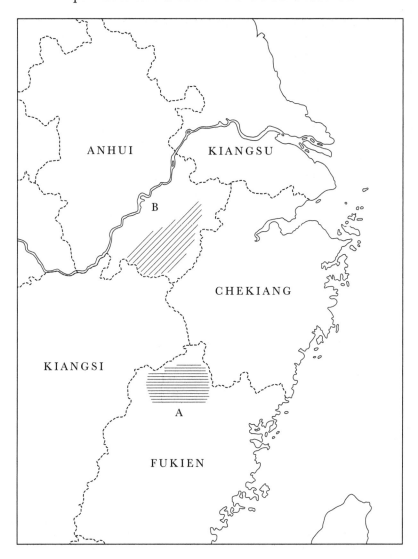

Fig 126. Map of Southeast China showing the *black tea* country of Wu-I in Fukien (A) and the green tea country of Sing-lo in Anhui (B). In the 18th and early 19th century, tea produced in these regions was transported southwards through Kiangsi all the way to Canton for shipment in the overseas trade.

But if Chinese red tea did not come into being until after 1850, what then was the *black tea* of the maritime trade for the preceding 150 years? During this period most of the tea exported to Europe came from two regions, green tea from the Sung-Lo 松蘿 hills in Anhui, and *black tea* from the Wu-I 武夷 mountains in Fukien (Figure 126).[80]

---

[80] Mui & Mui (1984), p. 4. Sung-lo was the home of the Hyson and Twankay green teas and Wu-I, the home of the Congu, Souchong and original Bohea black teas of maritime trade in the 18th and early 19th centuries.

As we have seen earlier (page 539), the principal tea made in the Wu-I region in late 17th century was the partly fermented Wu-I tea. Thus, the early *black tea* or Bohea tea of maritime trade had to be a semi-fermented tea, i.e. an oolong tea.

On the other hand, between +1786 and 1833 *black teas* from Fukien with such names as *congou* and *souchong* overtook Bohea in popularity in the export trade.[81] The Chinese records do not tell us how these teas were made.[82] Thus, the possibility that these were, in fact, fully fermented teas (i.e. *hung chha*) cannot be ruled out.

To resolve this issue we tried to find out how the *black tea* of maritime trade between +1700 and 1850 was processed. No such information could be obtained from Chinese sources. Since foreigners were not allowed to visit the tea growing areas of China in the 18th century, their records were of no help either. However, by early 1800s some westerners managed to penetrate the bamboo curtain and were able to provide accounts of how *black tea* was manufactured. The first is Samuel Ball.[83] Based on information he gathered from his Chinese contacts in Canton from 1804 to 1826, he found that the processing of the finest black *souchong* consisted of four steps: the leaves were (1) wilted and tossed about in air (2) inactivated by roasting (i.e. stir-frying) (3) rolled by hand, and (4) dried over a gentle fire. The second is

---

[81] Mui & Mui, *ibid.*, appendix 1, pp. 146–7, which is based on records of the East Indian Company. *Congou* is *kung fu* 工夫, which presumably means products processed with extra care. *Souchong* is *hsiao chung* 小種, a bushy variety of tea grown in the Wu-I region, cf. Ukers, (1935) I, p. 229. The best known *souchong* is *lapsang souchong* 正宗小種. According to Ukers (1935) II, p. 510, Bohea was originally applied to the best 'black tea, but subsequently to an inferior black', presumably because of the introduction of bogus Bohea teas from Kwangtung.

[82] Because these were called 'black tea' by foreign traders many Chinese scholars have automatically assumed that they were types of '*hung chha*', even though there is no record of how they were prepared in the Chinese literature. For example, Kung Chih (*1996*), p. 260, states that *hsiao chung hung chha* (小種紅茶) or Souchong black tea in tea trade terminology, was born in early 18th century in Thung-mu-tshun in the Wu-I region but provides no supporting documentation. Luo in Yamanishi ed. (1995), p. 431, also says that '*hsiao chung* 小種 black tea dates back to the 17th century'. There is an anecdote in circulation about the ascendence of red tea in northern Fukien which is cited by Kung Chih, *ibid.*, and the *CKCC*, p. 220. After the opening of the five treaty ports in 1842 a number of western tea trading firms established offices in Foochow to buy tea from producers in the hinterland. One night during the Taiping Rebellion a tea processing station in Chhung-an (in the Wu-I mountains) was billeted by rebel troops who used bags of freshly picked tea leaves as bedding. After they left the proprietor was greatly alarmed when he discovered that the leaves had turned dark and developed a special aroma. He could hardly wait to process the leaves, dry and ship them to Foochow. Expecting the worst, he was pleasantly surprised to learn that the foreigners actually liked this tea, and wanted more of the same. He made a small fortune, and received a standing order for the same type of tea year after year. He spread the word to tea growers in the Pai-lin, Than-yang and Chêng-ho areas, which soon became the premier red tea producers of the province.

[83] Ball (1948) who was an official of the East India Company from 1804 to 1826. A summary of his account by Mui & Mui (1984), p. 5, is abridged as follows:

Among the black teas, the finest quality of Souchong was made of large leaves picked from choice shrubs. Leaves that were of great succulency and extreme delicacy gathered in bright weather during the greatest heat of the day. They were then simply exposed to the air ... As soon as red spots were observed on the leaves or the edges began to turn red ... the leaves were roasted ... For inferior leaves exposure to the sun was sometimes necessary ... After roasting, all leaves were rolled by hand. In the treatment of fine quality teas both the roasting and rolling were repeated a number of times ... The final step was the drying of the rolled leaves in the drying implement. Built in the shape of an hourglass, the top had a diameter big enough to hold a bamboo tray-basket on which the leaves were spread. The drying took place over a well-regulated, slow-burning fire.

For further details cf. Ball (1848), chs. 6 and 7, pp. 103–53. Although Ball was not able to travel beyond the vicinity of Canton, he had access to fresh green leaves and was able to carry out experiments on how tea was processed using Chinese equipment.

Robert Fortune,[84] who, disguised as a Chinese, actually managed to visit the green tea country of Sung-Lo and the black tea country of Wu-I in the late 1840s. He confirmed Ball's observations, and showed conclusively that the plants in both the green tea plantations and black tea plantations belonged to the same species. Ball's and Fortune's accounts prove beyond a shadow of doubt that the pre-1850 *black tea of maritime commerce* had to be, in today's terminology, a partly fermented (i.e. oolong) tea, and not a fully fermented (i.e. red) tea.

Furthermore, Ball found that the pretreatment of the leaves before stir-frying (*chhao* 炒) consisted of three steps which he called: (a) *Leang Ching* (b) *To Ching* and (c) *Oc Ching*.[85] The *Leang Ching* step is literally that of cooling the leaves, which were strewn over bamboo trays and exposed to air in the open or a well-ventilated room. 'In this state they are kept until they begin to emit a slight degree of fragrance . . . *To Ching* signifies the tossing about the leaves with the hands.' This operation is repeated three to four hundred times. In *Oc Ching* the leaves were piled in a heap and covered with a piece of cloth. Ball stated:[86]

Lap Sing [Ball's Cantonese informant] says, 'It is only a common kind of tea that undergoes the process of *Oc Ching*, and which is consumed principally at Su-chao in Kiang-nan.' It is true, there is a particular kind of common tea called *Hong Cha*, or Red Tea, which I have seen and which is said to be made by a longer continuance of the process of *Oc Ching*, which is the tea he alludes to.

The statement is noteworthy for several reasons. First, it is remarkable that the term *hung chha*, denoting a product of China, should be recorded in print for the first time in English (in 1848) before it is seen in Chinese (in 1866). This is a tribute to the industry of the British merchants who were tireless in finding out and recording all that they could learn about the cultivation and processing of tea in China, in part, no doubt, to provide much needed information to support their fledgling tea industry in India and Ceylon. Secondly, it shows that a red tea (*hung chha*) was already in circulation when Ball finished his book (sometime before 1848). Since the process of *Oc Ching* had been performed before the leaves were fired and rolled, as the text would seem to indicate, we suspect the *hong chha* (red tea) that Ball saw was probably

---

[84] Fortune (1952) travelled to the tea countries in 1848 and 1849 disguised as a Chinese, staying in Chinese inns, eating Chinese food, drinking the local tea, saw the processing at first hand and was surely familiar with the properties of the export black tea. His account of the making of black tea, pp. 278–81, with minor variations, agrees very well with that of Ball's. He remarked, in summary, 'with reference to the leaves which are to be converted into black tea – 1st, that they are allowed to lie spread out in the factory after being gathered; 2nd, that they are tossed about until they become flaccid and then left in heaps; 3rd, that after being roasted for a few minutes and rolled, they are exposed for some hours in the air in a soft and moist state; and 4th, that they are at last dried slowly over charcoal fires.' His observations showed that the 'black tea' of the export trade was really a semi-fermented tea. He also demolished the notion that green tea was made from *Thea viridis* and black tea from *Thea bohea*.

[85] Ball (1848), pp. 109–14. It should be remembered that Ball's system of romanisation does not follow that of Wade-Giles (which had yet to be invented). After consulting *CKCC*, pp. 422–4, on the terms used in the pretreatment of tea leaves before 'firing' and 'rolling', and Cantonese-speaking friends on the pronunciation of the words, I have tentatively identified *Leang Ching* as 涼青, *To Ching* as 倒青 and *Oc Ching* as 漚青. These terms were presumably commonly used by the tea processors in Kuangtung. Other terms may have been preferred in other regions.

[86] Ball, *ibid.*, p. 114, footnote.

just a semi-fermented tea with a high degree of fermentation. Other accounts given by Ball, however, suggest that the Chinese had already achieved a true red tea process by 1848.[87] Finally, it suggests that the process for making red tea was undergoing rapid development in the 1840s and that its original home is Kuangtung and not Fukien as is commonly thought.[88] This explains why there is no early reference to *hung chha* in the gazetteers of Fukien, and why it was merchants from Kuangtung who, in the 1840s, taught the people in Chhung-yang county, Hupei, and Pa-ling county, Hunan, how to process red tea.[89] From these accounts it is clear that it was Oolong (i.e. partly fermented) tea and not Red (i.e. fully fermented) tea that was responsible for the unfavourable balance of payments against Britain in the Sino-British trade of the 18th and early 19th centuries that inexorably led to the coming of the Opium War, an event with momentous consequences for the unravelling of the modern history of China.

After 1850 semi-fermented tea was gradually displaced by 'red tea' in the *black tea* export trade to Europe. Before 1850 almost all the *black tea* exported came from Fukien. By 1880, Kiangsi, Hunan, Hupei and Anhui had joined Fukien as producers of the red tea exported under the name, *black tea*.[90] It was between 1850 and 1880 when it became necessary to distinguish between red, semi-fermented and green teas in commercial transactions and the terms *hung chha*, *wu lung* and *lu chha* came into general use. The years 1880–8 were the golden age of the Chinese red tea export trade. After 1888 the volume began to decline in the face of competition from the tea establishments of India and Ceylon.[91] Actually, tea plantations based on Chinese stock had been started in India in about 1835 and the first experimental batch of Indian tea had appeared in Britain in 1839. However, the Chinese imports did not do well on Indian soil, and the processing operation was plagued with

---

[87] This is suggested by other accounts of the making of tea recorded by Ball (1848). For example, on pp. 148–9, process 1, he describes how he saw a man from a village, about a day's journey from Canton, make a tea by wilting the leaves in the sun, rolling them by hand, exposing the twisted leaves again in the sun, repeating the rolling and sunning two more times, and drying them completely in the sun. He remarked that the product could command 3s 10d a pound (a very good price) at the tea market in London. On another occasion (p. 150, item 4) he saw the same process performed by the same operator except that the final drying was carried out by stir-frying in a wok. These processes are almost identical to those described for the making of red tea given earlier in the *Chung Chha Shuo Shih Thiao*, 1874 (note 76) and *Chhun Phu Sui Pi*, 1888 (note 78). It is clear that the Cantonese tea technicians with whom Ball was in contact, were on the verge of achieving a commercially successful 'red tea' process.

[88] Wu Chio-Nung (*1987*), pp. 90–1. The idea that Fukien is the original home of red tea is probably based on the common misconception that Bohea tea, which comes from Fukien and which has traditionally been rendered as 'black tea' in English, is a 'red tea'. In this connection it is interesting to note that when Fortune (1952), pp. 159–301, visited the Wu-I region in late spring 1849 he did not encounter any tea called *hung chha* in northern Fukien. It would thus appear that the 'red tea' process came to Fukien later, from another region, presumably Kuangtung.

[89] *Chhung Yang Hsien Chi* (Gazetteer of Chhung-yang County), cited above on p. 542, and footnote 71, and *Pa Ling Hsien Chi* (Gazetteer of the Pa-Ling County) cited above on p. 542 and footnote 72. It should be noted that *Hong Cha* and *Hung Chha* are merely different ways of romanising the word 紅茶.

[90] The Chinese could not call the new fully fermented tea 'black tea' i.e. *hei chha* 黑茶 since a *hei chha* already exists, cf. p. 551 below.

[91] According to Chhên Chhuan (*1984*), pp. 480–1, between 1871 and 1887 China exported 80 to 100 million kg of red tea per annum. By 1896 the amount had declined to 55 million kg. By 1917 it was down to 28 million kg and by 1932 to 9 million kg. The rise and decline of the tea trade in Fukien is described in detail in Gardella (1994).

labour problems. It was not until 1890 that the Indian tea industry became a viable and profitable enterprise. The eventual success was achieved with two innovations: first, the replacement of the Chinese stock by native varieties collected from the wild in Assam (or crosses between the Chinese and Assam varieties), and second, the streamlining and mechanisation of the processing system.[92] Soon India was able to displace China as the principal supplier of *black tea* (i.e. red tea in Chinese nomenclature) in international commerce.

What, then, was the type of tea made in the early years of the tea industry in India? Based on the limited information we have available, it seems likely to us that the early black teas of India, that is those produced from 1838 to 1860, were also a semi-fermented or oolong tea. Two pieces of evidence can be cited in support of this view. The first is the procedure introduced by Chinese technicians into Assam in the 1830s. The individual steps are as follows:[93]

(1) *Drying* – The leaves are exposed to the sun in large flat bamboo baskets.
(2) *Withering* – They are placed in the shade until they become soft and flaccid.
(3) *Panning* – They are stir-fried on a dull red pan.
(4) *Rolling* – They are rolled by hand.
(5) *Drying* – This is done in tall baskets shaped like an hourglass. The leaves are placed in the upper cone which is lined with paper. A charcoal stove is placed inside the lower cone.

The written record is further augmented by drawings of the implements used shown in Figure 127 (reproduced from Ukers, I, p. 464). The second is a description of how C. A. Bruce made the very first batch of tea grown in India in 1837,[94] 'He withered the leaf in the sun, rolled it by hand, and dried it over charcoal fires.' There was no allowance for a second fermentation after the leaves were rolled. These procedures are consistent with the Chinese method for processing oolong tea and significantly different from those used to process green tea or red (i.e. black) tea.

But by the time Indian tea became a significant threat to China's tea trade in the late 1880s, India was undoubtedly producing a red or fermented tea, although the British continued to call it *black tea*.[95] Presumably the Chinese red tea process was quickly transmitted to the British tea planters. How the transmission took place is probably buried in the archives of the tea plantations and factories in India between 1840 to 1880, and is well beyond the scope of this investigation.

---

[92] The saga of the rise of the tea industry in India is told in Ukers (1935) I, pp. 133–72. Cf. also pp. 467–74 for a description of tea processing machinery invented from 1854 to 1880. Today, it is generally assumed that all the tea cultivated in India is *Camellia sinensis* var. *assamica*. This is not quite true. According to Yamanishi *et al.* (1995), p. 390, 'in the famed Darjeeling area, where tea with exceptional flavour is produced, tea is produced from bushes originally derived from seed material from China'. Presumably, these are descendents of the plants that Robert Fortune, with enormous zeal and ingenuity, smuggled out of China in the late 1840s.
[93] Ukers *ibid.*, p. 466. See also Jacobson, J. I. L. L., *Handboek voor de kultuur en fabrikatie van thee*, Batavia, 1843, tr. 1873, cited in Ukers *ibid.*, pp. 465–6.
[94] Ukers *ibid.*, p. 146.
[95] Ukers *ibid.*, pp. 374–415, shows that the Indian 'black tea' of commerce made in the 1880s was a fully fermented tea i.e. *hung chha*.

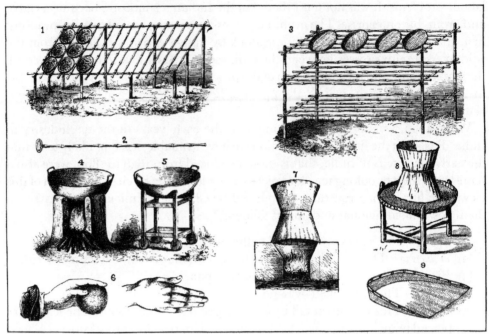

EARLY CHINESE APPARATUS FOR THE MANUFACTURE OF TEA, AND THE FIRST TO BE USED IN INDIA

Fig 127. Chinese style implements for processing tea in India, from Ukers (1935) 1, p. 464.

The history of the processing of tea in China may be summarised in the flow diagram shown in Figure 128. In the beginning, fresh leaves (or dried leaves) were used directly. As tea gained favour as a refreshing drink, efforts began to turn the leaves into a stable product that could be made into a 'tea drink' on demand. The result was the Han cake tea, which was refined into the Thang cake tea and then into the Sung cake tea. In the Yuan–Ming era, the technology went through a major change; the cake tea was displaced by processed leaves in loose form, which has remained as the most popular type of tea consumed in China today. So far, all the tea products were unfermented. During the Ming a partly fermented tea made in the Wu-I region of Fukien gradually gained national recognition. This Bohea tea (called 'black tea' by foreign traders) soon found favour in the maritime export trade. The flavour induced by 'fermentation' apparently appealed to the palate of the Western tea drinking public. Further exploitation of this trend led to the emergence of the fully fermented 'red tea' in the latter part of the 19th century.

(vi) *White, yellow, dark, compressed and scented teas*

In addition to green, red, and oolong we have so far considered, three other teas have also been processed from leaves in China, white tea (*pai chha* 白茶), yellow tea

40 FERMENTATIONS AND FOOD SCIENCE

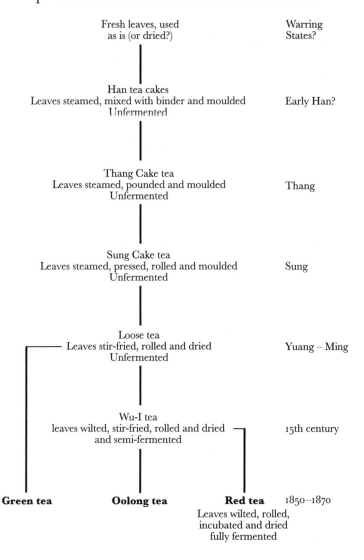

Fig 128. The genealogy of the major teas of China.

(*huang chha* 黃茶) and dark tea (*hei chha* 黑茶).[96] White tea is a speciality of Fukien and is made from young leaves rich in white hair which confers the tea its name. It is a lightly fermented tea. The leaves are wilted and dried in the sun without rolling. It probably originated during the Ming dynasty.[97] Yellow tea is a product of Anhui

---

[96] *Hei chha* (literally black tea) has been translated as 'dark tea' in English since 'black tea' is preempted by *hung chha*. For further information on the processing of these teas, see. Luo in Yamanishi (1995), pp. 421–6, and *CKCC*, pp. 236–48. Both white and yellow teas are elite teas of which only about 1,000 tons are made per year each.

[97] See Thien I-Heng, *Chu-Chhuan Hsiao Phin* (+1554), cited C&C, p. 130, says that 'the best tea is sunned and not stir-fried'. The same sentiment is expressed in Thu Lung, *Chha Shuo* (+1594), C&C, p. 134. According to Wen Lung, *Chha Chien* (1630), C&C, p. 174, 'the best tea is sunned, without firing or rolling'. These references show that white tea was already known by mid-Ming, cf. *CKCC*, p. 236, and Chhên Chhuan (*1984*), p. 198.

and Hupei. It is prepared in a manner similar to green tea except that after firing and rolling, the leaves are stacked and allowed to remain moist and warm for a significant length of time during which the leaves turn yellow through non-enzymatic oxidation. It is said to have been made since the Thang–Sung period.[98] Dark tea (*hei chha* 黑茶) is processed in the same way as green tea except that after rolling the leaves are piled for a long time in a warm and humid condition, before drying. During piling, growth of *Aspergillus glaucus*, *Saccharomyces* and other microorganisms occurs naturally, and the leaves become black.[99] It is believed that a prototype of dark tea was made during the Sung, but the term *hei chha* did not appear in print until the Ming.[100]

The processes for all the types of teas produced in China may be summarised as follows:[101]

Green tea: fresh leaf → firing → rolling → drying
Yellow tea: fresh leaf → firing → rolling → yellowing → drying
White tea: fresh leaf → withering → drying
Dark tea: fresh leaf → firing → rolling → piling → drying
Oolong tea: fresh leaf → withering (rotating) → firing → rolling → drying
Red tea: fresh leaf → withering (rotating) → rolling → fermenting → drying

The teas from all these processes are called raw tea (*mao chha* 毛茶) which are refined by sifting, cutting, firing, rerolling, selecting, drying, blending and packaging to give the finished products.[102] The finished teas may undergo further reprocessing to give compressed teas or scented teas. Compressed teas were, of course, the familiar cake tea (*ping chha* 餅茶 or *phien chha* 片茶) of the Thang and Sung. Although compressed teas were largely supplanted by loose teas during the Ming, they continued to be made in Szechuan and southwest China for export to the Tibetans and other pastoral peoples of Inner Asia.[103] Little information is available in the Chinese literature about the way they were produced in the Ming and Chhing, but

---

[98] *CKCC*, p. 238. The references are not entirely convincing since they do not state how the yellow tea was prepared. The yellowing step is known as *meng huang* 悶黃, cf. Yamanishi ed. (1995), p. 422.

[99] The piling step is known as *wu tui* 屋堆. In this case the 'firing' is intentionally made less severe so that some enzymatic activity remains. Thus, there was some 'fermentation' as well as microbial oxidation.

[100] According to *CKCC*, pp. 246–7, this type of tea was already made during the Sung in Szechuan. It was processed in brick form and used to exchange horses with Tibetans and other nomadic tribes. The term *hei chha* was first seen in a memorial from Chhen Chiang (+1524) about the use of this tea in the horse barter trade. For details on the tea–horse trade cf. Paul Smith (1991). For a history of dark tea in Hunan, cf. Shih Chao-Phêng & Huang Chien-An (*1991*) and for brick tea in Szechuan, Liu Chhin-Tsin (*1991*).

[101] After Luo in Yamanishi ed. (1995), p. 417 except that 'firing' is used instead of 'fixing'. 'Firing' in traditional Chinese processes in usually done by stir-frying, but it is probably done in a machine today. China currently produces about 540,000 tons of tea per year, distributed approximately as follows: green 60%, red 20%, dark >10%, oolong <10% white and yellow each <1%.

[102] This refining is called 'secondary processing' while the preparation of raw tea from leaves is 'primary processing', cf. Yamanishi ed. (1995), p. 416. For the refining of raw green into Hyson, Hyson Skin and Young Hyson for the export trade in the 18th and early 19th centuries cf. Mui & Mui (1984), pp. 6–8.

[103] The *Hsü Po Wu Chi*, c. +12th century cited in *CKCC*, p. 248, states that 'since the Thang the Western peoples have used *Phu* tea 西藩之用普茶，已自唐朝'. *Phu* tea is said to be a prototype of the famous *Phu-erh* 普洱 tea.

today they are manufactured mainly by the reprocessing of raw dark teas, and to a minor extent, of red and green teas. The teas are steamed, compressed and moulded into the desired shapes.[104] The best-known compressed tea is the renowned *phu-erh* 普洱 tea of Yunnan. It was already well known in the 18th century.[105]

Much more is known about the history of scented teas. In fact, the Tea Commissioner, Tshai Hsiang himself, remarked in about +1050 that

tea has its own scent, but for tribute tea it is customary to incorporate a little ambergris and gel to enhance its natural fragrance. When the people in Chien-an process tea for their own pleasure, no such additions are made so that the true aroma of tea is preserved. If while making the infusion, fruits and herbs are also used, the damage to the tea will be even greater. Such practices should be discouraged.[106] Tshai Hsiang's admonition, however, did not dampen the ardour of Sung tea connoisseurs to experiment with the incorporation of new and exotic scents to improve or alter the aroma and flavour to tea.

The *Chü Chia Pi Yung*, a Yuan food canon, gives a recipe for making scented tea by contacting processed tea leaves with Borneo camphor or musk perfume. The same procedure may be used to prepare teas scented with cassia, jasmine, orange and other flowers.[107] Details of these procedures are given in the *To Nêng Phi Shih*, +1370, as follows:[108]

Camphor tea, *Nao tzu chha* 腦子茶: finely grind a high quality tea. Wrap a piece of borneo camphor in thin paper and bury it in the tea. After one night the tea would have absorbed some of the aroma of camphor. How delightful!

Scented flower teas, *Hsün hua chha* 熏花茶: fabricate round pewter containers that sit on top of each other. The lowest tier is half-filled with good tea. Several tens of holes are drilled through the bottom of the middle tier. Lay on it a thin piece of paper, and fill it with flowers. Repeat the holes at the bottom of the top tier. Put a thin piece of paper on it and fill it loosely with tea. Cover it with its pewter cover. Seal the stacked containers with paper and let them stand overnight. [On the next day] replace the old flowers with new flowers. Perform this action three times. The tea will be ready.

A similar procedure is found in Chu Chhüan's *Chha Phu*, c. +1440:[109] scented teas are also discussed in the *Chha Phu*, +1539, of Chhien Chhun-Nien, which mentions the incubation of tea with orange peels, lotus bloom and the flowers of cassia, jasmine, rose, orchid, tangerine, gardenia and plum.[110] He recommends plucking the

---

[104] Yamanishi ed. (1995), p. 424, and *CKCC*, pp. 438–44.
[105] *Pên Tshao Kang Mu Shih I*, 18th century, p. 249, states that *phu-erh* tea is a product of *Phu-erh* prefecture in Yunnan. It is warming and fragrant. It dispels grease, improves digestion and dissolves phlegm.
[106] *Chha Lu*, by Tshai Hsiang, +1051, cf. C&C, p. 30. Cf. also *Hsuan Ho Pei Yuan Kung Chha Lu*, +1125, cited in C&C, p. 49.
[107] *CCPY*, p. 5.   [108] *To Nêng Phi Shih*, Shinoda & Seiichi (*1973*), p. 399; also cited in C&C, p. 290.
[109] *Chha Phu* of Chu Chhüan, C&C, p. 122, which says, 'To make flower scented teas: Any kind of flower may be used. When the flower blooms seal a two-level bamboo basket with paper. Place the tea leaves in the upper level and the flowers in the lower level. The seal must be air-tight. After one night replace the old flowers with fresh blooms. After a few days, the tea will be saturated with the fragrance of the flowers. The same method may be used to scent the tea with borneo camphor.'
[110] *Chha Phu* of Chhien Chhun-Nien, C&C, p. 126.

flowers when they are about half-way in bloom. Ideally, three parts of tea should be incubated with one part of flowers. One layer of flowers should be placed over one layer of tea, and the sequence repeated until the porcelain jar is full.

In spite of the objections of purists such as Thien I-Hêng, +1554, who disparaged teas scented with plum, chrysanthemum and jasmine flowers,[111] scented teas continued to gain acceptance during the Ming and Chhing. Today jasmine is one of the most popular teas of China. One of the best known jasmines is produced in the Foochow area in Fukien, where the soil and climate are particularly congenial to the cultivation of the jasmine shrub, *Jasminium sambac*, which we have already encountered in *SCC* Volume VI, Part 2 of this series. In fact, jasmine tea occupies such a unique position among Chinese teas, that it is called by a special name, *hsiang phien* 香片, or 'fragrant bits'.

Other well-known scented teas are orchid (*lan xiang* 蘭香), litchi fruit (*li chih* 荔枝) and rose congou teas. Jasmine and orchid are green teas, and litchi and rose congou are red (i.e. black) teas. Finally, there is one tea that contains no leaves of *Camellia sinensis*. That is chrysanthemum tea, which consists entirely of dried chrysanthemum flowers. It is much loved as a cooling drink in summer in South China.

(3) Tea Drinking and Health

The first cup caresses my dry lips and throat,
The second shatters the walls of my lonely sadness,
The third searches the dry rivulets of my soul to find the series of five thousand scrolls.
With the fourth the pain of past injustice vanishes through my pores.
The fifth purifies my flesh and bone.
With the sixth I am in touch with the immortals.
The seventh gives such pleasure I can hardly bear.
The fresh wind blows through my wings
As I make my way to Penglai.[1]

In this quotation from the poem 'Thanks to Imperial Censor Mêng for his Gift of Freshly Picked Tea', written in about +835, Lu Thung the Tea Doter expresses more eloquently than any other literary figure before or after him the lure and delight of drinking tea. If we were to paraphrase the verse at a more mundane level, we could say that tea quenches our thirst, raises our spirit, sharpens our mind, helps us to relax from the cares of the world, and bestows on us a general sense of well-being. But before we discuss how these perceptions came into being and what they really mean in terms of our physical and mental health, it will be helpful for us to review briefly how the Chinese prepared the tea product for drinking.

---

[111] *Chu Chhüan Hsiao Phin*, C&C, p. 131.
[1] C&C, p. 215; tr. Chow & Kramer (1990), pp. xiii and xiv. Phêng-lai, an island off the coast of Shantung, is said to be the home of the Immortals.

(i) *Making and drinking tea*

Even a casual perusal of the Tea Books of China would leave one with the impression that the authors were much more interested in the enjoyment of tea as a beverage than in its manufacture as a product. Tea drinking was the main event. The picking and processing of the leaves were merely prologues. A great deal more literary verbiage has been showered on the utensils and procedures used in preparing the beverage and the sources of water which would brew the best tea drink than on the steps involved in the processing of tea, the product. The making and drinking of tea have also fascinated Westerners who have found in these practices the key to the Chinese art and culture of tea.[2]

Our current objective is not to further expound the aesthetic and artistic merits of tea, but simply to ascertain how tea, the drink, has been prepared from tea, the product, through the ages in China. At the same time we would also like to find out how the physical and organoleptic character of the tea product have affected the technique for making the tea drink and vice versa. In the earliest days, as suggested in Kuo Pho's commentary (c. +300) to the *Erh Ya*, fresh leaves were plucked from the plant and boiled to make a decoction.[3] As soon as the value of tea as a medication or beverage was recognised, it was natural that attempts were made to preserve the leaves in such a way so that a drink could be prepared whenever and wherever it was needed. According to the *Thung Yüeh* (Contract with a Servant), c. −50, tea was already a processed product that could be shipped from place to place by the time of the Western Han. On the preparation of tea as a drink, the *Thung Yüeh* uses the term *phêng chha* 烹茶, which means literally cooking tea by boiling.[4] Presumably, this means that the tea leaves were boiled with water. In any case, we know from the anecdotes on Wei Yao 韋曜 and Wang Su 王肅 that tea was already drunk in copious amounts as a beverage during the Three Kingdoms and the North and South Dynasties.[5]

How was this tea drink prepared? A clue is provided by the *Hsin Hsiu Pên Tshao*, +569, which states that dogwood [seeds], scallion and ginger may be added to *ming* 茗 (tea) to make a drink.[6] The *Shih Liao Pên Tshao*, +670, further tells us that tea decoction 'makes a good medium for cooking congee', and that 'the best tea is made from fresh leaves. However, they can be steamed and pounded and then kept for later use'.[7] These references show that even at the beginning of the Thang, fresh leaves were still being used to make a drink, but they also affirm that tea leaves were processed to give a stable product. By the time Lu Yü's *Chha Ching*, +760, appeared on the scene, both the procedure for processing the tea product, and that for making the tea drink had been standardised.

---

[2] Blofeld (1985); Hardy (1979); Smith, M. (1986). Cf. also Okukura Kakuzo (1906) for a Japanese view on tea culture.
[3] Cf. p. 508 and *Erh Ya*, ch. 14, p. 40b. [4] P. 508 ns. 18 and 19. [5] Cf. pp. 509 and 511, and ns. 20 and 34.
[6] *HHPT*, ch. 13, p. 116. [7] *SLPT*, p. 23, item 44. Cf. also p. 512.

Table 54. *Implements used in making tea in Lu Yü's tea classic*

1. A stove made of brass, iron or clay in the shape of an ancient cauldron.
2. A bamboo basket 1 ft 2 in. high and 7 in. in diameter.
3. A six sided iron poker 1 ft long.
4. A pair of iron or brass tongs, 1 ft 3 in. long, for carrying hot charcoal.
5. A pot for boiling water, made of iron, pottery, or best of all, silver.
6. A stand for the boiler to rest on.
7. A long bamboo tong 1 ft 2 in. for baking tea cake.
8. A paper bag made of thick white paper.
9. A crusher, made of wood, for grinding tea cakes.
10. A sieve made of bamboo and covered with gauze.
11. A measure made of sea shell, bamboo, brass or iron.
12. A water tank, 1 *tou* size, made of wood and lined with lacquer.
13. A water strainer made of silk on a metal frame.
14. A ladle made of a gourd or wood.
15. A pair of sticks made of wood, 1 ft long, and tipped with silver.
16. A china salt-cellar 4 in. in diameter, with a bamboo spoon, 4 in. long and 1 in. wide.
17. A china jar (*shu yü* 熟盂) to hold boiled water (2 *shêng* in size) and act as a tea pot.
18. Tea bowls (*wan* 碗) from Yueh-chou 越州 or Ting-chou 鼎州.
19. A basket made of rushes to hold ten bowls.
20. A brush made of coir in the shape of a large writing brush.
21. A slop basin made of wood with a capacity of 8 *shêng*s of water to wash tea utensils.
22. A dustbin with a capacity of 5 *shêng*s for containings dregs from the tea.
23. Two 2 ft long dish towels for cleaning the utensils.
24. A cabinet for storing the implements.

We already know that the Thang tea product was in the form of cakes. The *Chha Ching* has a great deal to say about how an infusion was prepared from them and consumed as a drink. In chapter 4, it gives a detailed account of utensils used in the making of tea. A list of the implements is presented in Table 54. The functions of the different items are more or less self-explanatory. Chapter 5 is devoted to the technique of preparing the infusion, and chapter 6 the art of drinking it. The procedures for making and drinking tea in the *Chha Ching* are summarised below.

(1) Hold the tea cake with the tong (item 4, Table 54) and roast it over the fire (item 1).[8]
(2) Place the hot cake in a paper bag (item 8) to cool.
(3) Grind the cooled cake in a grinder (item 9).
(4) Pass the tea powder through a sieve (item 10)
(5) Heat water [in a pot] until it starts to boil vigorously.[9]

---

[8] It is not clear why the cake is first baked before it is used. Perhaps, the Thang cakes were never dried adequately, or protected from atmospheric moisture during storage. In time they would develop a mouldy odour. This could be eliminated by baking the cake. Presumably, by early Sung, the technique for drying and storing the tea cake had so improved that the roasting of the cake before its use was no longer necessary.
[9] It is curious that although chapter 5 in the *Chha Ching* discusses in detail the three stages of boiling of the water and the addition of a pinch of salt after the first boil, it says nothing about the container in which the water is boiled.

(6) Add boiled water to the ground tea in a tea-bowl (item 18). Or pour the water over the tea in the tea-bottle (item 17) and allow it to stand.[10]

(7) Keep the froth in the tea-bowl. It is the essence of the tea.

The *Chha Ching* further comments with undisguised disdain that 'sometimes scallion, ginger, jujube, orange peel and mint are boiled with the tea and the froth skimmed off. This type of tea is no better than slop water, yet the practice continues.' This may have been a remnant of the way tea was used as a medication in antiquity, since it is customary in Chinese medicine for several herbs to be boiled together to give the desired decoction. But it was obviously not the way that a proper Thang gentleman would wish to make his tea.

The popularity of tea continued to grow during the Sung. The method for making the tea infusion used by the Son of Heaven himself is recorded in the *Ta Kuan Chha Lun*, +1107. The utensils involved are listed in Table 55. The procedure is as follows:[11]

(1) Slice off the required amount of caked tea.
(2) Crush it into a fine powder.
(3) Pass the powder several times through a sieve.
(4) Place the powder in a *chien* 菱, the mixing and drinking bowl.
(5) Boil pure water in a *phing* 缾, a bottle-like kettle.
(6) Pour the right amount of boiling water into the bowl and stir the infusion with the bamboo whisk, *hsien* 筅.
(7) Drink the infusion from the bowl, but avoid the sediment.
(8) Pour more water on the sediment, whisk and drink the infusion and repeat six more times.

This procedure is quite similar to that described by Lu Yü. There are some minor differences. The cake is no longer roasted before use and the ground tea powder is infused with water for up to seven times.[12] We are told that the water is boiled in a bottle-like kettle, which, according to the Emperor, is best made of silver or gold. Most interesting of all, a whisk is used to stir up the infusion. Readers familiar with Japanese culture will recognise this practice as elements in the celebrated Japanese tea ceremony, *cha-no-yu*.[13] This procedure has survived in Japan although it has disappeared from China since the early Ming.

In this connection it is instructive to compare the Thang and Sung methods of making tea with drawings of tea utensils (Figure 129) published in the *Chha Chü Thu*

---

[10] Fu & Ouyang (*1983*), p. 34 and Ukers (1935) I, p. 19. The second method is based on a statement in chapter 6, Fu & Ouyang (*1983*), p. 38, Ukers *ibid.*, p. 20. The full quotation reads, 'As for cake tea, chip off a piece, roast it, grind it and place it in a bottle. Pour boiled water over it. This is called steeping the tea.'

[11] *Ta Kuan Chha Lun*, cf. C&C, pp. 45–6. The summary by Blofeld (1985), pp. 27–31, is most helpful.

[12] In the *Chha Lu* of Tshai Hsiang, +1051, C&C, pp. 31–2, an implement for baking the cake tea is still listed. On the other hand, a *chien* (tea-bowl) is already mentioned in the text, which recommends that it should be warmed before use. This is reminiscent of today's practice of warming the pot with boiling water before it is used for making tea.

[13] For further information on the Japanese tea ceremony cf. Chow & Kramer (1990), pp. 51–62; Hardy (1979), pp. 62–70; Ukers (1935) II, pp. 361–78.

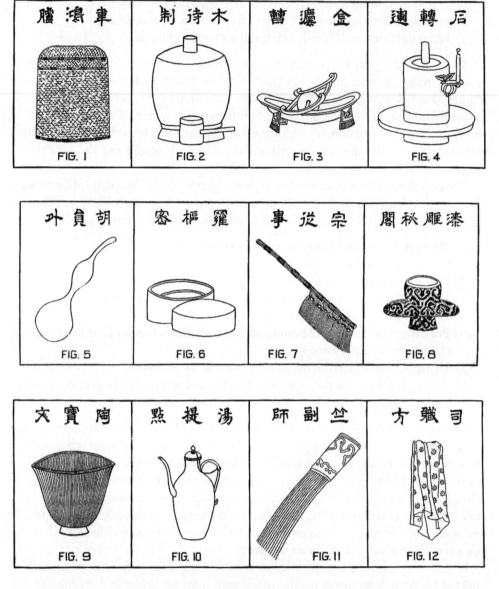

Fig 129. Tea utensils from *Chha Chü Thu Tsan* 茶具圖贊, pp. 5–27; a–l., +1269. Altogether twelve pictures, from Chhên & Chu (*1987*), pp. 96–119; Cf. also Ukers (1935) 1, pp. 16–18. (1) Bamboo basket for drying tea; (2) Wood anvil and iron mallet to mould the tea into cakes; (3) Iron grinding boat; (4) Stone grinding mill; (5) Gourd as ladle for measuring water; (6) Sieve to separate coarse from fine tea; (7) Brush for removal of dust; (8) Lacquer cup with holder; (9) Porcelain tea bowl; (10) Porcelain tea pot; (11) Bamboo brush for washing pots; (12) Towel for cleaning cups.

40 FERMENTATIONS AND FOOD SCIENCE

Table 55. *Tea utensils used by Emperor Hui Tsung*

1.	*Nien* 碾:	Crusher for powdering cake tea made of silver or cast iron.
2.	*Lo* 羅:	Sieve, should be finely meshed and very strong.
3.	*Chien* 盞:	A mixing and drinking bowl, should be deep, broad, and dark blue.
4.	*Hsien* 筅:	A whisk made of bamboo for mixing the tea infusion.
5.	*Phing* 餅:	A tall, narrow tea kettle with a long curved spout, should be made of gold or silver. The lip of the spout should be higher than the rest of the bottle so that when the water boils, none will be lost.

Tsan 茶具圖贊 (Tea Utensils with Illustrations) of +1269.[14] Although these drawings are often printed together with Lu Yü's *Tea Classic*, we believe it is more likely that they are representative of the type of tea utensils commonly seen during the Sung Dynasty.[15] Figures 129(1) and 129(2) are presumably used in the manufacture of cake tea, 129(3) and 129(4) in grinding the cake tea into a powder, and 129(5) to 129(6) in making the infusion. If we compare the utensils listed in Table 55 with these drawings, we can see that the *chien* (tea bowl) is shown in 129(9), the tall and narrow *phing* (tea-kettle) in 129(10) and the *hsien* (bamboo whisk) in 129(11). Since the Thang, tea-making utensils had apparently advanced from tools of daily living to finely crafted objects of art. We can just see Emperor Hui Tsung grinding a piece of his priceless tribute tea in his gold crusher, placing it in an exquisite porcelain bowl, pouring water onto it from his gold tea kettle, whisking the infusion with gusto and drinking it in solitary splendour. Little did he realise that his preoccupation with such indulgences to the detriment of the affairs of state would eventually cost him his throne.

It was Tshai Hsiang, the Tea Commissioner, who first introduced the term *tien chha* 點茶 for making tea. This term was used commonly during the Sung and even persisted in the Tea Books of the Ming Dynasty. As loose tea displaced cake tea during the Yuan, the method for making tea was further simplified during the Ming. Sung implements associated with the usage of cake tea, such as the crusher, sieve, and whisk were still listed in Chu Chhüan's *Chha Phu* of +1440, but they were nowhere to be seen in Chhien Chhun-Nien's book with the same name of +1539.[16] Listed in the latter work are only the tea-kettle (*chha phing* 茶瓶), ladle (*chha thiao* 茶銚) and tea bowl (*chha chien* 茶盞). It emphasises that in making a good infusion the tea should be washed once with hot water, and that the tea bowl should be warm.[17]

---

[14] Reproduced in C&C, pp. 95–120.
[15] These drawings are reproduced in Ukers (1935) 1, pp. 16–18, and Chow & Kramer (1990), pp. 8–9 and interpreted as utensils used in Lu Yü's time.
[16] Both books with the title *Chha Phu* are reproduced in C&C, pp. 120–8. The section of making tea in the latter work was copied *verbatim* into the Ming food canon *YCFSC* of +1591, pp. 25–8.
[17] The washing of the tea leaves is an integral part of the making of *kung fu* tea, cf. p. 540, and n. 59.

The most complete description of the making of tea in the Ming is found in the *Chha Shu* 茶疏 by Hsü Tzhu-Yü of +1597. The relevant passages have been translated into English by John Blofeld. We quote:[18]

Infusion: have the utensils ready at hand and make sure that they are perfectly clean. Set them out on the table, putting down teapot lid inner face upwards or laying it on a saucer. The inner face must not come into contact with the table, as the smell of lacquer or food remnants would spoil the taste of the tea. After boiling water has been poured into the pot (*hu* 壺), take some tea leaves in your hand and throw them in. Then replace the lid. Wait for as long as it takes to breathe in and out three times before pouring the tea into the cups, and then pour it straight back into the teapot so as to release the fragrance. After waiting for the space of another three breaths to let the leaves settle, pour out the tea for your guests. If this method is used, the tea will taste very fresh and its fragrance will be delicious. Its effect will be to promote well-being, banish weariness and raise your spirit.

Drinking: a pot of tea should not be replenished more than once. The first infusion will taste deliciously fresh; the second will have a sweet and pure taste, whereas a third will be insipid. So the quantity of water in the kettle should not be too much. However, rather than to have too little, there should be enough for some to be poured on the tea leaves after the second infusion, as it will continue to emit a pleasant aroma and can be used for cleaning the mouth after meals.

In addition to the *chha hu* 茶壺 (teapot), the *Chha Shu* also introduces us to the terms *chha ou* 茶甌 (tea cup) and *chha chu* (茶注) tea-kettle. Tea cups should be ceramic, and the tea-kettle silver or pewter. The same utensils are repeated in the *Chha Chieh* 茶解 (Tea Explanations), +1609, which prefers tea-kettle and teapots made of pottery to those made of tin. The terms *chha chu* and *chha ou* are no longer in use. They were replaced by *shui hu* 水壺 (or *kuo* 鍋) and *chha pei* 茶杯.[19] The term *tien chha* 點茶 was replaced by *phao chha* 泡茶.[20] Other than these changes in nomenclature the procedure for making tea has remained virtually unchanged from the Ming to the present day.

So far our discussion has been restricted to green tea. The mainstream tradition of making a beverage from green tea in China through the ages may be stated as follows:

(1) Earliest days: fresh leaves boiled in water;
(2) From −1st century to Thang: cake or loose tea boiled as needed;
(3) Thang: cake tea, roasted, ground and infused with boiling water;
(4) Sung: cake tea, ground and whipped in boiling water with a whisk;
(5) Ming to the present day: loose tea leaves infused with boiling water.

---

[18] Blofeld (1985), pp. 33–4. For the original text cf. C&C, pp. 154–6. At the end of the second quotation, Blofeld notes that the custom of cleansing the mouth with tea is still observed in North China after every meal. We do not know how true this statement is today (1995).
[19] *Chha pei* is seen in the *Chieh Chha Chien* 芥茶牋, by Fêng Kho-Ping 馮可賓, +1642, cf. C&C, p. 179.
[20] The term *phao chha* is first seen in the *Chha Khao* 茶考, +1593, by Chhên Shih 陳師, cf. C&C, p. 139; and the *Chha Lu* by Chang Yüan, +1595, cf. C&C, p. 142.

Because of the drastic treatment that the leaves received during processing, more of the essential oils in the leaves would be released in the Sung cake tea than in regular loose leaf teas. When whipped, the oils would form an emulsion and give the infusion a whitish milky appearance. Emperor Hui Tsung thus used the term *ju* 乳 (milk) to describe the first infusion. It was presumably for this reason that some of the best cake teas had names such as *shih ju* 石乳 (stony milk) and *pai ju* 白乳 (white milk).[21] For the same reason, during the Sung, dark ceramic bowls were preferred in order to accentuate the contrast. But during the Ming when cake teas had become rare and loose leaf tea was the norm, white became the preferred colour for tea bowls.

But what about oolong and red teas? When oolong appeared on the scene in the 17th century, it was considered simply as another type of green tea. It was probably in the 18th century that a special way of making oolong tea was developed, called *kung fu* tea.[22] The practice may have started among the tea gardens of the monasteries in the Wu-I mountains. In this procedure the tiny teapot and cups without handles are scalded by pouring boiling water over them (Figure 130). The pot is then more than half-filled with tea leaves. Boiling water is poured into the pot and is almost immediately drained off. This is called 'washing' the leaves. The pot is filled once again. After a minute, the infusion is dispensed into the little cups. The tea is savoured just like a liqueur before being drunk. Tea from the second infusion is considered the most flavourful. By the third infusion the aroma is all gone, but much flavour still remains. Green tea is sometimes also used in making *kung fu* tea but oolong is usually preferred.

No special way of making red tea has been described. As we have noted earlier, red tea was developed in response to the demands of foreign tea buyers. Most of the production was aimed at the export trade. The Chinese themselves consumed very little red tea. If they did drink any of it in the 19th century, it was presumably infused in the same way as green tea.[23]

The making of tea by infusion is not the only technique that has survived. The *Hsin Hsiu Pên Tshao* and the *Shih Liao Pên Tshao* have recommended using tea as a base for congee and boiling it with various additives such as onion, ginger, jujube, orange peel and peppermint.[24] These ancient procedures that Lu Yü regarded with such disdain have not all disappeared. In fact, during the Southern Sung, in his food canon *Shan Chia Chhing Kung* (Basic Needs for Rustic Living), Lin Hung complained that,

> Nowadays, people are not only careless about the quality of water used for making tea, but also oblivious to the lamentable effects the addition of salt and fruits would have on the infusion. Do they not know that it is scallion that dispels dizziness and plum that relieves fatigue? If one is not dizzy nor tired, what is the point of drinking tea?[25]

---

[21] *Sung Shih*, p. 4477.
[22] An excellent description of the making of *kung fu* tea is given in Blofeld (1985), pp. 133–40. Cf. also Chow & Kramer (1990), p. 37.
[23] Wang Shih-Hsiung, The *Sui Hsi Chü Yin Shih Phu*, 1861, p. 15, says, 'Tea that is red has gone through a "steamy" incubation and lost its freshness and purity. It cannot quench thirst. It is drunk for its own sake.' This suggests that the term 'red tea' (*hung chha* 紅茶) had not yet achieved general usage.
[24] Cf. p. 512.   [25] *SCCK*, pp. 109–10.

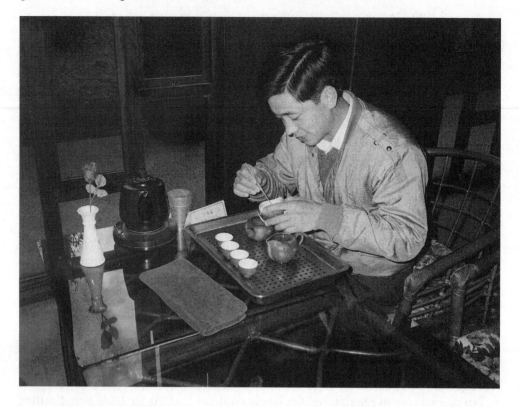

Fig 130. Preparation of *kung-fu* tea, Taipei, photograph HTH.

Several recipes for congees in which tea is boiled with substantial amounts of ingredients such as mung beans, rice, flowers or fruits of the matrimony vine, sesame, Szechuan peppercorn etc. are found in the Yuan food canon *Chü Chia Pi Yung*.[26] Furthermore, this work also contains two recipes for the making of tea with butter, which was and still is the way tea is drunk by the Mongolians and Tibetans. These recipes are copied in the *To Nêng Phi Shih* and *Chhü Hsien Shên Yin* of the Ming.[27] Indeed, Blofeld has noted that in Yunnan 'brick tea is actually boiled like soup together with such additives as sesame, ginger, the peel of citrus fruits and so on.'[28] He has also tasted tea brewed from freshly roasted Phu-erh tea from Yunnan and found it delicious.

(ii) *Effects of tea on health*

In his *Tea Classic* Lu Yü says, 'Birds, animals and humans all need food and drink to stay alive. The importance of drinks cannot be overstated. We drink water to quench our thirst, wine to drown our sorrow, and tea to stay awake.'[29] But tea was

[26] *CCPY*, pp. 5–8. [27] Similar recipes are also seen in the *I Ya II*, pp. 69–71. [28] Blofeld (1985), p. 140.
[29] *Chha Ching*, Fu & Ouyang eds. (1983), p. 37, tr. auct. adjv. Ukers (1935) I, p. 19.

renowned for other qualities besides the ability to keep one from falling asleep. In another part of the book Lu Yü declared: 'The nature of tea is cold. It is eminently suitable to serve as a drink for persons of self-reliance and good conduct. It should be drunk when one feels hot, thirsty, given to melancholia, aches in the head and eyes, fatigue of the four limbs, or pains in the hundred joints. Four to five sips of it and we would feel as if we had just tasted the life-giving dew of heaven.'[30] To this list of virtues, the foremost literary figure of the Sung Dynasty, Su Tung-Pho added, in +1083, the surprising discovery that tea helped to reduce tooth decay. He says:[31]

I have found it advantageous to rinse my mouth with strong tea after each meal. The oily heaviness thus dispelled, the stomach and spleen will feel no discomfort. The swirling of the tea will shrink and dislodge any meat particles that are stuck between the teeth. There is no need to use a tooth pick. The teeth will be strengthened. Dental disease will be reduced. The effect is achieved with teas of low and medium quality, since high quality teas are difficult to obtain.

These sentiments were reiterated by Chhien Chhun-Nien in the *Chha Phu* of +1539,[32]

Effect of drinking true tea: tea quenches thirst and aids digestion, dissolves phlegm and reduces sleep, promotes elimination of water, brightens the eyes and benefits the brain, dispels worries and counteracts fatty foods. It is something that one cannot do without every day. If for any reason one cannot drink tea, it would still be advisable to rinse one's mouth with strong tea. When worries are dispelled and fatty foods counteracted, the stomach and spleen would naturally function efficiently. Any meat particles caught between the teeth would be shrunk and washed away. No pricking is needed to remove them. The teeth would acquire the bitterness of tea and become stronger and firmer. Elements toxic to teeth would be nullified.

The verse on the effect of drinking tea in the celebrated poem of Lu Thung, the Tea Doter, +835, quoted earlier (page 554) may be regarded as a more poetic and romantic rendition of the above statements. But behind the literary imagery and poetic facade what did the mediaeval Chinese medical establishment actually think about the effect of tea drinking on the human constitution? For an answer to this question we need to look at what the *Pên Tshao* Pharmacopoeias have to say on this subject. The earliest record of this kind is in the *Hsin Hsiu Pên Tshao* of +659, which we have already quoted (page 512), but in view of its importance in this discussion, we need to quote it again.[33] '*Ming* (Tea): its taste is sweet and bitter. It is slightly cold and non-toxic. It alleviates boils and sores, promotes urination, dispels phlegm and relieves thirst. It keeps one awake.'

---

[30] *Chha Ching*, Fu & Ouyang eds. (1983), p. 5, tr. auct. adjv. Ukers (1935) I, p. 15 and Carpenter (1974), ch. 1. The last sentence in this paragraph is very difficult to translate. The original reads, *Liu szu wu chho yü thi hu kan lu khun hêng yeh* 柳四五啜與醍醐甘露抗衡也. Both Ukers and Carpenter have rendered *liu szu wu chho* as 'drink tea four to five times', while we prefer 'four to five sips' which is more faithful to the original.
[31] *Tung Pho Tsa Chi* 東坡雜記, cited in C&C, pp. 238–9.
[32] *Chha Phu* by Chhien Chhun-Nien, C&C, pp. 127–8; part of this passage is repeated in the *Chha Ching* by Chang Chhien-Tê, C&C, (*1981*), p. 146.
[33] *HHPT*, ch. 13, p. 116.

Similar statements are found in the *Shih Liao Pên Tshao*, +670, *Pên Tshao Shih I* (Supplement to the Pharmacopoeia), +725, *Chêng Lei Pên Tshao*, +1082, *Yin Shan Chêng Yao*, +1330, *Pên Tshao Phing Hui Ching Yao* (Essentials of the Pharmacopoeia), +1505 and the *Pên Tshao Kang Mu*, +1596.[34] The summation in the *Pên Tshao Kang Mu* reads as follows: '*Ming* (Tea) Leaf: it is bitter, sweet, slightly cold and non-toxic. It cures boils and sores, promotes urination, dispels phlegm and fever, quenches thirst, keeps one awake and heightens one's alertness and well-being. It also relieves flatulence and improves digestion.'

The *Pên Tshao Kang Mu* further mentions that tea is effective in controlling diarrhoea, heat-stroke and overindulgence in alcohol. The efficacy of specific types of tea and various prescriptions in curing or relieving some of these conditions are described in the *Pên Tshao Kang Mu Shih I*, +1780.[35] But not all the comments are in tea's favour. In the Southern Sung food canon *Shan Chia Chhing Kung* Lin Hung warns us,[36]

> Tea is a medication. When it is boiled the decoction will alleviate heartburn and facilitate digestion. If it is drunk as an infusion, it will impair the diaphragm and weaken the spleen and stomach. Some people mix old leaves with young leaves in processing the tea, and are too lazy to boil the product. The result will be doubly pernicious.

The celebrated *Yin Shih Hsü Chih*, a work on dietetics by Chia Ming at the end of the Yuan, says:[37]

> Tea tastes bitter and sweet. Its nature is cold, but the tea from Chieh is only mildly cold. Continuous drinking of tea will make one thin, induce loss of fat and diminish one's ability to fall asleep. Massive amounts of tea or drinking it after consumption of wine, will allow its coldness to penetrate the kidney and cause pain in the hip and bladder as well as swelling and rheumatism in the joints. It is particularly pernicious to add salt to tea or accompany it with salty food. This is like inviting a thief into one's home. One must not drink tea with an empty stomach. Taking tea while eating chives is likely to give one a bloated feeling.[38] Tea should be drunk when hot. Cold tea will aid the accumulation of phlegm. It is better to drink less of it, rather than more. Better yet! Don't drink it at all.

These discordant notes, however, had little influence on the continuing popularity of tea during the Ming and Chhing dynasties. The traditional view that tea is beneficial to one's health is still widely held in China today. How valid are the benefits claimed in the Chinese classical literature? If they are, what is their mechanism of action in the light of modern science? Before we can delve into these issues, it will be convenient for us to summarise and categorise the effects of tea cited in the above quotations in Table 56.[39] For our enlightenment and amusement we are also

---

[34] *SLPT*, p. 23, item 44; *PTSI* ard *CLPT* are cited in the *PTPHCY*, ch. 18, p. 508; *YSCY*, ch. 2, p. 58; *PTKM*, ch. 32, p. 1872.
[35] *PTKMSI*, pp. 249–60.   [36] *SCCK*, p. 109.
[37] *YSHC*, pp. 48–9. It is interesting that this passage is not included in the collectanea of tea literature compiled by C&C.
[38] The same observation is recorded in the *Ko Wu Tshu Than*, +980, p. 22.
[39] The total number of effects listed comes to nineteen. This is a simplification of the twenty-four effects listed by Lin Kan-Liang in *CKCC*, pp. 92–101.

Table 56. *Traditional Chinese views on the effects of tea*

1. Quenches thirst
2. Inhibits sleep, counteracts the effect of alcohol, and stimulates the mind
3. Cools the body, reduces fever, and relieves aches in the head and eyes
4. Induces urination, promotes bowel movement and slows down diarrhoea
5. Facilitates digestion, counteracts fatty foods and reduces flatulence
6. Dissolves phlegm, alleviates boils and sores, and lessens pains in the joints
7. Strengthens teeth
8. Endows one with a general sense of well-being
9. Promotes good health and longevity

reproducing in Table 57 an advertisement for tea, now in the British Museum, distributed by Thomas Garway in London in +1660.[40]

The first item in the Table 56 is that tea quenches thirst. This should come as no surprise since the infusion is, after all, mostly water. But this water is no ordinary water. It is boiled first before it is made into tea, and we now know that boiling kills any pathogens that may have been present. Tea is, therefore, a much safer drink than water itself, since one could be reasonably sure that it would be free from disease-carrying organisms. Indirectly, therefore, the custom of drinking tea was probably a great boon to public health in China. Every cup of tea drunk could mean one less cup of unsafe water consumed. There is, of course, also the likelihood that tea was actually preferred because it quenches thirst better than boiled water alone. But before we discuss this possibility, we need to know something about the chemistry and pharmacology of the constitutents in the tea leaf that are extracted into the infusion.

The three most important constituents of tea are the alkaloid caffeine, essential oils and polyphenols (previously but erroneously called tannins).[41] Caffeine is well known as a mild stimulant, analgesic and diuretic. It acts on the central nervous system, muscles (including cardiac muscles) and the kidney.[42] It increases mental alertness, facilitates muscular movements and improves the function of the kidney. The essential oils are responsible for the aroma of the tea.[43] Being volatile, they may be lost from the tea-product if it has been dried at too high a temperature or stored under improper conditions. The polyphenols are a group of complex phenolic compounds, chiefly flavanols commonly known as catechins. The major catechins are

---

[40] Taken from Ukers (1935) I, p. 39. The claims are similar to those found in *The Povey Manuscript*, an encomium on the Virtues of Tea, Ukers *ibid.*, p. 40, said to have been translated from the 'Chinese language', by Thomas Povey M. P. in +1686. Presumably, both the Garway Broadside and the Povey Manuscript are derived from the Tea Books of the Ming period.

[41] For a discussion of the chemistry of tea see Yamanishi ed. (1995), pp. 435–56. Cf. also Graham, H. N. (1992).

[42] Harler, C. R. in Ukers, W. (1935) I, pp. 538–43; Chow & Kramer (1990), ch. 9, pp. 91–9.

[43] The essential oils consist of a very complex mixture of hydrocarbons, alcohols, aldehydes, ketones etc. Among the components are linalool, geraniol and methyl salicylate, cf. LARC Monographs, Vol. 51 (1991), pp. 217–33.

Table 57. *Garway's First Tea Broadside (+1660)*

> It maketh the Body active and lusty.
> It helpeth the Headache, giddiness and heavyness thereof.
> It removeth the Obstructions of the Spleen.
> It is very good against the Stone and Gravel, cleansing the Kidneys and Uriters being drunk with Virgin's Honey instead of sugar.
> It taketh away the difficulty of breathing, opening Obstructions.
> It is good against Lipitude, Distillations and cleareth the Sight.
> It removeth Lassitude, and cleanseth and purifyeth adult Humours and hot Liver.
> It is good against Crudities, strengthening the weakness of the Ventricle or Stomack, causing good Appetite and Digestion, and particularly for Men of a corpulent Body, and such as are great eaters of Flesh.
> It vanquisheth heavy dreams, easeth the Brain, and strengtheneth the Memory.
> It overcometh superfluous Sleep, and prevents Sleepiness in general, a draught of the Infusion being taken, so that without trouble whole nights may be spent in study without hurt to the Body, in that it moderately heateth and bindeth the mouth of the Stomack.
> It prevents and cures aches, Surfets and Fevers, by infusing a fit quantity of the Leaf, thereby provoking a most gentle Vomit and breathing of the Pores, and hath been given with wonderful success.
> It (being prepared with Milk and Water) strengtheneth the inward parts, and prevents Consumptions, and powerfully assuageth the pains of the Bowels, or griping of the Guts and Looseness.
> It is good for Colds, Dropsies and Scurveys, if properly infused, purging the Blood by sweat and Urine, and expelleth infection.
> It drives away all pains in the Collick proceeding from Wind, and purgeth safely the Gall.
> And that the Vertues and excellencies of this Leaf and Drink are many and great is evident and manifest by the high esteem and use of it (especially of late years) among the Physitians and knowing men in France, Italy, Holland and other parts of Christendom.

(−)-epicatechin., (−)-epicatechin-3-gallate, (−)-epigallocatechin and (−)-epigal locatechin-3-gallate as well as (+)-catechin and (+)-gallocatechin. During 'fermentation' they are oxidised via tea polyphenol oxidase to theaflavins and thearubigins. Their formulae are shown in Figure 131. The reduction of catechins during processing of the three major types of teas is shown in Table 58. It shows that about 60% of the catechins in the fresh leaf is still present in the finished green tea, whereas hardly any catechins are left in the finished red or black tea. The level in oolong is in between the two. It is the presence of the catechins and oxidised polyphenols that gives the tea infusion its characteristic flavour and colour. It used to be thought that the polyphenols had a deleterious effect on the digestive system, but this prejudice has long since disappeared.[44] As it turned out, the polyphenols have become the

---

[44] Ukers (1935) I, p. 545, says, 'The consensus of medical opinion is that excessive tea drinking is bad, largely because of the effect of the tannin on the digestive tract, where its astringency may constipate the bowels, and its reduction in intestinal secretions may bring about indigestion.'

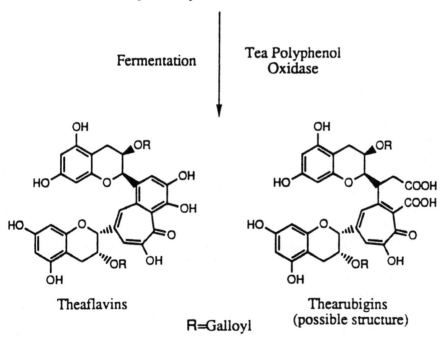

Fig 131. Chemical formulae for tea catechins and oxidised catechins.

Table 58. *Comparison of flavanol compounds in the three major teas*

Type	Catechin mg/g					
	EGC	GC	C & EP	ECG	EGCG	Total
Green						
Fresh leaf	18.99	8.11	11.01	88.58	31.09	158.38
Made tea	11.50	4.68	6.35	61.14	25.04	108.71
Reduced %	39.44	42.29	45.31	30.97	19.46	31.36
Oolong						
Fresh leaf	34.36	7.03	12.79	24.38	63.91	142.57
Made tea	5.00	3.69	4.01	8.15	16.91	37.76
Reduced %	85.49	47.51	68.64	66.57	73.54	73.51
Red						
Fresh leaf	29.00	11.05	7.34	18.74	81.84	147.93
Made tea	0.10	—	—	1.10	2.50	3.70
Reduced %	99.64	100.00	100.00	94.13	96.95	97.59

Adapted from Tei Yamanishi ed. (1995), p. 412, Table IV.2. We thank Luo Shaojun for correcting several errors in the original table and for providing the data on red tea.

EGC, (–)-epigallocatechin. GC, (+)-gallocatechin.
C, (+)-catechin. EP, (–)-epicatechin.
ECG, (–)-epicatechin-3-gallate. EGCG, (–)-epigallocatechin-3-gallate.

most intriguing component in tea in terms of their effect on human health and well-being.

The essential oils and polyphenols together are largely responsible for the immediate impression that one receives from a hot, steaming cup of tea – its aroma, colour and flavour. Since the Thang, the Chinese pharmacopoeias have stated repeatedly that tea is both bitter and sweet (*khu* 苦 and *kan* 甘). This would seem to be a contradiction, but once one has acquired an appreciation for green tea brewed in the traditional manner of the Ming tea books, the observation begins to make sense. When one first sips a freshly brewed infusion, one immediately feels a mild astringency developing around the mouth. The taste is deliciously fresh, grassy though slightly bitter. But after the tea is swallowed, there is a smoothness or sweetness that lingers on the tongue and in the throat. Presumably, this is the type of sensation that inspired Lu Thung to write almost twelve hundred years ago (page 554), 'The first cup caresses my dry lips and throat.' This is the direct effect of the essential oils and polyphenols.

But after the immediate reaction, it is the caffeine in the tea that provides the physical basis for the second and third lines of Lu Thung's poem, 'The second [cup] shatters the walls of my loneliness,' and 'The third searches the dry rivulets of my soul to find the stories of five thousand scrolls.' These are the sensations that makes tea truly 'the cup that cheers.' In fact, it is caffeine that accounts for most of the beneficial effects of tea listed in Table 56. It is the effect of caffeine on the central nervous system that keeps one awake, counteracts the numbing activity of alcohol and stimulates the mind (item 2). It is its stimulant activity that cools the body, reduces fever and relieves aches in the head and eyes (item 3) and lessens the pains in the joints. It is its effect on the kidney that promotes urination and controls diarrhoea (item 5), which in turn helps the body to eliminate toxic substances that may be circulating in the blood stream. Furthermore, caffeine stimulates movement of the intestines and the flow of digestive enzymes thus improving the function of the digestive system (item 4). It could also be the reason why a cup or two of tea after a heavy meal may help to dissipate the fat and the attendant feeling of discomfort.

On the other hand, for people who are highly sensitive to caffeine, it is the bane of an otherwise attractive beverage. In spite of the Garway Broadside (Table 57) to most people sleep can hardly ever be considered as superfluous. Nor 'can whole nights be spent in study without hurt to the Body'. By preventing one from falling asleep, caffeine robs the body of much needed rest and keeps it in a state of tension and nervousness for the next day. The half-life of caffeine in the blood stream is about six hours.[45] This means that in certain individuals the adverse effect of a cup of tea or coffee could last as long as twelve hours. Yet the lure of tea (and coffee) is so strong that a great deal of time and effort have been expended in the last half-century to remove caffeine from tea (and from coffee) with minimum damage to its organoleptic properties. Thanks to the wonders of modern chemical engineering, these efforts have met with considerable success. Decaffeinated tea and coffee are now commercially available and are a boon to legions of caffeine-sensitive lovers of tea and coffee. To those who do not suffer from a heightened sensitivity to caffeine such decaffeinated drinks may appear to be but a pale reflection of the genuine article. Indeed, one could ask, 'In terms of its effect on health, does it still make sense to drink tea after the caffeine is removed?'

The answer surprisingly seems to be 'yes'. The reason for this statement is found in the last three items in Table 56. We have not discussed them yet since we have so far concentrated only on the activity of caffeine. Item 7 is clear enough, but it is difficult to pin down what an increase in one's sense of well-being, general good health and longevity actually mean in practice. Yet research on the effect of tea on various aspects of human health and disease carried out in China, Japan, Britain and the United States in the last two decades have left us with the impression that these rather amorphous claims may actually have a reasonable scientific basis. Take first, for example, the case of strengthening the teeth (item 7).[46] It has now been found that tea contains a significant amount of fluoride. In areas where the water contains no fluoride, two to three cups of tea per day may supply enough fluoride to protect the tooth enamel from decay. But that is not all. The polyphenols in the tea may also inactivate the bacteria in the mouth and thus inhibit the formation of plaques and dental caries.[47]

---

[45] Chow & Kramer (1990), p. 95.
[46] Ibid., p. 108; cf. also Tshao Ching (*1992*), pp. 202–3, who has analysed the fluoride content of several varieties of Chinese tea.
[47] Otake *et al.* (1991), and Yu *et al.* (1992).

What is most exciting about the latest results is that it now appears quite possible that tea polyphenols, sometimes augmented by caffeine, can retard the onset of cancer and cardiovascular disease, the two most dreaded killers of the modern age.[48] If this view is correct then tea would confer on tea drinkers the blessings, that is a sense of well-being, good health and a long life, that are grouped in the last two items in Table 56. Two types of research have been and still are in progress. The first is epidemiological, which studies the effect of tea on specific illnesses among populations. The second is biochemical and pharmacological, which involves experiments carried out on laboratory animal, cell and organismal systems. It is beyond the scope of our survey to analyse the significance of these results. We can only say, in summary, that the jury on the epidemiological studies is still out, but the laboratory studies look highly encouraging. Tea infusions and extracts and isolated tea catechins have been shown to:

(1) Prevent cell mutations by increasing the faithfulness of DNA replications in healthy cells.[49]
(2) Inhibit the formation of cancer-causing chemicals such as nitrosamines.[50]
(3) Boost immunity by increasing the number of white blood cells.[51]
(4) Inhibit the formation of chemically induced tumour of the skin, oesophagus, forestomach, duodenum, small intestine, colon, pancreas, liver, pancreas and mammary gland.[52]
(5) Suppress the growth of implanted tumour cells.[53]
(6) Lower the cholesterol and triglyceride levels in the blood stream.[54]
(7) Reduce blood pressure and platelet aggregation.[55]
(8) Exhibit antibacterial and antiviral activities.[56]

These are the highlights of the activities of tea extracts and tea catechins that have been established under laboratory conditions. In each case one key reference is provided in the footnote for the convenience of readers who may wish to examine the original data themselves and pursue the matter further. It should be pointed out that the activities tend to be associated with the catechins rather than with oxidised catechins. Thus, in most cases, green or unfermented tea appears to be more effective than red (i.e. black) or fermented tea. Some of the observations may need further verification, and it remains to be seen whether any of these leads will actually result in applications of clinical importance. But one cannot help but feel that the story of the effect of tea on health is still evolving, and that the most exciting developments are yet to come.

---

[48] A popular account of these developments for the layman is provided by Chow & Kramer (1990), pp. 91–112. For a more academically oriented review of the latest results cf. Yang, C. S. & Wang Zhi-Yuan (1993), and Ho Chi-Tang *et al.* eds. (1994).
[49] Xu, Y, Ho, C. T., Amin, S. G., Han, C. & Chung, F. L. (1992).
[50] Nakamura & Kawabata, T, (1981). Cf. also Han & Xu (1990).
[51] Yan, Y. S. *et al.* cited in Yang & Wang (1993), p. 1046, ref. 126, and work at the Tea Research Institute in Hangchou, reported in Chow & Kramer (1990), pp. 102–3.
[52] Yang & Wang (1993), pp. 1042–3.   [53] Oguni *et al.* (1988).
[54] Ikeda *et al.* (1992).   [55] Sano *et al.* (1986).   [56] Toda *et al.* (1991).

(g) FOOD AND NUTRITIONAL DEFICIENCY DISEASES

If there is one central theme that runs through the Chinese literature on the relationship between diet and health it is the concept of '*I shih thung yuan* 醫食同源' or '*Yao shih thung yuan* 葯食同源,'[1] that is to say, 'medicine and food share a common origin.' Indeed, legend has it that both medicinal herbs and the five staple grains were discovered by the same person, the Heavenly Husbandman, the Emperor Shên Nung,[2] whom we have most recently encountered in the preceding chapter on the origin of tea. It was reasoned, therefore, that some foods should be able to serve as medicine and *vice versa*. This point of view has given rise to a tradition of medical practice known as *Shih Liao* 食療, diet therapy, in which foods are used as therapeutic agents for the treatment of diseases.[3] In fact, a *Shih I* 食醫 (Grand Dietitian) was a member of the group of medical officials listed in the *Chou Li* 周禮 (Rites of the Chou), c. −300, who looked after the health of the royal family at the Chou court.[4] A considerable body of literature has grown around this subject which we have reviewed earlier in Chapter (*b*). Furthermore, based on this theme, many common items of food, both raw and processed, have been routinely included in traditional pharmacopoeias, that is, the *Pên Tshao* 本草 corpus that has already been discussed at some length in a previous volume of this Series.[5]

Diet therapy is an integral part of traditional Chinese Medicine, which is based on principles that cannot yet be reconciled with the concepts of modern medicine. One cannot, therefore, adequately discuss diet therapy without first considering the theory behind Chinese Medicine, which is the responsibility of another Part of this series.[6] There is, however, one area in the practice of diet therapy that can be

---

[1] Cf., for example, Chiang Chhao (*1985*), p. 32, Chêng Chin-Shêng in Li Ching-Wei, Chhêng Chih-Fan *et al.*, eds. (*1987*), '*I Hsueh Shih*' in *Chung-kuo I-Hsüeh Pai Kho Chhuan Shu*, p. 70; Yang Ling-Ling (*1988*), p. 16, Cai Jinfeng (1988), p. 20, and Thao Wên-Thai (*1988*), p. 321. This concept is, however, also an integral part of the Western medical tradition, cf. Sigerist (1951), pp. 114–15. Guided by the humoral theory, foods have often been used by physicians in Greek, Roman, mediaeval Europe and as late as 18th-century America, to cure diseases, cf. Nutton (1993), pp. 281–91, Gerard's Herbal (Woodward 1985), Hippocrates (Chadwick & Mann 1950), Galen (Green 1951), and Estes (1990, 1996).
[2] For a quotation and discussion of the relevant passage which records this momentous discovery in the Book of the Prince of Huai Nan, the *Huai Nan Tzu* (c. −120), cf. Chapter (*b*), p. 8.
[3] Several books on the practice of diet therapy are now available in Western languages, for example, Flaws & Wolfe (1983), Henry C. Lu (1986), Cai Jinfeng (1988), Liu Jilin, ed. (1995) and Engelhardt & Hempen (1997).
[4] *Chou Li*, pp. 45–6. The group was headed by an Administrator of Medical Affairs, *I Shih* 醫師, who supervised a Grand Dietitian, a Grand Physician for Internal Illnesses, *Tsi I* 疾醫, a A Grand Physician for External Illnesses, *Yang I* 瘍醫 and a Grand Veterinarian, *Shou I* 獸醫. The term *Shih Liao* has been translated as 'Dietotherapy' by Cai Jinfeng (Tshai Ching-Feng) (1988), 'Food Cures' by Henry C. Lu (1986), and 'Dietary Therapy' by Liu Jilin ed. (1995).
[5] *SCC* Vol. VI, Pt 1, pp. 220–321.
[6] The concepts of *chhi*, Yin-Yang and Wu Hsing have been discussed in detail in Needham, *SCC* Vol. II, pp. 216–346. A simplified version of this discussion is found in Ho Peng Yoke (1985). For the principles of traditional Chinese medicine and diet therapy, cf. Vol. VI, Pt 6 on Medicine of this series, as well as Porkert (1974), Unschuld (1985) and Sivin (1987). Cf. also Beinfield & Korngold (1991) for a popular and readable account of the principles and practice of Chinese Medicine.

Table 59. Symbolic correlations of Wu-Hsing categories in heaven, earth and man

Wu-Hsing 五行	Wood 木	Fire 火	Earth 土	Metal 金	Water 水
Heaven 天時					
Position 方	East 東	South 南	Centre 中	West 西	North 北
Season 時	Spring 春	Summer 夏	Long Summer 長夏	Fall 秋	Winter 冬
Climate 氣	Wind 風	Hot 熱	Humid 濕	Dry 燥	Cold 寒
Planet 星	Jupiter	Mars	Saturn	Venus	Mercury
Number 數	3+5=8	2+5=7	5	4+5=9	1+5=6
Earth 地利					
Elements 品類	Wood	Fire	Earth	Metal	Water
Livestock 五畜	Chicken 雞	Sheep 羊	Ox 牛	Horse 馬	Swine 豕
Staple Grains 五穀	Wheat 麥	Panicum millet 黍	Setaria millet 稷	Rice 稻	Soybean 豆
Musical tones 五音	chio 角	chih 徵	kung 宮	shang 商	yü 羽
Colour 五色	Green 青	Red 紅	Yellow 黃	White 白	Black 黑
Flavour 五味	Sour 酸	Bitter 苦	Sweet 甘	Pungent 辛	Salty 鹹
Odour 五臭	Rancid 臊	Scorched 焦	Fragrant 香	Rotten 腥	Putrid 腐
Man 人和					
Viscera 五臟	Liver 肝	Heart 心	Spleen 脾	Lung 肺	Kidney 腎
Bowels 六腑	Gall bladder 膽	Small intestine 小腸	Stomach 胃	Large intestine 大腸	Bladder & 3 burning spaces 三焦
Orifices 九竅	Eyes 目	Ear 耳	Mouth 口	Nostrils 鼻	Lower orifices 二陰
Tissues 五體	Ligaments 筋	Pulse (blood) 脈	Muscle 肉	Skin & hair 皮毛	Bones 骨
Voice-sounds 五聲	Shout 呼	Laugh 笑	Sing 歌	Weep 哭	Groan 呻
Emotions 五志	Anger 怒	Joy 喜	Sympathy 思	Grief 憂	Fear 恐
Symptomology 病變	Clench fist/spasm 握	Depression 憂	Vomiting 噦	Coughing 咳	Trembling 慄

understood easily in terms of the science of nutrition, that is, the treatment of nutritional deficiency diseases. In fact, it was the topic that was originally intended to be the centrepiece of an extended chapter on Nutrition in the present volume. This plan is now obsolete, but we feel the discovery of nutritional deficiency diseases is of such historical and scientific interest that it should remain in this Section to remind us that processed foods with superior palatability may have an adverse as well as a salutary effect on the nutritional value of the original material.

For readers who possess only a passing acquaintance with the concept of Wu-Hsing 五行 (Five Elements or Phases) referred to in the discussion which follows, the correlations of Wu-Hsing categories with each other are presented in Table 59 for their convenience.

It is common knowledge today that the human body is able to produce from the foods it consumes almost all of the materials and chemicals it needs to maintain its anatomical integrity and physiological functions. But not all. In addition to certain key elements, our body is unable to synthesise a number of complex organic chemicals which are absolutely essential to its well-being. They must be obtained from the foods and drinks we consume. Deficiency in any one of these substances will result in illness or even death. These essential factors are of three types, vitamins, essential amino acids and trace elements.[7] The discovery, elucidation of the structure and function, and the synthesis of the vitamins are a major triumph of biochemistry in the first half of the twentieth century.

It is not surprising that among the most successful applications of diet therapy in China is the treatment of nutritional deficiency diseases. These are diseases, by their very nature, uniquely amenable to treatment by diet. Since the disease is caused by a lack of an essential element in the diet, it stands to reason that it can be reversed by restoring the missing element to the diet. The ancient and mediaeval Chinese, of course, had no way of knowing what was missing in the diet. But by exploiting the large numbers of materials at their disposal, by persistent trial and error, and by astute observation of the effects of substances ingested, they were able to recognise and treat the diseases which we now know as goitre, beriberi, and night blindness, as well as to try to control the crippling children's disease, rickets. Were they all simply empirical discoveries? Or were they in any way aided by the Wu-Hsing interactions that form the theoretical basis of diet therapy. Let us see what the historical background of the discoveries tells us. We shall consider each one of the four illnesses in turn.

(1) Goitre (*Ying* 癭)

Goitre is an enlargement of the thyroid gland on the front and sides of the neck.[8] It is caused by a deficiency of iodine which is an essential precursor for the synthesis of

---

[7] The vitamins are compounds that often function as coenzymes or hormones. The essential amino acids are those which the human cannot synthesise: leucine, isoleucine, lysine, methionine, phenylalanine, threonine, tryptophan and valine.

[8] For a brief history of goitre cf. Welbourn (1993), pp. 485–6 and 498–9 in Bynum & Porter (1993). Goitre was still common among primitive peoples such as the Neolithic tribes of the Mulia region of West New Guinea in the 1960s, cf. Gajdusek (1962).

the thyroid hormone thyroxine. The modern name for the disease is *Chia chuan hsien chung* 甲狀腺腫, that is, swelling of the thyroid, but in ancient and mediaeval China it was known as *ying* 癭. In fact the *Shuo Wên Chieh Tzu* (Analytical Dictionary of Characters), +121, defines *ying* as a tumour (*liu* 瘤) or swelling at the neck (*ying ching liu yeh* 癭頸瘤也).[9] The disease was probably recognised early in China, well before the start of the Warring States. The *Chuang Tzu* of the −4th century refers to a man who, in spite of a goitre (*ying*) as large as an earthenware jar, nevertheless won the favour of a feudal lord.[10] The *Lü Shih Chhun Chhiu* (Master Lü's Spring and Autumn Annals), −239, had this to say:[11]

In places where there is too much light (*chhing* 清) water there is much baldness (*thu* 禿) and goitre. In places where there is too much heavy (*chung* 重) water there will be a high incidence of swellings and dropsies which make it hard for people to walk. Where sweet (*kan* 甘) water abounds, the people will be healthy, but if there is much pungent (*hsin* 辛) water skin eruptions will be common. Finally where bitter (*khu* 苦) water is plentiful, there will be many people with ulcers and hunchbacks.

The quality of water was thus considered a major contributing factor in the occurrence of goitre. Other references tend to indicate that goitre is a disease prevalent in mountainous regions. The *Shan Hai Ching* 山海經 (Classic of Mountains and Seas), composed from Chou to Western Han, notes that[12] 'in the Mountain of the Celestial Emperor (*Thien ti chih shan* 天帝之山) ... is found a herbaceous plant with the appearance of mallow and the scent of selinum (*mi wu* 蘪蕪). It is called *tu hêng* 杜衡. It strengthens the health of horses and stops the growth of goitre.' The quotation implies that goitre is not uncommon in the *Thien ti* mountains and suggests that it can be cured by the herb *mi wu*. The *Po Wu Chih* states that[13] 'people living in the mountains often suffer from goitre (*shan chü chih min to ying* 山居之民多癭). The illness is caused by drinking water from springs that have become stagnant.' The commentary, however, disagrees, since the disease seems to occur only in mountains of South China and is hardly ever seen in similar terrain in the North. At any rate, the above references tend to indicate that goitre is the result of the peculiar water and soil conditions that exist in the regions where they are endemic.

A different view is expressed in The *Huai Nan Tzu* (Book of the Prince of Huai Nan), c. 120, which says,[14] 'When *chhi* is blocked by stress, goitre is stimulated (*hsien-tsu chhi to ying* 險阻氣多癭).' Kao Hsiu's commentary explains that when *chhi* in the throat is unable to move upwards or downwards, it will congeal and give rise to hard swellings, i.e. *ying*.

---

[9] *SWCT* (ch. 7B, p. 12), p. 154.
[10] *Chuang Tzu*, ch. 5, *Teh Chhung Fu* 德充符, Kuo Chhing-Fan ed. Chung Hua (1961), Vol. I, p. 216. The word *ying* 癭 is not translated in Watson tr. (1968), cf. p. 74, 'Mr. Pitcher-sized-Wen.'
[11] *LSCC*, ch. 3, segment 2, *Chin Shu* 盡數, p. 72. tr. Lu & Needham (1966), repr. in Needham *et al.* (1970), 'Proto-Endocrinology in Medieval China', p. 298, mod. auct.
[12] *Shan Hai Ching Chiao Chu*, ed. Yuan Kho (1980), ch. 2, p. 29.
[13] *Po Wu Chih*, ch. 1, *Wu fang chih min* 五方之民, p. 2b.
[14] *Huai Nan Tzu*, edn of *Kuang I Shu Chü*, Shanghai., ch. 4, *Chui Hsing Hsün* 墜形訓, p. 17.

Both points of view are expressed in the *Chu Ping Yuan Hou Lun* 諸病源候論 (Diseases and their Aetiology), +607, by Chhao Yuan-Fang. Chapter 31 tells us:[15]

Goitre is caused by the coagulation of the *chhi* of grief and anger. Others think it is due to the drinking of sandy water (*sha shui* 沙水). The sand particles follow the *chhi* 氣 into the pulse (blood). They attack the neck tissues and make them swell. At first the goitre may look like the stone of a fruit forming in the lower part of the neck. The skin remains loose and the goitre appears to hang freely. In those caused by emotional stress, the skin remains smooth and normal. In those caused by sandy water, the growth appears to float in a sac of skin. There are three kinds of goitre. The 'bloody' type can be punctured. The 'muscle' type can be surgically removed. The '*chhi* type' may be treated with needle.

The *Yang Shêng Fang* 養生方 (Prescriptions for Nurturing Life) says:[16] 'Mountainous regions where springs gush forth from the black earth are not places for a prolonged stay. Frequent consumption [of this water] will lead to goitre.'

Elsewhere the *Chu Ping Yuan Hou Lun* theorises that emotional stresses may disturb the *chhi* of kidney, causing it to rise. The kidney *chhi* then coagulates in the epiglottal region and gives rise to a goitrous swelling.[17] But long before they knew enough about the disease to indulge in such speculations, the Chinese had already learned how it could be reversed. The *Shên Nung Pên Tshao Ching* (Pharmacopoeia of the Heavenly Husbandman), −1st to +1st century, contains the first record we have on the use of seaweed for curing goitre:[18] 'Seaweed (*Hai tsao* 海藻): bitter in flavour and cold. It cures the *chhi* of goitre and tumours, and other swellings on the neck. It disperses *chhi* condensations.'

Hai tsao and another salt water alga, sweet tangle (*khun pu* 昆布) are listed in the *Ming I Pieh Lu*, +510.[19] Both are considered cures for goitre. But it is curious that in the *Shih Liao Pên Tshao*, +670, the anti-goitrous activity of neither material is mentioned.[20] The earliest prescription using seaweeds in goitre therapy is attributed to Ko Hung's *Chou Hou Pei Chi Fang* (Prescriptions for Emergencies), +340, which says:[21]

Wash off the salt on one catty of seaweed (*hai tsao*). Place it in a silk bag and soak it in two pints of clear wine (*chhing chiu* 清酒). In spring or summer after two days drink two *ko* 合 of the wine at a time until the wine is depleted on the third day. Repeat the soaking in another two pints of wine. Drink the wine as before. The residue is then dried and also taken internally.

This prescription is no longer extant in the surviving text of the *Chou Hou Pei Chi Fang*. It is preserved in the *Wai Thai Mi Yao*, +752, a collection of prescriptions from various sources which are themselves no longer extant. The latter also describes a

---

[15] *Chu Ping Yuan Hou Lun* (*CPYHL*), also known as *Tshao Shih Ping Yuan* 巢氏病源, Jenmin Weisheng, (*1992*), ch. 31, pp. 856–7. This passage is copied verbatim in the *Wai Thai Mi Yao*, +752. ch. 33 on goitre.
[16] Cited in *CPYHL*, ibid. This is probably the *Yang Shêng Fang* of Shangkuan I 上官翼 who lived in the early Thang. The book is no longer extant.
[17] *CPYHL*, ch. 39, p. 1135.
[18] *SNPTC*, Middle Class, Tshao Yuan-Yü, ed. (*1987*), p. 165. *Hai tsao* has been identified as *Sargassum siliquastrum*.
[19] *MIPL*, Shang Chih-Chün, ed. (*1986*), p. 157. *Kun pu* is *Laminaria saccharina*, usually called *hai tshai* 海菜 or *hai tai* 海帶, cf. Porter-Smith & Stuart, eds. (1973), pp. 23–4.
[20] *SLPT*, p. 10.
[21] This prescription is preserved in the *WTMY* ch. 33, *SKCS* **737**, p. 1 from which this quotation is taken.

second prescription by Ko Hung, in which he recommends the use of both seaweed and the sweet tangle *khun pu*. The ingredients are pulverised and mixed with a little honey to form pills the size of plum pits. These are then sucked in the mouth [like a hard candy].[22] In a recipe by the celebrated monk Shên Shih (深師) of the +5th century, the ingredients consist of seaweed, sweet tangle and nine other ingredients including clams and powdered shells of lamellibranch molluscs.[23] In another prescription Tshui Chih-Thi 崔知悌, c.+650,[24] distinguished between movable swellings, *chhi ying* 氣癭 and *shui ying* 水癭 that could be cured, and solid ones, *shih ying* 石癭, that could not.

Twenty-nine out of the thirty-six prescriptions for goitre collected in the *Wai Thai Mi Yao*, +752, utilise seaweeds as the active ingredient. It is clear that by the time of the Thang they had become the agent of choice in the treatment of goitre. This therapeutic measure remained popular through the ensuing centuries. *Hai tshai* 海菜 (sea vegetable) is mentioned as a cure for goitre in the Yuan nutritional classic, *Yin Shan Chêng Yao*, +1330. Four kinds of marine algae are described in the *Pên Tshao Kang Mu*, +1596, for use on goitre. But this is not the end of the story of goitre diet therapy. Six of the prescriptions preserved in the *Wai Thai Mi Yao* include an alternative agent for treating goitre, namely, the thyroid gland *yen* 厭 of domestic animals. This is an innovation that we did not see mentioned in the *Ming I Pieh Lu* or the early Thang *Shih Liao Pên Tshao*. Three of the prescriptions are attributed to *Ku Chin Lu Yen Fang* 古今錄驗方 (Well Tested Prescriptions, Old and New) written by Chên Li-Yen 甄立言, who lived in the +7th century;[25] One is ascribed to Shên Shih (+5th century) whom we have just met in the preceding paragraph, one to Sun Ssu-Mo's *Chhien Chin Yao Fang*, +659, and the last to Emperor Hsüan Tsung's *Kuang Chi Fang* 廣濟方, +723.

One of Chên Li-Yen's prescriptions calls for one hundred sheep thyroid glands to be washed in warm water, defatted, [chopped up] mixed with twenty skinless jujubes and made into pills. Another takes a single thyroid from a sheep, removes the fat and allows the patient to hold it in the mouth until it is sucked dry.[26] We first find the combined use of seaweed, thyroid gland and lamellibranch mollusc (*hai ko* 海蛤) in the prescription ascribed to *Chhien Chin Yao Fang*. The same triad is used in the prescription by Emperor Hsüan Tsung.[27]

Unlike seaweed, the thyroid gland was not an easily available food material. It is mentioned as a food in the *Mêng Liang Lu* of the Southern Sung, but is not seen in the

---

[22] This is also found in the *WTMY* ch. 33, *SKCS* **737**, p. 2.

[23] *Ibid.* According to the *MHS-CCM*, Medical Historiography Section of the Cyclopedia of Chinese Medicine (*1981*), p. 231, Shên Shih is a Buddhist monk–physician who lived during the North and South Dynasties. He authored a book of prescriptions known as *Shên Shih Fang* 深師方 (Prescriptions of Master Shên), said to consist of thirty chapters. It is no longer extant. Quotations from it may be seen in other works such as the *WTMY*.

[24] *WTMY*, ch. 33, 737–5. cf. *MHS-CCM*, p. 227 for further information on Tshui Chih-Thi.

[25] Cf. *MHS-CCM*, pp. 251–2 and Li Ching-Wei, Chhêng Chih-Fang *et al.* (*1987*), Medical History, Encyclopedia of Chinese Medicine, p. 123. Chên Li-Yen lived in the +7th century. In Needham *et al.* (1970), p. 301, the author of this work is erroneously identified as Chên Chhüan, who is actually Li-Yen's older brother. They are both celebrated physicians of early Thang. Chhüan is the author of several medical books, but it is Li-Yen who wrote the *Ku Ching Lu Yen Fang*.

[26] *WTMY*, pp. 737–4.   [27] *Ibid.*, pp. 737–4 and 737–6.

pharmaceutical and diet therapy literature until the *Pên Tshao Kang Mu* of +1596, which includes descriptions for pig, sheep and ox thyroid.²⁸ By then the anatomy of the thyroid was already well understood. Wang Hsi 王璽 in about +1475 tells us in the *I Lin Chi Yao* 醫林集要 (Collection on Medicinal) that 'the gland lies in front of the larynx and looks like a lump of flesh about the size of a jujube, flattish and of pink colour'. He further recommends that 'forty-nine thyroid glands should be heat dried, mixed with several herbs and ground to a powder, which is then taken with wine every night'. In another book of the same vintage, the *Hsing Lin Chê Yao* 杏林摘要 (Selections from the Apricot Forest), we find a similar prescription: 'seven pigs' thyroids are heated in wine, evaporated down and exposed to the dew in a water bottle overnight. They are then skewer-roasted and consumed.²⁸ᵃ The situation remained unchanged until the 20th century.

In the West, as summarised by Needham and Lu,²⁹ 'goitre was noted by Juvenal, Pliny and other Latin authors at least as early as the +1st century, and the thyroid gland itself was described by Galen in the +2nd century. After that there was no further advance until Roger of Palermo (Solerno?), about +1180, recommended the use of the ashes of seaweeds and sponges in cases of goitre – an empirical discovery . . . which continued in use and mentioned by medical writers long afterwards, as by Russell in +1755.' The next advance came when Thomas Linacre (+1460–1524), in +15th century, introduced the use of thyroid glands from healthy animals to treat the disease.

From at least the beginning of the 19th century onwards there was a tendency to blame the environment for the incidence of goitre, and Chatin in 1860 proved clearly the disease was correlated with a lack of iodine in the soil and water. In 1884 Schiff implanted thyroid gland into the abdomen of thyroidectomised animal and found that it would continue in good health. A dozen years later Baumann (1896) discovered the presence of iodine in the gland and about the same time Möbius (1891) ascribed exophthalmic goitre to the gland's hyperactivity.

Comparing the Western experience to the Chinese one, it is clear that while both cultures recognised the existence of goitre in antiquity, the Chinese were distinctly ahead of the West in the therapy of the disease. The Chinese were already using seaweeds before the time of Christ and thyroid glands by the time of Alexander of Tralles (+525–605). The corresponding dates in the West are +12th and +19th centuries. What led the Chinese to discover first seaweeds and later thyroid glands as cures for goitre? How could the discoveries have been made? Was it entirely empirical? We do not think so. In the Chinese view it is the proper balance of Yin and Yang principles that leads to good health. In the case of goitre it is obvious that there is an imbalance in the function of the body in general and of the thyroid gland in

---

²⁸, ²⁸ᵃ *MLL*, p. 137; *PTKM*, pp. 2708–9, 2739–40, 2758. The quotations from Wang Hsi and Wang Ying are cited on pp. 2708–9. See Needham *et al.* (1970), pp. 300–1.

²⁹ Needham *et al.* (1970), pp. 298 and 302. The 'Roger of Palermo' cited is probably 'Roger of Salerno'. For further details on the history of goitre in the West see Merke (1984), especially pp. 13, 17, 83–6, 92–100, 112–16.

particular. Let us put ourselves in the place of the ancient Chinese physicians and imagine what they would do.

In the first case, the ancients had long suspected that goitre had something to do with the condition of the soil and water in the mountainous regions where the disease is endemic. Obviously, there is either too much or little of a specific *chhi* 氣 in the diet that caused an imbalance in the function of the neck tissues. The opposite of mountain is the coastal plain. Indeed, there is no disease along the coast. Why not then bring something of the *chhi* in the diet of the coastal people to those suffering from goitre. Seaweed was a convenient choice, since it was plentiful and in a dried state readily portable.

In the second case, they might have asked what else they could do to restore the balance of function of the thyroid in a goitrous person. In a goitrous thyroid there is either too much or too little *chhi*. In a healthy thyroid the *chhi* is just right. If in the diseased state there is too little *chhi* it would make sense to feed healthy thyroids to the patient. But what if it is due to too much *chhi*? Would that not make the disease worse? The fact that the disease manifests itself as an enlargement of the thyroid may suggest that it is due to overactivity of the organ, the result of too much *chhi*. Needham and Lu have pointed out:[30] 'To have realised that the hyperplasia was essentially due to an incapacity (i.e. too little *chhi* or deficiency in a factor) would seem to have been an impossibly good guess – yet this is just what the old Chinese medical theorists achieved.' Perhaps, success in the use of seaweeds in reversing goitre suggests that the disease is caused by a deficiency in *chhi*. This gave the Chinese physicians the conviction to test the efficacy of the thyroid gland from animals. Let us quote Needham and Lu again:

When in 1890 Murray and others began to administer the thyroid or its extracts for myxoedema they acted following a course of reasoning based on physiological experiments of modern type on animals. Ko Hung and Chên Li-Yen[31] did not have this basis for their ideas, but that does not mean that they were destitute of theory; on the contrary, they reasoned out the matter with extreme acuity. Thyroid organotherapy must clearly be counted a signal success of mediaeval Chinese medicine.

(2) BERIBERI (*CHIAO-CHHI* 腳氣)

More than fifty years have elapsed since Lu and Needham submitted to ISIS their paper entitled 'A contribution to the History of Chinese Dietetics.'[32] This was the first time the history of the treatment of nutritional deficiency diseases in China was introduced to Western readers. A good part of the paper dealt with beriberi, a disease peculiar to the rice eating countries of East Asia, and quite unknown in Europe. In China it is called *Chiao chhi* 腳氣 (imbalance of *chhi* in the leg). Biochemical

---

[30] Needham *et al.*, *ibid.*, p. 302. [31] In the original text it is 'Ko Hung and Chên Chhüan'. Cf. n. 25 above.

[32] The paper was actually received by ISIS in 1939, but due to World War II it was not printed until April, 1951 (vol. 42, pp. 13–20). Although it attracted a considerable amount of attention at the time, apparently it made no lasting impression on historians of Western medicine. For example, the long history of beriberi in China is not mentioned at all in the account on beriberi in the article on 'Nutritional diseases' by Carpenter, K. J. in Bynum & Porter eds. (1993), pp. 472–4.

research in the early part of the 20th century has shown that it is caused by a deficiency of thiamin, or vitamin B_1 in the diet.[33] The disease was rare, perhaps even unknown, among the ancient Chinese. It has been suggested that the *chüeh* 厥 mentioned in the *Huang Ti Nei Ching Su Wên*,[34] the *wei chung* 微重 in the *Shih Ching* (Book of Odes),[35] the *wei pi* 委痺 in the *Huai Nan Tzu*,[36] and the *chhên ni chung chui* 沉溺重膇 in the *Tso Chuan*,[37] are all early terms that denote beriberi. But Li Thao has expressed doubt that the ailments referred to could be authentic cases of beriberi,[38] since the symptoms described are too general and ambiguous to permit a positive diagnosis of the disease. Moreover, in antiquity the population centre in China was in the North, and beriberi is a disease that first became prevalent in the South, among the rice eating population.

Beriberi was probably first noticed during the Han. The *Shih Chi* (Record of the Historian), c. −90, states that, 'in the *Chhu* 楚 and *Yüeh* 越 regions (i.e. Lower Yangtze Valley), the people boil seawater for salt and live mainly on rice and fish. The land is fertile, the food supply does not depend on trade and many suffer from a deterioration of the feet and leg' (*chu i ku chih yü* 足以故呰窳).[39] The illness was later called *huan fêng* 緩風 (slow wind) and, still later, *chiao jo* 腳弱 (weak legs).[40] Sun Ssu-Mo in the *Chhien Chin Yao Fang* (Thousand Golden Remedies), +659, has this to say on the origin of *chiao chhi*:[41]

I have examined the medical classics and found that weakness of the legs (*chiao jo*) is occasionally mentioned. But the ancients rarely suffered from this disease. It was only after the southern migration during the Yung Chia 永嘉 reign-period (+307–12) that it became prevalent among the officials and the gentry. In South of the Range and East of the River, Chih Fa-Tshun 支法存, Yang Tao-Jen 仰道人 and others investigated various remedies and became experts in curing the disease. Sufferers who survived the illness during the Chin reign all owe a debt of gratitude to these two pioneers.'

The text goes on to say that, 'In the Sung 宋 and Chhi 齊 period (+420 to +501) in South China, a Buddhist monk Shen Shih 深師 collected the formularies of Chih

---

[33] For a history of the discovery and elucidation of the structure and function of the vitamins see Harris, L. J. (1935). The fascinating story of Eijkman and his chickens in elucidating the cause of beriberi is an epic in the annals of nutritional science. Eijkman's report of his work in 1890 was recently reprinted in *Nutritional Reviews*, June (1990), *48* (6), pp. 242ff. The key role of Eijkman's study in the discovery of the vitamins is described by Pauling (1970), pp. 19–20. Cf. also Carpenter (1993), pp. 472–3, 476–7.

[34] *HTNCSW*, ch. 69, p. 538. Cf. for example, Wang & Wu (1932), p. 88 for the interpretation of *chüeh* as beriberi.

[35] *Shih Ching*, M 198. Chêng Hsüan's commentary interprets *wei chung* as 'legs swollen and ulcerated'.

[36] *Huai Nan Tzu*, cited in Li Thao (1936), p. 1030.

[37] *Tso Chuan*, Chhêng Kung 成公, 6th year, item 4, note 2.

[38] Li Thao (1936), p. 1010; idem (1940), repr. in Brothwell & Sandison (1940), pp. 417–22. Cf. also Hou Hsiang-Chhuan (1954), p. 16 and Thang Yü-Lin (1958), p. 54. *I Hsueh Shih yu Pao Chien Chu Chi* Vol. 2, No. 1.

[39] *Shih Chi*, ch. 129, *Huo Chih Lieh Chuan* 貨殖列傳.

[40] Cited in *Chiao-chhi Chih fa Tsung Yao* 腳氣治法總要 (Methods for Treating Chiao-chhi); cf. also Li Thao (1936).

[41] *CCYF*. ch. 7, p. 138. The text refers to the 5th and 6th centuries when China was divided into a Northern and a Southern empire each governed by a series of ruling houses. This is known as the Nan-Pei Chhao 南北朝 (North and South Dynasties) period. Sung 宋 and Chhi 齊 are the first and second Southern dynasties, while Wei 魏 and Chou 周 are the first and last of the Northern dynasties. For further information on Chih, Yang and Shên Shih see the *I Shuo* 醫說 by Chang Kao 張杲, +1224, ch. 1, *SKCS*, pp. 742–30. Yao Kung is Yao Sêng-Yuan 姚僧垣 and Hsu Wang is Hsü Chien 徐謇. Both are famous physicians of North China in the 6th century.

Fa-Tshun and Yang Tao-Jen into a book with thirty chapters, of which one, containing almost one hundred prescriptions, was devoted to treating weakness of the leg (i.e. beriberi). But the disease was not seen from the Chou to the Wei dynasties (*Chou Wei chih tai kai wu tshu ping* 周魏之代蓋無此病), +386 to +581, in North China. It is not mentioned in the well-known medical books of this period compiled by Doctors Yao Kung 姚公 and Hsü Wang 徐王.' It was, in fact, 'unknown West of the Gate and North of the river (*kuan hsi ho pei pu chien tzhu chi* 關西河北不見此疾).' These observations emphasise the fact that the illness was largely confined to South China during the period covered by the North and South Dynasties (i.e. c. +420 to +581).

At any rate, it would appear that by the late Han the famous physician Chang Chi 張機 was already familiar with the disease. In his book the *Chin Kuei Yao Lüeh* (Systematic Treasury of Medicine), c. +200, he gave two prescriptions for sufferers of *chiao chhi*.[42] One is a decoction of such familiar drugs as ephedra, paeonia, astragalus, glycyrrhiza and aconitum. The other is a bath of alum in rice water for soaking the affected leg. It was Ko Hung, in about +340, who first recognised that *chiao chhi* is a disease seen predominently in South China. He said,[43] '*Chiao chhi* originated South of the Range (*ling nan* 嶺南) and gradually moved to East of the River (*chiang tung* 江東), where it is still spreading. It may start as a slight pain and rheumatism [on the leg], a numbness around the shinbone, or a weakness in the leg that makes walking difficult.'

The first description of the pathology of *chiao chhi* is given in the *Chu Ping Yuan Hou Lun* (Diseases and their Aetiology) of +610. We quote:[44]

In the Eastern part of the Yangtse Valley and South China, the land is low and damp, and people are apt to be afflicted by this disease. It begins from lower extremities, usually as a little weakness in the legs, hence it is known as *chiao chhi*. It often sets in insidiously as a primary condition, but sometimes it follows other illnesses. At first the disease is so mild that the patient eats and works as usual and feels as strong as before. But upon closer observation, one finds numbness from the knees down to the feet. The skin over the affected area feels as if it is thickened. When the affected area is scratched lightly with the fingers, it feels as if the scratching is over the stockings and the patient may fail to notice it. The legs may become paretic and prone to formication. In some cases, the legs are so weak that the patient is unable to walk. There may be slight oedema, coldness, muscular discomfort and in severe cases complete paralysis.

The appetite may be good or lost. Nausea and vomiting may occur at the sight of food and such patients usually dislike to smell food. Some may feel distress in the lower abdomen which may ascend into the chest and bring about dyspnoea. There may be generalised uneasiness, high fever, headache, palpitation, slight photophobia and abdominal pains associated with diarrhoea. In some cases there may develop forgetfulness, blurring of vision, disorder in speech and even mental confusion.

When the cases are treated too late, the disease will spread upwards into the abdomen. When the abdomen is involved there may or may not be oedema. When oppression in the

---

[42] *CKYL*, ch. 1, pp. 13a, b.   [43] *CHPCF*, Renmin Weisheng reprint, ch. 3, sec. 21, p. 56.
[44] *CPYHL*, ch. 13, pp. 416, 413, tr. Kao Ching-Lang 高景朗 (1936). The arrangement in the Ting Kuang-Ti edition (1995) is slightly different from that quoted in Kao's translation.

chest and dyspnoea both develop, the patient will die. The unfavourable cases usually last less than one month, while mild ones may run a course from one to three months.

It is seen from this account that the presently understood forms of beriberi, the dry, wet and cardiac types were already recognised by the Chinese almost 1,400 years ago. The prescriptions developed by the Chin monks and others were collected and enlarged in the *Chhien Chin Yao Fang* and *Chhien Chin I Fang* of Sun Ssu-Mo in the 7th century. The latter gives twenty-one recipes for the treatment of *chiao chhi*. Most of them are preparations using various herbs.[45] To us the most interesting are two recipes that clearly fall into the diet therapy category. One uses pork liver as the main ingredient; the other uses rice bran (*ku pai phi* 穀白皮), which is incorporated into a congee made with regular rice. It is said that continuing intake of this bran congee will prevent the occurrence of beriberi. Unfortunately, this important prescription did not attract the attention that it deserved. In the special monograph on beriberi, the *Chiao Chhi Chih Fa Tsung Yao* 腳氣治法總要 (General Therapy of Beriberi), +1075, by Tung Chi 董及, forty-five recipes are presented, but none includes the use of liver or rice bran.[46] He recommends the use of many kinds of seeds which we now know to be rich sources of Vitamin B1.[47]

Even more tantalising is the observation by Chhên Tshang-Chhi in the *Pên Tshao Shih I* (+725), that 'prolonged consumption of rice weakens the body; feeding [polished] paddy or glutinous rice to young cats and dogs will so bend their legs that they will not be able to walk.'[48] Although this passage is repeated in subsequent pharmacopoeias, its far-reaching implication was never picked up and pursued. If someone had put together the significance of the two Thang discoveries, one, that rice bran cured beriberi, and two, that feeding polished rice to cats and dogs induced the illness, we would be talking today of Chhên Tshang-Chhi's cats rather than Eijkman's chickens. The true cause of beriberi would have been elucidated a millennium earlier.

Alas! It was not to be so. Both discoveries were ignored. Diet therapy for beriberi continued to move along in its traditional way as indicated in the *Yin Shan Chêng Yao*, +1330, which lists three recipes for curing *chiao chhi*.[49] All are to be taken on an empty stomach.

---

[45] *CCIF* (1965) reprint, ch. 17, no. 2, pp. 194–6.

[46] *Chiao Chhi Chih Fa Tsung Yao*, ch. 2, *SKCS* **738**, pp. 427–37. The author, Tung Chi, suffered from beriberi himself, and studied the disease for the rest of his life.

[47] After studying the Chinese herbal remedies for beriberi Yang & Read (1940) came to the following conclusion. 'Most of the seeds, especially that of plantain, contain significant quantities of vitamin B. The values of mulberry leaf, loquat leaf and carpenter weed are also high. The vitamin B content of barks and stem is low. Roots contain a moderate amount.'

[48] *PTSI*, cited in *CLPT*, ch. 26, item *Tao mi* 稻米, p. 3b. The original reads, 久食之令人身軟；黍米及糯食司小貓犬，令腳屈不能行。The importance of this observation, as well as that pertaining to the use of bran as a cure for beriberi, was also noted by Hou Hsiang-Chhuan (*1954*).

[49] *YSCY*, ch. 2, pp. 93–5. The same recipes are cited in Lu & Needham (1951), but they met with some difficulty in interpreting the text. (1) They could not identify the horse-tooth-vegetable, *ma-chhi-tshai* 馬齒菜. This should have been *ma-chhi-hsien* 馬齒莧, purselane. Li Shih-Chên has listed *ma-chhi-hsien* as one of the products used to treat beriberi, cf. *PTKM*, ch. 3, p. 195. (2) They list half a pound of adzuki beans. The original is two *ko* 合, i.e. two-tenths of a pint. (3) The original is bear meat. No doubt pork would work just as well, since pork liver, kidney and stomach are listed as effective against beriberi in the *PTKM*, p. 196.

(1) A rice congee with purslane.
(2) A carp cooked with adzuki beans (0.2 pt), tangerine peel (0.2 oz), peppercorn (0.2 oz) and grass seeds (0.2 oz).
(3) Bear's meat (one catty) cooked with fermented soybean, scallion, soy paste and five flavours.

In the *Pên Tshao Kang Mu* Li Shih-Chên gives a long list of 185 materials recommended for the treatment of beriberi.[50] Among them are many common food items such as Chinese cabbage, fish, chicken, milk (sheep and cow), pig's liver, kidney and stomach, radish, bamboo shoots and fermented soybeans. Wheat bran is mentioned, but not rice bran. Evidently, the key role of rice polishings in the aetiology of beriberi was never appreciated by the mediaeval Chinese physicians. They never suspected that the emergence of rice as the principal food grain had anything to do with the increasing occurrence of beriberi in South and Central China.

And yet Sun Ssu-Mo in the 7th century had pointed out that *chiao chhi* is a disease unknown to the ancients and that it started in South China and was spreading in the Eastern part of the [Yangtze] River Valley. We have remarked earlier on how during the Han dynasty, China began the process of dividing itself into two dietary zones, a Northern zone with wheat as the principal grain, and a Southern zone where the principal grain is rice.[51] This split was the natural outcome of the growing importance of wheat in the North and the increasing popularity of rice in South and Central China. It would seem that as the rice-eating population increased, so did the occurrence of beriberi. But then, as we have shown earlier, rice was hardly an unfamiliar commodity in ancient China. It is mentioned in the *Shih Ching* (−11th to −6th centuries) and ensconced as one of the five staple grains in the *Li Chi* and the *Chou Li*. It must have been consumed by a substantial portion of the populace even in the North. If so, why was beriberi rare or unknown?

Two answers are possible. The first is that beriberi was known but no one took the trouble to detail its symptoms. Since beriberi is an easily recognisable, conspicuously debilitating disease this seems highly unlikely. The second is to accept Sun Ssu-Mo's remarks at their face value. There was, indeed, little or no beriberi in antiquity. We are then left with a second puzzle. Why did beriberi suddenly appear during the late Han? This is an interesting epidemiological question. The answer, we submit, is because the rice consumed after the Han was no longer the same as the rice consumed before the Han. The pre-Han rice was effective in preventing beriberi, while the post-Han rice was not. How is that possible?

Earlier on we mentioned the refinements in milling technology during the Han that turned wheat from a coarse grain into an economical and versatile flour (*mien* 麵), amenable to being processed into a variety of attractive foods.[52] Similar refinements were no doubt also being applied to the processing of rice. Of these, the most important was the introduction of trip hammers activated by human foot

[50] *PTKM*, ch. 3, pp. 194–6. [51] Cf. p. 68 above. [52] Cf. pp. 463–6 above.

40 FERMENTATIONS AND FOOD SCIENCE 583

Fig 132. Trip hammers activated by human foot for polishing rice. *TKKW*, p. 91.

(Figure 132) or powered by animals or water-wheels (Figure 133) for the polishing of rice.[53] This made it possible to produce rice granules in quantity that were much more highly polished than those formerly prepared by the use of hand-held mortar and pestle. Polished rice cooks faster and has a softer mouth feel than brown or unpolished rice. It is organoleptically and gastronomically a superior product. Its popularity grew rapidly, and soon polished rice became the primary and preferred staple grain of South and Central China. Thus, we may presume that pre-Han rice

[53] The invention of the foot-activated trip hammer is described by Hêng Than 恒譚 (−25 to +56) in his *Hsin Lun* 新論, as follows, 'Many people benefited when Mi Hsi 宓犧 developed the pestle and mortar. Later, others added their ingenuity to utilise one's body weight and invented the foot-activated tilt hammer. The efficiency of the hand-held pestle was increased ten-fold. Still later, the efficiency was increased a hundred-fold when the tilt hammer was mechanically connected to the labour of donkeys, asses, ox and horse or to water power,' cf. Liu Hsien-Chou (1963), pp. 71–2. Cf. also *SCC* Vol. IV, Pt 2, pp. 174–83, 390–3 for further information on the pounding machinery used in the processing of grains in China.

碓水

Fig 133. A bank of trip hammers activated by water power. *TKKW*, p. 92.

was largely unpolished or lightly polished, while post-Han rice was highly polished, perhaps as highly polished as the rice we are accustomed to seeing today. We may even go a step further and say that unpolished or lightly polished rice prevents beriberi, but highly polished rice does not.[54]

---

[54] Fan Ka Wai (*1995*), pp. 168–9, has suggested another reason for the emergence of beriberi during the Chin and North/South Dynasties, the replacement of yoghurt drink by tea. Yoghurt drink would be a rich source of vitamin B1, while tea contains none of this vitamin. In our opinion, this factor is of minor importance, since yoghurt drink remained popular among the gentry throughout this period. The mechanisation of processing technology, which changed the character of the rice, is more likely to be the principal cause of the rise of beriberi. On this point we would like to inject a personal note. When I stayed in my ancestral village in Hothang (cf. author's note, pp. 1–4), in 1942 the staple food of the village people was coarse red rice that was very lightly polished. No incidence of beriberi was known. Yet among immigrants from our village who migrated in the 1930s to Situwan, a village in Perak, a state in northern Malaya, several cases of beriberi were known. Apparently, the rice from Thailand that the immigrants consumed was very highly polished. Because they were poor they could not afford to have other good sources of thiamine in their diet, and thus became susceptible to the disease.

As we now know, there is indeed an anti-beriberi factor, vitamin B1, i.e. thiamin, in rice. Unfortunately, it resides only in the outer aleurone layer, and is removed by the incessant grinding of the grains against each other when pounded by the triphammer. The factor is separated from the kernel and stays in the bran. And so, as more and more people began to consume highly polished rice, more and more cases of *chiao chhi* began to appear. In fact, what we have here is an early example of how the benefits of technology are often accompanied by unexpected, disastrous side effects that are not revealed until decades or centuries later. By the time of the Thang, even Northerners had started to eat rice (no doubt transported from the South) as the primary grain and had thus started to contract beriberi. The *Chhien Chin Yao Fang*, +659, says:[55]

In recent years even the official and scholarly class in the Central Kingdom (*chung kuo* 中國), who have never been to regions south of the Yangtse River, have suffered from this disease. It would seem that the air and the *chhi* 氣 of different regions have comingled, so that the speciality of one locality can now be found everywhere.

Sun Ssu-Mo and other mediaeval physicians were apparently aware that the disease was rare before the Han, that it was endemic only in the rice eating zone of the Empire, and that it could be prevented by eating a rice bran congee. Why then did they not make the connection between beriberi and the consumption of rice? Why did they not further pursue the efficacy of rice bran? One possible reason may be that the physicians were poorly informed about the technology of rice processing. They did not appreciate the difference in the extent of polishing between the rice of antiquity and the rice of their own time. The issue was further confused by the great variety of materials that could be used to help alleviate the debilitating condition of the disease, rice bran being only one of them. The prevailing theories of the cause of the disease were no help. Briefly, they are of two types:

(1) *Environmental*: the *Chu Ping Yuan Hou Lun* (Discourses on the Aetiology of Diseases), +610, suggests that *chiao chhi* is caused by *fêng tu* 風毒 (poisons in the wind).[56] This view is expanded in the *Chi Shêng Fang* 濟生方 (Prescriptions for Preserving Health) by Yen Yung-Ho 嚴用和, +1253, which blames windy chill and humid heat (*fêng han shu hsi* 風寒暑濕).[57] It is said that in the South and lower Yangtse valley there is much wind and humidity, which tends to rise from the ground and injure the leg, the body part closest to the earth.

(2) *Organic*: according to Su Chin the disease is caused by a deficiency in the kidney (*shên hsü* 腎虛).[58] This view is also echoed in the *Chi Shêng Fang*.[59] But there is no indication as to what caused the kidney to become deficient and what should be done to correct the deficiency.

These views are combined in a statement by Tung Chi in the *Chiao Chhi Chih Fa Tsung Yao*, +1075:[60]

---

[55] *CCYF*, ch. 7, pt 1 (*SKCS* ed. ch. 22).   [56] *CPYHL*, ch. 13, p. 413.   [57] *Chi Shêng Fang*, ch. 3, p. 101.
[58] Su Chin, cited in Li Thao (*1936*), possibly from 三家腳氣論.   [59] *Chi Shêng Fang*, ch. 3, p. 103.
[60] *Chiao Chhi Chih Fa Tsung Yao*, *SKCS*, 738, p. 417, tr. auct.

The illness is caused by the condensation of toxic wind and humid *chhi* in the liver, kidney and spleen resulting in obstruction of the circulatory system in the outer extremities (feet and fingers). Since the vital energy of the toxic wind emerges from the ground, the hot, the cold and the humid all rise and penetrate the covering of the feet. The impaired circulation gives rise to swelling, pain and weakness. So the illness is called *chiao chhi*.

These hypotheses have had little influence either in devising cures or in advancing an understanding of the disease. The successful treatment of *chiao chhi* must, therefore, be regarded as an entirely empirical discovery. Although in theory many materials were available for use as a cure for the disease, in practice they were probably too expensive for prolonged usage by a significant portion of the population. This would not have been the case had rice bran been clearly established and accepted as the agent of choice for the treatment of beriberi. As a result, beriberi remained a significant nutritional deficiency disease of the Chinese through the Sung, Ming, Yuan and Chhing until the twentieth century.[61]

(3) Night Blindness (*Chhüeh mu* 雀目)

Night blindness or nyctalopia is one of the early symptoms of a deficiency in vitamin A (*trans*-retinol), a substance essential for the functions of vision, growth and reproduction. It may be ingested as is from food, or indirectly as β-carotene, which is converted in the body to *trans*-retinol. Nyctalopia is mentioned by Galen in the +2nd century, but the disease attracted only minor attention in the West until William Briggs's study appeared in +1684.

Night blindness and related eye diseases, however, were well known in ancient China. According to the Wu-Hsing theory, the liver, gall bladder and eyes are all manifestations of the elemental phase, Wood. It is thus not surprising that the Chinese learned early on to use preparations of liver and gall bladder (and even eyes) to treat diseases of the eye. The *Shên Nung Pên Tshao Ching*, +100, states that[62] 'the gall bladder of the carp relieves burning pain and redness in the eyes (*chu chih mu jê chhi thung* 主治目熱赤痛), cures "green blindness" or amaurosis (*chhing mang* 青盲), and clarifies the vision (*ming mu* 明目).' Another effective agent for clarifying vision is the gall bladder of the dog. Similar statements are found in the *Ming I Pieh Lu* of +510.[63] For example, the gall bladder of the goat is said to cure green blindness, and the gall bladders of the goat and the chicken, to clarify vision. But of greater clinical significance is the observation that beef liver and rabbit liver promote a clearer vision. Thao Hung-Ching has also been quoted by Li Shih-Chên as saying that 'the eye of the rat improves light perception so much that one can see clearly enough to read books at night. This method is used by the astronomers.'[64]

---

[61] Kao Ching-Lan (1936) shows that beriberi was still a serious disease in Shanghai in the 1930s.
[62] *SNPTC*, Tshao Yuan-Yü (*1987*), pp. 278 (carp), 284 (dog). We should be reminded that some of the material included in this compilation may be as old as −3rd century.
[63] *MIPL*, reconstructed by Shan Chih-Chün (*1986*), pp. 79 (chicken), 172 (goat), 175 (beef) and 184 (rabbit). Cf. also Thang Yü-Lin (*1958*), p. 56.
[64] *PTKM*, ch. 51, p. 2904. The original reads, *Chu chih: ming mu, neng yeh tu shu, shu chia yung chih* 主治：明目，能夜讀書，術家用之。

We now know that vitamin A is essential for the proper functioning of the eyes. Shortage of vitamin A first leads to night blindness, and then progressively through xerophthalamia to total blindness.[65] Liver is a particularly rich dietary source of vitamin A. In this case the Wu-Hsing theory is perhaps fortuitously right on target. What is more remarkable is that as indicated in the *Ming I Pieh Lu* (c. +510), the early mediaeval Chinese already knew that certain vegetables, especially their seeds, such as *chhi* 薺 (shepherd's purse) and *wu chhing* 蕪菁 (Chinese turnip) also improved vision.[66] Green, leafy vegetables are now known as good sources of β-carotene, which is converted in the human body into vitamin A.

But what were the eye diseases that claimed the attention of the early mediaeval Chinese physicians? *Ming mu* 明目 simply means to clear or improve the vision when the vision of the patient, for various reasons, is not up to par. *Chhing mang* 青盲 denotes a gradual loss of sight while the physical appearance of the eye remains unimpaired. It may represent an advanced state of blurred vision akin to xerophthalmia or trachoma. The first mention of the term *chhüeh mu* 雀目 (sparrow eyes) is found in the *Chou Hou Pei Chi Fang*, +340, which devotes a whole subchapter to the treatment of eye diseases.[67] It occurs in a prescription aimed at curing this illness in which sheep's liver is a major ingredient. The prescription which follows, attributed to Mei Wen-Mei 梅文梅 (+7th century), uses slices of sheep's liver marinated in mild vinegar as a cure for the reduction of vision after dusk (*chih mu an huang hun pu chien wu che* 治目暗黃昏不見物者).[68] This may be regarded as another prescription for *chhüeh mu* 雀目 since 'reduction of vision after dusk' is merely another way of describing night blindness. In other parts of the chapter, the juice of canine gall bladder is dispensed as a drop for an itchy eye, and the liver of sheep is processed as a pill for patients with *chhing mang*. In the minds of the Chinese physicians, the efficacy of these prescriptions, serves to confirm their view that organs within the elemental sphere of Wood had a direct influence on the function of the eye.

Why is 'reduction of vision after dusk' called *chüeh mu* (sparrow eyes)? The answer is given in the *Chu Ping Yuan Hou Lun* (+610), which says,[69] 'Some people can see quite well in broad daylight, but cannot see anything after dark. This condition is commonly known as *chhüeh mu*, since they are like sparrows which are unable to see in dim light.'

Four prescriptions are listed. Three utilise traditional herbs such as *ti fu tzu* 地膚子 (Belvedere cypress seeds), *chüeh ming tzu* 決明子 (sickle senna seeds), *po pai phi* 柏白皮 (white cypress bark), *wu mei ju* 烏梅肉 (flesh of dark *mei* fruit), *hsi hsin* 細辛 (*Asarum sieboldi*) and *ti i tshao* 地衣草 (lichens). The fourth one is taken from the *Chhien Chin I*

---

[65] For a brief review of the effect of vitamin A deficiency on the eyes, cf. Oomen (1976).
[66] *MIPL*, ch. 1, p. 95.   [67] *CHPCF*, ch. 6, subch. 46, pp. 111–13.
[68] *Ibid.*, p. 112. There are two prescriptions that concern us on this page. The first one is as a cure for *chhüeh mu*. This could have been present in Ko Hung's original manuscript. The second one is and is quoted from the *Mei Shih Fang* 梅師方 (Master Mei's Prescriptions). It explains what *chhüeh mu* is, but it has to be a later interpolation since Mei Wen-Mei lived during the Sui (+589 to +618), almost three hundred years after Ko Hung.
[69] *CPYHL*, ch. 28, p. 790. The original reads, 人有晝而晴明，至瞑則不見物，世謂之雀目，言其如鳥雀瞑便無所見也. This statement is repeated verbatim in the *WTMY*, ch. 21, *SKCS* **736**, p. 696. There was apparently a general perception that birds such as the sparrow cannot see at night.

Fang (c. +682) and sliced pork liver is the only ingredient.[70] But the role of liver is given much greater prominence in the *Yin Hai Ching Wei* 銀海精微 (Essentials of the Silver Sea), a Sung monograph on ophthalmology. It explains that when the liver is injured by unhealthy influences, the Yin–Yang balance in the body is upset and the circulation of *chhi* and blood becomes retarded from dusk to dawn. This results in night blindness. Three medications are recommended as remedies: a powder made with pork liver, a pill containing the gall bladders of the bear, ox, fish and sheep, and a powder with the liver of bats as the chief ingredient.[71]

Similar views are expounded in the *I Hsüeh Chêng Chuan* 醫學正傳 of Yu Thuan (+16th century) and the *Chêng Chih Chun Shêng* 証治准繩 by Wang Khên-Thang (c. +1600).[72] The value of liver in the treatment of night blindness is reiterated in the *Phu Chi Fang* 普濟方 (Practical Prescriptions for Everyman), compiled by Prince Chu Hsiao 朱橚 of the Ming Dynasty in +1418. Of the twenty-six prescriptions collected in a supplement on night blindness to the chapter on eye diseases, fifteen include pork, beef or sheep's liver as the main ingredient. The others are said to serve as tonics to strengthen the vitality of the liver of the patient.[73]

Thus, we may say that night blindness has been recognised in China since at least the +3rd century. It was treated with reasonable success by the ingestion of the liver of domestic animals as well as by leafy green vegetables. Perhaps because of the routine consumption of dairy products, i.e. milk, butter and cheese, which are an excellent source of vitamin A, night blindness never achieved the same level of notoriety in Western Europe as it did in China until modern times.

(4) Rickets (*Kou lou* 佝僂)

So far we have dealt with three nutritional deficiency diseases which were successfully treated by dietary therapy. We now come to a disease which the Chinese had probably recognised no later than the +7th century, but for which no satisfactory cure was found, namely rickets. The disease is especially incident to children and is characterised by a softening of the bone structure resulting in deformities such as skull distortions, hunchbacks and bowlegs. It is caused by a deficiency in vitamin D (cholecalciferol), calcium and/or phosphate.[74] Strictly speaking, cholecalciferol is not a vitamin, since under normal conditions, in the presence of sunlight, the body can produce it from a plentiful metabolite, 7-dehydrocholesterol. But under

---

[70] According to Li Thao (*1936*), p. 1030, *ti fu tzu*, *hsi hsin*, *po phi* and *chüeh ming tzu* are all good sources of vitamin A. See also *Chhien Chin I Fang*, ch. 11, p. 132 for the prescription using pork liver.

[71] *Yin Hai Ching Wei*, Book I, repr. (*1954*), pp. 20b–21a. On another page, 17b, three herbal prescriptions are given as cures for a syndrome described as the lack of ability to see things after dusk. No liver or gall bladder are included in these prescriptions.

[72] Cited in Li Thao (*1936*), p. 1029 and (1967), p. 418.

[73] *Phu Chi Fang* (*1959*), supplement to ch. 83, pp. 843–7.

[74] Actually vitamin D occurs in two forms in foods: vitamin D_3 (cholecalciferol) in animal products and vitamin D_2 (ergosterol) in plants. It promotes absorption of calcium and phosphate in the intestine and mediates the mobilisation of calcium from bone.

crowded living conditions children, and even adults, may not receive adequate sunlight to produce sufficient vitamin D for their needs, and rickets ensues.

In the West rickets was mentioned in Galen (+2nd century), but a clear description of the disease did not appear until Francis Glisson in +1650. In China various defects in bone structure caused by a deficiency in vitamin D were noted, but they were treated as separate maladies that bore no connection with each other. Thus, in the chapter on children's ailments in the *Chu Ping Yuan Hou Lun* (+610), we see discussions of retarded emergence of teeth, lateness in walking, and three different types of deformed skull, *chiai lu* 解顱 (poor fit of skull bones), *sai han* 腮陷 (protrusion on forehead), and *sai thien* 腮填 (depression on forehead).[75] The aetiology of each illness is explained on the basis of the Wu-Hsing theory. For example, the cause of retarded growth of teeth is stated as follows:[76] 'Teeth are the resting place of bone tissue, which is nurtured by the marrow. If the *chhi* representing kidney that the child inherits is defective, then the marrow will be unable to nourish teeth and bone. Thus, growth of teeth will be slow.'

Another example is the explanation for the depression on the forehead:[77]

The depression shows an uneven forehead. The heat from the intestines rises and smothers the viscera. Heat in the viscera induces thirst and intake of water, which cools the bowels and promotes the elimination of excreta. The *chhi* of the blood is thus weakened. It is unable to ascend and fill the cranial cavity. To accommodate this deficiency a depression is formed.

The *Wai Thai Mi Yao* (+752) gives recipes for five ointments or poultices for external application on cranial deformities. Among the drugs used are *fang feng* 防風 (*Radix ledebouriellae*), *pai chi* 白及 (*Bletilla striata*), *hsiung chhiung* 芎藭 (Hemlock parsley), *hsi hsin* 細辛 (*Asarum sieboldi*), *kuei hsin* 桂心 (heart of cassia), *wu thou* 烏頭 (Wolfsbane), cypress seeds and dried ginger. The ingredients are finely ground and then mixed with milk, rich wine, or bone marrow before application.[78] There is no indication whether any of the treatments were effective.

The Sung pediatrics manual *Hsiao Erh Yao Chêng Chih Chüeh* 小兒藥証直訣 (Therapy of Childhood Diseases) by Chhien I 錢乙 (+1114) mentions *kuei hsiung* 龜胸 (turtle chest), *kuei pei* 龜背 (turtle back) and *hsing tzhu* 行遲 (lateness in walking) as notable illnesses of children. Their names are self-explanatory, but the author has more to say on the origin of these diseases:[79]

Turtle chest and turtle back. When the child is exposed to wind at birth, it hovers over the spine and penetrates the bone and marrow. This results in a turtle back. When the heat in the lungs overflow, it attacks the chest cavity and causes a turtle chest.

Lateness in walking: although the body grows [to normal size], the legs are too weak to support it for walking.

By this time it would appear that the Chinese physicians had begun to realise that these disparate symptoms were all manifestations of a basic malfunction in the

---

[75] *CPYHL*, ch. 48, items 109, 110, 111, 144, 147, pp. 1352–3 and 1374–5.
[76] *Ibid.*, p. 1374.   [77] *Ibid.*, p. 1353.   [78] *WTMY, SKCS*, 737, p. 489.
[79] *Hsiao Erh Yao Chêng Chih Chüeh*, cited in Li Thao (*1936*), p. 1034.

development of bone, but their attempts to explain their aetiology were like grasping straws in the wind. The Wu-Hsing theory failed to shed any light on them. The prescriptions they recommended were of little help. Indeed, the late Ming paediatrician, Wang Khên-Thang 王肯堂 in the paediatrics (*Yu Kho* 幼科) section of his *Chêng Chih Chun Shêng* 証治準繩 (Standard Methods for Diagnosis and Treatment), +1602, candidly admits that the disease cannot be reversed by any of the prevailing treatments.[80]

From what we know today of the natural distribution and physiological functions of vitamin D, calcium and phosphate, it is easy to see why rickets was a particularly recalcitrant disease for the Chinese medical practitioners to contain. Although the adoption of *tou fu* as a major component of the diet in Late Thang added a new source of calcium into the diet, calcium has remained the one macro-element that could be deficient among the Chinese.[81] The best dietary sources of vitamin D are animal products, molluscs, liver, milk, butter and eggs, which are by no means plentiful in a typical Chinese diet.[82] Plants, unfortunately, contain very little of this vitamin. Thus, most of the common herbs and plant products listed in the pharmacopoeias had little therapeutic value. Yeasts and moulds (mushrooms) are a rich source of ergosterol (vitamin D_2), but the Chinese had no inkling that the microbial *ferments* they so frequently used in the processing of fermented foods could have had a beneficial effect on this ailment. It is curious that the Chinese never prescribed pork or beef liver as part of the dietary therapy of illnesses of defective bone development, even though they were perfectly willing to use them as cures for beriberi and night blindness. One possibility is that while in the case of beriberi and night blindness a cure can be rapid or even instantaneous, a reversal of a bone deformation is a very slow process and may take several months to become visible. Perhaps liver was tried, but no one waited long enough for the beneficial effect to become manifest. Furthermore, if at the same time there is also a shortage of calcium and phosphate in the diet, the use of liver is unlikely to lead to a reversal of the disease.

As we have pointed out earlier, humans are not totally dependent on dietary sources of vitamin D. Vitamin D_3 is produced under the skin from cholesterol every time it is exposed to the sun.[83] In the course of their normal daily activities most individuals will receive enough sunshine to produce sufficient vitamin D for their metabolic needs. Provided they receive adequate levels of calcium, rickets will not develop. In the traditional family, however, women are often secluded inside the family chambers and are seldom seen outdoors. Since they are the primary

---

[80] *Chêng Chih Chun Shêng*, chs. 9 and 10.

[81] Both Chinese and Western nutritionists have expressed the opinion that calcium deficiency was a major drawback of the Chinese diet, cf. Adolph (1926), Wu Hsien (1928), Maynard & Swen (1937) and Snapper (1941). Cf. also Latourette (1957), p. 567 and Winfield (1948), p. 72.

[82] For analyses of the vitamin D content of foods available in Shanghai in the 1930s cf. Read, Lee & Chhêng (1937), *Shanghai Foods*, pp. 42–51 and 56–61. Vitamin D is found in some vegetables, e.g. carrots, mustard leaf and spinach and mushrooms (*Agaricus campestris* and *Auricularia auricula*), but the best sources are animal products.

[83] Hui, Y. H. (1985), p. 204. For a discussion of the role of vitamins in nutrition cf. pp. 156–223.

care-givers in the family, it is likely that some of the children under their charge will also spend little time exposed to the sun, thus running the risk of developing rickets. The cause of rickets is, therefore, both cultural and nutritional. It is caused as much by a defect in the life-style of the patient as by a deficiency in his diet. Thus the case of rickets represents not so much a failure of Chinese diet therapy and nutrition but rather a failure of Chinese society to appreciate the healthful effects of physical activity in an outdoor environment.

(h) REFLECTIONS AND EPILOGUE

In the *Shih Shih Wu Kuan* 食時五觀 (Five Points to Ponder at Meals), the Sung scholar Huang Thing-Chien 黃庭堅, +1045 to +1105, had this to say:[1]

It is well [for us] to reflect on the time and effort that have been spent before we sit down to enjoy a meal – all the exertions in farming, harvesting, processing and cooking, not to mention the slaughter of animal lives – that were expended to please our palate. The provision for one person requires the labours of ten. Those who can afford to stay home, without having to work for their sustenance, are merely depleting the resources gained through the toils of their ancestors. And we government officials are the privileged few that live off the sweat and blood of the common people.[2]

The passage reminds the scholar–official of the extensive and intensive labours that go into the making of every meal. It also reminds us that in the exploration we have just concluded we have concentrated on only one of the four tasks in the overall enterprise that culminates in a meal on the dining table, namely, food processing. We have traced the history of the major processing technologies that the Chinese have developed to make raw foods more palatable, enhance their digestibility, improve their keeping qualities and transform them into novel and attractive food products. We have also shown how diet has had an impact on the occurrence and cure of one specific type of diseases. As a parting note we now wish to offer a few reflections on what we have learned about the development of food technology in China and how processed foods have affected the health and well-being of the Chinese people.

(1) *The wonderful world of the grain moulds*

Even from a cursory inspection of the preceding chapters, the reader will have noticed the unusual preponderance of fermented foods in the Chinese dietary system. Further examination will show that many of the processes involved are mediated by a special *ferment* called *chhü* 麴, a culture of moulds, particularly species of *Aspergillus*, *Rhizopus* and *Mucor*, and yeast, mainly *Saccharomyces* spp., grown on a

---

[1] *Shih Shih Wu Kuan*, annotated by Thang Kên 唐艮 (*1987*) in *WSCKL, CKPJKCTK*, p. 64, tr. Lai, T. C. (1978), p. 86, mod. auct. The other four points are (1) Our worthiness to receive the gift of food (2) Balanced view of food (3) Therapeutic value of foods and (4) Food as a means to attain Tao.

[2] This is a fairly common theme in Chinese literature. For example, after witnessing the strenuous work of the field hands, Pai Chü-I was moved to say in the poem, 'Watching the Wheat-reapers',

> What deeds of merit have I done?
> I've neither farmed nor raised silkworms;
> My official's salary, three hundred piculs of rice,
> And at year's end there is surplus grain to eat.
> Thinking of this I feel guilty and ashamed;
> All day long I cannot keep it out of my mind.

The quotation is translated by Irving Y. Lo, in Liu & Lo eds. (1975), p. 202.

cooked grain medium. The enzymes present that are relevant to food processing include amylases that hydrolyse starch to sugars, proteinases that hydrolyse proteins to peptides and amino acids, pectinases that hydrolyse pectin to uronic acids, and lipases that hydrolyse fats to glycerol and fatty acids.

How did this unique product come about? It seems to us that the discovery of the *mould ferment* (*chhü* 麴 or *chiu yao* 酒藥) in the Neolithic period is the result of the happy conjunction of three factors, firstly, the nature of the ancient cereals cultivated by the Chinese, that is, rice and millets, secondly, the development of steaming as a preferred method for cooking such cereals, and thirdly, the kinds of fungal spores that were present in the environment.[3] It is the softness of the kernels of millets and rice that permits them to be steamed into individual granules of soft grain called *fan* 飯. It is the steaming process that turns the kernels of grain into fluffy yet dispersible granules of *fan* that are most pleasing to the human palate. It happens that the *fan* granules are also excellent substrates on which native fungal spores from the air can alight, germinate and proliferate, and it turns out that the kinds of fungi that thrive on them had the right complement of enzymes to function as a *ferment*.

As far as we know, the convergence of these distinctive factors occurred only in China, and in no other civilisation in the ancient world. This explains why the *mould ferment* was never developed in the early civilisations of the West. Although cereals (wheat and barley) were also cultivated by the ancient peoples of the Eastern Mediterranean, they were not cooked by steaming. They were usually ground into meal or flour and either turned into porridge or paste or baked into bread. None of these products provided the physical characteristics conducive to the growth of moulds that had the desired enzymatic properties.

Although the *mould ferment* was developed originally for making wine, further exploitation of its activities soon paved the way to the production of an array of fermented foods which have helped to shape the character and flavour of the Chinese diet and cuisine. Many of these are still familiar components of the dietary system today. An overview of the wonderful family of products derived from or inspired by the activities of the *ferment*, is presented in the chart shown in Figure 134. The *ferment* itself is placed in the centre with the products emanating from it in all directions. Included, are processes that do not currently utilise a *ferment* but may have done so in the past, such as the use of *moulded soybeans* in the making of fermented soyfoods, or the use of *fan* (cooked rice or millet granules) for the preparation of fish and vegetables pickles. The lines drawn to indicate the production of *li* 醴 (sweet liquor) and fermented meat and fish pastes are dotted to show that the products are now obsolete in China.[4] It is interesting to note that most of the end-products listed, wine, sweet liquor, vinegar, soy paste, fermented soybeans, meat and fish pastes, pickled meat and pickled vegetables were already known by the Han Dynasty. In addition, steamed bread (*chêng ping* 蒸餅), fermented flour paste (*mien chiang* 麵醬)

---

[3] Cf. pp. 161–2 and pp. 260–1 above.
[4] For the fate of *li* cf. pp. 262–3 and of fermented meat and fish pastes, pp. 388–9 above.

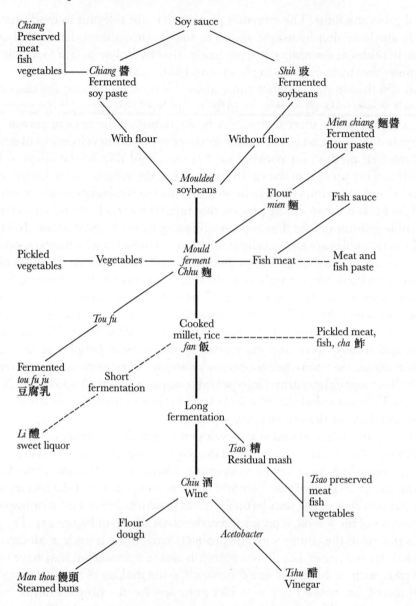

Fig 134. The Wonderful World of the Grain Moulds. Processed foods related to the *Mould Ferment – Chhü* 麴.

and soy sauce (*chiang yu* 醬油) began their development during the Han, although they did not achieve widespread utilisation until later.[5]

Only three new products were developed and added to this array after the Han, firstly, the application of *tsao* 糟 (wine residual mash) and *chiang* 醬 (soy paste)

---

[5] Steamed bread and flour paste during the Wei-Chin (+221 to +420), and soy sauce in Southern Sung (+1127 to +1279), cf. pp. 468–9 and pp. 358, 362.

Table 60. *The development interval of selected processed foods initiated during the Han*

Product	Initiation date	Utilisation date	Years in development
Distilled wine	+100 bronze still	+1300 *YSCY*	1200
Bean sprouts	−100 *SNPTC*	+1100 *TCMHL*	1200
Bean curd *tou fu*	+100 tomb mural	+900 *CIL*	800
Soy milk	+100 tomb mural	+1800 Fig. 71	1700
Soy sauce	−100 *52 Pingfang*	Southern Sung *SCCK*	1300
Filamentous noodles	+100 *Shih Ming*	+900 *TCMHL*	800

YSCY	*Yin Shan Chêng Yao.*	*52 Pingfang*	*Wu Shih Erh Ping Fang.*
SNPTC	*Shêng Nung Pên Tshao Ching.*	*SCCK*	*Shan Chia Chhing Kung.*
TCMHL	*Tung Ching Mêng Hua Lu.*	*Shih Ming*	*Shih Ming.*
CIL	*Chhing I Lu.*		

as preservatives to make preserved meat, fish and vegetables in the +6th century;[6] secondly, the making of fermented bean curd during the Ming;[7] and thirdly, the introduction of fish sauce probably in late Ming or early Chhing.[8] These observations all point to the Han as a remarkably productive and creative period in the utilisation and development of fermented foods in China.[9]

(2) *The uneven flow of food processing innovations*

The productivity during the Han, was not restricted just to fermented foods. Several other well-known processed foods also originated during the Han. They are listed together with soy sauce in Table 60. Except for soy milk, they all reached fruition as commercial products during the Sung–Yuan period.[10] In addition, two other innovations, red *ferment* and wheat gluten reached fruition, and one product, fermented bean curd, began its development during the Sung.[11] These activities mark the Sung as another period of high productivity in the history of processed foods in China.[12] What interests us most in Table 60, however, is what happened during

---

[6] These were established parts of the food system by the time of the *Chhi Min Yao Shu* (+6th century), cf. pp. 408–12.
[7] Cf. pp. 326–8.
[8] Cf. pp. 397–8. Although the fish sauce is not used widely in China, it is an important condiment in the cuisine of Fukien and Taiwan.
[9] Not shown in Figure 134 is the development of the red *ferment* during the Sung, which has given rise to a new series of traditional products, such as wines, vinegar, fermented *tou fu*, and meat, fish and vegetables preserved in *tsao* (wine residual mash), that have a unique appearance and a distinctive flavour, cf. pp. 192–202 above. Furthermore, the invention of the red *ferment* shows that organisms such as the *Monascus purpurea* can serve the functions of a *ferment* as well as the *Aspergillus*, *Rhizopus* and *Mucor* spp. found in the traditional *ferment*.
[10] The period covered is approximately +950 to +1300.
[11] Cf. pp. 192–203 for red *ferment*, pp. 487–502 for gluten and pp. 326–8 for fermented bean curd.
[12] The Sung is noted for the brilliance of its culinary arts. Freeman (1977), p. 144, has stated that the first 'cuisine' in the world emerged during the Sung, but we think he uses the word 'cuisine' in the sense that others would apply to *haute cuisine*, cf. Farb & Armelagos (1980), Goody (1982), Anderson (1988). On Sung cuisine cf. H. T. Huang, 'Sung Gastronomy – the Flowering of Chinese Cuisine' (in preparation).

the long gestation time between the initial discovery and its eventual utilisation. Unfortunately, there is usually little or no information in the literature to enlighten us on this point. The most rewarding example we have studied is perhaps distilled wine.[13] Archaeological evidence suggests that it was first prepared as a novelty, perhaps by alchemists, during the Eastern Han. Grape wine was possibly distilled during the Thang, and a distilled liquor was marketed in limited amounts during the Thang. But the still in use was expensive and inefficient. It was not until a new type of still was invented (during the Sung) and perfected (during the Yuan) that distilled wine became a viable commercial product. The complete process took about 1,200 years.

The reasons for the protracted development of this and other products in Table 60 can be stated briefly. For bean curd, a palatable product could not be achieved until it was understood that the soy milk had to be heated to boiling point before it was coagulated.[14] For soy milk itself, there was a problem with its beany flavour and its tendency to induce flatus. Perhaps the well-known Chinese aversion to dairy milk[15] spilled over to soy milk, and there was no interest in the product until dairy milk was reintroduced to China in the 18th and 19th centuries. Soy milk took on a new lease of life in late Chhing when it was found that prolonged heating improved its flavour as well as its digestibility.[16] In the case of bean sprouts and soy sauce, the technologies for making them were virtually in place at their date of initiation.[17] Their adoption required a change in attitude on the part of the consumer who became more receptive towards change during the Sung. Finally, in the case of the filamentous noodle, it was just one of a host of developing 'pastas' (*ping* 餅) that required a prolonged 'shake down' before a winner was finally selected.[18]

From the above examples we can see that innovation in processed foods was impeded by several factors. Firstly, there was a pervasive inertia towards change. Often there was no immediate advantage in making a change, as was the case with soy sauce, which stood for centuries, waiting as it were, in the shadow of its precursor, soy paste. Secondly, there were no institutionalised means to support a coordinated effort towards the development of a new technology. The only way such an effort could have been mounted was through the backing of the government. While the Emperors of China had always been supportive of improvements in agriculture, they were much less interested in the technology of food processing. Distilled wine could have received a measure of support from Taoist alchemists and bean curd from Buddhist monks, but it is impossible to substantiate the extent or value of

---

[13] Cf. pp. 203–32 above. The initiation date is inferred from the date of the Shanghai bronze still, identified as an Eastern Han (+25–220) object. Successful commercial production had to await the invention and adoption of the Chinese still, a remarkably simple and inexpensive device well suited for small scale production.

[14] For bean curd a satisfactory process had to await the discovery that boiling the soy milk was essential for the coagulation of a firm, resilient curd, cf. pp. 302–16.

[15] Cf. Hahn (1896), Laufer (1914–15), Creel (1937), p. 80, Harris (1985), pp. 130–53, Anderson (1988), p. 145, Simoons (1991), pp. 454ff. and H. T. Huang, 'Why the Chinese did not Develop Dairy Farming?' (paper in preparation). For the genetics of hypolactasia cf. Sahi (1994a) and Sahi (1994b) and the references cited, cf. especially Simoons (1954, 1970a, 1970b), Flatz & Rottawe (1973) and Flatz (1981).

[16] Cf. pp. 322–3.   [17] Bean sprouts, cf. pp. 295–8; soy sauce, cf. pp. 358–73.   [18] Cf. pp. 466–90.

this support.[19] Thirdly, developmental work was usually carried out by private entrepreneurs or employees of households for whom innovation was probably a peripheral activity. Most likely they were illiterate and unable to record their results in writing.[20] Fourthly, there was a tendency of the workers or their masters to keep their ideas and results secret, as for example, in the case of bean curd and distilled wine. There was no mechanism for the dissemination of ideas and results of experiments so that communication between groups interested in the same product was non-existent or incidental.[21]

These factors are a natural consequence of the social, political and cultural environment which govern the progress of all innovations.[22] The high productivity observed for the Han and Sung would suggest that these environments were more hospitable to innovations during these two periods than at other times in Chinese history.

(3) *The evolution of food technology in China*

Besides the family of fermented foods shown in Figure 134, there is another family of processed foods that is connected to the grinding activity of the rotary quern, which made the production of wheat flour and soy milk possible (lower part of Figure 135). These relationships tend to support George Basalla's evolutionary model for the development of technology, which is based on the thesis 'that each kind of made thing is not unique but is related to what has been made before'.[23] In other words, every artifact invented by man, no matter how novel in appearance, has had an antecedent in the existing technological order. The ultimate progenitor in the first family of foods is the *mould ferment*, and the second, the rotary quern. But what about the origin of the progenitors themselves? What are their antecedents?

---

[19] The alchemists were interested in the art of distillation, cf. *SCC* Vol. v, Pt 4, pp. 68–80, and the Shanghai still is believed to be an alchemical apparatus. Legend has it that the Prince of Huai Nan learned the art of making *tou fu* from the eight Taoist immortals, but there is no indication that the alchemists had ever tried to perfect the process. The Buddhists monks always had a particular interest in vegetarian substitutes of meat, and could have played a part in the development of *tou fu*, just as they did in the making of *mien chin* (gluten).

[20] Both male and female artisans were employed in the large households. Female maids can be identified in the famous Han kitchen scene from Chu-chhêng (Figure 25, p. 87). Although the artisans were probably illiterate, their masters were not. Chia Ssu-Hsieh, the author of the *Chhi Min Yao Shu*, is an example of a master who was knowledgable about what went on in the field and in the kitchen. But Tsui Hao's preface to his *Food Canon*, cf. p. 123, makes clear that even high-class ladies are often skilled in the culinary arts. It is thus not surprising that well-educated young women of good family were trained as chefs and became lady caterers (*chhu niang* 廚娘) who were much in demand among the upper echelon of society during the Sung, cf. Kao Yang (*1983*), pp. 45–56. They were greatly admired for their refined upbringing and technical skill and famous for the high price of their services. In order to impress his friends one Sung scholar, without careful thought, engaged a lady caterer to give a banquet. The banquet was a resounding success, but the bill almost drove the poor man into bankruptcy. The prominence of lady caterers in the Sung may be the reason why the author of the first Chinese cookbook, the *Wu shih chung khuei lu* (p. 127), is traditionally regarded as a woman even though nothing concrete is known about the author. Accordingly, we have translated the title as *Madam Wu's Recipe Book* and not *Wu's Recipe Book*.

[21] For example, the food artisans might not have been aware of the existence of bean sprouts as a drug, and the physicians were not interested in its value as a food.

[22] The factors that influence the invention, development and adoption of technology have been discussed by many authors, including Harrison (1954), Klemm (1964), Williams (1987), Sawers (1978), Basalla (1988), Pacey (1990), Mokyr (1990), Cardwell (1994) and Adams (1996). A succinct summary of these factors is found in Diamond (1996), pp. 249–51.

[23] Basalla (1988), p. 208. For another version of the evolution of technology from the point of view of an economic historian cf. Mokyr (1990), pp. 273–99.

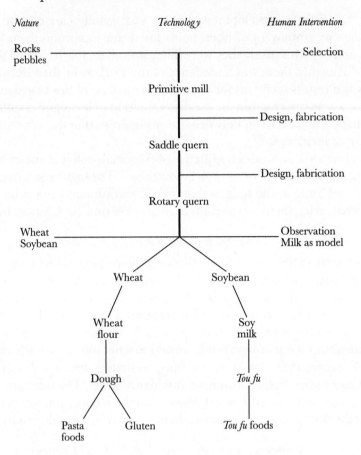

Fig 135. Evolutionary history of the rotary quern and its applications.

Friedrich Klemm, writing in 1957, began his history of western technology with this statement:[24] 'When man appeared at the dawn of history, he already had at his disposal a considerable treasury of technical instruments.'

In this case 'instruments' may be taken to mean products of technology in general. These 'instruments' were part of the foundation on which civilisations were built. Yet we know very little about their origin.[25] The application of Basalla's thesis to the products and processes discussed in this work may help us to understand how their primitive antecedents came into being and how they affected the development of the new technologies they inspired.

---

[24] Klemm (1964), p. 1.
[25] The issue of how prehistoric technologies developed is barely touched on by archaeologists. Nor is it one that has attracted much attention among historians of technology. Staudenmaier (1985 & 1990) has analysed the content of the articles that have appeared in the journal, *Technology and Culture* since its inception in 1958. None of the articles were concerned with the origin of technology in the prehistoric period. Cf. Kranzberg (1979) for a review of trends in the History and Philosophy of Technology as of 1973.

Let us take up the case of the rotary quern first.[26] Its antecedents are projected in Figure 135.[27] Its immediate antecedent is the saddle quern which was already well known during the Neolithic Age.[28] The antecedents to the saddle quern could be primitive mortars and grinding stones such as those found in the Nautufian (10,200 to 12,500 BP) and late Paleolithic (18,000 BP) age.[29] Their antecedents were probably a pair of stones of the proper size and shape that could be rubbed, rolled or banged against each other to crack cereal seeds or break the shells of nuts. Stones with the appropriate dimensions and surface smoothness could have easily been found along the banks of fast flowing streams. In other words, the distant prototype of the saddle quern was presumably selected from materials that were already in existence in nature. There are three factors in operation in the flow of events in this chart: natural resources, existing technology and human activity. In each step there is an interaction between human activity and at least one of the two remaining factors.

When we consider the case of the *mould ferment* in terms of Basalla's thesis, it is difficult to identify any existing model on which the idea of a *mould ferment* could have been based. How then was it related to other entities in the technological order? We have shown that it was the result of an accident, namely the prolonged exposure of *fan* (granules of cooked grain) to wild fungal spores in the air. In view of the critical role of *fan* in this discovery, we could say that the invention of the *mould ferment* owes its origin to the technique of steaming used in the making of *fan*. The Neolithic pottery steamer may, therefore, be considered as the antecedent which made possible the discovery of the *mould ferment*. But we can trace the lineage even further. The steamer is merely one of the myriad types of pottery vessels fabricated by the Neolithic Chinese. It is an artifact generated by the invention of pottery, which is itself the result of the application of fire by man on a naturally occurring material, clay. To summarise, we may depict all the steps involved in the evolution of the *mould ferment* in the chart presented in Figure 136. The same interactions between nature, technology and human activity shown in Figure 135 are also

---

[26] The rotary quern is a topic that has been treated extensively in *SCC* Volume IV, Pt 2, pp. 185–95. The invention of the rotary quern in China occurred in the late Warring States period (c. −300), slightly earlier than its appearance in Rome in the West. There was, however, a big difference between China and Rome in the way the wheat flour dough was fabricated and cooked. In the West, the dough, either leavened or unleavened was usually baked in ovens into bread. Ovens were uncommon in China and the dough was more often processed into various types of *ping* (pasta) and cooked by steaming, boiling or roasting on a hot surface.

[27] For foods derived from wheat flour, cf. pp. 462–97. Wheat flour was already a known entity before the Han since it could have been prepared by grinding wheat in a saddle quern, but it was the rotary quern that made it possible to produce it economically in quantity. The model for soy milk was dairy milk which was either known indigenously by the Shang period or introduced to the Chinese during the Han by the nomadic tribes of the north.

[28] *Khaifeng cultural relics bureau (1978)*, Plate No. 1; *Chung-kuo shê-huei kho-hsueh yuan (1982)*, Plate No. 2; and *idem*. (*1983*), Plate No. 1, show photographs of well-made saddle querns recovered from Neolithic sites (c. −6000) in Phei-li-kan 裴李崗 and Sha-wo-li 沙窩李. The saddle quern was used extensively in ancient Mesopotamia and Egypt for grinding flour to make bread.

[29] Cf. Valla (1995), p. 173 and Hole (1989), p. 102. Examples of grinding slabs dating from 10,000 to 14,000 BP are cited in Cohen (1977), p. 134.

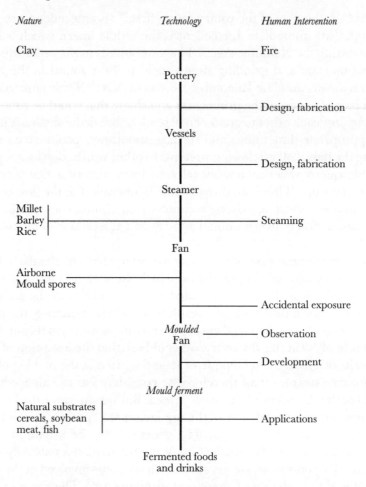

Fig 136. Evolutionary history of the *Mould ferment* and its applications.

involved in this scheme, except that in this case the most critical human intervention was the result of human imperfection.[30]

From the foregoing discussion it is clear that the pottery steamer and rotary quern are two key artifacts in the history of food processing in China. Both inventions had antecedents that reached far back to the dawn of the Neolithic Age. Each invention has given rise to a family of processed foods. But how do we account for the origin of the products that do not belong to these two families, such as bean

---

[30] The role of accidental discoveries in science as a result of human error is well known. Perhaps the most famous example in the present century is the discovery of penicillin when a petri dish of bacterial culture was accidentally exposed to air and left forgotten in the refrigerator for many days. The story is delightfully told in Macfarlane (1984). The key observation of a clear zone on a bed of bacteria was reported by Fleming in the *Brit. J. Exp. Path.*, 10, 226 (1929). Accidental discoveries played so important a role in her career that Levi-Montalcini (1988) titled her autobiography, 'In Praise of Imperfections.'

sprouts, malt sugar, vegetable oil and tea? The answers turn out to be fairly obvious. The antecedent to bean sprouts is sprouted grain.[31] Malt sugar was first discovered when someone attempted to make a congee (*chu* 粥) from rice or millet which had accidentally been allowed to sprout.[32] The antecedent to vegetable oil is animal fat, a key ingredient in the cookery of pre-Han China.[33] The discovery of tea was part of the massive prehistoric screening that tasted and tested all sorts of plant, animal and mineral products for possible use as food or medicine.[34] All in all, we can say that the origin of the diverse processed foods in the Chinese dietary system can be accounted for in terms of Basalla's theory of technological evolution.

(4) *Nature, technology and human intervention*

But we need to make a few adjustments to Basalla's theory. First of all, the key premise in Basalla's thesis is that 'each kind of made thing is not unique but is related to what has been made before'. If we go back far enough we are bound to encounter a man-made thing that had no prior man-made antecedent. In Figure 136, it is inconceivable that the first pottery pot ever made was in anyway related to another man-made object. But this does not mean that it did not have an antecedent. It could have followed the shape and form of an object that was already present in nature, such as the shell of a gourd, an animal pouch, or a large oyster shell. Any one of these objects could have been used by man as a container for liquids. In other words, its antecedent was a part of the natural rather than the man-made order.

Similarly, in the case of the rotary quern we can trace its history back to the Paleolithic grinding stones that had no man-made antecedent. They were also based on objects already in existence in nature. These examples indicate that when we come to an original artifact for which there is no man-made antecedent, the inventor may have had to draw his inspiration from things that were made by the forces of nature. Since the gourd, the shell of an oyster, or an animal pouch are objects derived from organic evolution, we could even say that the evolution of technology is an extension of organic evolution. In this case the evolution of technology is just one aspect of the streams of exogenetic evolution that have shaped the course of human culture and society.[35]

Secondly, as shown in Figures 135 and 136, each change in the scheme is the result of the interaction of human activity with either nature or technology or both. Human intervention is the primary agent for change. It may take the form of observation, design, fabrication, and development (i.e. experimentation and testing). But what are the driving forces behind human intervention? Why should early man have intervened at all? In the case of food and drink, surely one of the driving forces had to be necessity. This is a theme common in Chinese thought. For example, in the *Chou Li, Han Wên Chia* 周禮，含文嘉 (Rites of the Chou, Treasury of Cherished

---

[31] Cf. pp. 157–8, 457. [32] Cf. pp. 260–1. [33] Cf. pp. 28–32. [34] Cf. pp. 503–4.
[35] For a brief exposition of exogenetic heredity and evolution, cf. Medawar & Medawar (1983), pp. 94–7; Dawkins (1976, 1989), ch. 11, 'Memes: The New Replicators,' pp. 189–201; and Popper (1972) about World 3.

Literature), we read:[36] 'It was Sui Jên who first drilled wood to obtain fire and taught the people to cook food from raw materials in order that they might suffer no disease of the stomach, and to raise them above the level of the beasts.'

This sentiment is reiterated in the *Li Chi* and the *Han Fei Tzu*.[37] They illustrate the urgent need for Neolithic man to have a vessel not only for holding water, but also for cooking food, if he were to raise himself 'above the level of the beasts'. It was necessity that stimulated the invention of pottery. It was an invention that helped to secure the survival of the *sapiens* potential of *homo sapiens*.[38] It was only after the basic problem of securing an adequate supply of food was solved that the germ of civilisation could sprout and grow.

Basalla plays down the role of necessity in the evolution of technology. He says,[39] 'Instead of relying on nature directly for sustenance, we have devised the wholly unnecessary techniques of agriculture and cooking. They are unnecessary because plants and animals are able to grow and even thrive without human intervention, and because food need not be processed before it is fit for human consumption.' That is a rather simplistic and extreme statement. Many foods have to be processed before they are fit for human consumption.[40] Thus, in our view, agriculture, animal husbandry, processing and cooking were invented precisely because they were necessary to ensure that early man would have an adequate, wholesome and safe food supply, so that he could raise himself 'above the level of the beasts'.

To be sure, once the basic needs for survival are satisfied, the role of necessity in technological inventions may diminish in importance. It became just one of the factors that propelled the continuing evolution of technology. For example, in the case of pottery, once the technique of making a drinking bowl and the cooking pot was mastered, other factors began to influence the course of pottery design. Necessity had little to do with the fabrication of the myriad shapes and sizes of pottery artifacts that adorned the sites of all Neolithic cultures. And in the case of the pasta foods made from flour, it is difficult to see what necessity had to do with the varieties of *ping* that were fabricated and consumed. Perhaps, we can attribute the driving force behind these changes to human creativity. There is a basic urge to create new things to meet real or perceived challenges that cannot be denied.

---

[36] Cited in Needham *et al.* (1970), p. 364 and Thao Wên-Thai (*1983*), p. 2.

[37] *Li Chi*, ch. 9, *Li Yung*, p. 366. *Han Fei Tzu*, Section 49, tr. Watson (1963), p. 96.

[38] Chinese records have attributed the invention of pottery to such legendary figures as Huang Ti (c. −2697), the Yellow Emperor and Khun Wu 昆吾, an official of the Hsia dynasty (c. −2000). Based on archaeological records we know that these dates are much too late, cf. Thao Wên-Thai (*1983*), p. 9, and *LSCC*, (1983) edn, ch. 17, p. 515.

[39] Basalla (1988), p. 14; on p. 13 he quotes José Ortega y Gasset who defines 'technology as the production of the superfluous', and asserts that, 'Like the rest of the animal kingdom we, too, could have lived without fire and tools.' A biologist might argue that without the ability to invent and use tools and to form social groups for mutual protection, it is questionable that our distant ancestors, like their Neanderthal cousins, would have actually survived. Cf. also the criticism raised by Cardwell (1994), p. 499.

[40] For example, cassava contains a cyanogen that must be removed. Cereals have to be hulled to remove the indigestible husk. We have devoted a whole chapter to discuss ways to remove the protease inhibitors and flatus factors in soybean. Even milk has to be processed before it can be fed to people who are lactose intolerant. Fresh meat and fish spoil easily upon standing, and salting was used since prehistoric times to prolong their shelf life.

Thirdly, our study indicates that accidental discoveries play an important role in the development of processed foods. Besides the discovery of the *mould ferment*, other examples of accidental discoveries are the saccharifying activity of sprouted grain, vinegar from wine, raised bread and bean curd from soy milk. To these examples we must add the discovery of pottery, which probably came about accidentally when a fire was kept burning on a bed of clay.[41]

If technological evolution were, indeed, analogous to organic evolution, it would have to show in some important way a behaviour similar to that exhibited by organic evolution. For example, organic evolution does not advance at a constant rate along a straight line, but rather in fits and starts that go in different directions.[42] This appears to be the case with the evolution of Chinese processed foods. The practical result would be a family tree of products all linked to the primary invention such as the family of products linked to the *mould ferment* and another linked to the rotary quern. (Figures 134 and 135). The concept of fits and starts may also explain why there were periods of high productivity such as the Han and Sung, and periods of low or moderate productivity.

(5) *Effect of processed foods on health and nutrition*

In prehistoric China the most important grains were millets and rice. They were steamed to give fluffy granules called *fan*, which in turn were fermented into *chiu* or wine. When wheat and soybean were cultivated, they were also cooked into *fan*. The results were less than satisfactory. Since both were excellent crops for agronomic reasons, there was a strong incentive to find ways of turning wheat and soybean into foods that were more palatable than the *fan* that they generated. Thus, a majority of the processed foods we have discussed in this study, other than preserved meats, vegetables and fruits, deal with the processing of wheat and soybean. The products that developed are compared to those derived from rice and millet in Table 61. They fall into two groups: firstly, those with improved texture and digestibility, and secondly, those with improved flavour. A separate column is added to indicate the key intermediates used in the processing. Except for wine which is best known as an alcoholic drink, the products in the second group are familiar to us as flavouring agents.

---

[41] The development of pottery probably began in the Upper Paleolithic age. The earliest precursors of pottery are the fired clay figurines dated 27,000 BP found in Eastern Europe. They are 15,000 years earlier than the first pottery vessels of the Jomon culture in Japan. The earliest pottery pots could have been made by firing a basket heavily coated with clay. One Chinese legend holds that pottery was discovered as the result of the ancient *phao* 炮 technique of baking meat or fowl, in which the raw food is wrapped in leaves, daubed with clay and heated in the fire, in the manner still practised for making 'beggars chicken' (p. 73). One day the cast was left in the extra hot fire too long, so that when the clay was broken a section of it was found to serve perfectly as a water-proof bowl, cf. Ma Chien-Ying, *CKPJ*, **1995** (11), p. 13.

[42] A good analogy is given by Dawkins (1989), p. 223, who compares the progress of evolution with the journey of the ancient Israelites across the Sinai desert, a distance of about 200 miles. According to Exodus it took them forty years to complete the trip. If they had travelled at a uniform speed, they would have covered a distance of twenty-four yards a day. Obviously, they were travelling in fits and starts, perhaps camping at certain locations for long periods of time before they moved on to the next destination.

Table 61. *Summary of processed foods derived from grains*

Grain	Intermediate	Processed foods	
		Texture and digestibility	Flavour

Wheat	Wheat flour	Pasta family
		  Steamed bread, roasted
		  buns, wonton, *chiao-tzu*,
		  filamentous noodles
		Gluten
			Fermented wheat
			paste

Rice	Malted wheat or barley		Malt sugar
		Rice noodles	
Millet or Rice	*Moulded* wheat, barley or rice		Wine Vinegar

Soybean		Soy sprouts
		Tou fu family
		  Soft *tou fu*, hard *tou fu*,
		  tou fu skin, smoked *tou fu*	

	Soy milk		Fermented *tou fu*
	Moulded soybeans		Fermented soybeans, soy paste, soy sauce

The effects these products had on the Chinese dietary system and on the health of the people are twofold. The first is gastronomic. As shown in Table 61 wheat and soybean are processed into a series of foods with excellent texture and digestibility, namely pastas and *tou fu*, as well as unique condiments such as fermented soy sauce, soy paste and wheat paste. Similarly, rice was processed into noodles, wine and vinegar. These products have helped to provide a sound material base for the creation of one of the great cuisines of the world.[43]

The second is nutritional. In this case, the record is somewhat mixed. The traditional Chinese diet is often low in protein and calcium due to a shortage of meat and the absence of milk.[44] These deficiencies were to a large extent mitigated by the introduction of soyfoods.[45] Soybeans are rich in proteins, and bean curd, particularly

---

[43] For a comprehensive study of the efficacy of the traditional Chinese diet in the countryside in recent years, cf. Chen Junshi *et al.* (1990). The results indicate that under normal conditions of productivity the diet of the Chinese is adequate for the maintenance of health.

[44] This theme has been stressed by many observers of the Chinese food scene in the first part of the 20th century, cf. Mallory (1926), Adolph (1926 & 1946), Wu Hsien (1928), Maynard & Swen (1937), Read *et al.* (1937) and Snapper (1941), and echoed by Koo (1976). On the other hand, a high intake of protein is not necessarily always a blessing. High intake of proteins is known to inhibit calcium absorption, cf. Altchuler (1982). Osteoporosis is common in Europe and America but rare in East Asia. The Eskimos apparently have the world's highest incidence of osteoporosis despite having a high intake of dietary calcium (more than 2000 mg per day from fish bones). Their intake of proteins (250 to 400 g per day) is also the world's highest, cf. Mazess & Mather (1974).

[45] The availability of bean curd and other soyfoods presumably enabled Fortune (1857), pp. 42–3, to remark that 'the poorest classes in China seem to understand the art of preparing their food much better than the same classes at home', cited in Spence (1977), p. 267. Spence also lists the price of various foods in the early 18th century. The price of pork is 50 cash per catty, of bean curd 6 cash per catty.

when coagulated with calcium sulphate, is a rich source of calcium.[46] Bean sprouts provide water soluble vitamins in the winter when fresh vegetables are not available.[47] Gluten is another useful vegetarian source of protein.[48] Fermented foods are a major source of vitamin B12 for the poorer classes, since it is not present in plants.[49]

On the other hand, processing can also decrease the nutritional value of a food. This was what happened with beriberi, an illness unknown in ancient China. Beriberi became endemic after the Han when water power made it possible to polish rice to an extent that would have been impractical through the use of human or animal power. The polishing inadvertently removed vitamin B1, which prevents the occurrence of beriberi.[50] Yet highly polished rice is preferred by most people because it gives a softer and finer *fan*. Another example of an adverse effect is salted fish, formerly a major preserved fish product, but now valued mostly as a relish or condiment. Some types of salted fish are known to contain dimethyl-nitrosamine, a potent carcinogen that causes nasopharyngeal carcinoma, the incidence of which is particularly high among ethnic Chinese.[51]

Epilogue

In looking back on all that we have explored and discussed, we cannot help but be surprised by the transparent lack of interest the Chinese showed in following up valuable leads that could have raised the level of their technology and their understanding of the underlying phenomena. Three examples of such leads immediately come to mind. The first is Chia Ssu-Hsieh's observation that the activity of the loose *ferments* used in food processing resides in the yellow dust that covers the granules.[52] It is inexplicable to us that he did not go one step further and test the yellow dust on a new batch of substrate to see if it would grow to give more of the yellow dust. After all, he was the one who had previously determined that the ritual sacrifices and taboos associated with the making of *superior ferment* could be eliminated without harming the activity of the final product.[53] If he had followed his experimental bent

---

[46] The bioavailability of the calcium in bean curd compares very favourably with that in non-fat milk or Mozzarella cheese, cf. Poneros and Erdman (1988). Bean curd coagulated with calcium sulphate is a particularly rich source of calcium.

[47] Cf. pp. 295–9.   [48] Cf. pp. 497–502.

[49] This vitamin is produced only by microorganisms. Animals absorb it from foods contaminated with bacteria. Plants do not absorb it from the soil. This is one vitamin that is often lacking in a truly vegetarian diet, cf. Herbert (1976), pp. 193–4, Wang (1986), and Hesseltine & Wang (1986a), pp. 11 and 17.

[50] Cf. pp. 578–86.

[51] Cf. Fong & Chan (1972) and Henderson (1978). Another source of concern is the possible occurrence of aflatoxin, a potent carcinogen, in foods fermented by the grain moulds. *Aspergillus flavus*, a powerful producer of aflatoxin, is closely related to the *Aspergillus niger* and *Aspergillus oryzae* found in *chhü* 麴 (Chinese *ferment* or Japanese *koji*). This possibility has been thoroughly investigated, cf. Yokotsuka & Sasaki (1986) and Hesseltine & Wang (1986a), p. 21, and the results indicate that no detectable amounts of aflatoxin were produced during the production of fermented soyfoods.

[52] Cf. p. 335. Chia said, 'Do not winnow the yellow granules. The people of Chhi often winnow the granules in the wind to get rid of the yellow coat. This is a big mistake . . . it is the yellow coat that provides the bulk of the fermentative activity.'

[53] Cf. p. 171.

he might have become the founder of the science of mycology. At the very least, he would have found that the dust (i.e. spores) could be used as a seed to hasten the prepration of the next batch of *ferment*. This would have cut the time needed for the process from weeks to days.

The second is the failure of Chinese technologists to clearly distinguish between the saccharifying and fermenting functions of the mould *ferment* (*chhü* 麴). They already knew by the Han Dynasty that in the making of wine the *chhü* first converts steamed rice into a sugar solution (*li* 醴), and then ferments the sugar into wine (*chiu* 酒).[54] They also knew that a fermenting wine could be used as a leaven (*chiao* 酵) to make raised [steamed] bread.[55] The *Pei Shan Chiu Ching* (Wine Canon of North Hill), +1117, notes that *chiao* could be used to liven a sluggish fermentation and describes how it could be prepared from a fermenting must.[56] Even though these observations were well known, no one apparently realised that *chiao* is the principle in the *chhü* responsible for fermenting sugar into wine. As a result, no one ever tried to use *chiao* to ferment grape juice or honey into wine or mead.[57] If they had, the story of grape wine in China might have had a much happier ending.

The third is the linking of two key observations by the physicians of the Thang about the disease beriberi.[58] The first is a recipe by Sun Ssu-Mo in the *Chhien Chin I Fang* (+682), which recommends the incorporation of rice bran in a rice congee as a cure. The second is a statement by Chhên Tshang-Chhi in the *Pên Tshao Shih I* (+725) that 'prolonged consumption of rice weakens the body; feeding [polished] paddy or glutinous rice to young cats and dogs will so bend their legs that they will not be able to walk.' From our modern perspective, it is extraordinary that none of the Chinese physicians ever thought that the two observations might be connected. If they did, they might have induced the disease in cats or dogs by feeding them polished rice and cured it by adding rice bran to their diet. They would then have proved, as Eijkman did with his experimental chickens at the end of the 19th century, that beriberi is caused by the deficiency of a factor present in the bran fraction of the rice kernel. When the kernels are polished, the factor is removed with the bran. Therefore eating polished rice granules, when rice is the major component of the diet, therefore, induces beriberi. Ingesting rice bran cures it.

Why the Chinese missed these obvious opportunities to raise the level of their technology and to move from applied science to basic science, are matters of considerable interest in the larger context of why modern science did not arise in China. We would like to offer three observations.

---

[54] During the Han *li* (a sweet liquor) was prepared with the *ferment* in a one-night incubation; upon further incubation the sugar is fermented into alcohol, cf. pp. 262–3.

[55] Cf. p. 471. It is also noted here that the *Chhi Min Yao Shu*, +544, describes the preparation of a sourdough type of yeast leaven (*ping chiao* 餅酵) from rice.

[56] Cf. p. 184.

[57] In fact, even as late as the Ming (+1368 to +1644) grape wine and mead continued to be made with a *ferment*, cf. pp. 244 and 247.

[58] Cf. p. 581; Sun's recipe is in the *CCIF* repr. (1965), ch. 17, no. 2, pp. 194–6. Chhên's statement is preserved in *CLPT*, ch. 26, item on *tao mi* 稻米, p. 3b.

Firstly, the Chinese world view, based on the concept of Yin–Yang and Wu-Hsing tends to look at all phenomena in an organic or holistic way. It discourages the analytical approach, which involves the breaking down of a system into parts that can be controlled and tested experimentally.[59] In the case of the *ferment chhü*, the entity is regarded as an organic whole. Even when the evidence clearly points to two distinguishable activities, there is a reluctance to isolate and study them separately. And so, the Chinese continued the extravagant practice of using the *ferment* to make wine from grape juice and honey, when *chiao* (yeast) would have been a far better choice. In terms of beriberi, the physicians involved were so wrapped up in the Wu-Hsing theory of diseases, that they could not appreciate the connection between polished rice as a cause of beriberi and rice bran as a cure for it.[60]

Secondly, the stress on practical applications tends to divert attention away from any desire to gain an understanding of the intellectual basis of the technology. The wine technicians in the heartland of China were only interested in the fermentation of grains to wine. They had little interest in the fermentation of grape juice. In the case of beriberi the medical practitioners were interested mainly in curing the illness. They were actually quite successful in this task, and assumed that they already knew the cause of the disease.

Thirdly, there is a real question as to whether technical information was shared and disseminated in medieval China. Some of this inadequacy may have been deliberate. In wine fermentations very little detailed process information is recorded, especially quantitative data that would enable a material balance to be calculated. Such information was obviously available since wine had been produced commercially even before the Han dynasty. We cannot be sure that the technician carrying out a fermentation with grape juice actually knew about the existence of *chiao* (yeast) and how easily it could be prepared. Those who did know about *chiao* probably had no access to grape juice as a substrate. As for beriberi, we cannot be sure that Chhên Tshang-Chhi was aware of the existence of Sun Ssu-Mo's bran recipe. In fact, in Tung Chi's celebrated treatise on beriberi, *Chiao Chhi Chih Fa Tsung Yao*, +1075, neither Sun's recipe nor Chhên's observation is mentioned. It is possible that in this case what we consider as an opportunity missed, might have never existed during the Thang and Sung. But by the Yuan and Ming, all the key evidence should have been readily available to readers of the medical literature, yet the connection was not recognised.

This lack of efficient communication no doubt also contributed to the slow pace of development of some of the processed foods examined earlier. But it did not prevent the dissemination of the positive achievements through transmission, diffusion or migration to China's neighbours in East and Southeast Asia. Typical Chinese

---

[59] We do not wish to imply that the analytical approach is, in any way, superior to the holistic one. It all depends on the nature of the problem. In other situations, as in the application of Wu-Hsing to the treatment of chronic illnesses, the holistic approach is often more effective than the analytical.

[60] One might, in this case, say that they saw the wood but missed the trees.

processed foods such as *tou fu*, soy sauce, and noodles are well integrated into the diet of the peoples of Korea, Japan, Indochina and Indonesia.

The influence of Chinese processed foods was just beginning to be felt in the West in the latter part of the 20th century. But this influence may actually be much greater than meets the eye. The basic technology of the grain moulds or *chhü* 麴 spread from China to her neighbours in mediaeval times. When modern microbiology came to Japan and Southeast Asia in the late 19th century, European microbiologists began to take note of the interesting preparations of mixed cultures that were used by the 'natives' in the making of alcoholic drinks and condiments. Since then, Western and Japanese scientists have isolated many species of fungi from moulded grains and characterised the myriad enzymes they produce. Western microbiologists soon developed an interest in applying these organisms to their own fermentation processes. The Japanese pronunciation of the character 麴, *koji*, became so familiar in the West that it is listed in the *Oxford Dictionary of the English Language*.

The French were the first to develop a process based on the use of amylolytic organisms from the microbial *ferment* to convert starch to alcohol.[61] They called it the Amylomyces process (or Amylo process for short), but their attempt to commercialise it in the late 19th century ended in failure. Decades of further persistent efforts by Japanese and Western scientists and technologists eventually led to the successful development of several of the mould species as vehicles for the manufacture of enzymes used in the modern fermentation food processing industry. In fact, the production of microbial enzymes, such as amylases, proteinases, lipases and pectinases from these organisms, has become an industry in its own right in Europe, America and Japan. The enzymes are used widely in the manufacture of all manner of processed foods.[62] Thus, even though in the West today, where the average consumer has probably never heard of the word *chhü* or *koji*, enzymes derived from or inspired by it may be touching his/her life in unsuspecting ways. Indeed, this may happen each time he/she consumes a piece of cake, drinks a glass of clarified fruit juice, gulps down a tankard of beer, gobbles a slice of bread, eats a bowl of fast cooked oatmeal, inserts a slice of processed cheese in a sandwich, sprinkles grated Romano cheese on spaghetti or pours corn syrup on a pancake or waffle. These are examples of the fruits of technology transfer that the ancient Chinese could not have possibly dreamed of when they discovered a way to prepare a stable culture of grain moulds more than four thousand years ago.

---

[61] Lafar, F. (1903), pp. 94–8.

[62] Reed, ed. (1975) and Fogarty, ed. (1983) describe in detail the application of grain mould enzymes in food processing, e.g. pectinases in the clarification of fruit juices, proteases in the chill-proofing of beer, amylases in the manufacture of corn syrups and quick-cooking oat meal, amylases and proteases in milling and baking, and lactases in preparing low-lactose milk. Since 1983 microbial rennets from *Mucor*, *Aspergillus* and *Endothia* spp. have been commercialised for the making of cheese. It is estimated that about one-third of the rennets used in cheese making today are of microbial origin, cf. Carlile & Watkinson C. (1994), p. 389. *Mucor*-derived esterases are also being used to enhance the flavour of cheese, cf. Huang & Dooley (1976), and Carlile & Watkinson, *ibid*.

BIBLIOGRAPHIES

A   Chinese and Japanese books before +1800 and all gazetteers
B   Chinese and Japanese books and journal articles since +1800, excluding gazetters
C   Books and journal articles in Western languages

In Bibliography A there are two modifications of the Roman alphabetical sequence: transliterated *Chh-* comes after all other entries under *Ch-*, and transliterated *Hs-* comes after all other entries under *H-*. Thus *Chhên* comes after *Chung* and *Hsi* comes after *Huai*. This system applies only to the first word of the titles in Bibliography A, and does not apply to Bibliography B or Bibliography C.

When Chinese words in Bibliography C are romanised according to a different system (e.g. the Pinyin system) they remain unchanged. In Bibliographies B and C and in the text, the reference numbers are the years of publication. In Bibliography A, for references to translations of classical works that have been published many times in previous bibliographies, such as those made by Pfizmaier to parts of the *Chin Shu* and des Rotours to the *Chiu Thang Shu*, the old numerical key is retained and the reader is referred to a previous volume of the Series.

Korean books are included in Bibliography A. As in previous volumes reference numbers in italics imply that the work is in one of the East Asian languages.

ABBREVIATIONS

See also pages xxiii–xxiv immediately after the List of Contents for abbreviations of titles of books that are used frequently in the footnotes.

The current list applies to journals. Most of the western and some of the Chinese and Japanese references are cited only once or twice in the entire work, and are not abbreviated. In these cases the reader will have no difficulty identifying the name of the journal as listed in the entry in the bibliography.

AS/BIHP	Bulletin of the Institute of History and Philology, Academia Sinica, Taipei
BMFEA	Bulletin of the Museum of Far-Eastern Antiquities (Stockholm)
CHWSLT	*Chung-hua wên-shih lun-tshung* 中華文史論叢
CKKCSL	*Chung-kuo kho-chi shih-liao* 中國科技史料
CKPJ	*Chung-kuo phêng-jên* 中國烹飪
CKPJKCTK	*Chung-kuo phêng-jên ku chi tshung khan* 中國烹飪古籍叢刊
CKPJPKCS	*Chung-kuo phêng-jên pai-kho chhüan-shu* 中國烹飪百科全書
FMNHP/AS	Field Museum of Natural History (Chicago) Publications; Anthropological Series
HHTP	*Hua Hsüeh Thung Pao* 化學通報
KCSWC	*Kho-chi shih wên-chi* 科技史文集
KHSCK	*Kho-hsüeh shih chi khan* 科學史集刊
KK	*Khao-ku* 考古
KKHCK	*Khao ku hsüeh chi khan* 考古學集刊
KKHP	*Khao-ku hsüeh pao* 考古學報
KKYWW	*Khao-ku yü Wen-wu* 考古與文物
NSYC	*Nung-shi yen-chiu* 農史研究
NSYCCK	*Nung-shi yen-chiu chi-khan* 農史研究集刊
NYKK	*Nung yeh khao ku* 農業考古
PJSH	*Pheng Jen Shih Hua* (烹飪史話) see *Chung-kuo phêng-jen pien chi pu* (Editors of *CKPJ*)
SCYC	*Shih Chhien Yen Chiu* 史前研究
TJKHSYC	*Tzu-jang kho-hsüeh shih yen-chiu* 自然科學史研究
TLTC	*Ta lu tsa chi* 大陸雜誌
TP	*T'oung Pao*, Archives Concernant l'Histoire, les Langues, la Geographie, l'Ethnographie et les Arts de l'Asie Orientalie, Leiden
TPJMTH	*Tung-pei Jen-min Ta hsueh jên-wên ke-hsüeh hsüeh-pao* 東北人民大學人文科學學報
WW	*Wen-wu* 文物
WWTKTL	*Wên-wu tshan khao tzu liao* 文物參考資料
WWTLTK	*Wên-wu tzu-liao tshung-khan* 文物資料叢刊

A. CHINESE AND JAPANESE BOOKS BEFORE +1800

Each entry gives particulars in the following order:
  (a) title, alphabetically arranged, with characters;
  (b) alternative title, if any;
  (c) translation of title;
  (d) cross-reference to closely related book, if any;
  (e) dynasty;
  (f) date as accurate as possible;
  (g) name of author or editor, with characters;
  (h) title of other book, if the text of the work now exists only incorporated therein; or, in special cases, references to sinological studies of it;
  (i) references to translations, if any, given by the name of the translator in Bibliography C;
  (j) notice of any index or concordance to the book if such a work exists;
  (k) reference to the number of the book in the *Tao Tsang* catalogue of Wieger (6), if applicable;
  (l) reference to the number of the book in the *San Tsang* (Tripitaka) catalogues of Nanjio (1) and Takakusu & Watanabe, if applicable.

Words which assist in the translation of titles are added in round brackets.

Alternative titles or explanatory additions to the titles are added in square brackets.

It will be remembered that in Chinese indexes words beginning *Chh-* are all listed together after *Ch-*, and *Hs-* after *H-*, but that this applies to initial words of titles only. Where there are any differences between the entries in these bibliographies and those in earlier volumes, the information here given is to be taken as more correct.

ABBREVIATIONS

C/Han   Former Han.
E/Wei   Eastern Wei.
H/Han   Later Han.
H/Shu   Later Shu (Wu Tai).
H/Thang Later Thang (Wu Tai).
H/Chin   Later Chin (Wu Tai).
S/Han   Southern Han (Wu Tai).
S/Phing Southern Phing (Wu Tai).
J/Chin   Jurchen Chin.
L/Sung   Liu Sung.
N/Chou   Northern Chou.
N/Chhi   Northern Chhi.
N/Sung   Northern Sung (before the removal of the capital to Hangchow).
N/Wei   Northern Wei.
S/Chhi   Southern Chhi.
S/Sung   Southern Sung (after the removal of the capital to Hangchow).
W/Wei   Western Wei.

Chan Kuo Tshê 戰國策.
  Records of the Warring States [semi-fictional].
  Chhin.
  Writer unknown.

Chên Chih Chia I Ching 針灸甲乙經.
  Treatise on Acupuncture and Moxibustion.
  S/Kuo & Tsin, +256 to +282.
  Huang-fu Mi 皇甫謐.

Chêng Chih Chun Shêng 証治准繩.
  Standard Methods for Diagnosis and Treatment.
  Ming, +1602.
  Wang Khên-Thang 王肯堂.

Chêng-Ho Hsin-Hsiu Ching-Shih Chêng-Lei Pei-Yung Pên Tshao 政和新修經史證類備用本草.
  New Revision of the Classified and Consolidated Armamentarium Pharmacopoeia of the Chêng-Ho reign period.
  Sung, +1116; repr. +1143 (J/Chin).
  Thang Shen-Wei 唐慎微, ed. Tshao Hsiao-Chung 曹孝忠.

Chêng Lei Pên Tshao 證類本草.
  (*Chêng-Ho Hsin-Hsiu Ching-Shih Chêng-Lei Pei-Yung Pên Tshao*).
  Reorganised Pharmacopoeia.
  Northern Sung, +1108, enlarged +1116; re-edited in J/Chin, +1204, and definitively republished in Yuan, +1249; reprinted many times, e.g. in Ming, +1468.
  Original compiler Thang Shen-Wei 唐慎微, cf. Hummel (1941), Lung Po-chien (1957).

Chêng Tsu Thung 正字通.
  Orthography of Characters.
  Ming, +1627.
  Chang Tzu-Lieh 張字烈.

Chi Chiu Phien 急就篇.
  Handy Primer (Dictionary for Urgent Use).
  C/Han, between −48 and −33.
  Shih Yu 史游, with +7th-century commentary by Yen Shih-Ku 顏師古 and +13th-century commentary by Wang Ying-Lin 王應麟.

Chi Chung Chou Shu 汲冢周書.
  Same as *I Chou Shu*.

Chi Lê Pien 雞肋編.
  Miscellaneous Random Notes.
  Sung, +1133.
  Chuang Chho 莊綽.
  Textual refs. are to the Commercial Press 1920 reprint of the Han Fen Lou 涵芬樓 edn.

Chi Ni Tzu 計倪子 or *Fan Tzu Chi Jan* 范子計然.
  The Book of Master Chi Ni.
  Chou (Yueh), −4th century.
  Attributed to Fan Li 范蠡 recording the philosophy of his master Chi Ni.

Chi Shêng Fang 濟生方.
  Prescriptions for Preserving Health.
  Sung, +1253.
  Yen Yung-Ho 嚴用和.
  Renmin Weisheng (1956) edn.

Chi Yun 集韻.
  Rhyming Phonetic Dictionary [the *Kuang Yün* enlarged to include ancient and vernacular words].
  Sung, +1040.
  Ting Tu 丁度, reprinted 1983, Shanghai Classics Press.

Chiao Chhi Chih Fa Tsung Yao 腳氣治法總要.
  General Remedies for Beriberi.
  Sung, +1075.
  Tung Chi 董汲.
  SKCS, **738**, 417–37.

Chieh Chha Chien 芥茶牋.
  Chieh Tea Notes.
  Ming/Chhing, +1644.
  Fêng Kho-Ping 馮可賓.

Chien Chha Shui Chi 煎茶水記.
  Water for Making Tea.
  Thang, +825.
  Chang Yu-Hsin 張又新.

Chin Hua Chhung Pi Tan Ching Pi Chih 金華沖碧丹經秘旨.
  Confidential Instructions on the Manual of the Heaven-Piercing Golden Flower Elixir.
  Sung, +1225.
  Phêng Ssu 彭耜 & Mêng Hsü 孟煦 (pref. and ed. Mêng Hsü), TT/907.

Chin Kuei Yao Lüeh (Fang Lun) 金匱要略(方論).
  Systematic Treasury of Medicine.
  H/Han, originally c.+200, revised c. +300.
  Chang Chung-Ching 張仲景, edited and revised by Wang Shu-Ho 王叔和.

Chin Phing Mei Tzhu Hua 金瓶梅詞話 [better known as *Chin Phing Mei*]
  The Golden Lotus.
  Ming, date unknown, first printed +1610.
  Attributed to *Lan Ling hsiao hsiao shêng* 蘭陵笑笑生, identity unknown.
  Textual refs. are to the edn of Kuang-Chih Press, Hong Kong.

Chin Shih Khun Chhung Tshao Mu Chuang 金石昆蟲草木狀.
  Album (drawings and calligraphy) of Minerals, Insects and Plants.
  Ming, +1620.
  Wên Shu 文淑.
  MS at the National Palace Museum, Taipei.

Chin Shu 晉書.
  History of the Tsin Dynasty [+265 to +419].
  Thang, +635.
  Fang Hsüan-Ling 房玄齡.
  A few chs. tr. Pfizmaier (54–7); the astronomical ch. tr. Ho Ping-Yu (1). For translation of passages see the index of Frankel (1).
  Textual refs. are to the Chung Hua edition of 1974.

Ching Chhu Sui Shih Chi 荊楚歲時記.
  New Year Customs in the *Ching Chhu* Region (modern Hupei).
  Probably Liang, c. +550; perhaps early Sui, +610.
  Tsung Lin 宗懍, in *HWTS*.

Ching Chou Thu Ti Chi 荊州土地記.
  Record of the Ching-chou Region.
  Date unknown, possibly W/Tsin.
  Author unknown; no longer extant. Fragment in *CMYS*.

Ching Tien Shih Wên 經典釋文.
  Textual Criticism of the Classics.
  Sui, c. +600.
  Lu Tê-Ming 陸德明, tr. Egerton (1939).

Chiu Ching 酒經.
  Wine Canon.
  Sung, +1090.
  Su Shih (Tung-Pho) 蘇軾(東坡).

Chiu Hsiao Shih 酒小史.
  Mini-History of Wine.
  Sung.
  Sung Po-Jen 宋伯仁.

Chiu Huang Pên Tshao 救荒本草.
  Treatise on Wild Food Plants for use in Emergencies.
  Ming, +1406.
  Chu Hsiao 朱橚 (prince of the Ming), Chou Ting Wang 周定王.
  Incl. as chs. 46–59 of *Nung Chêng Chhuan Shu* (q.v.).

Chiu Kao 酒誥.
  Wine Edict.
  Chin, c. +310.
  Chiang Thung 江統.
  Cited in Hu Shan-Yüan (1935), p. 266, and Chang Yuan-Fêng (*1991*), p. 360.

Chiu Phu 酒譜.
  Wine Menu.
  Sung.
  Tou Phing 竇平.

Chiu Shih 酒史.
  History of Wine.
  Ming.
  Fêng Shih-Hua 馮時化.

Chiu Thang Shu 舊唐書.
  Old History of the Thang Dynasty [+618 to +906].
  Wu Tai, +945.
  Liu Hsü 劉昫.
  Cf. des Rotours (2), p. 64. For translation of passages cf. the index of Frankel (1).
  Textual refs. are to the Chung-Hua edition (1975).

Chou Hou Pei Chi Fang 肘後備急方.
  Prescriptions for Emergencies.
  Chin, +340.
  Ko Hung 葛洪.

Chou Li 周禮.
  Record of the Institutions (lit. Rites) of the Chou (Dynasty) [description of all government official posts and their duties].
  C/Han, containing some material from Late Chou, esp. *Khao Kung Chi* which may have come from the archives of the Chhi state.
  Compilers unknown, tr. E. Biot (1).
  Textual refs. are to the edn by Lin Yin (1985).

Chou Li Chu Shu 周禮注疏.
  The *Chou Li* with annotations and commentaries.
  Annotations by Chêng Hsüan 鄭玄, H/Han, and commentaries by Chia Kung-Yen 賈公彥, Thang, and edited by Huang Khan 黃侃.
  Classics Press, Shanghai, 1990.

Chu Chhüan Hsiao Phin 煮泉小品.
  Teas from Various Springs.
  Ming, +1554.
  Thien I-Hêng 田藝蘅.

Chu Fan Chih 諸番記.
  Records of Foreign Peoples and their Trade.

Sung, +1225.
Chao Ju-Kua 趙汝适.

Chu Ping Yuan Hou Lun 諸病源候論.
Treatise on Diseases and their Aetiology.
Sui, +610.
Chhao Yuan-Fang 巢元方.
Popularly known as the *Chhao shih chu ping yuan hou lun* 巢氏諸病源候論.
Textual refs. are to the *Chu ping yuan hou lun chiao chu* edn by Ting Kuang-Ti (1992).

Chu yü shan fang tsa pu 竹嶼山房雜部.
Miscellanies from the Bamboo Islet Studio.
Ming, c. +1500.
Sung Hsü's family, cf. *Sung Shih Tsun Shêng*. Cf. ch. 2, *Yang shêng pu* 養生部, SKCS, **871**, 129–49.

Chuang Tzu 莊子.
The Book of Master Chuang.
Chou, c. −290.
Tr. Legge, Feng Yu-Lan, Lin Yutang and Burton Watson.
Textual refs. are to Kuo Chhing-Fan ed. *Chuang Tzu Chi Shih*, Chung Hua, 1961.

Chung Chha Shuo Shih Thiao 種茶說十條.
Ten Talks on Tea.
Chhing, c. +1874.
Tsung Ching-Fan 宗景藩.

Chung Khuei Lu 中饋錄 (also known as *Wu shih Chung Khuei Lu* 吳氏中饋錄).
Madam Wu's Recipe Book.
Late Sung.
Author unknown, reputed to be a Madam Wu.
Textual refs. are to Sun Shih-Chêng and Thang Kên edn (1987), *CKPJKCTK*.

Chung Khuei Lu 中饋錄.
Book of Viands.
Chhing, c. late 19th century.
Tsêng I 曾懿.
Textual refs. are to Chhên Kuang-Hsin ed. (1984), *CKPJKCTK*.

Chü Chia Pi Yung Shih Lei Chhuan Chi 居家必用事類全集.
Essential Arts for Family Living.
Yuan.
Author unknown.
Textual refs. are to Chhiu Phang-Thung ed. (1986), *CKPJKCTK*.

Chü Lu 橘錄.
The Orange Record (Monograph on Citrus Horticulture).
Sung, +1178.
Han Yen-Chi 韓彥直.
Tr. Hagerty (1923) with Chiang Khang-Hu.

Chün Phu 菌譜.
A Monograph on Fungi.
Sung, +1245.
Chhên Jên-Yü 陳仁玉.

Chha Chieh 茶解.
Tea Explanations.
Ming, +1609.
Lo Lin 羅廩.

Chha Chien 茶箋.
Notes on Tea.
Ming, +1630.
Wên Lung 聞龍.

Chha Ching 茶經.
The Classic of Tea (Tea Canon).
Thang, c. +770.
Lu Yü 陸羽.
Tr. Carpenter (1974).
Textual refs. are to Fu Shu-Chhin & Ouyang-Hsün eds. (1983).

Chha Ching 茶經.
Tea Classic.
Ming, +1597.
Chang Chhien-Tê 張謙德.

Chha Chü Thu Tsan 茶具圖贊.
Tea Utensils with Illustrations.
Southern Sung, +1269.
Shên An Lao Jên 審安老人.

Chha Khao 茶考.
Comments on Tea.
Ming, +1593.
Chhên Shih 陳師.

Chha Lu 茶錄.
Tea Discourse.
Sung, c. +1060.
Tshai Hsiang 蔡襄 cf. *SF*.

Chha Lu 茶錄.
Tea Records.
Ming, +1595.
Chang Yuan 張源.

Chha Phu 茶譜.
Tea Compendium.
Wu Tai, +925.
Mao Wên-Hsi 毛文錫.

Chha Phu 茶譜.
Tea Discourse.
Ming, +1440.
Chu Chhüan 朱權.

Chha Phu 茶譜.
Tea Discourse.
Ming, +1539.
Chhien Chhun-Nien 錢椿年.

Chha Shih 茶史.
History of Tea.
Chhing, c. +1669.
Liu Yüan-Chang 劉源長.

Chha Shu 茶疏.
Tea Commentary.
Ming, +1597.
Hsü Tzhu-Yü 許次紓.

Chha Shuo 茶說.
Talks on Tea.
Ming, +1590.
Thu Lung 屠隆.

Chhang Ku Chi 昌谷集.
The Works of Li Hao.
Thang, +817.
Li Ho 李賀.
Now available as *Li Hao Shih Chi* 李賀詩集 (the Poems of Li Ho), Renmin Wenxue, 1959.

Chhêng Tsai Chi 誠齋集.
The Devout Vegetarian.

Sung, 12th century.
  Yang Wan-Li 楊萬里 (+1124–1192).
  Cf. *Tou Lu Tzu Jou Chuan* 豆盧子柔傳
    (Biography of *Jou* the son of *Tou Lu*), and
    subtitled '*Tou Fu*'.
Chhi Fa 七發.
  The Seven Stirrings [on food delicacies].
  C/Han, c. –141.
  Mei Chhên 枚乘.
  Cf. *Wên Hsüan*.
Chhi Min Yao Shu 齊民要術.
  Important Arts for the People's Welfare.
  N/Wei (& E/Wei or W/Wei) between +533 and +544.
  Chia Ssu-Hsieh 賈思勰.
  Partial translation Shih Shêng Shêng-Han (1958).
  Textual refs. are to 1982 edn of Miao chhi-Yü; cf.
    also 1957 edn of Shih Shêng-Han.
Chhieh Yün 切韻.
  Dictionary of Characters Arranged According to
    their Sounds when Split [rhyming phonetic
    dictionary: the title refers to the *fan:chhieh*
    method of 'spelling' Chinese characters].
    (Cf. Vol. 1, p. 33).
  Sui, +601.
  Lu Fa-Yen 陸法言.
  Now extant only in the *Kuang Yün* [q.v.] Teng &
    Biggerstaff (1936), p. 203.
Chhien Chin I Fang 千金翼方.
  Supplement to *A Thousand Golden Remedies*.
  Thang, +660.
  Sun Ssu-Mo 孫思邈.
  National Institute Traditional Chinese Medicine,
    Taipei, 1965.
Chhien Chin Yao Fang 千金要方.
  A Thousand Golden Remedies.
  Thang, between +150 and +659.
  Sun Ssu-Mo 孫思邈.
  Textual refs. are to ch. 26 (Food Therapy) are to
    Wu Shou-Chü ed. (1985), *Chhien Chin Shih Chi*,
    CKPJKCTK.
Chhien Chin Shih Chih 千金食治.
  Golden Dietary Remedies. Cf. *Chhien Chin Yao Fang*.
Chhien Han Shu 前漢書.
  History of the Former Han Dynasty (–206 to +24).
  H/Han, c. +100.
  Pan Ku 班固, and after his death, his sister Pan Chao 班昭.
  Partial tr. Dubs (2), Pfizmaier (32–4, 37–51), Wylie
    (2, 3, 10), Swann (1950).
  Yin-Tê Index, no. 36.
  Textual refs. are to the *pai na* or Chung Hua edn of
    1962.
Chhin Ting Ssu Khu Chuan Shu Tsung Mu Thi Yao 欽定
    四庫全書總目提要.
  Analytical Catalogue of the Complete *Ssu Khu
    Chhuan Shu* Library (made by Imperial order).
  Chhing, +1782.
  Ed. Chi Yün 紀昀.
  Yin-Tê Index, no. 7. Commonly known as *Ssu Khu
    Chuan Shu Tsung Mu Thi Yao*.

Chhing I Lu 清異錄.
  Anecdotes, Simple and Exotic.
  Wu Tai & Sung, c. +965.
  Thao Ku 陶穀.
  Textual refs. are to to the 1985 edn of the Food
    and Drink sections by Li I-Min *et al.*,
    CKPJKCTK.
Chhu Hsüeh Chi 初學紀.
  Primer of Learning.
  Thang, +700.
  Hsü Chien 徐堅.
Chhu Tzhu 楚辭.
  Elegies of *Chhu* State [Songs of the south].
  Chou, c. –300 (with Han additions).
  Chhü Yuan 屈原 (& Chia I 賈誼, Yen Chi 嚴忌,
    Sung Yü 宋玉 *et al.*).
  Tr. Hawkes (1959).
  Textual refs. are to Fu Hsi-Jen, ed. *Chhu Tzhu Tu
    Pên* (1976).
Chhuan Ming Lu 荈茗錄.
  Tea Record.
  Wu Tai, +960.
  Thao Ku 陶穀.
Chhun Chhiu 春秋.
  Spring and Autumn Annals.
  Chou: a chronicle of the State of Lu kept between
    –722 and –481.
  Cf. *Tso Chuan*, *Kungyang Chuan* and *Kuliang
    Chuan*.
Chhun Chhiu Fan Lu 春秋繁露.
  String of Pearls on the *Spring and Autumn Annals*.
  C/Han, c. –135.
  Tung Chung-Shu 董仲舒. Partial tr. Hughes,
    E. R. (1942), *Chinese Philosophy in Classical Times*,
    Dent, London.; Wieger (2).
  Chung-Fa Index no. 4.
Chhun Phu Sui Pi 純浦隨筆.
  Casual Notes on Local Products.
  Chhing, +1888.
  Yeh Jui-Yen 葉瑞延.
Chhü Hsien Shên Yin Shu 臞仙神隱書.
  Book of the Slender Hermit.
  Ming, +1440.
  Chu Chhüan 朱權.
  In Shinoda & Tanaka (1973), 406–51.
Chhü Pên Tshao 麴本草.
  Pandects of Ferment Cultures.
  Sung (or Ming)?
  Thien Hsi 田錫.
Chhü-Wei Chiu Wên 曲洧舊聞.
  Old Stories Heard in *Chhü-Wei* (modern Hsin-
    Cheng, Honan).
  Sung, +12th century.
  Chu Pien 朱弁.
Chhüan Fang Pei Tsu 全芳備祖.
  Complete Chronicle of Fragrances (Thesaurus of
    Botany).
  Sung, +1256.
  Chhên Ching-I 陳景沂.
  Agricultural Press, Peking, repr. 1982.
Chhüan Shang-ku San-tai Chhin Han San-kuo Liu Chhao Wên
    全上古三代秦漢三國六朝文.

Complete Collection of prose literature (including fragments) from remote antiquity through the Chhin and Han Dynasties, the Three Kingdoms, and the Six Dynasties.
Chhing, +1836.
Ed. Yen Kho-Chün 嚴可均, repr. Chung-Hua (1965).

Chhüan Thang Shih 全唐詩.
Complete Collection of Thang Poems.
Chhing, +1706.
Ed. Phêng Ting-Chhiu 彭定求 et al.
Chung Hua, Peking (1960).

Chhüan Thang Wên 全唐文.
Complete Collection of Thang Prose.

Chhün Fang Phu 群芳譜.
The Assembly of Perfumes.
Ming, +1630.
Wang Hsiang-Chin 王象晉.

Ekirinhon Setsuyoshu 易林本節用集.
Japanese Dictionary.
Japan, +1597.
Author unknown.

Engishiki 延喜式.
Codes of Laws and Conduct of Ryo.
Japan, +907.
Hujiwara-no-tokihira 藤原時平 et al.

Erh Ya 爾雅.
Literary Expositor (Dictionary).
Chou material; stabilised in Chhin or C/Han.
Compiler unknown. Enlarged and commented on by c. +300 by Kuo Pho 郭樸.
Yin-Tê Index no. Suppl. 18.

Erh Ya Chu Su 爾雅注疏.
Explanations of the Commentaries on the *Literary Expositor*.
Sung, c. +1000.
Hsing Ping 邢昺.

Erh Ya I 爾雅翼.
Wings for the *Literary Expositor*.
Sung, +1174.
Lo Yuan 羅願.

Fan Shêng-Chih Shu 范勝之書.
The Book of Fan Shêng-Chih (on Agriculture).
C/Han, −1st century.
Fan Shêng-Chih 范勝之.
Tr. Shih Shêng-Han (1959).
Textual refs. are to 1981 edn by Miao Chhi-Yü.

Fang Yen 方言.
Dictionary of Local Expressions.
C/Han, c. −15 (but much interpolated later).
Yang Hsiung 楊雄, *HWTS*.

Fêng Shih Wên Chien Lu 封氏聞見記.
Fêng's Record of what he saw and heard.
Thang, +8th century.
Fêng Yen 封演.

Fêng Su Thung I 風俗通義.
The Meaning of Populist Traditions and Customs.
H/Han +175.
Ying Shao 應劭.
Chung-Fa Index no. 3.

Fusō-ryakki 扶桑略記.
A Brief History of the Nation (from the reign-period of Emperor Jimmu to Emperor Horikawa).
Japan, +938.
Kō-en 皇圓.

Hai Wei So Yin 海外索隱.
Guide to the Delicacies of the Sea.
Ming, c. +1590.
Thu Pên-Tsun 屠本畯.

Han Chhang-Li Chi, 韓昌黎集.
Collected Works of Han Yü.
Thang, +768 to 824.
Kuo Hsüeh Chi Pen Tshung Shu 國學基本叢書.

Han Fei Tzu 韓非子.
The Book of Master Han Fei.
Chou, early −3rd century.
Han Fei 韓非.
Chu Tzu Pai Chia Tshung Shu 諸子百家叢書.
Classics Press, Shanghai (1989).

Han Shu 漢書 cf. *Chhien Han Shu*.

Hei Ta Shih Lüeh Chien Cheng 黑韃事略箋證.
Customs of the Black Tartars (Mongols).
Southern Sung, +1237.
Phêng Ta-Ya 彭大雅.
Annotated by Wang Kuo-Wei 王國維, Wen Tien Ko Press, Peking (1936).

Ho Han San Chhai Thu Huei 和漢三才圖會 Cf. *Wakan Sansaizue*.

Honzō Wamyō 本草和名.
Synonymic Materia Medica with Japanese Equivalents.
Japan, +918.
Fukane no Sukehito 深根輔仁.

Hou Han Shu 後漢書.
History of the Later Han Dynasty [+25 to +220].
L/Sung, +450.
Fan Yeh 范曄. The monograph chapters by Szuma Piao 司馬彪 (c. +305), with a commentary by Liu Chao 劉昭 (c. +570), who first incorporated it into the work.
A few chapters trs. Chavannes (6, 16); Pfizmaier (52, 53).
Yin-Tê Index, no. 41.

Hsi Yüan Lu 洗冤錄, also known as *Hsi Yüan Chi Lu* 洗冤集錄.
The Washing Away of Wrongs (or Instructions to Coroners).
Sung, +1247.
Sung Tzhu 宋慈, partially tr., A. H. Giles (1874).

Hu Ya 湖雅.
Lakeside Elegances.
Chhing, mid-19th century.
Wang Jih-Chen 汪日楨.
Ch. 8 reprinted in Shinoda & Tanaka (1973) II, pp. 497–516.

Hua Yang Kuo Chih 華陽國志.
Records of the Country South of Mount Hua [historical geography of Szechuan down to +138].

Chin, +347.
Chhang Chhü 常璩.
In *HWTS*.

Huai Nan Tzu 淮南子.
The Book of (the Prince of) Huai Nan [compendium of natural philosophy].
C/Han, c. −120.
Written by the group of scholars gathered by the Prince of Huai Nan, Liu An 劉安.
Partial trs. Morgan (1); Erkes (1); Hughes (1); Chatley (1); Wieger (2).
Chung-Fa Index no. 51.

Huang Ti Nei Ching, Ling Shu 黃帝內經靈樞.
The Yellow Emperor's Manual of Corporeal Medicine: the Vital Axis (medical physiology and anatomy).
Probably C/Han, c. −1st century.
Writers unknown; ed. Thang, +762, by Wang Ping 王冰.
Textual refs. are to edn in *SKCS*.

Huang Ti Nei Ching, Su Wên 黃帝內經素問.
The Yellow Emperor's Manual of Corporeal Medicine: Questions and Answers.
Chou, remodelled in Chhin and Han, reaching final form c. −2nd century.
Writers unknown. Ed. and comm. Thang (+762), Wang Ping 王冰; Sung (+1050), Lin I 林億.
Textual refs. are to 1959 edn by *Nanching Chung I Hsüeh Yuan*. (Nanking College of Chinese Medicine), repr. 1985.
Partial trs. Hübotter (1), chs. 4, 5, 10, 11, 21; Veith (1972); Chamfrault & Ung Kang-Sam (1).

Huang Ti Nei Ching Thai Su 黃帝內經太素.
The Yellow Emperor's Manual of Corporeal Medicine: The Great Innocence.
Chou, Chhin and Han, present form by −1st century, commented upon in Sui, +605 to +618.
Ed. and comm. Yang Shang-Shan 楊上善.

Hui Hui Yao Fang 回回藥方.
An Islamic Formulary.
Yuan, date not known.
Author unknown. Originally in 36 chapters. Only four are extant. Reproduced with articles on the Formulary and edited by Chiang Jun-Hsiang (*1996*)

Hung Lou Mêng 紅樓夢.
Dream of the Red Chamber.
Chhing, c. +1791.
Tshao Hsüeh-Chhin 曹雪芹.
Textual refs. are to the 1978 Chung-Hua edn.
Tr. Hawkes, Penguin Books (1973).

Hsi Ching Tsa Chi 西京雜記.
Miscellaneous Records of the Western Capital.
Liang or Chhen, mid +6th century.
Attributed to Liu Hsin 劉歆 (C/Han) or Ko Hung 葛洪 (Chin), but probably Wu Chün 吳均 (Liang).

Hsi Hu Lao Jen Fan Shêng Lu 西湖老人繁勝錄.
Memoir of the Old Man of the West Lake [in Hangchow].
Southern Sung.
Writer unknown.
One of the four works included in the 1982 edition of *Tung Ching Mêng Hua Lu*, Commerce Publishers Peking.

Hsi Yu Chi 西游記.
Journey to the West.
Ming, c. +1680.
Wu Chhêng-En 吳承恩.
Textual refs. are to 1974 edn of Chung-Hua, HK.

Hsi Yüan Lu 洗冤錄, also known as *Hsi Yüan Chi Lu* 洗冤集錄.
The Washing Away of Wrongs (or Instructions to Coroners).
Sung, +1247.
Sung Tzhu 宋慈, partially tr., A. H. Giles, *China Review, 1874*, 30, 92, 159 etc.

Hsia Hsiao Chêng 夏小正.
Lesser Annuary of the Hsia Dynasty.
Chou, between −7th and −4th century.
Writers unknown; incorporated in *Ta Tai Li Chi*, q.v.
Tr. Grynpas (1); R. Wilhelm (6); Soothill (5).

Hsiai Phu 蟹譜.
Monograph on Crabs.
Sung, +1060.
Fu Kung 傅肱 (+1060).

Hsiai Lo 蟹略.
Discourse on Crabs.
Sung, c. +1180.
Kao Shih-Sun 高似孫.

Hsiao Erh Yao Chêng Chih Chüeh 小兒葯証直訣.
Prescriptions for Children's Diseases.
Sung, +1114.
Chhien I 錢乙.

Hsien Chhin Ngo Chi 閒情偶寄.
Random Notes from a Leisurely Life.
Chhing, +1670.
Li Yü 李漁.
Ed. Yeh Ting-Kuo (1984) in the *CKPJKCTK* series.

Hsin Fêng Chiu Fa 新豐酒發.
The *Hsin Fêng* Wine Process.
Southern Sung.
Lin Hung 林洪.
Included in *Shan Chia Chhing Kung*.

Hsin Hsiu Pên Tshao 新修本草.
The Newly Improved Pharmacopoeia.
Thang, +650.
Ed. Su Ching (=Su Kung) 蘇敬(蘇恭) and a commission of twenty-two collaborators under the direction of Li Chi 李勣 & Yü Chih-Ning 于志寧, then of Chhangsun-Wu-Chi 長孫無忌.
The work was afterwards commonly but incorrectly known as *Thang Pên Tshao*. It was lost in China, apart from MS fragments in Tunhuang, but copied by a Japanese scholar in +731 and preserved in Japan though incompletely.
The name of the Japanese scholar was Tanabe Fubio 田邊史 and eleven chapters of the MS are preserved in the copy in the Ninnaji 仁和寺 in Kyoto.

Textual refs. are to the 1985 edn published by the Classics Press, Shanghai, with an introduction by Wu Tê-To based on the MS brought back from Japan in 1901 by the archaeologist Lo Chên-Yü 羅振玉.

Hsin Shu 新書.
  New Book.
  C/Han, c. −180.
  Chia I 賈誼.

Hsin Thang Shu 新唐書.
  New History of the Thang Dynasty [+618 to +906].
  Sung, +1061.
  Ouyang Hsiu 歐陽修 & Sung Chhi 宋祁.
  Cf. des Rotours (2), p. 56.
  Partial trs. des Rotours (1, 2); Pfizmaier (66–74). For translation of passages cf. the index of Franke (1).
  Yin-Tê Index no. 16.

Hsin Yuan Shih 新元史.
  New History of the Yuan.
  Cf. Kho Chao-Min 柯劭 (*1920*) in Bibliography B.

Hsing Yuan Lu 醒園錄.
  Memoir from the Garden of Awareness.
  Chhing, c. +1750.
  Li Hua-Nan 李化楠; compiled by his son Li Tiao-Yuan 李調元.
  Textual refs. are to the 1984 edn by Hou Han-Chhu & Hsiung Ssu Chi, in the *CKPJKCTK* series.

Hsü Chha Ching 續茶經.
  The Tea Classic – Continued.
  Chhing, +1734.
  Lu Thing-Tshan 陸廷燦.
  SKCS, **844**.

Hsü Hsiai Phu 續蟹譜.
  A Continuation Monograph on Crabs.
  Chhing.
  Chhu Jen-Huo 褚人獲.

Hsü Po Wu Chih 續博物志.
  A Continuation of the 'Records of the Investigation of Things'.
  Northern Sung.
  Li Shih 李石, *SKCS*, **1047**, 931–76.

Hsüan Ho Pei Yuan Kung Chha Lu 宣和北宛貢茶錄.
  Record of Pei-yuan Tribute Teas in the Hsüan Ho reign-period [from Chien-an in Fukien c. +1119–25].
  Sung, +1122–5.
  Hsiung Fan 熊蕃, enlarged by Hsiung Ko 熊克 in +1158.

Huai Hai Chi 淮海集.
  Collection from Huai Hai.
  Northern Sung, c. +1100.
  Chhin Kuan 秦觀, *SPTK*.

I Chien Chi 夷堅志.
  Strange Stories from *I-Chien*.
  Sung, c. +1170.
  Hung Mai 洪邁.
  Textual refs. are to Chung-Hua edn (1981), Peking.

I Ching 易經.
  The Book of Changes.
  Chou with C/Han additions.
  Compiler unknown.
  Tr. Legge (1899); Wilhelm (2); des Harlez (1).
  Yin-Tê Index, no. Suppl. 10.
  Textual refs. are to the *I Ching Yü Chieh* edited and annotated by Hsieh Ta-Huang (1959).

I Chou Shu 逸周書.
  Lost Records of the Chou Dynasty.
  Chou, −245 for such parts as are genuine. Found in the tomb of An Li Wang, a prince of the Wei state (r. −276 to −245) in +281.
  Writers unknown.

I Hsüeh Chêng Chuan 醫學正傳.
  Compendium of Medical Theories.
  Ming, +1515.
  Yü Thuan 虞摶.

I Li Chu Shu 儀禮注疏.
  The Book of Etiquette and Ceremonial, with Annotations and Commentaries.
  Chhin and Han, based on Chou material, some as old as the time of Confucius.
  Compilers unknown.
  Annot. by Chêng Hsüan (H/Han), comm. by Chia Kung-Yen (Thang).
  Classics Press, Shanghai, 1990. Tr. John Steele (1917).

I Mên Kuang Tu 夷門廣牘.
  Archives of the Hermit's Home.
  Ming, + 1598.
  Chou Lü-Ching 周履靖, *SKCS*.

I Ning Chou Chi 義寧州志.
  Gazetteer of the I-Ning Region.
  Chhing, +1873.

I Wên Lei Chü, Shi Wu Pu 藝文類聚, 食物部.
  An Encyclopedia of Art and Literature, Section on Food (ch. 72).
  Thang, c. +640.
  Ouyang Hsün 歐陽詢 Chung-Hua, Peking, 1965.

I Shuo 醫説.
  About Medicine.
  Sung, +1224.
  Chang Kao 張杲.

I Ya II 易牙遺意.
  Remnant Notions from *I Ya*.
  Yuan.
  Han I 韓奕.
  Ed. Chhiu Phang-Tung (1984) in the *CKPJKCTK* series.

Jih Chih Lu 日知錄.
  Daily Memoranda.
  Chhing, c. +1670.
  Ku Yen-Wu 顧炎武.

Ju Lin Wai Shih 儒林外史.
  Unofficial History of the Literati.
  Chhing, c. +1750; first printed 1803.
  Wu Ching-Tzu 吳敬梓.
  Tr. Yang Hsien-I & Gladys Yang (1957).
  Textual refs. are to the 1977 edn of Renmin Wenxue, Peking.

Kai yü chhung khao 陔餘叢考.
    Miscellanies beyond the Ledge.
    Chhing, +1790.
    Chao I 趙翼.
    Part of *Ou pei chhuang chi* 甌北全集 (Complete Works of Chao-I).

Khai Pao Pên Tshao 開寶本草.
    Of which the full name is *Khai Pao Hsin Hsiang-Ting Pên Tshao* 開寶新詳定本草.
    New and more detailed Pharmacopoeia of the Khai-Pao reign-period.
    Now known only in excerpts in later pharmacopoeias.

Khang-Hsi Tzu Tien 康熙字典.
    Imperial Dictionary of the Khang-Hsi reign period.
    Chhing, +1716.
    Ed. Chang Yü-Shu 張玉書.

Khuang Miao Chêng Shu 匡謬正俗.
    Corrections of Common Misconceptions.
    Thang, +651.
    Yen Shih-Ku 顏師古.

Ko Chih Ching Yuan 格致鏡原.
    Mirror of Scientific and Technological Origins.
    Chhing, +1717.
    Chhên Yuan-Lung 陳元龍.

Ko Wu Tshu Than 格物麤談.
    Simple Discourses on the Investigation of Things.
    Sung, c. +980.
    Attributed to Su Tung-Pho 蘇東坡.
    Actual writer (Lu) Tsan-Ning (錄)贊寧.
    With later additions, some concerning Su Tung-Pho.

Ku Chin Thu Shu Chi Chêng 古今圖書集成 Cf. *Thu Shu Chi Chêng*.

Ku Shih Khao 古史考.
    Investigation of Ancient History.
    San Kuo, c. 250.
    Chhiao Chou 譙周, now lost, fragments in *Thai Phing Yü Lan* etc.

Kua Shu Shu 瓜蔬書.
    Book of Melons and Vegetables.
    Ming, c. + mid 16th century.
    Wang Shih-Mou 王世懋.

Kuan Tzu 管子.
    The Book of Master Kuan.
    Chou and C/Han; perhaps mainly compiled in the Chi-Hsia Academy (late −4th century). In part from older materials.
    Attributed to Kuang Chung 管仲.

Kuang Chi Chu 觥記注.
    Notes on Wine Vessels.
    Sung.
    Chêng Hsiai 鄭獬.

Kuang Chih 廣志.
    Extensive Records of Remarkable Things.
    Chin, late +4th century.
    Kuo I-Kung 郭義恭.

Kuang-chou Chi 廣州記.
    Record of Kuangtung.
    Chin, +4th century.
    Phei Yüan 裴淵, now lost, fragment in *Thang Lei Han*.

Kuang Chün Phu 廣菌譜.
    Extensive Treatise on Fungi.
    Ming, c. +1550.
    Phan Chih-Heng 潘之恒.

Kuang Chhün Fang Phu 廣群芳譜.
    Monographs on Cultivated Plants, enlarged.
    Chhing, +1708.
    Wang Hao 王灝.

Kuang-tung Hsin Yü 廣東新語.
    New Talks about Kuangtung.
    Chhing, c. +1690.
    Chhü Ta-Chün 屈大均.

Kuang Ya 廣雅.
    Enlargement of the *Erh Ya*.
    San Kuo (Wei), +230.
    Chang I 張揖. *SKCS*, **221**, 425–69.

Kuang Yang Tsa Chi 廣陽雜記.
    Miscellanies about Kuang-Yang.
    Chhing, +1695.
    Liu Hsien-Thing 劉獻廷.

Kuang Yün 廣韻.
    Revision and Enlargement of the *Dictionary of Characters Arranged According to their Sounds when Split* [rhyming phonetic dictionary, based on, and including the *Chhieh Yün* and the *Thang Yün*, q.v.].
    Sung, +1011.
    Chhên Phêng-Nien 陳彭年 & Chhiu Yung 丘雍 et al.

Kuei-Hsin Tsa Shih 癸辛雜識.
    Miscellaneous Information from Kuei-Hsin Street (in Hanchow).
    Sung, late +13th century, perhaps not finished before +1308.
    Chou Mi 周蜜.

Kuei-Hsin Tsa Shih Hsü Chi 癸辛雜識續集.
    First Addendum to 'Miscellaneous Information from Kuei-Hsin Street'.
    Yuan, c. +1298.
    Chou Mi 周蜜.

Kuei-Hsin Tsa Shih Pieh Chi 癸辛雜識別集.
    Final Addendum to 'Miscellaneous Information from Kuei-Hsin Street'.
    Yuan, c. +1298.
    Chou Mi 周蜜.

Kuei Thien Lu 歸田錄.
    On Returning Home.
    Sung, +1067.
    Ouyang Hsiu 歐陽修.

Kui Thien Suo Chi 歸田瑣記.
    Miscellanies on Returning Home.
    Chhing, +1845.
    Liang Chang-Chü 梁章鉅.

Kuo Shu 果書.
    Discourse on Fruits.
    Ming, c. + mid-16th century.
    Wang Shih-Mou 王世懋.

Kuo Yü 國語.
    Discourses on the (ancient) states.

Late Chou, Chhin & C/Han containing material from ancient records.
Writers unknown.

Lan Shih Mi Chhang 蘭室秘藏.
Secrets of the Orchid Greenhouse.
J/Chin, printed +1276.
Li Kao 李杲, Renmin Weisheng, Peking, 1957.

Lao Hsüeh An Pi Chi 老學庵筆記.
Notes from the Hall of Learned Old Age.
Sung, c. +1190.
Lu Yu 陸游.

Lao Lao Hêng Yen: Chu Phu 老老恒言：粥譜.
Remarks of an Elder: Congee Menu.
Chhing, +1750.
Tshao Thing-Tung 曹庭棟.

Li Chi 禮記 [also known as *Hsiao Tai Li Chi*].
Record of Rites (compiled by Tai the Younger).
Ascr. C/Han, c. −70 to −50, but really H/Han, between +80 & +105, though the earliest pieces included may date from the time of Confucius (c. −5th century).
Attr., ed. Tai Shêng 戴聖, but actually ed., by Tshao Pao 曹褒.
Trs. Legge (1899); Couvreur (3); R. Wilhelm (6).
Yin-Tê Index, no. 27.

Li Chih Phu 荔枝譜.
A Treatise on the Litchi (*Nephelium litchi*).
Sung, +1059.
Tshai Hsiang 蔡襄.

Li Chih Phu 荔枝譜.
A Treatise on the Litchi (*Nephelium litchi*).
Ming, c. +1600.
Hsu Po 徐燉.

Li Chih Phu 荔枝譜.
A Treatise on the Litchi (*Nephelium litchi*).
Ming.
Sung Chüeh 宋珏.

Li Wei Han Wen Chia 禮緯含文嘉.
Apocryphal Treatise on Rites; Excellences of Cherished Literature.
Han, author unknown.
In ch. 17 of the *Ku Wei Shu* 古微書 (Minor Works of Antiquity), a Ming compilation, found in the *Shou Shan Ko Tshung Shu* 守山閣叢書, Vol. 18, reprinted 1922 by 上海博古暨齋.

Liang Shu 梁書.
History of the Liang.
Thang, +636.
Yao Ssu-Lien 姚思廉, Chung Hua, Peking, 1973.

Lieh Tzu 列子.
The Book of Master Lieh.
Chou, Warring States?
Attrib. Lieh Yu-Khou 列禦寇, probably compiled in +742 from Chin-Wei material.
Textual refs. are to *Lieh Tzu I Chu* 列子譯注 ed. & annot. by Yen Pei-Min 嚴北溟 & Yen Chieh 嚴捷, Classics Press, Shanghai, 1986.

Ling-Nan Li Chih Phu 嶺南荔枝譜.
Litchis South of the Range.
Chhing, +18th century.
Wu Ying-Ta 吳應達.

Ling Piao Lu I 嶺表錄異.
Strange Southern Ways of Men and Things [Special characteristics and natural history of Kuangtung].
Thang and Wu Tai, between 895 and 915.
Liu Hsün 劉恂.

Liu Mêng Tê Wên Chi 劉夢得文集.
The Complete Works of of Liu Mêng Tê.
Tang.
Liu Yü-Hsi 劉禹錫 (+772–842), *SPTK*.

Liu Shu Ku 六書故.
A History of the Six Graphs [the six principles of formation of the Chinese characters].
Sung, +1275; printed +1320.
Tai Thung 戴桐.

Lo-yang Chieh Lan Chi 洛陽伽藍記.
Description of the Buddhist Temples of Loyang.
N/Wei, c. +530.
Yang Hsüan-Chih 楊衒之.
Tr. Jenner (1981).
Textual refs. are to the *HWTS* edition.

Lü Shih Chhun Chhiu 呂氏春秋.
Master Lü's Spring and Autumn Annals [compendium of natural philosophy].
Chou (Chhin), −239.
Written by a group of scholars gathered by Lü Pu-Wei 呂不韋.
Tr. R. Wilhelm (3); Chung-Fa Index no. 3.
Textual refs. are to the 1985 edn by Lin Phin-Shih.

Lü Shih Chhun Chhiu Pên Wei Phien 呂氏春秋本味篇.
'The Root of Taste' chapter of the *Lü Shih Chhun Chhiu*.
Cf. *Chhiu Phang-Thung et al.* (1983).

Lun Yü 論語.
Conversations and Discourses [of Confucius]; Analects.
Chou (Lu), c. −465 to −450.
Compiled by disciples of Confucius (chs. 16, 17, 18 & 20 are later interpolations).
Tr. Legge (1861a); Lau (1979); Lyall (2); Waley (5); Kung Hung-Ming (1).
Yin-Tê Index, no. (suppl.) 16.

Ma-wang-tui chu chien 馬王堆竹簡.
Ma-wang-tui Inventory (of foods interred in Han tomb No. 1).
In *Hunan shêng po-wu kuan* (1973) 1, pp. 130–55.

Ma-wang-tui Han-mu po shu 馬王堆漢墓帛書（肆）.
Silk Manuscripts from Han tomb at Ma-wang-tui, Vol. 4.
Wen Wu, 1985.

Man'yoshu 万葉集.
A Collection of Early Japanese Poems.
Japan, +759.
Compiler unknown. English trans. by Pierson, J. L (1929–63).

Mao Shih Chêng I 毛詩正義.
The Standard *Book of Odes*.

Compiled by Mao Heng 毛亨, C/Han; annotated by Chêng Hsuan 鄭玄, H/Han and commented on by Khung Ying-Ta 孔穎達, Thang, +642.

Mao Shih Tshao Mu Niao Shou Chhung Yü Shu 毛詩草木鳥獸蟲魚疏.
An Elucidation of the Plants, Trees, Birds, Beasts, insects and Fishes mentioned in the *Book of Odes* edited by Mao (Heng and Mao Chhang).
San Kuo (Wu), +3rd century (c. +245).
Lu Chi 陸璣.

Mao Shih Tshao Mu Niao Shou Chhung Yü Shu Kuang Yao 毛詩草木鳥獸蟲魚疏廣要.
An Elaboration of the Essentials in the 'Elucidation of the Plants, Trees, Birds, Beasts, Insects and Fishes mentioned in the *Book of Odes* edited by Mao (Heng and Mao Chhang).'
Ming, +1639.
Mao Chin 毛晉.

Mêng Chhi Pi Than 夢溪筆談.
Dream Pool Essays.
Sung, +1086, last supplement dated +1091.
Shên Kua 沈括. Ed. Hu Tao-Ching (1956).

Mêng Liang Lu 夢粱錄.
Dreams of the Former Capital [description of Hangchow towards the end of the Sung].
Sung, +1275.
Wu Tzu-Mu 吳自牧.
Textual refs. are to the 1982 edn of Commerce Publishers, Peking.

Mêng Tzu 孟子.
The Book of Mencius.
Chou, c. −290.
Mêng Kho 孟軻.
Tr. Legge (1861b); Lau (1970); Lyall (1).
Yin-Tê Index, no. (suppl.) 17.

Min Chhan Lu I 閩產錄異.
Unusual Products of Fukien.
Chhing, +1886.
Kuo Po-Tshang 郭柏蒼.

Min Chung Hai Tsho Shu 閩中海錯疏.
Seafoods of the Fukien Region.
Ming, +1593.
Thu Pên-Tsun 屠本畯, later enlarged by Hsu Pen 徐勃.

Min Hsiao Chi 閩小記.
Lesser Min Records.
Chhing, +1655.
Chou Liang-Kung 周亮工.

Min Ta Chi 閩大記.
Greater Min Records.
Ming, +1585.
Wang Ying-Shan 王應山.

Min Tsa Chi 閩雜記.
Miscellaneous Notes on Fukien.
Chhing, 1857.
Shih Hung-Pao 施鴻保.

Ming I Pieh Lu 名醫別錄.
Informal Records of Famous Physicians.
Ascr. Liang, c. +510.
Attributed to Thao Hung-Ching 陶弘景.
This work was a disentanglement, made by other hands between +523 and +618 or +656, of the contributions of Li Tang-Chih (c. +225), Wu Phu (c. +235) and the commentaries of Thao Hung-Ching from the text of the *Shên Nung Pên Tshao Ching* itself. In other words it was the non-*Pên Ching* part of the *Pên Tshao Ching Chi Chu* (q.v.). It may or may not have included some of all of Thao Hung-Ching's commentaries. For centuries it has been extant only in quotations in the pharmaceutical histories.
Textual refs. are to the reconstruction by Shang Chih-Chün, Renmin Weisheng, Peking (1986).

Mo Chuang Man Lu 墨莊漫錄.
Random Notes from a Scholar's Cottage.
Southern Sung, c. +1131.
Chang Pang-Chi 張邦基.

Mo Ngo Hsiao Lu 墨娥小錄.
A Secretary's Commonplace Book.
Yuan/Ming, c. 14th century, 1st printed +1571 with Introduction by Wu Chi 吳繼.
Writer unknown, cf. Kuo Chêng-I (1979).

Mo Tzu 墨子.
The Book of Master Mo.
Chou, −4th century.
Mo Ti 墨翟.
Partial trs. Burton Watson (1963); tr. Mei Yi-Pao (1929); Forke (3).
Yin-Tê Index, no. (suppl.) 21.
Textual refs. are to Li Yü-Shu ed., *Mo Tzu ching chu ching I* (1974).

Mōshī Himbutsu Zukō 毛詩品物圖考.
Illustrated Study of the Plants and Animals in the *Book of Odes*.
Japan, +1785.
Subsequently reprinted in China without Katagana readings.
Oka Kōyoku 岡公翼.

Nan Fang Tshao Mu Chuang 南方草木狀.
A Flora of the Southern Region.
Chin, +304.
Chi (or Hsi) Han 稽含.
Textual refs. are to the edn and trans. of Li Hui-Lin (1979).

Nan-Yüeh Chih 南越誌.
Record of the Southern Yüeh.
North & South Dynasties, c. +460.
Shen Huai-Yüan 沈懷遠.

Nei Wai Shang Pien Kan Lun 內外傷辨感論.
Discourses on Internal and External Medicine.
J/Chin, +1231, printed in +1247.
Li Kao 李杲. Renmin Weisheng, 1957, Peking.

Nêng Kai Chai Man Lu 能改齊漫錄.
Random Records from the Corrigible Studio.
Sung, mid-12th century.
Wu Tshêng 吳曾.
Textual refs. are to Wang Jên-Hsiang ed. (1986) in the *CKPJKCTK* series.

Nung Chêng Chhüan Shu 農政全書.
    Complete Treatise on Agriculture.
    Ming, composed +1625 to +1628; printed in +1639.
    Hsü Kuang-Chhi 徐光啟.
    Textual refs. are to Shih Shêng-Han, ed. *Nung Chêng Chhüan Shu Chiao Chu* 農政全書校注 (1979), 3 vols.

Nung Sang Chi Yao 農桑輯要.
    Fundamentals of Agriculture and Sericulture.
    Yuan, +1273. Preface by Wang Phan 王磐.
    Imperially commissioned, and produced by the Agriculture Extension Bureau (Ssu Nung Ssu 司農司).
    Probable editor Mêng Chhi 孟祺.
    Probable later, ed. Chhang Shih-Wên 暢師文 (c. +1286) and Miao Hao-Chhien 苗好謙, +1339.
    Textual refs. are to photo reprint Shanghai Library, 1979.

Nung Sang I Shih Tsho Yao 農桑衣食撮要.
    Selected Essentials of Agriculture, Sericulture, Clothing and Food.
    Yuan, +1314 (again +1330).
    Lu Ming-Shan 魯明善.

Nung Shu 農書.
    Agricultural Treatise.
    Yuan, +1313.
    Wang Chên 王禎.
    Textual refs. are to Wang Yü-Hu, ed. *Wang Chên Nung Shu* (1981).

Oyudononoue no nikki 御湯殿上日記.
    Diaries at Court.
    Japan +1477 to +1637.
    Written by a series of minor court ladies.

Pai Chü-I Chi Chüan 白居易卷.
    Collected Works of Pai Chü-I.
    Thang.
    Pai Chü I (+772 to +846).

Pao Phu Tzu 抱朴子.
    Book of the Preservation-Solidity Master.
    Tsin, early +4th century, probably c. +320.
    Ko Hung 葛洪.
    Tr. Ware, *Alchemy, Medicine and Religion in the China of A.D. 320* (1966), *nei phien* ch. only.

Pao Shêng Yao Lu 寶生要錄.
    Essential Rules for a Healthy Life.
    Sung, date unknown.
    Pu Chhien-Kuan 蒲虔貫, *Shuo Fu*, ch. 84.

Pei Chi Chhien Chin Yao Fang 備急千金要方.
    The Thousand Golden Remedies for Use in Emergencies.
    Cf. *Chhien Chin Yao Fang*, of which it is the full title.

Pei Chhi Shu 北齊書.
    History of the Northern Chhi.
    Tang, c. +640.
    Li Pai-Yao 李百藥.

Pei Hu Lu 北戶錄.
    Record of the Northern Gate.
    Thang, c. +873.
    Tuan Kung-Lu 段公路.

Pei Shan Chiu Ching 北山酒經.
    Wine Canon of North Hill.
    Sung, +1117.
    Chu Kung 朱肱.
    SF, *SKCS* (844).

Pei Yuan Pieh Lu 北苑別錄.
    Pei-yuan Tea Record.
    Southern Sung, +1186.
    Chao Ju-Li 趙汝礪.

Pên Hsin Chai Su Shih Phu 本心齋素食譜.
    Vegetarian Recipes from the Pure Heart Studio.
    Sung/S.
    Chhên Ta-Sou 陳達叟.
    Textual refs. are to Wu Kuo-Tung & Yao Chêng-Chieh eds. (1987) in *CKPJKCTK* series.

Pên Tshao Kang Mu 本草綱目.
    The Great Pharmacopoeia [or Pandects of Natural History].
    Ming, +1596.
    Li Shih-Chên 李時珍.
    Paraphrased and abridged tr. Read and Collaborators (1–7).
    Textual refs. are to the 1975 edn published by Renmin Weisheng, Peking.

Pên Tshao Kang Mu Shih I 本草綱目拾遺.
    Supplemental Amplification of the *Pên Tshao Kang Mu*.
    Chhing, begun c. +1760; first prefaced +1765, prolegomena added 1780, last date in text 1803. First published 1871.
    Chao Hsüeh-Min 趙學敏.
    Textual refs. are to 1971 edn, Comm. Press, Hong Kong.

Pên Tshao Phin Hui Ching Yao 本草品彙精要.
    Essentials of the Pharmacopoeia Ranked according to Nature and Efficacy (Imperially commissioned).
    Ming, +1505.
    Liu Wên-Thai 劉文泰, Wang Phan 王槃 & Kao Thing-Ho 高廷和.
    Textual refs. are to 1982 edn Renmin Weisheng, Peking.

Pên Tshao Shih I 本草拾遺.
    A Supplement for the Pharmaceutical Natural Histories.
    Thang, c. +725.
    Chhên Tshang-Chhi 陳藏器.
    Now extant only in numerous quotations.

Pên Tshao Yen I 本草衍義.
    Dilations upon Pharmaceutical Natural Histories.
    Sung, pref. +1116; pr. +1119; repr. +1185, +1195.
    Khou Tsung-Shih 寇宗奭.

Phêng Lung Yeh Hua 蓬櫳夜話.
    Night Discourses from *Phêng Lung*.
    Ming, +1565.
    Chou Tan-Kuang 周旦光.
    Shuo Fu, vol. 18.

Phien Kho Yü Hsien Chi 片刻餘閒集.
: Notes from Moments of Leisure.
: Chhing, c.+1753.
: Liu Ching 劉靖.

Phin Chha Yao Lu 品茶要錄.
: A Compendium of Elite Teas.
: Northern Sung, +1075.
: Huang Ju 黃儒.

Phu Chi Fang 普濟方.
: Practical Prescriptions for Everyman.
: Ming, +1411.
: Chu Su 朱橚, who is a prince of the Imperial family.
: Renmin Weisheng, 1959, Peking.

Pi Hai Chi Yu 裨海紀游.
: Travels among the Islands of the Sea.
: Chhing, printed +1835.
: Yü Yung-Ho 郁永河.

Pi Shu Lu Hua 避暑錄話.
: Notes from a Summer Retreat.
: Quoted in Hu Shan-Yuan (1939).

Pi Wei Lun 脾胃論.
: Treatise on the Spleen-Stomach Network.
: J/Chin, +1249.
: Li Kao 李杲.

Pien Min Thu Tsuan 便民圖纂.
: Everyman's Handy Illustrated Compendium [or the Farmstead Manual].
: Ming, +1502; repr. +1552, +1593.
: Ed. Khuang Fan 鄺璠.

Po Wu Chih 博物志.
: Records of the Investigation of Things.
: Chin, c. +290.
: Chang Hua 張華.

Pu Chu Hsi Yüan Lu Chi Chêng 補注洗冤錄集證.
: The 'Washing of Wrongs' with Annotations and Amendments.
: Chhing, +1796.
: Wang Yu-Huai 王又槐 & Yuan Chhi-Hsin 沅其新 eds.

Ryō-no-gige 令義解.
: Detailed Regulations for the working of Ryo.
: Japan, +833.
: Kiyohara-no-Natuna *et al.* 清源夏野等.

Sandai-jituroku 三代實錄.
: Historical Records of Three Generations.
: Japan, +908.
: Hujiwara-no-tokihira *et al.* 藤原時平等.

San Fu Chüeh Lu 三輔決錄.
: A Considered Account of the Three Metropolitan Cities [Chhang-an, Feng-i and Fu-feng].
: H/Han, +153.
: Chao Chhi 趙岐.

San Kuo Chih 三國志.
: History of the Three Kingdoms [+220 to +280].
: Chin, c. +290.
: Chhên Shou 陳壽.
: Yin-Tê Index, no. 33.

San Tshai Thu Hui 三才圖會.
: Universal Encyclopedia.
: Ming, +1609.
: Wang Chhi 王圻.

Shan Chia Chhing Kung 山家清供.
: Basic Needs for Rustic Living.
: Sung/S.
: Lin Hung 林洪, ed. Wu Kê (1985), *CKPJKCTK* series.

Shan Fu Ching 膳夫經 [or *Shan Fu Ching Shou Lu* 膳夫經手錄].
: The Chef's Book.
: Thang, c. +857.
: Yang Yeh 楊曄, Shinoda & Seiichi eds. (1972), 113–16.

Shan Fu Lu 膳夫錄.
: The Chef's Manual.
: Sung/S.
: Chêng Wang 鄭望.
: Ed. by Tang Kên (1987) in the *CKPJKCTK* series.

Shan Hai Ching 山海經.
: Classic of the Mountains and Waters.
: Chou and C/Han.
: Writers unknown.
: Chung-Fa Index no. 9.

Shan Hai Ching Chiao Tsu 山海經校注.
: Ed. Yuan Kho 袁珂.
: Classics Press, Shanghai, *1980*.

Shang Lin Fu 上林賦.
: Ode to the Upper Forest.
: C/Han, c. −150.
: Ssuma Hsiang-ju 司馬相如.

Shên Nung Pên Tshao Ching 神農本草經.
: Classical Pharmacopoeia of the Heavenly Husbandman.
: C/Han, based on Chou and Chhin material, but not reaching final form until +2nd century.
: Writers unknown.
: Long lost as separate work, but the basis of all subsequent compendia of pharmaceutical natural history, in which it is constantly quoted. Reconstructed and annotated by many scholars. Textual refs. are to the edn by Tshao Yüan-Yü (1987).

Shêng Chi Tsung Lu (Shang, Hsia) 聖濟總錄.
: Imperial Medical Encyclopedia, I, II.
: Sung, c. +1115.
: Ed. Chao Chi 趙佶 (Emperor Hui Tsung), Shên Fu 申甫 *et al.*
: Renmin Weisheng, 1962, Peking.

Shih Chên Lu 食珍錄.
: Menu of Delectables.
: Sung/Northern.
: Yü Tshung 虞悰.
: Ed. Thang Kên (1987) in the *CKPJKCTK* series.

Shih Chi 史記.
: Records of the Historian.
: C/Han, c. −90 (first printed c. +1000)
: Ssuma Chhien 司馬遷, and his father Ssuma Than 司馬談.
: Partial trans. Yang & Yang (1979), Chavannes (1), Pfizmaier (13–36), Swann (1950) etc.
: Yin-Tê Index, no. 40.

Shih Ching 詩經.
    Book of Odes [ancient folksongs].
    Chou, −11th to −7th century.
    Writers and compilers unknown.
    Tr. Legge (1871); Waley (1937); Karlgren (1950).

Shih Ching 食經.
    Food Canon.
    Sui, c. +600.
    Hsien Fêng 謝諷, preserved in *Chhing I Lu*.

Shih Ching Pai Shu 詩經稗疏.
    Little Commentaries on *The Book of Odes*.
    Chhing, +1695.
    Wang Fu-Chih 王夫之.

Shih Hsien Hung Mi 食憲鴻秘.
    Guide to the Mysteries of Cuisine.
    Chhing, c. +1680.
    Attrib. To Wang Shih-Chên 王士禎, but more probably Chu I-Tsun 朱彝尊.
    Textual refs. are to Chhiu Phang-Thung ed. (1985) in the *CKPJKCTK* series.

Shih I Hsin Chien 食醫心鑑.
    Candid Views of a Nutritionist–Physician.
    Thang, Late.
    Tsan Yin 咎殷.

Shih Liao Pên Tshao 食療本草.
    Pandects of Diet Therapy [Materia Dietetica].
    Thang, c. +670.
    Mêng Shên 孟詵.
    Textual refs. are to the 1976 edn by Fan Feng-Yuan, Hsin Wen Feng Press, Taipei; and also to the Renmin Weisheng edition of 1984.

Shih Lin Kuang Chi 事林廣記.
    Record of Miscellanies.
    Southern Sung to Yuan, c. +1280.
    Chhên Yuan-Ching 陳元靚.
    In Shinoda & Tanaka (1973) I, 257–71.

Shih Liu Thang Phin 十六湯品.
    Sixteen Tea Decoctions.
    Thang, +900.
    Su I 蘇廙.

Shih Ming 釋名.
    Expositor of Names.
    H/Han, early +2nd century.
    Liu Hsi 劉熙, *HWTS*.

Shih Pên 世本.
    Book of Origins [imperial genealogies, family names and legendary inventors].
    C/Han (incorporating Chou material), −2nd century.
    Ed. Sung Chung (H/Han) 宋衷, *HWTS*.

Shih Pên Pa Chung 世本八種.
    Eight Types of Origins.
    Chhing.
    Ed. Chhin Chia-Mo 秦嘉謨.
    Comm. Press, Shanghai, 1957.

Shih Phu 食譜.
    List of Victuals.
    Thang, c. +710.
    Wei Chü-Yüan 韋巨源.

Shih Shih Wu Kuan 食時五觀.
    Five Aspects of Nutrition.
    Sung, +1090.
    Huang Thing-Chien 黃庭堅.
    Ed. Thang Kên 唐艮.

Shih Shuo Hsin Yü 世說新語.
    New Discourse on the Talk of the Times. [Notes on minor incidents from Han to Chin].
    Liu/Sung, c. +440, ed. +510.
    Liu I-Chhing 劉義慶, commentary by Liu Chün 劉峻.

Shih Wu Chi Yüan 事物紀原.
    The Origin of Events and Things.
    Northern Sung, c. +1080.
    Kao Chhêng 高承. In 惜陰軒叢書.

Shih Wu Pên Tshao 食物本草.
    Natural History of Foods [Materia Dietetica].
    Ming, +1641 (repr. from an earlier versions)
    Attributed to Li Kao 李杲 (J/Chin) or Wang Ying 王穎 (Ming); pref. by Li Shih-Chên, ed. Yao Kho-Chêng 姚可成.
    Textual refs. are to 1989 edn of Chêng Chin-Shêng *et al*.

Shih Wu Pên Tshao Hui Tsuan 食物本草會纂.
    New Compilation of the *Natural History of Foods*.
    Chhing, +1691.
    Shên Li-Lung 沈李龍.

Shōryōshū 性靈集.
    Japan, +811.
    Kŭkai 空海.

Shou Chhing Yang Lao Hsin Shu 授親養老新書.
    New Handbook on Care of the Elderly.
    Northern Sung, +1080; enlarged, Yuan, +1307.
    Original compiler, Chhen Chih 陳直; enlarged by Tsou Hsuan 鄒鉉.
    Chungkuo, Peking, 1986.

Shou Shih Thung Khao 授時通考.
    Compendium of Works and Days.
    Chhing, +1742.
    Compiled by imperial order under the direction of O-erh-thai 鄂爾泰.
    Textual refs. are all to the 1847 reprint of the original 1742 Palace edn.

Shu Ching 書經.
    Book of Documents [Historical Classic].
    The 29 'Chin Wen' chapters mainly Chou a few pieces possibly Shang; the 21 'Ku Wen' chapters a forgery by Mei Tsê 梅賾, c. +320, using fragments of genuine antiquity. Of the former, 13 are considered to go back to the −10th century, 10 to the −8th century and 6 not before the −5th century. Some scholars accept only 16 or 17 as pre-Confucian.
    Writers unknown.
    Tr. Medhurst (1846); Legge (1865b); Old (1904).

Shu Yüan Tsa Chi 菽園雜記.
    Miscellanies from the Soybean Garden.
    Ming, +15th century,
    Lu Yung 陸容 (+1436 to +1494).

Shuang Chhi Chi 雙溪集.
    Twin Rivulet Collection. Southern Sung, c. +1200.
    Wang Yen 王炎.

Shui Ching Chu 水經注.
    Commentary on the *Waterways Classic*.
    N/Wei, late +5th or early +6th century.
    Li Taao-Yuan 酈道元.

Shui Hu Chuan 水滸傳.
    The Water Margin.
    Ming, early.
    Shih Nai-An 施耐安.
    Tr. Buck, Shapiro etc. (1981).

Shun Thien Fu Chih 順天府志.
    Gazetteer of the *Shun-Thien* Prefecture.
    Chhing, 1884.
    Chou Chia-Mei 周家楣 *et al.*

Shuo Fu 說郛.
    Florilegium of Literature.
    Yuan, +1368.
    Ed., Thao Tsung-I 陶宗儀.

Shuo Wên.
    Cf. *Shuo Wên Chieh Tzu*

Shuo Wên Chieh Tzu 說文解字.
    Analytical Dictionary of Characters.
    H/Han, +121.
    Hsü Shên 許慎.
    Textual refs. are to the 1963 edn by Chung Hua, Peking.

Shuo Wên Chieh Tzu Yün Phu 說文解字韻譜.
    Rhyme Lists for the *Shuo Wên Chieh Tzu*.
    Chhing, c. +10th century.
    Hsü Hsüan 徐鉉 (+917 to +992).

Shuo Yuan 說苑.
    Garden of Discourses.
    Han, c. −20.
    Liu Hsiang 劉向.

Ssu Khu Chhüan Shu 四庫全書.
    Complete Library of the Four Categories.
    Chhing, +1782.
    Compiled by Yung Jung 永瑢 *et al.*
    Wên Yan Ge 文淵閣 version, reprinted Commercial Press, Taipei, 1986.

Ssu Khu Chhüan Shu Tsung Mu Thi Yao.
    Cf. *Chhin Ting Ssu Khu Chhüan Shu Tsung Mu Thi Yao.*

Ssu Min Yüeh Ling 四民月令.
    Monthly Ordinances for the Four Peoples (Scholars, farmers, artisans and merchants).
    H/Han, c. +160.
    Tshui Shih 崔寔.
    Textual refs. are to the 1981 edn by Miao Chhi-Yü.

Ssu Shih Tsuan Yao 四時纂要.
    Important Rules for the Four Seasons.
    Thang, probably late Thang.
    Han E 韓鄂.
    Textual refs. are to Miao Chhi-Yü edn (*1981*), *Ssu Shih Tsuan Yao Chiao Shih.*

Su Hsüeh Shih Chi 蘇學士集.
    The Collected Works of Su Shun-Chhing.
    Northern Sung, +1008 to +1048.
    Su Shun-Chhing 蘇舜欽, Chung Hua 1961 edn.

Su Shih Shuo Luo 素食說略.
    Cf. Hsüeh Pao-Chhen (1900).

Su Wên Chung Kung Ssu Pien Chu Chi Chhêng 蘇文忠公詩編注集成.
    Collected Poems of Su Shih (Tung-Pho).
    Ed., Wang Wên-Hao 王文浩, Chhing Dynasty.

Su Yung Chi 肅雝集.
    Collection from the Quiescent Marsh.
    Yuan, c. +1350.
    Chêng Yün-Tuan 鄭允端.

Sui Chien Lu 隨見錄.
    Record of Second Encounters.
    Chhing, c. +1734.
    Author unknown, no longer extant, cited in *Hsü Chha Ching*.

Sui Shih Kuang Chi 歲時廣記.
    Expanded Records of the New Year Season.
    Southern Sung.
    Chhên Yuan-Ching 陳元觀.

Sui Shu 隋書.
    History of the Sui Dynasty (+581 to +617).
    Thang, +636 (annals and biographies), +656 (monographs and bibliography).
    Wei Chêng 魏徵 *et al.*
    For translations of passages cf. the index of Franke (1).

Sui Yüan Shih Tan 隨園食單.
    Recipes from the Sui Garden.
    Chhing, +1790.
    Yüan Mei 袁枚.
    Textual refs. are to Chou San-Chin *et al.* eds. (1984), *CKPJKCTS.*

Sun Phu 筍譜.
    Treatise on Bamboo Shoots.
    Sung, c. +970.
    Tsan Ning 贊寧.

Sung Shih Tsun Shêng pu 宋氏尊生部.
    Sung Family's Guide to Living. [Part of *Chu yü shan fang tsa pu* 竹嶼山房雜部 'Miscellanies of the Bamboo Islet Studio.'].
    Ming, +1504.
    Sung Hsü 宋詡, *SKCS*, **871**, 282–328; or Shinoda & Seiichi eds. (*1973*), 452–86.

Sung Shih Yang Shêng pu 宋氏養生部.
    Sung Family Guide to Nourishing Life [Part of *Chu yü shan fang tsa pu* 竹嶼山房雜部 'Miscellanies of the Bamboo Islet Studio.'].
    Sung Hsü 宋詡, *SKCS*, **871**, 116–210.

Ta Kuan Chha Lun 大觀茶論.
    The Imperial Tea Book (Discourse on Tea of the *Ta Kuan* reign period, +1107–10).
    Sung, c. +1109.
    Chao Chi 趙佶 (Emperor Hui Tsung of the Sung).

Ta Kuang Yi Hui Yü Phien 大廣益會玉篇.
    Enlargement of the *Yü Phien* [cf. Yü Phien].
    Thang, enlargement by Sun Chhiang 孫強.
    Sung, re-edited by Chhên Phêng-Nien 陳彭年 *et al.*
    Peking, 1983, reprinted as *Sung Pên Yü Pien* 宋本玉篇.

Ta Tai Li Chi 大戴禮記.
    Record of Rites [compiled by Tai the Elder] (cf. *Li Chi, Hsiao Tai Li Chi*).
    Ascr. C/Han, c. −70 to −50, but really H/Han, +80 to +105.

BIBLIOGRAPHY A

Attrib. ed. Tai Tê 戴德, in fact, probably ed. Tshao Pao 曹褒.
Tr. Douglas (1); R. Wilhelm (6).

Ta Tsao Phu 打棗譜.
Monograph on Jujubes.
Yuan, early 14th century.
By Liu Kuan 柳貫.

Ta Yeh Shi I Chi 大業拾遺記 [also known as *Ta Yeh Tsa Chi* 大業雜記].
Miscellaneous Records of the Ta-Yeh Reign-Period (Sui +605 to +617).
Thang, c. +660 (as a supplement to the *Sui Shu*).
Tu Pao 杜寶.
Originally 10 chs., only one remains, in *TSCC*.

Tan Chhi Pu I 丹溪補遺.
Backup notions from the Elixir Brook.

Tan Fang Hsü Chih 丹房須知.
Indispensable Knowledge of the Chymical Elaboratory [with illustrations of apparatus].
Sung, +1163.
Wu Wu 吳悟, TT/893.

Tao Tê Ching 道德經.
The Canon of Reason and Virtue.
Chou, c. −5th century.
Attributed to Li Erh 李耳, better known as Lao Tzu 老子.
Numerous English translations.

Tao Tsang 道藏.
The Taoist Patrology [containing 1464 Taoist works].
All periods, but first collected in the Thang about +730, then again about +870 and definitively in +1019. First printed in the Sung (c. +1111 to +1117). Also printed in J/Chin (+1168 to +1191), Yuan (+1244), and Ming (+1445, 1598 and 1607).
Writers numerous. Index by Wieger (6), on which cf. Pelliot's review (58).
Yin-Tê Index, no. 25.

Thai Phing Huan Yü Chi 太平寰宇記.
Geographical Record of the Thai-Phing Reign-Period.
Sung, +976–83.
Yüeh Shi 樂史.

Thai Phing Huei Min Ho Chi Chü Fang 太平惠民和劑局方.
Standard Formularies of the Government Pharmacies in the Thai-Phing Reign-Period.
Sung, +1151.
Chhên Chhêng 陳承 *et al*. Of the *Thai phing huei min ho chi chü* 太平惠民和劑局 (Committee for Standard Formularies).

Thai Phing Kuang Chi 太平廣記.
Copious Collection by Imperial Solicitude in the Thai-Phing Reign-Period [anecdotes, stories, mirabilia and memorabilia].
Sung, +978.
Ed. Li Fang 李昉.

Thai Phing Shêng Huei Fang 太平聖惠方.
Beneficent Prescriptions of the Thai-Phing Reign-Period.
Sung, +992.
Wang Huai-Yin 王懷隱.

Thai Phing Yü Lan 太平御覽.
Imperial Encyclopedia of the Thai-Phing Reign-Period (Daily Readings for the Emperor of the Thai-Phing reign).
Sung, +983.
Ed. Li Fang 李昉.
Some chs. tr. Fitzmaier (84–106).
The Food and Drink sections have been edited and annotated by Wang Jen-Hsiang and published (1993) in the *CKPJKCTK* series, under the title *Thai Phing Yü Lan: Yin-shih pu*.

Thang Lei Han 唐類函.
A Thang Compendium.
Ming, +1618.
Yü An-Chhi 俞安期.

Thang Shih San Pai Shou 唐詩三百首.
Three Hundred Thang Poems.
Chhing, c. +1764.
Selected by Sun I 孫洙.

Thang Yü Lin 唐語林.
Forest of Mini-Discourses.
Sung, Northern.
Wang Tang 王讜.

Thang Yün 唐韻.
Thang Dictionary of Characters Arranged According to their Sounds [rhyming phonetic dictionary based on, and including, the *Chhieh Yün* q.v.].
Thang, +677, revised and republished +751.

Thang Yün Chêng 唐韻正.
Thang Rhyming Sounds.
Chhing, +1667.
Ku Yen-Wu 顧炎武.

Thiao Chhi Yü Yin 苕溪漁隱.
The Hermit Fisherman of *Thhiao Chhi*.
Sung, +1147.
Hu Tzu 胡仔.

Thiao Ting Chi 調鼎集.
The Harmonious Cauldron.
Chhing, +1760 to 1860.
Compiler unknown; MS in Peking Library; first time printed in 1986 edition by Hsing Po-Thao, *CKPJKCTS* series.

Thieh Wei Shan Tshung Than 鐵圍山叢談.
Collected Conversations at Iron-Fence Mountain.
Sung, +1115.
Tshai Thao 蔡絛.

Thien Kung Khai Wu 天工開物.
The Exploitation of the Works of Nature.
Ming, +1637.
Sung Ying-Hsing 宋應星.
Tr. Sun & Sun (1966) and Li Chhiao-Phing *et al*. (1980).

Thou Huang Tsa Lu 投荒雜錄.
Miscellaneous Jottings far from Home (lit. Records of one cast out in the wilderness).
Thang, c. +835.
Fang Chhien-Li 房千里.

Thu Shu Chi Chhêng 圖書集成.
Imperial Encyclopedia.
Chhing, +1726.

Index by L. Giles (1911), Chengtu 1985 edn.
Ed. Chhên Mêng-Lei 陳夢雷.

Thung Chih 通志.
Historical Collections.
Sung, c. +1150.
Chêng Chiao 鄭樵.

Thung Chih Lüeh 通志略.
Compendium of Information [part of *Thung Chih*, q.v.].

Thung Chün Yao Lu 桐君葯錄.
Master Thung's Herbal.
Probably Han, date unknown.
Sui Chih 隋志, now lost, fragment in *TPYL*, ch. 867.

Thung Su Pien 通俗編.
Origin of Common Expressions.
Chhing, +1751.
Chai Hao 翟灝, Commercial Press (*1959*).

Thung Yueh 僮約.
Contract with a Servant.
Han, −59.
Wang Pao 王褒. Reproduced in many collections, e.g. Chhen & Chu eds. (1981), pp. 202–3. Tr. Hsü Cho-Yun (1980), pp. 231–34.

Tou Kêng Fu 豆羹賦.
Ode to the Soybean Stew.
Chin, c. +3rd to 4th century.
Chang Han 張翰.

Tsêng Chhêng Li-chih Phu 增城荔枝譜.
Monograph on the Litchi of *Tsêng Chhêng* (in Kuangtung).
Sung, +1076.
Chang Tsung-Min 張宗閔.

Tshao Mu Tzu 草木子.
Herbs and Plants.
Ming, +1378.
Yeh Tzu-Chhi 葉子奇.

Tshung Hsün Chhi Yü 聰訓齊語.
Didactic Notes.
Chhing, +1697.
Chang Ying 張英.

Tso Chuan 左傳.
Master Tso-chhiu's Tradition of the *Chhun Chhiu* (Spring & Autumn Annals) [dealing with the period −722 to −453].
Late Chou, compiled from ancient written and oral traditions of several states between −430 and −250, but with additions and changes by Confucian scholars of the Chhin and Han, especially Liu Hsin. Greatest of the three commentaries on the *Chhun Chhiu*, the others being the *Kungyang Chuan* and the *Kuliang Chuan*, but unlike them, probably itself originally an independent book of history.
Attributed to Tshchhiu Ming 左邱明.
Tr. Couvreur (1); Legge ( ); Pfizmaier (1–12).
Textual refs. are to the *Chhun Chhiu Ching Chuan Yin Tê*. ed. by Hung Yeh et al. (*1983*).

Tsun Shêng Pa Chien 遵生八牋.
Eight Disquisitions for Healthy Living.
Ming, +1591.
Kao Lien 高濂.

The part dealing with food and drink (chs. 11–13) is published separately under the title *Yin Chuan Fu Shih Chien* (q.v.) in the *CKPJKCTS* series.

Tu Chhêng Chi Shêng 都城紀勝.
Wonders of the Capital [Hanchow].
Sung, +1235.
Mr Chao 趙氏 [Kuan Phu Nai Te Weng 灌圃耐得翁; the old gentleman of the water garden who achieved success through forebearance.]
Textual refs. are to the 1982 edn, Commerce Publishers Peking, *Tung-Ching Mêng Hua Lu*, etc.

Tung Ching Mêng Hua Lu 東京夢華錄.
Dreams of the Glories of the Eastern Capital [Khaifeng].
Northern Sung, +1148 (referring to the two decades which ended with the fall of the capital of the Northern Sung in +1126 and the completion of the move to Hangchow in +1135); first printed +1187.
Mêng yuan-Lao 孟元老.
Textual refs. are to the 1982 edn of Commercial Publishers together with four other titles of Sung reminiscence literature.

Tung Hsi Shih Chha Lu 東溪試茶錄.
Tea in Chien-an, Fukien.
Northern Sung, +1064.
Sun Tzu-An 孫子安.

Tung Kuan Han Chi 東觀漢記.
Eastern View of the Records of Han.
H/Han.
Started by Pan Ku 班固, continued by Liu Chên 劉珍 et al., completed c. +270.

Tuo Nêng Pi Shih 多能鄙事.
Routine Chores made Easy.
Ming, c. +1370.
Liu Chi 劉基.
In Shinoda & Tanaka (1973).

Tzu Lin 字林.
Forest of Characters.
Tsin, c. 4th century.
Lü Chhên 呂枕.
Preserved in *Tzu Lin Khao I* 字林.

Tzu Lin Khao I 字林考逸.
Investigation of Characters.
Chhing, c. +1780.
Jên Ta-Chhun 任大椿.

Utuho-monogatari 宇津保物語.
The History of Utuho.
Japan, +911 to +983.
Attributed to Minamoto-n-shitagō 源順.

Wai Thai Mi Yao 外台秘要.
Secret Formulary from the Outer Terrace.
Thang, +752.
Wang Thao 王燾.

Wakan Sansaizue 和漢三才圖會.
Collection of Japanese and Chinese Drawings of all Things.
Japan, +1711.
Terashima Ryoan.

Wamyō Ruizyūshō 和名類聚抄.
    General Encyclopedic Dictionary.
    Japan (Heian), +934.
    Minamoto no Shitagō 源順.
Wamyōshō 和名抄.
    Cf. *Wamyō Ruijūshō*.
Wang Tshao Thang Chha Shuo 王草堂茶説.
    Talks on Tea from the 'Wang' Grass Pavilion.
    Chhing, +1734.
    Author unknown, fragments found in the *Hsü Chha Ching*.
Wei Shêng Pao Chien 衛生寶鑑.
    Essential Precepts of Hygiene.
    Yuan.
    Lo Thien-I 羅天益, Renmin Weisheng, Peking, 1983.
Wei Shu 魏書.
    History of the Northern Wei Dynasty.
    N/Chhi, +554, revised +572.
    Wei Shou 魏收.
Wei Wang Tshao Mu Chih 魏王花木志.
    Flora of the King of Wei.
    Northern Wei, c. +515.
    Compiler unknown; only fragments have survived in other compilations.
Wen Hsüan 文選.
    Literary Selections.
    N&S Dynasties, c. +530.
    Liang Su-Thung 梁蕭統.
    Includes the poem '*Chhi-Fa*'.
Wu Chhao Hsiao Shuo 五朝小説.
    Collection of Small Talks from the Five Dynasties.
    Ming, date unknown.
    Compiler unknown, Ming print (University of Chicago Library).
Wu Chün Chi 吳郡志.
    Gazetteer of the Wu Region.
    Sung, +1192.
    Fan Chheng-Ta 范成大, re-ed. +1229.
Wu Chün Phu 吳菌譜.
    Mushrooms in the Wu Region.
    Chhing, +1683.
    Wu Lin 吳林.
Wu I Tsa Chi 武夷雜記.
    Random Notes on Wu-I.
    Ming, date not known.
    Wu Shih 吳拭.
Wu Lei Hsiang Kan Chih 物類相感志.
    On the Mutual Response of Things according to their Categories.
    Sung, c. +980.
    Attrib. wrongly to Su Tung-Pho; actual writer the monk Lu Tsang-Ning 錄贊寧.
Wu Li Hsiao Shih 物理小識.
    Mini-encyclopedia of the Principles of Things.
    Ming and Chhing, finished by +1643, sent to his son Fang Chung-Thung in +1650, printed +1664.
    Fang I-Chih 方以智.
Wu Shih Erh Ping Fang 五十二病方.
    Prescriptions for Fifty-Two Ailments.
    Chhin or C/Han, c. −200.
    Compiler unknown; discovered in Han Tomb No. 3 at Ma-wang-tui.
    Included in *Ma-wang-tui Han Mu Po Shu* (Silk Manuscripts from Han tombs at Ma-wang-tui), Vol. 4, Wen Wu, Peking, 1985. Tr. Harper (1982).
Wu shih Chung Khuei Lu 吳氏中饋錄.
    Cf. *Chung Khuei Lu* 中饋錄.
Wu Shih Pên Tshao 吳氏本草 also known as *Wu Phu Pên Tshao* 吳普本草.
    Wu's Pharmaceutical Natural History.
    San Kuo, Wei, c. +240.
    Wu Phu, 吳普.
    Textual refs. are to the reconstruction by Shang Chih-Chün et al. (1987), Renmin Weisheng, Peking.
Wu Yüan 物原.
    The Origin of Things.
    Ming, +15th century.
    Lo Chhi 羅頎.

Yang Hsiao Lu 養小錄.
    Guide to Nurturing Life.
    Chhing, +1698.
    Ku Chung 顧仲.
    Textual refs. are to Chhiu Phang-Thung tr. (1984), *CKPJKCTS*.
Yang Shêng Lun 養生論.
    Discourse on Nurturing Life.
    San-Kuo (Wei), 3rd century.
    Hsi Khang 稽康.
Yang Yü Ching 養魚經.
    Manual of Fish Culture.
    Chou, −5th century, author unknown.
    Attrib. to Thao Chu-Kung 陶朱公, i.e. Fan Li 范蠡.
Yang Yü Ching 養魚經 also known as *Chung Yü Ching* 種魚經.
    Monograph on Rearing Fish.
    Ming, c. +mid 16th century.
    Huang Shêng-Tshêng 黃省曾.
Yeh Chung Chi 鄴中記.
    Record of the Yeh Capital.
    Chin, Eastern.
    Lu Hui 陸翽, in *HWTS*.
Yeh Tshai Chien 野菜箋.
    Notes on Edible Wild Plants.
    Ming, c. +1600.
    Thu Pên Chun 屠本畯.
Yeh Tshai Po Lu 野菜博錄.
    Comprehensive Accounts of Wild Edible Plants.
    Ming, +1622.
    Pao Shan 鮑山.
Yen Thieh Lun 鹽鐵論.
    Discourses on Salt and Iron [record of the debate of −81 on state control of commerce and industry].
    C/Han, c. −80 to −60.
    Huan Khuan 桓寬.
Yen Tzu Chhun Chhiu 晏子春秋.
    Master Yen's Spring and Autumn Annals.

Chou, Chhin or C/Han proceeding from oral tradition but not stabilised before the −4th century.
Attributed to Yen Ying 晏嬰, but, in fact, a collection of stories about him.

Yin Chuan Fu Shih Chien 飲饌服食牋.
Compendium of Food and Drink.
Ming, +1591.
Kao Lien 高濂.
Textual refs. are to the 1985 edn by Thao Wên-Thai in *CKPJKCTS* series.

Yin Hai Ching Wei 銀海精微.
Essentials of the Silver Sea. [Treatise on Ophthalmology].
Sung, ascribed to Sun Ssu-Mou.
Author unknown repr. (*1954*) Ching Chang Book Co., Shanghai.

Yin Shan Chêng Yao 飲膳正要.
Principles of Correct Diet.
Yuan, +1330, reissued by imperial order in +1456.
Hu Ssu-Hui 忽思慧.
Cf. *CKPJKCTK* edn of 1988, annotated by Li Chhun-Fang; Peking. Textual refs. are to the Everyman Library edn, Commercial Press, Taipei, 1981.

Yin Shih Hsü Chih 飲食須知.
Essentials of Food and Drink.
Yuan/Ming, +1350.
Chia Ming 賈銘.

Yu Huan Chi Wên 游宦紀文.
Things Seen and Heard on Official Travel.
Sung, +1233.
Chang Shih-Nan 張世南.

Yu-Yang Tsa Tsu 酉陽雜俎.
Miscellany of the *Yu-Yang* Mountain cave [in S.E. Szechuan].
Thang, c. +860.
Tuan Chhêng-Shih 段成式.
See des Rotours (1), p. civ.

Yü Ching 芋經.
The Book of Taro.
Ming, c. +mid 16th century.
Huang Shêng-Tshêng 黃省曾.

Yü Han Shan Fang Chi I Shu 玉函山房集佚書.
Jade Box Mountain Studio Collection of Lost Books.
Chhing, 1853.
Ed. Ma Kuo-Han 馬國翰.

Yü Phien 玉篇.
The Jade Page Dictionary (Expanded *Shuo Wên Chieh Tzu*).
Liang, +543.
Ku Yeh-Wang 顧野王 cf. *Ta Kuang Yi Hui Yü Phien*.

Yü Phin 魚品.
Types of Edible Fish.
Ming, +16th century.
Tun Yuan Chu Shih 豚園居士.

Yü Shih Phi 玉食批.
The Imperial Food List.
Southern Sung.
Ssu Shan Nei Jên 司膳內人 (Identity unknown).
Ed. Thang Kên (1987), *CKPJKCTK* series.

Yuan Shih 元史.
History of the Yuan Dynasty [+1271 to +1368].
Ming, c. +1370.
Sung Lien 宋濂 et al. eds.
Yin-Tê Index, no. 35.

Yüan Chien Lei Han 淵鑑類函.
Mirror of the Infinite, a Classified Treasury [great encyclopaedia; the conflation of 4 Thang and 17 other encyclopaedias].
Chhing, presented +1701, printed +1710.
Ed. Chang Ying 張英 et al.

Yüeh Hsi Ngo Chi 粵西偶記.
Incidental Notes from Western Kuangtung
Chhing, +1705.
Liu Tsu-Fan 劉祚蕃.

Yung-Lo Ta Tien, I Hsüeh Chi 永樂大典, 醫學集.
The Great Collectanea of the Yung-Lo Reign, Medical Section.
Ming, +1408 by Imperial order.
Ed. Yao Kuang-Hsiao 姚廣孝, Chieh Chin 解縉 et al.
Renmin Weisheng, Peking, 1986.

Yün Lin Thang Yin-shih Chih-tu Chi 雲林堂飲食制度集.
Dietary System of the Cloud Forest Studio.
Yuan, +1360.
Ni Tsan 倪瓚.

B. CHINESE AND JAPANESE BOOKS AND JOURNAL ARTICLES SINCE 1800

Abe Koryu 阿部孤柳 &Tsuji Shigemitsu 辻重光 (*1974*).
　Tôfu no Hon 豆腐の本.
　The Book of Tofu.
　Shibata Shoten, Tokyo.
Adachi Iwao 足立嚴 (*1975*).
　Tabemono-denraisi たべもの伝來史.
　History of Food Transmissions [into Japan].
　Shibata Shoten, Tokyo.
Ah Ying 阿英 (*1990*).
　Man than hun thun 漫談餛飩.
　Random Talks on Wonton.
　CKPJ, **1990** (9), 9–11.
Aida Hiroshi, Ueda Seinosuke, Murata Marehisa & Watnabe Tadao, eds. (*1986*) 相田浩，上田誠之助，村田希久，渡辺忠雄.
　Ajia no muenhakkao daizushokuhin アジアの無鹽發酵大豆.
　Asian Symposium on Non-salted Soybean Fermentation.
　STEP, Tsukuba. Articles are both in Japanese and English.
Amano Motonosuke 天野元之助 (*1962*).
　Chūgoku nōgyōshi kenkyu 中國農業史研究.
　Researches into Chinese Agricultural History.
　Tokyo; 2nd expanded edn 1979.
An Chin-Huai 安金槐 (*1992*).
　Chung-kuo khao ku 中國考古.
　The Archaeology of China.
　Classics Press, Shanghai.
An Chin-Huai 安金槐 & Wang Yü-Kang 王興剛 (*1972*).
　Mi-hsien Ta-hu-thing Han-tai hua-hsiang shi-mu ho pi-hua mu 密縣打虎亭漢代畫像石墓和壁畫墓.
　Murals and Stone Engravings at Han tombs in Ta-hu-thing, Mi-hsien.
　WW, **1972** (10), 49–62.
Ang Tien-Sê (Hung Thien-Ssu) 洪天賜 (*1983*).
　Tou fu khao yuan 豆腐考源.
　On the Origin of Bean Curd.
　Papers on Chinese Studies (University of Malaysia), **1983** (2), 29–39.
Anhui shêng wên-wu kuan-li wei yuan hui 安徽省文物管理委員會 (*1959*).
　Ting-yuan hsien pa-wang-chuang ku-hua-hsiang shih-mu 定遠縣爐王莊古畫像石墓.
　Murals at a Stone Tomb at Pa-wang-chuang in Ting-yuan County.
　WW, **1959** (12), 43–6.
Chang Chi-Yuan 張集元 (*1989*).
　Chhung ku-ching li-shih khan liang-shih chia-kung ti fa-chang 從古今糧食看糧食加工的發展.
　The Development of Food Processing as seen from the Practice of Eating Granules.
　NYKK, **1989** (1), 281–3.

Chang Chung-Ko 張仲葛 (*1986*).
　Chung-kuo yang chi chien shih 中國養雞簡史.
　Brief History of Chicken Rearing in China.
　NYKK, **1986** (2), 279–82.
Chang Hou-Yung 張厚埔 (*1982*).
　Chhin-tu Hsien-yang i-hao chien-chu chi-chih khan Chhin-tai ti wei-shêng shê-shih 秦都咸陽一號建築基址看秦代衛生設施.
　Sanitary Facilities in the Chhin Period as Seen from the Foundation of Building No. 1 Excavated at the Chhin Capital of Hsien-yang.
　KKYWW, **1982** (5), 74–6.
Chang Hou-Yung 張厚埔 (*1987*).
　Yu Thang-mu chhu-thu ti shao-chiu pei khan wo-kuo shao-chiu chhu-hsien shih-chien 由唐墓出土的燒酒杯看我國燒酒出現時間.
　The Date of Appearance of Distilled Wine as Seen from Wine Cups Discovered in Thang Tombs.
　Shensi Chung I 陝西中醫, **1987** (8), 188–9.
Chang Kuang-Yuan 張光遠 (*1994*).
　Shang-ku shih-tai ti chêng-shih yong chhi 上古時代的蒸食用器.
　Utensils for Steaming Food in Ancient China.
　In Lin Chhing-Hu, ed. (*1994*), 167–214.
Chang Lien-Ming 張廉明 (*1987a*).
　Yu yen chiang chhu khao (1) 油鹽醬醋考 (1).
　Investigation of Oil, Salt, Soy Paste and Vinegar (I).
　CKPJ, **1987** (9), 7–8.
Chang Lien-Ming 張廉明 (*1987b*).
　Yu yen chiang chhu khao (2) 油鹽醬醋考 (2).
　Investigation of Oil, Salt, Soy Paste and Vinegar (II).
　CKPJ, **1987** (10), 7–9.
Chang Mêng-Lun 張孟倫 (*1988*).
　Han-Wei yin-shih khao 漢魏飲食考.
　Investigation of Food and Drink during the Han–Wei Era.
　Lanchou University Press.
Chang Phei-Yü 張培瑜 (*1988*).
　Yin-Shang Hsi-Chou shih-chhi chung-yuan wu-chhêng kho chien ti ji-shih 殷商西周時期中原五城可見的日食.
　Visible Solar Eclipses at Five Cities in Central China from the Yin–Shang period to the Western Chou Dynasty.
　TJKHSYC, **7** (3), 218–31.
Chang Tê-Shui 張德水 (*1994*).
　Yin-Shang chiu wên-hua chhu lun 殷商酒文化初論.
　Preliminary Discussion of the Culture of Wine during the Shang.
　Chung Yuan Wên Wu 中原文物, **1994** (3), 18–24.
Chang Thang-Hêng 張堂恒, Liu Chu-Shêng 劉祖生 & Liu Yüeh-Yun 劉岳耘 (*1994*).

Chha, Chha kho-hsüeh, Chha wên-hua 茶，茶科學，茶文化.
Tea: Its Science and Culture.
Liaoning Renmin, Shenyang.

Chang Tzu-Kao 張子高 (*1960*).
Lun wo-kuo niang-chiu chhi-yüan ti shih-tai wên-thi 論我國釀酒起源的時代問題.
Regarding the Date of Origin of Wine Fermentation in China.
Chhing-hua ta-hsüeh hsüeh-pao 清華大學學報, **7** (2), 31–3.

Chang-Tzu-Kao 張子高 (*1964a*).
Chung-kuo hua-hsüeh shih-kao (ku-tai chih pu) 中國化學史稿（古代之部）.
Draft on History of Chemistry in China (Part on Ancient China).
Science Press, Peking. Cf. *SCC*, v (4), 576.

Chang Tzu-Kao 張子高 (*1964b*).
Chung-kuo ku-tai hua-hsüeh shih 中國古代化學史.
A History of Chemistry in Ancient China.
Commercial Press, Peking.

Chang Wên-Hsü 張文緒 & Phei An-Phing 裴安平 (*1997*).
Li-hsien Mêng-hsi Pa-shih-tang chhu-thu tao-ku ti yen-chiu 澧縣夢溪八十璫出土稻穀的研究.
Investigation of the Rice Grains Discovered at Pa-shih-tang, Mêng-chhi, Li-hsien.
WW, **1997** (1), 36–41.

Chang Yüan-Fên 張遠芬, ed. (*1991*).
Chung-kuo chiu tien 中國酒典.
Literary Dictionary of Chinese Wines.
Kweichow Renmin.

Chao Chien-Min 趙建民 (*1985*).
Hsiang-yu than chhü 香油探趣.
Anecdotes on Sesame Oil.
CKPJ, **1985** (5), 36.

Chao Ho-Thao 趙和濤 (*1991*).
Wo-kuo chha-lei fa-chan yü yin-chha fang shih yen pien 我國茶類發展與飲茶方式演變.
The Development of Tea and Changing Styles of Tea Drinking in China.
NYKK, **1991** (2), 193–5.

Chao Hsüeh-Huei 趙學慧 (*1945*).
Szuchuan Chia-ting Chhü-chung-tzu yen-chiu 四川嘉定麴種子研究.
Investigation of the Microbes in the *Chhü Ferment* from Chiating, Szechuan.
Huang Hai 黃海, **7** (3), 27–38.

Chao Kang 趙岡 (*1988*, a,b).
Chung-kuo chu-yao nung tso-wu chien-shih (Shang, Hsia) 中國主要農作物簡史（上，下）.
A Brief History of the Major Crops of China (I & II).
TLTC, a. 1988, **76** (3), 129–39; b. 1988, **76** (4), 177–88.

Chao Khuang-Hua 趙匡華, ed. (*1985*).
Chung-kuo ku-tai hua-hsüeh shih yen-chiu 中國古代化學史研究.
Investigations on the History of Chemistry in China.
Peking University Press, Peking.

Chao Yung-Kuang 趙榮光 (*1990*).
Shih pu yen ching; khuai pu yen hsi 食不厭精，膾不厭細 'Grains cannot be too well cleaned, nor meat too finely sliced' [quotation from the *Lun Yü*].
CKPJ, **1990** (10), 20.

Chao Yung-Kuang 趙榮光 (*1995*).
Chao Yung-Kuang shih wên-hua lun chi 趙榮光食文化論集.
Collection of Essays on Food Culture by *Chao Yung-Kuang*.
Heilungkiang Renmin, Harbin.

Chao Yung-Kuang 趙榮光 (*1997*).
Chu yü Chung-hua min-chu yin-shih wên-hua 箸與中華民族飲食文化.
Chopsticks and Chinese Food Culture.
NYKK, **1997** (1), 225–35.

Chê-chiang shêng kung-yeh chhang 浙江省工業廠, Chekiang Provincial Industrial Bureau (*1958*).
Shao-hsing chiu niang tsao 紹興酒釀造.
The Manufacture of Shao-hsing Wine.
Light Industry Press, Peking, 1958.

Chê-chiang shêng po-wu-kuan 浙江省博物館, Chekiang Provincial Museum (*1960*).
Wu-hsing Chhien-shan-yang I-chih ti-i, -er-tzhu fa chüeh pao kao 吳興錢山漾遺址第一二次發掘報告.
Report of First and Second Archaeological Excavations at Chhien-shan-yang, near Wu-hsing.
KKHP, **1960** (2), 73–91.

Chê-chiang shêng po-wu-kuan 浙江省博物館 (*1978*).
Ho-mu-tu i-chih tung-chih wu i-chhang ti chien-ting yen-chiu 河姆渡遺址動植物遺存的鑒定研究.
A Study of the Animal and Plant Remains Unearthed at Ho-mu-tu.
KKHP, **1978** (1), 95–107.

Chê-chiang shêng wên-kuan-huei, po-wu-kuan 浙江省文管會，博物館, Chekiang Bureau of Cultural Relics and Provincial Museum, (*1976*).
Ho-mu-tu fa-hsien yüan-shih shê-hui chung-yao i-chih 河姆渡發現原始社會重要遺址.
Major Cultural Relics of a Primitive Society Discovered at Ho-mu-tu.
WW, **1976** (8), 6–14.

Chê-chiang shêng wên-kuan-huei, po-wu-kuan 浙江省文管會，博物館 (*1978*).
Ho-mu-tu i-chih ti-yi-chhi fa-chüeh pao kao 河姆渡遺址第一期發掘報告.
Report of the First Stage of Archaeological Excavation at Ho-mu-tu.
KKHP, **1978** (1), 39–94.

Chê-chiang shêng wên-kuan-huei 浙江省文管會 (*1960*).
Hangchou Shui-thien-fan i-chih fa-chüeh pao kao 杭州水田畈遺址發掘報告.
Report of Excavation of Archaeological Site at Shui-thien-fan, Hangchou.
KKHP, **1960** (2), 93–106.

Chêng Chin-Shêng 鄭金生, Liu Huei-Chên 李輝楨, Wang Li 王立 & Chang Thung-Chün 張同君, eds. (*1990*).
Shih Wu Pên Tshao 食物本草.
Pandects of Food Materials,

Chinese Medical Science and Technology Press, Peking, 1990.

Chêng-chou shih po-wu-kuan 鄭州市博物館 Chêng-chou City Municipal Museum (1973).
Chêng-chou Ta-ho-tshun Yang-shao wên-hua ti fang-chih i-chih 鄭州大河村仰韶文化的房基遺址.
Remains of Building Foundations of the Yang–Shao Period Discovered at Ta-ho-tshun near Chêngchou.
KK, **1973** (6), 330–6.

Chhang-chiang liu-yu ti-er-chhi wen-wu khao-ku kung-tso jen-yuan hsun-lian pan 長江流域第二期文物考古工作人員訓練班 Yangtzu Basin No. 2. Archaeology training team (1974).
Hupei Chiang-ling Fêng-huang-shan Hsi-Han-mu fa-chüeh chien pao 湖北江陵鳳凰山西漢墓發掘簡報.
Brief Report of Excavations at Western Han tombs at Fêng-huang Shan, near Chiang-ling, Hupei.
WW, **1974** (6), 41–53.

Chhên Chih-Ta 陳志達 (1985).
Shang tai wan-chhi ti chia-chhu ho chia-chhin. 商代晚期的家畜和家禽.
Domestic Animals and Fowls in the Late Shang.
NYKK, **1985** (2), 288–93.

Chhên Chu-Kuei 陳祖槼 & Chu Tzu-Chên 朱自振, eds. (1981).
Chung-kuo chha-yeh li-shih tzhu-liao hsüan-chi 中國茶葉歷史資料選輯.
Selected Historiographic Materials on the History of Tea in China.
Science Press, Peking.

Chhên Chhuan 陳椽 (1960, 1984).
Anhui Chha Ching 安徽茶經.
Book of Anhui Tea.
Anhui Science and Technology Press.

Chhên Chhuan 陳椽 (1984).
Chha yeh thung shih 茶葉通史.
A General History of Tea.
Agricultural Press, Peking.

Chhên Chhuan 陳椽 (1993).
Chung-kuo chha-yeh wai-hsiao shih 中國茶葉外銷史.
History of China's Tea Exports.
Pi-shan-yen Press, Taipei.

Chhên Chhuan 陳椽 (1994).
Lun chha yü wên-hua 論茶與文化.
On Tea and Culture.
Agricultural Press, Peking.

Chhên Ên-Chih 陳恩志 (1989).
Chung-kuo liu-pei thi phu-thung hsiao-mai tu-li chhi-yuan shuo 中國六倍體普通小麥獨立起源説.
The Indigenous Origin of Common Hexaploid Wheat in China.
NYKK, **1989** (1), 74–84.

Chhên Hsien-Ju 陳賢儒 (1955).
Kansu Lung-hsi hsien ti Sung mu. 甘肅隴西縣的宋墓.
The Sung Tomb at Lung-hsi County, Kansu.
WWTKTL, **1955** (9), 86–92.

Chhên Hsing-I 陳行一 (1995).
Wu-i chha shih wên-hsien tsa-hsieh 武夷茶史文獻雜纈.
Miscellaneous Notes on the Historiography of Wu-I Tea.
NYKK, **1995** (4), 202–5.

Chhên Hung 陳洪 (1987).
Hsien-Chhin shih-chhi ti lêng-chhang ho lêng-yin 先秦時期的冷藏和冷飲.
Cold Drinks and Cold Storage in the Pre-Chhin Period.
CKPJ, **1987** (7), 7–8.

Chhên Kuang-Hsin 陳光新 (1990).
Chung-kuo pheng-jen shih-hua 中國烹飪史話.
Remarks on the History of Chinese Cuisine.
Hupei Science and Technology Press.

Chhên Shao-Chün 陳紹軍 & Wu Chao-Su 吳兆蘇 (1994).
Chhung wo-kuo hsiao-mai, mien-shih chi chhi chia-kung kung-chü ti fa-chan li-shih shih than man-thou ti chhi-yüan wên-thi 從我國小麥麵食及其加工工具的發展歷史試談饅頭的起源問題.
The Origin of *Man-thou* (steamed buns) as seen from the History of Wheat, Pasta Foods and Processing Equipment in China.
NYKK, **1994** (1), 219–25.

Chhên Shao-Chün 陳紹軍 (1995a).
Hu-ping lai-yuan than-shih 胡餅來源探釋.
Investigation of the Origin of *Hu-ping*.
NYKK, **1995** (1), 260–3, 275.

Chhên Shao-Chün 陳紹軍 (1995b).
Man-thou chi hsiao-mien shih-phin chhi-yüan wên-thi ti tsai jên-shih 饅頭及酵麵食品起源問題的再認識.
Re-examination of the Origin of *Man-thou* and Leavened Pasta Foods.
NYKK, **1995** (2), 218–20.

Chhên Thao-Shêng 陳騊聲 (1979).
Chung-kuo wei-sheng-wu kung-yeh fa-chan shih 中國微生物工業發展史.
History of the Development of Microbiological Industry in China.
Light Industry Press, Peking, 1979.

Chhên Tsêng-Pi 陳增弼 (1982).
Lun Han-tai wu chuo 論漢代無桌.
Were Tables known during the Han?
KKYWW, **1982** (5), 91–7.

Chhên Tshun-Jên 陳存仁 (1978).
Tsin-tsin yu wei than 津津有味談.
Discourse on Food and Cuisine.
Commercial Press, Hong Kong.

Chhên Tsung-Mou 陳宗懋, ed. (1992).
Chung Kuo Chha Ching 中國茶經.
The Chinese Book of Tea.
Shanghai Wen Hua.

Chhên Wei-Min 陳偉明 (1989).
Thang-Sung shih-chhi yu, chiang-yu so-than 唐宋時期油，醬油瑣談.
A Brief Note on Oil and Soy Sauce during the Thang–Sung Period.
CKPJ, **1989** (11), 10.

Chhên Wên-Hua 陳文華 (1981).
Chung-kuo ku-tai nung-yeh kho-chih shih chiang-hua 中國古代農業科技史講話.

Talks on the History of Agricultural Science and Technology in China.
NYKK, **1981** (1), 114–24.

Chhên Wên-Hua 陳文華 (*1983*).
Chung-kuo ku-tai nung-yeh khao-ku tzu-liao so-yin (ssu) ti erh pien: shêng chhang kung chü 中國古代農業考古資料索引，四。第二篇。生產工具.
Index of Ancient Agricultural Materials found in China (4). Part 2, Production Implements.
NYKK, **1983** (1), 280–4.

Chhên Wên-Hua 陳文華 (*1987*).
Chung-kuo Han tai Chhang-chiang liu-yü ti shui-tao tsai-phei ho yu kuan nung chü ti chêng chiu. 中國漢代長江流域的水稻栽培和有關農具的成就.
Cultivation of Paddy Rice during the Han in the Yangtse Basin and Parallel Achievements in Agricultural Implements.
NYKK, **1987** (1), 90–114.

Chhên Wên-Hua 陳文華 (*1998*).
Hsiao-chung pan tou fu – kuan yü tou fu wên-thi ti ta-pien 小蔥拌豆腐－關於豆腐問題的答辯.
Scallion and *tou fu*, 'The Question of *tou fu*' Answered and Explained.
NYKK **1998** (3), 277–91.

Chhên Wên-Hua 陳文華 (*1989*).
Chung-kuo ku-tai nung-yeh khao-ku tzu-liao so-yin 中國古代農業考古資料索引.
Index of Ancient Agricultural Materials found in China (Summary).
NYKK, **1989** (1), 419–27.

Chhên Wên-Hua 陳文華 (*1991*).
Tou fu chhi yuan yü ho shih? 豆腐起源于何時?.
What is the Date of Origin of *Tou fu*?
NYKK, **1991** (1), 245–8.

Chhên Wên-Hua 陳文華 (*1995*).
Chung-kuo ku-tai nung-yeh khao-ku tzu-liao so-yin (shi wu) 中國古代農業考古資料索引，十五.
Index of Ancient Agricultural Materials found in China (15).
NYKK, **1995** (3), 303–11.

Chhêng Chien-Hua 程劍華 (*1984*).
Ku-tai nung-yeh yü chu-kuo I hsüeh ti shih-wu liao-fa (1) 古代農業與祖國醫學的食物療法(1).
Ancient Chinese Agriculture and Diet Therapy, Part 1.
NYKK, **1984** (2), 370–80.

Chhêng Chien-Hua 程劍華 (*1985*).
Ku-tai nung-yeh yü chu-kuo I hsüeh ti shih-wu liao-fa (2) 古代農業與祖國醫學的食物療法(2).
Ancient Chinese Agriculture and Diet Therapy, Pt 2.
NYKK, **1985** (1), 378–85.

Chhêng Chhing-Thai 成慶泰 (*1981*).
Shih-ching chung so chi-tsai ti yü-lei khao shih 詩經中所記載的魚類考釋.
Elucidation of the Fishes Cited in the Book of Odes.
Po Wu 博物, **1981** (4), 6–22.

Chhêng Pu-Khuei 程步奎 (*1983*).
Tou fu ti chhi-yüan 豆腐的起源.
The Origin of *Tou fu*.
CKPJ, **1983** (4); cf. also *CKPJ Editors* (1986), *Phêng Jên Shih Hua*, pp. 422–3.

Chheng-teh shih pi-shu shan chuang po-wu-kuan 承德市避暑山莊博物館 (*1980*).
Cf. Lin Yung-Kuei 林榮貴 (*1980*).

Chhêng Yao-Thien 程瑤田 (*1805*).
Chiu Ku Khao 九穀考.
Study of the Nine Grains.
Reprinted in Folk Literature Series, Taipei.

Chhi Hsia 漆俠 (*1987*).
Sung tai ching-chi shih, shang, hsia 宋代經濟史 (上，下).
Sung Economic History, I, II.
Shanghai Renmin, Shanghai.

Chhi Ssu-Ho 齊思和 *1981* (*1948*).
Mao-shih ku ming khao 毛詩穀名考.
An Inquiry into the Names of Grains in the Book of Odes.
Chung-kuo shih thang-so 中國史探索, 1981 (originally published in *Yenching hsueh-pao* 燕京學報 **36**, 276–88).

Chhien Mu 錢穆 (*1956*).
Chung-kuo ku-tai pei-fang nung tso-wu khao 中國古代北方農作物考.
Investigation of the Cultivated Grains of Ancient North China.
New Asia Journal, **1** (2), 1–27.

Chhing-lung hsien Ching-chan-tzu ta-tui ko-wei hui 青龍縣井丈子大隊革委會 (*1976*).
Hopei sheng Chhing-lung hsien chhu-thu chin-tai thung shao-chiu kuo 河北省青龍縣出土金代銅燒酒鍋.
A Chin Distillation Still Discovered in Chhing-lung County, Hopei.
WW, **1976** (9), 98–9.

Chhin-tu Hsien-yang khao-ku kung-tso chan 秦都咸陽考古工作站 (*1980*).
Chhin-tu Hsien-yang ti-i-hao kung-tien chien-chu i-chih chien-pao 秦都咸陽第一號宮殿建築遺址簡報.
Brief Report on the Remains of No. 1 Palace at the Old Chhin Capital of Hsien-yang.
WW **1980** (11), 12–24.

Chhin yung khao-ku tui 秦俑考古隊 (*1981*).
Lin-thung Cheng-chuang Chhin shih-liao chia-kung chhang i-chih thiao-chha chien-pao 臨潼鄭莊秦石料加工場遺址調查簡報.
A Brief Report on the Ruins of a Chhin Quarry at Lin-thung, Cheng-chuang.
KKyWW, **1981** (1), 39–43.

Chhing-lung hsien chin chang tzu ta tui kê-wei hui 河北省青龍縣井丈子大隊革委會 (*1976*).
Hopei shêng Chhing-lung hsien chhu-thu Chin-tai thung shao-chiu kuo 河北省青龍縣出土金代銅燒酒鍋.
A Chin Dynasty Bronze Distillation Still Discovered in Chhing-lung County, Hopei.
WW, **1976** (9), 98–9.

Chhiu Fêng 邱鋒 (*1982*).
Chung-kuo tan shui yü-yeh shih hua 中國淡水魚業史話.
Historical Notes on Fresh Water Fishery in China.
NYKK, **1982** (1), 152–7.

Chhiu Phang-Thung 邱龐同 (*1986a*).
Tzu liao chên kui, yao chuan ching mei – tu 'Chhi Min Yao Shu' ti pa, chiu chüan 資料珍貴，殽饌精美－讀〈齊民要術〉第八，九卷.

Chhiu Phang-Thung 邱龐同 (1986a).
 Invaluable Resources, Delectible Viands – Notes Upon reading Chs. 8 & 9 of the Chhi Min Yao Shu.
 PJSH, 502–8.
Chhiu Phang-Thung 邱龐同 (1986b).
 Shan fu ching shou lu 膳夫經手錄.
 About the Chef's Handbook.
 PJSH, 509–12.
Chhiu Phang-Thung 邱龐同 (1986c).
 Phêng-jen ho chhi, shu su yu liang – chien chieh 'Yün Lin Thang Yin Shih Chih Tu Chi' 烹飪和諧，蔬素尤良－簡介〈雲林堂飲食制度集〉.
 Harmony in Cookery, Best for Vegetarian fare – A Brief Introduction to the 'Dietary System of the Cloud Forest Studio'.
 PJSH, 516–20.
Chhiu Phang-Thung 邱龐同 (1986d).
 I Ya II 易牙遺壹.
 Remnant Notions from *I Ya*.
 PJSH, 525–9.
Chhiu Phang-Thung 邱龐同 (1986e).
 Yang Hsiao Lu 養小錄.
 Nurturing Life Booklet.
 PJSH, 535–8.
Chhiu Phang-Thung 邱龐同 (1986f).
 'Tiao Ting Chi' tso chieh hsien I〈調鼎集〉作者獻疑.
 Doubts on the Author of the 'Harmonious Cauldron'.
 PJSH, 550–3.
Chhiu Phang-Thung 邱龐同 (1987).
 Mi hsien 米線.
 Rice threads.
 CKPJ, **1987** (10), 9.
Chhiu Phang-Thung 邱龐同 (1988).
 Chung-kuo mien-thiao yüan-liu khao-shu 中國麵條源流考述.
 Exploration of the Origin of Filamentous Noodles in China.
 CKPJ, **1988** (7), 14–16.
Chhiu Phang-Thung 邱龐同 (1989a,b).
 Chung-kuo mien-tien shih, shang, hsia 中國麵點史，上，下.
 History of Pasta Foods in China, I, II.
 CKPJ, **1989** (5), 11–13 & (6), 8–10.
Chhiu Phang-Thung 邱龐同 (1989c).
 Chung-kuo phêng-jen wen-hsien kai shu 中國烹飪古籍概述.
 Outline of the Culinary Classics of China.
 Commercial Publishers, Peking.
Chhiu Phang-Thung 邱龐同 (1992a,b).
 Han-wei liu-chhao mian-tien yen-chiu, shang, hsia 漢魏六朝麵點研究，上，下.
 Investigation of Pasta Foods During the Han–Wei and Six Dynasties.
 CKPJ, **1992** (2), 12–13 & (3), 6–8.
Chhiu Phang-Thung 邱龐同, Wang Li-Chhi 王利器 & Wang Chêng-Min 王貞珉 (1983).
 Lü Shih Chhun Chhiu, Pên-wei phien 呂氏春秋本味篇.
 The *Pên Wei* Segment from Master Lü's Spring and Autumn Annals (with translation and annotations). *CKPJKCTK*.
Chhung-chhing shih po-wu-kuan 重慶市博物館
 Chungking Municipal Museum (1957).
 Szechuan Han Hua-hsiang chuan hsüan chi 四川漢畫像專選集.
 Selections of Han Brick Paintings from Szechuan.
 Wen Wu, Peking.
Chhü Chia-Wei 區嘉偉 & Fang Hsin-Fang 方心芳 (1935).
 Huang-hai Research Institute of Industrial Chemistry.
 Kung Yeh Chung Hsin 工業中心, 1935 (4), 373–6.
 Cited in Chhên Thao-Shêng (1979), p. 69.
Chhü Chih-Hsin 屈志信 (1959).
 Chih chhu kung-jên chi-pên chih shih 制醋工人基本知識.
 Basic Information for Vinegar Workers.
 Light Industry Press, Peking.
Chhü Wan-Li 屈萬里 (1959).
 Shih-ching shih I 詩經釋義.
 Glosses of the *Book of Odes*.
 Chung Hua Wên Hua, Taipei.
Chhü Wan-Li 屈萬里 (1969).
 Shang shu chin chu ching shih 尚書今注今釋.
 The Book of Documents: Annotations and Explanations.
 Commercial Press, Taipei.
Chia O 賈峨 (1998).
 Kuang-yü 'tou fu wên-thi' I-wên ti I-wên thi 關於〈豆腐問題〉一文的問題.
 The Question about 'The Question of *tou fu*'.
 NYKK **1998** (3), 267–76.
Chia Tsu-Chang 賈祖璋 & Chia Tsu-Shan 賈祖珊 (1936/1955/1956).
 Chung-kuo chih wu thu chien 中國植物圖鑑.
 Illustrated Dictionary of Chinese Flora [arranged on the Engler system]; 2602 entries.
 Chung-Hua, Peking, repr. 1955, 1956.
Chiang Chhao 姜超, ed. (1985).
 Shih-yung chung-i ying-yang hsüeh 實用中醫營養學.
 Practical Nutrition in Chinese Medicine.
 Liberation Army Press, Peking.
Chiang-hsi shêng po-wu-kuan 江西省博物館 (1978).
 Chiang-hsi Nanchang Tung-Han Tung-Wu mu 江西南昌東漢東吳墓.
 An Eastern-Wu Tomb of the Eastern Han Period [excavations of].
 KK, **1978** (3), 145–63.
Chiang Ju-Tao 姜汝燾 (1991).
 Tou chih-phin shêng-tshan chung ti chiao-ning chi-li 豆制品生產中的膠凝機理.
 The Mechanism of Coagulation in the Production of Soybean Products.
 CKPJ, **1981** (1), 21–2.
Chiang Jun-Hsiang 江潤祥, ed. (1996).
 Hui Hui Yao Fang 回回藥方.
 An Islamic Formulary (Original text and articles in both Chinese and English).
 Hong Kong.
Chiang Jun-Hsiang 江潤祥 & Kuan Phei-Shêng 關培生 (1991).
 Hsing Lin Shih Hua 杏林史話.
 Talks on the History of Chinese Medicine.
 Chinese University Press, Hong Kong.

Chiang-ling Fêng-huang-shan 167 hao Han-mu fa-chüeh chêng-li hsiao-chu 江陵鳳凰山一六七號漢墓發掘整理小組 (*1976*).
   Chiang-ling Fêng-huang-shan 167 hao Han-mu fa-chüeh chien-pao 江陵鳳凰山一六七號漢墓發掘簡報.
   Brief Report on the Excavation at Han Tomb No. 167 at Feng-huang-shan, Chiang-ling [in Hupei].
   WW, **1976** (10), 31–51.

Chiang Shih-Chhê 蔣師轍 (*1892*).
   Thai Yu Ji Chi 台游日記.
   Diary of Travels in Taiwan.
   Publisher not known, cited in Chhên & Chu (*1981*), 429.

Chiang-su hsin i-hsüeh yuan 江蘇新醫學院 (*1986*).
   Chung-I ta tzhu tien 中醫大辭典.
   Cyclopedia of Chinese Traditional Drugs.
   S & T Press, Shanghai. Textual refs. are to the miniaturised edition.

Chiang-su shêng po-wu kuan & Thai-chou hsien po-wu kuan Kiangsu Provincial Museum and Thai-chou County Museum (*1962*).
   Chiang-su Thai-chou Hsin-chuang han mu 江蘇泰州新莊漢墓.
   The Han Tomb at Hsin-chuang, Thai-chou, Kiangsu.
   KK, **1962** (10), 540–3.

Chiang Ying-Hsiang 江蔭香 (*1982*).
   Shih-Ching I-chu 詩經譯注.
   The Book of Odes: Translation and Commentaries.
   Chung Kuo, Peking, 1982. 1st pub. in 1934.

Chien ming Chung-kuo phêng-jên tzhu-tien pien-hsieh chu (editors) (*1987*).
   Chien ming Chung-kuo phêng-jên tzhu-tien 簡明中國烹飪辭典.
   Simplified Dictionary of Chinese Cookery.
   Shansi Renmin, Taiyuan.

Chih Tzu 知子 (*1986a*).
   Chung-kuo ku-tai tshan chha khao 中國古代餐叉考.
   An Inquiry into Dining Forks in Ancient China.
   CKPJ, **1986** (1), 16–19.

Chih Tzu 知子 (*1986b*).
   Wo-kuo wen-ming shih-tai ti tshan shi 我國文明時代的餐匙.
   Spoons and ladles in Chinese civilisation.
   CKPJ, **1986** (6), 21–2.

Chih Tzu 知子 (*1987*).
   Wo-chin shih-tai yin-shih phêng-jen fêng-shu ti shêng-tung thu-chüan 魏晉時代飲食烹飪風俗的生動圖卷.
   Album of Vivid Scenes of Dining and Cookery of the Wei-tsin Period.
   CKPJ, **1987** (11), 5–7.

Chih Tzu 知子 (*1989a*).
   Fên tshan yü ho tshan, Shang 分餐與合餐 (上).
   Individual and Communal Dining, Part I.
   CKPJ, **1989** (7), 8–9.

Chih Tzu 知子 (*1989b*).
   Fên tshan yü ho tshan, Hsia 分餐與合餐 (下).
   Individual and Communal Dining, Part II.
   CKPJ, **1989** (8), 6–7.

Chih Tzu 知子 (*1990*).
   Thung pi yü yin-shih phêng-jen 銅幣與飲食烹飪.
   Use of Copper Coins in Food and Cookery.
   CKPJ, **1990** (11), 12–13.

Chih-Wu Yen-chiu So 植物研究所 (Institute of Botany) (*1976*).
   Chung-kuo kao-têng chih-wu thu-chien 中國高等植物圖鑑，五卷.
   Iconographia Cormophytorum Sinicorum 5 Vols.
   Science Press, Peking.

Chi-nan chhêng Fêng-huang-shan 168 hao Han-mu fa-chüeh chêng-li tsu 紀南城鳳凰山一六八號漢墓發掘整理組 (*1975*).
   Hupei Chiang-ling Fêng-huang-shan 168 hao Han-mu fa-chüeh chien-pao 湖北江陵鳳凰山一六八號漢墓發掘簡報.
   A Brief Report of the Excavation at Han Tomb No. 168 at Fêng-huang-shan, Chiang-ling, Hupei.
   WW, **1975** (9), 1–28.

Chin Chiu-Kung 晉久工 (*1981*).
   Pai-chiu sheng-tshan wen-ta 白酒生產問答.
   The Production of Distilled Wine: Questions and Answers.
   Shansi Renmin Press.

Chou Chia-Hua 周嘉華 (*1988*).
   Su Shih pi-hsia ti chi-chung chiu chi chhi niang chiu chi-shu 蘇軾筆下的幾種酒及其釀酒技術.
   Wine Varieties and the Technology of their Fermentation as Described by Su Shih.
   TJKHSYC, **7** (1), 81–9.

Chou Hêng-Kang 周恒剛 (*1964*).
   Thang hua chhü 糖化麴.
   Saccharifying Ferments.
   Finance & Economics Press, Peking.

Chou Hêng-Kang 周恒剛 (*1982*).
   Pai chiu shêng-chhang kung-i hsüeh 白酒生產工藝學.
   Shansi Renmin, Taiyuan.

Chou Hsin-Chhun 周新春 (*1977*).
   Kao-liang chiu chih-chao kung-yeh 高粱酒制造工業.
   The Industrial Production of Sorghum Wine.
   Chhien Chhiu, Taipei.

Chou San-Chin 周三金 (*1986*).
   Tshung 'Sui Yuan Shih Tan' khan wo-kuo ku-tai ti phêng-tiao chi-shu 從〈隨源食單〉看我國古代的烹調技術.
   The Culinary Arts of Old China as seen from the 'Recipes from the Sui Garden'.
   PJSH, 539–45.

Chou Shih-Yung 周世榮 (*1979*).
   Kuang yü Chhang-sha Ma-wang-tui Han mu chung chien wên 楬 ti khao chêng 關於長沙馬王堆漢墓中簡文(楬)(檔)的考証.
   Evaluation of the character 楬 found on the bamboo slips at Han tombs in Ma-wang-tui, Changsha.
   Chha Yeh Thung Hsin 茶葉通訊 **1979** (3), 65.

Chou Shih-Yung 周世榮 (*1983*).
   Chhung Ma-wang-tui chhu-thu ku wên-tzu khan Han-tai nung-yeh kho-hsüeh 從馬王堆出土文字看漢代農業科學.

The State of Han Agricultural Science as seen from the Written Records Found in Ma-wang-tui.
NYKK, **1983** (1), 81–90.

Chou Shih-Yung 周世榮 (*1992*).
Tsai than Ma-wang-tui Han mu chung chien wên 再談馬王堆漢墓中簡文〈楢〉(櫃).
Re-evaluation of the character 〈楢〉 in the Ma-wang-tui Inventory.
Chha Yeh Thung Hsin **1992** (2), 200–3.

Chou Shu-Pin 周樹斌 (*1991*).
'*Shên-Nung te chha chieh tu' khao phing*〈神農得茶解毒〉考評.
Critique of the Statement 'Shên Nung found tea to counter poison'.
NYKK, **1991** (2), 196–200.

Chou Su-Phing 周蘇平 (*1985*).
Hsien-Chhin shih-chhi ti yü-yeh 先秦時期的漁業.
Fishery in the Pre-Chhin Period.
NYKK, **1985** (2), 164–70.

Chou Wên-Thang 周文棠 (*1995*).
Wang Pao Thung Yüeh chung thu fei chha ti khao cheng 王褒僮約中荼非茶的考証.
Evidence Indicating that the 'Thu' in Thung Yueh is not Tea.
NYKK **1995** (4), 181–3.

Chu Chhêng (or Shêng) 朱晟 (*1987*).
Pai-chiu (chêng-liu chiu) ti chhi-yüan 白酒 (蒸溜酒) 的起源.
The Origin of Distilled Wine in China.
CKPJ, **1987** (4), 12–13.

Chu Chung-Shêng 朱重聖 (*1985*).
Pei-Sung chha chih shêng-chhang yü ching yin 北宋茶之生產與經營.
The Production and Commerce of Tea during the Northern Sung.
Hsüeh Shêng Shu Chü, Taipei.

Chu Jui-Hsi 朱瑞熙 (*1994*).
〈*Mi-hsien*〉*khao* 米線考.
Note on 'Rice threads'.
CKPJ, **1994** (11), 37.

Chu Kho-Chên 竺可楨 (*1973*).
Chung-kuo chin wu-chhien nien lai chhi-hou pien-chhien ti chhu-pu yen-chiu 中國近五千年來氣候變遷的初步研究.
A Preliminary Study of Climatic Changes in China in the past Five Thousand Years.
Chung Kuo Kho Hsüeh, **1973** (2), 168–89.

Chu Tê-Hsi 朱德熙 & Chhiu Hsi-Kuei 裘錫圭 (*1980*).
Ma-wang-tui I-hao Han-mu chhien-tshê khao shih pu chêng 馬王堆一號漢墓遣策考釋補正.
Correction to the Interpretation of Bamboo Slips found in Han Tomb No. 1 at Ma-wang-tui.
Wên Shih, **1980** (10), 61–74.

Chu Wên-Hsin 朱文鑫 (*1934*).
Li tai jih shih khao 歷代日食考.
A Study of Eclipses Recorded in Chinese History.
Commercial Press, Shanghai.

Chuang Wan-Fang 莊晚芳 (*1981*).
Chha ti shih-yuan chi chhi yuan-chhang-ti wên-thi 茶的始原及其原產地問題.
The Earliest Use of Tea and its Original Habitat.
NYKK, **1981** (2), 134–9.

Chuang Wan-Fang 莊晚芳 (*1986*).
Kao-lu khhao-shih 皋蘆考釋.
Explanation of Kao-lu (as the Original Tea Tree, Found in Yunnan).
NYKK, **1986** (2), 369–71.

Chung I Ta Tzhu Tien pien chi wei-yuan hui 中醫大辭典編輯委員會 (*1981*).
Chung I Ta Tzhu Tien i shih wên hsien fêng chhê 中醫大辭典, 醫史文獻分冊.
Cyclopedia of Chinese Medicine, Section on Medical Historiography.
Renmin Weisheng, Peking.

Chung I-Yen 鐘益研 & Ling Hsiang 凌襄 (*1975*).
Wo-kuo hsien-i fa-hsien ti tsui-ku I-fang – po-shu〈*Wu Shih Erh Ping Fang*〉我國現已發現的最古醫方〈五十二病方〉.
The Oldest Pharmacopoeia Discovered in China – the Silk Manuscript 'Prescriptions for Fifty-Two Ailments'.
WW, **1979** (9), 49–60.

Chung-kuo chiu wên-hua yen-thao hui 中國酒文化研討會 (*1987*).
Cf. Li Yen-Heng, ed.

Chung-kuo i-hsüeh kho-hsüeh yuan wei-shêng yen-chiu so 中國醫學科學院衛生研究所 (*1963*).
Shih wu chhên-fen piao 食物成分表.
Composition of Foods.
Renmin Weisheng, Peking.

Chung-kuo kho-hsüeh yuan khao-ku yen-chiu so 中國科學院考古研究所 (*1959*).
Loyang Chung-chou-lu 洛陽中州路.
[Report of Excavations at] Loyang Chung-chou Lu.
Science Press, Peking.

Chung-kuo kho-hsüeh yuan khao-ku yen-chiu so shih yen shih 中國科學院考古研究所實驗室 (*1972*).
Fang-shê-hsing than-su tshê-ting nien-tai pao-kao.(erh) 放射性碳素測定年代報告(二).
Radiocarbon Determination of the Dates of Archaeological Artifacts (2).
KK, **1972** (5), 56–8.

Chung-kuo kho-hsüeh yuan khao-ku yen-chiu so shih yen shih 中國科學院考古研究所實驗室 (*1974*).
Fang-shê-hsing than-su tshê-ting nien-tai pao-kao.(san) 放射性碳素測定年代報告(三).
Radiocarbon Determination of the Dates of Archaeological Artifacts (3).
KK, **1974** (5), 333–8.

Chung-kuo kho-hsüeh yuan khao-ku yen-chiu so shih yen shih 中國科學院考古研究所實驗室 (*1977*).
Fang-shê-hsing than-su tshê-ting nien-tai pao-kao.(ssu) 放射性碳素測定年代報告(四).
Radiocarbon Determination of the Dates of Archaeological Artifacts (4).
KK, **1977** (3), 200–4.

Chung-kuo kho-hsüeh yuan wei-shêng-wu yen-chiu so 中國科學院微生物研究所 (*1975*).
〈*Chhi Min Yao Shu*〉*chung ti chih chhü niang chiu*〈齊民要術〉中的制麴釀酒.

The making of Ferments and Wine in the '*Chhi Min Yao Shu*'.
Acta Microbiol. Sinica, **15** (1), 1–4.
Chung-kuo kui-suan-yen hsüeh-hui 中國硅酸鹽學會 (1982).
Chung-kuo thao-tzhu shih 中國陶瓷史.
History of Ceramics in China.
Wen Wu, Peking.
Chung-kuo ku-tai hsiao-shuo pai-kho chhüan shu pien-chi wei-yuan hui 中國古代小說百科全書編輯委員會, Editorial Committee for the Encyclopedia of Novels (1993).
Chung-kuo ku-tai hsiao-shuo pai-kho chhüan shu 中國古代小說百科全書.
Encyclopedia of Novels (Small Talks) in Ancient China.
Science Press, Peking.
Chung-kuo ku-tai nung yeh kho-chi pien tsuan tsu 中國古代農業科技編纂組, (Editorial Committee for Ancient Chinese Agricultural Science and Technology) (1980).
Chung-kuo ku-tai nung yeh kho-chi 中國古代農業科技.
Agricultural Science and Technology in Ancient China.
Agricultural Press, Peking.
Chung-kuo nung-yeh pai-kho chhuan-shu 中國農業百科全書, Editorial Committee (1988).
Chung-kuo nung-yeh pai-kho chhuan-shu: chha-yeh chüan 中國農業百科全書, 茶業卷.
The Encyclopedia of Chinese Agriculture: Volume on Tea.
Agricultural Press, Peking.
Chung-kuo phêng-jen pai-kho chhüan shu pien wei huei 中國烹飪百科全書編委會 (1992).
Chung-kuo phêng-jen pai-kho chhüan shu 中國烹飪百科全書.
The Encyclopedia of Chinese Cuisine.
Encyclopedia Press, Peking.
Chung-kuo phêng-jen pien chi pu 中國烹飪編輯部 Editors of *CKPJ* (1986).
Phêng-jen shih hua 烹飪史話.
Historical Notes on Chinese Cuisine.
Commerce Publishers, Peking.
Chung-kuo shê-huei kho-hsüeh yuan, khao-ku yen-chiu-so & Hopei shêng wen-wu kuan li-chhu, Institute of Archaeology, CASS & Cultural Relics Bureau, Hopei (1980).
Man-chhêng Han-mu fa-chueh pao-kao, I, II 滿城漢墓發掘報告, I, II.
Report of the Excavations at Man-chhêng.
Wen Wu, Peking.
Chung-kuo shê-huei kho-hsüeh yuan, khao-ku yen-chiu-so Honan I-tui (1982).
1979 *nien phei-li-kan I-chih fa-chueh chien-pao* 1979 年裴李崗遺址發掘簡報.
Brief Report on the Phei-li-kan Site Excavated in 1979.
KK, **1982** (4), 337–40.
Chung-kuo shê-huei kho-hsüeh-yuan khao-ku yen-chiu so Honan I-tui 中國社會科學院考古研究所河南一隊, Honan Team I of the Institute of Archaeology, CASS (1983).
Honan Hsin-chêng Sha-wo-li hsin shih-chhi shih-tai i-chih 河南新鄭沙窩李新石器時代遺址.
Neolithic Remains at Sha-wo-li, Hsin-chêng County, Honan.
KK, **1983** (12), 1057–65.
Chung-kuo shih phing chhu pang shê 中國食品出版社 (1989).
Cf. Li Yin, ed. (1989).
Chung-kuo thu-nung yao-chih pien-chi wei-yuan hui 中國土農藥誌編輯委員會, Editorial Committee for the Repertorium (1959).
Chung-kuo thu-nung yao-chih 中國土農藥誌.
Repertorium of Plants used in Chinese Agricultural Chemistry.
Science Press, Peking.
Chung-yang Wei-shêng Yen-chiu Yuan 中央衛生研究院 (1952).
Shih-wu chhêng-feng piao 食物成分表.
Tables of Composition of Foods.
Commercial Press, Peking.

Fan Chia-Wei (Ka Wai) 范家偉 (1995).
Tung-Tsin tzu Sung-tai chiao-chhi ping chih than-thao 東晉至宋代腳氣病之探討.
Hsin Shih Hsueh 新史學, **6** (1), 155–77.
Fan Hsing-Chun 范行準 (1954).
Chung-kuo yü-fan I-hsüeh ssu-hsiang shih 中國預防醫學思想史.
History of the Ideas on Public Health in China.
Renmin Weisheng, Peking.
Fan Shu-Yün 樊樹雲 (1986).
Shih Ching Chhuan I Chu 詩經全譯注.
The *Shih Ching*: A Complete Translation [into Modern Chinese] with Annotations.
Heilungkiang Renmin, 1986.
Fang Chien 方健 (1996).
'*Phêng-chha ching chü*' ho '*Wu-tu mai chha*' khao pien 〈烹茶盡具〉和〈武都買茶〉考辨.
Textual Investigation of '*Phêng-chha ching chü*' and '*Wu-tu mai chha*' from the *Mock Slave Contract*.
NYKK, **1996** (2), 184–92 & 205.
Fang Chien 方健 (1997).
Kuan yü Ma-wang-tui Han-mu chhu-thu khao-pien erh-thi 關於馬王堆漢墓出土物考辨二題.
The Identity of Two Items Excavated from Han Tombs at Ma-wang-tui.
Chung-kuo li-shih ti-li lun tshun 中國歷史地理論叢, **1997** (1), 51–6, repr. *NYKK* **1997** (2), 212–13 & 258.
Fang Hao 方豪 (1987).
Chung-hsi chiao-thung shih, shang, hsia 中西交通史, 上, 下.
History of East–West Communications, I & II.
Yüeh Li Press, Shanghai.
Fang Hsin-Fang 方心芳 (1980).
Chhü-nieh niang chiu ti chhi-yüan yü fa-chan 麴糵釀酒的起源與發展.
Origin and Development of *Ferments* and the Wine Fermentation.
KCSWC, 1980 (4), 140–9.
Fang Hsin-Fang 方心芳 (1987).
Kuan yü Chung-kuo chêng-chiu chhi ti chhi-yüan 關於中國蒸酒器的起源.

Origin of Stills for Distilling Wine in China.
TJKHSYC, **6** (2), 131–4.

Fang Hsin-Fang 方心芳 (1989).
Tsai lun wo-kuo chhü-nieh niang chiu ti chhi-yuan yü fa-chang 再論我國麴糵釀酒的起源與發展.
Reexamination of the Origin and Development of *Ferments* and the Wine Fermentation.
In Li Ying, ed. (1989), 3 31.

Fang Hsin-Fang 方心芳 & Fang Wên-I 方聞一 (1993).
Chung-hua chiu wên-hua ti chhang-shih yü fa chan 中華酒文化的創始與發展.
The Origin and Development of Chinese Wine Culture.
In Wang Yen, Ho Thien-Cheng et al., eds. (1993), 103–8.

Fang Yang 方楊 (1964).
Wo-kuo niang-chiu tang shih yü Lung-shan wên-hua 我國釀酒當始於龍山文化.
Wine Fermentation in China began During the Lung-shan Cultural Period.
KK, **1964** (2), 94–7.

Fêng-huang Shan 167 hao Han-mu fa-chüeh chêng-li xiao chu 鳳凰山167號漢墓發掘整理小組, Excavation Team for Han Tomb No. 167 at Fêng-huang Shan (1976).
Chiang-ling Fêng-huang-shan i-liu-chhi hao Han-mu fa-chüeh chien pao 江陵鳳凰山一六七號漢墓簡報.
Brief Report of the Excavation of Han Tomb No. 167 at Fêng-huang Shan, Chiang-ling.
WW, **1976** (10), 31–48.

Fêng Lan-Chuang 馮蘭莊 (1960).
Chiang-yu shêng-chhan 醬油生產.
Production of Soy Sauce.
Light Industry Press, Peking, 1960.

Fu Hsi-Jên 傅錫壬 (1976).
Hsin I Chhu Tzhu tu pên 新譯楚辭讀本.
The *Chhu Tzhu* Reader, with new Annotations.
San Min, Taipei, 1976.

Fu Shu-Chhin 傅樹勤 & Ouyang-Hsün 歐陽勛 (1983).
Lu Yü Chha Ching I Chu 陸羽茶經譯注.
The Tea Canon of Lu Yü, Translation and Annotations.
Hupei Renmin, 1983.

Hayashi, Minao 林巳奈夫 (1975).
Kandai no Inshoku 漢代の飲食.
Food and Drink in the Han Dynasty.
Tōhō gakuhō, 1975, **48**, 1–98.

Ho Chhang-Hsiang 何昌祥 (1997).
Tshung mu-lan hua-shih lun chha-shu chhi-yuan ho yuan-chhan ti 從木蘭化石論茶樹起源和原產地.
The Evolution and Original Habitat of Tea as seen from Fossils of the Magnoliaceae.
NYKK, **1997** (2), 193–8.

Ho I 合毅 & Ting Wang-Phing 丁望平 (1986).
Wo-kuo hisien tshun tsui tsao ti shih liao chuan lun 我國現存最早的食療專論.
The Earliest Special Monographs Extant on Diet Therapy in China.
PJSH, 513–15.

Ho Ping-Ti 何炳棣 (1969).
Huang-thu yü Chung-kuo nung-yeh ti chhi-yüan 黃土與中國農業的起源.
Loess and the Origin of Agriculture in China.
Chinese University Press, Hong Kong.

Ho Suan-Chhuan 何雙全 (1986).
Chü-yen Han chien so chien Han tai nung tso-wu hsiao-khao 居延漢簡所見漢代農作物.
Han Crops as seen in Bamboo Slips found in Chü-yen.
NYKK, **1986** (2), 252–6.

Honan shêng po-wu-guan 河南省博物館 (1981).
Honan sheng wen-wu khao-ku kung-tso shan-shih nien 河南省文物考古工作三十年.
Thirty Years of Archaeological Investigation of Cultural Relics in Honan.
Wen Wu, Peking.

Honan shêng Hsin-yang ti-chhü wên kuan hui 河南省信陽地區文管會 etc. (1986).
Lo-shan Thien-hu Shang-Chou mu-ti 羅山天湖商周墓地.
The Shang and Chou Cemetery at Thien-hu, Lo-shan County.
KKHP, **1986** (2), 153–97.

Ho-nan shêng wên-wu yen-chiu so 河南省文物研究所 (1989).
Honan Wu-yang Chia-hu hsin shih chhi shih-tai i-chih ti er chih liu tzhu fa-chüeh chien-pao 河南省舞陽賈湖新石器時代遺址第二至六次發掘簡報.
Brief Report of the 2nd to 6th Excavations at the Neolithic Site at Chia-hu, Wu-yang, Honan.
WW, **1989** (1), 1–14.

Ho-nan wên-hua-chü wên-wu kung-tso tui 河南文化局文物工作隊, Honan Provincial Bureau of Cultural Relics (1956).
Chêng-chou ti-wu wên-wu chhü, ti-i hsiao-chhü fa-chüeh chien pao 鄭州第五文物區第一小區發掘簡報.
Brief Report of the Excavation at the First Subdivision of the Fifth Cultural Division at Chêng-chou.
WWTKTL, **1956** (5), 33–40.

Honan wên-wu yen-chiu-so 河南文物研究所, The Henan Provincial Institute of Archaeology (1993).
Mihsien Ta-hu-thing Han-mu 密縣打虎亭漢墓.
Han Dynasty Tombs at Ta-hu-thing Village in Mihsien County.
Wen Wu, Peking.

Hopei shêng khao-ku yen-chiu so 河北省文物研究所 (1985).
Kao-chhêng Thai-hsi Shang-tai i-chih 蒿城台西商代遺址.
The Shang Site at Thai-hsi, Kao-chhêng.
Wen Wu Press, 1985.

Ho-pei shêng wên-hua chü wên-wu kung-tso tui 河北省文化局文物工作隊, Hopei Provincial Cultural Relics Bureau (1965).
Hopei I-hsien Yen hsia-tu ku chhêng khan chha ho shih-chü 河北易縣燕下都故城勘察和試掘.
Preliminary Excavation and Examination of the Yen Capital at I-hsien, Hopei.
KKHP **1965** (1), 83–106.

Hou Hsiang-Chhuan 侯祥川 (1954).
Wo-kuo ku shu lun chiao chhi 我國古書論腳氣.

Discussion of Beriberi in Chinese Classics.
Chung-Hua I Shih Tsa Chih **1954** (1), 16–20.

Hsia Chih 夏志 (*1984*).
Chhao fa su yuan 炒法溯源.
The Origin of *Chhao* (Stir-Frying).
CKPJ, **1984** (1), 13–15.

Hsia Nai 夏鼐 (*1977*).
Than-14 tshê-ting nien-tai ho Chung-kuo shih chhien khao-ku-hsüeh 碳14測定年代和中國史前考古學.
Carbon 14 Dating and the Study of Chinese Prehistory.
KK, **1977** (4), 217–32.

Hsia Nai 夏鼐 (*1985*).
Chung-kuo wên-ming ti chhi-yuan 中國文明的起源.
The Origin of Chinese Civilisation.
Wen Wu, Peking.

Hsia Wei-Ying 夏緯英 (*1956*).
Lü Shih Chhun-Chhiu Shang-Nung têng shih-phien chiao-shih 呂氏春秋上農等四篇校釋.
An Analytic Study of the 'Exaltation of Agriculture' and Three Similar Chapters in *Master Lü's Spring and Autumn Annals*.
Chung Hua, Peking.

Hsia Wei-Ying 夏緯英 (*1979*).
Chou Li shu-chung yu-kuan nung-yeh thiao-wên ti chieh-shih 周禮書中有關農業條文的解釋.
Explanations of Passages Related to Agriculture from the *Chou Li*.
Agricultural Press, Peking.

Hsia Wei-Ying 夏緯英 (*1980*).
Tui 'Shih Ching' chung yu kuan nung shih chang chü ti chieh shih 對《詩經》中有關農事章句的解釋.
Explanations of Passages in the Shih Ching Related to Agriculture.
Agricultural Press, Peking.

Hsiang An-Chhiang 向安強 (*1993*).
Thung-thing hu chhü shih chhien nung-yeh chhu-than 洞庭湖區史前農業初探.
Preliminary Investigation of the Pre-Historic Agriculture in the Thung-thing Lake Region.
NYKK, **1993** (1), 23–43.

Hsiang Hsi 向熹 (*1986*).
Shih Ching Tzhu Tien 詩經辭典.
A Cyclopedia of Words in the Book of Odes.
Szechuan Renmin, Chengtu.

Hsieh Chhung-An 謝崇安 (*1985*).
Chung-kuo yüan-shih chhu-mu-yeh ti chhi-yüan ho fa-chan 中國原始畜牧業的起源和發展.
Origin and Development of Primitive Stock Farming in China.
NYKK, **1985** (1), 282–9.

Hsieh Chhung-An 謝崇安 (*1991*).
Phêng-thou-shan shih-chhien tao-tso I-chhan ti fa-hsien chi chhi i-i 彭頭山史前稻作遺存發現及其意義.
Prehistoric Remains of Rice at Phêng-thou-shan and their Significance.
NYKK, **1991** (1), 178–80.

Hsieh Kuo-Chên 謝國楨 (*1980*).
Ming tai shê-hui ching-chi shih-liao hsüan pien (Shang, Chung, Hsia) 明代社會經濟史料選編 (上，中，下).
Selections from the Social and Economic Historical Records of the Ming Dynasty (Parts I, II and III). Fukien Renmin, Foochow.

Hsieh Ta-Huang ed. 謝大荒 (*1959*).
I Ching Yü Chieh 易經語解.
The *Book of Changes* with Explanations.
Yuan-Tung, Taipei.

Hsin Shu-Chih 辛樹幟, revised by I Chhing-Hêng 伊欽恒 (*1983*).
Chung-kuo kuo shu shih yen-chiu 中國果樹史研究.
Studies on the History of Fruit Trees in China.
Agricultural Press, Peking, 2nd edn.

Hsing Hsiang-Chhên 邢湘臣 (*1989*).
Yü yeh yung ping hsiao shih 魚業用冰小史.
Mini-History of the Use of Ice in Fishery.
NYKK, **1989** (1), 298–9.

Hsing Hsiang-Chhên 邢湘臣 (*1997*).
Khuai tzu shih hua 筷子史話.
About the History of Chopsticks.
NYKK, **1997** (1), 236–40.

Hsing Jung-Chhuan 刑潤川 (*1982*).
Ku-tai niang-chiu chi-shu yü khao-ku fa-hsien 古代釀酒技術與考古發現.
Ancient Fermentation Technology and Archaeological Discoveries.
KCSWC, *1982* (9), 93–8.

Hsiung Kung-Tsê 熊公哲, ed. (*1975*).
Hsün Tzu Chin Chu Chin Shih 荀子今注今釋.
The Book of Master Hsün: Annotations and Translation.
Commercial Press, Taipei, 1975, rev. 1984.

Hsü Cho-Yün 許倬雲 (*1971*).
Liang Chou Nung-tso Chi-shu 兩周農作技術.
Agricultural Techniques of the Western and Eastern Chou.
AS/BIHP, 1971, *42* (4), 803–42.

Hsü Cho-Yün 許倬雲 (*1991*).
Chung-kuo wên-hua yü shih-chieh wên-hua 中國文化與世界文化.
Chinese Civilisation and World Civilisation.
Kweichow Renmin.

Hsü Cho-Yün 許倬雲 (*1993*).
Chung-kuo Chung-ku Shih-chhi Yin-shih Wen-hua ti Chuan-pien 中國中古時期飲食文化的轉變.
The Transformation of the Chinese Diet in Early Mediaeval Times.
The Second Symposium on Chinese Dietary Culture, Taipei, 1993.

Hsü Chu-Wên 徐祖文 (*1990*).
Ma nai chiu chhu than 馬奶酒初談.
Preliminary Discussion of Wine from Mare's milk.
CKPJ, **1990** (11), 17.

Hsü Hung-Shun 許洪順 (*1961*).
Huang-chiu shêng-chhang chi-shu kê-hsin 黃酒生產技術革新.
New Developments in the Fermentation of Yellow Wine.
Light Industry Press, Peking.

Hsü Hung-Shun 許洪順, Chou Chia-Hua 周嘉華 & Liu Chhang-Kuei 劉長貴 ed. (*1987*).
Huang Chiu Niang Tsao 黃酒釀造.

Manufacture of Yellow Wine.
Niang Chiu 釀酒, Harbin.
Hsü Tsung-Shu 徐中舒 (*1980*).
Han-yü ku wên-tzu tzu-hsing piao 漢語古字字形表.
The Configuration of Ancient Chinese Characters.
Chung Hua, Hong Kong.
Hsüeh Pao-Chhen 薛寶辰 (c. *1900*).
Su shih shuo lüeh 素食說略.
Brief Talks on a Vegetarian Diet.
Ed. and annot. by Wang Tsu-Hui (1984) in *CKPJKCTK* Series.
Hu Chia-Phêng 胡嘉鵬 (*1986*).
Yeh than tou-fu chhuan jih 也談豆腐傳日.
More about the Transmission of *Tou fu* to Japan.
in *PJSH*, 424–6.
Hu Chia-Phêng 胡嘉鵬 (*1989*).
Yeh than man-thou chhuan jih 也談饅頭傳日.
Further Remarks on the Transmission of *Man-thou* into Japan.
CKPJ, **1989** (8), 13.
Hu Chih-Hsiang 胡志祥 (*1990*).
Chhung fa ho chhu-chiu 舂法和杵臼.
Pounding with Mortar and Pestle.
CKPJ, **1990** (7), 13–14.
Hu Chih-Hsiang 胡志祥 (*1991a*).
Hsien-chhin chhi-fan pu yung chu, pi 先秦吃飯不用箸, 匕.
Chopsticks and Spoons were not used for Eating Cooked Grains in the Pre-Chhin Periods.
CKPJ, **1991** (3), 10–11.
Hu Chih-Hsiang 胡志祥 (*1991b*).
Chu-shih yü yin-liao 主食與飲料.
Primary Food and Drinks.
CKPJ, **1991** (4), 9–10.
Hu Chih-Hsiang 胡志祥 (*1994*).
Hsien-Chhin chu-shih phêng-shih fang-fa than chê 先秦主食烹食方法探折.
Exploration of Culinary Methods in the Pre-Chhin Period.
NYKK, **1994** (1), 214–18, 213.
Hu Chih-Hsiao 胡志孝 (*1984*).
Man thou man hua 饅頭漫話.
Random Talks on Steamed Buns.
CKPJ, **1984** (1), 18–19.
Hu Hou-Hsüan 胡厚宣 (*1944*).
Wu ting shih wu-chung chih-shih kho-tzhu khao 五丁時五種記事刻辭考.
Studies on Inscriptions on Five Events during the Wu Ting Period.
In *Chia-ku Hsüeh Shang Shih Lun-tshung*, 甲骨學商史論叢 Vol. I.
Chhi-lu University, Chengtu.
Hu Hou-Hsüan 胡厚宣 (*1944*).
Yin tai pu-kuei chih lai-yüan 殷代卜龜之來源.
The Origin of Oracle Bones of the Shang Dynasty.
In *Chia-ku Hsüeh Shang Shih Lun-tshung*, Vol. I.
Chhi-lu University, Chengtu, 1944.
Hu Hou-Hsüan 胡厚宣 (*1945*).
Pu-tzhu chung so chien chih Yin tai nung-yeh 卜辭中所見之殷代農業.

Yin Agriculture as seen in Oracle Bone Inscriptions.
In *Chia-ku Hsüeh Shang Shih Lun-tshung*, Vol. II.
Chhi-lu University, Chengtu.
Hu Hsi-Wên 胡錫文 (*1958*).
Mai, Shang-phien, Chung-kuo nung hsüeh I-chhan hsüan chi chia-lei ti-erh chung 麥, 上篇, 中國農學遺產選集, 甲類, 第二種.
Wheat and Barley, Chinese Agricultural Heritage Series, No. 2.
Chung Hua, Peking.
Hu Hsi-Wên 胡錫文 (*1981*).
Su, shu, chi, ku ming wu ti than-thao 粟黍稷古名物的探討.
An Exploration of the Identity of Ancient *Su*, *Shu* and *Chi*.
Agricultural Press, Peking.
Hu Hsin-Shêng 胡新生 (*1991*).
Than ku-tai jih-chhang shêng-huo chung ti 'San shih' hsi-kuan 談古代日常生活中的三食習慣.
On the Custom of Eating Three Meals a Day in Ancient China.
CKPJ, **1991** (3), 15–16.
Hu Shan-Yüan 胡山源 (*1939*).
Ku chin chiu shih 古今酒事.
On Wines, Ancient and Modern.
World Press, Shanghai.
Hu Tao-Ching 胡道靜 (*1963*).
Shih shu phien 釋菽篇.
Interpretation of the Term *Shu*.
CHWSLT, **1963** (3), 111–19.
Hu Tao-Ching 胡道靜, ed. (*1975*).
Hsin chiao chêng Mêng Chhi Pi Than 新校正夢溪筆談.
'Dream Pool Essays' ed. and annot.
Chung Hua, Hong Kong. 1st pub. in Shanghai, 1957.
Huang Chan-Yüeh 黃展岳 (*1956*).
I-chiu-wu-wu nien chhun Loyang Han-ho-nan hsien-chhêng tung-chhü fa-chüeh pao-kao 一九五五年春洛陽漢河南縣城東區發掘報告.
Report of the Excavations at the Eastern District of Loyang in 1955.
KKHP, **1956** (4), 21–54.
Huang Chan-Yüeh 黃展岳 (*1982*).
Han-tai jên ti yin-shih shêng-huo 漢代人的飲食生活.
Food and Drink During the Han Dynasty.
NYKK, **1982** (1), 71–80.
Huang Chin-Kuei 黃金貴 (*1993a*).
Shih shuo wo-kuo ku-tai ti shih-chih (shang) 試說我國古代的食制 (上).
About the Meal System in Ancient China (I).
CKPJ, **1993** (1), 10–12.
Huang Chin-Kuei 黃金貴 (*1993b*).
Shih shuo wo-kuo ku-tai ti shih-chih (hsia) 試說我國古代的食制 (下).
About the Meal System in Ancient China (II).
CKPJ, **1993** (2), 13–14.
Huang Chhi-Hsü 黃其煦 (*1982, 1983a,b*).
Huang-ho liu-yü hsin-shih-chhi shi-tai nung-kêng wên-hua chung ti tso-wu 黃河流域新石器時代農耕文化中的作物 (1, 2, 3).

Cultivated Crops in the Yellow River Basin during the Neolithic Cultural Period (1, 2, 3).
NYKK, 1. **1982** (2), 55–61; *2.* **1983** (1), 39–50; *3.* **1983** (2), 86–90.

Huang Chu-Liang 黃祖良 (*1986*).
'*Chung-kuo phêng-jen tien chi kou mu*' *pu-i*〈中國烹飪典籍鉤目〉補遺.
An Addendum to 'The Principal Culinary Works of China'.
PJSH, 493–5.

Huang Kuei-Shü 黃桂樞 (*1997*).
Shih-chiai chha wang – Yün-nan Chêng-yuan Chhien-chia-sai yeh-shêng ku chha-shu 世界茶王－雲南鎮沅千家寨野生古茶樹.
The King of Tea – Wild, Ancient Tea Trees in Chhien-chia-chai, Chên-yuan, Yunnan.
NYKK, **1997** (2), 202–3.

Huang Shih-Chien 黃時鑒 (*1988*).
A-la-chi yü Chung-kuo shao-chiu ti chhi-shih 阿剌吉與中國燒酒的起始.
A-la-chi and the Origin of Chinese Distilled Wine.
Wen Shih, **31**, 159–71.

Huang Shih-Chien 黃時鑒 (*1996*).
Chung-kuo shao-chiu ti chhi-shih yü chung-kuo chêng-liu chhi 中國燒酒的起始與中國蒸餾器.
Chinese Stills and the Origin of Distilled Wine in China.
Wên Shih, **41**, 141–52.

Huang Shih-Chien 黃時鑒, ed. (*1994*).
Chung-hsi kuang-shi shih nien piao 中西關系史年表.
Timetables on the History of East–West Relations.
Chechiang Renmin.

Huang Wên-Chhêng 黃文誠 (*1985*).
Fêng-mi niang chiu 蜂蜜釀酒.
Fermentation of Honey to Mead.
Agricultural Press, Peking.

Hu-nan nung-hsüeh-yuan 湖南農學院 (*1978*).
Changsha Ma-wang-tui i-hao Han-mu chhu-thu tung chi wu piao-pên ti yen-chiu 長沙馬王堆一號漢墓出土動植物標本的研究.
Investigation of Plant and Animal Specimens Excavated at Han Tomb No. 1 at Ma-wang-tui near Changsha.
Wên Wu, Peking.

Hu-nan shêng po-wu-kuan 湖南省博物館 (*1963*).
Changsha Sha-tzu-thang Hsi Han-mu Fa-chüeh Chien Pao 長沙砂子塘西漢墓發掘簡報.
Brief Report of the Excavation of a Western Han Tomb at Sha-tzu-thang, Changsha.
WW, **1963** (2), 13–24.

Hu-nan shêng po-wu-kuan 湖南省博物館 (Hunan Provincial Museum) (*1973*).
Changsha Ma-wang-tui I-hao Han-mu, shang, hsia 長沙馬王堆一號漢墓，上，下.
Han Tomb No.1 at Ma-wang-tui near Changsha, Vols. I and II.
Wên Wu, Peking (with an English abstract of the text).

Hu-nan shêng po-wu-kuan 湖南省博物館 (*1974*).
Changsha Ma-wang-tui erh san hao Han-mu fa-chüeh chien pao 長沙馬王堆二三號漢墓發掘簡報.
Brief Report of Excavations at Han Tombs Nos. 2 and 3 at Ma-wang-tui, near Changsha.
WW, **1974** (7), 39–48.

Hu-nan shêng po-wu-kuan 湖南省博物館 (*1984*).
Hunan Tzu-hsing Tung-Han mu 湖南資興東漢墓.
An Eastern Han Tomb at Tzu-hsing, Hunan.
KKHP, **1984** (1), 53–120.

Hu-nan shêng wên-wu-khao-ku yen-chiu so 湖南省文物考古研究所 (*1990*).
Hunan Li-hsien Phêng-thou-shan hsin-shih-chhi shih-tai chao-chhi I-chih fa-chüeh chien pao 湖南澧縣彭頭山新石器時代早期遺址發掘簡報.
A Brief Report of the Excavations of an Early Neolithic Site at Phêng-thou-shan, Li-hsien, Hunan.
WW, **1990** (8), 17–29.

Hu-nan shêng wên-wu-khao-ku yen-chiu so 湖南省文物考古研究所 (*1996*).
Hunan Li-hsien Mêng-chhi Pa-shih-tang hsin shih-chhi shih-tai tsao-chhi I-chi fa-chüeh chien-pao 湖南澧縣夢溪八十壋新石器時代早期遺址發掘簡報.
A Brief Report of the Excavation of an Early Neolithic Site at Pa-shih-tang, Mêng-chhi, Li-hsien, Hunan.
WW, **1996** (12), 26–39.

Hung Kuan-Chih 洪貫之 (*1954*).
Thang Hsien-Chhing 'Hsin Hsiu Pên Tshao' yao phin tshun-mu ti khao-chha 唐顯慶〈新修本草〉藥品存目的考察.
A study of the Preservation of the Index of the [lost] 'Newly Improved Pharmacopoeia'.
I Shih Tsa Chih, **6**, 239.

Hung Kuang-Chu 洪光住 (*1982*).
Tou chiang ho chiang yu shih yü wo-kuo 豆醬和醬油始於我國.
Soy Paste and Soy Sauce Originated in China.
KCSWC, **1982** (9), 99–102.

Hung Kuang-Chu 洪光住 (*1984a*).
Chung-kuo shih-phin kho-chi shih kao, shang tshê 中國食品科技史稿, 上冊.
Draft of a History of Foods in China, Vol. 1.
Commerce Publishers, Peking.

Hung Kuang-Chu 洪光住 (*1984b*).
Tou-fu shêng shih khao 豆腐身世考.
The Origin of *Tou fu*.
CKPJ, **1984** (2), 36–7.

Hung Kuang-Chu 洪光住 (*1985*).
〈*Chhi tzu mien*〉*ming shih chhu than*〈碁子麵〉名實初探.
Preliminary Investigation of the term 'Chess noodles'.
CKPJ, **1985** (5), 33.

Hung Kuang-Chu 洪光住 (*1987*).
Chung-kuo tou fu 中國豆腐.
Chinese Bean Curd.
Commercial Publishers, Peking.

Hung Po-Jên 洪卜仁 (*1983*).
Tou fu chhuan jih chê shui? 豆腐傳日者誰?
Who Transmitted *Tou fu* to Japan?
CKPJ, **1983** (8), 15.

Hung Yeh 洪業, Nie Chung-Chhi 聶崇岐, Li Shu-Chhun 李書春, & Ma Hsi-Yung 馬錫用 eds. (1983).
Chhun Chhiu Ching Chuan Yin-Tê 春秋經傳引得.
Index to the *Spring and Autumn Annals*.
Classics Press, Shanghai, repr. of Harvard-Yenching Index series.

Hu-pei shêng po wu-kuan 湖北省博物館 (1981).
Yunmeng ta-fen-thou I-hao Han mu 雲夢大墳頭一號漢墓.
Han Tomb No. 1 at Ta-fen-thou in Yunmeng.
WWTLTK, No. 4, 1–28.

Hu-pei shêng po wu-kuan Chiang-ling Chi-nan chhêng kung-tso chan 湖北省博物館江陵紀南城工作站 (1980).
I chiu chhi chiu nien Chi-nan chhêng ku-ching fa-chüeh chien-pao 一九七九年紀南城古井發掘簡報.
Brief Report on the Excavation of an Ancient Well at Chi-nan in 1977.
WW, **1980** (10), 42–9.

Ichino Naoko 市野尚子 & Takei Emiko 竹井惠美子 (1985).
Higashi Ajia no tofu zukuri 東アジアの豆腐づくり.
Tou-fu Making in East Asia.
In Ishige Naomichi, ed. (1985), 117–47.

I Min 益民 & Kuang Chün 光軍 (1984).
Shuo 'Khuai' 説〈膾〉About 'Khuai'.
CKPJ, **1984** (1), 7–8.

Ishige Naomichi 石毛直道, ed. (1985).
Ronshu higashi Ajia no shokuji bunka 論集東アジアの食事文化.
Essays on Food and Culture in East Asia.
Heibonsha Ltd, Tokyo.

Ishige Naomichi 石毛直道 & Kenneth Ruddle ケネス・ラドル (1985).
Shikara, gyoshōyu, narezushi 鹽辛，魚醬油，ナレズシ.
Fermented Salted Fish, Fish Sauce and Fish Pickle.
In Ishige Naomichi, ed. (1985), 177–242.

Ishige Naomichi 石毛直道, ed. (1989, 1990).
篠田統資料目錄 I, II *Catalogue of the Sinoda Document Collections at the National Museum of Ethnology*, Vols. I & II.
Bulletin of the National Museum of Ethnology, Special Issue. Nos. 8 & 10.

Ishige Naomichi 石毛直道 & Kenneth Ruddle ケネス・ラドル (1990).
Gyoshō-to-narezushi-no-kenkyū 魚醬とナレズシの研究.
Study of Fermented Fish Products.
Iwanami-shoten, Tokyo.

Ishige Naomichi 石毛直道, tr. Chao Yung-Kuang 趙榮光 (1995).
Yin-shih wên-ming lun 飲食文明論.
Essays on Dietary Culture.
Heilunkiang Science and Technology Press, Harbin.

Ishige Naomichi 石毛直道 (1995).
Bunkamensui gaku-kotahajime 文化麵類學ことはじめ.
World Culture of Noodles.
Kodansha, Tokyo.

Jên Jih-Hsin 任日新 (1981).
Shan-tung chu-chhêng Han-mu hua-hsiang shih 山東諸城漢墓畫像石.
Stone Mural in a Han Tomb in Chu-chhêng, Shantung.
WW, **1981** (10), 14–21.

Kan-su shêng po-wu-kuan 甘肅省博物館, Kansu Provincial Museum (1960).
Kansu Wu-wei Huang-niang-niang-thai i-chih fa-chüeh pao-kao 甘肅武威皇娘娘台遺址發掘報告.
Brief Report of the Excavation at Huang-niang-niang-thai, Wu-wei, Kansu Province.
KKHP, **1968** (2), 53–72.

Kansu wên-wu kuan-li wei-yuan-hui 甘肅文物管理委員會, Kansu Provincial Bureau of Cultural Relics (1959).
Chiu-chhüan Hsia-ho-chhing i-hao ho shih-pa-hao mu fa-chüeh chien pao 酒泉下河清一號和十八號墓發掘簡報.
Brief Report of Excavation of Tombs Nos. 1 and 18 at Hsia-ho-chhing, Chiu-chhüan.
WW, **1959**, *1959* (10), 71–6.

Kao Yang 高揚 (1983).
Ku-ching shih shih 古今食事.
Crown, Taipei.

Kawamura, Wataru 川村渉 & Tatsumi, Hamako 辰已浜子 (1972).
みその本 *Miso no hon*.
The Book of Miso.
Shibata shoten.

Kêng Chien-Thing 耿鑑庭 & Kêng Liu-Thung 耿劉同 (1980).
Chien-chên tung tu yü tou fu chhuan jih 鑒真東渡與豆腐傳日.
Kanshin's visit to Japan and the Transmission of *Tou-fu*.
CKPJ, **1980** (2), 55. repr. in *PJSH*, 420–1.

Kêng Hsüan 耿烜 (1974).
Shih-ching chung ti ching-chi chih-wu 詩經中的經濟植物.
The Economic Plants mentioned in the *Book of Odes*.
Commercial Press, Taipei, 1974.

Khai-fêng ti chhü wên-kuan hui, Khaifêng Cultural Relics Bureau (1978).
Honan Hsin-chêng Phei-li-kan hsin-shih-chhi shih-tai i-chih 河南新鄭裴李崗新石器時代遺址.
Phei-li-kan Neolithic Sites at Hsin-chêng, Honan.
KK, **1978** (2), 73–9.

Khao-ku yen-chiu-so 考古研究所, Institute of Archaeology, CASS. (1959).
Loyang Chung-chou Lu 洛陽中州路.
Report of Excavation at Chung-chou Lu, Loyang.
Science Press, Peking.

Khao-ku yen-chiu so 考古研究所, Institute of Archaeology, CASS (1975).
Ma-wang-tui er, san hao Han-mu fa-chüeh ti chu-yao shou-huo 馬王堆二，三號漢墓發掘的主要收獲.
Major Results of the Excavations of Han Tombs Nos. 2 & 3 at Ma-wang-tui.
KK, **1975** (1), 57–00.

Khao-ku yen-chiu-so 考古研究所 (*1980*).
: Man-chhêng Han-mu fa-chüeh pao kao 滿城漢墓發掘報告.
: Excavations of Han Tombs at Man-chhêng.
: Wên Wu, Peking.

Kho Chi-Chhêng 柯繼承 (*1985*).
: Lu chü khao pien 櫨橘考辨.
: Clarification of *Lu Chü* (loquat).
: *NYKK*, **1985** (2), 251–2.

Khuei Ming 夔明 (*1988*).
: Than 〈Mien chin〉 談〈麵筋〉 About Gluten.
: *CKPJ*, **1988** (9), 22.

Kikkoman K.K. (*1968*).
: Kikkoman shoyu shi キッコーマン醬油史.
: History of Kikkoman Soy Sauce [with Account of Early History of Soy Paste and Soy Sauce in Japan].
: Kikkoman KK, Noda, Japan.

Ko Chia-Chin 葛家瑾 (*1959*).
: Nan-ching hsi-hsia shan chi chhi fu-chin Han-mu chhing-li chien-pao 南京栖霞山及其附近漢墓清理簡報.
: Brief Report on the Han Tombs at Hsi-hsia Shan, Nanking, and Vicinity.
: *KK*, **1959** (1), 21–3.

Ku Fêng 顧風 (*1991*).
: Thang-tai tsai Chung-kuo chha wên-hua shih shang ti ti-wei 唐代在中國茶文化史上的地位.
: The place of Thang Dynasty in the History of Chinese Tea Culture.
: In Wang Chia-Yang, ed., 32–9.

Kuang-chou shih wên-guan hui 廣州市文管會, Kuangchou Municipal Cultural Relics Administration (*1961*).
: Kuangchou tung-chiao Sa-ho Han-mu fa-chüeh chien-pao 廣州東郊沙河漢墓發掘簡報.
: Brief Report of the Excavation of Han Tombs at Sa-ho, in the Eastern Suburb of Kuangchou.
: *WW*, *1961* (2), 54–7.

Kuang-si Chuang-chu tzu-chih chhü wen-wu kung-tso tui 廣西莊族自治區文物工作隊, Kuangsi Chuan Autonomous Region Cultural Relics Unit (*1978*).
: Kuangsi Kuei-hsien Lo-po-wan i-hao mu fa-chüeh chien pao 廣西貴縣羅泊灣一號墓發掘簡報.
: Brief Report on Tomb No. 1 at Lo-po-wan, Kuei-hsien County, Province of Kuangsi.
: *WW*, *1978* (9), 25–42.

Kuang-tung sheng po-wu-kuan 廣東省博物館, Kuangtung Provincial Museum (*1982*).
: Kuangtung Shih-hsing Tsin Thang fa-chüeh chien pao 廣東始興晉唐發掘簡報.
: Brief Report of Excavations of Tsin and Thang Sites at Shih-hsing, Kuangtung.
: Khao Ku Hsüeh Chi Khan 考古學集刊 **2** (Dec.1982), 113–33.

Kung Chih 鞏志 (*1996*).
: Hung-chha fa-hsiang ti Wu-i-shan Thung-mu-tshun 紅茶發祥地武夷山桐木村.
: Thung-mu-tshun in the Wu-I Mountains is the Original Home of Red Tea.
: *NYKK*, **1996** (4), 260–2.

Kung Chih 鞏志 & Yao Yüeh-Ming 姚月明 (*1995*).
: Chien chha shih wei 建茶史微.
: Brief History of Chien-chou (i.e. Northern Fukien) Tea.
: *NYKK*, **1995** (4), 192–201.

Kuo Chhing-Fan ed. 郭慶藩 (*1961*).
: Chuang Tzu Chi Shih 莊子輯釋.
: The Book of Master Chuang, ed. and annot.
: Chung Hua, Peking.

Kuo Po-Nan 郭伯南 (*1987a*).
: Tou fu ti chhi yüan yü tung-chhüan 豆腐的起源與東傳.
: The Origin and Eastward Transmission of Bean Curd.
: *NYKK*, **1987** (2), 373–7.

Kuo Po-Nan 郭伯南 (*1987b*).
: Tsai than tou fu ti chhi yuan yü tung-chhuan 再談豆腐的起源與東傳.
: More on the Origin of Bean Curd and its Transmission Eastwards.
: *CKPJ*, **1987** (3), 17–18.

Kuo Po-Nan 郭伯南 (*1989a*).
: Chiang, Shih, Chiang Yu 醬, 豉, 醬油.
: Fermented Soy Paste, Soy Beans and Soy Sauce.
: *CKPJ*, **1989** (2), 13–15.

Kuo Po-Nan 郭伯南 (*1989b*).
: Tshang ping yü ping tshang 藏冰與冰藏.
: Storage of Ice and Cold Storage.
: *CKPJ*, **1989** (8), 10–12.

Kuo Po-Nan 郭伯南 (*1989c, d, e*).
: Chha yü yin-chha (1, 2, 3) 茶與飲茶 (1, 2, 3).
: Tea and Drinking Tea (Parts 1, 2, 3).
: *CKPJ*, **1989** (9), 9–10; (10), 10–11; (11), 8–9.

Kweichou po-wu-guan 貴州博物館, Kweichow Provincial Museum (*1975*).
: Kweichou Hsing-I, Hsing-jen Han-mu 貴州興義, 興仁漢墓.
: Han Tombs at Hsing-I and Hsing-Jen, Kweichow.
: *WW*, **1975** (5), 20–35.

Lai Hsin-Hsia 來新夏 (*1991*).
: Hsüan chuan mo shang liu chhiung i, Chu yüeh tang chung kuen hsüeh hua 旋轉磨上流瓊液, 煮月鐺中滾雪花.
: 'Let the jade fluid flow from the quern, and the snowy flowers boil in the moon shaped pan'.
: *CKPJ*, **1991** (1), 13–15.

Li Chhang-Nien 李長年, ed. (*1958*).
: Tou Lei (Shang Phien) 豆類上編.
: Legumes (Part 1), Chinese Agricultural Heritage Series No. 4.
: Chung Hua, Peking.

Li Chhang-Nien 李長年 (*1982*).
: Lüeh shu wo-kuo ku-wu yüan-liu 略述我國穀物源流.
: The Origin of Grain Crops in China.
: *NSYC*, **1982** (2), 14–7.

Li Chhiao-Phing 李喬苹 (*1955*).
: Chung-kuo hua-hsüeh shih 中國化學史.
: History of Chemistry in China.
: Commercial Press, Changsha, 1940; 2nd edn Taipei, 1955.

Li Chhing-Chih 李清志 (1979).
Chin Shih Khun Chhung Tshao My Chuang, erh chhi chüan 金石昆蟲草木狀。二七卷.
Album of Minerals, Insects and Plants (Drawings and Calligraphy).
Bull. National Central Library, new **12** (1), 91–4.

Li Chien-Min 李建民 (1984).
Ta-wên-khou mu-tsung chhu-thu ti chiu-chhi 大汶口墓葬出土的酒器.
Wine Vessels Unearthed at Burial Sites at Ta-wên-khou.
KKYWW, **1984** (6), 64–8.

Li Chih-Chhao 李志超 & Kuang Tsêng-Chien 關增建 (1986).
Hsien tshun Han-tai chêng-liu chhi chhu-khao 現存漢代蒸餾器初考.
Preliminary Study of a Han Dynasty Still.
Science and Technology University of China, Mimeographed MS.

Li Ching-Wei 李經緯 & Chhêng Chih-Fan 程之范 et al., eds., Encyclopedia of Chinese Medicine, Editorial Committee (1987).
Chung-kuo I-hsüeh pai-kho chhüan shu: I-hsüeh shih 中國醫學百科全書：醫學史.
Encyclopedia of Chinese Medicine: History.
Science and Technology Press, Shanghai.

Li-chou i-chih lien-ho khao-ku fa-chueh tui 禮州遺址聯合考古發掘隊 (1980).
Sze-chhuan Hsi-chhang Li-chou fa-hsien ti Han-mu 四川西昌禮州發現的漢墓.
The Han Tomb Discovered at Li-chou, Hsi-chhang, Szechuan Province.
KK, **1980** (12). 1.

Li Fa-Lin 李發林 (1986).
Ku-tai hsüan chuan mo shih than 古代旋轉磨試探.
Tentative Discussion of Rotary Mills in Antiquity.
NYKK, **1986** (2), 146–67.

Li Fan 李璠 (1984).
Chung-kuo tsai phei chih wu fa chan shih 中國栽培植物發展史.
History of the Development of Cultivated Crops in China.
Science Press, Peking.

Li Fan 李璠, Li Ching-I 李敬儀, Lu Yeh 盧曄, Pai Phin 白品 & Chhêng Huan-Fang 程華芳 (1989).
Kansu shêng min-lo hsien Tung-hui-shan hsin shih chhi I chih ku nung-yeh I tshan hsin fa hsien 甘肅省民樂縣東灰山新石器遺址古農業遺存新發現.
New Discoveries of Ancient Agricultural Remains at a Neolithic Site in *Tung-hui* Hill in Min-lo County, Kansu.
NYKK, **1989** (2), 56–73.

Li Ho 栗禾 (1986).
〈*Chih*〉 *ti tshu than* 炙的趣談.
An Amusing Story about '*Chih*'.
CKPJ, **1986** (9), 19.

Li Hua-Jui 李華瑞 (1990).
Chung-kuo shao-chiu chhi-shih ti chêng-lun 中國燒酒起始的爭論.
The Controversy on the Origin of Distilled Wine in China.
Chung-kuo shih yen-chiu tung-thai 中國史研究動態, **1990** (8), 15–19.

Li Hua-Jui 李華瑞 (1995a).
Chung-hua chiu wên-hua 中華酒文化.
Wine in Chinese Culture.
Shansi Renmin, Taiyuan.

Li Hua-Jui 李華瑞 (1995b).
Sung tai chiu ti shêng chhan ho chêng chhüeh 宋代酒的生產和征榷.
The Production and Taxation of Wind During the Sung.
Hopei University Press, Baoding.

Li Kuang-Thao 李光濤, Ho Chhiang 何強 & Ho Shih-Hua 何仕華 (1997).
Yun-nan Lan-tshang hsien Mang-ching Ching-mai tsai-phei hsing ku cha-lin lueh khao 雲南瀾滄縣芒景景邁栽培型古茶林略考.
A Brief Investigation of the Ancient Domesticated Type of Tree Forest at Ching-mai, Mang-ching, Lan-tshang County, Yunnan.
NYKK, **1997** (2), 199–201.

Li Pin 李斌 (1991).
Shuo po tho 說餺飥.
About *Po-tho*.
CKPJ, **1991** (3), 14.

Li Pin 李斌 (1992).
Thang-Sung wên-hsien chung ti 'shao-chiu' shih fou chêng-liu chiu wên-thi 唐宋文獻中的燒酒是否蒸餾酒問題.
The Question of Whether the *Shao-chiu* Found in the Thang–Sung Literature is Distilled Wine.
CKKCSL, **13** (1), 78–83.

Li Thao 李濤 (1936).
Chung-kuo jen chhang huan ti chi chung ying yang pu chu ping chien khao 中國人常患的幾種營養不足病簡考.
Several Nutritional Deficiency Diseases Seen Among the Chinese.
Chung Hua I Hsüeh Tsa Chi, 中華醫史雜誌, **22** (11), 1027–38.

Li Thao 李濤 & Liu Szu-Chih 劉思職 (1953).
Shêng wu hua-hsüeh ti fa chan 生物化學的發展.
The Development of Biochemistry [in China].
Chung-hua I-shih tsa chih 中華醫史雜誌, **1953** (3), pp. 149–53.

Li Wên-Hsin 李文新 (1955).
Liao-yang fa-hsien ti san tso pi-hua ku-mu 遼陽發現的三座壁畫古墓.
Discovery of Murals in Three Ancient Tombs in Liao-yang.
WWTKTL, **1955** (5).

Li Yang-Sung 李仰松 (1962).
Tuei wo-kuo niang-chiu chhi-yüan ti than-thao 對我國釀酒起源的探討.
Origin of Wine Fermentation in China.
KK, **1962** (1), 41–4.

Li Yang-Sung 李仰松 (1993).
Wo-kuo ku-wu niang-chiu chhi-yuan hsin-lun 我國穀物釀酒起源新論.
A New Discussion on the Origin of the Fermentation of Grains to Wine in China.
KK, **1993** (6), 534–42.

Li Yen-Heng 李衍恒, ed. (1987).
   Chung-kuo chiu wen-hua yen-chiu wen-chi 中國酒文化文集.
   Essays on Chinese Wine Culture.
   Kuangtung Renmin 廣東人民.

Li Yen-Pang 李延邦 & Liu Ju-Wên 劉汝溫 (1982).
   Yu yung ya-ma shih lo 油用亞麻史略.
   The Use of Flax as a Source of Oil.
   NYKK, **1982** (2), 86–8.

Li Yin 黎瑩, ed. (1989).
   Chung-kuo chiu wen-hua ho Chung-kuo ming chiu 中國酒文化和中國名酒.
   Chinese Wine Culture and Famous Chinese Wines.
   China Food Press, 中國食品出版社.

Li Yü-Fang 李毓芳 (1986).
   Chhien than wo-kuo kao-liang ti tsai-phei shih-tai 淺談我國高粱的栽培時代.
   Brief Discourse on when Sorghum was first Cultivated in China.
   NYKK, **1986** (1), 267–70.

Li Yü-Shu 李漁叔, ed. (1974).
   Mo Tzu chin chu chin I 墨子今注今譯.
   The Book of Master Mo: Annotations and Translation.
   Commercial Press, Taipei, 1974.

Liang Chang-Chü 梁章鉅, (c. 1845).
   Kui Thien Suo Chi 歸田瑣記.
   Miscellanies on Returning Home.
   Private edn.

Liang Chhi-Chhao 梁啟超 (1925).
   Chung-kuo shih hsü lun 中國史敘論.
   Remarks on Chinese History.
   Ch. 34 in *Yin ping shih wên chi* 飲冰室文集.
   Collection from the Ice Drink Studio [Collected Works of Liang Chhi-Chhao].
   Commercial Press, Shanghai.

Liang Chhi-Chhao 梁啟超 (1955).
   Ku shu chên wei chi chhi nien-tai 古書真偽及其年代, pp. 109–17.

Liang-shi pu yu-chih chü chia kung chhu 糧食部油脂局加工處, Ministry of Grains, Bureau of Fats and Oil, Processing Bureau (1958).
   Thu-fa cha-yu shê-pei kê-hsin 土法榨油設備革新 (1).
   Improvement of Traditional Implements for Oilseed Pressing, No. 1.
   Light Industry Press, Peking.

Lin Chêng-Chhiu 林正秋 (1989).
   Chung-kuo man-thou chhuan ju Jih-pen than-shu 中國饅頭傳入日本探述.
   The Introduction of Chinese *Man-thou* into Japan.
   CKPJ, **1989** (1), 14–15.

Lin Chhing-Hu 林慶弧, ed. (1994).
   Ti-san chü Chung-kuo yin-shih wên-hua hsüeh-shu yen-thao hui lun-wên chi 第三屆中國飲食文化學術研討會論文集.
   Proceedings of the 3rd Symposium on Chinese Dietary Culture.
   Foundation of Chinese Dietary Culture, Taipei.

Lin Hao-Jan 林浩然, ed. (1985).
   Fên 粉.
   Rice Flour Pastry.
   Yin-shih Thien-ti Press 飲食天地出版社, Hong Kong.

Lin Nai-Shen 林乃燊 (1957).
   Chung-kuo ku-tai ti phêng-tiao ho yin-shih 中國古代的烹調和飲食.
   Cookery and Diet in Ancient China.
   Peiching ta-hsüeh hsüeh-pao 北京大學學報, **1957** (2), 131–44.

Lin Nai-Shen 林乃燊 (1986a).
   Tshung chia-ku wên khan wo-kuo yin-shih wên-hua ti yuan-liu 從甲骨文看我國飲食文化的源流.
   The Origin of Chinese Dietary Culture as seen from Oracle Bone Inscriptions.
   PJSH, 38–43.

Lin Nai-Shen 林乃燊 (1986b).
   Ku-tai ti shu-tshai shêng-tshan 古代的蔬菜生產.
   The Production of Vegetables in Ancient China.
   PJSH, 44–7.

Lin Nai-Shen 林乃燊 (1986c).
   Chung-kuo ku-tai ti phêng-thiao I-shu ho shih-liao wei-shêng 中國古代的烹調藝術和食療衛生.
   The Culinary Arts, Diet Therapy and Hygiene in Ancient China.
   PJSH, 48–54.

Lin Nai-Shen 林乃燊 (1986d).
   Chih-yen, hsiang liao, kuo-chiang, thang-lei ho niang-tsao yeh ti fa-shêng yü shih yung 制鹽，香料，果醬，糖類和釀造的發生與使用.
   The Discovery and Utilisation of Salt, Aromatics, Fruit Jam, Sugar and Fermentation Technology.
   PJSH, 55–7.

Lin Phin-Shih 林品石, ed. (1985).
   Lü shih chhun chhiu ching chu ching i 呂氏春秋今注今譯.
   Master Lü's Spring and Autumn Annals: Annotations and Translation.
   Commercial Press, Taipei.

Lin Yen-Ching 林衍經 (1987).
   Wu I niang chiu, chiu ju chhüan 無意釀酒酒如泉.
   The Accidental Brewing of Wine.
   CKPJ, **1987** (5), 17.

Lin Yin 林尹, ed. (1985).
   Chou Li chin chu chin i 周禮今注今譯.
   The Rites of the Chou: Annotations and Translation.
   Shu Mu Wên Hsien, Peking, (Originally published in Taipei).

Lin Yung-Kuei 林榮貴 (1980).
   Chin tai chêng-liu chhi khao luo 金代蒸餾器考略.
   Brief Investigation of the Chin Still.
   KK, **1980** (1), 466–71.

Ling Shun-Shêng 凌純聲 (1958).
   Chung-kuo chiu chih chhi-yuan 中國酒之起源.
   The Origin of Wine in China.
   AS/BIHP, **29**, 883–907.

Ling Shun-Shêng 凌純聲 (1957).
   Thai-phing yang chhü chüeh chiu wên-hua ti pi-chiao yen-chiu 太平洋區嚼酒文化的比較研究.
   Comparison of the Culture of Chewing Grains for Making Wine in the Pacific Region.
   Bull. Inst. Ethnology, Acad. Sinica, **5**.

Liu Chên-Ya 劉振亞 & Liu Phu-Yü 劉璞玉 (1982).
　　Wo-kuo ku-tai Huang-ho hsia-yu ti-yü kuo-shu ti fêng-pu yü pien-chhien 我國古代黃河下游地域果樹分布與變遷.
　　Ancient Distribution and Evolution of Fruit Trees in the Lower Huang Ho Basin.
　　NYKK, **1982** (1), 139–48.

Liu Chên-Ya 劉振亞 & Liu Phu-Yü 劉璞玉 (1994).
　　Wo-kuo ku-tai yin-liao yü lêng-liang shih-phin than yuan 我國古代飲料與冷涼食品探源.
　　The Origin of Cold Drinks and Chilled Foods in Ancient China.
　　NYKK, **1944** (3), 197–200.

Liu Chhang-Jun 劉昌潤 (1986).
　　'Han Fei-Tzu 韓非子'.
　　In Thao Wên-Thai *et al.* (1986).

Liu Chih-I 劉志一 (1994).
　　Kuan yü tao-tso nung-yeh chhi-yuan wên-thi ti thung-hsin 關於稻作農業起源問題的通訊.
　　Letters Discussing the Origin of Rice Cultivation.
　　NYKK, **1994** (3), 54–70.

Liu Chih-I 劉志一 (1996).
　　Yü-chhan-yen I-chih fa-chüeh ti wei-ta li-shi I-i 玉蟾岩遺址發掘的偉大意義.
　　The Great Significance of the Results of Excavations at Yü-chhan-yen, Hunan.
　　NYKK, **1996** (3), 95–8.

Liu Chin-Chin 劉勤晉 (1991).
　　Szechuan pien-chha yü tsang-chu chha wên-hua fa-tsan chhu khao 四川邊茶與藏族茶文化發展初考.
　　Brief Study of Szechuan Border-Tea and the Tea Culture of Tibetans.
　　In Wang Chia-Yang, ed., 105–12.

Liu Chih-Yuan 劉志遠, *et al.* eds. (1958).
　　Ssu-chhuan Han-tai hua-hsiang-chuan I-shu 四川漢代畫像磚藝術.
　　The Art of Han Brick Murals from Szechuan.
　　Chung-kuo ku-tien i-shu 中國古典藝術出版社, Peking.

Liu Chih-Yuan 劉志遠, *et al.*, eds. (1983).
　　Ssu-chhuan Han-tai hua-hsiang chuan yü Han-tai shê-hui 四川漢代畫像磚與漢代社會.
　　Han Society as Seen from the Han Brick Murals found in Szechuan.
　　Wen Wu, Peking.

Liu Chün-Shê 劉軍社 (1994a,b, 1995).
　　Hsien-Chhin jen ti yin-shih shêng-huo (I, II, III) 先秦人的飲食生活.
　　Lifestyle in Food and Drink before the Chhin (I, II, III).
　　NYKK, I: **1994** (1), 210–13; II: **1994** (3), 181–6, 193; III: **1995** (1), 231–6, 241.

Liu Hsien-Chou 劉仙洲 (1963).
　　Chung-kuo ku-tai numg-yeh chi-hsieh fa-ming shih 中國古代農業機械發明史.
　　History of Chinese Agro-Mechanical Inventions.
　　Science Press, Peking.

Liu Kuang-Ting 劉廣定 (1987).
　　Wo kuo chêng-liu chhi yü chêng-liu chiu ti wên-thi 我國蒸餾器與蒸餾酒的問題.
　　The Origin of Chinese Distilled Wine and Distillation Apparatus.
　　Kho Hsueh Shih Thung Hsin 科學史通訊, **6**.

Liu Kuang-Ting 劉廣定 undated MS.
　　Tsai than wo-kuo chêng-liu chiu ti shih-chhi 再談我國蒸餾酒的時期.
　　Private communication, received c. 1990.

Liu O 劉鶚 (1905).
　　Lao-Tshan Yu Chi 老殘遊紀.
　　Travels of Lao Can.
　　Textual refs. are to the 1958 edn, Taiping Press, Hong Kong.
　　Tr. by Yang & Yang, Panda Books, Peking, 1983.

Liu Po 劉波 (1958).
　　Wo-kuo ku-chi chung kuan yu chün lei ti chi shu 我國古籍中關於菌類的記述.
　　Accounts of Microbes in Classical Chinese Texts.
　　Sheng wu hsuen thung pao, **1958** (6), 19–22.

Liu Sung 劉松 & Yeh Ting-Kuo 葉定國 (1986).
　　Li Yü ho tha ti 'Hsien Chhin Ngo Chi' 李漁和他的〈閒情偶記〉.
　　Li Yü and his 'Random Notes from a Leisurely Life'.
　　PJSH, 530–4.

Liu Tsün-Ling 劉峻岭 (1985).
　　Mien chin 麵筋.
　　About Gluten.
　　CKPJ, **1985** (6), 32.

Liu Yü-Chhuan 劉毓瑔 (1960).
　　Shih Ching shi-tai chi shu pien 詩經時代稷黍辨.
　　The Identity of *Shu* and *Chi* in the Book of Odes.
　　NSYCCK, **1960**, 2, 38–47.

Lo Chih-Thêng 羅志騰 (1978, 1985).
　　Wo-kuo ku-tai ti niang-chiu fa-hsiao 我國古代的釀酒發酵.
　　Wine Fermentation in Ancient China.
　　Originally in *Hua-Hsüeh Thung Pao* 化學通報, **1978** (5).
　　In Chao Khuang-Hua, ed. (1985), 557–66.

Lou Tzu-Khuang 婁子匡, ed. (1950s).
　　Min-shu tshung shu chuang hao: yin-shih pu 民俗叢書專號：飲食部.
　　Folk Literature Series: Special Collection of Books on Food and Diet.
　　Chung-kuo Min Shu Hsüeh Huei, 中國民俗學會, Taipei.

Loyang fa-chüeh tui, Chung-kuo kho-hsüeh yuan khao-ku yen-chiu so 洛陽發掘隊，中國科學院考古研究所, Loyang Archaeology Team, IA/CAS, (1963).
　　Loyang hsi-chiao Han-mu fa-chüeh pao-kao 洛陽西郊漢墓發掘報告.
　　Excavations of Western Han Tombs in the Western Suburbs of Loyang.
　　KKHP, **1963** (1), 1–58 (includes an abstract in English).

Loyang khao-ku tuei 洛陽考古隊, Loyang District Archaeology Team (1959).
　　Loyang Shao-kou Han-mu 洛陽燒溝漢墓.
　　Han Tombs at Shao-kou, Loyang.
　　Science Press, Peking.

Loyang po-wu-kuan 洛陽博物館 (1980).
　　Loyang Hsi-kung Chhü Chang-Kuo Chhu-chhi Mu-tsang 洛陽西工區戰國初期墓葬.
　　Funeral Relics of the Early Warring States Period at Hsi-kung District in Loyang.

Loyang Municipal Museum.
WWTLTK (3), 118–20.
Lu Chao-Yin 盧兆蔭 & Chang Hsiao-Kuang 張孝光 (*1982*).
Man-chhêng Han-mu nung-chhi chhu i 滿城漢墓農器芻議.
Agricultural Implements in a Han Tomb in Man-chhêng.
NYKK **1982** (1), 90–6.
Lu Shu 盧蘇 (*1987*).
Pa-kung shan tou fu 八公山豆腐.
The *Tou fu* of the Mount of Eight Immortals.
NYKK, **1987** (2), 378–81.
Lu Wên-Yü 陸文郁 (*1957*).
Shih tshao mu chin shih 詩草木今釋.
A Modern Elucidation of the Plants and Trees Mentioned in the Book of Odes.
Renmin, Tientsin.
Lung Po-Chien 龍伯堅 (*1957*).
Hsien tshun pên tshao shu lu 現存本草書錄.
Bibliographical Study of all Extant Pharmacopoeias (from all Periods).
Renmin Weisheng, Peking.

Ma Chhêng-Yüan 馬承源, ed. (*1988*).
Chung-kuo chhing-thung chhi 中國青銅器.
The Chinese Bronzes.
Classics Press, Shanghai.
Ma Chhêng-Yüan 馬承源 (*1992*).
Han-tai chhing-thung chêng-liu chhi ti khao-ku khao-chha ho shih-yen 漢代青銅蒸餾器的考古考察和實驗.
Investigations of a Han Dynasty Still.
Bulletin of the Shanghai Museum 上海博物館集刊, **1992** (6), 174–83.
Ma Chi-Hsing 馬繼興 (*1985*).
Wo-kuo tsui-ku ti yao-chiu niang-chih-fang 我國最古的藥酒釀制方.
The Earliest Recipes for Medicated Wines in China.
Chao Khuang-Hua, ed. (*1985*), 570–4.
Ma Chi-Hsing 馬繼興 (*1990*).
Chung-i wên-hsien hsüeh 中醫文獻學.
Historiography of Chinese Medicine.
Shanghai Science and Technology Press.
Ma Chi-Hsing 馬繼興 (*1992*).
Ma-wang-tui ku i-shu khao-shih 馬王堆古醫書考釋.
Investigation of the Ancient Medical Texts from Ma-wang-tui.
Hunan Science and Technology Press.
Ma Chi-Hsing 馬繼興 & Li Hsüeh-Chhin 李學勤 (*1975*).
Wo-kuo i fa-hsien ti tsui ku i-fang po shu – 〈*Wu shih erh ping fang*〉 我國已發現的最古醫方帛書〈五十二病方〉.
The Oldest Pharmacopoeia Discovered in China – the Silk Manuscript – 'Prescriptions for Fifty-Two Ailments'.
WW, **1975** (9).
Ma Chien-Ying 馬健鷹 (*1989*).
Chuang-tzu yin-shih kuan chih wo-chien 莊子飲食觀之我見.
Interpretation of Master Chuang's Views on Food and Drink.
CKPJ, **1989** (7), 12–13.

Ma Chien-Ying 馬健鷹 (*1991*).
Yeh than 〈*Shih pu yen ching, khuai pu yan hsi*〉 也談〈食不厭精，膾不厭細〉.
More on 〈*Shih pu yen ching, khuai pu yen hsi*〉.
CKPJ, **1991** (4), 19.
Ma Chien-Ying 馬健鷹 (*1995*).
Yuan-shih shih-chhi phao-shih yü thao-chhi ti fa-ming 原始時期炮食與陶器的發明.
Phao-Roasting and the Invention of Pottery.
CKPJ, **1995** (11), 13.
Ma Shih-Chih 馬世之 (*1983*).
Lüeh lun Han-tu Hsin-chêng ti ti-hsia chien-chu chi lêng-tshang ching 略論韓都新鄭的地下建築及冷藏井.
Brief Discussion of the Underground Structure and Cold Storage at the Han Capital at Hsin-tsêng.
KKYWW, **1983** (1), 80–3.
Man-chhêng fa chüeh tui, Chung-kuo kho-hsüeh yuan khao-ku yen-chiu so, Man-chhêng Excavation Team, IA, CAS (*1972*).
Man-chhêng Han-mu fa-chüeh chi yao 滿城漢墓發掘紀要.
Principal Finds at the Excavation of the Han Tomb at Man-chhêng.
KK, **1972** (1), 8–28.
Ma-wang-tui Han-mu Po-shu 馬王堆漢墓帛書 (*1985*).
Ma-wang-tui Han mu po shu (*Ssu*) 馬王堆漢墓帛書 (肆).
Silk Manuscripts Discovered in Han Tombs at Ma-wang-tui (IV).
Wen Wu, Peking.
Mêng Nai-Chhang 孟乃昌 (*1984*).
Fêng chiu yüan-liu chhu than 汾酒源流初談.
Exploration of the Origin of *Fêng* Wine.
CKKCSL, **5** (4), 40–6.
Mêng Nai-Chhang 孟乃昌 (*1985*).
Chung-kuo chêng-liu chiu nien tai khao 中國蒸餾酒年代考.
Chronology of Distilled Wine in China.
CKKCSL, **6** (6), 31–7.
Mêng Ta-Phêng 門大鵬 (*1976*).
〈*Chhi Min Yao Shu*〉 *chung ti niang chhu* 〈齊民要術〉中的釀醋.
Acta Microbiologica Sinica, **16** (2), 98–101.
Mêng Ta-Phêng 門大鵬 & Chhêng Kuang-Sheng 程光勝 (*1976*).
Wo-kuo ku-tai jên shih ho li-yung wei-shêng-wu ti chhêng-chiu 我國古代認識和利用微生物的成就.
Knowledge and Utilisation of Microorganisms in Ancient China.
Wei shêng wu thung-pao 微生物通報, **5** (1), 39–41.
Miao Chhi-Yü 繆啟愉, ed. (*1981*).
Ssu Shih Tsuan Yao Chiao Shih 四時纂要校釋.
Important Rules for the Four Seasons: An Annotated Edition.
Agricultural Press, Peking.
Miao Chhi-Yü 繆啟愉, ed. (*1981*).
Ssu Min Yüeh Ling chi shih 四民月令輯釋.
Monthly Ordinances for the Four Peoples: An Annotated Edition.
Agricultural Press, Peking.

Miao Chhi-Yü 繆啟愉, ed. (*1982*).
    Chhi Min Yao Shu chiao shih 齊民要術校釋.
    Important Arts for the People's Welfare: An Annotated Edition.
    Agricultural Press, Peking, 1982.

Miao Chhi-Yü 繆啟愉 (*1984*).
    Liang shih shên mo? 粱是什麼? What is Liang?
    NYKK, **1984** (2), 289–93.

Miao Chhi-Yü 繆啟愉 (*1987*).
    〈*Chhi Min Yao Shu*〉 *chung li-yung wei-sheng-wu ti kho-hsueh chheng-chiu* 〈齊民要術〉中利用微生物 的科學成就.
    Scientific Basis of the Microbial Processes Recorded in the *Chhi Min Yao Shu*.
    Ku Chin Nung Yeh 古今農業 **1987** (1), 7–13.

Mu Kung 木公 (*1986*).
    Tien-hsin i tzhu hsiao khao 點心一詞小考.
    A Brief Note on 〈Tien-hsin〉.
    PJSH, 455–6.

Nakayama Tokiko 中山時子, ed. (*1988*).
    Chūgoku shoku bunka jiten 中國食事文化事典.
    Encyclopedia of Chinese Food Culture.
    Kodokawa Shoten, Tokyo.

Nan-ching chung-i hsüeh-yuan 南京中醫學院, Nanking College of Chinese Medicine (*1959*).
    Huang Ti Nei Ching Su Wên, I Shih 黃帝內經素問, 譯釋.
    The Yellow Emperor's Manual of Corporeal Medicine, Questions and Answers: Translation and Annotation.
    Science and Technology Press, Shanghai.

Nan-ching po-wu-kuan 南京博物館, Nanking Museum (*1979*).
    Chiang-su Hsü-i Tung-yang Han-mu 江蘇盱眙東陽漢墓.
    An Eastern Han Tomb at Tung-Yang, Hsü-I, in Kiangsu Province.
    KK, **1979** (5), 412–26.

Nankai Ta-hsueh 南開大學, Nankai University (*1973*).
    Kao Liang Chiu chih Chi Tsao 高粱酒之制造.
    The Fermentation of Sorghum to Wine.
    Report of Nankai University Inst. of Chemistry, 1973 (1), 15–40.

Nie Fêng-Chhiao 聶風喬 (*1985*).
    Khuei chih mi 葵之謎.
    The Riddle of Mallow.
    CKPJ, **1985** (10), 14.

Okura Nagatsune 大藏永常 (*1836*).
    Seiyū Roku 製油錄.
    On Manufacturing of Oil.
    Olearius Editions, tr. by Eiko Ariga and edited by Carter Litchfield (*1974*).

Ou Than-Sheng 歐潭生 (*1987*).
    San-chhien nien gu chiu chhu-thu chi 三千年古酒出土記.
    Recovery of a Three-Thousand-Year-Old Wine.
    In Li Yen-Heng, ed. 124–5.

Ou-yang Hsiao Thao 歐陽小桃 (*1991*).
    Wei-chin nan-pei chhao jen-shih yin-chha shu luo 魏晉南北朝人士飲茶述略.
    Brief Account of Tea Drinking in the Wei, Chin and North and South Dynasties.
    NYKK **1991** (2), 201–2, & 195.

Phan Chêng-Chung 潘振中 (*1988*).
    Hun thun so than 餛飩瑣談.
    Remarks on Wonton.
    CKPJ, **1988** (12), 13.

Phan Chi-Hsing 潘吉星 (*1993*).
    Thien-kung-kai-wu i chu 天工開物譯注.
    Thien-kung-kai-wu with Annotations and Translation.
    Classics Press, Shanghai.

Sakaguchi, Kinichiro 坂口謹一郎 (*1979*).
    Shoyu no ruutsu o saguru 醬油のルーツを探る.
    Searching for the Roots of Shoyu.
    Sekai 世界 Jan. No. *398*, 252–66.

Sasagawa Rinpu 世川臨風 & Adachi Isamu 足立勇 (*1942*).
    Kinsei-nihon shokumotushi 近世日本食物史.
    History of Recent Japanese Foods.
    Yūzankaku, Tokyo.

Sekine Shinryu 關根真隆 (*1969*).
    Narachō Shokuseikatu no Kenkyū 奈良朝食生活の研究.
    Food and Life-Style During the Nara Period.
    Yoshikawa Kōbunkami.

Shang Chih-Chün 尚志鈞 (*1981*).
    〈*Wu shih erh ping fang*〉 *yü* 〈*Shêng nung pên tshao ching*〉 〈五十二病方〉與〈神農本草經〉.
    Comparison of 'Prescriptions for 52 Ailments' with the 'Pharmacopoeia of the Heavenly Husbandman'.
    Monograph on the Medical Manuscripts of Ma-wang-tui. No. 2, 78–81.

Shang Ping-Ho 尚秉和 (*1991*).
    Li-tai shê hui feng-shu shih-wu khao 歷代社會風俗事物考.
    Social Artifacts and Customs through the Ages.
    Yüeh Lu Press, Changsha, 1st pub. Commercial Press, 1941.

Shansi nung-yeh kho-hsüeh yuan, 山西農業科學院 (*1977*).
    Ku tzu tsai phei chi shu 穀子栽培技術.
    The Technology of Millet Cultivation.
    Agricultural Press, Peking.

Shansi shêng wên-kuan-hui 山西省文管會, Shansi Bureau of Cultural Relics (*1959*).
    Hou-ma Tung-Chou shih-tai shao thao-yao chih fa-chüeh chi-yao 侯馬東周時代燒陶窯址發掘記要.
    Notes on the Excavation of a Pottery Kiln Site of the Eastern-Chou Period at Hou-ma.
    WW, **1959** (6); 45–6.

Shansi shêng wên-kuan-hui 山西省文管會 (*1960*).
    1959 Nien Hou-ma 'Niu-tshun ku chhêng' nan Tung-Chou i-chih fa-chüeh chien-pao 1959年侯馬〈牛村古城〉南東周遺址發掘簡報.
    A Brief Report of the 1959 Excavation at an Eastern-Chou Site South of *Niu Tshun* Old City in Hou-ma, Shansi.
    WW, **1960** (8, 9), 10–14.

Shantung shêng wen-wu khao-ku yen-chiu so 山東省文物考古研究所 (1991).
 Chü-hsien Ta-chu-chhun Ta-wên-khou wên-hua mu tsang 莒縣大朱家村大汶口文化墓葬.
 Burials of the Ta-wên-khou Culture from Ta-chu-chhun, Chü-hsien County.
 KKHP, **1991** (2), 167–206.

Shantung shêng wen-wu khao-ku yen-chiu so 山東省文物考古研究所 (1995).
 Shantung Tsao-chuang shi Chien-hsin I-chih ti-i, ti-erh tzhi fa-chüeh chien-pao 山東棗莊市建新遺址第一, 二次發掘簡報.
 A Brief Report on the First and Second Excavations at the Chien-Hsin Site, Tsao-chuang City, Shantung.
 KK, **1995** (1), 13–22.

Shantung ta-hsüeh li-shih hsi etc. 山東大學歷史系等 (1995).
 1984 nien chhiu Chi-nan ta-hsin-chuang I-chih shi-chüeh shu-yao 1984年秋濟南大辛莊遺址試掘述要.
 Test Excavations at the Ta-hsin-chuang Site, Chi-nan, Fall, 1984.
 WW, **1995** (6), 12–27.

Shen-chen po-wu-kuan 深圳博物館 (1987).
 Cf. Li Yen-Heng, ed. (1987).

Shensi sheng po-wu-kuan 陝西省博物館, Shensi Provincial Museum (1958).
 Shên-pei Tung-Han hua-hsiang shih-khê hsüan-chi 陝北東漢畫像石刻選集.
 Selections of Stone Engravings of the Eastern Han Period in Northern Shênsi.
 Wen Wu, Peking (1958).

Shensi shêng Yung-chhêng khao-ku tui 陝西省雍城考古隊 (1978).
 Shensi Fêng-hsiang Chhun-chhiu Chhin-kuo ling-yin i-chi fa-chüeh chien-pao 陝西鳳翔春秋秦國凌陰遺址發掘簡報.
 Brief Report of an Ice House of the Chhin Kingdom, Spring and Autumn Period, in Fêng-hsiang, Shensi.
 WW, **1978** (3), 43–7.

Shensi shêng wên-wu kuan-li wei-yuan hui 陝西省文物管理委員會 (1966).
 Chhin-tu Li-yang i-chih chhu-pu than-chi 秦都櫟陽遺址初步探記.
 Preliminary Report on the Ruins of the Chhin Capital, Li-yang.
 WW, **1966** (1), 10–18.

Shên Thao 沈濤 (1987).
 Chu Than 著探.
 About Chopsticks.
 CKPJ, **1987** (5), 9–11.

Shih Chao-Pêng 施兆鵬 & Huang Chien-An 黃建安 (1991).
 Hunan hei chha shih hua 湖南黑茶史話.
 Historical Notes on the Dark Tea of Hunan.
 In Wang Chia-Yang, ed., 89–96.

Shih Nieh-Shu 史念書 (1986).
 Kao-lu chha khao shu 皋蘆茶考述.
 Investigation of Kao-lu Tea.
 NYKK, **1986** (2), 360–8.

Shih Shêng-Han 石聲漢 (1957–1958).
 Chhi Min Yao Shu ching shih 齊民要術今釋.
 A Modern Translation and Annotation of the *Chhi Min Yao Shu*. 4 Vols.
 Science Press, Peking.

Shih Shêng-Han 石聲漢 (1957).
 Chhung 〈Chhi Min Yao Shu〉 khan Chung-kuo ku-tai ti nung-yeh kho-hsüeh chih-shih 從齊民要術看中國古代的農業科學知識.
 Ancient Chinese Scientific Knowledge of Agriculture as seen from the *Chhi Min Yao Shu*.
 Science Press, Peking.

Shih Shêng-Han (1963).
 Shih lun wo-kuo chhung 'hsi yü' yin ju ti chih-wu yü Chang Chhien ti kuan hsi 試論我國從西域引入的植物與張騫的關系.
 A Preliminary Study of the Relationship between Plants Transmitted from the Western Region and Chang Chhien.
 KHSCK, **5**, 16–33.

Shih Shêng-Han 石聲漢, ed. (1979).
 Nung Chêng Chhüan Shu, Chiao Chu 農政全書校注.
 The 'Complete Treatise of Agriculture' with Commentaries.
 Classics Press, Shanghai, 1979.

Shih Shêng-Han 石聲漢 (1980).
 Chung-kuo ku-tai nung-shu phing chieh 中國古代農書評介.
 A Critical Introduction to the Agricultural Treatises of Ancient China.
 Agricultural Press, Peking, 1980.

Shih Shêng-Han 石聲漢 (1984).
 Chhi Min Yao Shu (yin shih pu) 齊民要術(飲食部).
 The Food and Drink Section of the *Chhi Min Yao Shu*.
 Commercial Publishers, Peking, *CKPJKCTK* series.

Shinoda Osamu 篠田統 (1947).
 Pai kan chiu 白乾酒.
 White Dry Wine (Distilled Wine).
 Tr. Yü Chin-Jan 于景讓.
 TLTC, **14** (1), 418–21.

Shinoda Osamu 篠田統 (1951).
 Gokuku no Kigen 五穀の起源.
 Origin of the Five Grains [in East Asia].
 Shizen to Bunka 自然と文化, **2**, 37. Repr. in Shinoda tr. (1987), 3–32.

Shinoda Osamu 篠田統 (1952); tr.Yü Ching-Jang 于景讓 (1957).
 Cha khao: chung-kuo cha ti pien chhien 鮓考 – 中國的鮓的變遷.
 On *Cha* – the Evolution of *Cha* in China.
 TLTC, **15** (2), 39–44.

Shinoda Osamu 篠田統 (1957).
 Sushikou, sono 9 Sushi Nenpyo – Shina no bu 鮓考その9, 鮓年表 – シナの部.
 Study on Sushi, Part 9, Chronological Table for China.
 Seikatsubunka 生活文化研究 (6), Osakakugei University, 39–54.

Shinoda Osamu 篠田統 (1961).
: Sushikou, sono 10 Sushi Nenpyo – Nihon no bu 鮓考その10，鮓年表‐日本の部.
: Study on Sushi, Part 10, Chronological Table for Japan.
: Seikatsubunka 生活文化研究(10), Osakakugei University, 1–30.

Shinoda Osamu 篠田統 (1963).
: O tofu no hanashi お豆腐のはなし.
: Origin of Tou fu.
: Gaku-tano 樂味 1963 (6), 4–8.

Shinoda Osamu 篠田統 (1967).
: Nihonshu no Genryu 日本酒の源流.
: The Origin of Japanese Wine.
: Nihonminzuku to Nanpōbunka 日本民族と南方文化 552–74.
: The Japanese and Cultures of the Southern Areas, Committee for the Commemoration of Takeo Kanaseki's 70th Birthday, 1967.

Shinoda Osamu 篠田統 (1968).
: Tou fu kao 豆腐考.
: The Origin of Tou fu.
: Fuzoku 風俗, **8** (11), 30–6.
: Tr. Yü Chin-Jan 于景讓, TLTC 1971, **42** (6), 173–00.

Shinoda Osamu 篠田統 (1970).
: Kome no bunkashi 米の文化史.
: Cultural History of Rice.
: Shakaisisosha, Tokyo.

Shinoda Osamu 篠田統 (1972a).
: Chiu Sei no Saké 中世の酒.
: Wine in Mediaeval China.
: In Shinoda and Tanaka (1973), vol. I: 321–39 (522–41).

Shinoda Osamu 篠田統 (1972b).
: Soo-Gen Sakè than-shi 宋元酒造史.
: Wine Making During Sung and Yüan.
: In Shinoda and Tanaka (1973), vol. I: 279–329 (541–90).

Shinoda Osamu 篠田統 (1974).
: Chugoku Shokumotsu Shi 中國食物史.
: A History of Food in China.
: Shibata Shoten, Tokyo.

Shinoda Osamu 篠田統, tr. (1987a).
: Chung-kuo shih-wu shih yen-chiu 中國食物史研究.
: Studies on the History of Food in China.
: Tr. Kao Kuei-Lin 高桂林, Hsieh Lai-Yün 薛來運 & Sun Yin 孫音.
: Commercial Publishers, Peking.

Shinoda Osamu 篠田統 (1987b).
: Chung-shih shih-ching khao 中世食經考.
: In Shinoda, tr. (1987), 99–116.

Shinoda Osamu 篠田統 & Tanaka Seiichi 田中靜一, eds. (1972).
: Chugoku shokkei sosho (I) 中國食經叢書，上.
: A Collection of Chinese Food Classics, I.
: Shoseki Bunbutsu Ryutsukai, Tokyo.

Shinoda Osamu 篠田統 & Tanaka Seiichi 田中靜一, eds. (1973).
: Chugoku shokkei sosho (II) 中國食經叢書，下.
: A Collection of Chinese Food Classics, II.
: Shoseki Bunbutsu Ryutsukai, Tokyo.

Su Chao-Chhing 蘇兆慶, Chhang Hsing-Chao 常興照 & Chang An-Li 張安禮 (1989).
: Shantung Chü-hsien Ta-chu-chhun Ta-wen-khou wên-hua mu-ti chhing-li chien-pao. 山東莒縣大朱村大汶口文化墓地清理簡報.
: A Brief Report of the Re-examination of the Ta-wen-khou Cemetery at Ta-chu-chhun, Chu-hsien County, Shantung.
: SCYC, **1989**, 94–113.

Sugama Seinosuke 管間誠之助 (1993).
: Jih pên chêng-tsung shao-chiu te chhi-yuan yü fa-chang 日本正宗燒酒的起源與發展.
: Origin and Development of Typical Japanese Honkaku-Shocho.
: Wang & Ho, et al., eds. (1993), 117–28.

Ssu-chhuan shêng po-wu-kuan 四川省博物館 (1982).
: Kuang-yuan hsien wên-kuan su 廣元縣文管所.
: Szechuan Kuang-yuan shih-kho Sung mu chhing-li chien-pao 四川廣元石刻宋墓清里簡報.
: Brief Report on the Stone Murals of a Sung Tomb at Kuang-yuan, Szechuan.
: WW, **1982** (6), 53–61 & Pl. 7.

Ssu-chhuan shêng wên-wu khao-ku yen-chiu-so 四川省文物考古研究所 (1990).
: Ssu-chhuan Kuang-han hsien Lo-chhêng chên Sung mu chhing li chien pao 四川省廣漢縣雒城鎮宋墓清理簡報.
: A Brief Report on the Sung tomb at Luo-chhêng, Kuang-han County, Szechuan.
: KK, **1990** (2), 123–30.

Sui-hsien Lei-gu-tun yi-hao mu khao-ku tuei 隨縣雷鼓墩一號墓考古隊 (1979).
: Hupei Sui-hsien Tsêng-hou I mu fa-chüeh chien-pao 湖北隨縣曾侯乙墓發掘簡報.
: Brief Report of the Excavation of the Tomb of Marquis I in Sui-hsien, Hupei.
: WW, **1979** (7), 10–11.

Sun Chi 孫機 (1991).
: Han tai wu-tsu wen-hua tsu-liao thu-shuo 漢代物質文化資料圖說.
: Illustrated Compendium of Cultural Relics from the Han Dynasty.
: Wen Wu, Peking.

Sun Chi 孫機 (1996).
: Tou-fu wên-thi 豆腐問題.
: The question of Tou fu.
: In Sun Chi and Yang Hung, pp. 174–8, repr. in NYKK **1998** (3), 277–292–296.

Sun Chi 孫機 & Yang Hung 楊泓 (1996).
: Hsin-chang ti ching-chi 尋常的精致.
: Artistry in Ordinary Things.
: Liaoning Education Publisher, Shenyang.

Sun Chung-En 孫重恩 (1985).
: Chhung pa-chên ti chih-tso khan ku-tai ti phêng-tiao chi-i 從八珍的制作看古代的烹調技藝.
: Ancient Culinary Arts as Seen from the Recipes for Making the Eight Delicacies.
: CKPJ, **1985** (12), 20–3.

Sun Wên-Chhi 孫文奇 & Chu Chün-Po 朱君波 (1986).
: Yao-chiu yen fang hsüan 藥酒驗方選.

A Selection of Chinese Medicated Wines.
China Books, Kong Kong.

Sun Ying-Chhüan 孫穎川 & Fang Hsin-Fang 方心芳 (1934).
Fên chiu yüng shui chi chhi fa-hsiao pi chih fên-hsi 汾酒用水及其發酵秕之分析.
Analysis of the Effect of Water on the Progress of the Fermentation of Fên Wine.
Huang-Hai Reports, 1934 (8).

Sung Yü-Kho 宋玉珂 (1984).
Shih 'Kêng' 釋〈羹〉.
Explanation of 'Kêng'.
CKPJ, **1980** (10), 18–19.

Tai Fan-Chin 戴蕃瑨 (1985).
Ta-tou chi chhi tha liu-chung shih-yung tou-lei chih-wu lai-yüan ti than-thao 大豆及其他六種食用豆類植物來源的探討.
An Exploration of the Origin of Soybean and Six other Food Legumes.
J. Southwest Teacher's College, **1985** (3), 1–10.

Tai Ying-Hsin 戴應新 & Li Chung-Hsüan 李仲煊 (1983).
Shensi Sui-tê-hsien Yen-chia-chha Tung Han hua-hsiang shih mu 陝西綏德縣延家岔東漢畫像石墓.
Stone Mural at an Eastern Han Tomb at Yen-chia-chha, Sui-tê County, Shensi Province.
KK, **1983** (3), 233–7.

Tai Yung-Hsia 戴永夏 (1985).
Chiu shih so than 酒史瑣談.
Random Remarks on the History of Wine.
CKPJ, **1985** (5), 18–19.

Tai Yün 戴雲 (1994).
Thang-Sung yin-shih wên-hua yao-chih khao-shu 唐宋飲食文化要籍考述.
Major Classics of Dietary Culture During the Thang–Sung Era.
NYKK **1994** (1), 226–34.

Tamura, Heiji 田村平治 & Hirano, M. 平野正章 (1971).
Shoyu no hon しょうゆの本.
The Book of Soy Sauce.
Shibata shoten, Tokyo.

Tan Hsien-Chin 單先進 (1989).
Lüeh lun hsien Chhin shih-chhi ti ping chêng chi yu kuan yung ping ti chi-ko wên-thi 略論先秦時期的冰政暨有關用冰的幾個問題.
A Brief Discussion of the Storage and Utilisation of Ice During the Pre-Chhin Era.
NYKK, **1989** (1), 284–97.

Tanaka Seiichi 田中靜一 tr. (1991).
Chung-kuo yin-shih chhuan-ji Jipên shih 中國飲食傳入日本史.
History of the Transmission of Chinese Food and Drink into Japan.
Tr. Huo Fêng 藿風 & I Yung-Wên 伊永文, Heilungkiang Renmin Press, 1991.
Japanese original: *Ichiitaishui: Chungokuryouri-Denraisi*, Shibata shoten, Tokyo, 1987.

Tanaka Tan 田中談 (1985).
Kodai Chugoku Gazo no Kappo to Inshoku 古代中國畫像の割烹と飲食.
Diet and Cookery in Ancient China as Seen from Tomb Paintings.
In Ishige Naomichi, ed. (1985), 245–316.

Tanba Molotane 丹波元胤 (1819).
Chugoku Iseki Koh 中國醫籍考.
Synopsis of Chinese Medical Treatises.
Renmin Weisheng, Peking, 1900.

Tei Daisei 鄭大聲 (1981).
Chosen-shokumotosi 朝鮮食物誌.
The Food of Korea.
Shibata shoten, Tokyo.

Têng Kuang-Ming 鄧廣銘 (1986a).
Sung-tai mien-shih khao shih chih i 宋代麵食考釋之一.
Interpretation of Pasta Foods During the Sung (1).
CKPJ, **1986** (1), 5.

Têng Kuang-Ming 鄧廣銘 (1986b).
Sung-tai mien-shih khao shih chih erh 宋代麵食考釋之二.
Interpretation of Pasta Foods During the Sung.
CKPJ, **1986** (4), 7.

Than Chêng-Pi 譚正壁 (1981).
Chung-kuo wên-hsüeh chia ta tzhu tien 中國文學家大辭典.
Dictionary of Chinese Writers.
Shanghai Press, 1981 (originally pub. Kuang-ming, 1934).

Than Chhi-Kuang 譚旗光 (1987).
Thu-lu-fan chhu-thu ti Thang-tai nang, chiao-tzu, hun-thun ji hua-shih tien-hsin 吐魯番出土的唐代，[食囊]，餃子，餛飩及花式點心.
Thang Dynasty Flat Bread, Chiao-tzu, Wonton and Flower-Shaped Pastries from Turfan.
CKPJ, **1987** (11), 12–13.

Than Yao-Hui 檀耀輝 (1982).
Chian yen-tshai ti fa-hsiao chi-li chi chhi ying-yang chia-chi 醬腌菜的發酵機理及其營養價值.
Principles of the Fermentation of Vegetable Pickles and their Nutritional Value.
Paper presented at the *First National Symposium on Chiang-yen tshai*, Feb. 20, 1982.

Thang Huan 唐荒 (1988).
Fa hsiao mien shi yu ho shih? 發酵麵始于何時？
When was Raised Bread Invented?
CKPJ, **1988** (12), 16.

Thang Lan 唐蘭 (1980).
Changsha Ma-wang-tui Han Tai-hou chhi Hsin-Chui mu chhu thu sui tsang chien-tshê khao shih 長沙馬王堆漢忲侯妻辛追墓出土隨葬簡策考釋.
Elucidation of the Inscriptions on the Bamboo Slips Discovered in Tomb of Lady Hsin-Chui, Wife of the 3rd Marquis of Tai at Ma-wang-tui, near Changsha.
Wen Shih 文史, **1980** (10), 1–60.

Thang Yü-Lin 湯玉林 (1958).
Tsu-kuo ku-tai i-hsüeh tsai yin-shih, ying-yang, wei-shêng hsüeh fang-mien ti kung- hsien 祖國古代醫學在飲食營養偉生學方面的貢獻.
The Contributions of Ancient Chinese Medicine to Diet, Nutrition and Public Health.
I Hsueh Shih yü Pao Chien Chu Chi 醫學史與保建組職 **2**, (1), 54–8.

Thao Chen-Kang 陶振綱 & Chang Lien-Ming 張廉明 (1986).
Chung-kuo phêng-jen wên-hsien thi-yao 中國烹飪文獻提要.
Highlights of Chinese Culinary Literature.
Commercial Publishers, Peking.

Thao Wên-Thai 陶文台 (1983).
Chung-kuo phêng-jên shih lo 中國烹飪史咯.
A Brief History of Chinese Cuisine.
Kiangsu Science and Technology Press.

Thao Wên-Thai, et al. 陶文台等 (1983).
Hsien-Chhin phêng-jên shih-liao hsüan chu 先秦烹飪史料選注.
Selections of Passages of Culinary Interest in the Pre-Chhin Literature, with Annotations.
Commercial Publishers, Peking, *CKPJKCTK* series.

Thao Wên-Thai 陶文台 (1986a).
Chung-kuo phêng-jen tien chi kou mu 中國烹飪典籍鉤目.
The Principal Culinary Works of China.
PJSH, 487–92.

Thao Wên-Thai 陶文台 (1986b).
Liu shih-chi chung-kuo ti shih-phin pai-kho chhüan shu 六世紀中國的食品百科全書.
A 6th Century Chinese 'Food and Drink' Encyclopedia.
PJSH, 496–501.

Thao Wên-Thai 陶文台 (1986c).
Chhu Shan Mi Chi 〈Tiao Ting Ch〉 廚膳秘籍〈調鼎集〉.
The Secret Book of Culinary Delights – 'The Harmonious Cauldron.'
PJSH, 546–9.

Thao Wên-Thai 陶文台 (1988).
Chung-kuo phêng-jen kai lun 中國烹飪概論.
General Discourse on Chinese Cuisine.
Commercial Publishers, Peking.

Thu Ssu-Hua 屠思華 (1956).
Chiang-tu Feng-huang Ho-hsi Han-mu-kuo mu ti chhing-li 江都鳳凰河西漢木槨的清理.
Excavation of a Han tomb with a wooden coffin at Chiang-tu, Feng-huang, Ho-hsi.
KK, **1956** (1), 36–8.

Tsan Wei-Lian 昝維廉 (1982).
Chêng shih wo-kuo ku-tai ti wu ku 正視我國古代的五穀.
The Correct View of the Five Grains of Ancient China.
NYKK, **1982** (2), 44–9.

Tsêng Chao-Yü 曾昭燏, Chiang Pao-Keng 蔣寶庚 & Li Chung-I 黎忠義 (1956).
I-nan ku-hua hsiang shih mu fa chüeh pao-kao 沂南古畫像石墓發掘報告.
Report on the Murals in an Ancient Stone Tomb in I-nan.
Cultural Relics Bureau, Peking.

Tsêng Ching-Min 曾敬民 (1990).
Po I-Erh yü Chung kuo 波義耳與中國.
Robert Boyle and China.
CKKCSL, **1990**, *11* (3) 22–30.

Tsêng Tsung-Yeh 曾縱野 (1986a,b,c).
Wo-kuo niang-chiu than-yüan 1, 2, 3 我國釀酒探源 I, II, III.
The Origin of Wine Fermentation in China (Parts 1,2,3).
CKPJ, **1986** (3), 15–16; (4), 31–2; (6), 23–4.

Tsêng Tsung-Yeh 曾縱野 (1988).
Chung-kuo yin-chuan shih· ti I chüan 中國飲饌史：第一卷.
A History of Food and Drink in China, vol. I.
Commercial Publishers, Peking.

Tsêng Tzu-Fan 曾子凡 (1986).
Kuangchou hua / Phutung hua kou yü tuei-I shou tshê 廣州話／普通話口語對譯手冊.
A Handy Manual of Colloquial Cantonese and Phutunghua Equivalents.
English version tr. Lai, S. K.
Joint Publishing Co., Hong Kong.

Tshao Chin 曹進 (1991).
Hsiang-yin chha lueh khao 湘陰茶略考.
Brief Review of the Tea of Hsiang-yin County.
In Wang Chia-Yang, ed., 97–104.

Tshao Ching 曹進 (1992).
Chang-sha Ma-wang-tui I-hao Han mu ti ku chha khao-chêng chi chhi fang chhü i-i 長沙馬王堆一號漢墓的古茶考証及其防齲意義.
Occurrence of Tea at Han Tomb No. 1 at Ma-wang-tui and its Significance in the the Prevention of Tooth Decay.
NYKK, **1992** (2), pp. 200–3.

Tshao Lung-Kung 曹隆恭 (1983).
Kuan yü chung-kuo hsiao-mai ti chhi-yüan wên-thi 關於中國小麥的起源問題.
On the Origin of Wheat in China.
NYKK, **1983** (1), 19–24.

Tshao Yuan-Yü 曹元于 (1965).
Kuan yü Thang tai yu-mei-yu chêng-liu chiu ti wên-thi 關於唐代有沒有蒸餾酒的問題.
Was there Distilled Wine during the Thang?
KHSCK, **1963** (6), 24–8.

Tshao Yuan-Yü 曹元于 (1979).
Chung-kuo hua-hsueh shih-hua 中國化學史話.
Notes on the History of Chemistry in China.
Kiangsu Science and Technology Press.

Tshao Yuan-Yü 曹元于 (1985a).
Shao-chiu shih liao ti sou-chi he fen-hsi 燒酒史料的搜集和分析.
Assembly and Analysis of Historical Materials on Distilled Wine.
In Chao Kuang-Hua, ed. (1985), 550–6, original in *HHTP*, **1979**, (2).

Tshao Yüan-Yü 曹元于 (1985b).
Tou-fu chih tsao yüan liu khao 豆腐制造源流考.
Search for the Origin of *Tou fu*.
In Chao Khuang-Hua, ed. (1985), 622–8, originally *CKKCSL*, **1981** (4).

Tshao Yuan-Yü 曹元于, ed. (1987).
Pên Tshao Ching 本草經.
Annotated Edition of the *Shên Nung Pên Tshao Ching*.
Science and Technology Press, Shanghai.

Tsi-nan Fêng-huang shan 168 hao Han-mu fa-chüeh chêng-li-chu 紀南鳳凰山 168 號漢墓發掘整理組, Tsi-nan Fêng-huang Shan tomb No. 168 Excavation Unit, (*1975*).
 Hupei Chiang-ling Fêng-huang-shan i-liu-pa hao Han-mu fa-chüeh chien Pao 湖北江陵鳳凰山 168 號漢墓發掘簡報.
 Brief Report of the Excavation of Han Tomb No. 168 at Fêng-huang-shan, near Chiangling, Hupei.
 WW, **1975** (9), 1–8.

Tsou Shu-Wên 鄒樹文 (*1960*).
 Shih Ching shu chi pien 詩經黍稷辨.
 The Identity of *Shu* and *Chi* in the Book of Odes.
 NSYCCK, **1960** (2), 18–34.

Tuan Wên-Chieh 段文杰, ed. (*1957*).
 Yü Lin Khu 榆林窟.
 The Frescoes of Yü Lin Khu [i.e. Wan-fu-hsia, a Series of Cave Temples in Kansu].
 Tunhuang Research Institute, Peking.

Tung Chhu-Phing 董楚平 (*1986*).
 Chhu Tzhu I Chu 楚辭譯注.
 The *Chhu Tzhu* Translated (into Modern Chinese) and Annotated.
 Classics Press, Shanghai.

Tung Hsiao-Chüan 董曉娟 & Wên Wu 聞悟 (*1997*).
 Pu hsin-chang ti tou fu wên-thi 不尋常的豆腐問題.
 The Unusual Question of *Tou fu*.
 Kuangming jipao 光明日報 **1997**, Aug. 26. 5th printing, repr. in *NYKK*, **1998** (3), 273.

Ueda Seinosuke 上田誠之助 (*1992*).
 Nihonkodai no Sakezukuri o omou – Kuchikamizake to Lai 日本古代の酒づくりりを想 – 口嚼酒と醴.
 Discussion of the Origin of Ancient Japanese *saké* – Saké Brewing by Means of Chewing and Sprouting.
 Hakkkokogaku, **70**, 133–7.

Ueda Seinosuke 上田誠之助 (*1996*).
 Shitogi to Kodai no Sake 'づとき' 古代の酒.
 'Uncooked Rice Cake as an Offering to the God' and Ancient Saké Brewing.
 日本釀造協會誌 **91** (7), 498–501.

Wan Ling 萬陵 (*1990*).
 Shuo chêng-liu 說蒸餾.
 On Steam Cookery.
 CKPJ, **1990** (11), 14.

Wan Ling 萬陵 (*1991*).
 Shuo shih 說食.
 About Food and Eating.
 CKPJ, **1991** (5), 8.

Wang Chia-Yang 王家揚, ed. (*1991*).
 Chha ti li-shih yü wên-hua 茶的歷史與文化.
 The History and Culture of Tea.
 Hangchou.

Wang Chih-Tsun 王志俊 (*1994*).
 Shih-lun wo-kuo niang-chiu ti chhi-yuan 試論我國釀酒的起源.
 A Preliminary Discussion of the Origin of Wine in China.
 Wên Po 文博, **1994** (3), 17–21.

Wang Chin 王璡 (*1921*).
 Chung-kuo ku-tai chiu-ching fa-hsiao yeh chih i-pan 中國古代酒酒精發酵之一班.
 A brief Study of Alcoholic Fermentation in Ancient China.
 Kho Hsüeh 科學, **6** (3), 270–82.

Wang Chu-Lou 王竹樓 (*1986*).
 〈*Chung Khuai Lu*〉*ti tso-che Tseng I chi chhi tsuo che* 〈中饋錄〉的作者曾懿及其家世.
 The Background and Authorship of the Chung Khuai Lu.
 PJSH, 554–5.

Wang Fêng-Lin 汪玢玲, et al., eds. (*1994*).
 Chung-kuo gu-wen hsien ta tzhu tien, wen-hsüeh chüan 中國古文獻大辭典, 文學卷.
 Great Dictionary of Ancient Chinese Scholarly Works, Section on Literature.
 Encyclopedia Press, Peking.

Wang Hsüeh-Thai 王學泰 (*1985*).
 Wo-kuo ku-tai tso-tshan ti chu-yao shih-phing 我國古代佐餐的主要食品.
 The Main Entrées served in Ancient China.
 CKPJ, **1985** (10), 15–17.

Wang Jên-Hsiang 王仁湘 (*1985*).
 Ku-tai ti khuai-tzu chi chhi shih-yung 古代的筷子及其使用.
 Chopsticks of Antiquity and How they were Used.
 CKPJ **1985** (9), 20–1.

Wang Jên-Hsiang 王仁湘, ed. (*1993*).
 Thai Phing Yü Lan (*Yin shih pu*) 太平御覽〈飲食部〉.
 The Food and Drink Sections of the *Thai Phing Yü Lan* [edited and annotated].
 CKPJKCTK, Series.

Wang Jên-Hsing 王仁興 (*1985*).
 Chung Kuo Yin Shih Than Ku 中國飲食談古.
 Talks on Food and Drink in Ancient China.
 Light Industry Press, Peking.

Wang Jen-Hsing 王仁興 (*1987*).
 Chung-kuo ku-tai ming tshai 中國古代名菜.
 Famous Dishes of Ancient China.
 Food Press, Peking 1987.

Wang Ling 王玲 (*1991*).
 Liang Chin chih Thang-tai ti yin-chha chih fêng yü Chung-kuo chha wên-hua ti mêng-ya yü hsing-chheng 兩晉至唐代的飲茶之風與中國茶文化的萌芽與形成.
 The Practice of Tea-Drinking from Tsin to Thang and the Sprouting and Fruition of the Culture of Tea in China.
 In Wang Chia-Yang, ed., 20–31.

Wang Lung-Hsüeh 王龍學 (*1991*).
 Yeh shuo po tho 也說餺飥.
 Als about *Po-tho*.
 CKPJ, **1991** (11), 15–16.

Wang Mêng-Ou 王夢鷗, ed. (*1984*).
 Li Chi chin chu chin I 禮記今注今釋.
 The Record of Rites, with Commentaries and Translation.
 Commercial Press, Taipei.

Wang Sai-Shih 王賽時 (*1988*).
 Wo-kuo ku-tai ti lêng-yin yü ping-shih 我國古代的冷飲與冰食.

Consumption of Ice and Iced Foods in Ancient China.
CKPJ, **1983** (7), 17–18.

Wang Shang-Tien 王尚殿 (*1987*).
Chung-kuo shih-phing kung-yeh fa-chan chien shih 中國食品工業發展簡史.
A Brief History of the Development of the Food Processing Industry in China
Shansi Science and Technology Press, Thaiyuan.

Wang Shih-Hsiung 王士雄 (*1861*).
Sui hsi chü yin shih phu 隨息居飲食譜.
Dietetics from the Random Rest Studio.
Chou San-Chin 周三金, ed. (*1985*), *CKPJKCTK*.

Wang Shu-Ming 王樹明 (*1987a*).
Ling-yang-ho mu-ti chhu-i 陵陽河墓地雛議.
Preliminary Discussion of the Ling-yang-ho Cemetery.
SCYC, **1987** (3), 49–58.

Wang Shu-Ming 王樹明 (*1987b*).
Shantung Chü-hsien Ling-yang-ho Ta-wên-khou wên-hua mu-tsang fa-chüeh chien-pao 山東莒縣陵陽河大汶口文化墓葬發掘簡報.
A Brief Report on the Excavation of the *Ta*-wên-khou Burials at Ling-yang-ho, Chü-hsien County in Shantung. *SCYC*, **1987** (3), 62–82.

Wang Shu-Ming 王樹明 (*1987c*).
Ta-wen-khou wen-hua wan chih ti niang chiu 大汶口文化晚期的釀酒.
Fermentation in the Late Ta-wen-khou Period.
CKPJ, **1987** (9), 5–6.

Wang Tsêng-Hsin 王增新 (*1960*).
Liao-yang Phêng-thai-tzu er-hao pi-hua mu 遼陽捧台子二號壁畫墓.
Murals at Tomb No. 2 at Phêng-thai-tzu, Liao-yang.
KK, **1960** (1).

Wang Tzu-Hui 王子輝, ed. (*1984*).
Su shih shuo lüeh 素食說略.
Brief Talks on a Vegetarian Diet.
By Hsüeh Pao-Chhên 薛寶辰 (c. +1900).
CKPJKCTK.

Wang Tzu-Hui 王子輝 (*1988*).
Shuo hua su tshai 說話素菜.
Talks on Vegetarian Diets.
CKPJ, **1988** (6), 8–9.

Wang Wên-Chê 王文哲 (*1991*).
Hung yang min chu yin-shih wên-hua, chien-shê you chung-kuo thê-sê ti shih-phin kung-yeh 弘揚民族飲食文化,建設有中國特色的食品工業.
Propagate Chinese National Dietary Culture; Establish a Distinctively Chinese Food and Beverage Industry.
Introduction to Collected Papers of the First International Symposium on Chinese Dietary Culture. July 1991, Peking, pp. 1–8.

Wang Yen 王炎, Ho Thien-Cheng 何天正, eds. (*1993*).
Hui huang ti shih-chieh chiu wên-hua 輝煌的世界酒文化.
Glories of the World of Wine Culture, Proceedings of the First International Symposium on Alcoholic Beverages and Human Culture.
Chengtu Press, Chengtu.

Wang Yu-Phêng 王有鵬 (*1987, 1989*).
Wo-kuo chêng-liu chiu chhi yuan yü Tung-Han shuo 我國蒸餾酒起源與東漢說.
Distilled Wine may have Existed in China during the Eastern Han.
1st pub. in Li Yen-Heng, ed., 1925 & repr. in Li Ying ed. (*1989*), 277–82 but without illustrations.

Wang Yü-Hu 王毓瑚 (*1964, 1979*).
Chung-kuo nung-hsüeh shu lu 中國農學書錄.
Bibliography of Chinese Agriculture.
Agricultural Press, Peking; 2nd edn 1979.

Wang Yü-Hu 王毓瑚 (*1981a,b, 1982*).
Wo-kuo tzu-ku i-lai ti chung-yao nung tso-wu (shang, chung, hsia) 我國自古以來的重要農作物,上,中,下.
Major Crops of Ancient China (I, II, III).
NYKK, **1981**/1, 79–89; **1981**/2, 13–20; **1982**/1, 42–9.

Wei Chhün 衛群 (*1978*).
〈*Chhi Min Yao Shu*〉 *ho wo-kuo ku-tai wei-shêng-wu hsüeh* 〈齊民要術〉和我國古代衛生物學.
The '*Chhi Min Yao Shu*' and Ancient Chinese Microbiology.
Acta Microbiol. Sinica, **14** (2), 129–31.

Wei Ssu 衛斯 (*1986*).
Wo-kuo ku-tai ping-chên ti-wên chu-chhang chi-shu fang mien ti chung ta fa-hsien 我國古代冰鎮低溫貯藏技術方面的重大發現.
Important Discoveries Relating to Cold Storage in Ancient China.
NYKK, **1986** (1), 115–16 & 142.

Wên Ching-Ming 文景明 & Liu Ching-An 柳靜安 (*1989*).
Hsing-hua chhun yu chiu neng tsui jen 杏花村有酒能醉人.
Getting Drunk from the Wine from the Valley of Apricot Blossoms.
In Li Yin, ed. (*1989*), 289–93.

Wu Chhêng-Lo 吳承洛, 1937; revised by Chhêng Li-Chün 程麗濬 (*1957*).
Chung-kuo tu liang hêng shi 中國度量衡史.
History of Chinese Metrology (Weights and Measures).
Commercial Press, 1937, rev. 1957.

Wu Chhi-Chhang 吳其昌 (*1935*).
Chia-ku chin-wenchung su chien Yin tai nung-chia chhing hsing 甲骨金文中所見殷代農稼情形.
Yin Agriculture as Seen in Oracle Bones and Bronzes.
Commercial Press, Shanghai.

Wu Chia-Khuo 吳家闊 (*1993*).
Shên Nung fa-hsien chha-yeh chieh-tu ti min-chien chhuan-shuo 神農發現茶葉解毒的民間傳說.
The Folk Legend of Shên Nung's Discovery of the Detoxifying Action of Tea.
NYKK **1993** (2), 227–8, & 237.

Wu Chih-Sung 吳稚松 (*1993*).
Mien-chin ti hsing-chhêng chi-li yü mien-thuan ti hsing chih 麵筋的形成機理與麵團的性質.
The Mechanism of Gluten Formation and Properties of Gluten Dough.
CKPJ, **1993** (4), 14–16.

Wu Chio-Nung 吳覺農, ed. (1987).
  Chha Ching shu shih 茶經述釋.
  The 'Classic of Tea' (of Lu Yü) with Annotations and Explanations.
  Agricultural Press, Peking.

Wu Chio-Nung 吳覺農, ed. (1990).
  Chung-kuo ti-fang-chih chha-yeh li-shih tzu-liao hsuan-chi 中國地方志茶葉歷史資料選輯.
  Selected Historiographic Materials on Tea in the Local Gazetteers of China.
  Agricultural Press, Peking.

Wu Fêng 吳楓, *et al.*, eds. (1987).
  Chien ming Chung-kuo ku chi tzhu-tien 簡明中國古籍辭典.
  A Brief Cyclopedia of Ancient Chinese Classics.
  Chi-lin Wên Shih 吉林文史, Changchun.

Wu-hsi chhing-kung-yeh hsüeh-yuan 無錫輕工業學院 & Hopei chhing kung yeh hsüeh-yuan 河北輕工業學院, Wushih and Hopei Colleges of Light Industry (1964).
  Niang Chiu Kung I Hsüeh 釀酒工藝學.
  The Technology of Wine Fermentation.
  Finance and Economic Press, Peking.

Wu Hsien-Wên 吳獻文 (1949).
  Chi Yin-hsü chhu-thu chih yü ku 記殷墟出土之魚骨.
  Record of the Fish Bones Excavated from *Yin-hsü*.
  KKHP, **1949** (4), 139–43.

Wu Shih-Chhih 吳詩池 (1987).
  Chhung khao-ku tzu-liao khan wo-kuo shih-chhien ti yü-yeh shêng-chhang 從考古資料看我國史前的漁業生產.
  Prehistoric Fishery in China as seen in Archaeological Finds.
  NYKK, **1987** (1), 234–48.

Wu Tê-To 吳德鐸 (1966).
  Thang-Sung wen-hsien chung kuan-yü chêng-liu chiu yü chêng-liu chhi wên-thi 唐宋文獻中關於蒸餾酒與蒸餾器問題.
  On the Question of Liquor Distillation and Stills in the Literature of the Thang and Sung Periods.
  KHSCK, **1966** (9), 53–5.

Wu Tê-To 吳德鐸 (1982).
  Ho chia tshun chhu-thu i-yao wen-wu pu-chêng 何家村出土醫藥文物補証.
  Re-Evaluation of the Cultural Relics Discovered at Ho-chia-tshun.
  KK, **1982** (5), 528–31.

Wu Tê-To 吳德鐸 (1988a).
  Chieh khai shao-chiu chhi-yuan chih mi 解開燒酒起源之謎, Part 1.
  Unravelling the Puzzle of the Origin of Distilled Spirits in China.
  Min Pao Yueh Khan 民報月刊, **1988**, July, 84–9.

Wu Tê-To 吳德鐸 (1988b).
  Chieh khai shao-chiu chhi-yuan chih mi 解開燒酒起源之謎, Part 2.
  Unravelling the Puzzle of the Origin of Distilled Spirits in China.
  Min Pao Yueh Khan **1988**, August, 90–3.

Xi Zezong 席澤宗 (1994).
  Kho hsueh shih pa chiang 科學史八講.
  Eight Talks on the History of Science in China.
  Lien Ching, Taipei.

Yamasaki Hiyachi 山崎百治 (1945).
  Tōa Hakko Kagaku Ronshi 東亞發酵化學論考.
  Investigation of the Chemistry of East Asian Fermentations.
  Tokyo.

Yang Ching-Shêng 楊競生 (1980).
  Lun 〈Hsia Hsiao Chêng〉 chung ti liang yu chih wu 論〈夏小正〉中的糧油植物.
  On Cereal and Oilseed Crops in the *Hsia Hsiao Chêng*.
  In *Chung-kuo ku-tai nung yeh kho-chi pien tsuan tsu* 中國古代農業科技編纂組 (1980), 289–303.

Yang-chou shih po-wu kuan 揚州市博物館, Yangchou Municipal Museum (1980).
  Yang-chou tung fêng chiang wa chhang Han tai mu kuo mu chhü 揚州東風磚瓦廠漢代木槨慕群.
  Han Dynasty Graves with Wooden Coffins at the Tile and Brick Works in Yang-chou.
  KK, **1980** (5), 405–17.

Yang-chou shih po-wu-kuan 揚州市博物館 (1980).
  Yangchou hsi Han 'chhieh-mo-shu' mu kuo mu 楊州西漢（妾莫書）木槨墓.
  The 'Chhieh-mo-shu' Western Han Tomb at Yangchou.
  WW, **1980** (12): 1–6.

Yang-chou shih po-wu-kuan 揚州市博物館 (1988).
  Chiang-su Han-chou Yao-chuang 101 hao Hsi-Han mu 江蘇邗州姚莊101號西漢墓.
  Western Han Tomb No. 101 at Yao-chuang, Han-chou in Kiangsu.
  WW, **1988** (2), 19–43.

Yang Ling-Ling 楊玲玲 (1988).
  Chê-yang chhi tsuei pu, vols. 1, 2. 怎樣吃最補. 1, 2.
  How to Eat to get the Most Nourishment. 1, 2.
  Wen-ching, Taipei.

Yang Po-Chün 楊伯峻 & Hsü Thi 徐提 (1985).
  Chhun Chhiu Tso Chuan tzhu-tien 春秋左傳詞典.
  A Cyclopedia of Words in the *Chhun Chhiu Tso Chuan*.
  Chung Hua, Peking.

Yang Shu-Ai 楊淑媛, Thien Yüan-Lan 田元嵐 & Ting Shun-Xiao 丁純孝 (1989).
  Hsin pien ta tou shih phing 新編大豆食品.
  A New Compendium of Soyfoods.
  Commercial Publishers, Peking.

Yang Wên-Chhi 楊文騏 (1983).
  Chung-kuo win-shih wên-hua ho shih-phin kung-yeh fa-chan chien shih 中國飲食文化和食品工業發展簡史.
  A Brief History of the Development of Food Culture and Industry in China.
  Chung-kuo Chan-wang, Peking.

Yang Ya-Chhang 楊亞長 (1994).
  Pan-pho wên-hua hsien-min chih yin-shih khao-ku 半坡文化先民之飲食考古.
  The Archaeology of Food and Drink in the Culture of Pan-pho.
  KKYWW, **1994** (3), 63–72.

Yang Yin-Shêng 楊蔭深 (1986).
  Shih wu chang ku chhung than 事物掌故叢談.

Talks on Various Things and Events.
Shanghai Books, 1986.

Yen Kho-Chün 嚴可均, ed. (*1836*).
Chhüan Shang-ku San-tai Chhin-Han San-Kuo Liu-Chao Wên 全上古三代秦漢三國六朝文.
Complete Collection of Prose Literature from Remote Antiquity through the Chhin–Han Dynasties, the Three Kingdoms and the Six Dynasties.
Repr. 1956, Chung-Hua, Peking.

Yen Wên-Ming 嚴文明 (*1997*).
Wo-kuo tao-tso chhi-yuan yan-chiu ti hsin chin-chan 我國稻作起源研究的新進展.
New Developments in the Study of Rice Domestication in China.
KK, **1997** (9), 1–6.

Yo Hua-Ai 樂華愛 & Fang Hsin-Fang 方心芳 (*1959*).
News Reports on Moulds Occurring in Chinese Ferments.
Wei sheng wu hsueh thung-hsin 微生物學通訊 (*Microbiological News*).
1 (2), 86–9; (3), 151–63.

Yoshida, Shuji 吉田集而 (*1985*).
Minzokugaku kara mita muenhakkodaizu to sono shuhen 民族學から見た無鹽發酵大豆とその周邊.
Ethnological Studies on Non-Salted Soybean Fermentation.
In Aida Hiroshi, *et al.*, eds. (1985), 166–78.

Yu Hsiu-Ling 游修齡 (*1976*).
Tui Ho-mu-tu i-chih ti-si wên-hua tshêng chhu-thu tao-ku ho ku-sha ti chi tien khan-fa 對河姆渡遺址第四文化層出土稻谷和骨耜的幾點看法.
Views on the Remains of Rice and Bone Implements Excavated at Level 4 from the Archaeological site at Ho-mu-tu.
WW, **1976** (8), 20–3.

Yu Hsiu-Ling 游修齡 (*1984*).
Lun shu ho chi 論黍和稷.
On *Shu* amd *Chi*.
NYKK, **1984** (2), 277–88.

Yü Ching-Jang 余景讓 (*1956a,b*).
Shu, chi, su, liang yü kao-liang, I, II. 黍稷粟粱與高粱 I, II.
The Identity of *shu, chi, su, liang* and *kao-liang*, I, II.
TLTC, 1956, **13**, 67–76; 115–20.

Yü Ching-Jang 余景讓, tr. (*1957*).
'*Cha khao* 鮓考 – 中國的鮓的變遷'.
Cf. Shinoda Osamu (1957).

Yü Hua-Chhing 余華青 & Chang Thing-Hao 張廷皓 (*1980*).
Han-tai niang-chiu yeh than-thao 漢代釀酒業探討.
Investigation of the Fermentation Industry during the Han.
Li Shih Yen Chiu 歷史研究, **1980** (5), 99–116.

Yü Hsing-Wu 余省吾 (*1957*).
Shang-tai ti ku-lei tso-wu 商代的穀類作物.
Cereal Crops of the Shang Dynasty.
TPJMTH 東北人民大學人文科學學報, **1957** (1), 81–107.

Yü Hsing-Wu 余省吾 (*1972*).
Tshung chia-ku wên khan Shang-tai ti nung-thien khên-chih 從甲骨文看商代的農田墾殖.
Shang Land Clearance and Cultivation as seen in Oracle Bone Inscriptions.
KK, **1972** (4), 39.

Yü Ming-Hsien 禹明先 (*1993*).
Chhüan Chhien ti-chhü liang chien chiu-shih wên-wu khao shih 川黔地區兩件酒史文物考釋.
Investigation of two Wine Relics unearthed in Szechuan and Kweichow.
Wang & Ho *et al.* (1993), 331–5.

Yü Tê-Tsün 俞德浚 (*1979*).
Chung-kuo kuo shu fen lei hsüeh 中國果樹分類學.
Taxanomy of Fruit Trees in China.
Agricultural Press, Peking.

Yüan Han-Chhing 袁翰青 (*1954*).
Kuan yü 'shêng-wu hua-hsüeh ti fa-chang' i wen ti i-tien i-chien 關於〈生物化學的發展〉一文的一點意見.
A few Comments on 'The Development of Biochemistry [in China]'.
Chung-hua I-shih tsa chih 中華醫史雜誌, **1954** (1), 52–3.

Yüan Han-Chhing 袁翰青 (*1956*).
Chung-kuo hua-hsüeh shih lun wên chi 中國化學史論文集.
Essays on the History of Chemistry in China.
San Lien, Peking, 1956.

Yüan Han-Chhing 袁翰青 (*1981*).
Kuan yü tou-fu ti chhi-yuan wên-thi 關於豆腐的起源問題.
The Problem of the Origin of *Tou fu*.
CKKCSL, **1981** (2), 84–6.

Yüan Han-Chhing 袁翰青 (*1989*).
Niang chiu tsai wo-kuo ti chhi-yüan ho fa-chan 釀酒在我國的起源和發展.
The Origin and Development of Wine Fermentations in China.
In Li Yin, ed. (*1989*), 35–62; originally published in Yüan Han-Chhing (1956).

Yüan Kuo-Fan 袁國藩 (*1967*).
Shih-san shih-chi Meng-ku jen yin-chiu chih hsi-su i-li chi chhi yu kuan wên-thi 十三世紀蒙古人飲酒之習俗儀禮及其有關問題.
The Rites, Custom and Culture of Wine Drinking among the Mongols in the 13th Century.
TLTC, **34** (5), 14–18.

Yüan Thing-Tung 袁庭棟 (*1986*).
Wo kuo ho shih shih niu nai 我國何時食牛奶.
When did the Consumption of Milk begin in China
PJSH, 73–5.

Yün Fêng 耘楓 (*1987*).
Mien-thiao ku chhü 麵條古趣.
Ancient Anecdotes on Filamentous Noodles.
CKPJ, **1987** (11), 9.

Yün Fêng 雲峰 (*1989*).
Shih lun Hopei niang-chiu tzu-liao ti khao-ku fa-hsien yü wo-kuo niang-chiu ti chhi-yüan 試論河北釀酒資料的考古發現與我國釀酒的起源.
A Preliminary Discussion of the Discovery of Archaeological Remains of Fermentation Raw Materials and the Origin of Wine Fermentation in China.
In Li Yin, *et al.* eds., 259–70.

Yün-mêng hsien wên-wu kung-tso chu 雲夢縣文物工作組 (1981).
    Hu-pei Yün-mêng Shui-hu-ti Chhin-Han mu fa chüeh chien-pao 湖北雲夢睡虎地秦漢墓發掘簡報.
    Brief Report on the Excavation of a Chhin-Han Tomb at Shui-hu-ti, Yun-mêng, Hupei.
    KK, **1981** (1), 27–47.

Yün-mêng Shui-hu-ti Chhin-mu pien hsieh tsu 雲夢睡虎地秦漢墓編寫組 (1981).
    Yun-mêng Shui-hu-ti Chhin-mu 雲夢睡虎地秦漢墓.
    The Chhin Tomb at Shui-hu-ti, Yun-meng.
    Wen Wu, Peking.

Yunnan shêng wên wu kung-tso tui 雲南省文物工作隊 (1964).
    Yunnan Hsiang-yun Ta-pho-na mu kuo thung kuan mu chhing-li pao-kao 雲南祥云大波那木槨銅棺墓清理報告.
    Report of the Excavation of a Wooden Tomb with a Bronze Coffin at Ta-pho-na Hsiang-yün, Yunnan.
    KK, **1964** (12), 607–14.

Yünnan shêng wên-wu kung-tso tui 雲南省文物工作隊 (1983).
    Chhu-hsiung Wan-chia-chü ku-mu chhün fa chüeh pao-kao 楚雄萬家垻古墓發掘報告.
    Report of the Excavation of an Ancient Tomb at Wan-chia-chü, Chhu-hsiung.
    Yunnan Bureau of Cultural Relics, 1983.

C. BOOKS AND JOURNAL ARTICLES IN WESTERN LANGUAGES

ACHAYA, K. T. (1993). *Ghani: The Traditional Oilmill of India*, Olearius Editions, New Brunswich, NJ.
ADAMS, ROBERT McC. (1996). *Paths of Fire: An Anthropologist's Inquiry into Western Technology*, Princeton University Press, NJ.
ADOLPH, WILLIAM H. (1926). 'Chinese Foodstuffs: Composition and Nutritive Value', *Trans. Sc. Soc. China*, **4**, 11–32.
ADOLPH, WILLIAM H. (1946). 'Prewar Nutrition in Rural China', *J. Amer. Dietetic Assoc.*, **22**, 946–70.
AIDOO, K. E., HENDRY, R. & WOOD, B. J. B. (1982). 'Solid Substrate Fermentations', *Advances in Appl. Microbiol.*, **28**, 201–37.
AIDOO, K. E., SMITH, JOHN E. & WOOD, B. J. B. (1994). 'Industrial Aspects of Soy Sauce Fermentations using Aspergillus', in Powell, *et al.*, 155–68.
ALBRECHT, H. (1825). *Die vortheilhafteste Gewinnung des Oels; oder Anweisung, höchst möglichen Ölertrag aus öligen Samen und Früchten zu ziehen; nebst Anhang von den besten Vorschriften und Lehren über Aufbewarung, Reinigung und Nutzung aller fetten Oele*, Leipzig.
ALFORD, JEFFREY & DUGUID, NAOMI (1995). *Flatbreads and Flavors: A Baker's Atlas*, William Morrow & Co.
ALGAR, AYLA (1991). *Classical Turkish Cooking*, Harper Collins, New York.
ALTCHULER, S. I. (1982). 'Dietary Proteins and Calcium Loss: A Review', *Nutrition Research*, 2 (2), 193–200.
AN ZHIMIN (1989). 'Prehistoric Agriculture in China', in Harris & Hillman (1989), 643–9.
ANDERSON, E. N. (1988). *The Food of China*, Yale University Press, New Haven, CT.
ANDERSON, E. N. (1990). 'Up Against Famine: Chinese Diet in the Early Twentieth Century', *Crossroads*, **1** (1), 11–24.
ANDERSON, E. N. (1994). 'Food and Health at the Mongol Court', Kaplan & Whisenhunt, eds. (1994), 17–43.
ANDERSON, E. N. & ANDERSON, MARJA L. (1977). 'Modern China – South', in Chang, K. C. ed. (1977), 318–82.
ANDERSSEN, J. G. (1947). 'Prehistoric Sites in Honan', *BMFEA*, **1947** (19), 1ff.
ANDRÉ, JACQUES (1965). *Apicius: L'art culinaire De re coquinaria*, Librairie Klincksieck, Paris.
APICIUS (1977). *De Re Coquinaria*, 'Cooking and Dining in Imperial Rome', ed. and tr. Vehling, J. D., Dover, New York. Cf. also Edwards, John tr. & adapted (1984).
ARBERRY, J. A., tr. (1939). 'A Bagdad Cookery Book', *Islamic Culture*, **13**, 21–214.
ARNOTT, M. ed. (1976). *Gastronomy: The Anthropology of Foods and Food Habits*, Mouton, The Hague.
ASIAN ART MUSEUM OF SAN FRANCISCO (1975). *The Chinese Exhibition: The Exhibition of Archaeological Finds of the People's Republic of China*, June 28–Aug. 28, 1975, repr. Asian Art Museum of San Francisco.
ATHANAEUS (1927). *Deipnosophists*, Gulik, Charles Burton tr. Loeb Classical Library.
ATKINSON, R. W. (1881). 'On the Diastase of Koji', *Proc. Royal Soc.* (London), **32**: 299–332.
ATWELL, WILLIAM S. (1982). 'International Bullion Flows and the Chinese Economy, c. 1530–1650', *Past and Present*, No. 95, May, 1982.
BADLER, V. R. (1995). 'The Archaeological Evidence for Winemaking, Distribution and Consumption at Proto-historic Godin Tepe, Iran', in McGovern *et al.*, eds. (1995), 45–56.
BAGLEY, ROBERT W. (1988). 'Sacrificial Pits of the Shang Period at Sanxingdui in Guanghan County, Sichuan Province', *Arts Antiques*, **43**, 78–86.
BAGLEY, ROBERT W. (1990). 'A Shang City in Sichuan Province', *Orientation*, November **1990**, 52–67.
BALFOUR, EDWARD (1885). *The Cyclopaedia of India and of Eastern and Southern Asia*, 3 vols., Bernard Quarich, London.
BALL, SAMUEL (1848). *An Account of the Cultivation and Manufacture of Tea in China, derived from personal observations during an official residence in that country from 1805 to 1826*, Longman, Green, Brown & Longmans, London.
BAR-JOSEF, OFER (1992). 'The Neolithic Period', in Ben-Tor ed., 10–28.
BAR-YOSEF, OFER (1995a). 'Prehistoric Chronological Framework', in Levy ed. (1995), xiv–xvi.
BAR-YOSEL, OFER (1995b). 'Earliest Food Producers – Pre Pottery Neolithic (8000–5500)', in Levy ed. (1995), 190–201.
BARNETT, JAMES A. (1953). 'Cheese Biology', *Science News*, **29**, 60–83. Penguin Books.
BASALLA, GEORGE (1988). *The Evolution of Technology*, Cambridge University Press.
BEARD, JAMES (1973). *Beard on Bread*, Ballantine Books, New York.
BEAZLEY, MITCHELL (1985). *The World Atlas of Archaeology*, Mitchell Beazley International, London. Original French edition *Le grand atlas de l'archaéologie*, Encyclopedia Universalis, 1985.
BEDDOWS, C. G. & ARDESHIR, A. G. (1979). 'The Production of Soluble Fish Protein Solution for use in Fish Sauce Manufacture, 1 The Use of Added Enzymes', *J. Food. Technol.*, **14**, 603–12.
BEDINI, SILVIO (1994). *The Trail of Time*, Cambridge University Press.

BEINFIELD, HARRIET & KORNGOLD, EFREM (1991). *Between Heaven and Earth: A Guide to Chinese Medicine*, Ballantine Books, NY.
BELL, JOHN (1763). *A Journey from St. Petersburg to Pekin, 1719–22*, Edinburgh, 1965.
BENDA, I. (1982). 'Wine and Brandy', in Reed ed. (1982), 293–381.
BEN-TOR, AMNON, ed. (1992). *The Archaeology of Ancient Israel*, tr. R. Greenberg, Open University, Israel.
BERTUCCIOLI, G. (1956). 'A Note on Two Ming Manuscripts of the *Pên Tshao Phin Hui Ching Yao*', *Journal of Oriental Studies* (Hong Kong University), **3**, 63.
BIANCHINI, F. & CORBETTA, F. (1976). *The Complete Book of Fruits and Vegetables*, Crown, New York.
BIELENSTEIN, HANS (1980). *The Bureaucracy of the Han Times*, Cambridge University Press.
BIOT, E. tr. (1851) (1). *Le Tcheou-Li ou Tites des Tcheu*. 3 vols., Imp. Nat. Paris, 1851 (photo reproduction Wêntienko, Peiping, 1930).
BLOFELD, JOHN (1985). *The Chinese Art of Tea*, Allen Unwin, London.
BLUE, GREGORY (1990, 1991). 'Marco Polo's Pasta', in Hakim M. Said ed. (1990), 39–48, and in *Mediévales* (1991), **20**, 91–8 (in French).
BODDE, DERK (1942). 'Early References to Tea Drinking in China', *Amer. Oriental Soc.*, **62**, 74–6.
BOWRA, C. M. (1965). *Classical Greece*, Time-Life Books, New York.
BOYLE, ROBERT (1663). *Some Considerations touching the Usefulness of Experimental Natural Philosophy, proposed in a Familiar Discourse to a Friend, by way of Introduction to the Study of it* Hall and Davis, Oxford, 1663, 2nd edn 1664, in Thomas Birch ed. *The Works of Robert Boyle* Vol. 1, London, +1772; pp. 104–5 on wine in China.
BRAIDWOOD, LINDA S., BRAIDWOOD, ROBERT J., HOWE, BRUCE, REED, CHARLES A. & WATSON, PATTY JO (1983). *Prehistoric Archaeology along the Zagros Flanks*, Oriental Institute, University of Chicago Press.
BRAIDWOOD, R. J., SAUER, J. D., HELBAEK, H., MANGELSDORF, P. C., CUTLER, H. C., COON, C., LINTON, R., STEWART, J. & OPPENHEIM, L. (1953). 'Did Man once Live by Beer Alone?', *American Anthropologist*, **55**, 515.
BRAY, FRANCESCA (1984). *Science and Civilisation in China*. Vol. VI, Pt 2, *Agriculture*, Cambridge University Press.
BRAY, FRANCESCA (1986). *The Rice Economies*, Cambridge University Press.
BRAY, FRANCESCA (1993). 'Chinese Medicine', in Bynaum & Porter eds. (1993), 728–54.
BRETSCHNEIDER, E. (1892). *Botanicum Sinicum*, Vol. II, Pt II, *The Botany of the Chinese Classics*, Kelly & Walsh, Shanghai; also in *JRAS/NCB* (1893), **25**, 1–468.
BRILLAT-SAVARIN, JEAN A. (1926). *The Physiology of Taste*, Liveright, New York, 1926; originally published in 1825.
BROTHWELL, DON & SANDISON, A. T. eds. (1967). *Diseases in Antiquity*, Charles C. Thomas, Springfield, IL.
BROTHWELL, DON & BROTHWELL, PATRICIA (1969). *Food in Antiquity*, Thames & Hudson, London.
BROWN, SHANNON R. (1979). 'The Transfer of Technology to China in the Nineteenth Century: The Role of Direct Foreign Investment', *J. Econ. Hist.*, **39** (1), 181–97.
BROWN, SHANNON R. (1981). 'Cakes and Oil: Technology Transfer and Chinese Soybean Processing, 1860–95', *Comp. Studies in Society and History*, **23** (1), 449–63.
BRUCE, C. A. (1840). 'Report on the Manufacture of Tea, and on the Extent and Produce of the Tea Plantations of Assam', *Transactions of the Agricultural and Horticultural Society of India*, Calcutta, **7**, 1.
BUCHANAN, KEITH, FITZGERALD, CHARLES & RONAN, COLIN (1981). *China, the Land and the People: The History, the Art and the Science*, Crown, New York.
BUCK, JOHN LOSSING ed. (1937). *Land Utilization in China*, Council on Economic & Cultural Affairs, New York.
BUELL, PAUL D. (1986). 'The *Yin-shan-cheng-yao*, a Sino-Uighur Dietary: Synopsis, Problems, Prospects', in Unschuld ed. (1986), 109–27.
BUELL, PAUL D. (1990). 'Pleasing the Palate of the *Qan*: Changing Foodways of the Imperial Mongols', *Mongolian Studies*, **13**, 57–81.
BUGLIARELLO, GEORGE & DONER, DEAN B. eds. (1979). *The History and Philosophy of Technology*, University of Illinois Press.
BUONASSISI, VINCENZO (1985). *The Classic Book of Pasta*, tr. Evans, Elizabeth, Futura, London; first published MacDonald & Jane (1977). Original in Italian, *Il Codice della Pasta*, Rizzoli, Milano (1973).
BURKHILL, I. H. (1935). *A Dictionary of the Economic Products of the Malay Peninsula*, I & II, Crown Agents, London.
BYNUM, W. F., BROWNE, E. J. & PORTER, ROY (1981). *Dictionary of the History of Science*, Princeton University Press, NJ.
BYNUM, W. F. & PORTER, ROY (1993). *Companion Encyclopedia of the History of Medicine*, 1, Routledge, London & New York.
CAI, JINGFENG (1988). *Eating Your Way to Health: Dietotherapy in Traditional Chinese Medicine*, Foreign Language Press, Peking.
CALLOWAY, D. H., HICKEY, C. A. & MURPHY, E. L. (1971). 'Reduction of Intestinal Gas-forming Properties of Legumes by Traditional and Experimental Food Processing Methods', *J. Food Sc.*, **36**, 251.
CAMPBELL-PLATT, GEOFFREY (1987). *Fermented Foods of the World – A Dictionary and Guide*, Butterworths, London.
DE CANDOLLE, ALPHONSE (1884). *The Origin of Cultivated Plants*, Kegan Paul, London, repr. Hafner, NY, 1959, Tr. from the French edition, Geneva, 1983.
CARDWELL, DONALD (1994). *The Fontana History of Technology*, Fontana Press, London.

CARLILE, MICHAEL J. & WATKINSON, SARAH C. (1994). *The Fungi*, Academy Press, London.
CARPENTER, KENNETH J. (1993). 'Nutritional Diseases', in Bynum & Porter eds. (1993), 464–83.
CASTELLI, W. P. & ANDERSON, K. (1986). 'A Population at Risk', *Am. J. Med.*, **80** (Supplement 2A), 23–32.
CATER, C. M., CRAVEN, W. W., HORAN, F. E., LEWIS, C. J., MATTIL, K. F. & WILLIAMS, L. D. (1978). 'Oilseed Proteins', in Milner *et al.*, eds. (1978).
CATO THE ELDER (MARCUS PORCIUS CATO) (1934). *De Agri Cultura*, tr. Hooper, W. D. & Ash, H. B., Loeb Classical Library.
CAVALLI-SFORZA, L. LUCA, MENOZZI, PAOLO & PIAZZA, ALBERTO (1994). *The History and Geography of Human Genes*, Princeton University Press.
CERESA, MARCO (1996). 'Diffusion of Tea-Drinking Habit in Pre-Tang and Early Tang Period', *Asiatica Venetianna*, **1996** (1), 19–25.
CHADWICK, J. & MANN, W. N. (1950). *The Medical Works of Hippocrates*, Blackwell Science Publications, Oxford.
CHAKRAVARTY, TAPONATH N. (1959). *Food and Drink in Ancient Bengal*, Mukhopadhyay, Calcutta.
CHAMPION, PAUL (1866). 'Sur la fabrication du fromage de pois en Chine et au Japon', *Bulletin de la Societe d'Acclimatation*, **13**, 562–5.
CHAN, HARVEY T. ed. (1983). *Handbook of Tropical Foods*, Marcel Dekker, New York & Basel.
CHAN, SUCHENG (1986). *The Bittersweet Soil: The Chinese in California Agriculture, 1860–1910*, University of California Press, Berkeley, CA.
CHANG, K. C. (1973). 'Food and Food Vessels in Ancient China', *Transactions, 2nd Series, NY Acad, Sciences* **35**, 495–520.
CHANG, K. C. ed. (1977). *Food in Chinese Culture: Anthropological and Historical Perspectives*, Yale University Press, New Haven, CT.
CHANG, K. C. (1977a). 'Ancient China', in K. C. Chang ed. (1977), 25–52.
CHANG, KWANG-CHIH (1977b). *The Archaeology of Ancient China* 3rd edn, Yale University Press, New Haven, CT, 1977, earlier edn 1963, 1968.
CHANG, KWANG-CHIH (1980a). *Shang Civilisation*, Yale University Press, New Haven, CT.
CHANG, KWANG-CHIH (1980b). 'The Chinese Bronze Age: A Modern Synthesis', in Fong, Wen ed. (1980), pp. 35–50.
CHANG, KWANG-CHIH (1986). *The Archaeology of Ancient China* 4th edn, Yale University Press, New Haven, CT.
CHANG, TÊ-TZU (1983). 'The Origins and Early Culture of Cereal Grains and Food Legumes', in David Keightley ed. (1983), 65–94.
CHANG TSHO (1987). 'The Legend of Tea', in Cheung, ed. & tr. (1987), pp. 146–7.
CHAO, BUWEI YANG (1963). *How to Cook and Eat in Chinese*, Vintage Books, New York. First published in 1945.
CHEN JUNSHI, CAMPBELL, T. COLIN, LI JUNYAO & PETO, RICHARD (1990). *Diet, Life-Style and Mortality in China* (in English and Chinese), Oxford University Press, Cornell University Press, and People's Medical Publishing House.
CHEUNG, DOMINIC, ed. & tr. (1987). *The Isle Full of Noises; Modern Chinese Poetry from Taiwan*, Columbia University Press.
CHIAO, J. S. (1981). 'Modernisation of Traditional Chinese Fermented Foods and Beverages', *Adv. in Biotechnology*, **2**, 511–16.
CH'OE PU'S DIARY, ed. and tr., cf. JOHN MESKILL tr. (1965).
CHOW, KIT & KRAMER, IONE (1990). *All the Tea in China*, China Books, San Francisco.
CHURCH, ARTHUR H. (1886). *Food Grains of India*, Chapman & Hall, London, reprinted in India (1983) by Ajay Book Service, New Delhi.
CHURCH, C. G. (1924). 'Composition of the Chinese Jujube', *USDA Bulletin*, **1215**, 24–9.
CHURCHILL, AWNSHAM & CHURCHILL, JOHN (1704). *A Collection of Voyages and Travels* 4 vols., London.
CIVIL, M. (1964). 'A Hymn to the Beer Goddess and a Drinking Song', in *Studies Presented to A. L. Oppenheim, June 7, 1964*, University of Chicago Press, 67–89.
CLUTTON-BROCK, JULIET, ed. (1981). *The Walking Larder, Patterns of Domestication, Pastoralism and Predation*, Unwin Hyman, London.
COHEN, MARK (1977). *The Food Crisis in Prehistory: Overpopulation and the Origins of Agriculture*, Yale University Press, NH.
COLUMELLA, LUCIUS JUNIUS MODERATUS (1968). *De Re Rustica* Pt 2, tr. E. S. Foster and H. Heffnee, Loeb Classical Library, Harvard University Press.
COOPER, WILLIAM C. & N. SIVIN (1973). 'Man as a Medicine: Pharmacological and Ritual Aspects of Traditional Therapy using Drugs Derived from the Human Body', in Nakayama and Sivin eds. (1973), 203–772.
CORRAN, H. S. (1975). *A History of Brewing*, David & Charles, London.
COWAN, C. WESLEY & WATSON, PATTY JO eds. (1992). *The Origins of Agriculture*, Smithsonian Institution Press, Washington & London.
COYLE, L. PATRICK (1982). *The World Encyclopedia of Food*, Facts-on-File Inc., New York.
CRABTREE, PAM (1993). 'Early Animal Domestication in the Middle East and Europe', in Michael B. Schiffer (1993), pp. 201–45.

CRAIG, T. W. (1978). 'Proteins from Dairy Products', in Milner *et al.*, eds. (1978).
CRAWFORD, GARY W. (1992). 'Prehistoric Plant Domestication in East Asia', in Cowan & Watson (1992), 7–38.
CREEL, H. G. (1937). *The Birth of China: A Study of the Formative Period of Chinese Civilization*, Frederick Ungar, New York.
CROCE, JULIA DELLA (1987). *Pasta Classica: The Art of Italian Pasta Cooking*, Chronicle Books, San Francisco.
CROSBY, ALFRED W. (1972). *The Columbian Exchange: Biological and Cultural Consequences of 1492*, Greenwood Press, Westport, CT.
CROSBY, ALFRED W. (1986). *Ecological Imperialism*, Cambridge University Press.
CROSBY, ALFRED W. (1997). *The Measure of Reality: Quantification and Western Society 1250–1600*, Cambridge University Press.
CURRIER, R. L. (1966). 'The Hot–Cold Syndrome and Symbolic Balance in Mexican and Spanish–American Folk Medicine', *Ethnology*, **5**: 251–63.
CURTIS, ROBERT I. (1991). *Garum and Salsamentua*, E. J. Brill, Leiden.
DALBY, ANDREW (1995). *Siren Feasts*, Routledge, London & New York.
DALBY, ANDREW & GRAINGER, SALLY (1995). *The Classical Cookbook*, Getty Museum.
DANIELS, CHRISTIAN (1996). *Science and Civilisation in China*, Vol. VI, Pt 3. *Agro-Industries: Sugarcane Technology*, Cambridge University Press.
DARBY, WILLIAM J., GHALIOUNGIU, PAUL & GRIVETTI, LOUIS (1977). *Food: The Gift of Osiris* I, II, Academic Press, New York.
DAUMAS, MAURICE (1969–80). *A History of Technology & Invention*: Vol. I. The Origin of Technological Civilization to 1450, Vol. II. The First Stages of Mechanization 1450–1725, Vol. III. The Expansion of Mechanization 1725–1860, John Murray, Crown Publishers, London.
DAVIES, TENNY L. (1945). 'Introduction' to Huang and Chao tr. (1945), 'The Preparation of Ferments and Wines', 24–9.
DAVIS, JOHN FRANCIS (1836). *The Chinese: A General Description of China and its Inhabitants* Vol. II, New York.
DAWKINS, RICHARD (1976, 1989). *The Selfish Gene*, Oxford University Press.
DAWKINS, RICHARD (1989). *The Blind Watchmaker*, W. W. Norton, New York.
DIAMOND, JARED (1993). *The Third Chimpanzee*, Harper Perennial, New York.
DIAMOND, JARED (1997). *Guns, Germs and Steel: The Fates of Human Societies*, W. W. Norton, New York.
DJIEN, KO SWAN (1986). 'Some Microbiological Aspects of Tempe Starters', in Aida Hiroshi *et al.*, eds. (1986), 101–41. Cf. Bibliography B.
DOBSON, W. A. C. H. (1964). 'Linguistic Evidence and the Dating of the *Book of Songs*', *TP*, LT 322–34.
DRACHMANN, A. G. (1932). *Ancient Mills and Presses*, Levin & Munksgaard, Copenhagen.
DUDGEON, JOHN (1895). *The Beverages of the Chinese*, Tientsin.
DUMÉRIL, AUGUSTE (1859). 'Extraits des procès-verbaux des séances générales de la société. Séance du 4 Feb. 1859', *Bulletin de la Societe d'Acclimatation*, **6**, 86–110.
EBERHARD, W. (1933). 'Beiträge sur kosmologischen Spekulation Chinas in der Han Zeit', *Baessler Archiv*, **16**, 1ff.
EBERHARD, WOLFRAM (1971). *A History of China*, University of California Press, Berkeley, CA.
EDEN, T. (1976). 3rd edn. *Tea*, Longman, London. 1st published 1958, 2nd, edn, 1965.
EDWARDS, JOHN tr. and adapted (1984). *The Roman Cookery of Apicius*, Rider, London.
EGERTON, CLEMENT tr. (1939). *The Golden Lotus* [*Chin Phing Mei*], 4 vols., Routledge & Sons, London.
EIJKMAN, CHRISTIAAN (1890, repr. 1990). 'Report of the Investigation carried out in the Laboratory of Pathology and Bacteriology, Weltevreden, During the year 1889. VI. Polyneuritis in chickens', classical article, *Nutrition Reviews*, **48** (6), 242–3.
ELVIN, MARK (1973). *The Pattern of the Chinese Past*, Stanford University Press.
ENDO, A. (1979). 'Monacolin K, a new hypocholesterolemic agent produced by a *Monascus* species', *J. Antibiot.* (Tokyo), **32**, 852–4.
ENGELHARDT, U. & HEMPEN, C. H. (1997). *Chinesische Diätetik*, Urban and Schwarzenberg, München.
EPSTEIN, H. (1969). *Domestic Animals of China*, Commonwealth Agricultural Bureaux.
ESTES, J. WORTH (1990). *Dictionary of Protopharmacology: Therapeutic Practices, 1700–1850*, Science History Publications, USA.
ESTES, J. WORTH (1996). 'The Medical Properties of Food in the Eighteenth Century', *Journal of the History of Medicine and Allied Sciences*, **51** (2), 127–54.
EVANS, JOHN C. (1992). *Tea in China: The History of China's National Drink*, Greenwood Press, New York.
EVANS, L. T. & PEACOCK, W. J. eds. (1981). *Wheat Science – Today and Tomorrow*, Cambridge University Press.
FAGAN, BRIAN M. (1986). *People of the Earth*, Little Brown & Co., Boston.
FAIRBANK, WILMA (1972). *Adventures in Retrieval: Han Murals and Shang Bronze Molds*, Harvard University Press, Cambridge MA.
FARB, PETER & ARMELAGOS, GREGORY (1980). *Consuming Passions: The Anthropology of Eating*, Houghton Mifflin, Boston.
FARRINGTON, BENJAMIN (1955). *Greek Science: Its Meaning For Us*, Penguin Books, Middlesex.

Feldman, Moshe, Lupton, F. G. H. & Miller, T. E. (1995). 'Wheats', in Smartt & Simmonds, eds. (1995), 184–92, Longman, London.
Feng Gia-Fu & English, Jane (1972). *Tao Tê Ching – A New Translation*, Vintage Books, New York.
Finney, P. L. (1983). 'Effect of Germination on Cereal and Legume, Nutrient Changes and Food or Food Value: A Comprehensive Review', in Nozzollilo *et al.*, eds. (1983), 229–305.
Fisher, M. E. K, (1990). *The Art of Eating*, Collier Books, New York.
Flatz, G. (1981). 'Genetics of Lactose Digestion in Humans', *Adv. Hum. Genet.*, **16**, 1–77.
Flatz, G. & Rottauwe, H. W. (1973). 'Lactose nutrition and natural selection', *Lancet*, **2**, 76–7.
Flaws, Bob & Wolfe, Honora L. (1983). *Prince Wen Hui's Cook: Chinese Dietary Therapy*, Paradigm Publications, Brookline, MA.
Flon, Christine ed. (1985). *The World Atlas of Archaeology*, cf. Beazley, Mitchell.
Fogarty, W. M. ed. (1983). *Microbial Enzymes and Biotechnology*, Applied Science Publishers, London.
Fong, Wen ed. (1980). *The Great Bronze Age of China: An Exhibition from the People's Republic of China*, The Metropolitan Museum of Art, New York.
Fong, Y. Y. & Chan, W. C. (1973). 'Bacterial Production of Dimethy-Nitrosomaine in Salted Fish', *Nature*, **243**, 421–2.
Fontein, Jan & Wu Tung (1975). *Han and T'ang Murals*, Museum of Fine Arts, Boston.
Forbes, R. J. (1954). 'Chemical, Culinary and Cosmetic Arts', in Singer *et al.*, eds. (1954), 238–98.
Forbes, R. J. (1956). 'Food and Drink', in Singer *et al.*, eds. (1956), 103–46.
Forbes, R. J. (1957). 'Food and Drink', in Singer *et al.*, eds. (1957), 1–26.
Fortune, Robert (1845). Letter to Editor, The Gardener's Chronicle, 28 August, 1845.
Fortune, Robert (1847). *Three Years' Wanderings in the Northern Provinces of China, including a Visit to the Tea, Silk and Cotton Countries*, London.
Fortune, Robert (1852). *A Journey to the Tea Countries of China, including Sung-Lo and the Bohea Hills, with a short notice on the East India Company's Tea plantations in the Himalaya mountains*, John Murray, London. Repr. Mildmay Books, 1987.
Fortune, Robert (1853). *Two Visits to the Tea countries of China and the British Tea Plantations in the Himalayas; with a Narrative of Adventures, and a full Description of the Culture of the Tea Plant, the Agriculture, Horticulture and botany of China*. Vols. I & II, John Murray, London 1853 (3rd edn).
Fortune, Robert (1857). *A Resident Among the Chinese: Inland, On the Coast and at Sea*, John Murray, London.
Fotheringham, J. K. (1921). *Historical Eclipses*, Halley Lecture, Oxford (Abstr. *JBASA*, 1921, **32**, 197ff.).
Fowles, Gerry (1989). 'The Compleat Home Wine-maker', *New Scientist*, Sept. 2, 1989, 38–43.
Frake, Charles O. (1980). *Language and Cultural Description*, Stanford University Press.
Franke, Herbert (1970). 'Additional Notes on Non-Chinese Terms in the Yuan Imperial Dietary Compendium *yinshan zhenyao*', *Zentralasiatische Studien*, **4**, 7–16.
Franke, Otto (1930–53). *Geschichte d. Chinesischen Reiches* 5 vols., de Gruyter, Berlin, 1930–53.
Franklin, Benjamin (1770). 'Letter to John Bartram in Philadelphia, from London, dated January 11, 1770', in Smyth, Albert H. ed. (1907), Vol. V, 245–6.
Freeman, Michael (1977). 'Sung', in Chang, K. C. ed. (1977), 143–76.
Fu, Marilyn & Wen Fong (1973). *The Wilderness Colors of Tao-Chi*, Metropolitan Museum of Art, New York, 1973.
Fukushima, D. (1986). *Soy Sauce and Other Fermented Foods of Japan*, in Hesseltine & Wang eds., 121–49.
Gajdusek, D. C. (1962). 'Congenital Defects of the Central Nervous System Associated with Hyperendemic Goiter in a Neolithic Highland Society of Netherlands New Guinea', I. *Epidemiology, Pediatrics*, **25**, 345.
Galdston, I. ed. (1960). *Human Nutrition: Historic and Scientific*, International University Press, New York.
Gardella, Robert (1994). *Harvesting Mountains: Fujian and the China Tea Trade, 1757–1937*, University of California Press.
Gerard, John (1597). *Gerard's Herball*, Woodward, Marcus ed. (1985) based on the 1636 edition of Th. Thompson Crescent Books, New York.
Ghandi, M. K. (1948). *Ghandi's Autobiography: The Story of my Experiments with Truth*, Public Affairs Press, Washington DC.
Giles, H. A. (1874). 'The *Hsi Yüan Lu* or "Instruction to Coroners" [tr. From the Chinese]', *China Review*, **3**, 30, 92, 159, etc.
Giles, A. H. (1911). *An Alphabetical Index to the Chinese Encyclopaedia (Chhin Ting Ku Chin Thu Shu Chi Chêng)*, British Museum, London.
Godley, Michael R. (1986). 'Bacchus in the East: The Chinese Grape Wine Industry, 1892–1938', *Business History Review 60*, Autumn **1986**.
Gonen, Rivka (1992). 'The Chalcolithic Period', in Ben-Tor (1992), 40–80.
Goodrich, L. Carrington & Wilbur, C. Martin (1942). 'Additional Notes on Tea', *J. Amer. Oriental Soc.*, **62**, 195–7.
Goody, Jack. *Cooking, Cuisine and Class: A Study in Comparative Sociology*, Cambridge University Press, 1982.
Graham, A. C. (1989). *Disputers of the Tao*, Open Court, LaSalle, IL.

GRAHAM, HAROLD N. (1992). 'Green Tea Composition, Consumption, and Polyphenol chemistry', *Preventive Medicine*, **21**: 334–50.
GREEN, R. M. (1951). *A Translation of Galen's Hygiene (De sanitate tuenda)*, C. C. Thomas, Springfield, IL, pp. 210–11.
GRIGSON, CAROLINE (1995). 'Plough and Pasture in the Early Economy on the Southern Levant', in Levy (1995), 245–68.
GROFF, ELIZABETH (1919). 'Soy Sauce Manufacture in Kuangtung, China', *Philippine Journal of Science*, **15** (3), pp. 307–16.
GUPPY, H. B. (1884). Samshu-Brewing in North China, *Journ. North China Branch, Royal Asiatic Society*, **18**, 163–4.
GUTZLASS, CHARLES (1834). *Journey of Three Voyages along the Coast of China, in 1831, 1832 and 1833*, London.
HAAS, FRANÇOIS & HAAS, SHEILA S. 'The Origins of Mycobacterium Tuberculosis and the Notion of its Contagiousness', in Rom & Garay (1996), 3–19.
HAGERTY, M. J. with CHIANG KHANG-HU, tr. (1923). 'Han Yen-Chhi's *Chü Lu* (Monograph on the Oranges of Wên-Chou, Chekiang)' with introduction by P. Pelliot, *TP*, 1923, **22**, 63.
HAHN, EDUARD (1896). *Die Haustiere und ihre Beziehungen zur Wirtschaft des Menschen*, Dunker & Humblot, Leipzig.
DU HALDE, JEAN BAPTISTE (1736). *The General History of China* 4 vols., tr. from the French by R. Brookes, John Watts, London.
HAN, C. & XU, Y. (1990). 'The Effect of Chinese Tea on the Occurrence of Esophageal Tumors Induced by N-Nitrosomethyl-Benzylamine formed in Vivo', *Biomed. Environ. Sci.*, **3** (1), 35–42.
HARDY, SERENA (1979). *The Tea Book*, Whitlet Books, Weybridge.
HARLAN, J. R. (1971). 'Agricultural Origins: Centers and Non-Centers', *Science*, **174**, 468–74.
HARLAN, J. R. (1981). 'The Early History of Wheat – Earliest Traces to the Sack of Rome', in Evans and Peacock, eds. (1981), 1–20.
HARLAN, J. R. (1992). *Crops and Man* 2nd edn, Amer. Soc. Agronomy, Madison, WI.
HARLAN, J. R. (1995). 'Barley', in Smartt & Simmonds eds. (1995), 140–7.
HAROUTUNIAN, ARTO DER (1982). *Middle Eastern Cookery*, Pan Books, London.
HARPER, DONALD (1982). *The 'Wu shih erh ping fang' Translation and Prolegomena*, Ph.D. Thesis, University Microfilm International, Ann Arbor, MI, 1988.
HARPER, DONALD (1984). 'Gastronomy in Ancient China – Cooking for the Sage King', *Parabola*, **9** (4), 38–47.
HARRIS, DAVID R. & HILLMAN, GORDON C. eds. (1989). *Foraging and Farming: The Evolution of Plant Exploitation*, Unwin Hyman, London.
HARRIS, LESLIE J. (1935). *Vitamins*, Cambridge University Press.
HARRIS, MARVIN (1985). *Good to Eat: Riddles of Food and Culture*, Simon Schuster, NY.
HARRISON, H. S. (1954). 'Discovery, Invention and Diffusion', in Singer, Holmyard et al., eds. (1954), 58–84.
HART, D. V. (1069). 'Bisayan Filipino and Malay Humoral Pathologies', SE Asian Program *Data Paper 76*, Cornell University Press, Ithaca.
HARTMAN, LOUIS F. & OPPENHEIM, A. L. (1950). *On Beer and Brewing Techniques in Ancient Mesopotamia*, J. Amer. Oriental Soc., Suppl. 10, December, 1950.
HARTNER, W. (1935). 'Das Datum der *Shih-Ching* Finsternis', *TP*, **31**, 188.
AL-HASSAN, AHMAD, Y. & HILL, DONALD R. (1986). *Islamic Technology: An Illustrated History*, Cambridge University Press.
HAWKES, DAVID, tr. (1959, 1985). *Chhu Tzhu, The Songs of the South, An Ancient Chinese Anthology*, Penguin Books, 2nd edn 1985; 1st publ. Oxford, Clarendon Press, 1959.
HAWKES, J. G. (1983). *The Diversity of Crop Plants*, Harvard University Press, Cambridge, MA.
HEBER, DAVID, YIP, IAN, ASHLEY, JUDITH, M., ELASHOFF, DAVID, A. & GO, VAY LIANG W. (1999). 'Cholesterol-lowering effects of a proprietary Chinese red-yeast-rice dietary supplement', *Am. J. Clin. Nutr.*, **69**, 231–6.
HEGSTEAD, D. MARK et al. (ed.) (1976). *Present Knowledge in Nutrition*, The Nutrition Foundation, New York.
HEISER, CHARLES B. JR (1973). *Seed to Civilisation: The Story of Food*, Harvard University Press, Cambridge, MA.
HENDERSON, BRIAN E. (1978). 'Nasopharyngeal Cancer', in Kaplan & Tsuchitani eds. (1978), 83–100.
HERBERT, VICTOR (1976). Vitamin B_{12}, in Hegstead et al., eds. (1976), 191–203.
HERO (1893). *Mechanica*, Arabic edn and French trans. by Bernard Carra de Vaux, *J. Asia.*, neuvième série, **1893** (2).
HERON, CARL & EVERSHED, RICHARD P. (1993). 'The Analysis of Organic Residues and the Study of Pottery Use', in Schiffer ed. (1993), Vol. III, 247–84.
HESSE, BRIAN (1997). 'Animal Husbandry', in Meyers ed. (1997), 140–3.
HESSELTINE, C. W. (1965). 'A Millennium of Fungi, Food and Fermentation', *Mycologia*, **57**, 149–97.
HESSELTINE, C. W. (1983). 'Microbiology of Oriental Fermented Foods', *Ann. Rev. Microbiol.*, **37**, 575–601.
HESSELTINE, C. W. & WANG, H. L. eds. (1986). *Indigenous Fermented Food of Non-Western Origin*, Mycologia Memoir No. 11, J. Cramer, Berlin, Stuttgart.
HESSELTINE, C. W. & WANG, HUA L. (1986a). 'Food Fermentation Research and Development', in Hesseltine and Wang, eds. (1986), 9–21.
HESSELTINE, C. W. & WANG, H. L. (1986b). 'Glossary of Indigenous Fermented Foods', in Hesseltine & Wang, eds. (1986), 317–44.

Heywood, V. H. Consulting Editor (1985). *Flowering Plants of the World*, Equinox, Oxford; 1st publ. Oxford University Press (1978).
Hilbert, J. Raymond (1982). 'Beer', in Reed ed. (1982), 403–27.
Hillman, Gordan (1984). 'Traditional Husbandry and Processing of Archaic Cereals in Recent Times, The Operation, Products and Equipment which might feature in Sumerian Texts', Part I, *Bull. on Sumerian Agric.*, **1**, 114–52.
Hillman, Gordan (1985). 'Traditional Husbandry and Processing of Archaic Cereals in Recent Times, The Operation, Products and Equipment which might feature in Sumerian Texts', Part II, *Bull. on Sumerian Agric.*, **2**, 1–31.
Hillman, Howard (1981). *Kitchen Science: A Compendium of Information for Every Cook*, Houghton Mifflin, New York.
Hirayama, K. & Ogura, S. (1915). 'On the Eclipses recorded in the *Shu Ching* and *Shih Ching*', PPMST, 1915 (2nd series), **8**, 2.
Ho, Chi-Tang, Osawa, T., Huang, M. T. & Rosen, Robert, eds. (1994). *Food Phytochemicals for Cancer Prevention II: Teas, Spices and Herbs, ACS Symposium Series No. 547*, Amer. Chem. Soc., Washington, DC.
Ho Peng-Yoke (1955). 'Astronomy in the *Chin Shu* and *Sui Shu*', Inaugural Dissertation Singapore.
Ho Peng-Yoke (1985). *Li, Qi and Shu*, Hong Kong University Press.
Ho Ping-Ti (1955). 'The Introduction of American Food Plants into China', *American Anthropologist*, **57** (2), Pt 1, 191–201.
Ho Ping-Ti (1969). 'The Loess and the Origin of Agriculture in China', *Amer. Hist. Rev.*, **75** (1), 1–36.
Ho Ping-Ti (1975). *The Cradle of the East*, Chinese University of Hong Kong & University of Chicago Press.
Hodder, I, Isaac, G. & Hammond, N. eds. (1981). *Pattern of the Past*, Cambridge University Press.
Hole, Frank (1989). 'A Two-Part, Two-Stage Model of Domestication', in Clutton-Brock ed. (1989).
Homer tr. E. V. Rieu (1985). *The Iliad*, Penguin Books, Middlesex.
Homer tr. E. V. Rieu (1985). *The Odyssey*, Penguin Books, Middlesex.
Hommel, Rudolph, P. (1937). *China at Work*, John Day Co., New York.
Hopkins, David, C. (1997). 'Cereals', in Eric M. Meyers ed. (1997), pp. 479–81.
Howe, Bruce (1983). 'Karim Sharir', in Braidwood *et al.*, eds. (1983), 23–154.
Hsiung, Deh-Ta (1978). *The Home Book of Chinese Cookery*, Faber & Faber, London.
Hsu, Cho-yun (1978). 'Agricultural Intensification and Marketing: Agrarianism in the Han Dynasty', in Roy & Tsien (1978), 253–68.
Hsu, Cho-yün (1980). *Han Agriculture: The Formation of Early Chinese Agrarian Economy (206 BC to AD 220)*, University Washington Press, Seattle.
Hsu, Cho-yün & Linduff, K. M. (1988). *Western Chou Civilization*, Yale University Press, New Haven, CT.
Huang, H. T. (1982). 'Peregrinations with Joseph Needham in China, 1943–44', in Li *et al.*, eds. (1982), 39–75.
Huang, H. T. (1990). 'Han Gastronomy – Chinese Cuisine *in statu nascendi*', *Interdisciplinary Science Reviews*, **15** (2), 139–52.
Huang, H. T. & Dooley, J. G. (1976). 'Enhancement of Cheese Flavours with Microbial Esterases', *Biotech. & Bioeng.*, **18**, 909–19.
Huang, Ray (1988). *China: A Macro History*, M. E. Sharpe Inc., New York.
Huang Tzu-Chhing & Chao Yün-Tshung tr. (1945). ' "The Preparation of Ferments and Wines" by Chia Ssu-Hsieh of the Later Wei Dynasty', with an introduction by Tenney L. Davis. *Harvard J. Asiatic Studies*, 1945, **9**, 24–44.
Huc, M. (1855). *A Journey Through the Chinese Empire*, Vol. II, New York, 1855.
Huff, Toby E. (1993). *The Rise of Early Modern Science: Islam, China and the West*, Cambridge University Press.
Huff, Toby E. (1995). *The Rise of Early Modern Science*, Cambridge University Press.
Hui, Y. H. (1985). *Principles and Issues in Nutrition*, Wadsworth Health Sciences, Belmont, CA.
Hume, Edward H. (1949). *Doctors East, Doctors West*, George Allen & Unwin, London.
Hummel, A. W. (1941). 'The Printed Herbal of +1249', *ISIS*, 1941, **33**, 439.
Hymowitz, T. (1970). On the Domestication of the Soybean, *Econ. Bot.*, **23**, 408–21, 1970.
Hymowitz, T. (1976). 'Soybeans', in Simmonds ed. (1976), 159–62.
Hymowitz, T. & Harlan, J. R. (1983). 'The Introduction of the Soybean to North America by Samuel Bowen in 1765', *Econ. Botany*, **37** (4), 371–9.
IARC (1991). *Monographs on the Evaluation of Cancer Risks to Humans*, Vol. 51: 'Coffee, Tea, Mate, Methylxanthine and Methylglyoxal', WHO, 1991.
Ikeda, I., Imasato, Y., Sasaki, E., Nakayama, M., Nagao, H., Takeo, T., Yayabe, F. & Sugano, M., (1992). 'Tea Catechins Decrease Micellar-Solubility and Intestinal Absorption of Cholesterol in Rats', *Biochem. Biophys. Acta*, **1127**, 141–6.
The Indian Agriculturist, Calcutta (1882). 'The Japan Pea in India', December 1, 1882, 454–5.
Ishige Naomichi ed. (1988). *Catalog of the Sinoda Document Collections at the National Museum of Ethnology*, Vol. 1, *Bulletin of the National Museum of Ethnology* Special Issue, no. 8.

Ishige Naomichi (1990). 'Filamentous noodles (*miantiao*), its origin and dissemination', Paper presented at the 6th International Conference of the History of Science in China, Cambridge, England, Aug. 6, 1990.
Ishige Naomichi (1993). 'Cultural Aspects of Fermented Foods in East Asia', in Lee *et al.*, eds. (1993), 13–32.
Itoh, H., Tachi, H. & Kikuchi, S. (1993). 'Fish Fermentation Technology in Japan', in Lee *et al.*, eds. (1993), 177–86.
Jacob, H. E., tr. by Richard & Clara Winston (1944). *Six Thousand Years of Bread: Its Holy and Unholy History*, Greenwood Press, Westport, CT.
James, Peter & Thorpe, Nick (1994). *Ancient Inventions*, Ballantine Books, 1994.
Jeanes, Allene & Hodge, John eds. (1975). *Physiological Effects of Food Carbohydrates*, Am. Chem. Soc, Washington DC.
Jenner, W. J. F. tr. (1981). *Memories of Loyang* [trs. of *Lo-yang chieh Lan Chi*], Clarendon Press, Oxford.
Johnstone, Bob (1986). 'Japan Turns Soy Sauce into Biotechnology', *New Scientist*, Sept. 4, 1986, 38–40.
Kao Ching-Lang (1936). 'Infantile Beriberi in Shanghai', *Chinese. Med. J*, **50**, 324–40.
Kaplan, Henry S. & Tsuchitani, Patricia, J. eds. (1978). *Cancer in China*, Alan R. Liss, New York.
Kaplan, Edward H. & Whisenhunt eds. (1994). *Opuscula Altaica: Essays Presented in Honor of Henry Schwarz*, Bellingham, Western Washington University.
Kare, Morley R. & Maller, Owen, eds. (1977). *The Chemical Senses and Nutrition*, Academic Press, New York.
Karim, Mohamed Ismail Abdul (1993). 'Fermented Fish Products in Malaysia', in Lee *et al.*, eds. (1993), 95–106.
Karlgren, B. (1923). *Analytic Dictionary of Chinese and Sino-Japanese*, Paris.
Karlgren, B. tr. (1950). *The Book of Odes; Chinese Text, Transcription and Translation*, A reprint of the translation only from his papers in *BMFEA*, **16** and **17**.
Katz, Solomon (1987). News report in *Expedition* (University of Pennsylvania), March.
Katz, Solomon H. & Maytag, Fritz (1991). '*Brewing an Ancient Beer*', *Archaeology*, **44** (4): 24–33.
Keightley, David N. ed. (1983). *The Origins of Chinese Civilisation*, University California Press, Berkeley, CA.
Kemp, Barry ed. (1989). *Amarana Reports V*, Egypt Exploration Society, London.
Kêng Hsüan (1974). 'Economic Plants of Ancient North China as mentioned in the Book of Odes', *Econ. Bot.* **28** (4), 391–410.
Khader, Vijaya (1983). 'Nutritional Studies on Fermented, Germinated and Baked Soya Bean (*Glycine max*) Preparations', *J. Plant Foods*, **5**, 31–7.
King, Lester S. (1963). *The Growth of Medical Thought*, University of Chicago Press.
Klemm, Friedrich (1964). *History of Western Technology*, MIT Press, tr. Dorothea W. Singer. The original was published in 1959.
Knechtges, David (1986). 'A Literary Feast: Food in Early Chinese Literature', *J. Am. Oriental Soc.*, **106** (1), 49–63.
Knochel, Susanne (1993). 'Processing and Properties of North European Pickled Fish Products', in Lee *et al.*, eds. (1993), 213–29.
Ko, Swan Djien (1986). 'Some Microbiological Aspects of Tempe Starters', in Aida Hiroshi *et al.*, eds. (1986), 101–41.
Kodama, K. & Yoshizawa, R. (1977). 'Saké', in Rose ed., 423–75.
Kollipara, K. P., Singh, R. J. & Hymowitz, T. (1997). 'Phylogenetic and Genomic Relationships in the Genus *Glycine* Willd, based on sequences in the ITS region of nuclear DNA', *Genome*, **40**, 57–68.
Kondo, Hiroshi (1984). *SAKÉ A Drinker's Guide*, Kodansha, Tokyo.
Kong, Y. C. (1996). *Hui Hui Yao Fang – An Islamic Formulary*, Hong Kong cf. Chiang Jun-Hsiang (1996) in Bibliography B.
Koo, Linda (1976). 'Traditional Chinese Diet and its Relationship to Health', *Kroeher Anthropological Society Papers*, Vol. 48, pp. 116–47.
Kosikowski, Frank (1977). *Cheese and Fermented Milk Foods*, Edwards Brothers, Ann Arbor, MI.
Kranzberg, Melvin (1979). 'Introduction: Trends in the History and Philosophy of Technology', in Bugliarello and Doner eds., xiii–xxxi.
Kulp, Karel (1975). 'Carbohydrases', in Reed, ed. (1975).
Kunkee, Ralph E. & Goswell, Robin W. (1977). 'Table Wines', in Rose ed. (1977), 315–79.
Lafar, Franz tr. Salter, Charles C. T. (1903). *Technical Mycology: The Utilisation of Microorganisms in the Arts and Manufacture Vol. 2, Eumycetic Fermentation*, Charles Griffin & Co., London.
Lai, T. C. (1978). *Chinese Food for Thought*, Hong Kong Book Centre.
Lao, Yan-Shuan (1969). 'Notes on Non-Chinese Terms in the Yuan Imperial Dietary Compendium', *AS/BIHP*, **39**, 399–416.
Latham, Ronald tr. (1958). *The Travels of Marco Polo*, Penguin Books, 1958/82.
Latourette, Kenneth S. (1957). *The Chinese: Their History and Culture*, Macmillan, New York.
Lattimore, Owen (1988). *Inner Asian Frontiers of China*, Oxford University Press, 1st publ. 1940.
Lau, D. C., tr. (1970). *Mencius: Translation with an Introduction*. Penguin Books, Middlesex.
Lau, D. C., tr. (1979). *Confucius: The Analects*, Penguin Books, Middlesex.
Laudan, Rachel (1996). *The Food of Paradise*, University of Hawaii Press.

LAUFER, BERTHOLD (1914–15). 'Some Fundamental Ideas of Chinese Culture', *The Journal of Race Development*, **5**, pp. 160–74.
LAUFER, BERTHOLD (1919). *Sino-Iranica: Chinese Contributions to the History of Civilisation in Ancient Iran*, FMNHP/AS, **15**, no. 3 (Pub. No. 201).
LEACH, HENRY W. (1965). 'Gelatinization of Starch', in Whistler & Paschall eds. (1965), 289–306.
LEE, CHERL-HO (1993). 'Fish Fermentation Technology in Korea', in Lee *et al.*, eds. (1993), 187–201.
LEE, CHERL-HO, STEINKRAUS, KEITH H. & REILLY, P. J. ALAN eds. (1993). *Fish Fermentation Technology*, UN University Press, Tokyo, New York, Paris.
LEE, GARY (1974). *Chinese Tasty Tales Cook Book*, Chinese Treasure Productions, San Francisco.
LEE HYO-GEE (1996). 'History of Korean Alcoholic Drinks', *Koreana*, **1996**, pp. 4–9.
LEE SUNG WOO (1993). 'Cultural Aspects of Korean Fermented Marine Products in East Asia', in Lee *et al.*, eds. (1993), 33–43.
LEGGE, J. tr. (1861a). *Confucian Analects, The Great Learning and the Doctrine of the Mean, The Chinese Classics*, Vol. I, Hong Kong, Trubner, London, 1861.
LEGGE, J. tr. (1861b). *The Works of Mencius, The Chinese Classics*, Vol. II, Hong Kong, Trubner, London, 1861.
LEGGE, J. tr. (1871). *The She King* (The Book of Poetry), *The Chinese Classics*, Vol. IV, Pts 1 & 2, Lane Crawford, Hong Kong, 1871; Trubner, London, 1971.
LEGGE, J. tr. (1879). *The Texts of Confucianism, Pt I The Shu King, the Religious Portions* of the Shih King and the Hsiao King, Oxford.
LEGGE, J. tr. (1885a). *The Li Chi*, 2 vols. *The Texts of Confucianism, Pt III*, Oxford (*SBE* nos. 27, 28).
LEGGE, J. tr. (1885b). *The Shu Ching* (with Chinese text and notes), *The Chinese Classics*, Vol. III, Pts 1 & 2, Hong Kong, Trubner, London 1865.
LEGGE, J. tr. (1899). *The Texts of Confucianism*, Pt II *The Yi King* (*I Ching*), Oxford, 1899, (*SBE* no. 16).
LEICESTER, HENRY M. (1974). *Development of Biochemical Concepts from Ancient to Modern Times*, Harvard University Press, Cambridge, MA.
LEPKOVSKY, SAMUEL (1977). 'The Role of the Chemical Senses in Nutrition', in Kare & Maller eds. (1977), 413–57.
LEROI-GOURHAN, ANDRÉ (1969a). 'Primitive Societies', in Daumas ed. (1969–80) I, 18–58.
LEROI-GOURHAN, ANDRÉ (1969b). 'The first Agricultural Societies', in Daumas ed. (1969–80) I, 59–64.
LEVI-MONTALCINI, RITA (1988). *In Praise of Imperfection*, Basic Books, New York.
LEVY, THOMAS E. ed. (1995). *The Archaeology of Society in the Holy Land*, Facts-On-File, New York & Oxford.
LI CHIAO-PHING *et al.*, tr. (1980). *Thien Kung Khai Wu: Exploration of the Works of Nature*, China Academy, Taipei.
LI GUOHAO, ZHANG MENGWEN & CAO TIANQIN, eds. (1982). *Explorations in the History of Science and Technology in China*, Classics Press, Shanghai.
LI HUI-LIN (1969). 'The Vegetables of Ancient China', *Econ. Bot.* 1969, **23** (3), 253–60.
LI HUI-LIN (1974). 'An Archaeological and Historical Account of Cannabis in China', *Econ. Bot.*, 1974, **23** (4), 437–48.
LI HUI-LIN (1979). *A Fourth Century Flora of Southeast Asia*, Chinese University Press, Hong Kong.
LI HUI-LIN (1983). 'The Domestication of Plants in China, Ecogeographical Considerations', in Keightley ed. (1983), 21–64.
LI THAO (1940). 'Historical Notes on Some Vitamin-Deficiency Diseases in China', *CMJ*, **58**, 314. Repr. in Brothwell and Sandison (1967), 417–22.
LI XUEQIN, tr. by K. C. CHANG (1985). *Eastern Zhou and Qin Civilisations*, Yale University Press, New Haven, CT.
LIDDELL, CAROLINE & WEIR, ROBIN (1993). *Ices*, Hodder & Stoughton, London.
LIENER, IRVIN E. (1976). 'Legume Toxins in Relation to Protein Digestibility – A Review', *J. Food Sc.*, **41**, 1076–81.
LIN HSIANG-JU & TSUIFENG LIN (1969). *Chinese Gastronomy* 知味, Hastings House, New York.
LIU JILIN, PECK, G. *et al.* eds. (1995). *Chinese Dietary Therapy*, Churchill Livingstone, Edinburgh.
LIN YUTANG (1939). *My Country and My People*, John Day, New York, repr. Mei Ya Publ. Taipei, 1983.
LIN YUTANG (1948). *The Wisdom of Laotse*, The Modern Library, New York.
LITCHFIELD, CARTER ed. & ARIGA EIKO tr. (1974), original by OKURA NAGATSUNE (1836). *Seiyū Roku On Oil Manufacturing*, Olearius Editions, New Brunswick, NJ.
LIU, WU-CHI & LO, IRVING Y. (eds.) (1975). *Sunflower Splendour: Three Thousand Years of Chinese Poetry*, Indiana University Press, Bloomington.
LOEHR, MAX (1968). *Ritual Vessels of Bronze Age China*, The Asia Society, New York.
LOEWE, MICHAEL (1982). *Chinese Ideas of Life and Death: Faith, Myth and Reason in the Han Period*, Allen & Unwin, London.
LOUIE, AI-LING (1982). *Yeh Shen: A Cinderella Story from China*, Adapted from *Yu Yang Tsa Tsu* 西陽雜俎, Philomel Books, New York, 1982.
LU GWEI-DJEN & NEEDHAM, JOSEPH (1951). 'A Contribution to the History of Chinese dietetics', *Isis*, **42**, pp. 13–20.
LU GWEI-DJEN & NEEDHAM, JOSEPH (1966). 'Proto-Endocrinology in Medieval China', *Japanese Studies in the History of Science*, **5**, 150ff.

Lu Gwei-Djen & Needham, Joseph (1980). *Celestial Lancets, A History and Rationale of Acupuncture and Moxa*, Cambridge University Press.
Lu, Henry C. (1986). *Chinese System of Food Cures, Prevention & Remedies*, Sterling Publ. Co., New York.
Luo Shao-Jun (1995). 'Processing of Tea', in Yamanishi Tei *et al.* (1995), pp. 409–34.
Ma, Chengyuan (1980). 'The Splendour of Ancient Chinese Bronzes', in Wen Fong ed. (1980), 1–19.
Ma, Chengyuan (1986). *Ancient Chinese Bronzes*, Oxford University Press.
Mabesa, R. C. & Babaan, J. S. (1993). 'Fish Fermentation Technology in the Philippines', in Lee *et al.*, eds. (1993), 85–94.
Macfarlane, Gwyn (1984). *Alexander Fleming: The Man and the Myth*, Chatto & Windus, London. Reviewed by Max Perutz (1991), 149–63.
MacGowan, Dr (1871–2). 'On the mutton wine of the Mongols and analogous preparations of the Chinese', *JNCB/RAS.*, New Series VII (1871–2), 237–40.
Maciocia, Giovanni (1989). *The Foundations of Chinese Medicine: A Comprehensive Text for Acupuncturists and Herbalists*, Churchill Livingstone, London.
MacLeod, Anna M. (1977). 'Beer', in Rose ed. (1977), 43–137.
Madsen, W. (1955). 'Hot and Cold in the Universe of San Francisco Teeospa, Valley of Mexico', *J. Amer. Folklore*, **68**: 123–39.
Magno-Orejana, Florian (1983). 'Fermented Fish products', in Chan, Harvey ed. (1983), 255–95.
Mahdihassen, S. (1966). 'Alchemy and its Chinese Origin, as Revealed by the Etymology, Doctrines and Symbols', *Iqbal Review*, **1966**, 22ff.
Mair-Waldburg, H. ed. (1974). *Handbook of Cheese*, Volkswirtschaftlicher Verlag, GmbH, Kempten, Germany.
Mair-Waldburg, H. (1974a). 'On the history of cheesemaking in ancient times', in Mair-Waldburg ed. (1974).
Maitland, Derek (1982). *5000 Years of Tea – A Pictoral Companion*, Gallery Books, New York.
Mallory, Walter H. (1926). *China, Land of Famine*, American Geographic Society, New York.
Mallory, J. P. (1989). *In Search of the Indo-European*, Thames and Hudson, London.
Marks, Gil (1996). *The World of Jewish Cooking*, Simon & Schuster, New York.
Marsden, William tr. & ed. (1908). *The Travels of Marco Polo*, Dent, London.
Maytag, Fritz (1992). 'Sense and Nonsense about Beer', *ChemTech*. March, 1992, 138–41.
Maynard, Leonard A. & Swen Wen-yuh (1937). 'Nutrition', in Buck ed. (1937), 400–36.
Mazess, R. B. & Mather, W. (1974). 'Bone minerals Content of North Alaskan Eskimos', *Amer. J. Clinical Nutrition*, **27** (9), 916–25.
McGee, Harold (1984). *On Food and Cooking: The Science and Lore of the Kitchen*, Scribner's Sons, New York.
McGovern, Patrick E., Fleming, Stewart J. & Katz, Solomon H. eds. (1995). *The Origin and Ancient History of Wine*, Gordon and Breach, Amsterdam.
McGovern, P. E. & Michel, Rudolph H. (1995). 'The Analytical and Archaeological Challenge of Detecting Ancient Wine: Two Case Studies from the Ancient Near East', in McGovern *et al.*, eds. (1995), 57–65.
McGovern, P. E., Glusker, D. L., Exner, L. E. & Voigt, M. M. (1996). 'Neolithic Resinated Wine', *Nature*, **381**, 480–1.
McNeill, William H. (1991). 'American Food Crops in the Old World', in Viola & Margolis, eds. 1991, *Seeds of Change*, 42–xx.
Medawar, P. B. & Medawar, J. S. (1983). *Aristotle to Zoos, A Philosophical Dictionary of Biology*, Harvard University Press, Cambridge MA.
Medhurst, W. H. tr. (1846). *The Shoo King or the Historical Classic*, Mission Press, Shanghai.
Mei Yi-Pao tr. (1929). *The Ethical and Political Works of Mo Tzu*, Probsthain, London.
Merke, F. (1984). *History and Iconography of Endemic Goitre and Cretinism*, Hans Huber, Berne.
Meskill, John, tr. (1965). *Ch'oe Pu's Diary: A Record of Drifting Across the Sea*, with Introduction and Notes, University of Arizona Press, Tucson.
Meyers, Eric M. (ed.) (1997). *The Oxford Encyclopedia of Archaeology in the Near East*, Oxford University Press, New York & Oxford.
Michalowski, P. (1994). 'The Drinking Gods: Alcohol in Mesopotamian Ritual and Mythology', in Milano, ed., 27–44.
Michel, Rudolph, H., McGovern, Patrick E. & Badler, Virginia, R. (1992). 'Chemical Evidence for Ancient Beer', *Nature*, **360**, 24.
Milano, Lucio ed. (1994). *Drinking in Ancient Societies: History and Culture of Drinks in the Ancient Near East*, Papers of a Symposium held in Rome, Italy, May 17–19, 1990, Sargon srl, Padova.
Miller, Gloria Bley (1972). *The Thousand Recipe Chinese Cookbook*, Grosset & Dunlap, New York.
Miller, Naomi (1991). 'The Near East', in van Zeist *et al.*, eds. (1991), 133–60.
Miller, Naomi (1992). 'The Origin of Plant Cultivation in the Near East', in Cowan & Watson (1992), 39–58.
Milner, Max, Scrimshaw, N. S. & Wang, D. I. C. eds. (1978). *Protein Resources and Technology: Status and Research Needs*, AVI Publ. Co. Westport, CT, 1978.
Mizuno, Seiichi, tr. J. O. Gauntlett (1959). *Bronzes and Jades of Ancient China*, Nihon Keizei, Tokyo.
Mokyr, Joel (1990). *The Lever of Riches*, Oxford University Press.

MOLDENKE, H. N. & MOLDENKE, A. L. (1952). *Plants of the Bible*, Chronica Botanica, Waltham, MA [New series of plant science books, no. 28].
MONTGAUDRY, BARON DE (1855). 'Expériences faites pour l'acclimatation des semences importées en France par M. De Montigny', *Bulletin de la Societe d'Acclimatation*, **2** (1), 16–22.
MORITZ, L. A. (1958). *Grain Mills and Flour in Classical Antiquity*, Oxford University Press.
MOTE, FREDERICK W. (1977). 'Yuan and Ming', in Chang, K. C. ed. (1977), 195–257.
MOULE, A. C. & PELLIOT, PAUL (eds. & trs.). *The Description of the World*, Routledge & Kegan Paul, London.
MUI, HOH-CHEUNG & MUI, LORNA H (1984) *The Management of Monopoly: A Study of the English East India Company's Conduct of its Tea Trade, 1784–1833*, University British Columbia Press, Vancouver, Canada.
NAJOR, JULIA (1981). *Babylonian Cuisine: Chaldean Cookbook from the Middle East*, Vintage Press, New York.
NAKAMURA, M. & KAWABATA, T. (1981). 'Effect of Japanese Green Tea on Nitrosamine Formation in Vitro', *J. Food Sci.*, **46**, 306–7.
NAKAYAMA SHIGERU & SIVIN, NATHAN eds. (1973). *Chinese Science*, MIT Press, Cambridge, MA.
NAKAYAMA, TOMMY (1983). 'Tropical Fruit Wines', in Chan, Harvey ed. (1983), 537–53.
NASR, SEYYED HOSSEIN (1976). *Islamic Science: An Illustrated Study*, World Islam Festival Publ. Ltd.
NATHANAEL, W. R. N. (1954). 'The Manufacture and Characteristics of Ceylon Arrack', *Ceylon Coconut Quarterly*, **5** (2), 1–7.
NATIONAL PALACE MUSEUM (1980). *A City of Cathay*, National Palace Museum, Taipei.
NAVARETE, DOMINGO FERNANDEZ DE (1665). 'An account of the Empire of China, historical, political, moral and religious', in Awnsham Churchill and John Churchill (1744), *A Collection of Voyages and Travels*, Vol. 1 (of 4) tr. from Spanish, 3rd edn Churchills, London, ch. 13, pp. 251–2.
NEEDHAM, JOSEPH (1970). 'The Roles of China and Europe in the Development of Oecumenical Science', in Needham *et al.* (1970), *Clerks and Craftsmen in China and the West*, 396–418.
NEEDHAM, JOSEPH (1981). *Science in Traditional China, A Comparative Perspective*, Chinese University Press, Hong Kong.
NEEDHAM, JOSEPH & NEEDHAM, DOROTHY (1948). *Science Outpost, Papers of the Sino-British Science Co-operation Office, 1942–46*, The Pilot Press, London.
NEEDHAM, JOSEPH & WANG LING (1954). *Science and Civilisation in China*, Vol. I, *Introductory Orientations*, Cambridge University Press.
NEEDHAM, JOSEPH & WANG LING (1956). *Science and Civilisation in China*, Vol. II, *History of Scientific Thought*, Cambridge University Press.
NEEDHAM, JOSEPH & WANG LING (1969). *Science and Civilisation in China*, Vol. IV, Pt 2, *Mechanical Engineering*, Cambridge University Press.
NEEDHAM, JOSEPH, WANG LING, LU GWEI-DJEN & HO PING-YÜ (1970). *Clerks and Craftsmen in China and the West*, Cambridge University Press.
NEEDHAM, JOSEPH, HO PING-YÜ, LU GWEI-DJEN & SIVIN, NATHAN (1980). *Science and Civilisation in China*, Vol. V, Pt 4, *Spagyrical Discovery and Invention: Apparatus, Theories and Gifts*, Cambridge University Press.
NEEDHAM, JOSEPH, LU DWEI-DJEN & HUANG, HSING-TSUNG (1986). *Science and Civilisation in China*, Vol. VI, Pt 1, *Botany*, Cambridge University Press.
NEUBERGER, ALBERT, tr. (1930). *The Technical Arts and Sciences of the Ancients*, Brose, New York, 1930.
NEUMANN, H. (1994). *Beer as Means of Compensation for Work in Mesopotamia During the Ur III Period*, in Milano ed. (1994), 321–31.
NI MAOSHING tr. (1995). *The Yellow Emperor's Classic of Medicine*, Shambhala, Boston.
NICKEL, G. B. (1979). 'Vinegar', in Peppler & Perlman (1979), II, 155–72.
NODA, M., VAN, T. V., KUSAKABE, I. & K. MURAKAMI, K. (1982). 'Substrate Specificity and Salt Inhibition of Five Proteinases Isolated from the Pyloric Caeca and Stomach of Sardine', *J. Agric. Biol. Chem.*, **46** (6), 1565–9.
NORMILE, DENNIS (1997). 'Yangtze Seen as Earliest Rice Site', *Science*, **275**, Jan. 17, 309.
NOZZOLILLO, CONSTANCE, LEA, PETER J. & LOEWUS, FRANK A. eds. (1983). *Mobilization of Reserves in Germination*. Vol. XVII, Plenum Press, New York.
NUTTON, VIVIAN (1993). 'Humoralism', in Bynum & Porter eds. (1993), 281–91.
OGUNI, ITARO, NASU, KEIKO, YAMAMOTO, SHIGEHIRO & NOMURA, TAKEO (1988). 'On the Anti-Tumour Activity of Fresh Green Tea Leaf', *Agric. Biol. Chem.*, **52**, 1879–80.
OKUKURA KAKUZO (1906). *The Book of Tea*, Charles E. Tuttle (repr. 1956), Rutland VT.
OKURA NAGATSUNE tr. EIKO ARIGA, ed. CARTER LITCHFIELD (1974). *Seiyu Roku* 制油録 (On Oil Manufacturing), Olearius Editions, New Brunswick, NJ.
OLD, WALTER GORN (1904). *The Shu King or the Chinese Historical Classic*, Theosophical Publishing Society, London.
OLMO, HAROLD (1995). 'The Origin and Domestication of the *Vinifera* grape', in McGovern *et al.*, eds. (1995), 31–43.
OLSCHKI, LEONARDO (1960). *Marco Polo's Asia*, tr. J. A. Scott, University of California Press, Berkeley, CA.
ONIONS, C. T. ed. (1966). *The Oxford Dictionary of English Etymology*, Clarendon Press, Oxford.
OOMEN, H. A. P. C. (1976). 'Vitamin A Deficiency, Xerophthalmia and Blindness', in Hegstead *et al.*, eds. (1976), 73–81.
OSHIMA, KINTARO (1905). *A Digest of Japanese Investigations on the Nutrition of Man*, USDA Office of Experiment Stations, Bulletin No. 159.

OSHIMA, HARRY T. (1967). 'Food Consumption, Nutrition and Economic Development in Asian Countries', *Economic Development and Cultural Change*, vol. 15, no. 4, pp. 385–97. University of Chicago Press.
OTAKE, S., MAKIMURA, M., KUROKI, T., NISHIHARA, Y. & HIRASAWA, M. (1991). 'Anticaries Effects of Polyphenolic Compounds from Japanese Green Tea', *Caries Research*, **25** (6): 438–43.
PACEY, ARNOLD (1983). *The Culture of Technology*, MIT Press, Cambridge MA.
PACEY, ARNOLD (1990). *Technology in World Civilisation*, MIT Press, Cambridge, MA.
PAIGE, DAVID M. & BAYLESS, THEODORE M. eds. (1981). *Lactose Digestion: Clinical and Nutritional Implications*, Johns Hopkins University Press, Baltimore.
PANATI, CHARLES (1987). *Extraordinary Origins of Everyday Things*, Harper & Row, New York.
PANJABI, CAMELLIA (1995). *The Great Curries of India*, Simon & Schuster, New York.
PAPPAS, L. E. (1975). *Bread Making*, Nitty Gritty Productions, Concord, CA.
PARTINGTON, J. R. (1935). *Origin and Development of Applied Chemistry*, Longman, Green & Co., 1935.
PAULING, LINUS (1981). *Vitamin C, the Common Cold and the Flu*, Berkeley Books, originally published in 1970.
PEPPLER, HENRY J. & PERLMAN, DAVID, eds. (1979). *Microbial Technology* Vol. I *Microbial Processes*, & Vol. II *Fermentation Technology*, Academic Press, New York.
PERUTZ, MAX (1991). *Is Science Necessary?*, Oxford University Press.
PFIZMAIER, A., tr. from the Chin Shu, ref. 54–7 in *SCC* Vol. I, p. 288.
PHAFF, HERMAN J. & AMERINE, M. A. (1979). 'Wine', in Peppler & Perlman eds. (1979), II 132–52.
PHITHAKPOL, BULAN (1993). 'Fish Fermentation Technology in Thailand', in Lee *et al.* (1993), 155–66.
PIERSON, J. L. tr. (1929–63). *Man'yoshi*, E. J. Brill, Leiden. 20 vols.
PINKERTON, J. (1808–14). 'The Remarkable Travels of William de Rubriquis . . . into Tartary and China, 1253', in *A General Collection of Voyages and Travels*, Vol. VII.
PIRAZZOLI-T'SERSTEVENS, MICHELE (1985). 'A Second-Century Chinese Kitchen Scene', *Food & Foodways*, **1985** (1), 95–104.
PIRES-BIANCHI, MARIA DE L., CANDIDO SILVA, H. & POURCHET CAMPOS, M. A. (1983). 'Effect of Several Treatments on the Oligosaccharide Content of a Brazilian Soybean Variety', *J. Agric. Food Chem.* 1983, **31**, 1363–4.
PLATT, B. S. (1956). 'The Soya Bean in Human Nutrition', *Chemistry & Industry*, **1956**, 834–7.
PLINY THE ELDER (GAIUS PLINIUS SECUNDUS) (1938). *Natural History* (37 books, in ten vols.), tr. Rackham, H. Loeb Classical Library, Harvard University Press.
POLO, MARCO. *The Travels* cf. Ramusio (1583), Yule & Cordier (1903), Marsden (1908), Moule & Pelliot (1938), Latham (1958), Waugh (1984).
PONEROS, A. G. & ERDMAN, J. W. JR. (1988). 'Bioavailability of Calcium from Tofu, Tortillas, Nonfat Dry Milk and Mozzarella Cheese in Rats: Effect of Supplemental Ascorbic Acid', *J. Food Sc.*, **53** (1), 208–10, 230.
POPPER, KARL (1971, 1979). *Objective Knowledge*, Oxford University Press.
PORKERT, MANFRED (1974). *The Theoretical Foundations of Chinese Medicine*, MIT Press, Cambridge, MA.
PORTER-SMITH, F. & STUART, G. A. (1973). Chinese Medicinal Herbs, Georgetown Press, San Francisco, enlarged from Stuart, G. A. (1910).
POWELL, KEITH A., RENWICK, ANNABEL & PEBERDY, JOHN E. eds. (1994). *The Genus Aspergillus: From Taxonomy and Genetics to Industrial Application*, FEMS Symposium No. 69.
POWELL, M. A. (1994). 'Metron Ariston: Measure as a Tool for Studying Beer in Ancient Mesopotamia', in Milano, ed. (1994), 91–119.
PULLAR, PHILIPPA (1970). *Consuming Passions: Being an Historic Inquiry into Certain English Appetites*, Little, Brown & Co., Boston.
PULLEYBLANK, E. G. (1963). 'The Consonantal System of Old Chinese' (Pt 2), *Asia Major*, **9**, 205–65.
PULLEYBLANK, E. G. (1983). 'The Chinese and their Neighbours in Prehistoric and Early Historic Times', in Keightley, ed. (1983), 411–66.
PUTRO, SUMPENA (1993). 'Fish Fermentation Technology in Indonesia', in Lee *et al.*, eds. (1993), 107–28.
QIAN HAO, CHEN HEYI & RU SUICHU (1981). *Out of China's Earth*, H. N. Abrams, New York and China Pictoral, Peking.
QUILLER-COUCH, SIR ARTHUR ed. (1939). *The Oxford Book of English Verse, 1250–1918*, Oxford University Press. 1st publ. 1900; new edition 1939, repr. 1941.
QUINN, JOSEPH R. ed. (1973). *Medicine and Public Health in the People's Republic of China*, DHEW Publ. No (NIH) 73–67. Bethesda, MD.
RACKIS, JOSEPH J. (1975). 'Oligosaccharides of Food Legumes Alpha-Galactosidase Activity and the Flatus Problem', in Jeanes & Hodge eds. (1975), 207–22.
RACKIS, JOSEPH L. (1981). 'Flatulence caused by soya and its control through processing', *J. Am. Oil chem. Soc.*, **58** (3), 503–9.
RACKIS, JOSEPH L., GUMBBMANN, M. R. & LIENER, I. E. (1985). 'The USDA Trypsin Inhibitor Study. I. Background, Objectives and Procedural Details', in *Quality of Plant Foods in Human Nutrition*, **35**, 213–42, Martinus Nijhoff/Dr W. Junk Publishers, Dordrecht.
RAFFAEL, MICHAEL (1991). 'Vegetarian Pleasures', *High Life* (British Airways Magazine), July, 1991, 52–5.

RAMSEY, S. ROBERT (1987). *The Languages of China*, Princeton University Press, NJ.
RAMUSIO, GIOVANNI BATTISTA (ed. & tr.) (1583). *Delle Navigatione et Viaggi*, Vol. II, Venetia, 2nd edn.
RAWSON, JESSICA (1980). *Ancient China – Art and Archaeology*, British Museum Press, London.
RAWSON, JESSICA ed. (1996). *Mysteries of Ancient China – New Discoveries from the Early Dynasties*, British Museum Press, London (1996).
READ, BERNARD E. (with LIU JU-CHHIANG) (1936). *Chinese Medicinal Plants from the 'Pên Tshao Kang Mu' A.D. 1596 . . . a Botanical, Chemical and Pharmacological Reference List.* Publication of the Peking Natural History Bulletin, French Bookstore, 1936. (Chu. 10–37 of the *Pên Tshao Kang Mu*). Reviewed by W. T. Swingle ARLC/DO, 1937, 191. Expanded from an earlier version by Read in 1927 (Cf. Vol. VI, Pt 1, p. 646).
READ, BERNARD, LEE WEI-YING & CHHÊNG JIH-KUANG (1937). 'Shanghai Foods', *Chinese Medical Association Special Report No. 8*.
REED, GERALD ed. (1975). *Enzymes in Food Processing*, Academic Press, New York.
REED, GERALD ed. (1982). *Prescott and Duns's Industrial Microbiology*, 4th edn, AVI Publ. Co.
RIDDERVOLD, A. & ROPEID, A., eds. (1988). *Food Conservation*, Prospect Books, London.
RINDOS, DAVID (1989). 'Darwinism and its Role in the Explanation of Domestication', in Harris & Hillman eds. (1989), 27–41.
RITTER, DR (1874). *Tofu, Yuba, Ame, Mittheilungen der Deutschen Gesellschaft fuer Natur- und Verlkerkunde Ostasiens (Yokohama)*, **1** (5), 3–5.
ROAF, MICHAEL (1990). *Cultural Atlas of Mesopotamia and the Ancient Near East*, Facts-On-File, New York & Oxford.
ROCKHILL, WILLIAM tr. & ed. (1900), cf. Rubruck, William of.
RODEN, CLAUDIA (1968/1974). *A Book of Middle Eastern Food*, Vintage Books, New York.
RODEN, CLAUDIA. *The Food of Italy*, Arrow, London.
RODINSON, MAXIME (1949). 'Recherches sur les documents arabes relatifs à la cuisine', *Revue des Études Islamiques*, **1949**, 95–165.
ROM, WILLIAM M. & GARAY, STUART M. eds. (1996). *Tuberculosis*, Little, Brown & Co., Boston.
ROMBAUER, IRMA & BECKER, MARION R. (1975). *The Joy of Cooking*, Bobbs Merrill Co., Indianapolis, 1931, new edn. 1975.
RONAN, COLIN A. (1982). *Science: Its History and Development Among the World's Cultures*, Facts-On-File, Oxford.
ROOT, WAVERLEY (1971). *The Food of Italy*, Athenaeum, New York.
ROSE, A. H. ed. (1977). *Economic Microbiology. Vol. 1: Alcoholic Beverages*, Academic Press, New York.
ROSE, A. H. (1977). 'History and Scientific Basis for Alcoholic Beverage Production', in Rose ed. (1977), 1–37.
ROSENBERGER, BERNARD (1989). 'Les Pâtes dans le Monde Mussulman', in Sabban-Serventi *et al.*, eds. (1989), 77–98.
ROY, DAVID & TSIEN, T. H. eds. (1979). *Ancient China: Studies in Early Civilization*, Chinese University Press, Hong Kong.
RUBRUCK, WILLIAM OF (1808–14). *The Remarkable Travels of William of de Rubriequis into Tartary and China, 1253* in John Pinkerton ed. 1808–14, *A General Collection of Voyages and Travels*, Vol. VII, London, 1808–14.
RUBRUCK, WILLIAM OF (1900). *The Journey of William of Rubruck to the Eastern Parts of the world, 1253–55 as narrated by himself*, tr. From the Latin, and edited with an Introductory Notice by William W. Rockhill, Hakluyt Society, London, 1900.
RUMPHIUS, GEORGIUS EVERHARDUS (1747). Herbarium Amboinese, Vol. v. p. 388, Amstelaedami.
RUSSELL, KENNETH (1988). *After Eden*, BAR International Series, 391. Oxford.
SABBAN, FRANÇOISE (1983a). 'Cuisine à la cour de l'empereur de Chine: les aspects culinaires du Yinshan Zhengyao de Hu Sihui', *Médiévales*, **1983** (5), 32–56.
SABBAN, FRANÇOISE (1983b). 'Le système des cuissons dans la tradition culinaire chinoise', *Annales*, **1983** (2), 341–68.
SABBAN, FRANÇOISE (1986a). 'Un savoir-faire oublié: le travail du lait en Chine ancienne', *Zinbun: Memoirs of the Res. Inst. for Humanistic Studies, Kyoto University*, No. 21, 31–65.
SABBAN, F. (1986b). 'Court Cuisine in 14th century Imperial China: Some Culinary Aspects of Hu Sihui's *Yinshan Zhengyao*', *Food and Foodways*, **1**, pp. 161–96.
SABBAN, FRANÇOISE (1988a). 'Sucre candi et confiseries de Qinsai: L'essor du sucre de canne dans la Chine des Song (X^e–XIIIe siècle)', *Journal D'Agriculture Traditionelle et des Botanique Appliquée*, **35**, 195–214.
SABBAN, FRANÇOISE (1988b). 'Insights into the Problem of Preservation by Fermentation in 6th century China', in Riddervold & Ropeid (1988), 45–55.
SABBAN, FRANÇOISE (1990a). 'De la main à la pate: refléxion sur l'origine des pates alimentaires et les transformations du blé en Chine ancienne (IIIe av. J-C – Vi^e siècle ap. J-C)', *L'Homme*, **1990** (113), 102–37.
SABBAN, FRANÇOISE (1990b). 'Food Provisioning, the Treatment of Foodstuffs and other Culinary Aspects of the *Qimin yaoshu*', paper presented at the 6th ICHSC, Cambridge, Aug. 1990.
SABBAN, FRANÇOISE (1993). 'La viande en Chine: imaginaire et les usages culinaires', *Anthropozoologica*, **1993** (18), 79–90.
SABBAN-SERVENTI, FRANÇOISE *et al.*, eds. (1989). Contre Marco Polo: Une Histoire Comparée des Pâtes Alimentaires, *Médiévales*, **1989** (16–17), 27–100.

SABBAN-SERVENTI, FRANÇOISE (1989). 'Ravioli cristallins et Tagliatelle rouges: les Pâtes Chinoises entre XIIe et XIVe siècle', in Sabban et al., eds. (1989), 29–50.
SAGGS, H. W. F. (1965). *Everyday Life in Babylonia and Assyria*, Dorset Press, New York.
SAHI, T. (1994a). 'Hypolactasia and Lactase Persistence: Historical Review and the Terminology', *Scand. J. Gastroenterol*, Suppl. **202**, 1–7.
SAHI, T. (1994b). 'Genetics and Epidemiology of Adult-Type Hypolactasia', *Scand. J. Gastroenterol.*, Suppl. **202**, 7–20.
SAID, HAKIM M. ed. (1990). *Essays on Science: Felicitation Volume in Honour of Dr. Joseph Needham*, Hambard Foundation, Karachi.
SAMPSON, THEOS (1869). 'The Song of the Grape', *Notes and Queries on China and Japan*, Vol. III, 52. Cited in Schafer (1963), p. 144.
SAMUEL, DELWEN (1989). 'Their Stuff of Life: Initial Investigations on Ancient Egyptian Bread Making', in Kemp ed., 253–90.
SAMUEL, DELWEN (1993). 'Ancient Egyptian Cereal Processing: Beyond the Artistic Record', *Camb. Arch. J.*, **3** (2), 271–83.
SAMUEL, DELWEN (1996a). 'Archaeology of Ancient Egyptian Beer', *J. Am. Soc. Brew. Chem.*, **54** (1), 3–12.
SAMUEL, DELWEN (1996b). 'Investigations of Ancient Egyptian Baking and Brewing Methods by Correlative Microscopy', *Science*, *273*: 488–90.
SAMUEL, DELWEN & BOLT, PETER (1995). 'Rediscovering Ancient Egyptian Beer', *Brewer's Guardian*, **124**, Dec. 1995, 26–31.
SANO, M., TAKENAKA, Y., KOJIMA, R. et al. (1986). 'Effects of Pu-erh Tea on Lipid Metabolism in Rats', *Chem. Pharm. Bull* (Tokyo), **34**, 221–8.
SARIS, JOHN (+1613). 'The Voyage of Captain John Saris to Japan' Log of a trip to Japan, in Satow ed. (1900), Vol. v, Series 2, of the works issued by The Hakluyt Society, London.
SASSON, J. A. (1994). 'The Blood of grapes', in Milano ed., 1994, 399–419.
SAWERS, DAVID (1978). 'The Sources of Innovation', in Williams, Trevor ed., 1978, 27–47.
SCHAFER, EDWARD H. (1962). 'Eating Turtles in Ancient China', *J. Amer. Orient. Soc.*, **82**, 73.
SCHAFER, EDWARD H. (1963). *The Golden Peaches of Samarkand*, University of California Press, Berkeley, CA.
SCHAFER, EDWARD H. (1967). *The Vermillion Bird*, University of California Press, Berkeley, CA.
SCHAFER, EDWARD H. (1977). 'Thang', in Chang, K. C. ed. (1977), 87–140.
SCHIFFER, MICHAEL B. ed. (1993). *Archaeological Methods and Theory*, Vol. v. University of Arizona Press, Tucson, AZ.
SCHROEDER, C. A. & FLETCHER, W. A. (1967). 'The Chinese Gooseberry in New Zealand', *Econ. Bot.*, **21**, 81–92.
SHERRATT, ANDREW (1981). 'Plough and Pastorialism: Aspects of the Secondary Products Revolution', in Hodder et al., eds. (1981), 261–305.
SHIBA YOSHINOBU, tr. (1970). *Commerce and Society in Sung China*, abstr. tr. Mark Elvin, Centre for Chinese Studies, University of Michigan.
SHIBASAKI, K. & HESSELTINE, C. W. (1962). 'Miso Fermentation', *Econ. Bot.*, **16** (3), 180–95.
SHIH SHÊNG-HAN (1958). *A Preliminary Survey of the Book 'Ch'i Min Yao Shu'* – An Agricultural Encyclopedia of the 6th Century, Science Press, Peking.
SHIH, SHÊNG-HAN (1959). *Fan Shêng-Chih Shu*: An Agriculturist Book of China, written by Fan Shêng-Chih in the 1st century BC, Science Press, Peking.
SHILLINGLAW, C. A. (1957). Memorandum on a Visit to Noda Shoyu Co. Ltd, Private Communication, on file at the Needham Research Institute.
SHORT, THOMAS (1750). *Discourses on Tea, Sugar, Milk, Mead, Wines, Spirits, Punch, Tobacco etc. with Plain Rules for Gouty People*, London, cited in Ukers (1935), I, p. 29.
SHURTLEFF, WILLIAM & AYOYAGI, A. (1976). *The Book of Miso*, Ten Speed Press, Berkeley, CA.
SHURTLEFF, WILLIAM & AYOYAGI, A. (1975, 1979). *The Book of Tofu*, Ten Speed Press, Berkeley, CA.
SHURTLEFF, WILLIAM & AOYAGI, AKIKO (1980). 'In Search of the Real Tamari', *Soyfoods*, I (3), Summer, 20–4.
SHURTLEFF, WILLIAM & AOYAGI, AKIKO (1985). *History of Tempeh*, Soyfoods Centre, Lafayette, CA.
SHURTLEFF, WILLIAM & AOYAGI, AKIKO (1988a). *Bibliography of Soy Sauce*, Soyfoods Centre, Lafayette, CA.
SHURTLEFF, WILLIAM & AOYAGI, AKIKO (1988b). *Amazake and Amazake Frozen Desserts*, Soyfoods Centre, Lafayette, CA.
SIGERIST, HENRY E. (1951). *A History of Medicine*, I. *Primitive and Archaic Medicine*, Oxford University Press, New York.
SIMMONDS, N. W. ed. (1976). *Evolution of Crop Plants*, Longman, London.
SIMOONS, F. J. (1954). 'The non-milking area of Africa', *Anthropos*, **49**, 58–66.
SIMOONS, F. J. (1970a). 'Primary Adult Lactose Intolerance and the Milking Habit: A Problem in Biological and Cultural Interrelations. II. A Culture Historical Hypothesis', *Am. J. Dig. Dis.*, **15**, 695–710.
SIMOONS, F. J. (1970b). 'The Traditional Limits of Milk and Milk-Use in Southern Asia', *Anthropos*, **65**, 547–93.
SIMOONS, F. J. (1981). *Geographic Patterns of Primary Adult Lactose Malabsorption: A Further Interpretation of Evidence for the Old World*, in Paige & Bayless eds. (1981), pp. 23–48.
SIMOONS, F. J. (1991). *Food in China: A Historical and Cultural Inquiry*, CRC Press, Boca Raton.

SINGER, CHARLES (1941). *A Short History of Scientific Ideas – to 1900*, Oxford University Press.
SINGER, CHARLES, HOLMYARD, E. J. & HILL, A. R. eds. (1954). *A History of Technology, Vol. 1: From Early Times to Fall of Ancient Empires*, Oxford University Press, 1954.
SINGER, CHARLES, HOLMYARD, E. J., HILL, A. R. & WILLIAMS, TREVOR I. eds. (1956). *A History of Technology, Vol. 2: The Mediterranean Civilisation and the Middle Ages, c. 700 B.C. to c. A.D. 1500*, Oxford University Press.
SINGER, CHARLES, HOLMYARD, E. J., HILL, A. R. & WILLIAMS, TREVOR I. eds. (1957). *A History of Technology, Vol. 3: From the Renaissance to the Industrial Revolution: c. 1500 to c. 1750*, Oxford University Press.
SIVIN, NATHAN (1987). *Traditional Medicine in Contemporary China*, University of Michigan, Ann Arbor.
SMARTT, J. & SIMMONDS, N. W. eds. (1995). *The Evolution of Crop Plants*, Longman, London, 2nd edn of Simmonds (1976).
SMITH, BRUCE D. (1995). *The Emergence of Agriculture*, Scientific American Library. Reviewed by Wu Yao-Li 吳耀利, *NYKK*, **1997** (1), 48–50.
SMITH, MICHAEL (1986). *The Afternoon Tea Book*, Collier Books, New York.
SMITH, PAUL (1991). *Taxing Heaven's Storehouse: Horses, Bureaucrats and the Destruction of the Sichuan Tea Industry, 1074–1224*, Harvard University Press Cambridge, MA.
SMITH, R. C. E. & CHRISTIAN, D. (1984). *Bread and Salt: A Social and Economic History of Food and Drink in Russia*, Cambridge University Press.
SMYTH, ALBERT ed. (1907). *The Writings of Benjamin Franklin* 5 vols. The Macmillan Co., New York.
SNAPPER, ISADORE (1941). *Chinese Lessons to Western Medicine; A Contribution to Geographical Medicine from the Clinics of Peiping Union Medical College*, Interscience, New York.
SOEWITO, AUGUSTINA (1986). 'The cooking of Tempe – Indonesia', in Aida Hiroshi *et al.*, eds. (1986), 270–3. See Bibliography B.
SOSULSKI, F. W., ELKOWICZ, L. & REICHERT, R. D. (1982). 'Oligosaccharides in Eleven Legumes and their Air-Classified Proteins and Starch Fractions', *J. Food Sc.*, 1982, **47**, 498–502.
SOYER, ALEXIS (1853). *The Pantropheon or History of Food and its Preparation in Ancient Times*, Simpsin Marshall, London, 1853.
STAHEL, G. (1946). 'Foods from Fermented Soybeans as Prepared in the Netherland Indies. II. Tempe, a Tropical Staple', *J. N. Y. Bot. Gdn*, **47** (564), 285–96.
STAUDENMAIER, JOHN M. (1985). *Technology's Storytellers – Reweaving the Human Fabric*, MIT Press, Cambridge, MA.
STAUDENMAIER, JOHN M. (1990). 'Recent Trends in the History of Technology', *Am. Hist. Review*, **95** (3), 715–25.
STEELE, JOHN tr. (1917). *The I Li, Book of Etiquette and Ceremonial* (2 vols.), Probsthain & Co., London.
STEINKRAUS, KEITH H., *et al.*, eds. (1983). Handbook of Indigenous Fermented Foods, Marcel Dekker, Inc. New York.
STOCKWELL, FORSTER & TANG BOWEN eds. (1984). *Recent Advances in Chinese Archaeology*, Foreign Language Press, Beijing.
STOECKHARDT, ADOLPH & SENFF, EMANUEL (1872). 'Untersuchung von chinesesischen Oelbohnen', *Der Chemische Ackermann*, **18**, 122–5.
STOL, M. (1994). 'Beer in Neo-Babylonian Times', in Milano, ed., 1994, 155–83.
STUART, G. A. (1910). *Chinese Materia Medica, Vegetable Kingdom*, American Presbyterian Mission Press, Shanghai.
SUMIYOSHI, YASUO (1987). 'Present Status of *Shoyu* and *Miso* Industries in Japan', *Daizu Geppo* (Soybean Monthly News), No. 10/11, 19–28.
SUN, E-TU ZEN (1979). 'Chinese History of Technology: Some Points for Comparison with the West', in Bugliarello & Doner eds. 38–49.
SUN, E-TU ZEN & SUN, SHIOU-CHUAN, tr. (1966). *T'ien Kung K'ai Wu: Chinese Technology in the Seventeenth Century*, Pennsylvania State University Press, University Park and London.
SUPARMO & MARKAKIS, P. (1987). 'Tempeh Prepared from Germinated Soybeans', *J. Food Sci*, **52** (6), 1736–7.
SWANN, NANCY L. tr. (1950). *Food and Money in Ancient China: The Earliest Economic History of China to +25* (with tr. of *[Chhien] Han Shu*, ch. 24 and related texts, *[Chhien] Han Shu*, ch. 91, and *Shih Chi*, ch. 129), Princeton University Press, NJ.
TAKAMINE JOKICHI (1891). *Improvements in the Production of Alcoholic Ferments and of Fermented Liquids thereby*, British Patent 5,700, Oct. 17, 1891.
TAKAMINE JOKICHI (1894). *Alcohol-ferment mash*, US Patent 0,525,825, Sept. 11, 1984.
TANG, P. S. & CHANG, L. H. (1939). 'A Calculation of the Chinese Rural Diet from Crop Reports', *Chinese J. Physiol.*, **14**, 497–508.
TANNAHILL, REAY (1988). *Food in History*, 2nd edn, Crown Publ., New York, 1988 (1st edn 1973).
TAYLOR, F. SHERWOOD (1945). 'The Evolution of the Still', *Annals of Science*, **5** (3), 185–202.
TEMKIN, OWSEI (1960). 'Nutrition from Classical Antiquity to the Baroque', in Galdston, 1 (1960), 78–97.
TEMPLE, ROBERT K. G. (1986). *China, Land of Discovery and Invention*, Patrick Stephens, Wellingborough.
TENG, SSU-YÜ & BIGGERSTAFF, K. (1936). *An Annotated Bibliography of Selected Chinese Reference Works*, Harvard-Yenching Institute, Peiping (Yenching Journ. Chin. Studies, monograph. No. 12.).

TERAMOTO, Y., OKAMOTO, K., KAYASHIMA, S. & UEDA, S. (1993). 'Rice Wine Brewing with Sprouting Rice and Barley Malt', *J. Ferm. & Bioeng*, **75** (6), 460–2.
TEUBNER, CHRISTIAN, RIZZZI, SILVIO & TAN LEE LENG (1996). *The Pasta Bible*, Penguin Books, London.
TODA, M., OKUBO, S., HARA, Y. & SHIMAMURA, T. (1991). 'Antibacterial and Bacteriocidal Activities of Tea Extracts and Catechins Against Methicillin Resistant *Staphylococcus aureus*', *Nippon Saikingaku Zasshi*, **46** (5), 839–45 [in Japanese].
THOM, CHARLES & CHURCH, MARGARET B. (1926). *The Aspergilli*, Williams & Wilkins, Baltimore.
TRAUFFER, REGULA ed. (1997). *Manger en Chine*, Alimentarium Vevey.
TSIEN, TSUEN-HSUIN (1985). *Science and Civilisation in China*, Vol. V, Pt 1, Paper and Printing, Cambridge University Press.
TYN, MYO THANT (1993). 'Trends of Fermented Fish Technology in Burma', in Lee *et al.*, eds. (1993), 129–54.
UEDA, SEINOSUKE & TERAMOTO, YUJI (1995). 'Design of Microbial Processes and Manufacture Based on the Specialities and Traditions of a Region: A Kumamoto Case', *J. Ferm.& Bioeng.*, **80** (5), 522–7.
UKERS, WILLIAM H. (1935). *All About Tea*, Vols. I & II, The Tea & Coffee Trade Journal, New York.
UKERS, WILLIAM H. (1936). *The Romance of Tea*, Alfred Knopf, New York.
UNDERHILL, ANNE P. (forthcoming). 'Archaeological and Textual Evidence for the Production and Use of Alcohol in Ancient China', in Moskowitz, Marina ed. *Proceedings of the Ninth Yale-Smithsonian Seminar on the Material Culture of Alcoholic Beverages*.
UNITED STATES DEPARTMENT OF AGRICULTURE (1976). *Composition of Foods: Dairy and Egg Products: Raw, Processed and Prepared*, Agriculture Handbook, Nos. 8–1, USDA, Washington, DC, Revised Nov. 1976.
UNITED STATES DEPARTMENT OF AGRICULTURE (1986). *Composition of Foods: Legumes and Legume Products*, Agriculture Handbook, Nos. 8–16, USDA, Washington, DC, Revised Dec. 1986.
UNSCHULD, PAUL U. (1986a). *Medicine in China: A History of Ideas*, University of California Press, Berkeley, CA.
UNSCHULD, PAUL U. (1986b). *Medicine in China: A History of Pharmaceutics*, University of California Press, Berkeley, CA.
UNSCHULD, PAUL U. ed. (1986). *Approaches to Traditional Chinese Medical Literature*, Kluwer Acad. Publ., Dordrecht.
VALLA, FRANÇOIS (1995). 'The First Settled Societies – Natufian (12,500 – 10,200 BP)', in Thomas E. Levy ed. (1995), 170–85.
VAVILOV, N. I. (1931). 'The role of Central Asia in the Origin of Cultivated Plants', *Bull. Applied Bot., Genetics & Plant Breeding* (USSR), **26** (3), 3ff. (In Russian and English).
VAVILOV, N. I., tr. STARR, CHESTER (1949/50). *The Origin, Variation, Immunity, and Breeding of Cultivated Plants: Selected Writings, Chronica Botanica*, **13**, 1949/50, Waltham, MA.
VALVILOV, N. I., tr. LÖVE, DORIS (1992). *The Origin and Geography of Cultivated Plants*, Cambridge University Press.
VEHLING, J. D. ed. tr. (1977). *Apicius: Cookery and Dining in Imperial Rome*, Dover, New York, originally publ. in 1936.
VEITH, ILZA tr. (1972). *The Yellow Emperor's Classic of Internal Medicine*, University California Press, 1st publ. 1949.
VIOLA, HERMAN J. & MARGOLIS, CAROLYN, eds. (1991). *Seeds of Change*, Smithsonian Institution Books, Washington DC.
WAI, NGANSHOU (1929). 'New Species of Mono-Mucor, *Mucor sufu*, on Chinese Soybean Cheese', *Science*, **70**, 307–8.
WAI, NGANSHOU (1964). 'Soybean Cheese', *Bulletin Inst. Chem.*, Academia Sinica, Taipei, **1964** (9), 75–94.
WALEY, A. (1936). 'The Eclipse Poem [in the *Shih-Ching*] and its Group', *Thien Hsia Monthly*, **3**, 245.
WALEY, A. tr. (1937). *The Book of Songs* [translation of the *Shih Ching*], Allen & Unwin, London.
WALSHER, D. N., KRETCHMER, N. & BARNETT, H. L. eds. (1976). *Food, Man and Society*, Plenum Publ., New York.
WANG CHI-MIN & WU LIEN-Tê (1932). *History of Chinese Medicine*, National Quarantine Service, Shanghai, 2nd edn 1936.
WANG, HUA L. (1986). 'Nutritional Quality of Fermented Foods', in Hesseltine & Wang eds., 289–301.
WANG, H. L. & FANG, S. F. (1986). 'History of Chinese Fermented Foods', in Hesseltine & Wang eds., 23–35.
WANG, H. L. & HESSELTINE, C. W. (1965). 'Studies on Extracellular Proteolytic Enzymes of *Rhizopus oligosporus*', *Can. J. Microbiol.*, **11**, 727–32.
WANG, HWA L. & HESSELTINE, C. W. (1970). 'Sufu and Lao-Chao', *J. Agric. & Food Chem*, **18** (4), 572–5.
WANG, HUA L. & HESSELTINE, C. W. (1979). 'Mold Modified Foods', in Peppler & Perlman (1979) II, 96–129.
WANG ZHONGSHU, tr. K. C. CHANG *et al.* (1982). *Han Civilisation*, Yale University Press.
WARE, JAMES R. tr. & ed. (1981). *Alchemy, Medicine and Religion in the China of A.D. 320: The Nei Pien of Ko Hung*, Dover Publishers, New York, 1st published by M.I.T. Press, 1966.
WATKINS, RAY (1995). 'Cherry, Plum, Peach, Apricot and Almond', in Smartt & Simmonds eds. (1995).
WATSON, BURTON tr. (1963). *Basic Writings of Mo Tzu, Hsün Tzu and Han Fei Tzu*, Columbia University Press.
WATSON, BURTON tr. (1968). *The Complete Works of Chuang Tzu*, Columbia University Press.
WATSON, BURTON tr. & ed. (1984). *The Columbia Book of Chinese Poetry*, Columbia University Press.
WAUGH, TERESA (1984). *The Travels of Marco Polo*, tr. from the Italian work by Maria Bellonci, Sidgwick & Jackson, London, 1984.
WEATHERFORD, JACK (1988). *Indian Givers: How the Indians of the Americas Transformed the World*, Crown Publ., New York.

WEINHOLD, RUDOLF (1988). 'Baking, Brewing and Fermenting of the Grape-Must: Historical Proofs of their Connection', in Riddervold & Ropeid eds. (1988), 73–80.
WELBOURN, R. B. (1993). 'Endocrine Diseases', in Bynum & Porter, eds. (1993), 484–511.
WESTERMANN, D. H. & HUIGA, N. J. (1979). 'Beer Brewing', in Peppler and Perlman, eds. (1979), 2–36.
WHISTLER, ROY L. & PASCHALL, EUGENE F. eds. (1965). *Starch: Chemistry & Technology. I. Fundamental Aspects*, Academic Press, New York & London.
WHITE, LYNN JR (1976). 'Food and History', in Walsher *et al.*, eds. (1976), 12–30.
WILHELM, RICHARD tr. (1968, 3rd edn). *I Ching* (tr. English C. F. Baynes). Routledge & Kegan Paul, London.
WILLIAMS, TREVOR I. ed. (1978). *A History of Technology*, Vol. VI *The Twentieth Century: c. 1900 to c. 1950*, Pt 1. Clarendon Press, Oxford.
WILLIAMS, TREVOR I. (1987). *The History of Invention*, Facts-On-File, New York & Oxford.
WILSON, CHRISTINE S. (1975). 'Nutrition in Two Cultures: Mexican American and Malay Ways with Food', in Margaret Arnott, ed. (1975), 131–44.
WILSON, HILARY (1988). *Egyptian Food and Drink*, Shire Publications, Aylesbury, UK.
WINARNO, F. G. (1987). 'Tempe Making on Various Substrates – Including Unconventional Legumes', in Aida Hiroshi *et al.*, eds. (1987), 125–41. Cf. Bibliography B.
WINFIELD, GERALD F. (1948). *China: The Land and the People*, Wm Sloan Assoc. New York.
WINSLOW, C. E. A. & BELLINGER, R. R. (1945). 'Hippocratic and Galenic Concepts of Metabolism', *Bulletin of the History of Medicine*, **17**, 127–37.
WITTFOGEL, KARL (1960). 'Food and Society in China and India', in Galdston, ed. (1960), 61–77.
WITTWER, SYLVAN, YU YOUTAI, SUN HAN & WANG LIANZHENG (1987). *Feeding a Billion: Frontiers of Chinese Agriculture*, Michigan State University Press, East Lansing.
WOLFROM, M. L. & KHADEM, H. EL. (1965). 'Chemical Evidence for the Structure of Starch', in Whistler & Paschall eds. (1965), 251–74.
WOOD, ED. (1996a). *World Sourdoughs from Antiquity*, Ten Speed Press, Berkeley, CA.
WOOD, ED. (1996b). 'Bake like an Egyptian', *Modern Maturity* Sept.–Oct., 1996, pp. 66–7.
WOOD, FRANCES (1995, 1996). *Did Marco Polo go to China?*, Westview Press, Boulder, CO.
WOODWARD, NANCY H. (1980). *Teas of the World*, Collier Books, London.
WU HSIEN (1928). 'Nutritive Value of Chinese Foods', *Chinese J. Physiol. Report Series*, No. 1, 153.
WYLIE, A. (1867/1964). *Notes on Chinese Literature*, Shanghai, 1867; repr. Paragon, New York, 1964.
XING RUNCHUAN & TANG YUNMING (1984). 'Archaeological Evidence for Ancient Wine Making', in Stockwell & Tang, eds. (1984), 56–58.
XU, Y., HO, C. T., AMIN, S. G., HAN, C. & CHUNG, F. L. (1992). 'Inhibition of Tobacco-Specific Nitrosamine-Induced Lung Tumorigenesis in A/J Mice by Green Tea and its Major Polyphenol as Antioxidants', *Cancer Research*, **52** (14), 3875–9.
YAMANISHI, TEI ed. (1995). *Special Issue on Tea, Food Reviews International*, Vol. **11** (3), 1995, Marcel Dekker, New York.
YANG, C. S. & WANG ZHI-YUAN (1993). 'Tea and Cancer', *J. Natl Cancer Inst*, **85**: 1038–49.
YANG, E. F. & READ, B. E. (1940). 'Vitamin B content of Chinese plant Beriberi remedies', *Chinese J. Physiol.*, **15** (1), 9ff.
YANG XIANYI & GLADYS YANG, tr. (1957). *The Scholars* [*Ju Lin Wai Shih*], Foreign Language Press, Peking.
YANG XIANYI & GLADYS YANG, tr. (1979). *Selections from the Records of the Historian* by Szuma Chhien, Foreign Language Press, Peking.
YANG XIANYI, GLADYS YANG & HU SHIGUANG, tr. (1983). 'Selections from *The Book of Songs*', with Introduction by Yu Guanying, 'China's earliest Anthology of Poetry', Panda Books, Peking.
YEH, SAMUEL & CHOW, BACON (1973). 'Medicine and Public Health in the People's Republic of China', in Quinn ed. (1973), 215–39.
YOKOTSUKA, T. (1986). 'Soy sauce biochemistry', *Advances in Food Research*, **30**, 195–329.
YOKOTSUKA, T. & SASAKI, M. (1986). 'Risks of Mycotoxin in Fermented Foods', in Hesseltine & Wang, eds. (1986), 259–87.
YONG, F. M. & WOOD, B. J. B. (1974). 'Microbiology and Biochemistry of Soy Sauce Fermentation', *Advances in Applied Microbiology*, **17**, 157–94.
YU, H., OHO, T., TAGOMORI, T. & MORIOKA, T. (1992). 'Anticariogenic Effects of Green Tea', *Fukuoka Igaku Zasshi*, **83** (4): 174–80.
YULE, HENRY & CORDIER, HENRI (trs. & eds.) (1903–20). *The Book of Ser Marco Polo the Venetian Concerning the Kingdoms and Marvels of the East*, 3rd edn (3 vols., 1903–20), repr. in 2 vols., Murray, London, 1975.
YÜ GUANYING (1983). 'China's Earliest Anthology of Poetry', in Yang *et al.*, tr. (1983).
YÜ YING-SHIH (1977). 'Han', in Chang, K. C., ed. (1977), 53–83.
ZACCAGNINI, C. (1994). 'Breath of Life and Water to Drink', in Milano ed. (1994), 347–60.
ZEE, A. (1990). *Swallowing Clouds*, Simon & Schuster, New York.
VAN ZEIST, WILLEM, WASYLIKOWA, KRYSTYNA & BEHRE, KARL-ERNST eds. (1991). *Progress in Old World Paleoethnobotany*, A. A. Balkema, Rotterdam.

VAN ZEIST, WILLEM (1991). 'Economic Aspects', in van Zeist *et al.*, eds. (1991), 109–29.
VICKERY, KENTON F. (1936). 'Food in Early Greece', Illinois Studies in the Social Sciences, xx (2), University of Illinois Press.
ZENG ZIFAN, tr. S. K. LAI (1986). *Colloquial Cantonese and Putonghua Equivalents* (in both English and Chinese), Joint Publ. Co, Hong Kong.
ZHU, Y., LI C. L. & WANG, Y. Y. (1995). 'Effects of Xuezhikang on blood lipids and lipoprotein concentration of rabbits and quails with hyperlipidemia', *Chin J. Pharmacol.*, **30**, 4–8.
ZHU, Y., LI C. L. & WANG, Y. Y., ZHU J. S., CHANG J., KRITCHEVSKY, D. (1998). '*Monascus purpureus* (red yeast): a natural product that lowers blood cholesterol in animal models of hypercholesterolemia', *Nutr Research*, **18** (1), 71–81.
ZITO, R. (1994). 'Biochimica nutrizionale degli alimenti liquidi', in Milano, ed. (1994), 69–75.
ZOHARY, MICHAEL (1982). *Plants of the Bible*, Cambridge University Press.
ZOHARY, DANIEL & HOPF, MARIA (1993). *Domestication of Plants in the Old World*, 2nd edn Oxford University Press, 1st edn 1988.
ZOHARY, DANIEL (1995). 'The Domestication of the Grapevine *Vitis vinifera* L. in the Near East', in McGovern *et al.*, eds. (1995), 23–30.

GENERAL INDEX

abalone, mock 502
Abe Koryu & Tsuji Shigemitsu 318n
ablution, fingers 101, 105
Absidia 280
Abu Ali al-Husein Ibn Sina *see* Avicenna
Acanthobromo simoni 63
Acanthopanax spinosum (*wu chia phi*) 169n, 235
  bark 192
  acetic acid 283, 287, 291
Acetobacter 280, 283, 287, 291n, 594
  alcohol inhibition 288
  film formation 289
  oxygen requirement 288
  slime forming 289
Achaya, K. T. 452n, *454*
Achyranthes bidentata 234
acidulant 92
Acipenser sinensis 62
aconitum 580
Aconitum fischeri 235, 283n
Acorus calamus see Typha latifolia
Actinidia chinensis 52
Actinomucor 328
Adams, R. McC. 597n
Adolph, W. H. 590n, 604n
aflatoxin 605n
Afshana (Bukhara) 495
Agaricus campestris 145, 590n
agricultural treatises 143–4, 515
agriculture, origins in China 24–5, 27
Ah Ying 478n
Ai Kung 102n
Ajinomoto Library of Dietary Culture 473n
a-la-chi wine 204, 227, 229, 231
Albrecht, H. 452, *453*
albumins 465
alchemy
  distillation 226, 229, 596, 597n
  capability 207, 208
  Mount of the Eight Venerables 300
  Taoist 300n
alder wood 445
ale *see* beer
Aleurites fordii 441
Alexander of Tralles (+525–605) 577
alfalfa 193, 241
Algar, A. 496n, 497n
alkali 477
alkaline carbonates 477, 478, 483
All About Tea (Ukers) 542n
allec see garum
Allium bakeri *33*, 39
Allium fistulosum *33*, 38–9
Allium odorum 38
Allium ramosum *33*
Allium sativum 39

Allium scorodoprasum 39
almond 190, 437, 440
Altchuler, S. I. 604n
alum
  for fruit storage 428
  mountain 303, 305
  in rice water 580
  water 197, 201
Amano Motonosuke 19n, 463n
amaranthus oil 452
Amarna (Egypt) 271
amaurosis 586
amazaké (sweet wine) 263
ambix 207
amé (malt sugar) 263n
America
  ice house 430n
  microbial enzyme production 608
  osteoporosis 604n
  peanut introduction to China 456
  soybean oil 457n; starch crops 462; *tou fu* 328, 329
amino acids 593
  essential 573
Amoy 179n, 538, 397
amphora 275, *276*, 277
Amygda sinensis see Trionyx sinensis
amylase 154–5, 174, 177, 593, 608
  chhü ferment 279, 334
  exploitation by Chinese vintners 278
  heat resistance 260n
  malt or sprouting grains 154, 278–9
amyloglucosidase 279
Amylomyces (Amylo) process 155–7, 157, 159, 238, 608
  Taiwan aborigines 161
an (narrow platform bed) 113, 140
an (trays) 86, 87, *101*, 112–13
An Chin-Huai 305n, 306n
An Hua Hsien Chi (Gazetteer of the An-hua County; 1872) 543
An-i (Phing-yang) 243, 244n
Analects of Confucius, The *see Lun Yü*
analytical approach 607
Anchor Brewing Co. (San Francisco) 269n
ancient Egypt 242n
  beer 239, 258, 267; making 267, *268*, 269–72
  bread making *268*, 269n, 270, 463n
  grape wine 243n, 258
  kitchen 272n; olive oil 455n
  saddle quern 599n
  salted fish 396n; sausages 423n
  tandor oven, tomb mural of Dynasty XVIII 266n
  vinegar 283n; yeast 271n, 272n
Anderson, E. N. 10, 67n, 138n, 148, 149n, 245n, 252n, 258n, 259n, 270n, 293n, 298n, 328n, 467n, 595n, 596n

Anderson, M. L. 293n, 298n, 467n
Anderssen, J. G. 77
André, J. 497n
Ang Chin-Huai 430n
Ang Tian-Se 303n
angkak (Thailand, the Philippines) 396–7n
Anhui 191, *545*
  tea exports 548
  tea plantations 513
  yellow tea 551
Anhui shêng wên-wu ... (1959) 463n
animal bones 432
animal fat 28, 93, 437, 439–40, 456, 601
animal products, monographs 145–6
animals, land 55–61
anise 96
Anser 60
ant eggs 66
antecedents of objects 601
antifoam agents 313
anvil for tea processing *521*, 522, 558
Anyang, Shang ruins 61
ao (dry frying; parching) *70*, 72–3, 84–5, *88*, 89
Apicius de re Coquinaria (Cookery and Dining in Imperial Rome; +1st century) 400, 400n
apiculture 246n
Apium graveolens 36
apples, steeped in water 428
apricot, Chinese (Jaapanese) *44*, 46–8, 49, 284, 291n
  honey preserves 426; salting *418*
  smoked 425; sugared 426n
aqua ardens/vitae 203
aquaculture 62, 63
aquatic animals 61–6, *see also* fish
Arberry, A. J. 494n
arbor, grape vine 242n
Ardeshir, A. G. & Beddows, C. G. 383n, 398n
Arisaema consanguineum 187n
Armillaria matsutake 144
Aronson, S. 163n
arrak 245 *see also a-la-chi* wine
arrowroot 462, honey preserves 426
artemisia *170*, 173, 340, *tsu 406*
arum, serrated 187
Asana (Turfan; Sinkiang) 478
Asarum sieboldi 587, 589
ascorbic acid 298
ash, oil processing 455
asparagus 239
Aspergillus 5, 6n, 8, 167
  conditions for growth 277
  ferment 173, 194, 280, 282, 592, 595n
Aspergillus glaucus 552
Aspergillus niger 397n, 605n
Aspergillus oryzae 165, 377, 397n, 605n
Aspergillus terreus 202n
Assam (India) 549
  wild tea trees 503n
astralagus 580
Astronomica (Manilius; +1st century) 400n
aubergine *see* eggplant
Auricularia auriculae (wood ear) 42, 145, 590n

autolysis 380, 382–3, 386, 390, 392, *shiokara* 397
Autumn Festival, moon cakes 3
Averrhoa carambola 52n
Avicenna (+980 to 1037) 494–5, 496n
awl 520, *521*

back deformity, turtle 589
bacteria *see also* lactic acid bacteria
  acetic acid 283; in *ferment* 280
bacteriophage 251n
Badler, V. R. 153n, 258n
Bagdad, Islamic pasta recipes 494
bagels 472n
Bagley, R. W. 513n
baking, clay-wrapped 73, *88*, 89
Balfour, E. 536n
Bali, *tempeh* making 343–4
Ball, S. (official of East India Company in Canton from 1804 to 1826) 534n, 546, 547–8
balsam pear 95n
bamboo
  basket 195, 520, *521*, 539, *558*; container 80, 99
  leaves 187, 342n, 529, in *chiang* making 352
  mats 109, 187, *521*, 531; rod 520, *521*; rope 449, 450
  sieve 473, 558; sieve tube for soy sauce 364, 365, *367*
  splints *521*, 522; steamer 88–9, *90*, 230
  thatch 435; trays 197, 199, 520n, *521*, 526, 527;
  winnowing 473n; urn covers 364, 365–6, *367*
bamboo shoots *33*, 36, 43, 297
  beriberi 582
  cha 409; honey preserves 426
  sauce 358n; *tsao* preserves 409
  tsu pickles 379, 380, *406*
bamboo slips, Han Tomb No. 1 at Ma-wang-tui 25, *26*, 67, *70*, 121n
  fermented soy paste 346
  fermented soybeans 336
  wine 166n
  see also Han Tomb No. 1 (Ma-wang-tui)
Bamboosa 36
Ban Kao, Thailand 76n
'Banquet at Hung-Men' (mural, Western Han tomb near Loyang) 102–3
bappir (beer bread) 269, 271, 272
Bar-Josef, O. 258n
barley 19, *20*, 22, 25, 27
  beer development in West 258; bread 267
  chiang 346, 355, *356*
  distilled wine, wine mash for 204
  ferments 167; grinding 172, 262
  importance in Han dynasty 27; introduction in China 262
  kernel separation 265, 266; *koji* 376, 377
  malted 459; malting 262, *604*; malt sugar 461n
  meal 263, 266, 593; paste 278
  milling 593; moulded *604*; for *miso* 375
  parching 265; partially polished 285
  porridge 266n, 593; sprouting 163n, 457
  varieties 266; vinegar preparation 285–6
Barnett, J. A. 252
barrel, open-ended wooden 230–1
Basalla, G. 597, 598, 599, 601, 602

Basiodiomycetes 33
basket
    bamboo 195, 520, *521*, 539
    drying tea leaves 531, *558*
    rattan 162
Bastone, Ponvio (Italian soldier; 13th century) *495*
bean 23, *see also* soybeans
    aduki/adzuki 40n
    broad 297n; *chiang* 355, *356*, 373n; sprouts 297
    lesser 19n, *33*, 40; *chiang* 355, *356*; flavoured vinegar 286–7
    mung 190; noodles 491n, 492; tea congee 562; wax film 429n; red 43, 185, 235; wild *33*, 38, 520
bean sprouts 2, *595*, 596, 597n, *604 see also* soybeans
    antecedent 601; broad bean 297
    dietary 298; mung bean 296–7
    pharmaceutical 298; preparation 14
    vitamins 605
bean curd *see tou fu*
bear
    gall bladder 588
    meat 581n, 582
Beard, J. 466n
bed, narrow platform 113, 114, 115
Bedini, S. 533n
beef 57 *see also* cattle; ox
    chiang 381; *fu* 420n; *khuai* 70n; tallow 93, 440
beer 149
    ale 149, 259, 267
    bread 270, 272, 274
    brewing in West 275, 277
    containers for making 265
    Egyptian 239; Iberian 239
    making
        Egyptian models *268*
        fermentation 270, 271; saccharification 270, 271
        soaking raised bread 270
        origin and discovery in West 263–72
        porridge 266; bread 267–70; Sumerian 269; in Egypt 270–2
beeswax 187
Beijing 59n
Beinfield, H. & Korngold, E. 571n
Bell, J. 244n
Benda, I. 153n
Benzoni, Girolamo 154n
beriberi 573, 578–86, 605, 606
    diet therapy 581–2
    early records 579–80
    pathology 580–1
    rice, association with 578, 579, 581, 582, 605, 607 and
    bran 581, 585, 586, 606, 607
    spread to North 582, 585
    theories of cause 585–6; treatment 581, 586
betel nut 186
betle pepper 52n
beverages *see* drinks
Bianchini, F. & Corbetta, F. 258n
The Bible 257n
    Genesis 257n; Exodus 63n, 267n, 603n
    Leviticus 401; Numbers 91n; Samuel 456n

Bibliography of the History of the Former Han 124
Bibliography to the History of the *Sui* 124
Bielenstein, H. 248n
bihun (rice noodles, Indonesia & Malaysia) 492
bird's nest 146
bitter taste 91–2, 95
bittern 303, 305, 319, 329
blanching 407
blindness *see also* vision
    green 586
    night 573, 586–8
    total 587
Blofeld, J. 14n, 506n, 507n, 515n, 517n, 519n, 525n, 535n, 540n, 555, 557n, 560, 560n, 561n, 562
Blue, G. 467n, 493n, 494n, 496–7
Bo-hea (Wu-I) 538
Bodde, D. 507n, 509n
Bodhidharma (Indian price and Buddhist saint; +6th century) 506
boglu (malt) 269
boiler 76, 77, 78, 79, 80, 230
    bronze still 208, 209, *210–11*, 214
    steamer 444, 449
boiling 70, 71, 74n, 82, *88*, 89
    oil extraction 452
    pasta 599n
    soy milk, in making bean curd 303, 308–9, 311–12
Bombay mastic 235
bone deformities 588, 589
    reversal 590
Book of Odes *see Shih Ching*
Book of Songs (Waley's translation of the Book of Odes)
Book of Tofu (1979) 310n, 329
boorher (wooden half-barrel) 237n, 238
Bos taurus domesticus 57
Boswellia carterii 234n
Botanicum Sinicum (Bretschneider) 510n
bouza beer 161n, 267, 270
Bowen, Samuel (Savannah, Georgia) 292
bowl
    gold 106; origins 603n; shallow 100
    rice 99; tea 557, *558*, 559, 561
Bowra, C. M. 456n
box, China 305
Boyle, Robert (chemist; 17th century) 149
bracken *406*, *409*, 520
Braidwood, R. J. 265n
braising 84, *88*
bran
    gluten preparation 501
    rice 581, 585, 586, 606, 607
    vinegar mash 288
brandy 204, 244
Brassica alboglabra 35
Brassica campestris 35
Brassica cernua 31
Brassica chinensis 35, 38
Brassica chinensis var. *oleifera* 31
Brassica juncea 31
Brassica oleracea 31
Brassica rapa 31, *33*

brassicas 31, 34-5
Bray, F. 19, 24, 27n, 28n, 47n, 68n, 123n, 143n, 262n, 292n, 293n, 396n, 459n, 463n
bread 266, 463 *see also* dough; flour
   barley 267
   beer 270, 274; precursor 267, 272
   Egyptian models making *268*
   flat 265, 471n
   loaves for Egyptian beer 272
   origins in raised pasta 469
   raised 267, 269, 270, 274, 466, 603, 606
   steamed 282n, 469, 593n, 594n, *604*, 606
break (in yoghurt making) 251
breakfast, shredded meat or fish 424n
bream 61, *62*
Bretschneider, E. 39n, 507n, 510n
brewing *see* beer
bridegrooom gifts for bride 49
Briggs, W. (+1684) 586
Brillat-Savarin, J. A. 17n
brine 284n, 305 *see also* salt; salting
   soy sauce making 371
British Museum (London) 218n
bronze vessels *see also* still
   characters 18
   cold storage of food 430
   ritual 76n
   for serving food and drink 98, *99*, *100*
broth 89, 93 *see also* soup
   fermentable 267
   fermented 266
Brothwell, D. & Brothwell, P. 266n
Brothwell, D. & Sandison, A. 579n
Broussonetia papyrifera 190
Brown, S. R. 452n, 457n
Bruce, C. A. 503n, 549
brunch, Cantonese 485
buckwheat flour
   dough 488
   noodles 485, 488, 492
Buddhism, Chhan 506, 515
Buddhists/Buddhist monks
   gluten use 497, 502
   tea 515; tea making 561
   tou fu making 596, 597n, in Japan 318
   vegetarian recipes 323n
Buell, P. D. 138n
Bukhara 494, 495, 496, 496n
buns *see also* man-thou
   bright 3, 472n; roasted *604*
   steamed 282n, 469, 475-8, 490n, *594*, in Japan 492
Buonassisi, V. 497n
Burkhill, I. H. 342n
Burma
   fermented salted fish *395*, 396
   fish sauce 392
Burnt-wine method of Southern Tribal folk *see Nan fan shao chiu fa*
butcher 2, 3
   dog 59
   of King Hui of Liang 68

butter 358, 437
   iced 434
   tea 562
buttermilk 252, 253
Bynum, W. F. & Porter, R. 573n, 578

cabbage 3, 31, 38, 407, *409*, 582
   salting *418*
   seed oil, uses 450, yield 452
   tsu 405-6, *406*
cafes, tea 527, *528*
caffeine 565, 568-9, 570
Cai Jinfeng 571n
cake, pressed oilseed 450, *451*
cakes 463, 467
   little dragon 525; moon 3
   pan-frying 439; steamed 71
calcium 590
   absorption 604n
   bean curd 604-5
   deficiency 588, 590, 604; metabolism 588n
calcium sulphate 329n, 605
calendar, Chinese 171n
Cambodia
   fermented salted fish *395*, 396
   fish sauce 392
   rice noodles 492
Camellia sinensis 503, *504*
Camellia sinensis var. *assamica* 549n
camellia tea 553
Campbell-Platt, G. 249n
camphor 451, 553
camphorwood 444-5
Campsis grandiflora 234n
cancer 570, 605
Candida 378n
candles 29, 441
Candolle, A. de 44, 48
cane sugar 95, 426, 459-60, 461
   fruit preservation 426
Canis familiaris 58
Cannabis sativa 18, 25, 28
Canon of the Virtue of the Tao *see Tao Tê Ching*
Canton 59n
   fish sauce 397
   tea technicians 548n; tea trade *545*
Capsicum 96
carambola 52n
Carassius auratus 63
carbohydrates, hydrolysis 351
carbon dioxide 415, 465, 466
   raised dough 477
carbonates *see* alkaline carbonates
carcinogens 605
cardamon 239
cardiovascular disease 570
Cardwell, D. 597n, 602n
Caretta caretta olivacea 64-5
Carica papya 49
Carlile, M. J. & Watkinson, S. C. 608n
carotene 587
carp 61, *62*, 384-6, *385*, 582, 586

Carpenter, R. 510n, 515n, 519n, 520n, 522n, 563n, 578n, 579n
carpenter weed, vitamin B 581n
carrot 590n, *409*
casein hydrolysis 329
cassava 237, 602n
cassia *44*, 52, 91, 92, 96, 186 *see also* cinnamon
   heart of 589
   oil 214; tea 553
   wine 233, 237, medicated 238
Castanea mollissima *44*, 49
castor oil 451, 452, 456
catechins 536, 565–6, *567*, *568*
   pharmacological effects 570
Cater, C. M., Craven, W. W. & Horan, F. E. 292n
Cato the Elder (c. −180) 169, 437, *438*, 462
cats, polished rice feeding 581, 606
cattail *33*, 35, 379, 403, *406*
cattle 55, 57, 60, 61 *see also* beef; ox
cauldron 71, *79*, 80
   serving 98, 101
cauliflower, creamed 7
celery 379, *406*, *418*
   dryland 36
   Oriental *33*, 35
   wetland 35–6, 43
cellar, ice house 430–1, 434
Central Asia, grape wine 242
cereals 18, 154, 298n, 285, 379, 602n *see* rice, millet, barley, wheat
ceremonial offerings
   Chinese cherry 52; dog 59
   ritual baskets 468; wines 164
   see also rituals, wine
Ceresa, M. 513n
Ceylon, tea industry 547, 548
cha (Chinese hawthorn) *44*, 49
cha (deep frying) *88*, 89, 325
cha (jellyfish) 146
cha see meat & fish, pickle *594*
cha-no-yu (Japanese tea ceremony) 557
Chadwick, J. 571n
Chaenomeles sinensis *44*, 49
Chakravarty, T. N. 496n
Chamfrault, A. 121
Champion, P. 329n
chan (sturgeon) 62, 380
chan (thick gruel) 82
Chan Kuo Tshê (Records of the Warring States; Chhin) 155, 294
Chan, W. C. 605n
Chang Ao, King of Tsao 112
Chang Chhien (envoy; −2nd century) 39n, 240, 241
Chang Chhien-Tê (writer; 16th century) *516*, 563n
Chang Chi (physician; c. +200) 580
Chang Chung-Ko 60n
Chang Hou-Yung 203n, 223, 432n
Chang Hsiao-Kuang 314n
Chang I Lieh Chuan 294n
Chang, K. C. 18n, 32n, 59n, 76n, 80n, 98n, 102n, 104n, 111, 113, 148, 259n, 260n, 277n, 466n
Chang, Kwang-Chih 48n, 55n, 57n, 58n, 61n

Chang Lien-Ming 128, 147n, 358n
Chang Mêng-Lun 147, 169n, 337n
Chang Shih-Nan (writer; +13th century) 206
Chang Tê-Shui 151n
Chang Tê-Tzu 24n, 262n
Chang Thing-Hao 168n
chang tho (dough, held by hand) 499n
Chang Ting 255n
Chang Tsho 506n
Chang Tsung-Min (11th century) 144
Chang Tzu-Kao 155
Chang Wên-Hsü 27n
Chang Yu-Hsin (writer; +9th century) *516*
Chang Yuan (writer; +16th century) *516*, 530–1, 560
Chang Yuan-Fang 257n
Chao, Buwei Yang 88n, 91n, 148, 358n, 414n, 466n, 471n, 478n, 479n
Chao Chi (writer; +12th century) *516*
Chao Chün (poet, bibliophile; 17th century) *142*, 143
Chao Ho-Thao 506n
Chao Hun (Summons of the Soul; c. −300) 94, 433, 457
Chao Ju-Kua (writer; +1225) 206
Chao Ju-Li (writer; +12th century) *516*
Chao Kang 24n
Chao Khuang-Hua 206n
Chao Kung (Duke of State of Lu; −6th century) 102n, 430
Chao Ping 221
Chao Yao (Hunan), cassia from 52, 91
Chao Yün-Tshung 154n, 169n, 171n
Chao Yung-Kuang 69n, 104, 265n
char-siu-pao (Cantonese for pork-filled bun) 471n
charcoal 522, 523, 534n, 549
chê chiang (sugarcane juice) 457
chê chiang (yam juice) 457n
Chê-chiang shêng kung-yeh chhang (Manufacture of Shao-hsing Wine; 1958) 179n, 261n, 279n, 281n
Chê-chiang shêng po-wu-kuan (on Ho-mu-tu; 1978) 31n, 44n, 260n
Chê-chiang shêng wên-kuan-huei (on Shui-thien-fan, Hangchou; 1960; 1978) 44n
cheese 328–9
   comparison with *tou fu* 325, 328–30
   cottage 329; lactic 252; curd 255, 328n
   flavour enhancement 608n
   microbial rennet 608n
   nutrient composition 330
   preparation *325*; uses 329–30
Chekiang 326n
   Neolithic sites 30–1
   tea plantations 513
   tsao preserved duck eggs 413; *yü lu* 391
chên (Chinese hazelnut) *44*, 49, *51*
Chên Chhüan (physician; +7th century) 576n
Chen Junshi 604n
Chên Kua (*I Ching* hexagram) 157n
Chên Li-Yen (physician; +7th century) 576, 578
Chênchiang (Kiangsu province) 376n
chêng (steaming) 67, 70–1, *88*–9, 170, 205n, 521, *524*
chêng chhi shui (distilled water) 226
Chêng Chih Chun Shêng (Standard Methods for Diagnosis and Treatment; +1602) 588, 590

Chêng Chin-Shêng 571n
chêng chiu (steaming wine; distilled wine) 206–7, 225–6
Chêng Hsiai (author of *Kuang Chi Chu*; Sung) 132
Chêng Hsiung-Tuan (Yüan poetess) 303
Chêng Hsüan (bibliophile; +127–200) 19n, 21, 22–3, 27, 53, 102n, 115, 240n, 334, 429, 469n, 579n
    chhi and *hai* 379, 380; *chiang* 382, 390; *er* and *tzu* 468; *fu* and *hsi* 420; soup 83–4, 93; suckling pig 84; *tsu* 403, 404, 416
Chêng, I (author of *Chung Khuei Lu*) 132, 391
Chêng Lei Pên Tshao (Reorganised Pharmacopoeia; +1082) 135, 136, 137, 237, 290, 296, 460
    gluten 498n; milk products 254
    shih chih preparation 361n
    tea 512, 564
chêng ling chhi liu (steam distillation) 226n
chêng lung (bamboo steamer) 88–9, 90
chêng ping (steamed bread) 468, 475, 477, 593
Chêng-chou shih po-wu-kuan (on Yang-Shao Period remains near Chêngchou; 1973) 36n
Chêng-Ho (Fukien) 202, 544, 546n
Chengtu (Szechuan) 1, 5, 508n
    dining or drinking scene brick painting 113–14
cherry, Chinese 44, 51–2
chest deformity, turtle 589
chestnut, Chinese 40, 44, 49, 50, 427–8, 429, 457
chha see tea
Chha Chieh (Tea Explanations; +1609) 516, 518, 530, 531–2, 534n, 560
chha chien (tea bowl) 559
Chha Chien (Notes on Tea; +1630) 516, 518, 530, 532, 551n
Chha Ching (Classic of Tea; c. +770) 88n, 121n, 230, 506, 510, 514, 515–17, 520, 521, 523, 526, 529, 556, 557,
    effect on health 562–3
    harvesting and processing of tea 520–2
    preparation of tea from cake 556–7, 522n
    Shên Nung tea legend 506; water for tea 517, 556–7
Chha Ching (Tea Classic; +1597) 516, 518
chha chu (tea kettle) 560
Chha Chü Thu Tsan (Tea Utensils with Illustrations; +1269) 516, 517, 557, 558, 559
chha hu (teapot) 560
chha huang (tea yellow, steamed tea leaves) 524
Chha Khao (Comments on Tea; +1593) 516, 518, 538n, 560n
Chha Lu (Tea Discourse; +1051) 516, 517, 523, 534, 557n, 560n
    drying furnace 526; scented teas 553n
Chha Lu (Tea Records; +1595) 516, 518, 530–1
chha ou (tea cup) 560
chha pei (bamboo tray) 526; *chha pei* (tea cup) 560
chha phing (tea kettle) 559
Chha Phu (Tea Compendium; +925) 516, 517, 523
Chha Phu (Tea Discourse; +1440) 516, 518, 529–30, 553
    implements 559
Chha Phu (Tea Discourse; +1539) 516, 518, 530, 553
    drinking tea 563
Chha Phu (Tea Discourse; +1554) 530, 532
chha shên (Tea god) 515n; *chha shêng* (Tea sage) 515n
Chha Shih (History of Tea; +1669) 518
Chha Shu Chhüan Chi (Complete Collection of Tea Books; +1613) 515n

Chha Shu (Tea Books) 515, 516, 517–19
Chha Shu (Tea Commentary; +1597) 516, 518, 530, 531, 532n, 534n, 538; tea making 560
Chha Shuo (Talks on Tea; +1590) 516, 518, 530, 551n
chha szu (tea cafes) 527, 528
chha thiao (tea ladle) 559
chhai yu (rape seed oil) 358
chhan tou (broad bean) 297n
chhang (ritual wine) 156–7, 162, 183, 185, 232
chhang chhu (Chinese gooseberry) 52
Chhang Chi Chieh Ku 500
chhang huang (dried malt sugar) 457
Chhang Jên (Horticultural Officer) 53–4
Chhang Jên (Superintendant of the Imperial Gardens) 240n
Chhang Jen (Superintendant of Ritual Wines) 232
Chhang Ku Chi (Works of Li Hao; +817) 192n
chhang (ritual wine) 156–7, 162, 183, 185
    drug and herb addition 232
chhang shih (spoon) 106
chhang tshao (fragrant wine herb) 232–3
Chhang-chiang liu yü etc. (on Western Han tombs near Chiang-ling, Hupei; 1974) 103n, 104n, 112n
chhao (stir-frying; pan roasting) 88, 89, 170, 171
    soybeans 352
    tea 527, 530, 531, 539, 547
chhao hsia jên (stir fried shrimps) 497
chhao shou (wonton) 478n
Chhao Yuan-Fang 575
chhên (clam) 66
Chhên Chhuan 503n, 506n, 507n, 527n, 537n, 542, 548n, 551n
Chhên Chi (Imperial Tutor at Yuan court) 434
Chhên Chia-Hua 397n
Chhên Chih (author of *Shou Chhing Yang Lao Hsin Shu*; 11th century) 137
Chhên Chih-Ta 55n
Chhên Chin-Shi 225n
Chhên Chu-Kuei & Chu Tzu-Chên 509n, 510n, 515n, 516, 518, 519n, 522n, 523n, 525n, 526, 529n, 530n, 531n, 534n, 538n, 539n, 540n, 541n, 542n, 551n, 553n, 554n, 557n, 558, 559n, 560n, 563n, 564n
Chhên Hsien-Ju 88n
Chhên Hsing-I 538n
Chhên Hung 429n, 430n
Chhên Jên-Yu (writer; 13th century) 144
chhên mien (pulled noodle) 484–5
chhên ni chung chui (beriberi) 579
Chhên Shao-Chün 463n, 468n, 469n, 475
Chhên Shih 516, 560n
Chhên Sou-Ta (writer; Southern Sung) 301
Chhên Thao-Shêng 147, 155n, 159n, 169n, 179n, 182n, 186n, 353n, 365n
Chhên Ting (Ming Dynasty) 144n
Chhên Tsêng-Pi 112
Chhên Tshun-Jên 147
Chhên Tsung-Mao 503n
Chhên Tsung-Mou 517n, 518, 529n, 544
Chhên Wên-Hua 24n, 305n, 307, 309n, 310n, 314n, 332–3, 463n
Chhên Yüan-Ching (writer; 14th century) 421, 500n
chhêng (orange) 54

chhêng (stone or wood platform; anvil) *521*
Chhêng Chhing-Thai 61, *62*
Chhêng Chien-hua 298n
Chhêng Chih-Fan 571, 576n
Chhêng Kung 579n
Chhêng Li-Chün 247n
Chhêng Pu-Chi 305
Chhêng Ta-Chhang (writer; Sung) 199n
Chhêng Tsai Chi (Devout Vegetarian; +12th century) 301
Chhêng Tshang-Chhi (physician and writer; +8th century) 581, 606, 607
Chhêng Yao-Thien 20n, 24n
Chhêng-teh still (Hopei) 208–9, *210*
chhi (vital energy) 337n, 571n, 574
    effect on disease, bone deformities 589; goitre 574, 575, 578;
    imbalance in the leg 578; kidney 575; circulation 588
chhi (ant eggs) 66; *chhi* (awl) *521*
chhi (broth) 84; *chhi* (lettuce) *33*, 38
chhi (pickled chopped vegetables) 74n, 76, 402, 403, 404
chhi (receiver) 204; *chhi* (shepherd's purse) 587
chhi (small fish) 397–8n; *chhi* (vessel) 204
chhi hai (boneless meat paste) 379
Chhi Hsieh 441n
Chhi Min Yao Shu (Important Arts for the People's Welfare, +544) 6, 10, 16, 24n, 25n, 29, 29n, 31, 34, 35, 36, 38, 40, 43, 49, 52n, 63n, 69n, 73n, 75n, 96n, 158, 159, 161, 169, 241, 262, 279, 298, 382, 383, 393, 423, 428n
    dried meat and fish (*i yü, fu* and *hsi*) 383, 419–21
    ferment (*chhü*) 277–8
    for making wine (*chiu*) 169–80
    for food processing 335–6
    leaven 477, 606n
    fermented meat and fish 380–1, 390–1
    Fermented meat and fish paste (*chiang*) 381–3
    pickled fish (*cha*) 384–6, *tsao* preserved meat 413
    fermented soybean (*shih*) 336–40
    soybean paste (*chiang*) 347–51
    soy sauce precursors (*shih chih etc*.) 359–60)
    gluten (*po tho*) 498–9
    malt sugar (*i*) 58–9, 461
    milk products 248, 251–6,
    pasta (*ping*) 469–75
    steamed bun and wonton 475
    filamentous noodles (*mien*) 480–1
    pickled vegetables (*tsu*) 405–8
    starch preparation 462–2
    tou fu absence 300, 313–4
    tea 509–10
    vegetable oil 29, 31, 439–50
    vinegar 285–97
chhi pickles 379
chhi pien (*Acanthobromo simoni*) 63
Chhi Ssu-Ho 24n
chhi tsu (pickled vegetables) 379
chhi tzu mien (chessmen pasta) *470*, 472–3, 474, 482
chhi wines for sacrificial offerings 164
chhi ying (movable swellings) 576
chhi yu (fish sauce) *see yü lu*

Chhi Yüeh (Ode to the Seventh Month) 18, 19, 429
Chhi-men (Anwhui) *544*
chhieh (slicing) 75n
chhieh mien (sliced noodles) 3, 474, 484, 485, *486*
chhieh mien chu (sliced pasta congee) 474
chhien (poaching in water or broth) *70*, 74
chhien chang (pressed bean curd; thousand sheets) 320, *322*, *324*
Chhien Chhun-Nien (writer; +16th century) *516*, 518, 530, 553, 559, 563
Chhien Chin I Fang (Supplement to A Thousand Golden Remedies; +655) 16n, 136, 238, 240n
    beriberi 581, 606
    night blindness 587–8
Chhien Chin Pao Yao (A Thousand Golden Treasures; +1124) 371
Chhien Chin Shih Chih (Golden Dietary Remedies; +655) *135*, 136, 341, 460n
Chhien Ching Yao Fang (Thousand Golden Remedies; +150 to +659) 136, 136n
    beriberi 579, 581, 585; goitre 576
    grape 240n, malt sugar 460, *shih* 341
    soybeans 296
Chhien Han Shu (History of the Former Han Dynasty; c. +100) 59, 60n, 141, 315n, 353, 468n
    fermentation of wine 167–8; wine 233
    fermented soybeans 337
    gifts to Hsiung Nu 166
    Prefect of Mare Milkers 248
Chhien I (writer; +12th century) 205, 589
Chhien Lung (Chhing emperor) 329n
chhien tshêng man thou (thousand layer steam bun) 476n
chhih fan (to have a meal) 466
Chhih Ma (official) 56n
chhih tou (aduki bean) 40n
chhin (oriental celery) *33*, 35
chhin chiao (*Zanthoxylum*) 52
Chhin Kuan 225
Chhin-tu Hsien-yang khao-ku kung-tso chan (on Old Chhin Capital of Hsien-yang; 1980) 432n
chhing chiang (clarified sauce) 347n, 359, 362, 363
    condiment use *372*, 373
Chhing chiu (Chou wine, clear wine) *165*, 222
Chhing I Lu (Anecdotes, Simple and Exotic, +965) 126, *127*, 354, 387n, *409–10*, 411–12
    bone of wine 202
    milk products 255n
    pao-tzu 476; pasta 481
    red *ferment 193*; wine 180–1
    tea 517; *tou-fu* 300, 301, *595*
    tsao 390n, preserved crab 413n
chhing mang (green blindness; amaurosis) 586, 587
Chhing Ming Shang Ho Thu (City of Cathay; scroll painting +1736 and 1980) 527, *528*
Chhing-lung hsien ching chan tzu ta-tui ko-wei hui (1976) 208n, *209*, 213
Chhing-yang (southern Anhui) 300n
chhiu (Chinese hawthorn) 49
Chhiu Fêng 61n
Chhiu Phang-Thung 54n, 63n, 91n, 120, 123, 126n, 128, 131n, 132n, 147n, 148, 463n, 467n, 468n, 469n, 480n, 485n, 488n

chhiu yu (soy sauce harvested in autumn) 371, 373
Chhiung Kuan Hsien Sêng Chi (Collection of the Excellent Stone Master) 501n
chho shu (sip or suck soybean) 294, 316
chho tshai (pickled vegetable) 404
chhao tshai (wild bean) 38
Chhoe Pu (Korean official; 15th century) 329n
chhou (pull) 483
chhou tou fu kan (*tou fu* salted and fermented) 320
chhou yu (drawn sauce; soy sauce) 366, *372*
chhü (*ferment*) ancient character *see* ferment
chhu (kitchen) 432
chhu (vinegar) 354
Chhü Chia-Wei 186n
Chhü Chih-Hsin 287n, 288n
chhu chiu (mortar and pestle) *521*, 522
Chhü Hsien Shên Yin Shu (Notes of the Slender Hermit; +1440) *129*, 130, 254, 353, 357, 387, 414n, 419n
   dairy products 255n, *256*
   honeyed and sugared fruits 426n
   tsao-preserved crab 413n
Chhu Hsüeh Chi (Primer of Learning; +700) 22n, 192n
Chhu Jen-Huo (Chhing dynasty) 145
chhü mu (mother of ferment) 195, 200
chhu niang (lady caterers) 597n
chhü nieh interpretations 157–60
Chhü Pên Tshao (Pandects of Ferment Cultures; +990) 132–3, 194, 205, 224, 237
chhü shih (ferment power) 174, *175*, 177, 279
Chhü Ta-Chün (author of *Kuangtung Hsin Yü*; 17th century) 134
Chhu Tzhu (Elegies of the Chhu State; c. −300) 23, 43n, *117*, 118–19
   cooking, methods 67, *70*, 71, 72, 73; pots 76n, braising meat 84
   foods 94, 95; *erh* 468, cooled drink, five flavours 283n, *hai thun* 380; *tsao* 408; turtle 64–5
   sweeteners 457, malt sugar 158; honey 246
   tea 507; *lo* 250; wine 156, 157; herbal 233, 237
Chhü Wan-Li 157, 232n
Chhü Wei Chiu Wên (Old News from the Winding Wei River; +1127) 194
Chhü Yuan (principal author of *Chhu Tzhu*; c. −340 to 278) 118–19
Chhu-chou Archaeological Bureau (Anhui, Nanking), bronze still 209–10, 213
chhuan (bamboo strips) *521*, 522
chhüan (dog) 55; *chhuan* (hair pin) 522n
chhüan (iron mould for tea pressing) *521*
chhuan (old leaves of tea) 510n
chhuan hsiung (hemlock parsley) 186
Chhüan Jên (dog supervisor) 56n
Chhuan Ming Lu (Tea Record; +960) *516*, 517
Chhüan Shang-ku San-tai Chhin-Han San-kuo Liu-chao wen (1836) 240n
Chhüan Thang Wên 515n
chhuang (narrow platform bed) 113
chhüeh (sparrow) 380
chhüeh mu (sparrow eyes; night blindness) 586–8
chhui (blow fire) 80; *chhui fan* (cook rice) 80
chhun (alcoholic flavour or aroma) 260, 261

Chhün Fang Phu (An Assembly of Perfumes or Monographs on Cultivated Plants; +1630) 145, 456n
Chhun Phu Sui Pi (Casual Notes on Local Products; 1888) 543, 548n
chhung yang (9th day of 9th month) 235
Chhung Yang Hsien Chi (Gazetteer of Chhung-yang County; 1866) 542, 548n
Chhung-an (Wu-I mountains; Fukien) 540, 546n
chi (*Carassius auratus*) 63
chi (chicken) 55, 60 *see* chicken
chi (goblets) 166; *chi* (low table) 109, 114
chi (millet) 24, 27, 59, 140; *chi* (millet: for cooking) 18
chi (millet: foxtail) 18, 25; *chi* (millet: setaria) 19n, *20*, 23
chi (stools) 109, 111, 112, 140; *chi* (wild date) 48
Chi Chiu Pien (c. −40) 284, 336–7, 346, 457, 468, 354n
Chi I 156n, 250n
Chi Jên (Chicken Keeper) 60
Chi Lê Pien (+1133) 97, 440–1, 525n
chi mi (setaria millet, kernels) 459
chi ming chiu (cock crow wine) 189, 190
chi shao (pasteurised wine) 222
Chi Shêng Fang (Prescriptions for Preserving Health; +1253) 585
chi tzu mien (egg noodle) 491
chi tzu ping (egg pasta) *470*, 472
Chi-nan, capital of Chhu Kingdom 432, *433*
Chi-nan chhêng Fêng-huang-shan 168 etc. (on Han Tomb at Fêng-huang-shan, Chiang-ling; 1975) 101n, 102n, 103n
Chi-nan (capital of Shantung) 112, 151n
chia (chopsticks) 104, 105
chia (wine serving vessel) 99, *100*, 102, 155
chia chuan hsien chung (goitre) 574
Chia Ming (author of *Yin Shih Hsü Chi*; +14th century) 138, 564
Chia, O. 305n, 332
Chia Ssu-Hsieh (author of *Chhi Min Yao Shu*; +6th century) 123, 125, 169, 173, 251, 288, 336, 597n, 605
chia thu (bitter *thu*) 508
chia tshao (*Zingiber mioga*) 40
Chia Tsu-Chang & Chia Tsu-Shan 39n
chia yü (lucky fish) 61, *62*
chia yü (shelled or armoured fish) 65
Chia-hu (Honan), early rice culture 27n
Chia-yü-kuan tomb paintings (Wei-Chin period) 85
chiai lu (poor fit of skull bones) 589
chiang (fermented meat or fish paste) 380, 381–3, *381*, 390, 391 *see* meat, fish; fermented paste
chiang (fermented soy paste) 335, 342, 346–55, *356*, 357 *see* soy paste, fermented
chiang (savoury pastes and sauces) 95, 96, 333
   definitions 334–5
   Japanese sauces 374
   other types: wheat, elm nut, lesser bean, broad bean 355–6
chiang (water drained after boiling grain) 97, 187, 189
chiang (ginger) *33*, 40, 91
Chiang Chhao 571n
chiang chhing (clarified soy paste) 359, 362, 371, *372*, 408
chiang chih (juice of *chiang*) 371, *372*
chiang chün khuei (general's helmet vessel) 151, *152*, 153

chiang huang (yellow *chiang* ferment) 349, 352, 353, 363, 364, 371, 375
Chiang Jên (Superintendant of Beverages) 232n
Chiang Ju-Tao 310, 312
Chiang Jun-Hsiang 146n
Chiang Shih-Chhê (writer; 19th century) 541n
chiang sui (water extract of *chiang*) 371, *372*
chiang thun (river dolphin) 69
Chiang Thung (Chin dynasty) 161
chiang tshai (pickled vegetables) 412
Chiang Tzhu (Housing Superintendant) 112–13
chiang yen chha (ginger-salt tea) 519n
chiang yen tshai (pickled vegetables) 412
Chiang Ying-Hsiang 32n, 118
Chiang Ying-Shu 193n
chiang yu 4, 335, 357, 358, 389 *see* soy sauce
Chiang-hsi shêng po-wu-kuan (Eastern-Wu Tomb of the Eastern Han Period; 1978) 80n
Chiang-ling (Hupeh) 47, 52, 101, 102, 103
    utensils in Han tomb 105
    vessels for food and drink in Han tomb 101, 102, 103
chiang-lu method 414–15
Chiang-phu 54
Chiang-su hsin i-hsüeh yuan (Cyclopedia of Chinese Drugs; 1986) 232n, 234n, 298n
Chiangling Fêng-huang-shan 167 etc. (on Han Tomb No. 167 at Feng-huang-shan, Chian-ling; 1976) 101n
Chiangnan Beer Factory (Shanghai) 279n, *281*
chiao (drinking vessels) 102n
chiao (leaven pasta) *470*, 471, 476
chiao (yeast) 184, 187, 188–9, 251, 606 *see also* yeast
chiao chhi (imbalance of *chhi* in the leg) *see* beriberi
Chiao Chhi Chih Fa Tsung Yao (Methods for Treating Chiao-chhi; +1075) 579n, 581, 585–6, 607
chiao erh (leaven cakes) 478, 479
chiao i (jellied malt sugar) 460
Chiao, J. S. 377n
chiao jo (weakness of the legs) 579
chiao mai (buckwheat) 488n
chiao mu (fresh yeast) 189; *chiao tzu* (little leaven) 477
chiao tzu (ravioli like dumpling) 478–80, 490, *604*
chicken 2, 55, 56, 60, 61, *409*, 582
    lu method 414; mock 497–8, 502
    served to Son of Heaven 58
    steamed 71; *sung* (parched shredded chicken) 423
    tsao preserves 409
Chieh Chha Chien (Chieh Tea Notes; +1642) *516*, 518, 560n
chieh chhi (solar terms) 171n
Chieh district (Chekiang) 531
chieh keng (*Platycodon grandiflorum*) 235
chieh lan (brassica) 35
chieh tshai (mustard) 31
chieh tzu (Chinese mustard) 31
chien (cold storage urn) 430
chien (mixing and drinking bowl for tea) 557, 559
chien (shallow frying) 70, 72, 84, *88*, 89, 325
chien chha (grading tea) *524*, 525
Chien Chha Shui Chi (Water for Tea; +825) *516*, 517
chien fêng hsiao (fritters of glutinous rice) 126
Chien-an (Chien-an, Northwestern Fukien) 517, 523, 553

Chien-chhang hung chiu (red wine of Chien-chhang) 196
Chien-Chhang (Szechuan) 196n
Chien-Hsi (Northwestern Fukien) 523
Chien-Ou (Fukien) 202
chih see cha
chih (drinking vessels) 102n
chih (hard animal fat) 437
chih (roasting meat on a skewer) 70, 73, 85–6, *88*, 89
chih (stoup) 99, *100*, 103, 113
chih chü (raisin tree) *44*, 49
chih erh (mushrooms) *33*
Chih Fa-Tshun (physician; 4th century) 579–80
Chih Fang Shih (Director of Regions) 21
chih ju (steeped meat) 417
chih kao (sheep or ox tallow) 440
chih ma (fragrant or grease hemp) 440
Chih ti (Eastern Han Emperor; died +146) 469n
Chih Tzu 106n, 148
chih tzu (*Gardenia jasminoides*) 508
chih-erh (mushrooms) 40, 42
childbirth 234
chilli 96
chin (wild plant used as vegetable) 457n
Chin Hua Chhung Pi Tan Ching Pi Chih, +1225 207–8
Chin Kuei Yao Lüeh (Systematic Treasury of Medicine; +200) 580
chin ping (gold pasta) 468
Chin San Ssu (Gold Mountain Temple) 376
Chin Shan-Pao 265n
Chin Shih Khun Chhung Tshao Mu Chuan (Illustrations of Minerals, Insects, Herbs and Trees; +1620) 20, *142*, *143*
Chin Shu (History of the Chin Dynasty; +265 to +419) 295n, 358, 408
    Biography of Shih Chhung, *Lieh Chuang* 294n, 295n
Chin tomb (+1115 to +1234; Chhêng-teh, Hopei bronze sstill) 208–9, *210*
chin tou (soaking soybeans) 308
chin ying tzu (cherokee rose) 451n
chin-chü (cumquat) 54n
Chin-hua (Chekiang) 190n; ham 422
chin-tai-mien (ribbon noodles) 484
China at Work (Hommel) 215n
Chinese Gastronomy (1969) 148
Chinese quince 49
Ching Chhu Sui Shih Chih (New Year Customs in the *Ching Chhu* Region; c. +550) 294n, 316
    Ho Yen story 469n
Ching Chou Thu Ti Chi (Record of the Ching Chou Region) 510
Ching Hua Yüan (novel; 18th century) 323n
Chin Phing Mei (Golden Lotus; Ming novel) 498n
ching thao (Chinese cherry) 51
Ching Tien Shih Wên (Textual Criticism of the Classics; c. +600) 511
chiu (Chinese leek) *33*, 38, 40, *41*
chiu (wine from cooked grains) 96, 97, 149–50, 164, *594*, 606 *see* wine
    chhü nieh in making 157–60, 260
    chiu English term for 149–50
    Japanese 281–2
    origin in China 259–63, 274

Chiu Chêng (Wine Superintendant) 164, 232n
Chiu Ching (Wine Canon; +7th century) 180
Chiu Ching (Wine Canon; +1090) 132, *133*, 182
Chiu Hsi-Kuai 513n
Chiu Hsiao Shih (Mini-History of Wine; +1235) 133, 190n, 192, 247, 237, 194
Chiu Huan Pên Tshao (Treatise of Wild Plants for use in Emergencies; +1406) 144, 529n
Chiu Jên (Chief Wine Master) 164
Chiu Kao (Wine Edict; +300) 161, 257n
chiu ku (nine grains) 19n, *20*
Chiu Ku Khao (Study of the Nine Grains) 20n
chiu lao (thin fermensed mash) 155
chiu lu (dew of wine) 204
chiu mu (mother of wine; wine inoculum) *164*, 281
chiu niang (mother of wine) 281n
chiu pai (muddy wine) 161
Chiu Phu (Wine List; +7th Century) 132, *133*, 180
Chiu Shih (History of Wine; Ming) 133, 192n, 206n
Chiu Thang Shu (Old History of the Thang Dynasty; +945) 245, 255
chiu yao (wine medicament) 163, 183, 261, 281n, 593
chiu yüan (yeast culture) 282
chives 418, *379*, 403
Cho Ho-Thao 506n
cholecalciferol 588
cholesterol 330, 590
  lowering effect of red *ferment* 202
chopsticks 104, 105
  noodle eating with 482
chou (strong drink) 164
Chou Chia-Hua 237n, 245n, 246n, 247n
Chou Hêng-Kang 291n
Chou Hou Fang (Handbook of Medicines for Emergencies; +340) 134, *135*, 575 587
Chou Hsin-Chhun 179n, 279n
Chou Jên (Yoke-harness Superintendent) 57n
Chou, King (last ruler of Shang dynasty; −1100) 104
Chou Kung, Duke of Chou 119
Chou Li (The Rites of the Chou) 19, 20n, *21*, 23, 28, 49, 54, 55, 56, 57, 59, 60, 63, 65–6, 66n, 69, 84–591, 92, 96n, 115, *117*, 102, 119, 164, 232, 283n, 284, 333n, 334, 430, 416, 420, 424, 427, 507, 507n
  cooking methods 67, 70–4, 437
  dining furniture 109, 112–13
  fat for cooking 437
  Hai Jên (Superintendent of Fermented Victuals) 379–91, 402–4
  nutritionist-physician 137, 571
  raised pasta 469; serving vessels 99n, 100
  wine 156–7, 232
Chou Li, Han Wên Chia (The Rites of the Chou, Treasury of Cherished Literature) 601–2
Chou Mi (writer; c. +1300) 141, 245
Chou San-Chin 131n
Chou Shih-Yung 513n
Chou Shu-Pin 506n
Chou Su-Phing 63n
Chou Wên-Thang 508n
Chou wines; *Fa chhi, Li chhi, Ang chhi, Thi chhi & Chhên chhi; Shih chiu, Hsi chiu & Chhing chiu*

Chow, K. & Kramer, I. 554n, 557n, 559n, 561n, 565n, 569n, 570
chrysanthemum, tea 554, wine 235–6
Chrysanthemum morifolium 235
chu (candles) 29; *chu* (chopsticks) 104
chu (Chinese quince tree) 49
chu (cooking with water) 70, 71–2, *88*
chu (gruel; congee) 82, 260, 284
chu (millet) 24
chü (Chinese lettuce) 38
chü (female hemp plant) 28, 29
chü (raisin tree) 49
chü (sweet orange; tangerine) 54
chü chhang (ritual wine) 157, 232
Chu Chhêng 207n, 229n
Chü Chhü (nr Lake Thai; Kiangsu Province) 34n
Chu Chhüan (compiler of *Chhü Hsien Shên Yin Shu* and *Chha Phu*; +15th century) 130, *516*, 518
Chu Chhüan Hsiao Phin (Teas from Various Springs; +1554) *516*, 518, 551n
Chü Chia Pi Yung (Essential Arts for Family Living; c. +1300) *127*, 128, 130, 132n
  cha 386, 387–8; chiang 352, 355, 357, 386, 414
  distilled wine 227–8
  food preservation 409–10, 411n, 412n, 413, *418*, 419, 422, 423, 426, 428, 429
  gluten 500
  pasta: *man-thou & pao-tzu* 476, noodles 483
  milk products 254–6
  red *ferment* 194–7, 200–1
  shih 340, 360
  tea 542n; congee 562; scented 553, 562
  vinegar 290, 291n
chü chiang (*Piper nigrum*) 52n
Chu Chun-Po 234n, 237n, 238, 243n, 248n
Chu Chung-Shêng 527n
Chu Fan Chih (Records of Foreign Peoples and their Trade; +1225) 206
chu fu (chilled *fu*) 421
Chu Hsi (writer, poet; +1130 to +1200) 299n, 301, 402n
Chu Hsiao, Prince (esculentist movement in Ming dynasty; +15th century) 131, 588
chü i (yellow robe) 165
chu i chiang (chasing *i chiang*) 382–3, 397, 399
chu I wei chih yü (deterioration of the feet and leg) 579
Chu I-Tsun (author of *Shih Hsien Hung Mi*; 17th century) 131
Chu Jui-Hsi 488n
chu kao (lard) 437n
Chu Kho-Chên 48n
Chu Kung (author of *Pei Shan Chiu Ching*; +12th century) 132, 161, 183, 191, 193
chü leaves 348
Chü Lu (Monograph on Citrus; +1178) 144
Chü Lu (The Orange Record; +1178)
  fruit storage 428–9; honey preserved fruits 426
chu mien (cooking noodle) 484n
chü nü (ring pasta) 473
Chu Phu (Congee Menu; +1750) *135*, 139
Chu Pien (poet of Southern Sung) 34
chu ping (cooked pasta) 468, 469n

GENERAL INDEX

Chu Ping Yuan Hou Lun (Diseases and their Aetiology; +610) 575, 580, 585
  children's ailments 589
chu shih (primary food) 316
Chü Sung (Ode to the Orange) 43n
Chu Têh-Hsi 513n
Chu Têh-Jun (poet; +14th century) 229
Chu Wên-Hsin 118n
chu yu (lard) 358, 437n
chu yu (wine for *chhung yang* day) 235
chu yü (*Zanthoxylum ailanthoides*) 52n
Chu Yü Shan Fang Tsa Pu (Miscellanies from the Bamboo Islet Studio; c. +1500) 130, 485, 491
Chu-chhêng (Shantung), kitchen scene from Eastern Han tomb 86, 87, 332n, 597n
Chu-Ko Liang (general and strategist; Three Kingdom period +221–265) 34n
chuai (stretch) 483
chuan (duck) 84
chüan (pongee bag) 461
Chuang Chi-Yü 97n
Chuang Kung 102n
Chuang (Prince of State of Chêng) 83
Chuang Tzu (Book of Master Chuang; c. –290) 68, 140, 574
  hsi 420; *hun thun* 478
chüeh (lettuce) 405
chüeh (beriberi) 579
chüeh (bracken) 520n
chüeh (wine serving vessel) 99, *100*, 102
chüeh ming tzu (sickle senna seeds) 587
Chûjiruki (c. +1090) 491
Chuko Liang (general of Three Kingdoms period) 475
Chün Phu (Treatise on Fungi; +1245) 144
Chung Chha Shuo Shih Thiao (Ten Talks on Tea; c. 1874) 543, 548n
Chung Hang (eunuch at the Han court) 250n
Chung Hua 441n, 507n, 511n, 574n
Chung I-Yen & Ling Hsiang 246n, 295n
Chung Khuei Lu (Book of Viands; +1870) *129*, *132*, 391
Chung Kuo Chha Ching (Modern Chinese Classic of Tea; 1992) 503n, 506n, 515n, *516*, 520n
  black tea (red tea) 542n, *544*, 547n, 551n
  effects of tea 564n; reprocessed tea 553n
  varieties of tea 535n; yellow tea 552n
Chung Kuo Pheng Jen Shih Lo (1983) 147
Chung Kuo Shih Ching Tshung Shu 122, 147
Chung Kuo Shih Phin Kho Chih Shih Kao, Shang 147
Chung Kuo Shih Phin Kung Yeh Fa Chan Chien Shih 147
Chung Kuo Shih Wên Hua ho Shih Phin Kung Yeh Fa Chan Chien Shih (1983) 147
Chung Kuo Wei Sheng Wu Kung Yeh Fa Chan Shih 147
Chung Kuo Yin Chuan Shih ti-i-Chüan (1988) 147
Chung Kuo Yin Shih Than Ku (1985) 147
Chung yuan (Summer Festival) 296n
Chung-chhing shih po-wu-kuan 113n
Chung-kuo chha-yeh li-shih tzu-liao hsuan-chi (Selected Historiographic Materials on Tea in China; 1981) 518
Chung-kuo I-Hsüeh Pai Kho Chhuan Shu (1988) 571n
Chung-kuo kho-hsüeh yuan khao-ku yen-chiu so 108
Chung-kuo nung-yeh pai-kho chhuan shu 519

Chung-kuo phêng-jen pai-kho chhüan shu (1992) 79, 147n, 373n, 391, 413n, 419n, 421n, 423n, 427n, 472n, 479, 482
Chung-kuo pheng-jen wên-hsien kai shu 122n
Chung-kuo pheng-jen wên-hsien thi-yao 122n
Chung-kuo shê-huei kho-hsüeh yuan (1979) 146, *464*, 599n
Chung-kuo thu-nung yao-chih (1959) 39n
Chung-kuo ti-fang-chih chha-yeh li-shih tzu-liao hsuan-chi (Selected Historiographic Materials on Tea in the local Gazetteers of China; 1990) 518
Chung-yang Wei-sheng Yen-chiu Yuan (Tables of Composition of Foods; 1952) 179n
chuo (poaching) 70, 74, 82, *88*
chuo (table) 111
Church, A. H. 329n
Church, C. G. 49n
Cibot (Jesuit; +1780) 21
cinnabar 235, 207
Cinnamomum cassia *44*, 52, 235
cinnamon 52, 92, 186n, 233, 239 see also cassia
citron, *chiang* preserves 409, 412
Citrus grandis 54; *Citrus reticulata* 54
City of Cathay; scroll painting see *Chhing Ming Shang Ho Thu*
clams 66, 379, *409*, 576
climate 48; China 277; West 178
Clostridium 280
cloud-ear mushroom 144–5
clove 96, 186, 235
coats, lamb skin 57
cocklebur 29n, *170*, 184, 185, 335, 352, 437, 441
coconut
  flower wine 245; milk wine 245n; oil extraction 452
coffee, decaffeinated 569
Cohen, M. 265n, 599n
Coix lachryma 190
cold storage 8, 415, 429–36
Colocasia esculentum 42, 43
Columbus, Christopher (explorer; 15th century) 493
Columella (Roman author of +1st century) 398n
colza 31, *33*, 43, 403, 407, 587
  oil *440*; presing seeds 439
  salting *418*; *tsu* 405–6
communications in China 607
condenser 204, 208, 213, 231
confections 98, 460
Confucius 22, 99, 119
  chiang 390n
  drinks 96
  food habits 69n, 74n
  fu 420
  gourd dish 103
  meat preparation 69, 71
  sour taste agent 284
  wine 162–3
congee 2, 82, 232, 284, 294, 474, 601
  book on (*Chu Phu*, Congee Menu) 139
  heart of fermented bean 341
  discovery of *i* (malt sugar) and *li* (sweet liquor) 260, 601
  nieh, saccharifying activity of 260
  rice 260, 477
    bran 581, 585, 606; with purselane 582
  tea 512, 555, 561, 562; thin 97

GENERAL INDEX

congu tea 536n, *544*, 545n, 546 *see also kung fu* tea
  scented 554
controlled environment 415, 424; fruits 427–9
cooking 70–4, oils 439
Cooking Mallows (poem) 36n
cooking methods 16, 67, 70–6
  ancient compared with modern 88–9, *90*, 91
  with water *70*, 71–2
  see also baking; boiling; braising; frying; poaching; pot-roasting; roasting; steaming
cooking pot 76, 77–9, 80, 237n, 238, 260
  beer making 265, 266
  pottery 263, 265n
  tripod 260
cooler, bronze 430, *431*
copper carbonate, fruit storage 428
Cordier, H. 493n, 501n
corn, starch extraction 462
corpses, storage 430n
Corran, H. S. 266n, 267n, 272n, 275n
Corylus heterophylla 44, 49
cosmetics, starch 462
cotton seed oil 449, 450, 452, 456
couscous 465
crab 384, *381*, *409*
  chiang preserved 413–14
  tsao preserved *409*, 413
  monographs 145
crab apples 424, 437
cranberry 49
crane 61
Crataegus pinnatifida 44, 49
cream, clotted 256n, iced 434
Creel, H. G. 596n
Crete, olive oil press 437
crops, founder in prehistoric China 263
Crosby, A. W. 149n
crude solve 426
crullers, deep fried 2, 322, 471, 490n
crushing, oil seed 455
Ctenopharyngodon idellus 61n
Cucumis melo 33
cuisine 16, Sung 595n
culinary art 69, 140–8
Culter erythroopterus 62
cumquat 54n
cup-with-ears *100*, 103–4, 113, 114
cups, porcelain 223, tea *558*, 560, wine 223, 224
Curcuma aromatica 232
Curtis, R. I. 383n, 398n, 400n, 401
cutting and slicing meat 69
Cyclopedia of Chinese Medicine, Medical Historiography Section (1981) 576n
cypress 589; Belvedere 587; white 587
Cyprinus carpio 61n, *62*, 63

Daily Log of the Yü Thang Palace *see Oyudononoue no nikki*
dairy foods 124, 255, 315
  chilled 434
  Chinese diet 328; consumption 256
  koumiss, lo, su, thi hu 248–57
Dalby, A. 496n

Daniels, C. 7n, 426n, 437n, 459n
Darby, W. J., Ghalioungui, P. & Grivetti, L. 267n, 272n, 275n, 283n, 292n, 396n, 423n, 455n
dark dragon (oolong) tea 535–6
date *see also* jujube
  Chinese 44, 48–9, wild 48
  as sweetener 457
Davies, T. 267n
Davis, J. F. 149n
Davis, T. L. 153n
Dawkins, R. 601n, 603n
De Agri Cultura (Cato; c. –180) 437, 462
De alimentorum facultatibus (Galen) 400n
De Materia Medica (Discorides) 400n
decaffeinated drinks 569
decantation of wine 164, 179, 187; decanters 102
deer 60, 61; *ni* pastes 379 *see also* venison
Deir el-Bahri; XIth dynasty (ancient Egypt) *268*, 271
dental plaque and caries 569
Description of the World (Marco Polo) 493
di (screen) 113
Diamond, J. 55n, 597n
diet 604–5
  unbalanced 14
  vegetarian 497–8, 502, 605n
diet therapy 120, 121, 136, 571, 573
  nutritional deficiency diseases 573–91
Dietary Therapy (Liu Jilin; 1995) 571n
dietary zones of China 579n, 580, 582
dim-sum 485, 488
dimethyl-nitrosamine 605
dining furniture 109, 111–15
Diospyros kaki 44, 50, 51
Diospyros lotus 51
Director of Victuals for Mêng Shu kingdom 193
Discorea opposita 43
Discorides 400n
Discourses on Tea, Sugar, Milk, Mead, Wines, Spirits, Punch, Tobacco etc. with Plain Rules for Gouty People (Thomas Short; 1750) 538n
dish, serving 100–1
distillation *see also* wine, distilled
  alcohol 203, 204
  apparatus 207; flask 208; retort 227, 228
  essential oils 206, 214; grape wine 244; koumiss 217n, 231
  freezing 223
  steam 206, 213
    drug extraction 226; flower essences 227
  vessels 206
distillery 217; small 220
ditch, heating 520n, *521*, 522, 523, 526–7, 529
Divasament du Monde (Marco Polo) 493
Djien, Ko Swan 342n
Dobson, W. A. C. H. 18n
doburoku (fermented congee of Japan) 260n
doen jang (fermented soyfood) 378
dog 55–6, 58, 59–60, 60, 61
  butchers 59, 60; fat 59, 93
  gall bladder 586, 587
  polished rice feeding 581, 606
  served to Son of Heaven 58

dogwood 555
   seed in *cha* 385
dolphin, river 63
donkeys *448*
dough 3, 4, 265, 465, *594 see also* bread; flour; pasta
   addition of fermenting must 269, 282n, 469n
   gluten 466; kneading 465; noodle 3
   leaven 471, 477; plasticity 465
   raising 274, 465, 471n, 476, 477
Drachmann, A. G. 439n
Dragon Boat Festival 119
dragon's eye 238n
drinks 96–8 *see also* wine, tea
   decaffeinated 569
   fermented 266
   iced 433, 434
   storage vessels 102; serving vessels 98, *99*, 101–4
Drinking to the Eight Immortals (series of poems) 223
drinking vessels 223
   cups 102–3, 223
   Shang bronze 155
drugs
   steam distillation 226
   wine as vehicle 233
drunkenness 163
drying 415
   fruits 424–5
   meat and fish 419–24
   oil seed *455*
drying chamber *521*, 522, 530
duck 60, 61, 84
   eggs, salted 2; *tsao* preserves *409*, 413
   Peking 89; steamed 71
Dudgeon, J. 507n
Duméril, A. 329n
dumpling 480; with *tou fu* skin 323n

earth
   symbolic correlations *572 see also* Wu Hsing
earthenware 259
East India Company 546n
Eastern Han mural 2, 11, 597n
   dining or drinking scene from Chengtu, Szechuan 113–14
   kitchen scene from tomb in Chu-Chhêng (Shantung Province) 86, *87*, 332n, 597n
   tomb at Liao-yang, Liaoning 114–15
   tomb at Ta-hu-thing, Mi-hsien, *tou fu* making 306–7, 331–3
Eastern Han Tomb
   at Tung-Yang, Hsüi-I in Kiangsu Province 80n
      bronze *an* 112
   Liao-yang, Liaoning 114–15
      meal setup 113–15
   Ta-kua-liang 109, *110*
eating *see also* utensils
   rules for 16
Ebers Papyrus (XII to XIII dynasties) 271n
eclipse of the sun 118
Edwards, J. 400n
eel 62
Egerton, C. 498n

egg noodle 491
eggplant 342, *409*, *418*, *409*, 411, 412n
eggs 2, *409*, 413
Egyptian physicians 239
   see also ancient Egypt
Eight Delicacies of Chou cuisine 57, 69, 72, 73, 92, 354
Eijkman, C. 579n, 581, 606
Ekirinhon Setsuyoshu (Japanese dictionary; +1597) 376
elaphure, *ni* pastes 379
Eleocharis tuberosa 40
elm nut 337n, 354–6
Elopichthys bambusa 62, 63
Elvin, M. 440n, 441n
emmer wheat 267n, 269, 270, 271n
Encyclopedia of Chinese Medicine, Medical History (1987) 576n
Endo, A. 202n
Endothia microbial rennet *608*
Engelhardt, U. 263n, 571n
Engishiki (Codes of Laws and Conduct of Ryo; +907) 374n
enzymes 154, 380, 415
   autolysis 380, 395
   fungal (*chhü*) 608; visceral 498–9
   heat denaturation 537
ephedra 580
epicatechins 566, *567*, *568*
Epistula (Seneca; +1st century) 400n
Epsilon Tauri 85
Epstein, H. 55n
Equus caballus 55
Erdman, J. W. Jr. 605
ergosterol 588n, 590
erh (braising; pot-roasting) 84, *88*
erh (cooked food) 469
erh (fish or fish roe) 63
erh (food in ritual baskets) 467–8
erh pei (cup-with-ears; wine cups) *100*, 103–4, 166
Erh Ya I (Wings for the *Literary Expositor*; +1174) 294n
Erh Ya (Literary Expositor; –300) 35, 36, 39n, 140
   bitter vegetable 508
   cha 384, 396; *kêng* 83
   Chinese cherry 51; Chinese gooseberry 52; Chinese hawthorn 49
   tea 510, 555; yellow robe 165
Eriobotrya japonica 53
esculentist movement 131
Eskimos 604
essential oils
   distillation 214
   extraction 206, 229
   tea 561, 565, 568
esterase, *Mucor*-derived 608n
Estes, J. W. 571n
ethanol and ethylene, in fruit metabolism 429n
Eumsikbo (+1700s) 397
Euphorbia helioscopia 234n
Europe
   cheese use 329–30
   early pottery 603n
   humoral theory 571
   maritime trade with China 538

688   GENERAL INDEX

Europe (cont.)
  microbial enzyme production 608
  night blindness 588
  oil extraction process 455–7; oil press 452, *453*
  osteoporosis 604n
  sunflower oil introduced to China 456
  tou fu 329
Evans, J. C. 506n, 513n
evil spirits, warding off 235
evolution
  organic 601, 603
  technology 597–601
eye diseases 586, 587

fa chiu (regulation wine) 178, 181
fa hsiao (fermenting tea) *537*
fagara 44, 52, 92, 96, 327
Fairbank, W. *113*
fan (cooked grains, especially rice) 32, 68, 71, 80, 98, 99, 466, *594*
  cha making 384, 385, 388, 390, 393, 396
  exposure to airborne fungi 274, 599, *600*
  ferment making 61, 162, 174, 274, 593, 599, 603;
    red *ferment* 195
  making steamed rice 259, 260n
  malt sugar making 458
  polished rice 605
  vinegar making 285
  wine making 164, 175, 189; wine origins 161
fan (roasting) 70, 73, 85, *88*, 89
Fan Feng-Yuan 136n
Fan Hsing-Chun 97n
Fan Ka Wai 584n
Fan Khuai (late Chou and early Han period) 59
Fan Shêng-Chih Shu (Book of Fan Shêng-Chih; –1st century) 27, 27n, 31, *117*, 120
  gourd seeds 34; burning 437
  lesser bean 40; soybean 293n; taro 43
Fan Shu-Yün 75, 118
fan tou (cooked beans) 294
Fan Tzu Chi Jan (Book of Master Chi Ni; –4th century) 22
fan wan (rice bowl) 99
Fan Yen (Dictionary of Local Expressions; c. –15) 262
fang (bream) 61, *62*
fang (square drink storage vessel) 102
Fang Chhien-Li (writer; +9th century) 183, 205
Fang Chien 508n, 513n
fang fêng identified variously as *Siler divaricatum* 235, *Radix ledebouriellae* 589, or *Ledebouriella seseloides* 186
Fang Hsin-Fang 151, 155n, 156n, 157, 158, 159n, 160, 167, 179, 186n, 203n, 215n, *217*, *218*, 229, 231, 277n
Fang I-Chih (author of *Wu Li Hsiao Shih*; +17th century) 134
Fang Khuai 59n
Fang Shêng-Chi Shu (Book of Fan Shêng-Chih; –1st century) 143
Fang Yang 155
Fang Yen (Dictionary of Local Expressions; c. –15) 72, 140, 262, 300
  erh 468; *ferment* types 167n; *ping* 468, 478, 481n, 484;
  thang 457

Fansur, kingdom of 496
Farb, P. & Armelagos, G. 16n, 140n, 595n
farina granules 465n
fat 66
  animal 28, 93, 437, 439, 456, 601
  cooking use 28, 437
  hydrolysis 593
  fatty acids 593; esterification 378n
fei (radish) 32n, *33*, 34, 35
Feldman, M., Lupton, F. & Miller, T. E. 258n, 266n, 269n
fên (partially cooked grain) 175
fên (rice flour) 474
fen chiu (famous distilled spirit) 225n
fên phien (rice sheet) 488
fên ping (rice filaments) *470*, 473, 481
fên ssu (silky starch threads) 491n
fên ying (fine rice starch) 462, 485n
fêng (Chinese turnip; colza) 31, 32n, *33*, 34, *35*
Fêng Jên (Superintendent of Enclosures) 56n
Fêng Kho-Ping (writer; 17th century) *516*, 560n
Fêng Shih Wên Chien Lu (Fêng's Record of what he saw and heard; +8th century) 515n
Fêng Su Thung I 248n
fêng yü fa (air dried salted fish) 422
Fêng-hsiang (Yung-chêng, Shensi Province) 430, *432*
Fêng-huang Shan 167 etc. (on Han Tomb No. 167 at Fêng-huang Shan, Chiang-ling; 1976) 105n
Fengshien (Shansi) 216, *218*
fennel 96
  oil 214
ferment 5, 6n, 7, 8, 154, 405, 415
  Acetobacter 291n
  activity 174, 175, 177
  amylase 279, 334, 593; amyloglucosidase 279
  Aspergillus 592, 605n
  bacteria 280
  barley 167, 375
  brick 279n, 280, *281*
  cakes 167, 171, 173, 184
    incubation 173, 174, 184; wine preparation 174, 190
  cha making 387–8
  character 261
  chiang making, soy, wheat etc. 346–57; meat & fish 381–2, 390
  commodity value 169
  common 169, *170*, 171–4
    for autumn wine 169, *170*; for spring wine 176–7
  development 593–4, 599–600
  enzymes 593; etymology 261–2
  evolutionary history 599, *600*
ferments in food processing 335–6
  yellow coat 285n, 289, 335–6, 355, 605; yellow mould 335–6, 348, 360, 361, 362, 371
  glutinous rice 261n, 335–6, 339, 351
  yellow seed (*Huang tzu*) 350
fungal myceliae 162; population 279
grain moulds 166, 167, 280, 592, 595n *see Aspergillus, Rhizopus, Mucor, Monascus*
hai making 334, 380–1
herbal (medicated) 162, 183–6, 232, 238

GENERAL INDEX

incubation hut 171, 173–4, 178, 182
inoculum 161, 182n, 281
  made in modern China 280
  manufacture of standard 190
  medicament 232
  methods for making in
    early Mediaeval China 168–81; kings 171
    late Medieval and Pre-modern China 181–98
  microbial activity control 162
  millet 169, *170*, 262
  mixed cultures of organisms 280–1, 592, 595n
  moderate 169, *170*, 174
  mother 194–5, 197, 200
  moulded wheat 335, 336
  MWT Inventory 165, 166
  and *nieh* (sprouted grain) 156–60, 274
  origin and discovery 150–68, 257–74, 593
  power, *chhü shih* 174, 279
  products related to grain moulds 594
  protease 334
  red 3, 191, 192–203, 279, 290, 327, 595
    invention 194; making 194–5, 197–203; preservative 202
    temperature control 195, 197, 199, 201
    therapeutic properties 202–3
  Southeast Asia 395, 396–7
  soy (*chiang huang*) 339
  superior 169, 170–1, *172*, 174
    chiang making 348; ritual sacrifice 605; wine preparation 174–6
  technology dissemination 608
  temperature control 176
  tsu 406, 407; types 167, *185*; water 174, 184
  wheat 167, 169, *170*, 185, 190, 261, 277
    fu ju making 327
  white 290
  yeasts 261, 279, 280, 592
fermentation, alcoholic *see chiu*, wine, beer, *also* lactic acid, tea
  amphora 275–7
  by monkeys 245
  chewing cereals 154
  saccharification, distinction between 606
  technologies in East and West 272–8
  temperature 282n
fermenting activity (*chiao*) 188
fermentor
  jar 187
  temperature control 176, 177
Fernandez-Navarrete, Domingo (Friar; +1665) 319, 328
filtration, wine 164, 277
Finland, bread beer 270n
Finney, P. L. 298n
fire 602
  symbolic correlation 572
firewood 523
fish 60, 379, 400, 582
  armoured (*chia yü* turtle) 65
  cooking 71; steaming 71; frying 440
  dried, salted 420, 423
  fermented products in China 333–4, 359, 380–92

  cha (pickle) 384–8, 387–8, 390–1, *409–11*
  chiang (*paste*) 382–3, 387, 388–9
  chu i 382–3; *hai* 379; *hsi* 420; *i yü* 383, *395*
  fish sauce (*yü lu*) 395
  fermented products in East Asia 392–8
    distribution 393–5
    salted fish *yen hsin, shiokara* 392, 395
    fish paste *chiang* 392; fish sauce 395
    fish preserve *cha, narezushi* 392, 396
  fermented products in ancient Greece & Rome 398–402
    salsamentum 398
    garum, liquamen, muria and *allec* 399
    rise and decline 400–2
    medicine 400n
  finely shredded 423–4
  foodstuff category of ancient China 17
  frying 440
  long 61, *62*; lucky 61, *62*; mock 497–8
  oil 441
  preservation in ice 434, 435
  salted 2, 8, 420, 423, 602n, 605; wet salted 420
  storage 434n; transport 434
  tsao preserves *409*, 413
  literature 145–6
fish soup 71
Fishery Superintendent 63, 66
fishing 62–3
  early historic China 60
  implements 61
Fitzgerald, E. 150
five flavours (*wu wei*) 91–2, 283, 582
five grains (*wu ku*) 19, 20n, *21*, 22, 23, 55
  Chêng Hsüan's listing 27
five spice 95n
flag, sweet *see* cattail
flambeaux 29n, 34, 439; making 437
flask
  distillation 208
  pomegranate 207, 208
flatus production 293, 294, 298, 596
Flatz, G. 596n
flavanols 565–6, *567*, *568*
flavouring 16
  tea 519n, 557, 561, 562
Flaws, B. 571n
flax *see* hemp; linseed
flaying 68
Fleming, A. 600n
Fletcher, W. N. 52n
Flon, C. 76n, 86n
flour 265, 266, *594*
  elm-nut 355
  milling 582, 599
  paste 278, 594n
  sago palm 496
  sweet flour *chiang 356*, 357
  Western civilisations 593
  see also dough; milling; quern; wheat, flour
flour food 465, 466
flower essences, steam distillation 227
fluoride 569

Fogarty, W. M. 608n
Fong Wen 42, 76n, 98
Fong, Y. Y. 605n
Fontein, J. & Wu Tung 103n
Foochow (Fukien) 1
  fish sauce 397, 397-8n; soy sauce 365n, 366, 367; jasmine tea 554
  malt sugar making 461n; noodles 483n; oil press 450n
  tea trading 546n; *ting-pien-hu* 474n
food 98
  balanced consumption 17
  disease control and cures 14, 16; therapeutics 121
  implements for serving 104-7, *108*, 109
  imports 66
  and medicine have a common origin 134
  primary and supplemental 316
  shops 481-2
  texture 93
  uncooked 389n
  vessels for serving 98-101
food canons 121-2
  early mediaeval 124-6; late mediaeval 126-8; premodern 129-32
food preservation 402-15, 415-36
  chemical and physical methods 414, 415-36
  microbial action 405-7, 408, *409*-10, 411-12, 415
food processing 144, 592
  accidental discoveries 603
  innovations 595-7
  labour requirements 597
  necessity 602
  progenitors 597
  technology 7-8, 596-7
    evolution 597-601, 602
food resources 66-7, 122
  in ancient China 17-66
  imported 66
foodstuff categories in ancient China 17
Forbes, R. J. 239, 242n, 246n, 266n, 267, 272n, 275n, *276*, 437n, *438*
fork 104, 106-7, *108*, 109
  three-pronged 107, *108*
  (???) 181
  fou liang (floating log stage) 181
  fou phing mien (floating duckweed pasta) 481
Fowles, G. 258n
Frake, C. O. 260n
Frank, H. 496n
Franklin, B. 329n
Freeman, M. 255n, 354n, 436n, 595n
freezing, distillation 223
Frescoes of Yü Lin Khu 208
frog, *hai* pastes 379
fruit 2, 43-55, 379
  consumption 68
  foodstuff category of ancient China 17
  incubation in sugar 415
  preservation 424-9
    with sugar 460n
  sugar content 258

frying *see also* stir-frying
  deep *88*, 89, 325, 327
    pasta 471, 474
  dry 84
  oil 439, 441
  shallow 70, 72, 84, *88*, 89, 325
fu (delectable fish) 63
fu (cooking pot; boiler; wok) 76, 80, 82, 98, *521*
  bronze still 208, 209
fu (decayed; rotten) 302
fu (dried meat) 76, 419-22
fu (gluten) 500
fu (thick puree) 322
fu chin (gluten) 500-1
fu chu (bean curd bamboo; dried bean curd sticks) 323
Fu Hsi (Legendary Emperor) *15*
Fu Hsi-Jên 23n, 75, 84n, 119, 250n, 339n, 380n, 507n
fu ju (fermented bean curd) 7, 320, 321, *322*, 325-8, *594*
  organisms 328
Fu Kung (11th century) 145
fu shih (fermented wheat bran) 340n
fu shih (supplemental food) 316
Fu, Marilyn & Fong Wen *42*
Fu Shu-Chhin 515n, 520n, 522n, 557n, 562n, 563n
Fu Yun-Lung 136
fu yung tou fu (hibiscus *tou fu*) 324
Fu-ling, tea production 509
Fuga vermicularis (puffer fish) 145
Fujiwara Fuhito 374n
Fukien 5, 193, *545*
  bright bun 3, 472n
  emigration 492
  kung fu tea 540
  noodles 484, 490, 491; hung 3
  red *ferment* 202; red, i.e. yellow wine 151n, 190-1, 279
  red i.e. black tea 546n, 548
  rice noodles 490
  soy sauce making 365
  tea 543, 544
    exports 548
    plantations 513
  yü lu (fish sauce) 391, 595n
funeral rites of royal household 430n
  thu 507-8
funerals, *yü hsi* 420
fungi
  airborne spores 274, 277, 339, 593, *600*
  enzymes 608
  growth on *fan* 593, 599, *600*
  inoculum 168
  isolation from moulded grains 608
  Near East Neolithic Age 278
  red 191
  in *tempeh* 345
funnel, pottery for fermentation mash *152*, 153
furnace for tea drying *521*, 522, 523, 526-7, 529
  opening up 523
Fusō Ryakuki 375n

Gajdusek, D. C. 573n
Galen (+2nd century) 400n, 571n, 577, 586
  rickets 589

gall bladder 586, 587
Gallus gallus domesticus 60
Gami al-mufridat (Ibn al-Baytar) 494n
Gardella, R. 503n, 538n, 541n, 548n
Gardener's Chronicle 435
Gardenia jasminoides 508
garlic 39–40, *41*, 92
  chiang preserves 409; salting 418
garum (Greek and Roman condiment) 380, 383, 398–402
Garway broadside on tea (Thomas Garway; +1660) 565, *566*, 569
gasi (fermented congee of Philippines) 260n
Gaspar da Cruz, Father (Jesuit missionary; 16th century) 542
Gelsemium elegans 183
general's helmet vessel 151, *152*, 277n
Geoponica (+10th century), *garum* 398–9
geraniol 565n
Gerard's Herbal 571n
gerdeh (bagel with hole covered by a film of dough) 472n
getsu (sprouted rice) 166, 263
ghani (Indian oil mill) 452, *454*
ghee (Indian butter oil from goat's milk) 254
gifts of ice 430, 433, 434
ginger *33*, 40, 43, 91, 92, 96, 186
  chiang preserves *409*; *tsao* preserves *409*; *tsu* 380, *406*, 406n, 411, 412n
  dried 589; honey preserves 426
  medicated wine 235, 237
  salting *418*
  shih making 361
  tea flavouring 519n, 520, 555, 561, 562
  wild 40, 43
ginseng 186, 237
gliadin 465
Glisson, F. (+1650) 589
globulins 465
glucono-delta-lactone 305
glucose 154n
gluten 4, 8, 274, 465, 488, 497–502
  addition to flour 488
  chiang preserves *409*, 412
  cooked 501–2
  development 595, *598*, 604
  extraction 499
  honey soaked 500
  monks' use of 501
  protein source 466, 605
  recipes 501–2
  regard for in Sung 501
  rye 466
glutenin 465
glycerol 593
Glycine max 18, 25
glycyrrhiza 580
go (Japanese chess; Chinese *wei chhi*) 473n
Godley, M. R. 149n, 244n, 245n
goitre 573–8
  emotional stress 575
  mountain association 574, 575

seaweed for cure 575–6, 577, 578
soil 574, 575, 578
thyroid gland 578
  therapy 576–7
water 574, 575, 578
  quality 574, 575
West 577
Golingal Han tomb, Inner Mongolia 118
Goodrich, L. C. & Wilbur, C. M. 508n
Goody, J. 595n
goose 60, 61, 379, 409
gooseberry, Chinese (*kiwi*) 52
Goswell, R. W. 153n
gourd 29n, 33–4, 43
  dish 103
  seed, burning 29n, 34, 437; pounding 437
grading tea *524*, 525
Graham, H. N. 565n
grain 17–19, *20*, 21–5, *26*, 27–8
  amount digested by *ferment* 174
  conversion to alcohol 5, 154
  cooked (steamed) granules 68, 80, 83
  drinks derived from 96, 97
  fermentation 4, 272, *273*, 274–5, *276*, 277–8
  food resource 66, 98, 99
  foodstuff category of ancient China 17
  hulled 458; kernel separation 265; grinding 265
  making digestible 259
  moulds 8, 9, 593, *594*, 608
  parching 265, 270n
  soured 188; spent residues 278
  sprouted 154, 157, 158–9, 163, 601
    beer making 263
    grinding 274; *li* making 166
    malt sugar 457; porridge 266
    saccharifying activity 603
Grand Chef 60n, 71
Grand Dietician 57n, 59n, 60n, 137, 138, 334, 571
Grand Physician for External Illnesses 571n
Grand Physician for Internal Illnesses 571n
Grand Veterinarian 571n
granule food 465, 466
grape 53, 241
  cultivation 240
  storage controlled environment 427
  drying in honey 425
  grass dragon pearl 241; mare's treat 241–2
  Chang Chhien and seeds 240
  steeping in water 428
grape vines 241
  cultivation 242, 258
grape wine 153, 204, 205n, 240–45, 606
  distillation 244
  fermentation process 242–3
  ferment use 240, 242, 244
  Kansu & Shansi 241–2
  shelf life 240
  taxation 244
  Western origin 258, 274
  yeast *chiao* not used 244

Greece
  cookery 496n
  fermented fish sauce 380, 383, 398–402
  humoral theory 571
  medicated wine 238–9
  olive oil 437, 455n, 456
  yeast starter 243n
green meat (lü jou) 405, 414–15
green manure 345
Green, R. M. 571n
grid or grating (in a still) 208–13, *521*
Grigson, C. 258n
grinding stones 599, 601
Groff, E. 365, 366, *368–70*
gruel 82 see also congee
Guppy, H. B. 214
Gutzlaff, C. 244n
gypsum 310n, 319; powder 303

ha-la-huo wine 244
haemorrhoids, salve for 346n, 359
Hahn, E. 596n
hai (pickle) 333–4; (pickled minced meat) 47, 69, 76, 96; (paste) 379, 380, 381, 382, 390; *cha* 384
hai fu (sauce making by auto-digestion) 326n
Hai Jên (Superintendent of Meat Sauces) 333n, 334
hai tai (seaweed) 575n; *hai tsao* (seaweed) 575; *hai tshai* (sea vegetable) 575, 576
hai thun (suckling pig) 380
Hai Wei So Yin (Guide to the Delicacies of the Sea; +1590) 146
hai yen (sea salt) 305n
hair, oils for salving 439, 441
hairpin 522n
Halde, J. B. du (Jesuit scholar) 538n
ham 422
hamanatto (fermented savoury soybeans) 377
Han brick murals 217–21
  commentary 219
  interpretation 219
  tou fu making 331
  see also Eastern Han mural
Han Chhang-Li Chi (Collected Works of Han Yü; +768 to 824) 241n
han chhin (dryland celery) 36
han chü (cold pastry) 472
Han dynasty
  Chinese cuisine 16
  foods 19,
Han dynasty tombs
  Chi-nan (Shantung) 112; Golingal (Inner Mongolia) 112; Hupei 46; Kiangsu 46; Kuangsi 46; Kuangtung 46; Li-chou 80n; Liao-yang (Liaoning) 112, 114–15
  Loyang 102–3; Man-chhêng 146; Phêng-hsien (Szechuan) 112; Sa-ho (Kuangchou) 112n; Ta-fen-thou (Yunmeng) 93n
  cultivated grains 24–5, *26*
  drink storage vessels 102; pitchers 101; stoups 103; utensils 105

fruits 43, *44*, 54
  Chinese plum 46; loquat 53n; oranges 54; persimmon 51; pomelo 54; meal physical setup 111; monographs 146
  rotary mill models 314n; seasonings 93
  stoves 80, *81*; trays 112
  vegetables *33*, 43, *khuei* (mallows) 38
  wine 165–6
  see also Han Tomb No. 1; Han Tomb No. 3; *Hunan sheng po-wu-kuan*
Han Fei Tzu (Book of Master Han Fei; –3rd century) 85, 104n, 140
  flavours 283n
  vessels 602
Han Kuo Tshê (Record of the Han State) 294n
Han, O (writer; late Thang) 340, 351
Han Sen Tsai (nr Sian) Thang tomb 223
Han Shu see Chhien Han Shu
Han Shu I Wên Chi (Bibliography of the History of the Former Han) 124
han thao (Chinese cherry) 52
Han Tomb No. 1 (Ma-wang-tui) 19n, 24, 24–5, *26*, 33
  cooking methods 67, 68, *70*, 71
  dining furniture *101*, 112
  drink storage vessels 102; pitchers 101; stoups 103; *erh pei* 103
  fermented soy paste *chiang* 346; fermented soybeans *shih* 336
  ferments 165–6
  fish bones 63
  fish: *li, chi, chhi pien, yin ku, kan* and *kuei*
  food animal bones 61
    mammals: cattle, sheep, swine, dog, deer, and rabbit
    birds: chicken, duck, goose, pheasant, sparrow, crane
  fruits: apricots 48; Chinese apricot 47; Chinese date 48, 48–9; Chinese sand pear 50; Chinese strawberry 54; loquat 54; oranges 54; persimmon 51; pomelos 54
  monographs 146; *MWT Inventory* 121
  seasonings 93; sweeteners 92; utensils 105, 106
  wines 165–6
  see also bamboo slips, Han Tomb No. 1 at Ma-wang-tui; *Hunan nung-hsüeh-yuan*; *Ma-wang-tui Han-mu po-shu*
Han Tomb No. 3 (Ma-wang-tui) 67, *70*, 121
  fermented soy paste 346
  medicaments 134
  vinegar 284
  see also *Wu Shih Erh Ping Fang*
han tou (broad bean) 297
Han Wei Yin-Shih Khao (Investigation of Food and Drink during the Han-Wei Era) 147
Han Ye-Chih (12th century) 144
Han Yü (poet; +786 to +824) 241, 301
Hangchou (Southern Sung capital) 141, 183, 527
  dried salted fish 423
  gluten and pasta selling 500
  honeyed sweetmeats 426
  oil pressing establishments 441
  sausages 423
  tsao-preserved meat 413

hao yu (oyster sauce) 358n
Hardy, S. 506n, 534n, 555n, 557n
Harlan, J. R. 258n, 266n, 267n, 269n, 292n
Harler, C. R. 565n
harmonisation 404
  texture of food 93
harmonising agents 40, 91, 95–6
harmony concept 93, 94
Haroutunian, A. der 496n
Harper, D. 34n, 73n, 94n, 121, 138n, 284n, 315, 346n, 359n
Harris, L. J. 579n
Harris, M. 18n, 32n, 55n, 59n, 596n
Harrison, H. S. 597n
Hartman, L. F. 269n
Hartner, W. 118n
Al-Hassan, A. Y. 225n
Hawaii 412n
hawkers selling soy milk 323, iced drinks 434
Hawkes, D. 43n, 64n, 65n, 72n, 74n, 84n, 94n, 119, 156, 157n, 233n, 246, 250n, 283n, 380n, 433n, 457n, 507n
Hawkes, J. G. 258n
hawthorn, Chinese 44, 49
Hayashi Minao 334n, 336n, 353n
hazelnut, Chinese 44, 49, 51
health 603–5
heart disease 570
heat inactivation, of tea leaves 529–30
Heber, D. 202n
hei chha (dark tea) 535n, 548n, 551–2
Hei Tha Shih Lu (Brief Notes on the Black Tartars; c. +1237) 249n
Hemiculter leucisculus 62
hemlock parsley 186, 237, 589
hemorrhoids, treatment 73
hemp (*ma*) 28–30
  five grains 18, 20, 21–3, 25
  fibre and food 29
  fruit 31
  oil 28, 437, 439, 441, 472n
    cooking 439–40
    extraction 28, 29, 437
    fermentation jar sealing 440
    yield 452
  seed 28, 29, 437, 58
Hempen, C. H. 571n
henbane 234
Henderson, B. E. 605n
Hêng Than (writer; −25 to +56) 583n
Hêng-Ti (Emperor; +147–67) 221
Herbert, V. 605n
herbs
  ferment 162, 183, 185–6
  medicated wine 234, 235, 237
  red *ferment* 197
  steamed 71
Hero (*Mechanica*) 437, 439n
herring 401n
Hesseltine, C. W. 245n, 260n, 328n, 342n, 346n, 377n, 378n, 605n
hibiscus, flowers 301, oil 440–1, *tou fu* 324
Hilbert, J. R. 153n

Hill, D. R. 225
Hillman, G. 265n
Hillman, H. 69n
Hippocrates 400n, 571n
Hirayama, K. 118n
Hiroshi Kongo 167n
hishiho, hishio (Japanese savoury sauces) 374, 380
ho (grains) 19
ho (harmonizing agents) 40, 91
ho (lotus) 33, 36; *ho* (millet) 22; *ho* (spices) 17; *ho* (wine serving vessel) 99, 100, 102, 155
Ho Chhang-Hsiang 503n
Ho Chia Chhun (Sian) Thang tomb 208
ho chiang mien (rolled flat noodles) 484
ho fên (flat rice noodles) 490, 491
Ho Han San Tshai Thu-Huei 264
Ho, I. 136n
ho lo mien (noodles coagulated in boiling water) 488
ho lo mien (noodles made by dough squeezing through orifices; pressed noodles) 485, 491, 492
ho pei (without separation of the residues by filtration; muddy brew) 177, 181
Ho Peng Yoke 571n
ho ping (grub shaped pasta) 468, 472n
Ho Ping-Ti 23n, 24n, 262n, 277n, 456n
ho san (harmonise with rice crumbs) 93
Ho Suan-Chhuan 38n
ho thun (puffer fish) 145
Ho Tung (prefecture in southern Shansi) 159
ho tzu (cold pastry) 472n
Ho Yen (official at court of Wei Emperor Ming-ti) 469
ho-fên (thin rice noodle) 492
Ho-mu-tu (Chekiang) Neolithic site 44, 49, 58
  pottery steamer 76, 79
  rice husks 260
hoieu nood (dragon's eye) 238n
Hole, F. 265n, 599n
holistic approach 607
holly 452
Homer 57n, 150
Hommel, R. P. 215, 216, 222n, 231, 237, 449
Honan Provincial Institute of Archaeology 332n
Honan shêng Hsin-yang ti-chhü wên kuan hui (Shang and Chou Cemetery at Thien-hu, Lo-shan County; 1986) 153n
Honan shêng po wu kuan (Thirty Years of Archaeological Investigations in Honan; 1981) 305, 332n
Honan Shêng wên-hua-chü 108
Honan shêng wên-wu khao-ku yen-chiu-so 27n
Honan shêng wên-wu yen-chiu-so (on the Neolithic Site at Chia-hu, Honan; 1989) 27n, 331
honey 92, 95, 245–6, 457, 500
  fermentation 246–7, 282, 607
  pasta making 472, 473
  preservation of jujubes 49; preserved fruits 425–6
  sweetener 457; wine 8, 246–8; vinegar 287
Hopei 190
  dried jujubes 424n
Hopei shêng wên-hua chü wên-wu kung-tso tui (on Yen capital at I-hsien; 1965) 432n
Hopei sheng khao-ku yen-chiu so (Shang Site at Thai-hsi, Kao-chhêng; 1985) 151

Hopkins, D. C. 258n, 266n
hops 184, 185, 239
hor faan (Cantonese flat rice noodles) 490
Hordeum vulgare 19, 25
horn, drinking cups 102–3
horse 55–6, 61
    consumption of horseflesh 56
    domestication 55; trading 552n; uses 55
    see also donkeys
horticulture, Chou period 32
Hothang 1–2, 5, 6, 584n
Hou Han Shu (History of the Later Han Dynasty; +25–200) 52n, 141, 168
    distilled spirit 221?
    grape wine 242n
    lo 250
    pasta 468n
    soybeans 294
Hou Hsiang-Chhuan 579n, 581n
Hovenia dulcis 44, 49
Howe, B. 265n
Hsaio Erh Yao Chêng Chih Chüeh (Therapy of Childhood Diseases; +1114) 589
hsiao suan (rocambole) 39n
hsiao tshung (spring onion) 39
hsi (dried meat) 76, 419–22
hsi (*Caretta caretta olivacea*) 64–5
hsi (male hemp plant) 28
hsi (type of vinegar) 92, 284
hsi chiang (acid type of sauce) 334, 353
hsi chiang (freshly ground bean puree) 320
Hsi Ching Tsa Chi (Miscellaneous Records of the Western Capital; +6th century) 141, 235
hsi chhü (fine *ferment*) 477
hsi erh (*Xanthium strumarium*) 29n
hsi fan (thin or diluted *fan*, i.e. congee) 260n
hsi hsin (*Asarum sieboldi*) 587, 589
Hsi Hu Lao Jên Fan Shêng Lu (Memoir of the Old Man of West Lake; Southern Sung) 141, 325n, 479
hsi huan pin (ringlet pasta) 470, 472
Hsi Jên (Supervisor of Vinaigrette Viands) 92, 283
Hsi mien chin (washing gluten; +1325) 500
hsi shui hua (wet slippery noodle) 483n
Hsi Yu Chi (Journey to the West; +1570 and +1680) 44n, 501, *shih ping* 425n
Hsi Yüan Lu (Washing Away of Wrongs; +1247) 207n
Hsia Hsiao Chêng (Lesser Annuary of the Hsia Dynasty; –4th to –7th centuries) 48
Hsia Wei-Ying 20n, 24n, 27n, 119n, 402n
hsia yu (shrimp sauce) 358n, 391
Hsia-tu palace 432
Hsiai Lo (Discourse on Crabs; +1180) 145
Hsiai Phu (Monograph on Crabs; +1060) 145
hsiang (beef tallow flavour) 93
hsiang (dried salted fish) 423
Hsiang An-Chhiang 27n
hsiang chhang (sausages) 423
hsiang chün (shitake) 145
Hsiang Hsi 118
hsiang jih khuei (sun flower, sun-facing *khuei*) 38n
hsiang mo (fragrant ink) 207n
hsiang phien (fragrant bits), jasmine tea 554

Hsiang Thu (lord of Shang dynasty) 55
hsiang tshu (fragrant vinegar) 291
hsiang yu (sesame oil) 358
Hsiang-yin county (Hunan province) 519n
hsiao chiao (lesser leaven) 477
hsiao chung (souchong tea) 546n
Hsiao Jên (official) 56n
hsiao lung thuan (little dragon cake) 525
hsiao mai (lesser *mai*, or wheat) 19, 19n, 20, 25, 27, 169n
hsiao suan (rocambole) 39, *41*
hsiao thang (little sugar) 460
hsiao tou (lesser bean) 19n, *33*, 40, 355
hsieh (Chinese cherry) 51
hsieh (Chinese shallot) *33*, 39, *40*
Hsieh Chhung-An 27n, 55n, 60
Hsieh Fêng (writer; +589 to +618) 126, 254–5, 302, 354, 481
Hsieh Kuo-Chên 245n
Hsieh Ta-Huang 419n
hsien (bamboo whisk) 557, 559
hsien (rice) 197
hsien (salty) 91, *95*
hsien (steamer) 76, 77
Hsien Chhin Ngo Chi (Random Notes from a Leisurely Life; +1670) *129*, 131
    soybean sprouts 297
hsien mien (stretched noodles; thread noodles) 484n, 485, 491, 492
hsien thao (sacred peach, food of the immortals) 44
Hsien-yang, Chhin palace complex 432
hsin (mother of *ferment*) 197, 200
hsin (pungent) 91, 92–3, *95*, 96
hsin (sturgeon) 62
Hsin Fêng Chiu Fa (*Hsin Fêng* Wine Process; Southern Sung) 132, *133*, 188–9
Hsin Hsing 184n
Hsin Hsiu Pên Tshao (New Improved Pharmacopoeia; +659) 134, *135*, 136
    chiang use 354, 355
    fermented milk 253
    grape 240n, wine 242, 243
    malt sugar 460; mead 246
    shih use 341; soybeans 296
    tea 512, 555, 561, 563
    vinegar 290
Hsin Lun (–1st to +1st century) 583
hsin mao (day of the month) 117–18
Hsin Shu-Chih 44n, 48, 49n
hsin shui (rice wash water; trusty water?) 184
Hsin Thang Shu (New History of the Thang Dynasty; +618 to +906) 255n, 515n
    ice chambers 432
Hsin Yüan Shih (New History of the Yuan) 244
Hsin-tu (Szechuan), Han brick mural 218–21
hsing (almond) 437
hsing (apricot) *44*, 48
hsing (lard flavour) 93
hsing (malt sugar) 457
hsing (raw) 74n
Hsing Ching (Star Manual) 20n
Hsing Hsiang-Chhên 104n, 429n, 434n
Hsing Jung-Chhuan 151, 203n

hsing kêng (mixed soup) 83, 93
Hsing Lin Chê Yao (Selections from the Apricot Forest; 15th century) 577
hsing tzhu (lateness in walking) 588, 589
Hsing Yuan Lu (Memoir from the Garden of Awareness; +1750) *129*, 131–2, *256*, 358n, 388, 412, 419n, 427n
  chiang making 349n, 355, 357n, 373
  dried meat and fish 422
  orange storage 429
  processed fruits 427
  salting of vegetables *418*
  sausages 423; *sung* 423–4
  shih making 341, 342; soy condiments *372*; soy sauce 363–4, 365, 373, 399
  tou fu 324; *fu ju* making 327
  vinegar recipes 291
Hsing-chêng (Honan) 431
hsiu (confections) 98
Hsiu-shui (Kiangsi) 543n
Hsiung Deh-Ta 239n
hsiung chhiung (hemlock parsley) 589
Hsiung Fan (writer; +12th century) *516*
Hsiung Kho (writer; +12th century) *516*
Hsiung Nu (northern pastoral people) 166
hsü (tench) 61, *62*
Hsü Chha Ching (Tea Classic Continued; +1734) *516*, 518, 539n
Hsü Chien (physician; +6th century) 579n
Hsü Cho-yun 17n, 18n, 19n, 36n, 38n, 52n, 68n, 120, 347n, 359n, 508n
Hsu Cho-yun & Linduff, K. 18n
Hsü Chu-Wên 248n
Hsü Hsiai Phu (Continuation Monograph on Crabs; Chhing dynasty) 145
Hsü Hsüan 510–11
Hsü Hung-Shun 179n, 279n, 281n
Hsü Kuang-Chhi (17th century) 143, 443
Hsü Pei Shan Chiu Ching (Continuation to the Wine Canon of North Hill) 247
Hsü Pho (writer; 17th century) 144, 538
Hsü Po Wu Chih (Continuation of the 'Records of the Investigation of Things'; Northern Sung) 509, 552n
Hsü Tsung-Shu *276*, 277n
Hsü Tzhu-Yü (writer; +16th century) *516*, *530*, 531, 534, *538*
Hsü Wang (physician; Chou to Wei dynasties) 580
Hsuan Kung (Duke of Lu State; –605) 65n
Hsuan Kung (King of Chhi State; –290) 58
hsüan (large slices of meat) 69
hsüan chiu (water; primordal wine) 97, 301
Hsuan Ho Pei Yuan Kung Chha Lu (Pei-yuan Tribute Teas; +1121, +1125) *516*, 518, 523
  insignia *524*, 525, *526*
  scented teas 553
hsüan tshao mien (daylily pasta) 481
Hsüan Tsung (Thang Emperor; +723) 576
Hsüan-wei, ham 422n
hsüeh hsia kêng (snow and red cloud soup) 301
Hsüeh Pao-Chhên 485n, 502n
hsün tou fu (dried smoked *tou fu*) 320, 321, *322*, 325
Hsün Tzu (Book of Master Hsün; c. –240) 58

Hsüu Po 145
hu (drink storage vessel) *100*, 102, 151n
hu (garlic) 39, *41*
hu (gourd) 33–4
hu (large pieces of fish) 69
hu (vase) 215–16, 229
Hu Chia-Phêng 317n, 492n
hu chiao (black pepper) 52
Hu Chih-Hsiang 67n
Hu Chih-Hsiao 475n
Hu Hou-Hsüan 57n, 157n
Hu Hsi-Wên 23n
Hu Hsin-Shêng 115n
hu ma (sesame; sesame seeds) 29, 30, 31, 440, 468
hu ma yu (sesame oil) 439
hu mi ssu (koumiss) 248
hu pho thang (amber coloured malt sugar) 459
hu ping (flat bread; baked pasta) 468, 472, 475, 490n
Hu Shan-Yüan 257n
Hu Shih 148n
Hu Ssu-Hui (author of *Yin Shan Chêng Yao*; Grand Dietician of Mongol court; 14th century) 137, 227
Hu Tao-Ching 18n, 27n
hu tou (pea) 40
Hu Ya (Lakeside Elegance; c. 1850) 305, 319, 324, 325
hua chiao (fagara) 44, 52
hua chüan (flower roll) 476n
Hua Tho (physician; +208 to ?) 235
hua tshai (slippery vegetable) 38n
Hua Yang Kuo Chi (Records of the Country South of Mount Hua; +347) 508–9
  tea production 513
Huai Hai Chi (Collection from Huai Hai; c. +1100) 225n
Huai Nan Tzu (Book of the Prince of Huai-Nan; c. –120) 134, 140, 503, 506
  beriberi 579
  diet and health association 571n
  goitre 574
  tou fu 300
Huai-Nan (Anhui Province) 299, 300n
  Prince of (legendary inventor of *tou fu*) 299, 300, 337, 597
huan fêng (slow wind) alias for beriberi 579
Huang Chan-Yüeh *33*, *43*n, 46n, 48n, 51n, 53n, 54n, 93n, 305n, 463n
huang chêng (yellow mould *ferment*) 348–9, 350–1, 360–1
huang chha (yellow tea) 551
Huang Chhi-Hsü 23n
huang chhuan (Yellow Springs) 83
Huang Chien-An 552n
Huang Chin-Kuei 115n
huang chiu (yellow wine) 151n, 183, 190–1, 202
Huang Chu-Liang 128n
huang chüan (pharmaceutical bean sprout) 296
Huang, H. T. *78*, 595n, 596n
Huang, H. T. & Dooley, J. G. 608n
Huang Hai Research Laboratory 5
Huang Ho, Neolithic communities 61
huang i (yellow coat *ferment*) 289, 335, 336, 350, 355
Huang I-Chêng (writer; Ming) 498
Huang Ju (writer; +11th century) *516*
Huang Kuang-Chu 192n

Huang Kuei-Shu 503n
Huang, R. 221n
Huang Shan, monkeys 245
Huang Shêng-Tsheng (Ming dynasty) 145
Huang Shih-Chien 203n, 204n, 217n, 221n, 224, 225n, 229
Huang Thing-Chien (author of *Shih Shih Wu Kuan*; +1045 to 1105) 137, 592
Huang Ti (legendary Yellow Emperor; c. −2697) *15*, 80, 82, 259, 602n
Huang Ti Nei Ching Su Wên, I Shih (Translation and Annotation to the Yellow Emperor's Manual of Corporeal Medicine; 1981) 14, *15*, 17n, *117*, 120–1
　balanced consumption of foodstuffs 17
　beriberi 579
　five flavours 283n
　harmonising agents 91n
　staple grains 23n
　wine 161, 233
huang tou (yellow soybean substrate) 363
huang tou ya (soybean sprouts) 298
huang tzu (yellow moulded beans or yellow seed *ferment*) 340, 342, 349, 350, 352, 362, 371, 375
Huang Tzu-Chhing 154n, 169n, 171n, 174n
Huang Wên-Chhêng 247n
Huc, M. 149n, 244
Huei Hung (Buddhist monk +1071 to 1128) 375n
Hui, K. *90*
Hui (King of Liang; c. −290), butcher to 68
Hui Tsung (Emperor, supreme ruler of China +1101 to +1125) 517–18, 526–7
　first infusion of tea 561
　tea making 559
Hui, Y. H. 590n
human creativity 601–2
Humulus lupulus 185
hün (mushroom) 91
hun thun (wonton) 475, 478–80
hun-thun mien (wonton soup with noodles) 480
hun yu (animal oil) 437
Hunan dark tea 552n; tea exports 548; tea plantations 513
Hunan nung-hsüeh-yuan (Plant and Animal Specimens at Han Tomb No. 1 at Ma-wang-tui; 1978) 25n, 27n, 33n, 61, 61n, 63n, 121n *see also* Han Tomb No. 1 (Ma-wang-tui)
Hunan sheng po-wu-kuan (Han Tomb No. 1 at Ma-wang-tui Vols I & II, 1973) *26*, 33n, 121, 336n, 346 *see also* Han Tomb No. 1 (Ma-wang-tui)
Hunan sheng po-wu-kuan (on a Western Han Tomb at Sha-tzu-thang, Changsha; 1963) 25, *26*, 33n, 73n, 104n, 146n
　fermented soy paste *chiang* 346n, *shih* 336n
　tea? 513n
hung (granular wheat *ferment*) 285, 286
hung chha (red tea) 4, 8, 536, 541–2, 561n
　early reports 542–3, 547–8
　exports 541, 548
　Indian exports 549n
　processing 536, 544
　traders 546n

hung chhü (red *ferment*) 3, 191, 192–203
　making 195–8
　therapeutic properties 202
hung chiu (red wine) 191, 202
Hung Kuang-Chu 10, 147, 196n, 203n, 217n, 281n, 283n, 284n, 287, 288, 291n, 300–1, 302, 303n, *304*, 305, 313, 315n, 317n, 319n, 320n, 323n, 324n, 325n, 326n, 328, 332, 334n, 347n, 349, 358n, 359n, 360n, 364, *365*, *367*, 377n, 391–2, 403, 411, 419, 466n, 469n, 480n, 482n, 483n, 484n
Hung Lou Mêng (novel; Chhing) 323n, 324n
Hung Po-Jên 317n
hung shao (stew or red-cooking) 89, 414
hung ssu mien (silky red noodles) 484
hung tsao (red fermented mash) 202, 203
Hung Yeh 93n
hung-chha (black tea) 541n
hunting, early historic China 60
huo (braising; pot-roasting) 84, *88*
huo (cooking pot) 76, 80, 82, 98
　stir-roasting wheat 171
huo chhü (fire ferment) 191
huo chiu (fire wine) 204
Huo Fêng 374n, 377n, 491n
huo kang (fire urns) 532
huo kêng (soybean leaf soup) 294
huo pho chiu (fire-pressured wine) 206, 222
huo thui (ham) 422
Hupei
　Han tombs 46; tea exports 548; tea plantations 513; yellow tea 552
Hupei shêng po-wu-kuan (Han Tomb No. 1 at Ta-fen-thou in Yunmeng; 1981) 93n, 432n, *433*
hut, thatched for *ferment* cake incubation 171, 173–4, 182
Hymowitz, T. 292n
Hyoscymus niger 234
hypolactasia genetics 596n
Hypophthamichthys molitrix 62
Hyson tea 545n, 552n

i (coat or film) 286, 289
i (drink from plum juice) 97; *i* (honey cakes) 457n; *i* (Job's tears) 190
i (malt sugar) 92, 158, 160, 260, 263
　discovery 266, 274, 282
　making 457, 459
　references to 460
i (medical arts) 232; *i* (malt syrup) 262; *i* (rice cakes) 457n
i (thin congee) 97; *i* (water pitcher) 101, 105
i chhü (common ferment for autumn wine) 169, *170*
I Chien Chi (Strange Stories from *I-Chien*, c. 1170) 88n, 206, 225, 500
　steamer 230
I Chien San Chih 207n, 225n
I Chien Ting Chih 207n
I Ching (Book of Changes; Chou) 71, 140
　chhang ritual wine 156–7; *hsi* 419
I Chou Shu (Lost Records of the Chou Dynasty; c. −245) 23, 80, 82n
I Fang Lei Chi (Korean medical encyclopedia; +1443) 137

i fu mien (*I* family noodle) 491n
I Hsin Fang (Collection of Essential Prescriptions; +984) 136
I Hsüeh Chêng Chuan (Compendium of Medical Theories; +1515) 588
I Li (Ritual of Personal Conduct) 119, 140
I Li Chu Shu (*I Li* with Annotations and Commentaries; Chhin and Han) 83, 84n, 93, 107, 109; *chiang* 334
I Lin Chi Yao (Collection on Medicinal; +1475) 577
I, Marquis of (tomb, State of Tsêng, Sui-Hsien, Hupei) 106, 430, *431*
I Ning Chou Chhi (Gazetteer of the I-Ning Prefecture) 543
I people 382n
I Shih Thung Yuan (food and medicine have a common origin) 134
I Shuo (About Medicine; +1224) 579n
i thang (malt sugar) 460
I Ti (preparer of alcoholic drinks) 155, 161
i tsêng chêng chhü (distilling) 204
I Wên Lo (Bibliography of the *Thung Chih*; +1150) 125–6
I Wên Lei Chü (Encyclopedia of Art and Literature; +640) 240n
I Ya II (Remnant Notions from *I Ya*; Yuan) *127*, 128, 130, 411
   cha 387; *chiang* preparation 353, 355; *chiang yu* 363
   leaven recipe 476–7; *man-thou* 476; noodles 483
   red *ferment* 194, 196
   soy condiments *372*; soybean sprouts 297
   tea recipes 562n
   tou fu 324; soy milk 322
   vinegar recipes 290
   wine fermentation 189, 190n
I Yin (legendary master chef of Shang dynasty) 17, 31, 35
   cassia 52; fish 63; ginger 40
   harmonising agents 91, 94
   oranges 54; taro 43
i yü (wet salted fish) 383, 393, *395*, 420, 423
I Yung-Wên 374n, 377n, 491n
I-chang (Hupei) *544*
I-Ning Prefecture 543
Iacopone da Todi (ca. 1240–1306) 497
Ibn al-Baytar 494n
ice 433, 434
   cellars 430–1, 434; chambers 432–3; collection 430
   cutting 430; fish preservation 434
   gifts 430, 433, 434; price 434
   selling in Khaifeng market 434; storage 429, 430
   summer gifts 430; use 430, 433–4
   wells 431–2, *433*
ice box 430
ice house 429, 430
   construction *432*, 435; pits 430–1
   Fêng-hsiang (Yung-chêng, Shensi Province) 430, *432*
   Hsing-chêng (Honan) 431
   insulation 431, 435
   Montecello (Virginia, USA) 430n
   Ningpo region 435, *436*
   wells 431–2
ice-cream 434

Ichino Naoko 303n, 320n, *321*, 324n
Iconographia Cormophytorum Sinicorum (Institute of Botany; 1976) 39n
Idrisi (Arab traveller; 12th century) 494
Ikeda, I. 570n
Il Milione (Ramusio) 493
Iliad (Homer) 57n, 266n
Immortals, home of 554n
Imperial Academy of Medicine 143
imports of food 66
incantations for fermentation process 171
India
   British tea planters 549
   cookery 496n
   heating ditch for tea processing 520n
   labour problems 549
   oil mill (ghani) 452, *454*
   tea industry 503, 547, 548–9
   tea processing implements *550*
Indian Agriculturalist (Calcutta) (1882) 329n
Indochina, technology dissemination from China 608
Indonesia
   fermented salted fish *395*
   fermented soyfoods 378
   fish sauce 392
   rice noodles 492
   technology dissemination from China 608
   tempeh 342–5
   tou fu 328
Ingen (Zen master; +17th century) 317, 318
inoculum
   ferment 161, 182n, 281
   fermented soy paste 353
   koumiss 249
   lactic acid fermentation 254n
insignia template for tea *524*, 525, *526*
insulation, straw 431, 435
Internal Culinary Supervisor 56, 57n
International Conference (6th) on the History of Science in China (Cambridge; 1990) 209n, *212*, 305n
International Tea Committee 505
intestines, sausages 423
iodine deficiency 573–4
Ipomeia batatas 43
Iran 495
   wine (grape) process origination 153
Ishige Naomichi 10, 147, 345, 374n, 376n, 389n, 392, *393*, *394*, 395, 396, 397, 397n, 480n, 490, 491n, 492, 496n
Italy 493–7
   ice-cream 434
   noodles, arrival from China 493
   pasta 493–7, 495
Itoh, H. 383n, 392n, 397n
itriyya (filamentous pasta, macaroni) 494, 495, 496

Jabir (Arab chemist; 12th to 13th century) 225n
Jacob (Bible) 267n
Jacobson, J. I. L. L. 549n
James, P. 266n, 467n
jang (fermented soyfood) 378
jang ho (wild ginger) 380

jang ho (*Zingiber mioga*) 40, 43
Japan
    Buddhist monasteries 318
    fermentation components 282
    fermented salted fish *395*; fish sauce 392, 394; *hishiho* 380
    Jomon culture 603n
    microbial enzyme production 608
    noodles & pasta
        introduction 491; technology 492
    oil extraction process 455–7; oil press 452, *453*
    savoury sauces 374
    soy fermentation technology 374–8
    soybean curd technology 316, 317–19
    sprouted rice wine 163
    steamed bun (*man-thou*) 492
    sweet liquor (*amazake*) production 166
    tea ceremony 557
    technology dissemination from China 608
    tou fu 316
        characteristics 319; transmission from China 317–19
    wine making 263, 281n
Japanese scroll by Seibe Wake *15*
jar 348, *451*
    pointed bottom 275, *276*, 277
    shih incubation 339, 340, 341
    wide-mouthed 151, *152*, 277, 332
    see also pot; vessels
jars, starch preparation 461
jasmine tea 553, 554
Jasminium sambac (jasmine shrub) 554
Jatropha janipha 186
Jaubert, A. 494n
Jefferson, Thomas, ice house at Montecello (Virginia) 430n
jellyfish *Jeh* 146
jên (perilla seed) 439
Jên Jih-Hsin 86n
jen shên (ginseng) 186
Jên Ti 27n
Jenner, W. J. F. 253n, 511
jeot-cal (*chu-i* in Korea) 394–5
Jerome, St 401
jo (bamboo) 529
jo (reeds) 526
jo (roll) 529
jo nien (rolling) 537
Job's tears 190
Johnson, B. 150
Johnson, P. 495
Johnstone, B. 377n
Jomon culture (−5 th century), Japan 154, 603n
ju (meat; animal flesh) 17, 60, 380
ju (milky emulsion) 561
ju chiang (meat paste) 346, 359
ju chih (milk) 315
ju chiu (milk wine) 248
ju fu (milk curds; decayed milk) 254–6, 302
ju hsiang (Bombay mastic; *Boswellia carterii*?) 234n, 235
Jü Lin Wai Shih (Unofficial History of the Literati; c. +1750) 324n, 501

ju lo (drink from milk) 250
ju pei (oven) 532
ju ping (milk curds) 255, 256n
Judaic tradition 257
jujube *see also* date
    chiang making 349n
    drying 424
    monographs 145
    seeds 235
    steeped in water 428; stored in millet straw 429
    tea flavouring 561
    wine 245, 259n
Juvenal 577

Kua-chou area (Northwest China) 192
Kaifeng 527
    honeyed sweetmeats 426
    sausages 423
Kakushin (Buddhist monk; +13th century) 376, 377n
kál kuksu (sliced noodle, Korean) 492
Kalimantan, fish sauce 392
kalja (bread beer) 270n
kan (*Elopichthys bambusa*) 63
kan (orange) 54; *kan* (sweet taste) 91, *95*
kan chiao (dry yeast) 184
kan lan (*Brassica oleracea*) 31; *kan lan* (Chinese olives) 428
kan lo (dry yogurt or cheese) 252
Kan Lu Chiu Ching (Sweet Dew Wine Canon; Thang dynasty) 180
kan pi wêng tshai (drying pickling in a sealed jar) 417, 419
kan tang (wild pear) 49
kan tsao (drying by heat) 537
kan-cha (fruit of loquat/cumquat type) 54n
kang (wide-mouthed basin) 332n, 333
Kanshin (master Buddhist monk; +8th century) 317, 319
Kansu province 7, 241
    grape cultivation 240
    Neolithic site 23n
    tombs of Later Han, Wei and Chin 313n
Kansu shêng po-wu-kuan (on the Excavation at Huang-niang-niang-thai, Wu-wei, Kansu Province; 1968) *108*
Kansu shêng wên-wu kuan-li wei-yuan-hui (on Tombs 1 & 2 near Chiu-chhuan; 1959) *108*
kao (soft animal fat) 437
kao (cooked food of *erh* type) 468
kao (juice) *524*
kao (young lamb) 57
Kao Chan-Li (late Chou and early Han period) 59
Kao Chhuan 310n
kao chih (sheep tallow) 440
Kao Ching-Lang 125n, 580n, 586n
Kao Hsiu 159, 574
kao huan (ring pastry) *470*, 473
Kao Jui (prince in Northern Chhi; +550 to +557) 433
Kao Kuei-Lin, Hsieh Lai-Yung & Sun Yin 147
Kao Lien (writer; Ming dynasty) 130, 196
kao pên (*Lingusticum sinense*) 234
Kao Shih-Sun (12th century) 145
Kao Tsu, Emperor 112
Kao Yang 597n

kao yu (beef tallow) 440
Kao-Chhang or Kaaochhang (Sinkiang) 204, 242
    conquest (+640) 241, 243
kao-liang (sorghum) 20n, 23
Karim, Mohamed Ismail Abdul 392n
karintou (Japanese sweet pastry) 473
Karlgren, B. 19n, 32n, 39n, 67n, 74n, 82n, 118, 149n, 419n, 457n, 508n
Katz, S. 265n
Katz, S. & Maytag, F. 269n
kechap (fermented soyfoods in S.E. Asia) 378
Keeper of Victuals 28, 57n, 59
kêng (rice) 195, 197n
kêng (soup or stew) 34n, 71, 82–4, 89, 93
    cauldrons 98
    utensils 105
Kêng Chien-Thing 317n
Kêng Hsüan 18n, 19n, 24n
Kêng Liu-Thung 317n
kêng mi (rice) 204
ketchup 398n
kettle, tea 557, 559
Khader, V. 298n
Khai Pao Pên Tshao (New and More Detailed Pharmacopoeia of the Khai-Pao reign-period; +973) 52
khai pei (open up the furnace or tea work station) 523
Khaifeng (Northern Sung capital) 141, 296
    ice snow and cold water selling in market 434
Khang Hsi (Chhing emperor) 329
khao (roasting) *88*, 89
Khien Du 392n
Kho Chi-Chhêng 54n
Khou Tsung-Hsi (writer; +11–12th century) 300, 498
khu (wild rice) 20n
khu chiu (bitter wine or vinegar) 284, 285
khu kua (balsam pear) 95n
khu thu (bitter *thu*) 508
khu tshai (bitter vegetable) 507, 508, 512
khu tshai (sowthistle) 38
khuai (thin slices or strips of meat) 69, 70, 74–6, 420
khuai (thinly sliced fish) 384n
khuai tsu (chopsticks) 105
khuei (mallow) *33*, 36, *37*, 38, 93
Khuei Ming 499n, 500n
khun pu (sweet tangle) 575, 576
Khun Wu (legendary official of Hsia dynasty; c. −2000) 602n
Khung Ying-Ta 48n, 84, 111, 402n
Kiangsi 5
    tea plantations 513; tea trade *545*, 548
Kiangsu *545*
    Han tombs 46
kidney
    chhi 575
    deficiency 585
kieh chiap 398n
Kikkoman, K. K. 374n, 375n
Kikuchi, S. 383n, 392n, 397n
kimchi (Korean fermented fish product) 395
Kinzanji (Temple of the Gold Mountain) 376
Kitab al-Wusla ila I-Habib (+1230) 495

Kitan, honeyed sweetmeats 426
kitchen 432
kitchen scene, Eastern Han 86, *87*, 332n, 597n
kiwi fruit (Chinese gooseberry) 52
Klemm, F. 597n, 598
Knechtges, D. 7n, 94n, 96n, 402n, 404n, 469n, 472n, 478n
Knochel, S. 383n, 401n
Ko Chang-Kêng (Taoist monk) 501
Ko Chia-Chin 463n
Ko Chih Ching Yuan (Mirror of the Origin of Things; +1717) 498
Ko Hsuan (great-uncle of Ko Hung) 207
Ko Hung (alchemist and author; +4th century) 207n, 576, 578, 580, 587
ko phêng (culinary art) 69
Ko Wu Chhu Than (Simple Discourses on the Investigation of Things; c. +980) 295n, 564
Kodama, K. 167n, 279, 281n, 282n
koji (*chhü* in Japanese) 6n, 161n, 166–7, 263, 282, 380, 397n, 608
    Aspergillus 605n
    barley 376, 377; rice 377
    miso preparation 375, 376, 377–8
    soybean 377
Kollipara, K. P., Singh, R. J. & Hymowitz, T. 18n
Kondo, H. 161n
konton (flat piece of dough) 491
Korea
    fermented fish products 388, 394, 396–7
    fermented salted fish *395*; fish sauce 392
    fermented soyfoods 378
    noodles 492
    technology dissemination from China 608
    tou fu 318n
Kosikowski, F. 249n, 254n
kou (dog) 55
kou lou (rickets) 588–91
kou thu (dog butchers) 59
kou tzu wine *236*, 237
koumiss 248–50, 254
    alcohol content 249
    distillation 217n, 231
    inoculum 249
    making 249
Kranzberg, M. 598n
ku (grains) 17, 18, 259
ku (wine serving vessel) *99*, *100*, 102
Ku Chhi-Yuan 434n
Ku Chin Lu Yen Fang (Well Tested Prescriptions, Old and New; +7th century) 576
Ku Chin Thu Shu Chi Chhêng 221n
Ku Chu Shan (Kiangsu and Chekiang region) 517
Ku Fêng 515n
ku pai phi (rice bran) 581
ku ping (pressed oilseed cake) 450
Ku Shih Khao (Investigation of Ancient History; +250) 156, 260n
ku tzu (setaria millet) 27
Ku Yen-Wu (writer; 17th century) 512
Ku-chu-shan area (Chekiang) 522n
Ku-Thien (Fukien) 202

kua (melon) 33, *34*, 380
kua mien (hung noodles) 483n, 484–5, 485, *486*
Kua Shu Shu (Book of Melons and Vegetables; Ming dynasty) 145
kua-mien (hung noodle) 3, 484
kuan (bamboo splints; skewers) *521*, 522
kuan (roach) 61, *62*
kuan (urn) 151n, 153
kuan chhang (sausages) 423
Kuan Phei-Shêng 146n
Kuan Tzu (Book of Master Kuan; –4th century) 23n, 42, 140, 160
Kuang Chhün Fang Phu (Monographs on Cultivated Plants, Enlarged; +1708) 145
Kuang Chi Chu (Notes on Wine Vessels; Sung) 132, *133*
Kuang Chi Fang (+723) 576
Kuang Chih (Extensive Records of Remarkable Things; late +4th century) 509–10
Kuang Chün Phu (Extensive Treatise on Fungi; +1550) 144
Kuang Tsêng-Chien 209n
Kuang Wu (Emperor, founder of Eastern Han dynasty) 294n
Kuang Ya (Enlargement of the *Erh Ya*; +230) 478, 509
 tea 519–20
Kuang Yang Tsa Chi (Miscellanies about Kuang-Yang; +1695) 538
Kuang Yun (Revision and Enlargement of Dictionary of Characters Arranged According to their Sounds when Split; +1011) *167*, 265n
Kuang-chou Chi (Record of Kuangtung; +4th century) 511
Kuang-nan 441
kuang-ping (bright bun; half-sized bagel) 3, 472n
Kuangchou shi wen-kuan-hui (on Han Tombs at Sa-ho in the Eastern Suburb of Canton; 1961) 112n
Kuangsi 5, 46
Kuangtung 5, 7, 46
 emigration 492
 tea brokers 544; *kung fu* tea 540; red tea 548
 rice noodles (*ho fên*) 490
 sausages 423;
 soy sauce making 365–6, 368–70; fish sauce (*yü lu*) 391
Kuangtung Hsin Yü (New Talks about Kuangtung; +1690) 134
Kuangtung shêng po-wu-kuan (Chin & Thang sssites at Shih-hsing, Kuangtung, 1982) *108*
Kuanthou soy sauce of Fukien 365, *367*
Kublai Khan (Yuan emperor) 493
kudzu vine 462
kuei (cassia) *44*, 52, 91
kuei (iron mould for tea pressing) *521*
kuei (serving vessel) 98–9, *100*, 101
kuei (*Siniperca*) 63; *kuei* (tortoise) 66
kuei chih (*Cinnamomum cassia*) 235
kuei hsin (heart of cassia) 589
Kuei Hsin Tsa Shih (Miscellanies from the Kuei-Hsin Street; c. +1300) 245
kuei hsiung (turtle chest) 589
Kuei Jên (Tortoise Keeper) 66n
kuei pei (turtle back) 589
kuei wine 236
kui thiao (rice cake filaments) 490

Kui Thien Lu (On Returning Home; +1067) 525n
Kui Thien Suo Chi (Miscellanies on Returning Home; +1845) 540
kuki (fermented savoury soybeans) 377
Kulp, K. 278n
Kumamoto Prefecture (Japan) 163n
kung (palace) 432
kung (wine serving vessel) 99, *100*, 102
kung chhu (palace chef) 432
Kung Chih 538n, 542n, 546n
kung fu tea 540, 546n
 drinking 561; preparation *562*;
 washing of tea leaves 559n, 561
Kunkee, R. E. 153n
Kuntze, O. 503, *504*
kuo (cooking pot; boiler; wok) 82, 208, 215, 216–17, 229, 231
 tea processing 530, 534
kuo (fruit) 17, 43
Kuo Chhing-Fan 420n, 574n
kuo huang (firming the tea yellow) *524*
kuo lo (drink from fruits) 250
kuo lü (filtering) 308
kuo lu (tea called in Nan Yueh) 511n
Kuo Pho (c. +300) 508, 555
Kuo Phu 457
Kuo Po-Nan 148, 299n, 305, 305n, 317n, 358n, 359n, 360n, 375n, 429n, 430n, 434n
Kuo Shu (Discourse on Fruits; +mid-16th century) 145
Kuo Su (Fruits and Vegetables) 144n
kuo-thieh (pot stickers) 479
kvas (bread beer) 270n
Kwangsi, *yü lu* 391
Kweichow 5, 7
 distilleries 217

la (cured by drying) 422
la chha (cake tea; waxed cake tea) 527, 529
la chhang (sausages) 423
la jou (cured meat) 421
la mien (pull noodle) 484–5, *487*
la yüeh (twelfth month) 421
laap cheong (Cantonese sausages) 423
labour
 for food production 592
 ice collection and storage 430
 large households 597n
 tea production 517n, 522n, 523, 549
 women 597n
lactic acid bacteria 8, 96n, 249, 251n, 252n, 254n
 in fermented soyfoods 378n
lactic acid fermentation 252, 254, 383–6, 390–1, 393, 396, 408, 415
 inoculum 254n
 suan tshai 412; *tsu* 404
Lactobacillus 253, 280
lactose 249, 253, 322
 intolerance 602n
 malabsorption 257, 596n
ladle 85, 104, 105–6
 tea 559
Lafar, F. 155n, 608n

laganum (pastry) 497
Lagenaria leucantha 33, 34
lagman (filamentous noodles) 492
lai (or *rai* Japanese sweet liquor) 166
lai chhi (*tou fu*) 300
lai fu (radish) 35
Lai Hsin-Hsia 329n
Lai, T. C. 592n
lakare-no-lao (wine-making dregs) 162
lamb 58
  cha 409; *chiang* 381; *tsao* preserves 409
  meat bun 471; red pot-roast 193
  sacrificial ritual 58;
  skins 57; tallow 93
  young 57
Laminaria saccharina 575n
lamps *see also* flambeaux
  oil 441, 450
lan (bamboo basket) 521
lan (steeped pears) 428
lan xiang (orchid tea) 554
land animals 55–61
  domestication 55; food resources 66
lang tang (henbane) 234
lao (wine mixed with lees; muddy wine) 156, 161, 185, 233, 234, 260n
lao chao (fermented congee) 260n
lao chhên tshu (well-tempered vinegar) 291
lao chhou (mellow drawn soy sauce) 366
Lao Hsüeh An Pi Chi (Notes from the Learned Old Age Studio; +1190) 301, 500
Lao Tzu 60
lao wan (dumpling or wonton) 478
Laos
  fermented salted fish 395, 396; fish sauce 392
lard 59, 93, 358, 437n
lasagna 496–7
Latham, R. 243n, 493n, 501n
Latourette, K. S. 590n
Lattimore, O. 250n
Latuca denticulata 33, 38
Lau, D. C. 74n, 109n, 420n
Laudan, R. 412n
Laufer, B. 31n, 44, 204n, 596
Leang Ching (cooling the leaves) 547
leaven 606
  types 465, 470, 471, 476, 477–8
Leclerc L. 494
Ledebouriella seseloides 186
Lee, Arthur (Professor at Amoy University) 1
Lee, Cherl-Ho 392n
Lee, Cherl-Ho, Steinkraus & Reilly 383n, 388n,
Lee, G. 72n
Lee, H.-G. 154n
Lee, Sung Woo 392n, 395n, 397n
leek, Chinese 33, 38, 40, 41, 43, 92
Legenaria leucantha 29n
Legge, J. 19n, 20n, 21n, 28n, 32n, 39n, 47n, 49n, 52n, 56n, 57n, 59n, 68n, 69n, 71n, 72n, 73n, 74n, 80n, 82n, 83n, 84n, 96n, 99n, 102n, 103n, 109n, 118, 120, 149n, 154n, 156n, 157n, 163n, 164n, 259n, 283n, 284n, 334n, 354n, 403n, 457n

legumes
  germination 298n
  toxins 293; inactivation of 294
Lehrner, M. 272n
lei (drink storage vessel) 102
lei (pottery vessel) 151n
lei chiu (filtered wine) 166
lettuce 33, 38, 405
  salting 418
Levi-Montalcini, R. 600n
li (boiler) 76, 77, 80, 82, 98
li (carp) 61, 62, 63, 64
li (Chinese chestnut) 44, 49, 50
  storage 427–8
li (Chinese Japanese plum) 44, 45–6
li (mountain *suan*) 39n
li (pear) 44, 49–50
li (propriety as in *Li Chi*) 160
li (sweet weak wine) 97, 156, 157, 593, 594, 606
  association with soups and decoctions 161
  definition 264
  discovery 263, 266; discontinuation 274, 275
  making 158, 159, 160, 162, 166, 259, 260–1
  nieh for making 158; origin in China 259–63, 274;
  ritual purposes 164
Li, C. L. 202n
Li Chao (+9th century) 205
Li Chhang-Nien 21n, 22n, 23n, 294n, 295n, 316n
Li Chhi 248n
li chhi (*tou fu*) 300, 303
Li Chhi (Utensils of Rites) 157n
Li Chhiao-Phing 215n, 299n, 133, 154, 190, 197n, *444*, *445*, *446*, *447*
Li Chhing-Chi 143n
Li Chhun-Fang 138n
li chi (litchi) 144
Li Chi Phu (Monograph on Litchi; +1059) (Tshai Hsiang), dried litchi 425–6
Li Chi (Record of Rites; +1st century) 28, 117, 119–20
  bed 113; dining furniture 109, 111
  beef 57; dogflesh 59; game mammals and birds 60;
  horseflesh 56; mutton 57; pork 59; cooking
  methods 67, 70, 71, 72, 73, 74–5
  food, *fan*, *shan* & *hsiu* 98; braised meats 84; congee 82; *erh* 468; *kêng* 83, 93; soybean gruel 294; *shan* food 466; eight delicacies 354n
  drinks 97; drink storage vessels 102; fat for
  cooking 437; harmonising agents 91; starch 461;
  sweeteners 92, 457; vinegar 92, 284
  fermented meat or fish paste 380, 382, 387, 387n, 390
  fruit 43; Chinese cherry 52; raisin tree 49
  fu and *hsi* 420; *tsu* 403, 416
  grain preparation 259; meat cutting 69
  meal setup 114, 115; hand cleaning 105
  Ming Thang Wei (Hall of Distinction) 156n
  nieh (sprouted grain) 157, 160; rice 582; salt 92;
  wheat flour 68n
  serving implements 104, 107; chopsticks 104;
  spoon 85
  serving vessels 98, 99n, 100, 102; vessels 602

Li Chi (cont.)
  Son of Heaven 58, 60n
  vegetables 32–3; lesser bean 40; smartweed 92; mushrooms 40
  wine 150, 156, 157; medicated 232
  Yüeh Ling (Monthly Ordinances) 57, 163, 164, 165
  Yin-Tê index 120
Li Chien-Min 155n
li chih (litchi) 554
Li Chih Phu (Treatise on the Litchi; Ming) 144n
Li Chih-Chên 296n
Li Chih-Chhao 209n
Li Ching-Wei 571, 576n
Li Chung-Hsüan 109
Li Fa-Lin 314, 463n
Li Fan 19n, 23n, 30n, 44n, 47n, 48n, 49n, 50n, 265n, 507n
Li Hai Chi (c. +1400) 282n
li hing mui (salted *mei* fruit of Hawaii) 412n
Li Ho (Thang poet; +791–817) 85n, 192, 193
Li Hua-Jui 203n, 207, 221n, 225, 229n
Li Hua-Nan (writer; 18th century) 341
Li Hui-Lin 18n, 24n, 28n, 29n, 34n, 35n, 36n, 38, 40n, 49n, 51n, 52n, 183n
Li Ji-Hua (writer; +1565 to 1635) 325
Li Ju-Chên (writer of novels; +1763–1830) 323n
Li Kao (author of *Shih Wu Pên Tshao*; 17th century) 139
Li Kuang-Thao 503n
li lo (processed milk products) 255n
Li Pai (Thang poet) 180, 181, 223–4, 243n, 301
Li Pao (writer) 247n
Li Pin 203n, 221n, 222, 225, 499n
li shih (eating granule food) 68, 465, 466
Li Shih-Chên (author of the *Pên Tshao Kang Mu*; Ming) 24n, 38, 54n, 139, 204–5, 224, 227, 230, 231, 235, 241, 248, 297n, 299, 303, 305, 307, 309, 315, 323, 349, 341, 425n, 498, 501, 581n, 582, 586
Li Tao-Yuan (writer; +5th century) 432n
Li Thao 300n, 579n, 585n, 588n, 589n
Li Thiao-Yuan (scholar; theatre critic; 18th century) 131, 323n
Li Tshang (first Marquis of Tai, r. −193 to −186), tomb of wife 24–5, *26*
li tsu (pear pickle) *406*, 428
Li Wei Han Wen Chia (Apocryphal Treatise on Rites; Excellences of Cherished Literature; early Han) 97n
Li Wên-Hsin 114n
Li Yang-Sung 153n, 155
Li Yin 245n
Li Yü (author of *Hsien Chhin Ngo Chi*; 17th century) 131
Li Yü-Fang 23n
Li Yü-Hsi (Thang poet) 482n
Li Yüan-Tsung (Northern Chhi Dynasty) 240
Li Yün (Conveyance of Rights) 157n, 160n, 250n
Li-Chhi Phu (Monograph on Litchi; +1059) 144
Li-chi Lou Tui Chiu (Drinking at the Litchi Pavilion) 203n
Li-chou i-chih lien-ho-khao-ku fa-chüeh tui (Han tomb discovered at Li-chou; 1980) 80n
Li-I (nr Samarkand) 242n
Li-yang (Chhin capital) 463n

liang (millet or sorghum) 19n, 23, 59, 82
liang (tea from parched grain) 97
Liang Chang-Chü (writer; 19th century) 540
Liang Chhi-Chhao 17n, 18n
liang mi (large-grained setaria millet) 459
Liang Shu (History of the Liang; +635) 316n
Liang Wu-Ti (emperor r. +502 to +549) 498
Liang-chou tzhu (Song of Liang-chou; c. +713) 239
Liang-shi pu yu-chih chü chia kung chhu (Oil pressing improvements, 1958) 450n, *451*
liao (smartweed) *33*, 38, 92, 93, 190
Liao Shih (History of the Liao Dynasty) 426
Liao-yang Han tomb, Liaoning 112, 114–15
lichens 587
Lieh Chuan (Biography of Shih Chhung) 294n, 295n
Lieh Tzu (Book of Master Lieh; +742) 294, 315n
lien (lotus) 36; *lien ou* (lotus rhizome) 36
Liener, I. E. 294n
Ligusticum wallichii 286
lime, oil processing 455
Lin An (Hangchou) 301
Lin Chêng-Chhiu 492n
Lin Ching-Yin (traveller to Japan; +1349) 492
Lin Hsiang-Ju & Lin Tsuifeng 128, 131, 131n, 355n
Lin Hung (author of *Shan Chia Chhing Kung*; southern Sung) 132, 254, 301, 358
  pasta making 472n; tea 561, 564
Lin Ju-Lin 131
Lin Kan-Liang 564n
Lin Nai-Shên 32n, 43n, 148
Lin Phin-Shih 17n, 156
Lin Shun-Shêng 157n, 161–2
Lin Yen-Ching 245n
Lin Yin *22*, 57n, 119, 333n
Lin Yung-Kuei 208n, *210*
Lin Yutang 68n, 94, 127n
Lin-I Chi (Records of Champa; −2nd century) 54
Linacre, T. (+1460–1524) 577
Lindera strychnifolia 441
ling chiao (water caltrop) *33*, 40
Ling Hsiang 246n, 295n
Ling Jên (Superintendent of Ice Management) 430
Ling Phing-Shih 120
Ling Piao Lu I (Strange Southern Ways of Men and Things; +895) 183, 205n
  distilled wine 222; herbal *ferment* 232
Ling Shun-Shêng 154n, 156n, 157n, 161, 183n, 185n, 232, 233, 250n
Ling Thai (Magic Tower) 62–3
ling yin (ice house) 429
ling yu (soy sauce) 360n
Lingnan Li Chih Phu (Litchis South of the Range; Chhing Dynasty) 144
Linnaeus 539
linseed oil 450, 452, 456
lipases 593, 608
lipids, *tou fu* and cheese 330
lipoxidase 293n, 312, 322
liquaman see garum
Litchfield, C. *453*
litchi (*Litchi sinensis*) 144
  dried 425–6; tea 554; wine 245n

little dragon cake 525
liu (tumour) 574
Liu An (Prince of Huai-Nan; −179 to −122) 299–300, 303
Liu Chên-Ya 433n
Liu Chhang-Jun 85n
Liu Chhin-Tsin 552n
liu chhu (six livestock) 55, 57, 60
Liu Chhüan-Mei 225n
Liu Chi (author of *To Nêng Phi Shih*) 129–30
Liu Chih-I 27n
Liu Chih-Yuan 218n
Liu Ching (c. +1753) 542n
Liu Ching-An 225n
Liu Chün-Shê 67n
Liu Hsien-Chou 583n
Liu Hsun (writer; +9th century) 183, 205, 222
Liu Jilin 571n
liu ku (six grains) 20n
Liu Kuan (Yuan dynasty) 145
Liu Kuang-Ting 133n, 203n, 205n, 208n, 224n, 226n
Liu Lung Chuan (in Hou Han Shu) 168n
Liu Mêng Tê Wên Chi (Complete Works of Liu Mêng Tê; Tang) 242n
Liu Pai-Tuo (wine maker and producer of *pai tuo ferment*; +5th century) 170, 180
Liu Phu-Yü 433n
Liu, S. T. 486
liu shêng (six food animals for slaughter) 60
Liu Sung 131n
Liu Szu-Chih 300
Liu Tsün-Ling 501n
Liu Tsung-Yüan 301
Liu Wu-Chi & Lo, I. Y. 38n, 43n, 592n
Liu Yü-Chhuan 24n
Liu Yü-Hsi (poet; +772 to +842) 242, 523, 527
liver 586
    pig 588
    sheep's 587
    vitamin A source 587
livestock *see also* land animals
    keepers 56
lo (yoghurt drink, boiled/soured milk, koumiss?) 248, 250–1, 254, 315
    alcohol content 253
    boiling 316
    consistency and flavour 253
    dry (lactic cheese) 252
    methods for making 250–2
    starter 251–2; sour 253; strained 252; sweet 253
    waning of interest 259
lo (ancient fruit wine) 156, 157
Lo Chên-Yü 136
Lo Chhi (writer; +15th century) 299
lo chiang (yoghurt drink) 251n, 511
Lo Chih-Thêng 229n
lo fu mien (radish noodles) 482, 484
Lo Lin (writer; +17th century) 516, 530, 531–2
lo nu (yoghurt's slave; tea) 511, 513
lo po (radish) 35
Loehr, M. 76n
long fish 61, 62
loquat 53–4, 428; vitamin B content of leaf 581
lotus 33, 36, 43
    flower tea 553; honeyed seeds 426
    iced root strips 434
    leaves to make *cha* 385
lou chih (leaky pot) 511
Lou Tzu-Khuang 122
lovastatin 202–3
Loyang, Western Han tomb 102–3
Loyang Chieh Lan Chi (Description of the Buddhist Temples of Loyang; +547) 141, 180, 192
    grapes 240; ice chambers 432; *lo* 253n; tea 511
Loyang fa-chüeh tui (1963) 463n
Loyang khao-ku tuei (1959) 463n
Loyang po-wu-kuan 108
lu (furnace) 527; *lu* (pot roast) 88, 89; *lu* (salted liquor) 419; *lü* (green) 405
Lu Chao-Yin 314n
lü chha (green tea) 536, 541
    exports 548
lu fu (radish) 35
Lu Gwei-Djen 9, 138n, 246n, 574n, 577, 581
Lu, H. C. 571n
Lu Hui (writer; +4th century) 432n
lü ju (green meat from *lü ju fa*) 414–15
lü ju fa (recipe for pork chicken or duck) 414
lu lo (strained *lo*) 252
Lu Ming-Shan (Uighur folk hero; 14th century) 143
lu pha (radish) 35
Lu Pu-Wei (first prime minister to Chhin Shih Huang) 120
Lu Shan 305n
Lü Shih Chhun Chhiu (Master Lü's Spring and Autumn Annals; −239) 27, 34, 117, 139
    celery 35; ginger 40; oranges 54; harmonising agents 91, 94n; fish 63; *hai* 380; wine 156, 159
    goitre 574
Lu Shih Chhun Chhiu Pên Wei Phien (1983) 120
Lu Shu 300
Lu Thing-Tshan (writer; 18th century) 516, 539n
Lu Thung (Tea Doter; +835) 554, 563, 568
lü tou (mung bean) 298
lu tou (white sprouts from the mung bean) 296–7
Lu Wên-Yü 32n, 38n, 48n, 118
Lu Yü (author of *Chha Ching*; +8th century) 121n, 506
Lu Yu (Sung writer and poet; +1125–1210) 38n, 121n, 230, 301, 303, 482n, 506n
lu-chü (cumquat) 54n
lucky fish 61, 62
Lun Hêng (Discourse in the Balance; +82) 294n, 346–7
    I Chêng (part of *Lun Hêng*) 294n
Lun Yü (Analects of Confucius; −5th century) 19, 21n, 140
    chiang 334; *fu* 420
    fan (cooked grain) 80n; drinks 96
    meat preparation 69, 74n
    serving vessels 99, 103n
    vinegar 284, wine 163n
lung (drying basket) 531
lung (quern) 4
Lung Po-Chien 121, 139n
Lung-ching tea (Dragon well) 539
Lung-hsi (Kansu) 240

luo (snail) 66
Luo Shao-Jun 551n, 552n
luo suo (pasta nuggets) 470, 472, 473, 481
lupin 239
Luzon, fermented fish 395

ma (type of fish?) 380
ma (hemp) 18, 19n, 20, 22, 23, 25; cultivation 28, 29
ma (horse) 55
Ma Chhêng-Yüan 76n, 98n, 209n, 211, 213
Ma Chi-Hsing 234n
ma chiang (fish paste) 346
Ma Chien-Ying 69n, 603n
Ma Chih (official) 56n
ma fen (hemp fruit) 31
ma ju phu thao (mare teat grape) 241
ma lo (mare's milk *lo*) 252
Ma Shih-Chih 431n
ma thung (koumiss) 248
ma tzu (hemp seed) 439; *ma yu* (hemp seed oil) 437, 439–50
ma-chhi-hsien (purselane) 581n
Ma-wang-tui Han-mu po-shu (Silk Manuscripts Han Tombs at Ma-wang-tui; 1985) 121, 234, 238, 284
see also Mawangtui Chu Chien (MWT Inventory)
Mabesa, R. C. & Babaan, J. S. 392n, 396–7n
macaroni 494, 495; maccheroni 495
Macfarlane, G. 600n
McGee, H. 154n, 266n, 270n, 466n, 467n
McGovern, P. E. 153n, 258n, 272n
MacGowan, Dr. 231, 237, 250n
MacLeod, A. M. 153n, 267n, 282n
McNeill, W. H. 497n
macrobiotic movement 263
Madam Tshui's Food Canon 125
Magic Tower 62–3
magnesium chloride 305
magnesium sulphate 305
mai (barley or wheat) 18, 19, 27, 158, 262
mai chhü (wheat or barley *ferment*) 262
mai chiang (wheat *chiang*) 346, 355, 357
mai hun (moulded wheat *ferment*) 335, 336
mai nieh (malt) 186
mai ya thang (malt sugar) 2, 263n, 460–1, 461n
mai yu (water-based sauce of fermented wheat) 358n
Mair-Waldberg, H. 257n
Maitland, D. 506n
maize *see* corn
Malaysia 492
  fermented salted fish 395
  fermented soyfoods 378
mallet 451, 453
Mallory, J. P. 55n, 604n
mallow 33, 36, 37, 38, 43, 93, 379, 405–6, 409
malt 158, 186, 187, 458
  discovery 267; amylase 278, 459; cock crow wine 189;
  mashing 260n, 272, 274, 282; *sikhae* 397
  Sumerian beer brewing 269
malt sugar (maltose) 8, 92, 95, 263, 278–9, 459, 601, 604
  amber 458, 459, 460
  cereals, from 457–61; cloudy syrup 459
  colour 458, 459, 460; dark 458, 459; jellied 460

making 158, 160, 262, 457–61
  sweetener use 457; syrup collection 458
  vinegar making 290; white 458–9, 460
malt syrup 262, 459
maltase 278
malting, invention 266, 266n, 267
Malva verticillata 33, 36
Mampuku-ji temple (Kyoto) 317
man ching (Chinese turnip; colza) 31, 34–5, 439
Man-chhêng fa chüeh tui (1972) 463n
Man-chhêng Han tombs, monographs 146
man-thou (steamed bun) 282n, 469, 475–8, 490n, 594
  in Japan 492
mandrake 239
Manilius (Roman author of +1st century) 398n, 400
Mann, W. N. 571n
Man'yoshu (poem; +7th century) 374
Mao (Mao Heng) 17n, 116
mao chha (raw tea) 552
mao li (bean curd) 318
Mao Shih Chêng I (Standard *Book of Odes*; +642) 48n, 333n, 402n
Mao Shih Tshao Mu Niao Shou Chhung Yü Su (c. +254) 507
Mao Wên-Hsi (writer; +10th century) 516, 523
Marco Polo (world traveller +13th century) 59n, 243, 501n
  gluten 501; ice-cream 434
  noodles from China 467, 493–7
maritime trade 8, 538, 541, 545, 546, 547
Markakis, P. 298n
Marsden, W. 493n
mash, fermented 272n, 187, 196, 202, 458, 606
  pressing 187, 220
  residual 193, 594
  steam distillation 214, 229
  vinegar making 286, 288
  wine fermentation 182, 183, 187
Massachusetts Institute of Technology 14n
Materia Dietetica classics 134–9
mats 109, 113, 187, 521, 531
Mawangtui Chu Chien (MWT Inventory; −200) 117, 121
  pastes and pickles 379–80, 391, 396n
  tea 513?
  see also Hunan nung-hsüeh-yuan; Ma-wang-tui Han-mu po-shu
Maynard, L. A. & Swen Wen-Yuh 590n, 604n
maza (paste or porridge) 265
Mazess, R. B. & Mather, W. 604n
mead 246–7, 606n
  alcoholic content 247
  ferment use 246, 247; recipe 247; yeast 247
meal 111, 113–15
measurements, use in cookbooks 127
meat 379
  baking in clay 603; braising 84; pot roasting 84;
  lu method 414–15; finely shredded (*sung*) 423–4;
  sausages 423
  bear 581n, 582
  chiang 381–2, 381, 382n, 388–9; *chiang* preserves 413–14; *tsao* preserves 409, 413
  cutting 69; preparation 68–70; finely sliced (*khuai*) 70, 74–6, 420; minced 75; sauce for 70

dried (*fu* or *hsi*) 419–22
foodstuff category of ancient China 17
protein 334
salting 8, 602n; steeped 69n, 417
   see also other cooking methods
meat paste, fermented (*chiang*) 333–5, 346, 359, 593, 594
   decline in importance 388–90
   ferment use 380–2
   method 381–2, 383–4
   than, hai, hsi, ni, chiang 333–5, 379–80
Mechanica (Hero) 439, 439n
Medawar & Medawar 601n
medical arts 232
medical officers, Chou court 571
medicine, Chinese 571–3
meditation 506, 515
Megalobrama terminalis 62
mei (apricot) 284, 291n, 587
mei (Chinese or Japanese apricot) *44*, 46–8, 92, 412n, 434
Mei Jao-Chhen (poet; +1002 to 1060) 434
Mei Shi Fang (Master Mei's Prescriptions) 587
Mei Wen-Mei (prescriptions; +589 to +618) 587
meju (moulded beans) 375n
melon 33, *34*, 43, 342, *409*
   preservation 405, 425n; *tsao* preserves *409*; *tsu* 380, *406*, 406n, 411n, 412n
   salting *418*; sliced 342
   sweet, steeped in water 428
   writings 145
Mêng Chhi Pi Than (Dream Pool Essays; +1086) 441n, 500
meng huang (yellowing of tea) 552
Mêng Liang Lu (Dreaming of the City of Abundance *or* Dreams of the Former Capital; +1334) 141, 205, 354, 387
   burnt wine 224; *hsi* and *fu* 421–2; dried salted fish 423; *tsao*-preserved meat 413
   dairy products 256; gluten 500; honey 460; *tou fu* 301
   everyday necessities 436
   ice 434; snow 434; *lu* method 414
   noodles 483; pasta 476; wonton 479; *tien hsin* (dimsum) 485n
   oil pressing establishments 441
   shih 341; soy condiments *372*
   sugar 460; sugared fruits 426n
   tea houses 527
Mêng Nai-Chhang 203n, 223
Mêng Shên (of *Shih Liao Pên Tshao*; +7th century) 136, 205
Mêng Shu kingdom 193
Mêng Ta-Phêng 288n, 289
Mêng Tzu (Book of Mencius; c. −290) 21, 21n, 23n, 58, 75n, 99, 100, 140
   dining furniture 109
   food preparation 69, 71, 75n, 80n
   serving vessels 99n, 100n, 102n
Mêng Yuan-Lao (+12th century) 141, 296
menshhitshi (Tibetan noodle) 492
menthol, for fruit storage 428
mercury, distillation 207
Merke, F. 577n

Meskill, J. 329n
Mesopotamia
   alcoholic beverages 258; beer 258
   barley 267n; bread making 463
   brewing terms 269n
   grape fermentation 243n
   saddle quern 599n
   wine process origination 159, 258
metal, symbolic correlation *572*
methyl salicylate 565n
mi cooked grain for brewing 178n, 187n, 274
   for malt sugar making 457, 458, 459
mi (gruel) 82
mi (honey) 92, 246, 457, 460
mi chien (honey preserved fruits) 425–6
mi chiu (mead) 246–8
mi hou thao (Chinese gooseberry, kiwi fruit) 52
mi fên (refined starch, also rice flour) 461, 485, 488
mi-fên (rice flour noodle) 488, 491, 492
mi li (dried yogurt or cheese) 315n
mi lo (alcoholic drink from grains) 250
mi nieh (sprouted grain kernels) 166
mi wu (selinium) 574
mi-hsien or *mi-lan* (fine rice noodles) 488
Mi-hsien (Honan Province) *tou-fu* mural 305
Miao Chhi-Yü 23n, 29n, 31n, 75n, 120, 123n, 124, 169n, 170n, 171n, *172*, 179n, 187n, 285n, 314, 340n, 347n, 348n, 351n, 359n, 361n, 406n, 424n, 427n, 468n, 469n, 485n, 499n
Michaowski, P. 269n
Michel, R. H. 153n, 258n
microbes, on surface of rice grains 429n
microbial action 405–7, 408, *409–10*, 411–12, 415
Middle East, pasta 494
midzu-amé (Japanese malt syrup) 263
mien (wheat flour) 68, 314, 485
mien (filamentous noodles) 466, 480–3
mien or *mien fên* (wheat flour) 463, 466, 467, 473–4, 481–2, 582
mien chiang (wheat flour *chiang*) 355, 357, 373n, 593, *594*
mien chin (wheat gluten) 498, 499, 500 *see also* gluten
   mock chicken, fish and shrimp dishes 498
   preparation of 501
mien shih (flour or pasta food) 465, 466, 467n
mien shih (eating flour foods) 68 as opposed to *li shih* (eating granular foods)
mien shih tien (pasta shops) 482
mien thiao (filamentous pasta or noodle) 467, 480, 484, 492
mien-tien (pasta foods made from rice flour) 475n
Milano, L. 269n
milk 9, 322, 582
   consumption 256–7, 596
   curds 255; acid coagulated 302; *rennet* coagulated 329
   decayed (*ju fu*) 302; digestion 257, 322
   fermented products, *lo, su & thi-hu* 248–57
   honey soaked 500; mare's 248
   koumiss and other milk wines 248–50
   pasta making, in 472
   pastoral nomads 315
   processing 602n
   soured 252, 254, 255; defatted 253

mill *see also* milling; quern
  oilseed grinding 441, *446*, 452, *453*, *558*
  primitive *598*
  roller 439, 448, 449
  rotary 314, 463, *464*, *465*
Miller, G. B. 36n, 75n, 298n, 323, 479n
Miller, N. 258n, 266n
millet 18, 23, 27, 59, 140, *604*
  broomcorn 18, 25; foxtail 18, 25
  congee 260; gruel 82; cooked 80, 593, *594*
  crumbs for thickening 461; chaff 461
  cultivation 593
  dehusked grain 295
  distilled wine 204; medicated wines 232, 234
  ferment 262; making 161–2, 603; fermented drinks 274
  fermentation 153, 168, 169, *170*; founder crop 263
  panicum 19, 19n, *20*
    malt sugar making 459; staple grain *21*; wine 174–5
    served to Son of Heaven 58
  setaria 19, *20*, 23, 27
    malt sugar making 459; staple grain *21*; starch preparation 461
    vinegar making 286; wine 174, 178
  spiked 461
  sprouted 166n, 295, 601; sprouting 163n, 457
  steamed 274n, 278
  straw for fruit storage 429
  terminology 23–4, 459n
  wine making 18, 157, 174–8, 259
millet flour
  erh 468; pasta recipes *470*, 473; *tzu* 468
milling *see also* mill; quern
  technology 582–3
  wheat 68, 82, 463, *464*, 582
Min *see* Fukien
Min Chhan Lu I (Unusual Products of Fukien; 1886) 541
Min Chung Hai Tsho Shu (Seafoods of the Fukien Region; +1593) 145
Min Hsiao Chi (Lesser Min Records; +1655) 538
Min Shu Tshung Shu (Series on Folk Literature) 122, 132, 182n, 205n; tea 516
Min Ta Chi (Greater Min Records; +1582) 538
Min Tsa Chi (Miscellaneous Notes on Fukien; 1857) 540n
mincing 75n
ming (tea) 510–11, 555, 564
ming (young buds of tea) 510n
ming chha (tea) 509n; *ming chha* (loose tea) 529
ming chih (tea) 511
Ming I Pieh Lu (Informal Records of Famous Physicians; +510) 39n, 240n, 253, 296, 315n, 337, *341*, 354, 424
  eye diseases 586; vision 587; goitre 575
  tea 511; *i* (malt sugar) 460
ming mu (clarifying the vision) 586, 587
ming tshai (tea) 510n; *ming yin* (tea) 511
Ming-ti (Wei Emperor) 469
Miscanthus sacchariflorus 335
miso, *misho* (Japanese *chiang*) 362n, 374
  fermentation 377–8; preparation 374–5
  importance 375

Kinzanji (Golden Mountain Temple) 376
  modern technology 377
  soup 375–6
  types 377
Mizuno, S. 98n
mo (rotary quern) 448n
mo chha (loose tea) 529, 529n
Mo Chuang Man Lu (Random Notes from Scholar's Cottage; +1131) 246
Mo Ho, Biography of (in *Sui Shu*) 154n
mo ku chün (mushroom) 145
Mo Ngo Hsiao Lu (A Secretary's Common Placebook; c. 14th century) 196, 197, 200
Mo Tzu (Book of Master Mo; –4th century) 23n, 58, 140
  ping pastry 467
mo-tu (fermented soy paste) 347n
Mokyr, J. 597n
mola olearia (olive crushing mill) 439
Moldenke, H. N. & Moldenke, A. L. 39n
mollusc, lamellibranch 576
Momordica charantia 95n
monacolins 202–3
Monascus red mould (yeast) 191, 194, 197
  ferment 280, 595n; propagation 202
Mongols
  koumiss 248, 249; tea with butter 562
monkeys, Huang Shan 245
monographs, specific foods 144–6
monosodium glutamate (MSG) 96
Montgaudry, Baron de 329n
Moraceae 18
Mori Tateyaki 136
mortar
  primitive 599
  mortar and pestle *521*, 583
  primitive 599
  oil mill 452, *453*
  starch grinding 461
Mortitz, L. A. 265n
mortuary rituals, wine vessels 155n
Mōshi Himbutsu Zukō (Illustrated Study of the Plants and Animals in the *Book of Odes*; +1785) 37
  apricot *47*; Chinese chestnut *50*; Chinese hazelnut *51*; peach *45*; plum *46*; carp *64*; turtle *65*
mostardadi Cremona 412n
mota (bean curd) 318
Mote, F. W. 138, 434n
moto (yeast culture) 282
mou (barley) 262
mould, iron for tea pressing *521*, 522
moulding tea *521*, 522, *524*, 525, 553
moulds 280
  red 191, 194
  vitamin D_2 590
  see also ferment; grain moulds
Moule, A. C. 493n, 496n, 501n
mountains, goitre association 574, 575
Mrs Cheng's Soybean Products (Honolulu, HI) 310n, 311
mu erh (wood ear) 42, 144–5, 406n, 408, *409*
mu kua (Chinese quince; papaya) *44*, 49
Mu Kung 485n

Mu Ning (+8th century) 255, sons of 302n
Mucor 6n, 8
    conditions for growth 277
    esterases 608n
    ferment 280, 592, 595n
    fu ju making 327
    microbial rennet 608
mud fish 61, 62
mug *see* stoup
Mugil 61n
muginawa (noodles) 491
Mui, H.-C. & Mui, L. C. 534n, 545n, 546n, 552n
mulberry
mulberry leaves *170*, 190, 187, *190*, 342
    source of vitamin B 581
mulberry-fall wine 180
Mulia region (West New Guinea) 573n
mullet, *chiang 381*
mung beans 190, 322, 429
    flour for noodles 488; sprouts 296–7, 298
muria *see* garum
Murraya exotica 305
mushrooms *33*, 40, 42, 91, 92, 297
    chiang preserves *409*; *tsu* 412n
    cloud-ear or wood-ear 144–5
    pilaff 126; pine (*sung chün*) 144; shitake 145
    vitamin D_2 590; monographs 144–5
musk perfume 553
mussels *409*
must *see* mash
mustard 31, 35, 334, *406*, *409*, *418*, 590n
mustard green 405–6, *406*, *409*, *418*
mutton 57, 58; mutton wine 231, 237–8
Mylopharyngodon aethiops 61n
Myrica rubra 54
myrrh 239
myxoedema 578

naeng myon (Korean cold noodle) 492
nai (crab apple) 437
nai fu (dried sliced crab apple) 424
Najor, J. 496n
Nakamura, M. & Kawabata, T. 570n
Nakatsomi Sukeshige (Japanese court official; +12th c.) 317
Nakayama Tommy 245n
Nakayama Tokiko 305n
nan or *nang* (flat bread) 468n, 472n
nan (*Phoebe nanmu*) 48n
Nan fan shao chiu fa (Burnt-wine method of Southern Tribal folk) 227
Nan Fang Tshao Mu Chuang (Flora of the Southern Region; +304) 52n, 121n
    chhü nieh 160n
    herbal *ferment* 183, 185, 232
    rice flour *ferment* 261n
Nan Pu Hsin Shu (Sung) 205n
Nan-Yüeh Chih (Record of the Southern Yüeh; c. +460) 511
Nan-Yüeh Hsing Chi (Travels in Southern Yüeh; c. –175) 54
Nanching po-wu-kuan (Eastern Han Tomb at Tung-Yang, Hsüi-I in Kiangsu Province; 1973) 80n

Nanking *tsao* eggplant 411n
nao tzu chha (camphor tea) 553
narezushi (Japan) 389n, 392, 394, 412 *see also cha*
nasopharyngeal carcinoma 605
Nathanael, W. R. N. 245n
Nature's Laws 224
necessity as driving force in intervention 601
nectar, honey 245
Needham, J. 5, 6, 8, 46n, 54n, 68n, 97n, 130n, 134n, 136n, 137n, 138n, 139n, 140n, 141n, 143n, 144n, 146n, 192n, 203n, 204n, 206n, 208, 227n, 240n, 241n, 426n, 463n, 571n, 574n, 576n, 577, 578n, 581n, 602n
Needham Research Institute, Cambridge, England 495n
Nei Yung (Internal Culinary Supervisor) 56
Nelumbo nucifera 33, 36
Nêng Kai Chai Man Lu (Random Records from the Corrigible Studio; +12th century) *127*, 128, 180n, 181n, 341
    tien hsin (dim-sum) 485n
Neolithic Age
    alcoholic drink 160–1; origins of 257, 258
    fermentation 275
    grain preparation 259
    Near East 265n
    saccharification 275
    wine 155–6, 157
Neomeris phocaenoides 63
nervous disorder treatment 234–5
nest of silken threads 460
Neumann, H. 269n
nga chhoi (sprouted vegetable) 298
Ngo Huang Tou Shêng (Swan-like Yellow Bean Sprout) 296
Ni Tsan (Yuan landscape painter) 128, 362
niang (brew, ferment) 204, 240
Nichiren Shonin (Buddhist priest; +13th century) 317
Nickel, G. B. 283n, 287, 289n
nicotinic acid 298
nieh (sprouted cereal grain; malt) 157, 158, 159–60, 163
    discovery 263
    malt sugar making 295, 457, 458
    origin 260, *273*, 274
    technology migration 263
    yeast contamination 261
Nieh Chêng (late Chou and early Han period) 59
nieh chiu (sweet wine, made from sprouted grain) 166
nieh mi (sprouted grain kernel) 260, 295
nien (roller mill) 448n
nigari (bittern) 305
night blindness 573, 586–8
nine grains 20n
Nine Provinces of Ancient China 19, *22*
Ning-chou (Kiangsi) *544*
Ninghsia province, tombs of Later Han, Wei and Chin 313n
Ningpo 435, *436*
Ninkasi (Sumerian goddess) beer brewing 269n, 271, 272
niu (ox; cattle) 55, 57
niu hsi (*Achyranthes bidentata*) 234, *236*, 237
Niu Jên (Keeper of the Cattle) 57n

niu phi (clotted cream in sheets) 256n
niu yu (butter) 358, 437n
niu ju (cow's milk) 500
no mi (glutinous rice) 204
no mi kao (pastries made of glutinous rice) 2
Noah 257
Noda, M. 383n, 398, 398n
noodles 8, 463, 480–5, *486*–7, 488, *489*, 490–1
  buckwheat flour 485
  cutting 484, 485, *486*
  dissemination in Asia 491–7
  egg 491
  filamentous 463, 466–9, *470*, 471–5
    development *595*, 596
    Europe 494–7
  flat 481; hung 483n, 484–5, *486*; round 480–1
  instant 492
  Japan 488, 492; Japanese 'soba' 488
  Korea 492
  making 2, 3, 481, 482–4
  Marco Polo, *Travels of* 467, 493–7
  rice flour 485, 488, 490, 491, 604
  yam flour 483, 485, 500
Normile, D. 396n
nü chhü (glutinous rice *ferment*) 261n, 335–6, 339, 350–1, 405
nuan (warming wine) 222
numbanhchock (rice noodles, Caambodia) 492
Nung Chêng Chhuan Shu (Complete Treatise on Agriculture; +1639) 24n, 143, *443*
nung chiu (strong wine) 204
Nung Sang Chi Yao (Fundamental of Agriculture and Sericulture; +1273) 143
Nung Sang I Shih Tsho Yao (Selected Essentials of Agriculture, Sericulture, Clothing and Food; +1314) 24n, 143
  chiang preparation 352, *353*, 357n; *shih* preparation 342
  oil seed processing 441; red *ferment* 194
  tea origins 506n, processing 529n
  vinegar recipes 290
Nung Shu (Agricultural Treatise of Wang Chên +1313) 143, 488, 529, 532, 425
  loose tea 529; oil extraction 450
nutmeg 186, 237
nutrition 14, 603–5
nutritional deficiency diseases 9, 571, 573, 604
  beriberi 578–86
  diet therapy 573
  goitre 573–8
  night blindness 573, 586–8
  rickets 588–91
Nutritionist–Physician 57n
Nutton, V. 571n
nyctalopia *see* night blindness

oak 447
objects, antecedents 601
Odyssey (Homer) 57n
Oenanthe javanica *33*, 35
Oguni, I. 570n
Ogura, S. 118n

oil 2
  animal 437; vegetable 437, 441, 456–7
  aroma 93, cakes *451*
  cooking 439; evaluation 450
  extraction process comparison with Europe and Japan 455–7
  fire hazard 437; flambeaux making 29n, 34, 437, 439
  flavour 93; food resources 66; frying 439, 441
  hair salving 439, 441; lamps 441, 450
  secondary recovery 450; silk painting 437, 439
  uses 440–1, 450, 456; yield 449, 450–2
oil of citrus 206
oil cloth 437, 439, *521*, 522
oil press
  Chinese description 441, *442*, 443–50
  olive oil 437, *438*
  ram 447; southern style 450
  tea seed oil 4
  wedge 447
  Western 437, *438*, 439, *452*, *453*
oil pressing 29, 441, 443, 456
oilseeds 28–31, 379, 436–57
  boiling 452
  cake 450, *451*
  crops 456
  grinding 441, *446*; meal 449, 455
  oil extraction 28, 29, 31, 441, *442*, 443–50
  pressing 437, 440
Ojuyeonmunjang jeonsango (c. 1850) 397
Okimuru Imiki (poet; +686–707) 374
Okukura Kakuzo 555n
Okura Nagatsune 452, *453*
Oleschki L. 493
olive oil 437
  beam press 437, *438*; press 437; pressing 455
  roller mill 439
  significance in West 456
olives, Chinese 428
Olmo, H. 258n
onion 39n, 96
  tea flavouring 561
  see also scallion; spring onion
oolong tea 8, 535–41, *551*
  catechins 566; drinking 561
  exports 541, 548; Indian production 549
  kung fu tea making 561
  processing 536, 537, 552
Oomen, H. A. P. C. 587n
Ophiocephalus argus 62
Opium War 548
Oppenheim, A. L. 269n
oracle bones, Shang
  cattle records 57
  chhü nieh 159–60
  fermentation radical 277
  inscriptions 18
  plant pictographs 32
  ritual wine (*chhang*) 156, 157n, 232
orange *see also* tangerine
  golden 54; sweet 54–5
  tea flavouring 520, 553, 561
  wine 245; in *cha* 385; honey preserved 426

GENERAL INDEX

Orange Record, The see Chü Lu
orchard 32
orchid tea 553, 554
Ortega y Gasset, J. 602n
Oryza sativa 18, 25
Oshima, K. 263n
osteoporosis 604n
Otake, S. 569n
Ou Erh Thai (10th century) 144
Ou Than-Sheng 153n
ou thien (shallow pits) 31
Ou-Yang Hsiao Thao 506n
Ouyang Hsiu (writer; +1007 to +1072) 525, 525n
Ouyang-Hsün 515n, 520n, 522n, 557n, 562n, 563n
oven 266, 399n
    bright bun baking 3
    pasta cooking 472n, 474
    tandoor 3n, 266n, 468n, 472n
Ovis aries 57
Owen, T. 398n
ox 55, 57 *see also* beef; cattle
    keeper 56
    roller mill powering 449
    sacrifical ritual 58
oyster 66
    hai pastes 379; sauce 96, 358n
Oyudononoue no nikki (Daily Log of the Yü Thang Palace) 318

pa chên (eight delicacies) 57, 72, 354
pa chhi (*Smilex china*) 235
pa kung shan tou fu (*tou fu* of the Mount of Eight Venerables) 299, 305n
Pa Ling Hsien Chi (Gazetteer of the Pa-Ling County; Hunan, 1872) 542, 548, 548n
Pa-shih-tang (Hunan) 27n
Pacey, A. 597n
Pachomius, St 401
paeonia 580
pai chha (white tea) 550
pai chhü (white *ferment*) 290
pai chi (*Bletilla striata*) 589
pai chiu (mature wine) 165–6; (white wine) 191, 203
Pai Chü-I (Thang poet; +772 to +846) 36n, 160n, 180, 203, 205
    dairy products 257n; gift of ice 434
    grape wine 243n; *tou fu* 301
    wine warming 222
pai fu tzu (sweet cassava) 186
pai ho mien (lily pasta) 482
pai hsin (sturgeon) 62
pai hu ma (white sesame) 439n
pai i (white coat) 290
pai ju (white milk) 561
Pai Kêng (white soup) 84
pai lao chhü (moderate *ferment*) 169, 170, 174
pai mien (raised pasta) 470n
pai pi (bean curd) 318
pai ping (raised dough) 471
pai pu (white pellicles) 289–90
pai su (white perilla) 439n
pai tou fu kan (plain dried *tou fu*) 320

pai tshai (Chinese cabbage) 31, 35
pai tuo chhü (special cake *ferment*) 169, 170, 174, 180
pai yang wine 236
pai yeh (pressed bean curd) 320, 324
Pai-lin (Fukien) 544, 546n
Palermo (Sicily) 494
pan, for stir-frying oilseeds 448–9
Pan Ku (author of History of the Former Han) 233
Pan-pho (Sian, Shensi) 49, 76, 78, 79, 156, 276
Panati, C. 270n, 434n
Panax ginseng 196
pang (oyster) 66
Panicum miliaceum 18, 25
Panjabi, C. 496n
pao ching (treasure wells) 432
Pao Phu Tzu (Book of the Preservation-Solidarity Master; c. +320) 207n
pao-tzu (pork filled pasta; little wrap-around) 471n, 476
    leaven 477
Paonin tshu (Pao-ning vinegar) 291
papaya 49
    wine 245n
paper sizing 461n
paper-making transmission 496
Pappas, L. E. 270n
parching 84–5, 88, 89
parsley-seed 239
Partington, J. R. 258n, 267n, 269n, 272n, 283
pasta 463, 466–9, 468, 470, 471–5 *see also* dough
    cooking 471–4, 472, 474, 599n
    development 596, 598, 602, 604
    Italy 493–7, 495; Marco Polo legend 467, 494–7
    leaven 470, 471, 476; made with wine 468, 469
    making 469, 470, 471–5, 477
    raised 469, 470, 471
    sheets 490–1
    types 468, 469
pasteurisation of wine 187, 222, 225–6
pastoral nomad view of Han foods 315
pastry 467
    steamed 71
    sweet 473n
Pauling, L. 579n
peach 44, 45, 49
    goat's 52n; vinegar 285; wine 259n
peanuts
    oil 456–7; salting 418
pear 44, 49–50
    balsam 95n; Chinese sand 49, 50; white 50
    chiang preserves 409, 412; pickled 406n, 428; *tsu* 406
    storage 427, 429
    wine 245
peas 40, 297n
pectin hydrolysis 593
pectinase 593, 608
Pediococcus halophilus 378n
pei (drying process, for tea) 527, 531
    pei (heating ditch; furnace) 520n, 521, 522, 523, 529, 530
    pei (heat) 530; *pei* (oven) 532–3; *pei* (oven drying) 539
    pei chha (roasting tea) 524, 526

Pei Chhi Shu (History of the Northern Chhi Dynasty; +550–577 and +636) 240, 433n
pei ching khao ya (Peking duck) 89
Pei Hu Lu (Records of the Northern Gate; +875) 475, 478
pei lou (bamboo tray) 527
Pei Shan Chiu Ching (Wine Canon of North Hill; +1117) 132, 160n, 184n
  cooking wine 207n, 225; steaming wine 206
  fermentation process 186–8, 189, 191, 192, 193, 281
  fire-pressured wine 206, 222
  herbs in *ferment* 183
  mash pressing 220
  medicated wine 237, 238, 243
  origin of wine 161
  rice wine with grape juice 243
  yeast use 282; leaven 606
Pei Yuan Pieh Lu (Pei-yuan Tea Record; +1186) *516*, 517, 523
  drying of tea cakes 526
  loose tea 529
  small bud tea 525n
Pekochelys bibroni 65
pellicles in vinegar making, white 286, 289–90
Pelliot, P. 493n, 496n, 501n, 509n
Pelteobargurs fulvidraco 61n
pên chhü (common ferment) 169, *170*, 171–4
Pên Hsin Tsai Shu Shih Phu (Vegetarian Recipes from the Pure Heart Studio; Sung) *127*, 128
  tou fu 301
Pên Tshao catalogue of Lung Po-Chien 121
Pên Tshao Kang Mu (Great Pharmacopoeia; +1596) 16n, 24n, 38, 39n, 52n, 141
  beriberi 581n, 582; eye diseases 586; goitre 576, 577
  chhü (ferment) 158; *nieh* 158; leaven 606; red *ferment* 196, 200–1, 202
  cumquat 54n; dried jujubes 424n; dried litchi 425n; loquat 54n
  distilled wine 204, 230, 231n, 244n
  gluten 498; malt sugar 460; mead 247
  medicaments 134; rehmannia 499n
  medicated wine 235, 237, 238
  milk fermentation 248n, 254, 255n
  soybeans 296; soy sauce making 363; *shih* use 341n; *shih chih* making 361n; yellow coated bean 342
  tea 512, 564
  tou fu making 299, 303; curdling agents 315
  vinegar 290; bitter wine 284n
Pên Tshao Kang Mu Shi I (Supplemental Amplification of the *Pên Tshao Kang Mu*; +1765) 323, 536n, 540, 564
  Wu-I tea 540; *Phu-erh* tea 553n
Pen Tshao Phin Hui Ching Yao (Essentials of the Pharmacopoeia Ranked According to Nature and Efficacy; +1505) 141, 143
  chiang preparation 353, 357
  grape wine 240n
  malt sugar 460
  pasta 472n
  shih recipe 340, 341; *shih chih* making 360, 361
  soybeans 296; soy milk 322n; bean curd skin 320, 323
  tea 564

Pên Tshao Shih I (Supplement to the Natural Histories; +725) 226, 512, 564, 581, 606
Pen Tshao Thu Ching (Illustrated Pharmacopoeia; +1061) 296, 297n
Pên Tshao Yen I (Dilations of the Pharmacopoeia; +1116) 300, 301
Pên Wei (Root of Taste, *LSCC*) 17n, 40n, 52n, 54n, 120, 380n
penicillin discovery 600n
Penicillium 280
pepper *170*, 186, 235, 361, 582
  black 52, 96, 235, 257n
  Szechuan 95n, 96, 189, 562
peppermint as tea flavouring 520, 553, 561
Perdono, G. de 495n
perfume, steam distillation 214
Perilla frutescens 439n
Perilla ocymoides 439
perilla seed 439
  oil 440, 450, 452, 456
persimmon *44*, 50–1
  dried 425; wine 245n
pestle and mortar *521*, 583
  oil mill 452, *454*
petroleum 358
phan (serving dish; dining plate) 100–1, 105, 113
Phan Chên-Chung 478n
Phan Chi-Hsing 444n, 447n, 451n, 461n
Phan Chih-Heng (16th century) 144
phang phi (*Lindera strychnifolia*) seed oil 441
phao (clay-wrapped baking) 73, 85, *88*, 89, 603n
phao (steeping) 88
phao chha (making tea) 560
Phao Jên (Keeper of Victuals) 28, 57n, 59
phao tshai (pickled vegetable) of Szechuan 411
pharmacopoeias 16, 141, 143
  see also individual titles
Phaseolus calcaratus 33
  see also Vigna calcaratus
pheasant 59, 60, 61, *381*
phei (millet) 24
phei (muddy brew) 181
Phei An-Phing 27n
Phei-li-kang culture (–6th to –7th millennia) 48, 156, 260, 599
phêng (boiling) 70, 71, *88*
phêng (rack) *521*
Phêng Jên (Grand Chef) 71, 416
Phêng Jên Shih Hua (Talks on the History of Cuisine) 148, 485
Phêng Khuei (Cooking Mallows) 36n
Phêng Lung Yeh Hua (Night Discourses from *Phêng Lung*; +1565) 245, 325
phêng thu (prepare tea) 508
Phêng-hsien (Szechuan) Han brick mural 112, 217–21
Phêng-lai (Shantung) 554n
Phêng-thou-shan (Hunan) 27n
phi li (bamboo tray) 520n, *521*
phi-pha (loquat) 53–4
phiao (gourd dish) 103
phien chha (cake tea) 527, 552

Phien Kho Yü Hsien Chi (Notes from Moments of Leisure; c. +1753) 542
Philippines
  angkak 396n
  fermented salted fish 395; fish sauce 392
  tou fu 328
Phin Chha Yao Lu (Elite Teas Compendium; +1075) 516, 517
phing (bottle-like kettle) 557
Phing-yang prefecture 244
Phithakpol, B. 392n, 396n
Phoebe nanmu 48n
phosphate deficiency 588, 590
phou thou (fried pasta) 470n, 471
phu (bamboo rod) 521; *phu* (cattail) 33, 35
phu (vegetable garden; orchard) 32
Phu Chi Fang (Practical Prescriptions for Everyman; +1418) 588
phu hsiao (crude solve) 426
phu tea 552n said to be prototype of phu-erh
phu thao (grape) 240
phu thao chiu (grape wine) 243
phu-erh tea 535n, 552n, 553, 562
Phyllostachys 33, 36
pi (grid) 521
pi (spoon; ladle) 85, 105–6; *pi* (two-pronged fork) 107
pi ba (black pepper) 257n
Pi Shu Lu Hua (Notes from a Summer Retreat) 247n
Pichia yeast 280
picking tea 524–5, 529, 539
pickles 70, 74, 284n, 333, 594
  chhi 379
  fish 3, 8, 392, 393, 394, 397, 401n, 593, 594
  meat 3, 8, 47, 69, 76, 96, 379, 594
  pear 406, 428
  vegetables 2, 8, 74n, 76, 379–80, 390, 402–12, 414, 416–17
pieh (turtle) 64, 66
Pieh Jên (Shellfish Keeper) 66
pien (bamboo bowl) 100
pien (to burn husked rice to obtain the kernel) 265n
pien (ritual baskets) 468
Pien Chhüeh (physician of Spring and Autumn period) 234n
Pien Min Thu Tsuan (Everyman's Handy Illustrated Compendium; +1502) 129, 130, 143
  chiang preservation 414n
  fruit storage 428, 429
  honeyed and sugared fruits 426n
  red *ferment* 196
  tsao-preserved crab 413
pien shih (wonton) 478n
Pierson, J. L. 374n
pig 55, 58, 59, 60, 61 *see also* pork
  ham 422
  kidney 582; liver 581, 582, 588; stomach 582
  suckling 439
pin chhü (cake, brick ferment) 167
pine needles, to store oranges 429
pineapple wine 245n
ping (cake; biscuit; pastry) 467, 468

ping (dough products; pasta) 466–8, 470, 482, 596, 599n, 602
ping (noodles) 490
ping (wine medicament) 183
ping chha (cake tea) 522, 552
ping chiao (leaven pasta) 470, 471, 606n
ping ching (ice well) 432
Ping Fu (Ode to Pasta; +3rd century) 469, 475, 478
ping lo (iced yoghurt) 434
ping shih (ice chamber) 432
ping su (iced cream or butter) 434
52 *Ping-fang see Wu Shih Erh Ping Fang*
Pinkerton, J. 249n
Piper betle 52n
Piper nigrum 52
Pirazzoli-t'Serstevens, M. 86n
Pires-Bianchi, M. de L. 294n, 322n
Pisum sativum 40, 297n
pits, shallow 31
plantain seeds 581
plastics industry 378n
plate 99–100, 113
platform for tea processing 521, 522, 558
Platt, B. S. 293n, 298
Platycodon grandiflorum 235
Pliny the Elder (author, *Natural History*; +1st century) 168–9, 398n, 400, 577
  leaven making 471n
  medicated wines 239
  oil presses 437, 438, 439, 456n
plum, Chinese (Japanese) 44, 45–6, 428
  dried 425; sugared 426
  juice drink 97; wine 259n
  salting 418; sauce 96
  tea flavouring 553, 561
po (parching) 88 *see ao*
po chi (bamboo winnowing tray) 473n
po lo mêng chhing kao mien (steamed Brahmin light wheat cake) 126
po pai phi (white cypress bark) 587
po ping (rice pancakes) 474n
po tho (gluten) 499–500
po tho (thin drawn pasta) 470, 472, 491, 499
Po Wu Chi (Records of the Investigation of Things; +290) 233, 240n, 295, 509, 510, 574
  hemp seed oil 437
poaching 70, 74, 82, 88, 89
poison 451n
Polygonatum odoratum 234n
Polygonum 190
Polygonum hydropiper 33, 38
polyphenol oxidase 566, 567
polyphenols in tea 565–6, 568, 570
pomegranate 235, 428
pomelo 54–5
Pompey 437, 439
Poneros, A. G. 605n
pongee bag 461
Popper, K. 601n
poppy seed 322
porcelain cups 223

pork 58 *see also* pig
   cha 385, 386, 409; *chhi* pickles 379; cured 422
   salted 8; sausages 423; *sung* 423–4
   tsao preserves 409
Porkert, M. 571n
porridge 265, 266, fermenting 266
Porter-Smith, F. & Stuart, G. M. 39n, 49n, 52n, 54n, 234n, 296n, 337n, 354n, 575n
pot-roasting 84, *88*, 89
potato starch 462, 492
pots *see also* jar; vessels
   earthenware 259
   malt sugar making 458
pottery wares 76
   antecedents 601; evolution 602; invention 266, 603
   sealing jars with oil 440
   serving vessels *100*
   well collar 431, 432
   wine and beer storage 151, *152*, 153, 275
pounding tea 520, *521*, 522, 525
Povey Manuscript (Thomas Povey; +1686) 565n
Powell, M. 266n, 269n
prayers & incantations in fermentation process 171n
Prefect of Mare Milkers 248
Prescott & Dunn 249n
preservative 595
   red *ferment* 202
press *see also* oil press; wedge press
   beam 437, *438*, 456
   for bean curds 2
   mash 220
   screw 456
pressing *see also* oilseeds, pressing
   tea 520, *521*, 522, *524*, 525
prince's feather 417
protease 383n, 398
   ferment, in 334
   inhibitors 322, 345n, 602n
   suppression by soy proteins 293
protein
   coagulation 253
   deficiency 604
   hydrolysis 351, 593
   intake 604; meat 334
   wheat 465; gluten 605
proteinase 593, 608
Prunus armeniaca 44, 48
Prunus mume 44, 46
Prunus persica 44, *44*
Prunus pseudocerasus 44, 51
Prunus salicina 44, 45
Psephurus gladius 62
Pseudobagrus fulvidraco 62
pu (malt sugar) 458, 459
pu tho (dough, not held by hand) 499n
pu tho (pasta) 481
Pu Yang Fang (version of *Shih Liao Pên Tshao*) 136
puffer fish 145
Pulleyblank, E. G. 248n, 252n, 253n, 254
puls (paste or porridge) 265
pungent taste 91, 92–3
Purdue University 14n

purselane 581n, 582
Putro, S. 392n
Pyrus 49
Pyrus bretschneideri 44, 50
Pyrus cathayensis 49n
Pyrus communis 50
Pyrus pyrifolia 49, 50

Qian Hao, Chen Heyi & Ru Suichu 106n, 153n, 430n, *431*
quail 60
quern 463, 598
   antecedents 601
   rotary *303–4*, *464–5*, 600, 603
   development 597, *598*, 599
   oilseed extraction *446*, 448, 449
   saddle 463, *598*, 599
Quiller-Couch, A. 150n
quince, Chinese 44, 49, 426

rabbit 60, 61, *95*, 96, 379, *381*
rack for drying tea cakes *521*, 522
Rackham, H. 471n
Rackis, J. L., Gumbmann, M. R. & Liener, I. E. 293n
radish 32n, *33*, 35, 43, *406*, *409*, 411, *418*, 429, 582
Radix ledebouriellae 589
raffinose 293, 322n
rai (Japanese for *li*, sweet liquor or wine) 166, 263
rain garments 440
raisin tree 44, 49
ram, for oil press 447, 449, 450
Ramses II (Egyptian pharaoh) 455n
Ramusio, G. B. 493
rangaku (pharmaceutical extractor-still) 214–15
rape seed oil 358, 447, 448–9, 450
   extraction 452, *453*; importance 456; yield 451
rape-turnip seed 441
Raphanus sativus 33, 35
Rawson, J. 218n, *219*, 220, 221n, *465*
Read, B., Lee Wei-Ying 590n & Chhêng Jih-Kuang 590n
Read, B. E. 39n, 337n, 581n, 604n
red-cooking 414
Reed, G. 608n
reeds 526
   miscanthus 335, 336
   over-mats 109
refrigeration 429
rehmannia 257n, 499n; wine *236*, 237
rennet 257, 328, 329; microbial 608n
retort 227, *228*
Rheum officinale 235
Rhinogobius giurnius 62
Rhizopus 6n, 8, 167, 173, 194
   conditions for growth 277
   ferment 280, 592, 595n
Rhizopus oligosporus 345
riboflavin 298
rice 18, 25
   aleurone layer 585; polished 581, 582, 583, 584, 605, 606
   beriberi association 578, 579, 581, 582; bran 581, 585, 586, 606, 607

bowl 99, 101; eating 104–5
candy (Japanese *amè*) 263n
congee 260, 477, 585, 606; with purselane 582
cooked i.e. steamed (*fan*); cooking 260n; digestion 458
domestication 27, 166n, 396n; cultivation 593; wild 20n
ferment 261–2
    aboriginal tribes 161–2, 183, 190, 351, 603
    red, making 195–202; colour changes during 198
five staple grains *21*, 19, *20*, 22, 23; founder crop 263
in fruit storage 429; microbes on surface of grain 429n
glutinous *170*, 174, 178, 182, 183, 185
    distilled wine 204; mead making 247–8; vinegar 291
hulled 429n; husks 260; stalks 190
kêng 195; thickening 83, 93, 95; crumbs for thickening 93, 461
koji (Jaapanese *chhü*) 377
malt sugar preparation 457–8, 460
medicated wines 234
moulded for *miso* 375
noodles 488, 490; dumpling 119
roasting 265n; serving 99, 101
sheets to make noodles and pastry 488, 491
steeped in soured water 252, 471; sour steamed 187
sprouted 163n, 166, 262n, 263, 601
steamed 67–8, *164*, 166, 260, 265n, 274n, 278
    moulded by accident 166–7
tea congee 562
vinegar 189, 194, 290
vitamin B$_1$ 585
washing in making red *ferment* 197, *198*, 201
wine 178, 182, 187n
    making 161, 168, 182, 259, 282
yeast leaven 184, 188, 606n
see also ferment; starch substrate; steamer
rice flour 183, 261n
    batter 474n; dough 488; *erh* 468; cake 467; noodles 485, 491; pasta *470*, 471, 472n, 473, 474
    sifting for pasta making 474
    tzu 468; *yum cha* (tea brunch) 485, 488
rickets 573, 588–91
    cultural causes 590–1; West 589
Rieu, E. V. 266n
rishta (Islamic pasta recipe) 494, 495, 496
Ritter Dr. 329n
ritual sacrifice 605; in *kêng* 83
rituals
    ablution of fingers 101
    bronze vessels 76n
    mortuary 155n
    ox 58, sheep meat 57, 58; chicken 60
    see also ceremonial offerings; sacrificial offerings; slaughter
Rizzi, S. 493n
roach 61, *62*
Roaf, M. 258n
roasting 70, 73, 85, *88*, 89
    meat on a skewer *70*, 73, 85–6, *88*, 89
    oil seed *455*
    pasta 599n; tea *524*, 526–7
Robinson College (Cambridge) 329n

rocambole 39, 40, *41*, 92
Rockhill, W. 249n
Roden, C. 496n
Rodinson, M. 495n
roebuck, *ni* pastes 379
rolling tea *524*, 525, 529, 537, 552
    by foot *535*, 543
    by hand 532, 533, 534n, 546, 549
Rombauer, I. & Becker, M. R. 39n, 407n, 473n
Rome
    fermented fish sauce 380, 383, 398–402
    humoral theory 571
    olive oil use 456
    rennet 257n
    rotary quern 599n
    starch preparation 462
Root, W. 467n, 495
rope *451*
    bamboo *449*, 450; rattan 449; tree bark 522
Rosa lavigata 451n
rose, cherokee 451n; tea 553, 554
Rose, A. H. 153n
rose-water 206
Rosenberger, B. 494n, 496
Rottauwe, H. W. 596n
rudd 61, *62*
Ruddle, K. 389n, 392n, 397n
rue 239
Rumphius, G. E. (Dutch botanist; 18th century) 345
rush 507n, 338–9
rush wool 507n
Russell (+1755) 577
Russia, bread beer 270n
Rustichello of Pisa 493
rye 169n, gluten 466

Sabban, F. 10, 123n, 124n, 138n, 147, 252n, 254, 255n, 412n, 457n, 465n, 469n, 482n, 484n, 496n
saccharification 5, 154, 606
    agent 9, 157, 160n, 459
    beer making 270, 271; malt sugar production 457
    discovery 603; Neolithic Age 275
    temperature 282n; vinegar making 287, 288
    West and East 282
saccharifying activity 188, 260
Saccharomyces 254n, 280, 282, 552, 592
Saccharomyces cerevisiae 153n
Saccharomyces rouxii 378n
sacrificial offerings 49, 430
    poached meat 74n
    wine 156, 164, 263
    see also ceremonial offerings; rituals; slaughter, ritual
saffron 239
sago palm 496
Sahi, T. 596n
sai han (protrusion on forehead) 589
sai thien (depression on forehead) 589
Sakaguchi Kinichiro 362n, 377n
saké (Japanese rice wine) 167, 263, 279, 282
salads 407
saliva, amylase 154
salsamentum (Greek and Roman relish) 380, 383, 398

salt 2, 95–6, 390n, 396, 400
  addition to *chiang* 348, 352, 355
  autolysis 390, 396
  fermented fish 383, 385; *tsu* 405–8, *406*; in *fu* 421; incubation *385*, 408, 422
  leaven, used with 477
  making *chu i chiang* 383; *garum* 399; *hai* 380
  shih preparation 361; soy sauce making 371
  tea flavouring 519n, 556n
  yen tshai 412
salting 415 *see also* brine
  fish 605
  meat and fish preservation 602n
  vegetables 416–17, *418*, 419
salty taste 91, 95–6
samna (Middle-Eastern cookery) 254
Sampson, T. 242n
samshu distillation equipment 214
Samuel beer recipe 271, 272
Samuel, D. beer recipe 269n, 270–72, 278
san (rice crumbs) 93
san chha (loose tea) 527, 529n
San Fu Chüeh Lu (Considered Account of the Three Metropolitan Cities; c. +153) 337n
san huang (primordial sovereigns of Chinese prehistory) 503
San Kuo Chih (History of the Three Kingdoms; c. +290) 337n, 509; *mi chien* 426
San Tshai Thu Hui (Universal Encyclopedia; +1609) 204n
sand, fruit storage 428, 429
sandalwood 445, 447
sang (starch substrate) 385
Sano, M. 570n
Sanxindui (Kuang-han, Szechuan) 513n
sao (dog fat flavour) 93
Sapium sebiferum 441
sarano-lao (wine, Taiwan) 162
Sargassum siliquastrum 575n
Sargent, C. B. R. (Anglican Bishop of Fukien) 1
Saris, Captain John (+1613) 329n
Sasagawa, R. & Adachi, I. 377n
Sasaki, M. 605n
Sasson, J. A. 257n
saturation, with sugar 425–6
sauce 333
  soy, fresh drawn and mature drawn 366
  soy, harvested in autumn 371, 373
  ancient meat 333; fish 397–8
  savoury Japanese 374
sausages 2, 423
savoury flavour *95*
Sawers, D. 597n
scallion 301, 582 *see also* spring onion
  tea flavouring 520, 555, 561
Schafer, E. H. 64n, 241, 242n, 248n, 252n, 253n, 254, 255n, 257n, 400n
Schroeder, C. A. 52n
Scottish and Newcastle Brewery (Newcastle) 271
seafood monographs 145–6
seasonings 91–6
  in ancient and modern Chinese cookery 95–6

seaweed 406, *409,* 412; in goitre 575–8
seeds, vitamin B_1 source 581
Seibe Wake *15*
Seiyu Roku (On Oil Manufacturing; 1836) 452
Sekine, S. 374n, 375n
selinium 574
semolina 465n
sen mee (rice noodles, Thailand) 492
Seneca (Roman writer of +1st century) 400
Senff, E. 329n
senna, sickle 587
Serat Centini (1815) 345
serving
  dishes 100–1
  implements 104–7, *108*, 109
  vessels 98–101
sesame seeds 29–31, 96
  bun 490n
  puree 473
  tea congee 562; tea flavouring 519n, 562
  to store oranges 429, 429n
sesame oil 2, 189, 440, 441
  pan-frying cakes 439
  production 456; uses 450; yield 451
  oil extraction 441, 443, *444, 445, 446,* 452
  steaming 441, *444*
  stir-frying 441, 447, 448–9
Sesamum orientale 29
Setaria italica 18, 23, 24n, 25
seven necessities of life 354, 436–7, 527
sha (kill; digest) 174
sha chhing (heat inactivation of tea leaves) *537*
sha li (Chinese sand pear) 49, 50
sha ti (west) 170n
Sha-hsien (south of Wu-I region) 541
Sha-wo-li Neolithic site 599n
shad 434n
shai chhing (wilting tea leaves in the sun) 539
Shakespeare, W. 150
shallot, Chinese *33*, 39, 40, *41*, 92, *406*
shan (viands) 80n, 98
shan (lamb tallow flavour) 93
shan cha (Chinese hawthorn) 49
Shan Chia Chhing Kung (Basic Needs for Rustic Living; Southern Sung) *127*, 128, 132, 189, 417n, 472n, 482
  pu tho remedy for worms 499
  soy condiments *372*; soy paste 354; soy sauce 358, *595*
  soybean sprouts 296; tea 561, 564; *tou fu* 301
shan fan (mountain alum) 303, 305
Shan Fu (Grand Chef; Chief Steward) 20n, 60n, 115
Shan Fu Ching Shou Lu (Chef's Handbook; c. +857) 126, 481
Shan Fu Lu (Chef's Manual; Southern Sung) 126–7, 334n
Shan Hai Ching (Classic of Mountains and Seas; Chou to Western Han) 574
Shan prefecture 360
shan yao mien (Chinese yam noodle) 483n
shan yen (mountain salt) 305n
Shan-cha-kao (Chinese hawthorn jelly) 49n
Shang Chih-Chün 136, 460n, 575n, 586n
shang han (inducing sweat) 543

Shang Lin Fu (Ode to the Imperial Forest; −2nd century) 53, 54
Shang Ping-Ho 468n
Shanghai Museum, bronze still vessel 209, *211*, 212–14
Shangkuan I (writer; early Thang) 575n
Shansi
  buckwheat noodles 488
  distilleries 217; *shih ping* (persimmon cake) 425n
Shansi shêng wên-kuan-hui (bone forks 1959, 1960) *108*
Shantung 190, 244n, 245
  burial sites, Ta-wên-khou 255
  tombs of Later Han, Wei and Chin 313n
  fish sauce 392, 395
Shantung shêng wen-wu khao-ku yen-chiu so (Ta-wên-khou Culture from Ta-chu-chhun, Chü-hsien County; 1991) 155n
Shantung ta-hsüeh li-shih hsi (Test Excavations at the Ta-hsin-chuang Site, Chi-nan, Fall, 1984; 1995) 151n
shao (burnt; burn) 205, 222
shao chhu (burnt vinegar) 207n
shao chiu (distilled wine; burnt wine) 191, 203–5, 220
  development & commercialisation 226–31
  meaning of term 221–3
  Han potent wine 221
  Thang poets 222–4
  Sung period 225–6
shao fan (small distillery) 220
Shao Hsing wine of Chekiang 163, *164*, 179n
  alcohol content 279; process 163, 164, 183, 261n, 281–2
shao ping (roast pasta; sesame bun) 470n, 471, 474, 490n
Sharpe, A. 495n
shashlik 85
shê ma (hops) 185
sheep 55, 57–8, 60, 61 *see also* mutton
  keeper 56
  milking 251n
  tallow 440
Shellfish Keeper 66
shen (plant) 169n
Shên An Lao Jên (writer; +13th century) *516*
shên chhü (superior *ferment*) 169, 170–1, *172*, 174
  chiang preparation 348
  wine preparation 174–6
Shên Hsien Chuan (Lives of Holy Immortals; +4th century) 221
shên hsü (deficiency in the kidney) 585
Shên Kua (writer, scientist, 11th century) 441, 500
Shên Li-Lung (author of *Shih Wu Pên Tshao Hui Tsuan*) 139
Shên Nung (legendary emperor; Heavenly Husbandman) 5, 15, 134, 503, 506, 571
Shên Nung Pên Tshao Ching (Pharmacopoeia of the Heavenly Husbandman; +2nd century) 31, 38, 52, 141, 337n, 508
  eye diseases 586; goitre 575
  grape wine 240; honey 246
  medicaments 134; mushrooms 42; shellfish 66
  Shên Nung tea legend 506n
  soybeans 295; bean sprouts *595*
  wine as drug vehicle 233

Shên Nung Shih Ching (Food Canon of the Heavenly Husbandman) 121n, 124, 506n
Shên Shih Fang (Prescriptions of Master Shên, monk; +5th century) 576, 576n, 579–80
Shên Thao 104n
shêng chhou (fresh drawn sauce) 366
shêng huang chiang (raw yellow soy paste) 352
Shêng Min (Origin of the People; poem) 67, 68
Shêng Shan (raw meat strips) preparation 69n
Shensi sheng po-wu-kuan (Stone Engravings of the Eastern Han Period in Northern Shênsi; 1958) 109n, 110
Shensi
  buckwheat noodles 488
  fruit storage 429; *shih ping* (persimmon cake) 425n
Shensi shêng wên-wu-kuan (Preliminary Report on the Ruins of the Chhin Capital, Li-yang; 1966) 463n, *464*
Shensi shêng Yung-chhêng khao-ku tui (Ice House of the Chhin Kingdom, in Fêng-hsiang, Shensi; 1978) 430n
shepherd's purse 587
shi pei (insufficiently dried) 527
Shiba Yoshinobu 440n, 441n
shih (to eat) 294
shih (fermented soybeans) 4, 335, 336–42, 388–9, 391, 397
  condiment use 371, *372*, 373
  in *fu* 421; *hishio* 374
  importance 371, 373; process 376
  salty and bland 338–9, 361
  sauce from 358, 359, 362–3
  Thang recipe 361
shih (food) 98, 104
shih (persimmon) *44*, 50–1
shih (pig) 55
shih (reed over-mats) 109
shih (steamed rice or millet) 80, 83
Shih Chao-Phêng 552n
Shih Chêng Lu (Menu of Delectables; Northern Sung) 126–7
Shih Chhi (by Yen Tshan; Sung) 402n
shih chhing (clarified fermented soybeans) 359, 361, 371, *372*
Shih Chhü Chhi (Four Bend Brooks) 540
shih chi (foods to be avoided) 509
Shih Chi (Historical Records; c. −90) 103, 104, 141, 234n
  beriberi 579
  grape wine 240
  lo 250; pastoral nomad view of Han foods 315
  salt consumption 305n
  shih (fermented soybeans) 336–7
  soybean 294, 294n
  trays 112; wine 150–1, 159n
Shih Chi (magistrate of Chhing Yang) 300
Shih Chi Chuan (by Chu Hsi; Sung) 402n
shih chih (clarified fermented soybeans) 359–61, 371–3, 408
Shih Chih (diet therapy) 136
Shih Ching (Book of Odes; −11th to −7th century) 17n, 18, 19, 23, 24n, 96n, 116–18, 121
  apricots 48; Chinese apricots 48; Chinese date 48; Chinese gooseberry 52; fruit 43; peach 44, 45; pear 49; plum 45; raisin tree 49; grape 240n

Shih Ching (cont.)
  bamboo 36; dining furniture 109, 111; bed 113
  bitter spices 91-2
  chicken 60; goose 60; horse 55; ox 57; sheep 57; pig 58; dog 58, 59; game mammals and birds 60
  clan feast celebration 333
  commentaries & translations 118
  cooking methods 67, 70, 71, 73, 74; cooking pots 76n; cauldrons 98; stove 80; fishing 61, 62; turtle 64
  food preparation 67; *fu* 419; *hai* and *than* 380, 390; *tsu* 402, 416
  ice house 429
  pickle 333; sauce 333; soup 82
  serving vessels 99n, 100; spoon 85
  sunflower (wrong identification for the mallow) 38n
  sweeteners 92; malt sugar 158, 457
  tea 507, 513
  vegetables 32, 33; turnip 31; *thu* 507, 508
  wine 149n, 150, 156; *chhang* 157; medicated 232 importance 162, 163; wine serving vessels 102
Shih Ching Pai Shu (+1695) 32n
Shih Ching (Food Canon; c. +600) of Hsieh Fêng 126, 254-5, 302, pasta 481
Shih Ching (Food Canons or Classics) 147
Shih Hsien Hung Mi (Guide to the Mysteries of Cuisine; c. +1680) *129*, 130, 131, 290, 355, 357n, 373, 387, 414, *418*, 422-3, 476,
  dairy products *256*; soured milk 255-6
  food preservation methods 409-10, 412n
  fruit storage 428; honeyed and sugared fruits 426n
  shih stewed with *tou-fu-kan* 341
  soy condiments *372*; soy sauce 373
  tou fu 310, 324, 325, 326
  tsao-preserved crab 413
  vinegar recipes 290
shih hsin (heart of *shih*) 340n
shih hsin chu (heart of fermented bean congee) 341
Shih Hung-Pao (writer; 19th century) 540n, 541
Shih I (Nutritionist-Physician; Grand Dietician) 57n, 59n, 60n, 137, 334, 571
Shih I Hsin Chien (Candid Views of a Nutritionist-Physician; late Thang) *135*, 137, 334n
  noodles 481; wonton 478
Shih Jên (keeper of pigs) 56n
shih ju (tea infusion, stony milk) 561
Shih Liao (Diet Therapy tradition) 120, 121, 571
Shih Liao Fang (Diet Therapy Prescriptions) 138
Shih Liao Pên Tshao (Pandects of Diet Therapy; +670) 52, 116, *135*, 136, 141, 417
  dairy products *256*; fermented milk 253, 255, 302
  goitre 575
  malt sugar 460; mead 246; medicated wines 237
  mung bean sprouts 298
  shih use 341, *shih chih* making 360
  soybeans 296
  tea 512, 519, 520, 555, 561, 564
  vinegar 290

Shih Lin Kuang Chi (Record of Miscellanies; +1280) *127*, 128, 130, 189, 353, 355, 357n, 386-7, 411
  gluten 500, 501; *pu tho* 499-500
  soy condiments *372*
  vinegar recipes 290, 291n
shih liu (pomegranates) 428
shih liu kuan (pomegranate flask) 207
Shih Liu Thang Phin (Sixteen Tea Decoctions; +900) *516*, 517
shih mi (stone or wild honey) 246
Shih Ming (Expositor of Names; +2nd century) 113, 140, 300
  cha 384, 407, 408
  ferment types 167
  fu 302; *hai* 334; *lo* 250
  i (malt sugar) 457-8
  noodles (filamentous pasta) 480, 484, *595*
  ping (pasta), varieties of 468, 472n
  seed pounding 437
  shih (fermented soybeans) 337; *shih chih* 360
  steamed pasta (*man-thou*) 475
  tsu 404, 416
  wine making 158
Shih Pên (Book of Origins; -2nd century) 155, 156
  wine 161
Shih Pên Pa Chung (Eight Types of Origins) 55n
Shih Phu (List of Victuals; c. +710) 126
  dairy products *256*
  wonton 478
Shih Phu (recipe books) 122
shih ping (persimmon cake) 425
Shih Shêng-Han (1907-71) 5, 6, 9, 29n, 31n, 34n, 39n, 40n, 43n, 75, 120, 123, 124, 125, 154n, 159n, 171n, 251n, 252n, 285n, 286n, 289n, 293n, 338n, 427n, 429n, 437n, 458n, 461n
Shih Shih Wu Kuan (Five Aspects of Nutrition; +1090) *135*, 137, 582
Shih Shuo Hsin Yü (New Discourse on the Talk of the Times; c. +440) 72n, 337n
  Ho Yen story 469n
  shih chih 360
  tea 510
Shih Thao (painter; c. +1697) 42
shih tien (food shops) 481-2
Shih Wu Chi Yüan (The Origin of Things; c. +1080) 255n
Shih Wu Kan Chu (Notes on Miscellaneous Affairs; Ming) 498
Shih Wu Pên Tshao Hui Tsuan (New Compilation of the Natural History of Foods; +1691) *135*, 139
Shih Wu Pên Tshao (Natural History of Foods; +1000) *135*, 139
Shih Wu Pên Tshao (Natural History of Foods; +1641) 121-2, *135*, 139
Shih Yeh Thien I 507n
shih ying (solid swellings) 576
shih yu (petroleum) 358
shih yu (soy sauce, i.e. sauce from *shih*) 4, 335, 358, 371, *372*
shih yü (shad) 434n
Shih Yüeh Chih Chiao (Conjunction in the Tenth Month) 117
Shillinglaw, C. A. 378n

Shinodas Osamu (1899–1978) 9, 19n, 20n, 68n, 119n, 121n, 122, 125, 126n, 128, 136n, 147, 154n, 160n, 166n, 180n, 181n, 189n, 192n, 193n, 196n, 204n, 222n, 225n, 243n, 248n, 291n, 300n, 317, 318, 319n, 353n, 354n, 355n, 357n, 384, 387n, 388n, 389n, 396n, 419n, 423n, 426n, 429n, 463n, 481, 500n, 542n, 553n
shiokara (pickled fish) 392, 393, 394, 397, 401n
shish kebab 85
shitake 145
shitogi (raw rice flour cake) 261n
shoa ping (sesame bun) 471
Short, T. (1690–1772) 538
shortening 255
Shou Chhing Yang Lao Hsin Shu (New Handbook on Care of the Elderly; +1080 and +1307) *135*, 137
   dairy products 257n; milk drinking 256
   medicated wine 237
   shih use 341
Shou I (Grand Veterinarian) 571n
Shou Shih Thung Khao (Compendium of Work Days; +1742) 144
shoyu (Japanese soy sauce) 377
   fermentation 377–8
   Japanese process (modern) 377
   Japanese references 376
shrimp 384, *381*, 414
   mock 497–8, 502
   sauce (*hsia yu*) 96, 358n, 391
shu (broomcorn millet) 18, 25
shu (Chinese yam) 43
shu (glutinous millet) 204
shu (millet) 24
shu (millet for wine) 18
shu (panicum millet) 19n, *20*, 24, 27
   li making 159
shu (soybean) 18, 22, 23, 25
   recipes 295
Shu (Szechuan) 508
shu an (tray for books) 112
shu chi (stool for books) 112
shu chiang (fermented soy paste) 346
shu chiao (*Zanthoxylum*) 52, 235
shu chieh (oil-bearing vegetable) 31
Shu Ching (Book of Documents; −5th century) 47, 54, 140
   chhü discovery 157
   turtle 63–4
   vinegar 283
   wine 157, 160; medicated 232
Shu Hsi (writer; +3rd century) 469
shu huang chiang (ripe yellow soy paste) 352
Shu Kingdom (Szechuan) 475
shu mi (dehusked millet grain) 295
shu mi (panicum millet) 459
shu nieh (sprouted millet) 166n
shu pei (over-dried) 527
Shu Yuan Tsa Chi (Miscellanies from the Soy Garden +15c.) 105n
Shuang Chhi Chi (Double Rivulets Collection; +1200) 500n
Shui Ching Chu (Commentary on the Waterways Classic; +5th century) 180, 432

shui hu (tea kettle) 560
Shui Hu Chuan (novel; late Yüan to early Ming) 59n
shui o (water penance) 511, 513
shui su (water perilla) 439n
shui ya (water sprout) 525n
shui yin (drawn pasta) *470*, 472, 480, 481, 499
shui ying (movable swellings) 576
shui-hua-mien (wet slippery noodles) 483
shün hua chha (scented flower tea) 553
Shun (legendary emperor of China) 297n
Shun Thien Fu Chi (Gazetteer of the Shun-Thien Prefecture; 1884) 362
Shun Ti (Emperor; Yuan) 434
Shun-Yü, I (physician of Western Han) 234
shuo (ladle) 105
Shuo Fu (Florilegium of Literature; +1368) 329n, *516*
Shuo Wên Chieh Tzu (Analytical Dictionary of Characters; +121) 36, 140, 155, 300
   chiang (soy, meat or fish paste) 334
   Chinese gooseberry 52
   cooking methods 71, 72n, 74
   ferment chhü character 261; *ferment* types 167
   fermented soybeans (*shih*) 336
   fish roe 63
   fu (decayed) 302
   goitre 574
   lo (juice from milk) 250
   i (malt sugar) 458
   mien 466; *ping* 468; pastries: *erh* 468; *kao* 468; *tzu* 468
   serving vessels 99
   tea 510; *tsu* (pickle) 404; *yên* 404
   vinegar 284; wine making 157–8, 161, 165
Shurtleff, W. & Aoyagi, A. 10, 303n, 310, 317, 318n, 320n, 328n, 329n, 330n, 345n, 362n, 365n, 373n, 374n, 375n, 376n, 377n
sieve 473
   for *chiang* 364
   tea *558*
Sigerist, H. E. 571n
sikhae (Korea) 394, 397, 397n
Siler divaricatum 235
silk 166
   painting with oil 437, 439
Silk Road (Central Asia) 496
simmering 88 *see also* braising
Simoons, F. J. 67n, 148, 149n, 596n
Sinai desert 603n
Singh, R. J. 18n
Siniperca 63
Sinkiang 472n
   distilleries (pot or Chinese type) 217
Sino-British Science Cooperation Office (Chungking) 5
Situwan (Perak; Malaysia) 584n
Sivin, N. 571n
six livestock (food animals for slaughter) 55, 57, 60
skewers 73, 85–6, 107, *521*
skirret 239
skull deformities 589
slaughter, ritual 56, 57, 58
sleeping trough 443

smartweed *33*, 38, 92, 93, 184, 185
  flour *ferment* 190
  spirit of *ferment* 191
  thu 507, 508
Smilex china 235
Smith, B. 27, 258n, 396n
Smith, M. 503n, 555n
Smith, P. 519n, 552n
smoking, fruits 425
snails 66, 379, *409*
snake bite
  remedy 82; toxin extraction 207n
Snapper, I. 590n, 604n
snout fish 62, 380
snow 434
so mien (thread noodles; rope noodles) 483, 485, *488*, 491n
so ping (Chinese snacks) 491
so ping (filamentous noodles) 480, 481, 484
so ping (rope pasta) 468, 472n, 485, 499
soba (buckwheat noodles) 488, 492
sodium chloride 305 *see also* salt
Soewitao, A. 342n
soil & goitre 574, 575, 578
Solomon 57n
solve, crude 426
somen (fine noodles) 491–2
Son of Heaven, grains and meat served to 58
Sonchus arvensis 33, 38
Songs of the South *see Chhu Tzhu* (Elegies of the Chhu State)
sorghum 20n, 23, 179n
  vinegar 291; distilled wine 279n
sou (bad, mouldy) 161
sou ping (water-based pasta) 468
souchong tea 536n, 545n, 546
soup 71, 82, 84
  ancient grand 83, 84, 301; white (plain) 84
  fish 71; vegetable meat 84
  harmony 93
  pasta 469, 473
  serving dish 98, 101; utensils 105; *ting* 111
  texture 93; thickening 461
sour taste 91, 95
sourdough 270, 606n
soured preserves 412
Southeast Asia
  fermented fish products 388
  fish sauce 392
Southern Sung tombs 88n
  Kansu *90*
southernwood 239
sowthistle *33*, 38, 507, 508
Soxhlet extractor 213, 226
soy *see* soy milk, soy paste, soy sauce and soybeans, fermented
soy milk 7, 295, 299, 314, 321–2, *595*, 604
  cooking & boiling 309–10, 311–12, 313, 596
  defoamer 313
  derivation 320, *321*
  digestibility 322, 596; flavour 322, 596; palatability 315, 316, 322

process 303–4, 308, *598*
  grinding soaked beans to give *hsi chiang* (puree) 303, 308
  filtering puree as is (cold extraction) 303, 308
  heating the puree before filtration (hot extraction) 320, *321*
  prolonged heating 322n
  selling by hawkers 322, *323*
  spontaneous coagulation 315n
  uncooked 310–12, 333
soy paste, fermented (*chiang*) 295, 316, 346–7, 373n, 388–9, 408, *409–10*, 582, 594 *see also chiang*
  anaerobic fermentation, early 1st stage seen in *CMYS* 349–50
  aerobic fermentation, later 1st stage as in *SSCY* 351
  chiang huang (*chiang* substrate) 1st stage product 349, 351
  commodity value 353
  development 604
  herb and spice addition 352
  inoculum 353
  method for making 347–52
  moulded bean 349
  process for various types of *chiang* 356
  raw yellow 352; ripe yellow 352
  salt addition 348, 352, 355
  1st and 2nd stage fermentations (flow diagram) 349, *350*
  taboos 347n; uses 354
  wheat flour 347, 349
soy protein 312
  coagulation 310n
  digestion 293
  micelles 312, 329
soy sauce, fermented (*chiang yu*) 2, 4, 96, 295, 358–74
  boiler for soybean cooking *368*
  condiment use 371
  cylindrical bamboo filter to collect sauce
  fermentation process 363–70, flow diagram 364
    1st stage: making *moulded beans or chiang substrate* 363
    trays for surface culture of cooked beans *369*
    2nd second stage: incubation in brine to give soy sauce 364
  incubation urns *365–70*, 371
  Japanese *shoyu* 376, 377
  prototypes 359–362
  siphoning off liquid 366, *370*
  therapeutic use 371
  preservative (*chiang-lu* method) 414
soybeans 18, 19, *20*, 22, 25, 292–5, 389
  black 296
  congee 294, 295, 314, 316
  consumption 314
  cooked granules 294, 295, 314, 316
  cooking 294, 295, 308, 310n, *368*
    for *shih* 338, 340; *tempeh* 342; *chiang* 347; soy sauce 363
  cultivation 292, in West 292
  curd 2, 3, 295, 299–303, *304*, 305–33 *see tou-fu*
  defatted meal 292; dehulling 348, 349
  digestibility 293, 294, 604
  domestication 27–8; economic crop 292

soybeans, fermented (*shih*) 2, 4, 295, 336–46, 593, *594*, *604*
  beriberi 582
  chiang process difference from *shih* 349
  condiment use 341; as relish 364
  fungal growth 338, 340, 342
  herb and spice addition 340, 341, 342n
  hydrolysis 339, 345
  incubation jar 339, 340
  moulded 339, 342, 375, *594*
  pharmaceutical use 337
  preparation method 337–42, 345
    salty *yen shih* and bland *tan shih* 337, 338
  soup 375n; taste 339
  tempeh 342, *343–4*, *345*–6
  trade 337
  wheat flour addition 342
soybeans 18, 19, *20*, 22, 25, 292–5, 389
  black 296
  congee 294, 295, 314, 316
  consumption 314
  cooking: boiled gruel, *tou chu*; steamed granules *tou fan* 294, 295, 308, 310n, 314, 316
  cooking for making
    shih 338, 340; *tempeh* 342; *chiang* 347; soy sauce 363
  cultivation 292, in West 292
  curd 2, 3, 295, 299–303, *304*, 305–33 see *tou-fu*
  defatted meal 292; dehulling 348, 349, 352
  digestibility 293, 294, 604
  domestication 27–8; economic crop 292
  flakes 314; meal 314n
  flatus production 293, 294, 298, 596, 602n
  flavour, beany 293
  food value 292–3, 298; nutritional value 604–5
  health hazards of eating 293–5
  keeping in the dark 296n
  leaf soup 294
  milling in water to give puree *304*, 308, 314–15, *598*
  oil 363, 457n
    commercialisation 293
    extraction 28, 452
    source 292; uses 450
  oligosaccharides 322, 323, 324
  production in China 292n, 456
  protein toxin inactivation 294n, 298
  recipes 295
  rotary quern 597
  sprouts 295–8, 316
    food value 298
    pharmaceutical 297–8
  status, change from Han to Thang in 316
  tea flavouring 519n
  whey 308, 324
  yellow curls from 295
  see also bean sprouts; protease, inhibitors;
Spaghetti Museum, Pontedassio, Imperia, Italy 495
sparrow 61, 346, 380, 388
sparrow eyes 587
Spence, J. 604n
spices 16, 17, 52
  steam distillation 214

spinach 590n
spirits, distilled 204–31 *see also* wine, distilled
sponges for goitre treatment 577
spoons 85, 104, 105–6
  styles 106, *107*
spring onion *33*, 38–9, 40, *41*, 43, 92, 189 *see also* scallion
Squaliobarbus curriculus 61n, *62*
Ssu Khu Chhüan Shu (Complete Library of the Four Categories; +1782) 184n, 491n, 498n, 434n
Ssu Min Yüeh Ling (Monthly Ordinances for the Four Peoples; +160) 29, 36, 38, 39n, *117*, 120, 143
  Allium species 40; ginger 40; peas 40; taro 43
  chiang 380; fermented soy paste 347
  malt sugar preparation 457
  pasta cooking 468, 469
  seed pounding 437; flambeaux making 437
  soy condiments 372; sour condiments 284; *tsu* 404
Ssu Shih Tsuan Yao (Important Rules for the Four Seasons; late Thang) 143
  chiang 386; preparation 351, 352
  hsi and *fu* 421
  malt sugar 459
  pressing seeds for oil 440
  shih preparation 340, 341, 361–2
Ssu Wei (Four Taboos) 347n
Ssu-chhuan shên po-wu-kuan (on the Stone Murals of a Sung Tomb at Kuang-yuan, Szechuan; 1982) 88n
Ssu-chhuan shêng wên-wu khao-ku (on the Sung tomb at Luo-chhêng, Kuang-han County, Szechuan) 221n
Ssuma Chhien 104
stachyose 293, 322n
Stachys aspera 439n
Stahel, G. 342n
starch 8
  amylase, in human saliva 154
    sprouted grains (*nieh*) 154, 157, 260
    mould culltures (*chhü*) 154, 157, 261
  conversion to alcohol with *chhü* organisms 160–1, 608
  denaturation, gelatisation 259, 271
  enzymatic hydrolysis to sugars i.e. saccharification 154, 259–60, 269, 274n, 278, 593
  pasta making *470*, 473
  preparation 461–2
  uses 461n
starch substrate 408, *409–10*, 411, 415 *see also* rice
  pickled fish *cha* 385, 386; vegetables *tsu* 406
  salt 416
state dinners 430
statins 202–3
Staudenmaier, J. M. 598n
steam, distillation 206, 213
steamer 76, 77, *78*, *79*, 80, 206, 230, 259
  bamboo 88–9, *90*; wood 230
  bronze 226, 208–9 derivation of still
  malt sugar making 458
  packing materials 214
  pottery 230, 260, 263, 278, 599, 600
  sesame seeds 441, *444*, 449
  tea processing *521*

steaming 67, 70–1, 76, 82, 88–9
　oilseeds 441, *444*, 448, 449, *455*
　pasta 599n
　sesame seeds 441, *444*
　tea 230, 520, 521, *524*, 525, 533, 539, 553
Steele, J. 83n, 84n
steeping 74, *88*, 89, 428; steeped meat 417
Steinkraus, K. H. 249, 383n, 388n
stews 34n, 89
　serving dish 98, 101; thickening 461
stick, filial 349n
still, bronze 226, 596
　from Shanghai Museum 209, *211*, 212–14, 597n
　from Chin tomb 208–10
　from Chhu-Chou 213
　annular gutter 212, 213, 214, 216, 230, 231
　commercial production 230–1
　drain tube 231; spout 212
　fermented mash 229
　Hellenistic vase type 214–17, *217*, 226n
　Chinese (Mongolian) pot type 215, 216–17, *218*, 231
　traditional Szechuan 220
stir-frying *88*, 89, 90
　sesame seeds 441, *444*, 447, 448
　tea leaves 523, 527, 530–2, 533, 539, 546, 547, 552n
Stockhardt, A. 329n
Stol, M. 269n
stools 109, 112, 140
storage of fruits 424, 427–9
stoup *99*, *100*, 103
stove 80
　charcoal 522
　Eastern Han tombs in Sui-tê *110*
　tea processing *521*
　traditional 2n
straw
　insulation 431, 435
　millet for fruit storage 429
　oil processing 441, 449
strawberry, Chinese 54
　sugared 426
Streptococcus 253, 378n
strip, bamboo (in Thang tea processing) *521*
sturgeon 62, 63, 91, 380
su (clotted cream; butter) 248, 251, 253, 437n
　adulteration 255
　preparation 253–4
　use 255
su (setaria millet) 23, 24, 140
Su Chao-Chhing 277n
Su Chin 585n
Su Ching (Su Kung) (editor of *Hsin Hsiu Pên Tshao*; +7th century) 24, 134, 136n
Su Hsüeh Shih Chi (Collected Works of Su Shun-Chhing; +1008 to +1048) 225n
Su I (writer; +10th century) *516*
Su Shih Shuo Luo (Collected Poems of Su Shih; c. 1900) 193n, 484n, 502
Su Shun-Chhing 225
Su Shih (Tung-Pho, poet–scholar of Northern Sung; +1036 to +1101) 34n, 38n, 54n, 132, 160n, 182, 183, 186n, 482n

sweet tooth 500
　tea 563; wines 202, 206, 237, 243, 245, 246–7
su yu (vegetable oil) 437
suan (rocambole) 39, 40, *41*
suan (sour) 91, *95*, 283
suan chiang (soured water drained from boiled grain) 96n
Suan Kung 102n
suan lo (sour *lo*) 253
suan tshai (soured preserves) 412
suckling pig, *hai thun* 380
sucrose 459
Sudan, beer 267
suei (wild pear) 49
sugar *see also* malt sugar
　cane 95, 426, 459–60, 461
　content of fruit 258n
　fermentation 245–6
　fruit preservation 460n
　hydrolysis from starch 154
　incubation 415
　saturation 424, 425–6
sugarcane 457, 459, 461n
sui chhin (wetland celery) 35–6
Sui Chien Lu (Record of Second Encounters; +1734) 539
Sui Hsi Chü Yin Shih Phu (Dietetics from the Random Rest Studio; 1861) 324, 373, 391, 501, 561n
Sui Jên (discoverer of fire) 602
sui ping (bone marrow pasta) 468, *470*, 471
Sui Shih Kuang Chi (Expanded Records of the New Year Season; Southern Sung)
　ice 434n; *la jou* 421
Sui Shu Ching Chi Chih (Bibliography to the History of the Sui) 124
Sui Shu (History of the Sui Dynasty; +581 to +617) 154n
Sui Yuan Shih Tan (Recipes from the Sui Garden; +1790) *129*, 131, 132, 256, 388, 412
　bean flower (*tou hua* or *tou fu hu*) 323–4; bean sprouts 297
　chiang use 355, 357n, 373; culinary sauces 358n
　food preservation methods *409–10*, 412n
　lu method 414; *sung* 423
　man-thou 476; dumpling (*tien-pu-ling*) 480
　processed fruits 427n
　soy condiments *372*; soy sauce 373
　tea 540; *tou fu* 324, 325, 327
Sumerians
　beer making 267, 269
　vinegar 283n
Sumiyoshi, Y. *377n*
sun, eclipse 118
sun (bamboo shoot) *33*, 36, 380
Sun Chi 332, 333
Sun Chung-En 69n
Sun, E. T. Z. & Sun, S. C. 29n, 133, 154n, 160n, 190n, 197n, 447n, 450n, 452n, 460n
Sun I-Jang 85n
Sun Phu (Treatise on Bamboo Shoots; +970) 144
Sun Ssu-Mo (Thang physician, pharmacist) 136, 234–5, 247
　beriberi 579, 581, 582, 585, 606, 607
　goitre 576

Sun Wen-Chhi 234n, 237n, 238, 243n, 248n
Sun Yin-Hsing (writer; 17th century) 452
Sun Ying-Chhüan 179n
sun yu (bamboo shoot sauce) 358n
sunflower 38
  oil 456–7
sung (finely shredded meat or fish) 423–4
Sung Chüeh (Ming Dynasty) 144
sung chün (pine mushrooms) 144
Sung Hsü (writer; c. +1500) 485
sung (oil-bearing vegetable) 31
Sung Po-Jen (Sung writer) 247n
Sung Shih (History of the Sung) 206n
Sung Shih Tsun Shêng (Sung Family Guide to Cultivation of Life; +1504) *129*, 130, 189n, 353, 357
  red *ferment* 196
  vinegar recipes 290, 291n
  tea 527, 561n
Sung Tzu-An (writer; +11th century) *516*
Sung Ying-Hsin (author of *Thien Kung Khai Wu*; 17th century) 133, 158, 159, 190n, 202, 450
Sung Yü-Kho 84n
Sung-Chhi (Fukine) 202
Sung-lo or Sing-lo (Anhui), green tea country *545*, 545, 547
sunshine 590
Suparmo 298n
Sus scrofa domestica 58
Swan-like Yellow Bean Sprout *see Ngo Huang Tou Shêng*
Swann, N. L. 168n
Swatow, fish sauce 397
sweet liquor 97, 156, 157
sweet potato 2n, 21n, 43, 462
sweet rush 239
sweet tangle 575, 576
sweet taste 91, 95
sweeteners 92, 267
sweetmeats, honeyed 426
swellings 574, 576
swordbill fish 62
syrup 458–9
Szechuan 5, 46, 425
  brick murals 217–21
  brick tea 552n; tea cakes 522n; tea use 511, 513, 523
  distilleries 217
  phao tshai 411
Szechuan Provincial Museum (Chengtu) 218n
Szu Chi Yen (Superindendant of Stools and Mats) 109
Szuma Chhien 104

Ta Chao (Great Summons) 23n, 65n, 94–5, 250
ta chhü (great ferment) 204
ta chiao (great leaven) 476–7
ta huang (*Rheum officinale*) 235
ta huang tou (soybean) 297
ta kêng (grand soup) 301
ta khu (great bitter) 339
Ta Kuan Chha Lun (Imperial Tea Book; +1107) *516*, 517, 523
  drying tea 526–7
  picking tea 524
  preparation of tea drink 557
  pressing tea 525n

ta kuei (giant turtle) 64
ta ma yu (hemp oil) 440
ta mai (greater wheat; barley) 19, 25, 27, 204
ta mai chiang (barley *chiang*) 346
ta man hu (flat bread?) 468
ta suan (garlic) 39–40, 39n, *41*
Ta Tai Li Chi (Record of Rites of Tai the Elder; +80 to +105) 119–20
ta tao (Great Truth) 224
ta tou (soybean) 19n
ta tou huang chüan (yellow curls from the great bean; soybean sprouts) 295, 296, 297, 298
ta tsao (great jujube) 424n
Ta Tsao Phu (Monograph on Jujubes; Yuan dynasty) 145
ta tshung (spring onion) 39
ta tu (powerful action) 204
Ta Yü (Emperor; founder of Hsia dynasty) 155, 257
Ta-hu-thing (Mi-hsien)
  Eastern Han mural 11, 306–7, 331–3
  engraved stone slab 305, 307
  equipment in modern China 308, *309*
  Tomb No. 1 306, 307, 312
Ta-kua-liang tomb, Eastern Han (Sui-Tê county, Northern Shansi) 109, *110*
Ta-pho-na tomb (Tali, Yunnan; c. −600) 104
Ta-wên-khou excavations 155
Ta-Yüan (Ferghana) 240
table 111
  low 109, 114
Tachi, H. 383n, 392n, 397n
tagliatele press 497
tai (sour condiments) 284
Tai Fan-Chin 40n, 298n
Tai Shêng (compiler of *Li Chi*, Western Han period) 119
Tai Ying-Hsin 109
Tai Yün 128n
Tai Yung-Hsia 245n
Taipei (Taiwan) *81*, *562*
Taiping Rebellion 546n
Taiwan *81*, *562*
  Chi-I tea 541
  fish sauce 595
  kung fu tea 540
  sung, shredded meat or fish 424n
  tea processing 536n
  wine-making 161–2, 185, 233, from rice 278n
Takei Emiko 303n, 320n, *321*, 324n
Tale of Genji (Heian period) 375
Tali (Yunnan) 104
tallow, meat flavours 93
tallow tree 441, 448, 450
  oil yield 452
tamari (Japanese *shih yu*) 376, 377
Tamba no Yasuyori (author of *I Hsin Fang*) 136
tan (bamboo container) 80, 99, 101
Tan Chhi Pu I (Backup Notions from the Elixir Book) 202
tan chhü (red *ferment*) 197
Tan Fang Hsü Chih (Indispensable Knowledge of the Chymical Elaboratory; +1163) 207

Tan Hsien-Chin 429n
Tang Lang 346n
tan shên wine 235, 237
tan shih see soybean, fermented
Tan Tu-Kho 540
Tanaka Seiichi 147, 180n, 189n, 196, 291, 319n, 353n, 354n, 355n, 357n, 374n, 377n, 388n, 419n, 423n, 426n, 429n, 481n, 491n, 500n, 542n, 553n
Tanaka Tan 82n, 86n, 88n, 112n, 113n, 114n, 313n
tandoor (oven) 3n, 266n, 468n, 472n
tang fu (soybean curd, Japanese) 317
tangerine 54, 582 *see also* oranges
    peel 186n; tea 553
Tangshan (Hopei) 216
Tannahill, R. 248n, 249n, 265n, 266n, 267n, 270n, 328n, 396n, 467n
tannin *see* polyphenols
tao (rice) 18, 19n, *20*, 25
tao chha (pounding tea) 525
Tao Tê Ching (Canon of the Virtue of the Tao; −5th century) 60, 71, 140, 283n
tao tou (broad bean) 355
Tao Tsang (Taoist Patrology; +730 onwards) 207, 208
tapeh (fermented congee of Indonesia) 260n
taro 42–3, 145
taucheo or taucho (fermented soyfood in Indonesia) 378
taxation, grape wine 244
tea 2, 38, 503, *504–5*, 506–7, 601 509–10, 512
    antidote to toxins 506; aroma 568
    art of drinking 518
    Assam, wild trees 549, 503n
    bamboo basket 520, *521*, 539
    barter 515n
    black 4, 8, 535n, 536, 538, 541
        chemical components 567
        exports 545; red tea exports 548; Indian exports 549
        generic name for fermented teas 541
        manufacture 546–9; processing 541n, 547n
        maritime trade 8, 536, 541, 546, 547, 548, 550
        scented 554
        varieties 545n
    Bohea 538, 541, 542n, 545n, 548n
        trade 546
    boiling 561
    bowl 557, *558*, 559, 561
    brick 562
    brokers 543, 544; buying 508
    brunch, Cantonese (*yin chha* or *yum chha*) 485
    Buddhism 515
    with butter 562
    café or house 527, *528*; iced *mei* flower wine served in 434
    caffeine 565, 568–9
    cake 520, *521*, 529, 550, *551*
        baking 556
        compressed (*ping chha, phien chha*) 552
        infusion preparation 556–7
        Thang 519–23; Sung 523–8
        waxed 527
    cancer protection 570
    cardiovascular protection 570
    chemical constituents 565–6, *567*, 568
    Chiang-hsi-wu (Kiangsi dark) 542
    chrysanthemum 554
    coarse, loose & powdered 522
    colour 568
    commodity value 527
    compressed, for Tibetans & pastoral peoples 552–3
    congee 512, 555, 561, 562
    Congu 536n, *544*, 545n, 546, rose 554
    consumption 503, *505*
    cooling stir-fried leaves 532
    cultivation 513
    cup *554*, 560
    dark 535n, 551–2, 553
    decaffeinated 569
    drinking 540, 554, 555
    drying 531, 532–3, 539, 546
        by heat 537; in India 549
    essential oils 561, 565, 568
    etymology 507–13, *514*, 515
    European trade 538; exports 541; foreign buyers 544
    fermentation, enzymatic oxidation 536–9, 550, 552
    firing (stir-frying) or heat inactivation 552
    flavour 568; flavourings 519n, 557, 561, 562
    fluoride content 569
    from parched grain 97
    gardens 561
    grading *524*, 525
    green 4, 535, 538, 540, 543n, 544, *551*
        unfermented 535n; chemical components *567*
        exports 541, 545, 548
        kung fu making 561
        pharmacological effects 570
        preparing drink 557–61; drinking 568
        process 552; processing 533–4, 537, 539; reprocessing 553
        scented 554
        varieties 545n
    growing areas, Fukien 517
    harvesting 520, 525n
        implements 520–2, 523
        time 523–4
    health effects 562–6, *567*, 568–70
    Hyson 545n, 552n
    infusion 560, 561; *kung fu* 540
    insignia template *524*, 525, *526*
    international influence 503
    Japanese ceremony 557
    kettle 557, 559, 560
    labour force 517n, 522n, 523
    ladle 559
    loose 522–3, 527, 528–34, 540, 550, *551*, 552
        preparation of drink 559
        processing 529, 561
    Lung-ching 539, 540
    medicinal use 506, 512, 513, 557, 563–4, *565*
    medium bud 524
    mill, stone for grinding cake tea *558*
    mixed with plant products 513n
    moulding *521*, 522, *524*, 525, 553
    mouth cleansing after meals 560n

oolong 8, 535–41, 544, 546, *551*
  semi-fermented 548, 549; catechin level 566
  drinking 561; *kung fu* making 561
  exports 541, 548; early Indian production 549
  process 552; processing 537, 539
origins 503, 506
pan for heating 531, 532; panning 549
pharmacological effects 570
phu 552n, phu-erh 535n, 552n, 553
picking 524–5, 529, 539
piling 552
plant 508; plantations 513, 515
popularity 527
pot *558*
pounding 520, *521*, 522, 525
powdered 522
preparation 14, 518, 555–7, *558*, 559–62
pressing 520, *521*, 522, *524*, 525
processing 230, 519–554
  during Thang 519–23; Sung cake teas 523–8
  loose tea during & Ming 528–35; oolog tea 535–41; red tea as black tea of maritime trade 541–50
  primary & secondary processing 552n; raw and refined 552
production, annual 552n
purple shoot 522n, *524*
red 4, 8, 535n, 536, 541–50, *551*, 561n
  catechins 566
  commercialisation 544; replacement of black tea 548; consumption 561
  exported as *black tea* 548, 549; Indian exports 549
  process 552; processing 537, 543, 548; reprocessing 553
  trade names *544*
roasting *524*, 526–7
rolling *524*, 525, 529, 537, 552
  by foot *535*, 543
  by hand 532, 533, 534n, 546, 549
scented 552–4, 561
serving to guests 508
seven necessities of daily living 527
sleep effects 509, 510, 512, 520, 562–3, 564; meditation 515
small bud *524*, 525n
Souchong 536n, 545n, 546
sparrow tongue leaf 534
spread through China 513
steaming 230, 520, *521*, *524*, 525, 533, 539, 553
stimulant effect 569
stir-frying 523, 527, 530–2, 533, 539, 546, 547, 552n
storage 534
sun-drying 530, 533, 539, 543, 551n
tooth decay 563, 569
trade
  maritime 538, 541, 545, 546, 547
  with Tibet 552
tribute 517, 522n, 523, *524*, 538
  abolition 529
  insignia *524*, 525, *526*
  to Emperor's court 509
Twankay 545n

use of fresh leaves 508
utensils *556*, 557, *558*, 559, 560
votaries 540
water 523, *558*, 565
whisk 557
white 550–2; process 552
wild trees 503n
wilting 537, 539, 543, 546; withering 552
  India 549
Wu-I 538, 539, 540, 541, 542, 546
Yang-Hsien 539, 540
yellow *524*, 550–2
  process 552
see also polyphenols; tea seed oil
Tea Books 515, *516*, 523
Tea Sage 515n
tea seed oil 4, 450, 451
  poison 451n
teapot 560
technology 602n
  dissemination 607–8
  to West 608
  evolutionary model for development 597
  food processing 7–8, 596–7
    evolution 597–601
    instruments 598
  sharing 607
  understanding 607
teeth, retarded emergence 589
Tei Daisei 377n
tempeh 342, *343–4*, 345–6
  connection to China 345
temperature, fruit storage 427
Temple, R. K. G. 149n
tench 61, 62, *62*
têng (pottery shallow bowl) 100
Têng Kuang-Ming 482n
Têng Tao-Hsieh (Ming Dynasty) 144n
Teramoto, Y. 163n, 166n, 262n
Teubner, C., Rizzi, S. & Tan, L. L. 493n
textile desizing 260n
texture of food 93
tha (couch bed) 113, 114, 115
thai chha shu (tea trees) 510n
thai kêng (grand soup) 83; *thai ku chih kêng* (ancient grand soup) 84
Thai Phing Huan Yu Chi (Geographical Record of the Thai-Phing Reign Period; +976–83) 160n, 183
  conquest of Kao-chhang (+640) 241
  food storage wells 432–3
  grape wine 243; herbal *ferment* 232
  ice 433n; tea 509n, 510n, 511
  treasure wells 432–3
Thai Phing Hui Ming Ho Chi Chü Fang (Standard Formularies of the People's Pharmacy in the *Thai-Phing* reign-period; +1178) 207n
Thai Phing Yü Lan (Emperor's Daily Readings; +983) 126, 205n, 240n, 300
  death of Emperor Chih-ti 469n
  gluten 500; tea 510
Thai Phing Yü Lan: Yin-shih Pu (Food and Drink Sections of the *TPYL*) 126

thai tshai (tea) 510n
Thai Yu Ji Chi (Diary of Travels in Taiwan; 1892) 541n
Thai-hsi (Honan) 151, *152*, 153
Thai-hsi winery (Chengchou) 259n
Thai-shan (Mount Tai) 512
Thai-yuan (Shansi) 241, 242, 243, 244
Thailand
   fermented salted fish *395*, 396
   fermented soyfoods 378
   fish sauce 392, 397; rice noodles 492
than (fermented meat or fish sauce) 333–4, 346, 380, 390, 391
Than Chhi-Kuang 478n, *479*
than hai (juice of meat) 333n
Than Kung Hsia 294n
Than Yao-Hui 412n
Than-yang (Fukien) 544, 546n
thang (broth) 89
thang (cane sugar) 457, 459–60
thang (malt sugar) 158, 457, 458, 460
thang (sugared) 426
thang fu (bean curd, Japan) 318
Thang Huan 475n
Thang Kên 592n
Thang kuo (Chinese snacks) 491
Thang Lan 84n, 121n, 166n, 336n, 513n
Thang Lei Han (Thang Compendium; +1618) 511n
thang mi kuo shih (sugar and honey treated fruits) 426n
thang ping (pasta soup) 468, 469, 473, 475, 485
   boiled 490; lily pasta 482; precursor of noodles 480; wonton 478
thang pu (bean curd) 318
Thang Shih San Pai Shou (Three Hundred Thang Poems; +1764) 181n, 239n
Thang Shu (History of the Thang) 302, 362n
Thang Suan Phu (Monograph on Cane Sugar; +1154) 460
Thang tombs
   Han Sen Tsai (Sian) 223
   Ho Chia Chhun (Sian) 208
   Turfan (Sinkiang) 478, *479*
thang tshui mei (sugared crispy apricots) 426n
Thang Yü Lin (Forest of Mini-Discourses; Northern Sung) 434n
Thang Yü-Lin 579n, 586n
Thang Yün Chêng (Thang Rhyming Sounds; +1667) 512
'Thanks to Imperial Censor Mêng for his Gift of Freshly Picked Tea' (poem; +835) 554
thao (peach) 44, *45*
Thao Chên-Kang 128, 147n
Thao Hung-Ching (c. +500) 24n, 31, 134, 284n, 296, 297n, 341n
   vision 586
Thao Ku (writer; +10th century) 300, 354, 516n
Thao Wên-Thai 122, 123, 128n, 132n, 147, 148, 260n, 571n, 602n
Thea bohea 539, 547n; *Thea viridis* 539, 547n
theaflavins 536, 566, *567*; thearubigins 536, 566, *567*
thi an (whey) 241
thi-hu (clarified butter) 248n, 253, 254
thiamin deficiency 579

thiao (long fish) 61, *62*
thiao kêng (spoon) 106
Thiao Ting Chi (Harmonious Cauldron; +1760 to 1860) 132, 256n, 281n, 297, 391
   cha 388, 412; *chiang* 373, 387n; preservation 414
   food preservation methods *409–10*, 412; fruits steeped in water 428
   gluten 502; ham 422; *tsao*-preserved crab 413
   honeyed and sugared fruits 426–7n; orange storage 429
   kan pi wêng tshai 419; *lu* method 414; *man-thou* 476
   salting of vegetables *418*
   soy condiments *372*; soy sauce 373
   vinegar recipes 291, 291n
Thieh Wei Shan Tshung Than (Collected Conversations at Iron Fence Mountain; +1115) 206
Thien Hsi (writer; +940–1003) 205n
Thien I-Hêng (writer; +16th century) *516*, 530, 532, 539, 551n, 554
thien kua (musk melon) 33
Thien Kung Khai Wu (Exploitation of the Works of Nature; +1637) 4, 29, 133, 349n
   malt sugar 460
   oilseed processing 441n, *444*, *445*, 450–2
   red *ferment* preparation 197–203
   rice grinding *583*, *584*
   wine making 158, 160, 190–1
thien lo (sweet *lo*) 253
thien men tung wine 236
thien mi (sweet steamed rice) 187
thien mien chiang (sweet wheat *chiang*) 357, 412, 414
thien suan yü (sweet and sour fish) 497
Thien-hu-tshun (Honan) 153
thung ma chiu (mare's milk-wine) 248
tho (noodle) 468, 481n
Thorpe, N. 266n, 467n
Thou Huang Tsa Lu (Miscellaneous Jottings far from Home; +835) 160n, 183
thou ju (adding charges in fermentation process) 187n
thu (bitter herb; bitter vegetable) 91–2, 507, 510
thu (radish) 35; *thu* (smartweed) 457n, 507
thu (sowthistle) *33*, 38, 457n, 507
thu (tea) 38, 457n, 507, 509–10, 511, 512
Thu Lung (writer; +16th century) *516*, 530, 539, 551n
Thu Pên-Tsun (16th century) 145, 146
Thu Ssu-Hua 463n
thu su wine 235, 237
thu tshai (bitter vegetable) 508
thu tshao (bitter herb) 512
thuan hua pi lo (mushroom pilaff) 126
thun phi ping (suckling pig skin) 470n, 473–4
thung (milk) 315
thung chhing mo (copper carbonate) 428
Thung Chih (Historical Collections; +1150) 125
Thung Chün Yao Lu (Master Thung's Herbal; Han) 511
Thung region, tea 543
Thung Su Pien (Origin of Common Expressions; +1751) 354
thung tree 441, 448, 450
thung yu (tung oil) 358, 450
Thung Yüeh (Contract with a Servant; –59) 294, 508, 509

tea drinking 513, 555
tea processing 520n, 555
Thung-mu-tshun (Wu-I region) 546n
thyroid gland 573, 576–7
  myxoedema therapy 578
  see also goitre
thyroxine 574
ti fu tzu (Belvedere cypress seeds) 587
ti huang (rehmannia) 257n, 499
  wine 236, 237
ti i tshao (lichens) 587
ti lu (condensing drops; distillate) 204
ti tsieh (*Polygonatum odoratum*?) 234
ti-li (ground chestnut) 40
Tibet
  cake tea imports 552
  Chinese plum 46
  tea with butter 562
tien chha (making tea) 559, 560
tien chiang (coagulating soy milk) 308
tien-hsin (dots of heart's desire) or dimsum 485
Tientsin 244n
tiger-bone wine 237
tilt-hammer 441, 445 *see also* trip hammer
time measurement 533n
ting (cauldron) 71, 76, 79, 80, 82
  serving dish 98, *100*, 101, 111
Ting Wang-Phing 136n
ting-pien-hu 474n
tien pu ling (dumpling) 480
To Celia (poem) 150n
To Chhing (tossing the leaves) 547
To Nêng Phi Shih (Routine Chores Made Easy; +1370)
  129–30, 189n, 290, 291n, 411
  cha 387; *chiang* 386–7, 353, 357, 414n
  orange storage 429; honeyed and sugared fruits
    426n
  pasta 500n; noodles 483n
  sausages 423
  shih recipe 340; soy condiments *372*
  tea 542n, 562; scented 553
  tsao-preserved crab 413n; *yên* 419n
Toda, M. 570n
Todaiji Shosoin documents (+730–48) 374n
toddy 245n, Siamese 205
tofu no misozukè (tofu pickled in miso) 328
tomato sauce (& spaghetti) 497
tombs
  Chiang-hsi shêng po-wu-kuan (Eastern-Wu) 80n
  Chin (Chhêng-teh, Hopei) 208–9, *210*
  of Later Han, Wei and Chin 313n
  Marquis of I (Sui-Hsien, Hupei) 106
  paintings at Chia-yü-kuan (Wei-Chin period) 85
  Southern Sung 88n, *90*
  Ta-hu-thing (Mi-hsien) 306, 307, 312
  Ta-pho-na 104
  see also Han dynasty tombs; Thang tombs
tooth decay, tea 563, 569
torches 441 *see also* flambeaux
tortoise 66; Tortoise Keeper 66n
Torula 254n
tou (beans) 23

tou (shallow bowl) 100, 111, 402n
tou chiang 299, 308, 314, 315, 320, 321–2 *see* soy milk
tou chu (bean congee) 294, 295, 314, 316
tou fan (cooked or steamed bean granules) 294, 295,
  314, 316
tou fu (soybean curd) 2, 3, 11, 255–6, 295, 299–316,
  594
  archaeological evidence 305–10; process as
    described in Han tomb at Ta-hu-thing 307–9
  brain (*tou nao*) or flower (*tou hua*) 320, *321*
  Buddhist monasteries, Japan 318
  coagulation of soy milk with bittern, alum or vinegar
    303
    with calcium sulphate 329, 605
  calcium, as source of 590
  Chinese-style, pressed 317
  comparison with cheese 328–30
  deep fried 320, 321, *322*; frozen 321, *322*; smoked
    320, 321, *322*; skin 303, 320, *321*, *322*; honey
    soaked 500
  development *595, 598, 604*
  dishes 301; hibiscus 324; a la Su Tung-Pho 354
  drying 320, 321, *322*
  effect of heating of soy milk on coagulation 312n
  fermented 7, 320, 321, *322*, 326–8, *594, 604*
    organisms 328
  mock chicken, fish and shrimp dishes 498
  of the Mount of Eight Venerables 299
  nutrient composition 330
  origin 299–301, 302–3, *304*, 305–16
  pressed 320, *321*, *322*
  pressing 303n, *304*, 308–9, 311n, 319
  processing 320–1; products derived from 319–26,
    604
  proto 311, 312–13
  skin 303, 320, *321*, *322*, 323
  technology transmission to Japan 317–19
  traditional process for making 303–4, 305, 308–9
  use in Chinese cuisine 329–40
  use of term 302
tou fu chiang (soy milk before coagulation) 319, 321
tou fu fu (deep fried stinky dewatered *tou fu*) 326
tou fu hua (soft bean curd) 320, *321*, 323–4
tou fu i (bean curd skin) 320
tou fu ju (fermented bean curd) 413
  see also fu ju
tou fu kan (blocks of *tou fu* simmered in soy sauce; dried
  tou fu) 320n, 321, *322*, 325
tou fu nao (soft bean curd) 320, 323, 324
tou fu phao (deep fried *tou fu*; *tou fu* puff) 321, *322*, 325
tou fu phi (bean curd skin) 303, 320, *321*, *322*, 323
tou hua (freshyl coagulated bean curd) *322*, 323–4
tou huang (yellow coated, i.e. moulded bean) 342, 345
Tou Kêng Fu (Ode to the Soybean Stew; c. +3rd century)
  294n
Tou Lu Tzu Jou Chuan (Biography of *Jou* the son of *Tou
  Lu*; 12th century) 301
tou pan chiang (broad bean paste) 373n
tou ping (bean cake) 345
tou shih (fermented soybeans, i.e. *shih*) 340n, 364, 376n
tou ya (bean sprouts) 297, 298

tou yu (sauce from soybeans) 371, *372*
tou yu (soybean oil) 363
tou yu (*tamari* i.e. *shih yu*) 376
trace elements 573
trachoma 587
trade, maritime 18, 538, 541, 545, 546, 547
trans-retinol deficiency 586
Trapa bicornis 33, 40; *Trapa natans* 40
trays *101*, 112
   bamboo 197, 199, 520n, *521*, 526, 527
treacle 459
treasure wells 432–3
treaty ports (1842) 546n
tree bark rope 522
triglycerides 378n
Trionyx sinensis 64
trip hammer *445*, 448, 582–3, 585; tri-hammers *584*
tripe, *chhi* pickles 379
Triticum turgidum 19, 19n, 25
trough 187, 192
   roller mill 449
   sleeping, for oil 443
   starch preparation 461
trypsin inhibitor 293
Tsa Liao Fang (Miscellaneous Prescriptions; −200) 234
tsamba (paste or porridge) 266n, 275
Tsan Ning (monk; 10th century) 144
Tsan Wei-Lien 24n
Tsan Yin (author of *Shih I Hsin Chien*) 137
tsang (preservation) 405
tsao (Chinese date, jujube) *44*, 48–9, 424
tsao (coarse grain barley) 285
tsao (stove) 80, *81*, 82, *521*
tsao (wine fermentation mash or residues) 179, 183, 193, 204, 340, 415, 477, 594
   preservative for meat, fish & vegetables *406*, 407, 408, 411, 412 *594*, 595n
tsao fu (dried jujube slices) 424
tsao ju fu (*tou-fu* aged with a fermented mash) 326
tse chhi (*Euphorbia helioscopia*?) 234
tsêng (steamer) 76, 77, 98, 204, 458 215, 226, 229
Tsêng Chhêng Li Chi Phu (Treatise on the Litchis of Tsêng-Chhêng; +1076) 144
Tsêng Tsung-Yeh 147, 155n, 260n, 277
tshai (dishes, category of food) 80n, 466
tshai (to gather) 32; *tshai* (vegetables) 17, 32
Tshai Ching-Feng 571n
Tshai Hsiang (governor of Fukien, Tea Commissioner; 11th century) 144, *516*, 523, 525, 553, 557n
   dried litchi 425
   making tea (*tien chha*) 559; scented teas 553
   tea drying furnace 526; Wu-I tea 540
Tshai Thao (Writer; +12th century) 206
tshan (loose pasta) 470n, 473
tshao (grasses) 32; *tshao* (pan-fry) 439; *tshao* (trough) 187, 192
Tshao Chen Chuan (Biography of Tshao Chen) 337n
tshao chha (loose tea) 529n; *tshao chha* (rolling tea) *524*
tshao chhü (herbal *ferment*) 162, 183, 185, 232
Tshao Chih (poet, brother of Tshao Phi; +3rd century) 71–2, 360

Tshao Chin 519n
Tshao Ching 513n, 569n
tshao chü (grass tangerine) 385n; *tshao chü tzu* (grass tangerine seeds) 385
Tshao Fan (Ming Dynasty) 144n
tshao hao (prince's feather) 417
Tshao Hsüeh-Chhin 323n
tshao lung chu (grass dragon pearl grape) 241
Tshao Lung-Kung 463n
Tshao Mu Tzu (Herbs and Plants; +1378) 206n, 299
Tshao Phi (Emperor of Wei; c. +220) 71, 72, 240
Tshao Thing-Tung (author of *Chu Phu*) 139
Tshao Tshao (founder of Wei Kingdom) 168
Tshao Yuan-Yu 42n, 155n, 157n, 203, 205n, 206n, 207, 233n, 240n, 246n, 300n, 337n, 354n, 508n, 575n, 586n
tshe mien (thin noodles) 485
tsho (mincing) 75n
tshu (vinegar) 283, 284, 285, *594*
tshu mi (sour steamed rice) 187
tshuan (poaching) *88*, 89
tshuan (stove) 80
Tshui Chih-Thi (prescriptions c. +650) 576
Tshui Hao (scholar and prime minister in early Northern Wei Dynasty) 125, 597
tshui lü mien (silken green noodles) 484
Tshui Shih Ching (Madam Tshui's Food Canon) 125
tshung (spring onion) *33*, 38–9, 40, *41*
Tshung Hsün Chhi Yü (Didactic Notes; +1697) 538
Tsi I (Grand Physician for Internal Illnesses) 571n
Tsien Tsuen-Hsuin 38n, 461n, 496n
Tsin Tsin Yu Wei Than (Talks about Tasty Viands; 1978) 147
Tsan Ning (monk; Sung) 206n
tso (sour condiments) 284, 285
Tso Chuan (Master Tso's Commentaries on the Spring and Autumn Annals; −722 to −453) 49, 65, 71, 82, 93n, 140
   beriberi 579; *chhü* 157; medicated wine 232; dining furniture 109; ice collection storage and use 429–30; wine making 157n; wine serving vessels 102; Yellow Springs 83
Tsou Hsüan (author of *Shou Chhing Yang Lao Hsin Shu*) 137
Tsou Shu-Wên 24n
Tsou Yen (founder of Yin–Yang School of Philosophy) 119
tsu (pickled vegetables) *70*, 74, 76, 379, 380, 396n, 402–12, 414
   lactic acid fermentation 390; salting 416–17
tsu hsiao (*tsu*, as seasoning for meat) 405
tsu shun (purple shoot tea) 522n
Tsu Thing (statesman; −6th century) 441
Tsui Kuei-Thu (Thang period) 475n, 478
tsun (drink storage vessel) *100*, 102, 151n, 155
tsun (rudd) 61, *62*
Tsung Jen (official with *pi* fork) 107, 109
tsung tzu (rice dumpling) 119
tu (wild pear) 49
Tu Chhêng Chi Shêng (Wonders of the Capital [Hangchou]; +1235) 141, 325n
   salty fermented soybean soup 375n

Tu Fu (Thang poet) 180, 181, 223, 224, 301
tu hêng (herb) 574
Tu Khang or Shao Khang (Hsia ruler, preparer of alcoholic drinks) 155, 161
Tu Yu 82n
tuan (break in milk fermentation) 251
Tuan Wên-Chieh 208n
Tuan Wu (Dragon Boat Festival) 119, 235
tui (trip hammer) 448n
tumour 574
Tun Yuan Chu Shih (Resident of the Hidden Garden; Ming Dynasty) 145
Tun-huang (Kansu) 240
tung (fish), *chiang* 380
Tung Chhu-Phing 23n, 119, 233n, 250n, 457n
Tung Chi (writer; +11th century) 581, 585
Tung Ching Mêng Hua Lu (Dreams of the Glories of the Eastern Capital; +1148) 141, *256*
  deep fried foods 325; *shih* use 341; salty *shih* soup 375n
  honey 460; honeyed sweetmeats 426; sugar 460; ice selling 434
  pasta 476, 481–2, 490n; wonton 479
  sausages 423n; *hsi* and *fu* 421–2; soybean sprouts 296, *595*
Tung Cho Chuan (Biography of Tung Chuo) 337n
Tung Hsi Shih Chha Lu (Tea in Chien-an; +1064) *516*, 517, 523
Tung Hsiao-Chüan 332n
Tung Kuan Han Chi (Biography of Ming Kung) 294n
Tung Ma (Prefect of Mare Milkers) 248
tung oil 450, 452
Tung Pho Tsa Chi 563n
tung tou fu (frozen *tou fu*) 321, *322*, 324
Tung-Pho tou fu (*tou fu* á la Tung-Pho) 301, 354
Tung-Thing Chhun Sê (Spring Colours by the Tung-thing Lake; +11th century) 206, 225, 245
Tungting Lake 63n
Tunhuang manuscript 136
Turfan (Sinkiang) 204–5 see also Kao-chhang
  conquest 241, Thang tomb 478, *479*
turnip, Chinese 379, 403, *406*, 407 *409*
turnip seed oil 450, 451
turtle 63–6; giant 64; water 64
Twankay tea 534
Tyn, Myo Thant 392n
Typha latifolia 33
Tze Lu (follower of Confucius) 20, 22
Tzeng Mao-Chi 310n, 311n
tzu (cooked food) 469; *tzu* (rice and millet pastry) 467–8
tzu (jars of wine) 165; *tzu* (large chunks of meat) 69
tzu (meat process) 382n; *tzu* (steeped delicacy) 69; *tzu* (steeping) 74, *88*
tzu jang (Nature's Laws) 224
tzu su (purple perilla) 439
tzu wei (*Campsis grandiflora*?) 234

udon (sliced noodles) 491, *492*
Ueda Seinosuke 154n, 163n, 166, 261n, 263n
Uigher (Turfan, Sinkiang) 143, 242, 472n
Ukers, W. H. 504, 506n, 507n, 515n, 520n, 522n, 534, 535, 536, 537n, 538n, 541n, 542n, 543n, 544, 546n, 549n, *550*, 557n, *558*, 559n, 562n, 563n, 565n, 566n

Ulmus campestris 354n; *Ulmus macrocarpa* 337n, 354n
Underhill, A. P. 259n, 277n
Ung Kang-Sam 121
Unschuld, P. U. 571n
urn 151n, 153, 266
  cold storage of food 430, 432; tea storage 534
  fu ju making 326; soy sauce making 364, 365–6, *367*; malt sugar making 458; vinegar 285, 286, 288
  sealing 440
uronic acids 593
Urumqi, Sinkiang *90*
USDA Handbook *330*
Ushiyama Terui 473n
utensils 67
  eating 104; food serving 85, 104–7, *108*, 109
  Han tombs 105, *106*
  ice house well at Hsin-chêng 431–2
  tea *556*, 557, *558*, 559, 560

Vaccinium macrocarpus 49
Valla, F. 258n, 265n, 599n
Varicorhinus simus 62
vats 204
Vavilov, N. I. 66–7
veal 59
vegetable garden 32
vegetable oil 8, 437, 441, 601
vegetables 32–6, *17*, *37*, 38–40, *41*, *42–3*, 379
  food resources 66; *kêng* 83; mucilaginous 38, 93, 95
  oil-bearing 31
  pickled 74n, 76, 379–80, 390, 402–12, 414, 416–17, *594*
  preparation 68; preserved 595
  salting and pickling 416–17, *418*, 419
  vitamin D 590n
vegetarian diet 329n, 605n
  gluten use 497–8, 502
Vehling, J. D. 400n, 497n
Veith, I. 14n, *15*, 16, 92n, 121, 283n
venison 59, *381*, 70n, 379
vermicelli 495
vessels *see also* jars; pots
  general's helmet 151, *152*, 153, 277n
  necessity 602; for serving drinks 98, *99*, 101–4
  for serving food 98–101; wine 155n
Vettii, house of (Pompey) 437, *439*
viands 83, 84, 98
Vickery 437, 455
Vicia augustifolia (wild bean) 33
Vietnam
  fermented salted fish *395*, 396–7; fermented soyfoods 378
  fish sauce 392, 397; *tou fu* 328
Vigna angularis 40n
Vigna calcaratus 40 *see also Phaseolus calcaratus*
Vigna faba 297n; *Vigna mungo* var. *radiata* 298
vine, kudzu 462, matrimony 562 *see also* grape vine
vinegar 2, 92, 95, 283–91, *594*
  alcohol concentration, effect of 288
  ancient Egypt 283n; Sumerians 283n; China 291
  fragrant *hsiang tshu*, well-temmpered *lao chhên tshu*, *Pao-nin tshu* 291

vinegar (cont.)
　apricot 284, 291n; coat or film formation 286, 289
　condiment use 354, 362n
　discovery 603; early history 283–5
　incubation in 415, 416, *418*; with salt 417, *418*
　medicated 287; medicament use 290
　oxidation of alcohol to acetic acid 287–8
　preparation from: barley 285–6; clear wine 286; unfiltered wine 286; honey 287; malt sugar 290; flavoured with lesser beans 286–7; rice 290; soghum, glutinous rice and wheat 291
　saccharification 287, 288
　temperature control 286, 287
　use in making red *ferment* 197, 201; pickling and salting of vegetables 416–19;
　tsu 406, 407–8; *suan tshai* 412; *tou fu* 303
　urn 285, 286, 288
　water addition 288
vineyards 242 *see also* grape vines
Virgil 150, 246n
Visayas, fermented fish 395
visceral enzymes 398, 399
vision *see also* blindness
　clarifying 586
　loss 441, 587
　reduction after dusk 587
　vitamin A requirement 587
vitamin A
　deficiency 586, 587
　sources 588n
vitamin B_1
　anti-beriberi factor 585
　deficiency 579
　seed source 581
　yoghurt 584
vitamin B_{12} 605
vitamin C 248n
vitamin D 590
　deficiency 588
　sources, sunshine 590
　vegetables 590n
vitamin D_2 588n, 590
vitamin D_3 588n, 590
vitamins 573, 579n
　bean sprouts 605
viticulture 243
Vitis thunbergii 240n
Vitis vinifera 240, 242–3
　culture in West 258
Vogt, Kim 392n

Wai, N. 328n
Wai Thai Mi Yao (Important Prescriptions from a Distant Post; +752) *135*, 136, 226
　goitre 575–6; rickets 589; soy sauce 371
Wai-Huang county (north-eastern Honan) 301
Waley, A. 17n, 18n, 19n, 32n, *33*, 39n, 45n, 52n, 55n, 59, 61, 67n, 68n, 74n, 82n, 96n, 102n, 118, 118n, 149n, 156n, 246n, 406n, 419n, 429n, 457n, 507n
walking, lateness in children 588, 589

Wamyo-ruizyushu (General Encyclopedic Dictionary of Japanese language; +934) 375, 491
Wan Jên-Hsing 469n
wan tou (peas) 297n
Wan-hsien (Szechuan) 450
Wang Chên (14th century) 143, 425, 441, 450, 452, 488, 529, 532
Wang Chên Nung Shu (+1313) 529n, 531
　oil press *442*, 443
Wang Chhung (writer; +1st century) 347
Wang Chi (early Thang writer) 180, 243n
Wang Chi-Min & Wu Lien-Tê 579n
Wang Chih-Tsun 156n
Wang Chin (Thang writer) 163, 179n, 180, 181n
Wang Cho (writer; +12th century) 460
Wang Chu-Lou 132n
Wang, H. L. 260n, 328n, 342n, 345n, 346n, 377n, 378n, 605n
Wang Hai (predynastic ancestor of the Shang) 57
Wang Han (Thang poet; +8th century) 239, 241
Wang Hao (18th century) 145
Wang Hsi (writer; 15th century) 577
Wang Hsiang-Chin (17th century) 145, 456n
Wang Hsüeh-Thai 84n
Wang, I. 23, 339n
Wang Jên-Hsiang 104n, 148, 500n
Wang Jên-Hsing 63n, 75n, 147, 467n, 469n, 477, 480n, 491n
Wang Jih-Chên (writer; 1813–81) 305n, 324
Wang Khên-Thang (paediatrician; 17th century) 588, 590
Wang Khu (official of Kuangsi) 327
Wang Li-Chhi & Wang Chêng Min 54n, 63n, 91n
Wang Ling 515n
Wang Lung-Hsüeh 499n
Wang Mang (father-in-law of the Chin Emperor Ai-ti) 510n
Wang Mêng-Ou 60n, 69n, 70n, 75, 120, 163n
　tsu 403
Wang Sai-Shih 429n, 434n
Wang Shan-Tien 147, 359n, 362n, 377n, 425n, 426n, 429n, 462n, 506n
Wang Shih-Hsiung (author of *Sui Hsi Chü Yin Shih Phu*; mid-19th century) 319n, 324, 373, 391, 561n
Wang Shih-Mou (Ming dynasty) 145
Wang Shu-Ming 155
Wang Su 511, 555
Wang Thao (author of *Wai Thai Mi Yao*; +8th century) 136
Wang Tsêng-Hsin 115
Wang Tshan (poet; +177–217) 192
Wang Tshao Chha Shuo (Talks on Tea from the 'Wang' Grass Pavilion; +1734) 539
Wang Tung-Fêng 472n
Wang Wên-Tsê 467n
Wang, Y. Y. 202n
Wang Yên (poet; +1138–1218) 500–1
Wang Yü-Hu 24n, 38, 143n, 144n
Wang Yü-Kang 305n, 306n
Wang Yu-Phêng 203n, 218n, 219–20
Wang Yuan (Taoist holy man; +2nd century) 221
Wang Yüeh-Chêng 388n

Wang Yusheng *487*
Wang Zhi-Yuan & Yang, C. S. 570n
Wang-fu-hsia cave temples, Yu Lin Cave (+1032 to +1227; Ansi, Kansu) 208
water 97, 170, 174–5, 177n, 192, 184, 348, 197, 351, 461, 523
   boiled 97, 565; drinking 436; tea making 556ff
   distilled 226
   foodstuff category of ancient China 17
   frozen 436; iced 434
   goitre association 574, 575, 578; health 574; safety 565
   perfumed 206
   wine fermentation 174–5
   steep 187, 188; steeping 428
   symbolic correlation *572*
water caltrop *33*, 40, 43, 462
   steeped in water 428
water chestnut 40
water lily, root 462
water mallow 379, 403, *406*
water melon 245n
water penance 511, 513
water pepper *see* smartweed
water power *584*
water rushes 379
water sprout of small bud tea 525n
water wheel 4, 5n
watercress *406*
Watson, B. 68n, 181n, 420n, 478n, 574n, 602
wedge press 4, 443–50, *451*, 452, *453*
   development 456
   rapeseed oil 455
wei (swordbill sturgeon; snoutfish) 62, 63, 380
wei (wild bean) *33*, 38, 520n
wei chhi (Chinese chess) 473n
wei chiang (miso) 374
Wei Chü-Yuan (writer; c. +700) 126, 478
wei chung or *wei pi* (beriberi) 579
Wei Shu (History of the Northern Wei; +554) 125, 154n
Wei Ssu 429n
wei tiao (wilting of tea leaves) 537
Wei Tshê (Record of the Wei State) 155n
Wei Wang Tshao Mu Chih (Flora of the King of Wei; +515) 510, 511
Wei Yao (historian at court of Sun Hao, King of Wu; c. +290) 509, 513, 520n
   tea drinking 555
Weinhold, R. 272n
Welbourn, R. B. 573n
wells
   ice house 431–2
   treasure 432–3
wên (warming wine) 222
Wên Chhing-Ming 225n
Wên (Duke of Chin) 85
Wên (Emperor of Wei Dynasty) 204
Wên Liu Shih Chiu (A Question for Liu, No 19) 181n
Wên Lung (writer; 17th century) *516*, *530*, 532, 534, 551n
Wên Shu (Ming album painter; 17th century) 143

Wên-ti (Han emperor) 337
wêng (wide-mouthed jar for wine fermentations) 151, *152*, 332
West 263–72, *273*, 274, *275*; wine origin 153, 258
   Chinese processed foods 608
   climate 278
   fungal contamination of food 278
   goitre 577; rickets 589
   grain fermentation 272, *273*, 274–5, *276*, 277–8
   oilseed pressing 437, *438*, 439, 452, *453*, 455–7
   rotary quern 599n
   saccharification 282
   soy foods 292; soybean cultivation 292
   use of *ferment* organisms 608
   vegetarianism 329n
   see also America; ancient Egypt; Europe; Greece; Mesopotamia; Rome; tea, trade
wheat 25, 27, 58
   cream of 465n
   date of introduction in China 19n, 262
   digestibility 604
   durum 19n, 465n; emmer 267n, 269, 270, 271n
   ferment 167, 169, 185, 190, 261, 277; yellow coat 335
   fermented bran 340n
   five staple grains 19, *20*, 22, 23
   flour 2, 3, 4, 68, 82, 314
      components 465
      pasta recipes *470*, 474
      processing *463*, *464*, 465–6, 603–4, *604*
      sifting for pasta making 474
   chiang making 347, 349, 352, 355–7, 373n, 593
   shih making 342
   grinding & milling 68, 170, 172, 463, *464*, 593, 582, *598*, 599n, 582, *598*
   hulled type 266n
   importance in Han dynasty 27
   kernel separation 265, 266
   malting 262, *604*
   meal 263; heating 271–2; paste 278
   moulded *604*
   northern China 582
   parching 265; pastry frying 440; roasted 169, 170, 265
   rotary quern 597, *598*
   sprouted 271, 397; sprouting 163n, 457n, 458, 460
   staple grain *21*
   starch preparation 462
   stir-roasted 171, 172
   varieties 266
   vinegar 291
   see also dough; gluten
whey 241, 252; soybean 308
whisk, tea 557, 559
White, L. Jr. 401n
Wilderness Colors of Tao Chi 42
Wilhelm, R. 419n
Willia 280
William of Rubruck (traveller; c. +1253) 249
Williams, T. I. 597n
Wilson, H. 267n, 275n
Winarno, F. G. 342n

wine (*chiu*) 2, 7, 95, 96, 97, 149, *594*
   a-la-chi 227, 229, 230
   alcohol content 179, *180*, 279
   benefits 168; importance 162–3; overindulgence 169n
   bitter 284, 285; bitter flavour 91
   bone of 202;
   cassia 233; chrysanthemum 235; cinnamon 233; coconut flower 245; in *cha* making *385*; *chiang* making 382; *hai* making 380
   chewing method of making 154, 166
   Chou *165*; clear 222
   cock crow 189, 190; 'cock crowing' summer 178
   colour 439n; conversion to vinegar 283–7; cooling 187, 188
   cups 223, 224
   decantation 164, 179, 187; decanters 102
   distilled 7, 8, 10, 191, 203–31
      archaeological discoveries 208–9, *210–11*, *212–21*
      Chinese pot still for production 230–1
      development 226–7, *228*, 229–32, *595*, 596
      early evidence 204–8
      Han period 221
      Thang burnt wine 205, 206, 224, 229, 231
      taste of Sung burnt wine 224–5
      catching fire 225
   dregs, residues (*tsao*) 162, 182
   drug absorption 233; drug vehicle 233
   ferment 154, 157, 161, 349, 593; applications 606, 607
   fermentation 3, 168–283, 606
      early mediaeval period 168–81
      late mediaeval/premodern period 181–92
      residues residues 179, 340, 594
      vessels 151, *152*, 332–3
   filtration 164, 182
   fire-pressured, pasteurisation 206; or by steaming 88n, 206–7, 222, 225–6, 230
   flavouring 603, *604*; fragrant (ritual or medicated) 232; herbal 233–4
   frozen-out 223; high-potency 224, 226
   fruit 259; coconut milk 245n; jujubes 245, 259n; litchi 245n; nectar 245; orange 245; papaya 245n; peach 259n; pears 245; persimmon 245n; pineapple 245n; plum 259n; water melon 245n
   grape *see* grape wine
   honey 8, 246–8; iced *mei* flower 434
   Hsin Fêng process of Southern Sung 132, *133*, 188–9
   incubation with salt and vinegar *418*
   Job's tear 190; mung beans 190; millet 174
   medicament 18, 163, *164*, 182, 183
   medicated 8, 192, 232–9
      grape juice addition 243
      treatment of specific illnesses 234–5
   muddy 161; mulberry-fall 180; mutton 231, 237–8
   MWT Inventory 165–6
   mulled, warmed or heated 188, 221, 222; origins, from cereals 257–83
      in China 259–63; in West 263–72
   preincubation step 163; pressing 187
   primordal 97, 301; production 151; protective 235
   quality control 162–3
   red 8, 191, 192–203
      making 3, 163
   red *ferment* 196–7
   temperature control 176, 177
   regular strength 229; regulation 178, 190n
   rice 174, 604; ritual 156–7, 162, 183, 232, 263; sorghum 279n
   seaweed steeping for goitre treatment 575
   Shao Hsing of Chekiang 163, *164*
   superior *ferment* (*shên chhü*) 174–6
   social life of Shang 151; spring 176–7; stability 179–80, 188
   steeping herbs 235; storage 151, 176, 188
   technology, works on 132–4
   therapeutic value 233; *ti huang* 236, 237; tiger-bone 237
   traditional process 153; types 156, *165*
   vessels for mortuary rituals 155n
   vinegar 92, 283–91
   warding off evil spirits 235
   water collection 174–5, 177n
   white 191; yellow 151n, 190–1, 279
   yeast 184, 282
   yield 182, 183, 188
   see also fermentation, process; grape wine; *tsao*
Wine Superintendant 164, 232n
Wine and Vingar Bureau, Yuan Government 244
wine warmer, charcoal 221
winery attendant, accidental death 206–7, 225
Winfield, G. F. 590
winnowing
   bamboo tray 473n
   chiang process 348, 352
   shih process 335, 336, 340, 341, 361
winter melon 426, *418*
Wittwer, S. 59n, 456n
wok, tea processing 4, *521*, 530, 531; cooking pot 82, 230
Wolfe, H. L. 571n
wolf's bane 283, 589
women working in food preparation 597n
won-ton (cloud swallowing) 3, 475, 478–80, 490, *604*
wood, symbolic correlation *572*
Wood, E. 270n, 272n
wood ear 144–5, 406n, 408, *409*
Wood, F. 434n, 496n
Woodward, N. H. 571n
work station, opening up tea 523
wort 272, 459n
Wu region (Kiangsu) 181n
Wu Chao-Su 463n, 469n, 475
Wu Chhêng-Lo 170n, 173n, 247, 285n, 338n, 520n
wu chhi (5 pickled meat or vegetables) 379
Wu Chhi-Chhang 265n
wu chiu (tallow tree) 441
Wu Chi, Biography of 154n
wu chia phi (bark of *Acanthopanax spinosum*) 192, 235, 237; wine 236
Wu Chia-Khuo 506n
Wu Chih-Sung 466n
wu ching (Chinese turnip; colza) 31, 34, *35*, *587*
Wu Ching-Tzu (writer of *Ju Lin Wai Shih*; +1701 to 1754) 323n

Wu Chio-Nung 518, 542, 543, 548n
Wu Chün Chi (Gazetteer of the Wu Region; +1190) 434
Wu Chün Phu (Mushrooms in the Wu Region; +1683) 145
Wu Hêng 250n
wu hsiang (five spice) 95n
wu hsiang tou fu kan (*tou fu* cooked with five spices) 320, 325–6
Wu Hsien 590n, 604n
Wu Hsien-Wên 61n
Wu Hsing (Five Elements or Phases) 571n, 572, 586, 607
Wu I Shan Chih (Gazetteer of the Wu-I Hills; 1846) 542n
Wu I Tsa Chi (Random Notes on Wu-I; Ming) 540n
Wu Jui (Yüan dynasty) 24n
wu ku (five grains) 19, 20n, 21, 22, 23, 55
Wu Lei Hsiang Kan Chi (On the Mutual Response of Things according to their Categories; c. +980) 206, 225, 301
Wu Li Hsiao Shih (Mini Encyclopedia of the Principle of Things; +1644) 133–4, 204n
  tou fu making 305; vegetarian diet 299n
Wu li (must of Wu) 156n
Wu Lien-Tê 579n
Wu Lin (17th century) 145
Wu Lin Chiu Shih (Institutions and Customs of the Old Capital; +1270) 141, 476, 479
  gluten and pasta selling 500; honey 460; *hsi* and *fu* 421–2; *mi chien* 426
  sausages 423n; sugar 460
wu lung (oolong; semi-fermented tea) 541, 548
Wu Ma 56n
wu mei (dark dried apricot) 425; *wu mei ju* (flesh of dark *mei* fruit) 587
wu mei thang (sugared dark apricots) 426n; *wu mei tshu* (dark *mei* vinegar) 291n
Wu Shih Chung Khuei Lu (Madam Wu's recipe book; Southern Sung) 127, 128n, 130, 137n, 413–14
  author 597n; *cha* 387, 388; *chiang* 386; 354–5
  food preservation methods 409–10, 412
  fu and *cha* 422; *kan pi wêng tshai* 417, 419; *lu* 414n, 419
  noodles 483; salting of vegetables 418; *shih* recipes 340
  soy condiments 372; soy sauce 358, 359
Wu Shih Erh Ping Fang (Prescriptions for Fifty-two Ailments; c. –200) 67, 68, 117, 121
  cooking methods 70, 71, 72, 73, 74; beggar's chicken 73
  fermented soy paste 346; gruel 82; honey 246; medicaments 134
  soy sauce 371, 595; soybeans 295, 315n; vinegar 284, 290
Wu Shih (Ming writer) 540n
Wu Shih Pên Tshao (Wu's Pharmaceutical Natural History; c. +235) 296; *shih* use 337
Wu Shih-Chhi 61n
Wu Tê-To 10, 11, 136n, 203, 205, 206n, 207, 208, 209n, 221n, 225
wu thou (*Aconitum fischeri*; wolf's bane) 235, 283n, 589
wu tui (piling of tea) 552
Wu Tzu-Mu (writer; +14th century) 141, 301
wu wei (five flavours) 283; *wu yao* (drug) 441

Wu Ying-Ta (Chhing Dynasty) 144
Wu Yüan (Origin of Things; +15th century) 299
wu-chou la chu fa (The Wu-chou Method for Curing Pork) 422
Wu-hsi chhing-kung-yeh hsüeh-yuan (Technology of Wine Fermentation; 1964) 179n, 238n, 279n, 281n
Wu-I chha (Wu-I tea) 542n
Wu-I mountains (Fukien) 538, 545, 561
  see also tea, Wu-I
Wu-I region (northwestern Fukien) 541, 546, 547, 548n
  black tea 545, 550
Wu-ling prefecture 437
Wu-ti (Han emperor; –140–88) 382–3
Wu-yang (nr. Chengtu)
  buying tea 508; tea production 509
wu-yü (dark yam) 40
Wu-yüan (Kansu) 240
Wylie, A. 130

Xanthium strumarium 29n, 335
Xanthoxylum simulans 44
Xenocypris argentea 63
xerophthalmia 587
Xi Zeong 118n
Xing, R. & Tang, Y. 151n, 259n
Xu, Y., Ho, C. T., Amin, S. G., Han, C. & Chung, F. L. 570n

ya chha (bud tea) 529; *ya mien* (pressed noodle) 485, 489, 491
ya tou (bean sprouts) 296
ya tshai (mung bean sprouts) 298
ya yu (pressing seeds for oil) 440
ya yu chia (oil pressing houses) 29
Ya-la-chi chiu fu (Ode to Araki Wine; +1344) 229
yam, Chinese 40, 43, 457n
yam flour noodles 483, 485, 500
Yamanishi, T. 537n, 546n, 549n, 551n, 552n, 553n, 565n
Yamasaki Hiyachi 157n
yan (goose) 60
Yan, Y. S. 570n
yang (sheep) 55, 57–8
Yang Ching-Shêng 24n
Yang, E. F. 581n
Yang, G. 59n
Yang Hsiao Lu (Guide to Nurturing Life; +1698) 129, 130, 131, 342, 357n, 373, 387, 388, 414
  food preservation methods 409–10, 412n
  honeyed and sugared fruits 426n
  kan pi wêng tshai 419; *lu* method 414; milk products 256
  salting of vegetables 418; soy condiments 372
  soy sauce 363, 373
  sung 423; *tou fu* 324, 325; *tsao*-preserved crab 413
  vinegar recipes 290–1
Yang I (Grand Physician for External Illnesses) 571n
Yang Lao Fêng Chhing Shu (New Handbook on Care of the Elderly; +1080) 137
Yang Ling-Ling 571n
yang mei (Chinese strawberry) 54
Yang Phu (Szechuan) 40n, 91

Yang Shêng Fang (Prescriptions for Nurturing Life; early Thang) 575, 575n
Yang Sheng Lun (Discourse on Nurturing Life; c. +260) 295
Yang Shih-Chhang (Taoist priest; +11th century) 246
Yang Shu-Ai 310, 312
Yang Ssu (killed by steaming wine) 206–7, 225
Yang Tao-Jen (physician; 4th century) 579, 580
yang thao (Chinese gooseberry) 52
Yang Wan-Li (writer; +1127–1206) 301, 302, 434
Yang Wên-Chhi 147, 257n, 467n
Yang Ya-Chhang 18n, 71n
Yang Yeh 126
Yang Yü Ching (Manual of Fish Culture; Ming Dynasty) 63n, 145
Yang Yung-Tao (c. +1000) 134
Yang-chou shih po-wu-kuan (Chhieh-mo-shu Western Han Tomb at Yangchou; 1980) 80n
Yang-Hsien tea 539
Yang-shao culture (−5000 to −3000) 58, *276*, 277
Yang-sheng pu (Section on Nourishing Life of *Sung Shih Tsun Shêng*) 130n
Yang-Shêng-Fang (Manual for Nurturing Vitality; −200) 234
Yang-ti (Sui Emperor) 413n
Yangchou *tsao* melon 411n
Yangtze region 40, 522n, 579
 Chinese apricot growing 47
 fish 63; wetlands 35; Grand Canal 434
Yangtze River 63n
yao (to gently cook in hot meat or vegetable soup) 74
Yao Chih-Han (artist; Chhing) *323*
Yao Chiu Yen Fang Hsüan (1985) 238
Yao Kung (physician; Chou to Wei dynasties) 580
Yao (legendary emperor of China) 297n
Yao Sêng-Yuan (physician; +6th century) 579n
Yao Yüeh-Ming 538n
yeast 153n, 160n, 187–9, 194, 607
 alcohol production 282; beer making 270, 272, *273*, 274
 ancient Egypt 271n, 272n
 contamination of malted flour 270
 dough, contamination 269, starter 243n & leavening 465
 fermented soyfoods 378n
 grape wine 244; inoculum 168; leaven 606n
 harvesting from must 184; propagation of red 202
 in *chhü* (*ferment*) 261, 280, 282n, 592; with red *ferment* 196
 mead 247; residues in pottery vessels 151
 sensitivity to alcohol 279
 sprouted grain (*nieh*) contamination 159, 261
 use in China 282; vitamin D_2 590; wild 157, 242; wine & beer fermentations 184, 188, 189–90, 275
yeh chha (leaf or loose tea) 529, 529n
Yeh Chung Chi (Record of the *Yeh* region; c. +4th century) 432
Yeh Ting-Kuo 131n
Yeh Tshai Po Lu (Comprehensive Account of Edible Wild Plants; +1597) *129*, 130–1
Yeh Tzu-Chhi (writer; +14th century) 299
yeh-ko (*Gelsemium elegans*) 183

yellow coat (*huang i*) 285n, 289, 335, 336, 350, 355, 605
 bean 342, 345, 352; wheat 335
yellow jaw fish 62
Yellow River, water collection 192n
yellow sparrow, *cha 409*
Yellow Springs 83
yen (bamboo mats) 109; *yen* (oil cloth) *521*
yên (incubation in salt) 417, 419, 422; *yen* (salt) 92
yen (mud fish) 61, *62*; *yên* (steamer) 76, 77, *78*, *79*, 80
yen (thyroid gland) 576
yen chha (rolling tea) 525
yên chhü (covered *ferment*) 185
yen chih (salt and incubate) 402
Yen Chih-Tui 478
Yen Fan Lu 499n
yen hsin (fermented salted fish) 392
Yen Jên (Salt Administrator) 92
yên ju (salted meat) 417, 422
yen lu (bittern) 303, 305
yen shih (salty fermented soybean product) 337
 making 338–9, 340, 341, 351
yen shih thang (soup of salty fermented soybeans) 375n
Yen Shih-Ku (writer; 7th century) 59–60, 315, 346n, 354
yên tshai (salted vegetable) 412, 416, 417
Yen Tshan 402n
Yen Tzu (Chou Philosopher) 93
Yen Tzu Chhun Chhiu (Master Yen's Spring and Autumn Annals; −4th century) 54, 513
Yen Wên-Ming 27n
yên yü (fermented salted fish) 388n, 390, 393, 395, 417
Yen Yung-Ho (writer; +13th century) 585
Yenching (Peking) regulation wine 190n
Yenching sui shih chi (New Year Foods in Peking) 434n
yew nuts 301
yin (drink) 98; *yin chha* (drinking tea) 485
Yin Chuan Fu Shih Chien (Compendium of Food and Drink; +1591) *129*, 130, 132n, 372n, 355, 357n, 372n, 373 387, 414
 dairy products *256*; dried meat and fish 422
 food preservation methods 409–10, 411n, 412
 grape wine making 244; *lu* 419; mead 247; milk drinking 257
 noodles 483; red *ferment* 196; salting of vegetables *418*
 shih recipe 340; soy sauce 373; soybean sprouts 297
 tea 559n; *tou fu* 324; yeast use 244
Yin Hai Ching Wei (Essentials of the Silver Sea; Sung) 588
yin ku (*Xenocypris argentea*) 63
Yin people, wine for sacrificial ceremonies 156
Yin Shan Chêng Yao (Principles of Correct Diet; +1330) 38n, *135*, 137, 138, 341, 355, 357
 a-la-chi wine 204, 227, 229, 231
 beriberi 581–2; goitre 576
 dairy products *256*; malt sugar 460
 grape wine 244; medicated wine 237
 pasta, making 477; origin of pasta 496n
 man-thou 476; noodles 483, 484; *pao-tzu* 476
 red *ferment* 196
 tea 540n, 564
 wine distillation 227, *595*

Yin Shan Thai I (Grand Dietician) 138
yin shih (food and drink) 149
Yin Shih Chih Wei Fang (+17th century) 492
Yin Shih Hsü Chi (Essentials of Food and Drink; +1350) *135*, 138–9, *256*, *324*, *501*, *564*
yin-shih (food and drink) 503
Yin–Yang 571n, 607; balance 588
ying (bamboo basket) 521; ying (goitre) 574
ying ching liu yeh (swelling at the neck) 574
Ying Shao (author of *Fêng Su Thung I*) 248n
ying thao (Chinese cherry) *44*, 51–2
ying yü (indigenous grape) 240n
yoghurt 252
  drink 253n, 315, 511, 584n
  iced 434; liquid 253
  pastoral nomads 315
  slave of 511, 513
Yoke-harness Superintendent 57n
Yokotsuka, T. 377n, 605n
Yong, F. M. & Wood, B. J. B. 377n
Yoryo Ritsuryo (+718) 374n
Yoshida, Shjui 345
Yoshizawa R. 167n, 279, 281, 282n
yü (drying chamber) *521*, 522
yü (fish) 17, 60; *yü* (taro) 42–3
yü (serving vessel) 98–9, *100*, 101
yu (drink storage vessel) *100*, 102, 113, bronze 153; *yu* (fermentation) 277
yu (oil) 358, 437; *yu* (pomelo) 54; *yu* (sauce) 389
Yü Chêng 515n
yü chhang (ritual wine with herbs) 232, 235
yu chia (oil press) 441, *442*, 443–50
yü chiang (fish paste) 359, 387, 391–2, *395*
yü chiang chih (juice from fermented fish paste) 387, 397n
yü chin tshao (*Curcuma aromatica*; herb) 232
Yü Ching (Book of Taro; Ming dynasty) 145
Yü Ching-Jang 19n, 23n, 24n, 300n, 389n
yü fou liang (pearly floating logs) 181
Yü Guanying 18n
Yu, H. 569n
yü hsi (dried fish) 420
Yü Hsing-Wu 18n
Yü Hsiu-Ling 24n, 260n
Yü Hua-Chhing 168n
Yu Huan Chi Wên (Things Seen and Heard on Official Travel; +1233) 206
Yü Jên (Fishery Superintendent) 56n, 63, 157n
Yü Jên (Superintendant of Ritual Herbs) 232
yü jên (elm nuts) 355
yü jên chiang (elm nut *chiang*) 346, 355
Yü Kêng (grand soup) 84
Yu, King (Chou Dynasty) 117
Yu Lin frescoes (+1032 to +1227; Ansi, Kansu) 208
Yü Lin Khu (Kansu) 208
Yü Lin Wai Shih (novel; 18th century) 323n
yü lu (fish sauce) 389, 391–2, 392n, *395*, 397–8, 398n
Yü Ming-Hsien 220
Yü Phin (Types of Edible Fish; Ming dynasty) 145
yu ping (fried pasta) 490n
Yü Shih 56n
Yü Shih Phi (Imperial Food List; Southern Sung) 126–7, 128

Yu shui (river originating near Wu-ling) 437
Yu Tê-Tsun 54n
yu thiao (deep fried crullers) 2, 322, 471, 490n
yü thou chiang (fish head paste) 387
yu tou fu (deep fried blocks of *tou fu*) 320, 325
yu tsa kui (deep fried crullers) 2
yu tshai (brassica) 35
Yü Ying-Shih 293n, 305n, 463n
yü yüan so ping (rope pasta from the Chinese yam) 482
Yü-chhan-yen (Hunan) 27n
Yu-Yang Tsa Tsu (Miscellany of the *Yu-Yang* Mountain cave; +860) 121, 192
yüan (turtle) 65
Yuan Chieh (Thang poet) 222
Yüan Chien Lei Han (Mirror of the Infinite; +1701) 240n
Yüan chiu (Chou wine) *165*
Yüan Han-Chhing 155, 157n, 203, 259n, 265n, 300
Yuan Kho 574n
Yüan Kuo-Fan 248n, 249n
Yuan Mei (author of *Sui Yuan Shih Tan*, gastronome of Chhing dynasty) 131, 324, 355n, 480, 476, 373, 540
Yüan Shih (History of the Yuan; +1271 to +1368) 243, 244n
Yuang Hsuan-Chih 141
Yuasa (Wakayama prefecture) 376
Yueh Lo (South China) 91
Yueh Shi (writer; +10th century) 183
Yüeh-Yang (Hunan) 543n
Yule, H. 493n, 501n
yum cha (tea brunch) 485, 488
yün chiu (wine fermented with multiple batches of substrate) 166
Yün Fêng 151n, 153n, 155n, 156n, 480n
Yün Lin Thang Yin Shih Chih Tu Chi (Dietary System of the Cloud Forest Studio; +1360) *127*, 128, 189, *256*, 362, 411
  man-thou 476; noodles 484n
  soy condiments 372n; soy sauce making 362
yun thai (oil-bearing vegetable) 31
Yün-mêng (Hupei province) 35, 54, 93, 102, 103
  monographs 146
Yun-meng Suei-hu-ti Chhin-mu (Chhin Tomb at Shui-hu-tui, Yun-meng; 1981) 102n, 146n
yung (vats) 204
Yung Thao (+9th century) 205
Yung-Lo Ta Tien, I Hsüeh Chi (Great Collectanea of the Yung-Lo Region, Medical Section; +1408) 358n
Yunmeng hsien wên-wu (on a Chhin-Han Tomb at Shui-hu-ti, Yun-mêng, Hupei; 1981) *81*, 103n
Yunnan, Phu-erh tea 535n, 562
Yünnan, sesame 31

Zaccagnini, C. 269n
Zagros Mountains (Iran) 272
Zanthoxylum 52, 235; *Zanthoxylum ailanthoides* (pepper) *170*
Zee, A. 478n
Zen Buddhism 506, 515; Zen Temple Cuisine 317
Zhu, Y. 201n
Zingiber mioga 40, 43; *Zingiber officinale* 33, 40
zizania 409; *Zizania caduciflora* 20n
Zizyphus jujuba 44, 48; *Zizyphus spinosa* 48, 349n
Zohary, D. 258n, 265n, 266n
Zohary, M. 39n

TABLE OF CHINESE DYNASTIES

	夏	Hsia kingdom (legendary?)		c. −2000 to c. −1520
	商	Shang (Yin) kingdom		c. −1520 to c. −1030
	周	Chou dynasty (Feudal Age)	Early Chou period	c. −1030 to −722
			春秋 Chhun Chhiu period	−722 to −480
			戰國 Warring States (Chan Kuo) period	−480 to −221
First Unification	秦	Chhin dynasty		−221 to −207
	漢	Han dynasty	Chhien Han (Earlier or Western)	−202 to +9
			Hsin interregnum	+9 to +23
			Hou Han (Later or Eastern)	+25 to +220
First Partition	三國	San Kuo (Three Kingdoms period)		+221 to +265
			蜀 Shu (Han)	+221 to +264
			魏 Wei	+220 to +265
			吳 Wu	+222 to +280
Second Unification	晉	Chin dynasty	Western Chin	+265 to +317
			Eastern Chin	+317 to +420
	劉宋	(Liu) Sung dynasty		+420 to +479
Second Partition		Northern and Southern Dynasties (Nan Pei Chhao)	齊 Chhi dynasty	+479 to +502
			梁 Liang dynasty	+502 to +557
			陳 Chhên dynasty	+557 to +589
			魏 Northern (Thopa) Wei dynasty	+386 to +535
			Western (Thopa) Wei dynasty	+535 to +556
			Eastern (Thopa) Wei dynasty	+534 to +550
			北齊 Northern Chhi dynasty	+550 to +577
			北周 Northern Chou (Hsienpi) dynasty	+557 to +581
Third Unification	隋	Sui dynasty		+581 to +618
	唐	Thang dynasty		+618 to +906
Third Partition	五代	Wu Tai (Five Dynasty period)	(Later Liang, Later Thang (Turkic), Later Chin (Turkic), Later Han (Turkic) and Later Chou)	+907 to +960
			遼 Liao (Chhitan Tartar) dynasty	+907 to +1124
			West Liao dynasty (Qarā-Khiṭāi)	+1124 to +1211
			西夏 Hsi Hsia (Tangut Tibetan) state	+986 to +1227
Fourth Unification	宋	Sung dynasty		+960 to +1279
			宋 Northern Sung dynasty	+960 to +1126
			宋 Southern Sung dynasty	+1127 to +1279
			金 Chin (Jurchen Tartar) dynasty	+1115 to +1234
	元	Yüan (Mongol) dynasty		+1260 to +1368
	明	Ming dynasty		+1368 to +1644
	清	Chhing (Manchu) dynasty		+1644 to 1911
	民國	Republic		1912
	人民共和國	People's Republic		1949

N.B. When no modifying term in brackets is given, the ruling house was Han. During the Eastern Chin period there were no less than eighteen independent States (Hunnish, Tibetan, Hsienpi, Turkic, etc.) in the north. The term 'Liu chhao' (Six Dynasties) is often used by historians of literature. It refers to a succession of southern dynasties with their capital at Nanking from the beginning of the +3rd to the end of the +6th centuries, including (San Kuo) Wu, Chin, (Liu) Sung, Chhi, Liang and Chhen.

ROMANISATION CONVERSION TABLES

by Robin Brilliant

PINYIN/MODIFIED WADE-GILES

Pinyin	Modified Wade-Giles	Pinyin	Modified Wade-Giles	Pinyin	Modified Wade-Giles
a	a	chi	chhih	dui	tui
ai	ai	chong	chhung	dun	tun
an	an	chou	chhou	duo	to
ang	ang	chu	chhu	e	ê, o
ao	ao	chuai	chhuai	en	ên
ba	pa	chuan	chhuan	eng	êng
bai	pai	chuang	chhuang	er	êrh
ban	pan	chui	chhui	fa	fa
bang	pang	chun	chhun	fan	fan
bao	pao	chuo	chho	fang	fang
bei	pei	ci	tzhu	fei	fei
ben	pên	cong	tshung	fen	fên
beng	pêng	cou	tshou	feng	fêng
bi	pi	cu	tshu	fo	fo
bian	pien	cuan	tshuan	fou	fou
biao	piao	cui	tshui	fu	fu
bie	pieh	cun	tshun	ga	ka
bin	pin	cuo	tsho	gai	kai
bing	ping	da	ta	gan	kan
bo	po	dai	tai	gang	kang
bu	pu	dan	tan	gao	kao
ca	tsha	dang	tang	ge	ko
cai	tshai	dao	tao	gei	kei
can	tshan	de	tê	gen	kên
cang	tshang	dei	tei	geng	kêng
cao	tsho	den	tên	gong	kung
ce	tshê	deng	têng	gou	kou
cen	tshên	di	ti	gu	ku
ceng	tshêng	dian	tien	gua	kua
cha	chha	diao	tiao	guai	kuai
chai	chhai	die	dieh	guan	kuan
chan	chhan	ding	ting	guang	kuang
chang	chhang	diu	tiu	gui	kuei
chao	chhao	dong	tung	gun	kun
che	chhê	dou	tou	ha	ha
chen	chhên	du	tu	hai	hai
cheng	chhêng	duan	tuan	han	han

735

Pinyin	Modified Wade-Giles	Pinyin	Modified Wade-Giles	Pinyin	Modified Wade-Giles
hang	hang	kui	khuei	mu	mu
hao	hao	kun	khun	na	na
he	ho	kuo	khuo	nai	nai
hei	hei	la	la	nan	nan
hen	hên	lai	lai	nang	nang
heng	hêng	lan	lan	nao	nao
hong	hung	lang	lang	nei	nei
hou	hou	lao	lao	nen	nên
hu	hu	le	lê	neng	nêng
hua	hua	lei	lei	ng	ng
huai	huai	leng	lêng	ni	ni
huan	huan	li	li	nian	nien
huang	huang	lia	lia	niang	niang
hui	hui	lian	lien	niao	niao
hun	hun	liang	liang	nie	nieh
huo	huo	liao	liao	nin	nin
ji	chi	lie	lieh	ning	ning
jia	chia	lin	lin	niu	niu
jian	chien	ling	ling	nong	nung
jiang	chiang	liu	liu	nou	nou
jiao	chiao	lo	lo	nu	nu
jie	chieh	long	lung	nü	nü
jin	chin	lou	lou	nuan	nuan
jing	ching	lu	lu	nüe	nio
jiong	chiung	lü	lü	nuo	no
jiu	chiu	luan	luan	o	o, ê
ju	chü	lüe	lüeh	ou	ou
juan	chüan	lun	lun	pa	pha
jue	chüeh, chio	luo	lo	pai	phai
jun	chün	ma	ma	pan	phan
ka	kha	mai	mai	pang	phang
kai	khai	man	man	pao	phao
kan	khan	mang	mang	pei	phei
kang	khang	mao	mao	pen	phên
kao	khao	mei	mei	peng	phêng
ke	kho	men	mên	pi	phi
kei	khei	meng	mêng	pian	phien
ken	khên	mi	mi	piao	phiao
keng	khêng	mian	mien	pie	phieh
kong	khung	miao	miao	pin	phin
kou	khou	mie	mieh	ping	phing
ku	khu	min	min	po	pho
kua	khua	ming	ming	pou	phou
kuai	khuai	miu	miu	pu	phu
kuan	khuan	mo	mo	qi	chhi
kuang	khuang	mou	mou	qia	chhia

Pinyin	Modified Wade-Giles	Pinyin	Modified Wade-Giles	Pinyin	Modified Wade-Giles
qian	chhien	shu	shu	xian	hsien
qiang	chhiang	shua	shua	xiang	hsiang
qiao	chhiao	shuai	shuai	xiao	hsiao
qie	chhieh	shuan	shuan	xie	hsieh
qin	chhin	shuang	shuang	xin	hsin
qing	chhing	shui	shui	xing	hsing
qiong	chhiung	shun	shun	xiong	hsiung
qiu	chhiu	shuo	shuo	xiu	hsiu
qu	chhü	si	ssu	xu	hsü
quan	chhüan	song	sung	xuan	hsüan
que	chhüeh, chhio	sou	sou	xue	hsüeh, hsio
qun	chhün	su	su	xun	hsün
ran	jan	suan	suan	ya	ya
rang	jang	sui	sui	yan	yen
rao	jao	sun	sun	yang	yang
re	jê	suo	so	yao	yao
ren	jên	ta	tha	ye	yeh
reng	jêng	tai	thai	yi	i
ri	jih	tan	than	yin	yin
rong	jung	tang	thang	ying	ying
rou	jou	tao	thao	yo	yo
ru	ju	te	thê	yong	yung
rua	jua	teng	thêng	you	yu
ruan	juan	ti	thi	yu	yü
rui	jui	tian	thien	yuan	yüan
run	jun	tiao	thiao	yue	yüeh, yo
ruo	jo	tie	thieh	yun	yün
sa	sa	ting	thing	za	tsa
sai	sai	tong	thung	zai	tsai
san	san	tou	thou	zan	tsan
sang	sang	tu	thu	zang	tsang
sao	sao	tuan	thuan	zao	tsao
se	sê	tui	thui	ze	tsê
sen	sên	tun	thun	zei	tsei
seng	sêng	tuo	tho	zen	tsên
sha	sha	wa	wa	zeng	tsêng
shai	shai	wai	wai	zha	cha
shan	shan	wan	wan	zhai	chai
shang	shang	wang	wang	zhan	chan
shao	shao	wei	wei	zhang	chang
she	shê	wen	wên	zhao	chao
shei	shei	weng	wêng	zhe	chê
shen	shen	wo	wo	zhei	chei
sheng	shêng, sêng	wu	wu	zhen	chên
shi	shih	xi	hsi	zheng	chêng
shou	shou	xia	hsia	zhi	chih

Pinyin	Modified Wade-Giles	Pinyin	Modified Wade-Giles	Pinyin	Modified Wade-Giles
zhong	chung	zhuang	chuang	zou	tsou
zhou	chou	zhui	chui	zu	tsu
zhu	chu	zhun	chun	zuan	tsuan
zhua	chua	zhuo	cho	zui	tsui
zhuai	chuai	zi	tzu	zun	tsun
zhuan	chuan	zong	tsung	zuo	tso

MODIFIED WADE-GILES/PINYIN

Modified Wade-Giles	Pinyin	Modified Wade-Giles	Pinyin	Modified Wade-Giles	Pinyin
a	a	chhio	que	chua	zhua
ai	ai	chhiu	qiu	chuai	zhuai
an	an	chhiung	qiong	chuan	zhuan
ang	ang	chho	chuo	chuang	zhuang
ao	ao	chhou	chou	chui	zhui
cha	zha	chhu	chu	chun	zhun
chai	zhai	chhuai	chuai	chung	zhong
chan	zhan	chhuan	chuan	chü	ju
chang	zhang	chhuang	chuang	chüan	juan
chao	zhao	chhui	chui	chüeh	jue
chê	zhe	chhun	chun	chün	jun
chei	zhei	chhung	chong	ê	e, o
chên	zhen	chhü	qu	ên	en
chêng	zheng	chhüan	quan	êng	eng
chha	cha	chhüeh	que	êrh	er
chhai	chai	chhün	qun	fa	fa
chhan	chan	chi	ji	fan	fan
chhang	chang	chia	jia	fang	fang
chhao	chao	chiang	jiang	fei	fei
chhê	che	chiao	jiao	fên	fen
chhên	chen	chieh	jie	fêng	feng
chhêng	cheng	chien	jian	fo	fo
chhi	qi	chih	zhi	fou	fou
chhia	qia	chin	jin	fu	fu
chhiang	qiang	ching	jing	ha	ha
chhiao	qiao	chio	jue	hai	hai
chhieh	qie	chiu	jiu	han	han
chhien	qian	chiung	jiong	hang	hang
chhih	chi	cho	zhuo	hao	hao
chhin	qin	chou	zhou	hên	hen
chhing	qing	chu	zhu	hêng	heng

ROMANISATION CONVERSION TABLES

Modified Wade-Giles	Pinyin	Modified Wade-Giles	Pinyin	Modified Wade-Giles	Pinyin
ho	he	kao	gao	lieh	lie
hou	hou	kei	gei	lien	lian
hsi	xi	kên	gen	lin	lin
hsia	xia	kêng	geng	ling	ling
hsiang	xiang	kha	ka	liu	liu
hsiao	xiao	khai	kai	lo	luo, lo
hsieh	xie	khan	kan	lou	lou
hsien	xian	khang	kang	lu	lu
hsin	xin	khao	kao	luan	luan
hsing	xing	khei	kei	lun	lun
hsio	xue	khên	ken	lung	long
hsiu	xiu	khêng	keng	lü	lü
hsiung	xiong	kho	ke	lüeh	lüe
hsü	xu	khou	kou	ma	ma
hsüan	xuan	khu	ku	mai	mai
hsüeh	xue	khua	kua	man	man
hsün	xun	khuai	kuai	mang	mang
hu	hu	khuan	kuan	mao	mao
hua	hua	khuang	kuang	mei	mei
huai	huai	khuei	kui	mên	men
huan	huan	khun	kun	mêng	meng
huang	huang	khung	kong	mi	mi
hui	hui	khuo	kuo	miao	miao
hun	hun	ko	ge	mieh	mie
hung	hong	kou	gou	mien	mian
huo	huo	ku	gu	min	min
i	yi	kua	gua	ming	ming
jan	ran	kuai	guai	miu	miu
jang	rang	kuan	guan	mo	mo
jao	rao	kuang	guang	mou	mou
jê	re	kuei	gui	mu	mu
jên	ren	kun	gun	na	na
jêng	reng	kung	gong	nai	nai
jih	ri	kuo	guo	nan	nan
jo	ruo	la	la	nang	nang
jou	rou	lai	lai	nao	nao
ju	ru	lan	lan	nei	nei
jua	rua	lang	lang	nên	nen
juan	ruan	lao	lao	nêng	neng
jui	rui	lê	le	ni	ni
jun	run	lei	lei	niang	niang
jung	rong	lêng	leng	niao	niao
ka	ga	li	li	nieh	nie
kai	gai	lia	lia	nien	nian
kan	gan	liang	liang	nin	nin
kang	gang	liao	liao	ning	ning

Modified Wade-Giles	Pinyin	Modified Wade-Giles	Pinyin	Modified Wade-Giles	Pinyin
niu	niu	sang	sang	thê	te
no	nuo	sao	sao	thêng	teng
nou	nou	sê	se	thi	ti
nu	nu	sên	sen	thiao	tiao
nuan	nuan	sêng	seng, sheng	thieh	tie
nung	nong	sha	sha	thien	tian
nü	nü	shai	shai	thing	ting
o	e, o	shan	shan	tho	tuo
ong	weng	shang	shang	thou	tou
ou	ou	shao	shao	thu	tu
pa	ba	shê	she	thuan	tuan
pai	bai	shei	shei	thui	tui
pan	ban	shên	shen	thun	tun
pang	bang	shêng	sheng	thung	tong
pao	bao	shih	shi	ti	di
pei	bei	shou	shou	tiao	diao
pên	ben	shu	shu	tieh	die
pêng	beng	shua	shua	tien	dian
pha	pa	shuai	shuai	ting	ding
phai	pai	shuan	shuan	tiu	diu
phan	pan	shuang	shuang	to	duo
phang	pang	shui	shui	tou	dou
phao	pao	shun	shun	tsa	za
phei	pei	shuo	shuo	tsai	zai
phên	pen	so	suo	tsan	zan
phêng	peng	sou	sou	tsang	zang
phi	pi	ssu	si	tsao	zao
phiao	piao	su	su	tsê	ze
phieh	pie	suan	suan	tsei	zei
phien	pian	sui	sui	tsên	zen
phin	pin	sun	sun	tsêng	zeng
phing	ping	sung	song	tsha	ca
pho	po	ta	da	tshai	cai
phou	pou	tai	dai	tshan	can
phu	pu	tan	dan	tshang	cang
pi	bi	tang	dang	tshao	cao
piao	biao	tao	dao	tshê	ce
pieh	bie	tê	de	tshên	cen
pien	bian	tei	dei	tshêng	ceng
pin	bin	tên	den	tsho	cuo
ping	bing	têng	deng	tshou	cou
po	bo	tha	ta	tshu	cu
pu	bu	thai	tai	tshuan	cuan
sa	sa	than	tan	tshui	cui
sai	sai	thang	tang	tshun	cun
san	san	thao	tao	tshung	cong

ROMANISATION CONVERSION TABLES

Modified Wade-Giles	Pinyin	Modified Wade-Giles	Pinyin	Modified Wade-Giles	Pinyin
tso	zuo	tzhu	ci	yao	yao
tsou	zou	tzu	zi	yeh	ye
tsu	zu	wa	wa	yen	yan
tsuan	zuan	wai	wai	yin	yin
tsui	zui	wan	wan	ying	ying
tsun	zun	wang	wang	yo	yue, yo
tsung	zong	wei	wei	yu	you
tu	du	wên	wen	yung	yong
tuan	duan	wo	wo	yü	yu
tui	dui	wu	wu	yüan	yuan
tun	dun	ya	ya	yüeh	yue
tung	dong	yang	yang	yün	yun